RECENT DEVELOPMENTS IN THE SYNTHESIS AND APPLICATIONS OF PYRIDINES

RECENT DEVELOPMENTS IN THE SYNTHESIS AND APPLICATIONS OF PYRIDINES

Edited by

PARVESH SINGH
School of Chemistry and Physics, University of KwaZulu-Natal, Durban, South Africa

ELSEVIER

Elsevier

Radarweg 29, PO Box 211, 1000 AE Amsterdam, Netherlands
The Boulevard, Langford Lane, Kidlington, Oxford OX5 1GB, United Kingdom
50 Hampshire Street, 5th Floor, Cambridge, MA 02139, United States

ISBN: 978-0-323-91221-1

For Information on all Elsevier publications visit our website at
https://www.elsevier.com/books-and-journals

Publisher: Susan Dennis
Acquisitions Editor: Gabriela Capille
Editorial Project Manager: Czarina Mae Osuyos
Production Project Manager: Bharatwaj Varatharajan
Cover Designer: Christian J. Bilbow

Typeset by Aptara, New Delhi, India

Contents

Contributors xv

Preface xix

1. Role of pyridine and its privileged derivatives as anti-infective agents

Pankaj Sharma, Vaishali Suthar, Meenu Aggarwal, Rajeev Sing and K. Poonam Kumar

1.1 Introduction to infectious diseases 1
- 1.1.1 Origin/sources of human diseases 5
- 1.1.2 Stages of endemic human diseases 5
- 1.1.3 Favorable conditions responsible for infectious diseases 8
- 1.1.4 Re-emergence of infectious diseases 8
- 1.1.5 Antimicrobial resistance 9

1.2 Anti-infective agents 9

1.3 Pyridine & its derivatives 11

1.4 Synthesis of pyridine derivatives 20

1.5 Bioisosteres and pyridine bioisosterism 25
- 1.5.1 General classification of bioisosteres 28
- 1.5.2 Classification of bioisosteres based on functional group 30
- 1.5.3 Few examples of bioisosterism in drug design are given below 30

1.6 Conclusion 38

Acknowledgment 39

References 39

2. Pyridine-based polymers and derivatives: Synthesis and applications

Shagun Varshney and Nidhi Mishra

2.1 Introduction 43
- 2.1.1 History of pyridines 44
- 2.1.2 Ligand properties of pyridine 45
- 2.1.3 Transition metal complexes of pyridine 46
- 2.1.4 The pharmacological activity of pyridines 47

2.2 Properties of pyridine 48
- 2.2.1 Physical properties 48
- 2.2.2 Chemical properties 49

2.3 Pyridine-based polymers of heterocyclic compounds 52

2.4 Polymers of pyridine 52

2.5 Vinyl pyridine-based polymers 53
- 2.5.1 Polyvinyl pyridine-based polymers in light-emitting devices 54
- 2.5.2 Applications in metal ion adsorption 54

2.6 Applications of pyridine-based polymers 58
2.6.1 Ion-exchange resins 59
2.6.2 Photoresists 59
2.6.3 Heat-resisting materials 59
2.6.4 High-temperature electrolyte membranes 59
2.6.5 Redox polymers 60
2.6.6 Biological applications 62
2.7 Conclusions 66
References 66

3. Synthetic strategies of functionalized pyridines and their therapeutic potential as multifunctional anti-Alzheimer's agents

Jeelan Basha Shaik, Mohammad Khaja Mohinuddin Pinjari, Damu Amooru Gangaiah and Chinna Gangi Reddy Nallagondu

3.1 Introduction 69
3.2 Synthetic strategies of pyridine based heterocycles 70
3.2.1 Synthesis of pyridine nucleus 70
3.2.2 Synthesis of 2-aminopyridines and 2-amino-3-cyano pyridines 73
3.2.3 Synthesis of pyridinium salts 76
3.2.4 Synthesis of imidazopyridines 78
3.2.5 Synthesis of pyrazolopyridines 81
3.2.6 Synthesis of furopyridines 83
3.3 Therapeutic potential of pyridine scaffolds as anti-Alzheimer's agents 85
3.3.1 Cholinesterase inhibitors 86
3.3.2 Aβ aggregation inhibitors 89
3.3.3 Secretase Inhibitors 95
3.3.4 Tau aggregation inhibitors 98
3.3.5 Glycogen synthase kinase-3β inhibitors 100
3.3.6 MAO-B inhibitors 101
3.3.7 Neuroprotective agents 102
3.3.8 Adenosine A_1 and A_{2A} receptor antagonists 104
3.3.9 Selective muscarinic M1 subtype activation 105
3.3.10 Phosphodiesterase (PDE) inhibitor 106
3.3.11 Nicotinic acetylcholine receptor (nAChRs) ligands 106
3.3.12 Multitarget directed ligands (MTDLs) 107
3.4 Conclusions 110
References 112

4. Design, synthesis, and in vitro anticancer activity of thiophene substituted pyridine derivatives

Vinayak Adimule, Basappa C Yallur and Sheetal Batakurki

4.1 Introduction 127
4.2 Materials and characterization 129
4.2.1 Synthesis of 2-(5-bromothiophen-2-yl) pyridin-3-amine (Compound 2, Step 1) 130
4.2.2 Synthesis of 2-(5-aminothiophen-2-yl) pyridin-3-amine (Compound 3, Step 2) 130

4.2.3 Synthesis of substituted N-[2-(5-aminothiophen-2-yl) pyridin-3-yl] benzamide 4 (a–h) (Compound 4, Step 3) 130
4.2.4 Synthesis of substituted N-[5-(3-benzamidopyridin-2-yl) thiophen-2-yl] benzamide (Compound 5, Step 4) 130
4.3 Purification of the compounds 130
4.4 Analytical characterization 131
4.4.1 ^{1}H-NMR, ^{13}CMR, and LCMS spectroscopic characterization data 131
4.4.2 Substituted N-[2-(5-aminothiophen-2-yl) pyridin-3-yl] benzamide (Compound 4 (a–h)) 131
4.4.3 Substituted N-[5-(3-benzamidopyridin-2-yl)thiophen-2-yl]benzamide (Compound 5 (a–h)) 133
4.5 In vitro anticancer activity of the compounds (5 a–h) 135
4.5.1 MTT assay 135
4.6 Anticancer activity 135
4.6.1 Preparation of the cell culture and fixation 135
4.7 Biological properties 136
4.7.1 Cytotoxic activities 136
4.7.2 Molecular modeling and computational characteristics of the synthesized compounds 137
4.8 Conclusion 139
Acknowledgments 140
Authors Contributions 140
Conflict of Interest 140
Data Availability 140
Funding 140
References 140

5. The role of pyridine derivatives on the treatment of some complex diseases: A review

Xolani Henry Makhoba

5.1 Introduction 143
5.2 Pyridines in general 144
5.3 Synthesis of pyridines derivatives 145
5.4 Life cycle of the malaria parasites 147
5.5 Pyridine containing drugs for malaria treatment 147
5.6 Cancer 147
5.7 Diabetes 151
5.8 Conclusion and future perspectives 152
Acknowledgments 154
Conflict of Interest 154
References 154

6. Pyridines in Alzheimer's disease therapy: Recent trends and advancements

Puja Mishra, Souvik Basak, Arup Mukherjee and Balaram Ghosh

6.1 Introduction 159
6.1.1 Etiology of Alzheimer's disease-cholinergic hypothesis 160
6.1.2 β-amyloid hypothesis 160
6.1.3 BACE-1 (β-amyloid cleaving enzyme-1) role in AD 162
6.1.4 Metal chelators in AD 162

6.2 Pyridine and its role in prevention of Alzheimer's disease 163
6.2.1 Pyridine—a short synopsis on its chemistry 163
6.2.2 Role of pyridines in AChE (acetylcholine esterase) inhibition 164
6.2.3 Role of pyridine in preventing β-amyloid aggregation 165
6.2.4 Role of pyridine as BACE-1 (beta-site amyloid precursor protein cleaving enzyme) inhibitors 170
6.2.5 Role of pyridine as metal chelators 174
6.3 Synthetic routes of some pyridine derivatives used in AD 178
6.3.1 Synthesis of imidazo [1,5a] pyridine carboxylic acids derivative 178
6.3.2 2,6-disubstituted pyridine 181
6.3.3 Pyrrolo aminopyridine 181
6.3.4 Bis-1,2,4-triazole/thiosemicarbazide 181
6.4 Toxicological manifestations of pyridine 182
6.5 Molecular modeling and computational simulation of pyridine in AD 182
6.6 Pyridine in AD research (marketed drugs as well as preclinical trial drug having pyridine) 183
6.7 Conclusion 184
References 184

7. Pyridine derivatives as anti-Alzheimer agents

Babita Veer and Ram Singh

7.1 Introduction 189
7.2 Target-based evaluation of pyridine derivatives as anti-Alzheimer agents 190
7.2.1 Pyridine derivatives as cholinesterase inhibitors 190
7.2.2 Pyridine derivatives as anti-β amyloid aggregates 194
7.2.3 Pyridine derivatives as BACE1 inhibitor 196
7.2.4 Pyridine derivatives as metal chelators 199
7.2.5 Pyridine derivatives for miscellaneous targets 199
7.3 Summary 202
References 202

8. Role of pyridines as enzyme inhibitors in medicinal chemistry

Khalid Mohammed Khan, Syeda Shaista Gillani and Faiza Saleem

8.1 General introduction 207
8.2 Marketed drugs based on pyridine nucleus 208
8.3 Synthesis of pyridine derivatives 212
8.4 Medicinal importance of pyridine 214
8.4.1 Anticancer activity 215
8.4.2 Cholinesterase inhibition activity 221
8.4.3 Antidiabetic activity 225
8.4.4 Urease inhibitory activity 232
8.4.5 Antioxidant activities 234
8.4.6 Antiinflammatory activity 235
8.4.7 Antimicrobial activity 236
References 245

9. Contemporary development in the synthesis and biological applications of pyridine-based heterocyclic motifs

Nisheeth C. Desai, Jahnvi D. Monapara, Aratiba M. Jethawa and Unnat Pandit

9.1 Introduction 253
9.2 Synthesis of pyridine 254
9.2.1 Hantzsch pyridine synthesis 254
9.2.2 Baeyer pyridine synthesis 254
9.2.3 Kröhnke pyridine synthesis 255
9.2.4 Katrizky pyridine synthesis 256
9.3 Green approach for the synthesis of pyridine 256
9.3.1 Ultrasound-assisted green synthesis 256
9.3.2 Microwave-assisted synthesis of pyridine 259
9.4 Synthesis and antimicrobial activity of some pyridine derivatives 259
9.5 Synthesis and antitubercular activity of some pyridine derivatives 264
9.6 Synthesis and anticancer activity of some pyridine derivatives 275
9.7 Synthesis and anti-inflammatory activity of some pyridine derivatives 285
9.8 Synthesis and antimalarial activity of some pyridine derivatives 288
9.9 Synthesis and anti-infective activity of some pyridine derivatives 292
9.10 Conclusion 293
9.11 Graphical conclusion 294
Acknowledgments 295
References 295

10. Synthesis of pyridine derivatives using multicomponent reactions

Shah Imtiaz, Bhoomika Singh and Md. Musawwer Khan

10.1 Introduction 299
10.2 Importance of pyridine and its derivatives 300
10.3 Classification of pyridines 302
10.3.1 Aromatic 302
10.3.2 Nonaromatic 303
10.4 Multicomponent reactions: its importance and green features 305
10.5 Synthesis of simple pyridine derivatives using multicomponent reactions 306
10.5.1 Ammonium acetate as a source of nitrogen to pyridine ring 306
10.5.2 Amine group as a source of nitrogen to pyridine ring 313
10.5.3 Malononitrile as a source of nitrogen to pyridine ring 316
10.5.4 Miscellaneous 322
10.6 Conclusion 324
Acknowledgments 324
References 324

11. Recent green synthesis of pyridines and their fused systems catalyzed by nanocatalysts

Amira Elsayed Mahmoud Abdallah

11.1 Introduction 331

11.2 Recent green synthesis of polysubstituted pyridines by using heterogeneous nanocatalyst systems 332
11.2.1 Multicomponent synthesis of 2-amino-3-cyano pyridines by using various heterogeneous nanocatalysts 332
11.2.2 Different heterogeneous nanocatalysts catalyzed the multicomponent synthesis of 1,4-dihydropyridines (1,4-DHPs) 340
11.2.3 Preparation of 2,4,6-triarylpyridine derivatives via various heterogeneous nanocatalysts 346
11.2.4 Various heterogeneous nanocatalysts catalyzed the synthesis of pyridine dicarbonitriles 348
11.2.5 Nanocatalytic activity of Cu/NCNTs catalyst for the synthesis of pyridine derivatives 350
11.3 Recent green synthesis of some selected fused heterocyclic-pyridine derivatives by using heterogeneous nanocatalyst systems 350
11.3.1 Recent green synthesis of pyrazolo[3,4-*b*]pyridine derivatives under various nanocatalysts 352
11.3.2 Recent green synthesis of pyridopyrimidine derivatives by using different nanocatalysts 362
11.3.3 Eco-benign approach for the synthesis of thiazolo[4,5-b]pyridine-6- carbonitrile derivatives via MgO nanocatalyst 367
11.3.4 Synthesis of 3-iminoaryl-imidazo[1,2-*a*]pyridines via manganese Schiff-base complex immobilized on chitosan-coated Fe_3O_4 (Fe_3O_4@CSBMn) 367
11.4 Conclusion 368
References 368

12. Pyridine ring as an important scaffold in anticancer drugs

Amr Elagamy, Laila K. Elghoneimy and Reem K. Arafa

12.1 Introduction 375
12.2 Pyridine-containing APIs in clinical use for treatment of cancer 376
12.2.1 Abemaciclib 376
12.2.2 Abiraterone acetate 378
12.2.3 Acalabrutinib 379
12.2.4 Alpelisib 381
12.2.5 Apalutamide 382
12.2.6 Axitinib 383
12.2.7 Enasidenib 385
12.2.8 Imatinib 385
12.2.9 Ivosidenib 387
12.2.10 Neratinib 388
12.2.11 Nilotinib 388
12.2.12 Palbociclib 390
12.2.13 Pexidartinib 391
12.2.14 Regorafenib 393
12.2.15 Ribociclib 393
12.2.16 Selpercatinib 394
12.2.17 Sonidegib 395
12.2.18 Sorafenib 396
12.2.19 Vismodegib 398
References 399

13. Recent developments in the synthesis of pyridine analogues as a potent anti-Alzheimer's therapeutic leads

Aluru Rammohan, Baki Vijaya Bhaskar and Grigory V. Zyryanov

13.1 Introduction 411
13.2 Role of pyridine in drug discovery 414
13.3 Pyridine analogues as Alzheimer's disease (AD) drug agents 415
13.3.1 Simple pyridine analogues 416
13.3.2 Bipyridine analogues 419
13.3.3 Nicotine analogues 420
13.3.4 Chromanone and *iso*-chromanone cohesive pyridine analogues 421
13.3.5 Benzofuran–pyridine analogues 422
13.3.6 Imidazole–pyridine analogues 424
13.3.7 Pyrazole–pyridine analogues 425
13.3.8 Triazole–pyridine analogues 427
13.3.9 Carbazole–pyridine analogues 429
13.3.10 Tacrine analogues 430
13.4 Structure–activity relationship studies (SARs) 434
13.5 Clinical approaches 436
13.6 Summary 436
Acknowledgments 438
References 438

14. Pyridine-based probes and chemosensors

Pawan Kumar, Bindu Syal and Princy Gupta

14.1 Introduction 445
14.2 Pyridine-based colorimetric chemosensors 448
14.2.1 Pyridine–pyrazole chemosensors 448
14.2.2 Pyridine–phenol chemosensors 449
14.2.3 Pyridine–carbohydrazide chemosensors 452
14.2.4 Lansopyrazole chemosensors 453
14.2.5 Naphthyl–pyridine chemosensors 453
14.2.6 Pyridine–boron chemosensors 454
14.2.7 Hydrazinyl–pyridine chemosensors 454
14.2.8 Triarylpyridine chemosensors 455
14.2.9 Pyridine–Schiff base chemosensors 455
14.2.10 Pyridine–thiourea chemosensors 456
14.2.11 Pyridine–phthalimide chemosensors 457
14.2.12 Pyridine–hydrazone chemosensors 457
14.2.13 Dipicolinimidamide chemosensors 457
14.3 Pyridine-based fluorescent chemosensors 458
14.3.1 Pyridine–carboxamide chemosensors 458
14.3.2 Pyridine–rhodamine chemosensors 460
14.3.3 Pyridine–Schiff base chemosensors 463
14.3.4 Bi/terpyridine chemosensors 466
14.3.5 Pyridine-BODIPY (4,4-difluoro-4-borato-3a,4a-diaza-s-indacene) chemosensors 469
14.3.6 Pyridine–pyrazole chemosensors 470

14.3.7 Pyridine–imidazole chemosensors 474
14.3.8 Pyridine–thiazole chemosensors 476
14.3.9 Pyridine–hydrazone chemosensors 476
14.3.10 Pyridine–coumarin chemosensors 477
14.3.11 Pyridine-based other chemosensors 478
14.4 Pyridine-based dual-mode chemosensors 480
14.4.1 Pyridine–carboxamide chemosensors 480
14.4.2 Pyridine–rhodamine chemosensors 482
14.4.3 Pyridine–Schiff base chemosensors 485
14.4.4 Pyridine–BODIPOY chemosensors 486
14.4.5 Pyridine–imidazole chemosensors 487
14.4.6 Pyridine–thiazole chemosensors 489
14.4.7 Pyridine–amine chemosensors 490
14.4.8 Pyridine-based some other chemosensors 491
14.5 Conclusions and future prospects 493
Acknowledgments 493
References 493

15. Recent advances in catalytic synthesis of pyridine derivatives

Morteza Torabi, Meysam Yarie, Saeed Baghery and Mohammad Ali Zolfigol

15.1 Introduction 503
15.2 Basic catalysts 505
15.3 Acidic catalysts 513
15.4 Ionic liquids and molten salts 517
15.5 Transition metal catalysis 520
15.5.1 Cu catalysis 520
15.5.2 Fe catalysis 525
15.5.3 Pd catalysis 527
15.5.4 Ru catalysis 527
15.5.5 Ag catalysis 530
15.5.6 Rh catalysis 535
15.5.7 Ni catalysis 537
15.5.8 Sn catalysis 538
15.5.9 Co catalysis 539
15.5.10 Zr catalysis 540
15.5.11 Mn catalysis 540
15.5.12 Mg catalysis 541
15.5.13 Al catalysis 541
15.5.14 I_2 catalysis 542
15.6 Heterogeneous catalytic systems 547
15.6.1 MOF-based heterogeneous catalysts 547
15.6.2 Nanomagnetic catalysis 548
15.6.3 Silica-supported catalysts 559
15.7 Organocatalysis 561
15.7.1 Solid acids 563
15.7.2 Carbon nanotube-supported catalysts 563
15.7.3 Miscellaneous 564

15.8 Conclusion 569
Acknowledgments 569
References 570

16. Pyridine as a potent antimicrobial agent and its recent discoveries

Nitish Kumar, Harmandeep Kaur, Anchal Khanna, Komalpreet Kaur, Jatinder Vir Singh, Sarabjit Kaur, Preet Mohinder Singh Bedi, Balbir Singh

16.1 Introduction 581
16.2 Pyridine as antimicrobial agents 583
16.2.1 Compounds containing oxygen in the heterocyclic nucleus 583
16.2.2 Compounds containing nitrogen in the heterocyclic nucleus 584
16.2.3 Compounds containing more than one hetero atom in heterocyclic ring 589
16.2.4 Organometallic 596
16.2.5 Miscellaneous 597
16.3 Pharmacophoric features 599
16.4 Conclusion 601
16.5 Authors contribution 601
References 601

17. Synthesis of pyridine derivatives for diverse biological activity profiles: A review

Tejeswara Rao Allaka and Naresh Kumar Katari

17.1 Introduction 605
17.2 Synthetic strategy 607
17.3 Biological activity profiles 612
17.3.1 Antimicrobial activities 612
17.3.2 Anticancer activity 613
17.3.3 Antidiabetic activity 615
17.3.4 Antimalarial agents 616
17.3.5 Antituberculosis 616
17.3.6 Anti-inflammatory activity 617
17.3.7 Antihypertensive agents 617
17.3.8 Antiamoebic agents 618
17.3.9 Antiarhythmatic activity 618
17.3.10 Enzyme inhibition 619
17.4 Conclusion 619
Acknowledgment 620
References 620

Index 625

Contributors

Damu Amooru Gangaiah Department of Chemistry, Yogi Vemana University, Kadapa, Andhra Pradesh, India

Amira Elsayed Mahmoud Abdallah Chemistry Department, Faculty of Science, Helwan University Ain, Helwan, Cairo, Egypt

Vinayak Adimule Angadi Institute of Technology and Management (AITM), Belagavi, Karnataka, India

Meenu Aggarwal Department of Chemistry, Aggarwal College Ballabgarh, Faridabad, Haryana, India

Tejeswara Rao Allaka Centre for Chemical Sciences and Technology, Institute of Science and Technology, Jawaharlal Nehru Technological University Hyderabad, Hyderabad, Telangana, India

Reem K. Arafa Drug Design and Discovery Lab, Helmy Institute for Medical Sciences, Zewail City of Science and Technology, Cairo, Egypt; Biomedical Sciences Program, University of Science and Technology, Zewail City of Science and Technology, Cairo, Egypt

Saeed Baghery Department of Organic Chemistry, Faculty of Chemistry, Bu-Ali Sina University, Hamedan, Iran

Souvik Basak Dr. B.C. Roy College of Pharmacy & Allied Health Sciences, Durgapur, West Bengal, India

Sheetal Batakurki Department of Chemistry, M.S. Ramaiah University of Applied Sciences, Bangalore, Karnataka, India

Preet Mohinder Singh Bedi Department of Pharmaceutical Sciences, Guru Nanak Dev University, Amritsar, Punjab, India

Baki Vijaya Bhaskar Department of Biochemisrty, University of Nebraska, Lincoln, Nebraska, United States

Basappa C Yallur Department of Chemistry, M.S. Ramaiah Institute of Technology, Bangalore, Karnataka, India

Nisheeth C. Desai Division of Medicinal Chemistry, Department of Chemistry, Maharaja Krishnakumarsinhji Bhavnagar University, Bhavnagar, India

Amr Elagamy Drug Design and Discovery Lab, Helmy Institute for Medical Sciences, Zewail City of Science and Technology, Cairo, Egypt

Laila K. Elghoneimy Drug Design and Discovery Lab, Helmy Institute for Medical Sciences, Zewail City of Science and Technology, Cairo, Egypt

Balaram Ghosh Epigenetic Research Laboratory, Department of Pharmacy, Birla Institute of Technology and Science, Hyderabad, Telangana, India

Syeda Shaista Gillani Department of Chemistry, Lahore Garrison University, Lahore, Pakistan; School of Chemistry, University of the Punjab, Lahore, Pakistan

Princy Gupta Department of Chemistry and Chemical Sciences, Central University of Jammu, Jammu, J&K, India

Shah Imtiaz Department of Chemistry, Aligarh Muslim University, Aligarh, India

Aratiba M. Jethawa Division of Medicinal Chemistry, Department of Chemistry, Maharaja Krishnakumarsinhji Bhavnagar University, Bhavnagar, India

Naresh Kumar Katari Department of Chemistry, School of Science, GITAM Deemed to be University, Hyderabad, Telangana, India

Harmandeep Kaur Department of Pharmaceutical Sciences, Guru Nanak Dev University, Amritsar, Punjab, India

Komalpreet Kaur Department of Pharmaceutical Sciences, Guru Nanak Dev University, Amritsar, Punjab, India

Sarabjit Kaur Department of Pharmaceutical Sciences, Guru Nanak Dev University, Amritsar, Punjab, India

Khalid Mohammed Khan H.E.J. Research Institute of Chemistry, International Center for Chemical and Biological Sciences, University of Karachi, Karachi, Pakistan

Md. Musawwer Khan Department of Chemistry, Aligarh Muslim University, Aligarh, India

Anchal Khanna Department of Pharmaceutical Sciences, Guru Nanak Dev University, Amritsar, Punjab, India

K. Poonam Kumar Ipca Laboratories Limited, Vadodara, Gujarat, India

Pawan Kumar Department of Chemistry and Chemical Sciences, Central University of Jammu, Jammu, J&K, India

Nitish Kumar Department of Pharmaceutical Sciences, Guru Nanak Dev University, Amritsar, Punjab, India

Xolani Henry Makhoba Department of Biochemistry and Microbiology, University of Fort Hare, Alice, South Africa

Nidhi Mishra Department of Applied Sciences, Indian Institute of Information Technology Allahabad, Prayagraj, Uttar Pradesh, India

Puja Mishra Dr. B.C. Roy College of Pharmacy & Allied Health Sciences, Durgapur, West Bengal, India

Jahnvi D. Monapara Division of Medicinal Chemistry, Department of Chemistry, Maharaja Krishnakumarsinhji Bhavnagar University, Bhavnagar, India

Arup Mukherjee Department of Biotechnology, Maulana Abul Kalam Azad University of Technology, West Bengal, India

Chinna Gangi Reddy Nallagondu Department of Chemistry, Yogi Vemana University, Kadapa, Andhra Pradesh, India

Unnat Pandit Special Centre for Systems Medicine, Jawaharlal Nehru University, New Delhi, India

Mohammad Khaja Mohinuddin Pinjari Department of Chemistry, Yogi Vemana University, Kadapa, Andhra Pradesh, India

Aluru Rammohan Department Organic and Biomolecular Chemistry, Ural Federal University, Ekaterinburg, Russian Federation

Faiza Saleem H.E.J. Research Institute of Chemistry, International Center for Chemical and Biological Sciences, University of Karachi, Karachi, Pakistan

Jeelan Basha Shaik Department of Chemistry, Yogi Vemana University, Kadapa, Andhra Pradesh, India

Pankaj Sharma Department of Applied Chemistry, Faculty of Technology & Engineering, The Maharaja Sayajirao University of Baroda, Vadodara, Gujarat, India

Rajeev Singh Department of Chemistry, Atma Ram Sanatan Dharma College, University of Delhi, New Delhi, India

Ram Singh Department of Applied Chemistry, Delhi Technological University, Delhi, India

Bhoomika Singh Department of Chemistry, Aligarh Muslim University, Aligarh, India

Jatinder Vir Singh Department of Pharmaceutical Sciences, Guru Nanak Dev University, Amritsar, Punjab, India

Balbir Singh Department of Pharmaceutical Sciences, Guru Nanak Dev University, Amritsar, Punjab, India

Vaishali Suthar Department of Applied Chemistry, Faculty of Technology & Engineering, The Maharaja Sayajirao University of Baroda, Vadodara, Gujarat, India

Bindu Syal Department of Chemistry and Chemical Sciences, Central University of Jammu, Jammu, J&K, India

Morteza Torabi Department of Organic Chemistry, Faculty of Chemistry, Bu-Ali Sina University, Hamedan, Iran

Shagun Varshney Department of Applied Sciences, Indian Institute of Information Technology Allahabad, Prayagraj, Uttar Pradesh, India

Babita Veer Department of Applied Chemistry, Delhi Technological University, Delhi, India

Meysam Yarie Department of Organic Chemistry, Faculty of Chemistry, Bu-Ali Sina University, Hamedan, Iran

Mohammad Ali Zolfigol Department of Organic Chemistry, Faculty of Chemistry, Bu-Ali Sina University, Hamedan, Iran

Grigory V. Zyryanov Department Organic and Biomolecular Chemistry, Ural Federal University, Ekaterinburg, Russian Federation; I. Ya. Postovskiy Institute of Organic Synthesis, UB of the RAS, Yekaterinburg, Russian Federation

Preface

Pyridine is an excellent scaffold that represents the structural core of numerous biologically, chemically, and medicinally significant synthetic heterocycles and natural products. The inimitable physical and chemical properties of pyridine make it a compatible solvent with both water and other organic solvents to prepare a range of useful compounds, including dyes, vitamins, pesticides, adhesives, agrochemicals, and pharmaceutical agents. Moreover, the basic nature and unique electronic arrangement of the pyridine nucleus account for its wide exploitation in various chemical transformations leading to the generation of a plethora of heterocycles and metal complexes with remarkable applications in various sectors. The quantum of research in pyridine heterocycles has shown a tremendous increase in the last decade, and thus keeping oneself abreast of recent developments is rather challenging.

The book offers an insightful perspective on the recent developments in pyridine chemistry, and consists of 17 comprehensive reviews, contributed by distinguished researchers in this field. These reviews broadly discuss different conventional and green synthetic routes dedicated to the preparation of pyridine derivatives and their biological activities, particularly emphasizing their antimicrobial, anticancer, anti-Alzheimer potential, as well as their applications in chemosensors and nanocatalysts. Overall, the compiled information in the book provides a combined academic and industrial approach for readers who are interested in the synthesis and manufacture of pyridine derivatives, including those working in the medicinal, polymer, and material sciences.

I would like to take the opportunity to thank all authors in the book for supporting this effort that forms an updated base platform for pyridine synthesis and its applications in different areas. I would also like to express my sincere gratitude to the entire team of Elsevier Publisher, particularly the editorial project managers, Mr. Regine Gandullas and Ms. Veronica Santos, and for their regular support and timely production of the book.

Prof. Parvesh Singh
School of Chemistry and Physics,
University of Kwa-Zulu Natal,
Durban, South Africa

CHAPTER

1

Role of pyridine and its privileged derivatives as anti-infective agents

Pankaj Sharma[a], *Vaishali Suthar*[a], *Meenu Aggarwal*[b], *Rajeev Singh*[c] and *K. Poonam Kumar*[d]

[a]Department of Applied Chemistry, Faculty of Technology & Engineering, The Maharaja Sayajirao University of Baroda, Vadodara, Gujarat, India [b]Department of Chemistry, Aggarwal College Ballabgarh, Faridabad, Haryana, India [c]Department of Chemistry, Atma Ram Sanatan Dharma College, University of Delhi, New Delhi, India [d]Ipca Laboratories Limited, Vadodara, Gujarat, India

1.1 Introduction to infectious diseases

Unlike earlier, in the 21st century, the progress, economy, and political situation will depend upon the health of its citizen. Recently, due to Coronavirus disease 2019 (COVID-19), most of Asian, European, American, and African countries including developing as well as developed proved fail to manage the crisis generated by COVID-19. As a result, not only did millions lost their lives but it had a serious impact on the pace of development, economy, and political condition. In the context of global health, human populations have suffered for centuries from various infectious diseases which create serious issues like recession, poverty, and propensity of a government collapse of any nation. The statistics of World Health Organization (WHO) and Centers for Disease Control and prevention are an authentic sources and provide sufficient proof that despite the enormous progress in the pharmaceutical field, infectious diseases account for nearly one-third of global deaths (Fig. 1.1A–C).

As per the data and available reports, infectious diseases are the second leading cause of death and the disability-adjusted life years worldwide. Among various infectious diseases, acute lower respiratory tract infections, human immunodeficiency virus (HIV)/acquired immune deficiency syndrome (AIDS), malaria, tuberculosis (TB), and diarrheal diseases predominate the cause of death occurring worldwide [1]. Infectious diseases have also been found to be one of the leading causes of morbidity, mortality in men throughout the world. Infectious diseases are disorders caused by microorganisms such as bacteria, viruses, fungi, protozoa, helminths, and prion. The broader term used for them is "Human Pathogens." Many

Recent Developments in the Synthesis and Applications of Pyridines.
DOI: https://doi.org/10.1016/B978-0-323-91221-1.00016-6

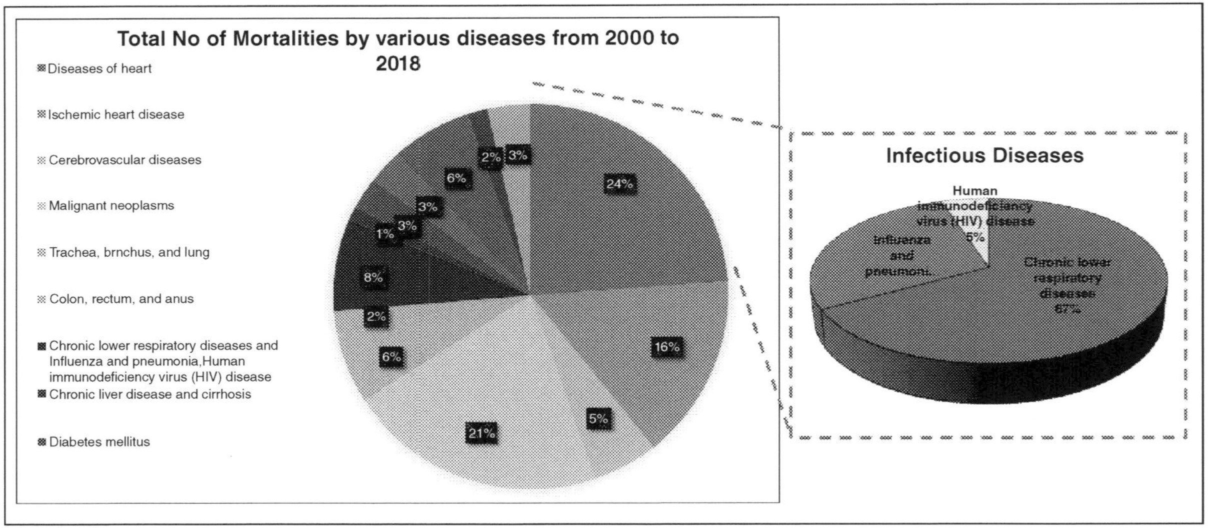

FIGURE 1.1 (A) An overview of the various diseases highlighted with infectious diseases as a I leading cause of deaths globally. Among all infectious diseases chronic lower respiratory infections are observed more commonly than pneumonia and HIV / AIDS. Data 2019, Source: Centers for disease control and prevention (CDC). (B) Number of deaths by various diseases Data 2019, Source: Centers for disease control and prevention (CDC). (C) Number of deaths by different infectious diseases Data 2019,Source: Centers for disease control and prevention (CDC).

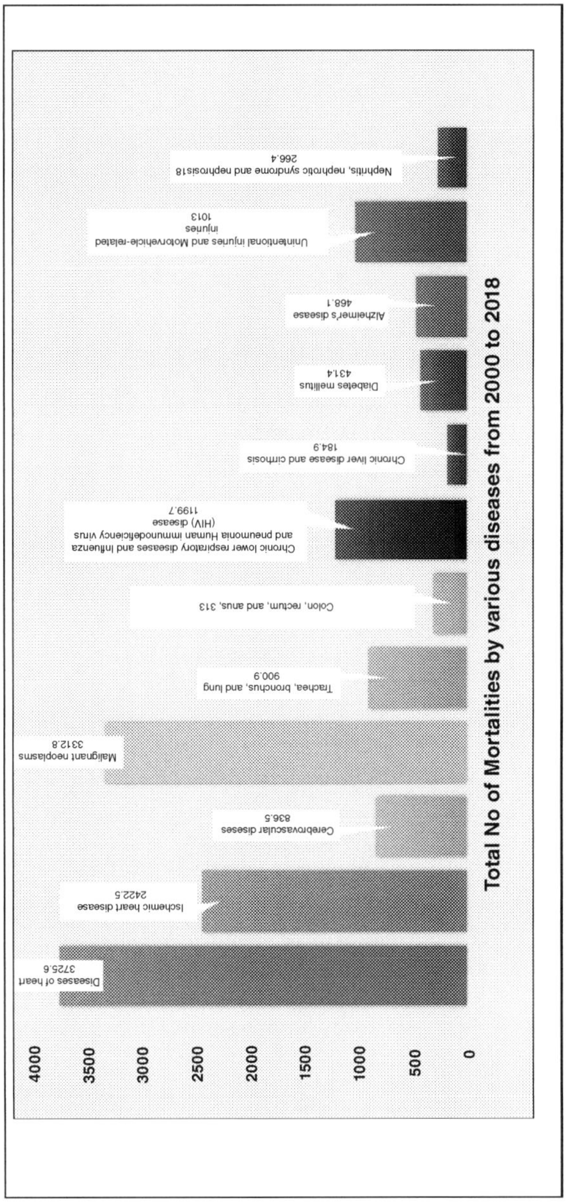

FIGURE 1.1, cont'd.

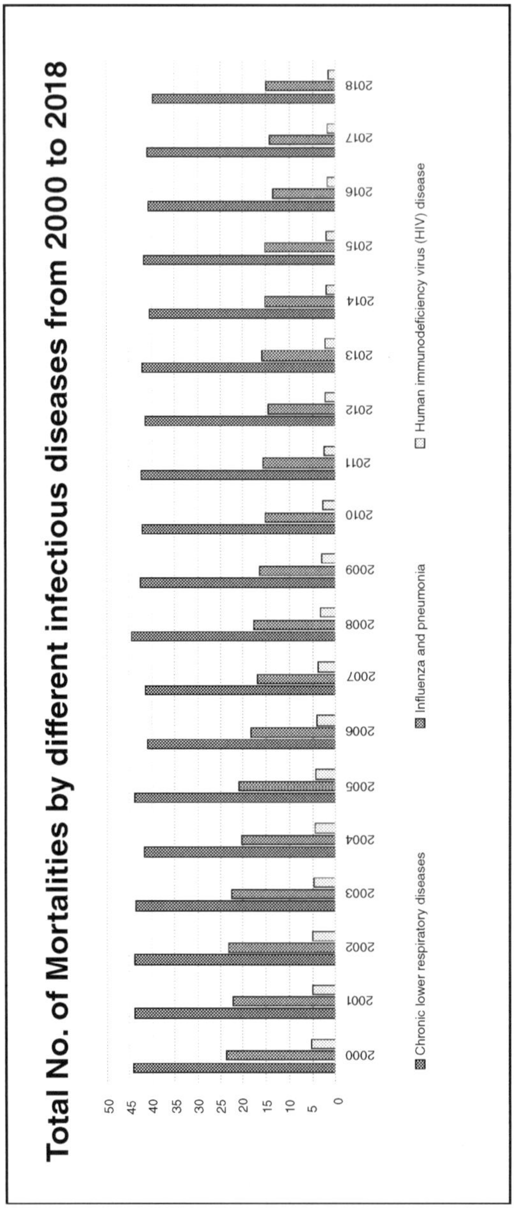

FIGURE 1.1, cont'd.

of these live in or on human bodies. Different types of human pathogens and their diseases are presented in Fig. 1.2.

Microbes or microorganisms are an essential component of the external and internal environment of the human species. Normally, these organisms are found to be harmless or in some cases even helpful. In normal circumstances, when the human immune system is fully functional, these organisms are not able to induce infection or say the disease symptoms may not develop. However, if the person's immune system is compromised due to any underlying health issues, an infection may be caused, and symptoms become visible. Diseases that affect men generally consist of community-acquired pneumonia, sinusitis, otitis media, influenza, hepatitis, and sexually transmitted diseases.

People in developing countries, particularly infants and children have borne the maximum burden of infectious diseases, mainly malaria, and diarrheal diseases. In developed countries, this is shared by indigenous and disadvantaged minorities [2].

1.1.1 Origin/sources of human diseases

An initial conceptualization has shown that the old world (Europe, Asia, and Africa) has been the center of the early emergence of infectious diseases, mainly from domestic animals. Infectious diseases also had major effects on the course of history, like, European Conquest of Native Americans, the failure of Napoleon's invasion of Russia, etc. [3,4], and the failure of French attempt to complete the construction of a Panama Canal [5]. Depending on the highest morbidity and mortality, and hence, of the highest historical and evolutionary significance. The disease's origin can be classified as depicted below:

1. ***Temperate/Tropical origin:*** Higher % of diseases are transferred by insect vectors in the tropics (8/10) than in the temperate regions (2/15) leading to long-lasting immunity, especially in the case of temperate. Temperate diseases are mostly acute as compared to slow, chronic, or latent diseases prevalent in tropical regions.
2. ***Pathogenic origin:*** Present data suggest that out of 15 diseases, 8 diseases in temperate regions have originated from domestic animals and later reached humans (diphtheria, influenza A, measles, mumps, pertussis, rotavirus, smallpox, and tuberculosis). Three originated from apes (hepatitis B) or rodents (plague, typhus) and the rest four (rubella, syphilis, tetanus, and typhoid) have been originated from unknown sources.

In tropical regions, diseases originating from domestic animals fell to 3 out of 10 while maximum tropic diseases were traced to wild nonhuman primate origins. The animal-derived human pathogens mostly arose from warm-blooded species, primarily mammals.

3. ***Geographic origin:*** The 25 major human infections have come from the old world as compared to 1, which originated in the new world.

1.1.2 Stages of endemic human diseases

In transformation of animal pathogens into a specialized pathogen of humans, sometimes four and sometimes five stages are involved which leads to human diseases. The five different stages of infective human diseases like staphylococcus foodborne illness, influenza, cholera, genital herpes, tetanus, syphilis, hepatitis B, AIDS, and leprosy are presented in Fig. 1.3 [3].

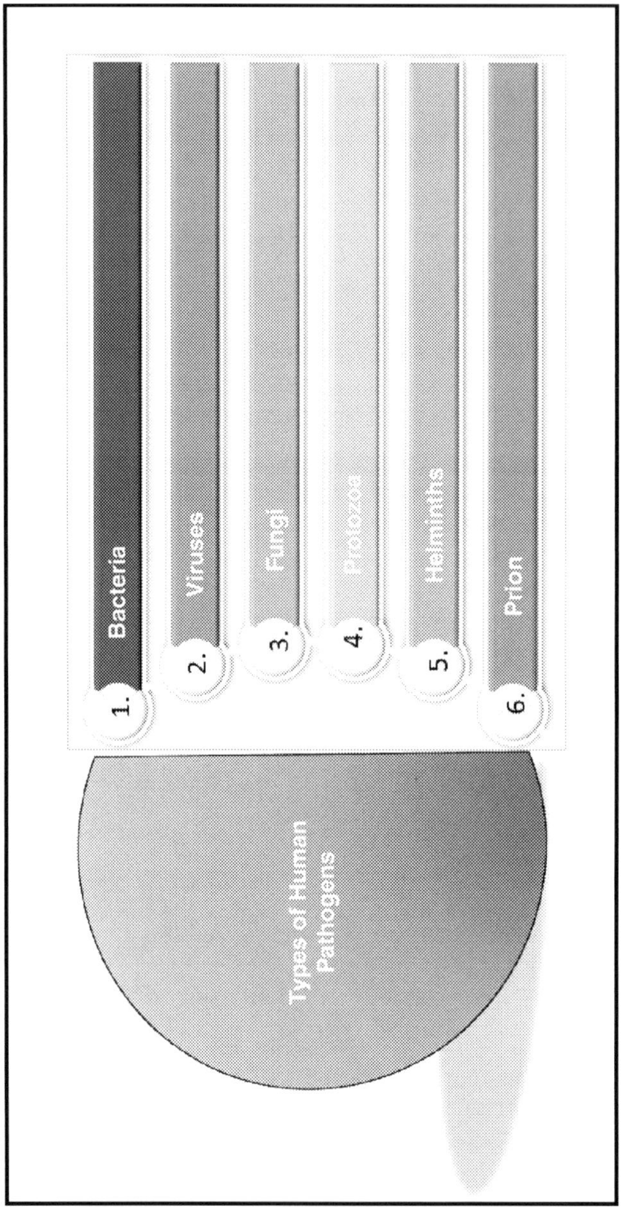

FIGURE 1.2 Different types of human pathogens.

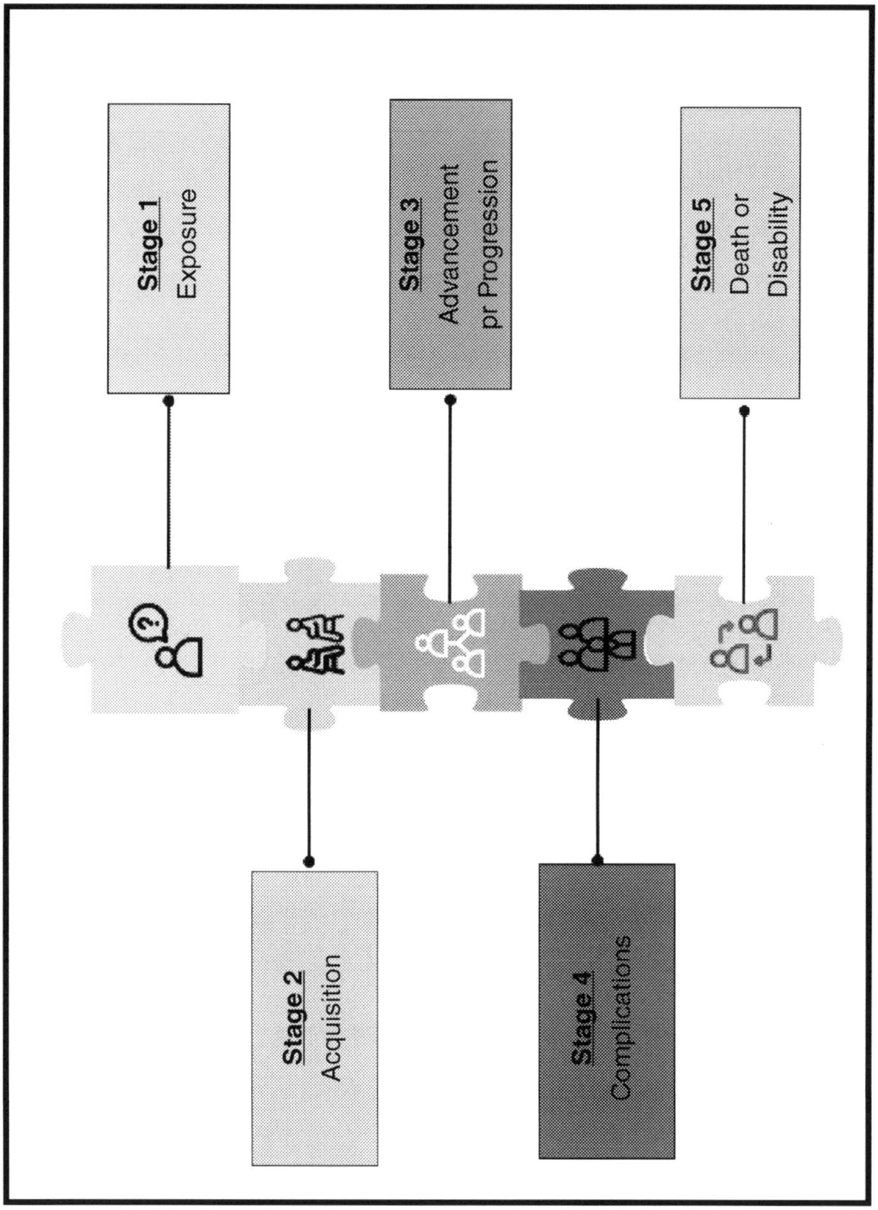

FIGURE 1.3 Different stages of endemic human diseases.

TABLE 1.1 Factors causing for emergence of diseases.

Sr. no.	Contributing factors	Infection caused
1.	Changes in agriculture	Argentine, Bolivian haemorrhagic fever
2.	Changes in rendering processes	Bovine spongiform encephalopathy (cattle)
3.	Transportation, travel, and migration; haemorrhagic fever urbanization	Dengue
4.	Transfusions, organ transplants, contaminated hypodermic apparatus, sexual transmission	Hepatitis B, C
5.	Migration to cities and travel; after introduction, sexual transmission, vertical spread from infected mother to child	HIV
6.	Conditions favoring mosquito vector	Malaria
7.	Unknown	Ebola, Marburg

These are as follows:

Stage 1 Exposure: Under natural conditions, a microbe present in animals has not been detected in humans. Example: malarial plasmodia, feline distemper.
Stage 2 Acquisition: Under natural conditions, a pathogen of animals has been transmitted from animals to humans (primary infection) but has not been transmitted between humans (secondary infection). Example: Anthrax and Nipah.
Stage 3 Advancement or Progression: Few cycles of secondary transmission of animal pathogens were observed between humans, leading to occasional human outbreaks by primary infection, which also dies out soon. Example: Ebola and monkeypox.
Stage 4 Complications: A natural (sylvatic) cycle of infections in humans by primary transmission from the animal host, with long sequences of secondary transmission between humans without the involvement of animal hosts. Example: Yellow fever and Dengue.
Stage 5 Death or Disability: Exclusive human pathogen. Example: Malaria, Syphilis, and Smallpox.

1.1.3 Favorable conditions responsible for infectious diseases

A lot of literature is available on favorable conditions required for a Stage 5 epidemic to continue. Briefly, the factors responsible for the emergence of diseases are discussed in Table 1.1 [4].

Wherever the disease originated, its emergence reaches a new height when it reaches a new population. Also, most emerging infections originate in one geographical area and then spreads to a new location.

1.1.4 Re-emergence of infectious diseases

Emerging infections (EIs) are the ones that have newly appeared in a population or have existed previously but are rapidly observed in the form of new infections or are spreading to new geographical location. The major cause for new EIs is a genetic mutation of microbes, genetic recombination of viruses, changes in the population of reservoir hosts or intermediate insect vectors, and transfer of microbes from animals to human [2]. Re-emergence is generally caused by microbial evolutionary vigor, zoonotic encounters, and environmental encroachment or may also be climate-related. Malaria is among the most important re-emerging disease worldwide, owing to mosquito resistance to chloroquine and mefloquine. TB is the second, cited as the most deadly re-emerging disease. The re-emergence of TB has been caused by immune deficiencies of people with AIDS. Opportunistic re-EIs in the form of immune deficiency disorders, associated with cancer therapy and immunosuppressive therapies have led to the global increase in the numbers of immunosuppressed people and hence, paving way for many opportunistic infections.

1.1.5 Antimicrobial resistance

Resistance to drugs by microbes is steadily leading to microbial and viral re-emergence on a wider scale has become a globally recognized problem. This is fueled by widespread and inappropriate use of antibiotics giving rise to resistant strains by mutation. Few examples include resistance to Sulpha drugs (the 1940s), Penicillin (1950s), Methicillin (1980s), and to Vancomycin in 2002. In addition, viral resistance is creating problems in the management of HIV-infected people. Therefore, the development of some effective novel antibacterial drugs has become the need of time. This goal can be successfully achieved with the joint venture of academia and industry, otherwise, it is going to be almost impossible to fight against these sudden epidemics for mankind [5].

1.2 Anti-infective agents

The top eight infectious diseases include AIDS, malaria, tuberculosis, and respiratory infections. The less developed countries are feeling the burden more due to the unavailability of drugs and the emergence of antimicrobial resistance (ABR). These infections can be broadly divided into six categories. Anti-infective agents are the ones that are used to prevent or treat infections which are caused by human pathogens. Anti-infective agents include antibacterial, antiviral, antifungal, antiprotozoal, antihelminths, and antiprion medications, as shown in Fig. 1.4.

(i) ***Bacterial infections:*** Ranges from minor such as skin or gastrointestinal infections, to serious and life-threatening, such as bloodstream infections or pneumonia.
(ii) ***Viral Infections:*** Ranges from common, self-resolving colds, to serious, long-term infections like hepatitis C.
(iii) ***Fungal infections:*** Ranges from athlete's foot or vaginal thrush to life-threatening, invasive infections in immunocompromised patients, such as those with cancer.
(iv) ***Protozoal infections:*** Caused by an organism living on or in a host and include malaria, toxoplasmosis, and intestinal worms.
(v) ***Helminth's infections:*** Caused by any parasite like tapeworms, flukes, and roundworm, and a part of the body get infected. Helminthiasis, also known as worm infection, is a very common example.

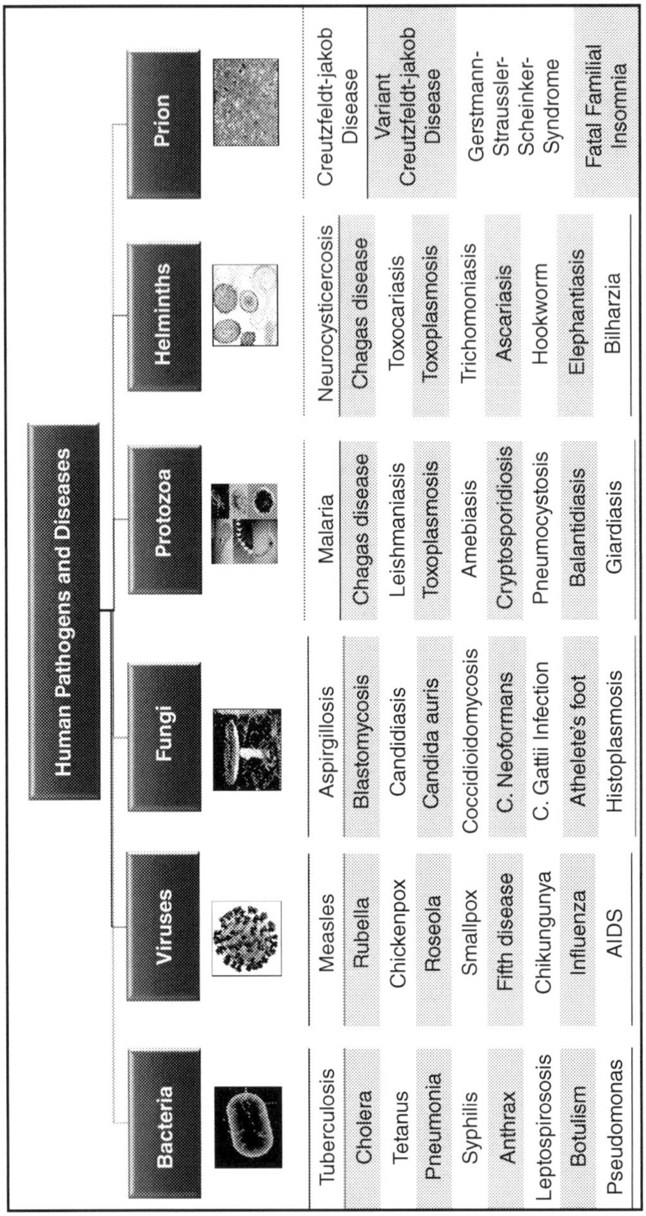

FIGURE 1.4 Different types of human pathogens and their diseases.

(vi) ***Prion Infections:*** The most common form of prion disease that affects humans is Creutzfeldt–Jakob disease. A prion is a type of protein that can trigger normal proteins in the brain to fold abnormally. Prion diseases can affect both humans and animals and are sometimes spread to humans by infected meat products.

Due to the evolution of multidrug resistance, research on new anti-infective agents has become an ongoing process and needs to be continued to explore all possible strategies. Of all the anti-infective drugs, antibacterial drugs have transformed the human ability to control and treat disease and in turn leading to a dramatic decrease in mortality and morbidity [5].

Antimicrobial agents can be classified into five different types based on their mode of action. These are cell wall synthesis inhibitors, protein synthesis inhibitors, essential metabolites inhibitors, nucleic acid replication and transcription inhibitors, and cell membrane synthesis inhibitors, as shown in Fig. 1.5 [6,7].

Among various types of available antibiotics, β-lactam antimicrobials are widely prescribed to treat serious infections even after 60 years of existence due to their excellent efficacy, safety, and tolerability profiles and they proved effective on Gram-positive and Gram-negative bacterial infections [8,9]. β-lactams inhibit bacterial cell wall synthesis and in some organism's trigger lysis of the cell wall. The cell wall inhibitors are further divided into subgroups as explained in Fig. 1.6.

Despite the important role played by antibiotics in exerting strong selective pressure on microorganisms, the incidence of multidrug resistance is continuously on the rise with a threat to return to the "preantibiotic" era. The resistance is affecting all antibiotic classes, particularly worrisome in the case of Gram-negative bacteria, for which treatment options are already limited.

Additionally, the availability of new antimicrobials for human consumption across the globe is lowering day by day due to "broken" economics of antibacterial research and development (R&D). No new class of antimicrobials has been developed after 2000 and the ones which entered the market before this time were modifications of existing molecules [10]. The development of new anti-infective agents has been affected due to the lack of success and low financial returns in bringing new antibacterial drugs to the market.

During the last decade, the focus has shifted from Methicillin-resistant Staphylococcus aureus to clostridium difficile and most recently to Gram-negative bacteria causing a further blow to antibacterial R&D. Changes in regulatory rules have also created uncertainty and additional financial risks for pharmaceutical industries.

To facilitate discovery, the scientific problems faced by pharmaceutical companies need to be addressed so that newer antimicrobials with novel modes of action can be delivered to treat the most urgent clinical challenges. However, in this gloom, natural product research still holds promise for providing new molecules which can act against multidrug-resistant strains.

1.3 Pyridine & its derivatives

For many centuries, humans were completely helpless against these epidemics until the establishment of germ theory and identification of specific microbes as causative agents of a wide variety of infectious diseases [2, 11,12]. One of the most significant medical achievements of the last century has been attributed to the discovery of antibiotics. Antibiotics are drugs

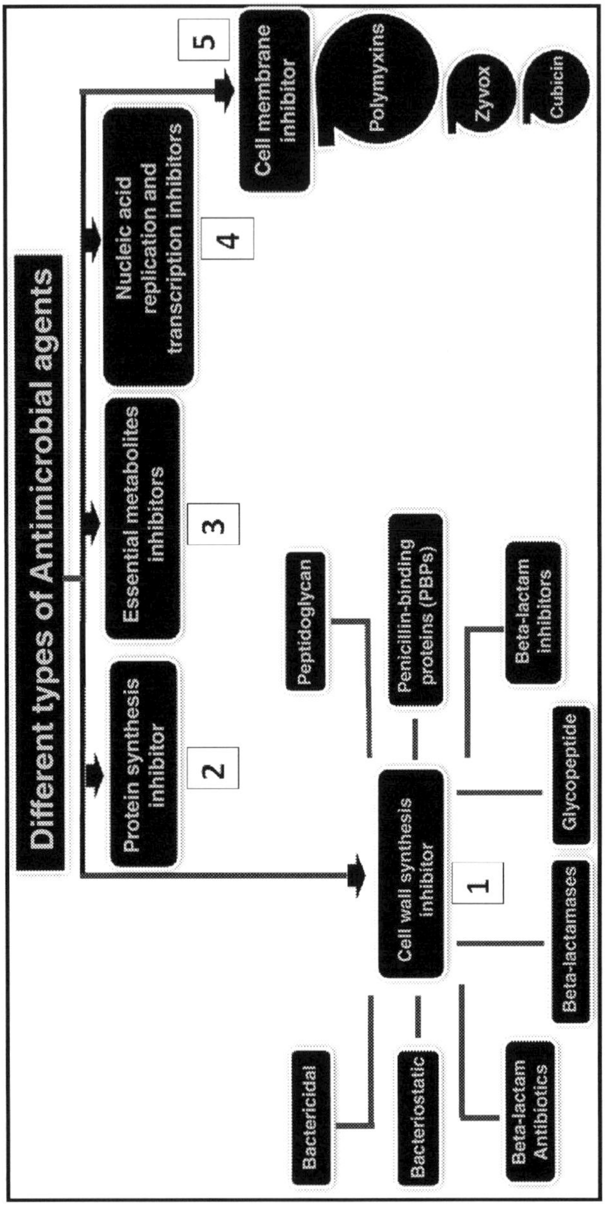

FIGURE 1.5 Different types of antimicrobial agents.

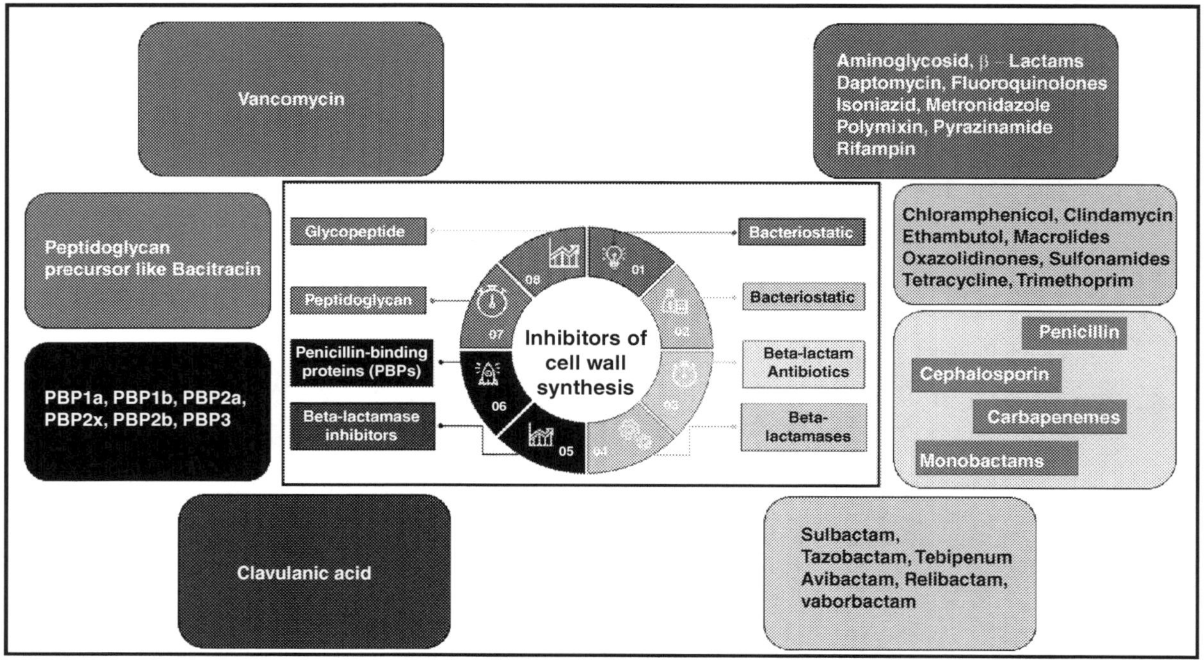

FIGURE 1.6 Classification of cell wall active antimicrobial agents.

that are used to treat bacterial infections. The first true antibiotic, Penicillin, was discovered in the year 1928 by Alexander Fleming and since then, millions of lives have been saved by a wide variety of antibiotics including their assistance in medical procedures like surgery and cancer chemotherapy [13]. The role of antibiotics in tackling a range of infections can be easily understood from Fig. 1.7.

Antibiotics, also known as antibacterial, have been proved as one most effective tools to control and treat bacterial infections. However, with the ready availability of antibiotics, the misuse and overuse of antibiotics also started leading to the emergence and spread of ABR in all parts of the world. This included hospital- and community-acquired infections, and thus jeopardizing the effectiveness of these potentially life-saving drugs. Recently, the spread of multidrug-resistant bacteria, and infections with no therapeutic options have been reported threatening the ability to treat bacterial infections in humans and animals. These reports are like a threat that we are progressing toward the return to the "pre-antibiotic" era [10]. According to the WHO's new global antimicrobial surveillance system, ABR currently exists among 500,000 people in 22 countries and kills thousands of individuals every year. The spread of ABR in the human population is rightly depicted in Fig. 1.8.

The ABR not only comprises a resistance to bacteria but also viruses, fungi, and parasites. This has created a need for new treatments, newer unexploited targets, and strategies for the next generation of antimicrobial drugs for combating the drug resistance and emerging pathogens of the 21st century (Fig. 1.9).

Among various available heterocyclic compounds, the furan and pyridine and their derivatives have gained maximum attention due to their structural analogues of quinolines and isoquinolines which possess amazing biological activities. Pyridine is a basic heterocyclic compound with the chemical formula C_5H_5N having the structure, as given in Scheme 1.1 [14].

Pyridine and pyridine-based compounds play diverse roles in organic chemistry. They are used as ligands, solvents, and catalysts for facilitating reactions, and as well as pyridine-based materials are valued for their optical and physical properties. Additionally, the high therapeutic properties of the pyridine related drugs have encouraged the medicinal chemists to synthesize a larger number of novel chemotherapeutic agents.

The first pyridine base Picoline, was isolated in 1846 by Anderson. Pyridine became an interesting target in 1930 with the importance of Niacin for the treatment of dermatitis and dementia. Niacin is pyridine derivative, as represented in Scheme 1.2 [15].

Six membered aromatic pyridine and its derivatives abundantly exist in nature, including vitamins such as niacin and vitamin B_6, coenzymes such as nicotinamide adenine dinucleotide, and alkaloids such as trigonelline. Pyridine moieties play critical role in medicinal chemistry as antimicrobial agents, antiviral agents, antioxidants, antidiabetic agents, antimalarial agents, anti-inflammatory agents, psychopharmacological antagonists, and antiamoebic agents [16–18].

Pyridine scaffolds have been utilized in various fields for the treatment of various ailments. The different pyridine scaffolds and their area of application are presented in Scheme 1.3 [19].

In the last few years, pyridine moieties clubbed with other heterocyclic ring systems such as pyrazole and pyrazoline have been reported as potential antimicrobial agents (Scheme 1.4) [20].

Synthesis and in vitro antimycobacterial and antifungal activities of various alkylthio pyridine derivatives bearing an alkylthio group in positions 2 or 4 as given by the general formula in Scheme 1.5 are continuously being explored by Klimesova et al. [21].

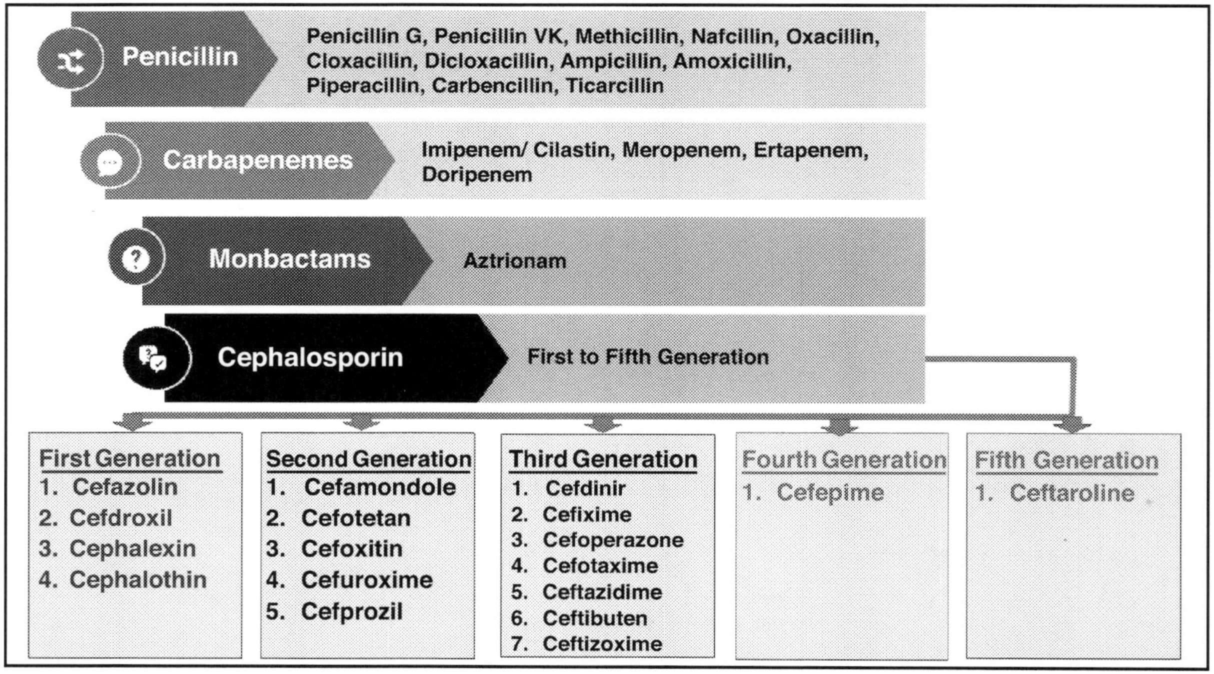

FIGURE 1.7 Classificationof beta lactam inhibitors antibiotics.

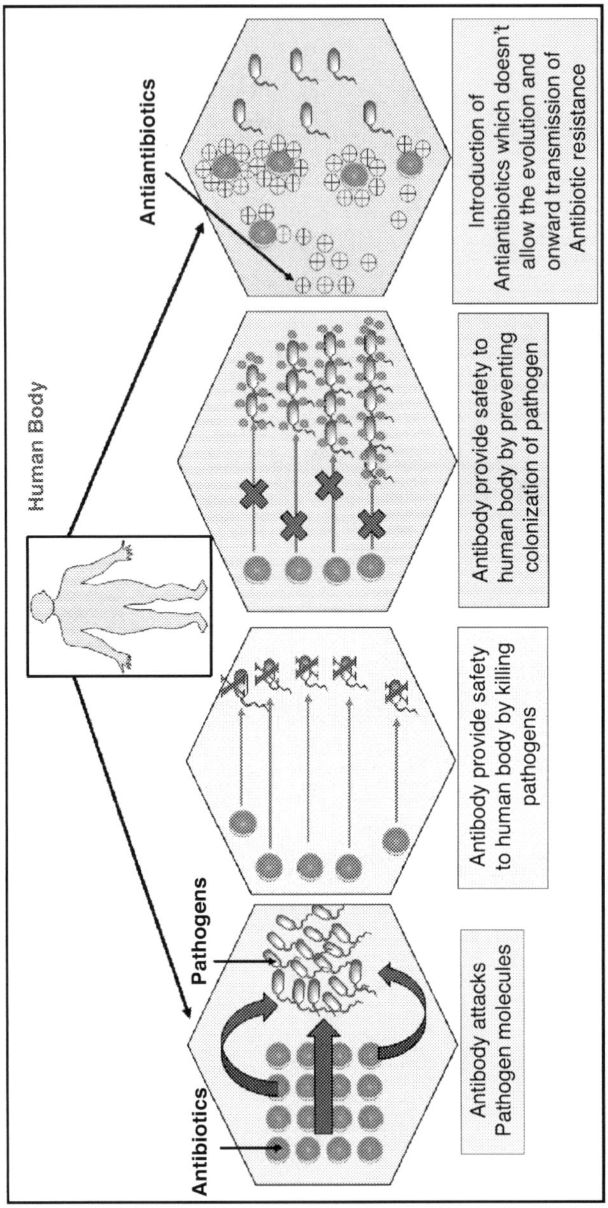

FIGURE 1.8 Mechanism of action of antibiotics in human body.

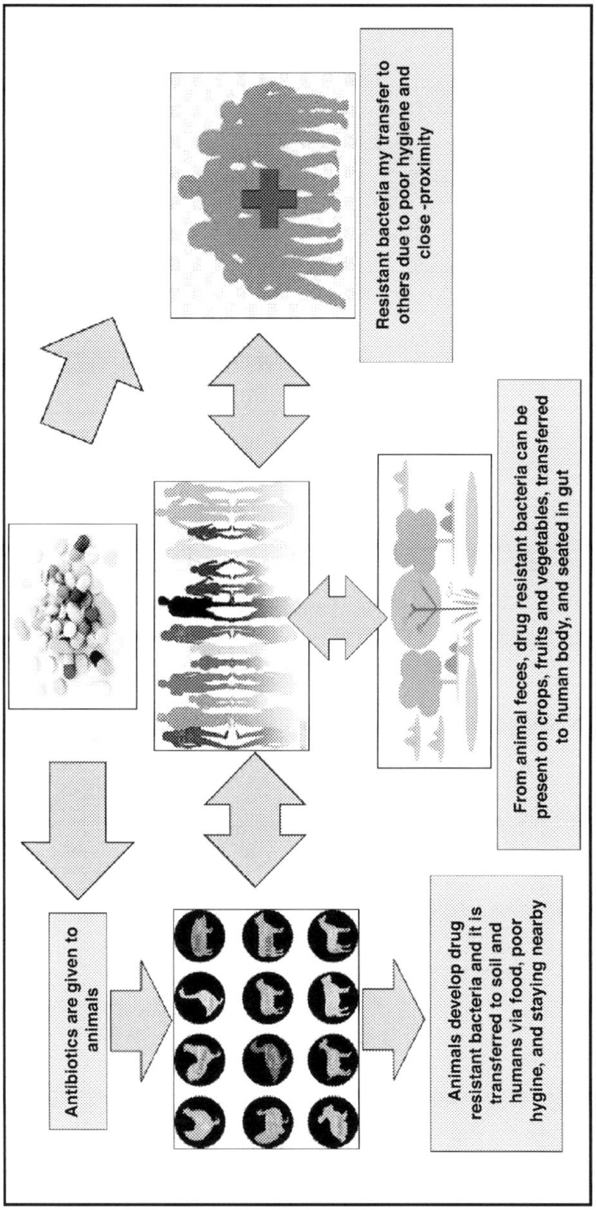

FIGURE 1.9 Process of development of antibiotic resistance.

Pyridine

SCHEME 1.1 Structure of Pyridine.

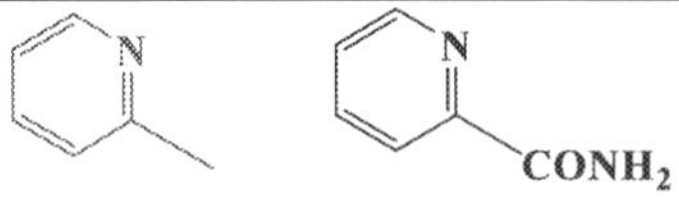

Picoline Niacin

SCHEME 1.2 Pyridine derivatives: Picoline and Niacin.

Name and Medicinal application	Name and Medicinal application
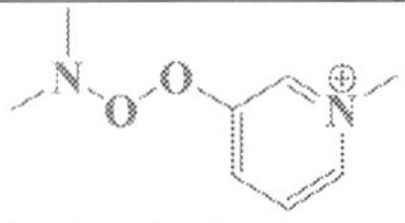 **Pyridostigmine** Anticholinesterase agent: used in myasthenia gravis	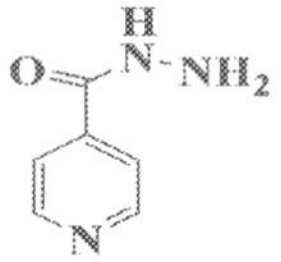 **Isoniazide** Antitubercular agent
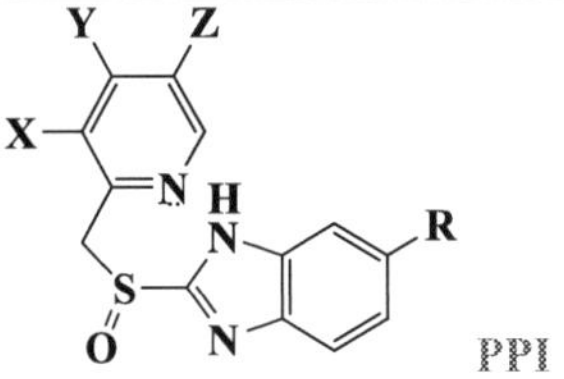 PPI **Omeprazole, Lansoprazole, Pantoprazole, Rabeprazole** Proton pump inhibitor: used in peptic ulcer.	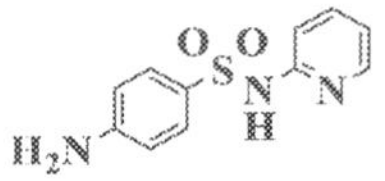 **Sulphapyridine, Sulfasalazine** Antibacterial agent

PPI	X	Y	Z	R
Omeprazole	CH_3-	CH_3O-	CH_3-	CH_3O-
Lansoprazole	CH_3-	CF_3-CH_2-O-	H-	H-
Pantoprazole	CH_3O-	CH_3O-	H-	CHF_2-CH-O-
Rabeprazole	CH_3-	CH_3O-CH_2-CH_2-CH_2-O-	H-	H-

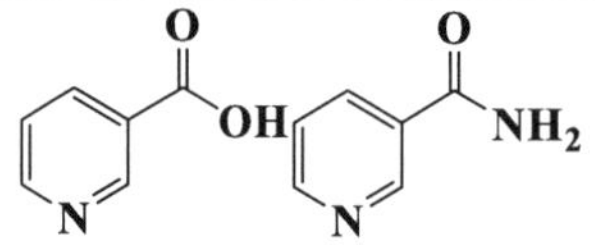

Nicotinic Acid **Nicotineamide**

Vitamin B_3

Niacin is collective name for these compounds. Helps with digestion and digestive system, also helps in processing Carbohydrates.

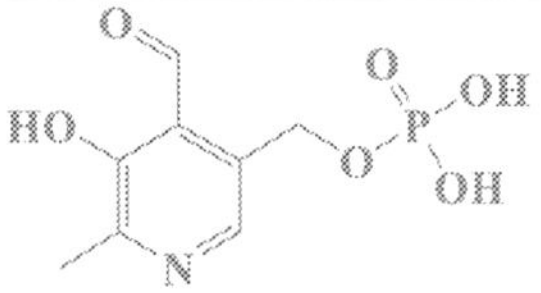

Pyridoxal Phosphate – Vitamin B_6

Active form in mammalian tissues. It helps in making some brain chemicals, needed for normal brain functioning, it also helps in making of RBCs and immune cells.

SCHEME 1.3 Pyridine derivatives and their medical applications.

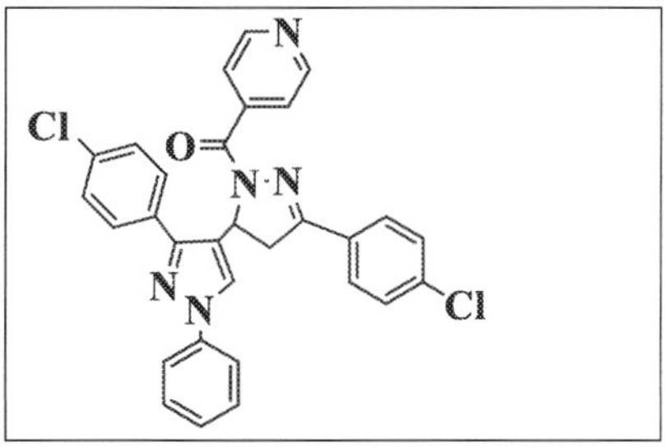

SCHEME 1.4 Pyridine derivatives clubbed with heterocyclic rings which are used as antimicrobial agents.

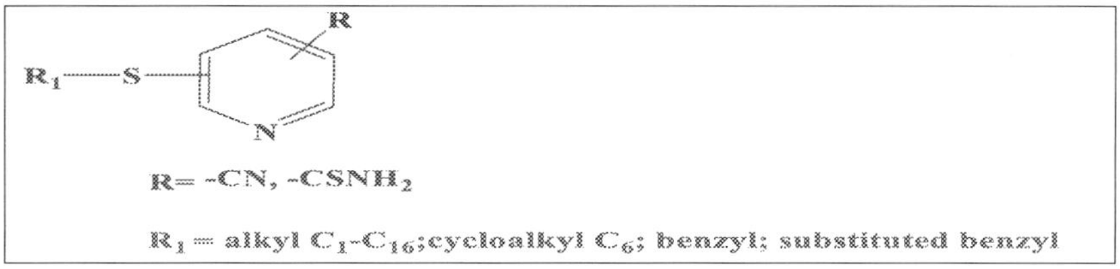

SCHEME 1.5 Alkylthio Pyridine derivatives which are used as antimycobacterial and antifungal.

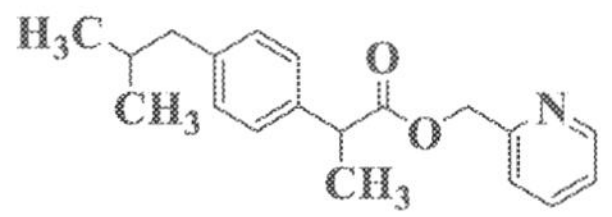 **Primeprofeir** Anti-inflammatory	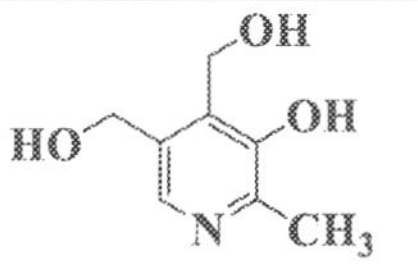 **Pyridoxine -Vitamin B_6** Balance Na and K salts, promotes RBC productions, essential for nervous system, helps immunity
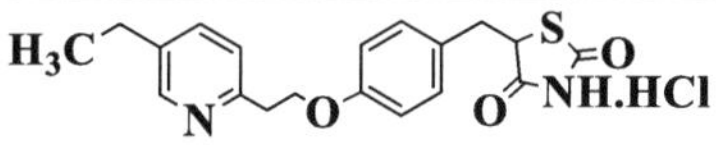 **Pioglitazone** Antidiabetic, Insulin enhancer	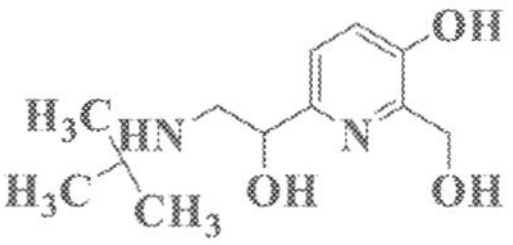 **Pirbuterol** Bronchodilator
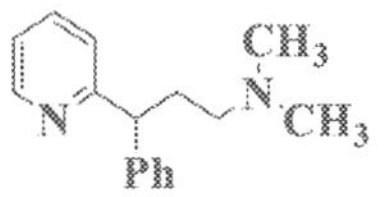 **Pheniramine** Antivitamin	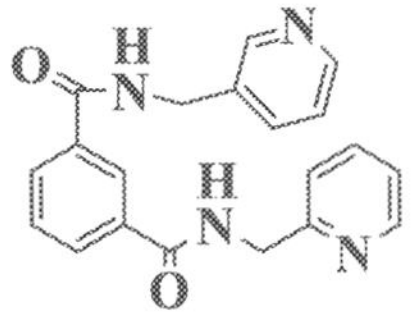 **Picotamide** Anticoagulant, Fibrinolytic

SCHEME 1.6 C-2 and C-6 substituted Pyridines and their medical applications.

A series of C-2 and C-6 substituted pyridines were synthesized and evaluated in vitro against Pseudomonas aeruginosa, Staphylococcus aureus, Streptococcus mutants, and Candida albicans by Almeida et al. are presented in Scheme 1.6 [16].

Pyridine derivatives containing novel hybrid sulphaguanidine moieties in a single molecular framework were evaluated in vitro by Ragab et al. for their antibacterial and antifungal

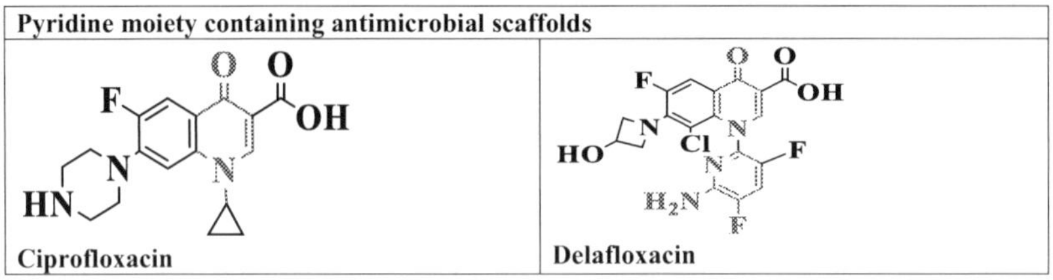

SCHEME 1.7 Pyridine moiety containing antimicrobial scaffolds.

activities, as well as DNA gyrase and dihydrofolate reductase enzymes with immunomodulatory potential are presented in Scheme 1.7 [22].

Series of 6-aryl-2-methylnicotinohydrazides, N′-arylidene-6-(4-bromophenyl)-2-methylnicotino hydrazides, and N′-(un/substituted 2-oxoindolin-3-ylidene)-6-(4-fluorophenyl)-2-methylnicotinohydrazides have been synthesized and evaluated for their potential in vitro antimycobacterial activity against M. tuberculosis. Isatin hydrazides were found to be more active than the parent hydrazides.

Pyridine derivatives like N-(pyridine-2-methyl)-2-(4-chlorophenyl)-3-methylbutanamide were determined to have good antifungal activities (Scheme 1.8) [23].

The Oxime derivatives of thiazolo[5,4-b] pyridine exhibit activity against influenza B Mass virus. The oxime derivatives of pyridine and naphthyridine have high activity against HIV. Some thiopyridine derivatives are antioxidant (SOD) in addition to their cytotoxic (DPPH) activities, these activities are quite attractive, especially compounds having Adamantane ring.

Additionally, pyridine derivatives containing thiazolidinones exhibit antidiabetic activities. Metal complexes of Cu (II) with Schiff base 2-[N-(apicolyl)-amino]-benzophenone on pyridine particularly brominated products have the highest cytotoxicity and show very good antitumor activity. Compounds of pyrazoline, pyridine, and pyrimidine coupled with indole functionality are found to have excellent activity against tumor cells.

Platinum complex with thiosemicarbazone of 2-acetyl pyridine and 4-acetyl pyridine exhibit excellent antiproliferative activity against human cells with an IC_{50} value in μM range best than commercial anti-tumor drugs cis-platin so these new compounds are considered as a potential anti-tumor agent.

The pyridine derivative, pyridine-2-carboxaldehydeisonicotinoylhydrazone (PCIH) makes a considerably stable complex with iron, the resulted compound is a chelator. The ligands derivatives independently are used as medicine in iron overload disease. Ligands equivalent activities as standard drug, desferrioxamine.

1.4 Synthesis of pyridine derivatives

Pyridine and its derivatives are an important class of compounds with varied biological applications in various fields. Pyridine nucleus is present in many naturally occurring molecules such as drugs, vitamins, food, flavorings, plants, dyes, rubber products, adhesives, insecticides, and herbicides. Pyridine nucleus is a principal feature of antimalarial drugs, as shown in Scheme 1.9 [24].

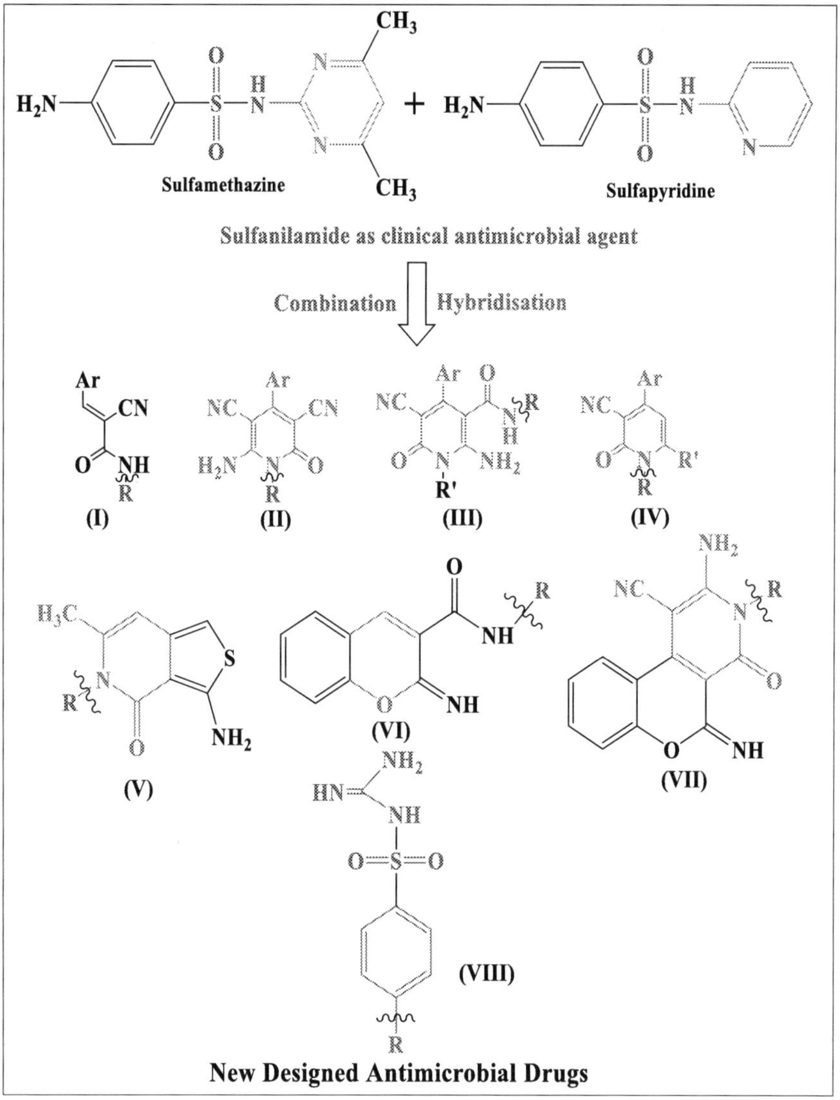

SCHEME 1.8 Pyridine derivatives: N-(pyridine-2-methyl)-2-(4-chlorophenyl)-3-methylbutanamide which shows good antifungal activities.

Imidazo1,2-pyridine scaffold is found in several marketed drug formulations, such as zolimidine (an antiulcer drug), zolpidem (a hypnotic drug), and alpidem (a nonsedative anxiolytic), as presented in Scheme 1.10 [25].

The synthesis of pyridine derivatives has been attempted by many scientific groups owing to the importance of this class of compounds and the incidence of multiple drug resistance in the antimicrobial field.

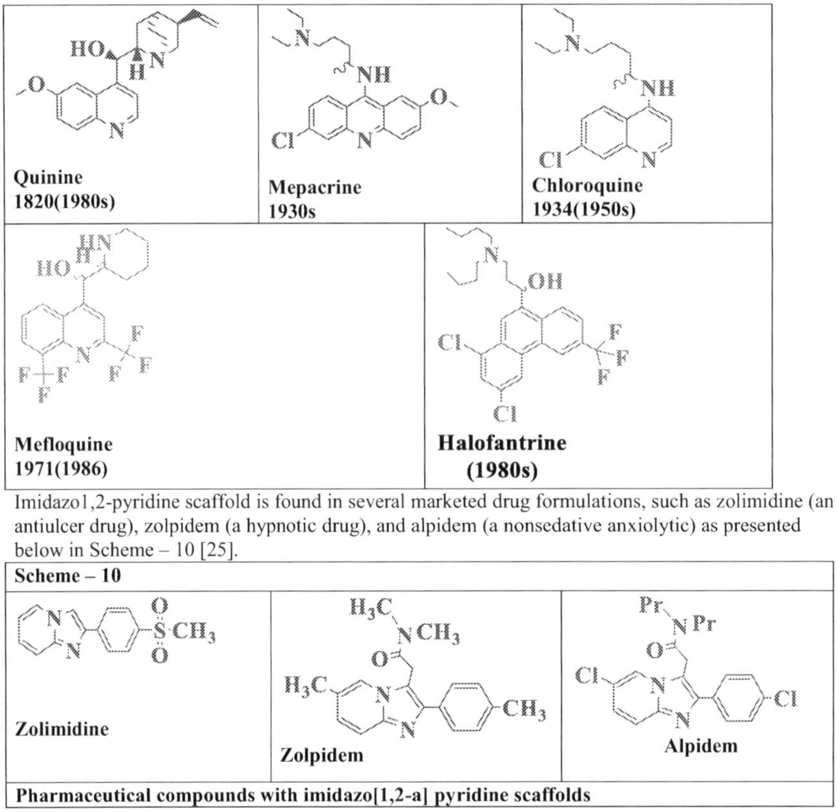

SCHEME 1.9 Pyridine derivatives and their applications in various medical areas.

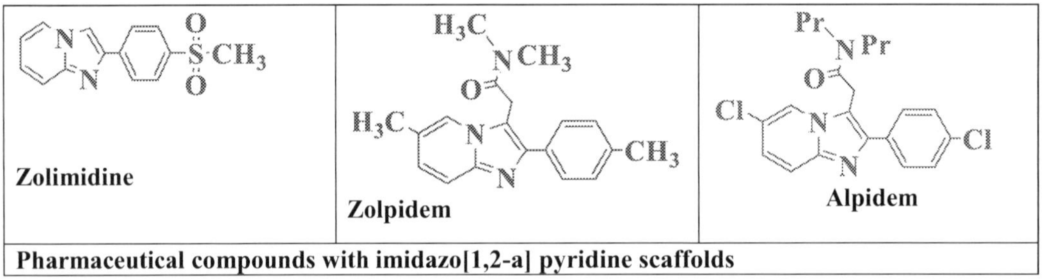

SCHEME 1.10 Imidazo1,2-pyridine scaffold and their medical applications.

One-pot synthesis of pyridine by NAP-MgO was first reported in 2010 with yields varying between 40% and 70%, as presented in Scheme 1.11 [26].

New derivatives were synthesized by reaction of acid chlorides with furfuryl amine, O-benzyl-hydroxylamine, aniline, or 4-bromo-aniline resulted in corresponding derivatives with yields between 48% and 86%, as presented in Scheme 1.12 [16].

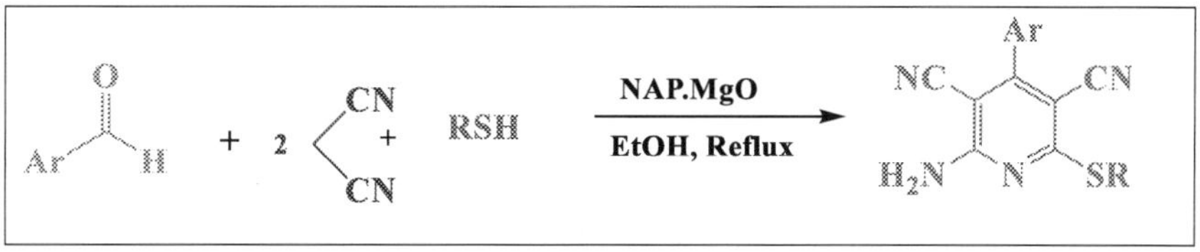

SCHEME 1.11 One-pot synthesis of pyridine by NAP-MgO.

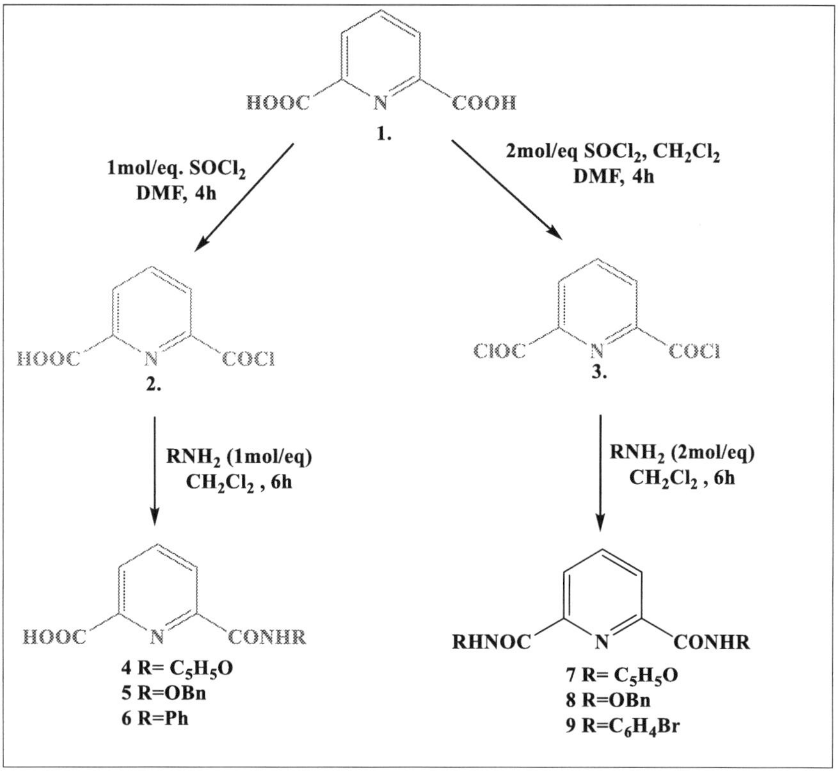

SCHEME 1.12 Synthesis of new derivatives of pyridine.

Chiral 4-(N, N-dimethylamino) pyridine derivatives were prepared through chemoenzymatic synthesis, as presented in Scheme 1.13 [27].

Imidazo[1,2-a] pyridines, a novel class of pharmaceutical compounds exhibiting a broad range of biological activities were prepared by utilizing 2-amino-5-bromo pyridine and 2-bromo-1-(3-nitrophenyl) ethenone, as presented in Scheme 1.14 [28].

Oligomers/polymers were synthesized from 3-aminopyridine and their antistaphylococcal activities were evaluated by Akgul et al., as presented in Scheme 1.15 [29].

Synthesis and evaluation of the antimalarial activity of new pyridine quinoline hybrid molecules against a chloroquine-susceptible strain of Plasmodium falciparum and as inhibitors of β-hematin formation are described by Kaushik et al.

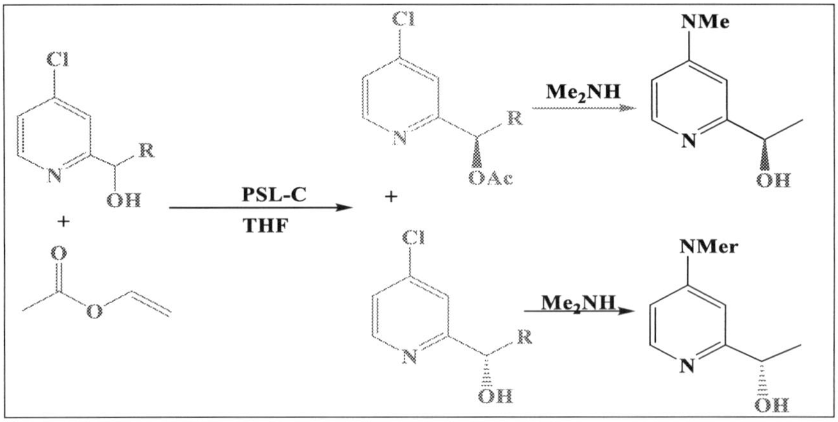

SCHEME 1.13 Synthesis of Chiral 4-(N, N-dimethylamino) pyridine derivatives.

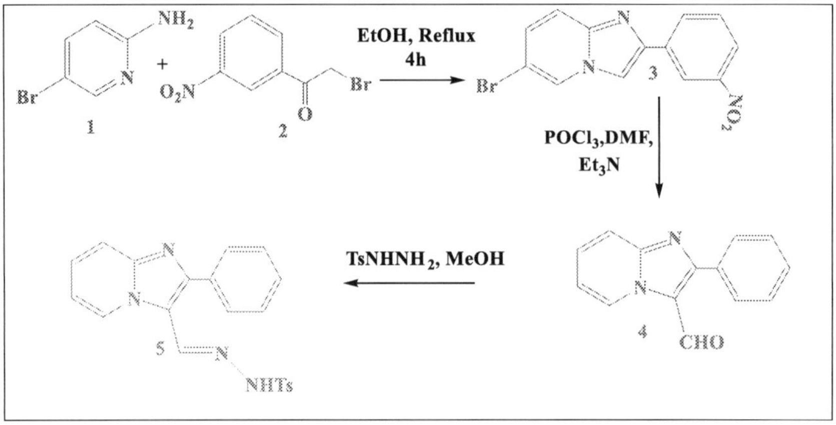

SCHEME 1.14 Synthesis of Imidazo[1,2-a] pyridines by utilizing 2-amino-5-bromo pyridine and 2-bromo-1-(3-nitrophenyl) ethenone.

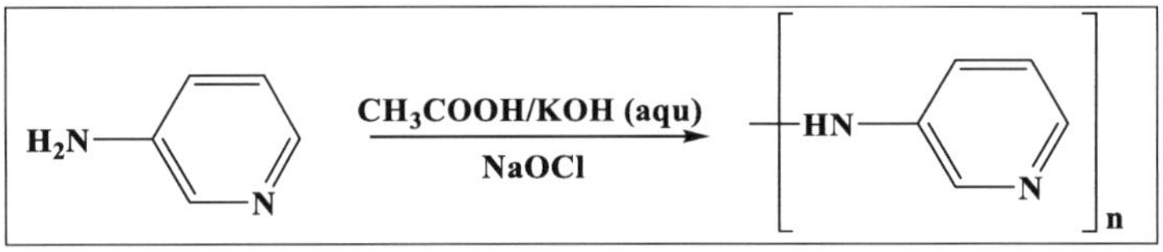

SCHEME 1.15 Synthesis of Oligomers/polymers from 3-aminopyridine.

Synthesis of Pyridine–Quinoline Hybrid is presented in Scheme 1.16 [30].

A set of pyridine derivatives bearing a substituted alkylthio chain or a piperidyl ring in positions 2 or 4 were synthesized and presented in Scheme 1.17 [31] and their antimycobacterial and antifungal activities were evaluated. The compounds were moderately active against both Mycobacterium tuberculosis and nontuberculous mycobacteria.

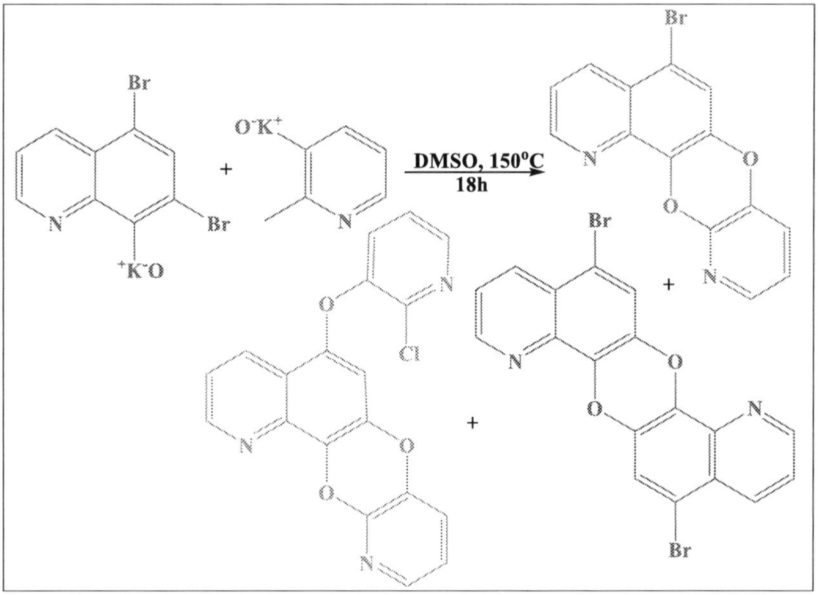

SCHEME 1.16 Synthesis of Pyridine–Quinoline Hybrid.

Recently, novel hybrid sulphaguanidine moieties bearing pyridine-2-one derivatives were prepared and their antimicrobial activity was evaluated against multidrug-resistant bacteria as dual DNA gyrase and DHFR inhibitors. Synthesis of the new pyridine and chromene-3-carboamyl derivatives containing the sulphaguanidine moiety is presented in Scheme 1.18 [23].

A series of nicotinic acid hydrazide derivatives with the aim of obtaining a new antimycobacterial agent were synthesized from ethyl 2-methyl-6-arylnicotinates. Synthesis of nicotinic acid hydrazides 4a-I is represented in Scheme 1.19 [32].

Synthesis of nicotinic acid hydrazides is summarized in detail in Scheme 1.20 [32].

In the recent past, imidazopyridine fragment-like bearing 3,4-difluorophenyl and phenyl groups at the 2-position of the imidazole region were prepared in Scheme 1.21 and their antitrypanosomal activities and cytotoxicity were evaluated [33].

1.5 Bioisosteres and pyridine bioisosterism

The concept of isostere was first formulated by Irving Langmuir in 1919, a little later in 1925 extended with Grimm's Hydride Displacement Law, which stated as "Atoms anywhere up to four places in the periodic system before an inert gas change their properties by uniting with one to four hydrogen atoms, in such a manner that the resulting combinations behave like pseudo atoms, which are similar to elements in the groups one to four places respectively, to their right." Lastly Erlenmeyer broadened Grimm's classification in 1932 by redefining Isosteres as atoms, ions, and molecules in which the peripheral layer of electrons

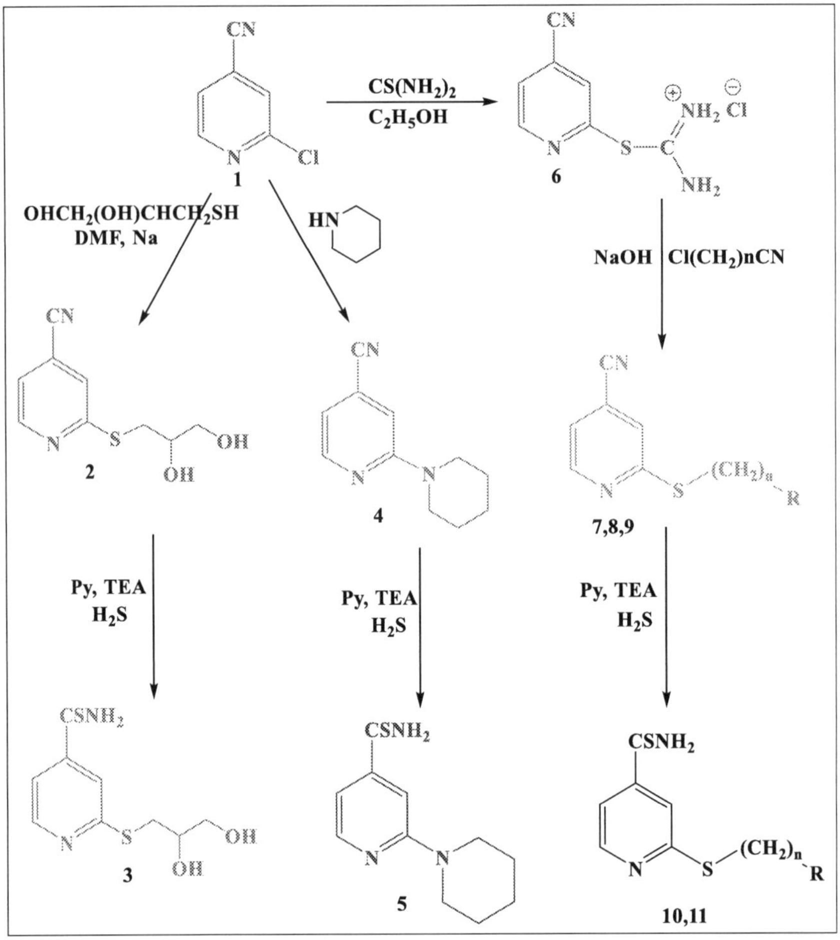

SCHEME 1.17 Synthesis of pyridine derivatives bearing a substituted alkylthio chain or a piperidyl ring in positions 2 or 4.

can be considered identical. Harris Friedman in 1950 introduced the term bioisostere for compounds inducing a similar biological effect. Parameters affected by bioisosteric replacements generally include size, conformation, inductive and mesomeric effects, polarizability, H-bond formation capacity, pKa, solubility, hydrophobicity, reactivity, and stability [34]. Furthermore, bioisosteres share analogous topology, volume, and electronic arrangements [35].

Today, in modern drug design, the concept of bioisosterism is used as a well-established and advanced technique that results greater selectivity, bioactivity, efficacy, potency, membrane permeability, biotransformation pathways, less side effects, decreased toxicity, improved pharmacokinetics (i.e., absorption, distribution, metabolism, and excretion), improved pharmacodynamic (i.e., receptor, enzyme, or channel level behavior) increased stability, simplified synthesis, patented lead compounds [34,35].

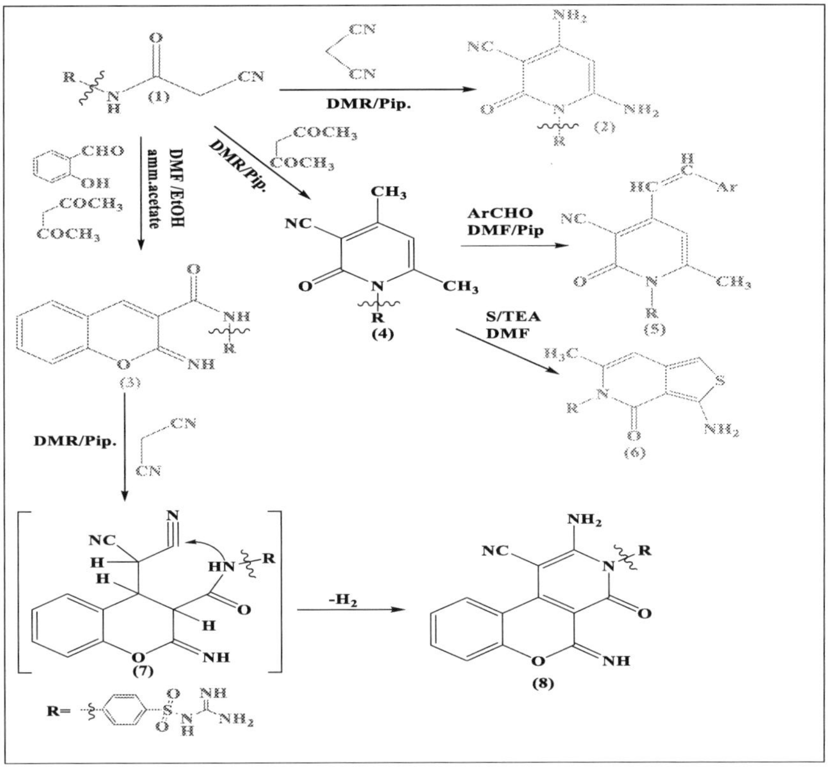

SCHEME 1.18 Synthesis of the new pyridine and chromene-3-carboamyl derivatives containing the sulphaguanidine moiety.

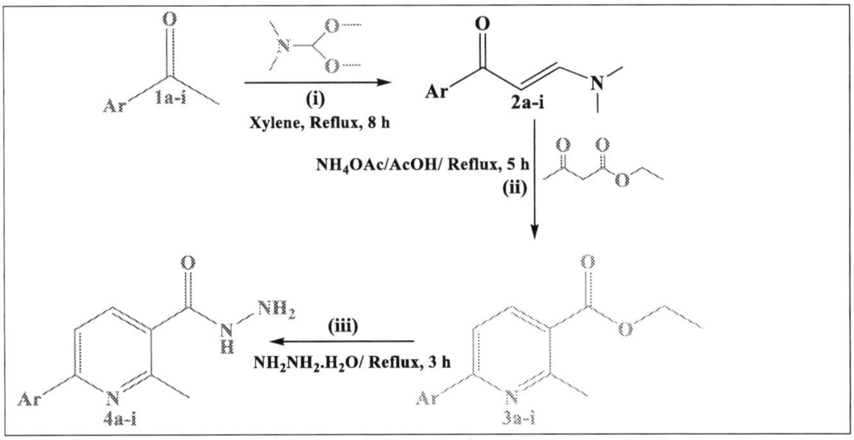

SCHEME 1.19 Synthesis of nicotinic acid hydrazides 4a-I.

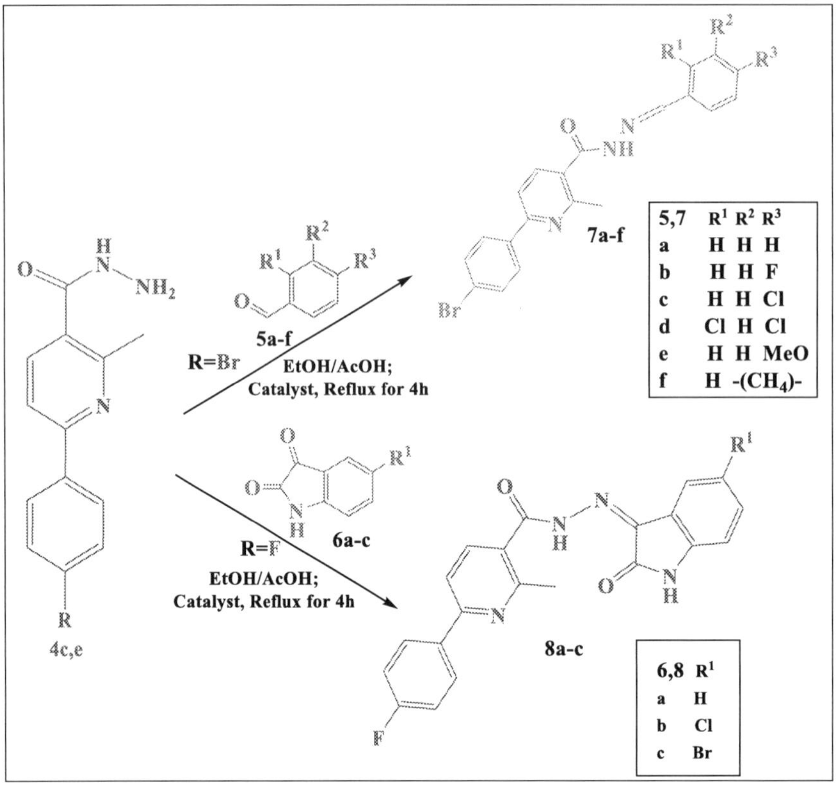

SCHEME 1.20 Synthesis of nicotinic acid hydrazides summary.

The main point of bioisosteric replacement is to alter the form of a biologically active molecule to improve its function.

1.5.1 General classification of bioisosteres

Bioisosteres are classified by Alfred Burger into two major classes that are classical and nonclassical. This classification is based on steric and electronic considerations.

1.5.1.1 Classical bioisostere

This group represents the exchange of structurally simple atoms or groups (Scheme 1.22) [34–36].

1.5.1.2 Nonclassical bioisostere

Are structurally distinct molecules, usually comprise a different number of atoms, and exhibit different steric and electronic properties (Scheme 1.23) [34–36].

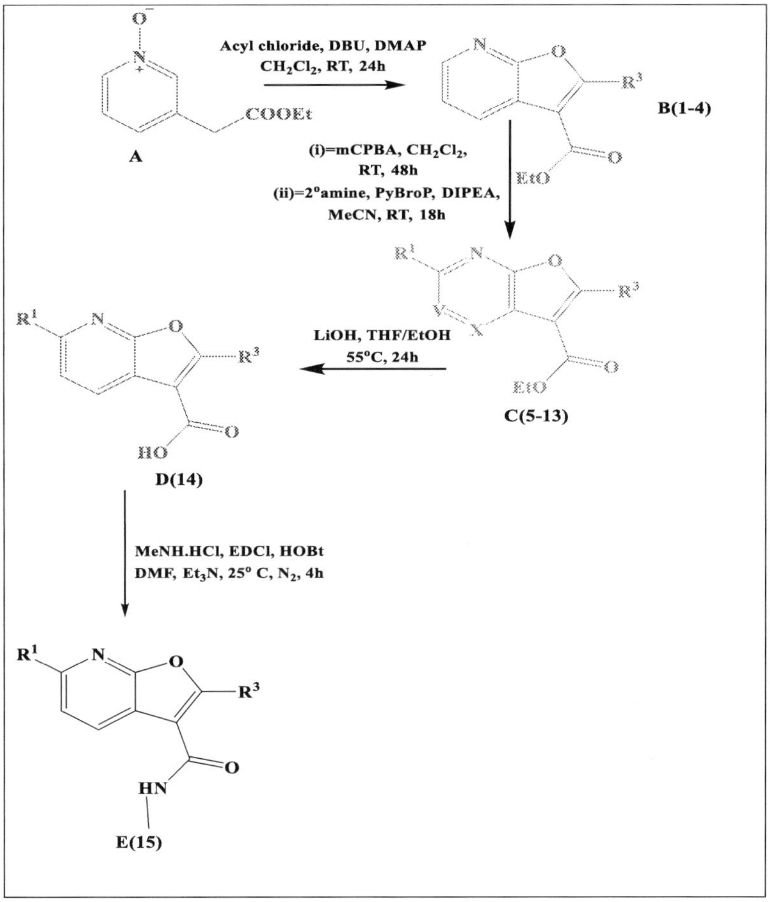

SCHEME 1.21 Synthesis of nicotinic acid hydrazides summary.

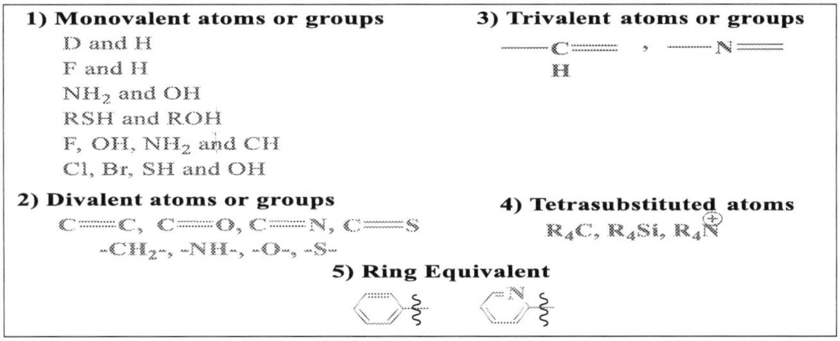

SCHEME 1.22 Classical bioisostere.

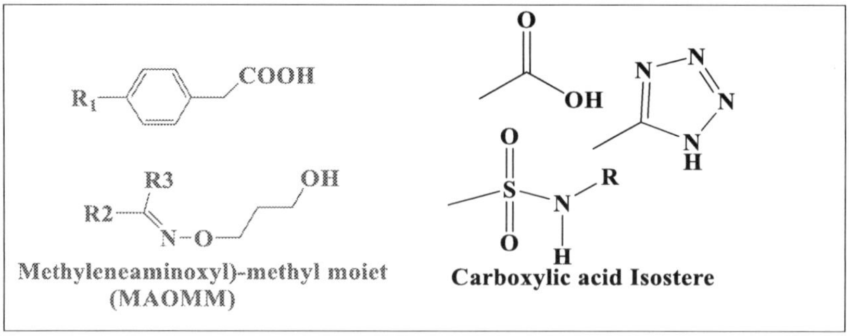

SCHEME 1.23 Nonclassical bioisostere.

1.5.2 Classification of bioisosteres based on functional group

Bioisosteric relationships among different functionalities are presented below (Scheme 1.24) [35].

1.5.3 Few examples of bioisosterism in drug design are given below

1.5.3.1 *Peptide and Dipeptide bioisosteres*

Peptide isosteres have been discussed in the designing of peptide hormones, neurotransmitters, and conformational restrictions of biologically active peptides. Peptide β-turn have been replaced by Ketovinyl and hydroxyethylidene dipeptide isosteres as well as heterocyclic and unnatural amino acid (Scheme 1.25) [37].

1.5.3.2 *Amide carbonyl group bioisosteres*

Replacement of N-acetylcysteamine side chain of thienamycin by 2- and 4-substituted quaternary pyridines or exchanging classical acylamino side chain in cephems by a 4-substituted quaternary pyridine give compounds having broad-spectrum antibacterial properties (Scheme 1.26) [37].

1.5.3.3 *Ketone carbonyl bioisosteres*

Active phosphonate analog of the antiherpetic arildone was prepared due to the isosteric similarity between ketone and phosphonate. Synthesis of diuretics is based on the bioisosteric relationship between benzoyl and the 1,2-benzisoxazoIe moiety related to (4-acy1phenoxy) acetic acids. A similar relationship can be discerned in neuroleptic and in benzisoxazole analogs of haloperidol and benperidol (Scheme 1.27) [37].

1.5.3.4 *Phosphate bioisosteres*

Phosphates play a critical role as structural elements of DNA and RNA. In bioisosteric replacement, the phosphate group provides several critical challenges as simple alkyl phosphates have two ionizations relevant at physiological pH. The simplest bioisosteric replacement is to replace phosphate with phosphonate, in which the ether oxygen is replaced by

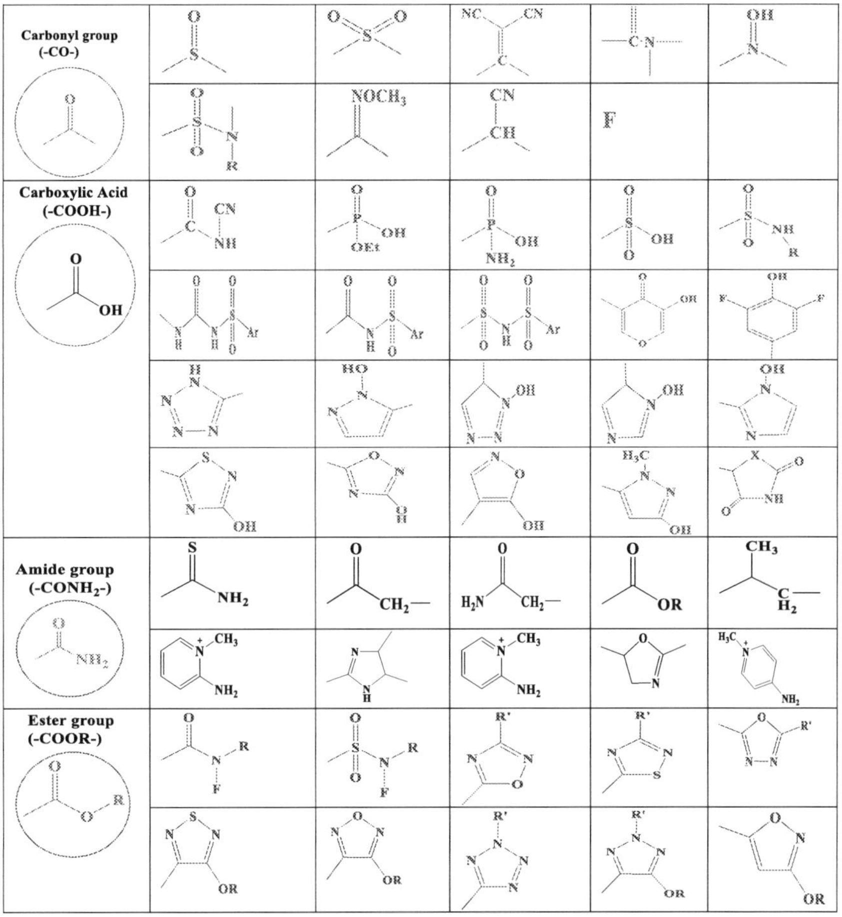

SCHEME 1.24 Classification of bioisosteres based on functional group.

carbon. The replacement of oxygen by carbon has also been explored in a series of analogues of adenosine triphosphate (ATP) by Blackburn et al. (Scheme 1.28) [37,38].

1.5.3.5 Phenol and catechol bioisosteres

The structural diversity, electronic properties, lipophilicity, and size of these functionalities varies widely, providing ample flexibility to overcome pharmacokinetic and toxicological limitations (Scheme 1.29) [37,39].

1.5.3.6 Urea and thiourea isosteres

Urea and thiourea represent well-established privileged structures in medicinal and synthetic chemistry. These structural motifs constitute a common framework of a variety of drugs and bioactive compounds. Cimetidine (a prototypical H-2 blocker/antiulcer) can be obtained

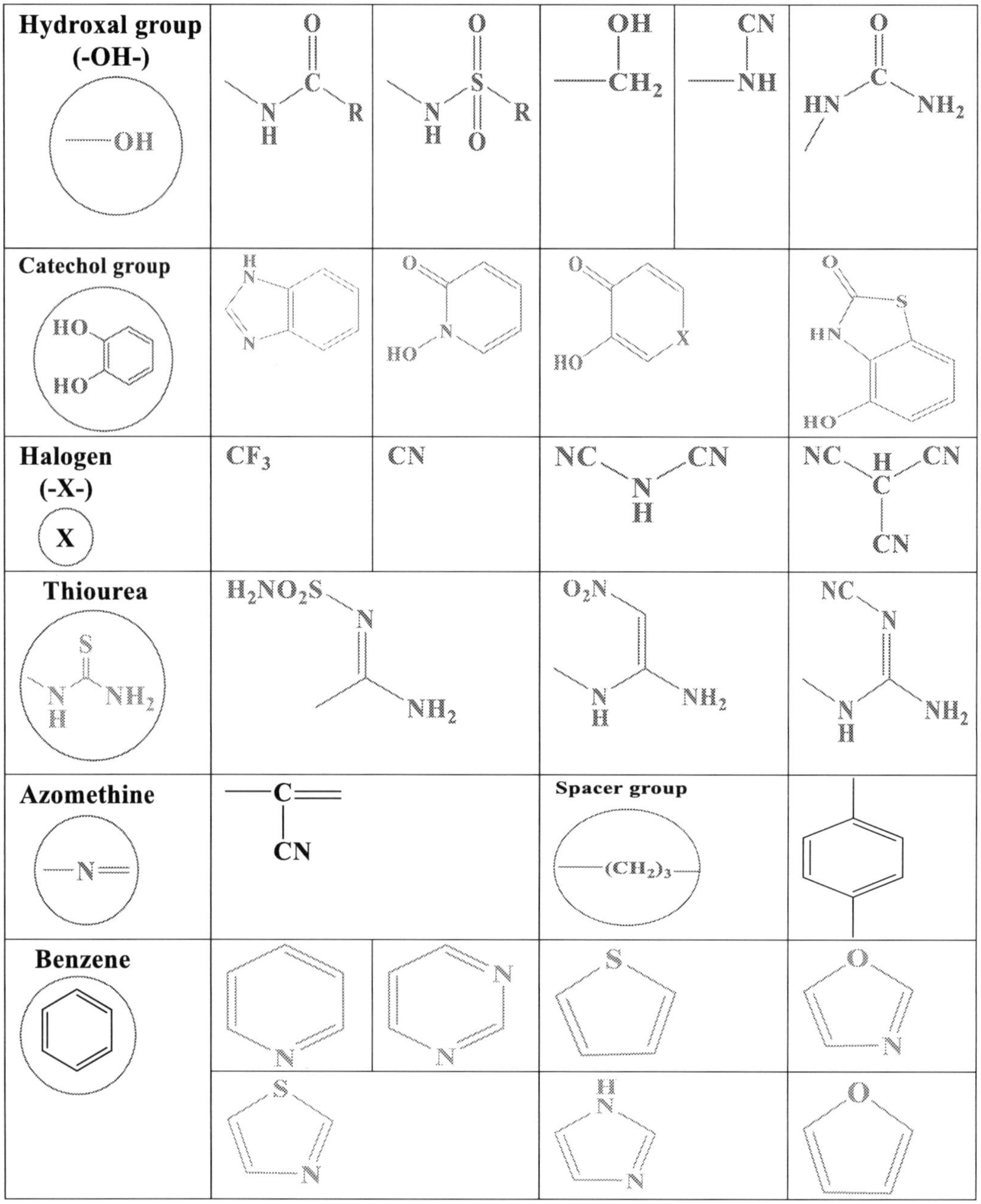

SCHEME 1.24, cont'd.

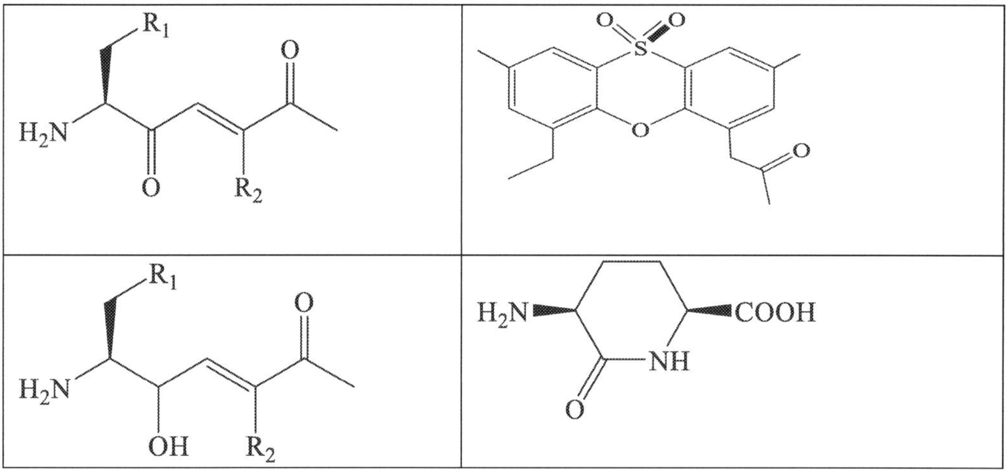

SCHEME 1.25 Peptide and Dipeptide bioisosteres.

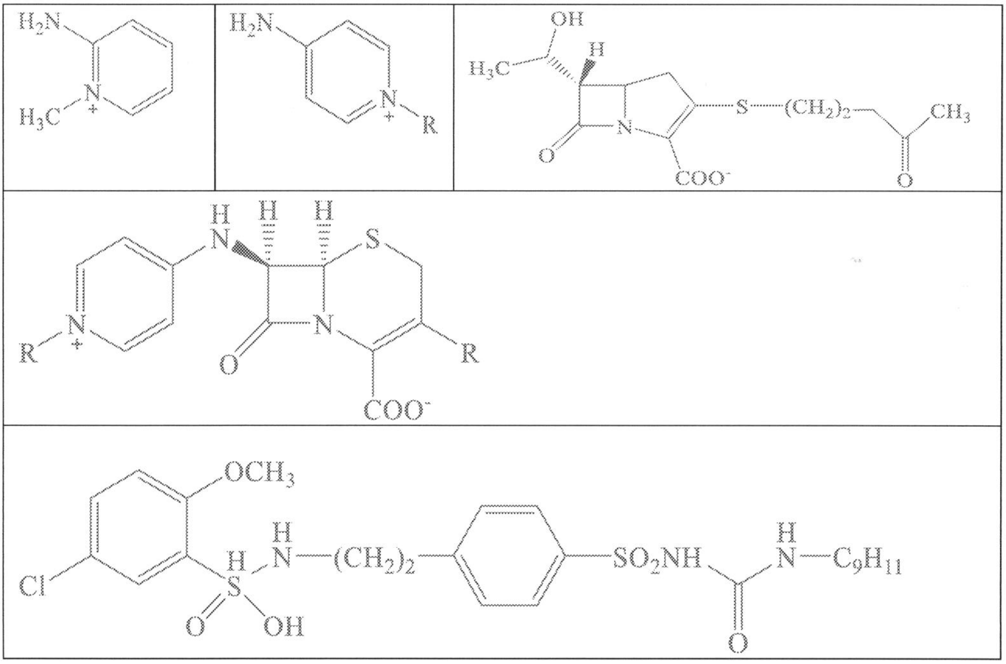

SCHEME 1.26 Amide carbonyl group bioisosteres.

from metiamide (H-2 blocker) by bioisosteric replacement of thiourea group with cyanoguanidine group and thus eliminate granulocytopenic toxicity associated with metiamide (Scheme 1.30) [40].

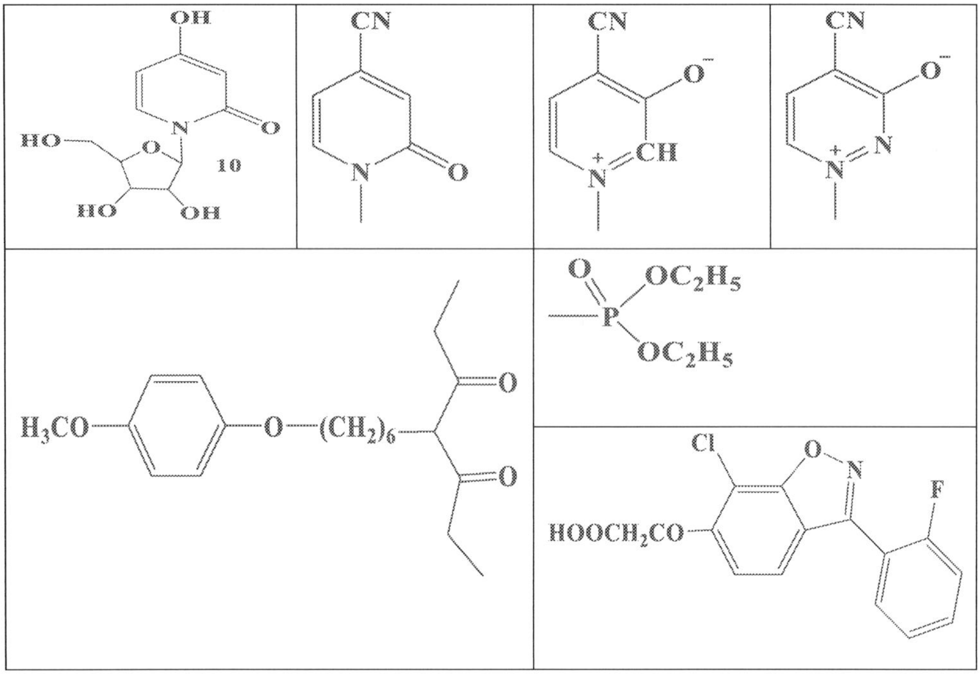

SCHEME 1.27 Ketone carbonyl bioisosteres.

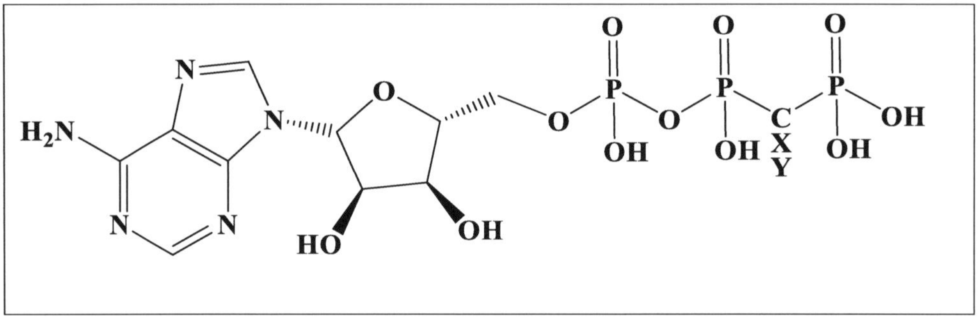

SCHEME 1.28 Phosphate bioisosteres.

1.5.3.7 Pyridine bioisosteres

The properties of pyridine include weak basicity, water solubility, in vivo/chemical stability, hydrogen bond-forming ability, and small molecular size, enabling them to be used as bioisosteres of amines, amides, heterocyclic rings containing some nitrogen atoms, and the benzene ring. Replacement of a portion of drugs with a pyridine moiety leads to improvement in their water solubility and hence, in the development of practical drugs that are suitable for an oral administration as well as an injectable formulation.

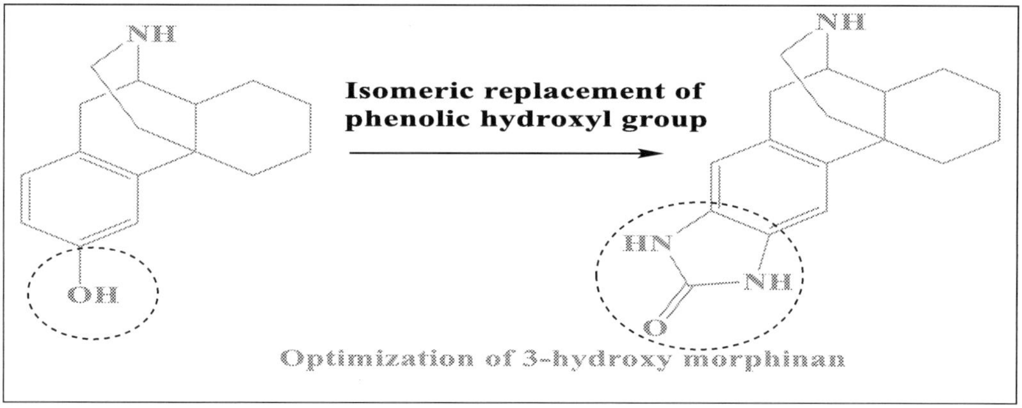

SCHEME 1.29 Phenol and catechol bioisosteres.

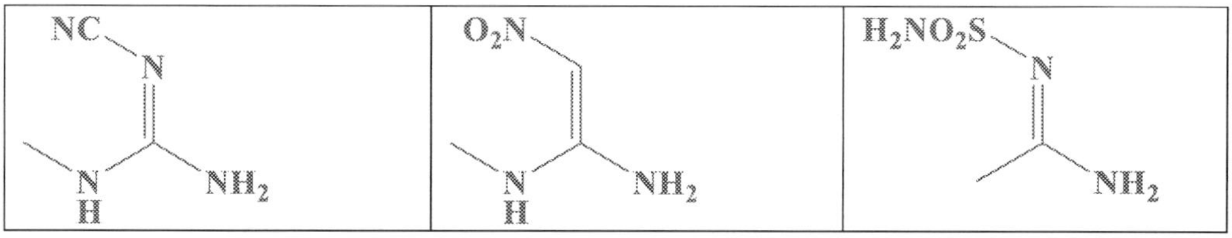

SCHEME 1.30 Urea and thiourea isosteres.

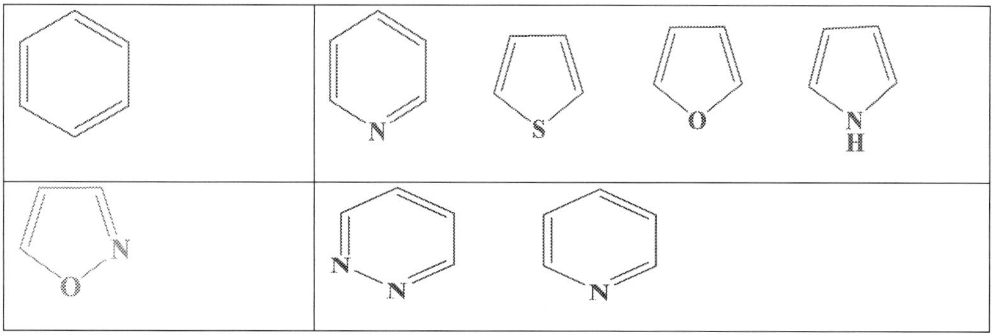

SCHEME 1.31 Pyridine bioisostere in COX-2 Inhibitors.

1.5.3.7.1 Pyridine bioisostere in COX-2 Inhibitors

COX-2 inhibitors are a subclass of nonsteroidal anti-inflammatory drugs which work by reducing the production of prostaglandins, chemicals that promote inflammation, pain, and fever. Pyridine ring has proved to be a good bioisostere of COX-2 inhibitors as well as Indanones (Scheme 1.31) [34].

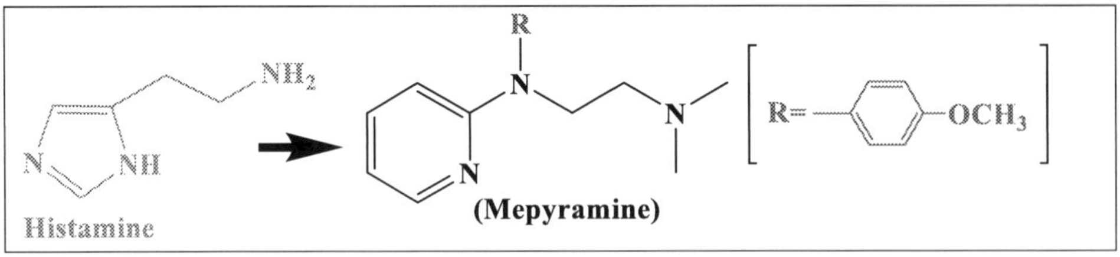

SCHEME 1.32 Pyridine ring as bioisostere of imidazole and benzene ring.

1.5.3.7.2 Pyridine ring as bioisostere of imidazole and benzene ring

Histamine contains an amino group and an imidazole ring, and hence, designing histamine receptor antagonists will require bioisosteres of the amino group and an imidazole ring. Mepyramine is an antihistamine agent used for the symptomatic treatment of allergy, hypersensitivity reactions, and pruritic skin disorders. Mepyramine, or pyrilamine, targets the H1 receptor. Mepyramine has a pyridine ring as a bioisostere of the imidazole ring (Scheme 1.32) [41].

1.5.3.7.3 Pyridine ring as bioisostere in MMP inhibitors

Matrix metalloproteinases (MMPs) are defined as calcium- and zinc-containing endopeptidases that have diverse roles in cell behaviors such as cell proliferation, migration, and differentiation. MMP family subtypes include MMP2, MMP3, and MMP9 which can degrade the extracellular matrix, resulting in the accelerated infiltration and migration of cancer cells. MMP subtype inhibitors have been reported for anticancer activity.

Prinomastat, a third-generation MMP inhibitor contains pyridine ring at the P1′ position, leading to improvement in the selectivity of MMP2 and MMP9 subtypes (Scheme 1.33) [41].

1.5.3.7.4 Pyridine ring as bioisostere in $BACE_1$ inhibitors

The most common cause of dementia is Alzheimer's disease (AD). AD is believed to be caused due to mutations of the gene encoding amyloid precursor protein (APP) or presenilin gene. The mutations in APP result in an increase in $A\beta$s, the main components of senile plaques in the brain. APPP produces $A\beta$s by two processing enzymes β-secretase (BACE1; β-site APP-cleaving enzyme 1) and γ-secretase, which are potential molecular targets for anti-AD drugs. Early BACE1 inhibitors were peptidomimetic with a substrate transition-state analog but associated with classical problems such as blood–brain barrier crossing, poor oral bioavailability, and susceptibility to P-glycoprotein transport. This led to a boom in research of nonpeptidic BACE1 inhibitors. Yoshio Hamada and the group reported a series of $BACE_1$ inhibitors possessing a pyridine scaffold. The replacement of the isophthalic scaffold of inhibitors with a pyridine dicarboxylic scaffold enabled the control of the conformation of inhibitors (Scheme 1.34) [41].

1.5.3.7.5 Pyridine ring as bioisostere in antitubercular drugs

Antitubercular therapy involves issues related to patient compliance and long-term therapy requiring the use of several drugs. Prodrug approach was initially tried to decrease problems with therapy adhesion mutual prodrugs. A combination of 4-amino-salicylic acid (PAS) with

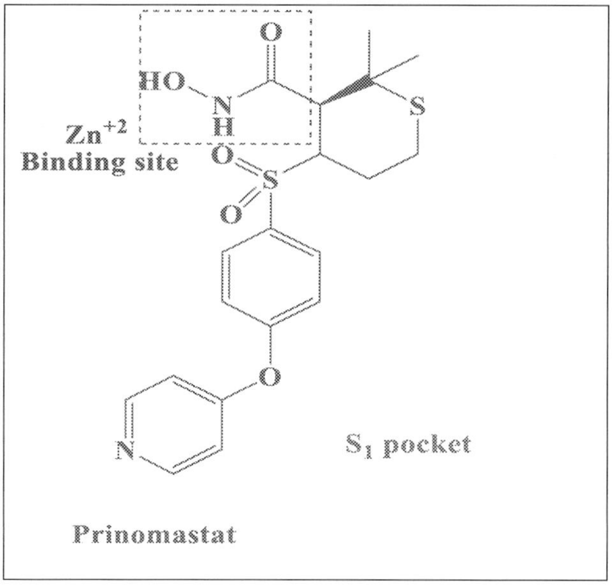

SCHEME 1.33 Pyridine ring as bioisostere in MMP inhibitors.

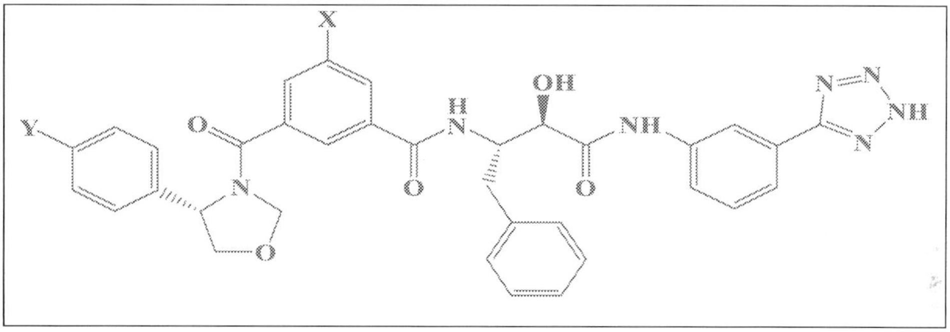

SCHEME 1.34 Pyridine ring as bioisostere in $BACE_1$ inhibitors.

isoniazid was reported (Scheme 1.35) [42]; however, the PAS treatment led to side effects such as gastrointestinal irritation effects and inadequate bioavailability due to rapid phase II metabolism. On the other hand, isoniazid gets rapidly absorbed after oral administration and quickly metabolizes to inactive products. Bioisosteres of isoniazid exploring the bioisosteric replacement of pyridine ring to imidazo[1,2-α]-pyridine can be found in the literature.

1.5.3.7.6 Pyridine bioisosteres in GluN2B subunit of NMDA receptor

Glutamate receptors are transmembrane proteins in neurons binding specifically the excitatory neurotransmitter (S)-glutamate. The glutamate receptors consist of three receptor types: the 2-amino-3-(3-hydroxy-5-methylisoxazol-4-yl) propanoic acid (AMPA), the kainate, and the N-methyl-D-aspartate (NMDA) receptor. The NDMA receptors in humans consist of seven units, termed GluN1, GluN2A–D and GluN3A-B. The negative allosteric modulators

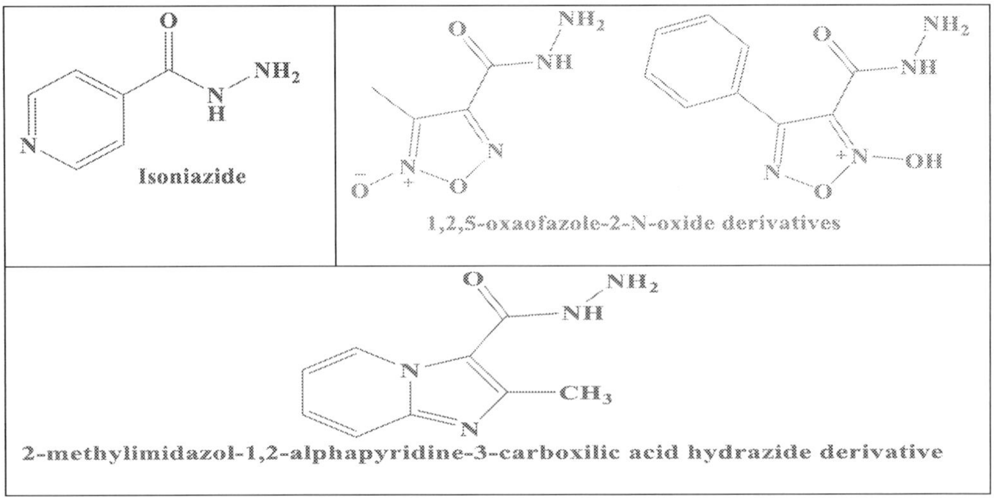

SCHEME 1.35 Pyridine ring as bioisostere in antitubercular drugs.

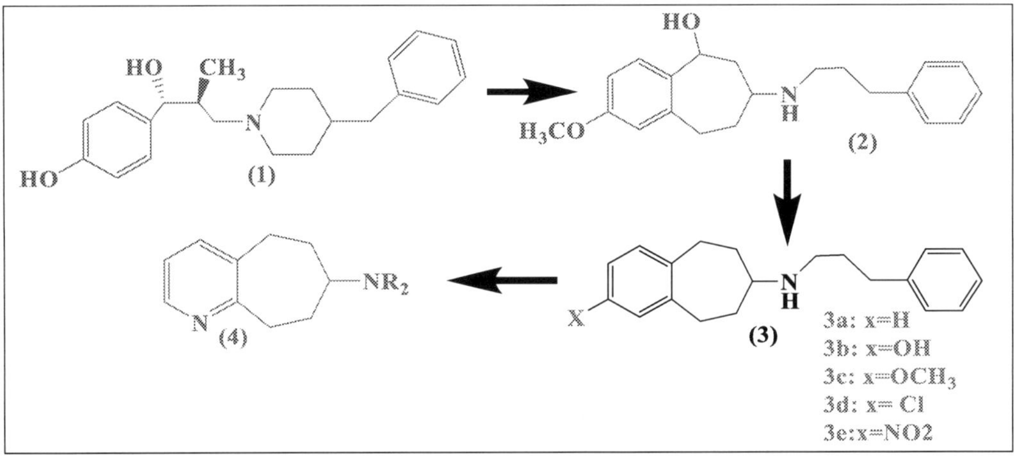

SCHEME 1.36 Pyridine bioisosteres in GluN2B subunit of NMDA receptor.

of GluN2B-NMDA receptors are being studied for the treatment of depression, cerebral ischemia, stroke, Parkinson's, Alzheimer's, and Huntington's disease. Recently, synthesis and pharmacological evaluation of pyridine bioisosteric GluN2B ligands of type 4 were reported by Bernhard Wünsch et al. (Scheme 1.36) [43].

1.6 Conclusion

In the 21st century, emerging and reemerging diseases will continue to challenge the medical practitioners as well as the worldwide ordinary population leading to an ever-increasing emphasis on infectious diseases. Globalization around the world has provided a route to an increased awareness of and commitment to addressing the terrible burden of infectious diseases in developing nations. Global health with an emphasis on infectious diseases is gradually assuming an important role in the foreign policies of all the nations. Drug discovery and development is a complex endeavor since new potential molecules need to meet acceptable pharmacological endpoints combined with favorable safety profiles.

Pyridine is an important class of compound present in many drugs, vitamins, food, flavorings, insecticides, and herbicides. Several important natural products like niconine, nicotine, and nucleic acid contain pyridine ring. Pyridine derivatives have occupied a unique place in the field of medicinal chemistry. The pharmacokinetic characteristics of lead molecules get improved due to the polar and ionizable aromatic nature of pyridine, and thus used as a remedy to optimize solubility and bioavailability parameters of proposed poorly soluble lead molecules.

Despite significant improvements in the delivery and efficacy of drug molecules, current challenges still include toxicity, poor biocompatibility, poor biodegradability, suboptimal targeting, and short circulation times. The concept of bioisosteres is a powerful and effective tool in drug design that has been applied to solve a wide range of problems encountered in drug discovery campaigns.

High solubility of pyridine in water has led to its applications as bioisosteres. Bioisosteres is one of the important roles of pyridines in novel drug design. The application of isosteric substitution of pyridine nucleus was explored in a fashion that focuses on the development of practical solutions to problems that are encountered in typical optimization campaigns. The role of pyridine bioisosteres to affect intrinsic potency and selectivity, influence conformation, solve problems associated with drug development, including receptor recognition, basicity, solubility, and toxicity was discussed to capture a spectrum of creative applications of pyridine in the design of drug candidates.

Finally, we can conclude that academia must play a lead role in gaining unexplored knowledge of bacterial physiology to enrich the field of antibiotic R&D to develop some new antibiotic drugs as per the list of bacteria published by WHO in a joint venture with small to medium-sized enterprises. This goal can be achieved by understanding the potential targets and gaining in-depth knowledge of membrane permeability barrier and multi-drug resistant properties achieved by Gram-negative bacteria. In finding the new drugs, the concept of bioisostere can be used as a tool to develop a drug of our choice as per our requirements.

Acknowledgment

All the authors are thankful to Mr. Kartik Mahesh Vithlani for his hard work and efforts in drawing structures of compounds in Chem Draw.

References

[1] A.S. Fauci, Infectious diseases: consideration for 21st century, Clin. Infect. Dis. 32 (2001) 675–685.

[2] D.M. Morens, G.K. Folkers, A.S. Fauci, The challenge of emerging and re-emerging infectious diseases, Nature 430 (2004) 242–249.

[3] N.D. Wolfe, C.P. Dunavan, Origins of major human infectious diseases, J. Diamond, Nature 447 (2007) 279–283.

[4] S.S. Morse, Plagues and Politics: Infectious Disease and Policy, Macmillan, A Division of Macmillan Publishers Limited, New York, United States, 2001 (Chapter 2).

[5] N. Jackson, L. Czaplewski, L.J.V. Piddock, Discovery and development of new antibacterial drugs: learning from experience? J. Antimicrob. Chemother. 73 (2018) 1452–1459.

[6] T. Yokota, Kinds of antimicrobial agents and their mode of actions, Nihon Rinsho 55 (5) (1997) 1155–1160 Japanese. PMID: 9155168.

[7] R. Koch, Untersuchungen über Bacterien. V. Die Aetiologie der Milzbrand-Krankheit, begründet auf die Entwicklungsgeschichte des Bacillus Anthracis, Beiträge zur Biologie der Pflanzen 2 (1876) 277–310.

[8] A.B. Jeon, A.O. Henao, D.F. Ackart, B.K. Podeli, J.M. Belardinelli, et al., 2-aminoimidazoles potentiate β-lactam antimicrobial activity against Mycobacterium tuberculosis by reducing β-lactamase secretion and increasing cell envelope permeability, PLOS One J. 12 (7) (2017) 1–29.

[9] T.V. Nguyen, M.S. Blachledge, D.F. Ackart, A.B. Jeon, A. Obregon-Henao, et al., The discovery of 2-aminobenzimidazoles that sensitize mycobacterium smegmatis and M. tuberculosis to β-Lactam antibiotics in a pattern distinct from β-lactamase inhibitor's, Angew. Chem. Int. Ed. Engl. 56 (14) (2017) 3940–3944.

[10] J. Rai, G.K. Randhawa, M. Kaur, Recent advances in antibacterial drugs, Int. J. Appl. Basic Res. 3 (1) (2013) 3–10.

[11] M. Worboys, Spreading Germs: Diseases, Theories, and Medical Practice in Britain, 1865–1900, Cambridge University Press, Cambridge, 2000.

[12] R. Porter, The Greatest Benefit to Mankind: A Medical History of Humanity from Antiquity to the Present, W. W. Norton and Co., London, 1997.

[13] S.Y. Tan, Y. Tatsumura, Alexander Fleming (1881-1955): discoverer of penicillin, Singap. Med. J. 56 (7) (2015) 366–367.

[14] S.N. Sirakanyan, A.A. Hovakimyan, A.S. Noravyan, Synthesis, transformations and biological properties of furo[2,3-b] pyridines, Russ. Chem. Rev. 84 (4) (2015) 441–454.

[15] A.A. Altaf, A. Shahzad, Z. Gul, N. Rasool, A. Badshah, B. Lal, E. Khan, A review on the medicinal importance of pyridine derivatives, J. Drug Des. Medic. Chem. 1 (1) (2015) 1–11.

[16] M.V. De Almeida, M.V.N Souza, N.R. Barbosa, F.P. Silva, et al., Synthesis and antimicrobial activity of pyridine derivatives substituted at C-2 and C-6 positions, Lett. Drug Des. Discov. 4 (2007) 149–153.

[17] J. Lazaar, A.S. Rebstock, F. Mongin, A. Godard, F. Trecourt, F. Marsais, G. Queguiner, Directed lithiation of unprotected pyridinecarboxylic acids: syntheses of halo derivatives, Tetrahedron 58 (2002) 6723–6728.

[18] M.C. Corredor, J.M.R. Mellado, On the electroreduction mechanism of 2-pyridinecarboxylic (picolinic) acid on mercury electrodes, Electrochim. Acta 49 (2004) 1843–1850.

[19] YC. Terrie, Recognizing and treating peptic ulcer disease, Pharm. Times 84 (7) (2018). https://www.pharmacytimes.com/view/recognizing-and-treating-peptic-ulcer-disease.

[20] N.C. Desai, D.V. Vaja, K.A. Jadeja, S.B. Joshi, V.M. Khedkar, Synthesis, biological evaluation and molecular docking study of pyrazole, pyrazoline clubbed pyridine as potential antimicrobial agents, anti-infective agents, SAR and QSAR in Environmental Research 18 (2020) 306–314.

[21] V. Klimesova, M. Svoboda, K. Waisser, J. Kaustova, V. Buchta, K. Kraluova, Synthesis of 2-benzylthiopyridine-4-carbothioamide derivatives and their antimicrobial, antifungal and photosynthesis-inhibiting activity, Eur. J. Med. Chem. 34 (1999) 433—440 V. Klimesova, P. Herzigova, K. Palat, Milos Machacek, J. Stolaríkova, H.M. Dahse, and U. Mollmann, The Synthesis and antimicrobial properties of 4-(substituted benzylsulfinyl)pyridine-2-crboxamides, ARKIVOC III (2012) 90-103.

[22] G. Thomas, M.S R.Groβ, Calcium channel modulation: ability to inhibit or promote calcium influx resides in the same dihydropyridine molecule, J. Cardiovasc. Pharmacol. 6 (6) (1984) 1170–1176.

[23] A. Ragab, S.A. Fouad, O.A. Abu Ali, E.M. Ahmed, A.M. Ali, A.A. Askar, Y.A. Ammar, Sulphaguanidine hybrid with some new pyridine-2-one derivatives: design, synthesis and antimicrobial activity against multidrug resistant bacteria as dual DNA gyrase and DHFR inhibitors, Antibiotics 10 (162) (2021) 1–31.

[24] E.G. Tse, M. Korsik, M.H. Todd, The past, present and future of antimalarial medicines, Malar. J. 18 (93) (2019) 1–21.

[25] Q. Cai, M. Liu, X.X B.Mao, F. Jia, Y. Zhu, A. Wu, Direct one-pot synthesis of zolimidine pharmaceutical drug and imidazole[1,2-a]pyridine derivatives via I_2/CuO-promoted tandem strategy, Chin. Chem. Lett. 26 (7) (2015) 881–884.

[26] M. Lakshmi Kantam, K. Mahendari, S. Bargava, One-pot, three component syntheses of highly substituted pyridines and 1,4-dihydropyridines by using nanocrystalline magnesium oxide, J. Chem. Sci. 122 (1) (2010) 63–69.

[27] E. Busto, V. Gotor-Fernandez, V. Gotor, Chemoenzymatic synthesis of chiral 4-(N,N-dimethylamino) pyridine derivatives, Tetrahedron Asymmetry 16 (2005) 3427–3435.

[28] S. Kona, R. Suresh Ravi, V.N.R. Chava, R. Sridhar Perali, A convenient synthesis of C-3-Aryloxymethyl Imidazo [1,2-a] pyridine derivatives, J. Chem. 2013 (2012) 1–6.

[29] C. Akgul, M. Ydirim, Molecular weight dependent antistaphylococcal activities of oligomers/polymers synthesized from 3-aminopyridine, J. Serb. Chem. Soc. 75 (9) (2010) 1203–1208.
[30] B.N. Acharya, E.D. Thavaselvam E. Kaushik, Synthesis, and antimalarial evaluation of novel pyridine quinoline hybrids, Med. Chem. Res. 17 (2008) 487–494.
[31] V. Klimesova, M. Svoboda, K. Waisser, M. Pour, J. Kaustova, New pyridine derivatives as potential antimicrobial agents, I L Farmaco 54 (1999) 666–672.
[32] W.M. Eldehna, M. Fares, M.M. Abdel-Aziz, H.A. Abdel Aziz, Design, synthesis and antitubercular activity of certain nicotinic acid hydrazides, Molecules 20 (2015) 8880 -8815.
[33] D.G. Silva, A. Junker, S.M.G. De Melo, F. Fumagalli, J.R. Gillespie, et al., Synthesis and structure-activity relationships of imidazopyridine/pyrimidine-and furopyridine-based anti-infective agents against trypanosomiases, Chem. Med. Chem. 16 (2021) 966–975.
[34] Y. Morita, Application of bioisosters in drug design, Literature Seminar 5 (7) (2012) 1–17. https://gousei.f.u-tokyo.ac.jp/seminar/pdf/Lit_Y_Morita_M1.pdf.
[35] G. Ali, F. Subhan, N. Islam, I. Khan, K. Rauf, et al., Input of isosteric and bioisosteric approach in drug design, J. Chem. Soc. Pak. 36 (2014) 1–20.
[36] P.I. Gaekwad, P.S. Gandhi, D.M. Jagdale, V.J. Kadam, The use of bioisosterism in drug design and molecular modification, Am. J. Pharm. Tech. Res. 2 (4) (2012) 1–24.
[37] C.A. Lipinsnski, Topics in chemistry and drug design-section VI, R. C. allen, Hoeschst-Roussel Pharmaceuticals Inc, Somervillr, New Jersey, Annu. Rep. Med. Chem. 21 (1986) 283–291.
[38] G. Michael Blackburnd, V. Kent, F. Kolkmann, Three New β, γ- methylene analogues of adenosine triphosphate, J.C.S. Chem. Comm. 1 (1981) 1188–1190.
[39] Z. Li, X.B. X.Bao, G. Zhang, J. Wang, Y.W. M.Zhu, C.S J.Shang, D. Zhang, Y. Wang, Design, syntheis, and biologicla evaluation of phenol bioisosteric analogues of 3-hydroxymorphinan, Sci. Rep. 9 (2019) 2247.
[40] R. Ronchetti, G. Moroni, A. Carotti, A. Gioiello, E. Camaioni, Recent advances in urea and thiourea-containing compounds: focus on innovative approaches in medicinal chemistry and organic synthesis, RSC Med. Chem. 12 (2021) 1046–1064.
[41] Y. Hamada, Role of pyridines in medicinal chemistry and design of BACE1 inhibitors possessing a pyridine scaffold; DOI:10.5772/intechopen.74719. Source: Intech Open.
[42] J. Leandro dos Santos, L. Antonio Dutra, T. Regina Ferreira de Melo, and C. Chin, New antitubercular drugs designed by molecular modification, (2012) DOI:10.5772/33169. Source: InTech.
[43] R. Zscherp, S. Baumeister, D. Schepmann, B. Wunsch, Pyridine bioisosters of potent GluN2B subunit containing NMDA receptor antagonists with benzo[7]annulene scaffold, Eur. J. Med. Chem. 61 (2018) 1–30.

CHAPTER

2

Pyridine-based polymers and derivatives: Synthesis and applications

Shagun Varshney and Nidhi Mishra

Department of Applied Sciences, Indian Institute of Information Technology Allahabad, Prayagraj, Uttar Pradesh, India

2.1 Introduction

Medicinal chemistry is a branch of science concerned with evaluating the impact of chemical compounds on the biological activity of living beings. It has evolved from an empirical approach including the organic synthesis of novel compounds based mainly on structural alteration and testing their biological activity. It is also concerned with the molecular level discovery, development, interpretation, and identification of physiologically active chemical mechanisms. Pyridine (C_5H_5N) is an aromatic molecule in which a ring shares all of the π-electrons, forming a constant circle of electrons in addition to the alternate double bonds shared by each atom on the circle. It is a heterocyclic molecule in conjunction with nitrogen, which aids in the formation of a tertiary amine by the alkylation and oxidation process. Since electrons are pulled more towards the electronegative nitrogen (lone pair electrons on the nitrogen) than other atoms in the ring, there is a small dipole on the ring due to amine. Pyridine exists as a clear liquid and tends to mix well with organic solvents and water. Several properties of pyridine enable it to be used in the production of a wide range of products, including pharmaceuticals, vitamins, food flavorings, insecticides, dyes, paints, adhesives, rubber products, nitrogen-containing plant products, and waterproofing fabrics. It is also used as a foundation agent for a variety of pharmaceuticals, agrochemicals due to its highly desirable characteristics. As a result, pyridine, its polymers, and its derivatives have a wide range of applications, particularly in the medical realm [1].

The pyridine-based compounds are frequently offered in the market as intermediates of chemicals for the production of finished consumer goods. Pyridine and its derivatives have been commercially important since the early 20th century, but especially during World War II

Recent Developments in the Synthesis and Applications of Pyridines.
DOI: https://doi.org/10.1016/B978-0-323-91221-1.00012-9

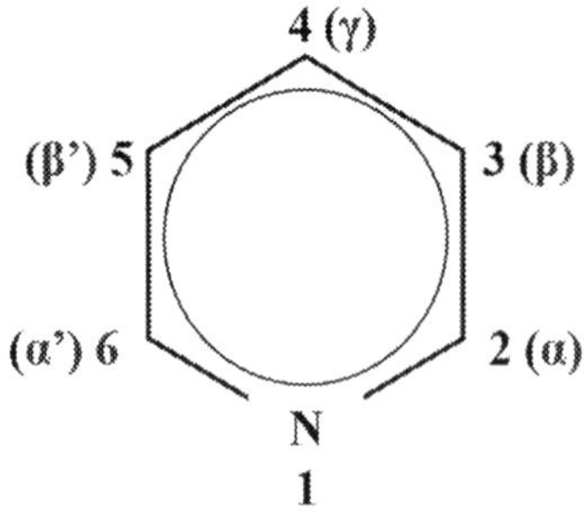

FIGURE 2.1 Chemical structure of pyridine.

and thereafter. In a marketplace where bioactivity is a significant parameter, such as pharmaceutical medications and products of agriculture, such as insecticides, herbicides, plant growth regulators, and fungicides. Pyridine has antihistaminic, antihypertensive, anticoagulant, antibacterial, anti-inflammatory, antifungal, antiviral, tuberculocidal, antidiabetic, and antimalarial properties. Polymers manufactured from pyridine-based monomers, for example, are marketed primarily for their physical qualities and functions, rather than for their bioactivity. Pyridine medications are being used to treat a wider range of conditions in clinical practice. At high doses, some pyridine medications may have a direct inhibitory effect on membranes, without interfering with sterols and sterol esters. Since bacteria have defied prevention or therapy longer than any other form of life, infectious microbial illness is still a worldwide concern. Antimicrobial resistance to β-lactam antibiotics, macrolides, quinolones, and vancomycin is becoming a major global concern. Pyridine and its derivatives are said to be physiologically and pharmacologically active and have been used to cure a variety of ailments [2].

The existence of a heterocyclic ring (six-membered) distinguishes pyridine from other chemicals. The valency of carbon atoms does not take up in constructing the rings are satisfied by atoms of hydrogen. Benzene and pyridine have identical atom arrangements, apart from that one of the hydrogen–carbon ring pairs have been substituted by an N-atom in pyridine. The numbering 1–6 can be used to identify pyridine ring substitutes, or the Greek letters (α, β), can be used to identify pyridine ring substitutes, as shown in Fig. 2.1.

These Greek letters are used to name mono-substituted pyridines and can also be used to indicate the place of the substituent relative to the ring N-atom. The para, meta, & ortho nomenclature, which is often used for di-substituted benzene, is not utilized to name pyridine-based compounds. The structures of pyridine derivatives are shown below in Fig. 2.2. (i) γ-picoline (ii), β-picoline (iii), α-picoline (iv), 3,5-lutidine (v), 2,6-lutidine, (vi) "5-ethyl-2-methylpyridine" (vii), and "2,4,6-collidine" are some economically significant alkyl pyridine compounds. In general, alkyl pyridines are precursors to a variety of different substituted pyridines employed in the industry. These alkyl pyridine-derived substituted pyridine compounds are often utilized as intermediates in the production of economically valuable end products.

2.1.1 History of pyridines

Pyridine is composed of two words: "*pyr*" means "fire," and "*idine*" means "aromatic base." Anderson discovered the first pyridine base i.e. picoline in 1846. The structure was separately identified by Wilhelm Körner in 1869 and James Dewar in 1871. Pyridine structure was thought

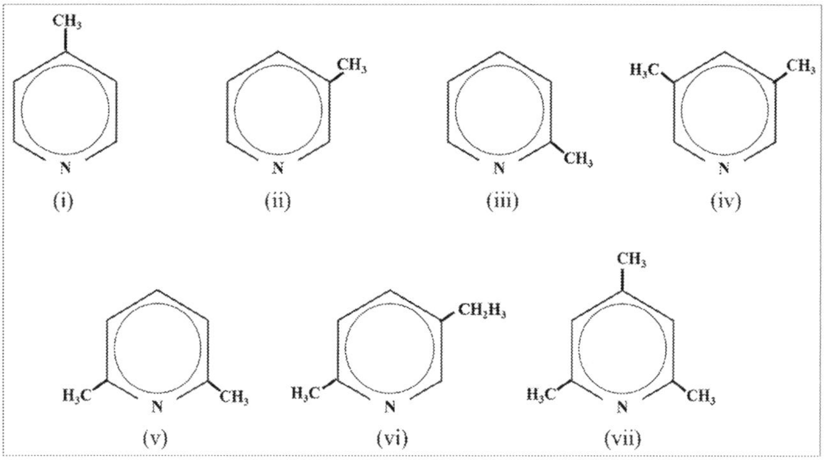

FIGURE 2.2 Chemical structures of (i) γ-picoline (ii) β-picoline (iii) α-picoline (iv) 3,5-lutidine (v) 2,6-lutidine (vi) 5-ethyl-2-methylpyridine, and (vii) 2,4,6-collidine.

to be similar to that of quinoline and naphthalene. Thomas Anderson was the first to discover pyridine by boiling animal bones at high temperatures. Pyridine was originally made from acetylene and hydrogen cyanide in 1876 [3]. It was determined that pyridine was generated from benzene and that its structure could be created by substituting a nitrogen atom for a CH moiety. William Ramsay synthesized pyridine in 1876 by mixing hydrogen cyanide and acetylene where the reaction took place in a high-temperature iron-tube furnace [4]. α-picoline, on the other hand, was the earliest C_5H_5N compound to be extracted in pure form, [5]. Surprisingly, in the 1940s, synthetic methods for pyridines were developed in response to market demand, rather than their separation from coal-tar sources. Pyridine is found in coal tar at a concentration of 0.01%. Pyridines have a chemistry that is quite distinct from that of benzenoids. It can go through reactions that only electron deficient compounds like benzoids can but they do not undergo some of the more common reactions like C-acylation and Friedel–Crafts alkylation.

2.1.2 Ligand properties of pyridine

Pyridine is a fairly strong ligand, as per the series of spectrochemical ligands included in crystal field theory (CFT). It displays ligand organization based on the splitting capacity of the d-orbital in metals [6]. This translates to significant pyridine lone pair electrostatic interactions with metal d-orbitals. Despite its neutrality, pyridine produces a considerable amount of d-orbital splitting, indicating a strong bonding contact with metal centers. In addition to the CFT, the valence bond theory views metal pyridine bonding as an overlap of sp2 lone pair orbital in pyridine with hybridized metal orbitals. Due to the differences in size, form, and energy of the combining orbitals, the amount of overlap in first transition metals is higher than in second and third transition metals. Apart from lone pair orbitals of nitrogen, the ring-electron can also bind to metal ions. Furthermore, antibonding orbitals which are delocalized can act as an electron density acceptor as shown in Fig. 2.3. As a result, pyridine can be bonded with metal ions present in abundance in the orbitals for better interactions.

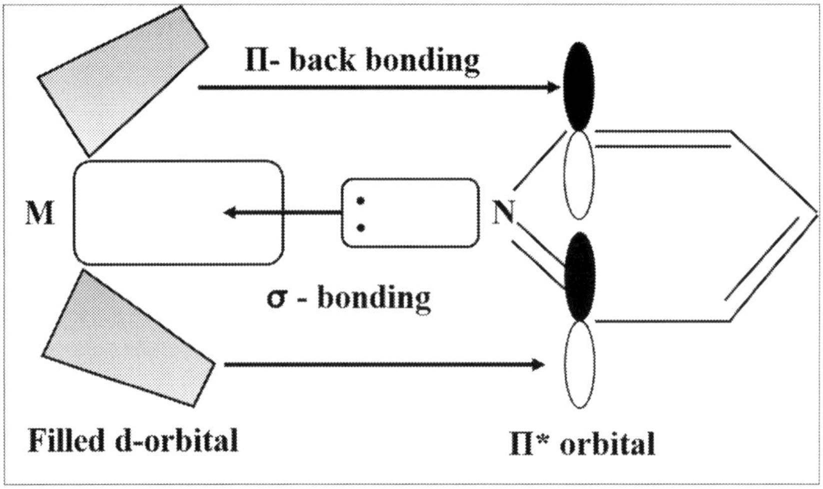

FIGURE 2.3 Bonding in pyridine.

2.1.3 Transition metal complexes of pyridine

The literature on pyridine transition metal compounds is extensive. Pyridine was discovered to coordinate with all transition metals, resulting in a wide range of metal complexes in various oxidation states. Although attempts have been made to include the growing number of pyridines into the metal coordination spheres including pyridine complexes like [*M*(py)4] *n*+ or [*M*(py)6] *n*+ (where *M* = transition metal) which are quite uncommon. Chemistry based on pyridine-metal combination uses pyridine and its derivatives to generate metal complexes comprising of bi-dentate or tri-dentate ligands [7]. Scandium and yttrium prefer to bind in a four-coordinated geometry to three pyridine units. The coordination number may vary depending on the properties of the binding ligands. Pyridine and thiocyanate ligands are also used to make five and six coordinated complexes. Picolinic acid, a pyridine derivative capable of functioning as a bidentate ligand, tends to form higher coordination number complexes. With differently substituted pyridines, a large variety of complexes are known. Sc and Y complexes in +1 and + 3 states are known. $[Sc(py)_3Cl]$ [8,9], $[Y(py)_3Cl]$ [7,10], $[Sc(py)_2(NCS)_3]$ [11], and $[Sc(py)_3(NCS)_3]$ [12] are examples of pyridine coordinated complexes. At room temperature, these complexes might be synthesized directly from their metal salts and pyridine. Some metal complexes of pyridine are shown in Fig. 2.4.

Pyridine complexes can also be synthesized by incorporating the inorganic salts of chromium, molybdenum, carbonyl and nitrosyl complexes, and tungsten as well. To maintain the stability of synthesized complexes, a neutral medium is considered the best option. These metals have a wide range of oxidation states (0–VI), which has resulted in a huge number of pyridine complexes. Pyridine complexes have the property to get oxidized to higher oxidation states, notably in the 0 and + 1 states which makes these complexes susceptible to air and moisture. In Cr and Mo, the common coordination number is still six, although it might be greater in tungsten pyridine complexes. Some metal-complexes of pyridine with Cr, W, and Mn are shown in Fig. 2.5.

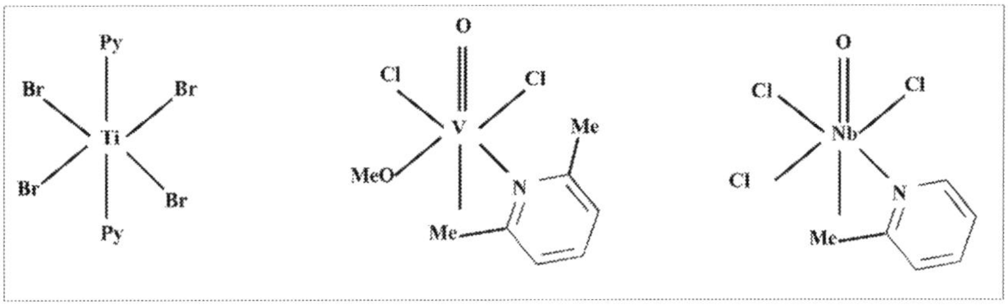

FIGURE 2.4 Pyridine-based metal complexes of Ti, V, and Nb.

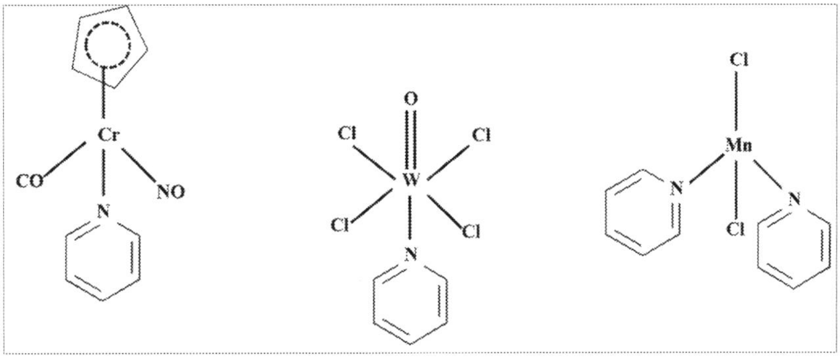

FIGURE 2.5 Pyridine-based metal complexes of Cr, W, and Mn.

In a variety of disciplines, transition metal pyridine compounds have proven their worth. Fig. 2.6 only mentions a few typical compounds and their uses. [$Ti(py)_xCl_y$] complexes of titanium pyridine have also been investigated as catalysts for alkene and alkyne polymerization. When combined with aluminum co-catalysts like $RAlCl_2$, R_2AlC, and R_3Al (R = alkyl groups), these titanium complexes are effective in the Ziegler–Natta catalytic process [7]. Olefin polymerization catalysts include pyridine complexes produced from VCl_3, VCl_4, and $VOCl_3$. Some applications of pyridine-based metal complexes are shown in Fig. 2.6.

2.1.4 The pharmacological activity of pyridines

Pyridines are well-known heterocyclic compounds that are used in a wide range of pharmacological-based applications due to their key properties. It is soluble in water and other polar solvents. Due to the presence of hydrogen atoms on either of the two nitrogen atoms, it exists in two equivalent tautomers. It is a highly polar molecule with an estimated dipole of 3.61 D and is completely water-soluble. With six electron pairs, including two from the protonated nitrogen atom and one each from the other four carbon atoms, it is said to be aromatic. Pyridine has a wide range of biological functions. Due to its biological activity and low toxicity, it has several advantages over traditional medicines, including a shorter time to market and a higher success rate in clinical trials. There is substantial antifungal action in the different substituted pyridines. Substituted pyridine derivatives with significant antiviral and antiasthma efficacy are now the top drugs on the market. While some modified pyridine is efficient as an antihypertensive, other derivatives are observed to have anti-inflammatory,

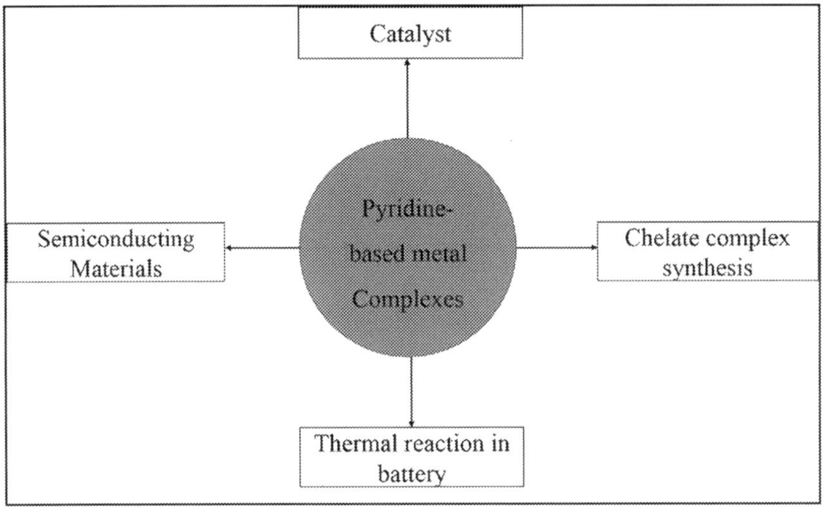

FIGURE 2.6 Application of metal pyridine complexes.

antiviral, as well as HIV-specific effects on the immune system. Many present organic-based medicines are expected to be replaced very soon by pyridine-based drugs. By using diverse research and development techniques, contemporary drug discovery will take pyridine-based medicines on a big scale for therapeutic application.

2.2 Properties of pyridine

2.2.1 Physical properties

Pyridine has a stable, cyclic, π-deficient, 6π-electron, aromatic structure with a nitrogen atom within the ring that gives it a range of physical characteristics. Since the ring carbons are less electronegative than the nitrogen in the ring, the six, four, and two-ring-C are more electropositive than would be anticipated based on benzenoid chemistry. This is because the aromatic π-electron system does not interact as the lone pair of electrons on the N-atom, pyridine compounds are described as π-deficient and weakly basic. Mostly nitrogen in pyridine atom undergoes protonation, alkylation, and acylation processes, which are a characteristic of weak, tertiary organic amines.

Liquid alkyl pyridines and pyridine are dipolar, aprotic solvents that are comparable to dimethylformamide or dimethyl sulfoxide. Most pyridines cannot be separated by simple distillation as they form large azeotropes with water. By distilling a small portion of water azeotrope, wet pyridines may also be dried quickly. The physical characteristics of pyridine vary from those of benzene, its homocyclic analog. The boiling point of pyridine is 115.3°C, which is 35.28°C higher than the boiling point of benzene (80.18°C vs 115.3°C). Contrary to benzene, it is miscible with H_2O in all concentrations at room temperature. Due to its larger dipole moment, pyridine has a higher boiling point and is more water-soluble than benzene. It is important to note that both compounds are aromatic, have similar resonance energies, and are miscible with other organic solvents as well. It is a weak organic (pKa = 5.22) base that acts

TABLE 2.1 Physicochemical characteristics of pyridine.

Physical properties	Values
Critical pressure, MPa	5.6
The critical temperature, 8°C	346.9
Enthalpy of vaporization, kJ/mol (at 258°C and 115.268°C)	40.1
	35.12
Enthalpy of fusion	8.2784
Enthalpy of formation	140.36
Heat capacity at 258°C, J/(Kmol)	78.22
Gibbs free energy of formation at 258°C, kJ/mol	190.47
Explosion limit, %	1.7–10.5
Ignition temperature, 8°C	551
Viscosity at 258°C, mPa ($\frac{1}{4}$ cP)	0.877
Surface tension, liquid at 258°C, mN/m($\frac{1}{4}$dyn/cm)	36.5
Thermal conductivity	0.166
Dielectric constant at 258°C	13.4

as an electron donor and a proton acceptor, and also as an aliphatic tertiary amine. Pyridine has lower basicity than most others. The physicochemical characteristics of pyridine are listed in Table 2.1.

2.2.2 Chemical properties

The presence of a basic N-atom in the ring, ring aromaticity, large dipole moment, activation of functional groups, the pi-deficient character of the ring, involved in the ring, easy polarizability of the π-electrons, and the existence of electron-deficient C-atoms at the γ- & α-locations are all factors that contribute to the chemical properties of pyridines. As far as chemistry is concerned, pyridines have two types of chemical reactions including those that occur at the ring atomic core and the one that occurs at the substituent atom.

2.2.2.1 Reactions occurring at ring atoms

Electrophiles may target ring nitrogen, ring carbons (less easily at ring carbons), and ring-atomic centers.

2.2.2.1.1. Electrophilic attack at nitrogen

Under moderate conditions, lone pair of pyridine interacts with protonic acids to produce Lewis acids and simple salts thereby producing coordination compounds (ix), transition metals to produce complex ions (x, xi), and electrophiles are shown in Fig. 2.7. Due to its moderate oxidizing tendency, pyridinium chlorochromate (xii) transforms –OH to carbonyl

(ix)

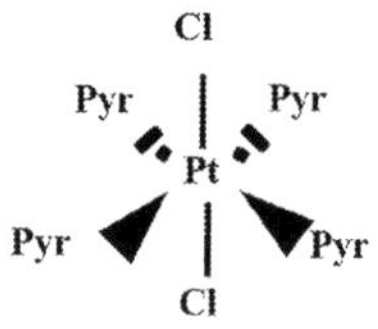

(x)

(xi)

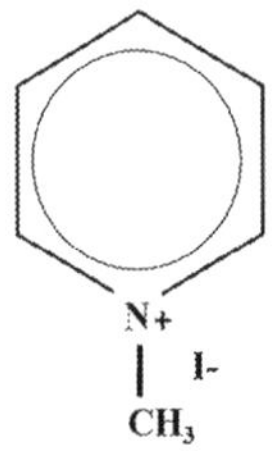

(xii)

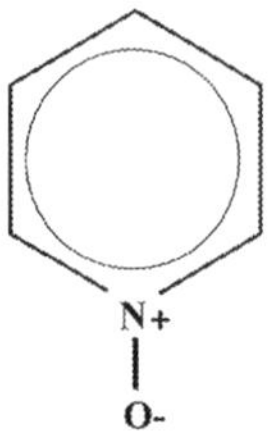

(xiii)

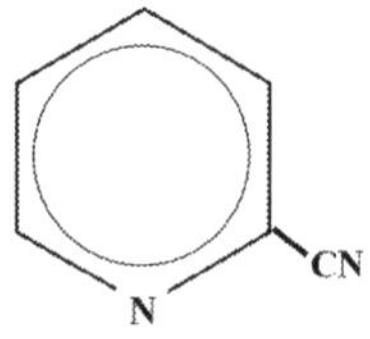

(xiv)

FIGURE 2.7 Depicting the electrophilic attack on nitrogen.

molecules. Reactive activated alkenes, alkyl halides, and halogen compounds generate quaternary pyridinium salts (xiii). Pyridine N-oxides are produced via oxidation with peracids (xiv) [13].

2.2.2.1.2. Electrophilic attack on carbon

Electrophiles have a difficulty in attacking the pyridine ring carbon atoms unless single or stronger electron-donating substituents are added to the ring [14].

2.2.2.2 Reactions of pyridine substituents

2.2.2.2.1 Carbon-based substituents

Carbanions are easily produced when alkyl carbons are joined at the 4- and 2-positions. The synthesis of 4- and 2-(phenyl propyl) pyridines, such as 4-(3-phenyl propyl) pyridine (xv), have been synthesized using this enhanced chemical reactivity, as shown in Fig. 2.8.

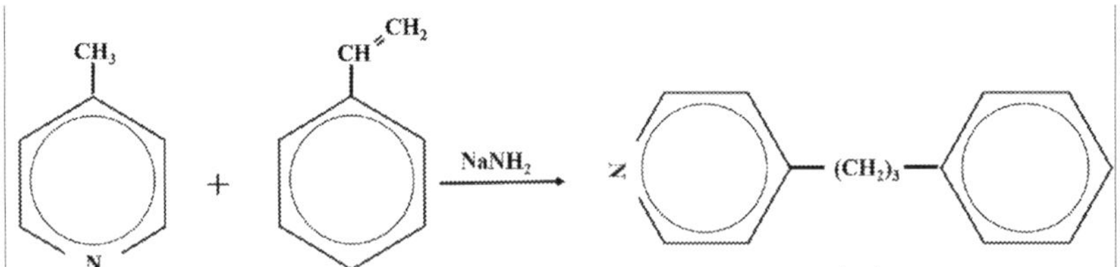

(xv)

2-ethanol pyridine (xvi) is formed by condensation of γ- or α-picoline with aqueous formaldehyde, followed by dehydration of -OH to produce "2-vinyl pyridine" (xvii). Many case studies are significant to the industry-related aspects in this regard [15].

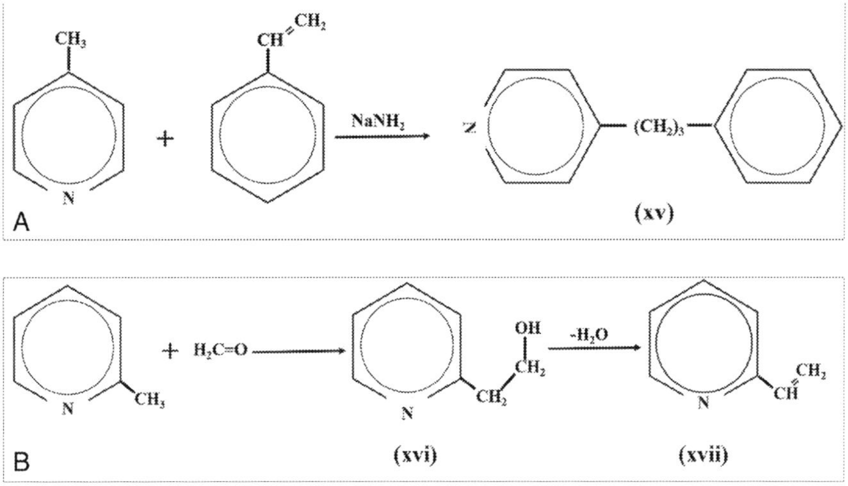

FIGURE 2.8 Different reactions of alkyl substituent.

2.2.2.2.2 Nitrogen substituents

2-aminopyridine and 4-aminopyridine react with cold HNO_3 to form insoluble nitramines. The nitro aminopyridines (with the -NO group mostly close to the -NH_2 group) are reorganized by nitramines intermolecularly when subjected to heating [16] are shown below.

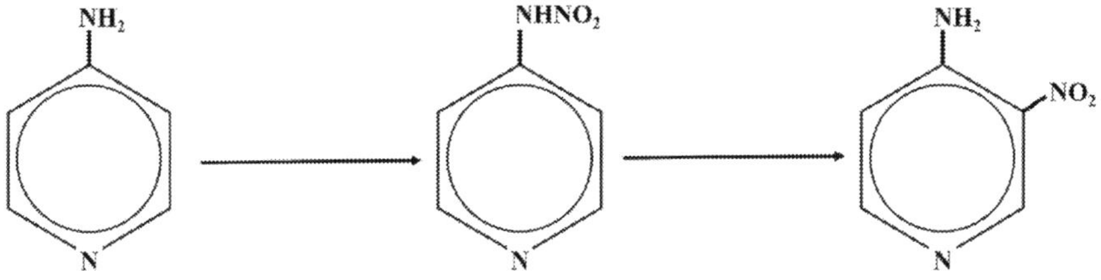

(xviii)

Sulphapyridine (xix) is produced by reacting 2-aminopyridine (xx) with N-acetylsulfanilyl chloride and then hydrolyzing it [17].

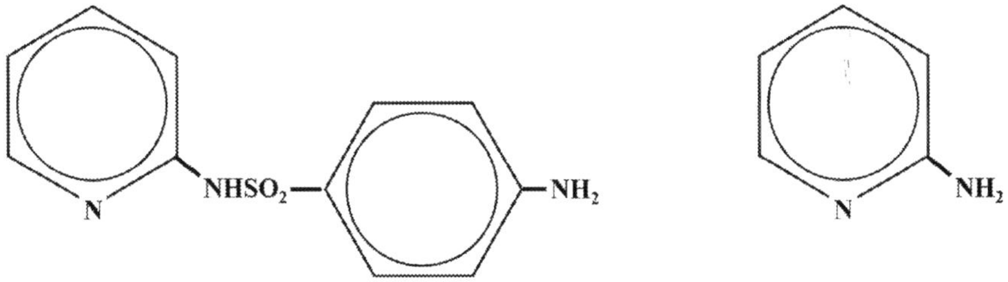

(xix) (xx)

The antibacterial agent nalidixic acid (xxvi) is obtained by reacting an alkoxy methyl-malonic ester with 2-amino-6-methylpyridine to produce the 1,8-naphthyridine carboxylic ester, which is then alkylated and hydrolyzed [18].

2.3 Pyridine-based polymers of heterocyclic compounds

The word "polymer" refers to materials that have more than five repeating units in them. Heterocyclic polymers cover a wide range of materials, from simple and uncomplicated linear polymers prepared from vinyl-substituted heterocyclic monomers to highly functionalized, crosslinked networks in which the heterocycle is generated during the polymerization or network formation reaction. In the main, side, or terminal chain of a heterocyclic polymer, there is a repeating unit containing a heterocycle. Heterocyclic "oligomers" or compounds with lower degrees of polymerization have been studied and their potential has also been investigated.

Compounds containing one or two heteroatoms are classified as heterocyclic compounds. Based on the method of production, heterocyclic polymers can be classified into three main categories:

a) Heterocyclic polymers through polymer modification.
b) Heterocycle-producing polymerizations.
c) Heterocyclic polymers through heterocyclic monomers [19].

2.4 Polymers of pyridine

Polymerization of vinyl pyridine monomers is the most frequent method for incorporating pyridine rings into polymers. 2-vinyl pyridine (xxi), 4-vinyl pyridine (xxii), and 2-methyl-5-vinyl pyridine (xxiii) are the most prevalent compounds and have gained economic importance (xxix). Poly (vinyl pyridine N-oxides) and poly (vinyl pyridine N-oxides) can be made by modifying the presynthesized polymer or poly (vinyl pyridine salts). For example, poly (vinyl pyridines) and its copolymers are used as adhesives, impact modifiers, antistat materials, conductive polymers, ion exchange resins, heavy metal chelating resins, dye receptor materials, catalyst materials, and more. The antisilicostic action of poly (vinyl pyridine N-oxides) is especially intriguing.

N, N′-dialkyl-4, 4′-bipyridinium salts (viologens) have been used in a variety of polymeric systems due to their low redox potential. Monoalkylated bipyridyl when reacts with vinyl benzyl chloride produces additional polymerizable viologens. Monomer (xxiv) and related monomers have been discovered to polymerize and copolymerize effectively in aqueous free radical polymerization. The absorbance maxima of the cation radical of these polymers were produced by a decrease in aqueous solution to 530 nm as compared to 600 nm for the monomers. UV radiation of a polymer sheet resulted in absorption at 600 nm. The hypsochromic shift seems to be caused by the involvement of adjacent groups on the polymer in an aqueous solution.

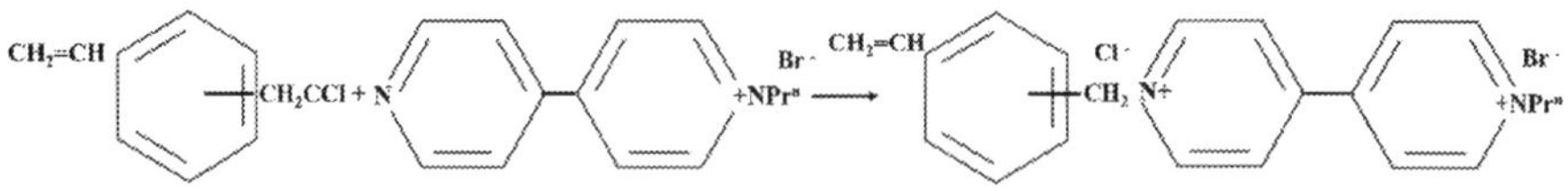

(xxiv)

Poly(viologens) (xxv) have been prepared directly using the Menschutkin reaction. When it is complexed with the "7,7′, 8,8′-tetracyanoquinodimethane" (TCNQ) radical anion. In addition, these polymers are extremely useful as redox and as conductive polymers.

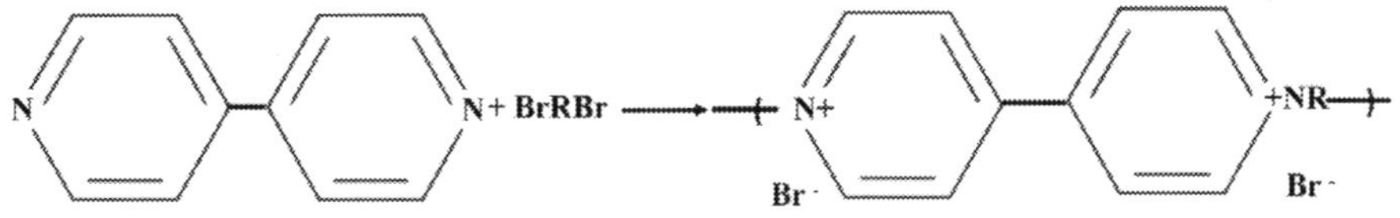

(xxv)

The related pyridine diesters or diacid chlorides have been used to make a variety of condensation polymers, for example, polyamides with the nucleus of pyridine in the backbone. Another method for integrating the nucleus of pyridine in a condensation framework is the Knoevenagel condensation. Poly (styryl pyridines) (xxvi) has been discovered to have excellent flame resistance and may be used in reinforced composites.

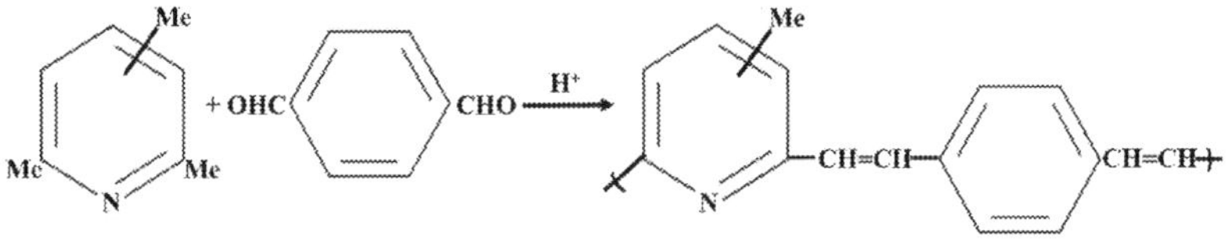

(xxvi)

Polymers that are solvable in a range of nonpolar solvents yet come to be insoluble when hydrolyzed have been created via an intriguing insertion process.

2.5 Vinyl pyridine-based polymers

A broad variety of polymers, organic synthesis, and catalytic processes employ polyvinyl pyridine (PVP) and its derivatives. For a range of chemicals and catalysts, PVP that has been cross-linked is well-known polymer support. Due to their interesting properties, such as ease of functionalization, high accessibility of functional groups, nonhygroscopicity, simplicity of synthesis, filtering, and swelling in various organic solvents, PVP is of great interest to researchers and scientists. According to different studies, PVP is used in many different scientific fields, and there are ways for increasing the catalytic efficacy of PVP-based reagents in organic synthesis. PVP has been utilized in different electrode designs for detecting humidity, metal ions, and other substances due to its electrical conductivity and excellent electrochemical properties. PVP-modified electrodes have also been used in electrochemical studies and several redox processes. Based on this background, PVP may be utilized in conjunction with other electroactive materials such as conductive polymers and inorganic materials to fabricate appropriate sensors for determining and monitoring solution compositions. It is worth noting that this study provides insights into innovative and convenient methods of producing PVP-based electrodes in the future, to provide better and superior electrochemical properties PVP, unlike polystyrene, is a true industrial polymer with a unit cost many times that of polystyrene. PVP is very resistant to electrophilic aromatic substitution due to its nucleophilic and weakly basic N-ring, although it has been reported to have a high chemical reactivity. The quaternization of PVP by nucleophilic attack on alkyl halides or protonation are two common processes.

PVP is a polar compound and may be anticipated to cooperate strongly with other polar molecules or may even serve as a suitable ligand for a range of metal ions. This unique reactivity is what makes PVP such an appealing material for a variety of specialty applications. The PVP-based polymers have been used as free radicals, as initiators, and also as noncrystallizable amorphous macromolecules. Treatment with aliphatic hydrocarbons at high temperatures (130–160°C) or simple heating in an inert environment at 140–160°C may also cause the amorphous polymer to crystallize. The crystal structure of polymers crystallized at high temperatures is usually different from that of polymers formed at low temperatures. In boiling aliphatic hydrocarbons and diethyl ether, the crystalline polymers of 2-vinyl pyridine are insoluble. Some polar solvents, such as methyl alcohol, dimethylformamide, chloroform, as well as boiling aromatic hydrocarbons, are soluble in them. The determination of melting temperatures (by polarized light microscope) of the crystalline acetone insoluble pyridine-based polymers gives the values generally between 190°C and 212°C [20]. There are some application-based properties of vinyl pyridine. Some of them are as follows.

2.5.1 Polyvinyl pyridine-based polymers in light-emitting devices

Pyridine-based polymers have been demonstrated to be potential candidates for light-radiating devices [21,22]. This feature is explained by the higher electron affinity of pyridine-based conjugated polymers. Due to this improvement in oxidation resistance, the polymer exhibits enhanced electron transport properties as well.

- The luminescence of various copolymers is determined by their internal photoluminescent quantum efficiencies.
- Relatively stable metals, such as Al, can be utilized as electron inserting species since pyridine-based polymers have a high electron affinity. Using poly (9-vinyl carbazole) as an electron-blocking/hole-transporting polymer, bilayer devices may be made with improved device efficiency due to charge confinement. There is an advantage in using pyridine-containing polymers because of their enhanced electron transport properties [23–25].

2.5.2 Applications in metal ion adsorption

With several research papers and patents detailing polymeric ligands based on partially or completely quaternized PVP, interest in selective ion-exchange resins, and chelate-forming resins has grown tremendously. Porous crosslinked VP-DVB resins quaternized by sulfuric acid reaction, for example, have been used to remove heavy metal ions such as Cr(V1) from wastewater [19] or gold from aqueous solution. A similar resin was also found to be efficient in uranium recovery, and a copolymer of 2-Me-5-VP and DVB was successfully utilized in the de-mercuration of effluent from mercury cells used in the electrolysis of brine. Although it was anticipated that the polymer-ligand chain would be maintained in the optimal conformation for the coordination sphere of a specific metal ion, the crosslinked polymer retains some flexibility and can adopt different conformations once the template ion is removed, resulting in a lack of complete ion selectivity.

2.5.2.1 Removal of organic contaminants

Vinyl pyridine-DVB copolymers have demonstrated remarkable promise for phenol removal from wastewater. Due to the presence of mildly basic pyridyl groups, these resins outperform the styrene-DVB copolymers and are used in several applications. Furthermore, the resins are less prone to fouling by inorganic salts in solution than more traditional anion exchange resins. The porous crosslinked PVP resins have high breakthrough capacities, and evolution of the absorbed phenols from the polymer can be accomplished by simple solvent extraction using methanol or acetone, or by heat treatment. PVP is interesting for usage as a reagent due to the presence of the reactive pyridyl ring on the repeating unit of the polymer. The electron-rich nitrogen may add to a variety of chemicals, resulting in novel reactive PVP resins. Reactivity of these chemicals is similar to but not identical to that of related low molecular weight pyridine derivatives, configurational effects, the local concentration of reactive species, ionic interactions, solubility, and so on. The porous crosslinked PVP resins have high breakthrough capacities, and the absorbed phenols may be eluted from the polymer by simple solvent extraction using methanol or acetone, or by heat treatment (Figs. 2.9–2.14).

PVP resins have also been utilized to remove aromatic compounds and carboxylic acids [26] from wastewater, while quaternized PVP resins were effective in adsorbing organic soaps and detergents. Furthermore, recent researches on crosslinked PVP or poly (VP-styrene) membranes have shown that they may be used to pervaporate aqueous amine solutions, with excellent efficiency in the separation of dipropyl amine from water. Researchers have also observed that pyridine-based polyurethanes (PUs) have differing degrees of resistance and durability against fungal biodegradation, depending on their structure and hard segment content. The different chemical structures of pyridine derivatives employed as chain-extenders have been discovered to impact the cohesion capacity of the hard segments, depending on their ability to create strong hydrogen bonds that inhibit the fungus from acting. In research reported by Orpera *et al.* [27], it has been revealed that PUs with 2-amino-3-hydroxy pyridine and 2,3-di-aminopyridine in the chain promote biodegradation by fungus *A. tenuissima*, whereas those with 2,3-di-hydroxypyridine and 3,4-di-aminopyridine in the chain demonstrate improved resistance to biodegradation. A. tenuissima is an appropriate fungal species for the breakdown of this type of PU. Thermomechanical studies by Oprea *et al.* [28] revealed poor molecular interconnectivity owing to the chemical structure, which eventually prevented pyridine nitrogen and other groups from PU urea chains from forming hydrogen bonds. Based on this behavior, these pyridine-based polymers have lower dielectric constants than other PUs produced from different pyridine derivatives even though they are dielectrically active themselves.

2.5.2.2 Oxidizing and reducing agents

PVP resins, as previously stated, offer good complexing capabilities, allowing them to be used in the removal of heavy metal ions. Due to their ability to form complex ions like Cr(V1) and the high reactivity of pyridine-based chromium (VI) oxidizing agents, polymeric reagents are more appealing than their low molecular weight equivalents. These polymers are electron-rich, making them suitable for the synthesis of complexes with diborane, as shown in Fig. 2.15. Such complexes have been made, and the PVP-borane reagent produced has been utilized to reduce carbonyl compounds.

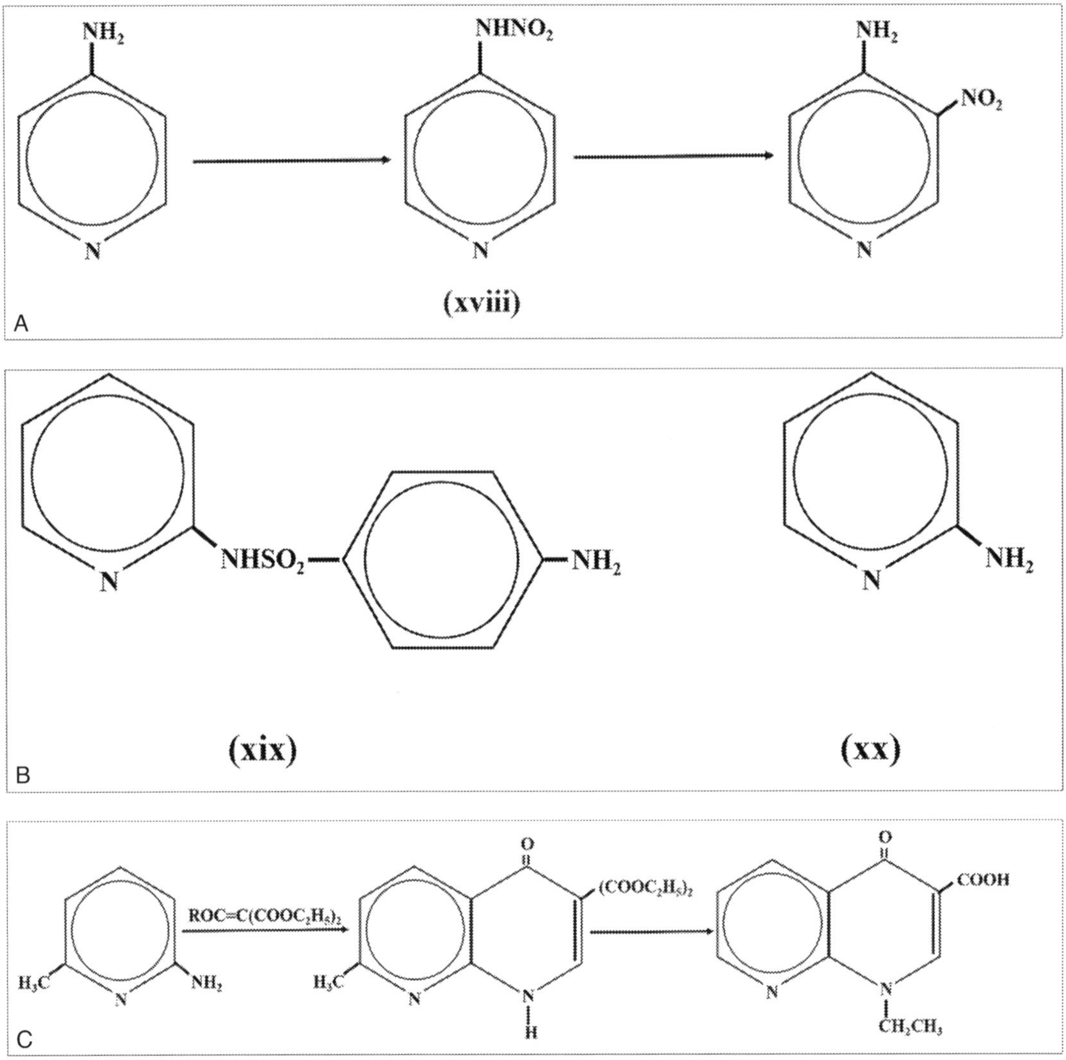

FIGURE 2.9 Different reactions of nitrogen substituent.

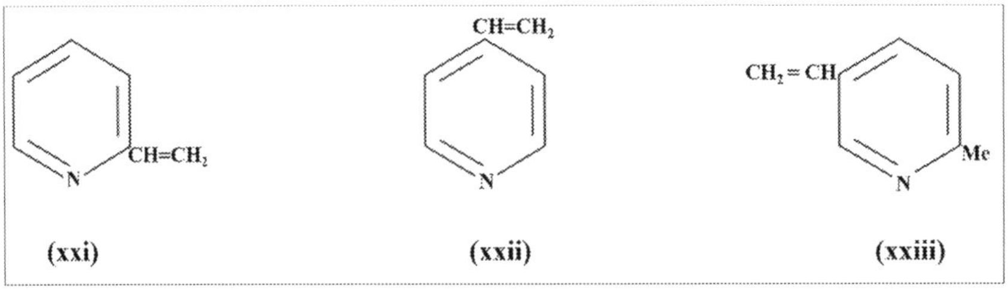

FIGURE 2.10 Different monomers of pyridines.

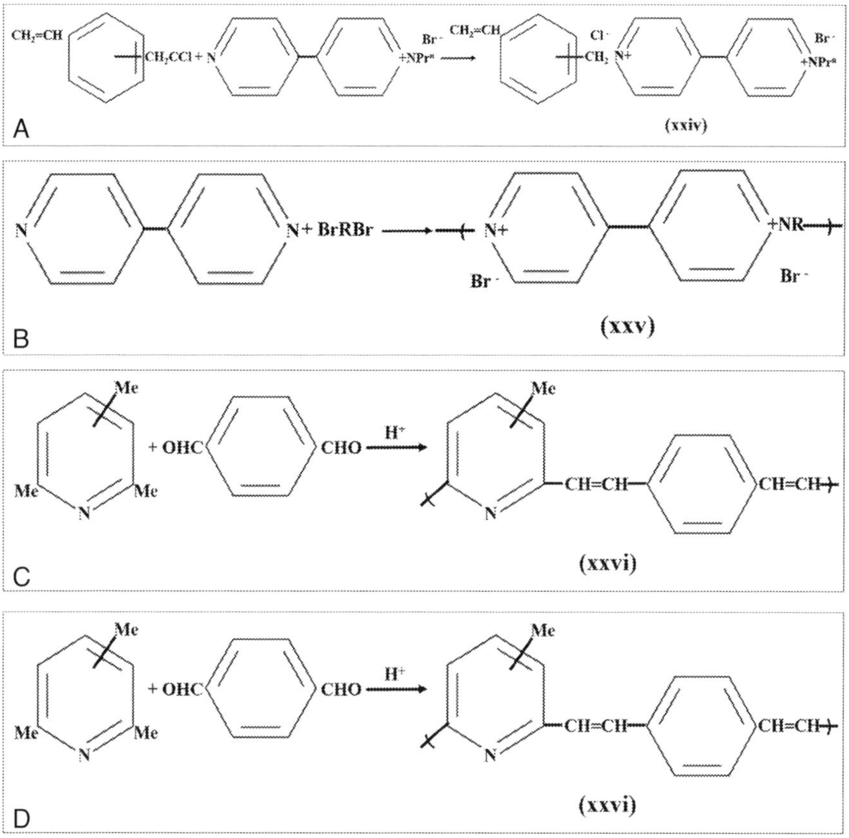

FIGURE 2.11 Synthesis of different pyridine polymers.

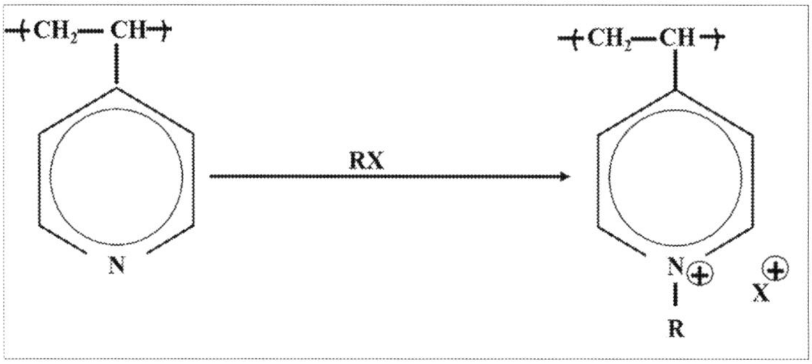

FIGURE 2.12 Reaction showing the nucleophilic attack.

2.5.2.3 *Pyridine-based urethanes*

PUs are one-of-a-kind polymeric materials having a diverse set of physical and chemical properties [29,30]. PUs may be utilized in a wide range of applications in contemporary

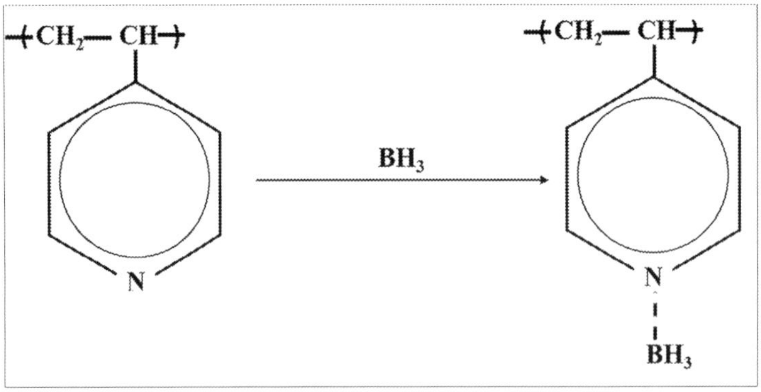

FIGURE 2.13 Pyridine-based reducing agent.

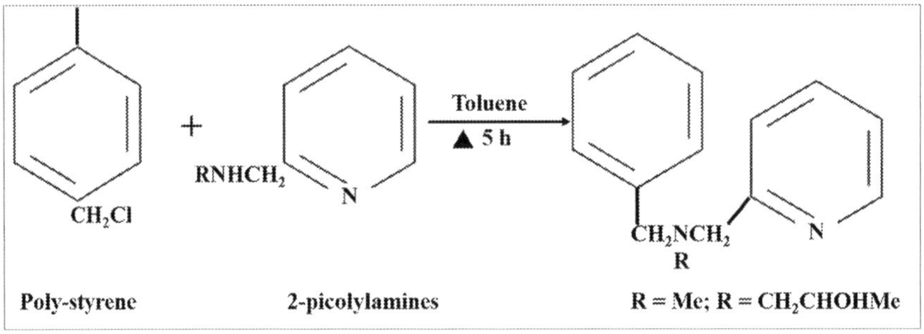

FIGURE 2.14 Ion exchange resin of pyridine.

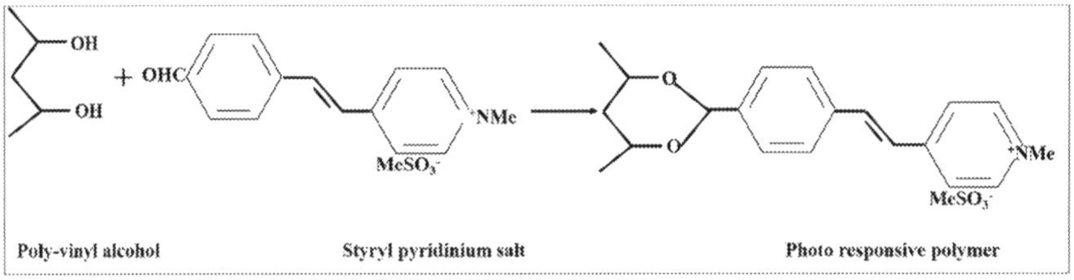

FIGURE 2.15 Synthesis of photoresponsive polymer.

technology, including automobiles, medicine, coatings, adhesives, fiber, foam, thermoplastic elastomers, and packaging [29]. Their characteristics are influenced by many variables and may be altered to a large extent by the appropriate component selection, composition, and preparation methods. There are a few options for incorporating functions into PU materials [31].

2.6 Applications of pyridine-based polymers

As modifying heterocycles, pyridine derivatives have garnered the greatest attention.

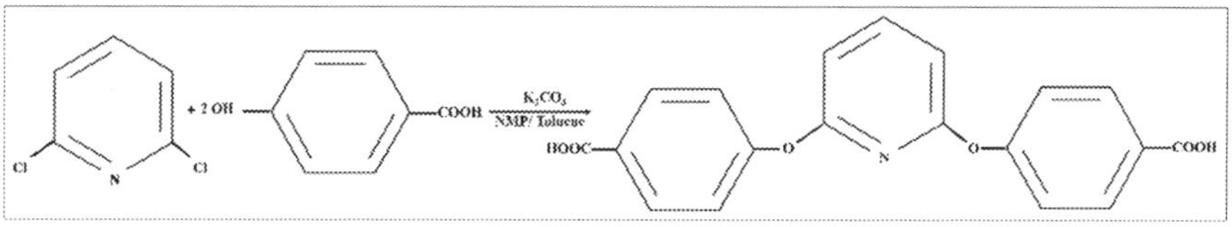

FIGURE 2.16 Diacid containing pyridine with improved solubility and high thermal stability [38].

2.6.1 Ion-exchange resins

The reaction of 2-picolylamines with chloromethylated poly (styrene) with produced resins is selective toward nickel and copper used for ion exchange. Due to its high selectivity in the existence of low pH and excess iron (III) ion, the resin can be well suited for nickel and copper recovery in hydrometallurgical processes. The use of "N-(hydroxyalkyl) polyamines" is ensured in even higher capacity resins. The addition of the "2-hydroxypropyl" group to resin results in a six-fold stronger adsorbent for Cu^{+2} ions, with an ultimate capacity of 40 g/L [32].

2.6.2 Photoresists

Derivatives of pyridine have also been used to make photoresists. The stilbazolium group can be attached when "poly (vinyl alcohol)" interacts with styryl pyridinium salt [33]. When irradiated in an aqueous solution, the polymer proceeds through cis-trans isomerization, but when radiated as a dry film, the stilbazolium groups dimerized and the polymer remains insoluble. A 1 mol percent stilbazolium group can be were loaded into the polymer resulted in a photoresponsive polymer that was 10 times more delicate than standard photoresists based on ammonium dichromate made this occurrence even more remarkable. The increased sensitivity was explained as the consequence of ionic group aggregation in the film state [34].

2.6.3 Heat-resisting materials

Researchers in polymer science have been studying heat-resistant or thermally stable polymers for more than 60 years, and their applications are still growing. Polyarylate, polyimide, polyamide, and their copolymers are the most common types of these polymers. Polyarylates are a kind of high-performance engineering plastic with outstanding thermal and mechanical characteristics with a wide range of commercial uses [35,36]. Their most typical applications include enameled wire coatings, hot melt adhesives, heat-resistant films, high-strength fibers, and printed circuit boards [37–40]. Optical activity polyesters are among those having unique characteristics that find use in certain optics-related applications. The production of optically active polymers has attracted interest because of their chiral character, which may play a significant role in the molecular organization and assembly necessary for optoelectronics-related applications (Figs. 2.16–2.26).

2.6.4 High-temperature electrolyte membranes

Due to its potential as high-energy-density power sources for both mobile and stationary applications, fuel cells based on solid polymer electrolytes have lately attracted a lot of

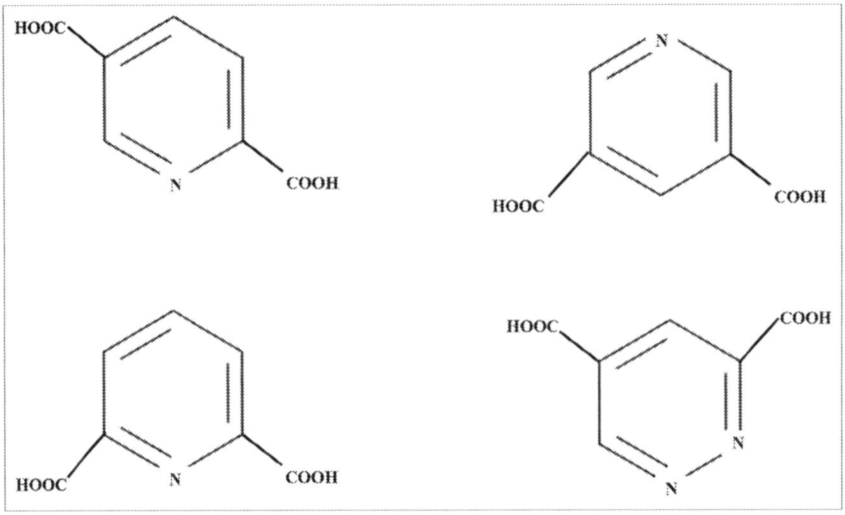

FIGURE 2.17 Pyridine-based compounds used for electrolyte membranes.

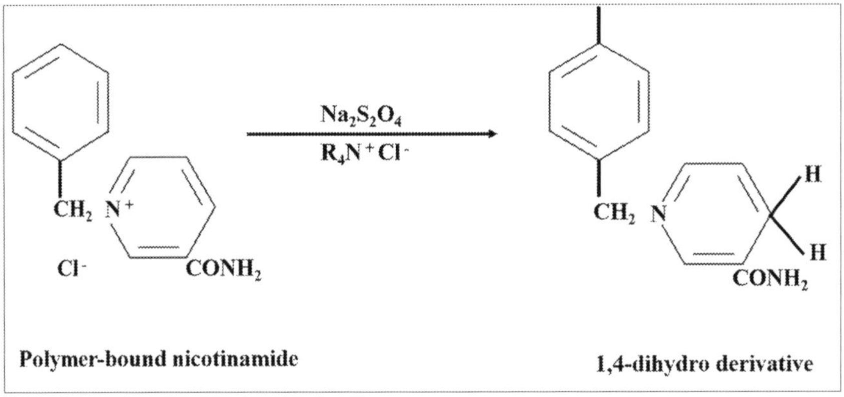

FIGURE 2.18 Synthesis of redox polymer.

attention [41]. There are several advantages of using high-temperature polymer electrolyte membranes that function at temperatures exceeding 120°C without humidification, including fast electrode kinetics, high tolerance to fuel impurities like carbon monoxide, and a simpler design [42].

2.6.5 Redox polymers

A quaternization process using chloromethylated polymer has been used to integrate nicotinamide (styrene). The 1,4-dihydro derivatives can be obtained by reducing polymer-bound nicotinamide with dithionite, to reduce a variety of substrates, including dyes like Tillman regent and malachite green, stable free radicals like l, l-diphenyl-2-picrylhydrazyl, and

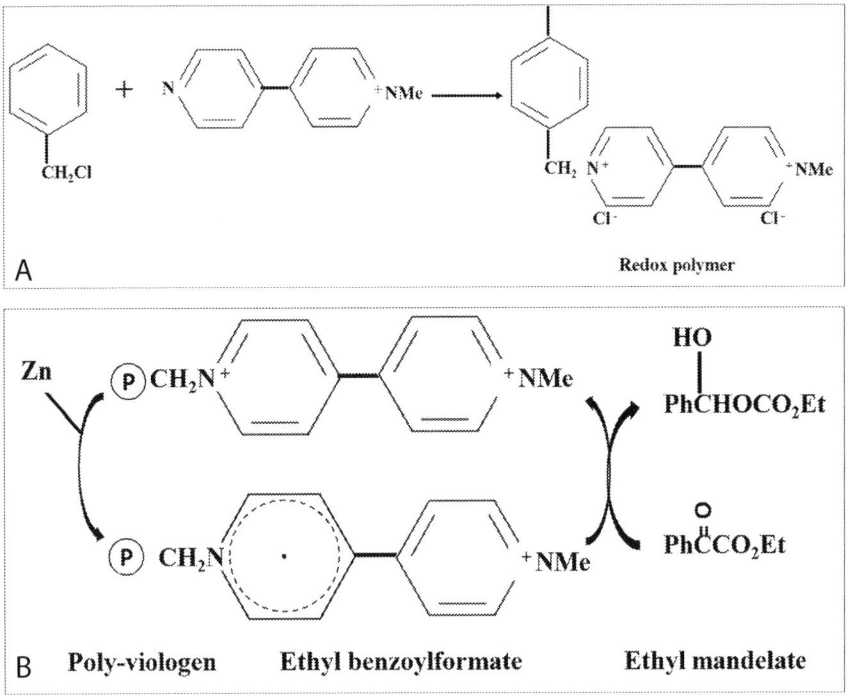

FIGURE 2.19 Effect of poly-viologen on redox polymer.

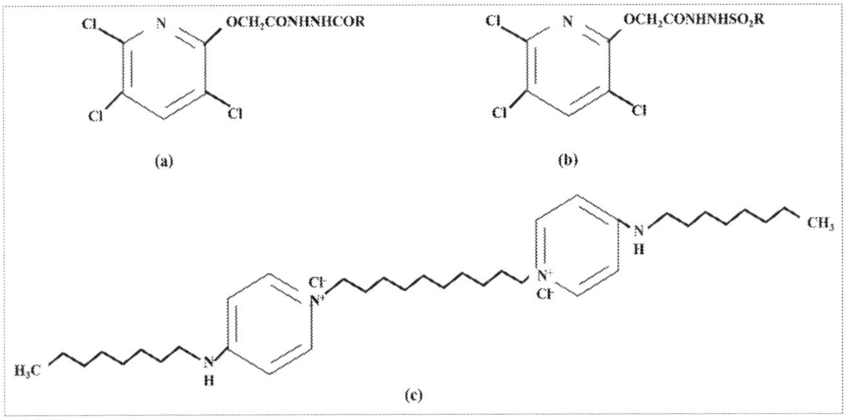

FIGURE 2.20 Chemical structure of (A) N-(3,5,6-Trichloro-2-pyridyloxyacetyl)-N′-aroylhydrazines and (B) N-(3,5,6-Trichloro-2-pyridyloxyacetyl)-N′-arylsulfonylhydr-azines, and (C) pyridine hybrid made up of hydrophilic and lipophilic components showing antimicrobial activity [44].

activated carbonyl compounds like quinone, choranil, and ninhydrin. Using these polymeric reducing agents is particularly appealing since the process generally consists of filtering the discarded polymeric reagent, draining the solvent to extract the reduced result, and replenishing the redox polymer by reduction with dithionite.

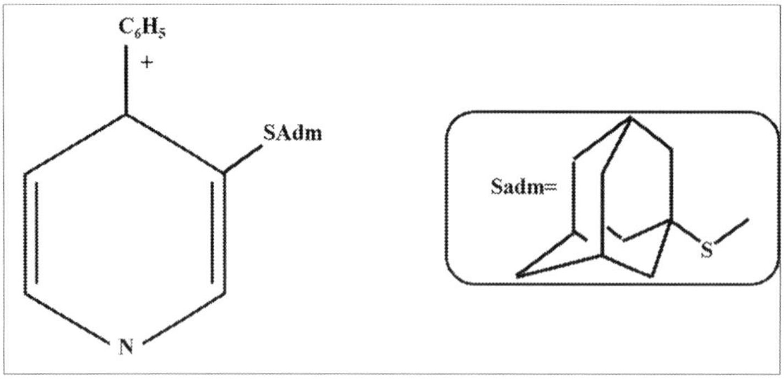

FIGURE 2.21 Pyridine compound as an antioxidant.

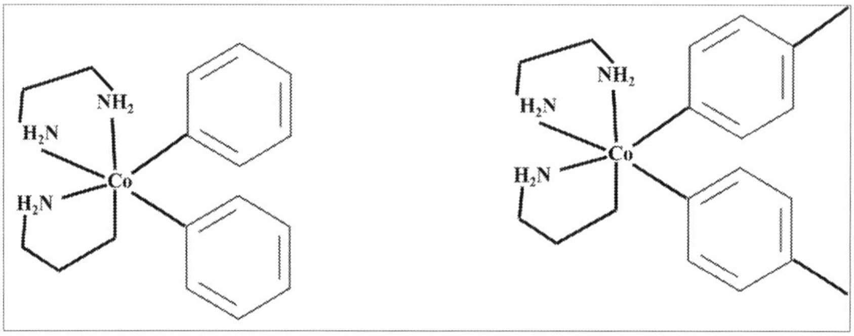

FIGURE 2.22 Heterolyptic complexes of cobalt (III) showing good anticancerous activity.

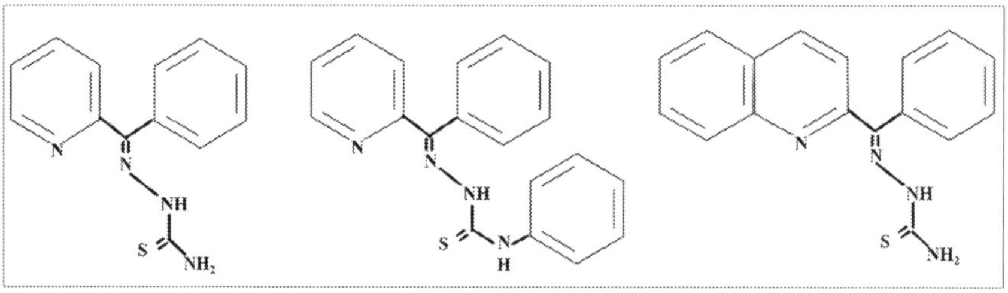

FIGURE 2.23 Pyridine-derived thiosemicarbazones as anticancer drugs.

2.6.6 Biological applications

2.6.6.1 Antimicrobial applications

Di-acyl hydrazine and acyl (aryl sulfonyl) hydrazine are pyridine derivatives with promising antibacterial properties against gram-negative bacteria *E. coli* and Gram-positive *S. albus*. These have also been put to the test as herbicides against *C. dactylon, C. rotundus, E. crusgalli, C. hirta*, and *T. indica*. The same chemicals have antifungal properties against *A. niger* and *A. niger Teniussiama* has also been studied using Griseofulvin as a standard [43].

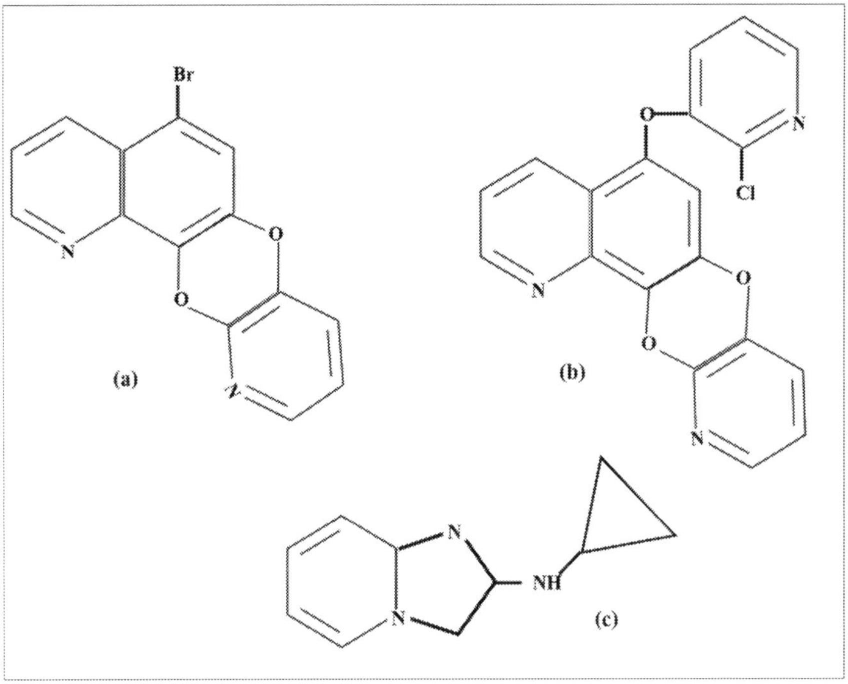

FIGURE 2.24 (A and B) Compounds showing antimalarial, and (C) anti-inflammatory property.

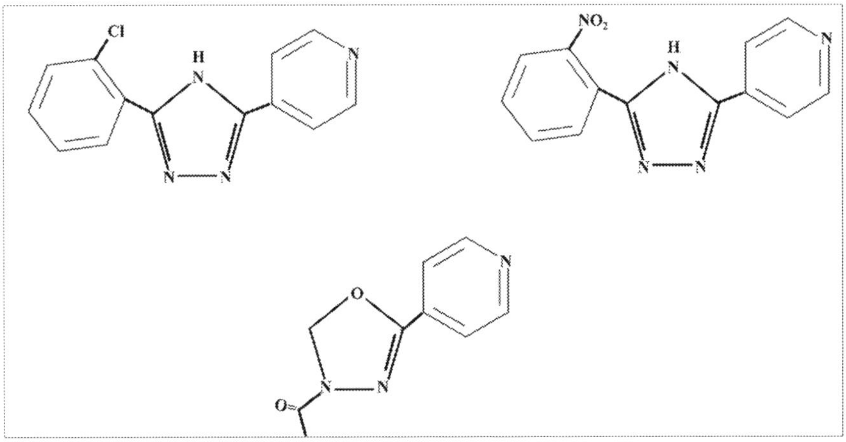

FIGURE 2.25 Pyridine derivatives showing analgesic property.

The antimicrobial activity of thienopyridine and other pyridine derivatives have also been demonstrated against Gram-positive bacteria S. aureus as well as the gram-negative *E. coli*, *P. vulgaris*, and *P. aeruginosa* [45]. Some Schiff base ligands, such as 1-phenyl-2,3-dimethyl-4-salicylalidene pyrazole-5-one, are simple to be synthesized and may be coupled with metals like Cu^{2+}, Ni^{2+}, Zn^{2+}, and Fe^{3+}. These ligands, both in their free and in metal complexes, have

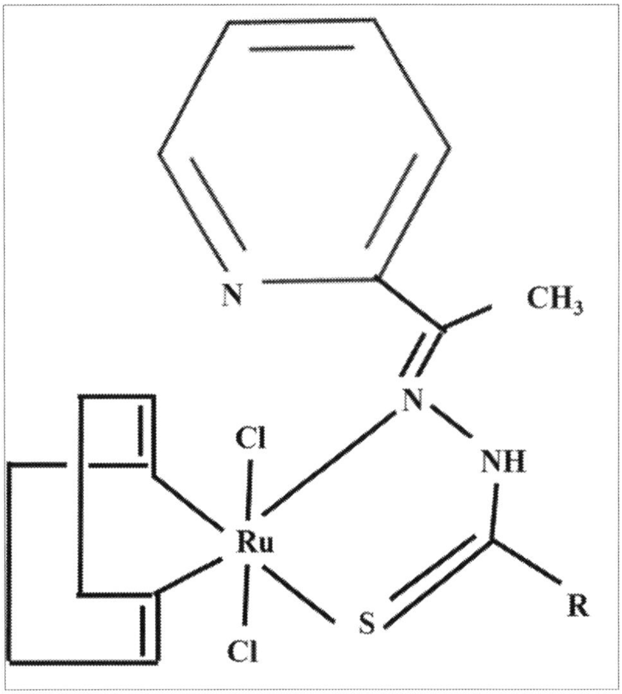

FIGURE 2.26 Pyridine derivatives showing antiamoebic property.

antibacterial and antifungal properties that are superior to several well-known antibiotics [46]. Synthetic compounds of pyridine with a benzothiazolylamino group substituted at position 2 and carboxylic functionality have been tested for biological activity. Experiments on their antibacterial properties have revealed that they exhibit weak to mild antimicrobial activity [47]. Substituents have a stronger impact on the biological activity of pyridine derivatives. As a conversion product of synthetic chalcone, pyrazoline, pyridine, and pyrimidine derivatives were produced. These conversion products were tested for anticancer as well as antimicrobial activity. According to the results, ciprofloxacin had moderate to high bactericidal effectiveness against Staphylococcus aureus and Pseudomonas aeruginosa. Almost all of the pyridine-based polymers were observed to have modest anti-Candida-Albicans action [48]. The antifungal activity of N-(pyridine-2-methyl)-2-(4-chlorophenyl)-3-methyl butanamide, a pyridine-containing molecule, is excellent [4]. The antibacterial, antifungal, and anti-inflammatory properties of the triazoles substituted pyridine derivatives have also been investigated. The results of the studies indicated that all of the compounds have moderate antibacterial and antifungal activity. It was discovered that compounds with free NH_2 at position 4 have the most antibacterial property. Triazoles ring system may be connected to the activities/efficacy against bacteria [49].

2.6.6.2 Antiviral applications

Influenza B-Mass virus: Thiazolo [50,51] pyridine Oxime derivatives have antiviral efficacy against the influenza B-Mass virus. The pyridine and naphthyridine oxime compounds exhibit strong anti-HIV activity. Oximes-Pyridine derivatives are utilized as antidotes for organophosphorus chemical poisoning cases [52]. Antiviral (HCV): Hydrazones of 3- and 4-acetyl pyridine, in combination with antitumor action, suppress HCV replication in both RNA (+) and RNA (-) strain [53]. Antiviral activity of Bi Pyridinyl derivatives complexed with ruthenium against hepatitis C-virus (HCV) [54] is also found effective to be used in medicinal research.

2.6.6.3 *Antioxidant applications*

Some thiopyridine derivatives have antioxidant as well as cytotoxic activity, and both properties are found to be appealing. The structure of compounds has a substantial influence on these activities. The most relevant descriptors for connecting the molecular structure of substances with their different SOD activities are dipole moment and electrophilic index, according to the QSAR research. Compounds having a high dipole moment and electrophilic index show increased antioxidant activity. Compounds having the lowest atomic polarizability are seen to have the highest cytotoxic activity [4].

2.6.6.4 *Anticancerous applications*

Cu (II) metal complexes with pyridine and Schiff base 2-[N-(apicolyl)-amino]-benzophenone, especially brominated products, exhibit a high degree of cytotoxicity and anticancer activity [55]. Pyazoline, pyridine, and pyrimidine compounds having indole functionality can be purified to a fair degree and can be expected to show good activity against tumor cells [48]. 2-Benzoxazolylhydrazon, 2-acetylpyridine derivatives, and 2-benzoxazolylpyridine derivatives exhibit antitumor action, and when the acetyl group is replaced with an acyl group. It has a good inhibitory effect against leukemia, colon, and ovarian cancer cell lines [56]. As a result of the high antitumor effect and minimal cytotoxicity, palladium and zinc complexes of 2-acetyl pyridine thiosemicarbazone can also be used as antitumor alternatives in near future [57]. Cd complexes containing 2,6-diacetyl pyridine also show antitumor action in the C6 glioma cell line, suggesting that they might be used to treat drug-resistant brain tumors [58].

2.6.6.5 *Antimalarial and analgesic applications*

Some heterocyclic compounds with pyridine have been synthesized and tested for analgesic efficacy. Compounds with pyridine nucleolus numbers below 38–40 exhibited comparable analgesic activity as compared to the conventional (pentazocine). On the heterocyclic ring, mecamylamine is observed to block the activity [59].

Hybrid molecules of pyridine-quinoline have been evaluated for their antimalarial efficacy against *Plasmodium falciparum* and the results showed that these molecules exhibited modest antimalarial efficacy. The findings also revealed that these molecules had weak antimalarial activity.

These chemicals suggest that they might be utilized as templates for developing new antimalarial drugs with enhanced activity. HPIA (haem polymerization inhibition activity) was also demonstrated by these compounds [60].

2.7 Conclusions

The chapter develops a strong foundation to understand the importance and various aspects about pyridine, its derivatives, and its polymers. Pyridine has a broad range of applications in almost every field of science and engineering. Pyridine-based polymers and derivatives have been widely exploited for their unique and interesting characteristics. Due to its chemical nature, pyridine has been studied to offer a wide range of properties which are highly useful and interesting for research-based works. Some effective substituted pyridine derivatives are now the top medication on the market and have significant biomedical applications possessing antiviral and antiasthmatic properties. A number of modified pyridine derivatives have been shown to have antihypertensive, antiasthma, antiviral, and particular anti-HIV effects as well. In the coming future, pyridine-based medicines are anticipated to replace many existing organic-based drugs.

References

[1] P.P. Pandey, Introductory chapter: pyridine, Pyridine, IntechOpen, 2018, doi:10.5772/intechopen.77969.
[2] P. Patil, et al., Pyridine and its biological activity: a review, Asian J. Res. Chem. 6 (10) (2013) 888–899.
[3] S. Shimizu, et al., Synthesis of pyridine bases: general methods and recent advances in gas phase synthesis over ZSM-5 zeolite, Catal. Surv. Asia 2 (1) (1998) 71–76.
[4] A.A. Altaf, et al., A review on the medicinal importance of pyridine derivatives, J. Drug Des. Med. Chem. 1 (1) (2015) 1–11.
[5] E.W. Warnhoff, When piperidine was a structural problem, Bull. Hist. Chem. 22 (1998) 29–34.
[6] J. Huheey, E. Keiter, R. Keiter, AppendixG: tanabe-sugano diagrams, Inorganic Chemistry: Principles of Structure and Reactivity, 4th ed., Harper Collins College Publishers, New York, NY, 1993, pp. A38–A39.
[7] S. Pal, Pyridine: A useful ligand in transition metal complexes, IntechOpen, Pyridine 57, 2018 doi: http://dx.doi.org/10.5772/intechopen.76986.
[8] N. Firsova, Yu.V. Kolodyazhnyi, O.A. Osipov, Pyridine Metal Complexes, Zh. Obshch. Khim 14, part 6A (1969) 2151–2154.
[9] S. Shibata, Spectro photometric determination of bromate withoff-arsanilic acid, Anal. Chim. Acta 28 (1963) 388–392.
[10] N. Dutt, A.S. Gupta, Chemistry of lanthanons, XLII preparation and characterization of isonicotine and picoline hydrazide complexes of lanthanons, Zeitschrift für Naturforschung B 30 (9-10) (1975) 769–772.
[11] M. Hniličková, L. Sommer, Pyridin-2-azo-1-dihydroxynaphthalin-2, 3-sulfonsäure-6 und Pyridin-2-azo-4-orcin, zwei neue chelatometrische Indicatoren, Fresenius' Zeitschrift für analytische Chemie 177 (6) (1960) 425–429.
[12] T. Sas, K. LN, A. NI, Scanduim isothiocyanate complexes with pyridine, Zh. Neorg. Khim. 16 (1) (1971) 87 -&.
[13] C. Hansch, et al., Chem-bioinformatics: comparative QSAR at the interface between chemistry and biology, Chem. Rev. 102 (3) (2002) 783–812.
[14] J.A. Joule, K. Mills, G.F. Smith, Heterocyclic Chemistry, CRC Press, London, 2020.
[15] L.W. Deady, M.R. Grimmett, C.H. Potts, Studies on the mechanism of the nitraminopyridine rearrangement, Tetrahedron 35 (24) (1979) 2895–2900.
[16] C. Giam, et al., Phthalic acid esters, Anthropogenic compounds, Springer, United States, 1984, pp. 67–142.
[17] G.Y. Lesher, et al., 1, 8-Naphthyridine derivatives. A new class of chemotherapeutic agents, J. Med. Chem. 5 (5) (1962) 1063–1065.
[18] J.K. Rasmussen, et al., Poly (2-imidazolin-5-ones)—A new class of heterocyclic polymers, J. Polym. Sci. Part A Polym. Chem. 24 (11) (1986) 2739–2747.
[19] J.M. Fréchet, M.V. de Meftahi, Poly (vinyl pyridine) s: Simple reactive polymers with multiple applications, Br. Polym. J. 16 (4) (1984) 193–198.
[20] D. Gebler, et al., Blue electroluminescent devices based on soluble poly (p-pyridine), J. Appl. Phys. 78 (6) (1995) 4264–4266.

[21] H. Wang, et al., Application of polyaniline (emeraldine base, EB) in polymer light-emitting devices, Synth. Met. 78 (1) (1996) 33–37.
[22] Y. Wang, A. Epstein, Interface control of light-emitting devices based on pyridine-containing conjugated polymers, Acc. Chem. Res. 32 (3) (1999) 217–224.
[23] C. Wang, et al., Tuning the optoelectronic properties of pyridine-containing polymers for light-emitting devices, Adv. Mater. 12 (3) (2000) 217–222.
[24] A.J. Epstein, et al., Pyridine-based conjugated polymers: Photophysical properties and light-emitting devices, Macromolecular Symposia, Wiley Online Library, United States, 1997.
[25] M. Szycher, Szycher's handbook of polyurethanes, CRC press, United States, 1999.
[26] X. Zhang, et al., Determination of ochratoxin A in raisins by SPE-HPLC-FLD, Chin. J. Anal. Lab. 31 (2012) 64–67.
[27] S. Oprea, et al., Biodegradation of pyridine-based polyether polyurethanes by the Alternaria tenuissima fungus, J. Appl. Polym. Sci. 135 (14) (2018) 46096.
[28] S. Oprea, V.O. Potolinca, Thermomechanical and dielectric properties of novel pyridine-based polyurethane urea elastomers, J. Elastomers Plast. 50 (3) (2018) 276–292.
[29] S. Gogolewski, Selected topics in biomedical polyurethanes. A review, Colloid. Polym. Sci. 267 (9) (1989) 757–785.
[30] K.R. Reddy, et al., Synthesis and characterization of pyridine-based polyurethanes, Des. Monomers Polym. 12 (2) (2009) 109–118.
[31] N. Kawabata, K. Ohira, Removal and recovery of organic pollutants from aquatic environment. 1. Vinylpyridine-divinylbenzene copolymer as a polymeric adsorbent for removal and recovery of phenol from aqueous solution, Environ. Sci. Technol. 13 (11) (1979) 1396–1402.
[32] S. Song, et al., Novel dual-mode photoresist based on decarboxylation by photogenerated base compound, Polym. Adv. Technol. 9 (6) (1998) 326–333.
[33] C.L. Aronson, et al., The effect of macromolecular architecture on functional group accessibility: hydrogen bonding in blends containing phenolic photoresist polymers, Polym. Bull. 53 (5) (2005) 413–424.
[34] B. Fravel, R. Murugan, and E. Scriven, *31.9 Product Class 9: Arenesulfonic Acids and Derivatives.* doi:10.1055/sos-SD-031-00615.
[35] P. CASSIDY, Thermally stable polymers: Syntheses and properties(Book), Marcel Dekker, Inc., New York, NY, 1980, p. 1980. Research supported by the Robert A. Welch Foundation 407.
[36] R. Seymour, G. Kirshenbaum, Durability (eventually for several thousand years), in: High Performance Polymers: Their Origin and Development: Proceedings of the Symposium on the History of High Performance Polymers at the American Chemical Society Meeting Held in New York, April 15–18, 1986, Springer, 1986.
[37] A.-C. Albertsson, I.K. Varma, Aliphatic polyesters: Synthesis, properties and applications, Degradable Aliphatic Polyesters, 2002, pp. 1–40.
[38] S. Mehdipour-Ataei, A. Mahmoodi, Heat-resistant pyridine-based poly (ether-ester) s: synthesis, characterization and properties, Chin. J. Polym. Sci. 31 (1) (2013) 171–178.
[39] S. Maiti, S. Das, Synthesis and properties of a new polyesterimide, Die Angewandte Makromolekulare Chemie: Applied Macromolecular Chemistry and Physics 86 (1) (1980) 181–191.
[40] S. Maiti, S. Das, Synthesis and properties of polyesterimides and their isomers, J. Appl. Polym. Sci. 26 (3) (1981) 957–978.
[41] B.C. Steele, A. Heinzel, Materials for fuel-cell technologies, Materials for Sustainable Energy: A Collection of Peer-Reviewed Research and Review Articles From Nature Publishing Group, World Scientific, London, 2011, pp. 224–231.
[42] C. Yang, et al., Approaches and technical challenges to high temperature operation of proton exchange membrane fuel cells, J. Power Sources 103 (1) (2001) 1–9.
[43] V. Chavan, et al., Synthesis, characterization, and biological activities of some 3, 5, 6-trichloropyridine derivatives, Chem. Heterocycl. Compd. 42 (5) (2006) 625–630.
[44] V. Savchenko, et al., Developing state-of-the-art antiseptics based on pyridine derivatives, Her. Russ. Acad. Sci. 80 (2) (2010) 149–154.
[45] V. Zav'yalova, A. Zubarev, A. Shestopalov, Synthesis and reactions of 3-acetyl-6-methyl-2-(methylthio) pyridine, Russ. Chem. Bull. 58 (9) (2009) 1939–1944.
[46] M.M. Mashaly, Z.H. Abd-Elwahab, A.A. Faheim, Preparation, spectral characterization and antimicrobial activities of schiff base complexes derived from 4-Aminoantipyrine. Mixed Ligand Complexes with 2-Aminopyridine, 8-Hydroxyquinoline and Oxalic Acid and Their Pyrolytical Products, J. Chin. Chem. Soc. 51 (5A) (2004) 901–915.

[47] N.B. Patel, S.N. Agravat, F.M. Shaikh, Synthesis and antimicrobial activity of new pyridine derivatives-I, Med. Chem. Res. 20 (7) (2011) 1033–1041.

[48] E. Nassar, Synthesis,(in vitro) antitumor and antimicrobial activity of some pyrazoline, pyridine, and pyrimidine derivatives linked to indole moiety, J. Am. Sci. 6 (8) (2010) 463–471.

[49] M. Nitin, et al., Synthesis, antimicrobial and anti-inflammatory activity of some 5-substituted-3-pyridine-1, 2, 4-triazoles, Int. J. PharmTech Res. 2 (4) (2010) 2450–2455.

[50] Y.-L. Jin, et al., Synthesis of a novel series of imidazo [1, 2-α] pyridines as acyl-CoA: cholesterol acyltransferase (ACAT) inhibitors, Bull. Korean Chem. Soc. 30 (6) (2009) 1297–1304.

[51] E. Busto, V. Gotor-Fernández, V. Gotor, Chemoenzymatic synthesis of chiral 4-(N, N-dimethylamino) pyridine derivatives, Tetrahedron Asymmetry 16 (20) (2005) 3427–3435.

[52] E. Abele, R. Abele, E. Lukevics, Pyridine oximes: Synthesis, reactions, and biological activity, Chem. Heterocycl. Compd. 39 (7) (2003) 825–865.

[53] S.A. El-Hawash, A.E. Abdel Wahab, M.A. El-Demellawy, Cyanoacetic acid hydrazones of 3-(and 4-) Acetylpyridine and some derived ring systems as potential antitumor and anti-HCV agents, Archiv der Pharmazie 339 (1) (2006) 14–23.

[54] M. Vrábel, et al., Purines bearing phenanthroline or bipyridine ligands and their RuII complexes in position 8 as model compounds for electrochemical DNA labeling–synthesis, crystal structure, electrochemistry, quantum chemical calculations, cytostatic and antiviral activity, Eur. J. Inorg. Chem. 2007 (12) (2007) 1752–1769.

[55] Y.-T. Wang, et al., New homochiral ferroelectric supramolecular networks of complexes constructed by chiral S-naproxen ligand, CrystEngComm 14 (10) (2012) 3802–3812.

[56] J. Easmon, et al., Synthesis, structure– activity relationships, and antitumor studies of 2-benzoxazolyl hydrazones derived from alpha-(N)-acyl heteroaromatics, J. Med. Chem. 49 (21) (2006) 6343–6350.

[57] D. Kovala-Demertzi, et al., Synthesis, characterization, crystal structures, in vitro and in vivo antitumor activity of palladium (II) and zinc (II) complexes with 2-formyl and 2-acetyl pyridine N (4)-1-(2-pyridyl)-piperazinyl thiosemicarbazone, Polyhedron 27 (13) (2008) 2731–2738.

[58] N.A. Illán-Cabeza, et al., New 2, 6-bis-[uracil-imino] ethylpyridine complexes containing the CdN6 core: Synthesis, crystal structures, luminescent properties and antiproliferative activity against C6 glioma cells, J. Inorg. Biochem. 103 (8) (2009) 1176–1184.

[59] G. Nigade, P. Chavan, M. Deodhar, Synthesis and analgesic activity of new pyridine-based heterocyclic derivatives, Med. Chem. Res. 21 (1) (2012) 27–37.

[60] B.N. Acharya, D. Thavaselvam, M.P. Kaushik, Synthesis and antimalarial evaluation of novel pyridine quinoline hybrids, Med. Chem. Res. 17 (8) (2008) 487–494.

CHAPTER

3

Synthetic strategies of functionalized pyridines and their therapeutic potential as multifunctional anti-Alzheimer's agents

Jeelan Basha Shaik, Mohammad Khaja Mohinuddin Pinjari, Damu Amooru Gangaiah and Chinna Gangi Reddy Nallagondu

Department of Chemistry, Yogi Vemana University, Kadapa, Andhra Pradesh, India

3.1 Introduction

Pyridine was first isolated in 1846 by Anderson from bone pyrolyzates and coal tar as picoline. Pyridine is not abundant in nature, except in the leaves and roots of belladonna (*Atropa belladonna*) and in marshmallow (*Althaea officinalis*). Trace amounts of pyridine is found in Beaufort cheese, black tea, sunflower honey, vaginal secretions, and saliva of those suffering from gingivitis. Further, the traces of pyridine are also found in roasted food items like potato chips, fried bacon, sukiyaki, fried chicken, roasted coffee, etc. Pyridines have a wide range of applications in several areas of chemistry as pyridine moiety is ubiquitous in natural products, pharmaceuticals, agrochemicals, and functional organic materials. Due to its weak basicity, the pyridine nucleus tends to enhance the water solubility of its analogs which has made this moiety as a significant part of many drugs and pesticides. This nucleus is present in more than 100 drugs with a wide spectrum of biological activities and this number is increasing day by day [1,2]. For example, omeprazole (proton pump inhibitor), lorlatinib (anaplastic lymphoma

DOI: https://doi.org/10.1016/B978-0-323-91221-1.00014-2

kinase inhibitor), ivosidenib (Isocitrate dehydrogenase 1 inhibitor), nevirapine, (HIV-1 reverse transcriptase inhibitor), abemaciclib (cyclin-dependent kinase 4/6 inhibitor), piroxicam (non-steroidal anti-inflammatory), apalutamide (androgen receptor antagonist), netupitant (NK-1 receptor antagonist), tripelenamine (antipruritic), delavirdine (non-nucleoside reverse transcriptase inhibitor), sulfapyridine (antibacterial), Loratadine (nonsedating histamine receptor-1 antagonist), cetylpyridinium chloride (antiseptic), zolimidine (gastroprotective), olprinone (cardiotonic agent), saripidem (sedative and anxiolytic), rifaximin (antibiotic), GSK812397 (CXC chemokine receptor 4), zolpidem, (insomnia), and alpidem (anxiolytic) are some of the representative drugs containing pyridine moiety with diverse applications (Fig. 3.1). Among pyridine analogues, 2-aminocyanopyridines, pyridinium derivatives, imidazopyridines, bis-imidazopyridines, pyrazolopyridines, and furopyridines are considered as highly privileged scaffolds as they are existing as core structure in many natural and bioactive compounds. Besides, pyridine and its derivatives have momentous applications as organic bases, ligands, catalysts, and directing groups in C–H activation reactions [3]. Consequently, this chapter mainly covered two sections highlighting (1) synthetic strategies of medicinally privileged pyridine-based heterocycles like 2-aminocyanopyridines, pyridinium derivatives, imidazopyridines, pyrazolopyridines, and furopyridines and (2) therapeutic potential of pyridine-based compounds as multifunctional anti-Alzheimer's agents.

3.2 Synthetic strategies of pyridine based heterocycles

3.2.1 Synthesis of pyridine nucleus

Owing to the diversity of pyridine analogs in the therapeutic response profile, several research groups have been focusing to develop synthetic methods for pyridine-based heterocycles. As a result, numerous synthetic strategies have been reported in the literature. Pyridine was first synthesized by William Ramsay in 1876 by passing a mixture of acetylene and HCN through a red-hot tube (Scheme 3.1) [4]. In 1881, Arthur Rudolf Hantzsch reported a multicomponent strategy for the synthesis of pyridine derivatives from formaldehyde, ethyl acetoacetate (2.0 equiv.), and ammonium acetate/ammonia (Scheme 3.2) [5]. Aleksei Chichibabin developed a scalable method for the synthesis of pyridine by the condensation reaction of aldehydes/ketones/α,β-unsaturated carbonyl compounds, or any combination of these compounds, in ammonia or ammonia derivatives (Scheme 3.3) [6,7]. Later, a huge number of improved protocols have been developed for the synthesis of pyridines. One of the most convenient synthetic routes is the [2+ 2+ 2] cycloaddition of alkynes and nitriles catalyzed by various transition-metals (Co, Rh, Fe, Ru, Ti, Zr/Ni, Zr/Cu, and Ta) (Scheme 3.4) [8]. He et al. reported an efficient method for the synthesis of highly substituted pyridines from α-amino allenes, aldehydes, and aryl halides in the presence of $MgSO_4$ and $Pd(PPh_3)_4$ (Scheme 3.5) [9]. A metal-free cascade reaction for the synthesis of pyridines from aldehydes, phosphorus ylides, and propargyl azide has also been developed (Scheme 3.6) [10]. Pyridines are readily prepared *via* [4+2] annulation reaction between propargylamine and electron-deficient alkynes catalyzed by AgOTf (Scheme 3.7) [11]. Recently, an easy and efficient protocol has been developed for the synthesis of highly substituted pyridines by Uredi et al. from α, $\beta-$unsaturated carbonyl compounds and propargylic amines (Scheme 3.8) (Uredi et al., [3]).

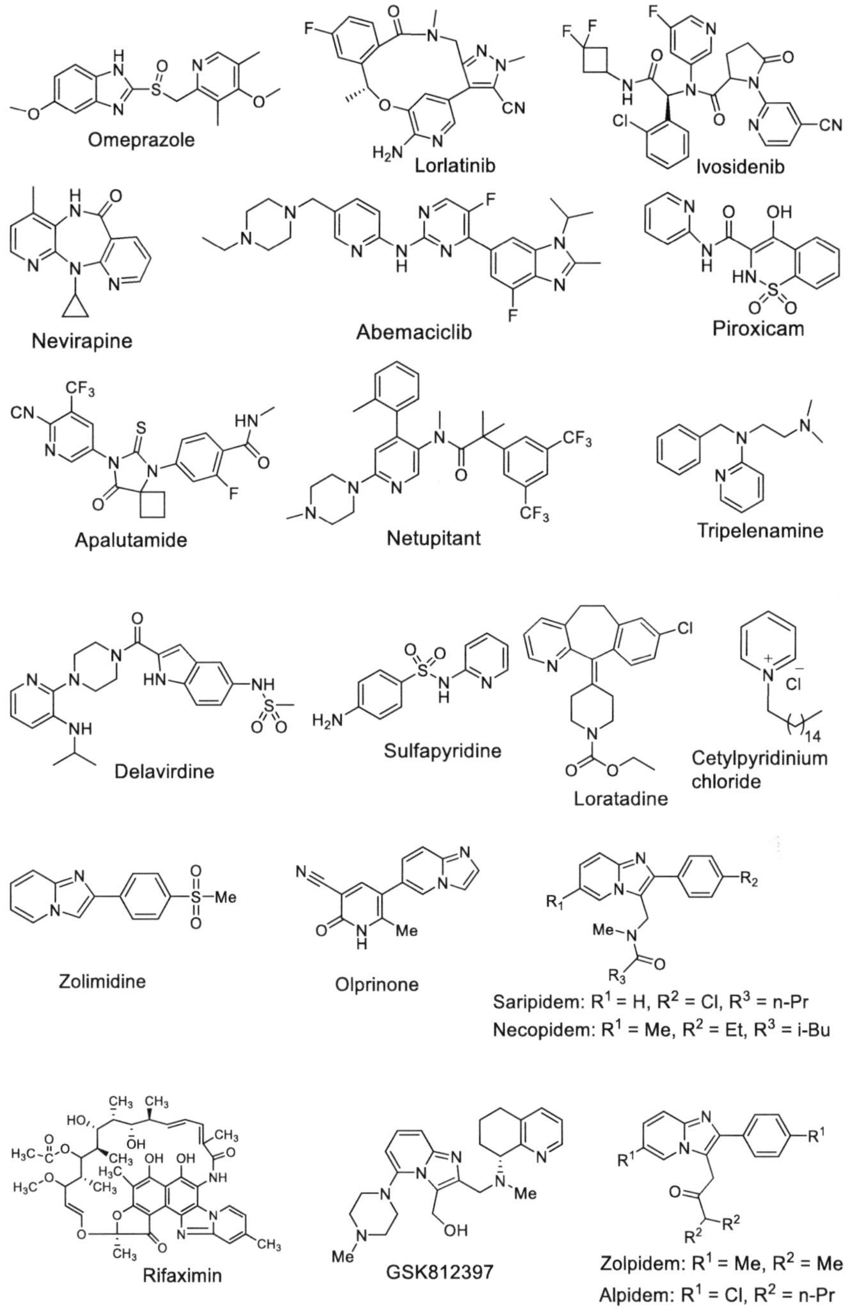

FIGURE 3.1 Representative examples of drugs with pyridine scaffolds.

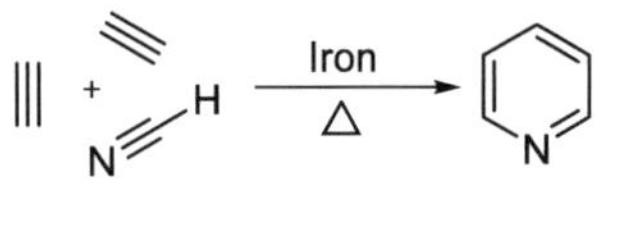

SCHEME 3.1 Ramsay pyridine synthesis.

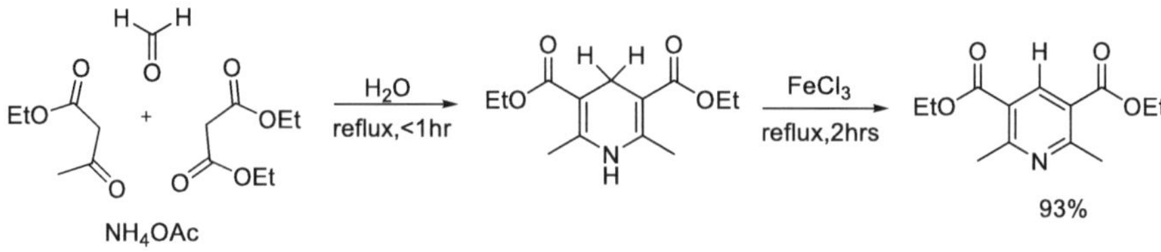

SCHEME 3.2 Hantzsch pyridine synthesis.

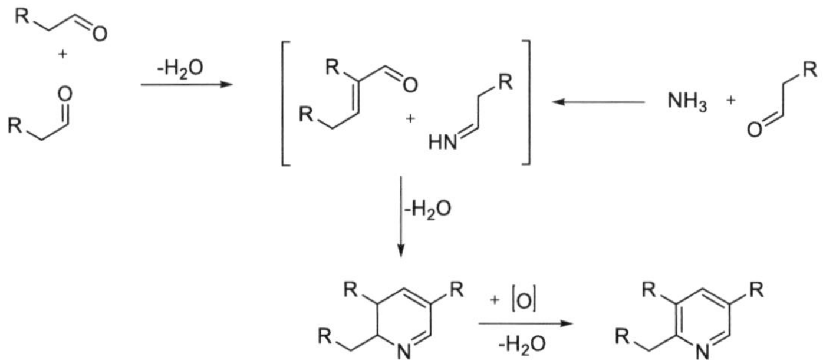

SCHEME 3.3 Chichibabin pyridine synthesis.

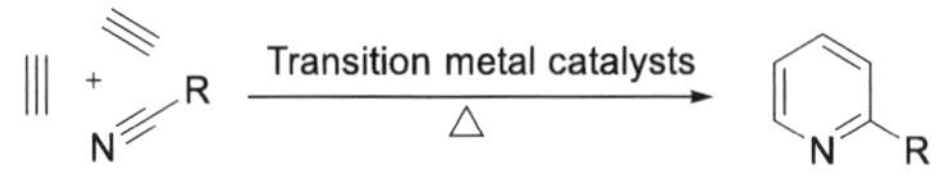

R = alkyl, vinyl, aryl
Transition-metals: Co, Rh, Fe, Ru, Ti, Zr/Ni, Zr/Cu and Ta)

SCHEME 3.4 Synthesis of pyridine from alkynes and nitriles.

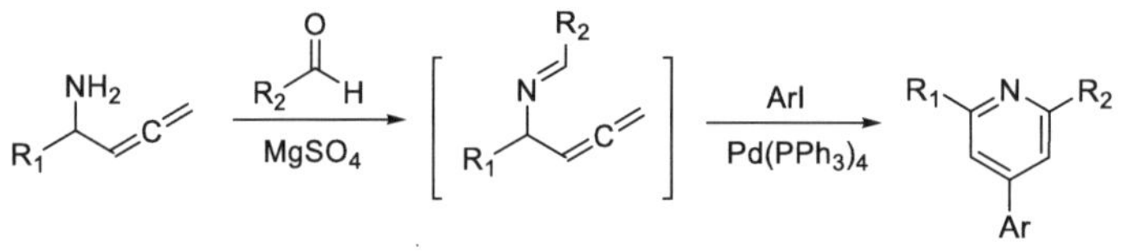

SCHEME 3.5 Preparation of pyridines from α-amino allenes, aldehydes, and aryl halides.

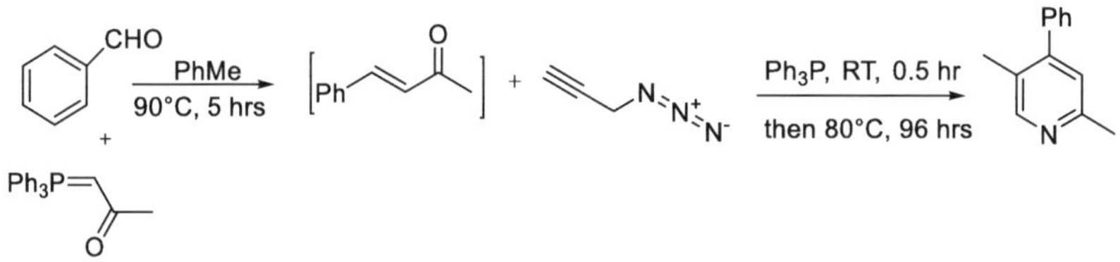

SCHEME 3.6 Synthesis of pyridines from aldehydes, phosphorus ylides, and propargyl azide.

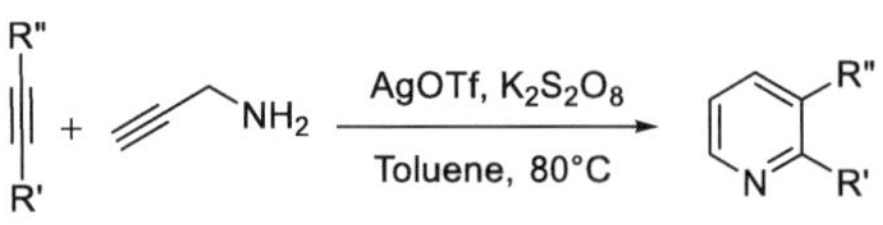

SCHEME 3.7 Synthesis of pyridines from propargylamine and alkynes.

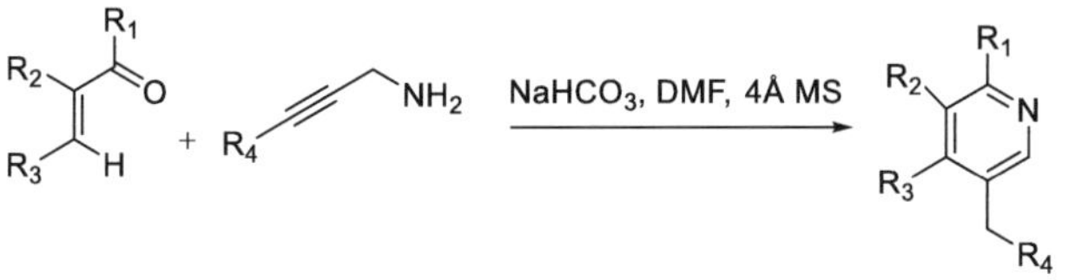

SCHEME 3.8 Synthesis of pyridines from α, β-unsaturated carbonyl compounds and propargylic amines.

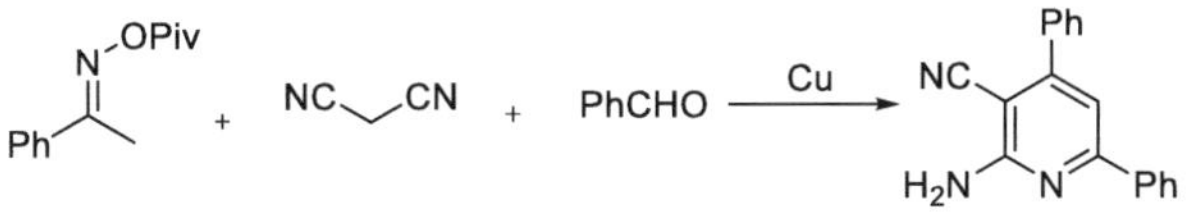

SCHEME 3.9 MCR synthesis of 2-amino-3-cyano pyridines from the reaction of O-pivaloyl acetophenone oxime, benzaldehyde and malononitrile.

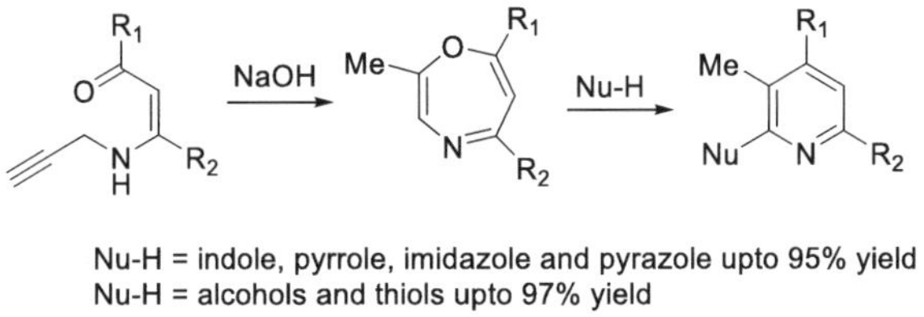

SCHEME 3.10 Synthesis of 2-(1-N-heteroaryl) pyridines from N-propargyl enaminones and N-heteroarenes.

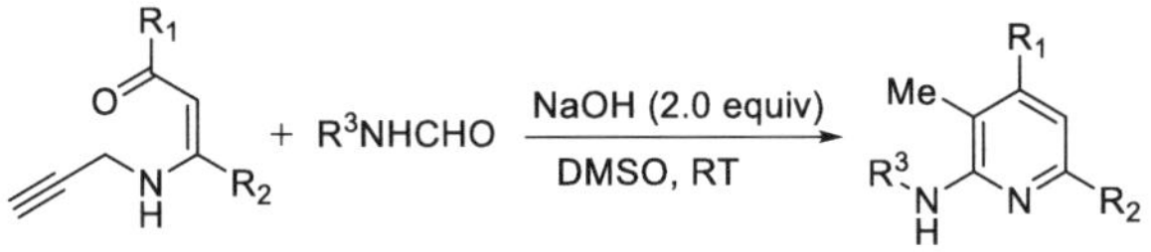

SCHEME 3.11 Synthesis of 2-aminopyridines from *N*-propargylic β-enaminones and formamides.

3.2.2 Synthesis of 2-aminopyridines and 2-amino-3-cyano pyridines

2-Aminopyridines and 2-amino-3-cyano pyridines represent an important structural motif in pharmaceuticals, natural products, and functional materials [12,13]. Besides, these derivatives are utilized as starting materials for the synthesis of a wide variety of bioactive pyridine-based heterocycles [14]. Therefore, substantial efforts have been devoted to develop a wide variety of methodologies for their synthesis. Copper-catalyzed multicomponent synthesis of 2-amino-3-cyano pyridines was achieved by the reaction of O-pivaloyl acetophenone oxime, benzaldehyde, and malononitrile (Scheme 3.9) [15]. 2-(1-N-heteroaryl)pyridines were synthesized from the reaction between N-propargyl enaminones which is generated *in situ* from propargyl amine and propynones *via* Michael addition, and N-heteroarenes which proceeds through the formation of 1,4-oxazepine as an intermediate (Scheme 3.10) [16]. Weng et al. reported a simple and highly efficient method for the synthesis of highly substituted 2-aminopyridines from the reaction of *N*-propargylic β-enaminones and formamides (Scheme 3.11) [17]. Further, several MCR strategies for the synthesis of 2-amino-3-cyanopyridines from readily available substituted benzaldehydes, acetophenones, malononitrile, and ammonium acetate under conventional/microwave/sonochemical conditions in the presence of various catalysts have been found in the literature (Schemes 3.12–3.18) [18–23]. This MCR proceeds through the Knoevenagel condensation, Michael addition,

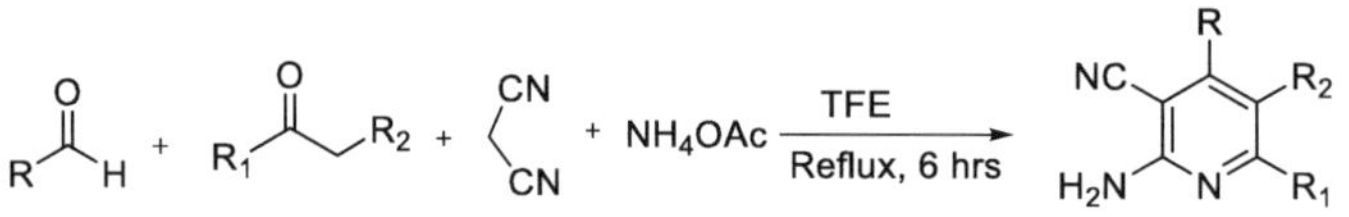

SCHEME 3.12 MCR synthesis of 2-amino-3-cyano pyridines.

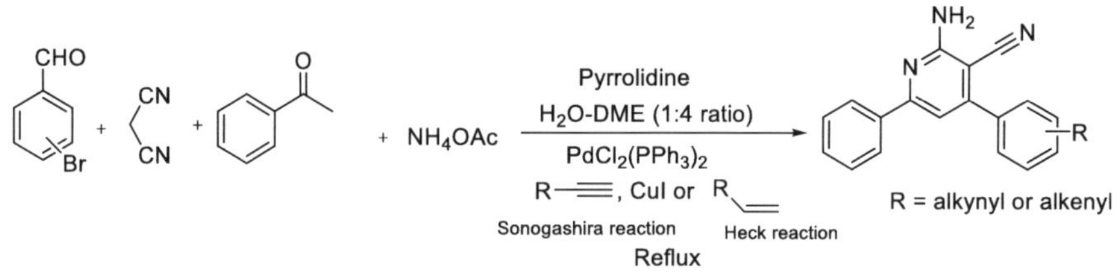

SCHEME 3.13 Five component strategy for the synthesis of 2-amino-3-cyanopyridines.

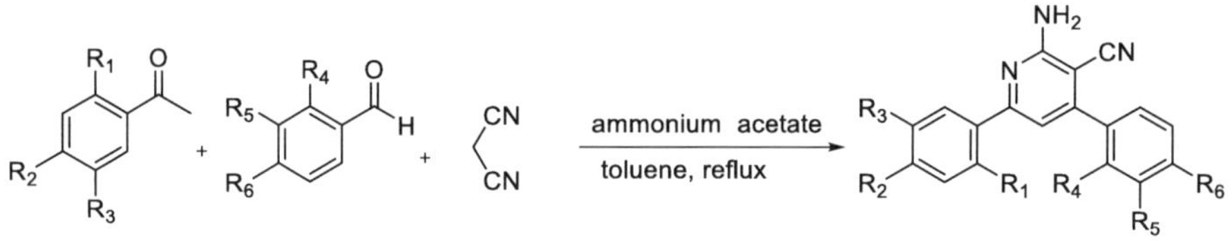

SCHEME 3.14 Four component synthesis of 2-amino-3-cyanopyridines.

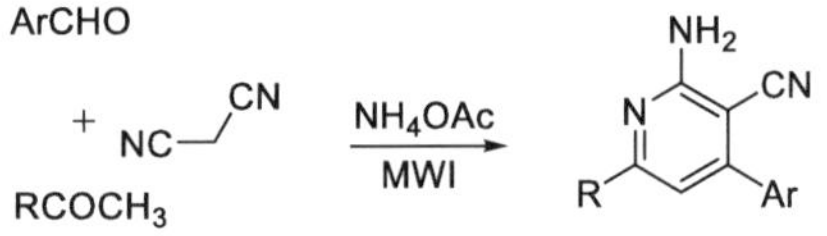

SCHEME 3.15 Microwave assisted MCR synthesis of 2-amino-3-cyanopyridines.

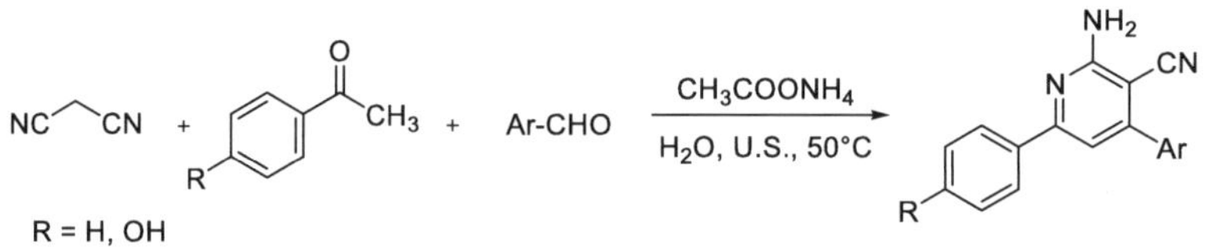

SCHEME 3.16 One-pot synthesis of 2-amino-3-cyanopyridines under ultrasonic irradiation.

cyclization, isomerization, and is followed by oxidative aromatization. Similarly, one of the important biologically active pyridine derivatives, 2-amino-3,5-dicarbonitrile-6-thio-pyridines were prepared by the standard MCR of aldehydes, malononitrile, and thiophenols under conventional/microwave/mechanochemical/sonochemical conditions (Schemes 3.19–3.23) [24–28], *via* the Knoevenagel condensation, Michael addition, intramolecular cyclization and oxidative aromatization.

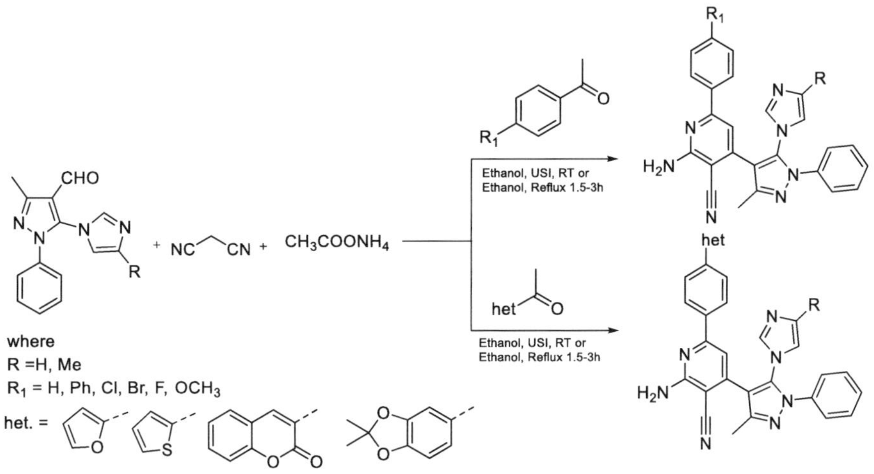

SCHEME 3.17 Ultrasound-assisted or conventional four-component synthesis of novel 2-amino-3-cyanopyridines.

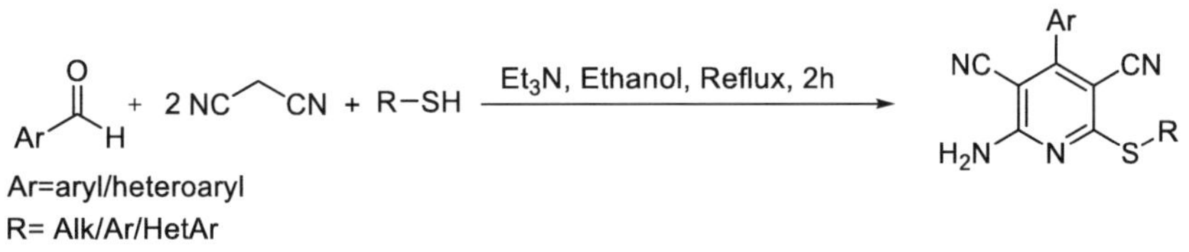

SCHEME 3.18 MCR synthesis of 2-amino-4-(hetero)aryl-3,5-dicarbonitrile-6-sulfanylpyridines.

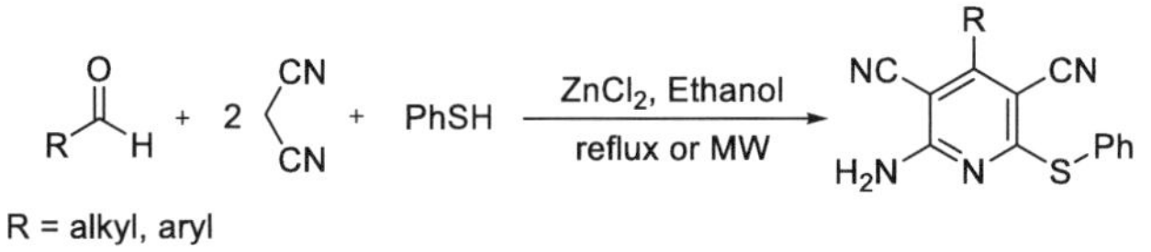

SCHEME 3.19 One-pot multicomponent synthesis of 2-amino-3,5-dicarbonitrile-6-thio-pyridines under conventional or microwave conditions.

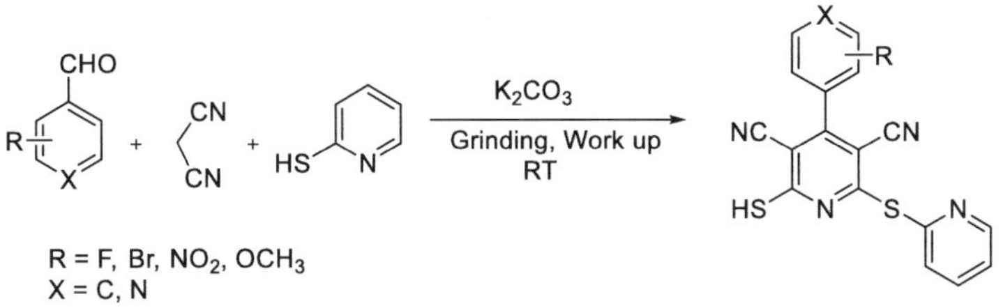

SCHEME 3.20 Mechanochemical MCR synthesis of 2-amino-3,5-dicarbonitrile-6-thio-pyridines.

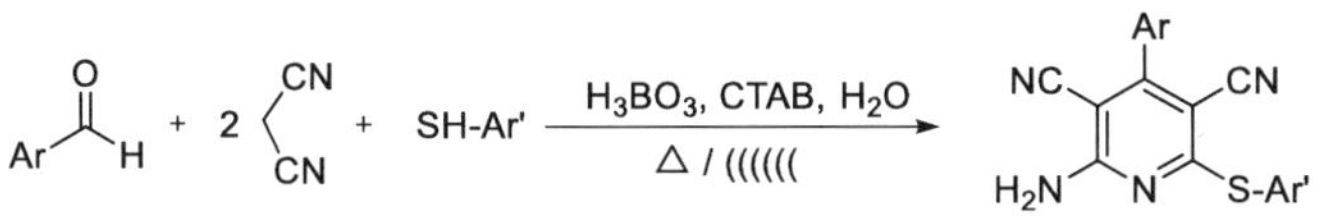

SCHEME 3.21 Three-component synthesis of 2-amino-3,5-dicarbonitrile-6-thio-pyridines under conventional or sonochemical conditions.

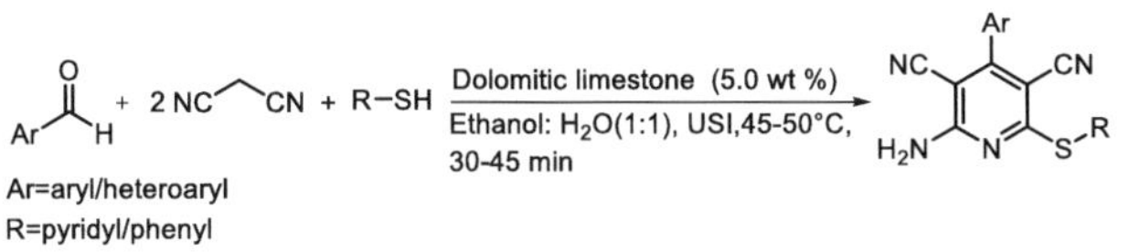

SCHEME 3.22 Ultrasound-assisted green MCR synthesis of 2-amino-4-(hetero)aryl-3,5-dicarbonitrile-6-sulfanylpyridines.

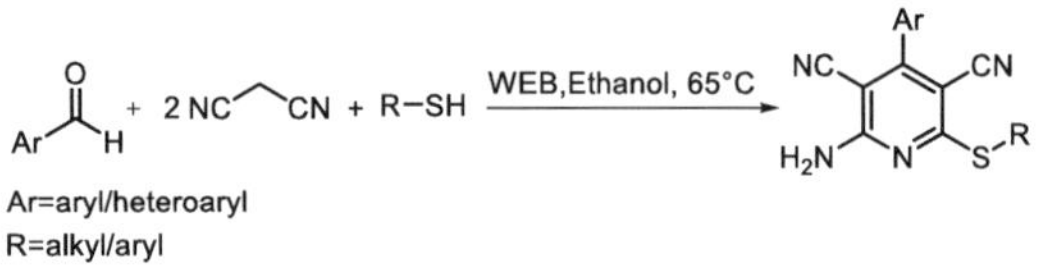

SCHEME 3.23 MCR synthesis of 2-amino-3,5-dicarbonitrile-6-thio-pyridines using water extract of banana.

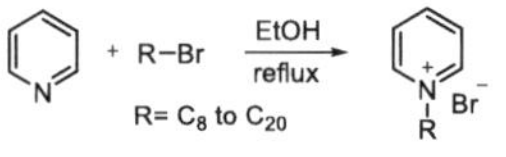

SCHEME 3.24 Preparation of pyridinium salts from pyridine with alkyl halides.

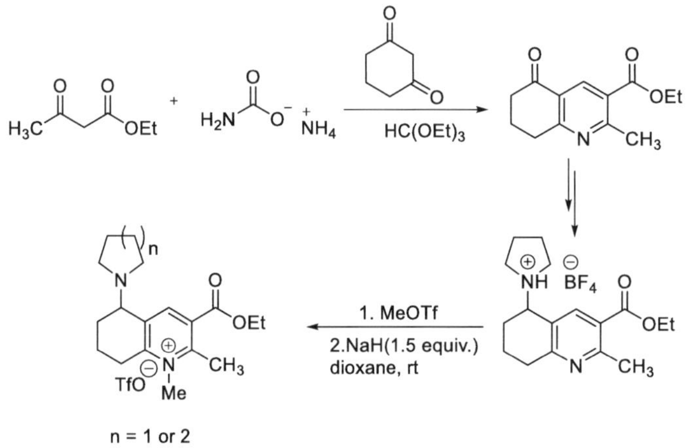

SCHEME 3.25 Synthesis of *N*-methylpyridinium salts.

3.2.3 Synthesis of pyridinium salts

For the past few decades, pyridinium salts have been considered as highly privileged scaffolds found in many bioactive compounds. Njaoaminiums and Pachychalines are the pyridinium-containing natural products. Pyridinium salts exist as liquids at room temperature, so-called "Pyridinium ionic liquids," such as 1-alkylpyridinium salts, are well known as potential solvents in synthesis and catalysis. These salts have also been utilized as key intermediates for the construction of pharmacologically active indole, piperidine, dihydro, tetrahydropyridine, containing natural products [29]. Pyridinium salts are easily synthesized by the S_N2 reaction of pyridine with alkyl halides in ethanol under reflux conditions (Scheme 3.24) [30]. *N*-methyl pyridinium salts are prepared by the reaction of ethyl acetoacetate, ammonium carbamate, 1,3-cyclohexanedione, ethylorthoformate, and methyl triflate in a step-wise manner, as shown in Scheme 3.25 [31]. An MCR strategy has been developed for the synthesis of highly substituted pyridinium salts by the reaction of α,β-unsaturated aldehydes/ketones, amines, and alkynes in the presence of rhodium catalyst (Scheme 3.26) [32] and found to proceed through in situ generated imine-assisted Rh(III)-catalyzed vinylic C–H activation. Recently, sustainable synthetic protocols have also been reported for the synthesis of pyridinium salts. For example, Kirchhecker et al. synthesized zwitterionic pyridinium salts by the condensation of aminoacids with biorenewable furfural (Scheme 3.27) [33], whereas Sowmiah et al. prepared *N*-alkyl pyridinium salts from

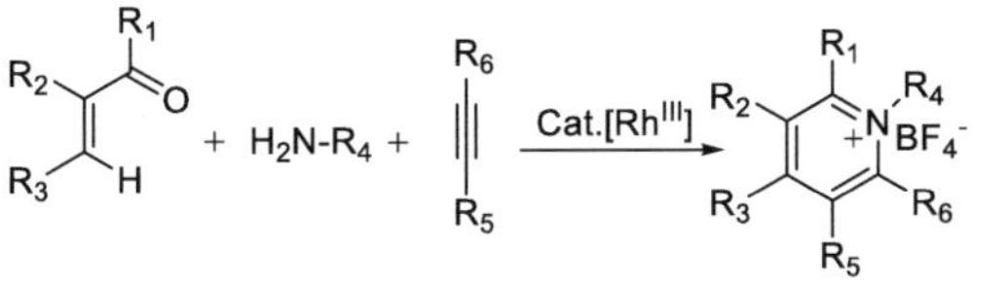

SCHEME 3.26 Rh(III) catalyzed route for highly substituted pyridinium salts.

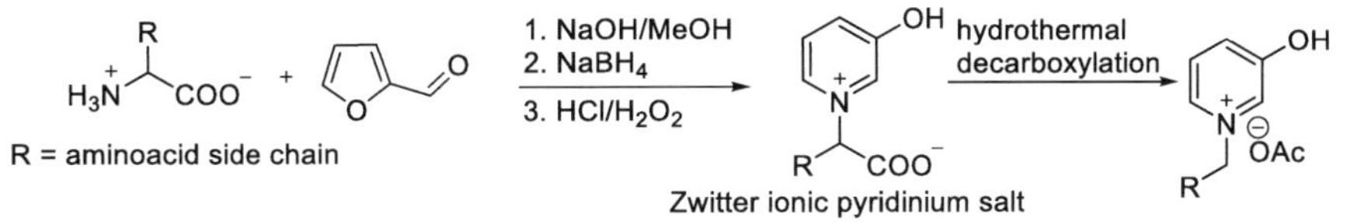

SCHEME 3.27 Synthesis of Zwitterionic pyridinium salts from furfural and aminoacids.

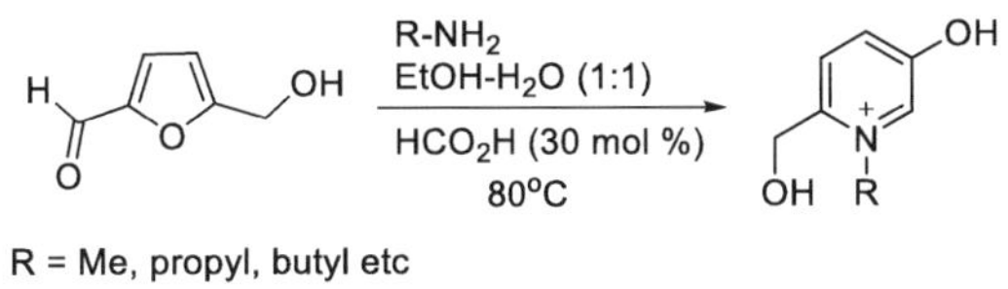

SCHEME 3.28 Synthesis of *N*-alkyl pyridinium salts from biomass derived 5-hydroxymethylfurfural.

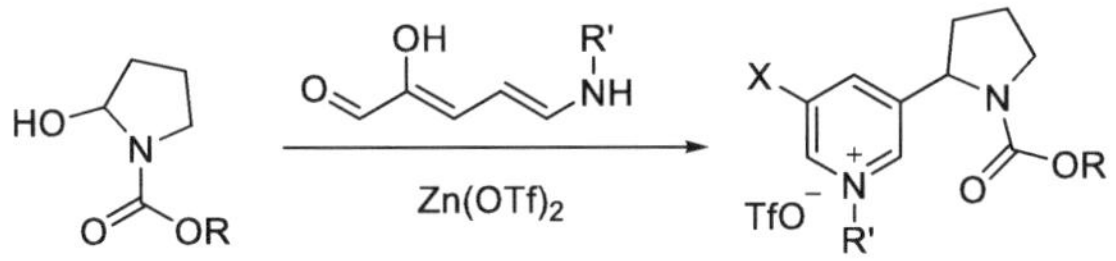

SCHEME 3.29 Synthesis of piperidine and pyrrolidine substituted pyridinium salts.

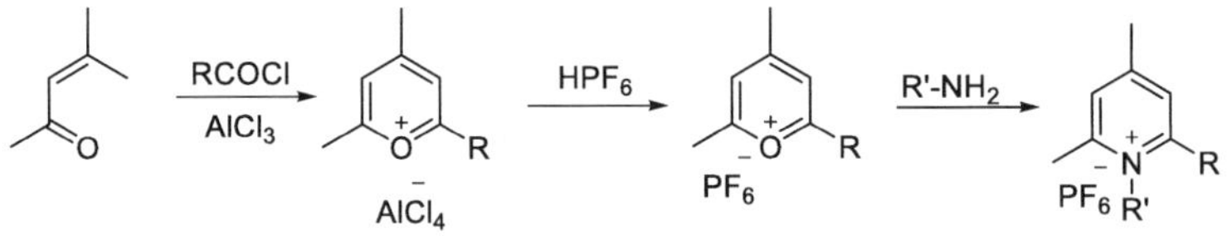

SCHEME 3.30 Synthesis of pyridinium salts from pyrylium salts.

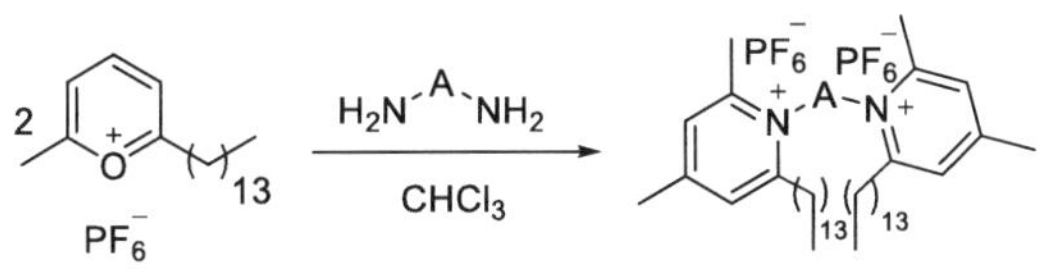

SCHEME 3.31 Synthesis of pyridinium salts from pyrylium salts.

the biomass-derived 5-hydroxymethylfurfural and various alkyl amines (Scheme 3.28) [34]. Pyridinium salts were synthesized by the reaction of 5-alkylaminopenta-2,4-dienals and the *N*-acyliminium ions using zinc triflate as catalyst (Scheme 3.29) [35]. A simple strategy has been demonstrated for the synthesis of pyridinium salts from highly reactive pyrylium salts and primary linear alkylamines (C10 to C18)/N-alkyldiimines (Schemes 3.30 and 3.31) [36–38]. Synthesis of 2,4-dioxochroman-pyridinium derivatives as potent anti-Alzheimer agents, was

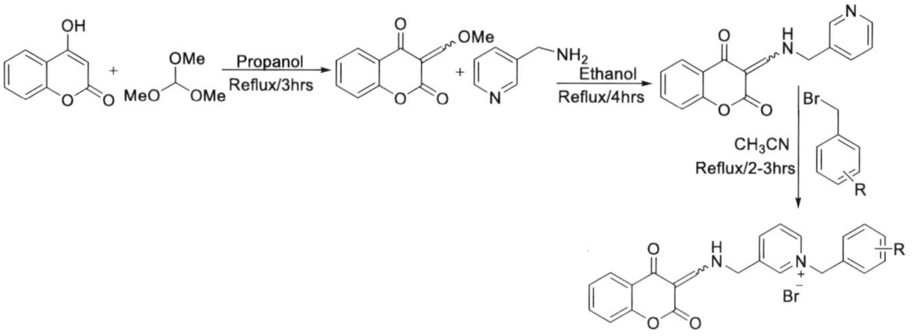

SCHEME 3.32 Synthesis of benzyl pyridinium-2,4-dioxochromans.

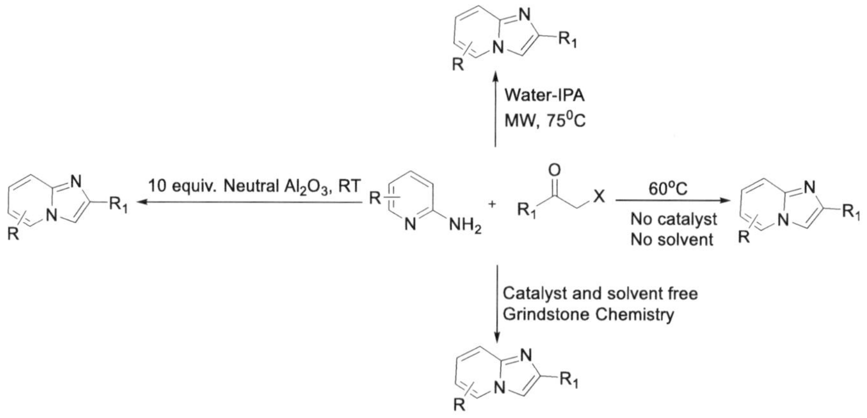

SCHEME 3.33 Synthesis of imidazopyridines from α-halo ketones.

achieved through a step-wise process from readily available starting materials, 4-hydroxy-2H-chromen-2-one, trimethoxymethane, pyridin-3-ylmethanamine, and various benzyl bromides (Scheme 3.32) [39].

3.2.4 Synthesis of imidazopyridines

Imidazopyridine where "the imidazole ring fused with the pyridine" is an important class of N-heterocycle. Among various imidazopyridines, the imidazo[1,2-a]pyridines have received considerable attention of the pharmaceutical industry owing to their wide range of biological activities. Moreover, these compounds are valuable building blocks in organometallic chemistry and materials science [40–43]. Hence, there is a continuous effort towards the development of new synthetic routes for the synthesis of imidazo[1,2-a]pyridines [44]. The most common method for the synthesis of imidazo[1,2-a]pyridines was the condensation reaction of α-haloketones/α-Diazoketones/α-tosyloxyketones with the 2-aminopyridines (Schemes 3.33–3.35) [45–50]. Gryko et al. reported a simple method for the synthesis of

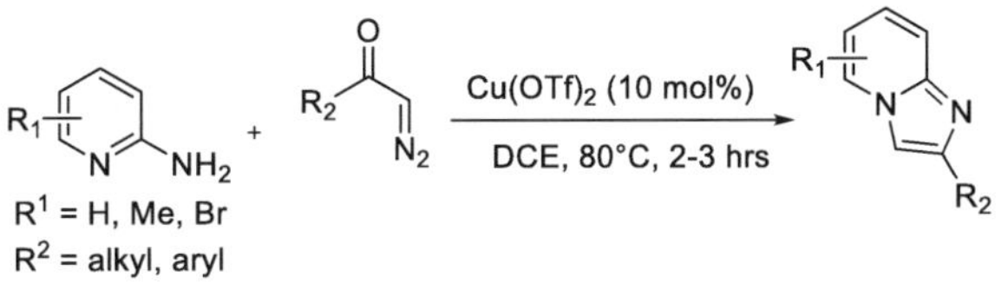

SCHEME 3.34 Synthesis of imidazopyridines from α-diazoketones.

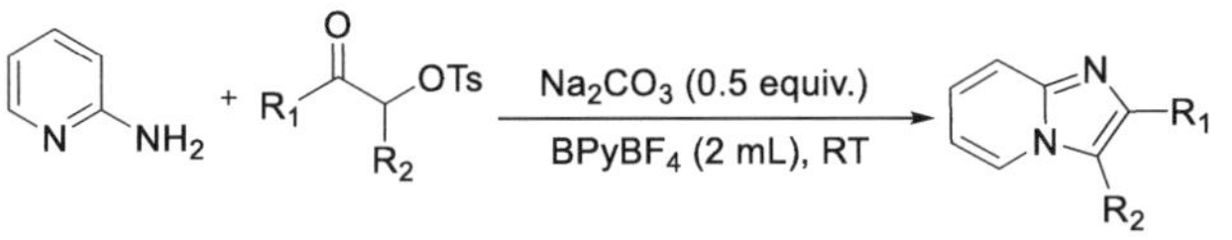

SCHEME 3.35 Synthesis of imidazopyridines from α-tosyloxyketones.

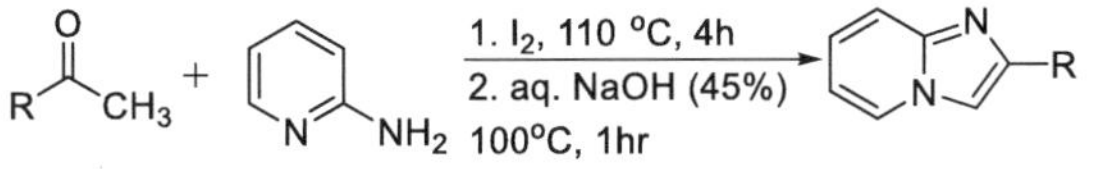

SCHEME 3.36 Synthesis of imidazo[1,2-*a*]pyridine derivatives from ketones and 2-aminopyridine.

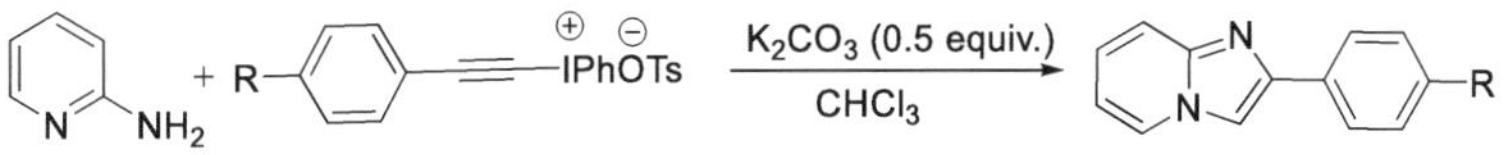

SCHEME 3.37 Synthesis of imidazopyridines from alkynyl(phenyl)iodonium salts and 2-aminopyridine.

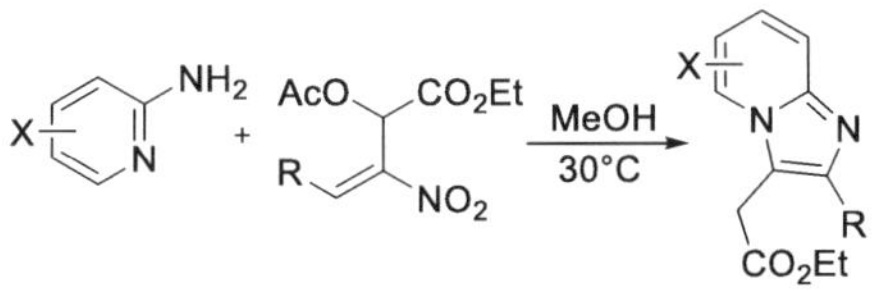

SCHEME 3.38 Synthesis of highly substituted imidazopyridines from MBH acetates of nitroalkenes and 2-aminopyridines.

imidazo[1,2-*a*]pyridines by treating 2-amino pyridines with ketones *via* Ortoleva–King reaction followed by ring closure (Scheme 3.36) [51]. Imidazopyridines were also prepared by the reaction of alkynyl(phenyl)iodonium salts with 2-aminopyridine in the presence of K_2CO_3 in $CHCl_3$ *via* [3,3]-sigmatropic rearrangement followed by intramolecular cyclization (Scheme 3.37) [52]. Nair et al. synthesized a highly substituted imidazo[1,2-*a*]pyridines through cascade inter–intra molecular double *aza*-Michael addition using Morita–Baylis–Hillman (MBH) acetates of nitroalkenes and 2-aminopyridines at room temperature in MeOH (Scheme 3.38) [53]. $FeCl_3$ catalyzed the synthesis of imidazo[1,2-*a*]pyridines by the reaction of nitroolefins and 2-aminopyridines *via* tandem Michael addition and subsequent intramolecular cyclization (Scheme 3.39) [54]. A 3-component strategy for the preparation of imidazo[1,2- *a*]pyridines from 2-aminopyridines, benzyl halides/benzyl tosylates, and isocyanides has been reported by Adib et al. (Scheme 3.40) [55]. Puthiaraj et al. developed a Cu(BDC)MOF (BDC: 1,4-benzenedicarboxylate) catalyzed 3-component synthesis of imidazo[1,2-*a*]pyridines from aldehydes, 2-aminopyridines and nitromethane (Scheme 3.41) [56] that proceeds *via* an intermolecular aza-Michael addition and subsequent intramolecular cyclization. A highly regioselective three-component synthesis of 2-alkoxy-3-arylimidazo[1,2-*a*]pyridines was attained

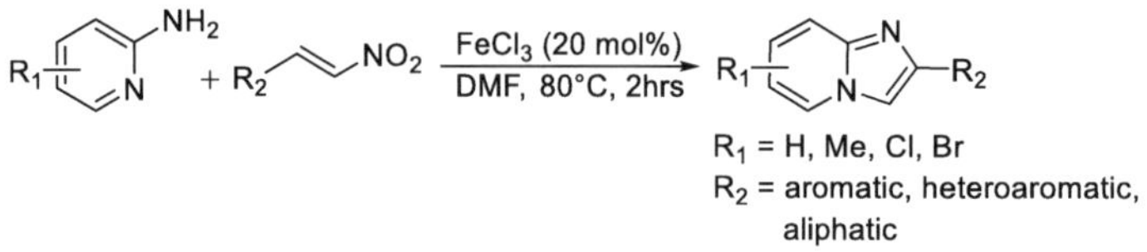

SCHEME 3.39 Synthesis of imidazopyridines from nitroolefins and 2-aminopyridines.

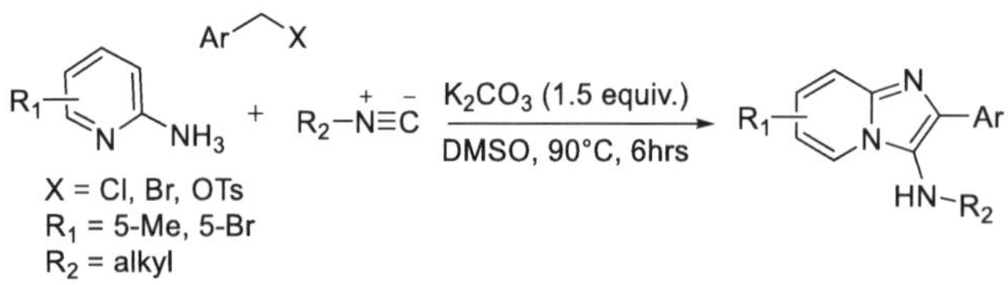

SCHEME 3.40 3-component synthesis of imidazo[1,2-*a*]pyridines.

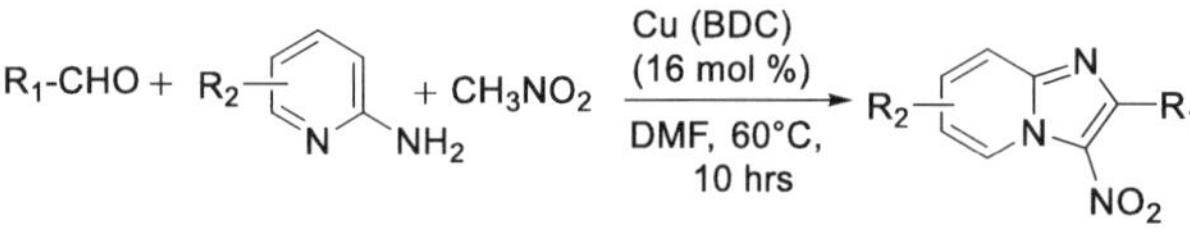

SCHEME 3.41 Copper-MOF-catalyzed 3-CR for imidazo[1,2-*a*]pyridines.

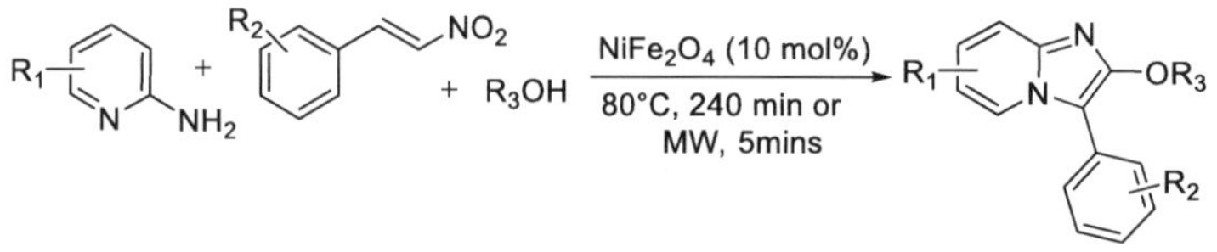

SCHEME 3.42 MCR synthesis of imidazopyridines from 2-aminopyridines, nitroolefins and alcohols.

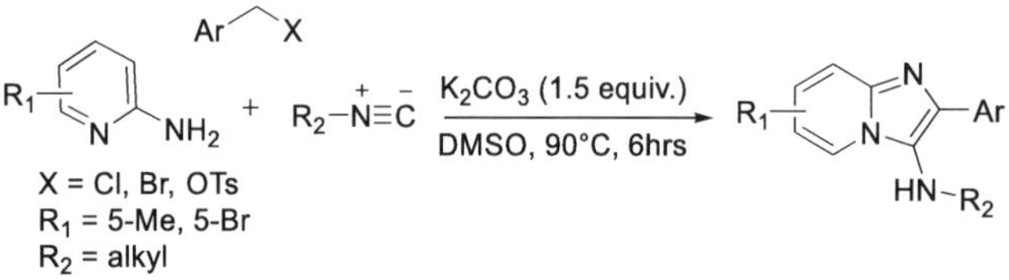

SCHEME 3.43 Copper catalyzed 4-CR for imidazo[1,2-*a*]pyridines.

from the reaction of 2-aminopyridines, nitroolefins, and alcohols catalyzed by nano-$NiFe_2O_4$ under microwave conditions (Scheme 3.42) [57]. A copper-catalyzed four-component reaction for the synthesis of imidazo[1,2-*a*]pyridines has been developed by Allahabadi et al. from the readily available 2-bromopyridine, sodium azide, aldehydes, and isocyanides (Scheme 3.43) [58]. An environmentally benevolent synthetic route has been reported by Swami et al. for the synthesis of biologically active pyrazole coupled imidazo[1,2-a]pyridines *via* a 3-component reaction of alkyl-4-formyl-1-phenyl-1H-pyrazole-3-carboxylate, 2-aminopyridine, and isocyanide in the presence of nano-crystalline ZnO in ethanol (Scheme 3.44) [59]. An environmentally benign strategy has been demonstrated for the synthesis of bis(imidazo[1,2-*a*]pyridin-3-yl)sulfanes from the reaction between imidazo[1,2-a]pyridines and sulfur powder under metal-free conditions. The reaction was found to proceed *via* oxidative dual C–H sulfenylation (Scheme 3.45) [60].

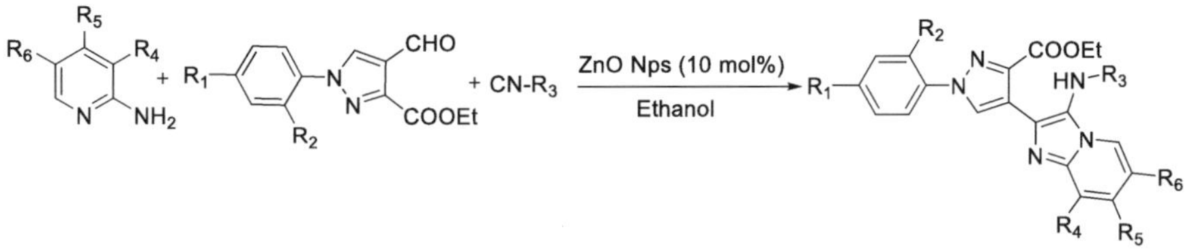

SCHEME 3.44 Synthesis of pyrazole coupled imidazo[1,2-a]pyridine derivatives.

SCHEME 3.45 Metal-free synthesis of bis(imidazo[1,2-a]pyridin-3-yl)sulfanes.

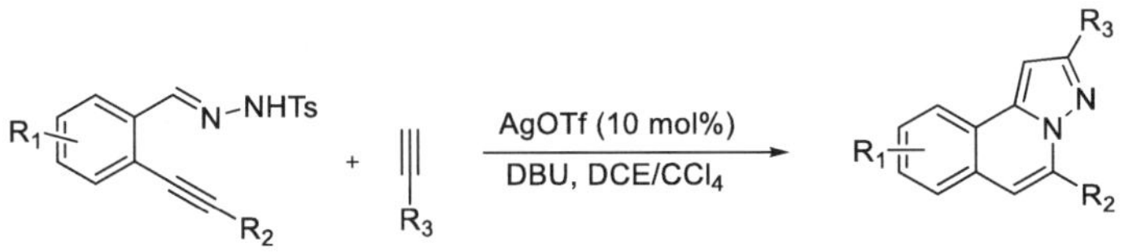

SCHEME 3.46 Preparation of pyrazolo[1,5-a]quinolines from N′-(2-alkynylbenzylidene)-hydrazide and various alkynes.

3.2.5 Synthesis of pyrazolopyridines

Pyrazolopyridines are nitrogen-containing fused heterocycles, with unique biological properties such as antidepressant, anti-inflammatory, antihyperglycemic, antitumor, antibacterial, anxiolytic, etc. Further, these derivatives are also used for the treatment of Alzheimer disease (AD), drugs addiction, and infertility [61–66]. Therefore, considerable efforts have been dedicated to the synthesis of these fused heterocyclic scaffolds. Pyrazolo[1,5-a]quinolines were prepared by the tandem reaction between N′-(2-alkynylbenzylidene)-hydrazides and various alkynes in the presence of silver triflate and DBU in DCE/CCl_4 under mild conditions (Scheme 3.46) [67]. Qin et al. established a multistep protocol for the preparation of pyrazolo[3,4-b]pyridines, as shown in Scheme 3.47 [68]. Pyrazolo[1,5-a]pyridines and 6-iodopyrazolo[1,5-a]pyridines have also been prepared by gold-catalyzed and iodine-mediated cyclization of enynylpyrazoles (Scheme 3.48) [69]. Pyrazolo[1,5-a]quinolines were synthesized through the combination of S_NAr substitution and Knoevenagel condensation reactions of 1H-pyrazoles and 2-fluorobenzaldehyde under the transition-metal-free conditions (Scheme 3.49) [70]. Pyrazolo[1,5-a]pyridines are also prepared by the thermal intramolecular cyclization of *N*-amino-2-alkynylpyridines in acetic acid (Scheme 3.50) [71]. Wang et al. reported a domino strategy for the synthesis of pyrazolo[1,5-a]pyridines by the reaction between ketones or aldehydes with (*E*)-4-bromobut-2-enenitrile in the presence of K_2CO_3 in DMF (Scheme 3.51) [72]. Pyrazolo[1,5-a]pyridines were synthesized by the reaction that involves [3+2] cycloaddition between benzonitriles and 2-substituted pyridines in the presence of $Cu(OAc)_2$/CuBr in DMSO (Scheme 3.52) [73]. Marjani et al. developed a green MCR strategy

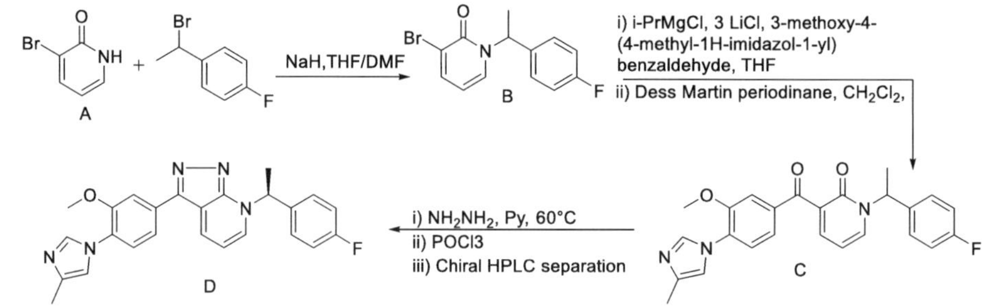

SCHEME 3.47 Preparation of pyrazolo[3,4-b]pyridine derivatives.

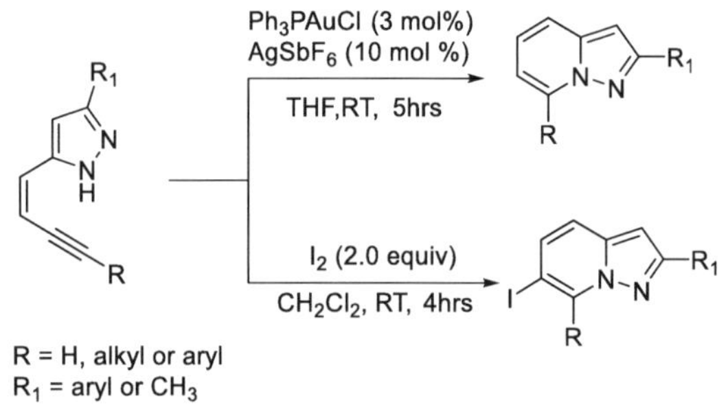

SCHEME 3.48 Synthesis of pyrazolo[1,5-a]pyridines and 6-iodopyrazolo[1,5-a]pyridines.

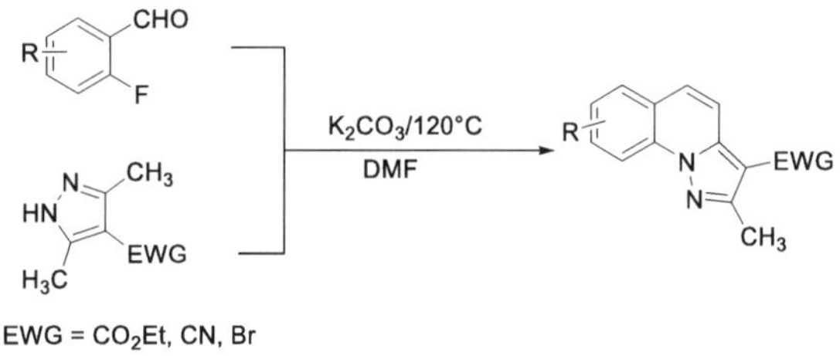

SCHEME 3.49 Synthesis of pyrazolo[1,5-a]quinolines from 1*H*-pyrazoles and 2-fluorobenzaldehyde.

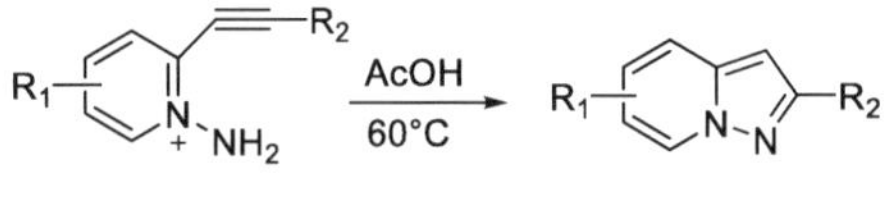

SCHEME 3.50 Synthesis of pyrazolo[1,5-a]pyridines *via* the thermal intramolecular cyclization of *N*-amino-2-alkynyl-pyridines.

for the synthesis of pyrazolo[3,4-b]pyridines by a 3-component reaction of arylglyoxals, 3-methyl-1-aryl-1H-pyrazol-5-amines and cyclic 1,3-dicarbonyl compounds in the presence of tetrapropylammonium bromide at 80°C in water which involves *via* Knoevenagel and Micheal reactions, subsequently intramolecular condensation, unexpected dearoylation and oxidation (Scheme 3.53) [74]. Aminochromane linked pyrazolo[3,4-b]pyridines have been synthesized by the 3-component reaction of (arylhydrazono)methyl-4H-chromen4-one, malononitrile, and primary amines in the presence of Et_3N in ethanol (Scheme 3.54) [75].

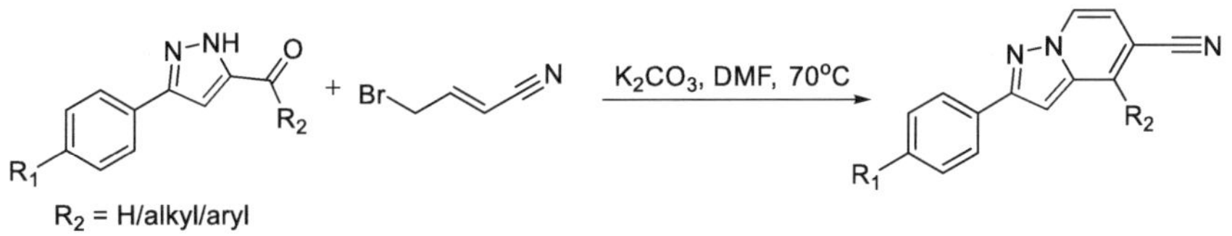

SCHEME 3.51 Synthesis of 5-cyanopyrazolo[1,5-a]pyridines from ketones or aldehydes and (*E*)-4-bromobut-2-enenitrile.

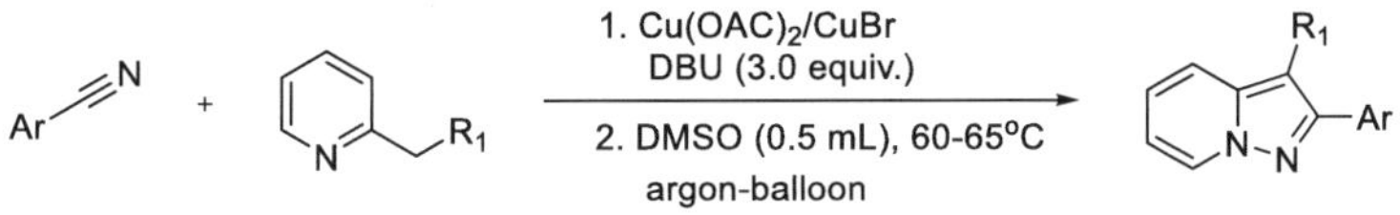

SCHEME 3.52 Synthesis of pyrazolo[1,5-a]pyridines *via* a [3+2] cycloaddition.

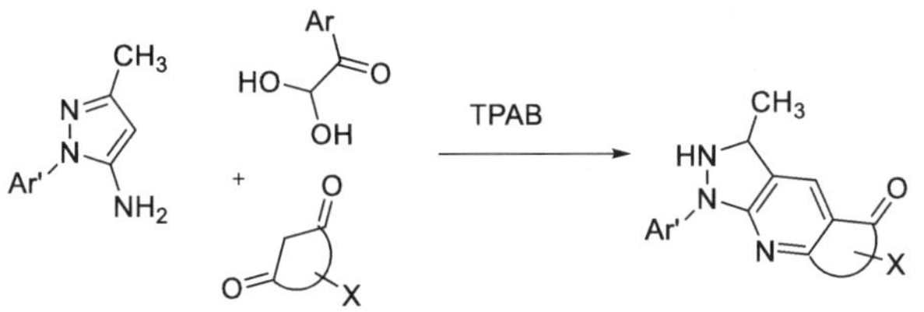

SCHEME 3.53 Three component synthesis pyrazolo[3,4-b]pyridines.

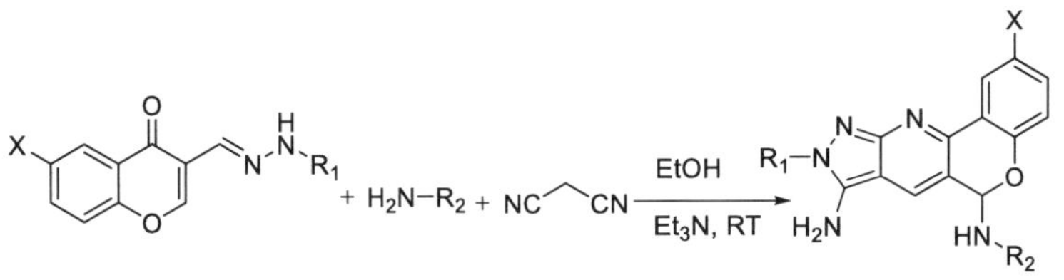

SCHEME 3.54 Three component synthesis of pyrazolopyridines.

3.2.6 Synthesis of furopyridines

Furopyridines are considered as privileged medicinal scaffolds because of its broad range of biological activities like anticonvulsant, antipsychotic, antiproliferative and anthelmintic, antianaphlactic activities [76–78]. Furopyridines are also used as calcium influx promoters [79], acetylcholinesterase inhibitors [80], and HIV-1 nonnucleoside reverse transcriptase inhibitor [81]. Therefore, significant efforts have been focused on the development of new synthetic strategies for furopyridines. Furo[2,3-*c*]pyridine was synthesized by Shiotani and Morita *via* a step-wise process (Scheme 3.55) [82]. Initially, the Knoevenagel condensation between 3-furaldehyde and malonic acid led to β-(3-furyl)acrylic acid, then transformed into an acylazide, which in turn undergo cyclization to give furo[2,3-*c*]pyridin-7(6*H*)-one. The obtained pyridinone on chlorination followed by replacement of the chlorine with hydrogen lead to the formation of target furo[2,3-c]pyridine. Furopyridines like furo[3,2-*b*]pyridines, furo[2,3-*b*]pyridines, and furo[2,3-*c*]pyridines were synthesized under mild conditions *via* the palladium-catalyzed cross-coupling of 1-alkynes with *o*-iodoacetoxy- or *o*-iodobenzyloxypyridines, followed by electrophilic cyclization by I_2 or by $PdCl_2$ under a balloon of carbon monoxide (Scheme 3.56) [83]. Cailly et al. developed a simple synthetic

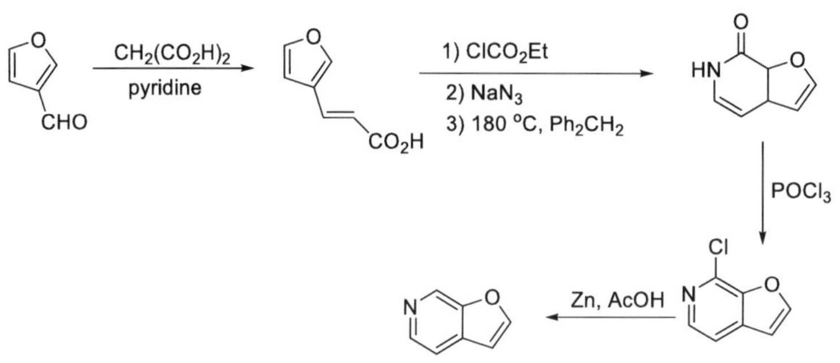

SCHEME 3.55 Synthesis of furo[2,3-*c*]pyridine.

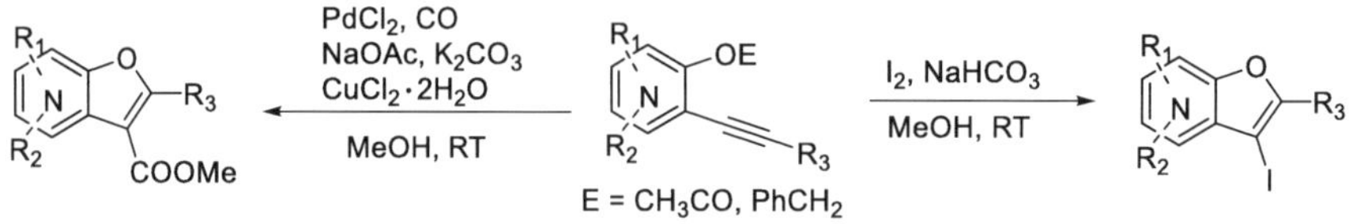

SCHEME 3.56 Synthesis of furopyridines from *o*-iodoacetoxy- or *o*-iodobenzyloxypyridines.

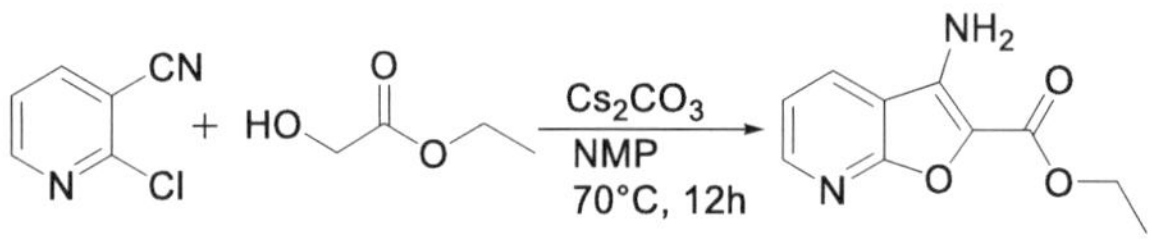

SCHEME 3.57 Synthesis of ethyl 3-amino-furopyridine-2-carboxylates form 1-chloro-2-cyanopyridines and ethyl glycolate.

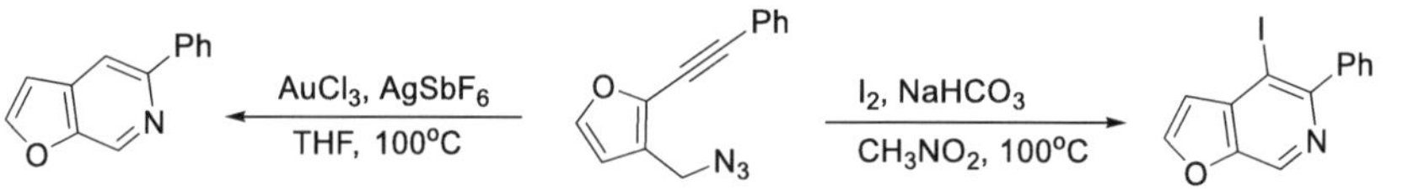

SCHEME 3.58 Synthesis of furo[2,3-*c*]pyridines from 3-(azidomethyl)-2-(phenylethynyl)furan.

route to ethyl 3-aminofuropyridine-2-carboxylates from 1-chloro-2-cyanopyridines and ethyl glycolate in the presence of Cs_2CO_3 and *N*-methylpyrrolidin-2-one (NMP) (Scheme 3.57) [84]. The furo[2,3-*c*]pyridines also synthesized from 3-(azidomethyl)-2-(phenylethynyl)furan in the presence of I_2 or gold *via* an electrophilic cyclization (Scheme 3.58) [85, 86]. Voluchnyuk et al. prepared a series of fused pyridine-4-carboxylic acids which includes furo[2,3-*b*]pyridines, thieno[2,3-*b*]pyridines, pyrazolo[3,4-*b*]pyridines, isoxazolo[5,4-*b*]pyridines, and pyrido[2,3-*d*]pyrimidines from the reaction of acyl pyruvates and electron-rich amino heterocycles (amino furans/amino thiophenes/amino pyrazoles/amino isoxazoles/amino pyridines) *via* a Combes-type reaction and subsequent ester hydrolysis (Scheme 3.59) [87]. Yan et al. reported an I_2-mediated oxidative tandem cyclization reaction of simple enaminones for the synthesis of furopyridines *via* construction of C–C/C–N/C–O bonds in a single-step operation (Scheme 3.60) [88]. 3-amino-N-phenylfuro[2,3-b]pyridine-2-carboxamides were prepared *via* a step-wise process (Scheme 3.61) in which 2-((3-cyanopyridin-2-yl)oxy)-N-phenylacetamides syn-

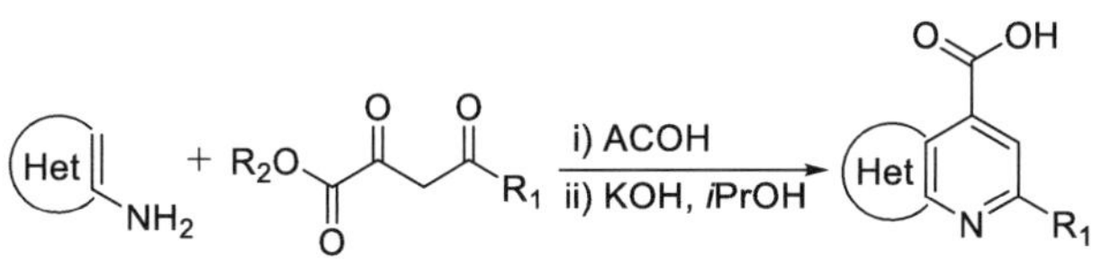

SCHEME 3.59 Synthesis of fused pyridine-4-carboxylic acids.

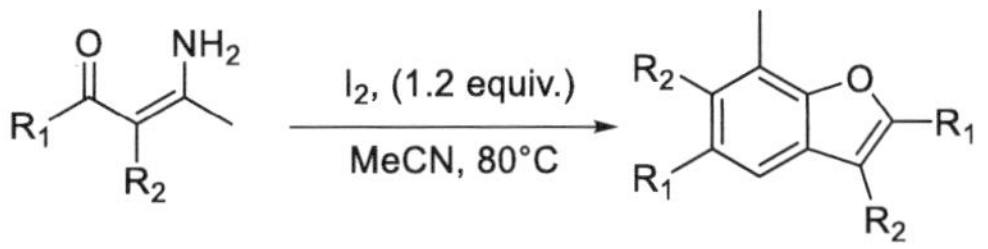

SCHEME 3.60 Synthesis of furopyridines from enaminones.

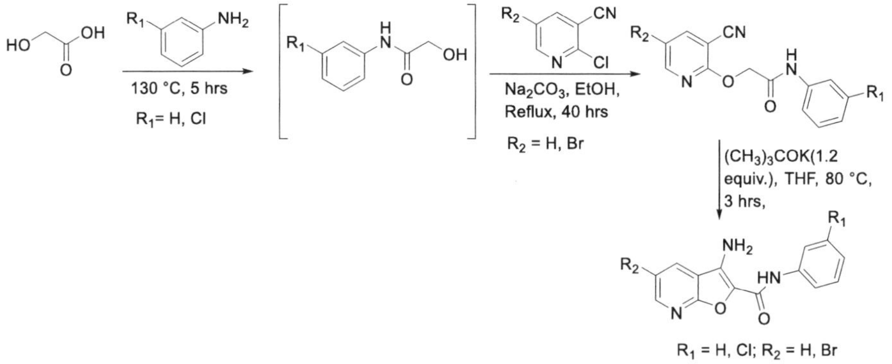

SCHEME 3.61 Synthesis of 3-amino-N-phenylfuro[2,3-b]pyridine-2-carboxamides.

thesized from the reaction of anilines, glycolic acid and cyanopyridines *via* a one-pot domino reaction followed by treatment with KOtBu in THF to afford furo[2,3-b]pyridines [89].

3.3 Therapeutic potential of pyridine scaffolds as anti-Alzheimer's agents

AD was first described as an "unusual disease of the cerebral cortex" in 1907 by Alois Alzheimer and is now deemed by World Health Organization as the most common cause of dementia [90]. AD is chronic and progressive neurodegenerative brain turmoil with the appearance of extracellular senile or neuritic plaques and intracellular neurofibrillary tangles in the brain [91]. This disorder is associated with silent, slow but sure, and conscious damage of the human brain that remains uncured and fatal. The first signals of AD are memory impairment and cognitive dysfunction, which progresses to memory loss, language problems, disorientation, loss of motivation, and behavioral impairment due to the destruction of neurons in the brain and causing enormous suffering to individuals, families, and society [92].

AD is categorized as sporadic AD (nonhereditary) and familial AD (FAD). Sporadic AD is the more common (95% of all cases) and less than 5% of cases reported are hereditary. About 4–8% of the elderly population of the world is affected by this disorder. It is estimated that AD strikes 5% of people over the age of 65. In Western countries, this number was 15% in the

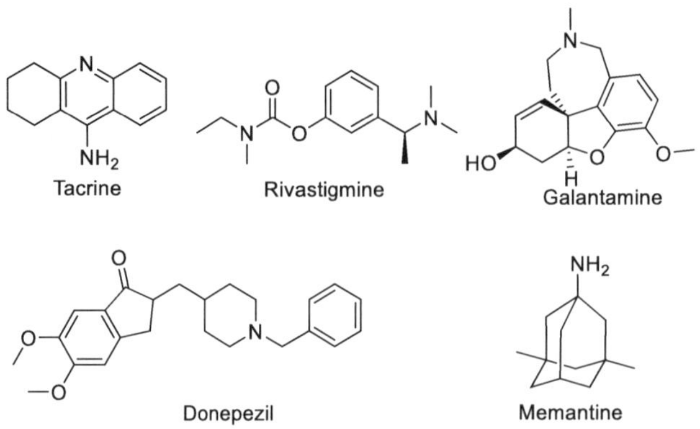

FIGURE 3.2 FDA approved medications against AD. *FDA,* Food and Drug Administration.

mid of 1990s; however, it is estimated to exceed 20% in 2020. According to World Alzheimer Report 2018 statistics, there are nearly 50 million AD patients worldwide in 2018, and the number is projected to be 75 million people by 2030 and 152 million by 2050 [93]. The increasing number of AD patients expected in the near future may produce a significant overload on the healthcare system. In this regard, AD has been recognized as a global public health priority by the World Health Organization due to the versatile and controversial health and economic issues [94]. Five medications have been approved as a treatment for AD: tacrine, rivastigmine, galantamine, donepezil, and memantine (Fig. 3.2). However, these agents can only provide symptomatic relief for AD patients [95].

The pathophysiology of AD is complex and the main histopathological hallmarks of AD are the presence of extracellular senile plaques caused by the deposits of amyloid-β (Aβ) peptide and the intracellular neurofibrillary tangles formed by abnormally phosphorylated Tau [96]. Several hypotheses have been proposed to explain the etiology of AD and these include the loss of cholinergic neuro transmission, excessive accumulation of Aβ aggregates, accumulation of Tau protein, oxidative stress, biometal dyshomeostasis, inflammatory processes, and disturbed mitochondrial transport to synaptic terminal.

Various pyridine-based compounds synthesized possessing anti-AD properties are classified into different groups based upon the type of the enzyme/receptor being inhibited by the molecule. Therefore, this chapter covered cholinesterase inhibitors, Aβ aggregation inhibitors, Secretase inhibitors, Tau aggregation inhibitors, glycogen synthase kinase-3β inhibitors, monoamine oxidase B inhibitors, neuroprotective agents, adenosine A_1 and A_{2A} receptor antagonists, selective muscarinic M1 subtype activation, phosphodiesterase inhibitor, nicotinic acetylcholine receptor, and multitarget directed ligands.

3.3.1 Cholinesterase inhibitors

In Alzheimer's disease (AD), cognitive impairment is due to loss of precholinergic cells and decreased access to synaptic acetylcholine. Therefore, one of the therapeutic approaches is to increase the level of acetylcholine (ACh) in the brain by blocking the cholinesterase (ChE) enzymes acetylcholinesterase (AChE) and butyrylcholinesterase (BChE) that are responsible for the ACh hydrolysis [97]. Furthermore, several lines of recent evidence have suggested that when the level of AChE gradually decreases, the level of BChE significantly increases in the hippocampus and temporal cortex of the brain of patients with AD. Accordingly, the concurrent inhibition of both AChE and BChE may improve AD signs and symptoms by increasing the synaptic levels of ACh [98]. Till to date, the cholinergic drugs: galantamine, donepezil, rivastigmine, and tacrine (Fig. 3.2) were regarded as the first-line pharmacotherapy for moderate AD [99].

The active site of AChE is a deep and narrow gorge mainly composed of two distinct binding sites: the catalytic (CAS) and the peripheral anionic site (PAS). The PAS has a crucial role in the polymerization of Aβ plaques and fibrils formation. Considering these aspects, dual AChE inhibitors that bind to both the CAS and PAS may simultaneously alleviate the cognitive deficit and prevent the assembly of Aβ-peptide. Despite the purely symptomatic mode of action of AChE inhibitors, they still represent the first-line treatment of AD [100].

Pyridine like tacrines that were designed by substitution of a benzene ring in tacrine by pyridine ring and annulated with saturated cyclohexane ring yielded anologues (**1**, **2**) with strong AChE inhibition capabilities [101]. A new group of tacrine analogues, modified at the aromatic ring A with a furan and a thiophene structural motive have been prepared and evaluated as AChEI/BuChEI. Interestingly, while the furan-like derivatives, 4-amino-2,3-diaryl-5,6,7,8-tetrahydrofuro[2,3-*b*]quinolines, and 4- amino-5,6,7,8,9-pentahydro-2,3-diphenylcyclohepta[*e*]furo[2,3-*b*]pyridine (**3**, **4**, **5**) are strong AChEIs and, **4** and **5** are selective regarding BChE, whereas the thieno derivatives are devoid of any biological activity [102]. Tacrine (THA) analogues, containing the azaheterocyclic pyrazolo-[4,3-*d*]pyridine system as isostere of the quinoline ring of THA (**6**, **7**) were identified as the most potent and selective AChE inhibitor and also presenting a selectivity index superior to that of THA [103]. Pyridonepezils like 2-amino-6-((2-(1-benzylpiperidin-4-yl)alkyl)amino)-4-phenylpyridine-3,5-dicarbonitriles (**8**), and 2 -amino-6-((2-(1-benzylpiperidin-4-yl)alkyl)amino)pyridine-3,5-dicarbonitriles (**9**) have been designed as hybrids resulting from a conjunctive approach that combines the N-benzylpiperidine moiety, present in donepezil, and the 2-amino-6-chloropyridine heterocyclic ring system, connected by an appropriate polymethylene linker [104]. Consequently, compound 2-amino-6-((3-(1-benzylpiperidin-4-yl)propyl)amino)pyridine-3,5-dicarbonitrile (**10**) is active at nano molar range, and found to be selective dual hAChE inhibitor. Good permeability values of pyridonepezils, as the known CNS drugs, pointing out that these molecules would cross the blood–brain barrier (BBB) by passive diffusion. Subsequently, 6-Chloro-pyridonepezils as hybrids of chloropyridine and donepezil by combining the N-benzylpiperidine moiety of donepezil with the 2-chloropyridine-3,5-dicarbonitrile heterocyclic ring system, *via* an appropriate polymethylene linker were designed. Particularly, a 6-chloro-pyridonepezil analog **11** is more selective for hAChE than for hBChE and equipotent to donepezil against hAChE [105]. The series of 5,6-dimethoxybenzofuranone derivatives bearing the benzyl pyridinium moiety inhibited AChE/BuChE activity in nanomolar range concentration. The 5,6-dimethoxybenzofuranone derivative containing methylbenzyl substituent on pyridine ring (**12**) was demonstrated to be more potent than their corresponding 6-ethoxy and 6-propoxy benzofuranone derivatives [106]. Mehdi Khoobi et al. described a series of fused coumarins namely 5-oxo-4,5-dihydropyrano[3,2-*c*]chromenes linked to *N*-benzylpyridinium scaffold at 3- or 4- positions as AChE inhibitors. The 1-(4-fluorobenzyl) derivative **13** from 4-pyridinium series showed the most potent anti-AChE activity and the highest AChE/BChE selectivity [107].

Benzofuran-based N-benzylpyridinium derivatives were assessed as novel AChE inhibitors, in particular, the N-(3,5-dimethylbenzyl) derivative (**14**) was the most active compound than donepezil. According to SAR studies, the presence of fluorine atom on benzyl pendent group and 4-nitro on the benzyl group had a positive effect on the inhibitory activity as these inhibitors were well-fitted in the active site of AChE being benzyl pyridinium part located in the CAS, while positively charged nitrogen at mid-gorge recognition site and benzofuran moiety in the PAS [108]. Manizheh Mostofi et al. reported a series of benzofuran-based chalconoids bearing 3-pyridinium and 4-pyridinium moiety as potent AChE inhibitors. Among them, 3-pyridinium derivative **15** consisting *N*-(2-bromobenzyl) moiety and 7-methoxy substituent on the benzofuran ring exhibited superior activity as potent as donepezil [109]. Pyridine 2,4,6-tricarbohydrazide derivatives were found to exhibit dual inhibitory activity against AChE and BChE. Moreover, analogs with halo substituents at C-4 of phenyl ring (**16, 17, 18**) were found to have significant inhibition property. As these congeners are too branched and not able to protrude deep into the active site of AChE, therefore, it could be bound mainly with the PAS at the entry to the gorge of AChE [110]. In 2017, Peglow et al. established a straightforward protocol to prepare bis(3-amino-2-pyridyl) diselenides and bis(2-pyridyl) diselenide derivatives and found that bis(3-amino-2-pyridyl) diselenide (**19**) significantly inhibited the AChE activity in the cerebral cortex at micromolar concentrations [111]. Hybridization of indanone and quinoline heterocyclic scaffolds resulted in a potent allosteric modulator of AChE that can target cholinergic and noncholinergic functions by fixing a specific AChE conformation [112]. The most active hybrid, olefin (**20**) had remarkable activity against hAChE in nanomolar range with a noncompetitive inhibition profile. According to Ulus et al. hybrid tacrine-sulfonamide derivative, **21** with 6 membered skeleton showed the highest inhibitory activity on AChE while analog **22** displayed the strongest inhibition of BChE [113]. Kumar et al. designed pyrimidine-containing molecules connected to cyanopyridine through piperazine linker using a multipronged approach employing computational, chemical, and biological approaches. As a result, an analog 2-(4-(6-(quinolin-8-yloxy)pyrimidin-4-yl)piperazin-1-yl) nicotinonitrile (**23**) having 8-hydroxy quinoline and 2-piperazine-3-cyano pyridine moieties has emerged as the most potent mixed type AChE inhibitor, and also as a stronger inhibitor for human AChE in neuronal cell extract *via* bivalent-binding mode [114]. A series of benzofuran-2-carboxamide-N-benzyl pyridinium halide derivatives were reported as new cholinesterase inhibitors. Amongst, the 2-fluoro-6-nitro containing derivative (**24**) was found to be a better BChE inhibitor rather than AChE as it strongly binds to the active site of BChE with hydrogen bonding, π-cation, and π–π interactions [115]. Coumarin-pyridinium hybrids, 1-(3-fluorobenzyl)-4-((2-oxo-2H-chromene-3-carboxamido)methyl) pyridinium bromide (**25**) was found to be the most active compound toward AChE, while analogs 1-(3-chlorobenzyl)-3-((2-oxo-2H-chromene-3-carboxamido)methyl)pyridinium bromide (**26**) and 1-(2,3-dichlorobenzyl)-3-((2-oxo-2H-chromene-3-carboxamido)methyl)pyridinium chloride (**27**) depicted the best BChE inhibitory activity [116]. In continuation, compounds with *N*-benzylpyridinium moiety with positively charged nitrogen at either 3- or 4- position linked to arylisoxazole carboxamide group were evaluated as competitive AChE and BChE inhibitors. Particularly, compound **28** possessing 2,4-dichloroaryl group on isoxazole ring was found to be the most potent AChE inhibitor whereas, compound **29** with phenyl group on isoxazole ring demonstrated the most promising inhibitory activity against BChE [117]. Ragab et al. designed 4-(chlorophenyl)tetrahydroquinoline derivatives to serve as tacrine analogs with lower hepatotoxicity. Consequently, 2-benzylamino derivative (**30**) had twice the inhibitory activity of tacrine with superior binding to the receptor over tacrine, promising drug-like characters, no hepatic injury and less GSH depletion than tacrine. Therefore, the hypothesis that the presence of a chloro group in tacrine related analogs results in encouraging anticholinesterase activity and reduced hepatotoxicity was supported [118]. The C-ring modifications of dehydroabietylamine, a natural AChE inhibitor yielded a set of 12-hydroxy-dehydroabietylamine derivatives, among which 12-hydroxy-*N*-(isonicotinoyl)dehydroabietylamine (**31**) emerged as selective and competitive BChE inhibitor [119]. While evaluated a series of benzyl pyridinium-2,4-dioxochroman derivatives as new anti-Alzheimer agents, two analogs 1-(2-chlorobenzyl)-3-((((2,4-dioxochroman-3-ylidene)methyl)amino)methyl)pyridinium bromide (**32**) and 1-(3,4-dichlorobenzyl)-3-((((2,4-dioxochroman-3-ylidene)methyl)amino)methyl)pyridin-1-ium bromide (**33**) were identified as the most potent inhibitors against AChE and BChE, respectively, through a mixed-type inhibition mode [39]. Four N-(4-methylpyridin-2-yl)thiophene-2-carboxamide analogs were identified as cholinesterase inhibitors. Compound with 3-chlorophenyl moiety (**34**) showed good BChE inhibition, however, compounds with 4-methyl ester on phenyl (**35**) and 4-methoxy phenyl moiety (**36**) emerged as a good inhibitor of eeAChE, but, high selectivity for BChE while compound with 3-acetyl phenyl group (**37**) as another good and selective inhibitor of eeAChE [120]. Bis(imino)pyridines bearing two *o*-pyridyl substituents (**38**) exerted AChE inhibition in the low micromolar range through binding to PAS [121] (Fig. 3.3).

3.3.2 Aβ aggregation inhibitors

The intracerebral accumulation of the amyloid β-peptide (Aβ) as senile plaques or vascular amyloid plays a key role in the pathogenesis of AD [122]. The high level Aβ peptides in the AD brain came from the abnormal proteolysis of amyloid precursor protein (APP) by β- and γ-secretase which gives $A\beta_{40}$ and $A\beta_{42}$ [123–125]. The accumulation of these peptides leads to the production of insoluble fibrils, which ultimately aggregate into the characteristic Aβ senile plaques and cause neuronal loss and cognitive impairment, increased oxidative stress which are responsible for the neuronal injury and death in AD [126]. Despite recent progress in symptomatic therapy with cholinergic drugs, an effective therapeutic approach aimed at halting and reversing amyloid formation and deposition that interferes directly with the neurodegenerative process in AD has gained much recent interest [127–129].

Michael et al. reported a series of dual-function triazole–pyridine ligands [4-(2-(4-(pyridin-2-yl)-1H-1,2,3-triazol-1-yl)ethyl)-morpholine (**39**) and 5-(4-(pyridin-2-yl)-1H-1,2,3-triazol-1-yl)pentan-1-amine (**40**) that interact with the Aβ peptide and modulate its aggregation. These compounds also found to exhibit an ability to limit metal-induced Aβ aggregation [130]. In 2014, Viayna et al. synthesized a family of rhein-huprine hybrids, which were expected to hit several AD-related targets, including Aβ and Tau aggregations [131]. Ex vivo studies with the lead **41** have shown a central soluble Aβ lowering effect, accompanied by an increase in the levels of mature APP and thus, **41** emerged as very promising disease-modifying anti-Alzheimer drug candidates. Kroth et al. developed 2,6-disubstituted pyridine derivatives composed of three 2,6-disubstituted pyridine units separated by at least one C2- (**42**) or C3- (**43**) linker as most potent inhibitor of Aβ aggregation by interacting with the β-sheet conformation of Aβ via donor–acceptor–donor hydrogen bond formation [132]. In 2017, Pandolfi et al.

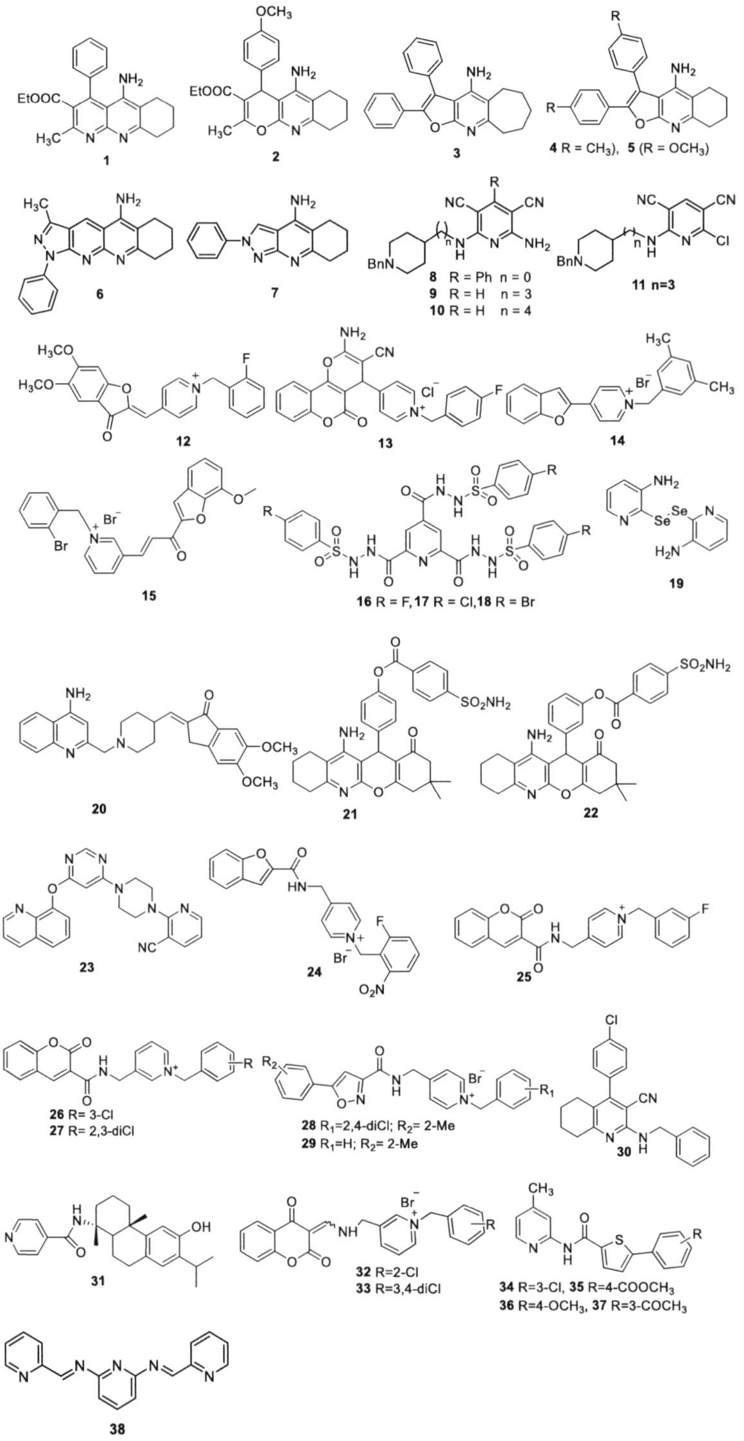

FIGURE 3.3 Pyridine based cholinesterase inhibitors.

synthesized the pyridine derivatives with carbamic or amidic functional groups where two aromatic regions linked by means of a flexible alkyl chain with variable length and tested toward $A\beta_{42}$ self-aggregation [133]. Among these, carbamates with unsubstituted phenyl ring and a five methylene spacer (**44**), and indole ring (**45**) were able to inhibit$A\beta_{42}$ self-aggregation with a moderate selectivity for EeAChE and quite low toxicity against human astrocytoma T67 and HeLa cell lines. Subsequently, Perez-Areales et al. have developed second-generation rhein–huprine hybrids by replacement of the chlorobenzene ring of the huprine Y moiety as potential multitarget anti-AD agents. The second-generation rhein–huprine hybrid **46** display or retain increased potencies as $A\beta_{42}$ and Tau antiaggregating agent [134]. A series of 2-(3-arylureido)pyridines and 2-(3-benzylureido)pyridines were synthesized and evaluated as potential modulators for amyloid beta (Aβ)-induced mitochondrial dysfunction in AD [135]. In the biological screening, 1-(3-(3-chlorobenzyloxy)pyridin-2-yl)-3-(2-fluorophenyl)ureaderivative (**47**) attained as a lead compound protecting neuronal cells against Aβ induced neurocytotoxicity and impairment of mitochondrial ATP production. A series of 1,2,3,4-tetrahydro-1-acridone analogues were evaluated as potential dual inhibitors for Aβ and Tau aggregation. N-methylation of the quinolone ring as in **48** effectively inhibited $A\beta_{1\text{-}42}$ aggregation and Tau aggregation and could permeate the BBB [136]. Mohamed Benchekroun et al. have reported (*E*)-*N*-benzyl-*N*-[2-(benzylamino)-2-oxoethyl]-3-(aryl)acrylamides (**49** and **50**) as potent anti-AD agents as they had excellent antioxidant, strong $A\beta_{1\text{-}40}$ self-aggregation, and significant neuroprotection effect against H_2O_2 induced cell death in SH-SY5Y cells [137]. As per Zhu et al., a pyridine amine derivative (PAT) (3-bis(pyridin-2-ylmethyl)aminomethyl-5-hydroxybenzyltriphenyl phosphonium bromide) can down-regulate the mRNA level of Aβ gene, reduce the expression of $A\beta_{42}$, Aβ-induced paralysis, and inhibit the AChE activity; protecting the mitochondrial oxidative damage and improve the cognition and memory, thus showing a promising prospect for the treatment of AD [138]. Ghotbi et al. evolved hybrids containing thiazole and pyridinium moieties as selective inhibitors of β-amyloid aggregation (Aβ). Amongst, (1-(2-bromobenzyl)-4-((4-oxo-4-((4-phenylthiazol-2-yl)amino)butanamido)methyl)pyridin-1-ium bromide**)** (**51**) shown to have more potency than donepezil, in preventing Aβ aggregation and very good neuroprotective activity against H_2O_2 induced oxidative stress [139] (Fig. 3.4).

3.3.2.1 *Amyloid-specific imaging agents*

Amyloid β plaques are the main cause of cerebral amyloid angiopathy (CAA) when they accumulate in the walls of cerebral capillaries and arteries [140]. In the past, the detection of these pathological abnormalities was only possible by postmortem investigation of AD brains using classic staining reagents for Aβ plaques including Congo Red (CR) and Thioflavin-T (ThT) [141]. To that end, detection of Aβ deposits with noninvasive techniques including positron emission tomography (PET) and single-photon emission computed tomography (SPECT) could provide a unique tool for *in vivo* monitoring of AD progression in the early stages in patients. Currently, Amyloid-imaging tracers could also facilitate the evaluation of the efficacy of antiamyloid therapies. Therefore, the development of amyloid-specific imaging agents could be potentially useful for early diagnosis and further neuropathogenesis studies of AD.

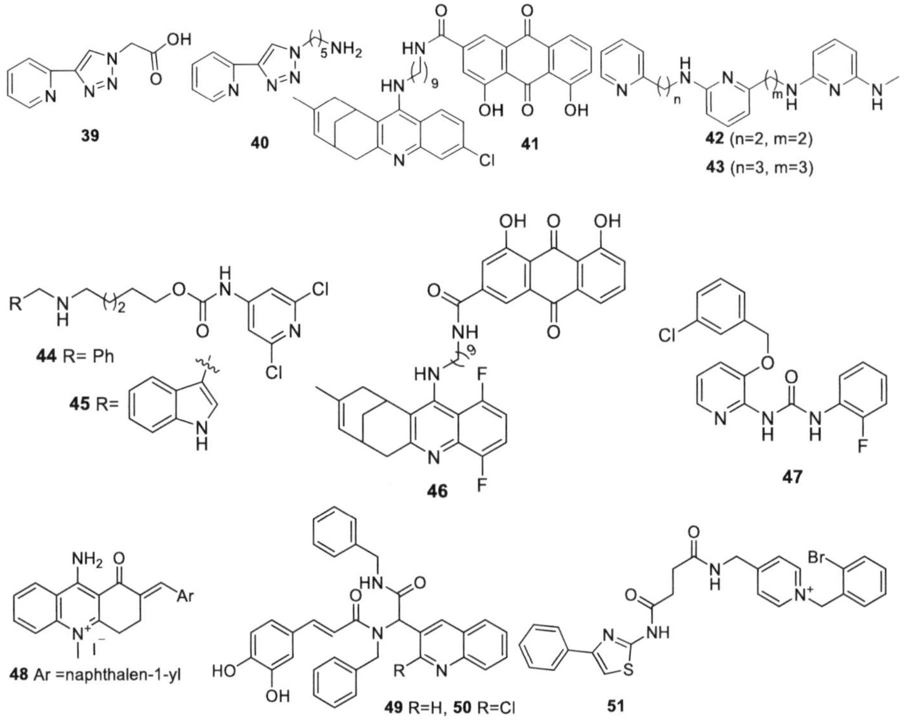

FIGURE 3.4 Pyridine based β-amyloid aggregation (Aβ) inhibitors.

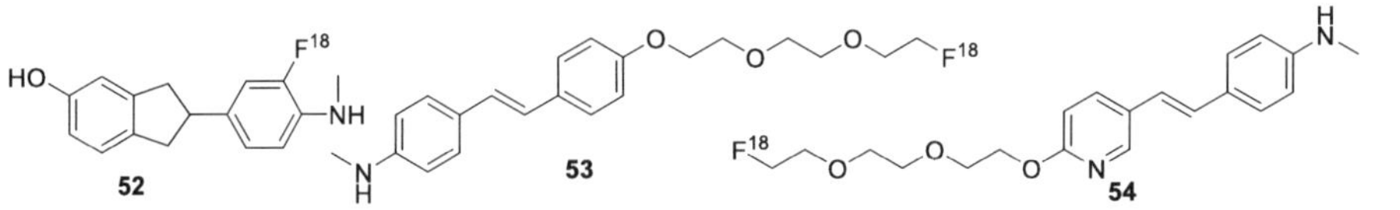

FIGURE 3.5 FDA approved probes for AD PET imaging. *FDA*, Food and Drug Administration; *PET*, positron emission tomography.

3.3.2.1.1 Positron emission tomography (PET)

The U.S. Food and Drug Administration approved a few of the probes for AD PET imaging, including [^{18}F]GE-067 [**52**], [^{18}F]BAY94-9172 [**53**], and [^{18}F]AV-45 [**54**] (Fig. 3.5). A series of β-amyloid (Aβ) aggregate-specific ligands, 2-(4′-dimethylaminophenyl)-6-iodoimidazo[1,2-*a*]pyridine, **55**(IMPY), and its related derivatives were prepared and assessed to have highly desirable properties: fast kinetics of high initial brain uptake and rapid wash out in the non-Aβ plaque-containing areas. Therefore, they are most likely to be successful as imaging agents for Aβ plaques in the brain [142]. A series of (*E*)-3-styrylpyridine derivatives as potential diagnostic imaging agents targeting Aβ plaques in AD was synthesized and examined. Among them, (*E*)-2-Bromo-5-(4-dimethylaminostyryl)pyridine (**56**) with a dimethylamino group shown to have the highest binding affinity, may be useful as a radioiodinated imaging agent for mapping Aβ plaques in the brains of patients with AD [143]. Cai et al. evaluated ^{18}F-labeled IMPY [6-iodo-2-(4¢-*N*,*N*-dimethylamino)phenylimidazo[1,2-*a*]pyridine] derivatives as agents for imaging β-amyloid plaque with PET. One of the two *N*-methyl groups of IMPY was substituted with either a 3-fluoropropyl (FPM-IMPY) or a 2-fluoroethyl (FEM-IMPY) group have favorable *in vivo* brain pharmacokinetics and a high specific labeling of β-amyloid plaques [144]. Later, 6-thiolato-substituted 2-(4′-*N*,*N*-dimethylamino)phenylimidazo[1,2-*a*]pyridines (*RS*-IMPYs) were synthesized as candidates for labeling with carbon-11 and imaging of Aβ plaques in living human brain using PET. Two MeSIMPY analogs [^{11}C] **57** and [^{11}C] **58** labeled with carbon-11 at its *S*- or *N*-methyl position had moderately high brain uptakes of radioactivity followed by rapid washout to low levels. [^{11}C] **57** also bound selectively to Aβ plaques in post mortem human AD brain [145]. Cai et al. evaluated ^{18}F-labeled IMPY [6-iodo-2-(4′-*N*,*N*-dimethylamino)phenylimidazo[1,2-*a*]pyridine] derivatives as agents for imaging β-amyloid plaque with PET. The two ^{18}F-labeled IMPY derivatives 2-fluoroethyl(FEM-IMPY) **59** and 3-fluoropropyl (FPM-IMPY) **60** reported entering the brain of normal mice readily and quickly with favorable pharmacokinetics with a moderate affinity for imaging β-amyloid plaques [145]. Seneca et al. reported a novel PET radio ligand, [^{11}C]MeS-IMPY([*S-methyl*-^{11}C]*N*,*N*-dimethyl-4-(6-(methylthio)imidazo[1,2-*a*]pyridine-2-yl)aniline) (**57**) for imaging β-amyloid plaques in patients with AD [146]. Two bifunctional small molecules, **61** and **62**, were prepared by Sarmad et al. based on a design strategy that integrates metal-binding properties into Aβ imaging agents. These bifunctional molecules were found to modulate the generation of Cu-triggered Aβ aggregates and promote their disaggregation and hence, considered as potential therapeutic agents in metal-ion chelation therapy. A series of ^{18}F styrylpyridine derivatives with high molecular weights for selectively targeting Aβ plaques in the blood vessels of the brain were developed by Zha et al. in 2011. The styrylpyridine derivative, **63**, displayed high binding affinities and specificity to Aβ plaques and therefore it can be potentially used as a biomarker and PET imaging agent in patients with CAA [147]. Yousefi et al. evaluated a series of novel phenyl-imidazo[1,2-a]pyridines as improved imaging agents for Aβ in AD and demonstrated that radioiodinated 2-(4′-bromophenyl)-6-iodoimidazo[1,2-a]pyridine (BrIMPY, **64**) possesses the desirable properties as a tracer suitable for *in vivo* and *in vitro* studies, with potential for PET and SPECT imaging of β-amyloid plaques in the living brain [148]. Harrison et al. investigated 5-fluoro-2-aryloxazolo-[5,4-b]pyridines as potential ^{18}F containing β-amyloid PET ligands. Analog **66** (MK-3328) was identified as a promising candidate, as it exhibited amyloid binding potency balanced with low levels of nonspecific binding. *In vivo*, [^{18}F] **65** demonstrated favorable kinetics, high brain uptake, and good washout in normal rhesus monkey PET imaging studies [149]. As dysregulated metal ions are linked to the aggregation of the amyloid-β (Aβ) peptide, the disruption of metal–Aβ interactions has become a viable strategy for AD therapeutic development. Considering this Jones et al. reported dual-function triazole–pyridine ligand namely [4-(2-(4-(pyridin-2-yl)-1H-1,2,3-triazol-1-yl)ethyl)-morpholine (**66**) to possess an ability to limit metal-induced Aβ aggregation [130]. Bandara et al. developed a series of bifunctional chelators (BFCs) that generated with triazacyclononane (TACN) and 2,11-diaza[3.3]-(2,6)pyridinophane (N4) macrocycles linked to 2-phenylbenzothiazole fragments that bind tightly to longer-lived and a strong chelator radionuclide ^{64}Cu as novel imaging agents for noninvasive PET imaging which would be

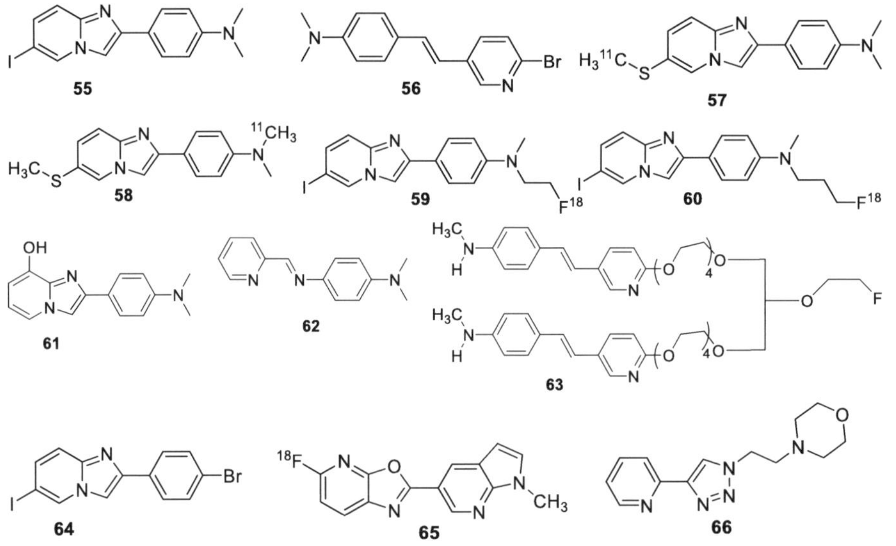

FIGURE 3.6 Pyridine-based PET imaging agents for Aβ plaques. *PET*, positron emission tomography.

advantageous for both diagnostic and drug development purposes. These ligands were found to exhibit low nanomolar affinity for Aβ aggregates *in vitro* as well as specific binding to amyloid plaques in the brain sections of AD transgenic mice [150] (Fig. 3.6).

3.3.2.1.2 Single photon emission computed tomography (SPECT) imaging agents

Zeng et al. synthesized and biologically evaluated the fluorinated imidazo[1,2-a]pyridine derivatives, FEPIP (6-(20-fluoroethyl)2-(40-dimethylamino)phenylimidazo[1,2-a]pyridine) (**67**) and FPPIP (6-(30-fluoropropyl)-2-(40-dimethylamino)phenylimidazo[1,2-a]pyridine) (**68**) as a potential SPECT agents for imaging Aβ plaques with high binding affinity [151]. Amongst several synthetic bis-pyridylethenylbenzenes, (E,E)-1,4-bis(4-pyridylethenyl)benzene (**69**) with Aβ binding properties, ability to pass the BBB, improved solubility, lower LogP values, accumulation of cerebral amyloid Aβ and fluorescent properties was identified as a novel backbone for the development of amyloid targeting PET or SPECT probes [152]. Okumura et al. developed 2-phenyl- and 2-pyridyl-imidazo[1,2-a]pyridine derivatives containing an alkoxy group or a five-membered aromatic heteroring system as novel SPECT imaging probes for Aβ plaques. The autoradiography results revealed that 1H-1,2,3-triazole derivatives [123I]**70** and [123I]**71**, and 1H-1,2,4-triazole derivative [1238 I]**72** displayed excellent binding affinities to Aβ plaques in the hippocampal region of an AD brain, exhibiting a higher selectivity than [12310I]IMPY and higher *in vivo* stability in normal rats [153]. Due to the unique characteristics of SPECT imaging with 99mTc-labeled radiopharmaceuticals such as simplicity, availability, and favorable physical properties, the development of 99mTc-labeled amyloid imaging agents for AD has attracted the researchers' attention. 2-Arylimidazo[2,1-b]benzothiazole (IBT) derivatives were synthesized as potential tridentate radiotracers for AD imaging purposes. Two of these ligands (**73, 74**) were successfully labeled with 99mTc

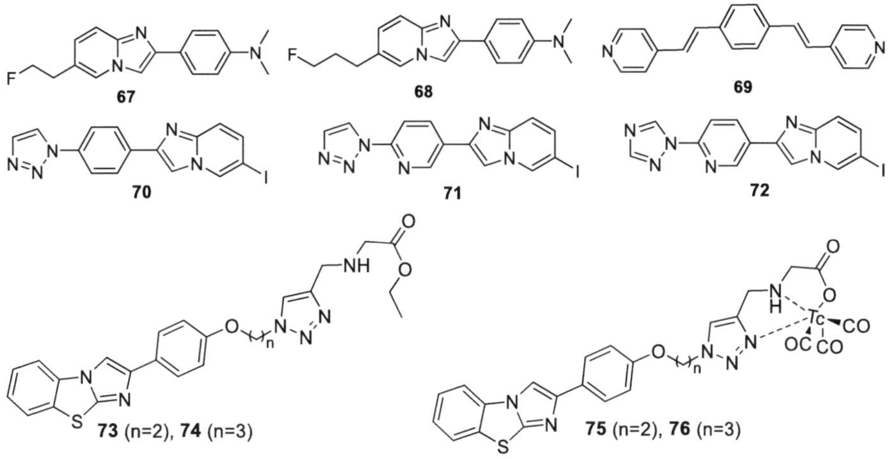

FIGURE 3.7 Pyridine-based SPECT imaging agents for Aβ plaques.

radionuclide at high radiochemical purity using fac-[99mTc$(CO)_3(H_2O)_3]^+$ synthon. [99mTc] **75** and [99mTc] **76** were evaluated as SPECT imaging agents for Aβ plaque in AD, since they had a suitable affinity toward Aβ aggregates [154] (Fig. 3.7).

3.3.3 Secretase Inhibitors

Conferring to the amyloid hypothesis, Aβ peptides are produced from APP across two sequential cleavages by β-secretase (BACE1) and γ-secretase. BACE1 cleaves APP into a C-terminal fragment (CTF) and a soluble APPβ fragment. The CTF further cleaved by γ-secretase to generate the Aβ_{42} fragment that subsequently aggregates to form plaques [155–157]. Therefore, lowering the Aβ_{42} levels in the brain by inhibiting the activity of β-secretase and γ-secretase represents a reasonable approach for evolving the disease-modifying drugs for AD.

3.3.3.1 *β−secretase (BACE1) inhibitor*

Further, BACE1 expression and activity are increased by oxidative stress cause metabolic impairment and apoptosis of neurons in AD [158]. Previously, it is reported that inhibitors with a peptidic and pseudopetidic structure show poor drug-like properties such as insufficient oral bioavailability, low serum half-life, or low BBB penetration [159]. Consequently, there remains a demand for small molecule BACE1 inhibitors. Around 10 inhibitors of BACE1 are now under clinical investigation in Phases I to III, and all of them are orally active small molecules. Hence, BACE1 inhibition holds tremendous potential as a therapeutic target for the development of novel small molecules for AD treatment.

Aminoimidazoles were reported as potent and selective human β-secretase inhibitors by Malamas et al. and in particular compound (**77**) as the most active analog, as it demonstrated low nano molar potency for BACE1 and selectivity for the other structurally related aspartyl proteases BACE2, cathepsin D, renin, and pepsin [160]. Later, Malamas et al. identified the small molecule di-substituted pyridinyl aminohydantoins as highly selective and potent BACE1 inhibitors. One of the more potent compounds **78** demonstrated low nanomolar potency for BACE1 and excellent selectivity toward the other structurally related aspartyl proteases BACE2, cathepsin D, pepsin, and renin [161]. Pyridinyl aminohydantoins were designed as highly potent BACE1 inhibitors by Zhou et al. and found that compound **79** had excellent potency in the nanomolar range against BACE1 [162]. The interaction between pyridine nitrogen and the tryptophan Trp76 was determined as a key feature in the S2′ region of the enzyme that contributed to increased potency. Peng et al. conducted a structure-based design and successfully produced a series of new nontacrine based novel framework [163]. From this series, the compound **80**, bearing a pyridinium functionality was identified as an excellent β-secretase inhibitor in addition to its inhibitory effects toward human BChE, and Aβ aggregation. A series of bicyclic aminoimidazoles was described as potent BACE-1 inhibitors. The interaction of aminoimidazole moiety of **81** with catalytic residues Asp32 and Asp228 of BACE-1 via hydrogen bonding was regarded as an important feature for achieving potent inhibition [164]. Structure-based design was developed by Allen A. Thomas et al. with substituted 1,3,4,4′,10,10′hexahydropyrano[4,3-b]chromene core to improve affinity for BACE1 and selectivity. Three different Asp-binding moieties: spirocyclic acyl guanidines, aminooxazolines, and aminothiazolines were examined in order to modulate potency, selectivity, efflux, and permeability to improve binding to both the S3 and S2′ sites of BACE1. As a result, compounds with an acyl guanidine moiety (**82**) from the series provided as most potent analogues with excellent selectivity for BACE1 versus CatD [165]. In the search of novel small compounds with potentially improved BACE1 inhibitory properties, Azimi et al. synthesized and evaluated the anti-BACE1activity of hybrid imidazopyridines containing phthalimide moieties [166]. The aminocyclohexyl derivatives with a methyl substituent at 6 or 7 positions of the imidazopyridine core (**83** and **84**) resulted in considerable improvement of BACE1 inhibitory potential. 4-Oxobenzo[*d*]1,2,3-triazin derivatives bearing pyridinium moiety were synthesized by Hosseini, et al. and screened against β-secretase for their inhibitory activity and found that the derivative 1-(2-nitrobenzyl)-3-((4-oxobenzo[d][1,2,3]triazin-3(4H)-yl)methyl)pyridin-1-ium (**85**) possessed inhibitory activity against β-secretase [167]. The discovery of BACE1 inhibitors with a 1-amino-3,4-dihydro-2,6-naphthyridine scaffold was achieved by K. Nakahara et al. [168]. In the design of compounds **86** and **87**, removing a structurally labile moiety and incorporating pyridine rings, increased their biochemical and cellular potency, along with reduced basicity on the amidine moiety and enhanced BACE1 inhibition activity. Moreover, the introduction of a fluorine atom on the pyridine culminated in **88** further enhanced the BACE1 inhibition activity, thus **88** become more potential than 86 and 87 (Fig. 3.8).

3.3.3.2 γ-secretase modulators (GSMs)

Though several γ-secretase inhibitors (GSIs) are currently in clinical trials, GSIs have been revealed to cause serious side effects in several clinical trials due to inhibition of Notch-processing [169, 170] and/or β-carboxy-terminal fragment (β-CTF) accumulation [171]. In contrast, γ-secretase modulators (GSMs) shift the APP cleavage site and selectively lower pathogenic Aβ_{42} levels without affecting Notch inhibition or β-CTF accumulation [172]. GSMs are therefore promising therapeutic agents for AD.

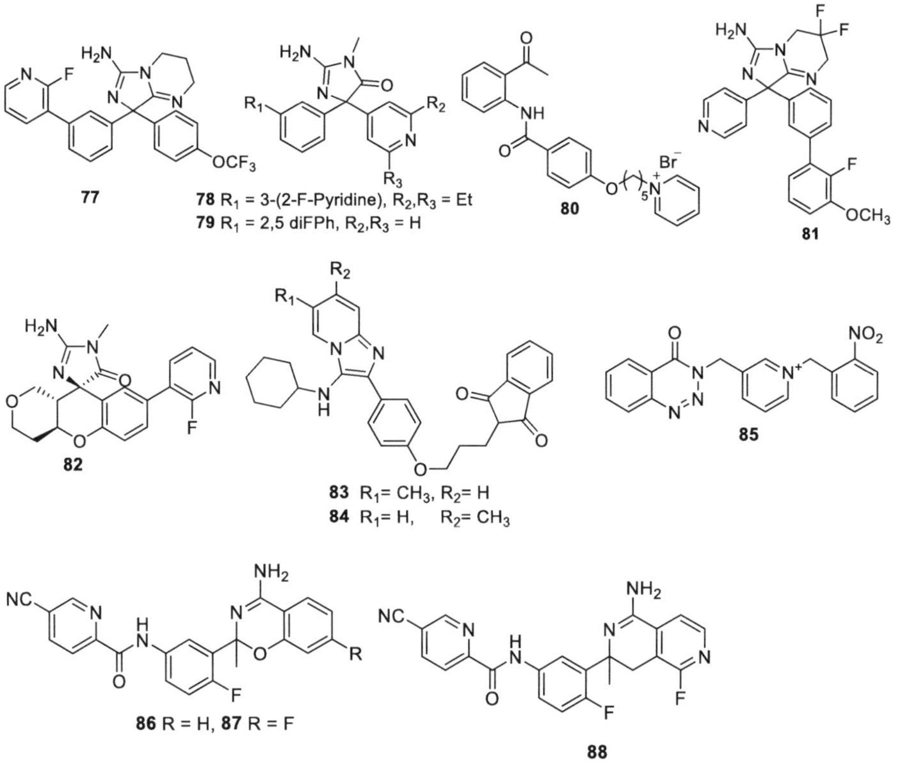

FIGURE 3.8 Pyridine-based BACE1 inhibitors.

A novel series of pyridines was discovered by Z. Wan et al. as potent γ-secretase modulators. The representative analog (**89**) N-((S)-1-(4-fluorophenyl)ethyl)-5-(3-methoxy-4-(4-methyl-1H-imidazol-1-yl)phenyl)pyridin-2-amine from the series displayed *in vivo* efficacy to inhibit $A\beta_{42}$ without altering Notch processing and had the mechanism of action consistent with that of GSMs [173]. J. Qin et al. discovered a series of pyrazolopyridines as potent GSMs that demonstrated good *in vitro* activity for reducing $A\beta_{42}$ production. An analog 5-chloro-7-((S)-1(3,4,5-trifluorophenyl)ethyl)-3-(3-methoxy-4-(4-methyl-1H-imidazol-1-yl)phenyl)-7Hpyrazolo[3,4-b]pyridine **90** has provided the greatest reduction of cerebrospinal fluid $A\beta_{42}$ in rats [68]. In another example, replacement of the methoxyphenyl ring with a methoxypyridyl ring in imidazole based GSMs has led to [2-(4-fluoro-phenyl)-1-methyl-1H-benzoimidazol-4-yl]-[6-methoxy-5-(4-methyl-imidazol-1-yl)-pyridin-2-yl]-amine (**91)** with an improved drug-like profile and enhanced *in vivo* activity, as it displayed the typical GSM profile by lowering $A\beta_{42}$ and $A\beta_{40}$ levels combined with an especially pronounced increase in $A\beta_{38}$ and $A\beta_{37}$ levels while leaving the total levels of amyloid peptides unchanged in mouse, rat, and dog [174]. Chen et al. reported 2-methylpyridine-based biaryl amides as GSMs. Lead optimization studies with the replacement of oxazole by 2-methylpyridine identified compound **92** as the most potent GSM as it selectively lower pathogenic $A\beta_{42}$ levels by shifting the enzyme cleavage sites without inhibiting γ-secretase activity and has no Notch-associated effects [175].

Through high throughput screening, R. Sekioka et al. discovered imidazopyridine derivatives as a new class of GSMs. The imidazopyridine derivative 5-[8-(benzyloxy)-2-methylimidazo[1,2-a]pyridin-3-yl]-2-ethylisoindolin-1-one (**93**) with the highest *in vitro* GSM activity and an acceptable pharmacokinetics profile that did not detectably inhibit CYP3A4 activity was considered as the most potent analog. Further, **93** demonstrated as a potent inhibitor of $A\beta_{42}$ and sufficiently penetrated the brain [176]. Later, Sekioka et al. also discovered a series of *N*-ethylpyridine-2-carboxamide derivatives as a novel scaffold for orally-active GSMs, and determined their structure-activity relationships by replacing isoindolinone moiety with a picolinamide moiety. As a result, compound 5-{8-[([1,1′-biphenyl]-4-yl)methoxy]-2-methylimidazo[1,2-*a*]pyridin-3-yl}-*N*ethylpyridine-2-carboxamide hydrogen chloride **94**, was determined as a promising analog as it had high in vitro GSM activity and was undetectable *in vitro* CYP3A4 inhibition with a sustained pharmacokinetic profile and significantly reduced brain $A\beta_{42}$ levels in mice [177]. Subsequently, the same group carried out an extensive study on GSMs and reported the imidazopyridine derivatives as valuable GSMs. With the engagement of pyridine-2-amide moiety by structural optimization of the biphenyl group, compounds greatly improved GSM activity and rat microsomal stability [177]. Specifically, 5-{8-[(3,4′-difluoro[1,1′-*biphenyl*]-4-yl)methoxy]-2-methylimidazo[1,2*a*]pyridin-3-yl}-*N*-methylpyridine-2-carboxamide (**95**) showed the highest *in vitro* GSM activity, undetectable CYP inhibition and exhibited undetectable inhibition of cytochrome p_{450} enzymes. A focused effort to improve the activity of aminothiazole-derived GSMs through the potential replacement of the fluorophenyl moiety and insertion of a methoxypyridine motif within the tetracyclic scaffold by Rynearson, et al. provided compounds with improved activity for arresting $A\beta_{42}$ production as well as improved properties, including solubility [178]. *In vivo* pharmacokinetic analysis demonstrated that these congeners are capable of crossing the BBB and accessing the therapeutic target. Methoxypyridine-derived compounds tetrahydroindazole **96** and cyclopentapyrazole **97** reduced $A\beta_{42}$ levels in the plasma and brain of mice (Fig. 3.9).

3.3.4 Tau aggregation inhibitors

Tau is an axonal protein known to form abnormal aggregates and is the biomarker of AD. Upon hyperphosphorylation, Tau disassembles from MTs and self-assembles to form NFTs which consist of paired helical filaments [179]. Metal-based therapeutics for inhibition of Tau aggregation is limited and rarely reported in contemporary science.

Gorantla et al. rationally designed molecular cobalt(II)-complexes (CBMCs) for effective inhibition of Tau and disaggregation of preformed Tau fibrils. The mechanistic studies revealed that CBMCs (**98**) play a dual role in causing disassembly of preformed aggregates as well as inhibition of complete Tau aggregation. CBMCs also prevented OA-induced toxicity in SH-SY5Y thus preventing cytotoxicity due to hyperphosphorylation of Tau. Hence, CBMCs have a potential for metal-based therapeutics for AD [180].

3.3.4.1 Tau PET tracer

PET is a powerful and noninvasive neuroimaging technique that could allow the longitudinal detection, characterization, and quantification of pathological patterns of NFTs and $A\beta$

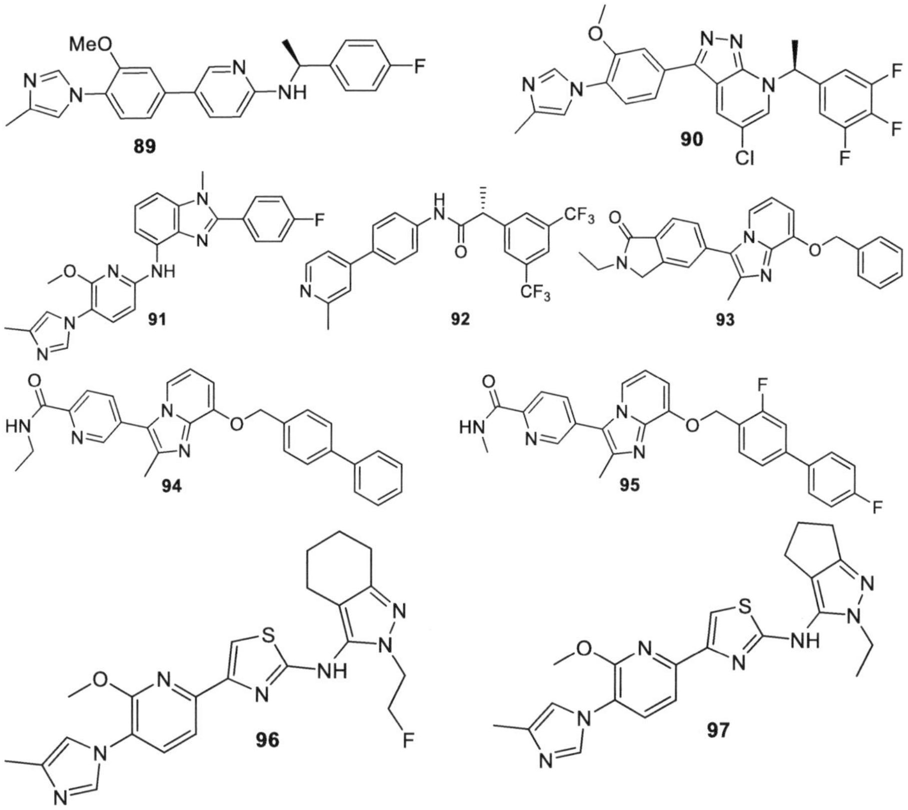

FIGURE 3.9 Pyridine-based γ-secretase modulators (GSMs).

deposition *in vivo*. Furthermore, the detection of NFTs at different stages of AD would allow a better patient selection for disease-modifying clinical studies, and would improve patient monitoring in such therapeutic trials to increase the success rate of AD drug development [181]. Several Tau PET tracers have been discovered and tested in humans so far, namely [^{18}F]FDDNP, [^{18}F]AV-680, [^{18}F]T808, [^{18}F]GTP-1, Flortaucipir, [^{18}F]AV-1451 formerly known as [^{18}F]T-807, [^{18}F]THK-5351, [^{18}F]RO6958948, [^{18}F]MK-6240, [^{18}F]PI-2620, [^{18}F]JNJ-067, and [^{18}F]APN-1607, and [^{18}F]PM-PBB3 (Fig. 3.10).

Gabellieri et al. identified the pyrrolo[2,3-b:4,5-c′]dipyridine core structure with high affinity for aggregated Tau as fused aromatic systems are likely required to interact with the β-sheet fibrillary aggregates present in Tau aggregates. Furthermore, the hydrogen bond donor (NH) found in the pyrrolo[2,3-b:4,5-c′] dipyridine core was masked with a methyl group to achieve high brain uptake. Consequently, 2-(4-(2-fluoroethoxy)piperidin-1-yl)-9-methyl-9H-pyrrolo[2,3-b:4,5-c′]dipyridine (PI- 2014, compound **99**) emerged as the most promising candidate, displaying high *in vitro* binding affinity and selectivity to neurofibrillary tangles. Fluorine-18 labeled compound **99** showed high brain uptake and rapid washout from the mouse brain with no observed bone uptake. Furthermore, compound **99** was able to detect Tau

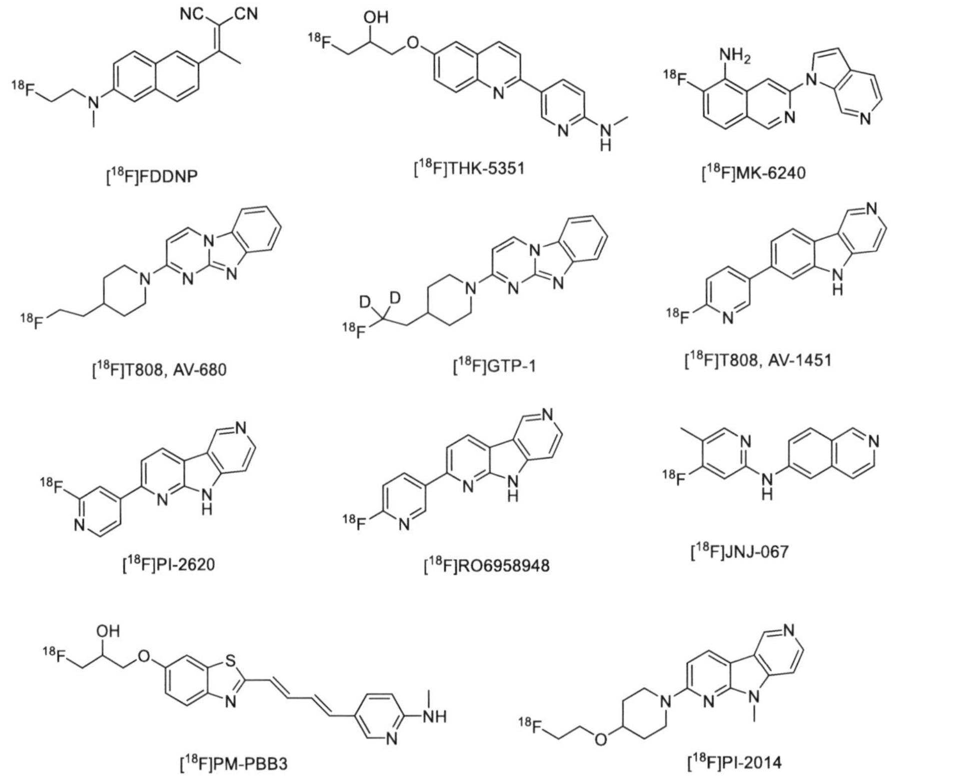

FIGURE 3.10 Structures of Tau PET ^{18}F-labeled tracers in clinical studies. *PET*, positron emission tomography.

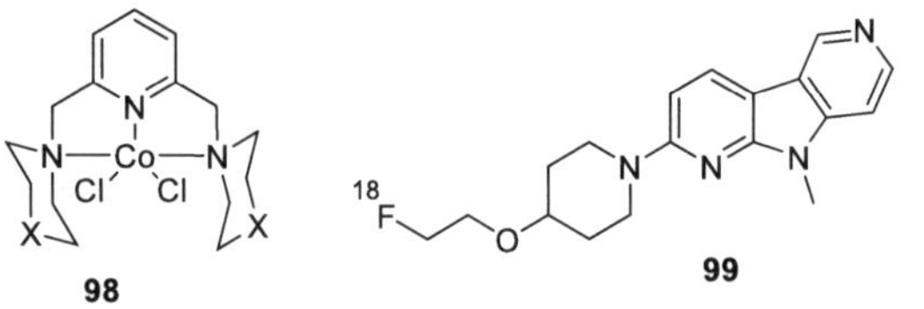

98 99

FIGURE 3.11 Pyridine-based Tau aggregation inhibitors and Tau PET tracers. *PET*, positron emission tomography.

aggregates in tauopathy brain sections from corticobasal degeneration, progressive supranuclear palsy, and Pick's disease donors [182](Fig. 3.11).

3.3.5 Glycogen synthase kinase-3*β* inhibitors

Glycogen synthase kinase-3*β*, also called Tau phosphorylating kinase, is a proline-directed serine/threonine kinase. GSK3*β* is abundant in the brain where it is localized primarily in neurons. GSK3*β* phosphorylates Tau, which leads to the formation of neurofibrillary tangles and subsequently neuronal death. It is also noted that abnormally increased activity of GSK3*β* is associated with a multitude of adverse events linked to microtubule dysfunction, neuritic

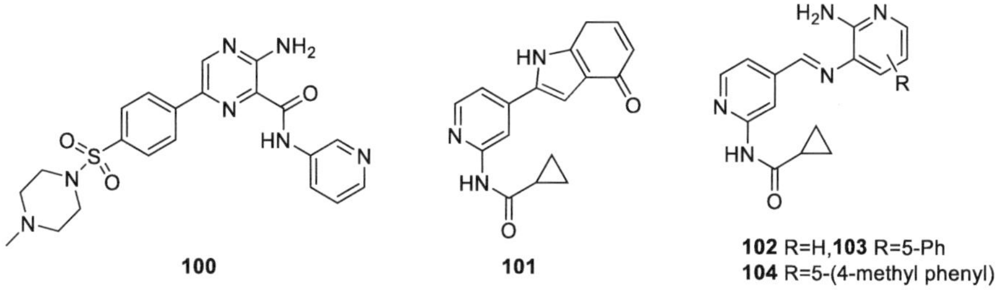

FIGURE 3.12 Pyridine-based GSK3β inhibitors.

dystrophy and cytoskeletal damage, cognitive deficits, and amyloid production. Given the significant role of GSK3β in a variety of effects linked to mechanism in AD and other CNS disorders, GSK3β inhibition using small molecule inhibitors for treating particular disease states will be worth testing in the clinic [183–185].

Stefan Berg et al. reported several pyrazine analogs as highly potent and selective GSK3β inhibitors showing cellular efficacy, good solubility, and permeability and thus predicting good bioavailability and brain penetrance. Several pyrazine analogues (**100**) are found suitable for testing of inhibition of Tau phosphorylation in the brain, and thus have therapeutic potential [186]. On the basis of the reported potent GSK-3β inhibitor, a pyrrolopyridinone derivative (**101**), a series of compounds containing the 2,3-diaminopyridine moiety was designed by keeping the N-(pyridin-2-yl)cyclopropane carboxamide portion of **101** to maintain the interaction with GSK-3b in the hinge region, replaced the pyrrolopyridinone core with a pyridine ring, and connected the two pyridine rings with an amide or an imine or a CH_2NH bridge and assessed as MTDLs against AD. Analogs **102, 103,** and **104** exhibited good Cu^{2+} induced Aβ aggregation inhibition, Cu^{2+}-Aβ complex disaggregation, ROS formation inhibition, and antioxidant activities, inhibition of Tau protein phosphorylation and protection of neuro cells against Cu^{2+}-A$\beta_{1\text{-}42}$ and H_2O_2-induced cell damage and ability to pass the BBB with drug-likeness properties and regarded as a good lead for the development of novel GSK-3β inhibitors targeting multifacets of AD [187] (Fig. 3.12).

3.3.6 MAO-B inhibitors

One of the several proteins that contribute to oxidative stress in AD is monoamine oxidase (MAO), a flavin adenine dinucleotide (FAD)-containing enzyme that metabolizes monoamine neurotransmitters and dietary amines by oxidative deamination, and thus plays an important role in regulating emotional and other brain functions. MAO-A and MAO-B are the two isoenzymes present in most mammalian tissues. The enzymatic activity of MAO-B is increased in the brains of patients with AD. This results in increased levels of oxidative deamination reaction products such as hydrogen peroxide, a source of hydroxyl radicals, aldehydes, and ammonia. These products contribute to oxidative stress, which enhances the neurodegeneration and synaptic dysfunction in AD [188, 189]. Human MAO-B inhibitors may also have neuroprotective and neurorestorative properties.

A series of (4-substituted-thiazol-2-yl)hydrazine derivatives demonstrated as privileged scaffold due to their *in vitro* hMAO inhibitory activity and selectivity in the nanomolar

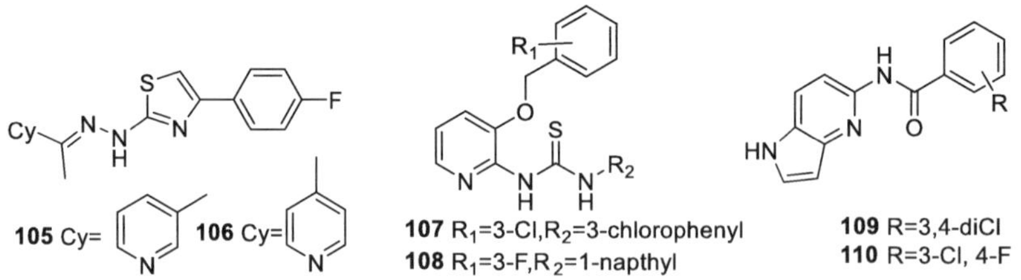

FIGURE 3.13 Pyridine-based GSK3β inhibitors.

range and hMAO-B selective inhibition higher than reference drugs. The structure–activity relationship of the different rings on the N1-hydrazine position indicated that a pyridine ring was preferred with the presence of electron-withdrawing substituents on the aryl group at C4 of the thiazole nucleus. The substituent on the α-carbon to the N1-hydrazine moiety (methyl or hydrogen) had a great influence on the activity and hMAO-B selectivity. Accordingly, the most active compounds, belonged to the 4-fluorophenyl and 2,4-difluorophenyl series and presented a pyridin-3-yl substituent on the N1-hydrazine and a methyl group on the α-carbon (**105, 106**) [190]. A series of 4-substituted-2-thiazolylhydrazone derivatives have been synthesized and tested *in vitro* for their human monoamine oxidase (hMAO) A and B inhibitory activity. The substitution at C4 of the thiazole ring was identified as an important feature to obtain highly potent and selective hMAO-B inhibition in the nanomolar range. Moreover, these derivatives were endowed with a reversible mechanism of enzyme inhibition. Insertion of phenyl or ethyl ester substituent at position R allowed for additional productive interactions and led to an enhancement in hMAO-B inhibition potency. MAO-B selectivity was further enhanced by the introduction of 3-/4-ethylpyridine moieties. Thiourea derivatives **107** and **108**, have been discovered as mitochondrial permeability transition pore (mPTP) modulators maintaining mitochondrial membrane potential in models of Aβ-induced mitochondrial membrane depolarization, and furthermore, protecting neuronal cells effectively against Aβ-induced cytotoxicity. They had a safe profile regarding ATP production and cell viability and exhibited significant neuroprotection [191]. A series of *N*-(1*H*-pyrrolo-pyridin-5-yl)benzamide derivatives were developed as BBB permeable MAO-B inhibitors. SAR analysis and structural optimization led to the identification of compounds with 3,4-Cl and 3-Cl, 4-F substituents (**109, 110**) within the series of (1*H*pyrrolo[3,2-*b*]pyridine-5-yl)benzamides as potent, selective, competitive and reversible MAO-B inhibitor with high oral absorption and brain penetration and improved physicochemical and drug-like properties required for CNS active drugs [192] (Fig. 3.13).

3.3.7 Neuroprotective agents

The mitochondrial Na^+/Ca^{2+} exchanger plays an important role in the control of cytosolic Ca^{2+} cycling in excitable cells, essential for the regulation of a plethora of Ca^{2+} -dependent physio-pathological events, such as apoptosis in the presence of a Ca^{2+} overload [193]. Although a minimal $[Ca^{2+}]c$ level is required to maintain neuronal viability, when the physiological $[Ca^{2+}]c$ level is highly altered, below or above a critical point, apoptosis is rapidly induced and the death of neurons occurs [194]. A mild sustained elevation of $[Ca^{2+}]c$ elicited by the Ca^{2+} promoters is associated with neuroprotection. The mitigation of the rate of mitochondrial Ca^{2+} efflux by the Na^+/Ca^{2+} exchanger (mNCX) blockers could afford neuroprotection against neurotoxicity elicited by cell Ca^{2+} overload [195].

Pyridothiazepines, analogs to a mNCX ligand CGP37157 (**111**) were prepared by replacing the benzene-fused ring with pyridine with the dual goal of supplying new mNCX ligands with an improved pharmacokinetic profile, as well as higher selectivity and potency. As expected, compound **112** disclosed the selective and marked mNCX blocking activity and a good neuroprotective profile.

The neuroinflammatory process in AD leads to activation of brain cells like microglia and astrocytes and also cytokines, chemokines, and the complement system, which result in neuronal dysfunction and brain cell death. The overexpression of cytokines such as interleukin (IL)-1β, tumor necrosis factor (TNF)-α, and IL-6 enhances APP production and the amyloidogenic processing of APP that leads to the formation of amyloid β_{42} (Aβ_{42}) peptide and hyperphosphorylation of Tau protein [196, 197]. Dehydroepiandrosterone (DHEA) and testosterone have been reported as neuroprotective steroids useful for the treatment of various neurodegenerative disorders. The 16-arylidene steroidal derivatives, 16-(4-pyridylidene) steroid **113,** and its 4-aza analogue **114** were found to be the most active neuroprotective agents in LPS-treated animal models and produced effects better than standard drug celecoxib and dexamethasone. These analogs improved LPS-induced learning, memory, and movement deficits in animal models and revealed suppression of oxidative and nitrosative stress, AChE activity, and reduction in TNF-α levels [198]. Subsequently, Ranjit Singh et al. identified 16,17-pyrazolinyl dehydroepiandrosterone (DHEA) analogues as neuroprotective agents using LPS-induced neuroinflammation animal models as they considerably improved the LPS-induced learning, memory and movement deficits in animal models. 16,17-Pyrazolinyl steroids substituted with a 4-pyridyl moiety at 5-position of heterocyclic ring **115, 116, 117** were found to be the most potent agents and produced neuroprotective effects better than standard drugs celecoxib and dexamethasone [199].

MPP^+ (1-methyl-4-phenylpyridinium) is a toxic cation that causes dopaminergic cell death after it is taken into the cell by the dopamine transporter. The production of free radicals following MPP^+ increases cell death, thereby inducing neuronal death and loss of dopaminergic neurons [200]. Jouha et al. synthesized pyrazolopyridine derivatives and a high percentage of neuroprotection against MPP^+-induced neurotoxicity in a human neuroblastoma cell line (SH-SY5Y cells) were noted for most of the analogs. Since, the neuroprotection of these compounds against MPP^+-induced apoptosis was found to associate with the regulation of pro- and anti-apoptotic proteins, oxidative stress, and down regulation of caspase-3 activation; it is evidenced that these pyrazolopyridine derivatives may have a role on dopaminergic neuroprotection *via* antiapoptotic pathways [201].

Pyridotacrines were designed to improve the pharmacological profile of the standard drug tacrine, concerns to its cholinergic and toxicity properties, as well as to discover activities on new biological targets involved in the origin and progression of AD. Consequently, ethyl 5-amino-2-methyl-6,7,8,9-tetrahydrobenzo[b] [1,8]naphthyridine-3-carboxylate (**118**), found to be a potent and selective AChEI compared with BChE, a very potent neuroprotective against

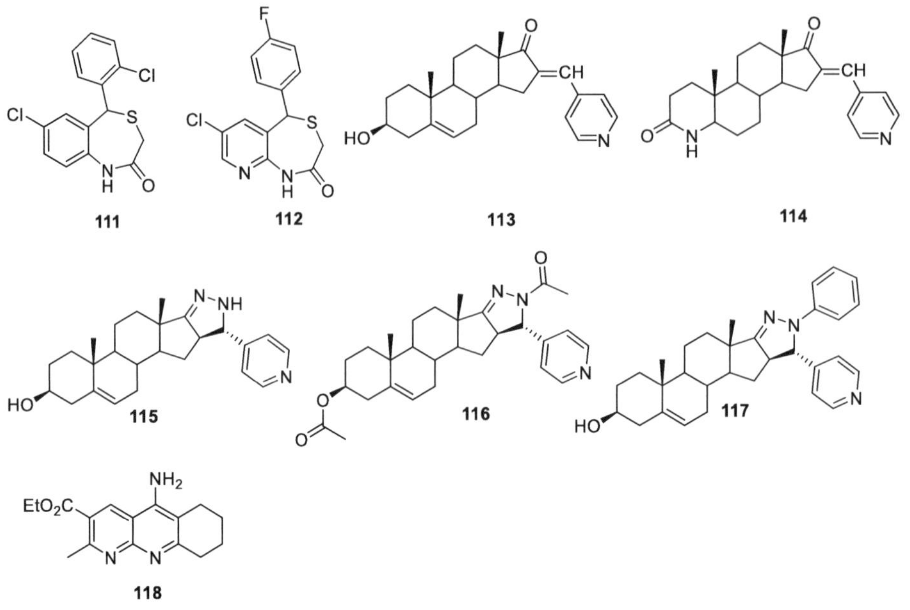

FIGURE 3.14 Pyridine-based neuroprotective agents.

diverse toxic insults, such as R/O, OA, Aβ, and glutamate overexposure and reduced the cognitive damage provoked by scopolamine, and the infarct volume provoked by photothrombosis [202] (Fig. 3.14).

3.3.8 Adenosine A_1 and A_{2A} receptor antagonists

Adenosine, a ubiquitous neuromodulator, exerts its effects through the activation of G protein-coupled receptors, subdivided into the four subtypes A_1, A_{2A}, A_{2B}, and A_3 [203]. The hA_1 AR subtype is the most abundant in the brain, sustained activation of which induces AMPA receptor endocytosis and the consequent, persistent synaptic depression may contribute to enhanced neuronal death. Thus it is traditionally considered a neuroprotective receptor due to its inhibitory effects [204]. The hA_{2A} AR subtype has widespread in the basal ganglia and blockade of this AR subtype exerts a protective effect in different models of cerebral ischemia and neurodegenerative disorders, such as PD or AD [205–207].

Falsini et al. studied an enlarged series of 1,2,4-triazolo[4,3-*a*]pyrazin-3-ones for their potential neuroprotective effect aiming to deepen SAR studies and to obtain hA_{2A}AR selective antagonists or dual-targeting hA_1 and hA_{2A} AR antagonists. It has been revealed that several analogs of this series possessed a nanomolar affinity for both hA_1 and hA_{2A} ARs and different degrees of selectivity versus the hA_3 AR. Specifically, compounds **119** and **120**, featuring a 2-phenyl ring and, respectively, a 4-nitro and 4-bromo substituent on the 6-phenyl moiety endowed with high affinity and a complete selectivity for the hA_{2A} AR and

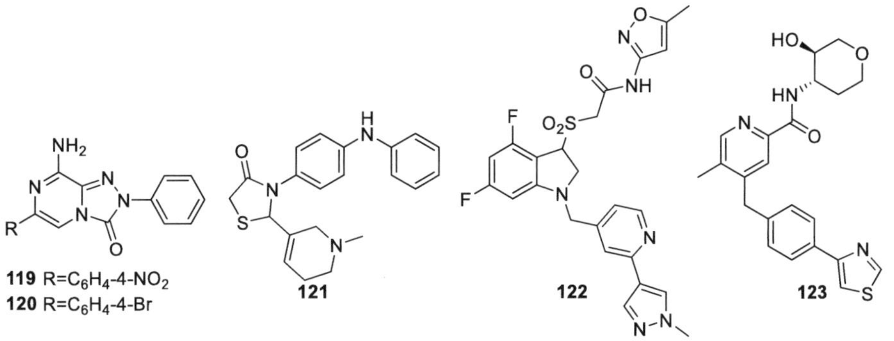

FIGURE 3.15 Pyridine-based hA AR antagonists and M1 PAMs.

demonstrated ability in preventing Aβ peptide-induced neurotoxicity in SH-SY5Y cells [208] (Fig. 3.15).

3.3.9 Selective muscarinic M1 subtype activation

The M1 muscarinic acetylcholine receptor is thought to play an important role in memory and cognition, making it a potential target for the treatment of AD and schizophrenia. Moreover, M1 interacts with BACE1 and regulates its proteosomal degradation, suggesting selective M1 activation could afford both palliative cognitive benefit as well as disease modification in AD. Selective activation of M1 with a positive allosteric modulator (PAM) has emerged as a new approach to achieve selective M1 activation [209–211].

Sadashiva et al. designed N-alkyl/aryl substituted thiazolidinone arecoline analogues to elucidate further SAR study on the chemistry and muscarinic receptor-binding efficacy of earlier reported arecoline thiazolidinone and morpholino arecoline analogues. Derivative **121** having diphenylamine moiety on nitrogen of thiazolidinone found to be the most potent with a significant affinity for the M1 receptor binding [212]. Further optimization of ML169 by James C. Tarr et al. led to identifying **122** as selective M1 Positive Allosteric Modulator as it maintained complete subtype selectivity for the M1 receptor over the other subtypes (M2–M5), displayed improved DMPK profiles, and potentiated the carbachol (CCh)-induced excitation in striatal MSNs, modest reductions in CYP450 inhibition at 2C9, 2D6, and 3A4 and significantly lowered cLogP and able to potentiate CCh-mediated nonamyloidogenic APPsα release, further strengthening the concept that M1 PAMs may afford a disease-modifying role in the treatment of AD [213]. Davoren et al. have identified a series of pyridones and pyridines as a new class of M1 PAMs and found that both series bind to the allosteric site of the M1 receptor in a similar fashion to each other. In the pyridine series, potent and selective M1 PAM-Agonist N-[(3R,4S)-3-hydroxytetrahydro-2H-pyran-4-yl]-5-methyl-4-[4-(1,3-thiazol-4-yl)benzyl]pyridine-2-carboxamide (PF-06767832) (**123**) has well-aligned physicochemical properties, good brain penetration, and pharmacokinetic properties. An extensive SAR study on the 4-position of the benzyl ring identified oxazole and thiazole rings as optimal substituents. Finally, Davoren et al. provided strong evidence that M1 activation contributes to the cholinergic liabilities that were previously attributed to activation of the M2 and M3 receptors [214] (Fig. 3.15).

3.3.10 Phosphodiesterase (PDE) inhibitor

Phosphodiesterases (PDEs) are a family of enzymes that catalyze the hydrolysis of the secondary signal messengers, cyclic adenosine monophosphate (cAMP), and cyclic guanosine monophosphate (cGMP). Specifically, Phosphodiesterase 5 (PDE5) hydrolyzes cGMP leading to increased levels of the cAMP response element-binding protein (CREB), a transcriptional factor involved with learning and memory processes. Therefore, the inhibition of PDE5, reduce PKG activity, which leads to an improvement of learning and memory in mouse models of AD [215–217]. Targeting the NO/cGMP/PKG/CREB signaling pathway can be achieved through the inhibition of phosphodiesterases (PDEs).

A series of novel PDE5Is with two scaffolds, 1,2,3,4-tetrahydrobenzo[b][1,6]naphthyridine and 2,3-dihydro-1H-pyrrolo[3,4-b]quinolin-1-one were developed aiming to improve the water solubility of reported PDE5I a quinoline analogue. Among them, compound **124**, 2-acetyl-10-((3-chloro-4-methoxybenzyl)amino)-1,2,3,4-tetrahydrobenzo[b][1,6]naphthyridine-8-carbonitrile was found to have excellent PDE5 potency and selectivity, with improved water solubility, as well as good efficacy in a mouse model of AD. It also increased cGMP levels in the hippocampus of mice and improved learning and memory deficits in a mouse model of AD. Moreover, the physicochemical and pharmacological properties of compound **124** were found to be optimal for a potential drug candidate [218].

Inhibition of PDE4 has been shown to have anti-inflammatory and antipsychotic effects in various preclinical models. In addition to inflammatory processes, PDE4 has been implicated in a number of other therapeutic roles, such as cancer, AD, addiction, and PD [219, 220]. Vadukoot et al. developed 1H-pyrrolo[2,3-b]pyridine-2-carboxamide derivatives as selective and potent PDE4 inhibitors. Analog with 3,3-difluoroazetidine moiety **125** is a PDE4 preferring inhibitor and exhibited acceptable *in vitro* ADME and significant inhibition of TNF-α release from macrophages exposed to pro-inflammatory stimuli, selectivity against a panel of CNS receptors and thus, represents an excellent lead for further optimization and preclinical testing in the setting of CNS diseases [221](Fig. 3.16).

3.3.11 Nicotinic acetylcholine receptor (nAChRs) ligands

Acetylcholine (ACh) was the first neurotransmitter discovered whose actions are mediated by two different cholinergic receptors, the metabotropic muscarinic acetylcholine receptors (mAChRs) and the ionotropic nicotinic acetylcholine receptors (nAChRs). At the presynaptic axon terminals, the nAChRs modulate the release of several neurotransmitters (acetylcholine, noradrenaline, dopamine, glutamate, and GABA) [222]. Since the activation of nAChRs by ACh can modulate a number of physiological processes, nAChRs are involved in several pathological conditions such as inflammation, cancer, and central nervous system disorders [223]. The $\alpha4\beta2$ type nicotinic acetylcholine receptors (nAChRs) are the most prominent subtype of nAChRs in mammalian brain. These receptors have been linked to many central nervous system disorders, including AD, PD, Tourette's syndrome, schizophrenia, and attention deficit hyperactivity disorder.

A series of 2-chloro-5-((1-methyl-2-(*S*)-pyrrolidinyl)methoxy)-3-(2-(4-pyridinyl)vinyl) pyridine was developed with the aim of improving lipophilicity and BBB penetration of 2-[^{18}F]fluoro-A-85380, **126**, a promising radiotracer for imaging the nicotinic acetylcholine

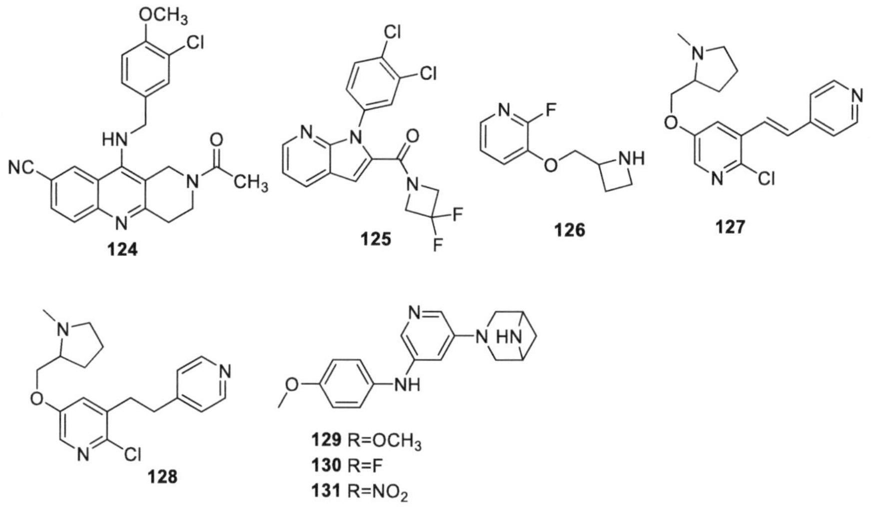

FIGURE 3.16 Pyridine-based PDE inhibitors and nAChRs ligands. *PDE*, Phosphodiesterase.

receptor (nAChR) by PET in humans. The N-methyl derivatives **127** and **128** demonstrated very high affinities at nAChRs and become targets for the development of $^{11}CH_3$-labeled derivatives as radiotracers for PET imaging of nAChRs [224]. Three series of derivatives: 3-(anilino)pyridines, 2-(anilino)pyridines and 6-(aniline)piperidines endowed with a diazabicyclo[3.1.1]heptane core have been synthesized and assessed for their binding affinity with nAChRs. Among the synthesized compounds, the most $\alpha4\beta2$ selective ligands resulted from 3-(5-anilinopyridin-3-yl)-3,6-diazabicyclo[3.1.1]heptane series. In particular, the aniline ring or its 4-OMe (**129**), 4-F (**130**) and 4-NO_2 (**131**), resulted the best substituents [225] (Fig. 3.16).

3.3.12 Multitarget directed ligands (MTDLs)

Based on the numerous AD-related targets in the disease network, multitarget design strategy is a crucial direction to seek for enhanced therapy, since; multitarget drugs have the ability to regulate more targets than single-target drugs, affecting the disease network with more potency. To this end, current efforts have been concentrating on multitarget directed ligand (MTDL) approaches against the complex pathologies of AD. In this concept, synthetic multifunctional hybrids designed based on the numerous AD-related targets could influence the different pharmacological targets and acquire a safer profile compared to single-targeted drugs [129, 226–228].

Samadi et al. described simple, and readily available 2-aminopyridine-3,5-dicarbonitriles and 2-chloropyridine-3,5-dicarbonitriles as multipotent agents [229]. These molecules were modest inhibitors of AChE and BChE in the micromolar range. Compound 2-Chloro-6-(piperidin-1-yl)pyridine-3,5-dicarbonitrile **132** was found to be a mixed-type inhibitor with the highest neuroprotection capability and hence, considered as attractive multipotent therapeutic molecules. A series of cyanopyridine–triazine hybrids were screened as multitargeted anti-AD agents by Rizzo et al. On the biological evaluation of homo- and heterodimer molecules referable to bis(7)-tacrine-derivatives, reported an analog N-(6-chloro-1,2,3,4-tetrahydroacridin-9-yl)-N′-(11H-indeno[1,2-b]quinolin-10-yl)heptane-1,7-diamine (**133**) of the series as potent MTDL. This compound disclosed the most potent AChE inhibition, good activity against BChE, a remarkable activity against BACE1 and capable of inhibiting the nonenzymatic function of AChE [230]. Xie et al. described a series of tacrine-coumarin hybrids as multifunctional candidates against AD as most of them exhibited a significant ability to inhibit ChEs and self-induced β-amyloid (Aβ) aggregation, and to act as metal chelators. Besides, compound 2-(4-(2-((4-methyl-2-oxo-2H-chromen-7-yl)oxy)ethyl)piperazin-1-yl)-N-(1,2,3,4-tetrahydroacridin-9-yl)acetamide **134** was indicated as a mixed-type inhibitor of AChE [231]. Further, Li et al. developed a series of tacrine-flavonoid hybrids as multifunctional ChE inhibitors against AD. These hybrids exhibited a significant ability to inhibit ChE and self-induced $A\beta_{1\text{-}42}$ aggregation. Among the series, compound **135** was particularly found highly potent and showed a balanced inhibitory profile against ChE and self-induced $A\beta_{1\text{-}42}$ aggregation. Analog **135** also showed excellent metal chelating property and low cell toxicity and hence, might be considered as a promising lead [232]. Bautista-Aguilera et al. developed the donepezil-pyridyl hybrids (DPHs) as multipotent cholinesterase (ChE) and monoamine oxidase (MAO) inhibitors. A donepezil-pyridyl hybrid **136** might be considered as a promising compound as it was found to be more potent for the inhibition of AChE and BChE than the reference with better drug-likeness profiles [233].

Maqbool et al. described cyanopyridine–triazinehybrids as powerful multifunctional agents, these compounds showed potent AChE inhibitory activity, good selectivity for AChE over BChE and potent radical scavenging activity. Specifically, compound 2-(4-(4-(3-chloro-4-fluorophenylamino-6-(3-(trifluoromethyl)phenylamino)-1,3,5-triazin-2-yl)piperazin-1-yl)nicotinonitrile (**137**) exhibited the highest anti-$A\beta_{1\text{–}42}$ aggregation, antioxidant, no cytotoxicity, and neuroprotection against $A\beta_{1\text{–}42}$-induced toxicity, which made these cyanopyridine–triazine hybrids powerful MTDL candidates against AD [234]. A series of 2,6-dichloro-4-aminopyridine derivatives with carbamic or amidic function where two aromatic regions linked by mean of a flexible alkyl chain with variable length has been reported to act as ChE inhibitors. The carbamates **138, 139,** and **140** were the most potent *h*AChE inhibitors with the mixed inhibition mechanism, able to inhibit *h*BChE and $A\beta_{42}$ self-aggregation, and possessed quite low toxicity against human astrocytoma T67 and HeLa cell lines, with the good predicted BBB permeation parameters [133]. Perez-Areales et al. have developed second-generation rhein–huprine hybrids as potential multitarget anti-AD agents. These compounds were designed by replacement of the chlorobenzene ring of the huprine moiety by other differently substituted benzene or heteroaromatic rings. These hybrids seem to display an interesting multitarget profile, in particular, analog **141** that lower hAChE and hBACE-1 activity, and had potent $A\beta_{42}$ and Tau antiaggregating and antioxidant activities. The rhein–huprine hybrids should also be able to readily penetrate the CNS [134]. A family of hybrids of pyridine carboxamide and tacrine were assessed as GSK-3β/AChE dual-target inhibitors. Among these hybrids, **142** had the most promising profile with nanomolar inhibition on both *h*AChE and *h*GSK-3β kinase activity, good inhibitory effect on β-amyloid self-aggregation and hyperphosphorylation of Tau protein and also significant amelioration of scopolamine induced cognitive impairment *in vivo* and less hepatotoxicity than tacrine [235]. A series of MTDLs was designed based on selective and potent hBChE inhibitor **143** to develop a series of highly merged chimeric multifunctional ligands against AD. The 8-HQ derivatives **144** and

145 showed multifunctional profile with balanced hBChE inhibition and radical scavenging, promising ion-chelating properties, and lowered redox cycling of chelated Cu^{2+} ions, also decreased intracellular levels of ROS after acute treatment with H_2O_2, and protected cells to some extent from toxic $A\beta_{1\text{-}42}$ species and also showed potential to cross the BBB [236]. CID 9998128 (N-(1H-indazol-5-yl)-2-(6-methylpyridin-2-yl) quinazolin-4-amine), which display high binding affinity to six targets including $A\beta$ fibril, peroxisome proliferator-activated receptor γ (PPAR γ), retinoic X receptor α (RXR α), β- and γ-secretases and AChE was further investigated by Nguyen Quoc Thai et al. and demonstrated that this compound strongly binds to both $A\beta_{42}$ fibrils and β-secretase with the van der Waals interactions over the electrostatic interactions in binding. Also inhibits the $A\beta_{42}$ fibrillization and is capable to clear $A\beta_{42}$ fibrils, decreases β-site APP cleaving enzyme (BACE-1) activity in the micromolar range [237]. According to Chalupova et al. tacrine-tryptophan heterodimers are multifunctional agents since most of the analogs significantly inhibited $A\beta_{42}$ self-aggregation and the hAChE-induced $A\beta_{40}$ aggregation and are predicted to cross BBB *via* passive diffusion and to exert moderate inhibition potency against neuronal nitric oxide. Out of these, compound **146** was found to demonstrate the highest levels of hAChE and hBChE inhibition compared to reference standards [238]. Umar et al. focused on evaluating the 4-substituted piperazine-pyrazolo-pyridin-3-yl-acetamide analogues *in vitro* against important biological targets of AD including, inhibition of AChE, $A\beta$ aggregation/disaggregation, antioxidant property to establish these compounds as multi target drug ligands [239]. The most potent molecule 2-chloro-N-(1H-pyrazolo[3,4-b]pyridin-3-yl)acetamide (**147**) exhibited excellent anti-AChE activity with mixed type inhibition, capable of inhibiting self-induced $A\beta$ aggregation, $A\beta$ disaggregation, antioxidation and metal-chelation activities. Yuying Fang et al. have proved the series of tetrahydroisoquinoline-benzimidazole hybrids as multifunctional agents against AD. Among them, compound **148** possessed significant antineuroinflammatory activity through inhibiting the expression and secretion of proinflammatory cytokines in BV2 cells, moderate hBACE1 inhibitory activity and potent neuroprotective effect by increasing GSH level and reducing ROS production and also cross the BBB [240]. Molecular hybrids of 2-pyridylpiperazine and 5-phenyl-1,3,4-oxadiazoles were identified as MTDLs against AD. Compound **149** containing 2,4-difluoro substitution at terminal phenyl ring considered as most potential lead with mixed-type inhibition of AChE, BChE and BACE1, significant displacement of propidium iodide from the PAS of hAChE, excellent BBB permeability and neuroprotective ability against SH-SY5Y neuroblastoma cell lines. Further, **149** also exhibited anti-$A\beta$ aggregation activity against self- and AChE-induced aggregation, signified learning and memory improvement in scopolamine- and $A\beta$-induced cognitive dysfunctions [241]. Naeimeh Salehi et al. reported benzylpyridinium-based benzoheterocycles *viz.* benzimidazole, benzoxazole or benzothiazole as potent AChE and BChE inhibitors. Analogs **150** and **151**, had potent anti-AChE activity than reference drug, in a competitive manner and also found to be inhibitors of $A\beta$ self-aggregation as well as AChE-induced $A\beta$ aggregation, significantly protect PC12 cells against H_2O_2-induced injury and showed no toxicity against HepG2 cells [242].

Haghighijoo et al. have developed a series of substituted benzyl-1*H*-1,2,3-triazol-4-yl-*N*-cyclohexylimidazo[1,2-*a*]pyridin-3-amine derivatives as MTDLs, since most of the compounds exhibited potent BACE1 and BChE inhibitory and antioxidant activities. In particular, Compounds **152** and **153**, bearing dichloro (2,3-diCl and 3,4-diCl) moieties on the benzyl pendant were the most active compounds against BACE1. In addition, 4-bromo derivatives **153** and **154** showed the highest BChE inhibitory, and antioxidant activities and metal chelation potential [243]. 1,3,4-Oxadiazole and 1,2,4-triazole conjugates were reported as promising leads for the amelioration of oxidative stress induced cognitive decline since compounds **155** and **156** with a pyridinyl substitution displayed significant activity toward AChE inhibition, decreased scopolamine-induced oxidative stress [244]. Jiang et al. synthesized a series of ChEs/GSK-3 dual-target inhibitors by connecting tetrahydroacridine scaffolds from three different sites of pyridothiazole core. One of the hybrids **157** has shown balanced *in vitro* activities like preferential AChE/GSK-3β inhibition in the low nanomolar range, high kinase selectivity for GSK-3, and a strong inhibition rate against DYRK1α and DYRK1β. In addition, **157** has shown an attractive BBB permeability profile, as well as antineuroinflammation in cells, ability to inhibit the phosphorylation of Tau protein, with no sign of toxicity, and also promising cognitive improvement in the scopolamine-induced cognitive deficit mice [245]. The tacrine analogs with various thieno[2,3-b]pyridine amines were evaluated for their anti-ChE activity. Among the series, 1,2,3,4,7,8,9,10-octahydrobenzo[4,5]thieno[2,3-b]quinolin-11-amine (**158**) showed a better AChE inhibitory activity than the BChE inhibitory activity, good activity against AChE-induced and self-induced Aβ aggregation, moderate BACE1 inhibitory activity and neuroprotectivity against Aβ-induced damage in PC12 cells together with lower toxicity than tacrine against the HepG2 cell line made it to consider as a promising lead compound [246]. G. Ghotbi et al. demonstrated that the hybrids containing thiazole-pyridinium structure may represent a useful multitargeted scaffold for the development of novel anti-AD agents due to their potency as selective inhibitors of AChE, and Aβ aggregation. Compounds **159** and **160** showed the best AChE inhibitory activities over BChE, with a mixed-type of inhibition mechanism, in addition to Aβ self-aggregation inhibitory effects stronger than donepezil and effective neuroprotective activity in H_2O_2-induced oxidative stress on PC12 cells and ability to pass BBB [139]. Martina Bortolami et al. designed a series of deferiprone derivatives possessing an arylalkylamine moiety connected *via* a flexible alkyl linker of variable length to a 3-hydroxy-4-pyridone fragment as potential multifunctional compounds for AD. Deferiprone moiety and 2-aminopyridine, 2-aminopyrimidine or 2,4-diaminopyrimidine groups have been incorporated into these compounds, to obtain molecules that potentially able to chelate biometals colocalized in Aβ [247](Fig. 3.17).

3.4 Conclusions

Pyridine-based heterocycles are particularly important scaffolds among other N-heterocycles in both synthetic organic and medicinal chemistry due to their unique properties like weak basicity, water solubility, *in vivo* chemical stability, hydrogen bond-forming ability, etc. Over the past few decades, enormous research has been conducted and reported several kinds of synthetic routes and biological activity studies of pyridine-based heterocycles. This chapter is an attempt by us to summarize the significant progress in both the synthetic strategies and multifunctional profile as anti-AD agents of pyridine-based scaffolds. The chapter begins with synthetic strategies of simple pyridine nucleus, 2-aminocyanopyridines, pyridinium salts, imidazopyridines, pyrazolopyridines, and furopyridines and proceeds to their therapeutic profile against target enzymes/receptors like AChE, BChE, Ab, Tau, BACE1, MAOB, PDE, nAChRs, hA_1 AR, hA_{2A} AR, etc., involved

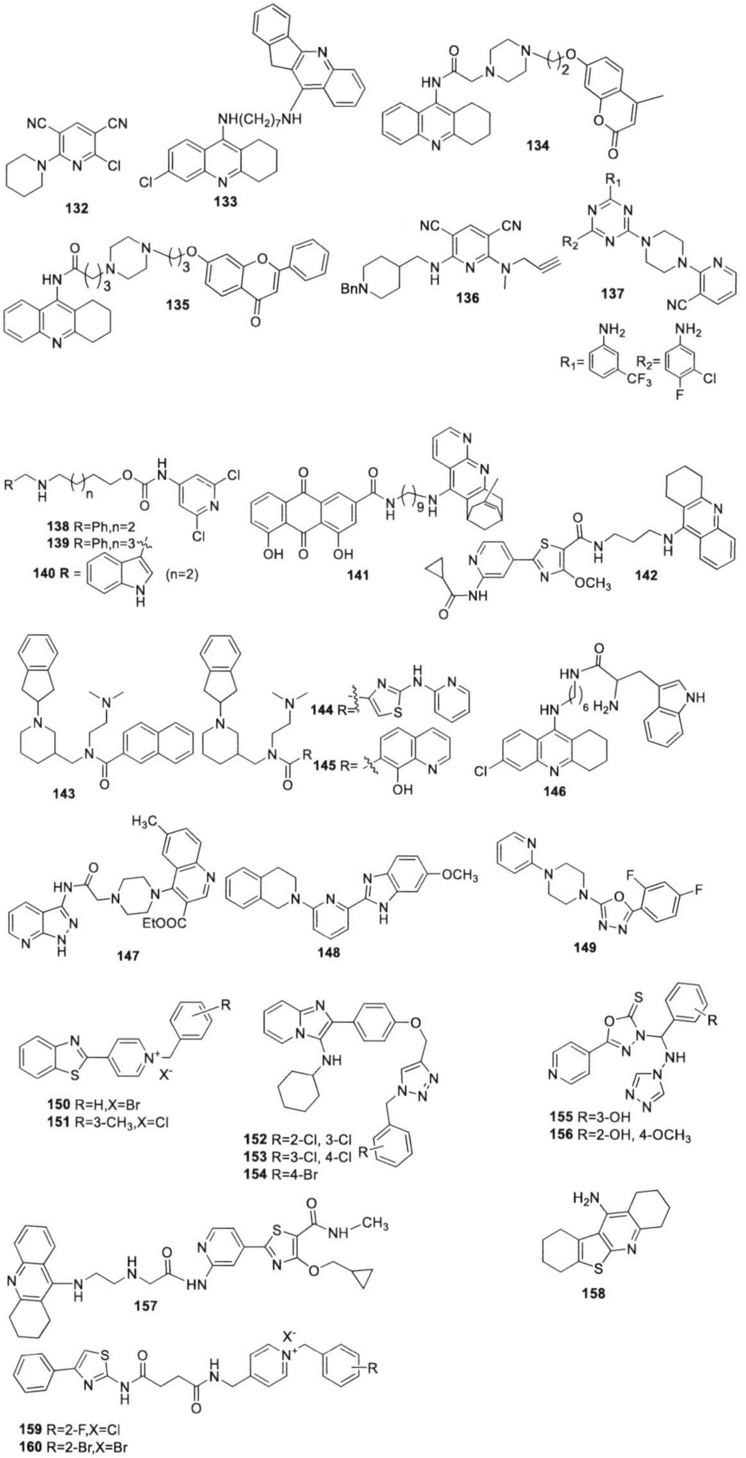

FIGURE 3.17 Pyridine-based MTDLs against AD.

in AD pathogenesis. Overall, the incorporation of pyridine scaffold into target analogs is endowed with multiple benefits like enhanced therapeutic potential, better pharmacokinetic properties, increased brain penetration, and improved physicochemical and drug-like properties required for CNS active drugs.

References

[1] M. Baumann, I.R. Baxendale, An overview of the synthetic routes to the best selling drugs containing 6 membered heterocycles, Beilstein J. Org.Chem. 9 (2013) 2265–2319.

[2] A.E. Goetz, N.K. Garg, Regioselective reactions of 3,4-pyridynes enabled by the aryne distortion model, Nat. Chem. 5 (2013) 54–60.

[3] D. Uredi, D.R. Motati, E.B. Watkins, A simple, tandem approach to the construction of pyridine derivatives under metal-free conditions: a one-step synthesis of the monoterpene natural product,(-)-actinidine, Chem. Commun. 55 (2019) 3270–3273.

[4] W. Ramsay, On picoline and its derivatives, Philos. Mag. 2 (1876) 269–281.

[5] A. Hantzsch, Condensationsprodukteaus aldehydammoniak und ketonartigen verbindungen, Chem. Ber. 14 (1881) 1637–1638.

[6] A.E. Chichibabin, On condensation of aldehydes with ammonia to make pyridines, J. für Praktische Chemie 107 (1924) 122.

[7] R.L. Frank, R.P. Seven, Pyridines. iv. a study of the Chichibabin synthesis, J. Am. Chem. Soc. 71 (1949) 2629–2635.

[8] J.A. Varela, C. Saa, Construction of pyridine rings by metal-mediated [2 2 2]-cycloaddition, Chem. Rev. 103 (2003) 3787–3802.

[9] Z. He, D. Dobrovolsky, P. Trinchera, A.K. Yudin, Synthesis of multi-substituted pyridines, Org. Lett. 15 (2013) 334–337.

[10] H. Wei, Y. Li, K. Xiao, B. Cheng, H. Wang, L. Hu, H. Zhai, Synthesis of polysubstituted pyridines *via* a one-pot metal-free strategy, Org. Lett. 17 (2015) 5974–5977.

[11] T.A. Nizami, R. Hua, Silver-catalyzed chemoselective annulation of propargyl amines with alkynes for access to pyridines and pyrroles, Tetrahedron 73 (2017) 6080–6084.

[12] A.S. Girgis, S.R. Tala, P.V. Oliferenko, A.A. Oliferenko, A.R. Katritzky, Computer-assisted rational design, synthesis, and bioassay of non-steroidal anti-inflammatory agents, Eur. J. Med. Chem. 50 (2012) 1–8.

[13] M. Mantri, O. de Graaf, J. van Veldhoven, A. Goblyos, J.K. von Frijtag Drabbe Kunzel, T. Mulder-Krieger, R. Link, H. de Vries, M.W. Beukers, J. Brussee, A.P. Ijzerman, 2-Amino-6-furan-2-yl-4-substituted nicotinonitriles as A2A adenosine receptor antagonists, J. Med. Chem. 51 (2008) 4449–4455.

[14] R. Kempe, The strained η^2-N_{Amido} N_{Pyridine} coordination of amino pyridinato ligands, Eur. J. Inorg. Chem. (2003) 791–803.

[15] Q. Wu, Y. Zhang, S. Cui, Divergent syntheses of 2-aminonicotinonitriles and pyrazolines by copper-catalyzed cyclization of oxime ester, Org. Lett. 16 (2014) 1350–1353.

[16] G. Cheng, Y. Weng, X. Yang, X. Cui, Base-promoted N-pyridylation of heteroarenes using *N*-propargyl enaminones as equivalents of pyridine scaffolds, Org. Lett. 17 (2015) 3790–3793.

[17] Y. Weng, C. Kuai, W. Lv, G. Cheng, Synthesis of 2-aminopyridines *via* a base-promoted cascade reaction of n-propargylic β-enaminones with formamides, J. Org. Chem. 83 (2018) 5002–5008.

[18] M.R. Bodireddy, N.C. Gangi Reddy, S.D. Kumar, Synthesis of alkynyl/alkenyl-substituted pyridine derivatives *via* heterocyclization and Pd-mediated Sonogashira/Heck coupling process in one-pot: a new MCR strategy, RSC Adv. 4 (2014) 17196–17205.

[19] P.N. Kalaria, S.P. Satasia, J.R. Avalani, D.K. Raval, Ultrasound-assisted one-pot four-component synthesis of novel 2-amino-3-cyanopyridine derivatives bearing 5-imidazopyrazole scaffold and their biological broadcast, Eur. J. Med. Chem. 83 (2014) 655–664.

[20] S. Khaksar, M. Yaghoobi, A concise and versatile synthesis of 2-amino-3-cyanopyridine derivatives in 2,2,2-trifluoroethanol, J. Fluorine Chem. 142 (2012) 41–44.

[21] J. Safari, S.H. Banitaba, S.D. Khalili, Ultrasound-promoted an efficient method for one-pot synthesis of 2-amino-4,6-diphenylnicotinonitriles in water: A rapid procedure without catalyst, Ultrason. Sonochem. 19 (2012) 1061–1069.

[22] F. Shi, S. Tu, F. Fang, T. Li, One-pot synthesis of 2-amino-3-cyanopyridine derivatives under microwave irradiation without solvent, ARKIVOC I (2005) 137–142.

[23] L. Xu, L. Shi, S. Qiu, S. Chen, M. Lin, Y. Xiang, C. Zhao, J. Zhu, L. Shen, Z. Zuo, Design, synthesis, and evaluation of cyanopyridines as anti-colorectal cancer agents via inhibiting stat3 pathway, Drug Des. Dev. Ther. 13 (2019) 3369–3381.

[24] N.M. Evdokimov, A.S. Kireev, A.A. Yakovenko, M.Y. Antipin, I.V. Magedov, A. Kormienko, One-step, three-component synthesis of pyridines and 1,4-dihydropyridines with manifold medicinal utility, Org. Lett. 8 (2006) 899–902.

[25] K. Godugu, V.D. Sri Yadala, M.K.M. Pinjari, T.R. Gundala, L.R. Sanapareddy, C.G.R. Nallagondu, Natural dolomitic limestone-catalyzed synthesis of benzimidazoles, dihydropyrimidinones, and highly substituted pyridines under ultrasound irradiation, Beilstein J. Org. Chem. 16 (2020) 1881–1900.

[26] P.V. Shinde, S.S. Sonar, B.B. Shingate, M.S. Shingare, Boric acid catalyzed convenient synthesis of 2-amino-3,5-dicarbonitrile-6-thio-pyridines in aqueous media, Tetrahedron Lett. 51 (2010) 1309–1312.

[27] M. Sridhar, B.C. Ramanaiah, C. Narsaiah, B. Mahesh, M. Kumaraswamy, K.K.R. MalluVishnu, M. Ankathi, P.Shanthan Rao, Novel $ZnCl_2$-catalyzed one-pot multi-component synthesis of 2-amino-3,5-dicarbonitrile-6-thio-pyridines, Tetrahedron Lett. 50 (2009) 3897–3900.

[28] L. Srinivasula Reddy, T. Ram Reddy, B. Mohan Reddy, A. Mahesh, Y. Lingappa, N.C. Gangi Reddy, An efficient green multi-component reaction strategy for the synthesis of highly functionalised pyridines and evaluation of their antibacterial activities, Chem. Pharm. Bull. 61 (2013) 1114–1120.

[29] S. Sowmiah, J.M.S.S. Esperança, L.P.N. Rebelo, C.A.M. Afonso, Pyridinium salts: from synthesis to reactivity and applications, Org. Chem. Front. 5 (2018) 453–493.

[30] J. Marek, P. Stodulka, J. Cabal, O. Soukup, M. Pohanka, J. Korabecny, K. Musilek, K. Kuca, Preparation of the pyridinium salts differing in the length of the*N*-alkyl substituent, Molecules 15 (2010) 1967–1972.

[31] J. Auth, P. Mauleon, A. Pfaltz, Synthesis of functionalized pyridinium salts bearing a free amino group, Arkivoc iii (2014) 154–169.

[32] C.Z. Luo, J. Jayakumar, P. Gandeepan, Y.C. Wu, C.H. Cheng, Rhodium(III)-catalyzed vinylic C–H activation: a direct route toward pyridinium salts, Org. Lett. 17 (2015) 924–927.

[33] S. Kirchhecker, S.T. Muller, S. Bake, M. Antonietti, A. Taubert, D. Esposito, Renewable pyridinium ionic liquids from the continuous hydrothermal decarboxylation of furfural-amino acid derived pyridinium zwitterions, Green Chem. 17 (2015) 4151–4156.

[34] S. Sowmiah, L.F. Veiros, J.M.S.S. Esperanca, L.P.N. Rebelo, C.A.M. Afonso, Organocatalyzed one-step synthesis of functionalized *N*-alkyl-pyridinium salts from biomass derived 5-hydroxymethylfurfural, Org. Lett. 17 (2015) 5244–5247.

[35] S. Peixoto, T.M. Nguyen, D. Crich, B. Delpech, C. Marazano, One-pot formation of piperidine- and pyrrolidine-substituted pyridinium salts via addition of 5-alkylaminopenta-2,4-dienals to *N*-acyliminium ions: application to the synthesis of (±)-nicotine and analogs, Org. Lett. 12 (2010) 4760–4763.

[36] M.A. Ilies, B.H. Johnson, F. Makori, A. Miller, W.A. Seitz, E.B. Thompson, A.T. Balaban, Pyridinium cationic lipids in gene delivery: an in vitro and in vivo comparison of transfection efficiency versus a tetraalkylammonium congener, Arch. Biochem. Biophys. 435 (2005) 217–226.

[37] M.A. Ilies, W.A. Seitz, M.T. Caproiu, M. Wentz, R.E. Garfield, A.T. Balaban, Pyridinium-based cationic lipids as gene-transfer agents, Eur. J. Org. Chem. 2003 (2003) 2645–2655.

[38] M.A. Ilies, W.A. Seitz, I. Ghiviriga, B.H. Johnson, A. Miller, E.B. Thompson, A.T. Balaban, Pyridinium cationic lipids in gene delivery: a structure activity correlation study, J. Med. Chem. 47 (2004) 3744–3754.

[39] M. Mollazadeh, M.M. Khanaposhtani, A. Zonouzi, H. Nadri, Z. Najafi, B. Larijani, M. Mahdavi, New benzyl pyridinium derivatives bearing 2,4-dioxochroman moiety as potent agents for treatment of Alzheimer's disease: design, synthesis, biological evaluation, and docking study, Bioorg. Chem. 87 (2019) 506–515.

[40] A. John, M.M. Shaikh, P. Ghosh, Palladium complexes of abnormal N-heterocyclic carbenes as pre catalysts for the much preferred Cu-free and amine-free Sonogashira coupling in air in a mixed-aqueous medium, Dalton Trans. 47 (2009) 10581–10591.

[41] G. Song, Y. Zhang, X. Li, Rhodium and Iridium complexes of abnormal N-heterocyclic carbenes derived from imidazo[1,2-*a*]pyridine, Organometallics 27 (2008) 1936–1943.

[42] J. Wan, C.J. Zheng, M.K. Fung, X.K. Liu, C.S. Lee, X.H. Zhang, Multifunctional electron-transporting indolizine derivatives for highly efficient blue fluorescence, orange phosphorescence host and two-color based white OLEDs, J. Mater. Chem. 22 (2012) 4502–4510.

[43] H. Shono, T. Ohkawa, H. Tomoda, T. Mutai, K. Araki, Fabrication of colorless organic materials exhibiting white luminescence using normal and excited-state intramolecular proton transfer processes, ACS Appl. Mater. Interfaces 3 (2011) 654–657.
[44] A. Hajra, A.K. Bagdi, S. Santra, K. Monir, Synthesis of imidazo[1,2-a]pyridines: a decade update, Chem. Commun. 51 (2015) 1555–1575.
[45] M. Fisher, A. Lusi, Imidazo[1,2-a]pyridine anthelmintic and antifungal agents, J. Med. Chem. 15 (1972) 982–985.
[46] R. Nishanth Rao, M.M. Balamurali, B. Maiti, T. Ranjit, C. Kaushik, Efficient access to Imidazo[1,2-*a*]pyridines/pyrazines/pyrimidines via catalyst-free annulation reaction under microwave irradiation in green solvent, ACS Comb. Sci. 20 (2018) 164–171.
[47] S. Ponnala, S.T.V.S.K. Kumar, B.A. Bhat, D.P. Sahu, Synthesis of bridgehead nitrogen heterocycles on a solid surface, Synthetic Commun. 35 (2005) 901–906.
[48] G. Kumar, N.C. Gangi Reddy, Solvent and catalyst-free synthesis of imidazo[1,2-a]pyridines by grindstone chemistry, J. Heterocyclic Chem. 58 (2021) 250–259.
[49] J.S. Yadav, B.V.S. Reddy, Y.G. Rao, M. Srinivas, A.V. Narsaiah, $Cu(OTf)_2$-catalyzed synthesis of imidazo[1,2-*a*]pyridines from α-diazoketones and 2-aminopyridines, Tetrahedron Lett. 48 (2007) 7717–7720.
[50] Y.-Y. Xie, Z.-C. Chen, Q.-G. Zheng, Organic reactions in ionic liquids: ionic liquid-accelerated cyclo condensation of α-tosyloxy ketones with 2-aminopyridine, Synthesis 11 (2002) 1505–1508.
[51] A.J. Stasyuk, M. Banasiewicz, M.K. Cyranski, D.T. Gryko, Imidazo[1,2-*a*]pyridines susceptible to excited state intramolecular proton transfer: one-pot synthesis via an Ortoleva–king reaction, J. Org. Chem. 77 (2012) 5552–5558.
[52] Z. Liu, Z.-C. Chen, Q.-G. Zheng, Hypervalent iodine in synthesis. 94. a facile synthesis of 2-substituted-imidazo[1,2-a]pyridines by cyclocondensation of alkynyl(phenyl) iodonium salts and 2-aminopyridine, Synthetic Commun 34 (2004) 361–367.
[53] D.K. Nair, S.M. Mobin, I.N.N. Namboothiri, Synthesis of imidazopyridines from the morita–baylis–hillman acetates of nitroalkenes and convenient access to alpidem and zolpidem, Org. Lett. 14 (2012) 4580–4583.
[54] S. Santra, A.K. Bagdi, A. Majee, A. Hajra, Iron(III)-catalyzed cascade reaction between nitroolefins and 2-aminopyridines: synthesis of imidazo[1,2-*a*]pyridines and easy access towards zolimidine, Adv. Synth. Catal. 355 (2013) 1065–1070.
[55] M. Adib, E. Sheikhi, N. Rezaei, One-pot synthesis of imidazo[1,2-*a*]pyridines from benzyl halides or benzyl tosylates, 2-aminopyridines and isocyanides, Tetrahedron Lett. 52 (2011) 3191–3194.
[56] P. Puthiaraj, A. Ramu, K. Pitchumani, Copper-based metal–organic frameworks as reusable heterogeneous catalysts for the one-pot syntheses of imidazo[1,2-*a*]pyridines, Asian J. Org. Chem. 3 (2014) 784–791.
[57] S. Payra, A. Saha, S. Banerjee, Nano-$NiFe_2O_4$ catalyzed microwave assisted one-pot regioselective synthesis of novel 2-alkoxyimidazo[1,2-*a*]pyridines under aerobic conditions, RSC Adv. 6 (2016) 12402–12407.
[58] E. Allahabadi, S. Ebrahimi, M. Soheilizad, M. Khoshneviszadeh, M. Mahdavi, Copper-catalyzed four- component synthesis of imidazo[1,2-*a*]pyridines *via* sequential reductive amination, condensation, and cycliza- tion, Tetrahedron Lett. 58 (2017) 121–124.
[59] S. Swami, N. Devi, A. Agarwala, V. Singh, R. Shrivastava, ZnO nanoparticles as reusable heterogeneous catalyst for efficient one pot three component synthesis of imidazo-fused polyheterocycles, Tetrahedron Lett. 57 (2016) 1346–1350.
[60] Z. Gan, X. Zhu, Q. Yan, X. Song, D. Yang, Oxidative dual C–H sulfenylation: a strategy for the synthesis of bis(imidazo[1,2-*a*]pyridin-3-yl)sulfanes under metal-free conditions using sulfur powder, Chin. Chem. Lett. 32 (2021) 1705–1708.
[61] K. Chavva, S. Pillalamarri, V. Banda, S. Gautham, J. Gaddamedi, P. Yedla, C.G. Kumar, N. Banda, Synthesis and biological evaluation of novel alkyl amide functionalized trifluoromethyl substituted pyrazolo[3,4-*b*]pyridine derivatives as potential anticancer agents, Bioorganic Med. Chem. Lett. 23 (2013) 5893–5895.
[62] L.R.S. Dias, M.B. Santos, S. de Albuquerque, H.C. Castro, A.M.T. de Souza, A.C.C. Freitas, M.A.V. DiVaio, L.M. Cabral, C.R. Rodrigues, Synthesis, in vitro evaluation, and SAR studies of a potential antichagasic 1*H*-pyrazolo[3,4-*b*]pyridine series, Bioorganic Med. Chem. 15 (2007) 211–219.
[63] E.E. Elboray, R. Grigg, C.W.G. Fishwick, C. Kilner, M.A.B. Sarker, M.F. Aly, H.H.T. Abbas, X Y–ZH compounds as potential 1,3-dipoles. Part 65: atom economic cascade synthesis of highly functionalized pyrimidinyl pyrrolidines, Tetrahedron 67 (2011) 5700–5710.

[64] T. El-Sayed Ali, Synthesis of some novel pyrazolo[3,4-*b*]pyridine and pyrazolo[3,4-*d*]pyrimidine derivatives bearing 5,6-diphenyl-1,2,4-triazine moiety as potential antimicrobial agents, Eur. J. Med. Chem. 44 (2009) 4385–4392.

[65] S. Huang, R. Lin, Y. Yu, Y. Lu, P.J. Connolly, G. Chiu, S. Li, S.L. Emanuel, S.A. Middleton, Synthesis of 3-(1*H*-benzimidazol-2-yl)-5-isoquinolin-4-ylpyrazolo[1,2-*b*]pyridine, a potent cyclin dependent kinase 1 (CDK1) inhibitor, Bioorganic Med. Chem. Lett. 17 (2007) 1243–1245.

[66] P. Nagender, G. Malla Reddy, R. Naresh Kumar, Y. Poornachandra, C. Ganesh Kumar, B. Narsaiah, Synthesis, cytotoxicity, antimicrobial and anti-biofilm activities of novel pyrazolo[3,4-*b*]pyridine and pyrimidine functionalized 1,2,3-triazole derivatives, Bioorganic Med. Chem. Lett. 24 (2014) 2905–2908.

[67] Z. Chen, X. Yang, J. Wu, AgOTf-catalyzed tandem reaction of n^{r}-(2-alkynylbenzylidene)hydrazide with alkyne, Chem. Commun. 23 (2009) 3469–3471.

[68] J. Qin, W. Zhou, X. Huang, P. Dhondi, A. Palani, R. Aslanian, Z. Zhu, W. Greenlee, M. Cohen-Williams, N. Jones, L. Hyde, L. Zhang, Discovery of a potent pyrazolopyridine series of γ -secretase modulators, ACS Med. Chem. Lett. 2 (2011) 471–476.

[69] H.C. Wu, C.H. Yang, L.C. Hwang, M. Wu, Au(I)-catalyzed and iodine-mediated cyclization of enynylpyrazoles to provide pyrazolo[1,5-*a*]pyridines, Org. Biomol. Chem. 10 (2012) 6640–6648.

[70] J.Y. Kato, H. Aoyama, T. Yokomatsu, Development of a new cascade reaction for convergent synthesis of pyrazolo[1,5-*a*]quinoline derivatives under transition-metal-free conditions, Org. Biomol. Chem. 11 (2013) 1171–1178.

[71] Y. Hoashi, T. Takafumi, K. Etsuo, K. Tatsuki, Synthesis of pyrazolo[1,5-*a*]pyridines by thermal intramolecular cyclization, Tetrahedron. Lett. 54 (2013) 2199–2202.

[72] B. Wang, F. Su, J. Jia, F. Wu, S.Y. Zhang, Y.Q. Ge, C. Wang, J.W. Wang, Synthesis of 5-cyanopyrazolo[1,5-*a*]pyridine derivatives via tandem reaction and their optical properties, Tetrahedron. Lett. 56 (2015) 425–429.

[73] C. Ravi, A. Qayum, D.C. Mohan, S.K. Singh, S. Adimurthy, Design, synthesis and cytotoxicity studies of novel pyrazolo[1, 5-a]pyridine derivatives, Eur. J. Med. Chem. 126 (2017) 277–285.

[74] A.P. Marjani, J. Khalafy, S. Akbarzadeh, Synthesis of pyrazolopyridine and pyrazoloquinoline derivatives by one-pot, three-component reactions of arylglyoxals, 3-methyl-1-aryl-1H-pyrazol-5-amines and cyclic 1,3-dicarbonyl compounds in the presence of tetrapropylammonium bromide, Green Process. Synthesis 8 (2019) 533–541.

[75] R. Navari, S. Balalaie, S. Mehrparvar, F. Darvish, F. Rominger, F. Hamdan, S. Mirzaie, Efficient synthesis of pyrazolopyridines containing a chromane backbone through domino reaction, Beilstein J. Org. Chem. 15 (2019) 874–880.

[76] E.G. Paronikyan, A.K. Oganisyan, A.S. Noravyan, R.G. Paronikyan, I.A. Dzhagatspanyan, Synthesis and anticonvulsant activity of pyrano [4′,3′:4,5]pyrido[2,3-b]furo[3,2-d]pyrimidine and pyrano[4′,3′:4,5]pyrido[2,3-b]furo[3,2-d]pyridine derivatives, Pharm. Chem. J. 36 (2002) 413–414.

[77] R.D. Bukoski, J. Bo, H. Xue, K. Bian, Antiproliferative and endothelium-dependent vasodilator properties of 1,3-dihydro-3- p-chlorophenyl-7-hydroxy-6-methylfuro[3,4-c] pyridine hydrochloride cicletanine, J. Pharmacol. Exp. Ther. 265 (1993) 30–35.

[78] P. Jeschke, A. Harder, W. Etzd, W. Gau, A. Goehrt, J. Benet-Buchholz, G. Thielking, Synthesis and anthelmintic activity of 7-substituted 3,4adimethyl-4a,5a,8a,8b-tetrahydro-6H-pyrrolo[3′,4′:4,5]furo[3,2-b]pyridine-6,8 (1H)-diones, Bioorg. Med. Chem. Lett. 15 (2005) 2375–2379.

[79] P. Gerster, C. Riegger, M. Fallert, Effects of 1,4,5,7-tetrahydro-furo[3,4-c]pyridine derivatives, a calcium influx promoter, on sodium-, calcium-, and magnesium-dependent bioelectrical activity of Purkinje fibers and papillary muscles, Arzneim. Forsch. 37 (1987) 309–315.

[80] J.L. Marco, M.C. Carreiras, Recent developments in the synthesis of acetylcholinesterase inhibitors, Mini-Rev. Med. Chem. 3 (2003) 518–524.

[81] D.G. Wishka, D.R. Graber, E.P. Seest, L.A. Dolak, F. Han, W. Watt, J. Morris, Stereoselective synthesis of furo[2,3-c]pyridine pyrimidine thioethers, a new class of potent HIV-1 non-nucleoside reverse transcriptase inhibitors, J. Org. Chem. 63 (1998) 7851–7859.

[82] S. Shiotani, H. Morita, I Furopyridines., Synthesis of furo[2,3-*c*]pyridine, J. Heterocycl. Chem. 19 (1982) 1207–1209.

[83] A. Arcadi, S. Cacchi, S. Di Giuseppe, G. Fabrizi, F. Marinelli, Electrophilic cyclization of o-acetoxy and o-benzyloxyalkynylpyridines: An easy entry into 2,3-disubstituted furopyridines, Org. Lett. 4 (2002) 2409–2412.

[84] T. Cailly, S. Lemaître, F. Fabis, S. Rault, Straightforward access to ethyl 3-aminofuropyridine-2-carboxylates from 1-Chloro-2-cyano- or 1-Hydroxy-2-cyano-Substituted Pyridines, Synthesis 20 (2007) 3247–3251.

[85] Z. Huo, Y. Yamamoto, Gold-catalyzed synthesis of isoquinolines via intramolecular cyclization of 2-alkynyl benzyl azides, Tetrahedron Lett. 50 (2009) 3651–3653.

[86] D. Fischer, H. Tomeba, N.K. Pahadi, N.T. Patil, Z. Huo, Y. Yamamoto, Iodine-mediated electrophilic cyclization of 2-alkynyl-1-methylene azide aromatics leading to highly substituted isoquinolines and its application to the synthesis of norchelerythrine, J. Am. Chem. Soc. 130 (2008) 15720–15725.

[87] D.M. Volochnyuk, S.V. Ryabukhin, A.S. Plaskon, Y.V. Dmytriv, O.O. Grygorenko, P.K. Mykhailiuk, D.G. Krotko, A. Pushechnikov, A.A. Tolmachev, Approach to the library of fused pyridine-4-carboxylic acids by combes-type reaction of acyl pyruvates and electron-rich amino heterocycles, J. Comb. Chem. 12 (2010) 510–517.

[88] R. Yan, X. Li, X. Yang, X. Kang, L. Xiang, G. Huang, A novel and one-pot method for the synthesis of substituted furopyridines: I_2-mediated oxidative reaction of enaminones via tandem cyclization under metal-free conditions, Chem. Commun. 51 (2015) 2573–2576.

[89] L. Zhang, Y. Liu, X. Li, Y. Guo, Z. Jiang, T. Jiao, J. Yang, Synthesis of a new amino-furopyridine-based compound as a novel fluorescent pH Sensor in aqueous solution, ACS Omega 6 (2021) 4806 4800.

[90] H. Hippius, G. Neundorfer, The discovery of Alzheimer's disease, Dialogues Clin. Neurosci. 5 (2003) 101–108.

[91] H. Forstl, A. Kurz, Clinical features of Alzheimer's disease, Eur. Arch. Psychiatry Clin. Neurosci. 249 (1999) 288–290.

[92] C.L. Masters, G. Simms, N.A. Weinman, G. Multhaup, B.L. McDonald, K. Beyreuther, Amyloid plaque core protein in alzheimer's disease and down syndrome, Proc. Natl. Acad. Sci. U S A 82 (1985) 4245–4249.

[93] Alzheimer Society: https://www.alzheimers.org.uk/. (accessed October 12, 2015).

[94] World Alzheimer Report 2016 sheet. https://www.alz.co.uk/research/(09.01.2017).

[95] D. Repantis, O. Laisney, I. Heuser, Acetylcholinesterase inhibitors and memantine for neuroenhancement in healthy individuals: a systematic review, Pharmacol. Res. 61 (2010) 473–481.

[96] M.I. Fernandez-Bachiller, C. Perez, N.E. Campillo, J.A. Paez, G.C. Gonzalez-Munoz, P. Usan, E. Garcia- Palomero, M.G. Lopez, M. Villarroya, A.G. Garcia, A. Martinez, M.I. Rodriguez-Franco, Tacrine-melatonin hybrids as multifunctional agents for Alzheimer's disease, with cholinergic, antioxidant, and neuroprotective properties, Chem. Med. Chem. 4 (2009) 828–841.

[97] R.T. Bartus, R.L. Dean 3rd, B. Beer, A.S. Lippa, The cholinergic hypothesis of geriatric memory dysfunction, Science 217 (1982) 408–414.

[98] A. Musial, M. Bajda, B. Malawska, Recent developments in cholinesterases inhibitors for Alzheimer's disease treatment, Curr. Med. Chem. 14 (2007) 2654–2679.

[99] M. Weinstock, The pharmacotherapy of alzheimer's disease based on the cholinergic hypothesis: an update, Neurodegeneration 4 (1995) 349–356.

[100] A. Martinez, A. Castro, Novel cholinesterase inhibitors as future effective drugs for the treatment of Alzheimer's disease, Expert Opin. Invest. Drugs 15 (2006) 1–12.

[101] J.L. Marco, C. Rios, M.C. Carreiras, J.E. Banos, A. Badi, N.M. Vivas, Synthesis and acetylcholinestarase/butyrylcholinestarase inhibition activity of new Tacrine-like analogues, Bioorg. Med. Chem. 9 (2001) 727–732.

[102] J.L. Marco, C. Ríos, M.C. Carreiras, J.E. Baños, A. Badi, N.M. Vivas, Synthesis and acetyl-cholinesterase/ butyrylcholinesteras inhibition activity of 4-Amino-2,3-diaryl-5,6,7,8-tetrahydrofuro(and thieno)[2,3-*b*]-quinolines, and 4-amino-5,6,7,8,9-pentahydro-2,3-diphenylcyclohepta[*e*]furo(and thieno)-[2,3- *b*]pyridines,, Arch. Pharm. Pharm. Med. Chem. 7 (2002) 347–353.

[103] E.J. Barreiro, C.A. Camara, H. Verli, L.B. Mas, N.G. Castro, W.M. Cintra, Y. Aracava, C.R. Rodrigues, C.A.M. Fraga, Design, synthesis, and pharmacological profile of novel fused pyrazolo[4,3-*d*]pyridine and pyrazolo[3,4-*b*][1,8]naphthyridine isosteres: a new class of potent and selective acetylcholinesterase inhibitors, J. Med. Chem. 46 (2003) 1144–1152.

[104] A. Samadi, M. Estrada, C. Pérez, M.I.R. Franco, I. Iriepa, I. Moraleda, M. Chioua, J.M. Contelles, Pyridonepezils, new dual AChE inhibitors as potential drugs for the treatment of Alzheimer's disease: synthesis, biological assessment, and molecular modeling, Eur. J. Med. Chem. 57 (2012) 296–301.

[105] A. Samadi, M.F. Revenga, C. Pérez, I. Iriepa, I. Moraleda, M.I. Rodríguez-Franco, J. Marco-Contelles, Synthesis, pharmacological assessment, and molecular modeling of 6-chloro-pyridonepezils: New dual AChE inhibitors as potential drugs for the treatment of Alzheimer's disease, Eur. J. Med. Chem. 67 (2013) 64–74.

[106] H. Nadri, M.P. Hamedani, A. Moradi, A. Sakhteman, A. Vahidi, V. Sheibani, A. Asadipour, N. Hosseinzadeh, M. Abdollahi, A. Shafiee, A. Foroumadi, 5,6-Dimethoxybenzofuran-3-one derivatives: a novel series of dual acetylcholinesterase/butyrylcholinesterase inhibitors bearing benzyl pyridinium moiety, DARU J. Pharm. Sci. 21 (2013) 15.

[107] M. Khoobi, M. Alipour, A. Sakhteman, H. Nadri, A. Moradi, M. Ghandi, S. Emami, A. Foroumadi, A. Shafiee, Design, synthesis, biological evaluation and docking study of 5-oxo-4,5-dihydropyrano[3,2-*c*]chromene derivatives as acetylcholinesterase and butyrylcholinesterase inhibitors, Eur. J. Med. Chem. 68 (2013) 260–269.

[108] F. Baharloo, M.H. Moslemin, H. Nadri, A. Asadipour, M. Mahdavi, S. Emami, L. Firoozpour, R. Mohebat, A. Shafiee, A. Foroumadi, Benzofuran-derived benzylpyridinium bromides as potent acetylcholinesterase inhibitors, Eur. J. Med. Chem. 93 (2015) 196–201.

[109] M. Mostofi, G.M. Ziarani, M. Mahdavi, A. Moradi, H. Nadri, S. Emami, H. Alinezhad, A. Foroumadi, A. Shafiee, Synthesis and structure-activity relationship study of benzofuran-based chalconoids bearing benzylpyridinium moiety as potent acetylcholinesterase inhibitors, Eur. J. Med. Chem. 103 (2015) 361–369.

[110] S. Riaz, I.U. Khan, M. Bajda, M. Ashraf, Q. ul-Ain, A. Shaukat, T. Ur Rehman, S. Mutahir, S. Hussain, G. Mustafa, M. Yar, Pyridine sulfonamide as a small key organic molecule for the potential treatment of type-II diabetes mellitus and Alzheimer's disease: In vitro studies against yeast α-glucosidase, acetyl- cholinesterase and butyrylcholinesterase, Bioorg. Chem. 63 (2015) 64–71.

[111] T.J. Peglow, R.F. Schumacher, R. Cargnelutti, A.S. Reis, C. Luchese, E.A. Wilhelm, G. Perin, Preparation of bis(2-pyridyl) diselenide derivatives: Synthesis of selenazolo[5,4-b]pyridines and unsymmetrical diorganyl selenides, and evaluation of antioxidant and anticholinesterasic activities, Tetrahedron Lett. 58 (2017) 3734–3738.

[112] T.P.C. Chierrito, S.P. Mantoani, C. Roca, C. Requena, V.S. Perez, W.O. Castillo, N.C.S. Moreira, C. Perez, E.T.S. Hojo, C.S. Takahashi, J.J. Enez-Barbero, F.J. Canada, N.E. Campillo, A. Martinez, I. Carvalho, From dual binding site acetylcholinesterase inhibitors to allosteric modulators: a new avenue for disease-modifying drugs in Alzheimer's disease, Eur. J. Med. Chem. 139 (2017) 773–791.

[113] R. Ulus, B.Z. Kurt, I. Gaziog˘ lu, M. Kaya, Microwave assisted synthesis of novel hybrid tacrine-sulfonamide derivatives and investigation of their antioxidant and anticholinesterase activities, Bioorg. Chem. 70 (2017) 245–255.

[114] J. Kumar, A. Gill, M. Shaikh, A. Singh, A. Shandilya, E. Jameel, N. Sharma, N. Mrinal, N. Hoda, B. Jayaram, Pyrimidine-triazolopyrimidine and pyrimidine-pyridine hybrids as potential acetylcholinesterase inhibitors for Alzheimer's disease, Chem. Select 3 (2018) 736–747.

[115] F. Abedinifar, S.M.F. Farnia, M. Mahdavi, H. Nadri, A. Moradi, J.B. Ghasemi, T.T. Küçükkılınç, L. Firoozpour, Foroumadi, synthesis and cholinesterase inhibitory activity of new 2-benzofuran carboxamide- benzylpyridinum salts, Bioorg. Chem. 80 (2018) 180–188.

[116] F. Vafadarnejad, M. Mahdavi, E.K. Razkenari, N. Edraki, B. Sameem, M. Khanavi, M. Saeedi, T. Akbarzadeh, Design and synthesis of novel coumarin-pyridinium hybrids: In vitro cholinesterase inhibitory activity, Bioorg. Chem. 77 (2018) 311–319.

[117] F. Vafadarnejad, E.K. Razkenari, B. Sameem, M. Saeedic, O. Firuzi, N. Edraki, M. Mahdavi, T. Akbarzadeh, Novel *N*-benzylpyridinium moiety linked to arylisoxazole derivatives as selective butyrylcholinesterase inhibitors: synthesis, biological evaluation, and docking study, Bioorg. Chem. 92 (2019) 103192.

[118] H.M. Ragab, M. Teleb, H.R. Haidar, N. Gouda, Chlorinated tacrine analogs: design, synthesis and biological evaluation of their anti-cholinesterase activity as potential treatment for Alzheimer's disease, Bioorg. Chem. 86 (2019) 557–568.

[119] A. Loesche, J. Wiemann, M. Rohmer, W. Brandt, R. Csuk, Novel 12-hydroxydehydroabietylamine derivatives act as potent and selective butyrylcholinesterase inhibitors, Bioorg. Chem. 90 (2019) 103092.

[120] M.D. Milošević, A.D. Marinković, P. Petrović, A. Klaus, M.G. Nikolić, N.Z. Prlainović, I.N. Cvijetić, Synthesis, in-vitro cholinesterase inhibition, in-vivo anticonvulsant activity and in-silico exploration of N-(4-methylpyridin-2-yl)thiophene-2-carboxamide analogs, Bioorg. Chem. 92 (2019) 103216.

[121] M.D. Milošević, A.D. Marinković, P. Petrović, A. Klaus, M.G. Nikolić, N.Ž. Prlainović, I.N. Cvijetić, Synthesis, characterization and SAR studies of bis(imino)pyridines as antioxidants, acetylcholinesterase inhibitors and antimicrobial agents, Bioorg. Chem. 102 (2020) 104073.

[122] D.J. Selkoe, The origins of Alzheimer's disease: A is for amyloid, JAMA 283 (2000) 1615–1617.

[123] F.M. LaFerla, K.N. Green, S. Oddo, Intracellular amyloid-beta in Alzheimer's disease, Nat. Rev. Neurosci. 8 (2007) 499–509.

[124] K.L. Viola, W.L. Klein, Amyloid beta oligomers in Alzheimer's disease pathogenesis, treatment, and diagnosis, Acta Neuropathol. 129 (2015) 183–206.

[125] M.P. Murphy, H.J. LeVine, Alzheimer's disease and the amyloid-β peptide 3rd, J. Alzheimer's Disease 19 (2010) 311–323.

[126] Y. Huang, L. Mucke, Alzheimer mechanisms and therapeutic strategies, Cell 148 (2012) 1204–1222.

[127] S.J. Basha, P.B. Kumar, P. Mohan, E. Siddhartha, D.M. Manidhar, A.D. Rao, V. Ramakrishna, A.G. Damu, Synthesis, biological evaluation and molecular docking of 8-imino-2-oxo-2H,8H-pyrano[2,3-f] chromene analogues: new dual AChE inhibitors as potential drugs for the treatment of Alzheimer's disease, Chem. Biol. Drug Des. 88 (2016) 43–53.

[128] S.J. Basha, P. Mohan, D.P. Yeggoni, Z.R. Babu, P.B. Kumar, A.D. Rao, R. Subramanyam, A.G. Damu, New flavone-cyanoacetamide hybrids with a combination of cholinergic, antioxidant, modulation of βamyloid aggregation, and neuroprotection properties as innovative multifunctional therapeutic candidates for Alzheimer's disease and unraveling their mechanism of action with acetylcholinesterase, Mol. Pharmaceutics 15 (2018) 2206–2223.

[129] S.J. Basha, D.P. Yeggoni, K.Y Rao, P. Mohan, Z.R Babu, K.K Viswanath, D.M. Manidhar, A.D Rao, R. Subramanyam, A.G. Damu, Synthesis and biological evaluation of flavone-8-acrylamide derivatives as potential multi-target-directed anti Alzheimer agents and investigation of binding mechanism with acetylcholinesterase, Bioorg. Chem. 88 (2019) 102960.

[130] M.R. Jones, E.L. Service, J.R. Thompson, M.C.P. Wang, I.J. Kimsey, A.S. DeToma, A. Ramamoorthy, M.H. Lim, T. Storr, Dual-function triazole–pyridine derivatives as inhibitors of metal-induced amyloid-β aggregation, Metallomics 4 (2012) 910–920.

[131] E. Viayna, I. Sola, M. Bartolini, A. De Simone, C. Tapia-Rojas, F.G. Serrano, R. Sabat_, e, J. Ju_arez-Jim_enez, B. P_erez, F.J. Luque, V. Andrisano, M.V. Clos, N.C., Inestrosa, Mu noz-Torrero, D. Synthesis and multitarget biological profiling of a novel family of rhein derivatives as disease-modifying anti-alzheimer agents, J. Med. Chem. 57 (2014) 2549–2567.

[132] H. Kroth, N. Sreenivasachary, A. Hamel, P. Benderitter, Y. Varisco, V. Giriens, P. Paganetti, W. Froestl, A. Pfeifer, A. Muhs, Synthesis and structure–activity relationship of 2,6-disubstituted pyridine derivatives as inhibitors of β-amyloid-42 aggregation, Bioorg. Med. Chem. Lett. 26 (2016) 3330–3335.

[133] F. Pandolfi, D. De Vita, M. Bortolami, A. Coluccia, R. Di Santo, R. Costi, V. Andrisano, F. Alabiso, C. Bergamini, R. Fato, M. Bartolini, L. Scipione, New pyridine derivatives as inhibitors of acetylcholinesterase and amyloid aggregation, Eur. J. Med. Chem. 141 (2017) 197–210.

[134] F.J. Perez-Areales, N. Betari, A. Viayna, C. Pont, A. Espargaro, M. Bartolini, A. De Simone, J.F. Rinaldi Alvarenga, B. Perez, R. Sabate, R.M. Lamuela-Raventos, V. Andrisano, F.J. Luque, D. Munoz-Torrero, Design, synthesis and multitarget biological profiling of second-generation antiAlzheimer rhein–huprine hybrids, Future Med. Chem. 9 (2017) 965–981.

[135] A. Elkamhawy, J-e. Park, A.H.E. Hassan, A. Nim Pae, J. Lee, B-G. Park, E. Joo Roh, Synthesis and evaluation of 2-(3-arylureido)pyridines and 2-(3-arylureido)pyrazines as potential modulators of aβ-induced mitochondrial dysfunction in Alzheimer's disease, Eur. J. Med. Chem. 144 (2018) 529–543.

[136] P. Lv, C.-L. Xia, N. Wang, Z.-Q. Liu, Z.-S. Huang, S.-L. Huang, Synthesis and evaluation of 1,2,3,4-tetrahydro-1-acridone analogues as potential dual inhibitors for amyloid-beta and tau aggregation, Bioorg. Med. Chem. 26 (2018) 4693–4705.

[137] M. Benchekroun, I.P. Angona, V. Luzet, H. Martin, M.J.O. Gasque, J.M. Contelles, L. Ismaili, Synthesis, antioxidant and Aβ anti-aggregation properties of new ferulic, caffeic and lipoic acid derivatives obtained by the UGI four-component reaction, Bioorg. Chem. 85 (2019) 221–228.

[138] Z. Zhu, T. Yang, L. Zhang, L. Liu, E. Yin, C. Zhang, Z. Guo, C. Xu, X. Wang, Inhibiting Aβ toxicity in Alzheimer's disease by a pyridine amine derivative, Eur. J. Med. Chem. 168 (2019) 330–339.

[139] G. Ghotbi, M. Mahdavi, Z. Najafi, F.H. Moghadam, M. Hamzeh-Mivehroud, S. Davaran, S. Dastmalchi, Design, synthesis, biological evaluation, and docking study of novel dualacting thiazole-pyridiniums inhibiting acetylcholinesterase and β-amyloid aggregation for Alzheimer's disease, Bioorg. Chem. 103 (2020) 104186.

[140] H. Jang, J.Y. Park, Y.K. Jang, H.J. Kim, J.S. Lee, D.L. Na, Y. Noh, S.N. Lockhart, J.K. Seong, S.W. Seo, Distinct amyloid distribution patterns in amyloid positive subcortical vascular cognitive impairment, Sci. Rep. 8 (2018) 16178.

[141] I. Maezawa, H.S. Hong, R. Liu, C.Y. Wu, R.H. Cheng, M.P. Kung, H.F. Kung, K.S. Lam, S. Oddo, F.M. Laferla, L.W. Jin, Congo red and thioflavin-T analogs detect A beta oligomers, J. Neurochem. 104 (2008) 457–468.

[142] Z.-P. Zhuang, M.-P. Kung, A. Wilson, C.-W. Lee, K. Plossl, C. Hou, D.M. Holtzman, H.F. Kung, Structure-activity relationship of imidazo[1,2-a]pyridines as ligands for detecting β-amyloid plaques in the brain, J. Med. Chem. 46 (2003) 237–243.

[143] M. Ono, M. Haratake, M. Nakayama, Y. Kaneko, K. Kawabata, H. Mori, M.-P. Kung, H.F. Kung, Synthesis and biological evaluation of (E)-3-styrylpyridine derivatives as amyloid imaging agents for Alzheimer's disease, Nucl. Med. Biol. 32 (2005) 329–335.

[144] L. Cai, F.T. Chin, V.W. Pike, H. Toyama, J.-S. Liow, S.S. Zoghbi, K. Modell, E. Briard, H.U. Shetty, K. Sinclair, S. Donohue, D. Tipre, M.-P. Kung, C. Dagostin, D.A. Widdowson, M. Green, W. Gao, M.M. Herman, M. Ichise, R.B. Inni, Synthesis and evaluation of two 18F-labeled 6-iodo-2-(4′-N,N-dimethylamino)phenylimidazo[1,2-a]pyridine derivatives as prospective radioligands for β-amyloid in Alzheimer's disease, J. Med. Chem. 47 (2004) 2208–2218.

[145] L. Cai, J.-S. Liow, S.S. Zoghbi, J. Cuevas, C. Baetas, J. Hong, H.U. Shetty, N.M. Seneca, A.K. Brown, R. Gladding, S.S. Temme, M.M. Herman, R.B. Innis, V.W. Pike, Synthesis and evaluation of N-methyl and S-methyl 11C-labeled 6-methylthio-2-(4r-N,N-dimethylamino)phenylimidazo[1,2-a]pyridines as radioligands for imaging β-amyloid plaques in Alzheimer's disease, J. Med. Chem. 51 (2008) 148–158.

[146] N. Seneca, L. Cai, J.-S. Liow, S.S. Zoghbi, R. Gladding, J.T. Little, P.S. Aisen, J. Hong, V.W. Pike, R.B. Innis, Low Retention of [S-methyl-11C]MeS-IMPY to b-amyloid Plaques in Patients with Alzheimer's Disease, Current Radiopharmaceuticals 2 (2009) 129–136.

[147] Z. Zha, S.R. Choi, K. Ploessl, B.P. Lieberman, W. Qu, F. Hefti, M. Mintun, D. Skovronsky, H.F. Kung, Multidentate 18F-Polypegylated Styrylpyridines as imaging agents for Aβ Plaques in Cerebral Amyloid Angiopathy (CAA), J. Med. Chem. 54 (2011) 8085–8098.

[148] B.H. Yousefi, A. Manook, B.V. Reutern, M. Schwaiger, A. Drzezga, H.-J. Westerb, G. Henriksen, Development of an improved radioiodinated 2-phenylimidazo[1,2-a]pyridine for non-invasive imaging of amyloid plaques, Med. Chem. Commun. 3 (2012) 775.

[149] S.T. Harrison, J. Mulhearn, S.E. Wolkenberg, P.J. Miller, S.S. O'Malley, Z. Zeng, D.L. Williams, J.E.D. Hostetler, S.S. Bohorquez, L. Gammage, H. Fan, C. Sur, J.C. Culberson, R.J. Hargreaves, J.J. Cook, G.D. Hartman, J.C. Barrow, Synthesis and evaluation of 5-Fluoro-2-aryloxazolo[5,4-b]pyridines as β-Amyloid PET ligands and identification of MK-3328, ACS Med. Chem. Lett. 2 (2011) 498–502.

[150] N. Bandara, A.K. Sharma, S. Krieger, J.W. Schultz, B.H. Han, B.E. Rogers, L.M. Mirica, Evaluation of ^{64}Cu-based radiopharmaceuticals that target Aβ peptide aggregates as diagnostic tools for Alzheimer's disease, J. Am. Chem. Soc. 139 (2017) 12550–12558.

[151] F. Zeng, J.A. Southerland, R.J. Voll, J.R. Votaw, L. Williams, B.J. Ciliax, A.I. Levey, M.M. Goodman, Synthesis and evaluation of two 18F-labeled imidazo[1,2-a]pyridine analogues as potential agents for imaging β-amyloid in Alzheimer's disease, Bioorg. Med. Chem. Lett. 16 (2006) 3015–3018.

[152] R.J.A. Nabuurs, V.V. Kapoerchan, A. Metaxas, M.Hafith S., M.M.Welling de Backer, W. Jiskoot, A.M.C.H. van den Nieuwendijk, A.D. Windhorst, H.S. Overkleeft, M.A. van Buchem, M. Overhand, L. van der Weerd, Bis-pyridylethenyl benzene as novel backbone for amyloid-β binding compounds, Bioorg. Med. Chem. 24 (2016) 6139–6148.

[153] Y. Okumura, Y. Maya, T. Onishi, Y. Shoyama, A. Izawa, D. Nakamura, S. Tanifuji, A. Tanaka, Y. Arano, H. Matsumoto, Design, synthesis, and preliminary evaluation of SPECT probes for imaging β-amyloid in Alzheimer's disease-affected brain, ACS Chem. Neurosci. 9 (2018) 1503–1514.

[154] S. Molavipordanjani, S. Emami, A. Mardanshahi, F.T. Amiri, Z. Noaparast, S.J. Hosseinimehr, Novel 99mTc-2-arylimidazo[2,1-b]benzothiazole derivatives as SPECT imaging agents for amyloid-β plaques, Eur. J. Med. Chem. 175 (2019) 149–161.

[155] M.E. Kennedy, A.W. Stamford, X. Chen, K. Cox, J.N. Cumming, M.F. Dockendorf, M. Egan, L. Ereshefsky, R.A. Hodgson, L.A. Hyde, The BACE1 inhibitor verubecestat (MK-8931) reduces CNS β-amyloid in animal models and in Alzheimer's disease patients, Sci. Transl. Med. 8 (2016) 363ra150 –363ra150.

[156] C. Tallon, M.H. Farah, Beta secretase activity in peripheral nerve regeneration, Neural Regener. Res. 12 (2017) 1565–1574.

[157] J.S. Tung, D.L. Davis, J.P. Anderson, D.E. Walker, S. Mamo, N. Jewett, R.K. Hom, S. Sinha, E.D. Thorsett, V. John, Design of substrate-based inhibitors of human β-secretase, J. Med. Chem. 45 (2002) 259–262.

[158] C.S. Kuruva, P.H. Reddy, Amyloid beta modulators and neuroprotection in Alzheimer's disease: a critical appraisal, Drug Discov. Today 22 (2017) 223–233.

[159] A. Iraji, M. Khoshneviszadeh, O. Firuzi, M. Khoshneviszadeh, N. Edraki, Novel small molecule therapeutic agents for Alzheimer disease: focusing on BACE1 and multi-target directed ligands, Bioorg. Chem. 97 (2020) 103649.

[160] M.S. Malamas, J. Erdei, I. Gunawan, K. Barnes, M. Johnson, Y. Hui, J. Turner, Y. Hu, E. Wagner, K. Fan, A. Olland, J. Bard, A.J. Robichaud, Aminoimidazoles as potent and selective human β-secretase (BACE1) inhibitors, J. Med. Chem. 52 (2009) 6314–6323.

[161] M.S. Malamas, K. Barnes, M. Johnson, Y. Hui, P. Zhou, J. Turner, Y. Hu, E. Wagner, K. Fan, R. Chopra, A. Olland, J. Bard, M. Pangalos, P. Reinhart, A.J. Robichaud, Di-substituted pyridinyl aminohydantoins as potent and highly selective human β-secretase (BACE1) inhibitors, Bioorg. Med. Chem. 18 (2010) 630–639.

[162] P. Zhou, Y. Li, Y. Fan, Z. Wang, R. Chopra, A. Olland, Y. Hu, R.L. Magolda, M. Pangalos, P.H. Reinhart, M.J. Turner, J. Bard, M.S. Malamas, A.J. Robichaud, Pyridinyl aminohydantoins as small molecule BACE1 inhibitors, Bioorg. Med. Chem. Lett. 20 (2010) 2326–2329.

[163] D.-Y. Peng, Q. Sun, X.-L. Zhu, H.-Y. Lin, Q. Chen, N.-X. Yu, W.-C. Yang, G.-F. Yang, Design, synthesis, and bioevaluation of benzamides: Novel acetylcholinesterase inhibitors with multi-functions on butylcholinesterase, Aβ aggregation, and β-secretase, Bioorg. Med. Chem. 20 (2012) 6739–6750.

[164] B.M. Swahn, J. Holenz, J. Kihlström, K. Kolmodin, J. Lindström, N. Plobeck, D. Rotticci, F. Sehgelmeble, M. Sundström, S.V. Berg, J. Fälting, B. Georgievska, S. Gustavsson, J. Neelissen, M. Ek, L.-L. Olsson, S. Berg, Aminoimidazoles as BACE-1 inhibitors: the challenge to achieve in vivo brain efficacy, Bioorg. Med. Chem. Lett. 22 (2012) 1854–1859.

[165] A.A. Thomas, K.W. Hunt, M. Volgraf, R.J. Watts, X. Liu, G. Vigers, D. Smith, D. Sammond, T.P. Tang, S.P. Rhodes, A.T. Metcalf, K.D. Brown, J.N. Otten, M. Burkard, A.A. Cox, M.K.G. Do, D. Dutcher, S. Rana, R.K. DeLisle, K. Regal, A.D. Wright, R. Groneberg, K.S. Levie, M. Siu, H.E. Purkey, J.P. Lyssikatos, I.W. Gunawardana, Discovery of 7–tetrahydropyran-2-yl chromans: β–Site amyloid precursor protein cleaving enzyme 1 (BACE1) inhibitors that reduce amyloid β–protein (Aβ) in the central nervous system, J. Med. Chem. 57 (2014) 878–902.

[166] S. Azimi, A. Zonouzi, O. Firuzi, A. Iraji, M. Saeedi, M. Mahdavi, N. Edraki, Discovery of imidazopyridines containing isoindoline-1,3-dione framework as a new class of BACE1 inhibitors: Design, synthesis and SAR analysis, Eur. J. Med. Chem. 138 (2017) 729–737.

[167] F. Hosseini, A. Ramazani, M.M. Khanaposhtani, M.B. Tehrani, H. Nadri, B. Larijani, M. Mahdavi, Design, synthesis, and biological evaluation of novel 4-oxobenzo[*d*]1,2,3-triazin-benzylpyridinum derivatives as potent anti-Alzheimer agents, Bioorg. Med. Chem. 27 (2019) 2914–2922.

[168] K. Nakahara, Y. Mitsuoka, S. Kasuya, T. Yamamoto, S. Yamamoto, H. Ito, Y. Kido, K. Kusakabe, Balancing potency and basicity by incorporating fluoropyridine moieties: Discovery of a 1-amino-3,4-dihydro-2,6-naphthyridine BACE1 inhibitor that affords robust and sustained central Aβ reduction, Eur. J. Med. Chem. 216 (2021) 113270.

[169] G.T. Wong, D. Manfra, F.M. Poulet, Q. Zhang, H. Josien, T. Bara, L. Engstrom, M.P. Ortiz, J.S. Fine, H.-J.J. Lee, L. Zhang, G.A. Higgins, E.M. Parker, Chronic Treatment with the γ -Secretase Inhibitor LY-411,575 Inhibits -Amyloid Peptide Production and Alters Lymphopoiesis and Intestinal Cell Differentiation, J. Biol. Chem. 279 (2004) 12876–12882.

[170] E.R. Siemers, J.F. Quinn, J. Kaye, M.R. Farlow, A. Porsteinsson, P. Tariot, P. Zoulnouni, J.E. Galvin, D.M. Holtzman, D.S. Knopman, J. Satterwhite, C. Gonzales, R.A. Dean, P.C. May, Effects of a γ -secretase inhibitor in a randomized study of patients with Alzheimer disease, Neurology 66 (2006) 602.

[171] Y. Mitani, J. Yarimizu, K. Saita, H. Uchino, H. Akashiba, Y. Shitaka, K. Ni, N. Matsuoka, Differential effects between γ -secretase inhibitors and modulators on cognitive function in amyloid precursor protein-transgenic and nontransgenic mice, J. Neuro. sci. 32 (2012) 2037–2050.

[172] S. Weggen, J.L. Eriksen, P. Das, S.A. Sagi, R. Wang, C.U. Pietrzik, K.A. Findlay, T.E. Smith, M.P. Murphy, T. Bulter, D.E. Kang, N.M. Sterlingk, T.E. Golde, E.H. Koo, A subset of NSAIDs lower amyloidogenic Aβ42 independently of cyclooxygenase activity, Nature 414 (2001) 212.

[173] Z. Wan, A. Hall, Y. Sang, J.-N. Xiang, E. Yang, B. Smith, D.C. Harrison, G. Yang, H. Yu, H.S. Price, J. Wang, J. Hawkins, L.-F. Lau, M.R. Johnson, T. Li, W. Zhao, W.L. Mitchell, X. Su, X. Zhang, Y. Zhou, Y. Jin, Z. Tong, Z. Cheng, I. Hussain, J.D. Elliott, Y. Mats, Pyridine-derived γ-secretase modulators, Bioorg. Med. Chem. Lett. 21 (2011) 4832–4835.

[174] F. Bischoff, D. Berthelot, M. De Cleyn, G. Macdonald, G. Minne, D. Oehlrich, S. Pieters, M. Surkyn, A.A. Trabanco, G. Tresadern, S. Van Brandt, I. Velter, M. Zaja, H. Borghys, C. Masungi, M. Mercken, H.J.M Gijsen, Design and synthesis of a novel series of bicyclic heterocycles as potent γ -secretase modulators, J. Med. Chem. 55 (2012) 9089–9106.

[175] J.J. Chen, W. Qian, K. Biswas, C. Yuan, A. Amegadzie, Q. Liu, T. Nixey, J. Zhu, M. Ncube, R.M. Rzasa, F.C.N. Chen, F. DeMorin, S. Rumfelt, C.M. Tegley, J.R. Allen, S. Hitchcock, R. Hungate, M.D. Bartberger, L. Zalameda, Y. Liu, J.D. McCarter, J. Zhang, L. Zhu, S. Babu-Khan, Y. Luo, J. Bradley, P.H. Wen, D.L. Reid, F. Koegler, C.D. Hickman, T.L. Correll, T. Williamson, S. Wood, Discovery of 2-methylpyridine-based biaryl amides as γ -secretase modulators for the treatment of Alzheimer's disease, Bioorg. Med. Chem. Lett. 23 (2013) 6447–6454.

[176] R. Sekioka, E. Honjo, S. Honda, H. Fuji, H. Akashiba, Y. Mitani, S. Yamasaki, Discovery of novel scaffolds for γ-secretase modulators without an arylimidazole moiety, Bioorg. Med. Chem. 26 (2018) 435–442.

[177] R. Sekioka, S. Honda, H. Akashiba, J. Yarimizu, Y. Mitani, S. Yamasaki, Optimization and biological evaluation of imidazopyridine derivatives as a novel scaffold for γ -secretase modulators with oral efficacy against cognitive deficits in Alzheimer's disease model mice, Bioorg. Med. Chem. 28 (2020) 115455.

[178] K.D. Rynearson, R.N. Buckle, R.J. Herr, N.J. Mayhew, X. Chen, W.D. Paquette, S.A. Sakwa, J. Yang, K.D. Barnes, P. Nguyen, W.C. Mobley, G. Johnson, J.H. Lin, R.E. Tanzi, S.L. Wagner, Design and synthesis of novel methoxypyridine-derived gamma-secretase modulators, Bioorg. Med. Chem. 28 (2020) 115734.

[179] K.S. Kosik, C.L. Joachim, D.J. Selkoe, Microtubule-associated protein tau (tau) is a major antigenic component of paired helical filaments in Alzheimer disease, Proc. Natl. Acad. Sci 83 (1986) 4044–4048.

[180] N.V. Gorantla, V.G. Landge, P.G. Nagaraju, P.C.G. Poornima, E. Balaraman, S. Chinnathambi, Molecular cobalt(II) complexes for tau polymerization in Alzheimer's disease, ACS Omega 4 (2019) 16702–16714.

[181] M. Shah, A.M. Catafau, Molecular imaging insights into neurodegeneration: focus on tau PET radiotracers, J. Nucl. Med. 55 (2014) 871–874.

[182] E. Gabellieri, F. Capotosti, J. Molette, N. Sreenivasachary, A. Mueller, M. Berndt, H. Schieferstein, T. Juergens, Y. Varisco, F. Oden, H.S. Willich, D. Hickman, L. Dinkelborg, A. Stephens, A. Pfeifer, H. Kroth, Discovery of 2-(4-(2-fluoroethoxy)piperidin-1-yl)-9-methyl-9Hpyrrolo[2,3-b:4,5-c′]dipyridine ([^{18}F]PI-2014) as PET tracer for the detection of pathological aggregated tau in Alzheimer's disease and other tauopathies, Eur. J. Med. Chem. 204 (2020) 112615.

[183] R.M. Kypta, Review: GSK3 inhibitors and their potential in the treatment of Alzheimer's disease, Expert Opin. Ther. Pat. 15 (2005) 1315–1331.

[184] K. Leroy, A. Boutajangout, M. Authelet, J.R. Woodget, B.H. Anderton, J.-P. Brion, The active form of glycogen synthase kinase-3β is associated with granulovacuolar degeneration in neurons in Alzheimer's disease, Acta Neuropathol. 103 (2002) 91–99.

[185] J.J. Pei, H. Braak, I. Grundke-Iqbal, B. Winblad, R.F. Cowburn, Distribution of active glycogen synthase kinase 3beta (GSK-3beta) in brains staged for Alzheimers's disease neurofibrillary changes, J. Neuropathol. Exp. Neurol. 58 (1999) 1010–1019.

[186] S. Berg, M. Bergh, S. Hellberg, K. Högdin, Y. Lo-Alfredsson, P. Söderman, S. von Berg, T. Weigelt, M. Ormö, Y. Xue, J. Tucker, J. Neelissen, E. Jerning, Y. Nilsson, R. Bhat, Discovery of novel potent and highly selective glycogen synthase kinase-3β (GSK3β) inhibitors for alzheimer's disease: design, synthesis, and characterization of pyrazines, J. Med. Chem. 55 (2012) 9107–9119.

[187] X.-L. Shi, J.-D. Wu, P. Liu, Z.-P. Liu, Synthesis and evaluation of novel GSK-3β inhibitors as multifunctional agents against Alzheimer's disease, Eur. J. Med. Chem. 167 (2019) 211–225.

[188] L. Emilsson, P. Saetre, J. Balciuniene, A. Castensson, N. Cairns, E.E. Jazin, Increased monoamine oxidase messenger RNA expression levels in frontal cortex of Alzheimer's disease patients, Neurosci. Lett. 326 (2002) 56–60.

[189] A.C. Tripathi, S. Upadhyay, S. Paliwal, S.K. Saraf, Privileged scaffolds as MAO inhibitors: retrospect and prospects, Eur. J. Med. Chem. 145 (2018) 445–497.

[190] D. Secci, A. Bolasco, S. Carradori, M. D'Ascenzio, R. Nescatelli, M. Yáñez, Recent advances in the development of selective human MAO-B inhibitors: (Hetero)arylidene-(4-substituted-thiazol-2-yl)hydrazines, Eur. J. Med. Chem. 58 (2012) 405–417.

[191] J.-e. Park, A. Elkamhawy, A.H.E. Hassan, A. Nim Pae, J. Lee, S. Paik, B.-G. Park, E. Joo Roh, Synthesis and evaluation of new pyridyl/pyrazinyl thiourea derivatives: Neuroprotection against amyloid-β-induced toxicity, Eur. J. Med. Chem. 141 (2017) 322–334.

[192] N.T. Tzvetkov, H.-G. Stammler, S. Hristova, A.G. Atanasov, L. Antonov, Pyrrolo-pyridin-5-yl)benzamides: BBB permeable monoamine oxidase B inhibitors with neuroprotective effect on cortical neurons, Eur. J. Med. Chem. 162 (2019) 793–809.
[193] J.L. Franklin, E.M. Johnson Jr, Elevated intracellular calcium blocks programmed neuronal death, Ann. N. Y. Acad. Sci. 747 (1994) 195–204.
[194] C. Orozco, C. de Los Rios, E. Arias, R. Leon, A.G. Garcia, J.L. Marco, M. Villarroya, M.G. Lopez, ITH4012 (ethyl 5-amino-6,7,8,9-tetrahydro-2-methyl-4-phenylbenzol[1,8]naphthyridine-3-carboxylate), a novel acetylcholinesterase inhibitor with "calcium promotor" and neuroprotective properties, J. Pharmacol. Exp. Ther. 310 (2004) 987–994.
[195] J. Egea, C. de los Rios, 1,8-Naphthyridine derivatives as cholinesterases inhibitors and cell Ca^{2+} regulators, a multitarget strategy for Alzheimer's disease, Curr. Top. Med. Chem. 11 (2011) 2807–2823.
[196] V.K. Ramanan, A.J. Saykin, Pathways to neurodegeneration: mechanistic insights from GWAS in Alzheimer's disease, Parkinson's disease, and related disorders, Am.J. Neurodegener. Dis. 2 (2013) 145–175.
[197] K.N. Shaw, S. Commins, S.M. O'Mara, Lipopolysaccharide causes deficits in spatial learning in the watermaze but not in BDNF expression in the rat dentate gyrus, Behav. Brain Res. 124 (2001) 47–54.
[198] R. Singh, R. Bansal, Investigations on 16-arylideno steroids as a new class of neuroprotective agents for the treatment of Alzheimer's and Parkinson's diseases, ACS Chem. Neurosci. 8 (2017) 186–200.
[199] R. Singh, S. Thota, R. Bansal, Studies on 16,17-pyrazoline substituted heterosteroids as anti-Alzheimer and anti-Parkinsonian agents using LPS induced neuroinflammation models of mice and rats, ACS Chem. Neurosci. 9 (2018) 272–283.
[200] S. Przedborski, M. Vila, The 1-methyl-4-phenyl-1,2,3,6-tetrahydropyridine mouse model: a tool to explore the pathogenesis of Parkinson's disease, Ann. N. Y. Acad. Sci. 991 (2003) 189–198.
[201] J. Jouha, M. Loubidi, J. Bouali, S. Hamri, A. Hafid, F. Suzenet, G. Guillaumet, T. Dagcı, M. Khouili, F. Aydın, L. Saso, G. Armagan, Synthesis of new heterocyclic compounds based on pyrazolopyridine scaffold and evaluation of their neuroprotective potential in MPP -induced neurodegeneration, Eur. J. Med. Chem. 129 (2017) 41–52.
[202] C. de los Ríos, J. Marco-Contelles, Tacrines for Alzheimer's disease therapy. III. The Pyridotacrines, Eur. J. Med. Chem. 166 (2019) 381–389.
[203] B.B. Fredholm, A.P. IJzerman, K.A. Jacobson, J. Linden, C.E. Muller, International union of Pharmacology LXXXI. Nomenclature and classification of adenosine receptors. An update, Pharmacol. Rev. 63 (2011) 1–34.
[204] R.A. Cunha, Neuroprotection by adenosine in the brain: from A1 receptor activation to A2A receptor blockade, Purinergic Signal 1 (2005) 111–134.
[205] R. Franco, G. Navarro, Adenosine A2A receptor antagonist in neurodegenerative diseases: huge potential and huge challenges, Front. Psychiatry 9 (2018) 68.
[206] M. Rivera-Oliver, M. Diaz-Rios, Using caffeine and other adenosine receptor antagonists and agonists as therapeutic tools against neurodegenerative diseases: a review, Life Sci. 101 (2014) 1–9.
[207] E. Faivre, J.E. Coelho, K. Zornbach, E. Malik, Y. Baqi, M. Schneider, L. Cellai, K. Carvalho, S. Sebda, M. Figeac, S. Eddarkaoui, R. Caillierez, Y. Chern, M. Heneka, N. Sergeant, C.E. Muller, A. Halle, L. Buee, L.V. Lopes, D. Blum, Beneficial effect of a selective adenosine A2A receptor antagonist in the APPswe/PS1dE9 mouse model of Alzheimer's disease, Front. Mol. Neurosci. 11 (2018) 235.
[208] M. Falsini, D. Catarzi, F. Varano, D. Dal Ben, G. Marucci, M. Buccioni, R. Volpini, L. Di Cesare Mannelli, C. Ghelardini, V. Colotta, Novel 8-amino-1,2,4-triazolo[4,3-a]pyrazin-3-one derivatives as potent human adenosine A1 and A2A receptor antagonists. Evaluation of their protective effect against β-amyloid-induced neurotoxicity in SH-SY5Y cells, Bioorg. Chem. 87 (2019) 380–394.
[209] D.J. Foster, D.L. Choi, P.J. Conn, J.M. Rook, Activation of M1 and M4 muscarinic receptors as potential treatments for Alzheimer's disease and schizophrenia, Neuropsychiatr. Dis. Treat. 10 (2014) 183–191.
[210] A.D. Korczyn, Muscarinic M(1) agonists in the treatment of alzheimer's disease, Expert Opin. Investig. Drugs 9 (2000) 2259–2267.
[211] A.C. Kruse, A.M. Ring, A. Manglik, J. Hu, K. Hu, K. Eitel, H. Hubner, E. Pardon, C. Valant, P.M. Sexton, A. Christopoulos, C.C. Felder, P. Gmeiner, J. Steyaert, W.I. Weis, K.C. Garcia, J. Wess, B.K. Kobilka, Activation and allosteric modulation of a muscarinic acetylcholine receptor, Nature 504 (2013) 101–106.
[212] C.T. Sadashiva, J.N. Narendra Sharath Chandra, C.V. Kavitha, A. Thimmegowda, M.N. Subhash, K.S. Rangappa, Synthesis and pharmacological evaluation of novel N-alkyl/aryl substituted thiazolidinone arecoline analogues as muscarinic receptor 1 agonist in Alzheimer's dementia models, Eur. J. Med. Chem. 44 (2009) 4848–4854.

[213] J.C. Tarr, M.L. Turlington, P.R. Reid, T.J. Utley, D.J. Sheffler, H.P. Cho, R. Klar, T. Pancani, M.T. Klein, T.M. Bridges, R.D. Morrison, A.L. Blobaum, Z. Xiang, J.S. Daniels, C.M. Niswender, P.J. Conn, M.R. Wood, C.W. Lindsley, Targeting selective activation of M1 for the treatment of Alzheimer's disease: further chemical optimization and pharmacological characterization of the M1 positive allosteric modulator ML169, ACS Chem. Neurosci. 3 (2012) 884–895.

[214] J.E. Davoren, C.-W. Lee, M. Garnsey, M.A. Brodney, J. Cordes, K. Dlugolenski, J.R. Edgerton, A.R. Harris, C.J. Helal, S. Jenkinson, G.W. Kauffman, T.P. Kenakin, J.T. Lazzaro, S.M. Lotarski, Y. Mao, D.M. Nason, C. Northcott, L. Nottebaum, S.V. O'Neil, B. Pettersen, M. Popiolek, V. Reinhart, R. Salomon-Ferrer, S.J. Steyn, D. Webb, L. Zhang, S. Grimwood, Discovery of the potent and selective M1 PAM-agonist N-[(3R,4S)-3-hydroxytetrahydro-2H-pyran-4-yl]-5-methyl-4-[4-(1,3-thiazol-4-yl)benzyl]pyridine-2-carboxamide (PF-06767832): evaluation of efficacy and cholinergic side effects, J. Med. Chem. 59 (2016) 6313–6328.

[215] A. Garcia-Osta, M. Cuadrado-Tejedor, C. Garcia-Barroso, J. Oyarzabal, R. Franco, Phosphodiesterases as therapeutic targets for Alzheimer's disease, ACS Chem. Neurosci. 3 (2012) 832–844.

[216] P.R. Heckman, C. Wouters, J. Prickaerts, Phosphodiesterase inhibitors as a target for cognition enhancement in aging and Alzheimer's disease: a translational overview, Curr. Pharm. Des. 21 (2014) 317–331.

[217] D. Puzzo, A. Staniszewski, S.X. Deng, L. Privitera, E. Leznik, S. Liu, H. Zhang, Y. Feng, A. Palmeri, D.W. Landry, O. Arancio, Phosphodiesterase 5 inhibition improves synaptic function, memory, and amyloid-beta load in an alzheimer's disease mouse model, J. Neurosci. 29 (2009) 8075–8086.

[218] J. Fiorito, J. Vendome, F. Saeed, A. Staniszewski, H. Zhang, S. Yan, S.-X. Deng, O. Arancio, D.W. Landry, Identification of a Novel 1,2,3,4-Tetrahydrobenzo[b][1,6]naphthyridine analogue as a potent phosphodiesterase 5 inhibitor with improved aqueous solubility for the treatment of Alzheimer's disease, J. Med. Chem. 60 (2017) 8858–8875.

[219] J. Prickaerts, P.R.A. Heckman, A. Blokland, Investigational phosphodiesterase inhibitors in phase I and phase II clinical trials for Alzheimer's disease, Expert Opin. Invest. Drugs 26 (2017) 1033–1048.

[220] Y. Wu, Z. Li, Y.-Y. Huang, D. Wu, H.-B. Luo, Novel phosphodiesterase inhibitors for cognitive improvement in Alzheimer's disease, J. Med. Chem. 61 (2018) 5467–5483.

[221] A.K. Vadukoot, S. Sharma, C.D. Aretz, S. Kumar, N. Gautam, Y. Alnouti, A.L. Aldrich, C.E. Heim, T. Kielian, C.R. Hopkins, Synthesis and SAR Studies of 1H-Pyrrolo[2,3-b]pyridine-2-carboxamides as Phosphodiesterase 4B (PDE4B) Inhibitors, ACS Med. Chem. Lett. 11 (2020) 1848–1854.

[222] E. Sher, Y. Chen, T.J. Sharples, L.M. Broad, G. Benedetti, R. Zwart, G.I. McPhie, K.H. Pearson, T. Baldwinson, G. De Filippi, Physiological roles of neuronal nicotinic receptor subtypes: new insights on the nicotinic modulation of neurotransmitter release, synaptic transmission and plasticity, Curr. Top. Med. Chem. 4 (2004) 283–297.

[223] M. Grupe, M. Grunnet, J.F. Bastlund, A.A. Jensen, Targeting $\alpha4\beta2$ nicotinic acetylcholine receptors in central nervous system disorders: perspectives on positive allosteric modulation as a therapeutic approach, Basic Clin. Pharmacol. Toxicol. 116 (2015) 187–200.

[224] L.L. Brown, S. Kulkarni, O.A. Pavlova, A.O. Koren, A.G. Mukhin, A.H. Newman, A.G. Horti, Synthesis and evaluation of a novel series of 2-chloro-5-((1-methyl-2-(s)-pyrrolidinyl)methoxy)-3-(2-(4-pyridinyl)vinyl)pyridine analogues as potential positron emission tomography imaging agents for nicotinic acetylcholine receptors, J. Med. Chem. 45 (2002) 2841–2849.

[225] F. Deligia, G. Murineddu, C. Gotti, G. Ragusa, F. Fasoli, M. Sciaccaluga, S. Plutino, S. Fucile, G. Loriga, B. Asproni, G.A. Pinna, Pyridinyl- and pyridazinyl-3,6-diazabicyclo[3.1.1]heptane-anilines: novel selective ligands with subnanomolar affinity for $\alpha4\beta2$ nACh receptors, Eur. J. Med. Chem. 152 (2018) 401–416.

[226] M. Rosini, E. Simoni, R. Caporaso, A. Minarini, Multitarget strategies in Alzheimer's disease: benefits and challenges on the road to therapeutics, Future Med. Chem. 8 (2016) 697–711.

[227] J.B. Shaik, B.K. Palaka, M. Penumala, K.V. Kotapati, S.R. Devineni, S. Eadlapalli, M.M. Darla, D.R. Ampasala, R. Vadde, G.D. Amooru, Synthesis pharmacological assessment, molecular modeling and *insilico* studies of fused tricyclic coumarin derivatives as a new family of multifunctional anti-Alzheimer agents, Eur. J. Med. Chem. 107 (2016) 219–232.

[228] M.J. Oset-Gasque, J. Marco-Contelles, Alzheimer's Disease, the "one molecule, one-target" paradigm, and the multitarget directed ligand approach, ACS Chem. Neurosci. 9 (2018) 401–403.

[229] A. Samadi, J. Marco-Contelles, E. Soriano, M. Álvarez-Pérez, M. Chioua, A. Romero, L. González-Lafuente, L. Gandía, J.M. Roda, M.G. López, M. Villarroya, A.G. García, C. De los Ríos, Multipotent drugs with cholinergic and neuroprotective properties for the treatment of Alzheimer and neuronal vascular diseases. I. Synthesis, biological assessment, and molecular modeling of simple and readily available 2-aminopyridine-, and 2-chloropyridine-3,5-dicarbonitriles, Bioorg. Med. Chem. 18 (2010) 5861–5872.

[230] S. Rizzo, A. Bisi, M. Bartolini, F. Mancini, F. Belluti, S. Gobbi, V. Andrisano, A. Rampa, Multi-target strategy to address Alzheimer's disease: Design, synthesis and biological evaluation of new tacrine-based dimers, Eur. J. Med. Chem. 46 (2011) 4336–4343.

[231] S.-S. Xie, X.-B. Wang, J.-Y. Li, L. Yang, L.-Y. Kong, Design, synthesis and evaluation of novel tacrine-coumarin hybrids as multifunctional cholinesterase inhibitors against Alzheimer's disease, Eur. J. Med. Chem. 64 (2013) 540–553.

[232] S.-Y. Li, X.-B. Wang, S.-S. Xie, N. Jiang, K.D.G. Wang, H.-Q. Yao, H.-B. Sun, L.-Y. Kong, Multifunctional tacrine-flavonoid hybrids with cholinergic, β-amyloid-reducing, and metal chelating properties for the treatment of Alzheimer's disease, Eur. J. Med. Chem. 69 (2013) 632–646.

[233] O.M. Bautista-Aguilera, G. Esteban, M. Chioua, K. Nikolic, D. Agbaba, I. Moraleda, I. Iriepa, E. Soriano, A. Samadi, M. Unzeta, J. Marco-Contelles, Multipotent cholinesterase/monoamine oxidase inhibitors for the treatment of Alzheimer's disease: design, synthesis, biochemical evaluation, ADMET, molecular modeling, and QSAR analysis of novel donepezil-pyridyl hybrids, Drug Des. Dev. Ther. 8 (2014) 1893–1910.

[234] M. Maqbool, A. Manral, E. Jameel, J. Kumar, V. Saini, A. Shandilya, M. Tiwari, N. Hoda, B. Jayaram, Development of cyanopyridine–triazine hybrids as lead multitarget anti-Alzheimer agents, Bioorg. Med. Chem. 24 (2016) 2777–2788.

[235] X.Y. Jiang, T.K. Chen, J.T. Zhou, S.Y. He, H.Y. Yang, Y. Chen, W. Qu, F. Feng, H.P. Sun, Dual GSK-3β/AChE inhibitors as a new strategy for multitargeting anti-Alzheimer's disease drug discovery, ACS Med. Chem. Lett. 9 (2018) 171–176.

[236] D. Knez, N. Coquelle, A. Pislar, S. Zakelj, M. Jukic, M. Sova, J. Mravljak, F. Nachon, X. Brazzolotto, J. Kos, J.-P. Colletier, S. Gobec, Multi-target-directed ligands for treating Alzheimer's disease: butyrylcholinesterase inhibitors displaying antioxidant and neuroprotective activities, Eur. J. Med. Chem. 156 (2018) 598–617.

[237] N.Q. Thai, Z. Bednarikova, M. Gancar, H.Q. Linh, C.K. Hu, M.S. Li, Z. Gazova, Compound CID 9998128 is a potential multi-target drug for Alzheimer's disease, ACS Chem. Neurosci. 9 (2018) 2588–2598.

[238] K. Chalupova, J. Korabecny, M. Bartolini, B. Monti, D. Lamba, R. Caliandro, A. Pesaresi, X. Brazzolotto, A.-J. Gastellier, F. Nachon, J. Pejchal, M. Jarosova, V. Hepnarova, D. Jun, M. Hrabinova, R. Dolezal, J.Z. Karasova, M. Mzik, Z. Kristofikova, J. Misik, L. Muckova, P. Jost, O. Soukup, M. Benkova, V. Setnicka, L. Habartova, M. Chvojkova, L. Kleteckova, K. Vales, E. Mezeiova, E. Uliassi, M. Valis, E. Nepovimova, M.L Bolognesi, K. Kuca, Novel tacrine-tryptophan hybrids: multi-target directed ligands as potential treatment for Alzheimer's disease, Eur. J. Med. Chem. 168 (2019) 491–514.

[239] T. Umar, S. Shalini, M.K. Raza, S. Gusain, J. Kumar, P. Seth, M. Tiwari, N. Hoda, A multifunctional therapeutic approach: synthesis, biological evaluation, crystal structure and molecular docking of diversified 1Hpyrazolo[3,4-b]pyridine derivatives against Alzheimer's disease, Eur. J. Med. Chem. 175 (2019) 2–19.

[240] Y. Fang, H. Zhou, Q. Gu, J. Xu, Synthesis and evaluation of tetrahydroisoquinoline-benzimidazole hybrids as multifunctional agents for the treatment of Alzheimer's disease, Eur. J. Med. Chem. 167 (2019) 133–145.

[241] A. Tripathi, P.K. Choubey, P. Sharma, A. Seth, P.N. Tripathi, M.K. Tripathi, S.K. Prajapati, S. Krishnamurthy, S.K. Shrivastava, Design and development of molecular hybrids of 2-pyridylpiperazine and 5-phenyl-1,3,4-oxadiazoles as potential multifunctional agents to treat Alzheimer's disease, Eur. J. Med. Chem. 183 (2019) 111707.

[242] N. Salehi, B.B.F. Mirjalili, H. Nadri, Z. Abdolahi, H. Forootanfar, A.S. Kermani, T.T. Kucukkılınc, B. Ayazgok, S. Emami, I. Haririan, M. Sharifzadeh, A. Foroumadii, M. Khoobik, Synthesis and biological evaluation of new *N*-benzylpyridinium-based benzo heterocycles as potential anti-Alzheimer's agents, Bioorg. Chem. 83 (2019) 559–568.

[243] Z. Haghighijoo, S. Akrami, M. Saeedi, A. Zonouzi, A. Iraji, B. Larijani, H. Fakherzadeh, F. Sharifi, S.M. Arzaghi, M. Mahdavi, N. Edraki, N-Cyclohexylimidazo[1,2-*a*]pyridine derivatives as multi-target-directed ligands for treatment of Alzheimer's disease, Bioorg. Chem. 103 (2020) 104146.

[244] A. Jain, P. Piplani, Design, synthesis and biological evaluation of triazole-oxadiazole conjugates for the management of cognitive dysfunction, Bioorg. Chem. 103 (2020) 104151.

[245] X.Y. Jiang, J. Zhou, Y. Wang, L. Chen, Y. Duan, J. Huang, C. Liu, Y. Chen, W. Liu, H. Sun, F. Feng, W. Qu, Rational design and biological evaluation of a new class of thiazolopyridyl tetrahydroacridines as cholinesterase and GSK-3 dual inhibitors for Alzheimer's disease, Eur. J. Med. Chem. 207 (2020) 112751.

[246] M. Saeedi, M. Safavi, E. Allahabadi, A. Rastegari, R. Hariri, S. Jafari, S.N.A. Bukhari, S.S. Mirfazli, O. Firuzi, N. Edraki, M. Mahdavi, T. Akbarzadeh, Thieno[2,3-b]pyridine amines: Synthesis and evaluation of tacrine analogs against biological activities related to Alzheimer's disease, Arch. Pharm. 353 (2020) e2000101.

[247] M. Bortolami, F. Pandolfi, D. De Vita, C. Carafa, A. Messore, R. Di Santo, M. Feroci, R. Costi, I. Chiarotto, D. Bagetta, S. Alcaro, M. Colone, A. Stringaro, L. Scipione, New deferiprone derivatives as multi-functional cholinesterase inhibitors: design, synthesis and in vitro evaluation, Eur. J. Med. Chem. 198 (2020) 112350.

CHAPTER

4

Design, synthesis, and in vitro anticancer activity of thiophene substituted pyridine derivatives

Vinayak Adimule[a], *Basappa C Yallur*[b] *and Sheetal Batakurki*[c]

[a] Angadi Institute of Technology and Management (AITM), Belagavi, Karnataka, India [b] Department of Chemistry, M.S. Ramaiah Institute of Technology, Bangalore, Karnataka, India [c] Department of Chemistry, M.S. Ramaiah University of Applied Sciences, Bangalore, Karnataka, India

4.1 Introduction

In recent years, rapid treatment of cancer biology involves multimodel such as chemotherapy, radiotherapy, and extensive surgical processes involved in chemotherapy. The limitations arise out of the existing agents, their side effects, and the cancer cells acquire resistance to the inhibition. The need for better anticancer agents with a prominent therapeutic profile over the years has established a strong link between inflammation diseases and cancer [1,2]. Inflammation is one of the key components in cancer diseases and [3] many compounds bearing anti-inflammatory properties are also treated for the treatment of anticancer. For the treatment of pain and inflammation in anticancer therapy, many of the nonsteroidal anti-inflammatory heterocycles are used [4]. Anti-inflammatory compounds such as ibuprofen, aspirin, and phenyl butazole are associated with several side effects when consumed for a long period of time. Extensive clinical trials are needed to produce novel anticancer molecules with enhanced efficacy and better safety profile. Important heterocyclic compounds mimicking the bioactive molecules attracted synthetic chemists to synthesize a large number of heterocyclic compounds with novel chemotherapeutic properties. In recent years, pyridine and its various derivatives attracted much attention due to its diversified medicinal properties [5]. The pyridine substitution at the second and third positions attracted further interest due to their broad spectrum of biological activities such as anticancer [6], antifungal [7], anti-inflammatory [8],

Recent Developments in the Synthesis and Applications of Pyridines.
DOI: https://doi.org/10.1016/B978-0-323-91221-1.00008-7

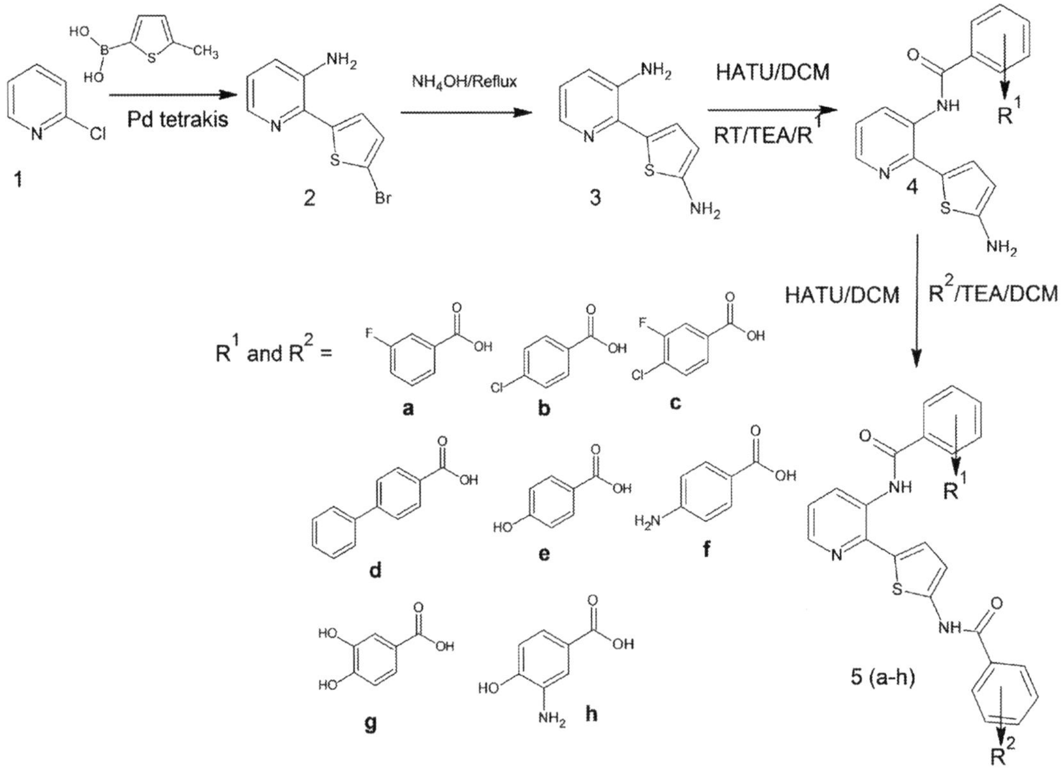

FIGURE 4.1 Schematic synthetic pathway of the thiophene substituted pyridine derivatives.

antidepressant [9], and antioxidant [10]. Moreover, pyridine nucleus is prevalent in numerous natural products and is extremely important in biochemical systems. Most of the pyridine derivatives are widely used as bactericides and anticancer agents [11]. The anticancer activities carried out in the present study also involve computational studies [12]. In recent years, the probability of the understanding of the various reaction intermediates, their binding energy, bond length, dipole moment, and HOMO and LUMO configurations [13]. The application of the semiempirical methods for the calculation of novel heterocyclic compounds has been evaluated [14].

In the many areas of the research to understand the chemical structural interaction with the bioactive compounds, SAR (structural activity relationship) becomes increasingly helpful. SAR has become an important tool to understand the biochemical interaction (heterocyclic interaction with the receptor) and to produce therapeutically important compounds (Fig. 4.1).

In the present investigation, we have developed novel thiophene containing amide derivatives of pyridine analogues **4 (a–h)** and **5 (a–h)** to test their in vitro anticancer properties. In total, 16 derivatives have been synthesized with almost 98% purity and tested against *HeLa, MCF-7, Hep G2* cell lines using docetaxel as a standard drug. The synthesized derivatives are characterized by ^{1}H-NMR, ^{13} C-NMR, LCMS, and purified by column chromatography using silica gel of 60–120 mesh. The IC_{50} values (inhibitory concentration of the compounds at 50) of the **5a, 5b,** and **5c** derivatives containing active –F (Fluorine), –Cl (Chlorine), –F, Cl (fluorine

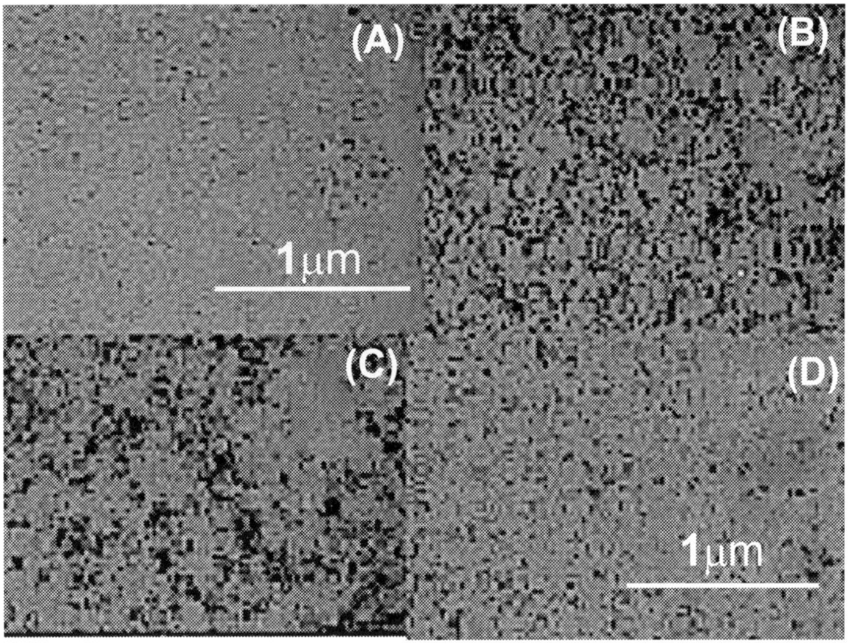

FIGURE 4.2 Cytotoxicity images of the compounds a. (A) against *HepG2* cell line b. (B) against *HepG2* c. (C) against *HepG2* and (D) reference standard drug.

and chlorine) groups were found to be 0.1298±0.098, 0.1456±0.001, and 0.1647±0.013 µM, respectively, against *HepG2* cell lines. The present research leads to the exploration of the new research in the area of pyridine containing thiophene derivatives for anticancer therapy and can help other researchers for the development of new and more potent anticancer molecules (Fig. 4.2).

4.2 Materials and characterization

The chemicals and reagents for the synthesis are 2-methyl thiophene-boronic acid (97.9% purity, Sigma-Aldrich, India), Tetrakis triphenyl palladium (0) (98% purity, Alfa Aesar, India), Triethyl amine (99.6% purity, S-d fine chemicals Ltd, India), HATU (99.5% purity, Sigma-Aldrich, India), Dichloromethane (DCM) (98.8% purity, spectrochem Ltd, India), NH_4OH (36% aqueous solution, S-d fine chemicals Ltd, India), The different acids (**a–h**) are obtained from Sigma-Aldrich and Spectrochem Ltd, India with average purity ranging in between 97.8% and 99.5%. All the other chemicals and reagents were used without any further purification. Melting points (m.p.) of the compounds are uncorrected, ^{1}H-NMR and ^{13} CNMR of the compounds recorded on a broker 400 Hz spectrophotometer using TMS (trimethyl silane) as the internal standard. Chemical shifts are reported in (ppm) s = singlet, bs = broad singlet, *m* = multiplet, *t* = triplet. Purity of the chemical compounds analysed using Agilent make ultivo triple quadra pole LC–MS instrument fitted with vortex collision cells. Elemental analysis was carried out using CHN analyzer Eltra elemental analyser.

4.2.1 Synthesis of 2-(5-bromothiophen-2-yl) pyridin-3-amine (Compound 2, Step 1)

In a 100 mL round bottom (RB) flask fitted with a reflux condenser, 2-chloropyridine (1 mole), K_2CO_3 (2–3 moles), 5-methyl thiophene-2-boronic acid (1.2 mole), tetrakis triphenyl palladium (0) (0.005 mole), Ethanol (250 mL) were refluxed for overnight. TLC was monitored for the completeness of the reaction, solvent was removed and residue was treated with ice-cold water, crude product was extracted with ethyl acetate (25 mL × 3) times, washed with brine, cold water, and the product was further purified by column-chromatography, silica gel 100–200 mesh; Yield = 4.5 g ; m.p. = 160.8°C; Color- off white solid,

4.2.2 Synthesis of 2-(5-aminothiophen-2-yl) pyridin-3-amine (Compound 3, Step 2)

Step 1 product was refluxed with passing NH_3 (g) in water for 2–3 h, the product was separated by adding methanol, the precipitate was filtered, washed with cold water, dried in an oven at 100–200°C, Yield = 4.16 g; m.p. = 157.8°C; Color- white colored solid.

4.2.3 Synthesis of substituted N-[2-(5-aminothiophen-2-yl) pyridin-3-yl] benzamide 4 (a–h) (Compound 4, Step 3)

In a 100 mL RB flask fitted with nitrogen, Step 3 (1 mole) compound was dissolved in dichloromethane (DCM) (10–20 mL), triethyl amine (2 mole), HATU (0.5–0.8 mole), and different acids 4a–4h were treated and stirred for overnight. After completion of the reaction monitored by TLC, solvent was evaporated, product was extracted in ethyl acetate (5 mL × 4) times, purified by column chromatography using silica gel 60–120 mesh, The table gives the % yield, color, m.p. of the various derivatives of **4** (**a–h**). Eluent for the system is ethyl acetate and hexane (30 : 70)

4.2.4 Synthesis of substituted N-[5-(3-benzamidopyridin-2-yl) thiophen-2-yl] benzamide (Compound 5, Step 4)

Step 3 compound (1mole) was dissolved in 10–20 mL of DCM, fitted with nitrogen, different acid derivatives **4** (**a–h**) (1.2 mole) were added under stirring at room temperature. HATU (1.5 mole) was added and stirred for overnight. TLC was monitored for the completeness of the reaction, after completion, reaction mixture was evaporated, product was precipitated by adding cold water and white-colored product separated out was extracted in ethyl acetate (5–10 mL), ethyl acetate layer was washed with brine, cold water, and purified by column-chromatography using silica gel 100–200 mesh, eluent (ethyl acetate : hexane : 50: 50), yield = 3.75 g; Color- Pale Yellow; m.p.- 175.8°C.

4.3 Purification of the compounds

All the derivatives **4** (**a–h**) and **5** (**a–h**) were purified by column chromatography using silica gel 100–200 mesh for close proximate impurities along with the main derivative and 60–120 silica gel mesh for other derivatives. Eluent started generally with hexane and slowly increased the polarity by the addition of ethyl acetate and increased the ethyl acetate content up to 80%. The eluted fractions were concentrated under a vacuum, the solid was obtained by the addition of n-pentane. Ultrasonicated the solid for 5 min, filtered and air-dried [15].

4.4 Analytical characterization

4.4.1 ^{1}H-NMR, ^{13}CMR, and LCMS spectroscopic characterization data

4.4.1.1 2-(5-Bromothiophen-2-yl) Pyridin-3-Amine (Compound 2)

Appearance: Off-white colored solid; M.P- 147.2°C; Yield- 56%; LCMS purity 98.2%; ^{1}H NMR: δ 7.05 (1H, d, J = 8.3 Hz), 7.21–7.34 (3H), 7.25 (dd, J = 8.5, 4.7 Hz), 7.32 (d, J = 8.3 Hz), 7.29 (dd, J = 8.5, 1.9 Hz, 8.65 (1H, dd, J = 4.7, 1.9 Hz).

4.4.1.2 2-(5-aminothiophen-2-yl) pyridin-3-amine (Compound 3)

Appearance: White colored solid; M.P- 137.2°C; Yield- 66%; LCMS- 98. 1%; ^{1}H NMR: δ 6.67 (1H, d, J = 8.3 Hz), 6.80 (1H, dd, J = 8.6, 4.7 Hz), 7.24–7.30 (2H, 7.27 (dd, J = 8.6, 1.9 Hz), 7.26 (d, J = 8.3 Hz)), 8.53 (1H, dd, J = 4.7, 1.9 Hz).

4.4.2 Substituted N-[2-(5-aminothiophen-2-yl) pyridin-3-yl] benzamide (Compound 4 (a–h))

4.4.2.1 N-[2-(5-aminothiophen-2-yl) pyridin-3-yl]-3-fluorobenzamide (4a)

Appearance: Off-white colored solid; M.P- 171.5°C; Yield- 56%; LCMS purity 98.9%; ^{1}H NMR: δ 7.01 (1H, ddd, J = 8.4, 1.7, 1.5 Hz), 7.16 (1H, dd, J = 7.9, 4.7 Hz), 7.31 (1H, ddd, J = 8.4, 8.2, 0.5 Hz), 7.50–7.59 (2H, 7.54 (ddd, J = 8.2, 1.7, 1.5 Hz), 7.56 (d, J = 8.6 Hz)), 7.72 (1H, td, J = 1.7, 0.5 Hz), 7.78 (1H, d, J = 8.6 Hz), 8.03 (1H, dd, J = 7.9, 1.9 Hz), 8.65 (1H, dd, J = 4.7, 1.9 Hz); ^{13}C NMR: δ 107.3 (1C, s), 114.5 (1C, s), 115.0 (1C, s), 120.6 (1C, s), 121.3 (1C, s), 127.0 (1C, s), 127.8 (1C, s), 130.2 (1C, s), 135.9 (1C, s), 137.3 (1C, s), 144.9 (1C, s), 149.1 (1C, s), 151.5 (1C, s), 161.2 (1C, s), 163.8 (1C, s), 164.8 (1C, s).

4.4.2.2 N-[2-(5-aminothiophen-2-yl) pyridin-3-yl]-4-chlorobenzamide (4b)

Appearance: Pale colored solid; M.P- 161.2°C; Yield- 63%; LCMS purity 98.5%; ^{1}H NMR: δ 7.16 (1H, dd, J = 7.9, 4.7 Hz), 7.42 (2H, ddd, J = 8.1, 1.6, 0.5 Hz), 7.56 (1H, d, J = 8.6 Hz), 7.72–7.81 (3H, 7.75 (ddd, J = 8.1, 1.5, 0.5 Hz), 7.78 (d, J = 8.6 Hz)), 8.03 (1H, dd, J = 7.9, 1.9 Hz), 8.65 (1H, dd, J = 4.7, 1.9 Hz); ^{13}C NMR: δ 107.3 (1C, s), 120.6 (1C, s), 121.3 (1C, s), 127.0 (1C, s), 128.3 (2C, s), 128.7 (2C, s), 133.6 (1C, s), 133.7 (1C, s), 135.9 (1C, s), 144.9 (1C, s), 149.1 (1C, s), 151.5 (1C, s), 163.8 (1C, s), 164.8 (1C, s).

4.4.2.3 N-[2-(5-aminothiophen-2-yl) pyridin-3-yl]-3-chloro-2-fluorobenzamide (4c)

Appearance: Yellow colored solid; M.P- 187.6°C; Yield- 56%; LCMS purity 99.2%; ^{1}H NMR: δ 7.16 (1H, dd, J = 7.9, 4.7 Hz), 7.24 (1H, dd, J = 8.3, 1.6 Hz), 7.40 (1H, dd, J = 8.3, 0.5 Hz), 7.56 (1H, d, J = 8.6 Hz), 7.60 (1H, dd, J = 1.6, 0.5 Hz), 7.78 (1H, d, J = 8.6 Hz), 8.03 (1H, dd, J = 7.9, 1.9 Hz), 8.65 (1H, dd, J = 4.7, 1.9 Hz); ^{13}C NMR: δ 107.3 (1C, s), 115.1 (1C, s), 120.6 (1C, s), 121.1

(1C, s), 121.3 (1C, s), 126.3 (1C, s), 127.0 (1C, s), 128.3 (1C, s), 135.9 (1C, s), 137.3 (1C, s), 144.9 (1C, s), 149.1 (1C, s), 151.5 (1C, s), 156.3 (1C, s), 163.8 (1C, s), 164.8 (1C, s).

4.4.2.4 N-[2-(5-aminothiophen-2-yl)pyridin-3-yl][1,1′-biphenyl]-4-carboxamide (4d)

Appearance: Off-white colored solid; M.P- 154.6°C; Yield- 76%; LCMS purity 98.5%; ^{1}H NMR: δ 7.16 (1H, dd, J = 7.9, 4.7 Hz), 7.29–7.50 (5H, 7.32 (ddd, J = 9.0, 1.5, 0.5 Hz), 7.36 (tt, J = 7.5, 1.6 Hz), 7.45 (dddd, J = 7.9, 7.5, 1.7, 0.4 Hz)), 7.54–7.61 (3H, 7.58 (dddd, J = 7.9, 1.6, 1.2, 0.4 Hz), 7.56 (d, J = 8.6 Hz)), 7.66 (2H, ddd, J = 9.0, 1.9, 0.5 Hz), 7.78 (1H, d, J = 8.6 Hz), 8.04 (1H, dd, J = 7.9, 1.9 Hz), 8.66 (1H, dd, J = 4.7, 1.9 Hz); ^{13}C NMR: δ 107.3 (1C, s), 120.6 (1C, s), 121.3 (1C, s), 127.0 (1C, s), 127.2 (2C, s), 127.7–127.8 (3C, 127.7 (s), 127.8 (s)), 128.0 (2C, s), 128.4 (2C, s), 133.6 (1C, s), 135.9 (1C, s), 138.9–139.0 (2C, 138.9 (s), 138.9 (s)), 144.9 (1C, s), 149.1 (1C, s), 151.5 (1C, s), 163.8 (1C, s), 164.8 (1C, s).

4.4.2.5 N-[2-(5-aminothiophen-2-yl) pyridin-3-yl]-4-hydroxybenzamide (4e)

Appearance: Pale yellowish white colored solid; M.P- 194.6°C; Yield- 29%; LCMS purity 98.2%; ^{1}H NMR: δ 6.67 (2H, ddd, J = 8.8, 2.6, 0.6 Hz), 7.16 (1H, dd, J = 7.9, 4.7 Hz), 7.28 (2H, ddd, J = 8.8, 1.7, 0.6 Hz), 7.56 (1H, d, J = 8.6 Hz), 7.78 (1H, d, J = 8.6 Hz), 8.03 (1H, dd, J = 7.9, 1.9 Hz), 8.65 (1H, dd, J = 4.7, 1.9 Hz); ^{13}C NMR: δ 107.3 (1C, s), 115.7 (2C, s), 120.6 (1C, s), 121.3 (1C, s), 127.0 (1C, s), 129.6 (2C, s), 133.6 (1C, s), 135.9 (1C, s), 144.9 (1C, s), 149.1 (1C, s), 151.5 (1C, s), 157.4 (1C, s), 163.8 (1C, s), 164.8 (1C, s).

4.4.2.6 4-amino-N-[2-(5-aminothiophen-2-yl)pyridin-3-yl]benzamide (4f)

Appearance: White colored solid; M.P- 184.6°C; Yield- 82%; LCMS purity 98.6%; ^{1}H NMR: δ 6.70 (2H, ddd, J = 8.7, 1.8, 0.5 Hz), 7.16 (1H, dd, J = 7.9, 4.7 Hz), 7.43 (2H, ddd, J = 8.7, 1.4, 0.5 Hz), 7.56 (1H, d, J = 8.6 Hz), 7.78 (1H, d, J = 8.6 Hz), 8.03 (1H, dd, J = 7.9, 1.9 Hz), 8.65 (1H, dd, J = 4.7, 1.9 Hz); ^{13}C NMR: δ 107.3 (1C, s), 114.3 (2C, s), 120.6 (1C, s), 121.3 (1C, s), 127.0 (1C, s), 129.0 (2C, s), 133.6 (1C, s), 135.9 (1C, s), 144.9 (1C, s), 148.4 (1C, s), 149.1 (1C, s), 151.5 (1C, s), 163.8 (1C, s), 164.8 (1C, s).

4.4.2.7 N-[2-(5-aminothiophen-2-yl) pyridin-3-yl]-3, 4-dihydroxybenzamide (4g)

Appearance: Off-white colored solid; M.P- 168.2°C; Yield- 84%; LCMS purity 97.2%; ^{1}H NMR: δ 6.75 (1H, dd, J = 8.7, 0.5 Hz), 6.84 (1H, dd, J = 2.5, 0.5 Hz), 7.16 (1H, dd, J = 7.9, 4.7 Hz), 7.29 (1H, dd, J = 8.7, 2.5 Hz), 7.56 (1H, d, J = 8.6 Hz), 7.78 (1H, d, J = 8.6 Hz), 8.03 (1H, dd, J = 7.9, 1.9 Hz), 8.65 (1H, dd, J = 4.7, 1.9 Hz); ^{13}C NMR: δ 107.3 (1C, s), 110.6 (1C, s), 115.8 (1C, s), 120.6 (1C, s), 121.3 (1C, s), 127.0 (1C, s), 129.6 (1C, s), 134.1 (1C, s), 135.9 (1C, s), 144.9 (1C, s), 145.8 (1C, s), 146.3 (1C, s), 149.1 (1C, s), 151.5 (1C, s), 163.8 (1C, s), 164.8 (1C, s).

4.4.2.8 3-amino-N-[2-(5-aminothiophen-2-yl) pyridin-3-yl]-4-hydroxybenzamide (4f)

Appearance: White colored solid; M.P- 138.9°C; Yield- 74%; LCMS purity 98.8%; ^{1}H NMR: δ 6.75 (1H, dd, J = 8.6, 2.7 Hz), 6.83 (1H, dd, J = 2.7, 0.4 Hz), 7.10–7.20 (2H, 7.16 (dd, J = 7.9, 4.7 Hz), 7.13 (dd, J = 8.6, 0.4 Hz)), 7.56 (1H, d, J = 8.6 Hz), 7.78 (1H, d, J = 8.6 Hz), 8.03 (1H, dd, J = 7.9, 1.9 Hz), 8.65 (1H, dd, J = 4.7, 1.9 Hz); ^{13}C NMR: δ 107.3 (1C, s), 113.0 (1C, s), 117.6

(1C, s), 120.6 (1C, s), 121.3 (1C, s), 127.0 (1C, s), 127.5 (1C, s), 129.6 (1C, s), 135.9 (1C, s), 136.1 (1C, s), 144.0 (1C, s), 144.9 (1C, s), 149.1 (1C, s), 151.5 (1C, s), 163.8 (1C, s), 164.8 (1C, s).

4.4.3 Substituted N-[5-(3-benzamidopyridin-2-yl)thiophen-2-yl]benzamide (Compound 5 (a–h))

4.4.3.1 *3-fluoro-N-{5-[3-(3-fluorobenzamido) pyridin-2-yl]thiophen-2-yl}benzamide (5a)*

Appearance: Off-white colored solid; M.P- 196.9°C; Yield- 38%; LCMS purity 98.9%; ^{1}H NMR: δ 6.99–7.08 (2H, 7.04 (ddd, J = 8.3, 7.7, 1.4 Hz), 7.03 (ddd, J = 8.3, 1.6, 0.5 Hz), 7.19–7.36 (2H, 7.31 (ddd, J = 8.1, 7.5, 1.3 Hz), 7.24 (ddd, J = 7.9, 7.7, 1.6 Hz)), 7.43–7.62 (4H, 7.46 (dd, J = 8.1, 4.7 Hz), 7.54 (ddd, J = 8.3, 7.5, 1.5 Hz), 7.59 (d, J = 8.8 Hz), 7.48 (ddd, J = 8.3, 1.3, 0.5 Hz)), 7.83–7.92 (2H, 7.89 (ddd, J = 8.1, 1.5, 0.5 Hz), 7.86 (d, J = 8.8 Hz)), 8.09 (1H, ddd, J = 7.9, 1.4, 0.5 Hz), 8.30 (1H, dd, J = 8.1, 1.9 Hz), 8.66 (1H, dd, J = 4.7, 1.9 Hz); ^{13}C NMR: δ 111.1 (1C, s), 114.5 (1C, s), 115.0 (1C, s), 115.4 (2C, s), 120.6 (1C, s), 121.3 (1C, s), 127.0 (1C, s), 127.8 (1C, s), 129.4 (2C, s), 130.2 (1C, s), 133.6 (1C, s), 135.9 (1C, s), 137.3 (1C, s), 144.9 (1C, s), 145.5 (1C, s), 149.1 (1C, s), 151.5 (1C, s), 161.2 (1C, s), 162.5 (1C, s), 164.8 (1C, s), 165.7 (1C, s).

4.4.3.2 *3-chloro-N-{5-[3-(3-chlorobenzamido) pyridin-2-yl] thiophen-2-yl}benzamide (5b)*

Appearance: Pale yellowish red colored solid; M.P- 185.9°C; Yield- 75%; LCMS purity 98.9%; ^{1}H NMR: δ 7.17–7.33 (2H, 7.22 (ddd, J = 8.0, 7.7, 1.4 Hz), 7.29 (ddd, J = 8.0, 7.7, 1.6 Hz)), 7.35–7.62 (6H, 7.46 (dd, J = 8.1, 4.7 Hz), 7.46 (ddd, J = 8.0, 1.6, 0.5 Hz), 7.52 (ddd, J = 8.4, 7.6, 1.4 Hz), 7.60 (d, J = 8.8 Hz), 7.53 (ddd, J = 8.4, 1.5, 0.5 Hz), 7.40 (ddd, J = 8.1, 7.6, 1.5 Hz)), 7.78 (1H, ddd, J = 8.0, 1.4, 0.5 Hz), 7.86 (1H, d, J = 8.8 Hz), 7.94 (1H, ddd, J = 8.1, 1.4, 0.5 Hz), 8.30 (1H, dd, J = 8.1, 1.9 Hz), 8.66 (1H, dd, J = 4.7, 1.9 Hz); ^{13}C NMR: δ 111.1 (1C, s), 120.6 (1C, s), 121.3 (1C, s), 126.8 (1C, s), 127.0–127.0 (2C, 127.0 (s), 127.0 (s)), 127.8 (1C, s), 128.3 (2C, s), 128.6–128.8 (3C, 128.7 (s), 128.7 (s)), 130.4 (1C, s), 133.6 (1C, s), 133.7 (1C, s), 134.1 (1C, s), 135.9 (1C, s), 144.9 (1C, s), 145.5 (1C, s), 149.1 (1C, s), 151.5 (1C, s), 164.8 (1C, s), 165.7 (1C, s).

4.4.3.3 *4-chloro-N-{5-[3-(4-chloro-3-fluorobenzamido)pyridin-2-yl]thiophen-2-yl}-3-fluorobenzamide (5c)*

Appearance: Reddish colored solid; M.P- 168.7°C; Yield- 71%; LCMS purity 97.9%; ^{1}H NMR: δ 6.82 (1H, dd, J = 8.4, 1.7 Hz), 7.22–7.39 (3H, 7.34 (t, J = 8.3 Hz), 7.30 (dd, J = 7.6, 1.4 Hz), 7.25 (dd, J = 8.3, 1.7 Hz)), 7.43–7.57 (2H, 7.46 (dd, J = 8.1, 4.7 Hz), 7.52 (dd, J = 7.8, 7.6 Hz)), 7.60 (1H, d, J = 8.8 Hz), 7.86 (1H, d, J = 8.8 Hz), 7.95 (1H, dd, J = 7.8, 1.4 Hz), 8.30 (1H, dd, J = 8.1, 1.9 Hz), 8.66 (1H, dd, J = 4.7, 1.9 Hz); ^{13}C NMR: δ 111.1 (1C, s), 115.1 (1C, s), 120.6 (1C, s), 121.1 (1C, s), 121.3 (1C, s), 121.6 (1C, s), 122.3 (1C, s), 124.6 (1C, s), 126.3 (1C, s), 127.0 (1C, s), 128.3 (1C, s), 128.7 (1C, s), 131.8 (1C, s), 135.9 (1C, s), 137.3 (1C, s), 144.9 (1C, s), 145.5 (1C, s), 149.1 (1C, s), 151.5 (1C, s), 156.3 (1C, s), 160.1 (1C, s), 164.8 (1C, s), 165.7 (1C, s).

4.4.3.4 *4-diphenyl-N-{5-[3-(4-chloro-3-fluorobenzamido)pyridin-2-yl]thiophen-2-yl}-3-fluorobenzamide (5d)*

Appearance: Red colored solid; M.P- 198.7°C; Yield- 82%; LCMS purity 97.9%; ^{1}H NMR: δ 7.29–7.52 (7H, 7.46 (dd, J = 8.1, 4.7 Hz), 7.49 (tdd, J = 7.2, 1.7, 1.2 Hz), 7.45 (dddd, J = 7.9, 7.5,

1.7, 0.4 Hz), 7.36 (tt, *J* = 7.5, 1.6 Hz), 7.32 (ddd, *J* = 9.0, 1.5, 0.5 Hz)), 7.44 (2H, dddd, *J* = 8.1, 7.2, 1.5, 0.5 Hz), 7.54–7.62 (3H, 7.60 (d, *J* = 8.8 Hz), 7.58 (dddd, *J* = 7.9, 1.6, 1.2, 0.4 Hz)), 7.63–7.82 (8H, 7.70 (ddd, *J* = 8.7, 1.7, 0.4 Hz), 7.74 (dddd, *J* = 8.1, 1.5, 1.2, 0.5 Hz), 7.79 (ddd, *J* = 8.7, 1.8, 0.5 Hz), 7.66 (ddd, *J* = 9.0, 1.9, 0.5 Hz)), 7.86 (1H, d, *J* = 8.8 Hz), 8.31 (1H, dd, *J* = 8.1, 1.9 Hz), 8.66 (1H, dd, *J* = 4.7, 1.9 Hz); ^{13}C NMR: δ 111.1 (1C, s), 120.6 (1C, s), 121.3 (1C, s), 126.9–127.0 (2C, 126.9 (s), 127.0 (s)), 127.4–127.5 (2C, 127.4 (s), 127.5 (s)), 127.7–127.9 (9C, 127.7 (s), 127.7 (s), 127.7 (s), 127.7 (s), 127.8 (s), 127.8 (s), 127.8 (s)), 128.0 (1C, s), 128.3–128.5 (6C, 128.4 (s), 128.4 (s), 128.4 (s), 128.4 (s)), 128.4 (s), 128.4 (s)), 134.0–134.3 (2C, 134.1 (s), 134.2 (s)), 135.1 (1C, s), 135.9 (1C, s), 138.9 (1C, s), 139.2 (1C, s), 144.9 (1C, s), 145.5 (1C, s), 149.1 (1C, s), 151.5 (1C, s), 164.8 (1C, s), 165.7 (1C, s).

4.4.3.5 *4-hydroxy-N-{5-[3-(4-hydroxybenzamido)pyridin-2-yl]thiophen-2-yl}benzamide (5e)*

Appearance: Red colored solid; M.P- 174.9°C; Yield- 76%; LCMS purity 99.2%; ^{1}H NMR: δ 6.67 (2H, ddd, *J* = 8.8, 2.6, 0.6 Hz), 6.88 (2H, ddd, *J* = 8.6, 1.2, 0.4 Hz), 7.28 (2H, ddd, *J* = 8.8, 1.7, 0.6 Hz), 7.46 (1H, dd, *J* = 8.1, 4.7 Hz), 7.59 (1H, d, *J* = 8.8 Hz), 7.86 (1H, d, *J* = 8.8 Hz), 8.04 (2H, ddd, *J* = 8.6, 1.8, 0.4 Hz), 8.30 (1H, dd, *J* = 8.1, 1.9 Hz), 8.66 (1H, dd, *J* = 4.7, 1.9 Hz); ^{13}C NMR: δ 111.1 (1C, s), 114.5–114.6 (2C, 114.5 (s), 114.6 (s), 115.1 (1C, s), 116.8 (1C, s), 120.6 (1C, s), 121.3 (1C, s), 127.0 (1C, s), 127.8 (1C, s), 128.1 (1C, s), 128.4 (1C, s), 129.4 (1C, s), 129.6 (1C, s), 134.1 (1C, s), 135.9 (1C, s), 144.9 (1C, s), 145.5 (1C, s), 149.1 (1C, s), 151.5 (1C, s), 154.4 (1C, s), 159.8 (1C, s), 164.8 (1C, s), 165.7 (1C, s).

4.4.3.6 *4-amino-N-{5-[3-(4-aminobenzamido)pyridin-2-yl]thiophen-2-yl}benzamide (5f)*

Appearance: Red colored solid; M.P- 185.9°C; Yield- 82%; LCMS purity 98.8%; ^{1}H NMR: δ 6.70 (2H, ddd, *J* = 8.7, 1.8, 0.5 Hz), 6.87 (2H, ddd, *J* = 8.6, 1.1, 0.4 Hz), 7.40–7.51 (5H, 7.48 (ddd, *J* = 8.6, 1.7, 0.4 Hz), 7.43 (ddd, *J* = 8.7, 1.4, 0.5 Hz), 7.46 (dd, *J* = 8.1, 4.7 Hz)), 7.59 (1H, d, *J* = 8.8 Hz), 7.86 (1H, d, *J* = 8.8 Hz), 8.30 (1H, dd, *J* = 8.1, 1.9 Hz), 8.66 (1H, dd, *J* = 4.7, 1.9 Hz); ^{13}C NMR: δ 111.1 (1C, s), 113.0–113.0 (2C, 113.0 (s), 113.0 (s)), 113.5–113.6 (2C, 113.6 (s), 113.6 (s)), 120.6 (1C, s), 121.3 (1C, s), 127.0 (1C, s), 127.4–127.5 (2C, 127.5 (s), 127.5 (s)), 127.8–127.9 (2C, 127.8 (s), 127.8 (s)), 128.9–129.1 (2C, 129.0 (s), 129.0 (s)), 135.9 (1C, s), 144.9 (1C, s), 145.5 (1C, s), 146.5–146.7 (2C, 146.6 (s), 146.6 (s)), 149.1 (1C, s), 151.5 (1C, s), 164.8 (1C, s), 165.7 (1C, s).

4.4.3.7 *N-{5-[3-(3,4-dihydroxybenzamido)pyridin-2-yl]thiophen-2-yl}-3,4dihydroxybenzamide (5g)*

Appearance: Red colored solid; M.P- 198.9°C; Yield- 69%; LCMS purity 98.9%; ^{1}H NMR: δ 6.75 (1H, dd, *J* = 8.6, 2.7 Hz), 6.83 (1H, dd, *J* = 2.7, 0.4 Hz), 7.10–7.16 (2H, 7.13 (dd, *J* = 8.6, 0.4 Hz), 7.13 (dd, *J* = 8.8, 0.5 Hz)), 7.43–7.52 (2H, 7.51 (dd, *J* = 1.7, 0.5 Hz), 7.46 (dd, *J* = 8.1, 4.7 Hz)), 7.59 (1H, d, *J* = 8.8 Hz), 7.77 (1H, dd, *J* = 8.8, 1.7 Hz), 7.86 (1H, d, *J* = 8.8 Hz), 8.30 (1H, dd, *J* = 8.1, 1.9 Hz), 8.66 (1H, dd, *J* = 4.7, 1.9 Hz); ^{13}C NMR: δ 110.6 (1C, s), 111.1 (1C, s), 115.1 (1C, s), 115.8 (1C, s), 118.7 (1C, s), 120.6 (1C, s), 121.3 (1C, s), 127.0 (1C, s), 128.1 (1C, s), 129.5–129.7 (2C, 129.6 (s), 129.6 (s)), 134.1 (1C, s), 135.9 (1C, s), 144.9 (1C, s), 145.4–145.7 (2C, 145.5 (s), 145.6 (s)), 145.8 (1C, s), 146.3 (1C, s), 146.8 (1C, s), 149.1 (1C, s), 151.5 (1C, s), 164.8 (1C, s), 165.7 (1C, s).

4.4.3.8 N-{5-[3-(3, 4-dihydroxy diamine benzamido) pyridin-2-yl] thiophen-2-yl}-3,4dihydroxybenzamide (5h)

Appearance: Yellowish red colored solid; M.P- 201.8°C; Yield- 76%; LCMS purity 98.9%; ^{1}H NMR: δ 6.75 (1H, dd, J = 8.7, 0.5 Hz), 6.83–6.90 (2H, 6.84 (dd, J = 2.5, 0.5 Hz), 6.88 (dd, J = 8.4, 0.5 Hz)), 7.29 (1H, dd, J = 8.7, 2.5 Hz), 7.46 (1H, dd, J = 8.1, 4.7 Hz), 7.55–7.68 (3H, 7.65 (dd, J = 8.4, 1.7 Hz), 7.59 (d, J = 8.8 Hz), 7.56 (dd, J = 1.7, 0.5 Hz)), 7.86 (1H, d, J = 8.8 Hz), 8.30 (1H, dd, J = 8.1, 1.9 Hz), 8.66 (1H, dd, J = 4.7, 1.9 Hz); ^{13}C NMR: δ 111.1 (1C, s), 113.0 (1C, s), 115.1 (1C, s), 117.6 (1C, s), 120.4 (1C, s), 120.6 (1C, s), 121.3 (1C, s), 127.0 (1C, s), 127.5 (1C, s), 127.9 (1C, s), 129.5–129.7 (2C, 129.6 (s), 129.6 (s)), 135.9 (1C, s), 136.1 (1C, s), 141.3 (1C, s), 143.7 (1C, s), 144.0 (1C, s), 144.9 (1C, s), 145.5 (1C, s), 149.1 (1C, s), 151.5 (1C, s), 164.8 (1C, s), 165.7 (1C, s).

4.5 In vitro anticancer activity of the compounds (5 a–h)

4.5.1 MTT assay

IC_{50}—The inhibitory concentration of the compounds at 50%; docetaxel used standard drug used in the MTT assay.

4.6 Anticancer activity

4.6.1 Preparation of the cell culture and fixation

The entire cell lines were grown in RPMI 1640 medium containing 10% of fetalbovinserum. Added with antibiotics (100 units/mL penicillin, 100 μg/mL streptomycin, L-glutamine 0.03%, w/v and sodium bicarbonate 2.2%, w/v). All the cell lines were kept in a 100% humidified atmosphere incubated with 10% of CO_2 at 37 °C for 24 h. Before the addition of synthesized thiophene substituted pyridine amide derivatives (**4a–4h**) and (**5a–5h**) after 24 h, the aliquots of 10 μL of the test compounds (100, 250, 500, and maximum of 1000 μg/mL) of different dilutions were added to the appropriate microtiter plate containing the 90 μL of cells. Subcultures were performed with 0.05% trypsin and 0.02% EDTA in phosphate-buffered saline solution [16–19].

MTT assay: The MTT assay was carried out according to the method set out by Mossman. Cells lines were plated in 96-well microtiter plates at a density of 1×10^4 cells/well. After 24 h, the culture medium was replaced with 200 μL RPMI 1640 medium supplemented with 10% fetal bovine serum-containing varying concentrations. The final concentration of solvent was less than 0.1% in the cell culture medium. The culture solutions were removed and replaced with 90 μL of culture medium. Ten microliters of sterile-filtered MTT (Sigma, USA) solution (5 mg/mL) suspended in PBS (pH = 7.4) was added to each well to achieve a final concentration of 0.5 mg MTT/mL. The cells were then incubated at 37°C for 4 h. After the medium and unreacted dye was removed, 200 μL of DMSO was added to each well. The absorbance at 490 nm of the dissolved solution was measured using a Bio-Rad 680 microplate reader (BIORAD, USA). The relative cell viability (%) of the control wells containing cell culture medium without the tested compound was calculated by dividing the absorbance of treated cells by that of the controls in each experiment [20–25]. The IC_{50} was calculated by using statistical software [26].

TABLE 4.1 In vitro anticancer activity data (IC $_{50}$ (μM) of the synthesised thiophene substituted pyridine derivatives.

Compounds	*HeLa* (μM)	*MCF-7* (μM)	*HepG2* (μM)
4a	1.214 ± 0.003	1.221 ± 0.036	1.214 ± 0.073
4b	1.014 ± 0.028	1.117 ± 0.013	1.347 ± 0.024
4c	1.589 ± 0.053	1.489 ± 0.098	1.547 ± 0.052
4d	1.894 ± 0.074	1.745 ± 0.078	1.632 ± 0.056
4e	1.478 ± 0.086	1.821 ± 0.005	1.987 ± 0.087
4f	1.214 ± 0.091	1.214 ± 0.050	1.114 ± 0.009
4g	1.001 ± 0.014	1.006 ± 0.027	1.008 ± 0.039
4h	1.104 ± 0.023	1.108 ± 0.039	1.109 ± 0.043
5a	0.235 ± 0.045	0.315 ± 0.063	0.129 ± 0.098[a]
5b	0.248 ± 0.032	0.417 ± 0.034	0.145 ± 0.001[a]
5c	0.258 ± 0.093	0.247 ± 0.028	0.164 ± 0.013[a]
5d	0.114 ± 0.049	0.156 ± 0.052	0.247 ± 0.035
5e	0.258 ± 0.086	0.427 ± 0.022	0.258 ± 0.011
5f	0.236 ± 0.029	0.415 ± 0.047	0.237 ± 0.014
5g	0.269 ± 0.059	0.417 ± 0.031	0.200 ± 0.013
5h	0.285 ± 0.021	0.358 ± 0.058	0.213 ± 0.013
Docetaxel	0.2200 ± 0.013	0.2100 ± 0.008	0.2250 ± 0.086

[a] Highest value of Cytotoxic of compounds.

4.7 Biological properties

4.7.1 Cytotoxic activities

In attempts to find new anticancer compounds, we reported synthesis of various thiophene substituted pyridine derivatives. In vitro cytotoxicity assays of these compounds were performed on human liver cancer cell line (*HepG2*) by MTT assay method. The results are expressed as log IC_{50} (inhibitory concentration at 50%). Using sigma plot software average value of the two consecutive determinations was performed with the help of sigmoidal concentration curve fittings and the various values are represented in Tables 4.1–4.3. The present investigation involves a series of amide compounds containing thiophene substitutions and tested against *HeLa, MCF-7,* and *HepG2* cell lines. The standard reference drug docetaxel was used and the compounds **5a, 5b,** and **5c** containing active –F group, –Cl, and –Cl, F exhibited better cytotoxicity against *HepG2* cell lines with log IC_{50} values of 0.1298±0.098, 0.1456±0.001, and 0.1647±0.013 μM, respectively. The standard drug docetaxel exhibited in vitro cytotoxicity of 0.2250 μM against *HepG2* cell lines. Rest other derivatives showed good cytotoxicity against

TABLE 4.2 Cytotoxic assay of the in vitro synthesized compounds.

Compound number	IC $_{50}$ (μM) *HepG2*	μ	Log P
2	1.478 ± 0.073	1.64	1.22
3	1.346 ± 0.073	1.04	1.11
4a	1.214 ± 0.073	1.34	1.68
4b	1.347 ± 0.024	1.87	1.96
4c	1.547 ± 0.052	1.96	1.87
4d	1.632 ± 0.056	1.95	1.96
4e	1.987 ± 0.087	1.96	2.02
4f	1.114 ± 0.009	2.14	2.08
4g	1.008±0.039	2.19	2.12
4h	1.109 ± 0.043	2.64	2.38
5a	0.129 ± 0.098[a]	3.87	3.49[a]
5b	0.145 ± 0.001[a]	3.89	3.89[a]
5c	0.164 ± 0.013[a]	3.64	3.22[a]
5d	0.247 ± 0.035	2.67	2.52
5e	0.258 ± 0.011	2.12	2.42
5f	0.237 ± 0.014	2.14	2.02
5g	0.200 ± 0.013	2.19	2.00
5h	0.213 ± 0.013	2.29	2.05
Docetaxel	0.2250 ± 0.086	3.16	3.04

[a] Highest value of solubility.

tested cell lines as compared with docetaxel. The increase in the solubility of the diamide derivatives containing F, Cl, and F along with Cl also contributed to the increased in vitro cytotoxicity. The role of increase in the Log P values for the functional groups responsible for the anticancer activity and the effect of –OH, $-NH_2$, –OH NH_2 groups in the derivatives **5e, 5f, 5g,** and **5h** caused increased binding coefficient in the synthesized derivatives.

4.7.2 Molecular modeling and computational characteristics of the synthesized compounds

Crystal structure, molecular conformation, and energy levels of the thiophene substituted pyridine derivatives obtained by the semiempirical PM3 model using Hyper Chem 7.5 software. PM3 parameters are used to evaluate the most stable structure of the synthesized local environment of the compounds. Using PM3 parameters the geometrical parameters were calculated and the values of optimized dipole moment (μ), total energy (E_T), binding

TABLE 4.3 Cytotoxicity assay of the active in vitro compounds against *HepG2* cell line and reference standard drug.

Compound number	IC $_{50}$ (μm) (*HepG2*)	μ	Log P
5a	0.1298 ± 0.098	3.12	2.41
5b	0.1456 ± 0.001	2.89	2.05
5c	0.1647 ± 0.013	1.47	2.08
Docetaxel[a]	**0.2250** ± 0.001	**1.49**	**5.12**

[a] Reference standard drug.

TABLE 4.4 Energetic properties of the synthesized compounds as calculated from PM3 method.

Compound	Total energy (E_T)	Binding energy (E_B)	Electronic energy (E_E)	Dipole moment (μ)
(k cal/mole)	(k cal/mole)	(k cal/mole)		
4a	−71464.5	−3719.1	−421612.8	1.64
4b	−78452.5	−3124.1	−41568.8	1.08
4c	−65871.5	−2471.1	−356987.8	1.14
4d	−55874.5	−2894.1	−475896.8	1.28
4e	−69874.5	−3547.1	−38947.8	2.64
4f	−45789.5	−3496.1	−48569.8	2.19
4g	−58742.5	−6987.1	−658741.8	2.89
4h	−65874.5	−3696.1	−547821.8	2.84
5a	−91464.5[a]	−4719.1	−721612.8	3.64[a]
5b	−92586.5	−4897.1	−721628.8	3.54
5c	−93698.5	−4874.1	−721587.8	3.44
5d	−74587.5	−3587.1	−69874.8	2.28
5e	−84715.5	−3614.1	−65247.8	2.96
5f	−74125.5	−3547.1	−471528.8	2.87
5g	−58473.5	−3698.1	−569874.8	2.34
5h	−6325.5	−2578.1	−584711.8	2.45

[a] Highest binding energy and dipole moment of the compound.

energy (E_B), electronic energy (E_E) were tabulated in Table 4.4. From the PM3 parameters, the highest occupied energy levels of thiophene substituted pyridine derivatives (E_{HOMO}), lowest unoccupied energy levels (E_{LUMO}), and difference in the energy levels (ΔE) are evaluated. In the first step, all the geometrics of the synthesized compounds were calculated in the gaseous phase (0.01 kcal/mole/A°) and later the descriptions of the PM3 parameters are optimized.

Mulliken electronegativity (χ), chemical potential (Pi), global softness (S), global harness (η), and global electrophilicity (Ψ) [27–30] can be calculated using the below equations

$$\chi = -\frac{1}{2}(E_{\mathrm{LUMO}} - E_{\mathrm{HOMO}}) \tag{4.1}$$

$$\mathrm{Pi} = -\chi \tag{4.2}$$

$$\eta = \frac{1}{2}(E_{\mathrm{LUMO}} - E_{\mathrm{HOMO}}) \tag{4.3}$$

$$S = \frac{1}{2\eta} \tag{4.4}$$

$$\Psi = \frac{Pi^2}{2\eta} \tag{4.5}$$

Recently, quantum-chemical calculation methods have become available to provide a powerful approach for crystal structure prediction [31,32]. Molecular stability and reactivity depend on the hardness and softness properties. Soft molecules are reactive since they can easily transfer electrons during the reaction and hard molecules are unable to facilitate the electrons. The rate of the reaction of soft molecules increases and the electron transfer or rearrangement inside the molecular structure also increases. When the system acquires an additional electronic charge from the reaction environment, the reactivity index also increases. The direction of the electron transfer is measured from the electronic chemical potential (Pi) of the molecule. Therefore, LUMO and HOMO are the quantum chemical parameters to measure the electronic chemical potential. Frontier molecular orbitals (FMO) are responsible for the strong interaction of the molecules during the course of the reaction. However, the HOMO orbital is involved in the transfer of electrons since it has the highest molecular orbital and FMO responsible for molecular stability. LUMO orbital accepts the transferred electrons. The energies of the HOMO and LUMO are negative in the present study which indicates the stability of the thiophene substituted pyridine molecules.

Anticancer activity (Table 4.1)**:** The synthesized novel thiophene substituted pyridine derivatives **4a–4h** and **5a–5h** were tested for in vitro anticancer activity [33–35] on three leukemic cell lines namely (1) human cervical carcinoma cell line (*HeLa*), (2) human liver carcinoma cell lines (*HepG2*), and (3) human breast carcinoma cell lines (*MCF7*). The results were expressed in the form of IC_{50} (Inhibitory concentration of the compound at 50%). Most of the compounds in this series have showed better cytotoxic effect on these three carcinoma cell lines. Compounds **5(a), 5(b),** and **5(c)** whose IC_{50} as compared to the standard docetaxel. The IC_{50} values of the **5a, 5b,** and **5c** derivatives containing active –F (Fluoro), –Cl (Chloro), –F, Cl (Fluoro and Chloro) groups were found to be 0.1298 ± 0.098, 0.1456 ± 0.001, and 0.1647 ± 0.013μM, respectively, against *HepG2* cell lines. Rest other compounds showed moderate activity against all the treated cell lines at 24 h. Thiophene substituted pyridine derivatives are promising candidates for anticancer treatment.

4.8 Conclusion

In the present research work, novel series of the thiophene substituted pyridine amide derivatives were synthesized and tested for in vitro anticancer activity against *HeLa, HepG2* and *MCF-7* cell lines using MTT assay method. The novel series of thiophene substituted pyridine derivatives were characterized by ^{1}H-NMR, ^{13}C-NMR, LCMS spectroscopic analyses. The docetaxel was used as a reference standard drug and log IC50 of the compounds **5a, 5b,** and **5c** showed remarkable properties against *HepG2* cell lines with the IC_{50} values are 0.1298 $\pm$ 0.098, 0.1456 $\pm$ 0.001, and 0.1647 $\pm$ 0.013μM, respectively. The in vivo anticancer activity (MTT assay method) was performed by standard methodologies. As regards the relationships between the structure of the heterocyclic scaffold and the detected biological activities, it showed varied biological activity. Probably in this case the presence of different substituents caused a certain change of activity. It is interesting to note that in general the compounds having the F, Cl groups at ortho and para position of the phenyl ring has exhibited potent activity compared to other substitutions. In conclusion, compounds **5a, 5b,** and **5c** emerged as lead compounds in the liver anticancer treatment. To support the solid-state structure, molecular modeling optimization and electronic parameters have been calculated using PM3 method. SAR studies suggested an increase in the Log P (solubility) values which is in good agreement with the calculated dipole moment and total energy possessed by the synthesized molecules. In general, it can be suggested that molecules that meet lipophilicity, and low dipole moment exhibited good anticancer activities. The above results revealed that substitutions like F, Cl which have greater solubility and are partly involved in the electron-donating, and may act as useful leads for biologically active drug development in the future.

Acknowledgments

All the authors are thankful to Mangalore University for analytical characterization of the samples.

Authors Contributions

Dr. Vinayak Adimule contributed for the SAR studies, synthesis of the compounds, and wrote the paper. Dr. Basappa C. Yallur contributed for the preparation and outline of the manuscript, characterization of the synthesized compounds, and partly by analysis of the compounds. Dr. Sheetal Batakurki involved in the in vitro anticancer activity of the synthesized compounds, assimilation of the data, and partly by manuscript preparation.

Conflict of Interest

All the authors declare that they do not have any conflict of interest.

Data Availability

All the data have been presented in the manuscript. Analytical data can be obtained from the corresponding author on request.

Funding

There is no funding from any institution or any organization.

References

[1] F. Balkwill, A. Mantovani., Inflammation and cancer: back to Virchow? Lancet 357 (9255) (2001) 539–545.

[2] J.K. Kundu, Y.J. Surh, Inflammation: gearing the journey to cancer, Mutat. Res./Rev. Mutat. Res. 659 (1-2) (2008) 15–30, doi:10.1016/j.mrrev.2008.03.002.

[3] E.R. Rayburn, S.J. Ezell, R. Zhang, Anti-inflammatory agents for cancer therapy, Mol. Cell. Pharm. 1 (1) (2009) 29–43.
[4] S.E. Abbas, F.M. Awadallah, N.A. Ibrahin, E.G. Said, G.M. Kamel, New quinazolinone–pyrimidine hybrids: synthesis, anti-inflammatory, and ulcerogenicity studies, Eur. J. Med. Chem. 53 (2012) 141–149, doi:10.1016/j.ejmech.2012.03.050.
[5] F.E. Silverstein, G. Faich, J.L. Goldstein, L.S. Simon, T. Pincus, A. Whelton, R. Makuch, G. Eisen, 5.NM Agrawal, W.F. Stenson, A.M. Burr, Gastrointestinal toxicity with celecoxib vs nonsteroidal anti-inflammatory drugs for osteoarthritis and rheumatoid arthritis: the CLASS study: a randomized controlled trial, JAMA 284 (10) (2000) 1247–1255, doi:10.1001/jama.284.10.1247.
[6] A. Gangjee, Y. Zhu, SF. Queener, 6-Substituted 2, 4-diaminopyrido [3, 2-d] pyrimidine analogues of piritrexim as inhibitors of dihydrofolate reductase from rat liver, pneumocystis carinii, and toxoplasma gondii and as antitumor agents, J. Med. Chem. 41 (23) (1998) 4533–4541, doi:10.1021/jm980206z.
[7] J.L. Greene Jr, A.M. Williams, J.A. Montgomery, Vitamin B6 Analogues. III. Some 5-Aminomethyl and 5-thiomethyl derivatives of pyridoxine and 4-desoxypyridoxine1, J. Med. Chem. 7 (1) (1964) 20–23.
[8] M.J. Gil, M.A. Manu, C. Arteaga, M. Migliaccio, I. Encio, A. Gonzalez, V. Martinez-Merino, Synthesis and cytotoxic activity of N-(2-pyridylsulfenyl) urea derivatives. A new class of potential antineoplastic agents, Bioorg. Med. Chem. Lett. 9 (16) (1999) 2321–2324.
[9] M. Sathish, J. Chetna, N. Hari Krishna, N. Shankaraiah, A. Alarifi, A. Kamal, Iron-mediated one-pot synthesis of 3, 5-diarylpyridines from β-nitrostyrenes, J. Org. Chem. 81 (5) (2016) 2159–2165.
[10] G. Semple, B.M. Andersson, V. Chhajlani, J. Georgsson, M.J. Johansson, Å. Rosenquist, L. Swanson, Synthesis and biological activity of kappa opioid receptor agonists. Part 2: Preparation of 3-aryl-2-pyridone analogues generated by solution-and solid-phase parallel synthesis methods, Bioorg. Med. Chem. Lett. 13 (6) (2003) 1141–1145.
[11] S. Bondock, T. Naser, YA. Ammar, Synthesis of some new 2-(3-pyridyl)-4, 5-disubstituted thiazoles as potent antimicrobial agents, Eur. J. Med. Chem. 62 (2013) 270–279.
[12] JA. Sordo, The configuration interaction expansion is the exact solution of the electronic Schrödinger equation, J. Chem. Educ. 83 (3) (2006) 480.
[13] Y.D. Scherson, S.J. Aboud, J. Wilcox, BJ. Cantwell, Surface structure and reactivity of rhodium oxide, J. Phys. Chem. C 115 (22) (2011) 11036–11044.
[14] JJ. Stewart, MOPAC Ver. 6.00 (QCPE# 455, VAX version), received as Ver. 6.01 (JCPE P049) for the UNIX-Sun SPARCstation version by Kazuhiro Nishina. (b) Stewart, JJPJ, Comput. Chem. 10 (1989) 209.
[15] E. Skau, Purification and physical properties of organic compounds. IX. Some binary freezing point diagrams and a study of their ideality, J. Phys. Chem. 39 (6) (2002) 761–768.
[16] D.A. Medvetz, K.M. Hindi, M.J. Panzner, A.J. Ditto, Y.H. Yun, WJ. Youngs, Anticancer activity of Ag (I) N-heterocyclic carbene complexes derived from 4, 5-dichloro-1H-imidazole, Met. Based Drugs (2008) 384010, doi:10.1155/2008/384010.
[17] P. Jänicke, C. Lennicke, A. Meister, B. Seliger, L.A. Wessjohann, GN. Kaluđerović, Fluorescent spherical mesoporous silica nanoparticles loaded with emodin: synthesis, cellular uptake and anticancer activity, Mater. Sci. Eng.: C 119 (2021) 111619.
[18] L. Ma, Y. Meng, C. Tu, X. Cao, H. Wang, Y. Li, S. Man, J. Zhou, M. Li, Z. Liu, Y. Su, A cardiac glycoside HTF-1 isolated from helleborus thibetanus franch displays potent in vitro anti-cancer activity via caspase-9, MAPK and PI3K-Akt-mTOR pathways, Eur. J. Med. Chem. 158 (2018) 743–752.
[19] T.S. Reddy, S.H. Privér, N. Mirzadeh, SK. Bhargava, Synthesis of gold (I) phosphine complexes containing the 2-BrC6F4PPh2 ligand: evaluation of anticancer activity in 2D and 3D spheroidal models of HeLa cancer cells, Eur. J. Med. Chem. 145 (2018) 291–301.
[20] R.H. Wang, J. Bai, J. Deng, C.J. Fang, X. Chen, TAT-modified gold nanoparticle carrier with enhanced anticancer activity and size effect on overcoming multidrug resistance, ACS Appl. Mater. Interfaces 9 (7) (2017) 5828–5837.
[21] S. Wang, X. Han, L. Zhang, Y. Zhang, H. Li, Y. Jiao, Whole peptidoglycan extracts from the lactobacillus paracasei subsp. Paracasei M5 strain exert anticancer activity in vitro, Biomed. Res. Int. (2018), doi:10.1155/2018/2871710.
[22] P. Tavolaro, S. Catalano, A. Tavolaro, Anticancer activity modulation of an innovative solid formulation of extra virgin olive oil by cultured zeolite scaffolds, Food Chem. Toxicol. 124 (2019) 139–150.
[23] E. Vidhya, S. Vijayakumar, S. Prathipkumar, PK. Praseetha, Green way biosynthesis: characterization, antimicrobial and anticancer activity of ZnO nanoparticles, Gene Reports 20 (2020) 100688.

[24] R. Kumar, A. Chauhan, BK. Kuanr, A robust in vitro anticancer activity via magnetic hyperthermia mediated by colloidally stabilized mesoporous silica encapsulated La $_{0.7}$ Sr0. $_3MnO_3$ core-shell structures, Colloids Surf. A 615 (2021) 126212.

[25] G. Bamias, I. Delladetsima, M. Perdiki, S.I. Siakavellas, D. Goukos, G.V. Papatheodoridis, G.L. Daikos, H. Gogas, Immunological characteristics of colitis associated with anti-CTLA-4 antibody therapy, Cancer Invest. 35 (7) (2017) 443–455.

[26] Y. Sun, J. Zhang, J. Zhou, Z. Huang, H. Hu, M. Qiao, X. Zhao, D. Chen, Synergistic effect of cucurbitacin B in combination with curcumin via enhancing apoptosis induction and reversing multidrug resistance in human hepatoma cells, Eur. J. Pharmacol. 768 (2015) 28–40.

[27] RG. Pearson, Absolute electronegativity and hardness: applications to organic chemistry, J. Org. Chem. 54 (6) (1989) 1423–1430.

[28] P. Geerlings, F. De Proft, W. Langenaeker, Conceptual density functional theory, Chem. Rev. 103 (5) (2003) 1793–1874.

[29] R.G. Parr, L.V. Szentpály, S. Liu, Electrophilicity Index, J. Am. Chem. Soc. 121 (9) (1999) 1922–1924, doi:10.1021/ja983494x.

[30] P.K. Chattaraj, S. Giri, Stability, reactivity, and aromaticity of compounds of a multivalent superatom, J. Phys. Chem. A 111 (43) (2007) 11116–11121.

[31] T. Soman, Quantum-chemical studies on hexaazaisowurtzitanes, J. Phys. Chem. A 114 (1) (2010) 498–503.

[32] N.N. Ji, Z.Q. Shi, R.G. Zhao, Z.B. Zheng, ZF. Li, Synthesis, crystal structure and quantum chemistry of a novel schiff base N-(2, 4-Dinitro-phenyl)-N′-(1-phenyl-ethylidene)-hydrazine, Bull. Korean Chem. Soc. 31 (4) (2010) 881–886.

[33] B.S. Chhikara, N.S. Jean, D. Mandal, A. Kumar, K. Parang, Fatty acyl amide derivatives of doxorubicin: synthesis and in vitro anticancer activities, Eur. J. Med. Chem. 46 (6) (2011) 2037–2042.

[34] M. Jung, N. Park, H.I. Moon, Y. Lee, W.Y. Chung, KK. Park, Synthesis and anticancer activity of novel amide derivatives of non-acetal deoxoartemisinin, Bioorg. Med. Chem. Lett. 19 (22) (2009) 6303–6306.

[35] H. Elamari, R. Slimi, G.G. Chabot, L. Quentin, D. Scherman, C. Girard, Synthesis and in vitro evaluation of potential anticancer activity of mono-and bis-1, 2, 3-triazole derivatives of bis-alkynes, Eur. J. Med. Chem. 60 (2013) 360–364.

CHAPTER

5

The role of pyridine derivatives on the treatment of some complex diseases: A review

Xolani Henry Makhoba

Department of Biochemistry and Microbiology, University of Fort Hare, Alice, South Africa

5.1 Introduction

The ever-increasing numbers of reported cases of malaria, cancer, and diabetes are a major concern in high-earning, middle and poor countries. For example, 229 million cases of malaria were reported in 2020 by the World Health Organization (WHO) [1]. The current coronavirus pandemic has made things even worse, as the focus has diverted to curb its spread and to urgently come up with an effective treatment. The growing number of reported cases of resistance to the currently available treatment of malaria in some parts of the world is also a major problem. There is no available effective treatment in the market for malaria. Therefore, there is an urgent need to develop alternative drugs for malaria. There are various types *Plasmodium species*, namely: *P. vivax, P. malariea, P.yoellii, P.berghei, P. knowlesi, P. ovale, P. chabaudi, P. gallinaceum*, and *P. falciparum. Plasmodium falciparum* is responsible for most cases of malaria [2]. With both pregnant women and children under the age of 5 are the most vulnerable to this disease because of their compromised immunity. *P. falciparum* survives between two hosts, namely the mosquito vector and humans. During its blood meal, the female *Anopheles* mosquito transfers the *P. falciparum* parasite to the human host. The life cycle of this parasite has been widely reported, thus most studies have reported different strategies to block its entry to the human host [3–5]. For example, a single dose has been the only treatment available for a number of years, but recent reports have proposed combination therapy. Artemisinin combination therapy is the most reported and promising approach for malarial management. However, even this approach is not effective in reducing or eliminating the growing numbers of malaria cases. Therefore, most scientists and scholars including pharmaceutical industries are working overtime to develop urgent treatment of malaria to save lives [6].

Recent Developments in the Synthesis and Applications of Pyridines.
DOI: https://doi.org/10.1016/B978-0-323-91221-1.00015-4

Another global health disease that affects and threatens human life is cancer, which falls under complex diseases. It is regarded as the second cause of death after cardiovascular disease worldwide. The uncontrolled growth of cells in the human body is one of the most challenging problems [7–9]. It results in the development of what is referred to as cancer. For example, statistics show that in 2018 as reported by WHO, at least 18.1 million people around the world had cancer, with 9.6 million people succumbing to it. These numbers are expected to double by 2040. The most affected will be the low and middle-income countries. Cancer is the cause of almost 30% of all hasty deaths from noncommunicable diseases among adults aged 30 to 69. Among all, the most frequently diagnosed cancer is lung (11.6% of all cases), followed by female breast (11.6%) and colorectal cancers (10.2%). Lung cancer is the foremost cause of death from cancer (18.4% of all deaths), followed by colorectal (9.2%) and stomach cancers (8.2%). The most usual cause of cancer is tobacco usage, which accounts for 25% of all cancer deaths around globally. Hence, cancer is a serious issue hampering human health [9–14]. Various new anticancer agents are being developed every now and then. Still, their toxicity profile has restricted their clinical use as anticancer agents. Therefore, further exploration of new chemotherapeutic agents with great efficacy and least adverse effects is critically essential for medicinal chemists.

Another complex disease that affects human life is diabetes, also known as the silent killer. For example, about 422 million people worldwide were reported to have diabetes, the majority living in low and middle-income countries, and 1.6 million deaths are directly attributed to it each year. Both the number of cases and the prevalence of diabetes have been steadily increasing over the past few decades [15]. All these growing numbers of reported cases of malaria, cancer, and diabetes and other complex diseases, suggests the need to come up with alternative and innovative treatment approaches to eliminate some of these complex diseases to prevent loss of life, both in developed and underdeveloped countries [16–23]. Pyridines have been suggested as the promising tools for drug delivery and the treatment of various complex diseases. Briefly, pyridines are an important entity in drug discovery field, they have been used in studies of various complex diseases such as malaria, cancer, and diabetes. They are regarded as one of the promising agents in pharmaceutical industry and research in general. In this review, we look at the past, present, and future perspectives of pyridine and its derivatives for the treatment of some complex diseases [24–29].

5.2 Pyridines in general

Pyridines are N-heterocyclic compounds that play a vital role in the field of science such as medicinal chemistry, drug development, and are found in several existing drugs that comprise heterocyclic rings (Fig. 5.1). The six-membered compound pyridine is used in many applications as it is an important solvent, reagent, and precursor in agrochemicals and pharmaceuticals [30]. The growing concern of drug resistance in current drugs available in the markets requires the development of novel chemotherapeutic agents for various complex diseases. Many pyridine scaffold derivatives play a major role in biological activities such as antimicrobial, antiviral, antioxidant, antidiabetic, anticancer, antimalarial, analgesic, anti-inflammatory activities, psychopharmacological antagonistic, antiamoebic agents, and antithrombotic activity [9]. Pyridines can be classified into different categories such as pyridine-containing

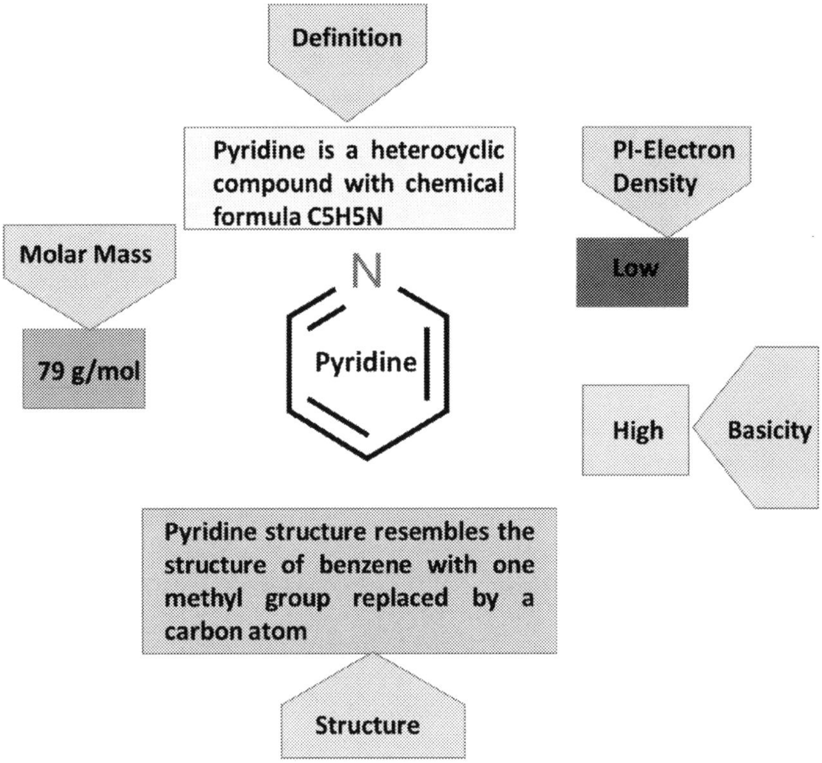

FIGURE 5.1 Summary of pyridine.

heterocycles and pyridine fused rings. Pyridine and pyridine-fused ring systems are ubiquitous in medicinal research and demonstrate such diverse pharmacological benefits as anticonvulsant therapies, treatment for fungal and bacterial infections as well as chemotherapy agents.

5.3 Synthesis of pyridines derivatives

Pyridine ring is regarded as the simplest heteroaromatic compound, yet it plays an important role as a scaffold in drug development. In addition, pyridines form one of the classes of heterocyclic compounds whose presence in various natural compounds and pharmaceuticals as well as their potent bioactivities have increased interest both in academia and pharmaceutical industry (Table 5.1). Certainly, pyridines have been studied for many decades, therefore, they have been used in many subdivisions of chemistry, such as catalysis, drug design, molecular recognition, and material science. It is important to note that most pyridine derivatives have an essential role namely: hypnotic and sedative, HIV antiviral, bone calcium regulator, cholesterol and triglyceride regulator, antidiabetic, antihistaminic,

TABLE 5.1 Some pyridine-containing compounds.

Drugs	Chemical structure	Role
Pyridoxal phosphate	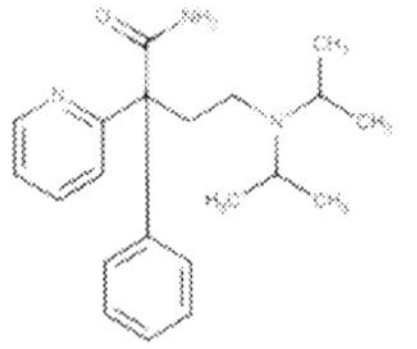	A vitamin available in many formulations to correct vitamin B6 deficiency
Pyridoxal	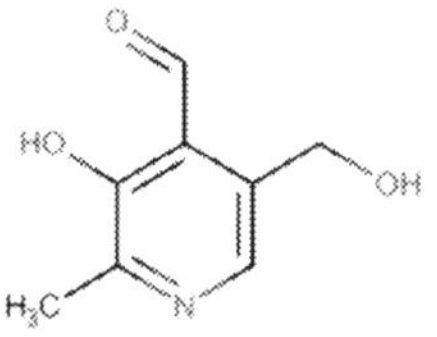	Pyridoxal is one of the natural forms available of vitamin B6; therefore, it is used for nutritional supplementation and for treating dietary shortage or imbalances
Pyridoxine	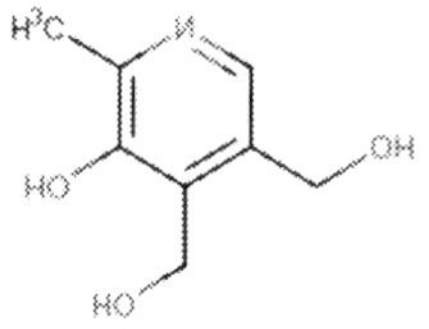	A vitamin used to correct vitamin B6 deficiency and to treat nausea during pregnancy
Nicotine	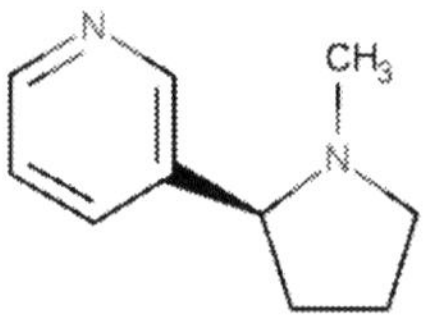	A stimulatory alkaloid found in tobacco products that is often used for the relief of nicotine withdrawal symptoms and as an aid to smoking cessation
Ticlopidine	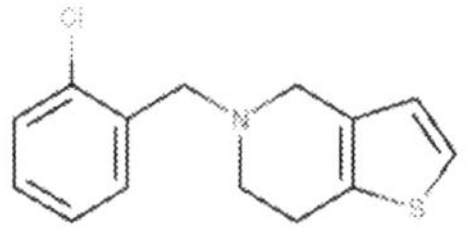	A platelet aggregation inhibitor used in the prevention of conditions associated with thrombi, such as stroke and transient isehemic attacks (TIA)
Pantoprazole	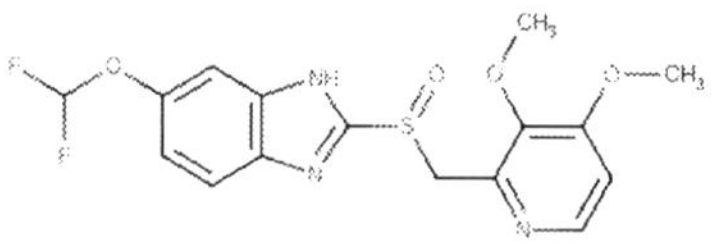	A proton pump inhibitor used to treat erosive esophagitis, gastric acid hypersecretion, and to promote healing of tissue damage caused by gastric acid
Indinavir	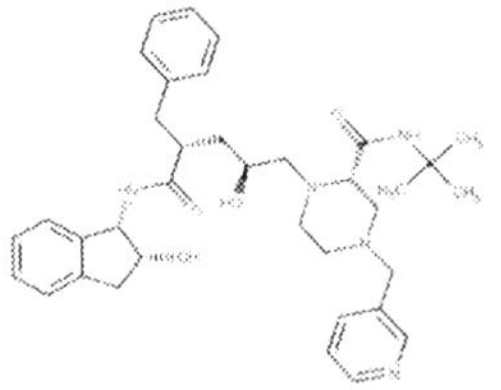	A protease inhibitor used to treat HIV infection

antiulcerate, antineoplastic, and anticancer activities [30–32]. This nitrogen ring compound is also found in various vital compounds such as vitamins. Pyridines are also a crucial ubiquitous redox system such as NADP/NADPH, alkaloids (nicotine), pyridine therefore play an important role in many biological activities. Utmost, pyridine-based alkaloid natural products are derivatives of nicotinic acid, also known as vitamin B3 and niacin. Furthermore, pyridines are the backbone during the preparation of conjugated polymers and functional materials which are used in light-emitting devices.

5.4 Life cycle of the malaria parasites

The *Plasmodium* parasite survives in two hosts, the mosquito vector and human host. The anopheles' female mosquito is found in muddy areas that have a temperature of approximately 22°C. When the mosquito takes a blood meal in the human, the parasites get transferred into the human host [33]. This marks the life cycle of the parasite. Inside the human host the parasite undergoes the developmental, proliferation, and growth phase, as shown in Fig. 5.2.

5.5 Pyridine containing drugs for malaria treatment

Malaria is one of the major causes of death worldwide especially in Sub-Saharan Africa, with the number of cases increasing in the past decades due to resistance that continues to be a growing concern in the current treatment. The malaria parasite that results in high morbidity and mortality is caused by *Plasmodium falciparum*. The *P. falciparum* parasite survives between the mosquito vector and human host. The flexibility of this parasite poses a major threat to human life [31]. As a result, there is an ongoing search for effective drugs against malaria. Table 5.2 gives a summary of some FDA-approved drugs for the treatment of malaria. However, there have been reported cases of drug resistance to the currently available drugs in the market in some parts of Africa, Asia and other areas with reported cases of malaria. This, therefore, requires the need for the development of new drugs that have a pyridine as a promising tool for malaria treatment [32].

5.6 Cancer

Cancer is a major cause of death. As a result, it causes severe economic burden throughout the world. There are many contributing factors to the development of cancer in humans. These include genetic, environmental, and lifestyle changes. It is reported that between 5% and 10% of many cancer-associated cases are due to genetic mutations inherited from a parent. Whereas the other 90%–95% are initiated by environmental and lifestyle factors [34,35]. The lifestyle factors often linked with cancer development include unhealthy diets (such as processed foods, red meat), cigarette smoking, environmental carcinogens, ultraviolet (UV) exposure, stress, obesity, and physical inactivity. Fig. 5.3 summarizes how a healthy cell develops into healthy tissues while some cells form or become cancer cells.

Doll and Peto in 1981 reported that diet alone contributes to ~30%–35% of cancer-related demises. Whereas, Bray and colleagues in 2012 estimated that the global burden of new cancer

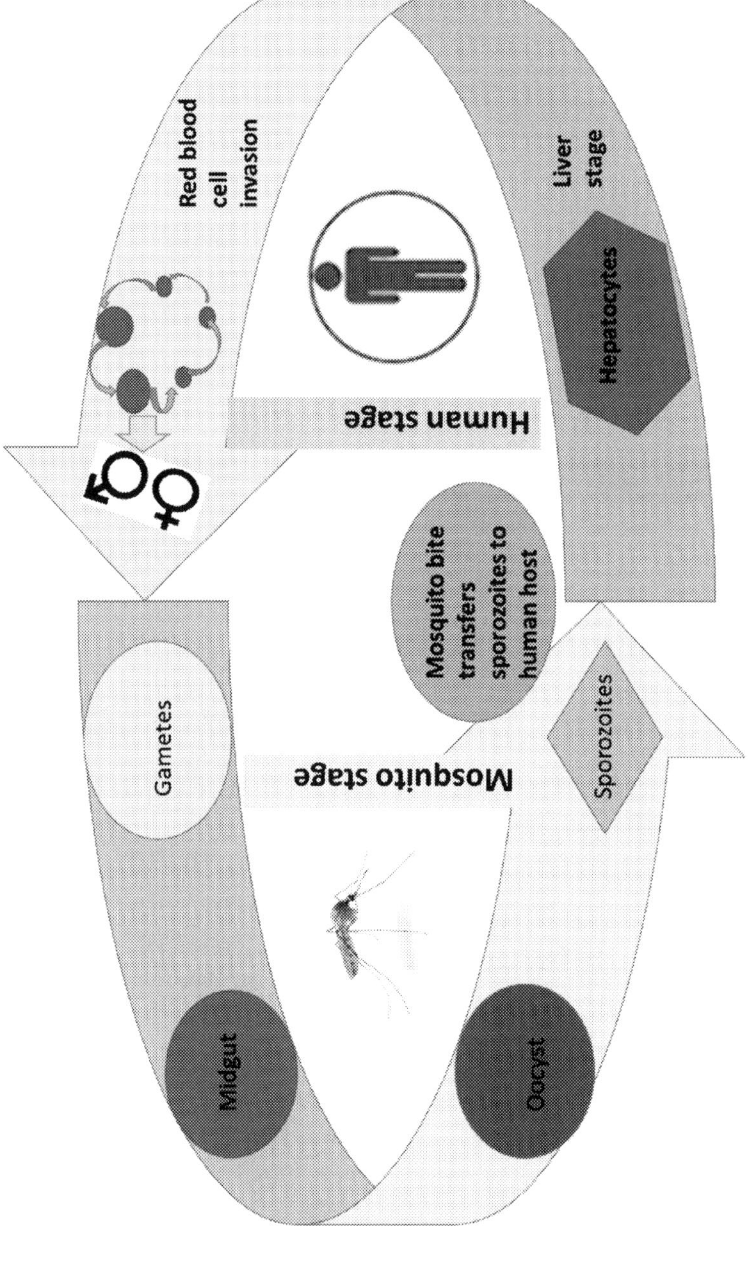

FIGURE 5.2 Summary of the parasite life cycle between mosquito and human host.

TABLE 5.2 Drugs containing six-ring structure tested against some plasmodium species.

Species name	Pyridine containing compound tested against some plasmodium species	Role
Berghei and falciparum	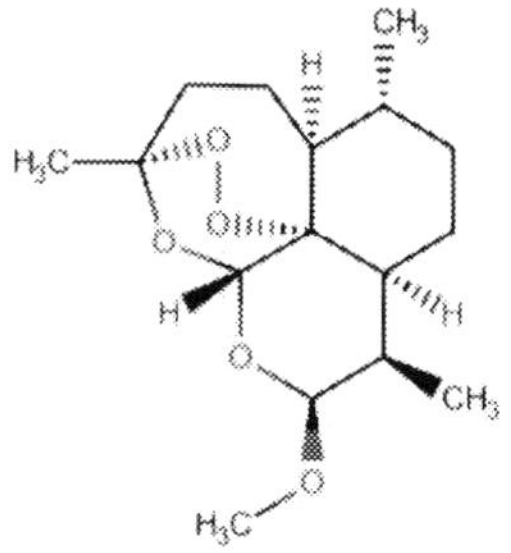	**Artemether** is an antimalarial agent used in combination with lumefantrine for the treatment of acute uncomplicated malaria caused by *berghei* and *falciparum*
malariea	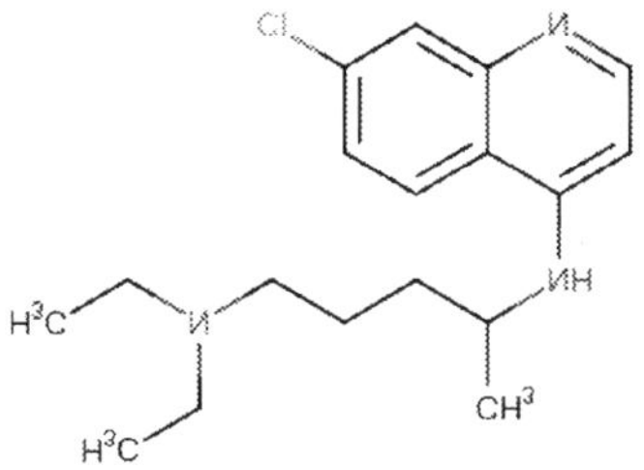	**Chloroquine** is an antimalarial drug used to treat susceptible infections with *P. vivax, P. malariae, P. ovale,* and *P. falciparum*. It is also used for the second line treatment for rheumatoid arthritis
yoellii	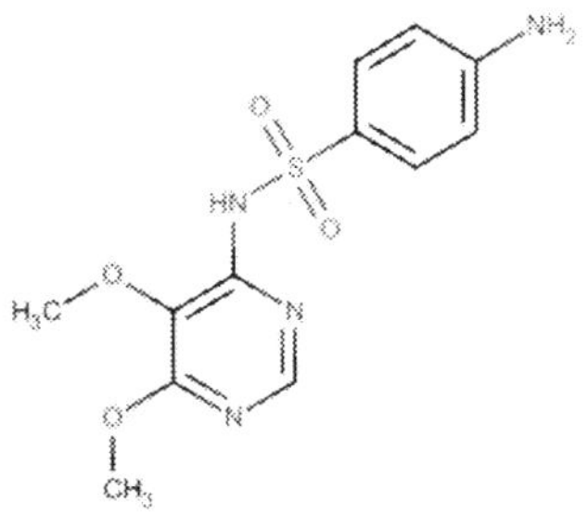	**Sulfadoxine** is a long acting sulfonamide used for the treatment or prevention of malaria
knowles	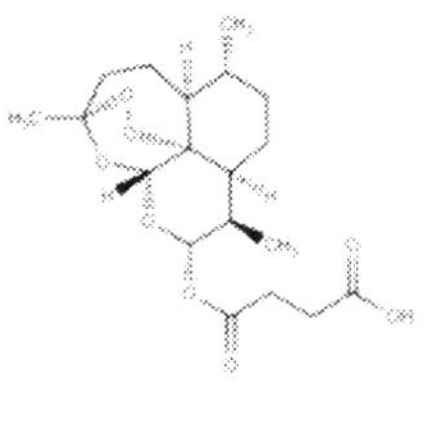	**Artesunate** is an artemesinin derivative indicated for the initial treatment of severe malaria
falciparum	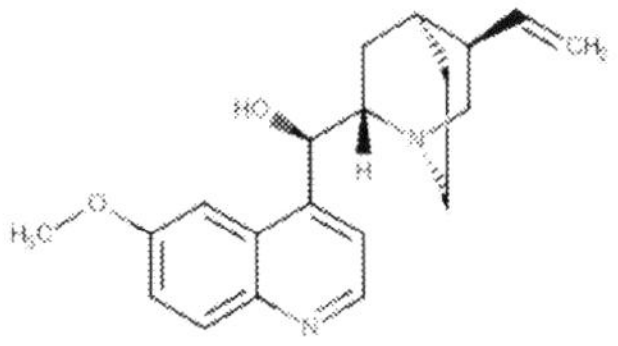	**Quinine** is an alkaloid used to treat uncomplicated *Plasmodium falciparum* malaria

(*continued on next page*)

TABLE 5.2 Drugs containing six-ring structure tested against some plasmodium species—cont'd

Species name	Pyridine containing compound tested against some plasmodium species	Role
Vivax and falciparum	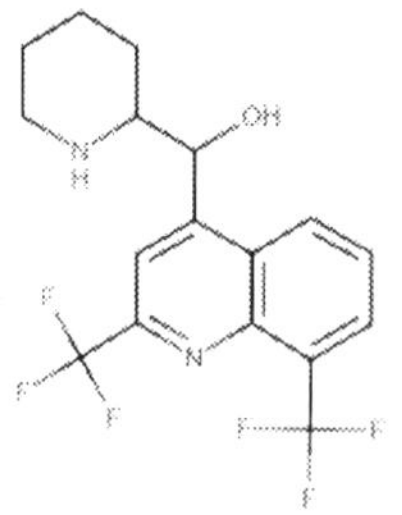	**Mefloquine** is an antimalarial agent used in the prophylaxis and treatment of malaria caused by *Plasmodium falciparum* and *Plasmodium vivax*
Ovale	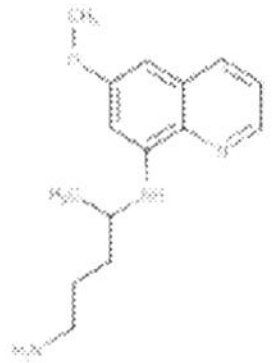	**Primaquine** is an antimalarial indicated to prevent relapse of *ovale* and *vivax* malaria

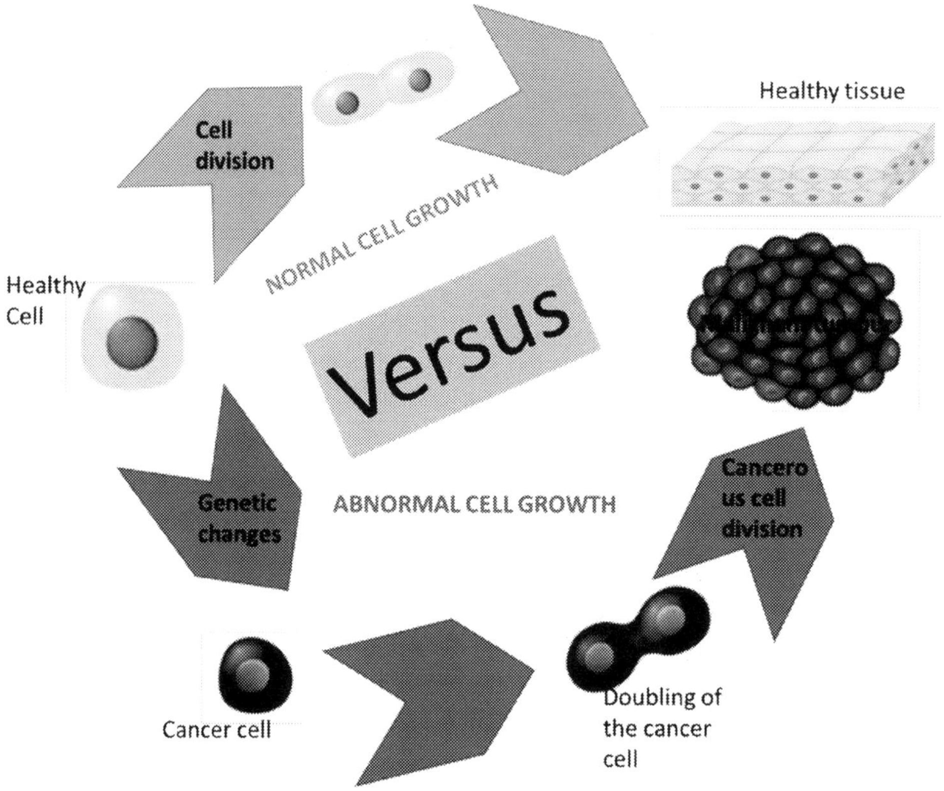

FIGURE 5.3 Normal and abnormal cell growth.

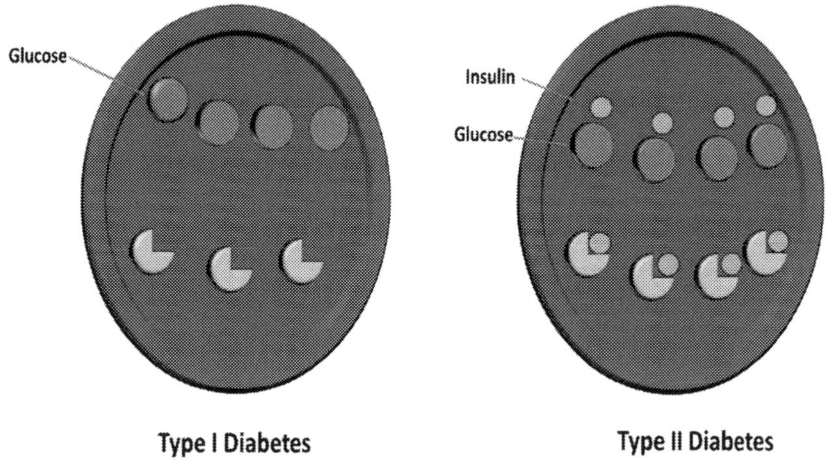

FIGURE 5.4 Difference between Type I and Type II diabetes.

incidences is likely to increase from 12.7 million in 2008 to 20.2 million by 2030. Various dietary agents and "whole foods" have potent cancer-preventive properties, but to date, only a few of these dietary agents have shown efficacy in human intervention trials [36]. The effectiveness of cancer prevention depends on the identification of potential risk factors and understanding how, at the molecular level, these factors trigger cancer initiation and progression. A better understanding of the signaling pathways and molecular events involved in cancer initiation can be acquired by following up patients with a higher risk for a specific type of cancer. Approximately 40%–50% of cancers can be reduced if the existing knowledge about potential risk factors is taken into account by public health strategists [37]. Table 5.3 depicts some pyridine-containing drugs that have been tested so far against various types of cancer.

5.7 Diabetes

There are currently at least 415 million cases of diabetic or infected people worldwide, based on the International Diabetes Federation report of 2015. This number is expected to continue increasing by 2040. More than 90% of diabetic patients in the world are reported to have type diabetes mellitus (T2DM). The economic burdens that are due to diabetes continue to increase in underdeveloped and well-developed countries. T2DM represents a major global health issue and the incidence of disease increases with various genetic and other associated factors such as age, obesity, stress, diet, ethnicity, lack of exercise and inflammation. The burden of type 2 diabetes and its major complications are rising worldwide [28,29]. In short, Type 1 diabetes is an autoimmune disease—the body's immune system attacks the cells in the pancreas that make the hormone insulin. Whereas the cause of type 2 diabetes is multifactorial, people inherit genes that make them susceptible to type 2, but lifestyle factors, like obesity and inactivity, are also important. In type 2 diabetes, at least in the early stages, there is enough insulin, but the body becomes resistant to it (Fig. 5.4). Table 5.4 shows different pyridine-containing compounds that have been used to treat either type 1 or type 2 diabetes [15,27–29,36].

TABLE 5.3 Drugs containing pyridine ring structure tested against cancer.

Type of cancer	Tested against some type of cancer	Target
Breast, ovarian, pancreatic and other types	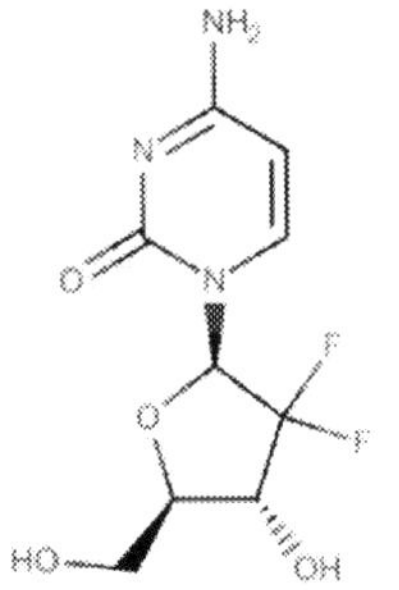	**Gemcitabine** is a nucleoside metabolic inhibitor used as adjunct therapy in the treatment of certain types of ovarian cancer, nonsmall cell lung carcinoma, metastatic breast cancer, and as a single agent for pancreatic cancer
Lung cancer	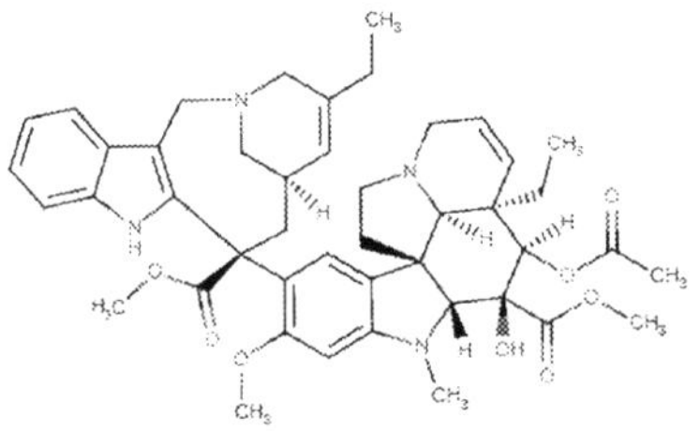	**Vinorelbine** is a vinca alkaloid used in the treatment of metastatic non-small cell lung carcinoma (NSLC) and in conjunction with other drugs in locally advanced NSCLC
Pancreatic	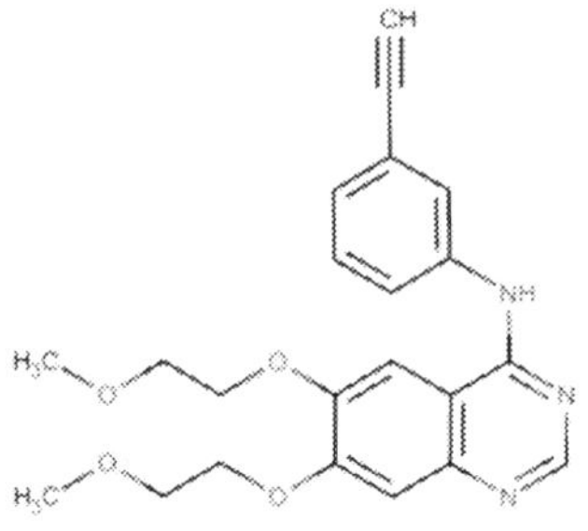	**Erlotinib** is an EGFR tyrosine kinase inhibitor used to treat certain small cell lung cancers or advanced metastatic pancreatic cancers
Prostate	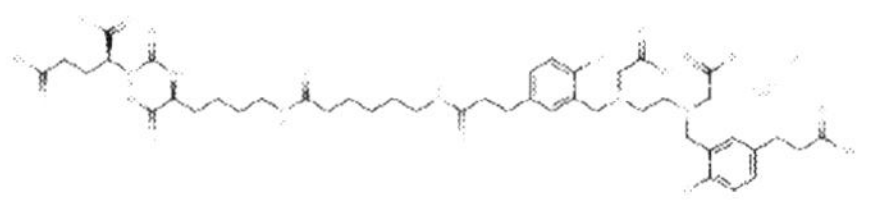	**Ga 68 PSMA-11** is a radiopharmaceutical agent used in the diagnosis of prostate-specific membrane antigen (PSMA) positive lesions in male patients during positron emission tomography

5.8 Conclusion and future perspectives

The increase in drug resistance to most complex diseases poses a threat to human life. Thus ongoing search for alternative treatment is justified. Pyridine, as many studies have shown, is a promising tool to deliver innovative treatment for diseases such as cancer, diabetes, and malaria. It is important to note this with the current pandemic of coronavirus; pyridine

TABLE 5.4 Compounds containing pyridine tested against diabetes.

Type of Diabetes	Chemical structure	Role
Type II	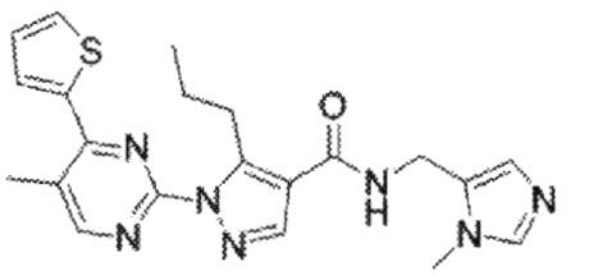	**Insulin aspart** is a rapid-acting form of insulin used for glycemic control in type 1 and type 2 diabetes mellitus
Non-Insulin dependent	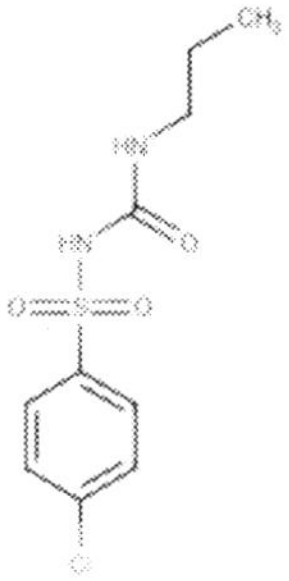	**Chlorpropamide** is a sulfonylurea used in the treatment of noninsulin dependent diabetes mellitus
DM	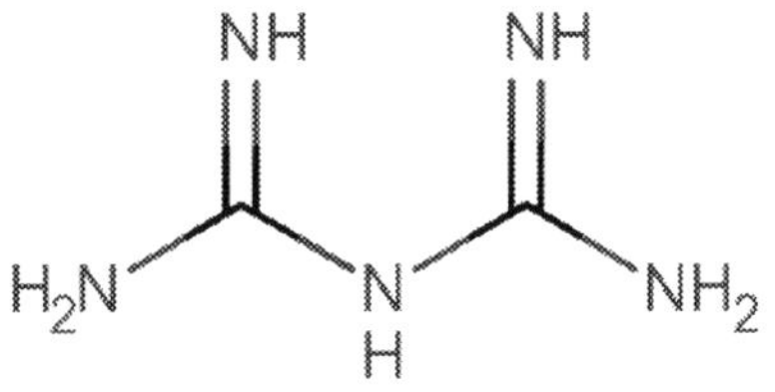	Biguanide has been investigated for the treatment of diabetes mellitus
Glycaemic control	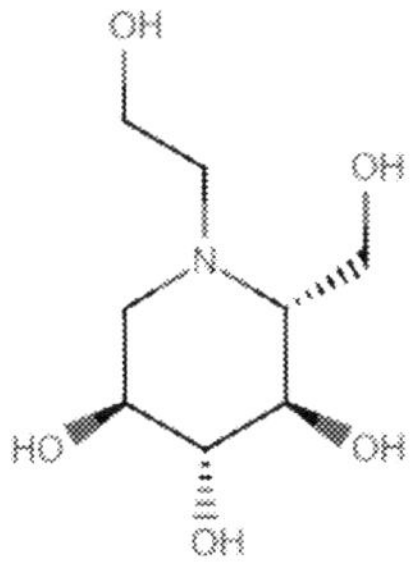	**Miglitol** is an oral alpha-glucosidase inhibitor used to improve glycemic control by delaying the digestion of carbohydrates
Type II	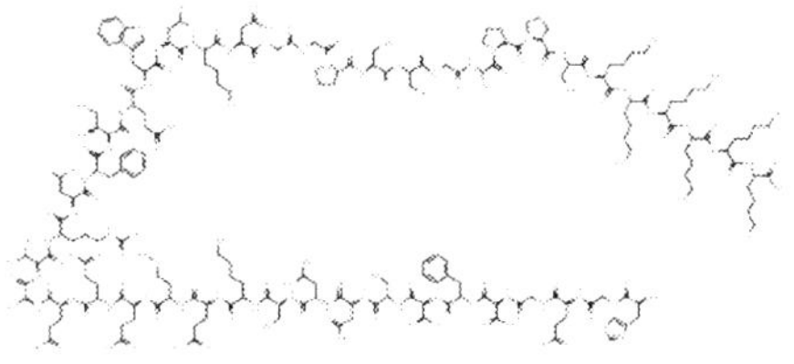	**Lixisenatide** is a GLP-1 receptor agonist used for the management of type 2 diabetes mellitus

(*continued on next page*)

TABLE 5.4 Compounds containing pyridine tested against diabetes—cont'd

Type of Diabetes	Chemical structure	Role
Type II	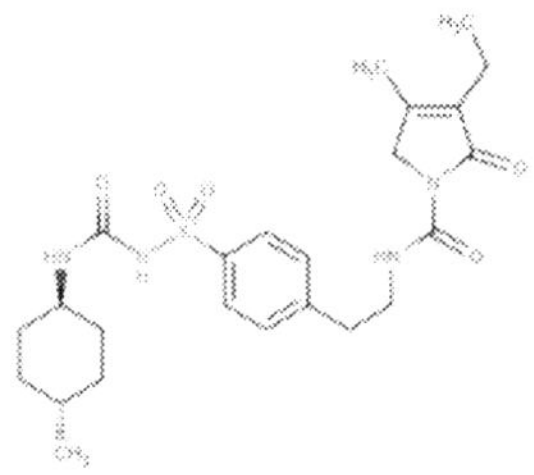	**Glimepiride** is a sulfonylurea drug used to treat type 2 diabetes mellitus

derivatives could also play a major role to bring an effective treatment. Besides the fact that COVID-19 somehow has taken a spotlight, therefore the fight against the aforementioned complex diseases has somehow slowed down. Therefore, it is important to continue looking for alternative treatment for various complex diseases with pyridine at our disposal, we are guaranteed to deliver innovative drugs.

Acknowledgments

The author wishes to thank the University of Fort Hare for support, Dr P. Motsweni from the University of Pretoria and Miss Sesethu Godlo for their technical support I would also like to thank the South African Medical Research Council (SA-MRC, PA19) and the University of Fort Hare, SEED grant (C415) for fincial support to this work.

Conflict of Interest

There is no conflict of interest among authors.

References

[1] World Health Organization (WHO), 2020
[2] Z.A. Damanhouri, A. Ahmad, A review on therapeutic potential of Piper nigrum L. black pepper): the king of spices, Med. Aromat. Plants 3 (2014) 161. http://doi.org/10.4172/2167-0412.1000161.
[3] N. Basilico, E. Pagini, D. Monti, P. Olliaro, D. Taramelli, A microtitre-based method for measuring the haem polymerization inhibitory activity (HPIA) of antimalarial drugs, J. Antimicrob. Chemother. 42 (1998) 55–60.
[4] P.G. Bray, S.R. Hawley, M. Mungthin, SA. Ward, Physicochemical properties correlated with drug resistance and the reversal of drug resistance in *Plasmodium falciparum*, Mol. Pharmacol. 50 (1996) 1559–1566.
[5] A. Dorn, R. Stoffel, H. Matile, A. Bubendorf, R. Ridley, Malarial haemozoin/b-haematin supports haem polymerization in the absence of protein, Nature 374 (1995) 269–271.
[6] A. Dorn, S.R. Vippagunta, H. Matile, C. Jacquet, J.L. Vennerstrom, RG. Ridley, An assessment of drughaematin binding as a mechanism for inhibition of haematin polymerisation by quinolone antimalarials, Biochem. Pharmacol. 55 (1998) 727–736.
[7] H. Sakurai, K. Tsuchiya, M. Nukatsuka, Insulin-like effect of vanadyl ion on streptozotocininduced diabetic rats, J. Endocrinol. 126 (1990) 451–459.
[8] S. Abdelhamed, S. Yokoyama, A. Refaat, K. Ogura, H. Yagita, S. Awale, Piperine enhances the efficacy of TRAIL-based therapy for triplenegative breast cancer cells, Anticancer Res. 34 (2014) 1893–1899.

[9] M.R. Abid, S. Guo, T. Minami, K.C. Spokes, K. Ueki, C. Skurk, Vascular endothelial growth factor activates PI3K/Akt/forkhead signaling in endothelial cells, Arterioscler. Thromb. Vasc. Biol. 24 (2004) 294–300. http://doi.org/10.1161/01.ATV.0000110502.10593.06.
[10] S.V. Ambudkar, C. Kimchi-Sarfaty, Z.E. Sauna, M.M. Gottesman, P-glycoprotein: from genomics to mechanism, Oncogene 22 (2003) 7468–7485. http://doi.org/10.1038/sj.onc.1206948.
[11] P. Anand, A.B. Kunnumakkara, C. Sundaram, K.B. Harikumar, S.T. Tharakan, O.S. Lai, Cancer is a preventable disease that requiresmajor lifestyle changes, Pharm. Res. 25 (2008) 2097–2116. http://doi.org/10.1007/s11095-008-9661-9.
[12] U. Andergassen, A.C. Kölbl, J.N. Mumm, S. Mahner, U. Jeschke, Triple-negative breast cancer: new therapeutic options via signalling transduction cascades, Oncol. Rep. 37 (2017) 3055–3060. http://doi.org/10.3892/or.2017.5512.
[13] C.K. Atal, R.K. Dubey, J. Singh, Biochemical basis of enhanced drug bioavailability by piperine: evidence that piperine is a potent inhibitor of drug metabolism, J. Pharmacol. Exp. Ther. 232 (1985) 258–262.
[14] N. Atal, K.L. Bedi, Bioenhancers: revolutionary concept to market, J.Ayurveda Integr. Med. 1 (2010) 96–99. http://doi.org/10.4103/0975-9476.65073.
[15] Y.G. Chen, P. Li, P. Li, R. Yan, X.Q. Zhang, Y. Wang, X.T. Zhang, W.C. Ye, QW. Zhang, α-Glucosidase inhibitory effect and simultaneous quantification of three major flavonoid glycosides in Microctis folium, Molecules 18 (2013) 4221–4232.
[16] Y. Wu, Y. Ding, Y. Tanaka, W. Zhang, Risk factors contributing to type 2 diabetes and recent advances in the treatment and prevention, Int. J. Med. Sci. 11 (2014) 1185–1200.
[17] U. Jung, M.-S. Choi, Obesity and its metabolic complications: the role of adipokines and the relationship between obesity, inflammation, insulin resistance, dyslipidemia and nonalcoholic fatty liver disease, Int. J. Mol. Sci. 15 (2014) 6184–6223.
[18] A. Das, S. Mukhopadhyay, The evil axis of obesity, inflammation and type-2 diabetes, Endocr. Metab. Immune Disord. Drug Targets 11 (2011) 23–31.
[19] M.Y. Donath, S.E. Shoelson, Type 2 diabetes as an inflammatory disease, Nat. Rev. Immunol. 11 (2011) 98–107.
[20] S. Mohammed, A. Yaqub, A. Nicholas, W. Arastus, M. Muhammad, S. Abdullahi, Review on diabetes, synthetic drugs and glycemic effects of medicinal plants, J. Med. Plants Res. 7 (2013) 2628–2637.
[21] D. Patel, S. Prasad, R. Kumar, S. Hemalatha, An overview on antidiabetic medicinal plants having insulin mimetic property, Asian Pac. J. Trop. Biomed. 2 (2012) 320–330.
[22] P. Arulselvan, H.A.A. Ghofar, G. Karthivashan, M.F.A. Halim, M.S.A. Ghafar, S. Fakurazi, Antidiabetic therapeutics from natural source: a systematic review, Biomed. Prev. Nutr. 4 (2014) 607–617.
[23] X. Wang, W. Bao, J. Liu, Y.Y. Ouyang, D. Wang, S. Rong, X. Xiao, Z.L. Shan, Y. Zhang, P. Yao, et al., Inflammatory markers and risk of type 2 diabetes: a systematic review and meta-analysis, Diabetes Care. 36 (2013) 166–175.
[24] A.B. Goldfine, V. Fonseca, K.A. Jablonski, L. Pyle, M.A. Staten, S.E. Shoelson, The effects of salsalate on glycemic control in patients with type 2 diabetes: a randomized trial, Ann. Intern. Med. 152 (2010) 346–357.
[25] N. Asano, K. Oseki, E. Tomioka, H. Kizu, K. Matsui, N-containing sugars from Morus alba and their glycosidase inhibitory activities, Carbohydr. Res. 259 (1994) 243–255.
[26] C. Brownson, A. Hipkiss, Carnosine reacts with a glycated protein, Free Radic Biol Med 28 (2000) 1564–1570.
[27] L. Bukowski, R. Kaliszan, Imidazo [4, 5-b] pyridine derivatives of potential tuberculostatic activity, II: synthesis and bioactivity of designed and some other 2-cyanomethylimidazo [4, 5-b] pyridine derivatives, Arch. Pharm. 324 (1991) 537–542.
[28] A.E. Butler, J. Janson, S. Bonner-Weir, R. Ritzel, R.A. Rizza, PC. Butler, Beta-cell deficit and increased beta-cell apoptosis in humans with type 2 diabetes, Diabetes 52 (2003) 102–110.
[29] M.J. Charron, R.A. Dubin, Michels CAStructural and functional analysis of the MAL1 locus of Saccharomyces cerevisiae, Mol. Cell. Biol. 6 (1986) 3891–3899.
[30] H. Sakurai, A new concept: the use of vanadium complexes in the treatment of diabetes mellitus, Chem. Rec. 2 (2002) 237–248.
[31] H. Sakurai, Y. Fujisawa, S. Fujimoto, Role of vanadium in treating diabetes, J. Trace Elem. Exp. Med. 12 (1999) 393–401.
[32] H. Sakurai, Y. Kojima, Y. Yoshikawa, Antidiabetic vanadium(IV) and zinc(II) complexes, Coord. Chem. Rev. 226 (2002) 187–198.

[33] T.J. Egan, HM. Marques, The role of haem in the activity of chloroquine and related antimalarial drugs, Coord. Chem. Rev. 190–192 (1999) 493–517.

[34] F. Bray, A. Jemal, N. Grey, J. Ferlay, D. Forman, Global cancer transitions according to the human development index (2008-2030): a population-based study, Lancet Oncol. 13 (2012) 790–801. http://doi.org/10.1016/S1470-2045(12)70211-5.

[35] S.C. Casey, A. Amedei, K. Aquilano, A.S. Azmi, F. Benencia, D. Bhakta, Cancer prevention and therapy through the modulation ofthe tumor microenvironment, Semin. Cancer Biol. 35 (Suppl.) (2015) S199–S223. http://doi.org/10.1016/j.semcancer.

[36] R.L. Clark, A.A. Pessolano, T.Y. Shen, D.P. Jacobus, H. Jones, V.J. Lotti, LM. Flataker, Synthesis and analgesic activity of 1, 3-dihydro-3-(substituted phenyl) imidazo [4, 5-b] pyridin-2-ones and 3-(substituted phenyl)-1, 2, 3-triazolo [4, 5-b] pyridines, J. Med. Chem. (2) (1978) 965–978.

[37] G. Danaei, M.M. Finucane, Y. Lu, G.M. Singh, M.J. Cowan, C.J. Paciorek, J.K. Lin, F. Farzadfar, Y.H. Khang, G.A. Stevens, M. Rao, M.K. Ali, L.M. Riley, C.A. Robinson, M. Ezzati, Global burden of metabolic risk factors of chronic diseases collaborating group (blood glucose), Lancet 378 (2011) 31–40.

[38] E.M. Duffy, WL. Jorgensen, Prediction of properties from simulations: free energies of solvation in hexadecane, octanol, and water, J Am Chem 122 (2000) 2878–2888.

[39] S.H. Hsiao, L.H. Liao, P.N. Cheng, TJ. Wu, Hepatotoxicity associated with acarbose therapy, Ann. Pharmacother. 40 (2006) 151–154.

[40] P. Hollander, Safety profile of acarbose, an α-glucosidase inhibitor, Drugs 44 (1992) 47–53.

41 M.J. Humphries, K. Matsumoto, S.L. White, K. Olden, Inhibition of experimental metastasis by castanospermine in mice: blockage of two distinct stages of tumor colonization by oligosaccharide processing inhibitors, Cancer Res. 46 (1986) 5215–5222.

[41] S. Imran, M. Taha, N.H. Ismail, S. Fayyaz, K.M. Khan, Choudhary MISynthesis, biological evaluation, and docking studies of novel thiourea derivatives of bisindolylmethane as carbonic anhydrase II inhibitor, Bioorg. Chem. 62 (2015) 83–93 a.

[43] S. Imran, M. Taha, N.H. Ismail, S.M. Kashif, F. Rahim, W. Jamil, M. Hariono, M. Yusuf, Wahab HSynthesis of novel flavone hydrazones: in-vitro evaluation of α-glucosidase inhibition, QSAR analysis and docking studies, Eur. J. Med. Chem. 105 (2015) 156–170 b.

[44] S. Imran, M. Taha, N.H. Ismail, S.M. Kashif, F. Rahim, W. Jamil, H. Wahab, KM. Khan, Synthesis, in vitro and docking studies of new flavone ethers as α-glucosidase inhibitors, Chem. Biol. Drug Des. 87 (2015) 361–373 c.

[45] S. Kumar, S. Narwal, V. Kumar, O. Prakash, α-glucosidase inhibitors from plants: a natural approach to treat diabetes, Pharmacogn. Rev. 5 (2011) 19–29.

[46] P. Lavanya, M. Suresh, Y. Kotaiah, N. Harikrishna, CV. Rao, Synthesis, antibacterial, antifungal and antioxidant activity studies on 6-bromo-2-substitutedphenyl-1H-imidazo [4, 5-b] pyridine, Asian J. Pharm Clin. Res. (4) (2011) 69–73.

[47] P. Lefebvre, A. Scheen, The use of acarbose in the prevention and treatment of hypoglycaemia, Eur. J. Clin. Invest. 24 (1994) 40–44.

[48] M. Halberstam, N. Cohen, Shlimovich POral vanadyl sulfate improves insulin sensitivity in NIDDM but not in obese nondiabetic subjects, Diabetes 45 (5) (1996) 659–666.

[49] N. Cohen, M. Halberstam, P. Shlimovich, Oral vanadyl sulfate improves hepatic and peripheral insulin sensitivity in patients with non-insulindependent diabetes mellitus, J. Clin. Invest. 95 (6) (1995) 2501–2509.

[50] K. Cusi, S. Cukier, R.A. DeFronzo, Vanadyl sulfate improves hepatic and muscle insulin sensitivity in type 2 diabetes, J. Clin. Endocrinol. Metabolism 86 (3) (2009) 1410–1417.

[51] H. Sakurai, H. Sano, T. Takino, A new type of orally active insulin-mimetic vanadyl complex: bis(1-oxy-2-pyridinethiolato)oxovanadium(IV) withVO(S2O2) coordination mode, Chem. Lett. 9 (1999) 913–914.

[52] H. Sakurai, H. Sano, T. Takino, An orally active antidiabetic vanadyl complex, bis(1-oxy-2-pyridinethiolato)oxovanadium(IV), with VO(S2O2) coordination mode, in vitro and in vivo evaluationsin rats, J. Inorg. Biochem. 80 (2000) 99–105.

[53] A. Bachi, I. Dalle-Donne, A. Scaloni, Redox proteomics: chemical principles, methodological approaches and biological/biomedical promises, Chem. Rev. 113 (2013) 596–698. http://doi.org/10.1021/cr300073p.

[54] R. Baena Ruiz, P. Salinas Hernández, Cancer chemoprevention by dietary phytochemicals: epidemiological evidence, Maturitas 94 (2016) 13–19. http://doi.org/10.1016/j.maturitas.2016.08.004.

[55] M. Balduyck, F. Zerimech, V. Gouyer, R. Lemaire, B. Hemon, G. Grard, Specific expression of matrix metalloproteinases 1, 3, 9 and 13 associated with invasiveness of breast cancer cells in vitro, Clin. Exp. Metastasis 18 (2000) 171–178. http://doi.org/10.1023/A:1006762425323.

[56] R. Baskar, J. Dai, N. Wenlong, R. Yeo, K.W. Yeoh, Biological response of cancer cells to radiation treatment, Front. Mol. Biosci. 1 (2014) 24. http://doi.org/10.3389/fmolb.2014.00024.

[57] R.K. Bhardwaj, H. Glaeser, L. Becquemont, U. Klotz, S.K. Gupta, M.F. Fromm, Piperine, a major constituent of black pepper, inhibitshuman P-glycoprotein and CYP3A4, J. Pharmacol. Exp. Ther. 302 (2002) 645–650. http://doi.org/10.1124/jpet.102.034728.

[58] C. Braicu, N. Mehterov, B. Vladimirov, V. Sarafian, S.M. Nabavi, A.G. Atanasov, Nutrigenomics in cancer: revisiting the effects of natural compounds, Semin. Cancer Biol. 46 (2017) 84–106. http://doi.org/10.1016/j.semcancer.2017.06.011.

[59] C.L. Chaffer, B.P. San Juan, E. Lim, R.A. Weinberg, EMT, cell plasticity and metastasis, Cancer Metastasis Rev. 35 (2016) 645–654. http://doi.org/10.1007/s10555-016-9648-7.

[60] L. Chen, A. Malhotra, Combination approach: the futureof the war against cancer, Cell Biochem. Biophys. 72 (2015) 637–641. http://doi.org/10.1007/s12013-015-0549-0.

[61] K. Collins, T. Jacks, N.P. Pavletich, The cell cycle and cancer, Proc.Natl. Acad. Sci. U.S.A. 94 (1997) 2776–2778. http://doi.org/10.1073/pnas.94.7.2776.

[62] G.M. Cooper, The Development and Causes of Cancer. The Cell: A Molecular Approach, Sinauer Associates, Sunderland, MA, 2000.

[63] G.A. Cordell, Introduction to Alkaloids: A Biogenetic Approach, JohnWiley & Sons, New York, NY, 1981.

[64] T.J. Egan, W.W. Mavuso, D.C. Ross, HM. Marques, Thermodynamic factors controlling the interaction of quinoline antimalarial drugs with ferriprotoporphyrin IX, J. Inorg. Biochem. 68 (1997) 137–145.

[65] D.A. Fidock, T. Nomura, A.K. Talley, R.A. Cooper, S.M. Dzekunov, M.T. Ferdig, L.M.B. Ursos, A.B.S. Sidhu, K.W. Naude, K.W. Deitsch, X. Su, J.C. Wootton, P.D. Roepe, TE. Wellems, Mutations in the P. falciparum digestive food vacuole transmembrane protein PfCRT and evidence for their role in chloroquine resistance, Mol. Cell 6 (2000) 861–871.

[66] D.E. Goldberg, AFG. Slater, The pathway of hemoglobin degradation in malaria parasites, Parasitol. Today 8 (1992) 280–283.

[67] S.R. Hawley, P.G. Bray, M. Mungthin, J.D Atkinson, P.M. O'Neill, S.A. Ward, Relationship between antimalarial drug activity, accumulation, and inhibition of heme polymerization in Plasmodium falciparum in vitro, Antimicrob. Agent Chemother. 42 (1998) 682–686.

[68] S. Hawley, P.C. Bray, P.M. O'neill, B.K. Park, S.A. Ward, The role of drug accumulation in 4-aminoquinoline antimalarial potency: the role of structural substitution and physicochemical properties, Biochem. Pharmacol. 52 (1996) 723–733.

[69] J.H. Holland. Adaption in natural and artificial system. Ann Arbor, University of Michigan Press Pagola S, Stephens PW, Bohle DS, Kosar AD, Madsen SK.The structure of malaria pigment bhaematin. Nature 404: (2000) 307–310

[70] C. Portela, C.M.M. Afonso, M.M. Pinto, MJ. Ramos Receptor–drug association studies in the inhibition of the hematin aggregation process of malaria. FEBS Lett. 27435:217–222

71 I.W. Sherman, Amino acid metabolism and protein synthesis in malarial parasites, WHO Bull 55 (2003) 265–276.

[71] I. Solmonov, I. Osipova, Y. Feldman, C. Baehtz, K. Kjaer, I.K. Robinson, G.T. Webster, D. McNaughton, B.R. Wood, I. Weissbuch, Leiserowitz, Crystal nucleation, growth and morphology of the synthetic malaria pigment b-hematin and the effect thereon by quinoline additives: the malaria pigment as a target of various antimalarial drugs, J. Am. Chem. Soc. 129 (2007) 2615–2627.

[73] W. Trager, J.B. Jensen, Human malaria parasite in continuous culture, Science 193 (1976) 673–675.

CHAPTER

6

Pyridines in Alzheimer's disease therapy: Recent trends and advancements

Puja Mishra[a], *Souvik Basak*[a], *Arup Mukherjee*[b] *and Balaram Ghosh*[c]

[a]Dr. B.C. Roy College of Pharmacy & Allied Health Sciences, Durgapur, West Bengal, India [b]Department of Biotechnology, Maulana Abul Kalam Azad University of Technology, West Bengal, India [c]Epigenetic Research Laboratory, Department of Pharmacy, Birla Institute of Technology and Science, Hyderabad, Telangana, India

6.1 Introduction

Alzheimer's is named after a German Psychiatrist Alois Alzheimer while performing a histopathological study of his patient (Auguste Deter) brain, who died suffering from dementia, known as a physical disease that affects the brain [1]. Further, this was defined as a progressive neurodegenerative disorder, characterized by gradual loss of cholinergic neurons and accumulation of β-amyloid protein in the brain areas like the cortex and hippocampus. The disease onset starts with short-term memory impairment that gradually progresses to complete loss of cognitive function, weak performance of activities of daily life, loss of logical thoughts, emotional disturbances and erratic mood fluctuations, paranoia and hallucinations, and ultimately death [2,3].

The main causes of Alzheimer's are age-related factors, genetic, mutation, health, lifestyle, and environmental factors. However, the exact cause of Alzheimer is not very clearly understood. Early onset of Alzheimer may be due to genetic mutation. Late-onset Alzheimer may be due to several changes in the brain which occur in decades. Mostly this disease is seen in the elderly due to atrophy (shrinking) in various regions of the brain. The increased levels of metals such as Fe, Al, and Hg in the brain generate unstable reactive species like free radicals, increased lipid peroxidation and unsaturated fatty acid content, DNA oxidation, and decreased cytochrome C oxidase. Mutations in critical regions of amyloid precursor

Recent Developments in the Synthesis and Applications of Pyridines.
DOI: https://doi.org/10.1016/B978-0-323-91221-1.00001-4

protein (APP) generate Amyloid-β, which is believed to be the main cause of Alzheimer's disease [3–5].

6.1.1 Etiology of Alzheimer's disease-cholinergic hypothesis

Hippocampus and associated parts of the brain have a large number of cholinergic neurons. Loss of such neurons is also a cause of Alzheimer's disease. Various neurotransmitters govern brain function, and the one which is most associated with cognition and memory is Acetylcholine (ACh). Acetylcholinesterase **(AChE)**, an enzyme that plays a key role in the termination of signaling at the cholinergic synaptic cleft by rapid hydrolysis of Ach. AChE is the target of the first-generation drugs for the treatment of Alzheimer's disease (AD) [6].

Acetylcholinesterase is an active enzyme that has a very high catalytic activity and rapidly hydrolyses ACh to give choline and acetic acid. The dominant toxicity of organophosphate poison is attributed primarily to its potent inhibition of acetylcholinesterase. Crystal structure of Torpedo californica acetylcholinesterase forms lead to visualization, for the first time in the research field and is shown to have a binding pocket for acetylcholine.

The crystal structure of human AChE shows a narrow gorge of 20 Å in depth, which contains the Catalytic Anionic Site (CAS) with four important regions anionic site, esteratic site, oxyanion hole, and an acyl pocket. The active site of AchE consists of two sites which are an anionic site, having Trp86 residue, and an esteratic site having residues like Ser203, Glu334, and His447. Trp 86 recognizes and binds to several active-site-directed inhibitors. Oxyanion hole and esteratic site are essential for substrate selectivity. To the top of the gorge, there is peripheral anionic site (PAS) which is about 15Å from CAS region and contains a significant residue Trp 279. Inside the gorge, the Phe330 acts as a swinging gate that keeps changing its conformation so that a molecule can interact with Trp 86 residues. The residues Trp279 and Phe330 are present in AChE conferring selectivity for ACh hydrolysis, but they are absent in Butyrylcholinesterase (BChE). The following figure (Fig. 6.1) depicts the outline of important structural features of AChE enzyme [7,8].

The ester link of the Acetylcholine (Ach) enzyme is cleaved by Acetylcholinesterase (AchE) which converts ACh into choline and acetate (Fig. 6.2) Individuals who suffer from Alzheimer's have numerous senile plaques and neuro fibrillary tangles (NFTs) characterized by neural cell loss and vascular damage. β-Amyloid (protein fragment) that builds the connection between the nerve cells, are the cause of the above condition.

6.1.2 β-amyloid hypothesis

The β-amyloid is a byproduct of the protein APP whose function is believed to be involved in neuronal degradation. On the surface of neurons, there is a transmembrane protein called the APP that is cleaved by a series of enzymes called secretase. In normal physiology, α-secretase cleaves APP to give α-APP and Aβ units [4]. The Aβ unit is cleaved by γ-secretase to give Aβ-40 which contains 40 amino acid-containing residues. In the case of Alzheimer disease, in the initial stage, instead of α-secretase, an abnormal β-secretase cleaves APP followed by γ-secretase, and produces amyloid-β protein plaques and the fiber tangles of tau protein building inside the cells [9]. These plaques and tangles tend to develop in a predictable pattern

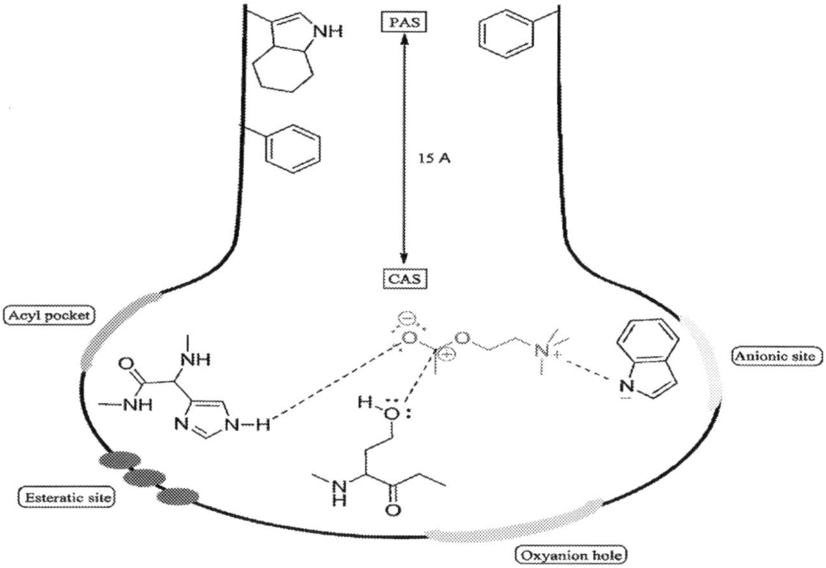

FIGURE 6.1 Schematic representation of Active site Gorge of Acetylcholinesterase. Redrawn based on reference [8].

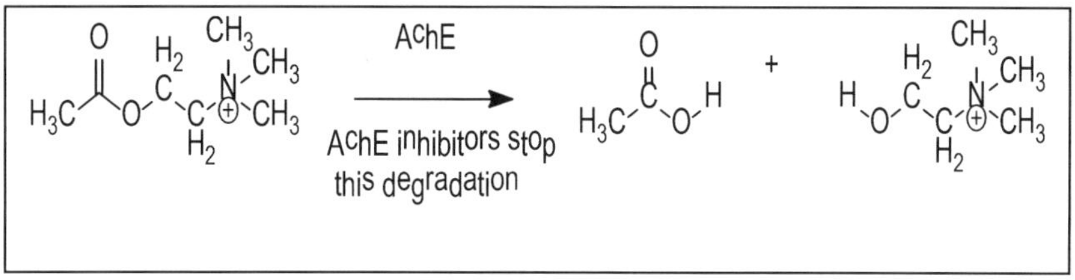

FIGURE 6.2 Breakdown of Ach by AchE enzyme into choline and acetic acid.

and far more numbers in the brain areas which are important for memory in persons suffering from Alzheimer's disease. The pathophysiology of this disease is related to the injury and death of neurons starting from the hippocampus that ultimately extends to the entire region of the brain [3,10]. To explain the pathophysiology there is the various hypothesis that need to be discussed (Fig. 6.3).

Amyloid-β (Aβ) is a protein having 36–42 amino acids that forms amyloid plaques in the brain of Alzheimer patients. They predominantly exist in two isoforms Aβ-40 and Aβ-42, and Aβ-42 is the major amyloidogenic form of the peptide. Aβ-42 has N-terminal domain and a flexible C-terminal prion forming domain. The peptide bears 16 hydrophilic residues (Asp 1-Lys 16) and the rest of the peptide (Leu 17-Val-40) is hydrophobic in nature. An earlier report has elicited that within this peptide, the amino acid residues forming KLVFFAE fragment (Lys 16-Glu 22) forms the core for fibril formation. After the nucleation, oligomers extend to form larger aggregates in the salt bridge region (Glu 22-Gly 29). Aβ-42 dimer gets stabilized by the salt bridge and HHQK fragment site contains His-13, His 14 residues which bring conformational change of Aβ from α-helix to β-sheet structure [11,12].

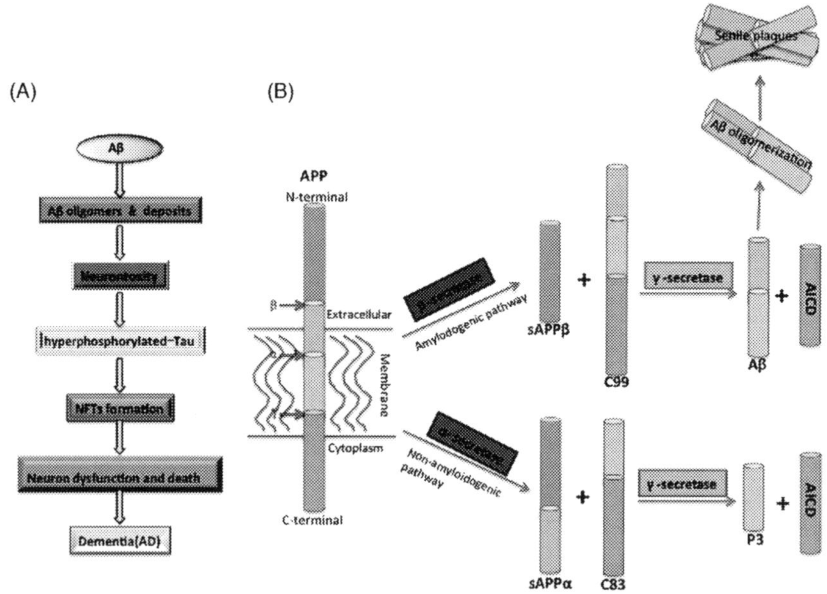

FIGURE 6.3 Elucidation of Amyloid β hypothesis pathway. (A) The toxic transformation and deposition of the associated hippocampus-located neurofibrils. (B) The enzymatic degradation of Amyloid precursor protein (APP) to shorter fragments nucleating for future fibrillation. However, the α / γ- secretase mediated pathway has been found to elicit shorter fibrils eventually getting eliminated from brain. Reprinted with Creative Commons License (CC-BY) Permission from [56].

6.1.3 BACE-1 (β-amyloid cleaving enzyme-1) role in AD

BACE-1 also forms a major target area for AD therapy. BACE-1 leads to abnormal cleavage of APP proteins thus such abnormal accumulation of Aβ can be stopped by BACE-1 inhibitors. Elaborated study of BACE-1 enzyme elucidates the presence of two catalytic active site aspartate residues Asp32 and Asp228 which make a strong salt bridge with the amide ring of common BACE1 inhibitors (Fig. 6.4). BACE-1 contains several subpockets like flap, a flexible antiparallel β-hairpin loop that controls substrate lining of the S2 pocket [13–15]. The β-hairpin loop is essential to develop selective BACE-1 inhibitors as the substrates of this loop are different in BACE-2. In order to reach S1 pocket, alpha quaternary center to amide is essential. S3 pockets can be reached by amide linker where –NH forms interactions with Gly230. S3 pocket is present in lining with 10s loop [13,16,17]. Various BACE-1 drugs had been developed and terminated after drug design due to less efficacy and high toxicity.

6.1.4 Metal chelators in AD

Amyloid β (Aβ) plaques show metals deposition in brain areas that are bound to Aβ species and are involved in the pathological pathways of AD. Metal cations like Zn^{2+}, Cu^{2+}, Fe^{2+} are present in 1, 400, and 1mM, respectively. These metal cations bind mostly to the three histidine residues (His6, His13, and His14) that ease amyloid aggregation. Redox metal ions bind to

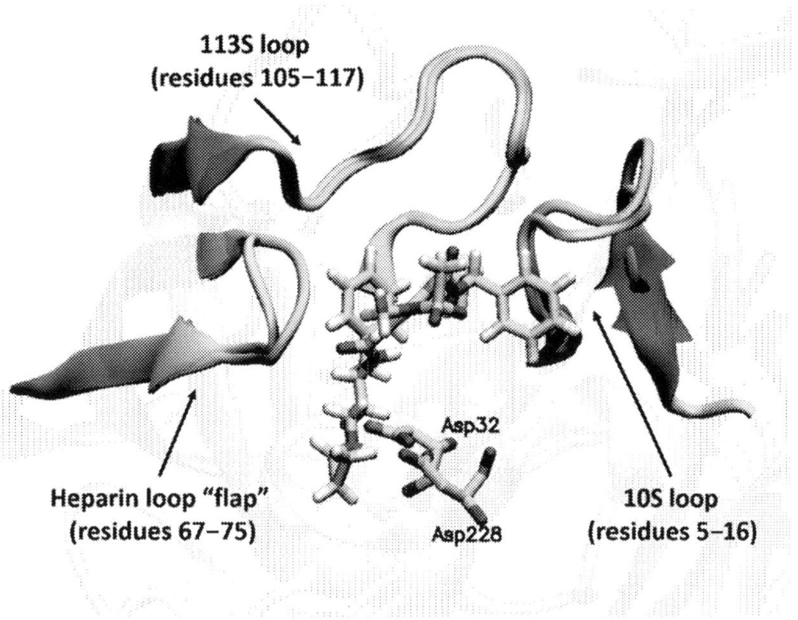

FIGURE 6.4 Representation of BACE-1 peptide with catalytic dyad (Asp32 and Asp228) together with loop 10S and 113S which hold the shape and diameter of the binding site with additional interaction with ligands (in cases) to aid in the seating of the latter in the catalytic (active) site. The flap is also shown accordingly. Reprinted with Creative Conmmons License (CC-BY) Permission from [57].

peptide to form reactive oxygen species which are belied to be linked to neurodegeneration [5,18,19].

6.2 Pyridine and its role in prevention of Alzheimer's disease

Pyridine, due to its unique chemical nature of exerting hydrophobic interactions together with its polar nitrogen terminal has been studied comprehensively as a scaffold against AD.

6.2.1 Pyridine—a short synopsis on its chemistry

Pyridine (C_5H_5N) is an aromatic compound where all the pi electrons are shared by a ring. It forms one continuous circle of electrons besides the alternate double bonds shared by every atom on the circle. Pyridine is a unique type with nitrogen on the ring to provide a tertiary amine by undergoing reactions such as alkylation and oxidation. Nitrogen is responsible for the slight dipole on the ring because electrons are drawn more toward the nitrogen being electronegative (lone pair electrons on the nitrogen) than other atoms in the ring. The ^1H nuclear magnetic radiation (^1H-NMR) shows three signals at the ortho, meta, and para positions on the molecule in respect of three different chemical shifts. These chemical shifts are the result of the different electron densities for each of these atoms. As a result, this is not very stable as other aromatic compounds.

Pyridine, a liquid similar to water, can easily mix with water and other organic solvents. This property is useful for making various products such as medicines, vitamins, food flavorings, pesticides, paints, dyes, rubber products, adhesives, waterproofing fabrics, and nitrogen-containing plant products. This nature of pyridine further makes it to be used as a precursor for many agrochemicals and pharmaceuticals. Hence, pyridine and its derivatives have significant applications in various fields, especially in the medicinal area. All these properties of pyridine make it worthwhile to have a full overview of pyridine and its derivatives with recent researches in one place for potential researchers.

It is the electronegative nature of pyridine in the backdrop responsible for the formation of all its derivatives. The synthesis of derivatives using pyridine has biological activities and vast applications.

6.2.2 Role of pyridines in AChE (acetylcholine esterase) inhibition

6.2.2.1 Pyridine 2,4,6-tricarbohydrazide

Acetylcholine esterase (AChE) activity is seen in various molecules like tris sulphonamide groups which are surrounded by pyridine group. The compound *N′,N″,N‴*-(Pyridine-2,4,6-tricarbonyl)tribenzenesulfonohydrazide) showed significant results with various targets like α-glucosidase (IC_{50} 32.2 $\pm$ 0.3 μM), acetyl choline esterase (AChE, IC_{50} 50.2 $\pm$ 0.8 μM) and butylcholine esterase (BChE, IC_{50} 43.8 $\pm$ 0.8 μM)[20]. Fluorobenzene substituted derivatives like *N′,N″,N‴*-(Pyridine-2,4,6-tricarbonyl)tris(4-fluorobenzenesulfonohydrazide) bearing 4-flouro benzyl group was found to be the most active (IC_{50} 25.6 $\pm$ 0.2 μM) against α-glucosidase, and it displayed weak inhibition activities against BChE and AChE. Thus halogen substituted tricarbohydrazides were found to be more active as compared to the unsubstituted ones.

6.2.2.2 Tacrine-pyrazolo [3,4-b]pyridine

Tacrine stacked with pyrazolo[3,4-b] pyridine moiety and separated by six-carbon spacer, was the most AChE inhibitor (IC_{50} 0.125 μM). Dual targeted drug was designed using tacrine and pyrazolo[3,4b] pyridine moiety as biologically active compounds against AChE and phoshodiesterase4 (PDE4). Moreover, compound having six carbon spacer between tacrine and pyrozolo[3,4b] pyridine provided a desired balance of butyl cholinesterase (BChE), AChE and PDE4 inhibition activities (IC_{50} 0.125, 0.449, and 0.271 μM) [21]. Thus the length of the carbon chain is an essential factor for its inhibitory activity.

6.2.2.3 Biphenyl imidazo [1,2-a]pyridine

Imidazo [1,2a]pyridine is a nitrogen-bearing fused heterocycle that has varied uses in clinical fields. 2-arylimidazo [1,2a]pyridinium salts like 2-(1,1′-biphenyl-4-yl)-6-methylimidazo[1,2a] pyridine show comparable inhibitory activity against AChE (IC_{50} 0.2-50 μM). *N*-benzylidene imidazo[1,2a]pyridine [22–25] with halogen substitution also act as potent AChE inhibitor. Adamantyl, phenyl, and biphenyl moieties were used as side chains and their activities were compared by replacing various substituents. Derivatives with biphenyl side chains show greater inhibitory activity against AChE, and phenyl side chains are essential for BChE activity. Adamantyl side chain containing imidazo[1,2a] pyridine(s) exhibit moderate to weak AChE inhibitory activity. Methyl side chains show weak inhibition against both AChE and BChE. They reach the catalytic active site of AChE enzyme and peripheral active gorge and also active site regions by hydrophobic interactions

6.2.2.4 Pyrimidine-triazolopyrimidine and pyrimidine-pyridine hybrids

Molecules show greater $\pi-\pi$ interactions with AChE enzyme when bulkier groups are substituted thus more and more aromatic groups were incorporated and the planar structure of pyrimidines ensured best fitting to the enzyme gorge. Pyridine stacked with nitrile groups reach to the catalytic active triad. 2-(4-(6-(quinolin-8-yloxy)pyrimidin-4-yl) piperazin-1-yl)nicotinonitrile shows IC_{50} 36 nM [26] and comparable to the standard drugs donepezil and tacrine. Substitution in pyrimidine ring show greater inhibitory activity. Inhibitory effect of the compound was strong for human AChE in neuronal cell extract. The 8-hydroxyquinoline moiety on 4,6-dichloro pyrimidine favors greater interactions with close residues in the active sites of the enzyme which in turn strongly increases the inhibitory activity of the drug. Enzyme inhibition study revealed by Lineweaver-Burk reciprocal plots revealed it to be mixed-type inhibitor with increasing slope and intercept. Acute toxicity study also reveals them as nontoxic and noncarcinogenic [27]. For a snapshot of such inhibitors, Table 6.1 may be referred.

6.2.3 Role of pyridine in preventing β-amyloid aggregation

6.2.3.1 2,3-dihydrothiazolo-2-pyridone

The ring-fused 2,3-dihydrothiazolo-2-pyridone unit has received considerable attention as a scaffold, which had been modified to demonstrate varied biological activities. Structures like (R)-Methyl 10-cyclopropyl-5-oxo-7,9-diphenyl-3,5-dihydro-2Hthiazolo[2,3-g][1,7]naphthyridine-3-carboxylate show rigidification of the bicyclic 2-pyridone, by decoration with sterically demanding aryl groups. Nitrogen heterocycle, (R)-Methyl 7-(4-Nitrophenyl)-5-oxo-10-phenyl-3,5-dihydro-2Hthiazolo[2,3-g][1,7]naphthyridine-3-carboxylate[28]offers a synthesis route of modulators of amyloid formation. Particular substitution pattern shows activity against specific bacterial infections, while others modulate the formation of amyloids associated with neurological disorders. Nitrene insertion approach for the facile synthesis of fluorescent multi ring-fused 2-pyridone poly heterocycle was under investigation. The benzoquinoline and benzothienopyridine annulated polyheterocycles modulated α-synuclein amyloid formation, while indole annulated 2- pyridones were ineffective. Multicomponent coupling reactions of bicyclic 6-amino-2-pyridones could allow a rapid construction of analogues containing a highly functionalized pyridine-fused tricyclic central fragment with the same hydrogen bond donor–acceptor configurations as in the previously reported polyheterocycles. In particular, transition-metal-catalyzed three-component couplings of amines, aldehydes, and alkynes (A3) and Lewis acid-catalyzed three-component Povarov reactions have been widely used for the construction of highly functionalized, biologically important heterocycles [28].

The preliminary in vitro biological evaluation suggested that 8 of the 32 tricyclic 2-pyridones tested were able to bind Aβ and α-synuclein amyloid fibrils. Annulation of

TABLE 6.1 Summary of role of pyridine structures used in AChE inhibitors.

Comp. Sr. No.	IUPAC name	Structure	Activity	References
(1)	N′, N″, N‴-(Pyridine-2,4,6-tricarbonyl)tris (4chlorobenzenesulfonohy-drazide)	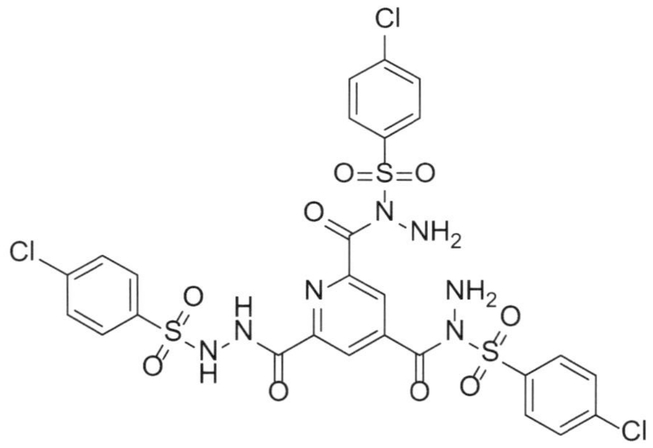	(IC_{50} = 32.2 ± 0.3μM), (AChE) (IC_{50} 50.2 ± 0.8 μM) (BChE)	[22]
(2)	N′,N″,N‴-(Pyridine-2,4,6-tricarbonyl)tris (4-fluorobenzene-sulfonohydrazide)	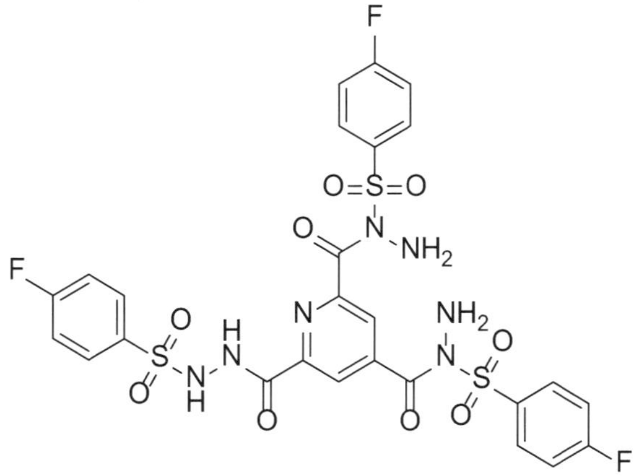	(IC_{50} = 25.6 ± 0.2 μM) Weak against AChE and BChE	[22]

(continued on next page)

(3)	pyrazolo[3,4-b] pyridine	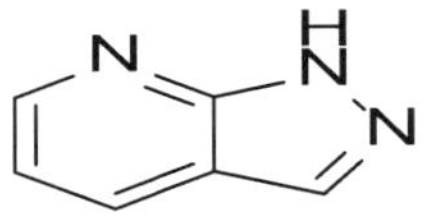	–	[23]
(4)	2-(1,1′-biphenyl-4-yl)-6-methylimidazo[1,2a] pyridine	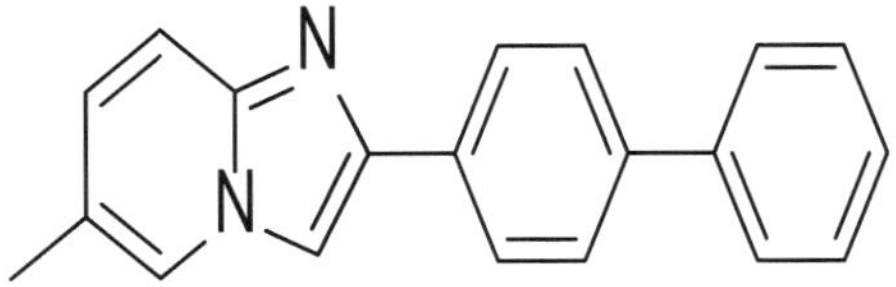	$IC_{50} = 0.2\text{-}50\ \mu M$	[24]
(5)	2-(4-(6-(quinolin-8-yloxy)pyrimidin-4-yl) piperazin-1-yl)nicotinonitrile	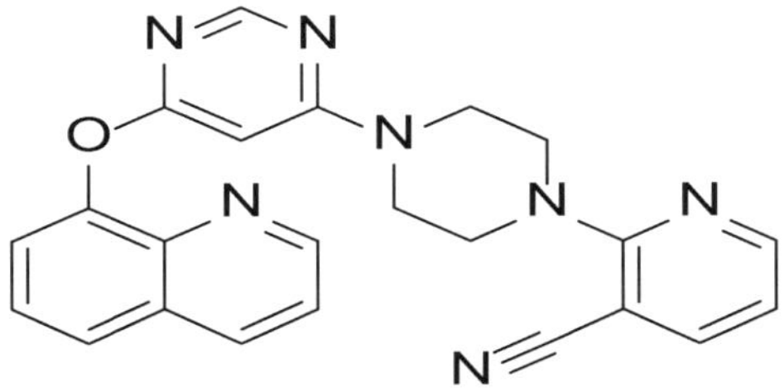	$IC_{50} = 36\ nM$	[28]

bicyclic 2-pyridones with functionalized pyridine rings, therefore, provides a straightforward synthetic route for novel tricyclic 2-pyridone peptidomimetics capable of binding to amyloid fibrils.

6.2.3.2 *Imidazo [1, 2-a] pyridine*

AD is the most prevalent neurodegenerative disorder among elderly people. Neuritic plaques containing β-amyloid (Aβ) peptides and NFTs in postmortem brain are the two pathological hallmarks characteristic of AD and provide the basis for the definitive, albeit postmortem, diagnosis of AD. While there are no definitive treatments available to affect a cure of AD, much recent interest has been given to the development of antiamyloid therapies aimed at halting and/or reversing amyloid formation and deposition. Therefore, the development of amyloid-specific imaging agents could be potentially useful in improving diagnosis by identifying patients likely to have AD for early diagnosis and further monitoring the progression of the disease. Efficacy of antiamyloid therapies currently under investigation could be evaluated using various amyloid-imaging tracers.

Development of Aβ aggregate specific imaging agents is often based on highly conjugated dyes, like congo red. Thioflavin T and S as well as congo red have been used in fluorescent tagging but these are heavier molecules that cannot penetrate brain regions. Imaging agents for Aβ aggregates must be small molecule-based and labeled with suitable isotopes for photoinduced electron transfer (PET) and single-photon emission computed tomography (SPECT) imaging. ^{123}I is used for SPECT and ^{11}C, ^{18}F are used for PET [29].

Thus, to achieve a high brain penetration, it is judicious to use small, neutral, and lipophilic compounds. Most of the active imaging compounds contain either an *N,N′*-dimethyl amino phenyl or *N*-methyl amino- group on one end of the molecule. In the search for novel ligands for Aβ aggregates based on the benzothiazole ring system, it was observed that imidazo[1,2-a]pyridine derivatives containing the desired *N,N′*-dimethylaminophenyl group, such as (IMPY), have binding properties similar to those of the other iodinated benzothiazole ligands. IMPY ([^{123}I] 6-iodo-2-(4′-*N,N′*-dimethylamino) phenyl imidazo[1,2-a] pyridine) has recently been reported as a potential fluorescence imaging agent for Aβ plaques with high binding affinity for preformed synthetic Aβ-40 aggregates and human AD cortical homogenates.

In vitro binding assay of IMPY, iodinated Thioflavin T ([^{125}I] TZDM) binding to Aβ aggregates, show excellent binding affinity (around $K_i = 15.0 \pm 5.0$ nM). A comparable binding affinity value was obtained for the corresponding bromo derivative of IMPY shows lesser inhibitory constant values of ($K_i = 10.3 \pm 1.2$ nM)

Removing halogen substitution in the 6-position show a drastic reduction in the binding affinity ($K_i > 2000$ nM) and methyl substitution also show reduced inhibition ($K_i > 242$nM). It is also worth noting that the 6-(*N,N′*-dimethlyamino)-2-(4′-bromophenyl) or 6-methyl derivative, showed low binding affinities (K_i=339 and 638 nM, respectively). The overall molecular size of (IMPY) is similar to that of ([^{125}I] TZDM) and the imidazo [1, 2-a] pyridine compounds show strong absorbance, with λ_{max} in the range of 320 -360 nm[32–34].

The brain sections of a 16-month-old mouse were labeled with ([^{125}I] IMPY). A biodistribution study in normal mice after an iv injection shows that ([^{125}I] IMPY) exhibited excellent brain uptake (2.9% ID/organ at 2 min) and fast removal (0.2% ID/organ at 1 h), which is essential for Aβ plaque imaging agents. The log P values determining the distribution coefficient of ([^{125}I] IMPY) in 1-octanol buffer which is comparable to that of ([^{125}I] TZDM). The most important property of this pyridine containing ligand, ([^{125}I] IMPY) is the faster rate of washout from the normal brain. As expected, the Aβ plaques in the brain of a mouse can be distinctly visualized and detected in vivo by using ([^{125}I] IMPY). In summary, this novel imidazo [1,2-a]pyridine derivative is a radioiodinated compound showing excellent in vitro binding and in vivo biodistribution data. Imidazo [1,2a] pyridine has desirable properties to be successful imaging agent such as enhanced initial brain uptake and fast washout from the brain areas.

6.2.3.3 Fluoroethyl and fluoropropyl Imidazo [1,2-a]pyridine

As part of an effort to develop ^{18}F-labeled tracers for PET imaging of Aβ plaques, two fluorinated analogues of IMPY were found to be effective against AD. The iodo group of IMPY is replaced with fluoroethyl to form FEPIP (6-(2′-fluoroethyl)-2-(4′ –dimethyl amino) phenyl imidazo [1, 2-a] pyridine) or fluoropropyl that forms FPPIP (6-(3′-fluoropropyl)-2-(4′ -dimethylamino) phenylimidazo [1, 2-a] pyridine).

The radiolabeling of [^{18}F] FPPIP [30] was accomplished by treatment of tosylate precursor with anhydrous [^{18}F] KF-K_{222} in acetonitrile for 10 min at 90°C. The entire procedure required almost 105 min from the end of bombardment. In vitro binding affinities of FEPIP and FPPIP to β-amyloid were determined via the binding competition with [^{125}I] IMPY using human AD cortical tissues by quantitative autoradiography. The results demonstrate that IMPY displayed a high binding affinity with $K_i = 10.3$ nM and FPPIP competed well with [^{125}I] IMPY, showing a moderate affinity ($K_i = 48.3$ nM). Decreasing side chain length in FEPIP reduced the binding affinity significantly ($K_i = 177$ nM). Partition coefficients of FEPIP and FPPIP were measured between 1-octanol and phosphate buffer at pH 7.4, and log P values were recorded. Moderate lipophilicity was displayed by FEPIP and FPPIP and the log P values are 2.82 and 2.84 which is in the optimal range (1–3) for brain permeability [30].

Further the micro PET imaging study of [^{18}F] FPPIP to assess brain penetration of the radiolabeled derivatives was performed in rhesus monkey that displayed no amyloid deposits in its brain. FPPIP easily penetrates the blood brain barrier after intravenous injection. Time activity curves of brain regions show peak value (SUV) of 1.5–2.8 at 9 min. Moreover, fast nonspecific binding clearance was observed with the radioactivity ratios of peak to 105 min to 1.9, and 2.0, 2.6 in the frontal cortex, white matter, and cerebellum which are devoid of specific binding sites. No bone toxicity in the skull was found after the intravenous administration of [^{18}F] FPPIP to the monkey, indicative of low in vivo defluorination. Thus, two new Aβ plaque ligands, FEPIP and FPPIP, were known to penetrate brain regions easily and [^{18}F] FPPIP is a promising candidate as PET imaging agent for Aβ, based on its favorable characteristics in a rhesus monkey.

6.2.3.4 Azaindolizinone derivative, ZSET845 and ZSET1446

Newly synthesized azaindolizinone derivative, spiro [imidazo [1, 2-a] pyridine-3, 2-indan]-2(3H)-one (ZSET1446), were assessed in rats with learning deficits induced by Aβ-40. The intracerebroventricular infusion of Aβ-40 caused impairments in short-term memory in a water-maze task, spontaneous alternation behavior in a Y-maze task and retention of passive-avoidance learning. Oral administration of ZSET1446 at the dose range of 0.01 to 1 mg/kg ameliorated Aβ-40 induced learning impairment in water-maze, Y-maze, and passive-avoidance tasks. ZSET1446 reversed the decrease of choline esterase transferase (ChAT) activity in the brain regions like hippocampus, glutathione like immuno reactivity in the cortex in Aβ-40 treated rats. Additionally, ZSET1446 showed ameliorative effects on learning at the dose of 0.001 to 0.1 mg/kg in a passive avoidance task. ZSET1446 may be a potential candidate for development as a therapeutic agent for AD [25].

6.2.3.5 *Carbamate and amide derivative of pyridine*

Previous studies by different research groups have validated that the 2-methoxyphenyl residue is a suitable fragment to achieve good AChE inhibition. Group insertion of methoxy phenyl and indole moiety was followed for hydrophobic and $\pi-\pi$ interactions with PAS regions. Two bulkier groups, namely 4-phenoxybenzyl groups and naphthyl also provided greater interactions. The carbamate derivative 6-(benzyl amino)hexyl (2,6-dichloropyridin-4-yl)carbamate was observed to be most potent AChE inhibitor ($IC_{50} = 0.153 \pm 0.016$ μM) while the carbamate 5-(((1H-indol-3-yl)methyl)amino)pentyl (2,6-dichloropyridin-4-yl)carbamate was the most potent inhibitor of butylcholine esterase (BChE) ($IC_{50} = 0.828 \pm 0.067$μM). A molecular docking study indicated that the carbamate 5-(benzyl amino) pentyl (2,6-dichloropyridin-4-yl)carbamate could bind to AChE by interacting with both CAS(catalytic active site) and PAS(peripheral active site), in agreement with the mixed inhibition mechanism. Further, synthesized lead was able to inhibit Aβ42 self-aggregation and possess low toxicity against human cell lines [27]. The pyridine derivatives used as Amyloid-β inhibitors have been summarized in Table 6.2.

6.2.4 Role of pyridine as BACE-1 (beta-site amyloid precursor protein cleaving enzyme) inhibitors

6.2.4.1 *Amino oxazoline xanthenes*

BACE-1 can be regarded as a potential target for the development of anti-Alzheimer drug. Various scientists have described that 2-amino xanthenes heterocycle scaffold is essential for hydrogen-bonding interactions with Asp residues of the protease enzyme. The 4-aza group substitution in xanthenes increases BACE-1 potency of BACE-1. Xanthene chain is substituted in both ends to improve pharmacokinetic profiles and BBB penetration compared to earlier inhibitors. Replacement of the side chain of amino xanthene ring with 2- fluoro-3-pyridyl ring exhibited excellent BACE1 inhibition potency in the enzymatic assays. After obtaining promising results with 2-amino xanthene [31], 5-amino xanthene was also designed and synthesized but such compounds showed reduced BACE-1 potency. Substitution like methoxy, neopentyloxy, and dihydropyran side chains reduces BACE-1 potency significantly.

Among various substitutions, a compound AMG-8718 ((S)-7-(2-fluoro-3-pyridinyl)-3-((3- methyloxetan-3-yl)ethynyl)-5′H-spiro[chromeno[2,3-b]pyridine-5,4′-oxazol]-2′-amine) has methyl oxetane and 2-fluoro-3-pyridinyl substitutions in opposite ends of the xanthene ring show greater permeability and high potency toward BACE-1($IC_{50} = 0.7 \pm 0.3$ nM) [31]. 2-fluoro-3-pyridinyl group is essential for Ser229 interactions in the S3 pocket of BACE-1, oxetane oxygen interacts with Arg128 and nitrogen in xanthene ring show hydrogen bonding with Trp76. Suitably AMG-8718 show enhanced ligand-receptor binding and potency. Male Sprague Dawley rats were given AMG-8718 by oral gavage and CSF fluids were analyzed for Aβ concentrations in brain regions. After 4h intervals CSF fluids were examined and found

TABLE 6.2 Summary of pyridine structures used as $A\beta$ amyloid aggregation inhibition.

Comp. Sr. No.	IUPAC name	Structure	Activity	References
(1)	(R)-Methyl 10-Cyclopropyl-5-oxo-7,9-diphenyl-3,5-dihydro-2H thiazolo[2,3-g][1,7]naphthyridine-3-carboxylate	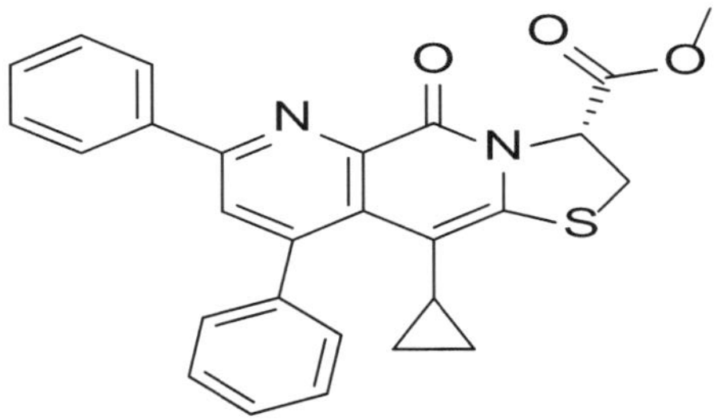	–	[30]
(2)	(R)-Methyl 7-(4-Nitrophenyl)-5-oxo-10-phenyl-3,5-dihydro-2Hthiazolo[2,3-g][1,7]naphthyridine-3-carboxylate	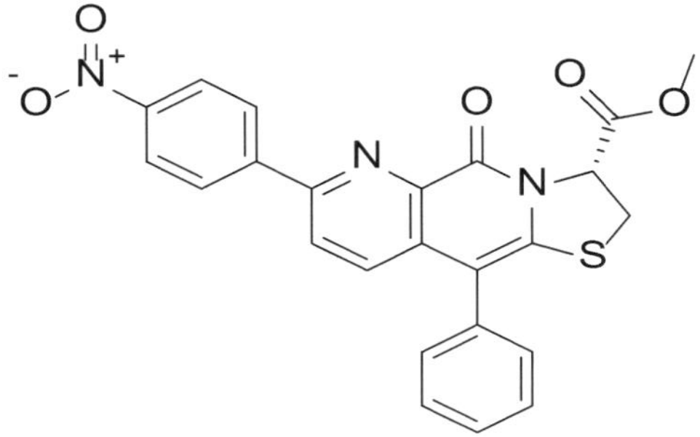	–	[30]
(3)	6-iodo-2-(4′-N,N′-dimethylamino) phenyl imidazo[1,2-a] pyridine **(IMPY)**	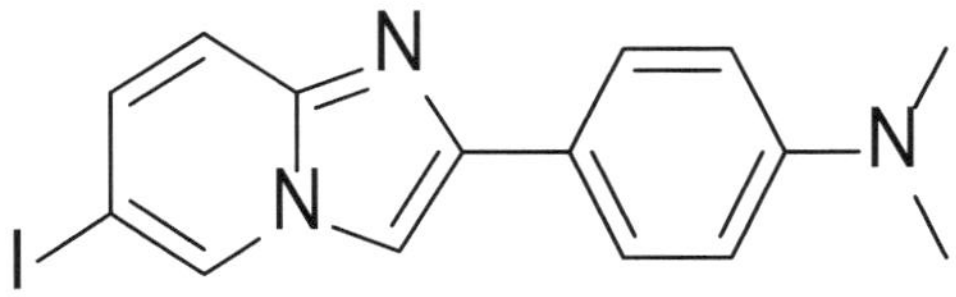	$K_i = 15.0 \pm 5.0$ nM Excellent brain uptake (2.9% ID/organ at 2 min) fast removal (0.2% ID/organ at 1 h)	[32,33]
(4)	6-(N,N′-dimethlyamino)-2-(4′-bromophenyl) phenyl imidazo[1,2-a] pyridine	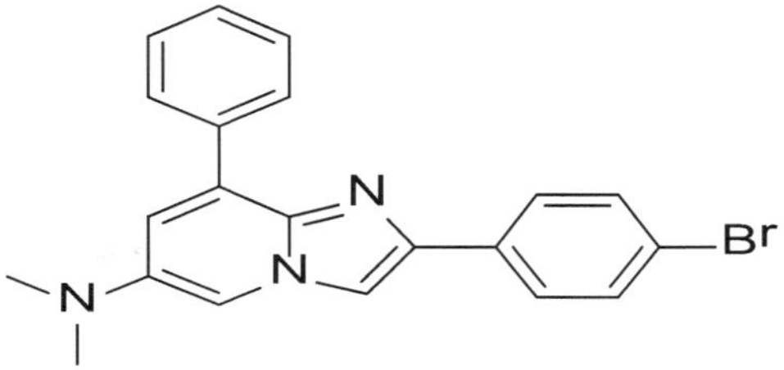	$K_i = 339$ nM	[31]

(continued on next page)

TABLE 6.2 Summary of pyridine structures used as Aβ amyloid aggregation inhibition—cont'd

Comp. Sr. No.	IUPAC name	Structure	Activity	References
(5)	(6-(2′-fluoroethyl)-2-(4′ –dimethyl amino) phenyl imidazo [1, 2-a] pyridine) **(FEPIP)**	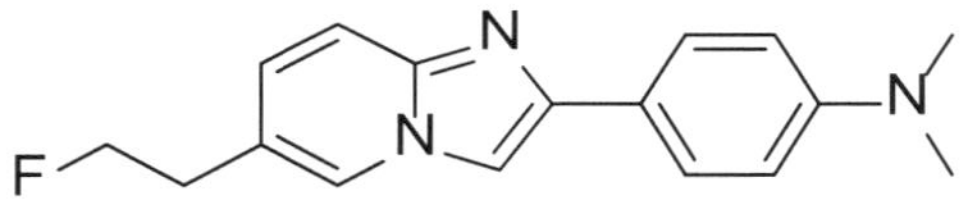	$K_i = 177$ nM	[35]
(6)	(6-(3′-fluoropropyl)-2-(4′ -dimethylamino) phenylimidazo [1, 2-a] pyridine) **(FPPIP)**	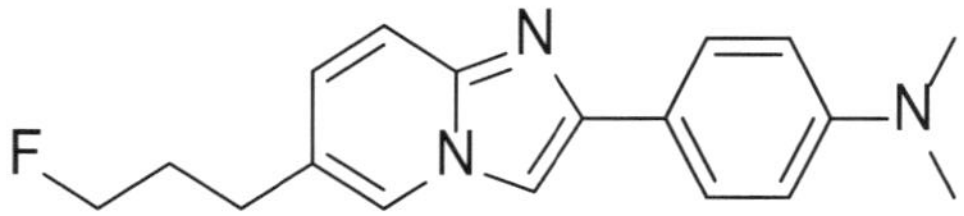	$K_i = 48.3$ nM	[35]
(7)	spiro [imidazo [1, 2-a] pyridine-3, 2-indan]-2(3H)-one **(ZSET1446)**	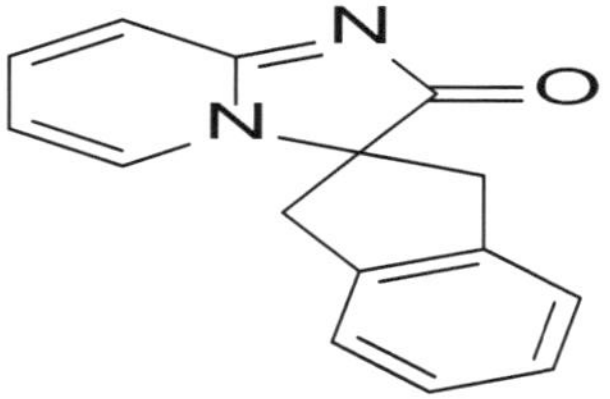	Ameliorative effects on learning at the dose of 0.001 to 0.1 mg/kg	[27]
(8)	6-(benzyl amino)hexyl (2,6-dichloropyridin-4-yl)carbamate	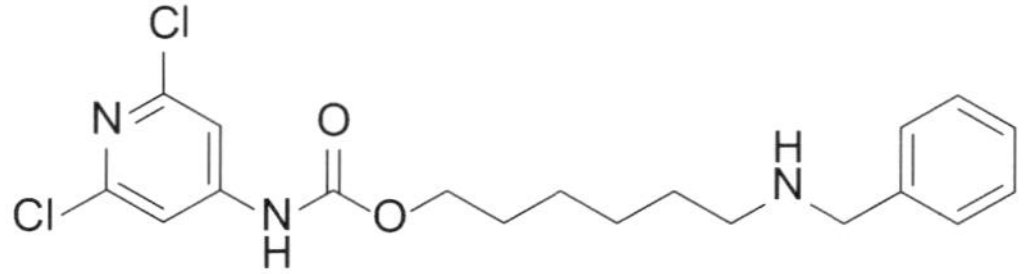	($IC_{50} = 0.153 \pm 0.016$ μM) for AChE	[29]
(9)	5-(((1H-indol-3-yl)methyl)amino)pentyl (2,6-dichloropyridin-4-yl)carbamate	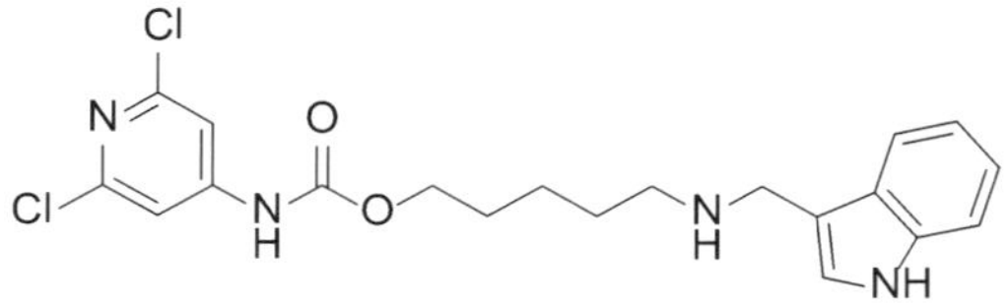	$IC_{50} = 0.828 \pm 0.067$μM	[29]

the reduction of 42% Aβ in CSF fluids. The compound showed moderate pharmacokinetic behavior by gradual total clearance and moderate half-life (5h–8h).

6.2.4.2 Iminothiadiazine dioxide [MK-8931(Verubecestat), JNJ-54861911(Atabecestat)]

During optimization of a preclinical candidate, AMG-8718 showed a balanced profile in permeability and BACE-1 potency. A series of patent compounds [MK-8931(Verubecestat), JNJ-54861911(Atabecestat), CNP-520(Umibecestat)] with pyridine moieties were developed with both BACE-1 and BACE-2 selectivity. MK-8931 (*N*-((5R)-3- amino-5,6-dihydro-2,5-dimethyl-1,1-dioxido-2H-1,2,4-thiadiazin-5-yl)-4-fluorophenyl)-5-fluoro-2-pyridine carboxamide) was highly selective for BACE-1 but failed in phase-3 clinical trials due to its efficacy issues and adverse effects [32,33].

6.2.4.3 Imidazo[1,2a]pyridine, AZD-3293 (Lanabecestat)

Another isoindole-based structure AZD-3293 (Lanabecestat) led to the development of multifaceted drug which could target both BACE-1 and Choline esterase inhibition. AZD-3293 showed moderate potency against BACE-1 (IC_{50} =2.84 μM) [34].

6.2.4.4 Disubstituted pyridinyl amino hydantoin

Small molecules containing amino hydantoins were substituted with aromatic ring in tetrahydro regions of hydantoin to get more compact structures. These substituted hydantoin were more potent (IC_{50} = 3.4 μM) compared to unsubstituted molecules. Further, it was observed that disubstituted pyridinyl amino hydantoin increased ligand affinity toward BACE-1 enzyme (IC_{50} =20 nM) to 1900 fold.

BACE-1selectivity was enhanced by closely studying the difference of BACE-1 and BACE-2 enzyme. The catalytic sites of both the enzymes are similar and a difference occurs in S2 pocket when Pro70 in BACE-1 is replaced by Lys86 of BACE-2[15]. Thus above enzyme study enabled to find enzyme specific ligand. The pyridine ring held an important role for entry into S3 pocket and hydrophobic interaction with Ser229 with water residues in the catalytic site of enzyme. Ligand exhibited greater potency ($IC_{50} = 0.06 \pm 0.01$ μM) [35] due to the Vander Waals interaction, water-bridge interactions and hydrogen bonding with Trp 76. However, BACE-1 selectivity got improved by substituting the ortho carbon of pyridine ring. *O*-methyl pyridine ring modification designed lead exhibited two-fold change in BACE-1 potency ($IC_{50} = 0.1 \pm 0.002$ μM). Thus, pyridine-based analogs declared more potency and selectivity toward BACE-1 in FRET (fluorescence resonance energy transfer) assay. The pyridines revealed promising results in cell line studies. In addition 2,6-diethyl pyridyl analog namely, (5S)-2-amino-5-(2,6- diethylpyridin-4-yl)-3-methyl-5-[3-(pyrimidin-5-yl)phenyl]- 3,5-dihydro-4H-imidazol-4-one show nanomolar potency (IC_{50} = 10nM) and 670 fold greater selectivity toward BACE-1. Cell line study of the above compound was satisfactory (EC_{50} = 130nM) in the enzyme-linked immunosorbent assay (ELISA) [35].

6.2.4.5 Substituted chelidamic acids

Analogs comprising of 2,6-pyridine dicarboxylic scaffold targets Arg235 which is active site of BACE-1 located between cleft and flap region. The S2 pocket-forming domain of BACE-1 is narrow thus isophthalic scaffold also referred to as substituted chelidamic acids was designed. Elaborated study of trans-membrane aspartic protease, BACE-1 enzyme envisaged sites β-site of APP. The β-site cleavage refers to DLVFFAE [14] fragment of which the first four amino acids are referred to as P1, P2, P3, P4 amino acids on the basis of which several peptidomimetic ligands were designed [22]. Phenyl norstatine or ((2R,3S)-3-amino-2-hydroxy-4- phenyl-butyric acid) was substituted at P1 position, chelidamic acids were replaced at P2, and glu bioisostere was added to P4 position to yield a new molecule. Such molecules include KMI-429 highlights two carboxylic acids at P1 position and are highly potent against BACE-1 ($IC_{50} =$ 3.9 nM). Carboxylic rings are replaced by tetrazolyl ring in KMI-429 [36] distinctly potent against BACE- ($IC_{50} =$ 1.2 nM).Halogens like chlorine (Table 6.3), bromine, and iodine substituted chelidamic acids (namely KMI-1283, KMI-1303, KMI-1302) were highly potent against BACE-1 with IC_{50} values of 13nM, 9nM, 10nM, respectively. Electron-rich halogenated groups increased lipophilicity and bioavailability. Coulomb's force interaction was observed between electron-poor guanidine and electron-rich halogens present in the above compounds [37].

6.2.5 Role of pyridine as metal chelators

6.2.5.1 *Aroyl hydrazones*

Endogenous metal ions like Zn^{2+}, Cu^{2+} may accord to Aβ accumulation in AD brains by cross-linkages of histidine residues. Metal protein attenuating compounds (MPAC) are mostly composed of hydroxyquinoline which delays or prevents the progression of AD. Metal chelators like 8-hydroxyquinoline-2-carboxaldehyde isonicotyl hydrazones reduce the accumulation and toxicity of Aβ oligomers. These chelators must bound to Zn^{2+}, Cu^{2+} ions and reduce the toxicity [5,19].

Thus, series of metal chelators using hydrazide and pyridine 2-carboxaldehyde formed pyridine-2-carboxaldehyde isonicotyl hydrazone (HPCIH) [38]. Such hydrazone formed co-ordinated complex with divalent cations like Zn^{2+} and Cu^{2+} in a diverse way to form [Zn (HPCIH) $SO_4]_n$ and $Cu(PCIH)_2$. HPCIH was found to be stable in methanol and DMSO solutions and exhibit dissociation constant K_d in the range of 1μM to 20 μM. Their partition coefficients or log P values were nearly 0.91 ± 0.1, adequate to cross the blood brain barrier. Electron spray ionization mass spectrometry (ESI–MS) shows changes in spectra on the addition of Zn^{2+} mostly focusing on metal coordination sites like Asp1, His6, His13, His14 residues. HPCIH has a greater affinity for Zn^{2+} and can efficiently compete with Aβ for Zn^{2+}.

6.2.5.2 *IMPY*

6-iodo-2-(4-dimethylamino)phenylimidazo [1,2-a]pyridine or IMPY derivatives have been earlier discussed as Aβ aggregation inhibitors. These novel compounds were aimed to be designed as bifunctional small molecules which reduce both Aβ accumulation and metal coupled Aβ induced toxicity. IMPY derivatives like 6-(6-iodoimidazo[1,2-a]pyridin-2-yl)-N,N-dimethylpyridin-3-amine and 6-(imidazo[1,2-a]pyridin-2-yl)-*N*,*N*-dimethylpyridin-3-amine [39] contains two nitrogen donor atoms which could regulate Cu^{2+} induced toxicity. The above derivatives were confirmed for metal chelation by high-resolution 2D NMR spectroscopy. Metal triggered inhibition was controlled by newly designed bifunctional molecules [40].

TABLE 6.3 Summary of pyridine structures used as BACE-1 inhibitors.

Comp. Sr. No.	IUPAC name	Structure	Activity	References
(1)	((S)-7-(2-fluoro-3-pyridinyl)-3-((3-methyloxetan-3-yl) ethynyl)-5′H-spiro[chromeno[2,3-b]pyridine-5,4′-oxazol]-2′-amine) **(AMG-8718)**	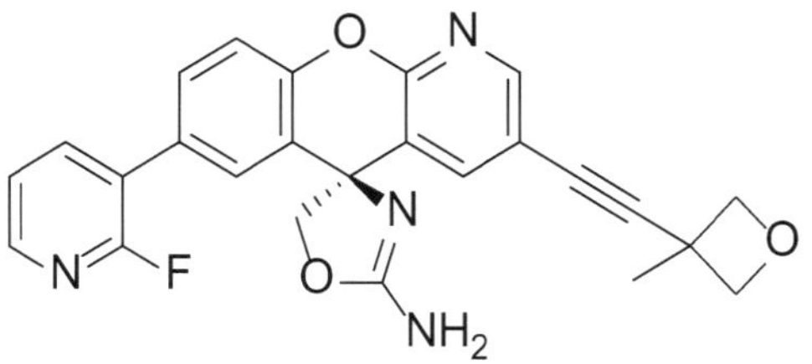	(IC_{50}= 0.7 ± 0.3 nM)	[36]
(2)	N-[3-[(5R)-3-Amino-5,6-dihydro-2,5-dimethyl-1,1-dioxido-2H-1,2,4-thiadiazin-5-yl]-4-fluorophenyl]-5-fluoro-2-pyridinecarboxamide **(MK-8931)**	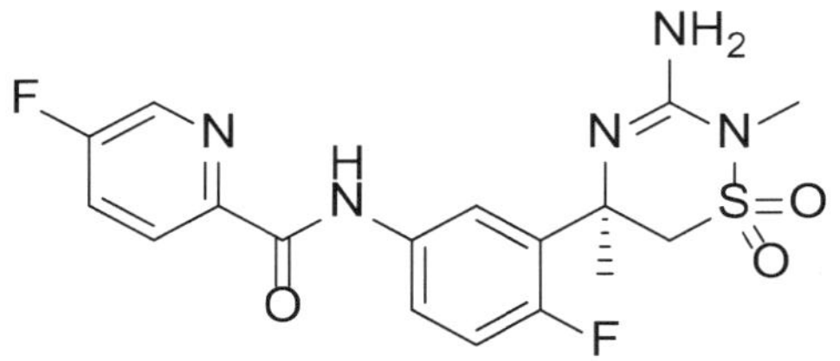	Phase 3 clinical trials	[37,38]
(3)	4-Methoxy-5″-methyl-6′-[5-(prop-1-yn-1-yl)pyridin-3-yl]-3′H-dispiro[cyclohexane-1,2′-indene-1′,2″-imidazole]-4″-amine **(AZD-3293)**	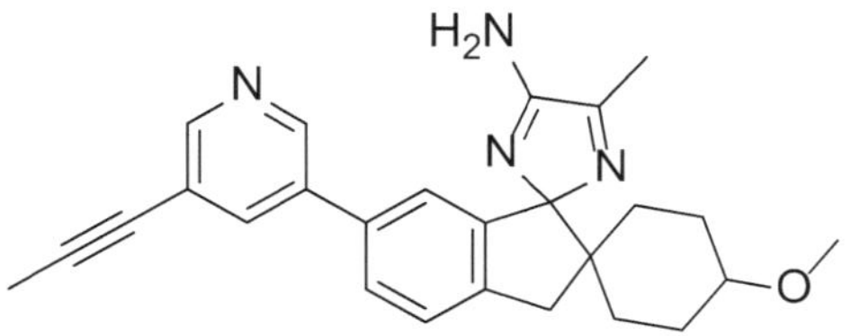	(IC_{50} =2.84 μM)	[39]

(continued on next page)

TABLE 6.3 Summary of pyridine structures used as BACE-1 inhibitors—cont'd

Comp. Sr. No.	IUPAC name	Structure	Activity	References
(4)	(5S)-2-amino-5-(2,6-diethylpyridin-4-yl)-3-methyl-5-[3-(pyrimidin-5-yl)phenyl]-3,5-dihydro-4H-imidazol-4-one	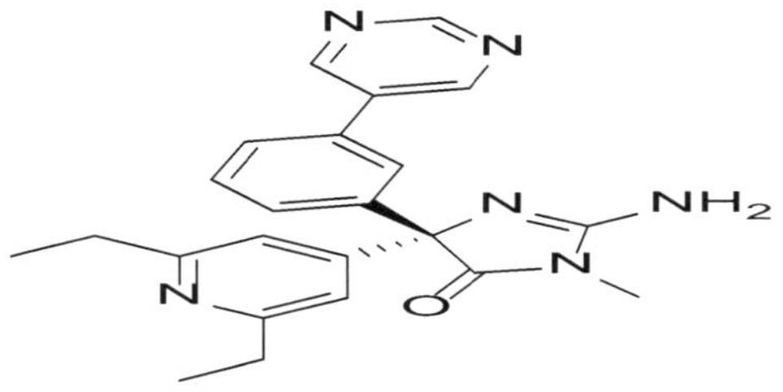	($IC_{50} = 10$ nM)	[40]
(5)	**(KMI-429)**	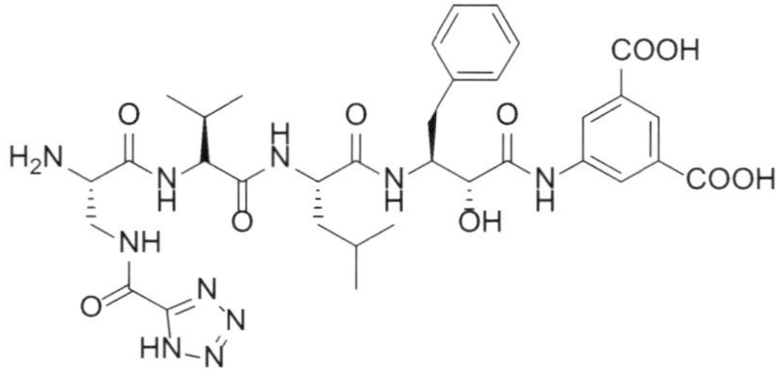	($IC_{50} = 3.9$ nM)	[41]
(6)	**(KMI-1283)**	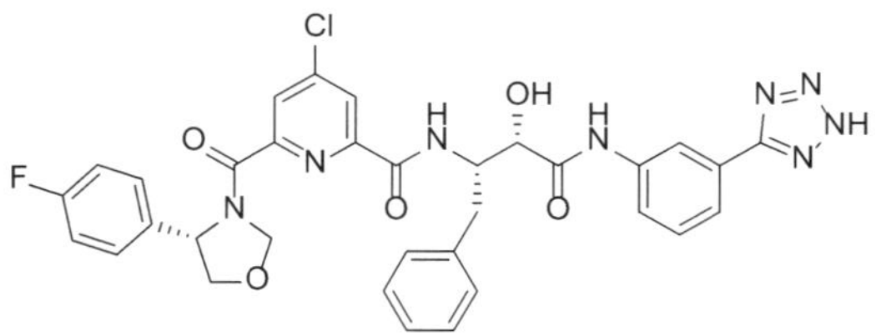	($IC_{50} = 13$ nM)	[41]

(continued on next page)

(7)	**(KMI-1303)**	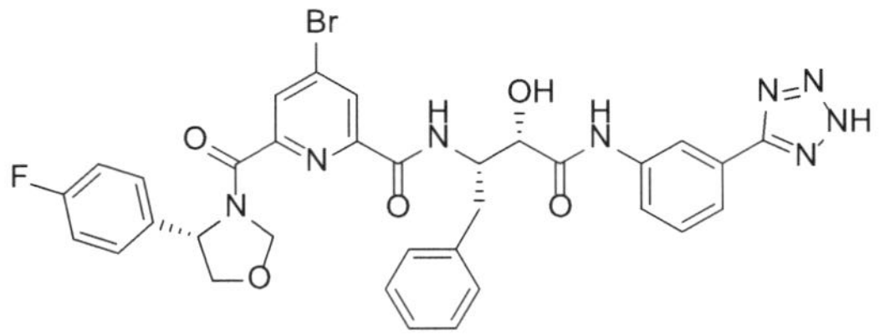	(IC_{50} = 9 nM)	[41]
(8)	**(KMI-1302)**	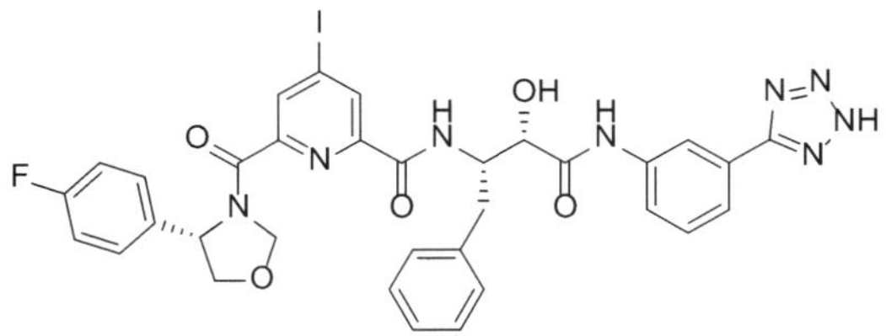	(IC_{50} = 10 nM)	[41,43]

6.2.5.3 DMAP and ENDIP (N′, N′–bis (pyridine-2-yl-methyl)ethane-1,2-diamine)

The derivatives containing pyridine were investigated as metal chelators in AD. DMAP contains two dissociating protons at pyridine and secondary amine group. Metal-binding ability was examined by pH potentiometry, EPR spectroscopy with Zn^{2+} and Cu^{2+}. ENDIP proves to be a better MPAC as compared to DMAP which forms a very stable ML type complex at physiological pH. Their dissociation constants (K_d) with Zn^{2+} and Cu^{2+} is 0.6mM and 3.98pM, respectively. DLS measurements and fluorescence results confronts that ENDIP displaced Aβ from their metal complexes [41].

6.2.5.4 Pyridine amine derivative (PAT)

Metal chelators are also referred to as potential anti-Alzheimer's agents. PAT forms coordinate complexes by introducing donor atoms like N, N, O responsible for antioxidant activity. PAT interacts with metals like Cu^{2+} and Zn^{2+} and disaggregates metal-related Aβ aggregates.

3-bis (pyridin-2-ylmethyl) aminomethyl-5-hydroxybenzyltriphenylphosphonium bromide (PAT), protects the cell against oxidative damage due to the presence of phenolic group and triphenyl phosphonium cation (TPP^+) [42] moiety mediated mitochondrial toxicity. PAT inhibited self-assembly of Aβ fibrils and metal-induced Aβ aggregation and transiently reduced metal oxidation. Cognitive and memory abilities of AD mice were improved significantly by PAT introduction. The chelating region in PAT is bis(2-pyridylmethyl)amine binds to fluorescent compound Thioflavin T and was observed that it decreases metal-induced aggregation. PAT shows potent activity in Zn^{2+} induced aggregation ($IC_{50} = 27.42 \pm 1.5$ mM). Aβ forms self-aggregates and leads to toxicity when it transforms from α-helix to β-helix. PAT is known to prevent self-aggregates of Aβ. ($IC_{50} = 12.5 \pm 1.3$ mM). The fluorescence intensity decreased when PAT was incorporated with Aβ which indicates Aβ fibril reduction [43]. Circular dichroism spectroscopy investigated conformational changes of Aβ from α-helix to β-helix and a positive peak in broader areas was observed with Aβ incubated with Zn^{2+}. Pyridine derivatives as metal chelators to combat AD have been demonstrated in Table 6.4.

6.3 Synthetic routes of some pyridine derivatives used in AD

Various pyridine derivatives were discussed on the basis of their major nucleus and their role as AChE inhibitors, Aβ inhibitors, BACE-1 inhibitors, and metal chelators. Thus, pyridine nucleus finds greater application against AD. Various synthetic procedures of important nucleus were discussed in order to give a vivid idea of such compounds.

6.3.1 Synthesis of imidazo [1,5a] pyridine carboxylic acids derivative

Ethyl-2-pyridinyl acetate was used as starting material and treated with sodium nitrite and glacial acetic acid to yield Schiff base substituted pyridine. The product in the reported pathway was hydrogenated in the presence of Pd to form amino ester. Amino ester resulted in an intermediate with butyryl chloride and final cyclization of the intermediate product was achieved using thionyl chloride. Imidazo [1,5a] pyridine [24] carboxylic acids were finally obtained after series of above-mentioned steps. Hydrolysis of ester was carried out to give

TABLE 6.4 Summary of pyridine structures used as metal chelators.

Comp. Sr. No.	IUPAC name/compound code	Structure	Activity	References
(1)	*N*-[(E)-pyridin-2-ylmethylideneamino]pyridine-4-carboxamide	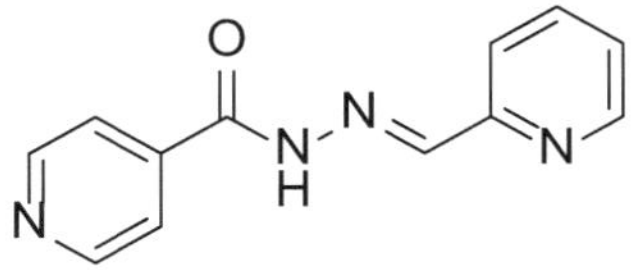	1–20 μM	[44]
(2)	6-(6-iodoimidazo[1,2-a]pyridin-2-yl)-*N*,*N*-dimethyl pyridin-3-amine	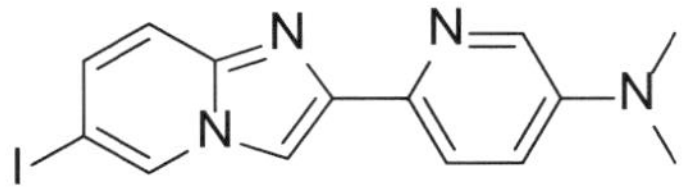	–	[33]
(3)	6-(imidazo[1,2-a]pyridin-2-yl)-*N*,*N*-dimethylpyridin-3-amine	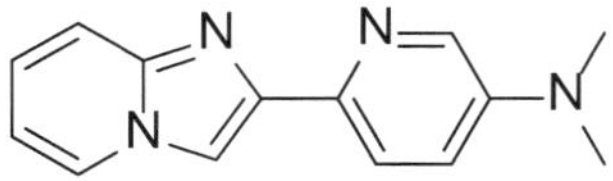	–	[46]
(4)	(N*N'*, *N'* –bis (pyridine-2-yl-methyl)ethane-1,2-diamine)(ENDIP)	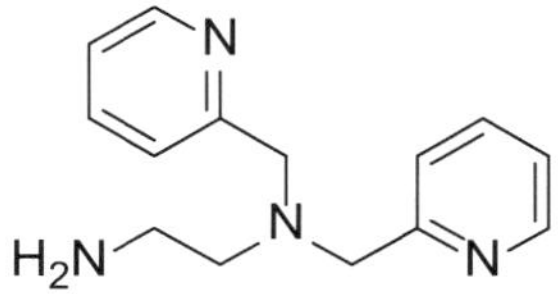	$K_d = 0.6$ mM	[46]

(continued on next page)

TABLE 6.4 Summary of pyridine structures used as metal chelators—cont'd

Comp. Sr. No.	IUPAC name/compound code	Structure	Activity	References
(5)	dimethyl aminopyridine (DMAP)	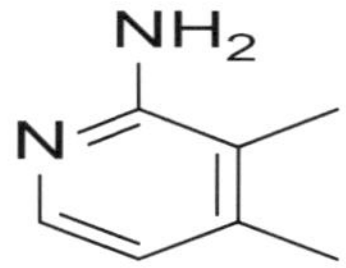	$K_d = 3.98$ pM	[46]
(6)	3-bis (pyridin-2-ylmethyl) aminomethyl-5-hydroxybenzyltriphenylphosphonium bromide	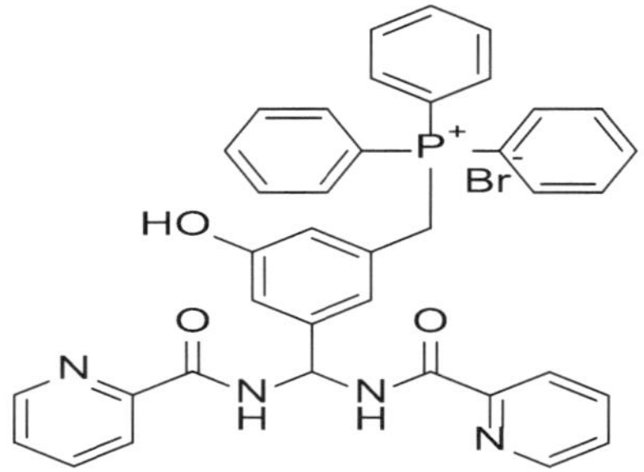	($IC_{50} = 27.42 \pm 1.5$ mM)	[47]

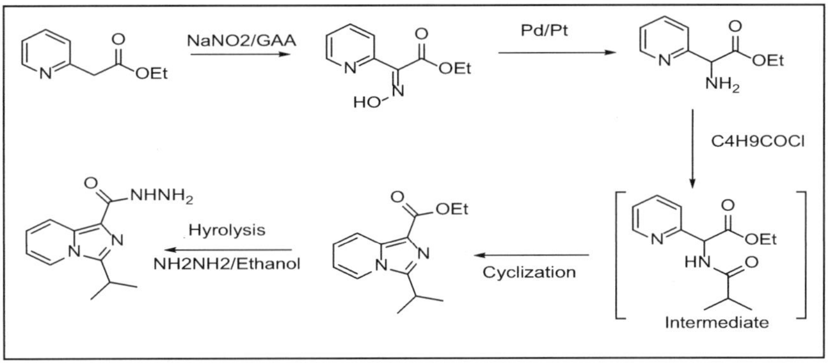

FIGURE 6.5 Synthetic scheme of imidazo [1,5a] pyridine carboxylic acid derivative [24].

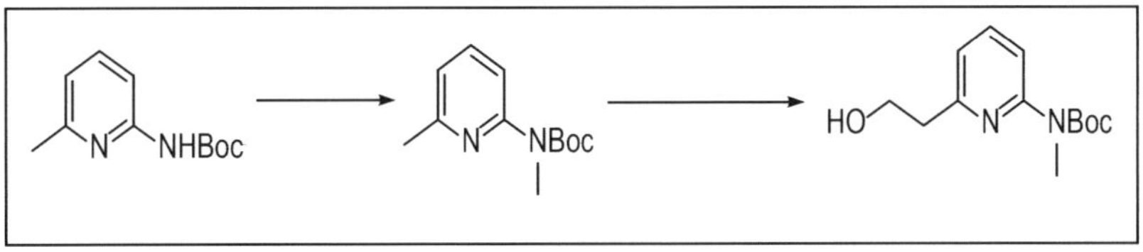

FIGURE 6.6 Synthetic scheme of 2,6-disubstituted pyridine derivatives [44].

acids and hydrazine hydrate was finally treated to give hydrazone imidazo [1,5a] pyridine acids (Fig. 6.5).

6.3.2 2,6-disubstituted pyridine

2-amino-boc (ter-butyloxy carbonyl) derived pyridine derivatives were alkylated with CH_3I using NaH. Further lithium diisopropyl amine (LDA) was treated at -78 °C in acetic acid and methanol corresponds to alcohol. Such protecting groups are substituted with 2,6-disubstituted pyridine amine and carbon linkers are attached on both sides [44] (Fig. 6.6).

6.3.3 Pyrrolo aminopyridine

Pyrrolo aminopyridine derivatives were aminated using HOSA (hydroxylamine-o-sulphonic acids). $OHSO_2ONH_2$ and condensation reaction takes place with 4-pyridinyl chloride to give pyrrolo aminopyridine [45] (Fig. 6.7).

6.3.4 Bis-1,2,4-triazole/thiosemicarbazide

0.02 mol of pyridine-2,5-dicarboxylic acid was refluxed for 24 h using 100 mL methanol solvent, 1.8 mL of conc. H_2SO_4. Hydrazine hydrate formed hydrazone complex, pyridine-2,5-dicarbohydrazide with dimethyl2,5-pyridinedicarboxylate (Fig. 6.8). Pyridine-2,

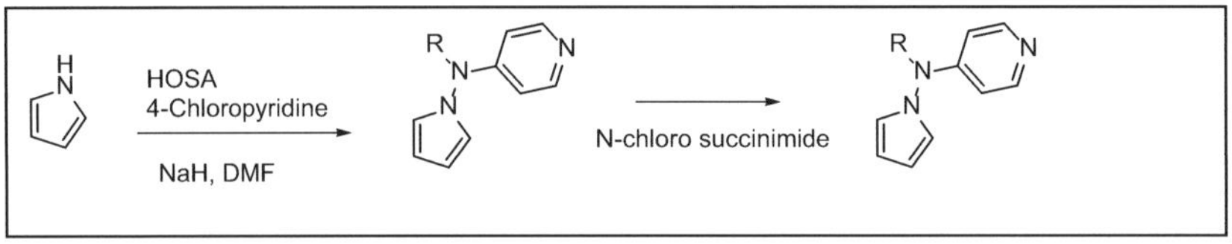

FIGURE 6.7 Synthetic scheme of pyrrolo aminopyridine [58].

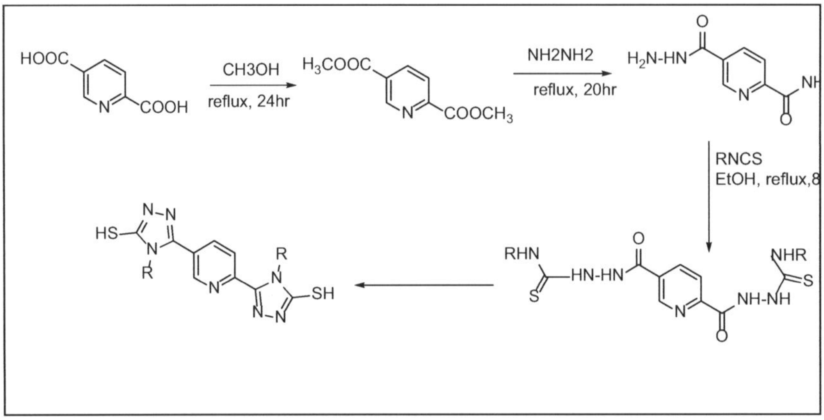

FIGURE 6.8 Synthetic scheme of bis-1,2,4-triazole/thiosemicarbazide [59].

5-dicarbohydrazide was refluxed with aryl isothiocyanate for 3 to 8 h formed thiosemicarbazides. Intramolecular cyclization formed bis-1, 2, mercapto triazole [22].

6.4 Toxicological manifestations of pyridine

Pyridine is a substance which has odor and leads to sensitivity in odor. The common toxicity observed in pyridine is myosis, salivation, nasal secretion, apnoea, micturition. Pyridine itself is a toxic compound and a dosage of 880 mg/kg body weight may lead to death due to cardiac failure. At a dose of 880 mg/kg body weight, there is a steep decline in blood pressure and a marked increase in heart rate.

In a decreased dose of 176 and 440 mg/kg body weight there is a fall in blood pressure along with an increased heart rate from 83 to 160 heartbeats per minute. The respiratory rate was not affected greatly but there was a slight decreased rate in respiration.

6.5 Molecular modeling and computational simulation of pyridine in AD

The probable binding conformations of imidazo[1,2a] pyridine and protein were observed by molecular modeling studies [46]. Protein stability study by recording root mean square deviation (RMSD) at 20ns time interval was noted. Molecular dynamic (MD) simulations were carried out using Maestro 11.1.011. The OPLS-2005 molecular mechanics force field at

TABLE 6.5 Pyridine scaffold in clinical trials as BACE-1 inhibitors [34].

Sr. No.	Drug Candidate	Developers	Class	Clinical trial phases	Withdrawing status
1.	AMG-8718	Amgen	Hydroxylamine(HEA) based inhibitors	Phase 1	–
2.	CTS-21166	Astellas	Spiro-1,3-oxazolidine	Phase 1	Terminated
3.	LY2886721	Bristol Myers Squibb(BMS)	Dihydro-1,3-thiazine	Phase 1	Terminated due to toxicity
4.	E-2609 (elenbecestat)	Eisai and Biogen	Amino-dihydrothiazine	Phase 3	Terminated due to their less efficacy
5.	MK-8931 (verubecestat)	Merck Sharp and Dohme (MSD)	Iminothiadiazine dioxide	Phase 3	Terminated due to their less efficacy
6.	LY-2886721	Eli lily	1,4-diazine oxide	Phase 2	Terminated due to toxicity
7.	JNJ-54861911 (atabecestat)	Jassen	1,4-oxazine	Phase 2	Terminated due to toxicity

1.01325 bar pressure and the trajectory interval was 4.8ps time. However standard temperature pressure was maintained at 300K and was used for performing the 1 bar pressure. The pyridine and the protein molecule were optimized before MD study [47].

The MD simulations postulated probable binding conformation of the docked complex by providing system changes like temperature, pressure, and water molecules similar to that present in our body. The RMSD value obtained was less than 3Å which proves the dynamic stability of the protein with and without pyridine ligand. Physiologically similar environments were provided by MD simulations. The simulation study gives an elaborated idea of binding conformations for the docked complex.

6.6 Pyridine in AD research (marketed drugs as well as preclinical trial drug having pyridine)

Various pyridine-containing fused rings with heterocyclic groups like imidazo, pyrazole formed the central core of study against AD. Imidazo [1,2a] pyridine [22] and pyrazolo [3,4b] [46] are promising candidates and are in widely in research especially IMPY [29,40,48,49] (6-iodo-2-(4-diethylamino) phenyl imidazo [1, 2-a] pyridine. Substitution in their central core forms a large library of compound which targets Aβ aggregates. However, these drugs are not reported in clinical trial phases. Candidate compound targeting BACE-1 [13,22,50]/BACE-2 exhibit positive inhibition and could reach mostly Phase2/3 clinical trials. Although most of compounds under trials were terminated due to their efficacy and toxicity. The molecules containing pyridine scaffold under clinical trial are discussed along with their current clinical phase status in Table 6.5.

6.7 Conclusion

Pyridine is a colorless liquid with an unpleasant odor. Pyridine is used as a good solvent and is formed from the degradation of various natural materials. Various molecules of pyridine and their applications were discussed. The above discussion implies pyridine to be a potent molecule for AD treatment as they show a broad inhibiting strategy. Various molecules containing pyridine are in clinical trials against AD.

Thus, pyridine can be used to design synthetic heterocyclic molecules containing pyridine analogs for multitargeted therapy for AD. Pyridine molecules can be a promising lead for neurodegeneration if their toxicity and efficacy were modified by changing the substitution. Additionally, pyridine show wide use in AChE inhibitors [9,26,27,31,51–53], Aβ inhibitors [2,5,19,35,54], BACE-1(lead under clinical trial), and metal chelators [5,19,31,38,41,43,55].

Pyridine leads to liver toxicity and few analogs are terminated in clinical trials. Thus, suitable molecules with enhanced efficiency, lesser toxicity, and greater potency should be designed to obtain promising drug-like candidates for future use.

References

[1] H. Konno, T. Sato, Y. Saito, I. Sakamoto, K. Akaji, Synthesis and evaluation of aminopyridine derivatives as potential BACE1 inhibitors, Bioorganic Med. Chem. Lett. 25 (22) (2015) 5127–5132. https://doi.org/10.1016/j.bmcl.2015.10.007.

[2] V.A. Online, A.P. Piccionello, C. Marino, M.G. Ortore, P. Picone, S. Vilasi, M. Carlo S. Di, Buscemi, D. Bulone RSC advances. doi:doi:10.1039/C4RA13556C.

[3] B.De Strooper. Amyloid-beta precursor protein processing in neurodegeneration. 2004, 1, 582–588. doi:doi:10.1016/j.conb.2004.08.001.

[4] G.F. Chen, T.H. Xu, Y. Yan, Y.R. Zhou, Y. Jiang, K. Melcher, H.E. Xu, Amyloid beta: structure, biology and structure-based therapeutic development, Acta Pharmacol. Sin 38 (9) (2017) 1205–1235. https://doi.org/10.1038/aps.2017.28.

[5] S. Bandyopadhyay, X. Huang, D.K. Lahiri, J.T. Rogers, Novel drug targets based on metallobiology of Alzheimer's disease, Expert Opin. Ther. Targets 14 (11) (2010) 1177–1197. https://doi.org/10.1517/14728222.2010.525352.

[6] P. Mishra, P. Sharma, P.N. Tripathi, S.K. Gupta, P. Srivastava, A. Seth, A. Tripathi, S. Krishnamurthy, S.K. Shrivastava, Design and development of 1,3,4-oxadiazole derivatives as potential inhibitors of acetylcholinesterase to ameliorate scopolamine-induced cognitive dysfunctions, Bioorg. Chem. 89 (2019), doi:10.1016/j.bioorg.2019.103025.

[7] G. Kryger, I. Silman, J.L. Sussman, Structure of acetylcholinesterase complexed with E2020 (Ariceptρ): implications for the design of new anti-alzheimer drugs, Structure 7 (3) (1999) 297–307. https://doi.org/10.1016/S0969-2126(99)80040-9.

[8] T. Silva, J. Reis, J. Teixeira, F. Borges, Alzheimer's disease, enzyme targets and drug discovery struggles: from natural products to drug prototypes, Ageing Res. Rev. 15 (1) (2014) 116–145. https://doi.org/10.1016/j.arr.2014.03.008.

[9] M. Hernández-Rodríguez, J. Correa-Basurto, F. Martínez-Ramos, I.I. Padilla-Martínez, C.G. Benítez-Cardoza, E. Mera-Jiménez, M.C. Rosales-Hernández, Design of multi-target compounds as AChE, BACE1, and Amyloid-B1-42 oligomerization inhibitors: in silico and in vitro studies, J. Alzheimer's Dis. 41 (4) (2014) 1073–1085. https://doi.org/10.3233/JAD-140471.

[10] C. Zhang, A. Browne, J.R. Divito, J.A. Stevenson, D. Romano Amyloid- β production via cleavage of amyloid- β protein precursor is modulated by cell density. 2010, 22, 683–694. https://doi.org/10.3233/JAD-2010-100816.

[11] H. Ramshini, M. mohammad-zadeh, A. Ebrahim-Habibi, Inhibition of amyloid fibril formation and cytotoxicity by a chemical analog of curcumin as a stable Inhibitor, Int. J. Biol. Macromol. 78 (2015) 396–404. https://doi.org/10.1016/j.ijbiomac.2015.04.038.

[12] X. Han, G. He, Toward a rational design to regulate β-amyloid fibrillation for Alzheimer's disease treatment, ACS Chem. Neurosci. 9 (2) (2018) 198–210. https://doi.org/10.1021/acschemneuro.7b00477.

[13] H.M. Kumalo, M.E.A Soliman, Comparative molecular dynamics study on BACE1 and BACE2 flap flexibility, J. Recept. Signal Transduct. 36 (5) (2016) 505–514. https://doi.org/10.3109/10799893.2015.1130058.

[14] H.M. Kumalo, S. Bhakat, M.E. Soliman, Investigation of flap flexibility of β-secretase using molecular dynamic simulations, J. Biomol. Struct. Dyn. 34 (5) (2016) 1008–1019. https://doi.org/10.1080/07391102.2015.1064831.

[15] H. Shimizu, A. Tosaki, K. Kaneko, T. Hisano, T. Sakurai, N. Nukina, Crystal structure of an active form of BACE1, an enzyme responsible for amyloid β protein production, Mol. Cell. Biol. 28 (11) (2008) 3663–3671. https://doi.org/10.1128/mcb.02185-07.

[16] S. Patel, L. Vuillard, A. Cleasby, C.W. Murray, J. Yon, Technology, A. Apo and inhibitor complex structures of BACE (b -Secretase). 2004, 407–416. 10.1016/j.jmb.2004.08.018.

[17] Y.J. Wu, J. Guernon, F. Yang, L. Snyder, J. Shi, A. McClure, R. Rajamani, H. Park, A. Ng, H. Lewis, C.Y. Chang, D. Camac, J.H. Toyn, M.K. Ahlijanian, C.F. Albright, J.E. Macor, L.A. Thompson, Targeting the BACE1 Active site flap leads to a potent inhibitor that elicits robust brain Aβ reduction in rodents, ACS Med. Chem. Lett. 7 (3) (2016) 271–276. https://doi.org/10.1021/acsmedchemlett.5b00432.

[18] N.-, Curcumin N. Ahsan, S. Mishra, M.K. Jain, A. Surolia, S. Gupta Modulate toxicity of wild type and. 2015. 10.1038/srep09862.

[19] S. Chan, S. Kantham, V.M. Rao, M.K. Palanivelu, H.L. Pham, P.N. Shaw, R.P. McGeary, B.P. Ross, Metal chelation, radical scavenging and inhibition of Aβ42 fibrillation by food constituents in relation to Alzheimer's disease, Food Chem. 199 (2016) 14–24. https://doi.org/10.1016/j.foodchem.2015.11.118.

[20] S. Riaz, I.U. Khan, M. Bajda, M. Ashraf, A. Qurat-Ul-Ain, Shaukat, T.U. Rehman, S. Mutahir, S. Hussain, G. Mustafa, M. Yar, Pyridine sulfonamide as a small key organic molecule for the potential treatment of Type-II diabetes mellitus and Alzheimer's disease: in vitro studies against Yeast α-glucosidase, acetylcholinesterase and butyrylcholinesterase, Bioorg. Chem. 63 (2015) 64–71. https://doi.org/10.1016/j.bioorg.2015.09.008.

[21] T. Pan, S. Xie, Y. Zhou, J. Hu, H. Luo, X. Li, L. Huang, Dual functional cholinesterase and PDE4D inhibitors for the treatment of Alzheimer's disease: design, synthesis and evaluation of tacrine-pyrazolo[3,4-b]pyridine hybrids, Bioorganic Med. Chem. Lett. 29 (16) (2019) 2150–2152. https://doi.org/10.1016/j.bmcl.2019.06.056.

[22] Z. Haghighijoo, S. Akrami, M. Saeedi, A. Zonouzi, A. Iraji, B. Larijani, H. Fakherzadeh, F. Sharifi, S.M. Arzaghi, M. Mahdavi, N. Edraki, N-Cyclohexylimidazo[1,2-a]pyridine derivatives as multi-target-directed ligands for treatment of Alzheimer's disease, Bioorg. Chem. 103 (July) (2020) 104146. https://doi.org/10.1016/j.bioorg.2020.104146.

[23] D. Karlsson, A. Fallarero, G. Brunhofer, P. Guzik, M. Prinz, U. Holzgrabe, T. Erker, P. Vuorela, Identification and characterization of diarylimidazoles as hybrid inhibitors of butyrylcholinesterase and amyloid beta fibril formation, Eur. J. Pharm. Sci. 45 (1–2) (2012) 169–183. https://doi.org/10.1016/j.ejps.2011.11.004.

[24] R. Nirogi, A.R. Mohammed, A.K. Shinde, N. Bogaraju, S.R. Gagginapalli, S.R. Ravella, L. Kota, G. Bhyrapuneni, N.R. Muddana, V. Benade, R.C. Palacharla, P. Jayarajan, R. Subramanian, V.K. Goyal, Synthesis and SAR of Imidazo[1,5-a[pyridine derivatives as 5-HT4 receptor partial agonists for the treatment of cognitive disorders associated with Alzheimer's disease, Eur. J. Med. Chem. 103 (2015) 289–301. https://doi.org/10.1016/j.ejmech.2015.08.051.

[25] Y. Yamaguchi, H. Miyashita, H. Tsunekawa, A. Mouri, H.C. Kim, K. Saito, T. Matsuno, S. Kawashima, T. Nabeshima, Effects of a novel cognitive enhancer, spiro[Imidazo-[1,2-a]Pyridine-3,2- Indan]-2(3H)-One (ZSET1446), on learning impairments induced by amyloid-B1-40 in the rat, J. Pharmacol. Exp. Ther. 317 (3) (2006) 1079–1087. https://doi.org/10.1124/jpet.105.098640.

[26] J. Kumar, A. Gill, M. Shaikh, A. Singh, A. Shandilya, E. Jameel, N. Sharma, N. Mrinal, N. Hoda, B. Jayaram, Pyrimidine-triazolopyrimidine and pyrimidine-pyridine hybrids as potential acetylcholinesterase inhibitors for Alzheimer's disease, ChemistrySelect 3 (2) (2018) 736–747. https://doi.org/10.1002/slct.201702599.

[27] F. Pandolfi, D. De Vita, M. Bortolami, A. Coluccia, R. Di Santo, R. Costi, V. Andrisano, F. Alabiso, C. Bergamini, R. Fato, M. Bartolini, L. Scipione, New pyridine derivatives as inhibitors of acetylcholinesterase and amyloid aggregation, Eur. J. Med. Chem. 141 (2017) 197–210. https://doi.org/10.1016/j.ejmech.2017.09.022.

[28] P. Singh, D.E. Adolfsson, J. Ådén, A.G. Cairns, C. Bartens, K. Brännström, A. Olofsson, F. Almqvist, Pyridine-fused 2-pyridones via povarov and A3 reactions: rapid generation of highly functionalized tricyclic heterocycles capable of amyloid fibril binding, J. Org. Chem. 84 (7) (2019) 3887–3903. https://doi.org/10.1021/acs.joc.8b03015.

[29] C.J. Chen, K. Bando, H. Ashino, K. Taguchi, H. Shiraishi, K. Shima, O. Fujimoto, C. Kitamura, S. Matsushima, K. Uchida, Y. Nakahara, H. Kasahara, T. Minamizawa, C. Jiang, M.R. Zhang, M. Ono, M. Tokunaga, T. Suhara, M. Higuchi, K. Yamada, B. Ji, In vivo SPECT imaging of amyloid-β deposition with radioiodinated imidazo[1, 2-α]pyridine derivative DRM106 in a mouse model of Alzheimer's disease, J. Nucl. Med. 56 (1) (2015) 120–126. https://doi.org/10.2967/jnumed.114.146944.

[30] F. Zeng, J.A. Southerland, R.J. Voll, J.R. Votaw, L. Williams, B.J. Ciliax, A.I. Levey, M.M. Goodman, Synthesis and evaluation of two 18F-labeled imidazo[1,2-a]pyridine analogues as potential agents for imaging β-amyloid in Alzheimer's disease, Bioorganic Med. Chem. Lett. 16 (11) (2006) 3015–3018. https://doi.org/10.1016/j.bmcl.2006.02.055.

[31] M. Saeedi, A. Rastegari, R. Hariri, S.S. Mirfazli, M. Mahdavi, N. Edraki, O. Firuzi, T. Akbarzadeh, Design and synthesis of novel arylisoxazole-chromenone carboxamides: investigation of biological activities associated with Alzheimer's disease, Chem. Biodivers. 17 (5) (2020), doi:10.1002/cbdv.201900746.

[32] M.E. Kennedy, A.W. Stamford, X. Chen, K. Cox, J.N. Cumming, M.F. Dockendorf, M. Egan, L. Ereshefsky, R.A. Hodgson, L.A. Hyde, S. Jhee, H.J. Kleijn, R. Kuvelkar, W. Li, B.A. Mattson, H. Mei, J. Palcza, J.D. Scott, M. Tanen, M.D. Troyer, J.L. Tseng, J.A. Stone, E.M. Parker, M.S. Forman, The BACE1 inhibitor verubecestat (MK-8931) reduces CNS b -amyloid in animal models and in Alzheimer's disease patients, Sci. Transl. Med. 8 (363) (2016) 1–14 Nov 2363ra150. doi:10.1126/scitranslmed.aad9704 .

[33] J.D. Scott, S.W. Li, A.P.J. Brunskill, X. Chen, K. Cox, J.N. Cumming, M. Forman, E.J. Gilbert, R.A. Hodgson, L.A. Hyde, Q. Jiang, U. Iserloh, I. Kazakevich, R. Kuvelkar, H. Mei, J. Meredith, J. Misiaszek, P. Orth, L.M. Rossiter, M. Slater, J. Stone, C.O. Strickland, J.H. Voigt, G. Wang, H. Wang, Y. Wu, W.J. Greenlee, E.M. Parker, M.E. Kennedy, A.W. Stamford, Discovery of the 3-Imino-1,2,4-thiadiazinane 1,1-dioxide derivative verubecestat (MK-8931)-A β-site amyloid precursor protein cleaving enzyme 1 inhibitor for the treatment of Alzheimer's disease, J. Med. Chem. 59 (23) (2016) 10435–10450. https://doi.org/10.1021/acs.jmedchem.6b00307.

[34] F. Rombouts, K. Kusakabe, C.C. Hsiao, H.J.M Gijsen, Small-molecule BACE1 inhibitors: a patent literature review (2011 to 2020), Expert Opin. Ther. Pat. 31 (1) (2021) 25–52. https://doi.org/10.1080/13543776.2021.1832463.

[35] M.S. Malamas, K. Barnes, M. Johnson, Y. Hui, P. Zhou, J. Turner, Y. Hu, E. Wagner, K. Fan, R. Chopra, A. Olland, J. Bard, M. Pangalos, P. Reinhart, A.J. Robichaud, Di-substituted pyridinyl aminohydantoins as potent and highly selective human β-secretase (BACE1) inhibitors, Bioorganic Med. Chem. 18 (2) (2010) 630–639. https://doi.org/10.1016/j.bmc.2009.12.007.

[36] T. Kimura, D. Shuto, Y. Hamada, N. Igawa, S. Kasai, P. Liu, K. Hidaka, T. Hamada, Y. Hayashi, Y. Kiso, Design and synthesis of highly active Alzheimer's β-secretase (BACE1) inhibitors, KMI-420 and KMI-429, with enhanced chemical stability, Bioorganic Med. Chem. Lett. 15 (1) (2005) 211–215. https://doi.org/10.1016/j.bmcl.2004.09.090.

[37] Y. Hamada, H. Ohta, N. Miyamoto, D. Sarma, T. Hamada, T. Nakanishi, M. Yamasaki, A. Yamani, S. Ishiura, Y. Kiso, Significance of interactions of BACE1-Arg235 with Its ligands and design of BACE1 inhibitors with P2 pyridine scaffold, Bioorganic Med. Chem. Lett. 19 (9) (2009) 2435–2439. https://doi.org/10.1016/j.bmcl.2009.03.049.

[38] D.S. Cukierman, E. Accardo, R.G. Gomes, A. De Falco, M.C. Miotto, M.C.R. Freitas, M. Lanznaster, C.O. Fernández, N.A. Rey, Aroylhydrazones constitute a promising class of 'Metal-Protein Attenuating Compounds' for the treatment of Alzheimer's disease: a proof-of-concept based on the study of the interactions between Zinc(II) and pyridine-2-carboxaldehyde isonicotinoyl hydrazon, J. Biol. Inorg. Chem. 23 (8) (2018) 1227–1241. https://doi.org/10.1007/s00775-018-1606-0.

[39] I.F. Sørensen, S. Purup, M. Ehrich, Modulation of Neurotoxicant-induced increases in intracellular calcium by phytoestrogens differ for amyloid beta peptide (Aβ) and 1-Methyl-4-Phenyl-Pyridine (MPP+), J. Appl. Toxicol. 29 (1) (2009) 84–89. https://doi.org/10.1002/jat.1376.

[40] J.S. Choi, J.J. Braymer, S.K. Park, S. Mustafa, J. Chae, M.H. Lim, Synthesis and characterization of IMPY derivatives that regulate metal-induced amyloid-β aggregation, Metallomics 3 (3) (2011) 284–291. https://doi.org/10.1039/c0mt00077a.

[41] A. Lakatos, É. Zsigó, D. Hollender, N.V. Nagy, L. Fülöp, D. Simon, Z. Bozsó, T. Kiss, Two pyridine derivatives as potential Cu(Ii) and Zn(Ii) chelators in therapy for Alzheimer's disease, Dalt. Trans. 39 (5) (2010) 1302–1315. https://doi.org/10.1039/b916366b.

[42] Z. Zhu, T. Yang, L. Zhang, L. Liu, E. Yin, C. Zhang, Z. Guo, C. Xu, X. Wang, Inhibiting Aβ toxicity in Alzheimer's disease by a pyridine amine derivative, Eur. J. Med. Chem. 168 (2019) 330–339. https://doi.org/10.1016/j.ejmech.2019.02.052.

[43] M. Asadbegi, A. Shamloo, Identification of a Novel Multifunctional Ligand for simultaneous inhibition of amyloid-beta (Aβ42) and chelation of zinc metal ion, ACS Chem. Neurosci. 10 (11) (2019) 4619–4632. https://doi.org/10.1021/acschemneuro.9b00468.

[44] H. Kroth, N. Sreenivasachary, A. Hamel, P. Benderitter, Y. Varisco, V. Giriens, P. Paganetti, W. Froestl, A. Pfeifer, A. Muhs, Synthesis and structure–activity relationship of 2,6-disubstituted pyridine derivatives as inhibitors of β-Amyloid-42 aggregation, Bioorganic Med. Chem. Lett. 26 (14) (2016) 3330–3335. https://doi.org/10.1016/j.bmcl.2016.05.040.

[45] L. Davis, G.E. Olsen, J.T. Klein, K.J. Kapples, F.P. Huger, C.P. Smith, W.W. Petko, M. Cornfeldt, R.C. Effland, ChemInform abstract: substituted (Pyrroloamino)pyridines: potential agents for the treatment of Alzheimer's disease, ChemInform 27 (19) (2010), doi:10.1002/chin.199619156.

[46] T. Umar, S. Shalini, M.K. Raza, S. Gusain, J. Kumar, P. Seth, M. Tiwari, N. Hoda, A multifunctional therapeutic approach: synthesis, biological evaluation, crystal structure and molecular docking of diversified 1H-Pyrazolo[3,4-b]pyridine derivatives against Alzheimer's disease, Eur. J. Med. Chem. 175 (2019) 2–19. https://doi.org/10.1016/j.ejmech.2019.04.038.

[47] O. Delgado, F. Delgado, J.A. Vega, A.A. Trabanco, N-Bridged 5,6-bicyclic pyridines: recent applications in central nervous system disorders, Eur. J. Med. Chem. 97 (2015) 719–731. https://doi.org/10.1016/j.ejmech.2014.12.034.

[48] Z.P. Zhuang, M.P. Kung, A. Wilson, C.W. Lee, K. Plössl, C. Hou, D.M. Holtzman, H.F. Kung, Structure-activity relationship of Imidazo[1,2-a]pyridines as ligands for detecting β-amyloid plaques in the brain, J. Med. Chem. 46 (2) (2003) 237–243. https://doi.org/10.1021/jm020351j.

[49] L. Cai, F.T. Chin, V.W. Pike, H. Toyama, J.S. Liow, S.S. Zoghbi, K. Modell, E. Briard, H.U. Shetty, K. Sinclair, S. Donohue, D. Tipre, M.P. Kung, C. Dagostin, D.A. Widdowson, M. Green, W. Gao, M.M. Herman, M. Ichise, R.B. Innis, Synthesis and evaluation of two 18F-labeled 6-Iodo-2-(4′ -N,N-Dimethylamino)Phenylimidazo[1,2-a]pyridine derivatives as prospective radioligands for β-amyloid in Alzheimer's disease, J. Med. Chem. 47 (9) (2004) 2208–2218. https://doi.org/10.1021/jm030477w.

[50] Y. Hamada, N. Miyamoto, Y. Kiso, Novel β-amyloid aggregation inhibitors possessing a turn mimic, Bioorganic Med. Chem. Lett. 25 (7) (2015) 1572–1576. https://doi.org/10.1016/j.bmcl.2015.02.016.

[51] M. Awasthi, A.K. Upadhyay, S. Singh, V.P. Pandey, U.N. Dwivedi, Terpenoids as promising therapeutic molecules against alzheimer's disease: amyloid beta- and acetylcholinesterase-directed pharmacokinetic and molecular docking analyses, Mol. Simul. 44 (1) (2018) 1–11. https://doi.org/10.1080/08927022.2017.1334880.

[52] M. Kliachyna, G. Santoni, V. Nussbaum, J. Renou, B. Sanson, J.P. Colletier, M. Arboléas, M. Loiodice, M. Weik, L. Jean, P.Y. Renard, F. Nachon, R.D Baati, Synthesis and biological evaluation of novel tetrahydroacridine pyridine- aldoxime and -amidoxime hybrids as efficient uncharged reactivators of nerve agent-inhibited human acetylcholinesterase, Eur. J. Med. Chem. 78 (2014) 455–467. https://doi.org/10.1016/j.ejmech.2014.03.044.

[53] T. Arslan, M. Buğrahan Ceylan, H. Baş, Z. Biyiklioglu, M. Senturk, Design, synthesis, characterization of peripherally tetra-pyridine-triazole-substituted phthalocyanines and their inhibitory effects on cholinesterases (AChE/BChE) and carbonic anhydrases (HCA I, II and IX), Dalt. Trans. 49 (1) (2019) 203–209. https://doi.org/10.1039/c9dt03897c.

[54] S.T. Ngo, M.S. Li, Top-leads from natural products for treatment of Alzheimer's disease: docking and molecular dynamics study, Mol. Simul. 39 (4) (2013) 279–291. https://doi.org/10.1080/08927022.2012.718769.

[55] C.B. Jalkute, S.H. Barage, M.J. Dhanavade, K.D. Sonawane, Molecular dynamics simulation and molecular docking studies of angiotensin converting enzyme with inhibitor lisinopril and amyloid beta peptide, Protein J. 32 (5) (2013) 356–364. https://doi.org/10.1007/s10930-013-9492-3.

[56] W. Chen, Y. Wang β-amyloid : the key peptide in the pathogenesis of Alzheimer ' s disease. 2015, *6* (September), 1–9. https://doi.org/10.3389/fphar.2015.00221.

[57] V.D. Mouchlis, G. Melagraki, L.C. Zacharia, A. Afantitis Computer - Aided Drug Design of β - Secretase , γ - secretase and anti - tau inhibitors for the discovery of novel Alzheimer ' s therapeutics. 2020. 10.3390/ijms21030703.

[58] J.P. Raval, T.N. Akhaja, D.M. Jaspara, K.N. Myangar, N.H. Patel, Synthesis and in vitro antibacterial activity of new oxoethylthio-1,3,4-oxadiazole derivatives, J. Saudi Chem. Soc. 18 (2) (2014) 101–106. https://doi.org/10.1016/j.jscs.2011.05.019.

[59] N. Bulut, U.M. Kocyigit, I.H. Gecibesler, T. Dastan, H. Karci, P. Taslimi, S. Durna Dastan, I. Gulcin, A. Cetin, Synthesis of some novel pyridine compounds containing Bis-1,2,4-triazole/thiosemicarbazide moiety and investigation of their antioxidant properties, carbonic anhydrase, and acetylcholinesterase enzymes inhibition profiles, J. Biochem. Mol. Toxicol. 32 (1) (2018) 2–11. https://doi.org/10.1002/jbt.22006.

CHAPTER

7

Pyridine derivatives as anti-Alzheimer agents

Babita Veer and Ram Singh

Department of Applied Chemistry, Delhi Technological University, Delhi, India

7.1 Introduction

Pyridine derivatives are well-defined as a six-membered heterocyclic ring with five carbon atoms and one nitrogen atom. Pyridine itself is the parent compound. The substituents in pyridine derivatives are shown either by numbers or by the Greek letters [1]. Pyridine derivatives are rare in nature but can be found in a few vitamins, amino acids, alkaloids, and cofactors. Pyridoxine (vitamin B_6), niacin (vitamin B_3), nicotine, nicotinic acid, nicotyrine, pyridoxal, anabasine, and pyridoxamine are the important pyridine derivatives found in nature [2,3] N(5-amino-5-carboxypenty1) pyridinium chloride and anabilysine are the amino acids responsible in crosslinking of proteins by glutaraldehyde [3]. Two rare amino acids, desmosine and isodesmosine, containing the pyridine moiety, were isolated from the protein elastin [3].

The importance of pyridine moiety in the pharmaceutical world can be judged by the fact that more than 7000 drugs with pyridine moiety exist [2]. For illustration, piroxicam is an NSAID to relieve pain, stiffness, and tenderness caused in arthritis; doxylamine is an antihistamine used for the short-term treatment of insomnia, upper respiratory infections and soft allergy, ABT-594 is an analgesic, niflumic acid is an analgesic and antirheumatic drug, Omeprazole, Netupitant an antiemetic drug, Food and Drugs Administration (FDA) approved drugs for cancer such as Abemaciclib (2015), Lorlatinib (2018), Apalutamide (2018), and Ivosidenib (2019), bay 60-5521 is an anticholesteryl ester transfer protein, lavendamycin is an antiproliferative and many more drugs are there in the market having pyridine ring (Fig. 7.1) [2,4,5]. The basic nature of pyridine is the probable reason for its medicinal properties [3]. The importance of pyridine derivatives in CNS-related disease was noted when vitamin B_3 was found effective in treating dementia [6–8]. Numerous pyridine alkaloids from various plants, fungi, amphibian, marine sources, and bacteria show CNS activity [9]. Other than pharmaceutical applications, pyridine derivatives are used for agrochemical applications like herbicides, fungicides, or bactericides [2].

Recent Developments in the Synthesis and Applications of Pyridines.
DOI: https://doi.org/10.1016/B978-0-323-91221-1.00017-8

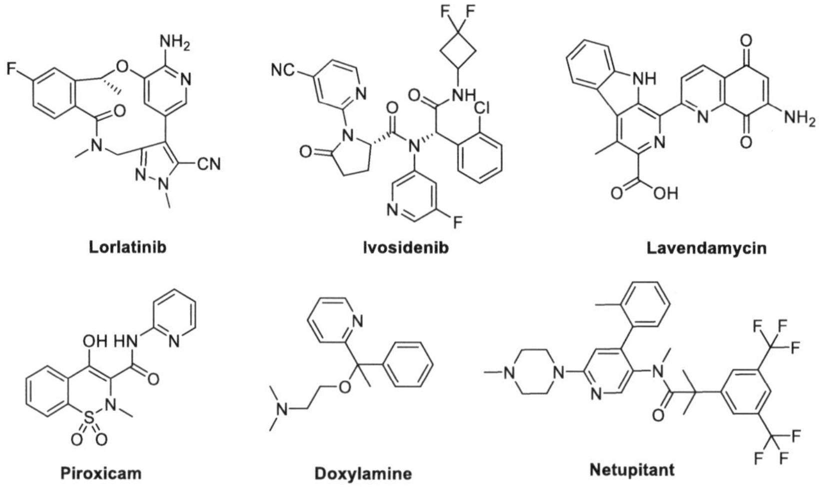

FIGURE 7.1 Drugs containing pyridine moiety.

Alzheimer's disease (AD) was introduced in 1906 by Dr. Alois Alzheimer on the basis of the autopsied brain of one of his patients who suffered memory loss [10,11]. AD is a neurodegenerative disease characterized by progressive loss of brain cells causing memory loss and cognitive decline [12,13]. It is the predominant form of dementia in the aged population and affects approximately 30–40 million people worldwide and is the third foremost cause of death in developed countries [14]. Various pathologies for AD are oxidative stress, deposition of beta-amyloid plaques and hyperphosphorylated tau proteins, decline in acetylcholine, dyshomeostatis of biometals, and various others [12,15,16]. The current pharmacotherapies provide only temporary relief and are not able to cure the disease [12]. The FDA approved drugs are donepezil, tacrine, rivastigmine, galantamine, and memantine (Fig. 7.2) [14]. This chapter discusses the synthesis of pyridine derivatives and their biological evaluation toward different targets of AD as cholinesterase inhibitors, anti-β amyloid aggregates, BACE1 inhibitor, and other miscellaneous targets.

7.2 Target-based evaluation of pyridine derivatives as anti-Alzheimer agents

7.2.1 Pyridine derivatives as cholinesterase inhibitors

2-Aminopyridine-3,5-dicarbonitriles and 2-chloropyridine-3,5-dicarbonitriles derivatives, synthesized from 2-amino-6-chloropyridine-3,5-dicarbonitrile and 2-amino-6-chloro-4-phenylpyridine-3,5-dicarbonitrile were found to be inhibitors for acetylcholinesterase (AChE) and butyrylcholinesterase (BChE). Compound (**1**) inhibited AChE with K_i value of 6.33 μM [17]. WO 2009/051922 Al patent reports (+)-isopropyl 2-methoxyethyl 4-(2-chloro-3-cyano-phenyl)-l,4-dihydro-2,6- dimethyl-pyridine-3,5-dicarboxylate (**2**) as a potential candidate for cholinesterase inhibition [18]. Another patent WO 2014/114742 Al, reported compounds

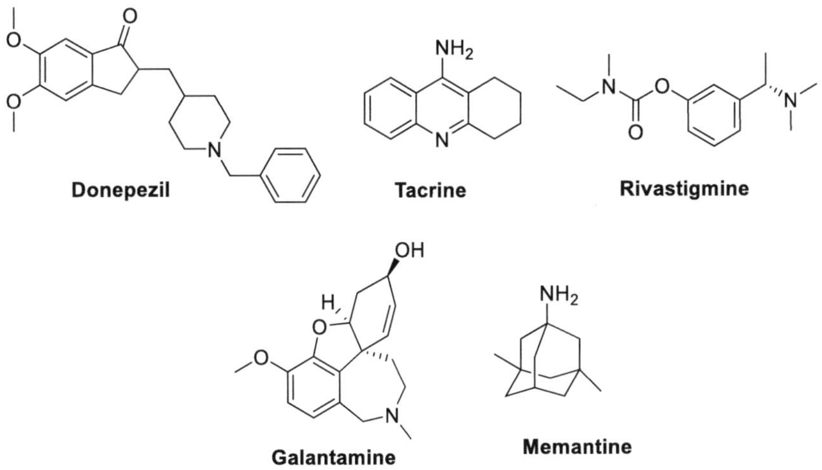

FIGURE 7.2 FDA-approved drugs for Alzheimer's disease.

of formula (**3**) as potent anticholinesterase agents [19]. A new series of pyridonepezils which are hybrids of N-benzylpiperidine moiety in donepezil and 2-amino-6-chloropyridine heterocyclic ringsystem, connected by an appropriate polymethylene linker. 2-amino-6-((3-(1-benzylpiperidin-4-yl)propyl)amino)pyridine-3,5-dicarbonitrile (**4**) was most potent AChE inhibitor with IC_{50} value 9.4 ± 0.4 nM. Compound (**4**) was 1.4-fold times more active than donepezil [20]. A novel series of pyridine 2,4,6-tricarbohydrazide were evaluated for ChE inhibition. Compound (**5**) inhibited AChE with IC_{50} value 50.2 ± 0.8 μM and BChE with IC_{50} 43.8 ± 0.8 μM [21].

A series of benzofuranone based derivatives having pyridinium moiety was reported as potent dual inhibitor for AChE and BChE. Compound (**6**) was potent with IC_{50} value 52 ± 6.38 nM as acetylcholinesterase inhibitor. The methylbenzyl substituent on pyridine ring was more potent than corresponding 6-ethoxy and 6-propoxy derivatives [22]. Phthalimide-based molecules containing a substituted N-benzylpyridinium residue were evaluated for ChE inhibition. 2-fluorobenzylpyridinium derivative (**7**) inhibited AChE and BChE with IC_{50} values of 0.77 and 8.71 μM [23]. A new series of benzylpyridinium-based benzoheterocycles were evaluated for ChE. The compounds (**8**) and (**9**) had cholinesterase inhibition comparable to the standard drug donepezil. Compound (**8**) inhibited in competitive manner and occupied the active site in the vicinity of catalytic triad [24]. Novel multitarget natural product-pyridoxine based triazole derivatives were synthesized and evaluated for AD. Compound (**10**) was the most potent AChE inhibitor [25]. The new series of edaravone derivatives (**11**) consisting of N-benzyl pyridinium moieties was synthesized and was found to be dual-site binding inhibitors of AChE. The molecular docking results showed that there are interactions with conserved amino acids Trp279 in PAS and Trp84 in CAS. The IC_{50} value ranges from 1.2–4.6 μM [26]. Abdpour et al. reported a series of 7-hydroxy-chromone derivatives bearing pyridine moiety. Most of the compounds showed IC_{50} values for AChE inhibition between 9.8–0.71 μM and for BChE inhibition between 1.9–0.006 μM. Compound (**12**) and (**13**) were most potent for AChE with IC_{50} value 0.71 μM and BChE with IC_{50} value 0.006 μM, respectively.

The pyridinium salt has a strong binding affinity toward catalytic active site (CAS) of AChE via π-stacking and charge interactions. The substitution of ethyl group at para position of pyridinium ring and 3- or 4- carbon linker length enhanced the ChE inhibition [27].

Hybrids of cyanopyridine-triazine were tested for inhibition of ChE and it was found that compounds were effective against ChE. Compounds (**14**) and (**15**) were most potent with IC_{50} values 0.059 and 0.080 μM for AChE and showed selectivity over BChE [28]. Bis(2-pyridyl) diselenide derivatives (**16**) were evaluated for AChE inhibition [29]. A new series of pyridine derivatives with carbamic or amidic function was synthesized and evaluated for cholinesterase inhibitors. The carbamate (**17**) was most potent AChE inhibitor with IC_{50} value 0.153 ± 0.016 μM and carbamate (**18**) was most potent BChE inhibitor with IC_{50} value 0.828 ± 0.067 μM [30]. A new series of (3-hydroxy-4-pyridinone)-benzofuran were evaluated for AChE inhibition. O-benzyl-hydroxypyridinone hybrids (**19**) containing a 2-methylene linker showed inhibitory activity similar to donepezil [31]. A novel class of glycogen synthase kinase-3β (GSK-3β) and AChE inhibitor was developed. Compound (**20**) was the most potent inhibitor with IC_{50} value 6.5 nM for AChE and 66 nM for GSK-3β kinase activity [32]. Few compounds were designed based on 2,3,4,9-tetrahydro-1H-carbazole (**21**) having benzyl pyridine moieties and were evaluated for BChE. The compound (**6i**) was most potent with IC_{50} value 0.088 ± 0.0009 μM [33].

The conjugates of 9-amino-1,2,3,4-tetrahydroacridine and 5,6-dichloronicotinic acid linked with different linkers were evaluated as potent ChE inhibitors. The compound (**22**) with IC_{50} value 1.02 nM had a high potential for inhibition of ChE [34]. Hybrid molecules of pyrimidine derivatives and triazolopyrimidine showed good inhibition activity against AChE. Compound 2-(4-(6-(quinolin-8-yloxy)pyrimidin-4-yl) piperazin-1-yl)nicotinonitrile (**23**) with IC_{50} value of 36 nM was most potent [35]. Analogues of 4-aminopyridine were evaluated for AChE inhibition. Compounds (**24**) and (**25**) showed noncompetitive activity comparable to the standard drug rivastigmine [36]. One of the research groups reported diversely functionalized tacrines by replacing benzene ring of tacrine with five or six-membered aromatic rings. To obtain a multitarget directed ligands, new analogues were designed on the basis of structures of tacrine and 1,4-dihydropyridines by replacing benzene ring of tacrine with pyridine ring. Ethyl 5-amino-2-methyl-6,7,8,9-tetrahydrobenzo[b][1,8]naphthyridine-3-carboxylate (**26**) had pharmacological properties beyond just inhibiting the ChE. The replacement of the benzene ring by pyridines prevented nondesired metabolism and the methyl group behaved as a metabolic blocker which redirected the oxidation metabolism leading to less toxic species [37]. New series of thienopyridine-tacrine analogues were evaluated for AChE inhibition activity and compound (**27**) was most potent with IC_{50} value of 172 nM [38]. Novel series of tacrine-pyrazolo[3,4-b]pyridine hybrids were evaluated for ChE and compound (**28**) which is tacrine linked with pyrazolo[3,4-b]pyridine moiety by a six-carbon spacer was the most potent AChE inhibitor [39]. Another group reported some thieno[2,3-b]pyridine amine derivatives for ChE inhibition as safer analogues of tacrine. Compounds (**29**) and (**30**) were the most potent inhibitors with IC_{50} values of 1.55 and 0.23 μM, respectively [40].

A series of new pyridine-2,6-dicarboxamide containing sulfonamide groups (**31**) were evaluated for cholinesterase (ChE) inhibition and showed IC_{50} values in the ranges 98.4–197.5 nM against acetylcholinesterase (AChE), and 82.2–172.7 nM against butyrylcholinesterase (BuChE). The results were comparable with the standard drug rivastigmine [41]. A series of hybrids of 5-phenyl-1,3,4-oxadiazoles and 2-pyridylpiperazine were synthesized. Compound (**32**) bearing 2,4-difluoro substitution at phenyl ring showed best results with

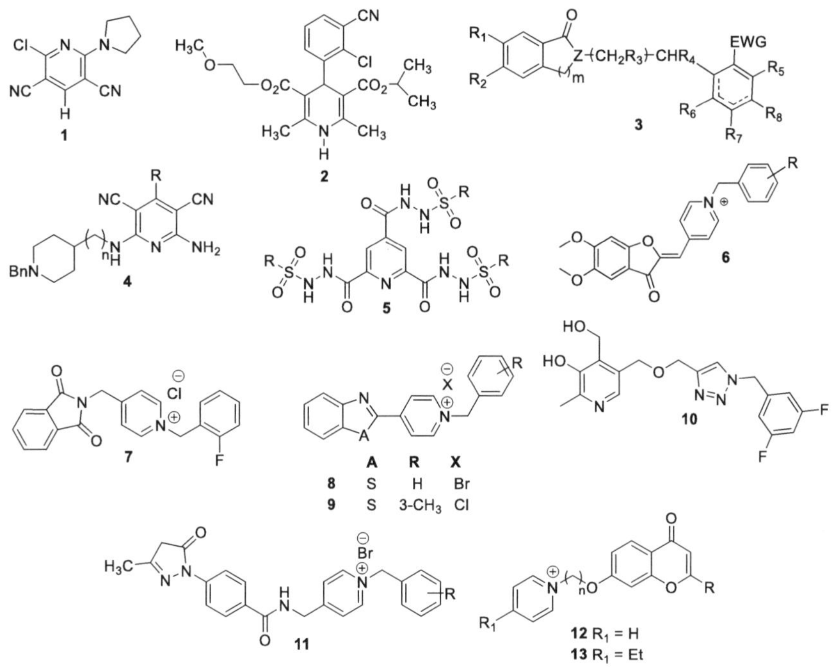

FIGURE 7.3 Cholinesterase inhibitors.

IC_{50} value 0.054 μM for AChE inhibition and 0.787 μM for BChE inhibition [42]. A new series of 2-(piperazin-1-yl)-N-(1H-pyrazolo[3,4-b]pyridin-3-yl)acetamides (**33–35**) were evaluated for ChE inhibition. Most potent compound (**34**) had IC_{50} value 4.8 nM for AChE inhibition [43]. 3-bis(pyridin-2-ylmethyl)aminomethyl-5-hydroxybenzyltriphenylphosphonium bromide (PAT) (**36**) was reported to inhibit AChE [44]. Two novel amide-based transition metal Zn^{2+} complexes AAZ7 (**37**) and AAZ8 (**38**) attached with pyridine moiety were found to be potent ChE inhibitors. The complex AAZ8 (**38**) was more potent with IC_{50} values of 14 μg/mL and 18μg/mL as AChE and BChE inhibitors respectively [45]. The significance of pyridine and sultones in medicinal chemistry prompted Zhang et al. to synthesize novel pyridine-containing sultones which were evaluated for cholinesterase (ChE) inhibition. The structure–activity relationship (SAR) exhibited that fused pyridine-containing sultones increased the AChE inhibition. Compound (**39**) (4-(4-chlorophenyl)-2,2-dioxide-3,4,5,6-tetrahydro-1,2-oxathiino[5,6-h]quinoline) was a selective AChE inhibitor with IC_{50} value 8.93 μM. It was a reversible and noncompetitive AChE inhibitor and showed non-toxic neuroprotective activity [46]. One of the research groups reported pyridine amine derivatives with phenyl and benzyl moieties for AChE inhibition. The compounds with electron withdrawing group showed better inhibition as compared to electron-donating groups. The analogues with heterocycles such as pyridine and pyrimidine (**40–42**) showed better activities with IC_{50} values of 18.77± 1.60, 14.22 ± 0.95, 21.24 ± 1.09 μM, respectively [47]. The chemical structures for the cholinesterase inhibitors (**1–42**) are given in Figs. 7.3–7.5.

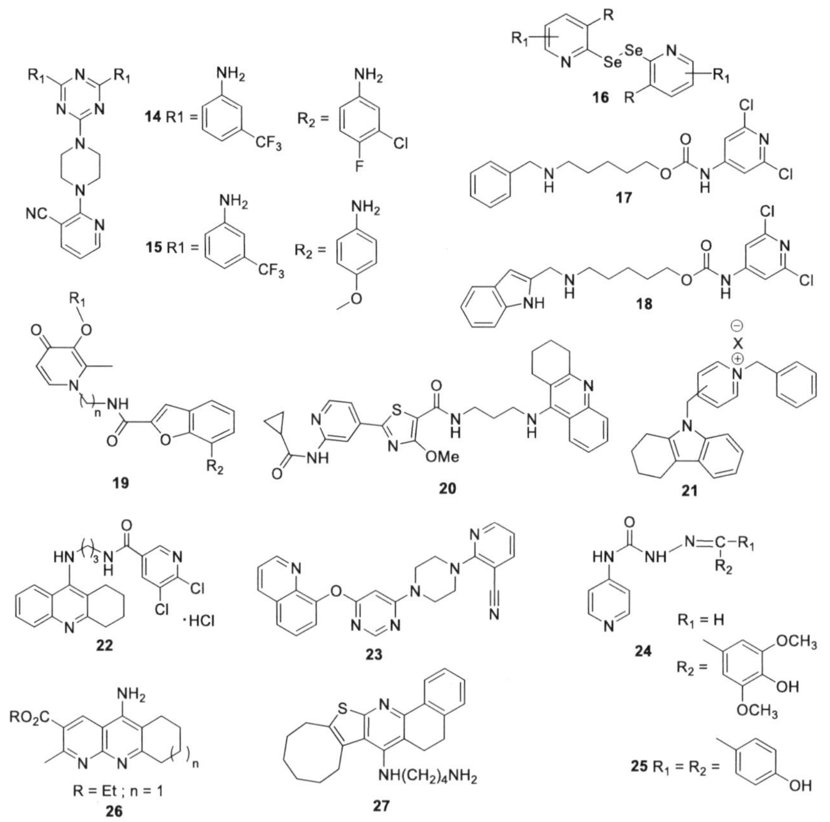

FIGURE 7.4 Cholinesterase inhibitors.

7.2.2 Pyridine derivatives as anti-*β* amyloid aggregates

New compounds pyridothieno-pyrimidines, pyridothienotriazines, and pyridothienoxazines (**43–48**) were synthesized and were evaluated for anti-Alzheimer activity [48]. The significant role of dysregulated metal ions in the aggregation of the amyloid-*β* (A*β*) peptide makes disruption of metal-A*β* interactions an important target for AD. The Cu and Fe metals catalyze the generation of reactive oxygen species which leads to neurotoxicity. Therefore, Jones et al. developed a new series of dual-function triazole-pyridine ligands [4-(2-(4-(pyridin-2-yl)-1H-1,2,3-triazol-1-yl)ethyl)-morpholine (**49**), 3-(4-(pyridin-2-yl)-1H-1,2,3-triazol-1-yl)propan-1-ol (**50**), 2-(4-(pyridin-2-yl)-1H-1,2,3-triazol-1-yl)acetic acid (**51**), and 5-(4-(pyridin-2-yl)-1H-1,2,3-triazol-1-yl)pentan-1-amine (**52**)] that interact with the A*β* peptide and modulates its aggregation [49]. Pyclen (**53**) was reported as a capable molecule for preventing as well as disrupting Cu^{2+}-induced A*β*1-40 aggregation. The pyridine backbone of pyclen is responsible for antioxidant capacity [50]. *β*-Amyloid is generated by sequential cleavage of amyloid precursor protein (APP) by *γ*-secretases. Inhibition of *γ*-secretases is one of the approaches for reducing *β*-amyloid levels. As *γ*-secretase targets many substrates, while

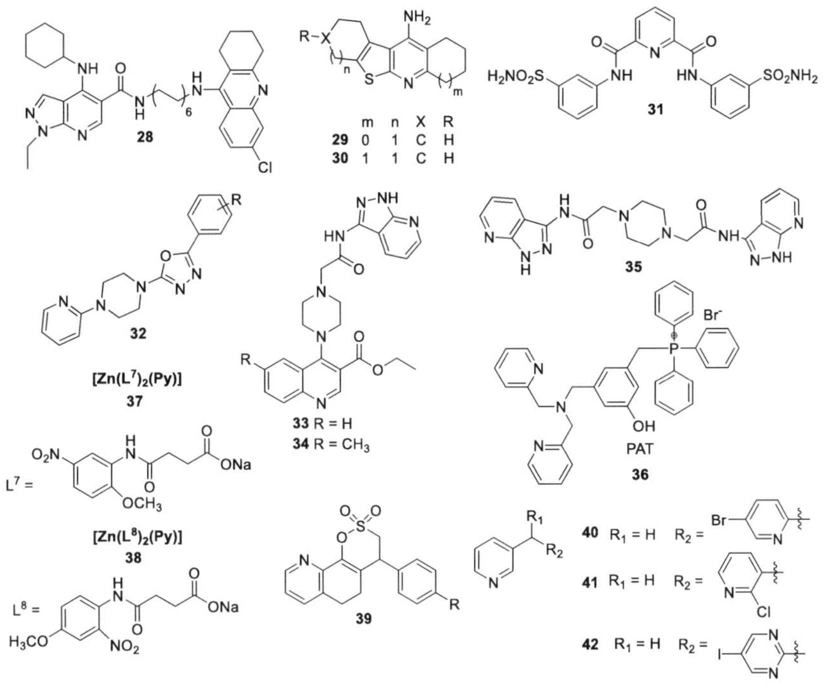

FIGURE 7.5 Cholinesterase inhibitors.

designing γ-secretase modulator (GSM), one has to be very careful. Huang et al. reported in-vivo Aβ-lowering by pyridazine (**54**) and pyridine-derived GSM (**55**). Both the molecules decreased Aβ40 and Aβ42 productions and have exceptional oral bioavailability and good brain permeability [51].

In another study, sulfonamide was optimized as lead for GSM. Compound (**56**) improved cell potency and decreased the Aβ42 levels in cerebrospinal fluid on oral administration [52]. Trigonelline (**57**) is a quaternary base pyridine alkaloid that is presumed to show anti-Alzheimer's activity as molecular modeling shows that the affinity of trigonelline to the Aβ (1–42) peptide is high and similar to the standard drug cotinine for AD. Trigonelline is related to nicotine which was considered for study as anti-AD drug candidate. It is assumed that the way cotinine and nicotine bind the segment of Aβ between amino acids 1–28 when folded in α-helical conformation inhibiting the conformational change from α-helical to β-sheet conformation, similar effect is of trigonelline on the helical Aβ (1–42) structure making it a possible drug candidate for AD [53]. Novel 5,6,7,8-tetrahydro[1,2,4]triazolo[4,3-a]pyridine derivatives were reported to be GSMs. Compound ((R)-**58**) decreased the Aβ42 and had high oral bioavailability and good blood-brain barrier permeability [54].

Hybrids of cyanopyridine-triazine can reduce H_2O_2-induced oxidative stress and Aβ_{1-42} induced cytotoxicity. Compounds (**14**) and (**15**) exhibited the highest anti-Aβ_{1-42} aggregation potential with IC_{50} values 10.1 and 10.9 μM, respectively [28].

A new series of (3-hydroxy-4-pyridinone)-benzofuran (**19**) were evaluated for inhibition of amyloid peptide (A*β*) aggregation. The free-hydroxypyridinone hybrids showed the highest inhibition of A*β* aggregation and metal chelating property [31]. The conjugates of 9-amino-1,2,3,4-tetrahydroacridine and 5,6-dichloronicotinic acid linked with different linkers (**22**) were evaluated for A*β* self-induced aggregation. The compound (**22**) in different concentrations had moderate activity against A*β* aggregation. At 50 μM and 5 μM concentrations, the inhibition of A*β* aggregation was 46.63% and 19.41%, respectively [34]. The compound (**21**) inhibited self- and AChE-induced A*β* peptide aggregation at concentration of 10 μM and 100 μM, respectively [33]. A series of 2-(3-arylureido) pyridines and 2-(3-benzylureido) pyridines were evaluated as modulators for A*β*-induced mitochondrial dysfunction in AD. Forty-one small molecules were evaluated for A*β*-induced mitochondrial permeability transition pore opening, out which six hit compounds were identified. Compound (**59**) was the lead compound protecting neuronal cells against 67% of neurotoxicity and 43% suppression of mitochondrial ATP production induced by 5 μM [56]. The compound (**20**) shows inhibitory effect on A*β* self-aggregation with an inhibition rate of 46% at 20 μM [32].

The compounds (**8**) and (**9**), benzylpyridinium-based benzoheterocycles inhibited A*β* self-aggregation as well as AChE-induced A*β* aggregation. They also protect PC12 cells against H_2O_2-induced injury and are nontoxic against HepG2 cells [24]. A new series of 2-(piperazin-1-yl)-N-(1H-pyrazolo[3,4-b]pyridin-3-yl)acetamides showed inhibition of self-mediated A*β* aggregation and Cu(II)-mediated A*β* aggregation. The highest inhibition percentage was 81.65%. Pyrazolopyridine, quinoline, pyridine, and triazolopyrimidine scaffolds were used in combinations. Due to the planar structure of pyrazolopyridine, more favorable intercalation between A*β* fibrils and strong inhibition of aggregation was observed [43]. Compound (**32**) exhibited anti-A*β* aggregation activity in self and AChE-induced thioflavin T assay [42]. Compounds (**12**) and (**13**) showed effective neuroprotection on H_2O_2- and A*β*-induced neurotoxicity in PC12 cells which was more than standard drugs [27]. Thieno[2,3-b]pyridine amine derivatives, compounds (**29**) and (**30**) showed inhibition against A*β*-aggregation and *β*-secretase 1 and neuroprotection and cytotoxicity against HepG2 cells [40]. The chemical structures for the *β*-Amyloid aggregation inhibitors are given in Figs. 7.6 and 7.7.

7.2.3 Pyridine derivatives as BACE1 inhibitor

A series of novel pyridinyl aminohydantoins were evaluated for BACE1 inhibition. Compound (**60**) was the most potent BACE1 inhibitor with IC_{50} value of 20 nM. The high potency of the compound was the result of interaction of pyridine nitrogen with tryptophan (Trp)76 [56]. A series of aminoisoindoles for BACE1 inhibition and discovery of ((S)-**61**) (AZD3839), a potent anti-Alzheimer's drug candidate was reported. The compounds ((S)-**62**) and ((R)-**63**) showed BACE1 inhibition with IC_{50} values 8.6 nM and 0.16 nM, respectively [57]. A hybrid of rhein-huprine (**64–67**) was developed by modification of huprine aromatic ring. Replacement by thieno[3,2-e]pyridine or [1,8]-naphthyridine system resulted in decreased but still potent inhibition of AChE and BACE1 [59]. 2,3,4,9-tetrahydro-1H-carbazole having benzyl pyridine moieties (**21**) were evaluated for BACE1 inhibition and compound (**21**) was most effective [33]. Compound CID 9998128 (**68**) was reported as a potential multitarget drug for AD. The

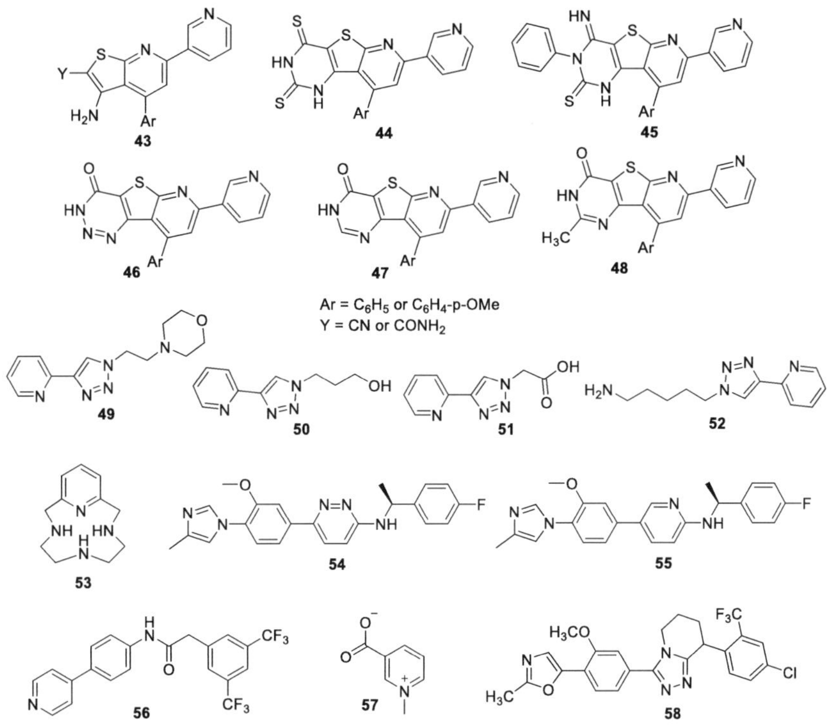

FIGURE 7.6 β-Amyloid aggregation inhibitors.

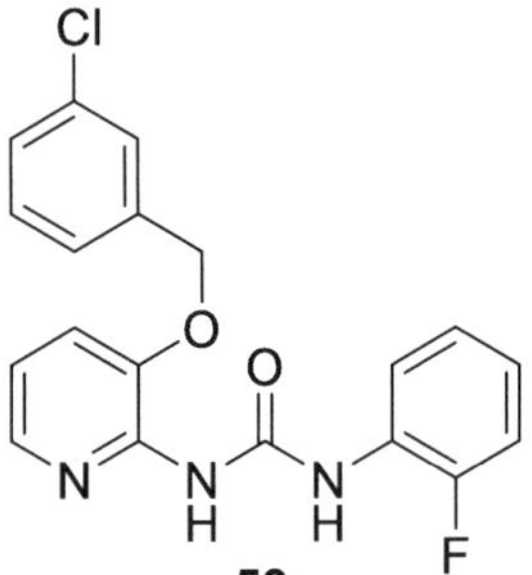

59

FIGURE 7.7 β-Amyloid aggregation inhibitor.

compound shows strong binding with both Aβ42 fibrils and β-secretase. It inhibited Aβ42 amyloid fibrillization and was capable to clear Aβ42 fibrils. It decreased the BACE1 activity with EC_{50} value in the micromolar range [59].

Hybrids of tetrahydroisoquinoline-benzimidazole were evaluated for inhibition of neuroinflammation and BACE1. Compound (**69**) had moderate BACE1 inhibition activity, that is, 65.7% inhibition at 20 μM [60]. Hybrids of 5-phenyl-1,3,4-oxadiazoles and 2-pyridylpiperazine

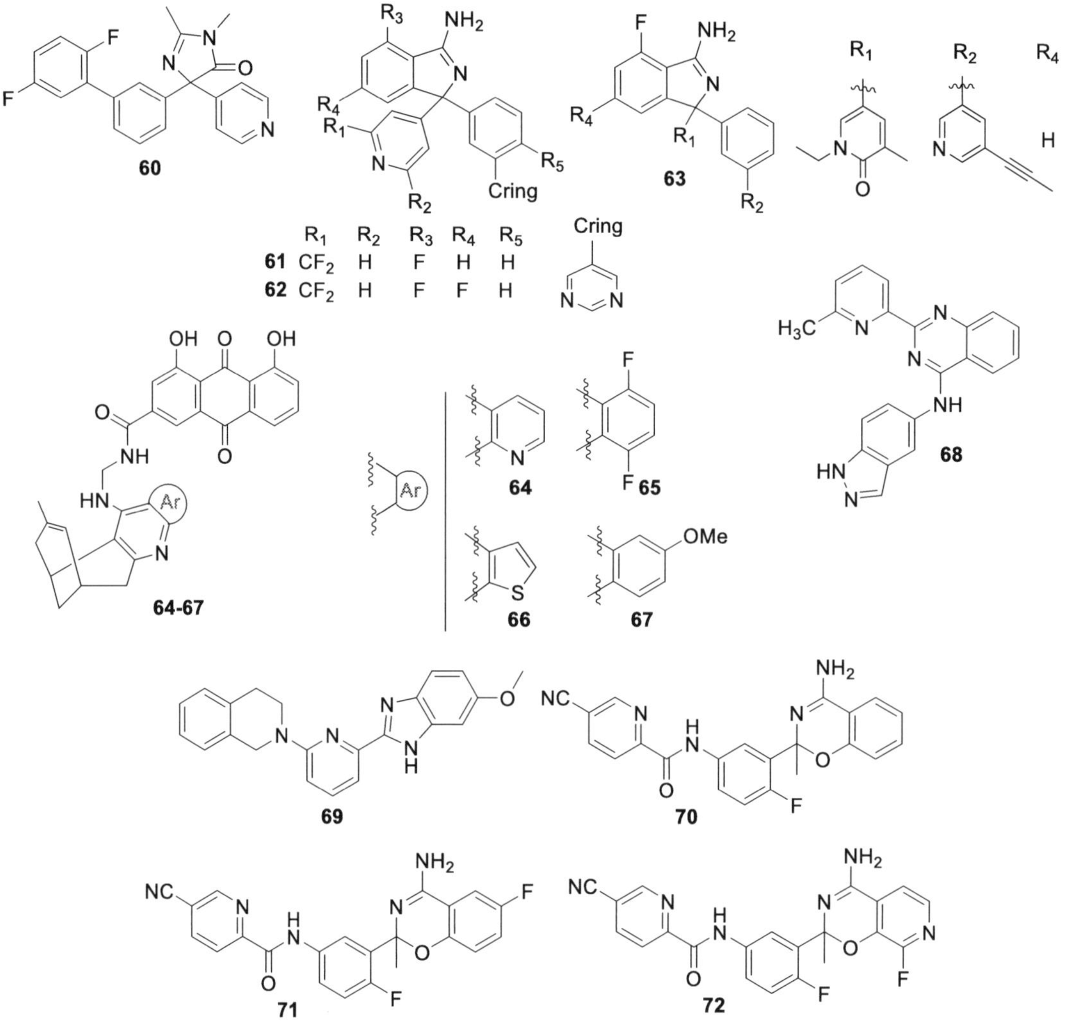

FIGURE 7.8 BACE1 inhibitors.

showed inhibition against *β*-site APP cleaving enzyme 1 (BACE-1) with IC_{50} value 0.098 μM. 1-amino-3,4-dihydro-2,6-naphthyridine scaffold was reported as BACE-1 inhibitors. The authors designed naphthyridine-based compounds by using the previously reported benzene-fused amidines (**70**) and (**71**). By removing the labile N, O-acetal substructure and replacing the phenyl moiety on the amidine with different substituted pyridine rings, basicity was controlled which led to increased biochemical and cellular potency. Compound (**72**) having fluorine atom substituted at 4-position on pyridine ring was the most potent inhibitor [61]. The chemical structures for the BACE1 inhibitors are given in Fig. 7.8.

7.2.4 Pyridine derivatives as metal chelators

Two bifunctional small molecules (**73**) and (**74**) were designed to target divalent metal ions involved in Aβ aggregation. Both the compounds were evaluated for disaggregation of Cu^{2+} triggered Aβ aggregates. Both the compounds showed low toxicity as compared to clinically available clioquinol. Compound (**74**) showed no toxicity up to 200 μM, thus making it more suitable drug candidate [62]. Another research reported bifunctional small molecules, N-(pyridin-2-ylmethyl)aniline (**75**) and N1,N1-dimethyl-N4-(pyridin-2-ylmethyl)benzene-1,4-diamine (**76**) which modulates the metal-induced Aβ aggregation and neurotoxicity [63].

Two pyridine derivatives, N-methyl-1-(pyridine-2-yl)-methaneamine (DMAP) (**77**) and N^1,N^2-bis(pyridine-2-yl-methyl)ethane-1,2-diamine (ENDIP) (**78**) effectively modulates aggregated amyloid-β peptides (Aβ) for both Cu(II) and Zn(II) and prevents metal-ion induced Aβ aggregation and resolubilises amyloid precipitates. DMAP was weak metal binder in comparison to ENDIP [64]. Glycosides of 3-hydroxy-4-pyridinones (**79**) were reported as metal chelators which resolubilizes metal-Aβ1-40 aggregates with zinc and copper [65].

2,6-Disubstituted pyridine derivatives were developed which interacted with β-sheet conformation of Aβ through donor–acceptor–donor hydrogen bond formation. The most potent inhibition of Aβ aggregation was shown by compounds having three 2,6-disubstituted pyridine (**80**) units separated by at least one C2- or C3-linker [66]. A pyridine amine derivative, 3-bis(pyridin-2-ylmethyl)aminomethyl-5-hydroxybenzyltriphenylphosphonium bromide (PAT) (**36**) was evaluated for inhibition of metal- and self-induced Aβ aggregation. It was reported that PAT forms stable metal complexes by including N,N,O donor atoms and shows antioxidant property due to phenolic group. The compound interacts with Cu^{2+} and Zn^{2+} to disaggregate the metal-associated Aβ aggregates [44]. Natural product-pyridoxine based triazole derivatives were found to possess antioxidant and metal chelation properties. Pyridoxine is in phase III of clinical trials for reducing homocysteine levels in AD patients. A high level of homocysteine is responsible for beta-amyloid plaques deposition. The directly attached hydroxyl group to the pyridine ring is responsible for metal chelation and antioxidant properties [25].

Ruthenium complexes (**81**) containing pyridine-based ligands were evaluated for modulating the metal-ion induced Aβ aggregation. The complexes having terminal primary amines were more effective in modulating the aggregation of Aβ and diminishing its cytotoxicity. The secondary amine of 4-methylamino pyridine and the carbonyl of 4-pyridinecarboxaldehyde were active followed by a series of compounds having hydrophobic alkyl group, which was further followed by polar hydroxy-containing compounds. The a morpholino and unsubstituted pyridine were the least active complexes. The absence of H-bond donating ability in morpholino-containing complex was responsible for poor performance. Polar substituent had great impact on performance, especially N–H hydrogen-bonding substituents led to dramatic decrease in aggregates [67]. The chemical structures for inhibitors of Aβ by metal chelators are given in Fig. 7.9.

7.2.5 Pyridine derivatives for miscellaneous targets

The derivatives of oxazolopyridine and thiazolopyridine (**82**) were investigated for MAO-B inhibition. The SAR study showed that the piperidino group substituted to the

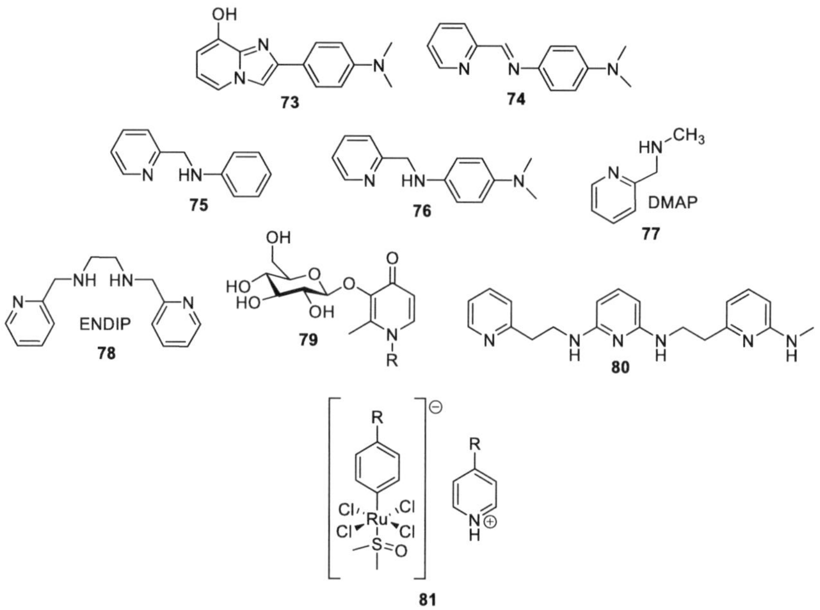

FIGURE 7.9 Inhibitors of Aβ by metal chelators.

oxazolopyridine core structure gave the best results. The IC_{50} value ranged between 267.1 and 889.5 nM. When the core structure oxazolopyridine was replaced by thiazolopyridine, the activity remarkably improved with the most potent derivative having IC_{50} value 26.5 nM [68]. A series of (pyrrolo-pyridin-5-yl) benzamides were developed as reversible MAO-B inhibitors. Compounds (**83**) inhibited MAO-B with IC_{50} value 1.11 nM, Ki = 0.56 nM and (**84**) had IC_{50} value 3.27 nM, Ki = 1.45 nM [69]. Hydroxypyridinone and coumarin hybrids were developed and evaluated for MAO-B inhibition. Compound (**85**) was most potent inhibitor with IC_{50} value 14.7 nM [70]. The new series of N-pyridyl-hydrazone derivatives (**86**) were checked for their potency against MAO. The compounds showed low inhibitory effects toward both isoforms of MAO [71].

6-Methyl-N-[3-[[3-(1-methylethoxy)propyl]carbamoyl]-1H-pyrazol-4-yl]pyridine-3-carboxamide (**87**) was evaluated for inhibition of tau phosphorylation and reduced the amount of tau aggregation. The compound is selective inhibitor of GSK-3 with IC_{50} value of 2 nM [72].

Several 3-amino-N-4-arylmethylenes, N-formamides, ethyl imidoformate, N,N-dimethyl-N-imidoformamides, N-acetyl-N-acetamides, 1,3,4-oxadiazole-2-thiol, pyrazolothienopyridines, 2-[(3,5-dimethyl-1H-pyrazol-1-yl)carbonyl]-4-aryl-6-pyridin-3-ylthieno[2,3-b]pyridin-3-amines, 2-carbonyl-5-methyl-2,4-dihydro-3H-pyrazol-3-ones and 2-carbonyl-pyrazolidine-3,5-diones (**88**) were evaluated as anti-Alzheimer's agents [73]. The compound with formula (**89**) is novel pyridine derivative as GSM [74]. Spiro[imidazo[1,2-a]pyridine-3,2-indan]-2(3H)-one coded as ST101 (previously coded as ZSET1446) was reported to improve cognition in CNS by targeting T-type voltage-gated calcium channels. 0.01 to 100 nM ST101 was tested on rat somatosensory cortical and hippocampal slices. 0.1 nM ST101

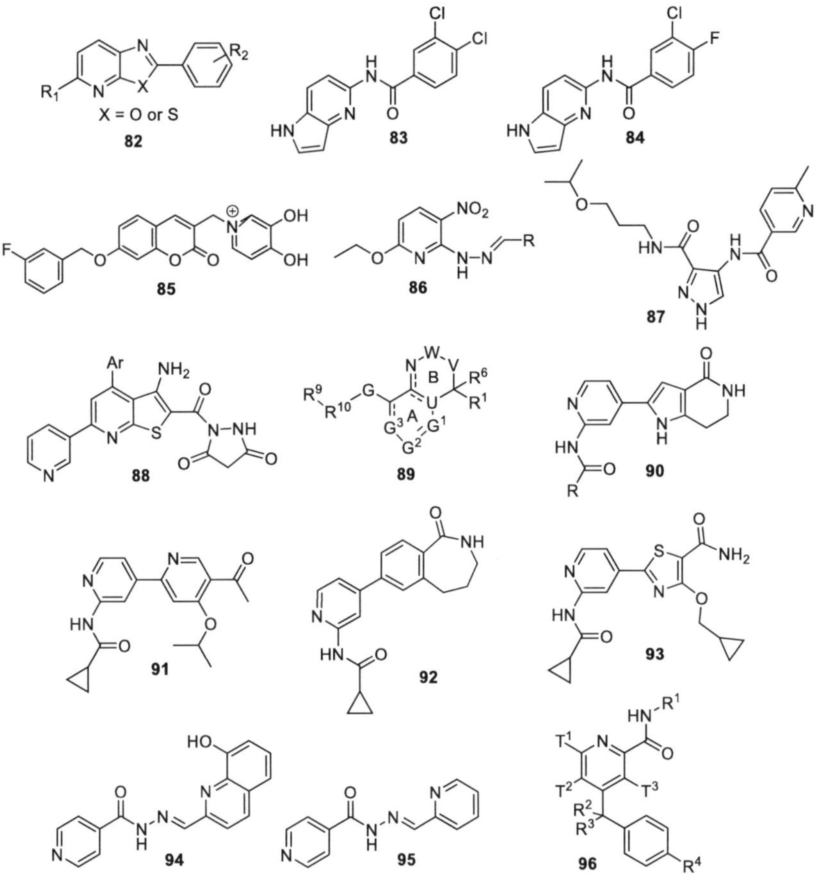

FIGURE 7.10 Inhibitors for miscellaneous targets.

was maximum effective concentration for increasing calcium/calmodulin-dependent protein kinase II (CaMKII) autophosphorylation in cortical slices but it did not affect the hippocampal protein kinase Ca autophosphorylation in slice preparations [75].

Glycogen synthase kinase-3 (GSK-3) are anticipated as crucial in neurodegenerative disease pathogenesis. A series of new pyrrolopyridinone (**90**), pyridylpyridines (**91**), phenyl- and fused-phenylpyridines (**92**), and thiazolylpyridine (**93**) derivatives were designed with appropriate central spacer and evaluated for GSK-3 inhibition. Compound (**93**) from thiazolylpyridine series was most potent with IC_{50} value 0.29 nM [76]. The selective muscarinic M1 subtype activation can be an effective strategy for AD patients. A series of M1-selective pyridine and pyridine amides are designed and it was reported that compound (**38**) is the most potent M1 selective positive allosteric modulator [77]. Hybrid of aroylhydrazone (**94, 95**) containing the 8-hydroxyquinoline group INHHQ as potent Metal-protein attenuating compounds (MPACs) were evaluated. The compound was remarkable as a ligand and efficiently compete with $A\beta(1–40)$ for Zn^{2+} ions [78]. Compound (**96**) is the novel pyridine derivatives for muscarinic M1 receptor modulators to treat AD [79]. The chemical structures of inhibitors (**82–96**) are given in Fig. 7.10.

7.3 Summary

AD is a neurodegenerative disorder that troubles mostly old-age people, however, in recent times, the people with age groups lower than 50 are also getting affected. The etiology of AD has been an exciting area of research; still, an evidence-based conclusive pathway is a matter of research. Some of the targets exploited for the AD treatment include shortage of choline neurotransmitters, accumulation of β-amyloid, oxidative stress, MAO B inhibition, etc.

Pyridine scaffold has been helping medicinal chemistry researchers for years in the development of different drugs. The synthesis of pyridine derivatives has been easily achieved by routine chemical reactions. This chapter discussed the synthesis of pyridine derivatives and their biological evaluation toward different targets of AD as cholinesterase inhibitors, anti-β amyloid aggregates, BACE1 inhibitor, and other miscellaneous targets.

References

[1] E. Scriven, R. Murugan, Pyridine and pyridine derivatives, Kirk-Othmer Encyclopedia of Chemical Technology, vol. 20, John Wiley & Sons, Inc., New York, 2005, pp. 1–53.
[2] C. Gonzalez-Bello, L. Castedo, Six-membered heterocycles: pyridines, Modern Heterocyclic Chem. 3 (2011) 1431–1525.
[3] G.R. Newkome, Pyridine and its derivatives: part five, in: A. Weissbergkr, E.C. Taylor (Eds.), The Chemistry of Heterocyclic Compounds: A Series of Monographs, First edition, John Wiley and Sons, New York, 1985, pp. 3–33.
[4] E. Khan, Pyridine derivatives as biologically active precursors; organics and selected coordination complexes, Chem. Select 6 (13) (2021) 3041–3064.
[5] M.V.K. Reddy, K.Y. Rao, G. Anusha, G.M. Kumar, A.G. Damu, K.R. Reddy, N.P. Shetti, T.M. Aminabhavi, P.V.G. Reddy, In-vitro evaluation of antioxidant and anticholinesterase activities of novel pyridine, quinoxaline and s-triazine derivatives, Environ. Res. 199 (2021) 111320.
[6] J.E. Prousky, Treating dementia with vitamin B3 and NADH, J. Orthomolecular Med. 26 (4) (2011) 163.
[7] K. Rajakumar, Pellagra in the United States: a historical perspective, South. Med. J. 93 (3) (2000) 272–277.
[8] R.H. Major, Classic Description of Disease, 3rd ed, Charles C. Thomas Publisher, Springfield, IL, 1978, pp. 607–612.
[9] S.X. Lin, M.A. Curtis, J. Sperry, Pyridine alkaloids with activity in the central nervous system, Bioorg. Med. Chem. 28 (24) (2020) 115820.
[10] B. Yang, Z.A. Xia, B. Zhong, X. Xiong, C. Sheng, Y. Wang, W. Gong, Y. Cao, Z. Wang, W. Peng, Distinct hippocampal expression profiles of long non-coding RNAs in an Alzheimer's disease model, Mol. Neurobiol. 54 (2017) 4833–4846.
[11] R.J. Bermudez, Alzheimer's disease: critical notes on the history of a medical concept, Arch. Med. Res. 43 (2012) 595–599.
[12] A. Kumar, C.M. Nisha, C. Silakari, I. Sharma, K. Anusha, N. Gupta, P. Nair, T. Tripathi, A. Kumar, Current and novel therapeutic molecules and targets in Alzheimer's disease, J. Formos. Med. Assoc. 115 (1) (2016) 3–10.
[13] D.K. Lahiri, M.R. Farlow, N.H. Greig, K. Sambamurti, Current drug targets for Alzheimer's disease treatment, Drug Dev. Res. 56 (2002) 267–281.
[14] Y. Biran, C.L. Masters, K.J. Barnham, A.I. Bush, P.A. Adlard, Pharmacotherapeutic targets in Alzheimer's disease, J. Cell. Mol. Med. 13 (1) (2009) 61–86.
[15] J.M. Long, D.M. Holtzman, Alzheimer disease: an update on pathobiology and treatment strategies, Cell 179 (2) (2019) 312–339.
[16] C. Reitz, C. Brayne, R. Mayeux, Epidemiology of Alzheimer disease, Nature Rev. Neurol. 7 (2011) 137–152.
[17] A. Samadi, J. Marco-Contelles, E. Soriano, M. Álvarez-Pérez, M. Chioua, A. Romero, L. González-Lafuente, L. Gandía, J.M. Roda, M.G. López, M. Villarroya, A.G. García, C. de los Ríos, Multipotent drugs with cholinergic and neuroprotective properties for the treatment of Alzheimer and neuronal vascular diseases. I. Synthesis, biological assessment, and molecular modeling of simple and readily available 2-aminopyridine-, and 2-chloropyridine-3,5-dicarbonitriles, Bioorganic Medicinal Chem. 18 (2010) 5861–5872.

[18] S.R. Murray, Methods of treating alzheimer's disease with (+) - isopropyl 2-methoxyethyl4-(2-chloro-3- cyanophenyl) -i, 4-dihydro-2, 6-dimethyl-pyridine-3, 5-dicarboxylate and a cholinesterase inhibitor, WO 2009/051922 Al, Memory Pharmaceuticals Corporation, 2009.

[19] F. Marsais, V. Levacher, C. Papamicael, P. Bohn, L. Peauger, V. Gembus, N. Le Fur, M.L. Dumartin-Lepine, Oxidisable pyridine derivatives, their preparation and use as anti-alzheimer agents, WO 2014/114742 Al, INSA (institut national des sciences appliquees) de rouen [FR/FR]; Avenue de l'Universite, centre national de la recherche scientifique (CNRS) [FR/FR], Uniyersite derouen [FR/FR], VFP therapies, WO 2014/114742 Al, 2014.

[20] A. Samadi, M. Estrada, C. Pérez, M.I. Rodríguez-Franco, I. Iriepa, I. Moraleda, M. Chioua, J. Marco-Contelles, Pyridonepezils, new dual AChE inhibitors as potential drugs for the treatment of Alzheimer's disease: synthesis, biological assessment, and molecular modeling, Eur. J. Med. Chem. 57 (2012) 296–301.

[21] S. Riaz, I.U. Khan, M. Bajda, Md. Ashraf, A.S Qurat-ul-Ain, T.U. Rehman, S. Mutahir, S. Hussain, G. Mustafa, Md. Yar, Pyridine sulfonamide as a small key organic molecule for the potential treatment of type-II diabetes mellitus and Alzheimer's disease: In vitro studies against yeast α-glucosidase, acetylcholinesterase and butyrylcholinesterase, Bioorg. Chem. 63 (2015) 64–71.

[22] H. Nadri, M. Pirali-Hamedani, A. Moradi, A. Sakhteman, A. Vahidi, V. Sheibani, A. Asadipour, N. Hosseinzadeh, Md. Abdollahi, A. Shafiee, A. Foroumadi, 5,6-Dimethoxybenzofuran-3-one derivatives: a novel series of dual Acetylcholinesterase/Butyrylcholinesterase inhibitors bearing benzyl pyridinium moiety, DARU J. Pharmaceutical Sci. 21 (2013) 15.

[23] M. Saeedi, M. Golipoor, Md. Mahdavi, A. Moradi, H. Nadri, S. Emami, A. Foroumadi, A. Shafiee, Phthalimide-derived N-Benzylpyridinium halides targeting cholinesterases: synthesis and bioactivity of new potential anti-Alzheimer's disease agents, Arch. Pharm. Chem. Life Sci. 349 (2016) 293–301.

[24] N. Salehi, B.B.F. Mirjalili, H. Nadri, Z. Abdolahi, H. Forootanfar, A. Samzadeh-Kermani, T.T. Küçükkılınç, B. Ayazgok, S. Emami, I. Haririan, Md. Sharifzadeh, A. Foroumadi, M. Khoobi, Synthesis and biological evaluation of new N-benzylpyridinium-based benzoheterocycles as potential anti-Alzheimer's agents, Bioorg. Chem. 83 (2019) 559–568.

[25] T. Pal, S. Bhimaneni, A. Sharma, S.J.S. Flora, Design, synthesis, biological evaluation and molecular docking study of novel pyridoxine-triazoles as anti-Alzheimer's agents, RSC Adv. 10 (2020) 26006–26021.

[26] L.S. Zondagh, S.F. Malan, J. Joubert, Design, synthesis and biological evaluation of edaravone derivatives bearing the N-benzyl pyridinium moiety as multifunctional anti-Alzheimer's agents, J. Enzyme Inhib. Med. Chem. 35 (1) (2020) 1596–1605.

[27] S. Abdpour, L. Jalili-Baleh, H. Nadri, H. Forootanfar, S.N.A. Bukhari, A. Ramazani, S.E.S. Ebrahimi, A. Foroumadi, M. Khoobi, Chromone derivatives bearing pyridinium moiety as multi-target-directed ligands against Alzheimer's disease, Bioorg. Chem. 110 (2021) 104750.

[28] M. Maqbool, A. Manral, E. Jameel, J. Kumar, V. Saini, A. Shandilya, M. Tiwari, N. Hoda, B. Jayaram, Development of cyanopyridine-triazine hybrids as lead multitarget anti-Alzheimer agents, Bioorg. Med. Chem. 24 (12) (2016) 2777–2788.

[29] T.J. Peglow, R.F. Schumacher, R. Cargnelutti, A.S. Reis, C. Luchese, E.A. Wilhelm, G. Perin, Preparation of bis(2-pyridyl) diselenide derivatives: Synthesis of selenazolo[5,4-b]pyridines and unsymmetrical diorganyl selenides, and evaluation of antioxidant and anticholinesterasic activities, Tetrahedron Lett. 58 (38) (2017) 3734–3738.

[30] F. Pandolfi, D.D. Vita, M. Bortolami, A. Coluccia, R.D. Santo, R. Costi, V. Andrisano, F. Alabiso, C. Bergamini, R. Fato, M. Bartolini, L. Scipione, New pyridine derivatives as inhibitors of acetylcholinesterase and amyloid aggregation, Eur. J. Med. Chem. 141 (2017) 197–210.

[31] A. Hiremathad, K. Chand, L. Tolayan, R.S.K Rajeshwari, A.R. Esteves, S.M. Cardoso, S. Chaves, M.A. Santos, Hydroxypyridinone-benzofuran hybrids with potential protective roles for Alzheimer´s disease therapy, J. Inorg. Biochem. 179 (2018) 82–96.

[32] X.Y. Jiang, T.K. Chen, J.T. Zhou, S.Y. He, H.Y. Yang, Y. Chen, W. Qu, F. Feng, H.P. Sun, Dual GSK-3β/AChE inhibitors as a new strategy for multitargeting anti-Alzheimer's disease drug discovery, ACS Med. Chem. Lett. 9 (3) (2018) 171–176.

[33] R. Ghobadian, Md. Mahdavi, H. Nadri, A. Moradi, N. Edraki, T. Akbarzadeh, Md. Sharifzadeh, S.N.A. Bukhari, M. Amini, Novel tetrahydrocarbazole benzyl pyridine hybrids as potent and selective butryl cholinesterase inhibitors with neuroprotective and β-secretase inhibition activities, Eur. J. Med. Chem. 155 (2018) 49–60.

[34] K. Czarnecka, N. Chufarova, K. Halczuk, K. Maciejewska, M. Girek, R. Skibiński, J. Jończyk, M. Bajda, J. Kabziński, I. Majsterek, P. Szymański, Tetrahydroacridine derivatives with dichloronicotinic acid moiety as attractive, multipotent agents for Alzheimer's disease treatment, Eur. J. Med. Chem. 145 (2018) 760–769.
[35] J. Kumar, A. Gill, M. Shaikh, A. Singh, A. Shandilya, E. Jameel, N. Sharma, N. Mrinal, N. Hoda, B. Jayaram, Pyrimidine-triazolopyrimidine and pyrimidine-pyridine hybrids as potential acetylcholinesterase inhibitors for Alzheimer's disease, Chem. Select 3 (2) (2018) 736–747.
[36] S.K. Singh, S.K. Sinha, M.K. Shirsat, Design, synthesis and evaluation of 4-aminopyridine analogues as cholinesterase inhibitors for management of Alzheimer's diseases, Indian J. Pharmaceutical Education Res. 52 (4) (2018) 644–654.
[37] C. de los Ríos, J. Marco-Contelles, Tacrines for Alzheimer's disease therapy. III. The PyridoTacrines, Eur. J. Med. Chem. 166 (2019) 381–389.
[38] A. El-Malah, A.I.Y. Abouelatta, Z. Mahmoud, H.H. Salem, New cyclooctathienopyridine derivatives in the aim of discovering better Anti-Alzheimer's agents, J. Mol. Struct. 1196 (2019) 162–168.
[39] T. Pan, S. Xie, Y. Zhou, J. Hu, H. Luo, X. Li, L. Huang, Dual functional cholinesterase and PDE4D inhibitors for the treatment of Alzheimer's disease: Design, synthesis and evaluation of tacrine-pyrazolo[3,4-b]pyridine hybrids, Bioorg. Med. Chem. Lett. 29 (16) (2019) 2150–2152.
[40] M. Saeedi, M. Safavi, E. Allahabadi, A. Rastegari, R. Hariri, S. Jafari, S.N.A. Bukhari, S.S. Mirfazli, O. Firuzi, N. Edraki, Md. Mahdavi, T. Akbarzadeh, Thieno[2,3-b]pyridine amines: synthesis and evaluation of tacrine analogues against biological activities related to Alzheimer's disease, Arch. Pharm. 353 (10) (2020) 2000101.
[41] N. Stellenboom, A.R. Baykan, Synthesis and enzyme inhibitory activity of novel pyridine-2,6-dicarboxamides bearing primary sulfonamide groups, Russ. J. Org. Chem. 55 (2019) 1951–1956.
[42] A. Tripathi, P.K. Choubey, P. Sharma, A. Seth, P.N. Tripathi, M.K. Tripathi, S.K. Prajapati, S. Krishnamurthy, S.K. Shrivastava, Design and development of molecular hybrids of 2-pyridylpiperazine and 5-phenyl-1,3,4-oxadiazoles as potential multifunctional agents to treat Alzheimer's disease, Eur. J. Med. Chem. 183 (2019) 111707.
[43] T. Umar, S. Shalini, Md.K. Raza, S. Gusain, J. Kumar, P. Seth, M. Tiwari, N. Hoda, A multifunctional therapeutic approach: synthesis, biological evaluation, crystal structure and molecular docking of diversified 1H-pyrazolo[3,4-b]pyridine derivatives against Alzheimer's disease, Eur. J. Med. Chem. 175 (2019) 2–19.
[44] Z. Zhu, T. Yang, L. Zhang, L. Liu, E. Yin, C. Zhang, Z. Guo, C. Xu, X. Wang, Inhibiting Aβ toxicity in Alzheimer's disease by a pyridine amine derivative, Eur. J. Med. Chem. 168 (2019) 330–339.
[45] R. Zafar, H. Naureen, Md. Zubair, K. Shahid, Md.S. Jan, S. Akhtar, H. Ahmad, W. Waseem, A. Haider, S. Ali, Md. Tariq, A. Sadiq, Prospective application of two new pyridine-based Zinc (II) amide carboxylate in management of Alzheimer's disease: synthesis, characterization, computational and in vitro approaches, Drug Design, Develop. Therapy 15 (2021) 2679–2694.
[46] H. Zhang, C. Wu, X. Chen, Z. Zhang, X. Jiang, H.L. Qin, W. Tang, Novel pyridine-containing sultones: structure-activity relationship and biological evaluation as selective AChE inhibitors for the treatment of Alzheimer's disease, ChemMedChem 16 (20) (2021) 3189–3200.
[47] M.V.K. Reddy, K.Y. Rao, G. Anusha, G.M. Kumar, A.G. Damu, K.R. Reddy, N.P. Shetti, T.M. Aminabhavi, P.V.G. Reddy, *In-vitro* evaluation of antioxidant and anticholinesterase activities of novel pyridine, quinoxaline and s-triazine derivatives, Environ. Res. 199 (2021) 111320.
[48] F.A. Attaby, A.M. Abdel-Fattah, L.M. Shaif, Md.M. Elsayed, Reactions, anti-Alzheimer and anti COX-2 activities of the newly synthesized 2-substituted thienopyridines, Curr. Org. Chem. 13 (16) (2009) 1654–1663.
[49] M.R. Jones, E.L. Service, J.R. Thompson, M.C. Wang, I.J. Kimsey, A.S. DeToma, A. Ramamoorthy, M.H. Lim, T. Storr, Dual-function triazole-pyridine derivatives as inhibitors of metal-induced amyloid-β aggregation, Metallomics 4 (9) (2012) 910–920.
[50] K.M. Lincoln, T.E. Richardson, L. Rutter, P. Gonzalez, J.W. Simpkins, K.N. Green, An N-Heterocyclic amine chelate capable of antioxidant capacity and amyloid disaggregation, ACS Chem. Neurosci. 3 (11) (2012) 919–927.
[51] Y. Huang, T. Li, A. Eatherton, W.L. Mitchell, N. Rong, L. Ye, X.J. Yang, S. Jin, Y. Ding, J. Zhang, Y. Li, Y. Wu, Y. Jin, Y. Sang, Z. Cheng, E.R. Browne, D.C. Harrison, I. Hussain, Z. Wan, A. Hall, L.F. Lau, Y. Matsuoka, Orally bioavailable and brain-penetrant pyridazine and pyridine-derived γ-secretase modulators reduced amyloidogenic Aβ peptides in vivo, Neuropharmacology 70 (2013) 278–286.
[52] J.J. Chen, W. Qian, K. Biswas, C. Yuan, A. Amegadzie, Q. Liu, T. Nixey, J. Zhu, M. Ncube, R.M. Rzasa, F. Chavez, N. Chen, F. DeMorin, S. Rumfelt, C.M. Tegley, J.R. Allen, S. Hitchcock, R. Hungate, M.D. Bartberger, L. Zalameda, Y. Liu, J.D. McCarter, J. Zhang, L. Zhu, S. Babu-Khan, Y. Luo, J. Bradley, P.H. Wen, D.L. Reid, F. Koegler, C. Dean,

D. Hickman, T.L. Correll, T. Williamson, S. Wood, Discovery of 2-methylpyridine-based biaryl amides as γ-secretase modulators for the treatment of Alzheimer's disease, Bioorg. Med. Chem. Lett. 23 (23) (2013) 6447–6454.

[53] J. Makowska, D. Szczesny, A. Lichucka, A. Giełdoń, L. Chmurzyński, R. Kaliszan, Preliminary studies on trigonelline as potential anti-Alzheimer disease agent: determination by hydrophilic interaction liquid chromatography and modeling of interactions with beta-amyloid, J. Chromatogr. B 968 (2014) 101–104.

[54] T. Takai, Y. Hoashi, Y. Tomata, S. Morimoto, M. Nakamura, T. Watanabe, T. Igari, T. Koike, Discovery of novel 5,6,7,8-tetrahydro[1,2,4]triazolo[4,3-a]pyridine derivatives as γ-secretase modulators, Bioorg. Med. Chem. Lett. 25 (19) (2015) 4245–4249.

[55] A. Elkamhawy, J. Park, A.H.E. Hassan, A.N. Pae, J. Lee, B.G. Park, E.J. Roh, Synthesis and evaluation of 2-(3-arylureido)pyridines and 2-(3-arylureido)pyrazines as potential modulators of Aβ-induced mitochondrial dysfunction in Alzheimer's disease, Eur. J. Med. Chem. 144 (2018) 529–543.

[56] P. Zhou, Y. Li, Yi Fan, Z. Wang, R. Chopra, A. Olland, Y. Hu, R.L. Magolda, M. Pangalos, P.H. Reinhart, M.J. Turner, J. Bard, M.S. Malamas, A.J. Robichaud, Pyridinyl aminohydantoins as small molecule BACE1 inhibitors, Bioorg. Med. Chem. Lett. 20 (7) (2010) 2326–2329.

[57] B.M. Swahn, K. Kolmodin, S. Karlström, S. von Berg, P. Söderman, J. Holenz, S. Berg, J. Lindström, M. Sundström, D. Turek, J. Kihlström, C. Slivo, L. Andersson, D. Pyring, D. Rotticci, L. Öhberg, A. Kers, K. Bogar, M. Bergh, L.L. Olsson, J. Janson, S. Eketjäll, B. Georgievska, F. Jeppsson, J. Fälting, Design and synthesis of β-site amyloid precursor protein cleaving enzyme (BACE1) inhibitors with in vivo brain reduction of β-Amyloid peptides, J. Med. Chem. 55 (21) (2012) 9346–9361.

[58] F.J. Pérez-Areales, N. Betari, A. Viayna, C. Pont, A. Espargaró, M. Bartolini, A. De Simone, J.F. Rinaldi Alvarenga, B. Pérez, R. Sabate, R.M. Lamuela-Raventós, V. Andrisano, F.J. Luque, D. Muñoz-Torrero, Design, synthesis and multitarget biological profiling of second-generation anti-Alzheimer rhein-huprine hybrids, Future Medicinal Chem. 9 (10) (2017) 965–981.

[59] N.Q. Thai, Z. Bednarikova, M. Gancar, H.Q. Linh, C.K. Hu, M.S. Li, Z. Gazova, Compound CID 9998128 is a potential multitarget drug for Alzheimer's disease, ACS Chem. Neurosci. 9 (11) (2018) 2588–2598.

[60] Y. Fang, H. Zhou, Q. Gu, J. Xu, Synthesis and evaluation of tetrahydroisoquinoline-benzimidazole hybrids as multifunctional agents for the treatment of Alzheimer's disease, Eur. J. Med. Chem. 167 (2019) 133–145.

[61] K. Nakahara, Y. Mitsuoka, S. Kasuya, T. Yamamoto, S. Yamamoto, H. Ito, Y. Kido, K. Kusakabe, Balancing potency and basicity by incorporating fluoropyridine moieties: discovery of a 1-amino-3,4-dihydro-2,6-naphthyridine BACE1 inhibitor that affords robust and sustained central Aβ reduction, Eur. J. Med. Chem. 216 (2021) 113270.

[62] S.S. Hindo, A.M. Mancino, J.J. Braymer, Y. Liu, S. Vivekanandan, A. Ramamoorthy, M.H. Lim, Small molecule modulators of copper-induced Aβ aggregation, J. Am. Chem. Soc. 131 (46) (2009) 16663–16665.

[63] J.S. Choi, J.J. Braymer, R.P.R. Nanga, A. Ramamoorthy, M.H. Lim, Design of small molecules that target metal-Aβ species and regulate metal-induced Aβ aggregation and neurotoxicity, Proc. Natl Acad. Sci. 107 (51) (2010) 21990–21995.

[64] A. Lakatos, É. Zsigó, D. Hollender, N.V. Nagy, L. Fülöp, D. Simon, Z. Bozsód, T. Kiss, Two pyridine derivatives as potential Cu(ii) and Zn(ii) chelators in therapy for Alzheimer's disease, Dalton Trans. 39 (2010) 1302–1315.

[65] D.E. Green, M.L. Bowen, L.E. Scott, T. Storr, M. Merkel, K. Böhmerle, K.H. Thompson, B.O. Patrick, H.J. Schugar, C. Orvig, In vitro studies of 3-hydroxy-4-pyridinones and their glycosylated derivatives as potential agents for Alzheimer's disease, Dalton Trans. 39 (2010) 1604–1615.

[66] H. Kroth, N. Sreenivasachary, A. Hamel, P. Benderitter, Y. Varisco, V. Giriens, P. Paganetti, W. Froestl, A. Pfeifer, A. Muhs, Synthesis and structure-activity relationship of 2,6-disubstituted pyridine derivatives as inhibitors of β-amyloid-42 aggregation, Bioorg. Med. Chem. Lett. 26 (14) (2016) 3330–3335.

[67] B.J. Wall, M.F. Will, G.K. Yawson, P.J. Bothwell, D.C. Platt, C.F. Apuzzo, M.A. Jones, G.M. Ferrence, M.I. Webb, Importance of hydrogen bonding: Structure-activity relationships of Ruthenium(III) complexes with pyridine-based ligands for Alzheimer's disease therapy, J. Med. Chem. 64 (14) (2021) 10124–10138.

[68] H.R. Park, J. Kim, T. Kim, S. Jo, M. Yeom, B. Moon, I.H. Choo, J. Lee, E.J. Lim, K.D. Park, S.J. Min, G. Nam, G. Keum, C.J. Lee, H. Choo, Oxazolopyridines and thiazolopyridines as monoamine oxidase B inhibitors for the treatment of Parkinson's disease, Bioorg. Med. Chem. 21 (17) (2013) 5480–5487.

[69] N.T. Tzvetkov, H.G. Stammler, S. Hristova, A.G. Atanasov, L. Antonov, Pyrrolo-pyridin-5-yl) benzamides: BBB permeable monoamine oxidase B inhibitors with neuroprotective effect on cortical neurons, Eur. J. Med. Chem. 162 (2019) 793–809.

[70] C. Zhang, K. Yang, S. Yu, J. Su, S. Yuan, J. Han, Y. Chen, J. Gu, T. Zhou, R. Bai, Y. Xie, Design, synthesis and biological evaluation of hydroxypyridinone-coumarin hybrids as multimodal monoamine oxidase B inhibitors and iron chelates against Alzheimer's disease, Eur. J. Med. Chem. 180 (2019) 367–382.

[71] G. Turan-Zitouni, W. Hussein, B.N. Sağlık, A. Tabbi, B. Korkut, Design, synthesis and biological evaluation of novel N-pyridyl-hydrazone derivatives as potential monoamine oxidase (MAO) inhibitors, Molecules 23 (1) (2018) 113.

[72] Y. Uno, H. Iwashita, T. Tsukamoto, N. Uchiyama, T. Kawamoto, M. Kori, A. Nakanishi, Efficacy of a novel, orally active GSK-3 inhibitor 6-Methyl-N-[3-[[3-(1-methylethoxy) propyl] carbamoyl]-1H-pyrazol-4-yl] pyridine-3-carboxamide in tau transgenic mice, Brain Res. 1296 (2009) 148–163.

[73] F.A. Attaby, A.M. Abdel-Fattah, L.M. Shaif, Md.M. Elsayed, Anti-Alzheimer and anti-COX-2 activities of the newly synthesized 2,3′-bipyridine derivatives (II), Phosphorus, Sulfur, Silicon 185 (2010) 668–679.

[74] X. Huang, W.G. Z.Zhu, A. Palani, R.G. Aslanian, Gamma secretase modulators, WO 2010/147975 Al, Schering corporation, 2010.

[75] S. Moriguchi, N. Shioda, Y. Yamamoto, H. Tagashira, K. Fukunaga, The T-type voltage-gated calcium channel as a molecular target of the novel cognitive enhancer ST101: enhancement of long-term potentiation and CaMKII autophosphorylation in rat cortical slices, J. Neurochem. 121 (1) (2012) 44–53.

[76] P. Sivaprakasam, X. Han, R.L. Civiello, S. Jacutin-Porte, K. Kish, M. Pokross, H.A. Lewis, N. Ahmed, N. Szapiel, J.A. Newitt, E.T. Baldwin, H. Xiao, C.M. Krause, H. Park, M. Nophsker, J.S. Lippy, C.R. Burton, D.R. Langley, J.E. Macor, G.M. Dubowchik, Discovery of new acylaminopyridines as GSK-3 inhibitors by a structure guided in-depth exploration of chemical space around a pyrrolopyridinone core, Bioorg. Med. Chem. Lett. 25 (9) (2015) 1856–1863.

[77] J.E. Davoren, C.W. Lee, M. Garnsey, M.A. Brodney, J. Cordes, K. Dlugolenski, J.R. Edgerton, A.R. Harris, C.J. Helal, S. Jenkinson, G.W. Kauffman, T.P. Kenakin, J.T. Lazzaro, S.M. Lotarski, Y. Mao, D.M. Nason, C. Northcott, L. Nottebaum, S.V. O'Neil, B. Pettersen, M. Popiolek, V. Reinhart, R. Salomon-Ferrer, S.J. Steyn, D. Webb, L. Zhang, S. Grimwood, Discovery of the potent and selective M1 PAM-agonist N-[(3R,4S)-3-Hydroxytetrahydro-2H-pyran-4-yl]-5-methyl-4-[4-(1,3-thiazol-4-yl)benzyl]pyridine-2-carboxamide (PF-06767832): evaluation of efficacy and cholinergic side effects, J. Med. Chem. 59 (13) (2016) 6313–6328.

[78] D.S. Cukierman, E. Accardo, R.G. Gomes, A. De Falco, M.C. Miotto, M.CR. Freitas, M. Lanznaster, C.O. Fernández, N.A. Rey, Aroylhydrazones constitute a promising class of 'metal-protein attenuating compounds' for the treatment of Alzheimer's disease: a proof-of-concept based on the study of the interactions between zinc(II) and pyridine-2-carboxaldehyde isonicotinoyl hydrazone, J. Biol. Inorg. Chem. 23 (2018) 1227–1241.

[79] M.A. Brodney, J.E. Davoren, M.R. Garnsey, L. Zhang, S.V. O'neil, pyridine derivatives as muscarinic ml receptor positive allosteric modulators, WO 2016/009297 Al, PFIZER INC. 2016.

CHAPTER

8

Role of pyridines as enzyme inhibitors in medicinal chemistry

Khalid Mohammed Khan[a], *Syeda Shaista Gillani*[b,c] and *Faiza Saleem*[a]

[a] H.E.J. Research Institute of Chemistry, International Center for Chemical and Biological Sciences, University of Karachi, Karachi, Pakistan [b] Department of Chemistry, Lahore Garrison University, Lahore, Pakistan [c] School of Chemistry, University of the Punjab, Lahore, Pakistan

8.1 General introduction

Pyridine is an organic monocyclic mancude azaarene with the chemical formula C_5H_5N that consists of a benzene ring where one –CH group is replaced by a nitrogen atom Fig. 8.1. It is a weakly alkaline, highly flammable, water-miscible liquid that appears colorless to yellow with 0.978 g/cm^3 density and 115.2–115.3°C boiling point. Pyridine also has a characteristic pungent nauseous fish-like odor, toxic by inhalation as well as ingestion, and upon combustion, it produces toxic nitrogen oxides [1,2].

In 1849, the Scottish chemist, Thomas Anderson, discovered pyridine by extracting pyridine from oil obtained from animal bones. The new molecule was very flammable, so it was called pyridine, from the Greek word "pyr" meaning fire and "idine" meaning nitrogen-containing organic compound. Anderson's method of isolating pyridine was based on the extraction of coal tar and was poor until the 1920s, when Aleksey Chichibabin, a Russian chemist, discovered a new method to synthesize it from ammonia, acetaldehyde, and formaldehyde [3]. The cyclic nature of pyridine was identified by Dewar and Korner in 1869. It plays an important role in catalyzing both biological and chemical systems. It has the conjugated six π electrons like benzene which are delocalized about the heterocycle. The molecule is essentially planar and follows the Hückel aromaticity rule. Moreover, pyridine is a class of both synthetically and naturally occurring heterocyclic compounds and its nucleus is a well-studied six-membered heterocycle with a wide-ranging therapeutic use. Pyridine is a useful precursor for the synthesis of other compounds. Its major role is in the herbicides production and also provides a building block for some pesticides, preservatives, and fungicidal agents.

Recent Developments in the Synthesis and Applications of Pyridines.
DOI: https://doi.org/10.1016/B978-0-323-91221-1.00010-5

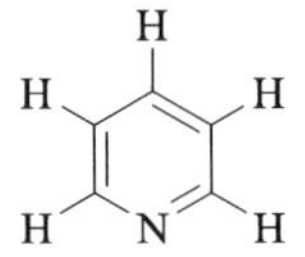

FIGURE 8.1 Structure of pyridine.

Pure pyridine has a variety of uses as an excellent solvent, primarily in antibiotics extraction, dehydrochlorination as well as acylation reactions, and commercially used in both dye and rubber industries [3–5].

Pyridine exhibits unique chemistry as a reagent, substrate, and ligands. In synthetic organic chemistry, pyridine has been hypothesized to play a significant role in C–H activation/oxidation as a functional group. Pyridines play countless roles in the supramolecular chemistry field either in organic polymers of metals as ligand or in nanodevices that control major electronic stations [6]. The exclusive pyridine reactivity is related to all fields dealing with heterocyclic chemistry. Pyridines form N-alkyl pyridinium salts from the respective alkyl-containing compounds *via* removing counterparts such as sulfate, halide, and other ions by S_N2 fashion. Pyridine reacts exothermally with dimethyl sulfate or iodomethane. Pyridine reactions are slower with alkyl halides except for methyl, and thus perform under reflux in a solvent with high dielectric constant such as acetonitrile which helps in ion formation. Instead, an extremely active alkylating agent, for example, alkyl triflate can be used [7].

The pyridine nucleus is usually present in numerous natural products and is important in biological systems. Pyridine molecules are frequently used in drugs due to their properties for example basicity, water solubility, small molecular size, stability, and their hydrogen bond-forming ability. Substituted pyridine derivatives are reported as anticancer, anti-inflammatory, antifungal, antiproliferative, antitumor, antimicrobial, antitubercular, anti-HIV, and enzyme inhibitors. Our lives are dependent on pyridine comprising cofactors which are essential nutrients. This blockbuster heterocycle is present in the important vitamins and in over-the-counter marketed drugs such as nexium, takepron, singulair, and actos with a business of over $20 billion last year (Fig. 8.2). However, nature is constantly inspiring us with new structures as the industry demonstrates its ingenuity in synthesizing new azine comprising compounds [8,9].

8.2 Marketed drugs based on pyridine nucleus

Some pyridine-based drugs are listed below.

Pyridoxine

Our body requires pyridoxine (vitamin B_6: dietary supplement) to treat pyridoxine deficiency, pyridoxine-dependent epilepsy, and certain metabolic abnormalities. It is used to produce amino acids, lipids, and carbohydrates (Fig. 8.3) [10].

Nicotine

Nicotine is an alkaloid naturally obtained from the family Solanaceae. Nicotine is extremely addictive and is one of the most abused drugs. The important medical use of nicotine is to eradicate smoking (Fig. 8.4) [11].

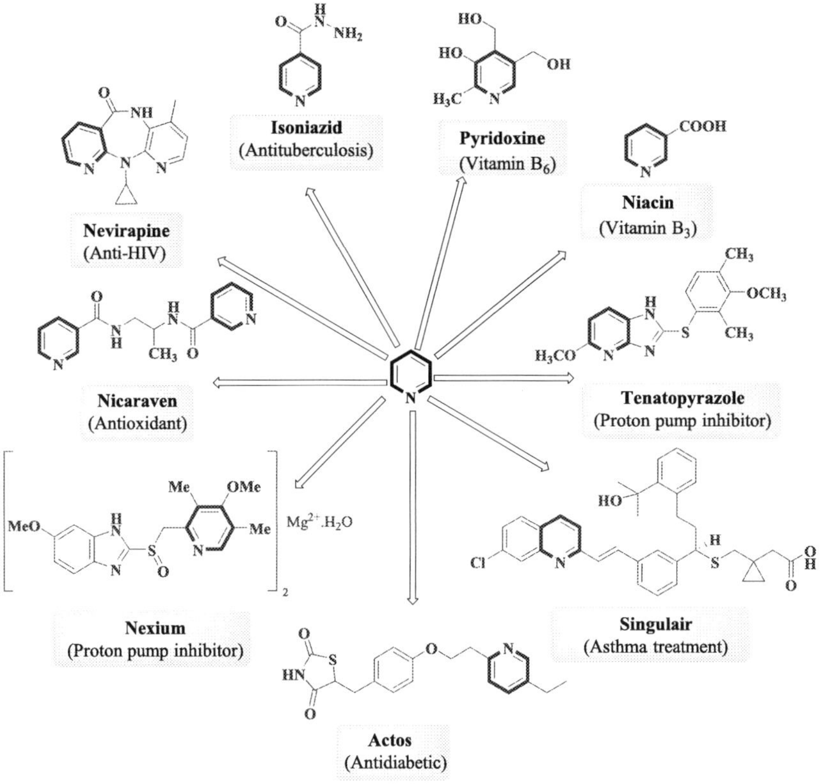

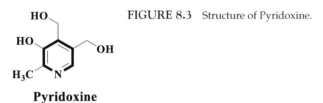

FIGURE 8.2 Some structure of pyridine-based marketed drugs.

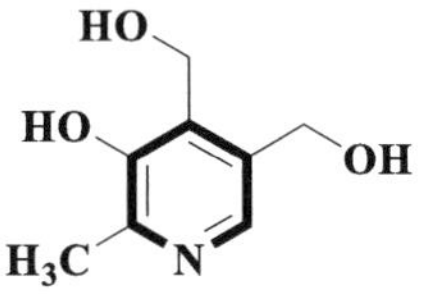

Pyridoxine

FIGURE 8.3 Structure of Pyridoxine.

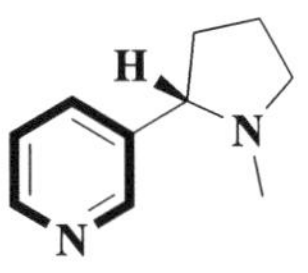

Nicotine

FIGURE 8.4 Structure of Nicotine.

Niacin

Niacin (Vitamin B_3) and its derivatives are broadly used for the synthesis of ion exchange resins and new polymeric materials, in agriculture, medicine, and food industry as supplements to control the growth of a plant. Niacin is taken as a supplement, also produced in the

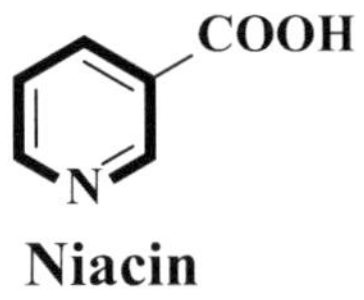

FIGURE 8.5 Structure of Niacin.

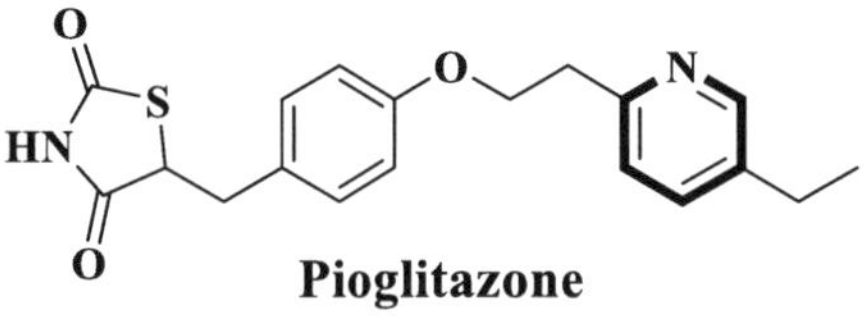

FIGURE 8.6 Structure of Pioglitazone.

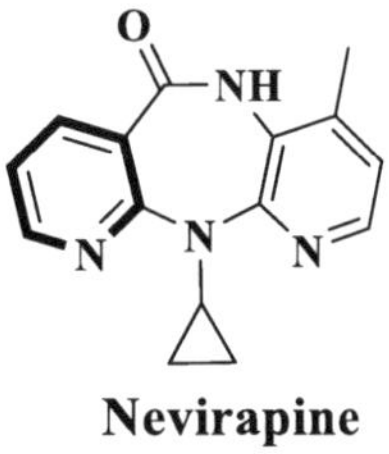

FIGURE 8.7 Structure of Nevirapine.

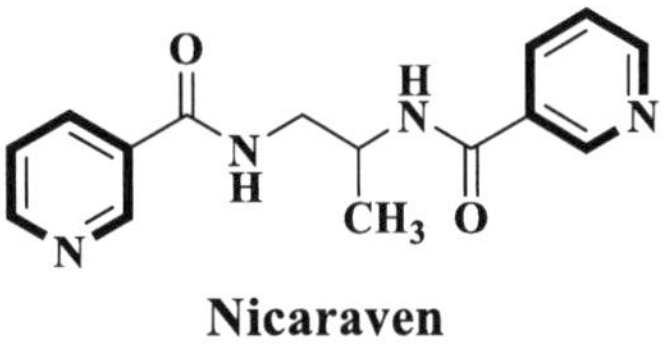

FIGURE 8.8 Structure of Nicaraven.

body from tryptophan, which is found in protein-containing food. It is mainly used to treat niacin deficiency (pellagra) (Fig. 8.5) [12].

Pioglitazone

Pioglitazone is a combination of two drugs, pioglitazone, and glimepiride. It is a thiazolidinedione-type, also called "glitazones" a diabetes drug. The patients having type-2 diabetes, use it to regulate high sugar levels of blood and increases the exposure of tissues to insulin (Fig. 8.6). [13]

Nevirapine

Nevirapine (Viramune) works against HIV, particularly HIV-1 in combination with antiretroviral drugs. It is thereby used to retard the immune system damage and also decreases the risk of developing AIDS and HIV viral loads (Fig. 8.7) [14].

Nicaraven

Nicaraven is a drug used for the treatment of cerebral stroke including subarachnoid hemorrhage. It is an antioxidant a specific radical scavenger for hydroxyl radicals, to protect against ischemia-reperfusion injury in several organs like kidney, brain, heart, and liver (Fig. 8.8) [15].

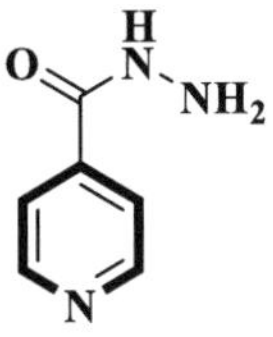

Isoniazid

FIGURE 8.9 Structure of Isoniazid.

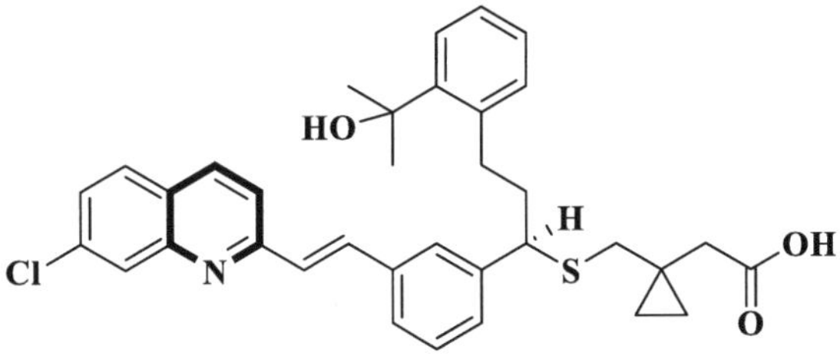

Singulair

FIGURE 8.10 Structure of Singulair.

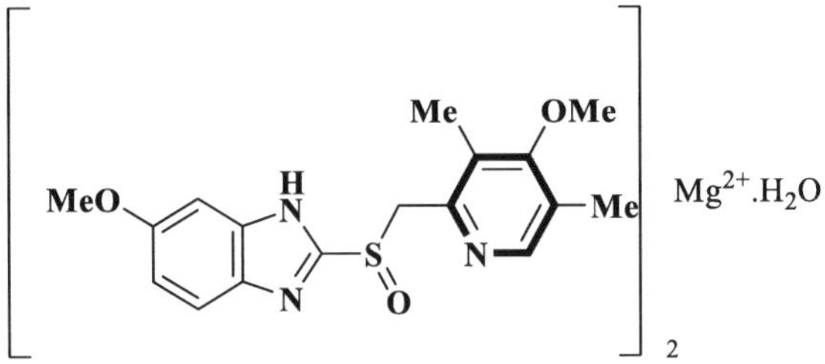

Nexium

FIGURE 8.11 Structure of Nexium.

Isoniazid

Isoniazid is used in combination with other drugs for the treatment and prevention of tuberculosis (TB). It is an antibiotic used only for the treatment of bacterial infections for their growth prevention (Fig. 8.9) [16].

Singulair

Singulair is used to block substances called leukotrienes in the body and to relieve asthma symptoms (such as wheezing and shortness of breath). It is used to prevent breathing problems (bronchospasm) during exercise. This helps to reduce the number of times you need to use a rapid relief inhaler (Fig. 8.10) [17].

Nexium

Nexium (esomeprazole magnesium) is a proton pump inhibitor that decreases the amount of acid produced in the stomach during ulceration. Nexium is used to treat symptoms of gastroesophageal reflux disease and other conditions related to excess stomach acid, such as Zollinger–Ellison syndrome, and to promote the healing of damaged esophagus due to stomach acid (Fig. 8.11) [18,19].

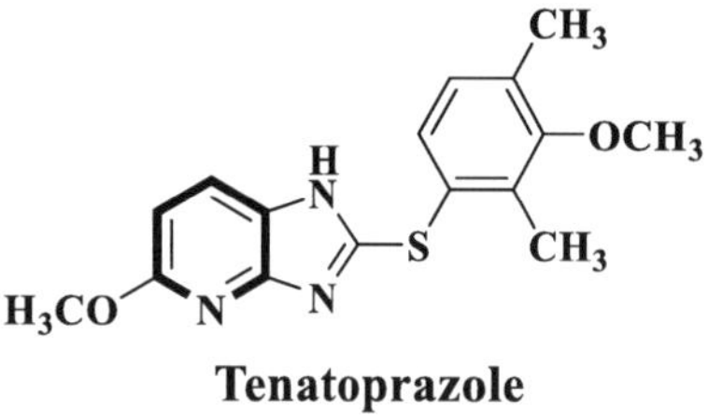

FIGURE 8.12 Structure of Tenatoprazole.

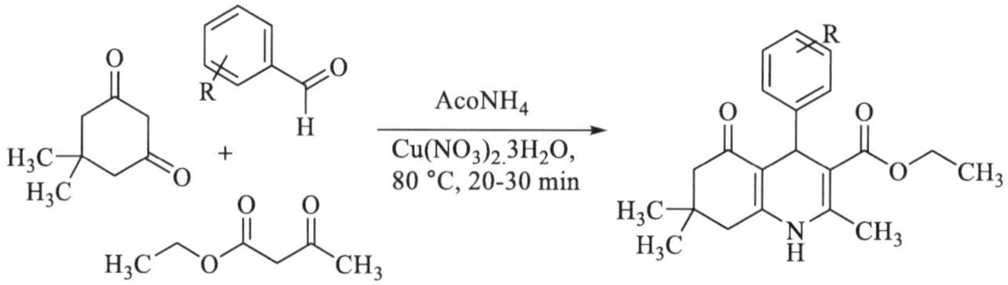

SCHEME 8.1 Hantzsch synthesis of dihydropyridines.

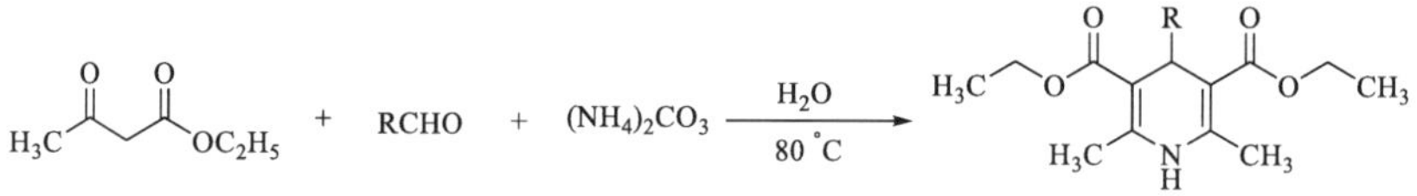

SCHEME 8.2 Synthesis of 1,4-dihydropyridines.

Tenatoprazole

Tenatoprazole is a novel proton pump inhibitor whose plasma half-life is seven times longer. Proton pump inhibitors control stomach acidity during the day than at night, when nocturnal acid breakthroughs can occur (Fig. 8.12) [20].

8.3 Synthesis of pyridine derivatives

Few synthetic methods are discussed below for the synthesis of pyridine and its synthetic derivatives.

One-pot synthesis of pyridine derivatives by modifying the Hantzsch synthesis protocol are reported in literature recently. By reacting ammonium acetate, different aryl aldehydes ethyl acetoacetate, and dimedone in the presence of $Cu(NO_3)_2.3H_2O$ as catalyst under solvent-free conditions, pyridine derivatives were obtained (Scheme 8.1) [21]. By using ammonium carbonate instead of ammonium acetate and dimedone, 1,4-dihydropyridines were obtained in good yields (Scheme 8.2) [22].

2-Aminocarbonitrile pyridines were also obtained by one-pot multicomponent reaction (Scheme 8.3) [23].

Pyridinedicarbonitriles were obtained by reacting various aryl aldehydes and malononitrile in the presence of a base to form the acrylonitrile intermediate which was then treated with malononitrile and thiophenol in basic conditions to give the final compound (Scheme 8.4) [24].

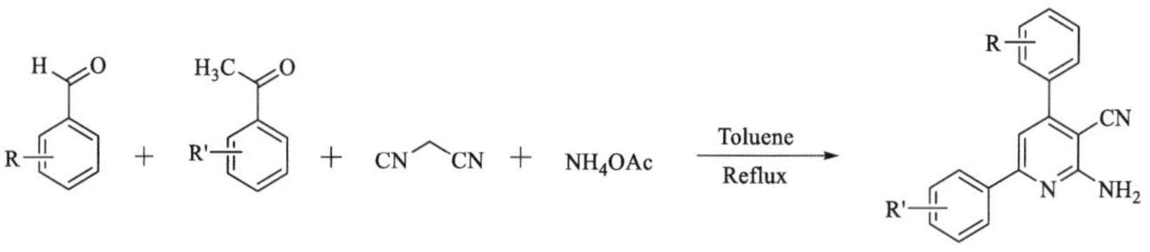

SCHEME 8.3 Synthesis of 2-aminocarbonitrile pyridines.

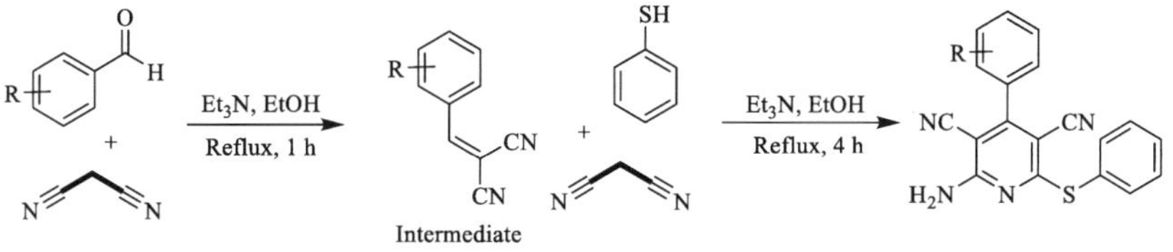

SCHEME 8.4 Synthesis of pyridinedicarbonitriles.

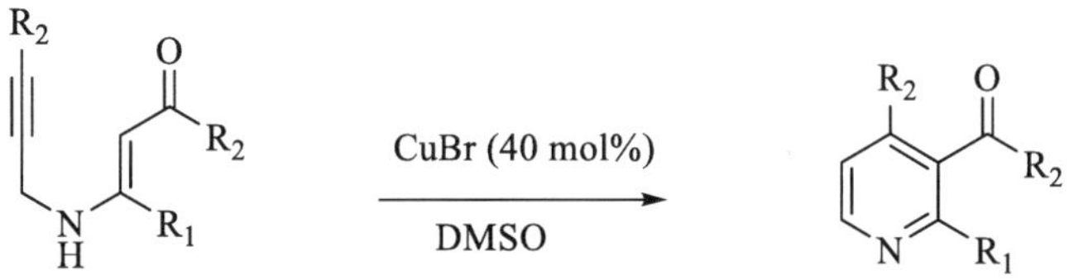

SCHEME 8.5 Synthesis of pyridines by *N*-propargylic-β-enaminones.

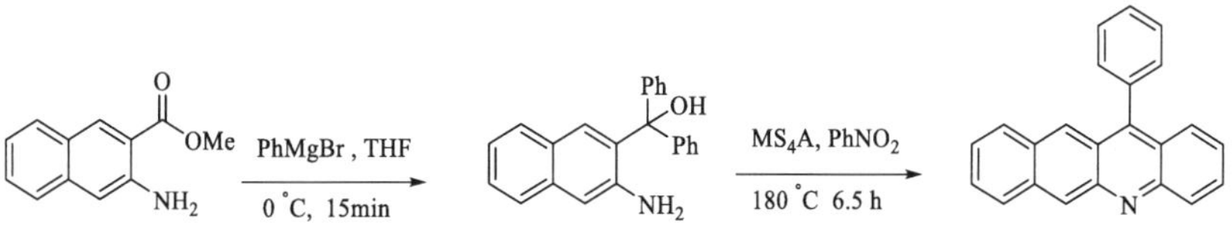

SCHEME 8.6 Synthesis of acridines.

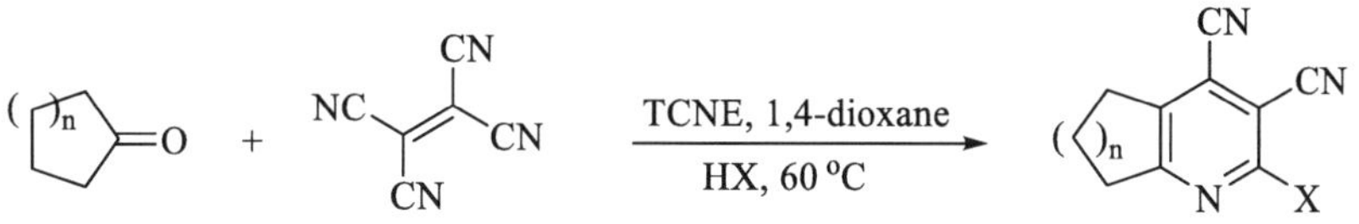

SCHEME 8.7 Synthesis of 2-halocycloalkylpyridine-3,4-dicarbonitriles.

N-Propargylic-β-enaminones afforded numerous pyridine compounds by cyclization catalyzed by CuBr in DMSO in an inert atmosphere (Scheme 8.5) [8].

Naphthalene-3-amino ester reacted with phenyl magnesium bromide *via* Grignard reaction to yield triaryl carbinol which when cyclized provided pyridine compounds (Scheme 8.6) [25].

Tetracyanoethylene, cycloalkanone, and hydrogen halide give 2-halocycloalkylpyridine-3,4-dicarbonitriles *via* coupling reaction (Scheme 8.7) [26].

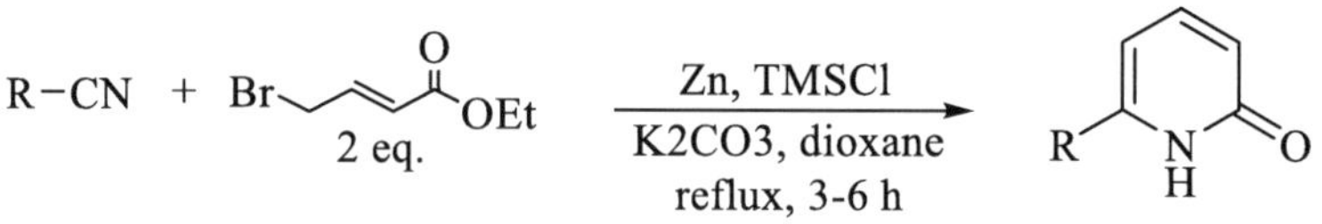

SCHEME 8.8 Synthesis of 2-pyridones.

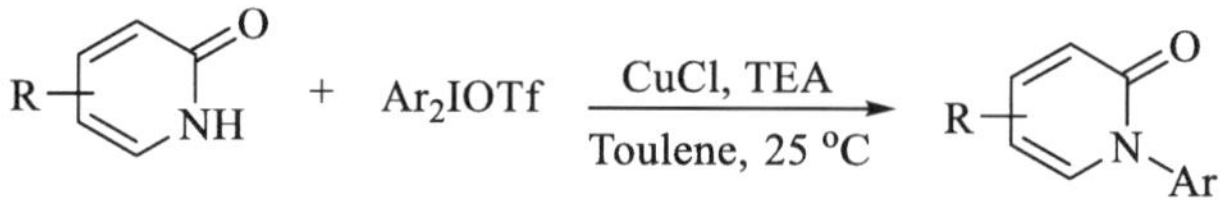

SCHEME 8.9 Synthesis of N-arylated-2-pyridones.

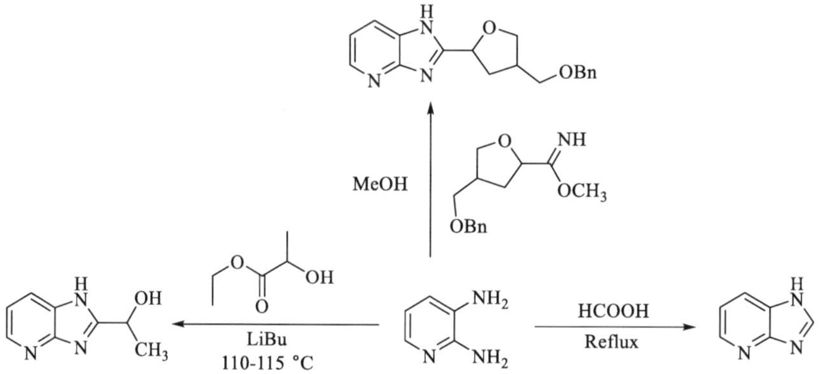

SCHEME 8.10 Synthesis of imidazopyridines.

Pyridone-2-ones were synthesized by treating ethy-4-bromobut-2-enoate, substituted nitriles, trimethylsilyl chloride (3 mol%), in the presence of zinc catalyst and potassium carbonate for 30 min (Scheme 8.8) [27].

N-Arylation of 2-pyridone with diaryliodonium salts is copper-catalyzed (0.1 eq) at room temperature to produce *N*-arylpyridine-2-ones (Scheme 8.9) [28].

Synthesis of imidazopyridines was known since 2000 to date. Imidazopyridines consist of various isomeric forms. Imidazopyridine-based drugs, such as bamaluzole and tenatoprazole, are available on the market. Several imidazopyridine derivatives are used clinically to treat many diseases. By using 2,3-diaminopyridine, variety of imidazopyridine derivatives could be synthesized as demonstrated in the following Scheme 8.10 [29–32].

8.4 Medicinal importance of pyridine

Pyridine is a heterocycle present in numerous pharmaceuticals and natural products. It imparts a significant character to bioactive natural substances such as nicotine, vitamin B_6, and coenzymes (NADP, NADPH). These heterocycles have found many pharmaceutical uses such

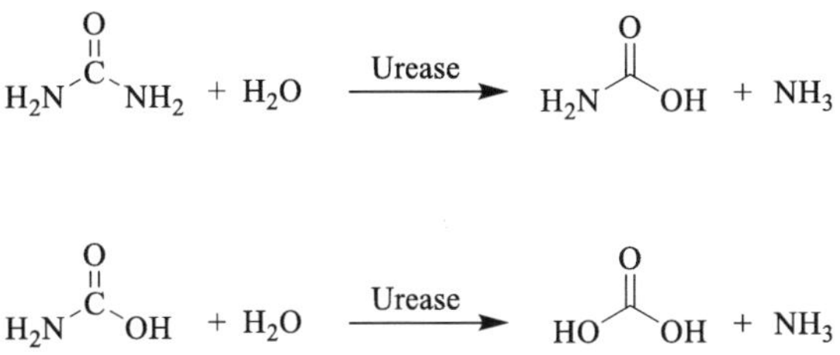

$$H_2N-\overset{\overset{O}{\|}}{C}-NH_2 + H_2O \xrightarrow{\text{Urease}} H_2N-\overset{\overset{O}{\|}}{C}-OH + NH_3$$

$$H_2N-\overset{\overset{O}{\|}}{C}-OH + H_2O \xrightarrow{\text{Urease}} HO-\overset{\overset{O}{\|}}{C}-OH + NH_3$$

SCHEME 8.11 Hydrolysis of urea.

as antidepressant, antidiabetic, anticancer, antioxidant, anti-Alzheimer, anti-inflammatory, antifungal, antiviral, anti-HIV, etc. Remarkably, right now up to hundred marketed drugs have this privileged pharmacophore having high demand in the market. Prominent drugs among them are lavendamycin (anticancer), Bay 60-5521 (CETP-inhibitors), and streptonigrin (anticancer) [33].

8.4.1 Anticancer activity

Cancer is unregulated and anomalous multiplication of cells that arise from cells of a particular organ. Cancer is a common name for a collection of more than 100 diseases. However, many types of cancer start due to the uncontrolled growth of abnormal cells. Cancer cells generate their own blood supply, separate from the organ where they originated, and then proliferate to other body organs (Scheme 8.11). Morbidity and mortality increase primarily due to organ damage instigated by local growth and metastases to distant anatomic areas [34,35]. Cancer remains a leading health problem in fact the first or second common reason of mortality in 91 countries before the age of 70 and accounting for 9.6 million deaths in 2018 worldwide [36]. Kinase inhibitors are the most successful group of medicines as anticancer drugs throughout the last three decades and half a dozen of them has the pyridine nucleus [36,37]. Some current studies have concentrated on searching new drugs especially small molecules and new treatments that can target very precisely the signaling paths in cancer cells. Additionally, current concerns for the novel anticancer drugs development may be partly due to the emergence of multidrug resistance and side effects that pose a serious risk to public health. Therefore, the development of novel and effective treatments is a very important goal, and most of the efforts of researchers in this area are for the designing of new efficacious agents. Some significant anticancer agents are reported to have pyridine nuclei that can be explored for in vivo and clinical studies in the future [38]. Pyridine has a broad anticancer profile, including numerous targets which can disseminate enormous scientific knowledge for the development and designing of new pyridine derivatives. Numerous pyridine compounds are published to inhibit human carbonic anhydrase, tubulin polymerization, kinase, topoisomerase enzyme, androgen receptors, and many other targets to control the global health issue of cancer. The thoroughgoing declaration of SAR/molecular docking will surely help the scientific society to produce effective and potent drugs with excellent pharmacological activity [39,40].

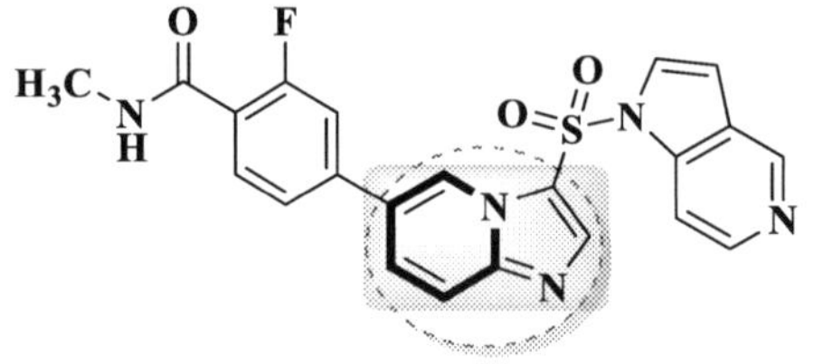

FIGURE 8.13 Most active imidazopyridine Compound 1.

Imidazopyridines

Abnormal activation of c-Met imparts a crucial role in the formation, progress, and propagation of cancer as well as in the progression of anticancer drugs resistance. In 1984, c-Met was discovered and is a part of receptor tyrosine kinases (RTKs) subfamily and makes a heterodimer by a disulfide bond that connects a membrane associating β-chain and a small extracellular α chain. Once the c-Met binds to its natural ligand, its kinase phosphorylation activity initiates and activates a series of signal pathways. Hence, c-Met has gained attention as a potential target for cancer treatment. A new series of compounds having imidazopyridine scaffold was discovered as c-Met inhibitor. Potential inhibitors were recognized through complete optimizations along with cellular and enzymatic bioassays. A selectively active compound was further explored for its impact on c-Met signaling, cell propagation, and cell dispersion in vitro. The most active compound **1** exhibited c-Met kinase inhibition with an IC_{50} value of 12.8 nmol/L which showed 78-fold more activity than 16 types of tyrosine kinases in a dose-dependent manner. Compound **1** inhibited the c-Met phosphorylation and its significant downstream ERK and Akt signaling pathways in c-Met abnormal human EBC-1 cancer cells. In 12 cancer cell lines of humans having various c-Met activation levels, compound **1** effectively stopped the cell proliferation instigated by c-Met. Moreover, dose depending compound **1** reduced c-Met driven cell dispersion of MDCK cells (Fig. 8.13) [41].

Polycyclic pyridine compounds

Antiproliferative potential of new synthetic polycyclic pyridine derivatives was tested in vitro against HeLa, Me180, and ZR751 human cancer cell lines. Compounds **2** and **3** were found to have good cytotoxicity in the cervical (Me180) and breast cancer cell line (ZR751), respectively, and some other compounds also exhibited moderate potential (Fig. 8.14) [42].

Pyrazolopyridines

Anaplastic lymphoma kinase (ALK), was first known as nucleophosmin (NPM)-ALK fusion protein belongs to the tyrosine kinase insulin receptor superfamily. ALK gene reordering outcome is numerous ALK fusion oncogenes. Abnormalities of ALK gene have been observed in various human cancers which include initiating mutations, for example, rearrangement of F1174L/R1275Q, and amplification of gene. ALK has been documented as a favorable molecular target for NSCLC targeted therapy. SAR study of pyrazolopyridines was performed to overcome ALK-L1196M mutation-induced crizotinib resistance, and compound **4** was

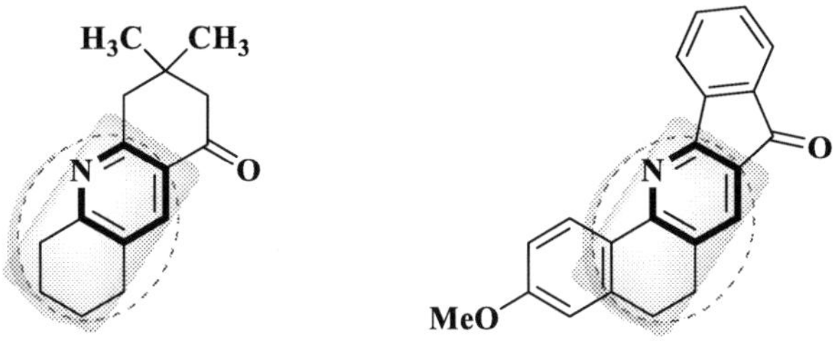

2 3

FIGURE 8.14 Most active polycyclic pyridine Compounds **2** and **3**.

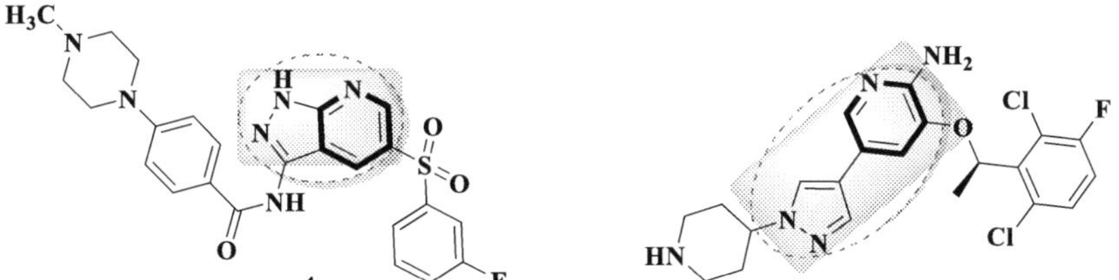

4 Crizotinib

FIGURE 8.15 Most active pyrazolopyridine Compound **4** and approved anticancer drug crizotinib.

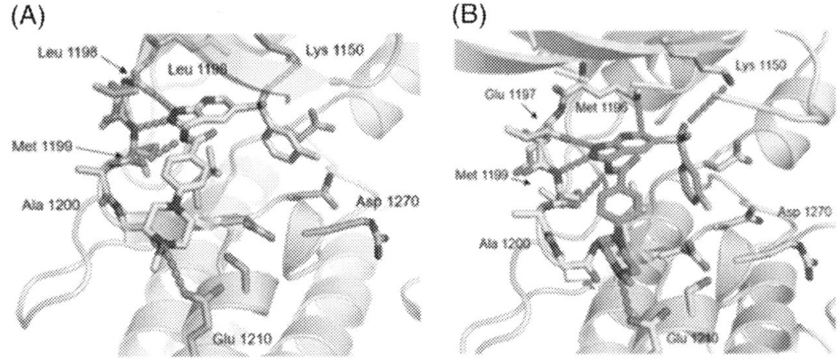

FIGURE 8.16 Docking interaction of **4** in kinase domain of ALK (A) WT and (B) L1196M on the basis of an X-ray crystal structure.

identified as a novel L1196M inhibitor. Molecule **4** showed excellent activities (<0.5nM of IC_{50}) against ALK-L1196M and ALK-wt. Additionally, **4** is an exceptionally active ROS1 inhibitor (<0.5nM of IC_{50}) and over c-Met shows incredible selectivity. Furthermore, **4** by blocking the signaling of ALK and apoptosis strongly decreases the cells proliferation of ALK-L1196M-Ba/F3 and H2228. The molecular docking results disclose that, in comparison to crizotinib, **4** involves in a promising interaction in ALK-L1196M kinase domain with M1196 and hydrogen bonding interactions with E1210 and K1150. This work has delivered a valuable perception to design the potent ALK gatekeeper mutant inhibitors (Figs 8.15 and 8.16) [43].

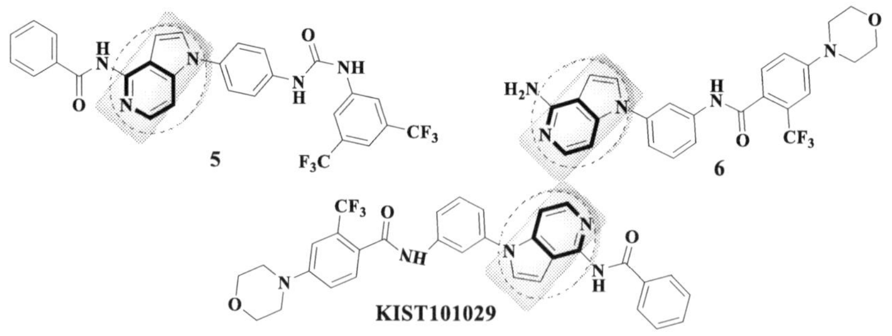

FIGURE 8.17 Most active pyrrolopyridine Compounds **5**, **6** and lead compound KIST101029.

Pyrrolopyridines

FMS, kinase, or CSF-1R (colony-stimulating factor-1 receptor) belongs to the type III RTKs family. It works in conjunction with IL-34 or CSF-1, the resulting signals causes proliferation, and persistence of the macrophage-lineage cells. Over-expression of FMS kinases is observed in numerous cancer types (e.g., breast, ovarian, and prostate) and in inflammatory diseases (such as rheumatoid arthritis). Therefore, FMS inhibitors may be potential drug candidates to treat these disorders. Pyrrolopyridine compounds were analyzed to find their inhibitory potential against FMS kinase. Compounds **5** (IC_{50} = 60 nM) and **6** (IC_{50} = 30 nM) were the most potent among the whole tested library of compounds. These compounds were almost 2–3 times more active than the standard, KIST101029 (IC_{50} = 96 nM), respectively. Compound **6** was examined over 40 kinases which also includes FMS and found to be selective for FMS kinase. It was additionally investigated against BMDM (bone marrow-derived macrophages) and exhibited an IC_{50} value of 84 nM which was 2-fold more active than KIST101029 with an IC_{50} value of 195 nM. Antiproliferative activity of derivative **6** was further investigated against two prostates, five breast, and six ovarian cancer cell lines with IC_{50} values ranging from 0.15–1.78 μM. It also retains the discrimination for cancer cells over fibroblasts (Fig. 8.17) [44].

Thienopyridines

Pim kinases belong to primarily active serine/threonine kinases. The word pim is derived from the recognition of gene pim-1 which is the site of proviral insertion in T-cell lymphoma. Three pim kinases are known [pim-1, -2, and -3] which are taking part in various processes such as apoptosis, metabolism, migration, differentiation, proliferation, and cell survival. Furthermore, pim-1 has been demonstrated to work as a control for cell cycle development in the G2/M and G1/S transitions. Over-expressed pim-1 and 2 were involved in hematologic cancers such as lymphoma, murine leukemia, and acute myeloid leukemia, as well as solid cancers such as pancreas, bladder, colon, and prostate cancers. Thus, targeting pim kinases could be a successful and promising approach to combat cancer. Three series of thienopyridines having benzoyl or amide groups at position-2 were prepared as inhibitors of pim-1. All the synthetic derivatives were evaluated for inhibitory activity against pim-1 enzyme. Two derivatives **7** and **8** exhibited pim-1 inhibition potential with IC_{50} values of 12.7 and 35.7 μM, respectively. The most potent derivatives were studied on five different cell lines

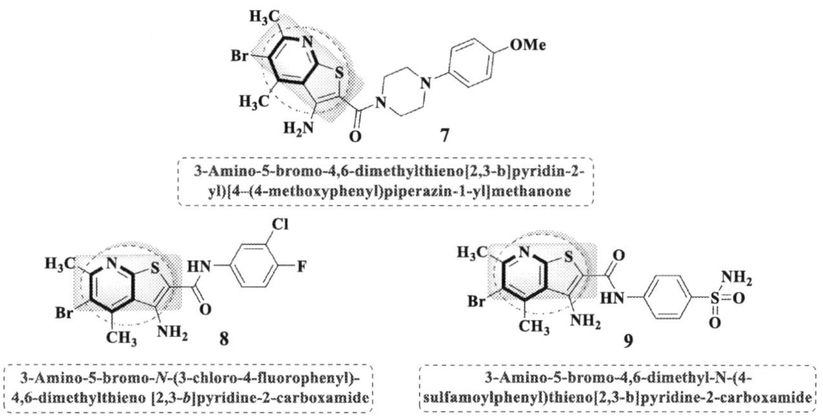

FIGURE 8.18 Most active thienopyridine derivatives **7**, **8**, and **9**.

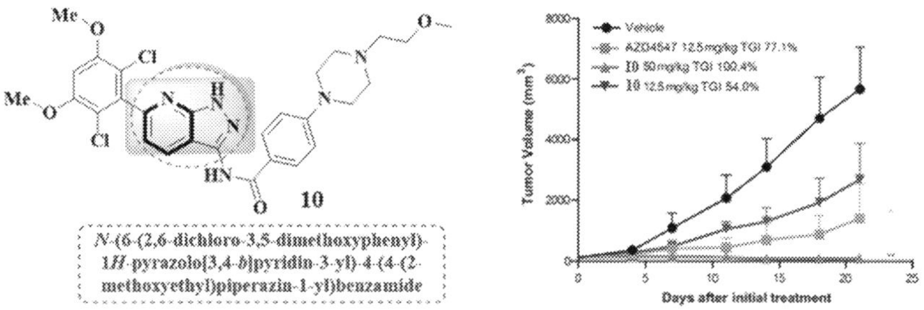

FIGURE 8.19 Structure and antitumor activity of derivative **10** in the H1581 xenograft model.

[MCF7, A549, HCT116, HEPG2 and PC3] for their cytotoxic activities. Compound **9** showed potent cytotoxicity on nearly all the tested cell lines (Fig. 8.18) [45].

Pyrazolopyridines

Fibroblast growth factors (FGFs) along with their receptors control various biological functions, such as angiogenesis, wound healing, tissue repair, and embryogenesis. FGFR upon connecting with FGF, go through dimerization followed by autophosphorylation which activates downstream signaling pathways, for example, PLCγ and MAPK pathways. These FGFR cascades confer significant parts in cellular activities, including migration, differentiation, propagation, and persistence, making FGFR signaling vulnerable to destruction by cancer cells. Abnormalities of FGFR signaling have been known in clinical samples of breast, lung, bladder cancers, etc. Hence, we documented the preparation, and bio-evaluation of a new series of 1*H*-pyrazolopyridine compounds which can selectively inhibit FGFR kinase. Due to its exceptional in vitro potential, derivative **10** was selected for in vivo studies and displayed substantial antitumor activity. These studies show that **10** may be a potential molecule in further drug discovery (Figs 8.19 and 8.20) [46].

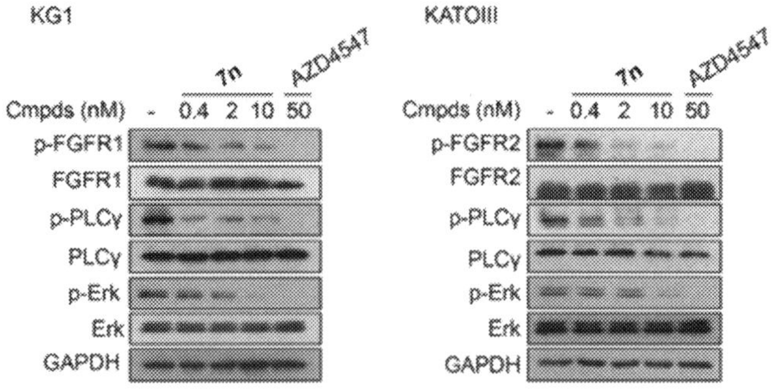

FIGURE 8.20 FGFR downstream signaling and phosphorylation inhibition by derivative **10** in KATOIII and KG1cells.

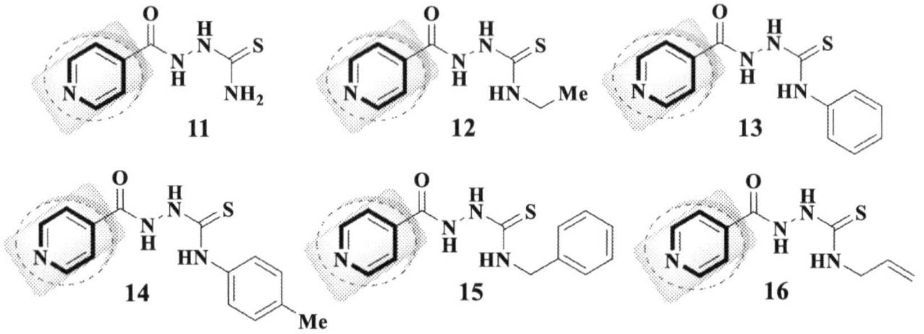

FIGURE 8.21 Active pyridine hydrazinecarbothioamide derivatives **11–16**.

Carbonic anhydrase inhibition

Carbonic anhydrase is a group of metalloenzymes which act as catalysts in the quick change of carbon dioxide (CO_2) into proton (H^+), and bicarbonate (HCO_3^-) ions as well as participated in the bio-catalyzed mineralization. CA isoforms are present in various tissues that involve in numerous essential processes such as respiration, bone resorption, pH balance, ureagenesis, ion transport, electrolyte secretion, lipogenesis, and gluconeogenesis. Many isozymes of CA that participated in these progressions are central targets that are likely to be activated/inhibited to treat wide-ranging diseases such as obesity, edema, cancer, osteoporosis, and epilepsy [47–49].

Pyridine hydrazine carbothioamide derivatives

Pyridine hydrazine carbothioamides libraries having tolyl, ethyl, benzyl, allyl, phenyl groups were prepared and studied as active inhibitors of three-members of pH regulating carbonic anhydrase enzyme family. The activatory and inhibitory potential of derivatives against the transmembrane, tumor-associated hCA IX, and the cytosolic human isoforms hCA I and hCA II were studied by a hydrase bioassay having CO_2 substrate. Most of the derivatives explored here show inhibitory constants for the three isozymes in low micromolar or nanomolar range. KI values for hCA I were ranging from 34–871 nM and molecules **11–16** exhibited exciting activation of the hCA II with KA values in the range of 0.81–12.5 μM. These derivatives were also considered for in silico studies at the binding sites of three target enzymes. The different inhibitory mechanism of these compounds and their good inhibitory activities for these isozymes have gained great attention for this class for the designing of potent CA inhibitors (Figs 8.21 and 8.22) [50].

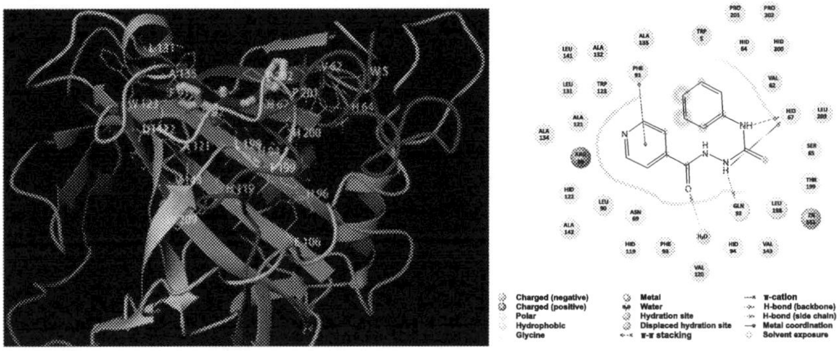

FIGURE 8.22 Docking study of derivative **13** in the hCA-I binding pocket.

17 (*Z*)-5-((*Z*)-3,4-Dihydroxybenzylidene)-2-(phenylimino)-3-(pyridin-4-ylmethyl)thiazolidin-4-one

18 (*Z*)-5-((*Z*)-4-Nitrobenzylidene)-2-(phenylimino)-3-(pyridin-4-ylmethyl)thiazolidin-4-one

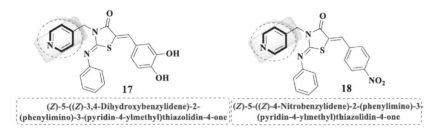

FIGURE 8.23 Structure of most active Compounds **17** and **18**.

Pyridine thiazolidinone derivatives

In search of potent CA IX inhibitors, a library of thiazolidinone-based pyridines was prepared and confirmed by several spectroscopic techniques. Fluorescence binding studies measured the binding affinity of all derivatives and esterase assay was performed for CA IX enzyme inhibition activity. Compounds **17** and 18 were found to be significantly active against the CA IX isoform having IC_{50} values 1.84 μM and 1.61 μM, respectively. Compounds **17** and **18** showed significant binding-affinity for CA IX with their KD values 2.32 μM and 11.21 μM, respectively. In silico studies showed that derivatives 17 and **18** bind effectively in the active site of CA IX by making good van der Waals interactions and H-bonds. All derivatives were evaluated for anticancer potential in vitro and show that compounds **17** and **18** showed significant activity for HepG-2 and MCF-7 cell lines. These results propose that derivatives **17** and **18** can be further explored as significant compounds for anticancer drug development (Figs 8.23 and 8.24) [51].

8.4.2 Cholinesterase inhibition activity

Alzheimer disease (AD) is the most fatal neurodegenerative disease among all types of dementia. Symptoms that appear in AD patients are memory loss, confusion, cognitive impairment and with the progression of the disease patients also lost the ability to perform daily activities independently. This problem is a gradual progression of brain degeneration

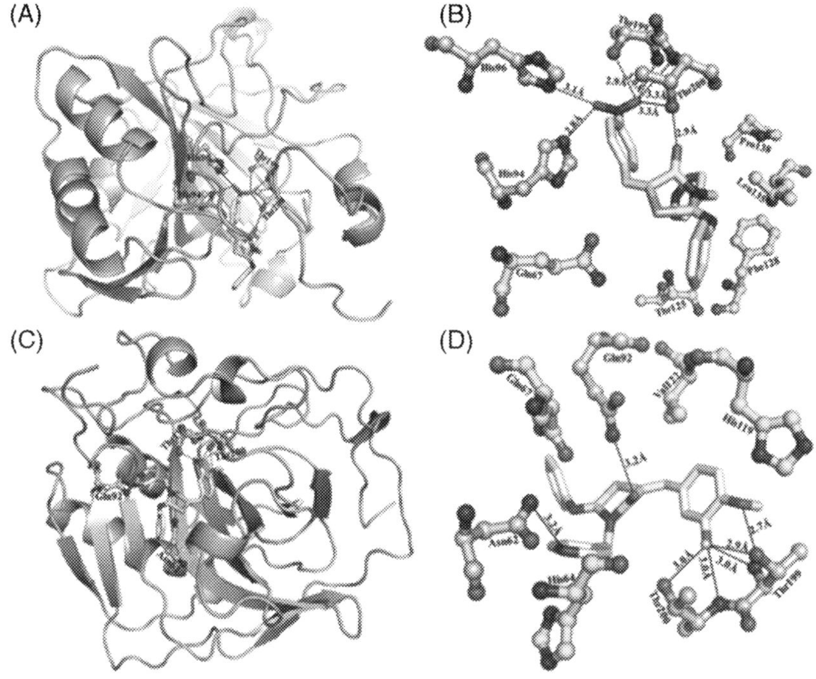

FIGURE 8.24 Docking studies: (**A,C**) cartoon view of Compounds **18** and **17** docked with CA IX. (**B,D**) Interaction of Compounds **18** and **17** with the active site residues.

and is very common in people who are middle-aged and older. Among the many causes of AD, hyperactivity of acetylcholinesterase (AChE) and butyrylcholinesterase (BChE) is considered to be a major cause of brain dysfunction [52,53]. The World Alzheimer's Report 2016 reported that 47 million people are affected from dementia and it is expected that there will be more than 132 million by 2050, if effective drugs to inhibit AChE and BChE are not developed to treat the progression of the disease [54]. Huge financial resources and human, materials are devoted each year for clinical study of AD and its treatment. Therefore, due to the complexity and severity of AD, some drugs have been used clinically to treat this disease [55,56]. AD has been put in the loop of top 10 causes of death worldwide. The brain impairments are initiated due to the loss of cholinergic neurons which decreases the acetylcholine (ACh) neurotransmitter in the brain. Cholinesterase enzymes, AChE and BChE, are mainly involved in the cholinergic cells damage, based on the so-called cholinergic hypothesis. Thus, dysfunctioning of the brain can be improved by restoring acetylcholine levels [57].

In recent reported treatments, inhibition of cholinesterases, AChE and BChE, is of considerable interest in the treatment of AD. The main etiology of cognitive dysfunction in the brain is caused by both AChE and BChE enzymes. The available marketed drugs include galantamine (Nivalin), tacrine (Cognex), rivastigmine (Exelon), and donepezil (Aricept) as cholinesterases inhibitors momentarily reverse the symptoms of AD in the initial stages (Fig. 8.25) [58]. The US Food and Drug Administration (FDA) approved cholinesterase inhibitors (ChEIs) (donepezil, galantamine, and rivastigmine) beneficial for mild to moderate AD. Still, they have several

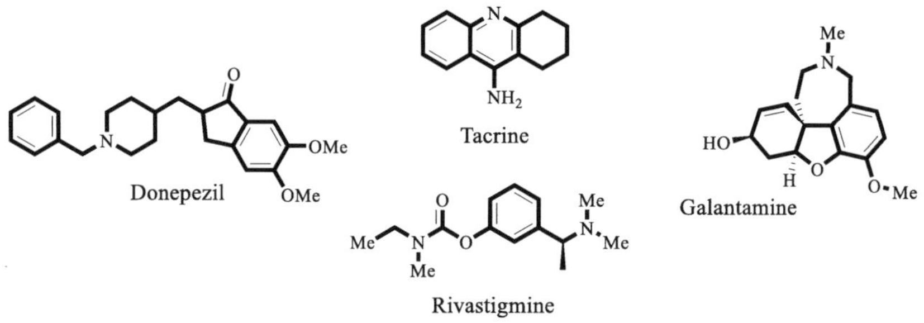

FIGURE 8.25 Clinical drugs to treat Alzheimer's disease (AD).

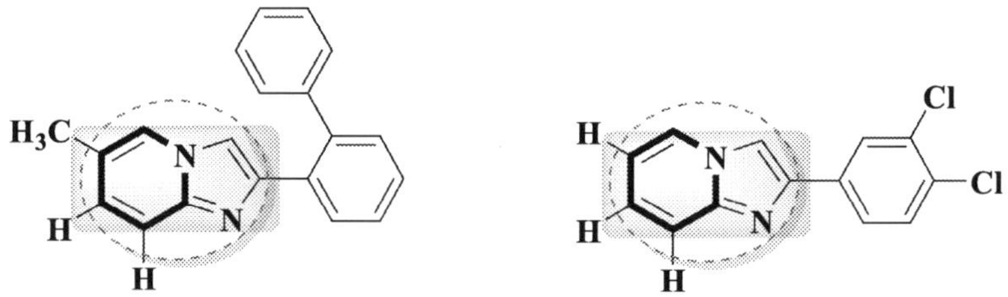

19 2-([1,1'-Biphenyl]-4-yl)-6-methylimidazo[1,2-*a*]pyridine

20 2-(3,4-Dichlorophenyl)imidazo[1,2-*a*]pyridine

FIGURE 8.26 Most active imidazopyridine Compounds **19** and **20**.

unwanted side effects and their benefits are only symptomatic rather than curative [59–61]. Therefore, to combat AD beneficial next generation ChEIs with better efficacy are needed to develop new effective drugs.

Imidazopyridines

Imidazopyridines are important clinically to treat circulatory and heart failures although for pharmaceutical applications various compounds are developing gradually. Imidazopyridine-based compounds bearing biphenyl residue are inhibiting AChE potentially. Derivative **19** which has a side chain of biphenyl and substituted at the R_4 position with methyl group on the imidazopyridine ring displayed good AChE inhibitory activity with an IC_{50} value of 79 μM. Furthermore, among the series imidazopyridines having a side chain of phenyl group display enhanced BChE inhibition potential. Compound **20** having 3,4-dichlorophenyl group and unsubstituted imidazopyridine ring seems to be the best inhibitor of BChE enzyme with an IC_{50} value of 65 μM with effective selectivity. The inhibition potential of most active compounds was also confirmed by molecular docking studies. The results revealed that AChE's outer anionic sites and BChE's acyl pocket were the dominated interaction sites for the respective inhibitors (Fig. 8.26) [62].

Pyridine alkaloids

An evergreen midsize tree *Senna multijuga* is extensively distributed in diverse Brazilian and African regions. Studies of the chemical properties of roots, seeds, and leaves have revealed

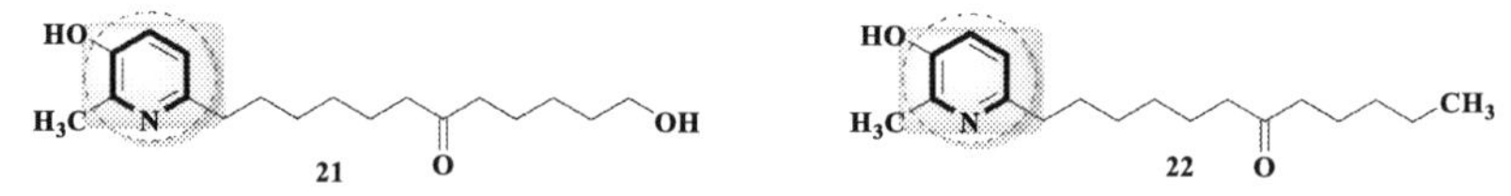

1-Hydroxy-12-(5-hydroxy-6-methylpyridin-2-yl)dodecan-6-one

12-(5-Hydroxy-6-methylpyridin-2-yl)dodecan-6-one

FIGURE 8.27 Pyridine alkaloids **21** and **22**.

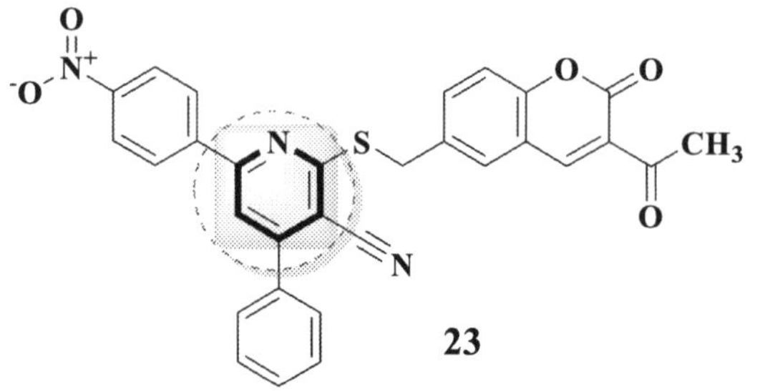

2-(((3-Acetyl-2-oxo-2*H*-chromen-6-yl)methyl)thio)-6-(4-nitrophenyl)-4-phenylnicotinonitrile

FIGURE 8.28 Nicotinonitrile coumarin hybrid **23**.

the presence of polysaccharides, flavonoids, chromones, and anthraquinones. However, few types of the genus *Senna* have been published as the main phenol compounds source with various pharmacological and biological properties, only *S. multijuga* seeds are utilized in the treatment of skin and eye infections. Although, for the secondary metabolites isolation no phytochemical study has been performed previously on this species. In this research work, two new alkaloids having pyridine moiety, 12′-hydroxy-7′-multijuguinone **21** and 7′-multijuguinone **22** were obtained from the leaves of *S. multijuga* along with the flavonoid rutin (Fig. 8.27). The structures of the new pyridine alkaloids were confirmed by interpreting the spectroscopic data. Both compounds displayed good in vitro AChE inhibition potential when compared with the standard physostigmine [63].

Nicotinonitrile coumarin hybrids

To increase the cholinergic neurotransmission by reducing the synaptic acetylcholine degradation, a new nicotinonitrile coumarin series were synthesized as potential inhibitors of acetylcholinesterase. The new hybrid molecules were synthesized by using pyridinethiones as initial precursors. The in vitro AChE inhibitory potential was observed for the new nicotinonitrile coumarin hybrid compounds in comparison with the reference drug donepezil having IC_{50} value of 14 nM. Coumarin compound **23** connected to 6-(4-nitrophenyl)-4-phenylnicotinonitrile, exhibited potent inhibitory activity than the standard donepezil with an IC_{50} value of 13 nM. Moreover, in vitro cytotoxicity of derivatives has been observed against many eukaryotic cells. Additionally, the docking study showed effective interaction between AChE and nicotinonitrile coumarin compounds (Figs 8.28 and 8.29) [64].

Pyridine and *N*-benzylpiperidines

Some new pyridine-based AChE inhibitors, *N*-benzylpiperidine, and acylhydrazone fragments were synthesized in a research work. The best-obtained structure was the compound

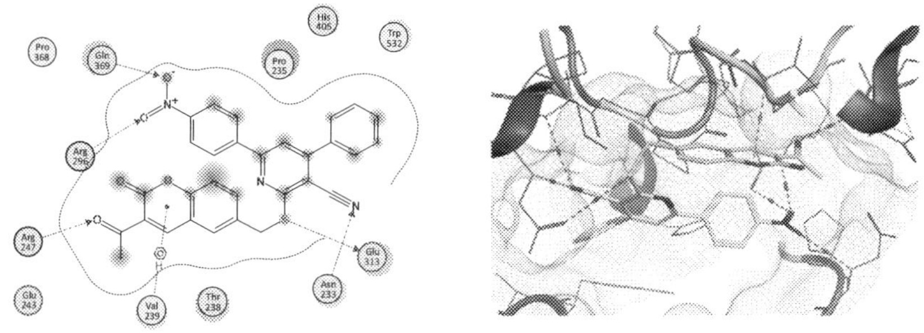

FIGURE 8.29 The 2D and 3D ligand interactions of 23 with AChE.

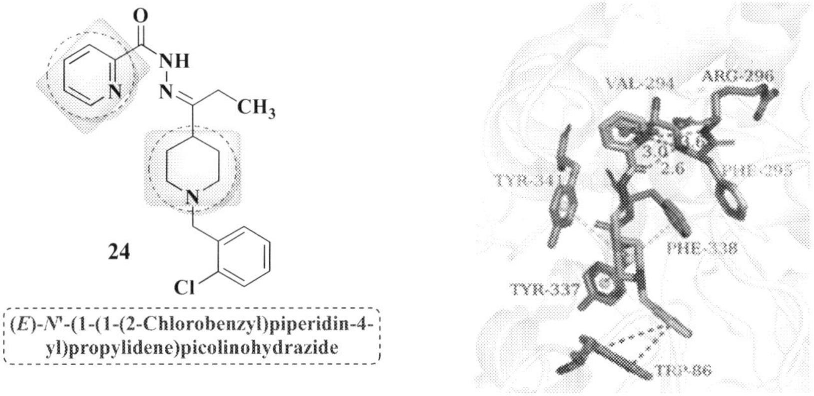

FIGURE 8.30 The key interaction analysis of representative Compounds **24** and AChE.

24 exhibiting AChE inhibition with IC_{50} value of 6.62 nM, while showed nearly no inhibitory potential for BChE. PAMPA permeability calculation and ADMET evaluation displayed good drug-like characteristics. The enhanced inhibitory activity with an alkyl chain substituted intermediate shows a new binding mechanism of AChE with the inhibitor. This outcome displays new perceptions into the mechanism of binding and is also supportive for the discovery of potent new AChE inhibitors with enhanced activity (Fig. 8.30) [65].

8.4.3 Antidiabetic activity

Diabetes is a long-lasting endocrinal metabolic disease growing rapidly. A high level of sugar for a long period is the symptom of the disease. Hyperglycemia is indicated by increased hunger, thirst, and repeated urination [66]. Glucose is a key source of energy that is produced in the body. In the body, the glucose is controlled by two main hormones glucagon and insulin which work oppositely to lower or raise the glucose level in blood. Both hormones are originated and released by the pancreas. Hyperglycemia triggers the insulin secretion which is released by the cells of pancreatic islets of Langerhans that stimulates glucose absorption by the cell. Conferring to its pathogenesis, diabetes is primarily divided into types I and II

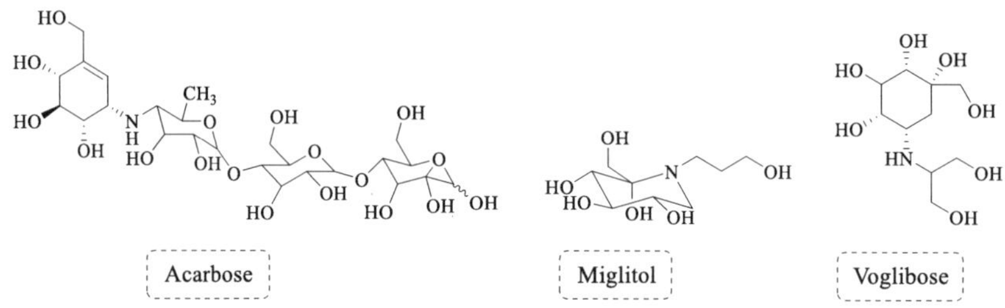

FIGURE 8.31 Clinical drugs to treat type II diabetes mellitus.

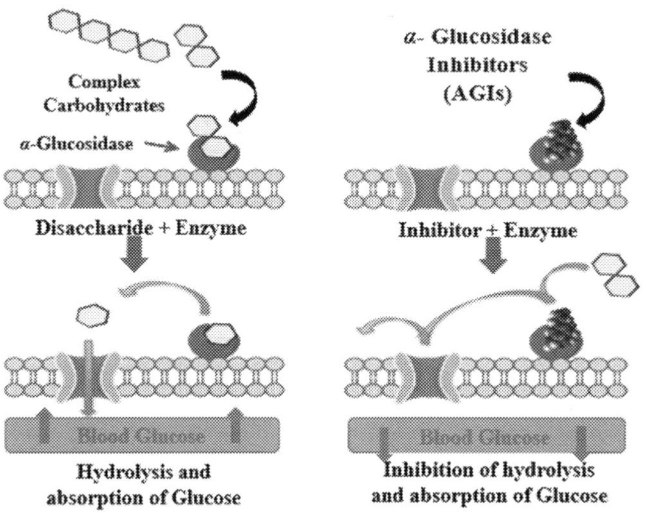

FIGURE 8.32 Mechanism of hydrolysis and inhibition of hydrolysis of carbohydrates.

diabetes. Type I is generally due to the less production of insulin. In type II, cells do not respond properly to insulin. Gestational diabetes happens during pregnancy and diminishes after birth. Carbolytic enzymes hydrolyze the carbohydrates resulting in the condition called postprandial hyperglycemia (PPHG) [67,68].

8.4.3.1 *α-Glucosidase and α-amylase inhibition*

α-Glucosidase and α-amylase enzymes are present in microorganisms, plants, animals, and humans. They catalyze the release of glucose in the small intestine by digesting carbohydrates [69]. The worldwide prevalence of diabetes mellitus has encouraged the medical community to find several new therapeutic methods to control the disease threat. One of the methods is by regulating the PPHG [70]. α-Glucosidase inhibitors are identified as effective agents for regulating PPHG. Acarbose, voglibose, and miglitol are the frequently used drugs for the treatment of diabetes by impeding the α-glucosidase enzyme (Figs 8.31and 8.32) [71,72]. To avoid the adverse effects of these drugs they are recommended with other antidiabetic drugs.

Furopyridinediones

Discovery of various groups of α-glucosidase inhibitors has been under consideration. In this regard, an assorted library of furopyridinediones with various substitutions was found to

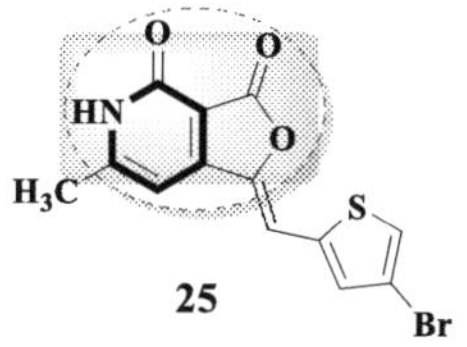

FIGURE 8.33 Most active furopyridinedione Compound **25**.

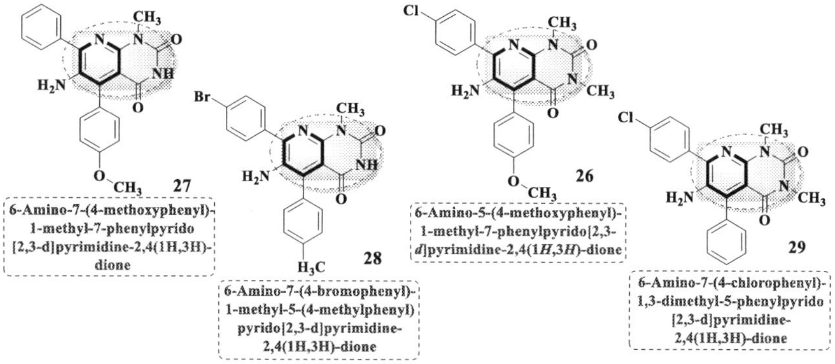

FIGURE 8.34 Most active pyrido pyrimidine Compound **26, 27, 28**, and **29**.

be active inhibitors of α-glucosidase by using an intuitive scaffold hopping approach. These derivatives were evaluated against α-glucosidase while acarbose was used as the reference compound. Among the tested derivatives **25** emerged as a lead compound having IC_{50} value of 0.24 μM. Lineweaver–Burk study of **25** showed it to be a mixed inhibitor. In silico studies and molecular dynamic simulation experiments were performed against α-glucosidase homology model to know the binding mechanism of **25** with the enzyme (Fig. 8.33) [73].

Pyridopyrimidines

A new library of pyridopyrimidine-2,4-diones was synthesized through an efficient and facile reaction from 6-amiouracils and α-azidochalcones (Fig. 8.34). Synthetic derivatives were tested for the inhibition of α-glucosidase and all compounds demonstrated exceptional in vitro inhibition activity against α-glucosidase with IC_{50} values ranging from 78 ± 2.0 to 252 ± 1.0 μM. The most potent derivative **26** was about 10-fold more active than the standard drug acarbose (IC_{50} = 750 ± 1.5 μM). Kinetic evaluation of derivative **26** displayed that it follows competitive inhibition mode for α-glucosidase enzyme. Molecular docking studies of the potent derivatives **26, 27, 28,** and **29** were also carried out (Fig. 8.35) [74].

Dihydropyridines

1,4-Dihydropyridine derivatives were obtained in good yields through Hantzsch reaction and tested for their inhibitory potential against α-glucosidase enzyme. Many compounds displayed effective inhibition potential against α-glucosidase and displayed IC_{50} values ranging from 35.0–273.7 μM in comparison with acarbose (IC_{50} = 937 ± 1.60 μM) the standard

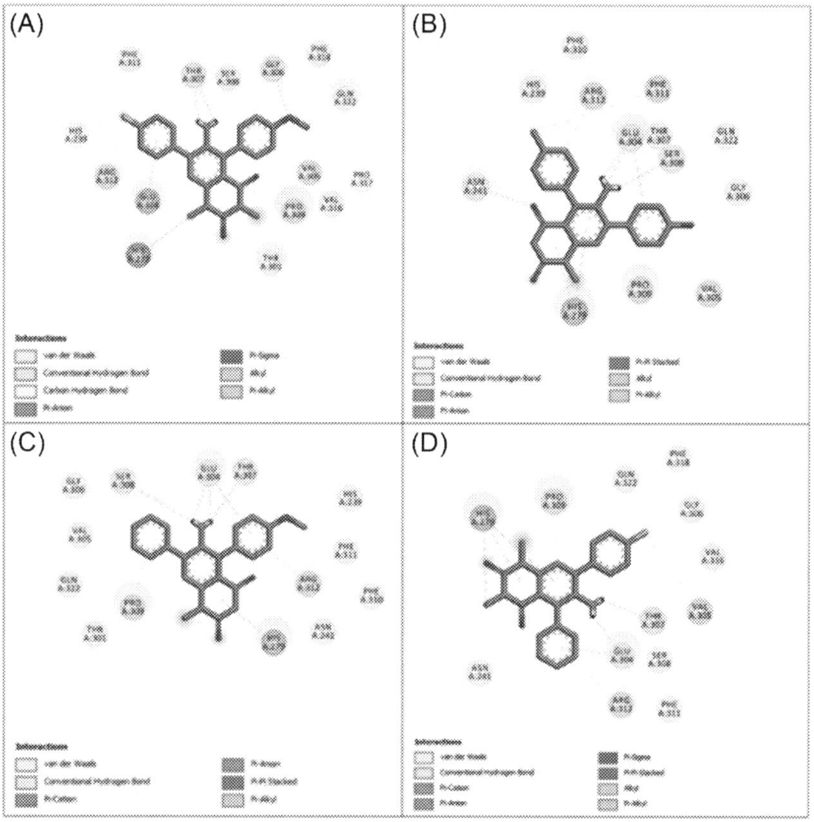

FIGURE 8.35 The predicted interaction mechanism of compounds (A) **26**, (B) **28**, (C) **27**, and (D) **29** in the enzyme's active site.

drug. The kinetic studies on highly active derivatives **30** and **31** demonstrate their inhibitory mechanism and dissociation constants Ki. Derivative **30** followed noncompetitive mode of inhibition with K_i value 25.0 ± 0.06 while derivative **31** was found to be a competitive inhibitor with K_i value 66.0 ± 0.07 μM (Fig. 8.36) [22].

Recently, a new library of dihydropyridines was synthesized by using one-pot multicomponents approach and evaluated against α-amylase and α-glucosidase in vitro. The synthetic compounds displayed good inhibition against α-amylase enzyme in the range of IC_{50} value 2.21 ± 0.06–9.97 ± 0.08 μM in comparison to the standard acarbose ($IC_{50} = 2.01 \pm 0.1$ μM) and against α-glucosidase in the range of $IC_{50} = 2.31 \pm 0.09$–9.9 ± 0.1 Mm in comparison to the acarbose ($IC_{50} = 2.07 \pm 0.1$ μM), respectively. To confirm the binding interaction mode of synthetic compounds with enzyme's active sites, in silico studies were also carried out (Figs 8.37–8.39) [21].

Pyridine chalcones

Exploring new drug-like compounds to treat diabetes due to its high prevalence has become a major global concern. In a research work, pyridine chalcones were prepared and tested for

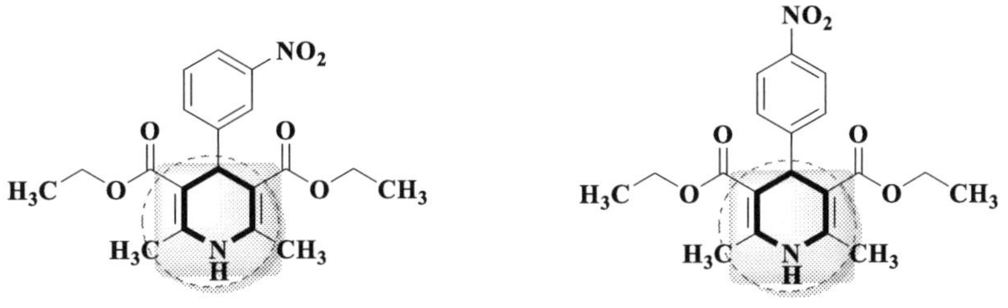

30

Diethyl 2,6-dimethyl-4-(3-nitrophenyl)-1,4-dihydropyridine-3,5-dicarboxylate

31

diethyl 2,6-dimethyl-4-(4-nitrophenyl)-1,4-dihydropyridine-3,5-dicarboxylate

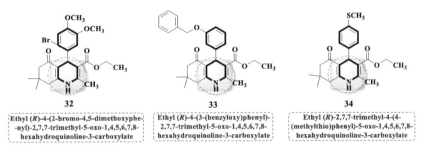

FIGURE 8.36 Most active dihydropyridine Compounds **30** and **31**.

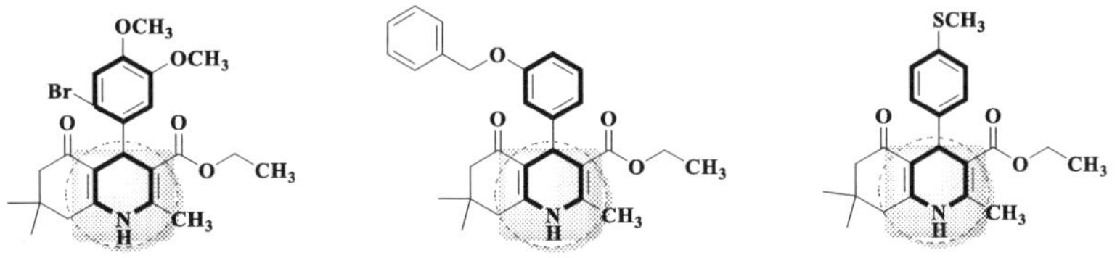

32

Ethyl (*R*)-4-(2-bromo-4,5-dimethoxyphe--nyl)-2,7,7-trimethyl-5-oxo-1,4,5,6,7,8-hexahydroquinoline-3-carboxylate

33

Ethyl (*R*)-4-(3-(benzyloxy)phenyl)-2,7,7-trimethyl-5-oxo-1,4,5,6,7,8-hexahydroquinoline-3-carboxylate

34

Ethyl (*R*)-2,7,7-trimethyl-4-(4-(methylthio)phenyl)-5-oxo-1,4,5,6,7,8-hexahydroquinoline-3-carboxylate

FIGURE 8.37 Most active dihydropyridine Compounds **32, 33**, and **34**.

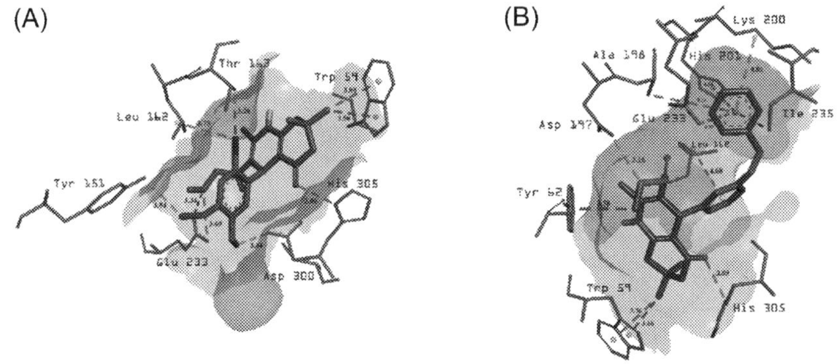

FIGURE 8.38 Docked poses of most active derivatives with the α-amylase (A) Compound **32** and (B) Compound **33**.

their antihyperglycemic potential. Three derivatives **35, 36**, and **37** showed good inhibition potential against α-amylase and α-glucosidase enzymes, respectively, when compared with the standard drug acarbose (Fig. 8.40). The kinetic studies of derivatives were also performed to find their mode of action and they displayed a competitive type of inhibition mechanism

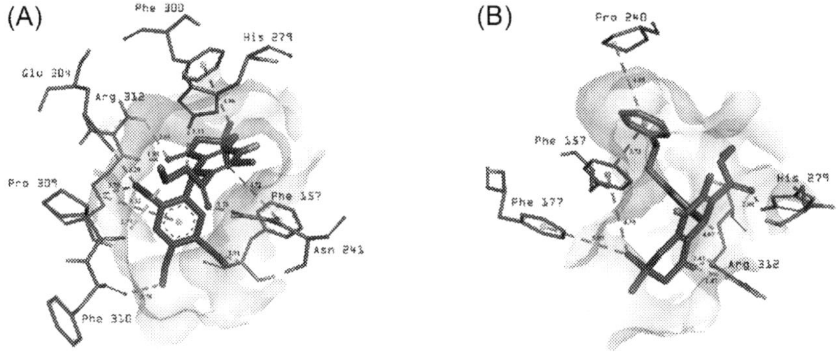

FIGURE 8.39 Docked poses of most active derivatives with α-glucosidase (A) Compound **32** and (B) Compound **33**.

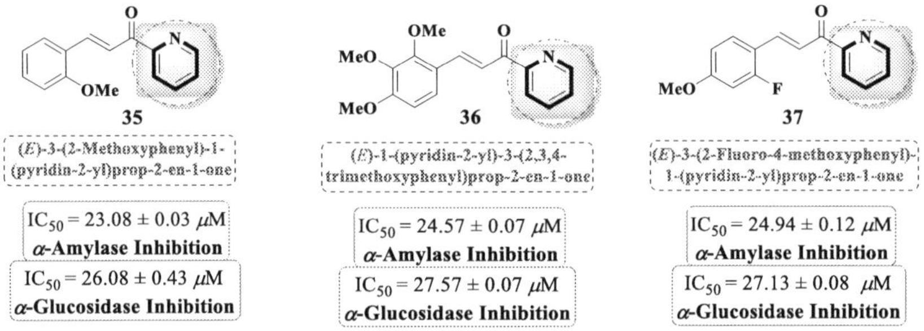

FIGURE 8.40 Most active pyridine chalcone Compounds **35, 36**, and **37**.

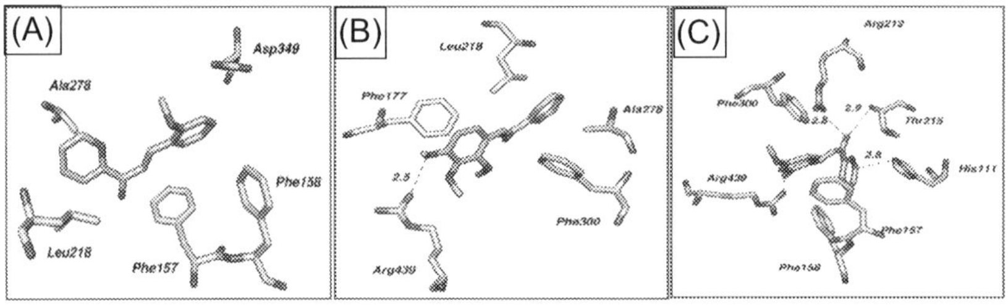

FIGURE 8.41 Predicted docked pose of active Compounds **35(A), 36(B)**, and **37(C)** in the pocket of α-glucosidase.

for both enzymes. The molecular docking studies have confirmed these results and indicated that these derivatives are involved in several interactions within the enzyme's active site (Figs 8.41 and 8.42) [75].

Pyridinecarbonitriles

Pyridinedicarbonitriles were prepared and screened for their α-glucosidase inhibition in vitro, in silico, and kinetics studies. Numerous derivatives exhibited many folds better inhibitory activity as compared to the standard acarbose ($IC_{50} = 750 \pm 10$ μM). Excitingly, derivative **38**

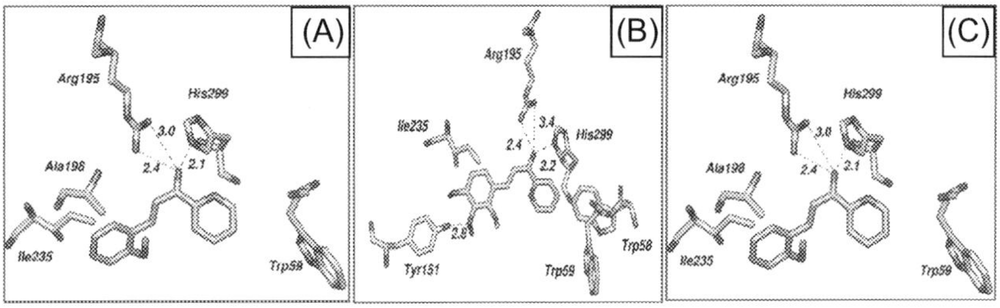

FIGURE 8.42 Predicted docked pose of active compounds **35(A), 36(B)**, and **37(C)** in the pocket of α-amylase.

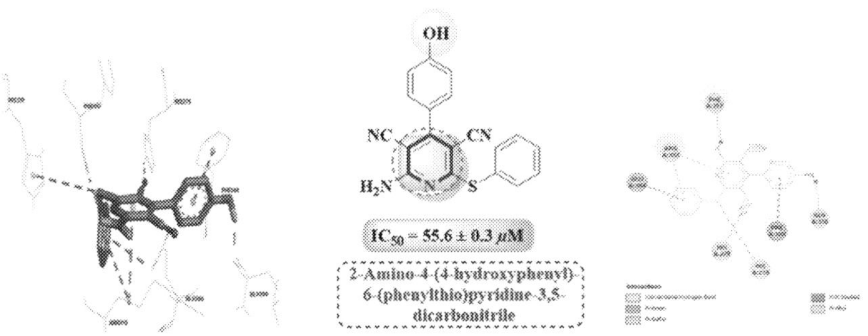

FIGURE 8.43 Docked pose of Compound **38** in the α-glucosidase's active site.

($IC_{50} = 55.6 \pm 0.3\ \mu M$) showed thirteen-fold better inhibitory potential than the standard. Most active compound **38** showed inhibitory mode of competitive type in kinetic studies. Docking studies revealed the interaction mechanism of compound **38** with the enzyme's active site (Figs 8.43 and 8.44) [24].

8.4.3.2 Aldose reductase inhibitors

In polyol pathway, the first enzyme is aldose reductase (AR) which act as a catalyst in the synthesis of sorbitol by reducing glucose. Sorbitol can be in turn converted to fructose by polyol dehydrogenase. Certainly, the negligible level of sorbitol is present in various tissues and the AR attraction for glucose is very less than the hexokinase. Therefore, it is possible that phosphorylation of glucose takes place in the cell which decreases the formation of sorbitol. Upon changing this condition such as diabetes mellitus, production of sorbitol increased, and deposited in numerous human tissues such as lens, retina, peripheral nerves, and kidney. These variations can be amended by preventing A, in the meantime aldose reductase inhibitors (ARIs) can treat diabetic problems directly since they are independent of the control of blood sugar levels therefore, they do not increase the danger of hypoglycemia [76–78]. These interpretations have stimulated great concern in developing ARIs, with rigid spiro hydantoins, for example, Sorbinil [79], and carboxylic acid group-containing, such as tolrestat [80], alrestatin [81], and zopolrestat [82]. In these compounds, the placement of an acidic proton in an appropriate spatial connection with a planar aromatic 50 group, seems to be important for the enzyme inhibition [81,82].

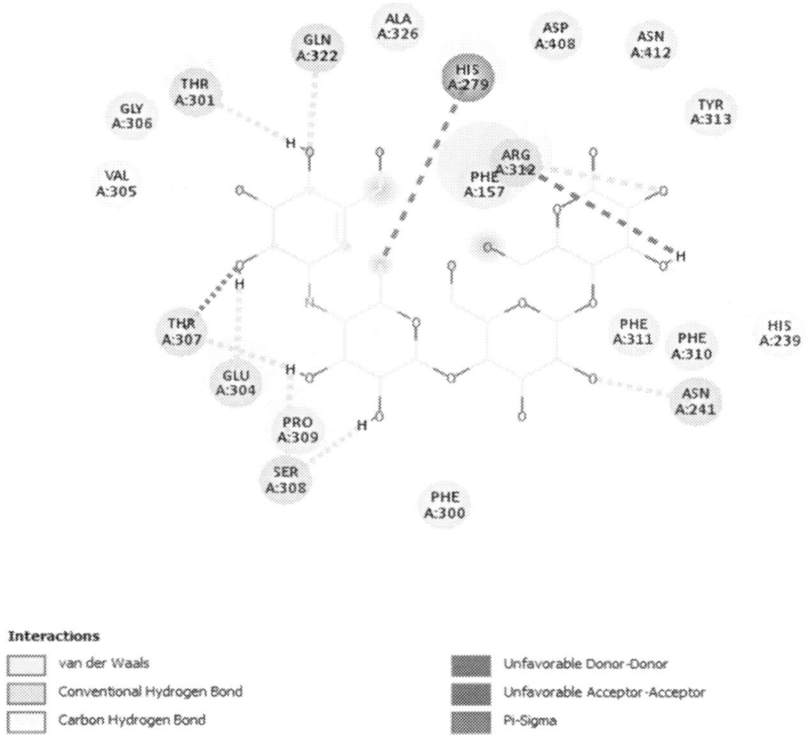

FIGURE 8.44 Docked pose of acarbose in the α-glucosidase's active site.

Pyrrolo[3,4-*c*]pyridines

A series of pyrrolopyridine alkanoic acids were prepared and their inhibition potential for AR was studied on lens enzyme of rat in vitro. The carboxylic acid analogs were found to be highly active inhibitors than the *iso*-propionic, and propionic analogs. Compounds **39** and **40** were found as the most potent in vitro and thus also studied in vivo as glutathione lens depletion inhibitors in galactosemic rats. None of the derivatives was succeeded in retaining the glutathione level in rat lens which predicts the possible complications of ocular metabolism and bioavailability. The inhibitory potential of derivatives **39** and **41** was also evaluated by considering their electronic and conformational features which was determined by theoretical calculations (Fig. 8.45) [83].

8.4.4 Urease inhibitory activity

Urease is a nickel-containing metallo-enzyme which takes part in its catalytic effects and is widely present in various species such as invertebrates, multicellular plants, and unicellular bacteria [84]. Hydrolysis of urea in CO_2 and ammonia is catalyzed by urease which increases the alkalinity of urine and supports the increasing concentration of struvite and calcium phosphate in crystalline form and therefore leads to the stone formation [85]. An increase in

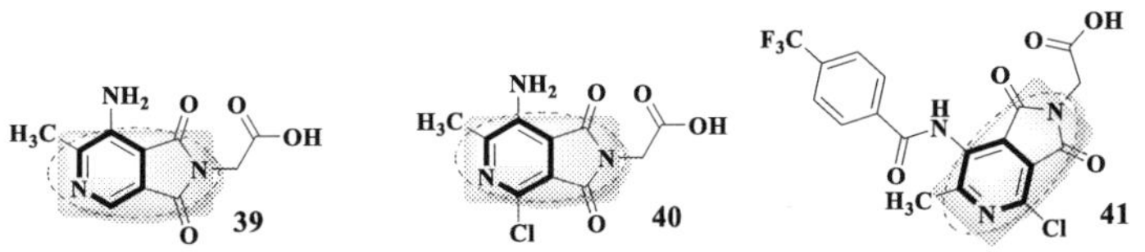

2-(7-Amino-6-methyl-1,3-dioxo-1,3-dihydro-2*H*-pyrrolo[3,4-*c*]pyridin-2-yl)acetic acid

2-(7-Amino-4-chloro-6-methyl-1,3-dioxo-1,3-dihydro-2*H*-pyrrolo[3,4-*c*]pyridin-2-yl)acetic acid

2-(4-Chloro-6-methyl-1,3-dioxo-7-(4-(trifluoromethyl)benzamido)-1,3-dihydro-2*H*-pyrrolo[3,4-*c*]pyridin-2-yl)acetic acid

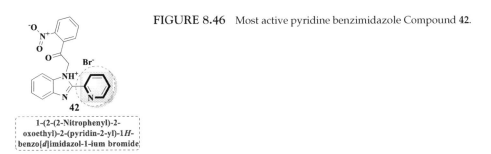

FIGURE 8.45 Most active pyrrolo[3,4-c]pyridines **39, 40,** and **41**.

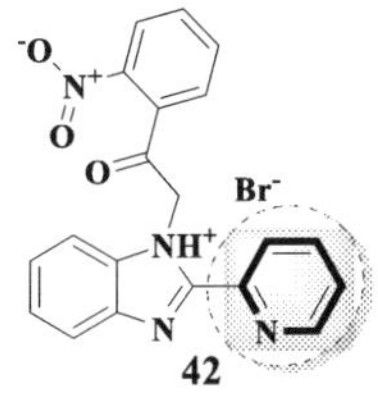

1-(2-(2-Nitrophenyl)-2-oxoethyl)-2-(pyridin-2-yl)-1*H*-benzo[*d*]imidazol-1-ium bromide

FIGURE 8.46 Most active pyridine benzimidazole Compound **42**.

the basic pH also helps in the survival and progression of *Helicobacter pylori* (HP), a bacterium that is responsible for causing several pathological disorders such as hepatic coma, formation of infectious urinary stones, ulceration, and pyelonephritis [86].

Continuous exposure to ammonia during the hydrolysis reaction causes several metabolic problems and damage to the gastrointestinal tract epithelium. Therefore, more active urease inhibitors are of great importance in medicinal, agricultural, and environmental sectors due to the severe effects of urease. Kidney stones are dissolved by urease inhibitors and stop crystals formation in urine [87,88]. Hundreds of potent urease inhibitors have been discovered which include polyphenols, hydroxamate, oxadiazole, phosphor amide, biscoumarin, dihydropyrimidine, thiourea, and urea derivatives [89]. Organometallic derivatives are also commonly used for the urease inhibition. In 1983, acetohydroxamic acid (AHA) has been approved by US FDA as the standard drug used clinically to combat the urease overexpression and the related complications [90]. Consequently, it is important to explore potential, selective, and novel inhibitors with substantial stability and bioavailability.

Pyridine benzimidazole compounds

Pyridyl benzimidazole derivatives with varyingly substituted phenacyl group were prepared and assessed for their urease inhibitory potential. All derivatives revealed inhibition of urease with IC_{50} values between 19.22 and 77.31 μM. Analog **42** ($IC_{50} = 19.22 \pm 0.49$ μM) displayed urease inhibition comparable to the standard thiourea with IC_{50} value of 21.00 ± 0.01 μM and showed twofolds greater activity than the standard AHA with IC_{50} value 42.00 ± 1.26 μM, respectively (Fig. 8.46). Additionally, few other derivatives ($IC_{50} = 21.55 \pm 0.36$ μM - 41.22 ± 0.42 μM), also demonstrated good activities when compared with the standards. The substituent's size, their mesomeric effect, and their occurrence on phenyl group actually controls the inhibition potential [91].

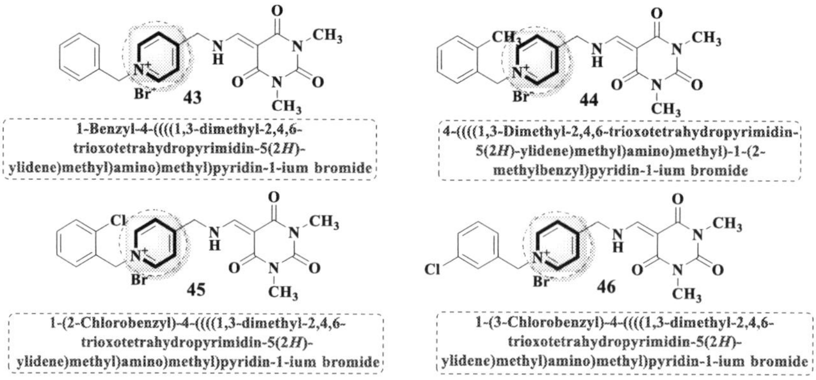

FIGURE 8.47 Most active *N,N*-dimethylbarbituric-pyridinium derivatives.

N,N-dimethylbarbituric-pyridinium derivatives

A new library of barbituric pyridinium derivatives was synthesized and studied as inhibitors of HP. All synthetic compounds ($IC_{50} = 10.37 \pm 1.0$–77.52 ± 2.7 μM) were more active against urease than the standard inhibitors hydroxyurea having an IC_{50} value of 100.00 ± 0.2 μm and thiourea having IC_{50} value 22.0 ± 0.03 μM these results indicated that derivatives **43, 44, 45,** and **46** were found to be more active than thiourea (Fig. 8.47). In silico study of most active derivatives was also performed. Furthermore, the drug-like characteristics of the synthetic derivatives were determined on the basis of Lipinski rule and other filters [92].

8.4.5 Antioxidant activities

The balance between the production of reactive oxygen species (ROS) and antioxidants can be determined either by overproducing ROS or by losing the antioxidant defense system [93]. This disbalance is termed as oxidative stress, and the ROS in excess can damage cellular DNA, lipids, and proteins which leads to the alteration and inhibition of their regular functioning [94]. The finding of ROS effects on the cell has aroused great attention for the discovery of antioxidants. There are a lot of advantageous effects of antioxidants on different conditions like cancer, ischemia-reperfusion, and numerous kinds of inflammation, where ROS impart a central role in the disease onset [95]. Many synthetic antioxidants such as edaravone [96] and ebselen [97] are used for neurologic recovery. Edaravone was the first drug developed for the treatment of acute cerebral infarctions that acts as a free radical scavenger while ebselen is still in clinical trials for acute cerebral infarction as well as to treat free radical cornea injury. Antioxidants are used as drugs and as dietary supplements ingredients to maintain health and for diseases prevention. Antioxidants also found their application as food additives, for example, TBHQ (tertiary-butylhydroquinone), BHT (butylated hydroxytoluene), OG (octyl gallate), BHA (butylated hydroxyanisole), EQ (ethoxyquin), and PG (propyl gallate). However, some of these compounds have higher concerns about possible health problems which they may promote [98,99]. For these reasons, the continuous discovery of new synthetic products with antioxidant activity is of primary importance as they may substitute with advantage the currently used compounds.

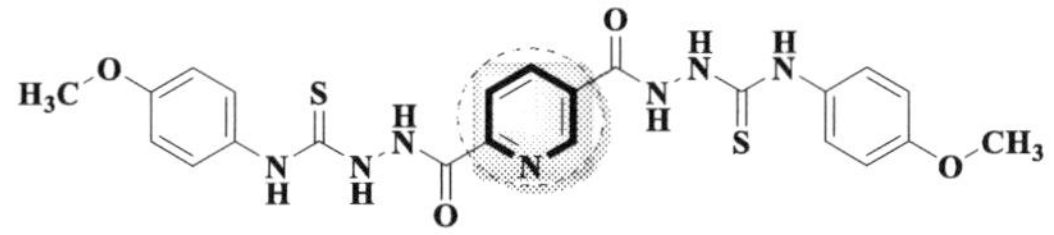

47

2,2'-(Pyridine-2,5-diyldicarbonyl)*bis*[*N*-(*p*-methoxyphenyl)hydrazinecarbothioamide]

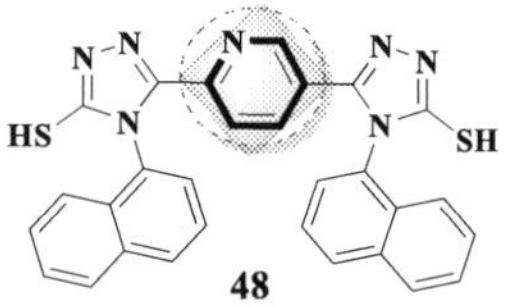

48

5,5'-Pyridine-2,5-diyl*bis*[4-(1-naphtyl)-4*H*-1,2,4-triazole-3-thiol]

FIGURE 8.48 Most active pyridine triazole Compounds **47** and **48**.

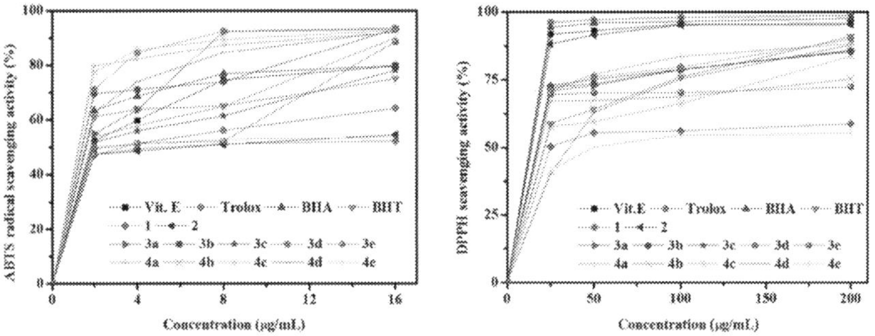

FIGURE 8.49 DPPH and ABTS scavenging activities of the derivatives **47** (3c), **48** (4d).

Pyridine *Bis*-1,2,4-triazole/thiosemicarbazide derivatives

A novel library of 1,2,4-triazole-3-thiol and thiosemicarbazide was prepared and studied for their bioactivities. A synthesis of dimethyl pyridine-2,5-dicarboxylate was achieved by the reaction commercially available pyridine-2,5-dicarboxylic acid with methanol in the presence of a catalytic amount of mineral acid. The dimethyl pyridine-2,5-dicarboxylate on reaction with $N_2H_4.H_2O$ afforded pyridine dicarbohydrazide in good yield. Pyridine dicarbohydrazide on reaction with aryl/alkyl isothiocyanates under reflux for 3–8 h furnished 1,4-disubstituted thiosemicarbazide intermediates. Intramolecular cyclization of these intermediates in the presence of base yielded the 4,5-disubstituted *bis*-mercaptotriazoles in 85%–95% yield. Among the synthetic compounds, **47** exhibited high activity with a 72.93% value against DPPH at the 25 μg/mL concentration (Figs 8.48 and 8.49) [100].

8.4.6 Antiinflammatory activity

Inflammation is a response of body to defective tissue homeostasis in which the immune cells of the body and defense chemicals like leukocytes, fluid, and plasma proteins enter the defective tissues that lead to vascular permeability and vasodilation [101,102]. Inflammation can be either chronic or acute. Acute inflammation disappears after a few days or hours while chronic inflammation can take months or years, even after the initial trigger has disappeared. Diseases associated with chronic inflammation are diabetes, cancer, Alzheimer's disease, asthma, and heart disease. Nonsteroidal anti-inflammatory drugs (NSAID) are effective for inflammation and as pain relievers which effect by suppressing cyclooxygenase (COX) enzymes [103]. There are two isoforms of COX enzymes, constitutive COX-1, and inducible COX-2.

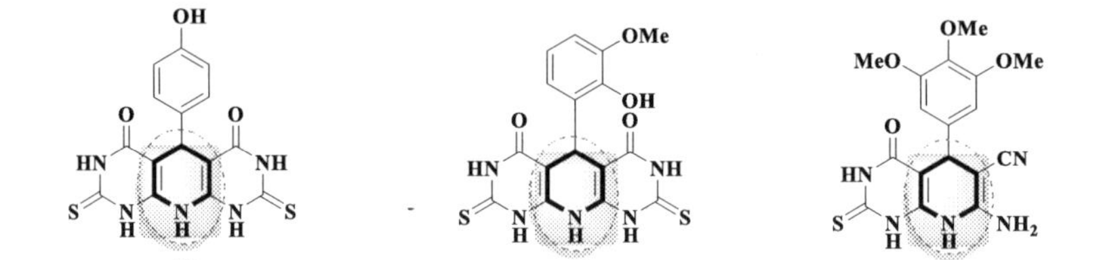

49 5-(4-Hydroxyphenyl)-2,8-dithioxo-2,3,5,8,9,10-hexahydropyrido[2,3-*d*:6,5-*d*′]dipyrimidine-4,6(1*H*,7*H*)-dione

50 5-(2-Hydroxy-3-methoxyphenyl)-2,8-dithioxo-2,3,5,8,9,10-hexahydropyrido[2,3-*d*:6,5-*d*′]dipyrimidine-4,6(1*H*,7*H*)-dione

51 7-Amino-4-oxo-2-thioxo-5-(3,4,5-trimethoxyphenyl)-1,2,3,4,5,8-hexahydropyrido[2,3-*d*]pyrimidine-6-carbonitrile

FIGURE 8.50 Most active pyrimidine pyridine Compounds **49, 50,** and **51**.

The constitutive COX-1 plays many roles in platelet aggregation, vascular homeostasis, and protection of gastric mucosa, while COX-2 is associated with prostaglandins which supports the growth of inflammation and pain [104]. The suppression of two isoenzymes is the main cause of gastrointestinal side effects caused by the traditional NSAID administration [104,105]. NSAIDs inhibit the synthesis of prostaglandin through nonselective COX enzymes inhibition but as a side effect, they cause ulceration, mucosal damage, and ulcer complication. Many selective COX-2 inhibitors have been reported and marketed as parecoxib sodium, rofecoxib, valdecoxib, and celecoxib [106].

Pyrimidine–pyridine hybrids

A variety of compounds containing pyridodipyrimidine, tetrahydropyridopyrimidine, and hexahydro-1*H*-pyrimidine were prepared. The anti-inflammatory activity of these derivatives was evaluated. The tetrahydropyridopyrimidine compound **51** revealed better inhibition of edema compared to celecoxib. Furthermore, derivatives **49, 50** and **51** displayed improve inhibition of COX-2 with IC_{50} value of 0.25–0.89 μM compared to celecoxib (IC_{50} = 1.11 μM) (Fig. 8.50). In silico studies were also carried out on the most potent candidates to study their mode of action as lead candidates to discover other anti-inflammatory agents (Fig. 8.51) [104].

Pyridine and thiazole-based hydrazides

A new class of derivatives was synthesized by linking –C(O)–NH– with thiazole and pyridine and confirmed by different spectral techniques. All derivatives were tested for their anti-inflammatory activities. The in vitro activity of these derivatives was evaluated by the bovine serum albumin denaturation method and exhibited inhibition with IC_{50} values ranging from 46.29–100.60 μg/mL. Off all the tested derivatives, **53** possess the maximum IC_{50} value and derivative **52** possess the lowest IC_{50} value (Fig. 8.52). Additionally, docking studies of the active derivatives showed a good docking score and better interaction with the enzyme. In silico studies also revealed their inhibitory potential. Derivative **54** appeared as a notable bioactive compound among the synthetic compounds [107].

8.4.7 Antimicrobial activity

8.4.7.1 Antifungal activity

Infectious fungal diseases are very common and can be classified into deep and superficial infections depending on the localization of infection. Recently, the fungal diseases are

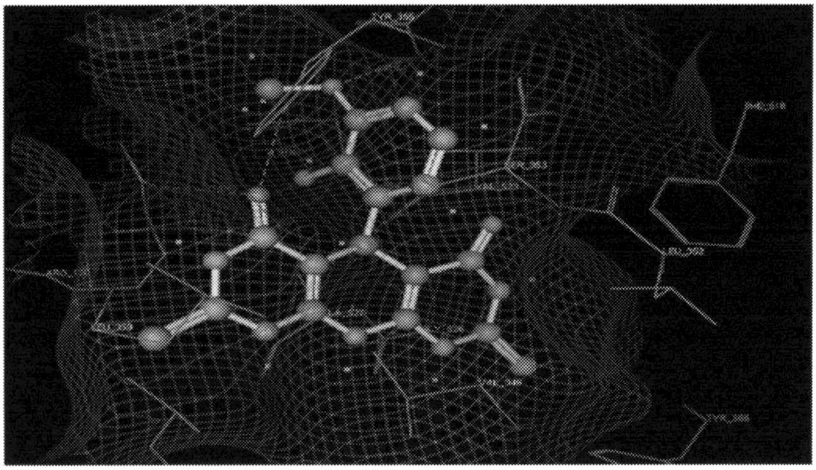

FIGURE 8.51 Interaction of Compound **50** inside COX-2 active site.

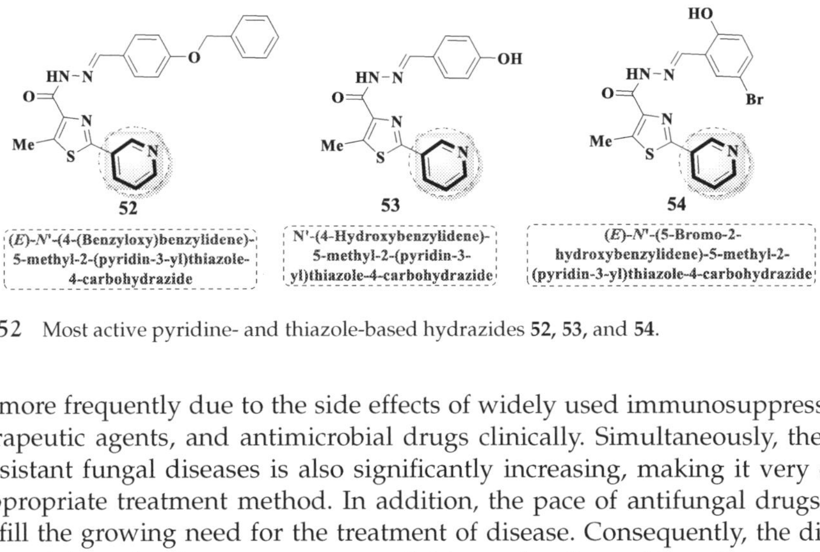

FIGURE 8.52 Most active pyridine- and thiazole-based hydrazides **52, 53,** and **54**.

occurring more frequently due to the side effects of widely used immunosuppressive drugs, chemotherapeutic agents, and antimicrobial drugs clinically. Simultaneously, the incidence of drug-resistant fungal diseases is also significantly increasing, making it very difficult to find an appropriate treatment method. In addition, the pace of antifungal drugs discovery cannot fulfill the growing need for the treatment of disease. Consequently, the discovery of new antifungal drugs has become more essential to resolve the problematic fungal resistance [108]. Currently, the antifungal drugs are mainly designed and developed by interfering the ergosterol and glucan biosynthesis. The synthesis of glucan is majorly affected by inhibitors and some antibiotics which have commendable clinical significance, but they are experiencing few problematic conditions such as incapability to manage oral intake, tedious synthesis, and high price [109]. However, most of the inhibitors of the ergosterol synthase don't face these difficulties, they can approach the particular targets to accomplish the treatment purpose. The ergosterol synthesis mechanism was further investigated which explained that the rate-limiting enzymes 14α-demethylase (CYP51) and squalene cyclooxygenase (SE) have crucial importance in this process. If the activity of these enzymes is blocked, the synthesis of

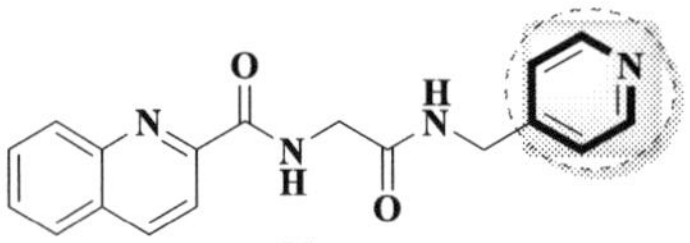

55

N-(2-Oxo-2-((pyridin-4-ylmethyl) amino)ethyl)quinoline-2-carboxamide

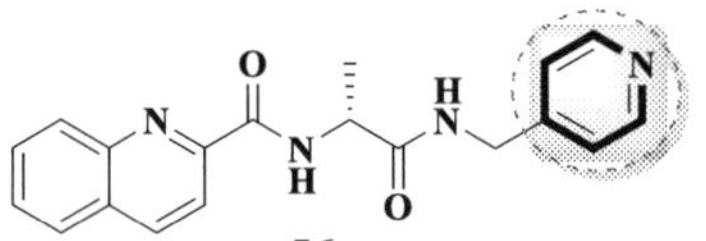

56

(R)-N-(1-Oxo-1-((pyridin-4-ylmethyl)amino) propan-2-yl)quinoline-2-carboxamide

FIGURE 8.53 Most active pyridine- and thiazole-based hydrazides **55** and **56**.

ergosterol can be inhibited, and the cells can accumulate the upstream components. Ultimately, it leads to the death of fungal cells. These single-target enzyme inhibitors, such as thioguanides, azoles, and acrylamines have been broadly and successfully used in clinic. Succinate dehydrogenase (SDH) also has been reported as the most important targets to discover more fungicides. By today, 23 SDH inhibitor fungicides have been commercially approved for the defense of plants since the first discovery of carboxin in 1966, and widely used to kill destructive fungi in plant. Conversely, the phenomenon of drug resistance and noxious side effects is persistently developing along with the long-term usage of these inhibitors. These disadvantages can be dodge by using dual- or multi-target drugs [110–112].

Amide–pyridines

On the basis of investigation of the 14α-demethylase (CYP51) and squalene cyclooxygenase (SE) inhibitors, pharmacophore feature as well as the dual-target active sites. A library of compounds having amide–pyridine skeletons was prepared for the treatment of increasing fungal infections enduring drug-resistant. These derivatives were evaluated in vitro and demonstrated that they exhibited a certain extent of antifungal activities. The most effective derivatives **55, 56** with MIC values ranging from 0.125–2 μg/mL possess a broad-spectrum antifungal activity and showed exceptional inhibition potential against drug-resistant pathogenic fungi (Fig. 8.53). Primary investigation of inhibition mechanism showed that the derivative **56** might impart an antifungal role by impeding the activity of CYP51 and SE. Remarkably derivatives did not demonstrate the genotoxicity through plasmid binding assay. Finally, the molecular docking study, ADME/T calculations and the designing of 3DQSAR model were completed. These outcomes can provide the pathway for further optimization of the lead molecules [113].

Pyridine Sulfides

Since the SDH is the most important target for the discovery of fungicide. Thus, 20 different pyridine sulfides comprising heterocyclic amide for their use as antifungal agent was planned, prepared, and confirmed by ^{1}H-NMR, ^{13}C-NMR, and HRMS. The target molecule **57** was evaluated in vitro against antifungal activity and exhibited effective inhibition for the pathogens with EC_{50} values of 5–39 ~g/mL. The in vivo studies revealed that the derivative **57** could efficiently stop *Botrytis cinerea* from affecting cucumber and tomato leaves with the blocking rates of 50% and 67%, respectively. The SDH enzyme assay, mitochondrial membrane potential detection, and the docking studies demonstrated that the mode of action of the derivative **57** and the related interactions with the targeted enzyme may be parallel to the fluopyram the standard fungicidal (Fig. 8.54). These results predicted that the derivative **57** could be explored for further studies as a potential SDH inhibitor [114].

Pyridine-grafted chitosan derivatives

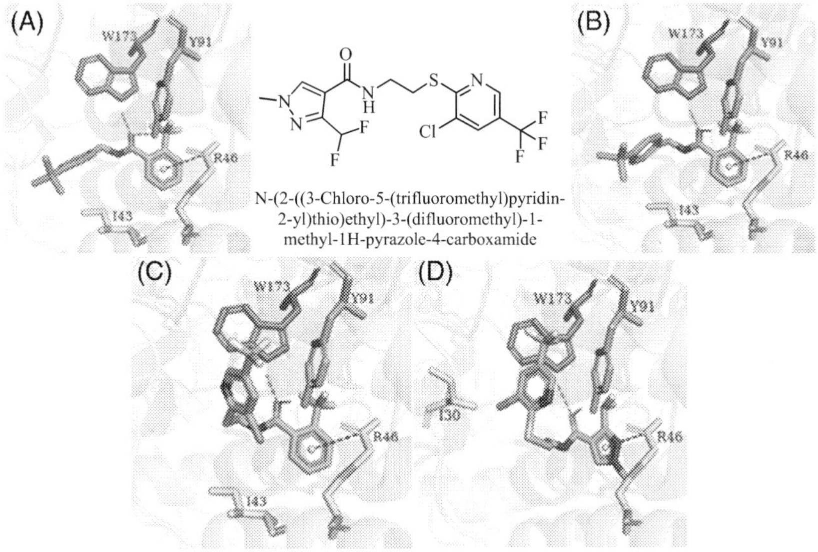

FIGURE 8.54 The reported interactions of ligand in 4YTP (A), and the binding of ligand (B), fluopyram (C), derivative **57** (D) to porcine heart mitochondrial SDH.

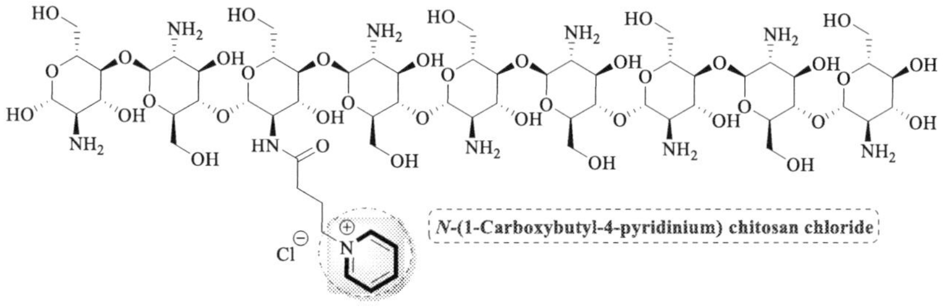

FIGURE 8.55 Structure of pyridine-grafted chitosan derivative.

By nucleophilic substitution reaction pyridine nucleus was incorporated into chitosan to synthesize pyridine chitosan (Fig. 8.55). The resulting hybrid was screened for its activity against fungi, on the basis of impeding the spore germination and mycelial growth. The results revealed that hybrid showed improved fungicidal activity compared to pristine chitosan. The values of the minimum concentration of pyridine chitosan for fungicidal activity and the minimal inhibitory concentration against *Fulvia fulva* were 1 mg/mL and 0.13 mg/mL, respectively, while the resultant values against *B. cinerea* were 4 mg/mL and 0.13 mg/mL, respectively. Extreme morphological variations in *B. cinerea* were observed after treating with pyridine chitosan which indicated that pyridine chitosan hybrid could deform and damage the fungal hyphae structure and successively prevent the growth of strain. Acute toxicity test of hybrid compound demonstrated that it is a nontoxic compound. These results are helpful

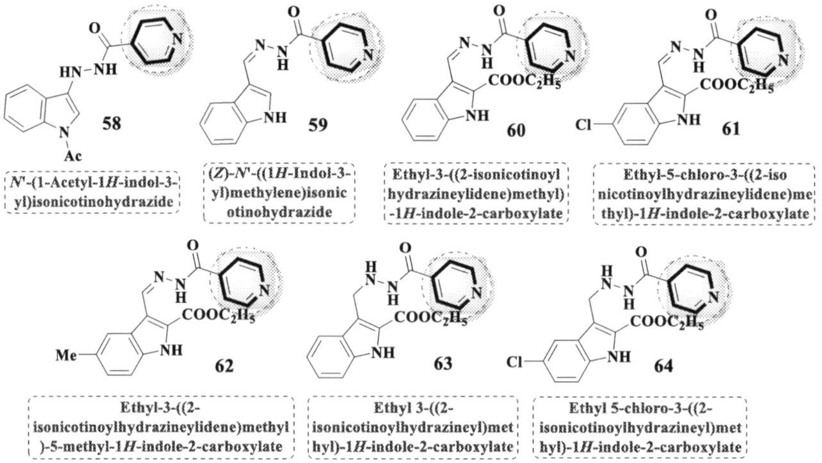

FIGURE 8.56 Most active indole–pyridine derived hydrazides **58–64**.

for evaluating the potency of chitosan analog and for finding new fungicidal with chitosan in the food industry [115].

8.4.7.2 Antituberculosis activity

TB is a foremost infectious disease in the world. Roughly one-third population is estimated to be affected with the *Mycobacterium tuberculosis*. Each year approximately 8 million people are infected with contagious, active pulmonary TB, and above 1.6 million people die due to this infection. *M. tuberculosis* is genetically resistant to numerous drugs therefore, the anti-TB chemotherapy treatment suggested by the WHO is based on the combination of four/five first-line drugs, including isoniazid (INH). However, TB-HIV coinfections, multidrug resistant TB and extremely drug-resistant TB are quickly spreading, thus suggesting to pay serious attention to discover and develop new anti-TB agents [116–118].

Indole–pyridines

Antimycobacterial activity of a library of simple synthetic indole and pyridine hybrids joined by hydrazides and hydrazine, were evaluated in vitro. The hybrids are analogs of substituted indole-3-carboxaldehydes or 1-acetylindoxyl chained by a hydrazine group with 4- and 3-pyridines, 1-oxide 4- and 3-pyridine carbohydrazides. Fifteen most potential derivatives displayed MICs values against an INH-sensitive strain of *M. tuberculosis* H37Rv equal to that of INH (0.05–21 g/mL). Five derivatives exhibited considerable potential against the INH-resistant *M. tuberculosis* CN-40 clinical isolate (MICs: 2–51 g/mL), making them countable for further in vivo experiments. Therefore, these experiments have revealed seven top 7 compounds **58–64** examined with MIC value of $\geq$ 0.2 μg/mL which exhibit the anti-TB potential of INH (Fig. 8.56). The antimycobacterial activity of these compounds was more than ethambutol the standard drug (MIC = 0.5 μg/mL) [118].

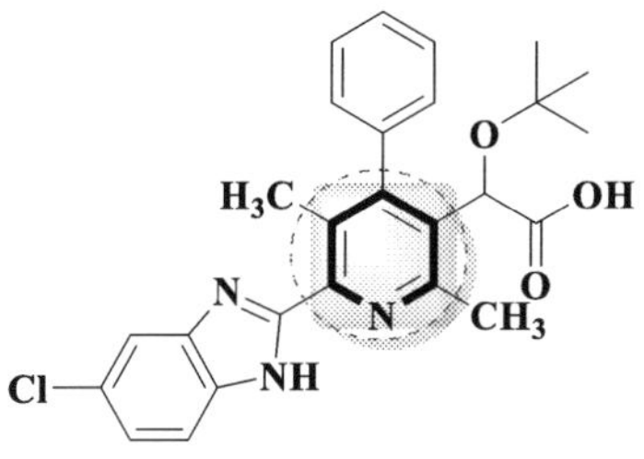

(-)-KF116

FIGURE 8.57 Structure of pyridine-based allosteric integrase inhibitors.

8.4.7.3 *Antiviral activity*

8.4.7.3.1 Anti-HIV activity

Healthy cells change into factories of virus when genetic code of HIV-1 enters into human genomes. For this purpose, the virus uses a multifunctional HIV-1 integrase (IN) enzyme. Integrase strand-transfer inhibitors (INSTIs) block the activity of HIV-1 enzyme during frontline treatments. These inhibitors have supported to improve the HIV/AIDS treatment by blocking the viral protein function during the initial stage of disease and stop the viral cDNA incorporation into human chromosomes thus defending the cells from more infections. Nevertheless, the incidence of drug resistance still remains a severe problem. As the virus develops, the integrase protein changes its shape, significantly drop the effectiveness of the existing therapies [118–120]. Clinical uses of first generation INSTIs elvitegravir (EVG) and raltegravir (RAL) leads to the progression of drug-resistant phenotypes comprising IN substitutions in the locality of the inhibitor interaction sites. Second generation bictegravir (BIC), dolutegravir (DTG), and cabotegravir (CAB) face considerably greater genetic pressure for the progression of drug resistance and generate complex resistance pathways with IN substitutions in both the vicinity of and substantially detached from inhibitor binding sites [121,122]. One approach to face this situation is the development of other treatments that can destroy the viruses that are drug-resistant by pointing to various areas of the integrase protein. It will be more difficult for the virus to develop the correct variation combinations to resist two or more therapies at once [123].

Benzimidazole–pyridine compounds

Allosteric HIV-1 integrase (IN) inhibitors (ALLINIs) are a new group of antiretroviral drugs that disturb appropriate viral development by starting hyper-multimerization of IN. Herein, it is reported that pyridine-based ALLINI KF116 are prominently selective for IN tetramers versus lower-order protein oligomers (Fig. 8.57). Therefore, very active (-)-KF116 enantiomer was synthesized which exhibited EC_{50} of ~7 nM against wild-type HIV-1 and about 10 times greater potential against dolutegravir resistant mutant virus proposing greater clinical significance for accompanying dolutegravir treatment with pyridine-based ALLINIs [123].

8.4.7.3.2 SARS coronavirus (SARS-CoV) inhibitory activity

SARS coronavirus (SARS-CoV) infection can cause a severe pneumonia. The SARS coronavirus was first appeared in Asia in February 2003 and blowout rapidly, followed by over 8000 patients with positive test and 774 deaths. China reports the first SARS-CoV-2 infection

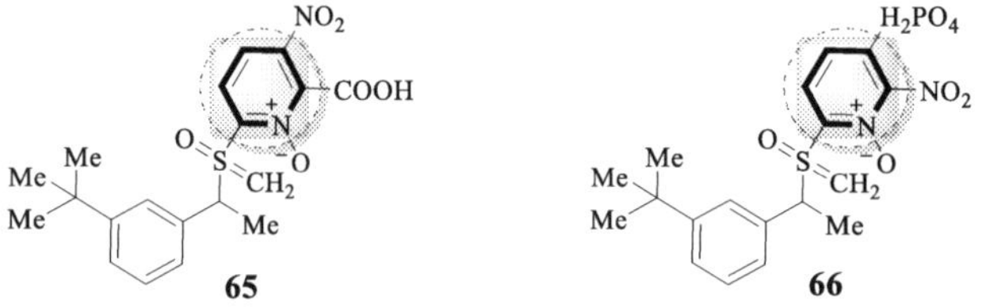

FIGURE 8.58 Structure of most active pyridine-*N*-oxide compounds **65** and **66**.

in December 2019 that was 80% identical with SARS-CoV. As a result, the major pandemic coronavirus that causes COVID-19, impose a significant risk to health, economy, and social stability around the world. It becomes a major global challenge for all mankind. In April 2020, over 2.4 million people were infected and approximately over 166 thousand people have died. Until an effective and safe vaccine is found that can take months or years to develop, newly exposed antiviral or repurposed drugs are used clinically and are found to be more useful and practical for treating severe conditions and decreasing the risk of disease. In this evolving health situation, chloroquine (CQ) and hydroxychloroquine (HCQ), are undergoing clinical trials known for their antimalarial potential. These repurposing drugs have displayed clear effectiveness to treat COVID-19 [124–127]. The gene sequence of SARS-CoV-2, was released on January 11, 2020, initiating strong global R&D activity for the development of vaccine against COVID-19 [128]. Currently, many vaccines are in use that are also listed in WHO Emergency Use Listing such as Pfizer/BioNtech Comirnaty, SII/Covishield, AstraZeneca/AZD1222, Janssen/Ad26.COV 2.S, Moderna (mRNA 1273), Sinopharm and Sinovac COVID-19 vaccines [129–130].

Pyridine *N*-oxide compounds

SARSCoV-2 is the leading challenge the world is experiencing today. In this work, the pyridine *N*-oxide compounds were explored as the potential class of human SARS inhibitors for CoV-2. The consistent CoMSIA developed model of 110 pyridine *N*-oxide based antiviral molecules, displayed $Q^2 = 0.54$ and $r^2_{ext} = 0{:}71$. The molecular surflex-docking was used to find the crystal structure of CoV-2 main protease 3CLpro (PDB: 6LU7) and two potentially used antiviral molecules, namely HCQ and CQ. The molecular surflex-docking between pyridine *N*-oxide, HCQ, CQ, antiviral molecules, and the crystal structure of COVID-19 main protease presented that compound **65** is stabilized by hydrogen bonds with CYS A:145, GLY A: 143, THR A:26 residues (Figs 8.58 and 8.59). The found free energy affinity and ADMET properties show that among the series, the new antiviral compound **65** could be an effective antiviral drug inhibitor against COVID-19. The inhibition potential of pyridine *N*-oxide derivatives against CoV-2 was compared with the potential of the two most common antiviral drugs, namely HCQ and CQ. DFT method was also used to describe the reactive sites of pyridine *N*-oxide derivatives as well as CQ and HCQ [127].

Hydrazone-pyridine compound

Hydrazinyl pyridine was synthesized and characterized by single-crystal X-ray crystallography and quantum theory to confirm the experimental data. The geometrical parameters, UV-visible spectrum, and vibrational frequencies were gained from the optimized structure of the CCPEHP compound and compared with the experimental results. The experimental vibrational frequency values were very close to the calculated values. Hirschfeld Surface analysis verified that significant interactions were the C ···· H/H ····C 20.3%, Cl ···H/H ···Cl 20.7% and H ···H 31.2% interactions. Stabilization energy values between

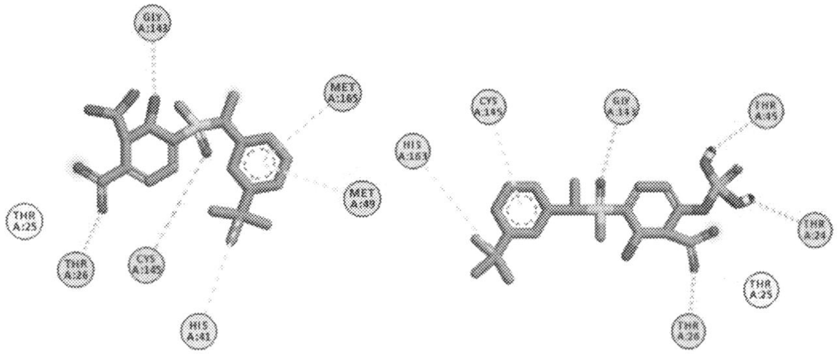

FIGURE 8.59 Different types of interactions of pyridine *N*-oxide Compounds **65** and **66** with Sars-Covid-19 receptor PDB:6LU7.

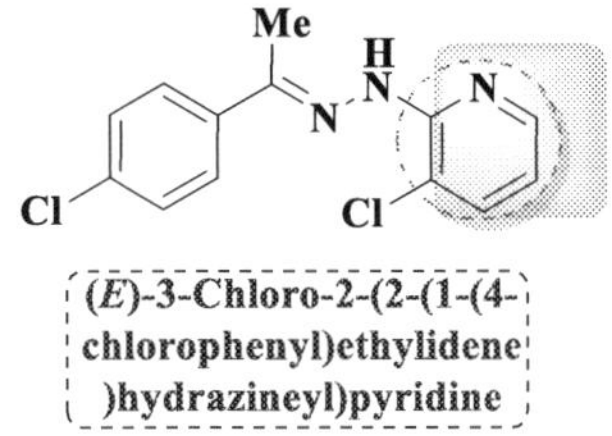

FIGURE 8.60 Structure of 3-chloro-2-{(2E) −2-[1-(4-chlorophenyl)ethylidene]hydrazinyl} pyridine (CCPEHP).

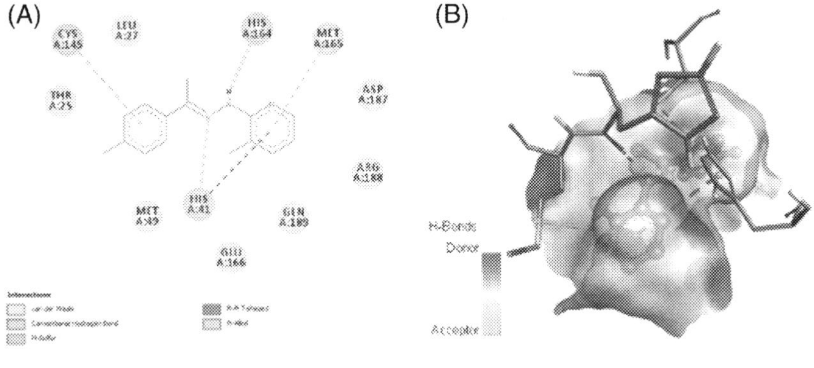

FIGURE 8.61 (A) and (B) Visualize the 2D and 3D interactions between CCPEHP and COVID-19 M^{pro}.

donor-acceptor orbitals and hyperconjugation interactions were obtained by NBO orbital analysis. The reactive zones of CCPEHP were found by molecular electrostatic potential results in the nitrogen atoms N8, N9, and N11 were nucleophilic zones. Frontier molecular orbitals and molecular interaction ability-related properties were projected. The results of docking, spectroscopic, and theoretical techniques give improved data about inhibitor-receptor interactions. Molecular docking analysis, showed that CCPEHP compound was fixed actively into the COVID-19 M^{pro} sites and displayed hydrogen bond, π–π T-Shaped, and π-alkyl type interactions. CCPEHP docked effectively with the amino acid molecule of the main protease of SARS-CoV2 with a high negative binding affinity. Finally, it is expected that the effective in silico results achieved in the CCPEHP molecule will be further confirmed by in vitro and in vivo studies (Figs 8.60 and 8.61) [131].

Mo(VI)-pyridine complex

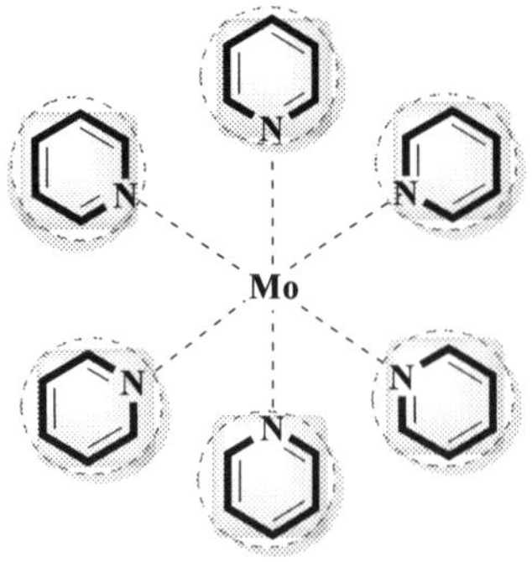

FIGURE 8.62 Structure of Mo(VI)-pyridine complex.

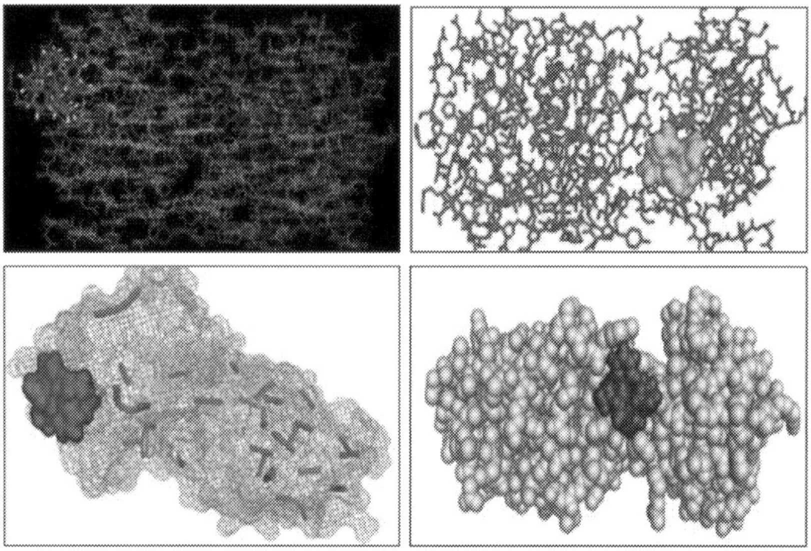

FIGURE 8.63 Molecular docking images of the complex with COVID-19 spike Protein (PDB No.: 6LU7).

The straightforward and high yield wet laboratory strategy has been used to form the combination of *N*-contributor heterocyclic ligand (Pyridine) functionalized Mo(VI) metal complex. The complex shows extremely good potential and steric energy of 35.66 and 35.684 kcal/mole, separately. The UV-visible electronic absorption behavior of the metal complex in a neat aqueous solvent and the interaction with BSA shows that the absorption quenching mechanism, a combined quenching process, dynamic, and static quenching, occurs at low complex concentration. The change in absorption values redshift 308 nm to 317 nm indicates that the metal complex binds to BSA mainly by hydrogen bonds and Van der Waals interactions. The absorption confirms the anchoring ability of our complex in a protein environment also the strong covalent binding may also be formed between them. The heterocyclic ligand-based metal complex showed the best binding energy of -5.5 kcal/mol with covid-19 spike proteins (6LU7). The strong anchoring of the complex on the covid-19 spike protein gives more details about M-protease and can assist in the inhibition of COVID-19 spread (Figs 8.62 and 8.63). This makes them potential lead compounds for advancement into competitors against the SARS-COV-2 treatment [132].

References

[1] S. Shimizu, N. Watanabe, T. Kataoka, T. Shoji, N. Abe, S. Morishita, H. Ichimura, Pyridine and pyridine derivatives, Ullmann's Encyclopedia Ind. Chem. 30 (2000) 557–586.

[2] G.L. Kennedy Jr., M.S. Dabt, Patty's Toxicology, Sixth Edition, John Wiley & Sons, Inc., Hoboken, New Jersey, United States, 2012 Volume 2 Bingham, E. and Cohrssen, B. editors http://doi.org/10.1002/0471435139.tox060.pub2.

[3] Y. Pawar, A. Nikum, A review on bioactive nitrogenous compounds, Research Interventions and Advancements in Plant Sciences, Bhumi Publishing, Nigave Khalasa, Kolhapur 416207, Maharashtra, India, 2020, p. 353.

[4] D. Smith, Compounds containing six-membered rings with one nitrogen atom; pyridine and its derivatives, Rodd's Chem. Carbon Compounds 4 (1964) 27–226.

[5] J. Elvidge, L. Jackman, Studies of aromaticity by nuclear magnetic resonance spectroscopy. Part I. 2-Pyridones and related systems, J. Chemical Society (Resumed) 181 (1961) 859–866.

[6] P.E. Alford, Six-membered ring systems: pyridines and benzo derivatives, Prog. Heterocycl. Chem. 22 (2011) 349–391.

[7] A.R. Katritzky, C.A. Ramsden, J.A. Joule, V.V. Zhdankin, 3.2-Reactivity of six membered rings, Handbook of Heterocyclic Chemistry, Elsevier, Amsterdam, Netherlands, 2010, pp. 242–382.

[8] E. Vessally, A. Hosseinian, L. Edjlali, A. Bekhradnia, M.D. Esrafili, New page to access pyridine derivatives: Synthesis from *N*-propargylamines, RSC Adv. 6 (75) (2016) 71662–71675.

[9] P. Thirumurugan, S. Mahalaxmi, P.T. Perumal, Synthesis and anti-inflammatory activity of 3-indolyl pyridine derivatives through one-pot multi component reaction, J. Chem. Sci. 122 (6) (2010) 819–832.

[10] R.J. Ramos, M. Albersen, E. Vringer, M. Bosma, S. Zwakenberg, F. Zwartkruis, N.M. Verhoeven-Duif, Discovery of pyridoxal reductase activity as part of human vitamin B_6 metabolism, Biochimica et Biophysica Acta -General Subjects 1863 (6) (2019) 1088–1097.

[11] R.M. Bastle, N.A. Peartree, J. Goenaga, K.N. Hatch, A. Henricks, S. Scott, L.E. Hood, J.L. Neisewander, Immediate early gene expression reveals interactions between social and nicotine rewards on brain activity in adolescent male rats, Behav. Brain Res. 313 (2016) 244–254.

[12] V.S. Kamanna, M.L. Kashyap, Mechanism of action of niacin, Am. J. Cardiol. 101 (8) (2008) S20–S26.

[13] R.H. Meymeh, E. Wooltorton, Diabetes drug pioglitazone (Actos): risk of fracture, Can. Med. Assoc. J. 177 (7) (2007) 723–724.

[14] S.W.E.-D.N.S. Study, Extended-dose nevirapine to 6 weeks of age for infants to prevent HIV transmission *via* breastfeeding in Ethiopia, India, and Uganda: an analysis of three randomised controlled trials, Lancet North Am. Ed. 372 (9635) (2008) 300–313.

[15] H. Ali, O. Galal, Y. Urata, S. Goto, C.-Y. Guo, L. Luo, E. Abdelrahim, Y. Ono, E. Mostafa, T.-S. Li, The potential benefits of nicaraven to protect against radiation-induced injury in hematopoietic stem/progenitor cells with relative low dose exposures, Biochem. Biophys. Res. Commun. 452 (3) (2014) 548–553.

[16] T.R. Sterling, M.E. Villarino, A.S. Borisov, N. Shang, F. Gordin, E.E. Bliven-Sizemore, J. Hackman, C.D. Hamilton, D. Menzies, A. Kerrigan, S.E. Weis, R.E Chaisson, Three months of rifapentine and isoniazid for latent tuberculosis infection, N. Engl. J. Med. 365 (23) (2011) 2155–2166.

[17] S. Amlani, T. Nadarajah, R.A. McIvor, Montelukast for the treatment of asthma in the adult population, Expert Opin. Pharmacother. 12 (13) (2011) 2119–2128.

[18] D.E. Baker, Esomeprazole magnesium (Nexium), Rev. Gastroenterol. Disord. 1 (1) (2001) 32–41.

[19] S. Halder, M.L. Shuma, A.K.L. Kabir, A.S.S. Rouf, In vitro release kinetics study of esomeprazole magnesium trihydrate tablet available in Bangladesh and comparison with the originator brand (Nexium®), Stamford J. Pharmaceutical Sci. 4 (1) (2011) 79–83.

[20] S.R. Dhaneshwar, V.N. Jagtap, Development and validation of stability indicating RP-HPLC-PDA method for tenatoprazole and its application for formulation analysis and dissolution study, Am. J. Analytical Chem. 2 (02) (2011) 126.

[21] H. Yousuf, S. Shamim, K.M. Khan, S. Chigurupati, S. Hameed, M.N. Khan, M. Taha, M. Arfeen, Dihydropyridines as potential α-amylase and α-glucosidase inhibitors: synthesis, in vitro and *in silico* studies, Bioorg. Chem. 96 (2020) 103581.
[22] H. Niaz, H. Kashtoh, J.A. Khan, A. Khan, M.T. Alam, K.M. Khan, S. Perveen, M.I. Choudhary, Synthesis of diethyl 4-substituted-2,6-dimethyl-1,4-dihydropyridine-3,5-dicarboxylates as a new series of inhibitors against yeast α-glucosidase, Eur. J. Med. Chem. 95 (2015) 199–209.
[23] J. Liu, D. Zuo, T. Jing, M. Guo, L. Xing, W. Zhang, D. Zhang, Synthesis, biological evaluation and molecular modeling of imidazo [1,2-*a*] pyridine derivatives as potent antitubulin agents, Bioorg. Med. Chem. 25 (15) (2017) 4088–4099.
[24] M. Ali, K.M. Khan, M. Mahdavi, A. Jabbar, S. Shamim, U. Salar, M. Taha, S. Perveen, B. Larijani, M.A. Faramarzi, Synthesis, in vitro and *in silico* screening of 2-amino-4-aryl-6-(phenylthio) pyridine-3, 5-dicarbonitriles as novel α-glucosidase inhibitors, Bioorg. Chem. 100 (2020) 103879.
[25] K. Torikai, R. Koga, X. Liu, K. Umehara, T. Kitano, K. Watanabe, Y. Shimohigashi, Design and synthesis of benzoacridines as estrogenic and anti-estrogenic agents, Bioorg. Med. Chem. 25 (20) (2017) 5216–5237.
[26] O.V. Ershov, V.N. Maksimova, K.V. Lipin, M.Y. Belikov, M.Y. Ievlev, V.A. Tafeenko, O.E. Nasakin, Regiospecific synthesis of *gem*-dinitro derivatives of 2-halogenocycloalka [*b*] pyridine-3,4-dicarbonitriles, Tetrahedron 71 (39) (2015) 7445–7450.
[27] H.S.P. Rao, N. Muthanna, A.H. Padder, Vinylogous blaise reaction: conceptually new synthesis of pyridin-2-ones, Synlett 29 (12) (2018) 1649–1653.
[28] S.H. Jung, D.B. Sung, C.H. Park, W.S. Kim, Copper-catalyzed *N*-arylation of 2-pyridones employing diaryliodonium salts at room temperature, J. Org. Chem. 81 (17) (2016) 7717–7724.
[29] A.C. Foster, J.A. Kemp, Glutamate-and GABA-based CNS therapeutics, Curr. Opin. Pharmacol. 6 (1) (2006) 7–17.
[30] M. Bamford, 3 H^+/K^+ ATPase inhibitors in the treatment of acid-related disorders, Prog. Med. Chem. 47 (2009) 75–162.
[31] M. Dowsett, D. Smithers, J. Moore, P.F. Trunet, R.C. Coombes, T.J. Powles, R. Rubens, I.E. Smith, Endocrine changes with the aromatase inhibitor fadrozole hydrochloride in breast cancer, Eur. J. Cancer 30 (10) (1994) 1453–1458.
[32] H. Mikashima, K. Goto, Inhibitory effect of 2-(4-(2-imidazo (1,2-*a*) pyridyl) phenyl) propionic acid (miroprofen) on platelet aggregation and prostaglandin I2 generation (author's transl), Yakugaku Zasshi 102 (1) (1982) 99–103.
[33] L. Goswami, S. Gogoi, J. Gogoi, R.K. Boruah, R.C. Boruah, P. Gogoi, Facile diversity-oriented synthesis of polycyclic pyridines and their cytotoxicity effects in human cancer cell lines, ACS Combinatorial Sci. 18 (5) (2016) 253–261.
[34] M. Mareel, A. Leroy, Clinical, cellular, and molecular aspects of cancer invasion, Physiol. Rev. 83 (2) (2003) 337–376.
[35] F.W. Hartel, S. de Coronado, R. Dionne, G. Fragoso, J. Golbeck, Modeling a description logic vocabulary for cancer research, J. Biomed. Inform. 38 (2) (2005) 114–129.
[36] T. Damghani, F. Moosavi, M. Khoshneviszadeh, M. Mortazavi, S. Pirhadi, Z. Kayani, L. Saso, N. Edraki, O. Firuzi, Imidazopyridine hydrazone derivatives exert antiproliferative effect on lung and pancreatic cancer cells and potentially inhibit receptor tyrosine kinases including c-Met, Sci. Rep. 11 (1) (2021) 1–20.
[37] M. Radi, E. Dreassi, C. Brullo, E. Crespan, C. Tintori, V. Bernardo, M. Valoti, C. Zamperini, H. Daigl, F. Musumeci, F. Carraro, Design, synthesis, biological activity, and ADME properties of pyrazolo [3,4-*d*] pyrimidines active in hypoxic human leukemia cells: a lead optimization study, J. Med. Chem. 54 (8) (2011) 2610–2626.
[38] I. Abbas, S. Gomha, M. Elaasser, M. Bauomi, Synthesis and biological evaluation of new pyridines containing imidazole moiety as antimicrobial and anticancer agents, TUrk. J. Chem. 39 (2) (2015) 334–346.
[39] A. Omar, Review article; anticancer activities of some fused heterocyclic moieties containing nitrogen and/or sulfur heteroatoms, Al-Azhar J. Pharmaceutical Sci. 62 (2) (2020) 39–54.
[40] E.A. Mohamed, N.S. Ismail, M. Hagras, H. Refaat, Medicinal attributes of pyridine scaffold as anticancer targeting agents, Fut. J. Pharma. Sci. 7 (1) (2021) 1–17.
[41] T.C. Liu, X. Peng, Y.C. Ma, Y.C. Ji, D.Q. Chen, M.Y. Zheng, D.M. Zhao, M.S. Cheng, M.Y. Geng, J.K. Shen, J. Ai, Discovery of a new series of imidazo [1,2-*a*] pyridine compounds as selective c-Met inhibitors, Acta Pharmacol. Sin. 37 (5) (2016) 698–707.

[42] L. Goswami, S. Gogoi, J. Gogoi, R.K. Boruah, R.C. Boruah, P. Gogoi, Facile diversity-oriented synthesis of polycyclic pyridines and their cytotoxicity effects in human cancer cell lines, ACS Combinatorial Sci. 18 (5) (2016) 253–261.
[43] Y. Nam, D. Hwang, N. Kim, H.S. Seo, K.B. Selim, T. Sim, Identification of 1*H*-pyrazolo [3,4-*b*] pyridine derivatives as potent ALK-L1196M inhibitors, J. Enzyme Inhib. Med. Chem. 34 (1) (2019) 1426–1438.
[44] M.I. El-Gamal, C.H. Oh, Pyrrolo [3,2-*c*] pyridine derivatives with potential inhibitory effect against FMS kinase: In vitro biological studies, J. Enzyme Inhib. Med. Chem. 33 (1) (2018) 1160–1166.
[45] B.H. Naguib, H.B. El-Nassan, Synthesis of new thieno [2,3-*b*] pyridine derivatives as pim-1 inhibitors, J. Enzyme Inhib. Med. Chem. 31 (6) (2016) 1718–1725.
[46] B. Zhao, Y. Li, P. Xu, Y. Dai, C. Luo, Y. Sun, J. Ai, M. Geng, W. Duan, Discovery of substituted 1-*H*-pyrazolo [3,4-*b*] pyridine derivatives as potent and selective FGFR kinase inhibitors, ACS Med. Chem. Lett. 7 (6) (2016) 629–634.
[47] F. Erdemir, D. Barut Celepci, A. Aktas, Y. Gok, R. Kaya, P. Taslimi, Y. Demir, I. Gulcin, Novel 2-aminopyridine liganded Pd(II) *N*-heterocyclic carbene complexes: synthesis, characterization, crystal structure and bioactivity properties, Bioorg. Chem. 91 (2019) 103134.
[48] S. Imran, M. Taha, N.H. Ismail, S. Fayyaz, K.M. Khan, M.I. Choudhary, Synthesis, biological evaluation, and docking studies of novel thiourea derivatives of bisindolylmethane as carbonic anhydrase II inhibitor, Bioorg. Chem. 62 (2015) 83–93.
[49] F. Turkan, A. Cetin, P. Taslimi, H.S. Karaman, I. Gulcin, Synthesis, characterization, molecular docking and biological activities of novel pyrazoline derivatives, Arch. Pharm. 352 (6) (2019) 1800359.
[50] S. Isık, D. Vullo, S. Durdagi, D. Ekinci, M. Senturk, A. Çetin, E. Senturk, C.T. Supuran, Interaction of carbonic anhydrase isozymes I, II, and IX with some pyridine and phenol hydrazinecarbothioamide derivatives, Bioorg. Med. Chem. Lett. 25 (23) (2015) 5636–5641.
[51] M.F. Ansari, D. Idrees, M.I. Hassan, K. Ahmad, F. Avecilla, A. Azam, Design, synthesis and biological evaluation of novel pyridine-thiazolidinone derivatives as anticancer agents: targeting human carbonic anhydrase IX, Eur. J. Med. Chem. 144 (2018) 544–556.
[52] H. Amat-ur-Rasool, M. Ahmed, S. Hasnain, W.G. Carter, Anti-cholinesterase combination drug therapy as a potential treatment for Alzheimer's disease, Brain Sci. 11 (2) (2021) 184.
[53] N.L. Batsch, M.S. Mittelman, World Alzheimer Report 2012. Overcoming the Stigma of Dementia, Alzheimer's Disease International, Alzheimer's Disease International (ADI, London, United Kingdom, 2012.
[54] P.T. Nelson, W.X. Wang, B.W Rajeev, MicroRNAs (miRNAs) in neurodegenerative diseases, Brain Pathol. 18 (1) (2008) 130–138.
[55] M. Tamilselvan, T. Tamilanban, V. Chitra, Unfolding remedial targets for Alzheimer's disease, Res. J. Pharmacy Technol. 13 (6) (2020) 3021–3027.
[56] K. Iqbal, I. Grundke-Iqbal, Alzheimer disease is multifactorial and heterogeneous, Neurobiol. Aging 6 (21) (2000) 901–902.
[57] F.H. Darras, S. Pockes, G. Huang, S. Wehle, A. Strasser, H.J. Wittmann, M. Nimczick, C.A. Sotriffer, M. Decker, Synthesis, biological evaluation, and computational studies of tri-and tetracyclic nitrogen-bridgehead compounds as potent dual-acting AChE inhibitors and hH3 receptor antagonists, ACS Chem. Neurosci. 5 (3) (2014) 225–242.
[58] P.T. Francis, A.M. Palmer, M. Snape, G.K. Wilcock, The cholinergic hypothesis of Alzheimer's disease: a review of progress, J. Neurol., Neurosurgery Psychiatry 66 (2) (1999) 137–147.
[59] M.W. Jann, Rivastigmine, a new-generation cholinesterase inhibitor for the treatment of Alzheimer's disease, Pharmacotherapy 20 (1) (2000) 1–12.
[60] R. Leon, A.G. Garcia, J. Marco-Contelles, Recent advances in the multitarget-directed ligands approach for the treatment of Alzheimer's disease, Med. Res. Rev. 33 (1) (2013) 139–189.
[61] A. Hasan, K.M. Khan, M. Sher, G.M. Maharvi, S.A. Nawaz, M.I. Choudhary, Atta-ur-Rahman, C.T Supuran, Synthesis and inhibitory potential towards acetylcholinesterase, butyrylcholinesterase and lipoxygenase of some variably substituted chalcones, J. Enzyme Inhib. Med. Chem. 20 (2005) 41–47.
[62] H.C. Kwong, C.C. Kumar, S.H. Mah, Y.L. Mah, T.S. Chia, C.K. Quah, G.K. Lim, S. Chandraju, Crystal correlation of heterocyclic imidazo [1,2-*a*] pyridine analogues and their anticholinesterase potential evaluation, Sci. Rep. 9 (1) (2019) 1–15.

[63] M.A. Serrano, M. Pivatto, W. Francisco, A. Danuello, L.O. Regasini, E.M. Lopes, M.N. Lopes, M.C. Young, V.S. Bolzani, Acetylcholinesterase inhibitory pyridine alkaloids of the leaves of *Senna multijuga*, J. Nat. Prod. 73 (3) (2010) 482–484.

[64] S.M. Sanad, A.E. Mekky, Novel nicotinonitrile-coumarin hybrids as potential acetylcholinesterase inhibitors: design, synthesis, in vitro and *in silico* studies, J. Iran. Chem. Soc. 18 (1) (2021) 213–224.

[65] Y. Zhou, W. Sun, J. Peng, H. Yan, L. Zhang, X. Liu, Z. Zuo, Design, synthesis and biological evaluation of novel copper-chelating acetylcholinesterase inhibitors with pyridine and *N*-benzylpiperidine fragments, Bioorg. Chem. 93 (2019) 103322.

[66] M. Solangi, K.M. Khan, F. Saleem, S. Hameed, J. Iqbal, Z. Shafique, U. Qureshi, Z. Ul-Haq, M. Taha, S. Perveen, Indole acrylonitriles as potential anti-hyperglycemic agents: Synthesis, α-glucosidase inhibitory activity and molecular docking studies, Bioorg. Med. Chem. 28 (21) (2020) 115605.

[67] S. Hameed, F. Seraj, R. Rafique, S. Chigurupati, A. Wadood, A.U. Rehman, V. Venugopal, U. Salar, M. Taha, K.M. Khan, Synthesis of benzotriazoles derivatives and their dual potential as α-amylase and α-glucosidase inhibitors in vitro: Structure-activity relationship, molecular docking, and kinetic studies, Eur. J. Med. Chem. 183 (2019) 111677.

[68] R. Rafique, K.M. Khan, K. Arshia, S. Chigurupati, A. Wadood, A.U. Rehman, A. Karunanidhi, S. Hameed, M. Taha, M. al-Rashida, Synthesis of new indazole based dual inhibitors of α-glucosidase and α-amylase enzymes, their in vitro, *in silico* and kinetics studies, Bioorg. Chem. 94 (2019) 103195.

[69] M. Taha, M.S. Baharudin, N.H. Ismail, M. Selvaraj, U. Salar, K.A. Alkadi, K.M. Khan, Synthesis and *in silico* studies of novel sulfonamides having oxadiazole ring: As β-glucuronidase inhibitors, Bioorg. Chem. 71 (2017) 86–96.

[70] M. Khan, A. Alam, K.M. Khan, U. Salar, S. Chigurupati, A. Wadood, F. Ali, J.I. Mohammad, M. Riaz, S. Perveen, Flurbiprofen derivatives as novel α-amylase inhibitors: biology-oriented drug synthesis (BIODS), in vitro, and *in silico* evaluation, Bioorg. Chem. 81 (2018) 157–167.

[71] N.P. Malik, M. Naz, U. Ashiq, R.A. Jamal, S. Gul, F. Saleem, K.M. Khan, S. Yousuf, Oxamide derivatives as potent α-glucosidase inhibitors: design, synthesis, in vitro inhibitory screening and *in silico* docking studies, Chemistry Select 6 (28) (2021) 7188–7201.

[72] P. Singh, R.H. Jayaramaiah, S.B. Agawane, G. Vannuruswamy, A.M. Korwar, A. Anand, V.S. Dhaygude, M.L. Shaikh, R.S. Joshi, R. Boppana, M.J. Kulkarni, Potential dual role of eugenol in inhibiting advanced glycation end products in diabetes: proteomic and mechanistic insights, Sci. Rep. 6 (1) (2016) 1–13.

[73] C. Bathula, R. Mamidala, C. Thulluri, R. Agarwal, K.K. Jha, P. Munshi, U. Adepally, A. Singh, M.T. Chary, S. Sen, Substituted furopyridinediones as novel inhibitors of α-glucosidase, RSC Adv. 5 (110) (2015) 90374–90385.

[74] M. Adib, F. Peytam, M. Rahmanian-Jazi, S. Mahernia, H.R. Bijanzadeh, M. Jahani, M. Mohammadi-Khanaposhtani, S. Imanparast, M.A. Faramarzi, M. Mahdavi, B. Larijani, New 6-amino-pyrido [2,3-*d*] pyrimidine-2, 4-diones as novel agents to treat type 2 diabetes: a simple and efficient synthesis, α-glucosidase inhibition, molecular modeling and kinetic study, Eur. J. Med. Chem. 155 (2018) 353–363.

[75] F. Saleem, K.M. Khan, S. Chigurupati, M. Solangi, A.R. Nemala, M. Mushtaq, Z. Ul-Haq, M. Taha, S. Perveen, Synthesis of azachalcones, their α-amylase, α-glucosidase inhibitory activities, kinetics, and molecular docking studies, Bioorg. Chem. 106 (2021) 104489.

[76] D.R. Tomlinson, E.J. Stevens, L.T. Diemel, Aldose reductase inhibitors and their potential for the treatment of diabetic complications, Trends Pharmacol. Sci. 15 (8) (1994) 293–297.

[77] J.M. Petrash, All in the family: aldose reductase and closely related aldo-keto reductases, Cellular Molecular Life Sci. 61 (7) (2004) 737–749.

[78] C. Yabe-Nishimura, Aldose reductase in glucose toxicity: a potential target for the prevention of diabetic complications, Pharmacol. Rev. 50 (1) (1998) 21–34.

[79] Q. Huang, Q. Liu, D. Ouyang, Sorbinil, an aldose reductase inhibitor, in fighting against diabetic complications, Med. Chem. 15 (1) (2019) 3–7.

[80] C. Zhang, Z. Min, X. Liu, C. Wang, Z. Wang, J. Shen, W. Tang, X. Zhang, D. Liu, X. Xu, Tolrestat acts atypically as a competitive inhibitor of the thermostable aldo-keto reductase Tm1743 from Thermotoga maritima, FEBS Lett. 594 (3) (2020) 564–580.

[81] M.N. Marinov, S.M. Bakalova, R.Y. Prodanova, N.V. Markova, Conformational and spectral properties of newly synthesized compounds obtained by reaction of alrestatin with 3-aminocycloalkanespiro-5-hydantoins, Bulg. Chem. Commun. 49 (2017) 146–152.

[82] P.B. Inskeep, R.A. Ronfeld, M.J. Peterson, N. Gerber, Pharmacokinetics of the aldose reductase inhibitor, zopolrestat, in humans, J. Clinical Pharmacol. 34 (7) (1994) 760–766.
[83] A. Da Settimo, G. Primofiore, F. Da Settimo, F. Simorini, C. La Motta, A. Martinelli, E. Boldrini, Synthesis of pyrrolo [3,4-*c*] pyridine derivatives possessing an acid group and their in vitro and *in vivo* evaluation as aldose reductase inhibitors, Eur. J. Med. Chem. 31 (1) (1996) 49–58.
[84] N.E. Dixon, C. Gazzola, R.L. Blakeley, B. Zerner, Jack bean urease (EC 3.5. 1.5). Metalloenzyme. Simple biological role for nickel, J. Am. Chem. Soc. 97 (1975) 4131–4133.
[85] M. Biglar, H. Sufi, K. Bagherzadeh, M. Amanlou, F. Mojab, Screening of 20 commonly used Iranian traditional medicinal plants against urease, Iranian J. Pharmaceutical Res. 13 (2014) 195.
[86] X.D. Yu, R.B. Zheng, J.H. Xie, J.Y. Su, X.Q. Huang, Y.H. Wang, Y.F. Zheng, Z.Z. Mo, D.W. Wu, Y.E. Liang, Biological evaluation and molecular docking of baicalin and scutellarin as *Helicobacter pylori* urease inhibitors, J. Ethnopharmacol. 162 (2015) 69–78.
[87] M.A. Abdullah, G.E.-D.A. Abuo-Rahma, E.-S.M. Abdelhafez, H.A. Hassan, R.M.A. El-Baky, Design, synthesis, molecular docking, anti-*Proteus mirabilis* and urease inhibition of new fluoroquinolone carboxylic acid derivatives, Bioorg. Chem. 70 (2017) 1–11.
[88] S. Vassiliou, A. Grabowiecka, P. Kosikowska, A. Yiotakis, P. Kafarski, L. Berlicki, Design, synthesis, and evaluation of novel organophosphorus inhibitors of bacterial ureases, J. Med. Chem. 51 (2008) 5736–5744.
[89] K.M. Khan, F. Naz, M. Taha, A. Khan, S. Perveen, M. Choudhary, W. Voelter, Synthesis and in vitro urease inhibitory activity of *N,N*′-disubstituted thioureas, Eur. J. Med. Chem. 74 (2014) 314–323.
[90] P. Kosikowska, Ł. Berlicki, Urease inhibitors as potential drugs for gastric and urinary tract infections: a patent review, Expert Opin. Ther. Pat. 21 (2011) 945–957.
[91] Z.S. Saify, A. Kamil, S. Akhtar, M. Taha, A. Khan, K.M. Khan, S. Jahan, F. Rahim, S. Perveen, M.I. Choudhary, 2-(2′-Pyridyl) benzimidazole derivatives and their urease inhibitory activity, Med. Chem. Res. 23 (10) (2014) 4447–4454.
[92] M. Biglar, R. Mirzazadeh, M. Asadi, S. Sepehri, Y. Valizadeh, Y. Sarrafi, M. Amanlou, B. Larijani, M. Mohammadi-Khanaposhtani, M. Mahdavi, Novel *N,N*-dimethylbarbituric-pyridinium derivatives as potent urease inhibitors: Synthesis, in vitro, and in silico studies, Bioorg. Chem. 95 (2020) 103529.
[93] B. Halliwell, Reactive species and antioxidants. Redox biology is a fundamental theme of aerobic life, Plant Physiol. 141 (2) (2006) 312–322.
[94] M. Valko, D. Leibfritz, J. Moncol, M.T. Cronin, M. Mazur, J. Telser, Free radicals and antioxidants in normal physiological functions and human disease, Int. J. Biochem. Cell Biol. 39 (1) (2007) 44–84.
[95] B. Halliwell, The wanderings of a free radical, Free Radical Biol. Med. 46 (5) (2009) 531–542.
[96] K. Toyoda, K. Fujii, M. Kamouchi, H. Nakane, S. Arihiro, Y. Okada, S. Ibayashi, M. Iida, Free radical scavenger, edaravone, in stroke with internal carotid artery occlusion, J. Neurol. Sci. 221 (1-2) (2004) 11–17.
[97] A.R. Green, T. Ashwood, T. Odergren, D.M Jackson, Nitrones as neuroprotective agents in cerebral ischemia, with particular reference to NXY-059, Pharmacol. Ther. 100 (3) (2003) 195–214.
[98] M. Carocho, I.C. Ferreira, A review on antioxidants, prooxidants and related controversy: natural and synthetic compounds, screening and analysis methodologies and future perspectives, Food Chem. Toxicol. 51 (2013) 15–25.
[99] R.C. Calhelha, D. Peixoto, M. Vilas Boas, M.J.R. Queiroz, I.C. Ferreira, Antioxidant activity of aminodiarylamines in the thieno [3,2-*b*] pyridine series: radical scavenging activity, lipid peroxidation inhibition and redox profile, J. Enzyme Inhib. Med. Chem. 29 (3) (2014) 311–316.
[100] N. Bulut, U.M. Kocyigit, I.H. Gecibesler, T. Dastan, H. Karci, P. Taslimi, S. Durna Dastan, I. Gulcin, A. Cetin, Synthesis of some novel pyridine compounds containing *bis*-1,2,4-triazole/thiosemicarbazide moiety and investigation of their antioxidant properties, carbonic anhydrase, and acetylcholinesterase enzymes inhibition profiles, J. Biochem. Mol. Toxicol. 32 (1) (2018) 22006.
[101] R. Shaykhiev, R. Bals, Interactions between epithelial cells and leukocytes in immunity and tissue homeostasis, J. Leukocyte Biol. 82 (1) (2007) 1–15.
[102] L.A. Abdulkhaleq, M.A. Assi, R. Abdullah, M. Zamri-Saad, Y.H. Taufiq-Yap, M.N.M. Hezmee, The crucial roles of inflammatory mediators in inflammation: a review, Veterinary World 11 (5) (2018) 627.
[103] M.A. Ingersoll, A.M. Platt, S. Potteaux, G.J. Randolph, Monocyte trafficking in acute and chronic inflammation, Trends Immunol. 32 (10) (2011) 470–477.

[104] M.A. Abdelgawad, R.B. Bakr, A.A. Azouz, Novel pyrimidine-pyridine hybrids: synthesis, cyclooxygenase inhibition, anti-inflammatory activity and ulcerogenic liability, Bioorg. Chem. 77 (2018) 339–348.
[105] C. Nathan, A. Ding, Nonresolving inflammation, Cell 140 (6) (2010) 871–882.
[106] L.W. Mohamed, M.A. Shaaban, A.F. Zaher, S.M. Alhamaky, A.M. Elsahar, Synthesis of new pyrazoles and pyrozolo [3,4-*b*] pyridines as anti-inflammatory agents by inhibition of COX-2 enzyme, Bioorg. Chem. 83 (2019) 47–54.
[107] V. Kamat, R. Santosh, B. Poojary, S.P. Nayak, B.K. Kumar, M. Sankaranarayanan, Faheem, S. Khanapure, D.A. Barretto, S.K Vootla, Pyridine-and thiazole-based hydrazides with promising anti-inflammatory and antimicrobial activities along with their *in silico* studies, ACS Omega 5 (39) (2020) 25228–25239.
[108] A. Jain, S. Jain, S. Rawat, Emerging fungal infections among children: A review on its clinical manifestations, diagnosis, and prevention, J. Pharmacy Bioallied Sci. 2 (4) (2010) 314.
[109] J. Guitard, M.D. Tabone, Y. Senghor, C. Cros, D. Moissenet, K. Markowicz, N. Valin, G. Leverger, C. Hennequin, Detection of β-D-glucan for the diagnosis of invasive fungal infection in children with hematological malignancy, J. Infect. 73 (6) (2016) 607–615.
[110] D. Zhao, S. Zhao, L. Zhao, X. Zhang, P. Wei, C. Liu, C. Hao, B. Sun, X. Su, M. Cheng, Discovery of biphenyl imidazole derivatives as potent antifungal agents: design, synthesis, and structure-activity relationship studies, Bioorg. Med. Chem. 25 (2) (2017) 750–758.
[111] Y. Dong, M. Liu, J. Wang, Z. Ding, B. Sun, Construction of antifungal dual-target (SE, CYP51) pharmacophore models and the discovery of novel antifungal inhibitors, RSC Adv. 9 (45) (2019) 26302–26314.
[112] L. Xiong, Y.Q. Shen, L.N. Jiang, X.L. Zhu, W.C. Yang, W. Huang, G.F. Yang, Succinate dehydrogenase: an ideal target for fungicide discovery, Discovery and Synthesis of Crop Protection Products, American Chemical Society, Washington, United States, 2015, pp. 175–194.
[113] B. Sun, Y. Dong, K. Lei, J. Wang, L. Zhao, M. Liu, Design, synthesis and biological evaluation of amide-pyridine derivatives as novel dual-target (SE, CYP51) antifungal inhibitors, Bioorg. Med. Chem. 27 (12) (2019) 2427–2437.
[114] X. Hua, W. Liu, Y. Su, X. Liu, J. Liu, N. Liu, G. Wang, X. Jiao, X. Fan, C. Xue, Y. Liu, Studies on the novel pyridine sulfide containing SDH based heterocyclic amide fungicide, Pest Manage. Sci. 76 (7) (2020) 2368–2378.
[115] R. Jia, Y. Duan, Q. Fang, X. Wang, J. Huang, Pyridine-grafted chitosan derivative as an antifungal agent, Food Chem. 196 (2016) 381–387.
[116] R. Diel, R. Loddenkemper, K. Meywald-Walter, S. Niemann, A. Nienhaus, Predictive value of a whole blood IFN-γ assay for the development of active tuberculosis disease after recent infection with *Mycobacterium tuberculosis*, Am. J. Respir. Crit. Care Med. 177 (10) (2008) 1164–1170.
[117] J.C. Brust, M. Lygizos, K. Chaiyachati, M. Scott, T.L. van der Merwe, A.P. Moll, X. Li, M. Loveday, S.A. Bamber, U.G. Lalloo, G.H. Friedland, Culture conversion among HIV co-infected multidrug-resistant tuberculosis patients in Tugela Ferry, South Africa, PLoS One 6 (1) (2011) e15841.
[118] V. Velezheva, P. Brennan, P. Ivanov, A. Kornienko, S. Lyubimov, K. Kazarian, B. Nikonenko, K. Majorov, A. Apt, Synthesis and antituberculosis activity of indole–pyridine derived hydrazides, hydrazide–hydrazones, and thiosemicarbazones, Bioorg. Med. Chem. Lett. 26 (3) (2016) 978–985.
[119] D.J. Hazuda, HIV integrase as a target for antiretroviral therapy, Current Opinion HIV AIDS 7 (5) (2012) 383–389.
[120] E. Thierry, E. Deprez, O. Delelis, Different pathways leading to integrase inhibitors resistance, Frontiers in Microbiology 7 (2017) 2165.
[121] M. Oliveira, R.I. Ibanescu, K. Anstett, T. Mésplède, J.P. Routy, M.A. Robbins, B.G. Brenner, Selective resistance profiles emerging in patient-derived clinical isolates with cabotegravir, bictegravir, dolutegravir, and elvitegravir, Retrovirology 15 (1) (2018) 1–14.
[122] K. Maeda, D. Das, T. Kobayakawa, H. Tamamura, H. Takeuchi, Discovery and development of anti-HIV therapeutic agents: progress towards improved HIV medication, Curr. Top. Med. Chem. 19 (18) (2019) 1621–1649.
[123] P.C. Koneru, A.C. Francis, N. Deng, S.V. Rebensburg, A.C. Hoyte, J. Lindenberger, D. Adu-Ampratwum, R.C. Larue, M.F. Wempe, A.N. Engelman, D. Lyumkis, HIV-1 integrase tetramers are the antiviral target of pyridine-based allosteric integrase inhibitors, Elife 8 (2019) 46344.
[124] K.W. Tsang, P.L. Ho, G.C. Ooi, W.K. Yee, T. Wang, M. Chan-Yeung, W.K. Lam, W.H. Seto, L.Y. Yam, T.M. Cheung, P.C. Wong, A cluster of cases of severe acute respiratory syndrome in Hong Kong, N. Engl. J. Med. 348 (20) (2003) 1977–1985.

[125] A. Ghaleb, A. Aouidate, M. Bouachrine, T. Lakhlifi, A. Sbai, *In silico* exploration of aryl halides analogues as checkpoint kinase 1 inhibitors by using 3D QSAR, molecular docking study, and ADMET screening, Adv. Pharmaceutical Bull. 9 (1) (2019) 84.

[126] R.M. Vivanco-Hidalgo, I. Molina, E. Martinez, R. Roman-Viñas, A. Sánchez-Montalvá, J. Fibla, C. Pontes, C.V. Muñoz, Real-World Data Working Group, Incidence of COVID-19 in patients exposed to chloroquine and hydroxychloroquine: results from a population-based prospective cohort in Catalonia, Spain, 2020, Eurosurveillance 26 (9) (2021) 2001202.

[127] A. Ghaleb, A. Aouidate, H.B.E. Ayouchia, M. Aarjane, H. Anane, S.E. Stiriba, *In silico* molecular investigations of pyridine *N*-Oxide compounds as potential inhibitors of SARS-CoV-2: 3D QSAR, molecular docking modeling, and ADMET screening, J. Biomol. Struct. Dyn. 40 (1) (2020) 1–11.

[128] Z. Andreadakis, A. Kumar, R.G. Román, S. Tollefsen, M. Saville, S. Mayhew, The COVID-19 vaccine development landscape, Nature Rev. Drug Discovery 19 (5) (2020) 305–306.

[129] M. Fabiani, M. Ramigni, V. Gobbetto, A. Mateo-Urdiales, P. Pezzotti, C. Piovesan, Effectiveness of the Comirnaty (BNT162b2, BioNTech/Pfizer) vaccine in preventing SARS-CoV-2 infection among healthcare workers, Treviso province, Veneto region, Italy, 27 December 2020 to 24 March 2021, Eurosurveillance 26 (17) (2021) 2100420.

[130] N. Ghiasi, R. Valizadeh, M. Arabsorkhi, T.S. Hoseyni, K. Esfandiari, T. Sadighpour, H.R. Jahantigh, Efficacy and side effects of Sputnik V, Sinopharm and AstraZeneca vaccines to stop COVID-19; a review and discussion, Immunopathologia Persa 7 (2) (2021) e31.

[131] T. Topal, Y. Zorlu, N. Karapınar, Synthesis, X-ray crystal structure, IR and Raman spectroscopic analysis, quantum chemical computational and molecular docking studies on hydrazone-pyridine compound: as an insight into the inhibitor capacity of main protease of SARS-CoV2, J. Mol. Struct. 1239 (2021) 130514.

[132] M. Santhiya, T.L. Pushparaj, Synthesis of smart Mo (VI)-pyridine complex for targeting amino acids on M-protease of COVID-19 virus, Synthesis 4 (2) (2021) 113–120.

CHAPTER

9

Contemporary development in the synthesis and biological applications of pyridine-based heterocyclic motifs

Nisheeth C. Desai[a], *Jahnvi D. Monapara*[a], *Aratiba M. Jethawa*[a] *and Unnat Pandit*[b]

[a]Division of Medicinal Chemistry, Department of Chemistry, Maharaja Krishnakumarsinhji Bhavnagar University, Bhavnagar, India [b]Special Centre for Systems Medicine, Jawaharlal Nehru University, New Delhi, India

9.1 Introduction

Pyridine derives its name from a Greek word. Pyr stands for fire while idine stands for aromatic bases. Pyridine, with the molecular formula C_5H_5N, is a basic heterocyclic molecule that was first isolated from picoline by Anderson in 1846 [1].

Pyridine is used in a variety of industries, including medicinal chemistry, agrochemicals, organic chemistry, metal complexes, polymers, food flavorings, paints, dyes, insecticides, pesticides, adhesives, and rubber products, waterproofing fabrics, nitrogen-containing plant products, catalysts, as well as others [2–7]. The replacement of a pyridine moiety for a component of a drug can improve its water solubility, allowing for the development of practical drugs appropriate for orally administered or injectable formulations. This approach may also be applied to the prodrug strategy. Pyridines have a significant function in drug design through the bioisostere. Because of the aforementioned properties of pyridines, they can be used as a bioisostere of amines, amides, heterocyclic rings containing some nitrogen atoms, and the benzene ring. The use of pyridines to replace a portion of lead compounds is an important tool in the development of powerful drugs. Furthermore, replacing a component of a pharmacological molecule with pyridines may influence selectivity toward subtypes of a target biomolecule, such as mepyramine, a histamine H1 receptor antagonist.

Recent Developments in the Synthesis and Applications of Pyridines.
DOI: https://doi.org/10.1016/B978-0-323-91221-1.00007-5

Over 60 pyridine-containing drugs have been approved by the FDA, making pyridine a privileged scaffold in medicinal chemistry and aroused the research interest in this versatile molecule. Pennington and Moustakas published an excellent review on "The Necessary Nitrogen Atom" in 2017 that discussed the role of pyridine as a bioisostere for the phenyl fragment (dubbed as the N scan SAR strategy) [8]. When a phenyl ring is replaced with a pyridine, a nitrogen atom acts as a hydrogen bond acceptor and may establish contact with the target, generating a hydrogen bond as well as a minor shift in pi-stacking. As a result, binding affinity and biological potency may improve. Pyridine and its derivatives have been used as antimicrobial, antitubercular, anticancer, antiviral, antimalarial, antidiabetic, antioxidant, analgesic, anti-inflammatory, and anti-infection agents in pharmaceutical chemistry due to their water solubility, basicity, stability, hydrogen bond-forming ability, lower molecular weight, and availability in many natural products [9–17]. Pyridine can be found in a variety of natural products, including vitamins such as niacin (Vitamin B_3) and pyridoxin (Vitamin B_6), alkaloids such as anabasine and ricinine, and co-enzymes such as nicotinamide adenine dinucleotide and nicotinamide adenine dinucleotide phosphate.

As a solvent, base, ligand, functional group, and molecular substrates, pyridine scaffolds have established a purpose in approaching all aspects of organic and medical chemistry. Pyridine is a valuable pharmacophore in medicinal chemistry and an important functionality for organic chemists as a structural element. Pyridine is a six-membered ring that varies from benzene by having a nitrogen atom in place of the –CH– (*N* atom) (Fig. 9.1). Pyridine ring is stabilized by resonance (Fig. 9.2). Pyridine was the quintessential π-deficient heterocycle, emphasizing distinct chemistry as both a substrate and a reagent. In the context of medicinal chemistry, replacing a CH group on a phenyl ring with *N* atom may affect a drug's molecular and physicochemical properties, as well as intra- and inter-molecular interactions, which can lead to enhanced pharmacological profiles. Pyridine scaffolds can be found in a variety of applications, most notably in the medical field. Fig. 9.3 shows commercially available drugs with various therapeutic purposes that contain the pyridine structural unit. These will be helpful in the research and design of novel pyridine-containing lead compounds. The FDA authorized medicines containing pyridine units in 2020, and the bulk of them were anticancer therapies, thus they had a favorable pharmacological profile (Fig. 9.4).

Several pyridine hybrids, involving one or more heterocycles with diverse pharmacological profiles, were conceptualized and developed by scientists, and their synthesis and biological activity were documented.

9.2 Synthesis of pyridine

9.2.1 Hantzsch pyridine synthesis

Arthur Rudolf Hantzsch described the first synthetic process for pyridine in 1881 [18]. An aldehyde - **2**, active methylene compounds - **3**, and ammonia - **4** are refluxed for about an hour in this procedure. The resultant 1,4-dihydropyridine - **5** is then oxidized to produce pyridine derivatives - **6** (Scheme 9.1).

9.2.2 Baeyer pyridine synthesis

The reaction of pyrylium salts - **8** with ammonia or ammonia solution to form pyridine derivatives - **9** is known as the Baeyer pyridine Synthesis. The pyrylium salts - **8** were formed

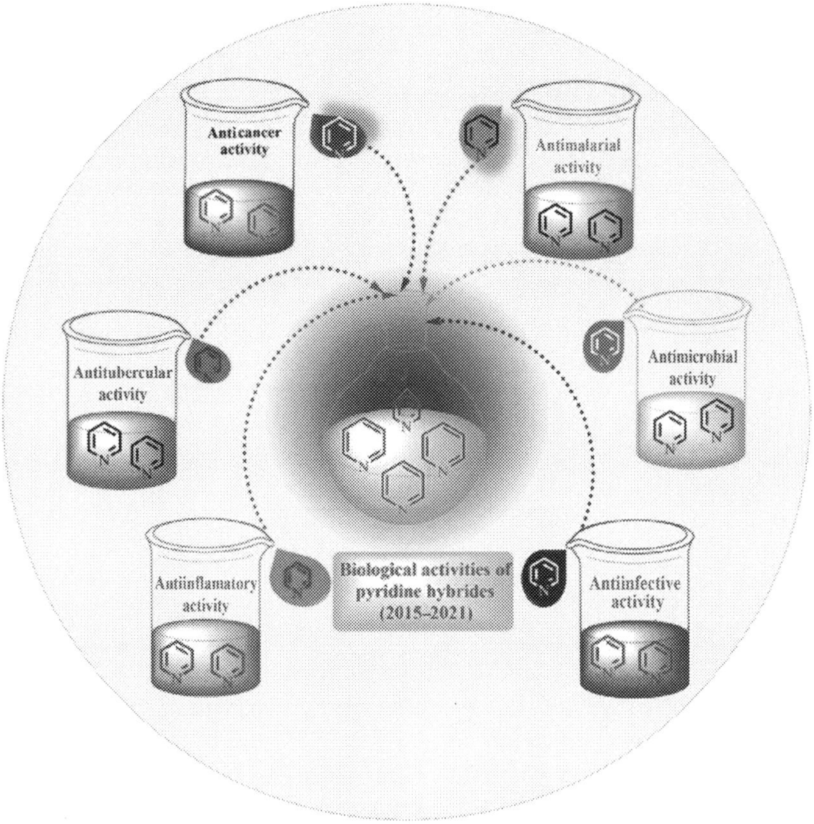

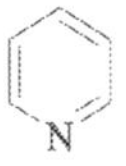

1

FIGURE 9.1 Structure of pyridine.

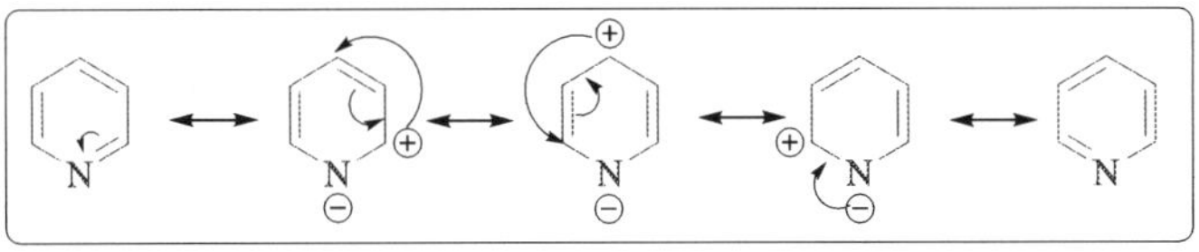

FIGURE 9.2 Resonance in pyridine.

by alkylation of pyran-4-one derivatives - **7** with dimethylsulfate followed by treatment of reaction mixture with perchloric acid [19] (Scheme 9.2).

9.2.3 Kröhnke pyridine synthesis

Kröhnke pyridine synthesis is the reaction of -pyridinium methyl ketone - **10** with, α, β-unsaturated carbonyl compounds - **11** utilizing the Michael reaction, which produces

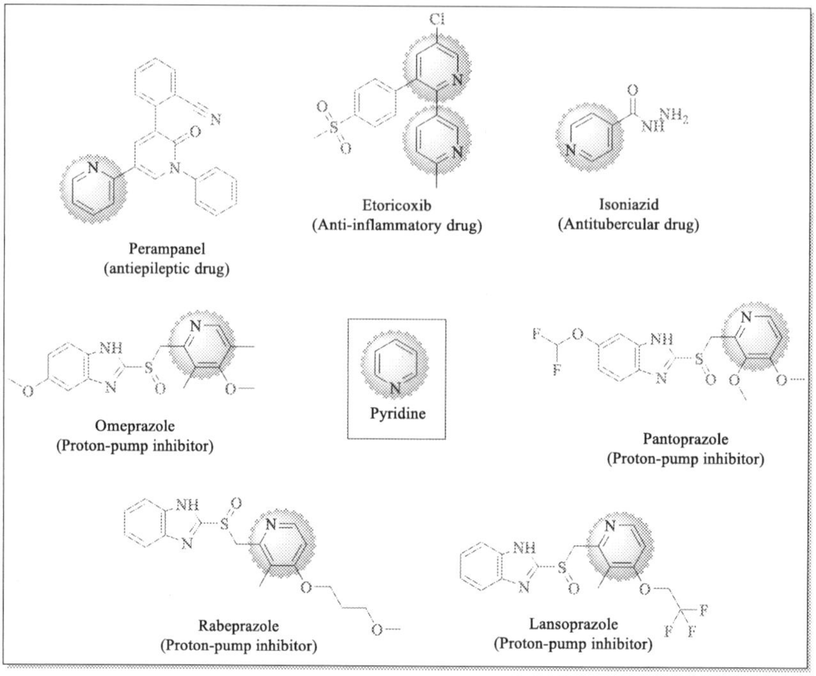

FIGURE 9.3 Pyridine and pyridine-containing medicines.

2,4,6-trisubstituted pyridines - **13** when treated with ammonium acetate. In mild conditions, excellent yields of the products were achieved [20] (Scheme 9.3).

9.2.4 Katrizky pyridine synthesis

The Katrizky pyridine Synthesis is similar to Kröhnke pyridine synthesis because it involves the Michael addition. The Michael addition of α-substituted ketones - **14** with α, β-unsaturated carbonyl compounds - **15** in the presence of ammonium acetate forms substituted pyridines - **16** is known as Katrizky pyridine Synthesis [19] (Scheme 9.4).

9.3 Green approach for the synthesis of pyridine

9.3.1 Ultrasound-assisted green synthesis

Rocío Gámez-Montaño et al. [21] reported the ultrasound-assisted green synthesis of bound type bis-heterocyclic carbazolyl imidazo[1,2-*a*]pyridines using ammonium chloride (10 mol%) as a catalyst in excellent yields (90%–96%) using Groebke-Blackburn-Bienayme reaction (GBBR). The synthetic pathway is described in Scheme 9.5. The incorporation of using 2-amino-pyridines - **18**, substituted carbazole aldehyde - **17**, and isocyanides - **19** in presence of NH_4Cl as a green catalyst and EtOH as a green solvent produced bound type bis-heterocycles

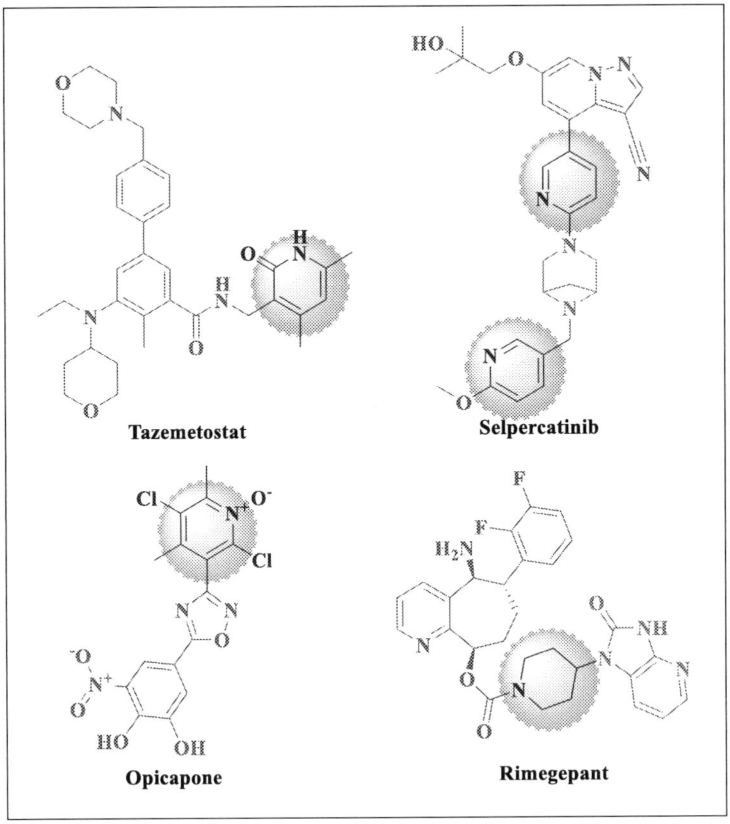

FIGURE 9.4 Pyridine-containing medicines approved in 2020.

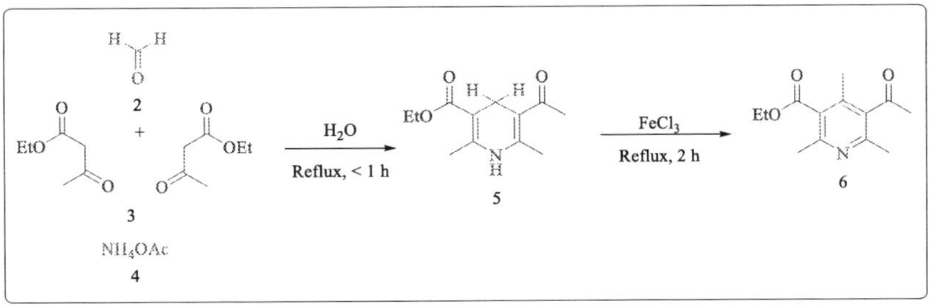

SCHEME 9.1 Hantzsch pyridine synthesis.

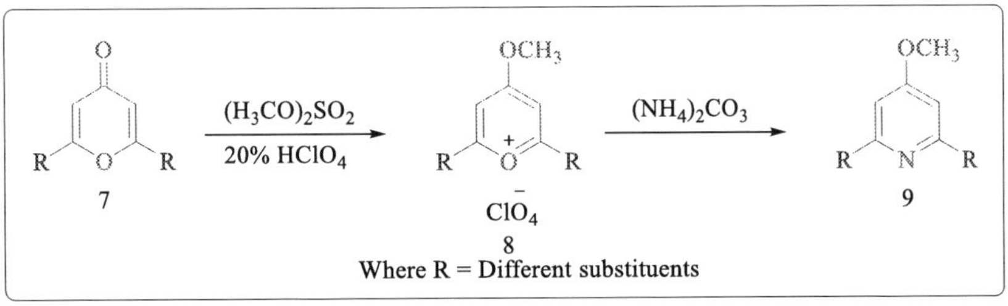

SCHEME 9.2 Baeyer pyridine synthesis.

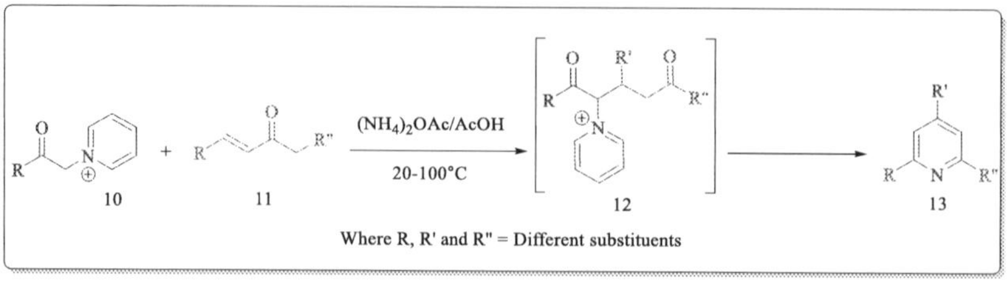

SCHEME 9.3 Kröhnke pyridine synthesis.

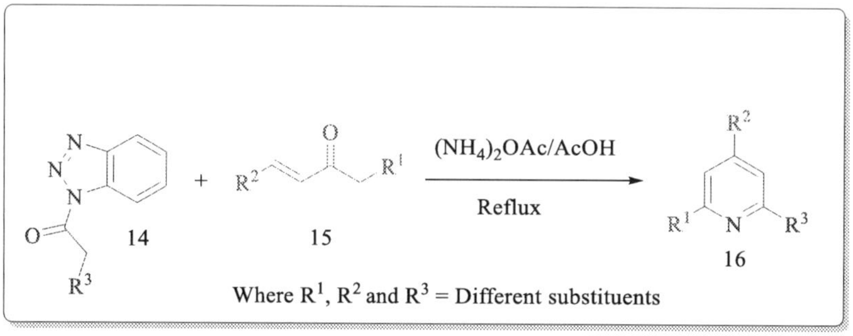

SCHEME 9.4 Katrizky pyridine synthesis.

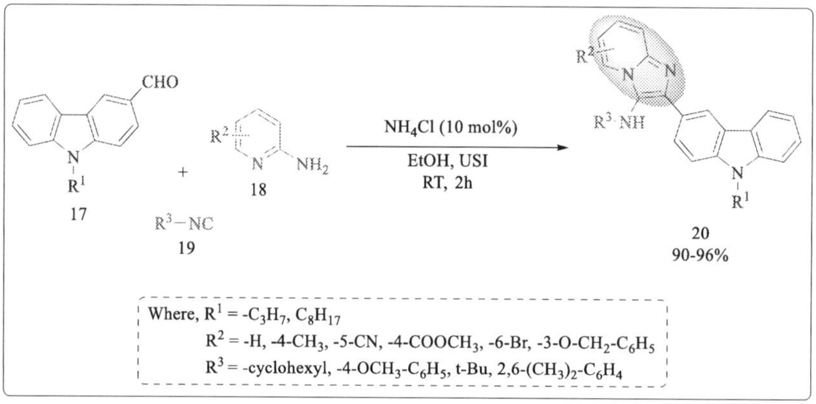

SCHEME 9.5 Synthesis of 3-amino fused imidazo[1,2-*a*]pyridines bis-heterocycles - 20 *via* green GBBR.

containing aromatic polyheterocycles at the C-3 position of imidazole - **20** using ultrasound-assisted Groebke-Blackburn-Bienayme reaction. This reported method has some advantages like green solvent and catalyst, short reaction time, improved yields, eco-friendliness, and operation simplicity. It also includes the formation of *bis*-heterocycles in a one-pot reaction.

Fourteen novel dihydropyrazolo[3,4-*b*]pyridine scaffolds were prepared by Kerru et al. [22] using gadolinium oxide (Gd_2O_3/ZrO_2) loaded zirconia under green conditions and evaluated for anticancer activity. An equimolar mixture of 5-methyl-1*H*-pyrazol-3-amine - **21**, benzyl acetoacetate - **22**, substituted aldehydes - **23,** and catalyst (Gd_2O_3/ZrO_2) was stirred

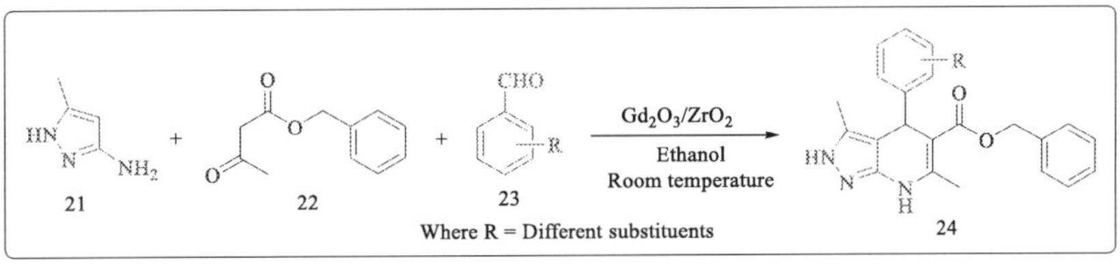

SCHEME 9.6 Multicomponent synthetic route for dihydropyrazolo [3,4-*b*]pyridines - 24.

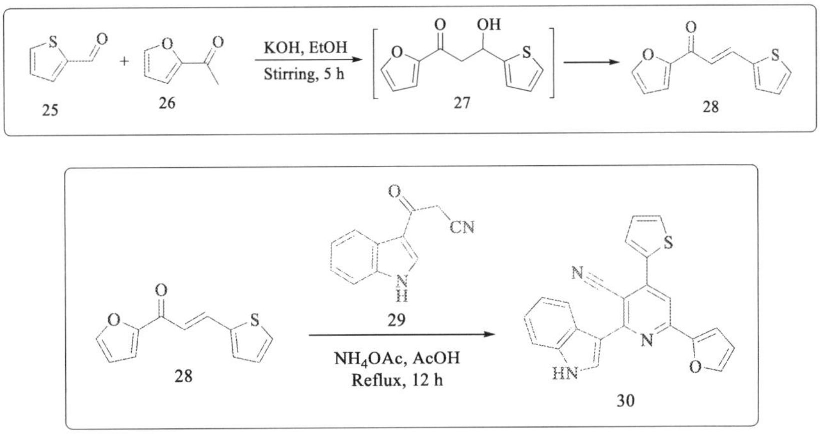

SCHEME 9.7 Synthetic route for 6-(furan-2-yl)-2-(*1H*-indol-3-yl)-4-(thiophen-2-yl) nicotinonitrile - 30.

at room temperature for 25–30 min to form benzyl 3,6-dimethyl-4-phenyl-4,7-dihydro-2*H*-pyrazolo[3,4-*b*]pyridine-5-carboxylates - **24** using absolute ethanol as a solvent. This synthesis incorporated green protocols like rapid synthesis, recyclability, excellent yields, green solvent, simple workup, and no use of column chromatography (Scheme 9.6).

9.3.2 Microwave-assisted synthesis of pyridine

Radwan et al. [23] reported microwave-assisted synthesis of some new heterocyclic ring systems incorporated pyridine moiety starting from 1-(furan-2-yl)-3-(thiophen-2-yl) chalcone. The reaction of 3-cyanoacetylindole – **29,** and 1-(furan-2-yl)-3-(thiophen-2-yl) chalcone - **28** in the presence of ammonium acetate produces 6-(furan-2-yl)-2-(*1H*-indol-3-yl)-4-(thiophen-2-yl)nicotinonitrile - **30** (Scheme 9.7).

9.4 Synthesis and antimicrobial activity of some pyridine derivatives

Desai et al. [24] synthesized 4-thiazolidinone derivatives based on pyridine and quinazoline in 2021. Compound - **33** is synthesized by refluxing 3-amino-2-phenylquinazolin-4(3*H*)-one - **32** and 4-pyridine carboxaldehyde - **31** at 80°C for 3 h with acetic acid as a catalyst and

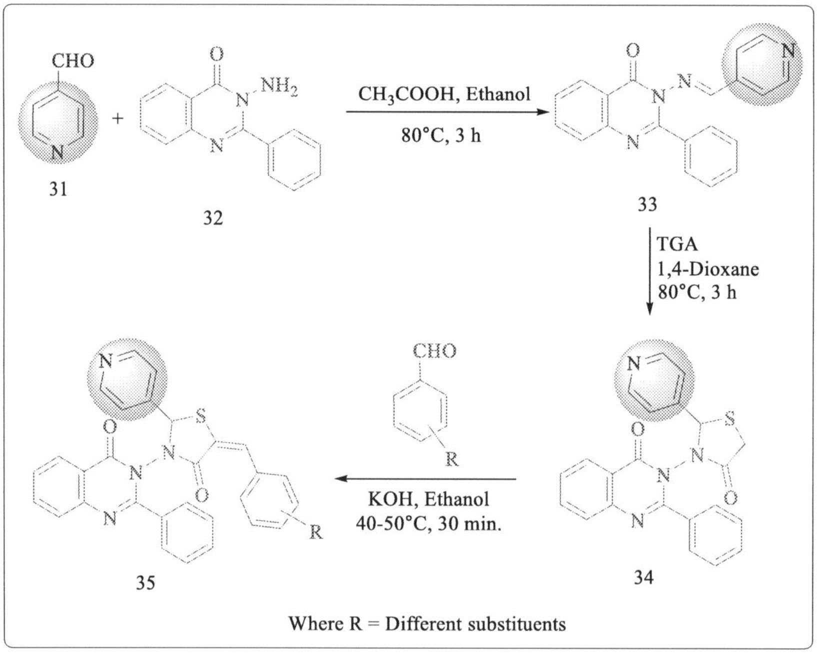

SCHEME 9.8 Synthetic route for pyridine and quinazoline-based 4-thiazolidinone derivatives -35.

ethanol as a solvent at 80°C. Compound 3-(4-oxo-2-phenylquinazolin-3(4*H*)-yl)-2-(pyridin-4-yl)thiazolidin-4-one), **34** is obtained by treating compound - **33** with $ZnCl_2$ and thioglycolic acid at 110°C for 6 h. Compound - **34** was then treated with various benzaldehydes to produce 5-arylidene-3-(4-oxo-2-phenylquinazolin-3(4*H*)-yl)-2-(pyridin-4-yl)thiazolidin-4-ones - **35**, which are pyridine and quinazoline based 4-thiazolidinone derivatives (Scheme 9.8).

The antibacterial activity of the synthesized compounds was evaluated using MIC values against various bacterial and fungus species. Compounds **36–38** had antibacterial activity against diverse bacterial strains in the range of 12.5 to 62.5 μg/mL. Compounds **39–41** had antifungal efficacy against diverse fungal strains with MICs of 100 μg/mL. Antibacterial activity was superior in compounds with electron-withdrawing groups, while antifungal activity was higher in compounds with electron-donating groups (Fig. 9.5). Positive controls included gentamicin, chloramphenicol, ciprofloxacin, and norfloxacin against diverse bacterial strains were used. Positive controls against fungal strains included nystatin and griseofulvin were used. With high binding affinities, the most active compounds were fit into the binding site of DNA gyrase.

Kamat et al. [25] prepared pyridine and thiazole-based hydrazides with promising antimicrobial activities. Compound - **43**, pyridine-3-carbothiamide was obtained from, 3-cyanopyridine - **42** by heating it with P_4S_{10}. The reaction of compound - **43** with ethyl-2-chloroacetoacetate at 70°C using ethanol as solvent provide compound **- 44**, ethyl-5methyl-2-(pyridine-3-yl)thiazole-4-carboxylate. Compound - **45**, 5-methyl-2-(pyridine-3-yl)thiazole-4-carbohydrazide was obtained from the compound - **44** by refluxing it with hydrazine hydrate

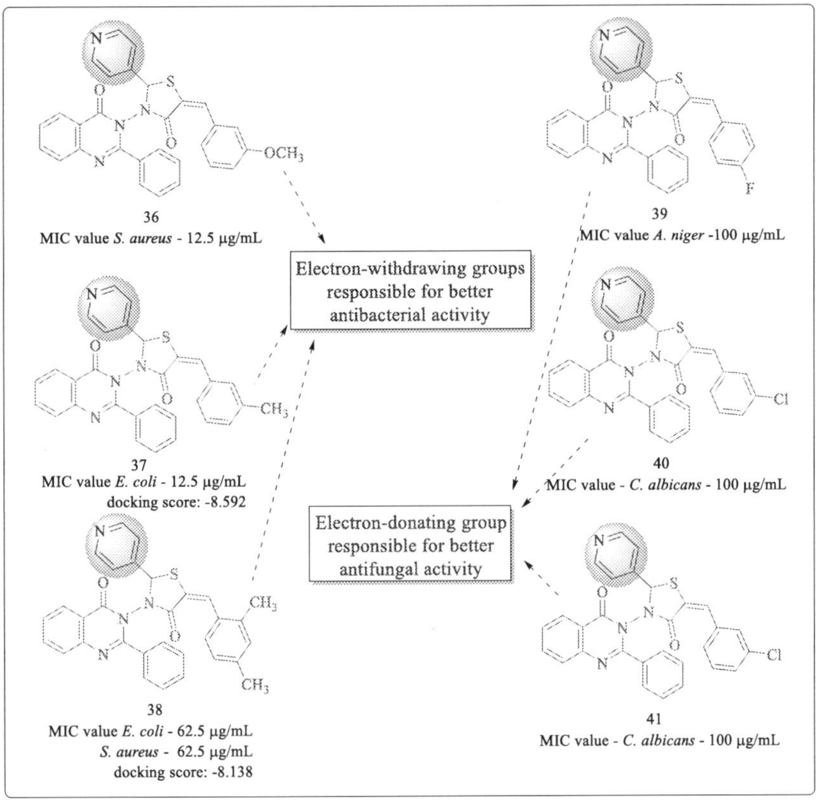

FIGURE 9.5 Most active molecules from pyridine and quinazoline-based 4-thiazolidinone derivatives.

at 70°C using ethanol as solvent. Pyridine and thiazole-based hydrazides - **46** were obtained by condensing compound - **45** with different aldehydes at 70°C using ethanol as solvent (Scheme 9.9).

The synthesized compounds were screened for antimicrobial activity using tetracyclin and streptomycin as antibacterial medicines and fluconazole and nystatin as antifungal medicines. The results of antimicrobial screening were promising. Compounds **47–51** possessed promising antimicrobial activities while Compound **49** having bromo substitution at the fifth position and hydroxyl substitution at the second position showed comparable MIC value with standard drugs against different bacterial and fungal strains. The variation of the position of the substituents and the introduction of the heterocyclic ring were responsible for the activity of the synthesized compounds (Fig. 9.6).

Abdel Hamid et al. [26] prepared 1,2,3-triazole and pyridine-based hybrid compounds as potential antimicrobial agents. The compound - **53**, 1-(4-azidophenyl)ethenone was prepared by diazotization of *p*-aminoacetophenone - **52** followed by treatment of sodium azide. Further catalytic [3+2] cycloaddition in presence of K_2CO_3 afford compound - **54**, 1-(4-(4-acetyl-5-methyl-1*H*-1,2,3-triazol-1-yl)phenyl)ethan-1-one. The reaction of the compound - **54**

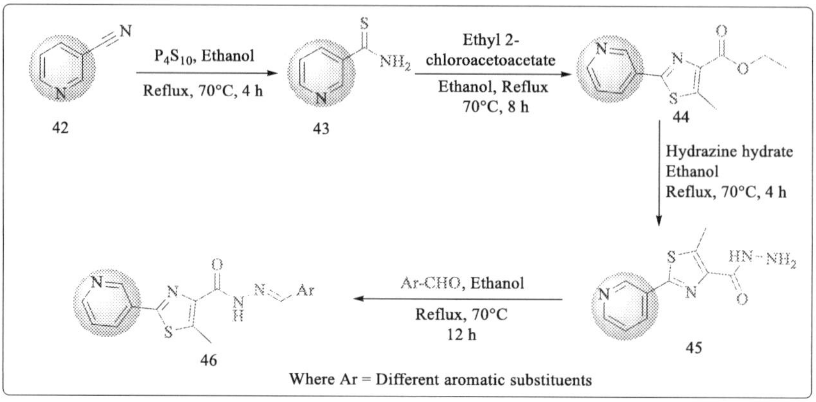

SCHEME 9.9 Synthetic route for pyridine and thiazole-based hydrazides 45.

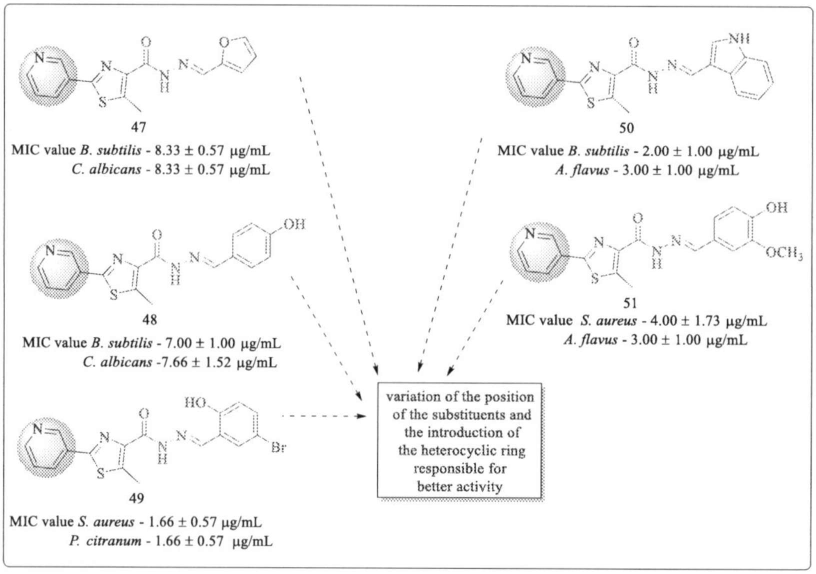

FIGURE 9.6 Most active molecules from pyridine and thiazole-based hydrazides.

with different aldehydes and ethyl cyanoacetate provides 1,2,3-triazole and pyridine-based hybrids - **55** (Scheme 9.10).

The synthesized compounds were screened for their *in vitro* antibacterial activity against Gram-positive bacteria like *Staphylococcus aureus* ATCC 6538 and *Micrococci luteus* ATCC 10240, and Gram-negative bacteria *Pseudomonas aeruginosa* ATCC 9027 and *Escherichia coli* ATCC 10536 using cefoxamide as a positive control for bacteria. Nystatin was used as a positive control for the antifungal screening of synthesized compounds against *Candida albicans* ATCC 10231 and *Aspergillus niger* ATCC 16404. The antimicrobial screening was carried out

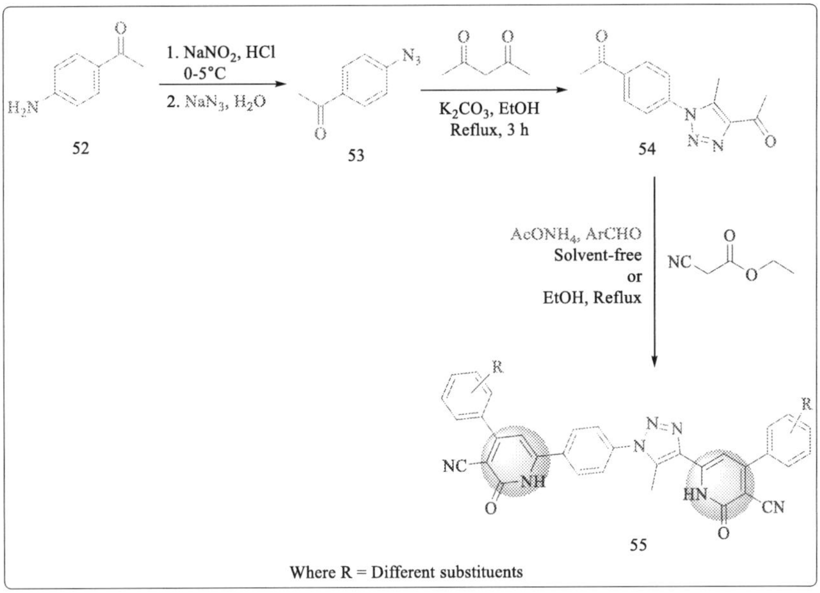

SCHEME 9.10 Synthetic route for 1,2,3-triazole and pyridine-based hybrid compounds 55.

using the agar-diffusion method. The synthesized Compounds **56, 57,** and **58** possessed better antimicrobial activities than the standard medicines (Fig. 9.7).

Al-Tel and Al-Qawasmeh [27] prepared imidazo[1,2-*a*]pyridine derivatives as potential antimicrobial agents. Compound - **60** 5-carboxy-2-aminopyridine was condensed with p-substituted benzaldehydes - **61** using $Sc(OTf)_3$ as the catalyst at room temperature for 45 min. The further reaction with phenyl isocyanide - **59** produced imidazopyridines **- 62**. Compound - **62** then treated with substituted phenylene diamines, TBTU (O-(benzotriazol-1-yl)-*N,N,N′,N′*-tetramethyluronium tetrafluoroborate) and DIPEA (*N,N*-diisopropylethylamine) in DMF at 0°C. The product formed was then treated with acetic acid for 3 h to get compound - **63**. Compound - **62** treated with substituted α-bromoacetophenones, DIPEA in DMF at 0°C. The product formed then condensed using AcO_2NH_4 and acetic acid for 4 h to get compound - **64**. Compound - **62** treated with benzohydrazides, TBTU, and DIPEA in DMF at 0°C. The product formed then treated with $POCl_3$ to get compound - **65** (Scheme 9.11).

The synthesized compounds were screened for their antimicrobial activities against various Gram-positive bacteria like *S. aureus, E. faecalis,* and *B. megaterium* and Gram-negative bacteria like *E. coli, P. aeruginosa,* and *E. aerogenes* using cefixime, amoxicillin, and vancomycin as positive controls. Compounds containing electron-withdrawing groups like bromo, chloro, fluoro, and nitro were most prominent. Compounds **66–69** possessed excellent activity against *S. aureus* with MIC of 0.51–10.32 μg/mL (Fig. 9.8).

Desai et al. [28] developed hybrid oxazine clubbed pyridine scaffolds as potential antimicrobial agents. Compound **71**, 1-((1-(4-(2*H*-benzo[*e*][1,3]oxazin-3(4*H*)-yl)phenyl) ethylidene)amino)-6-amino-4-(4-chlorophenyl)-2-oxo-1,2-dihydropyridine-3,5-dicarbonitrile was obtained by refluxing compound - **70**, *N′*-(1-(4-(2*H*-benzo[*e*][1,3]oxazin-3(4*H*)-yl)phenyl)ethylidene)-2-cyanoacetohydrazide with 2-(4-chlorobenzylidene)malononitrile

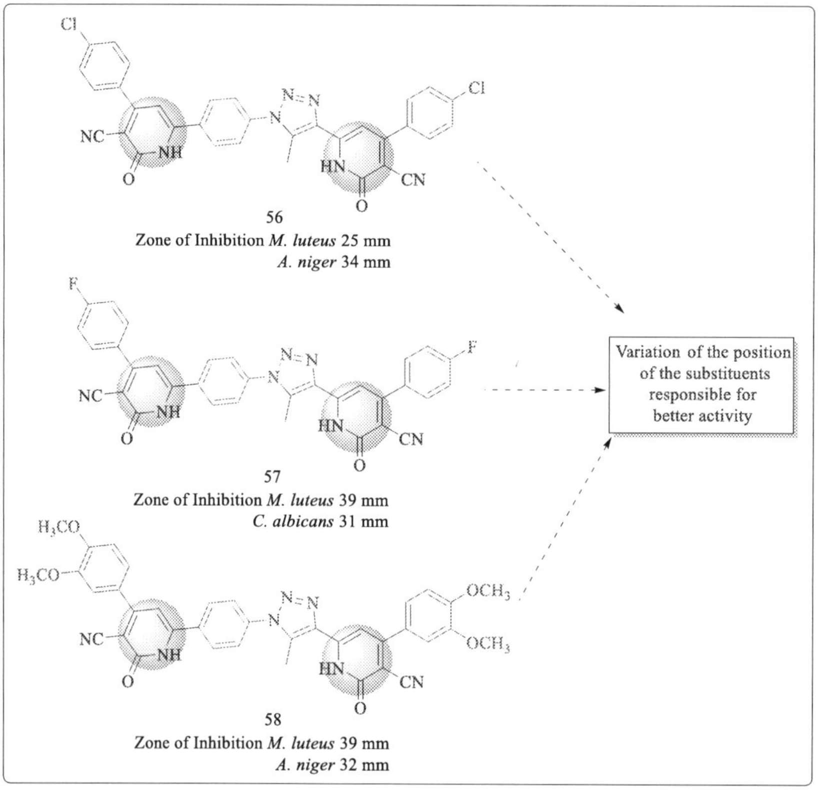

FIGURE 9.7 Most active molecules from 1,2,3-triazole and pyridine-based hybrid compounds.

using piperidine as catalyst and ethanol as solvent. Compound - **71** was refluxed with substituted benzaldehydes for 8–10 h using ethanol as solvent to furnish compound - **72** (Scheme 9.12).

The synthesized derivatives were screened for *in vitro* antimicrobial activities against various bacterial and fungal strains. The results of activity indicated that the synthesized derivatives **73–80** containing different electron-donating and withdrawing functional groups were prominent. These compounds possessed antibacterial activity and antifungal activity with a range of MIC of 25–100 μg/mL (Fig. 9.9).

9.5 Synthesis and antitubercular activity of some pyridine derivatives

In 2019, Bodige et al. [29] prepared pyrrolo[3,2-*b*]pyridine-3-carboxamide linked 2-methoxypyridine derivatives. Compound - **82**, 2-cyano-2-(4-methyl-3-nitropyridin-2-yl) acetate was prepared by refluxing compound - **81**, 2-chloro-4-methyl-3-nitropyridine with ethyl-2-cyanoacetate in the presence of *t*-BuOK for 6 h using isopropanol as solvent. The reduction of nitro group of compound - **82** by Pd/C in EtOH, followed by intramolecular

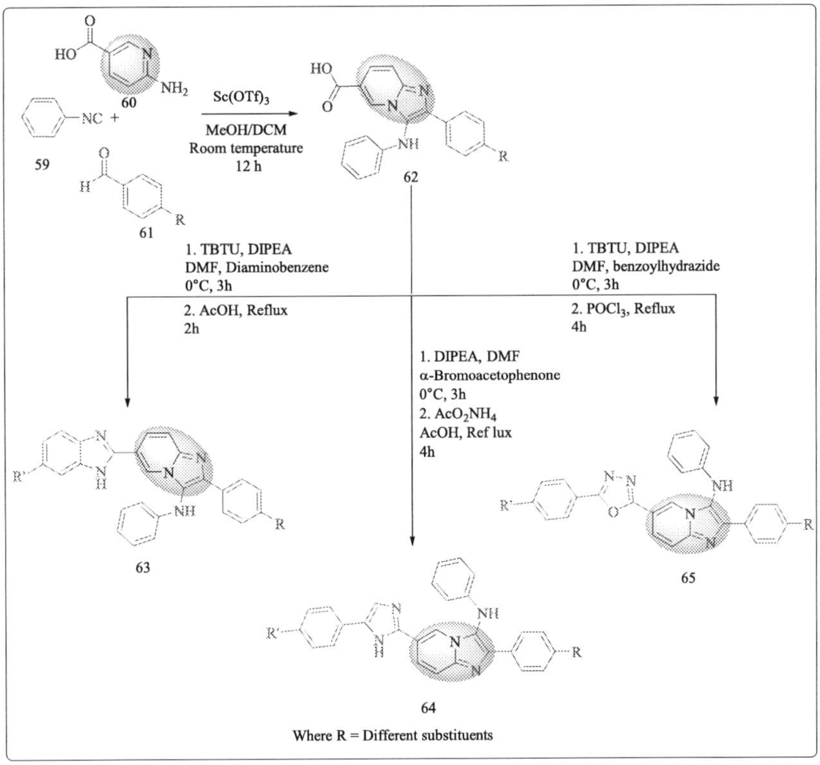

SCHEME 9.11 Synthetic route for imidazo[1,2-*a*]pyridine derivatives - 64.

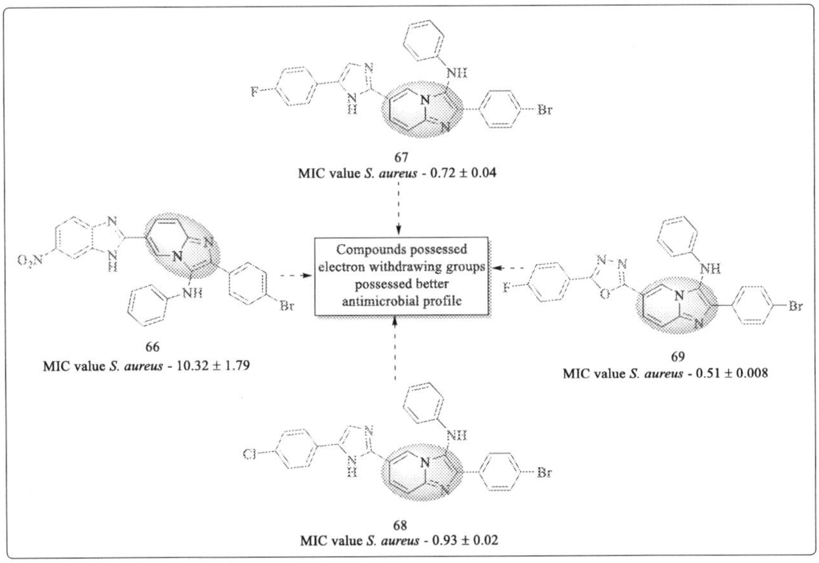

FIGURE 9.8 Most active molecules from imidazo[1,2-*a*]pyridine derivatives.

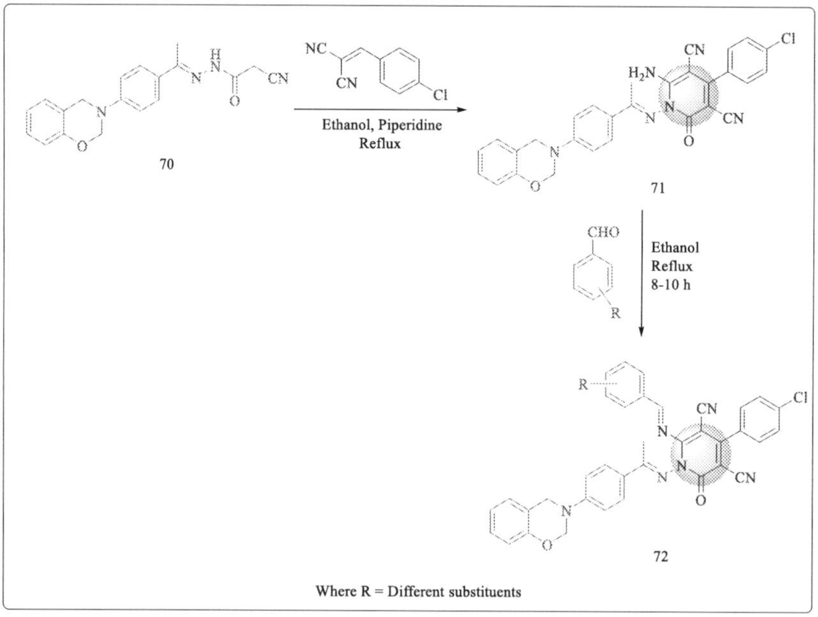

SCHEME 9.12 Synthetic route for hybrid oxazine clubbed pyridine scaffolds - 72.

cyclization produced 4-azaindole ester derivative, **83**, ethyl 7-methyl-1*H*-pyrrolo[3,2-*b*]pyridine-3-carboxylate. Compound - **83** was further reacted with 4-(bromomethyl)-2-methoxypyridine in the presence of K_2CO_3 to get compound - **84**, ethyl 1-((2-methoxypyridin-4-yl)methyl)-7-methyl-1*H*-pyrrolo[3,2-*b*]pyridine-3-carboxylate. Compound - **84** was hydrolyzed to an acid - **85**, 1-((2-methoxypyridin-4-yl)methyl)-7-methyl-1*H*-pyrrolo[3,2-*b*]pyridine-3-carboxylic acid using $LiOH.H_2O$. The pyrrolo[3,2-*b*]pyridine-3-carboxamide linked 2-methoxypyridine derivatives - **86** were prepared from compound - **85** by reacting with different amines in HATU/DIPEA condition (Scheme 9.13).

The synthesized derivatives were tested for antitubercular activity against *Mycobacterium tuberculosis* using microplate alamar blue assay method (MABA method). The antitubercular activity was determined in terms of MIC values against *M. tuberculosis*. Pyrazinamide was used as a positive control against *M. tuberculosis*. Compounds **87, 88**, and **89** possessed excellent antitubercular activity against *M. tuberculosis* with MIC of 3.12 µg/mL. Compound with aryl substitution on the fourth position possessed excellent activity. Compounds **91** and **92** bearing di and trifluoro ethyl side chain possessed better antitubercular activity (MIC = 6.25 µg/mL) than the Compound **90** (MIC = 12.5 µg/mL), containing ethyl side chain (Fig. 9.10). The tested compounds exhibited essential docking interactions against DprE1. The tested compounds possessed promising *in silico* predictions of toxicities, drug likeness, and drug score profiles.

Ambhore et al. [30] prepared pyridine based 1,3,4-oxadiazole embedded hydrazinecarbothioamide derivatives as potent antitubercular agents. Compound - **94**, 5-(pyridine-4-yl)-1,3,4-oxadiazole-2(3*H*)-thione was prepared by reacting isoniazid, **93** with carbon disulfide

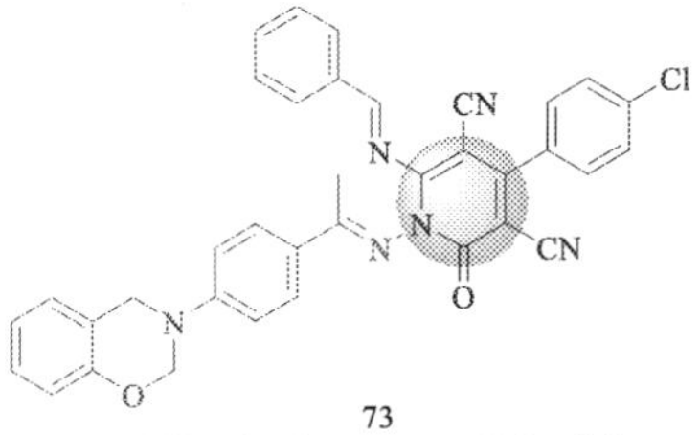

73
MIC value *S. pyogenes* - 62.5 μg/mL

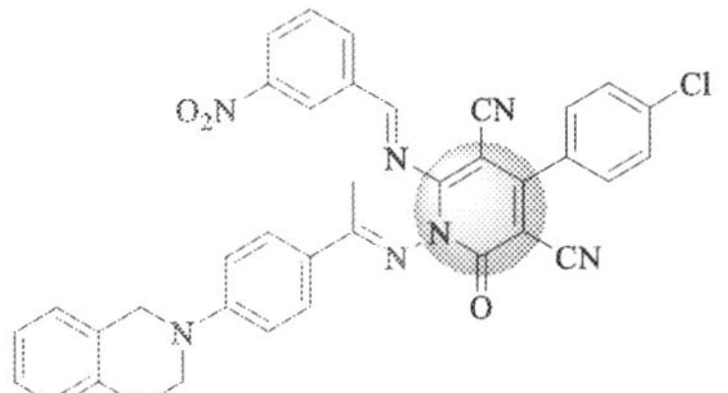

77
MIC value *P. aeruginosa* - 100 μg/mL

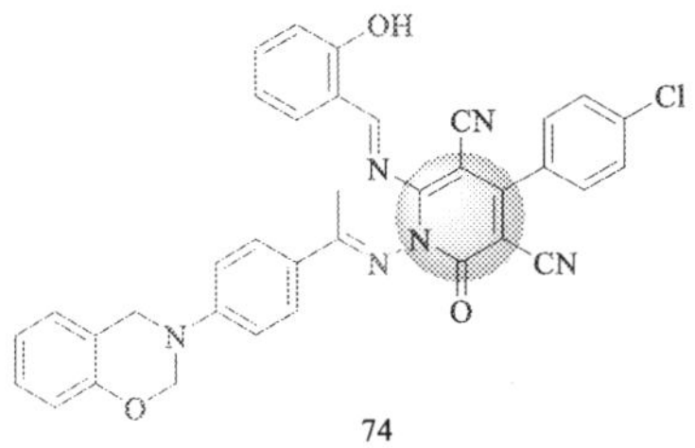

74
MIC value *S. aureus* - 50 μg/mL
P. aeruginosa - 25 μg/mL

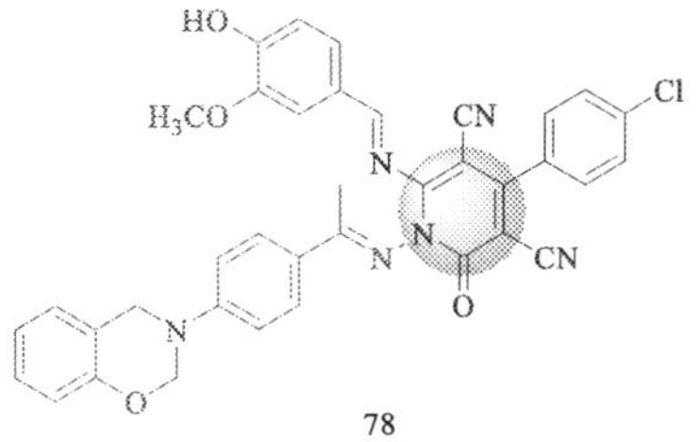

78
MIC value *E. coli* - 25 μg/mL
A. clavatus - 25 μg/mL

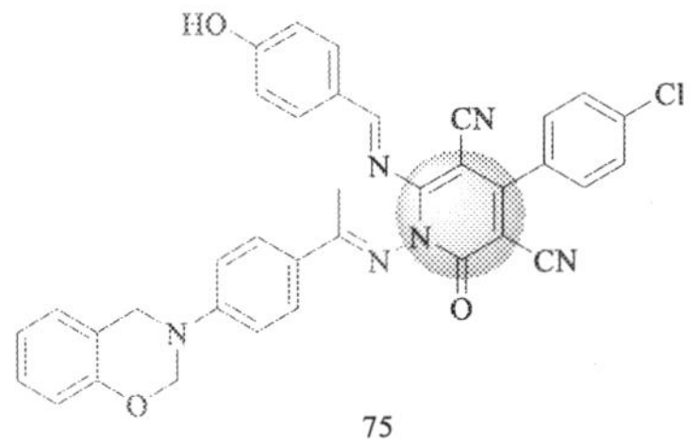

75
MIC value *E. coli* - 100 μg/mL

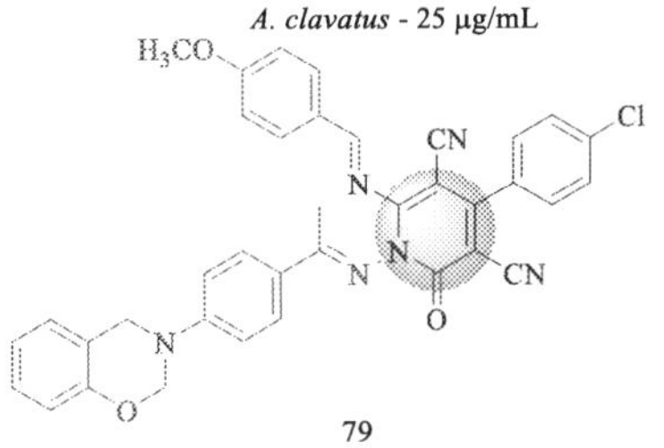

79
MIC value *C. albicans* - 50 μg/mL
A. niger - 100 μg/mL
A. clavatus - 100 μg/mL

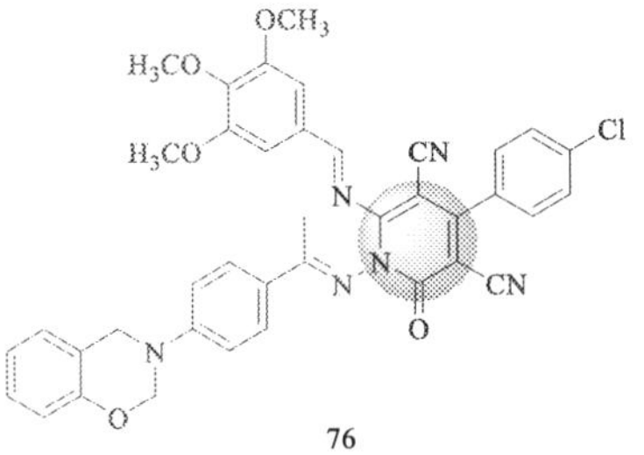

76
MIC value *A. niger* - 62.5 μg/mL
A. clavatus - 100 μg/mL

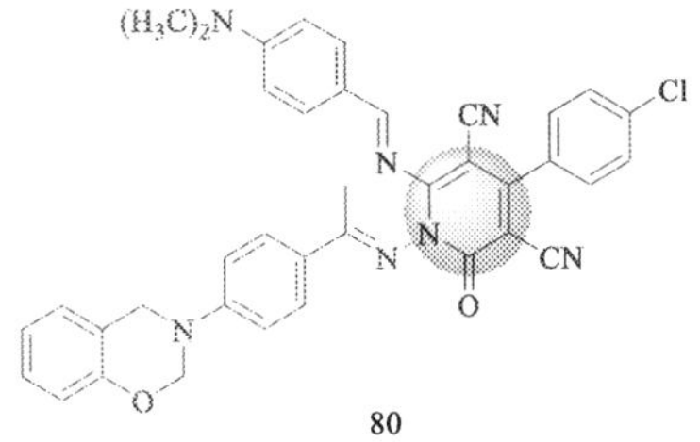

80
MIC value *C. albicans* - 100 μg/mL

FIGURE 9.9 Most active molecules from hybrid oxazine clubbed pyridine scaffolds.

and alcoholic KOH solution. 1-(substituted phenyl)-2((5-(pyridin-4-yl))-1,3,4-oxadiazole-2-yl) thio)ethanones, **95** were formed by reacting compound - **94** with different phenacyl bromides and BEC (S-(2-boronoethyl)-L-cysteine) (10 wt %) in PEG-400 (polyethylene glycol 400) at 70–80°C. Equimolar quantity of 1-(substituted phenyl)-2-((5-(pyridin-4-yl)-1,3,4-oxadiazol-2-yl)thio)ethanones, **95** and thiosemicarbazide were dissolved in PEG-400 and stirred at 70–80°C

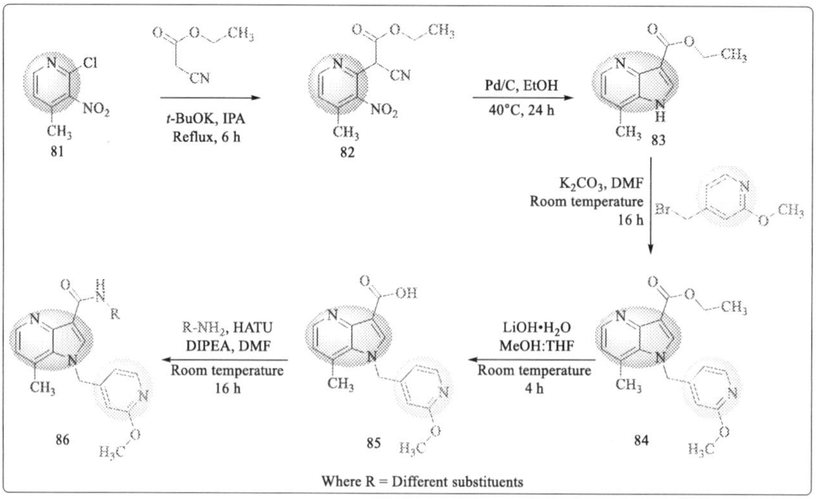

SCHEME 9.13 Synthetic route for pyrrolo[3,2-*b*]pyridine-3-carboxamide linked 2-methoxypyridine derivatives - 86.

for 2–3 h using acetic acid as catalyst provides Compound **96**, 2-(1-(substitutedphenyl)-2-((5-(pyridin-4-yl)-1,3,4-oxadiazol-2-yl)thio)ethylidene) hydrazinecarbothioamides (Scheme 9.14).

The 2-(1-(substitutedphenyl)-2-((5-(pyridin-4-yl)-1,3,4-oxadiazol-2-yl)thio)ethylidene) hydrazinecarbothioamide derivatives were screened for antimycobacterial activity against multidrug resistance *M. tuberculosis* strain (MTCC 300) using Resazurin microtiter assay. The synthesized derivatives **97, 98, 99,** and **100** were found most active against multidrug resistance *M. tuberculosis* strain (MTCC 300) with MIC of 3.90 μg/mL when compared to the standard drugs rifampicin (MIC = 0.24 μg/mL) and isoniazid (MIC = 0.48 μg/mL). Compound **101** also possessed good antitubercular activity against the tested strain with MIC value of 7.81 μg/mL. According to the activity data, Compound **98** with an electron-withdrawing -NO_2 group on the third position has exceptional activity, while Compound **101** with an electron-withdrawing –NO_2 group on the fourth position has good activity (Fig. 9.11).

Reddyrajula and Dalimba [31] prepared several imidazo[1,2-*a*]pyridine hybrids and screened for antitubercular activity. The condensation of 2-aminopyridine, **102** and bromo pyruvic acid in ethanol at 80°C produced compound - **103**, imidazo[1,2-*a*]pyridine-2-carboxylic acid. The condensation of compound - **103** with prop-2-yn-1-amine through the classical Hexafluorophosphate Azabenzotriazole Tetramethyl Uronium (HATU)-mediated reaction gives compound - **104**, *N*-(prop-2-yn-1-yl)imidazo[1,2-*a*]pyridine-2-carboxamide. Further, the reaction of the compound - **104** with different benzyl bromides, $CuSO_4$ and sodium ascorbate produced compound - **105** (Scheme 9.15). The reaction of compound - **102**, 2-aminopyridine with ethyl 2-chloroacetoacetate gives compound - **106**, ethyl 2-methylimidazo[1,2-a]pyridine-3-carboxylate which on further hydrolysis by lithium hydroxide produced compound - **107**, 2-methylimidazo[1,2-*a*]pyridine-3-carboxylic acid. The reaction of it with prop-2-yn-1-amine by HATU-mediated coupling afforded compound - **108**, 2-methyl-*N*-(prop-2-yn-1-yl)imidazo[1,2-*a*]pyridine-3-carboxamide. Further, the reaction

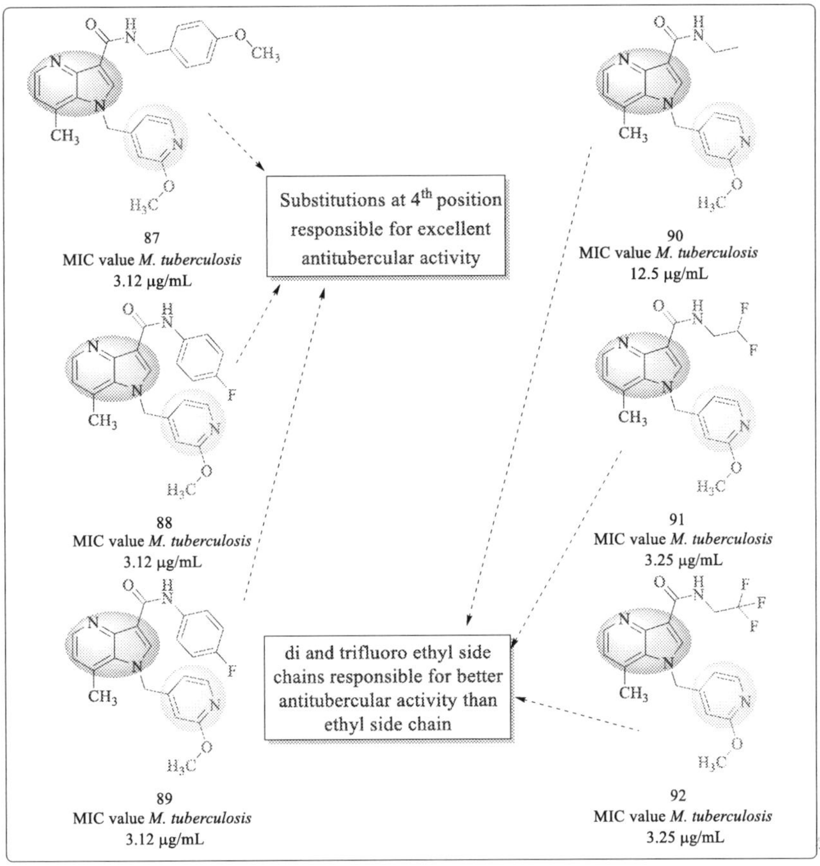

FIGURE 9.10 Most active molecules from pyrrolo[3,2-*b*]pyridine-3-carboxamide linked 2-methoxypyridine derivatives.

of Compound 3 with different benzyl bromides, $CuSO_4$ and sodium ascorbate produced *N*-((1-aryl-1*H*-1,2,3-triazol-4-yl)methyl)-2-methylimidazo[1,2-*a*]pyridine-3-carboxamide derivatives, **109** (Scheme 9.16). The synthetic pathway for the compounds - **114** was described in Scheme 9.17.

The synthesized imidazo[1,2-*a*]pyridine hybrids were screened for antitubercular activity against *M. tuberculosis* $H_{37}Rv$ (ATCC27294) strain using the MABA method. Among the synthesized derivatives total of sixteen derivatives possessed the most potent antitubercular activity with an MIC of 1.56 μg/mL which showed more potency than the MIC of standard drug zolpidem, pyrazinamide, and ciprofloxacin (MIC = 3.12 μg/mL) and streptomycin (MIC = 6.25 μg/mL) (Figs 9.12–9.14).

Desai et al. [32] prepared pyridine clubbed 1,3,4-oxadiazoles as potential antitubercular agents. The synthesis of final compounds was described in Scheme 9.18. Compound - **131**, 4-fluoroacetophenone was reacted with isoniazid, **132** and it formed compound - **133**, *N*′-(1-(4-fluorophenyl)ethylidene)isonicotinohydrazide. Compound - **133** on cyclization with

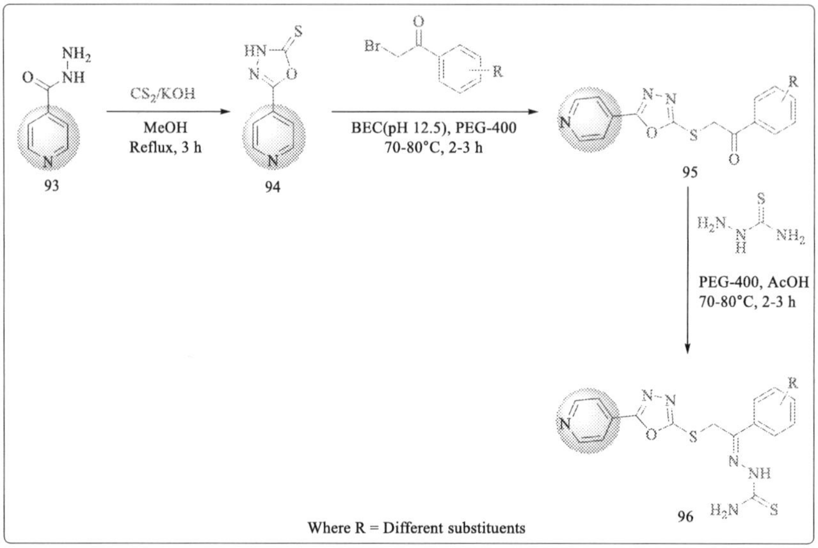

SCHEME 9.14 Synthetic route for pyridine-based 1,3,4-oxadiazole embedded hydrazinecarbothioamide derivatives 96.

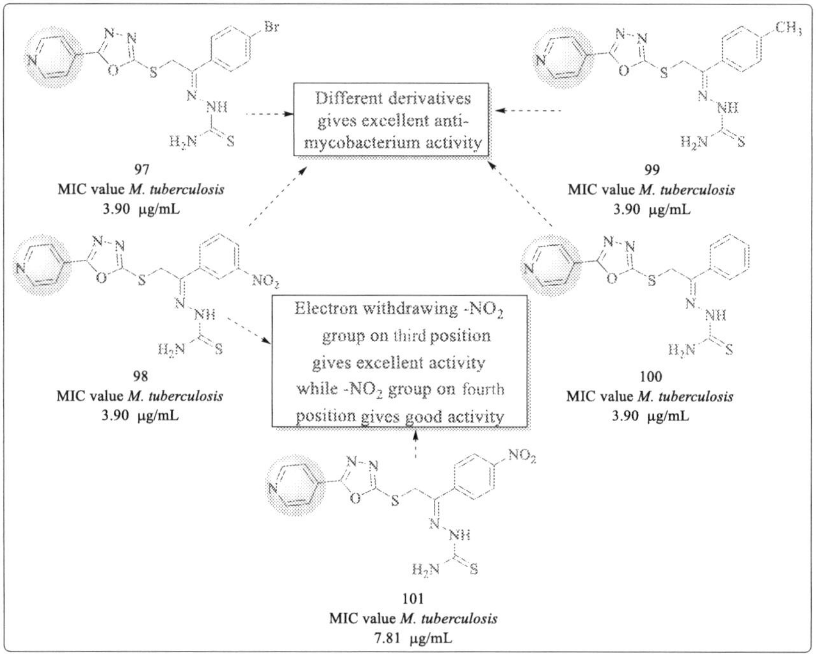

FIGURE 9.11 Most active molecules from pyridine based 1,3,4-oxadiazole embedded hydrazinecarbothioamide derivatives.

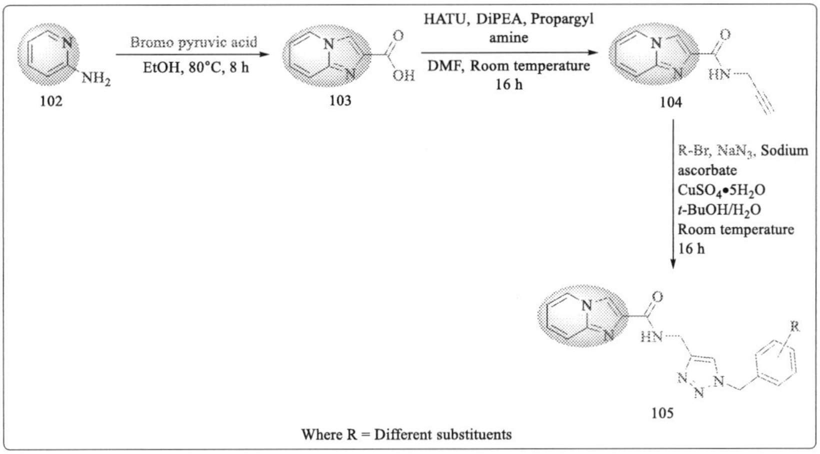

SCHEME 9.15 Synthetic route imidazo[1,2-*a*]pyridine hybrids 105.

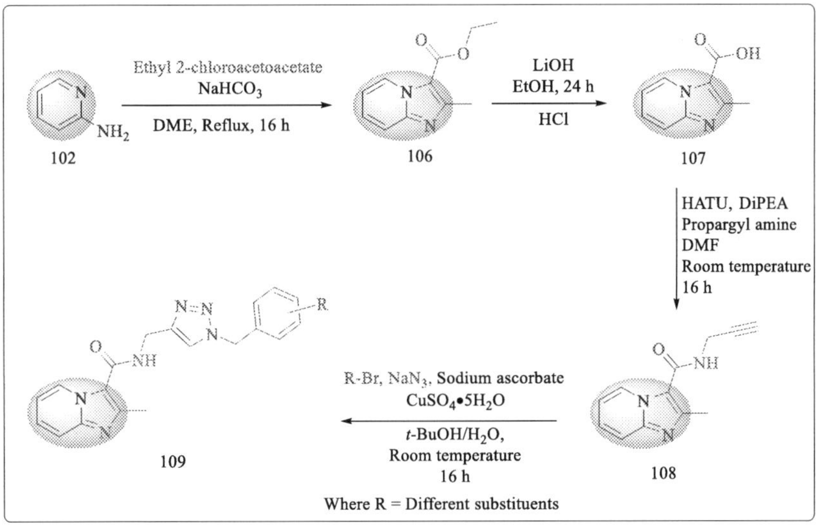

SCHEME 9.16 Synthetic route imidazo[1,2-*a*]pyridine hybrids 109.

acetic anhydride formed compound - **134**, 1-(2-(4-fluorophenyl)-2-methyl-5-(pyridin-4-yl)-1,3,4-oxadiazol-3(2*H*)-yl)ethenone. Compound - **134** reacted with different benzohydrazides along with the catalytic amount of anhydrous ZnCl2 in ethanol (99.9%) produced final compounds - **135**.

The synthesized compounds were screened for their *in vitro* antitubercular activity against *M. tuberculosis* $H_{37}Ra$ and *Mycobacterium bovis* BCG by MTT assay using rifampicin and isoniazid as references. The Compounds **136, 137,** and **139** possessed prominent *in vitro* antitubercular activity against active and dormant strains (Fig. 9.15). Furthermore, the selectivity

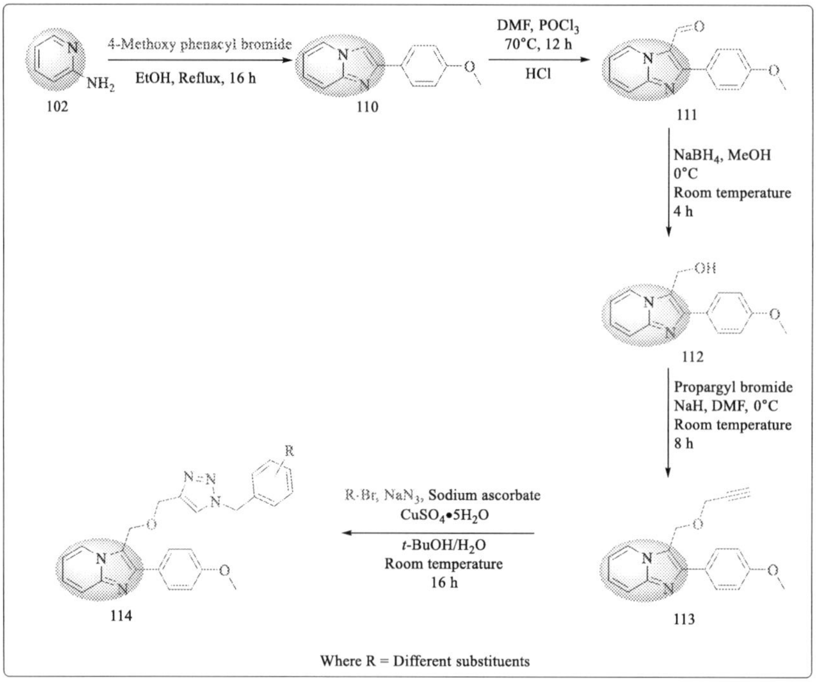

SCHEME 9.17 Synthetic route imidazo[1,2-*a*]pyridine hybrids 114.

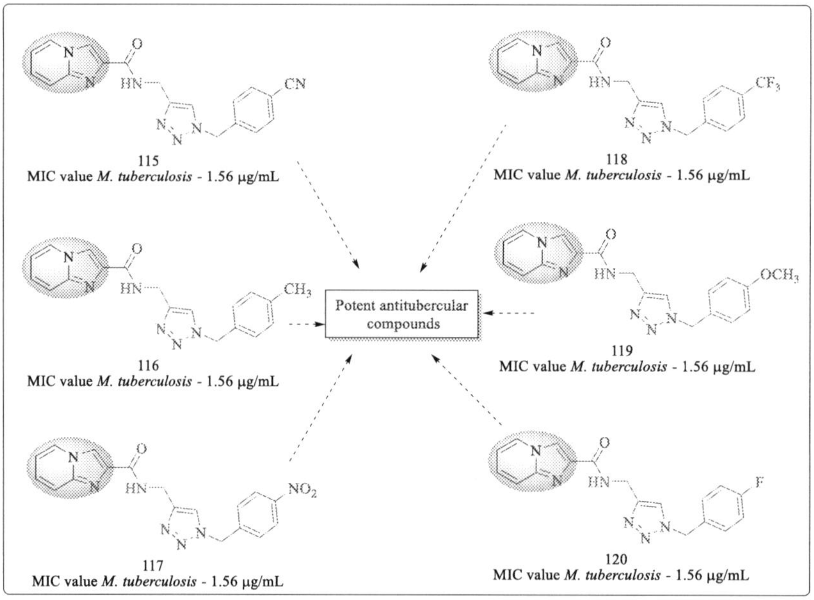

FIGURE 9.12 Most active molecules from imidazo[1,2-*a*]pyridine hybrids 105.

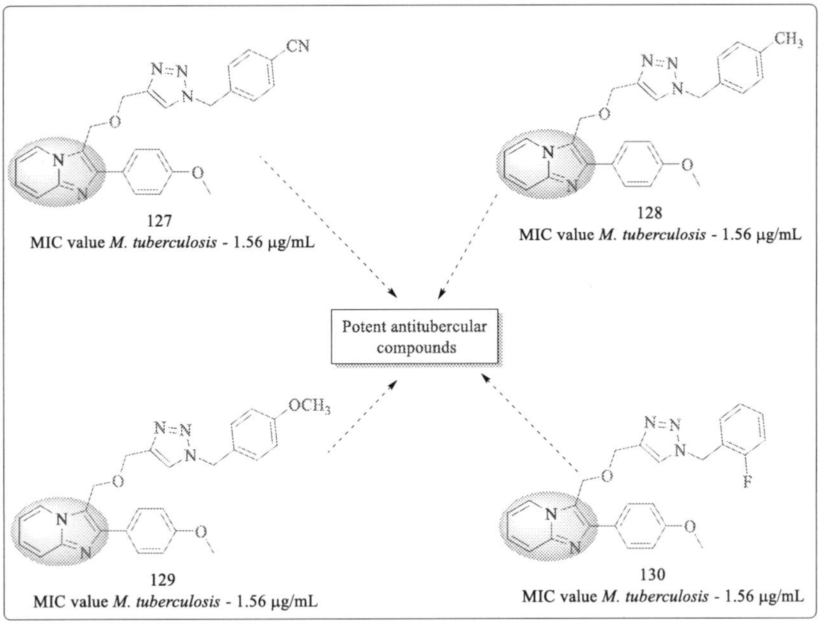

FIGURE 9.13 Most active molecules from imidazo[1,2-*a*]pyridine hybrids 114.

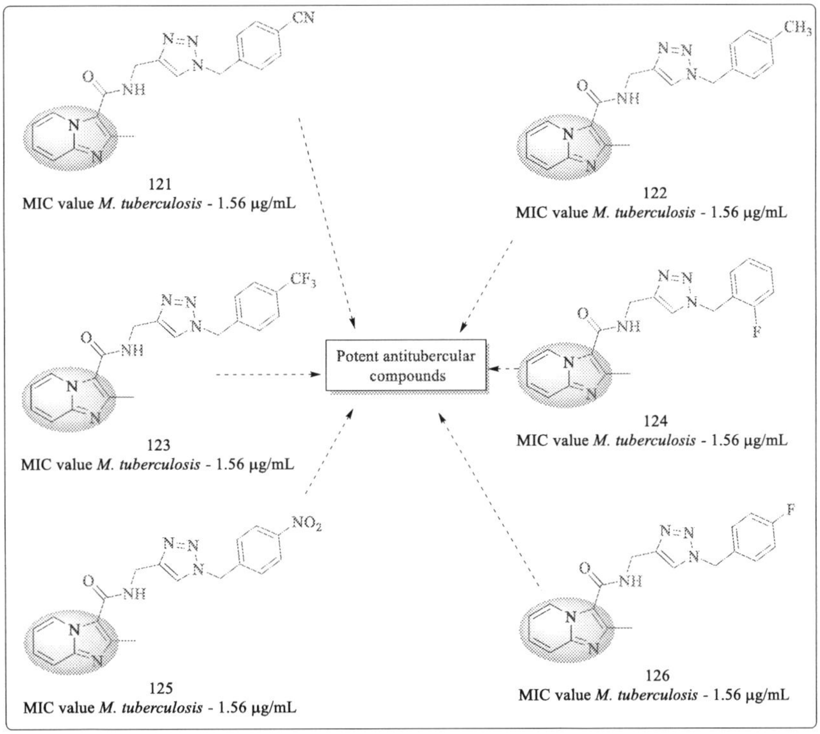

FIGURE 9.14 Most active molecules from imidazo[1,2-*a*]pyridine hybrids 109.

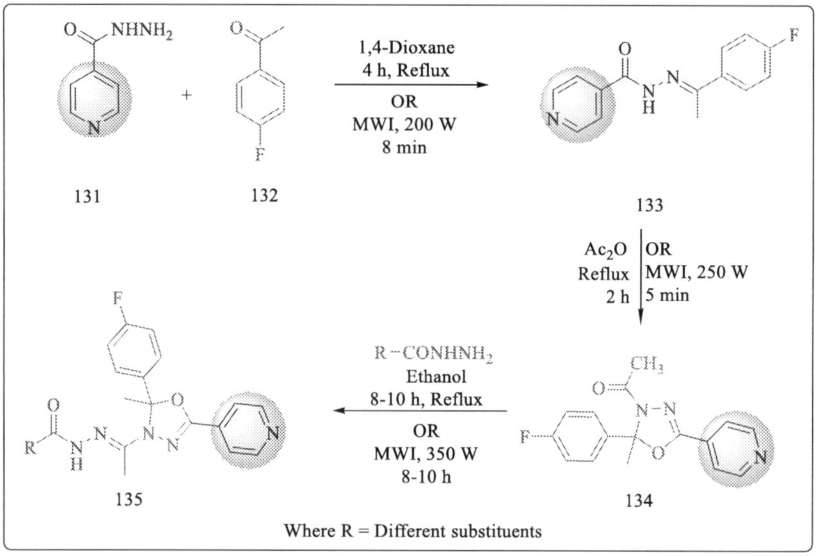

SCHEME 9.18 Synthetic route for imidazo[1,2-*a*]pyridine hybrids 135.

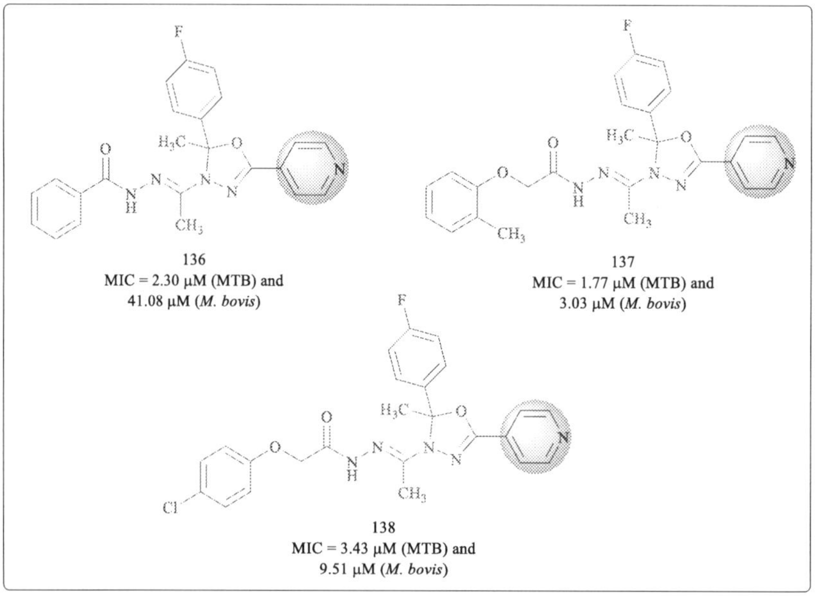

FIGURE 9.15 Most active molecules from imidazo[1,2-*a*]pyridine hybrids.

index of synthesized compounds toward the human cell line against BCG is high which indicates a prominent antitubercular agent.

Desai et al. [33] developed indole and pyridine-based 1,3,4-oxadiazole derivatives as potential antitubercular agents. The synthesis of final compounds were described in

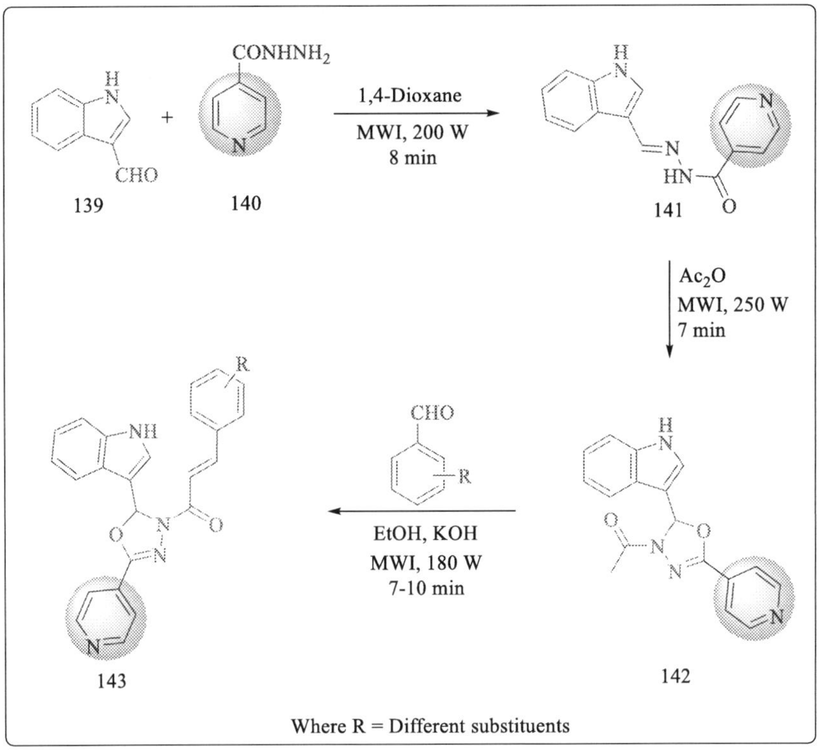

SCHEME 9.19 Synthetic route for indole and pyridine-based 1,3,4-oxadiazole derivatives 143.

Scheme 9.19. Compound - **139**, Indole-3-carbaldehyde was reacted with isoniazid, **140** and produced compound - **141**, *N′*-((1*H*-indol-3-yl)methylene)isonicotinohydrazide. Compound - **141** on cyclization with acetic anhydride formed compound - **142**, 1-(2-(1*H*-indol-3-yl)-5-(pyridin-4-yl)-1,3,4-oxadiazol-3(2*H*)-yl)ethenone which on further reaction with different aldehydes formed final compounds - **143**.

The synthesized derivatives screened for their *in vitro* antitubercular activity against *Mycobacterium bovis* BCG and *M. tuberculosis* $H_{37}Ra$ using XTT Reduction Menadione assay (XRMA) and NR (Nitrate reductase) assay respectively. Rifampicin and isoniazid were used as references. The synthesized Compounds **144, 145, 146**, and **147** possessed prominent *in vitro* antitubercular activity at active and dormant states against *Mycobacterium bovis* BCG (Fig. 9.16).

9.6 Synthesis and anticancer activity of some pyridine derivatives

Rashdan et al. [34] prepared various substituted nicotine analogs or 1,2,3-triazolyl-pyridine hybrids from the triazole derivative, **148** by a one-pot three-component reaction. The one-pot reaction of 1-(1-(4-bromophenyl)-5-methyl-1*H*-1,2,3-triazol-4-yl)ethenone, **148** with different aldehydes, **149** and malanonitrile, **150** in the presence of ammonium acetate catalyst

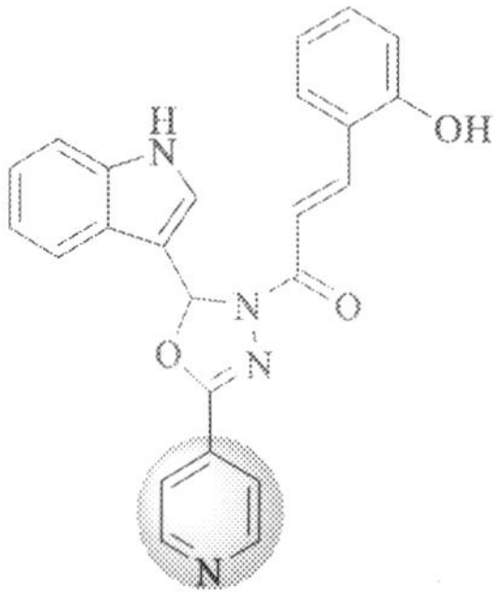

144
MIC (*M. bovis*)
0.94 μM (Active)
1.19 μM (Dormant)

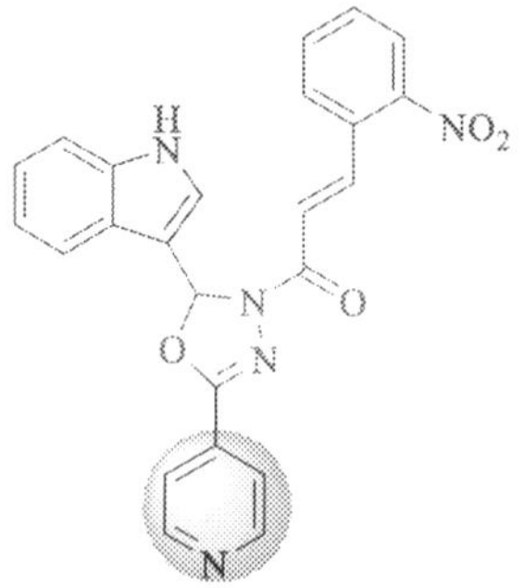

145
MIC (*M. bovis*)
1.03 μM (Active)
0.85 μM (Dormant)

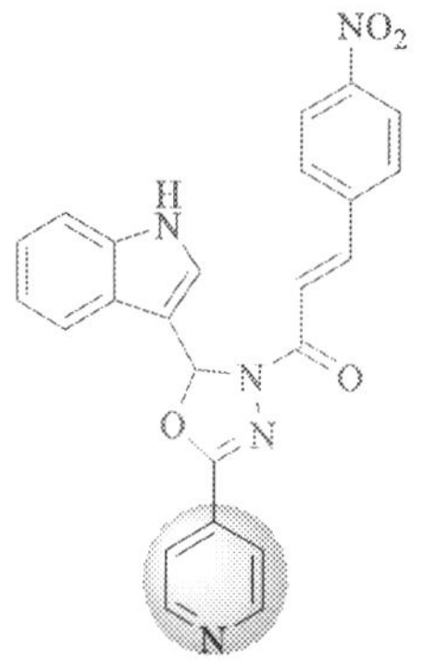

146
MIC (*M. bovis*)
2.50 μM (Active)
2.37 μM (Dormant)

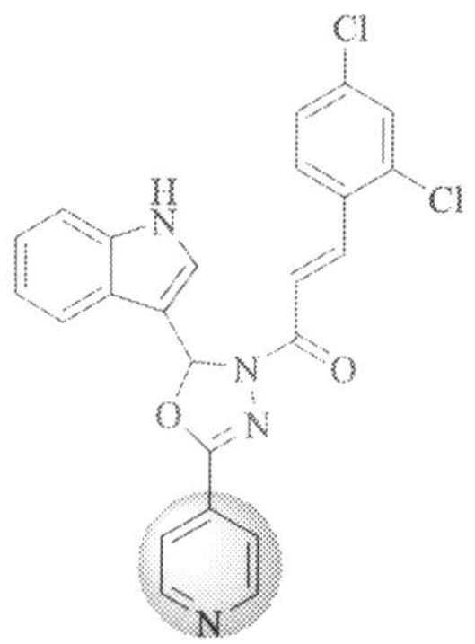

147
MIC (*M. bovis*)
5.17 μM (Active)
4.97 μM (Dormant)

FIGURE 9.16 Most active molecules from indole and pyridine-based 1,3,4-oxadiazole derivatives.

produced substituted 2-amino-6-(1-(4-bromophenyl)-5-methyl-1*H*-1,2,3-triazol-4-yl)nicotine nitriles (**151–157**) (Scheme 9.20).

The one-pot reaction of 1-(1-(4-bromophenyl)-5-methyl-1*H*-1,2,3-triazol-4-yl)ethenone, **148** with different aldehydes, **149** and, ethyl cyanoacetate, **158** in the presence of ammonium acetate catalyst provide substituted 2-amino-6-(1-(4-bromophenyl)-5-methyl-1*H*-1,2,3-triazol-4-yl)nicotine nitriles (**159–165**) (Scheme 9.21).

The synthesized compounds were evaluated for anticancer activity against HepG2 (hepatoblastoma) and BALB/3T3 (murine fibroblast) cell lines using doxorubicin (IC_{50} = 3.56 ± 0.46 μg/mL) as standard drug and MTT assay. The synthesized compounds possessed

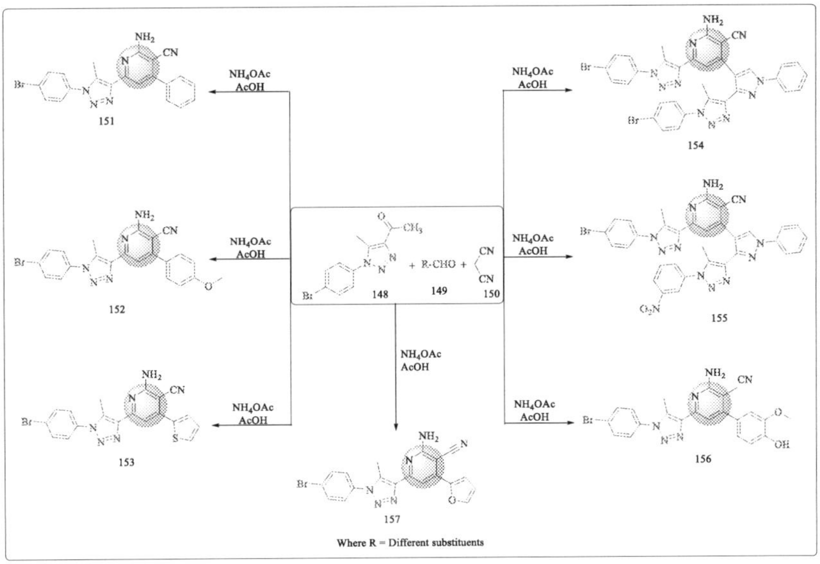

SCHEME 9.20 Synthetic route for substituted nicotine analogs or 1,2,3-triazolyl-pyridine hybrids 151–157.

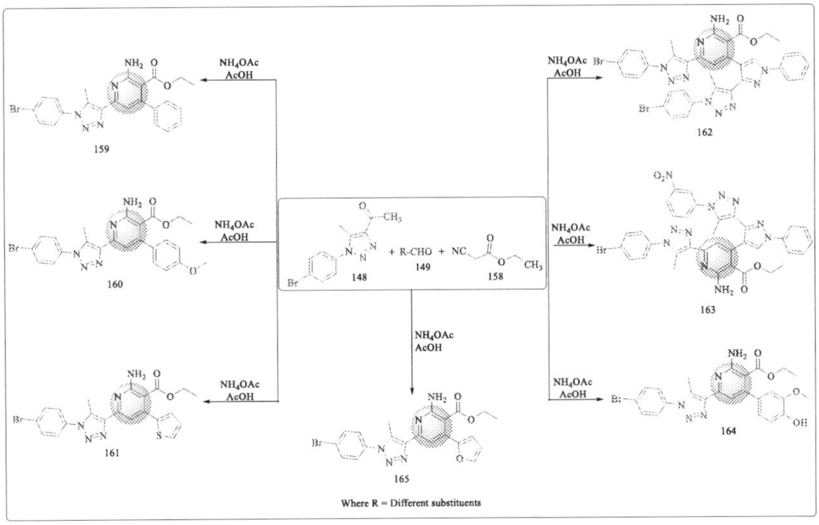

SCHEME 9.21 Synthetic route for substituted nicotine analogs or 1,2,3-triazolyl-pyridine hybrids 159–165.

promising anticancer activity against HepG2 compared with doxorubicin. Compound **154** and **155** possessed excellent anticancer activity against HepG2 with IC_{50} value of 0.64 ± 0.14 μg/mL and 1.08 ± 0.94 μg/mL, respectively. Compound **152** and **162** possessed better anticancer activity against HepG2 having IC_{50} values of 2.34 ± 0.16 μg/mL and 6.24 ± 0.71 μg/mL.

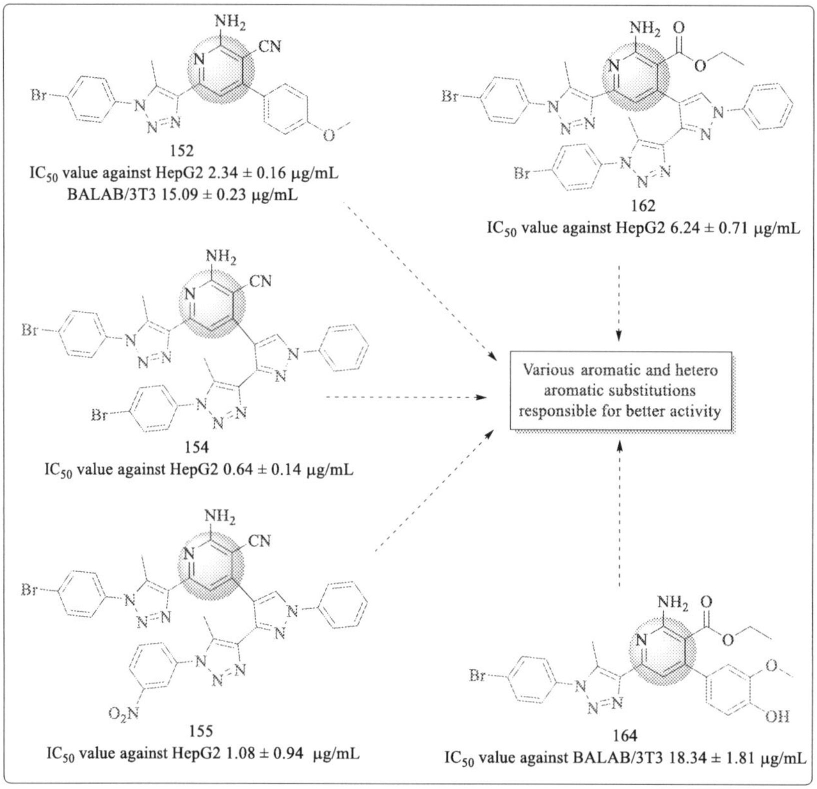

FIGURE 9.17 Most active molecules from substituted nicotine analogs or 1,2,3-triazolyl-pyridine hybrids.

Compound **152** and **164** possessed good anticancer activity against BALAB/3T3 with IC_{50} values of 15.09 ± 0.23 μg/mL and 18.34 ± 1.81μg/mL, respectively (Fig. 9.17).

Abdelaziz et al. [35] prepared several pyridine and thieno[2,3-b] pyridine derivatives as anticancer PIM-1 kinase inhibitors. The synthesis of pyridine derivatives was described in Scheme 9.22. Compound - **167**, hydroquinone diacetate was prepared by acetylation of dihydroquinone, **166**. Fries rearrangement of the compound - **167** produced 2,5-dihydroxy acetophenone, **168.** The one-pot three-component reaction of the compound - **168,** substituted aldehydes and ethyl cyanoacetate using ammonium acetate catalyst produced 3-cyanopyridones, **169**. The one-pot three-component reaction of the compound - **168**, substituted aldehydes and malanonitrile using ammonium acetate catalyst produced 2-amino-3-cyanopyridines, **170**. The reaction of compound - **168** with potassium carbonate and methyl iodide formed compound - **171**, 2-hydroxy-5-methoxyacetophenone which on esterification by benzoyl chloride in the presence of pyridine gives compound - **172**, 2-benzoyloxy-5-methoxyacetophenone. The one-pot three-component reaction of the compound - **172**, substituted aldehydes and ethyl cyanoacetate using ammonium acetate catalyst produced pyridine-2-one derivatives, **173**. Compound - **174** was prepared by methylation of the compound - **168** using methyl sulphate. Further reaction of the compound - **174** with substituted aldehydes and

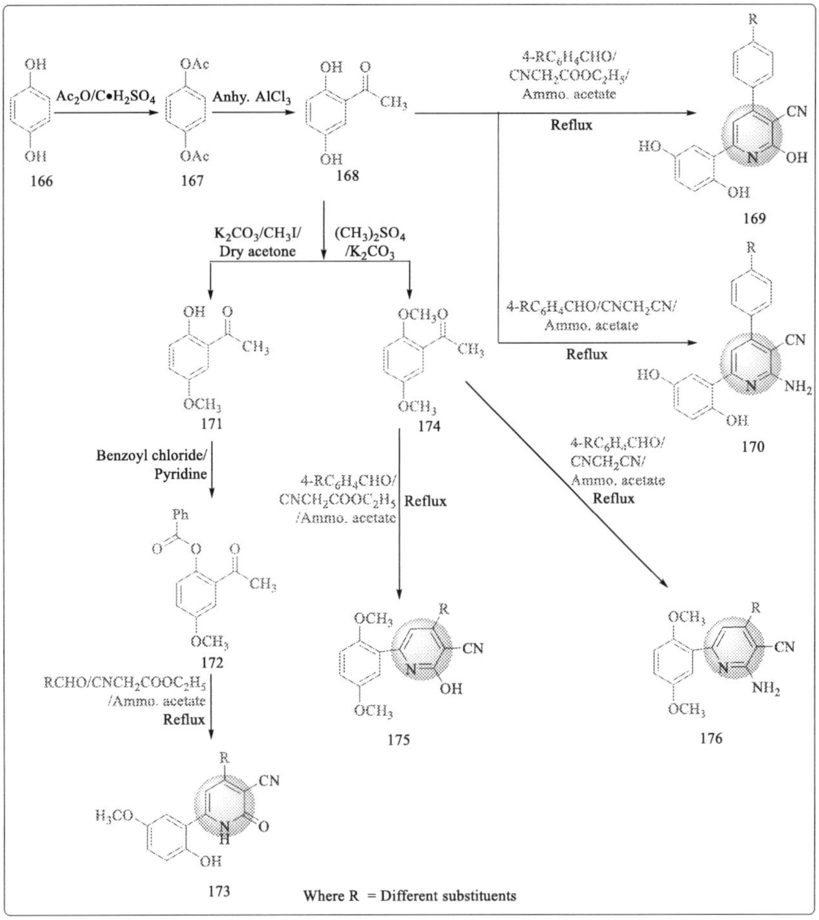

SCHEME 9.22 Synthetic route for pyridine derivatives.

ethyl cyanoacetate using ammonium acetate catalyst produced compound - **175**. Compound - **176** were prepared by the same method using malanonitrile in lieu of ethyl cyanoacetate.

The synthesis of pyridine and thieno[2,3-*b*] pyridine derivatives is described in Scheme 9.23. Compound - **177** was prepared by reaction of the compound - **174** with different aldehydes and cyano thioacetamide in the presence of ammonium acetate catalyst. Compound - **177** on reaction with chloroacetone or phenacyl bromide derivatives and anhydrous potassium carbonate formed compound - **178**. Compound - **178** on cyclization by in ethanolic sodium ethoxide solution formed compound - **179**. S-alkylation of the compound - **177** with alkyl halides and anhydrous potassium carbonate in acetone formed compound - **180**.

The antitumor activity of the synthesized pyridine and thieno[2,3-*b*] pyridine derivatives was tested against sixty cancer cell lines. With the exception of six cell lines, Compound **182** had a strong growth inhibition percentage and GI_{50} values in the range of 0.302 to 3.57 μM against the tested cell lines. The synthesized Compound **181** had the highest inhibitory

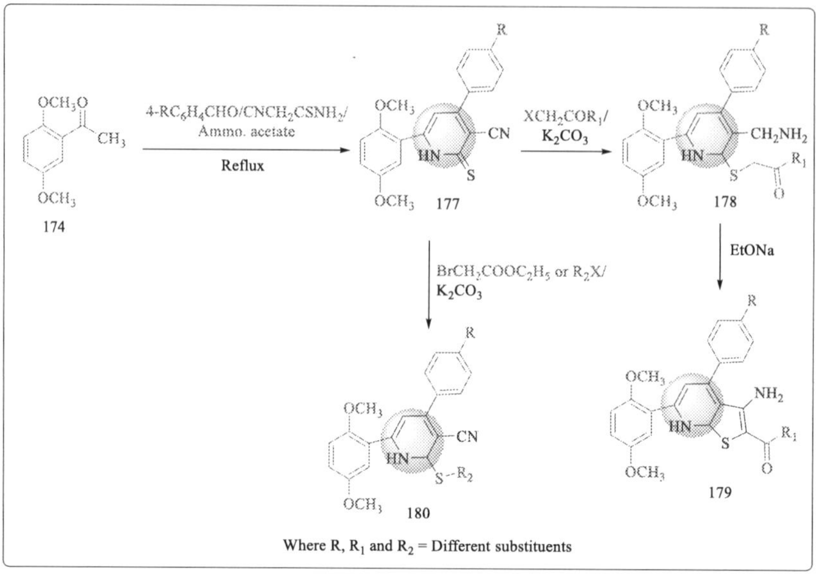

SCHEME 9.23 Synthetic route for pyridine and thieno[2,3-*b*] pyridine derivatives.

action against PIM-1 kinase, with an IC_{50} of 0.019 μM, which was comparable to the conventional drug staurosporin. (IC_{50} = 0.013 μM) (Fig. 9.18).

Şenkardeş et al. [36] developed 2,6-disubstituted pyridine hydrazones as potential anticancer agents. Compound - **187**, dimethyl pyridine-2,6-dicarboxylate was obtained by esterification of compound - **186**, pyridine-2,6-dicarboxylic acid with methanol and concentrated sulfuric acid. Further reaction of the compound - **187** with hydrazine hydrate formed compound - **188**, pyridine-2,6-dicarbohydrazide. The condensation of compound - **188** with different aldehydes in the presence of acetic acid formed 2,6-disubstituted pyridine hydrazones, **189** (Scheme 9.24).

The synthesized compounds were screened for *in vitro* anticancer activity against human cancer cell lines by using MTT assay. Compounds **190** and **192** possessed excellent anticancer activity with IC_{50} values of 6.78 μM and 8.88 μM against HT-29 Human colon cancer cell line, respectively, as a comparison of the standard drug paclitaxel (IC_{50} = 0.35 μM). Compounds **191** possessed excellent anticancer activity with IC_{50} value of 8.26 μM against ISH Ishikawa human endometrial cancer cell line in comparison to standard drug paclitaxel (IC_{50} = 1.025 μM) (Fig. 9.19).

Saleh et al. [37] prepared pyridine-derived VEGFR-2 inhibitors. The reaction of *o*-bromobenzaldehyde, **193** with 2-cyanoacetohydrazide, **195** using acetic acid catalyst in the presence of ethanol formed Schiff base N′-(2-bromobenzylidene)-2-cyanoacetohydrazide, **196**. Compound - **196** on reaction with malanonitrile in the presence of pyridine formed compound - **197**, 1,6-diamino-4-(2-bromophenyl)-2-oxo-1,2-dihydropyridine-3,5-dicarbonitrile. Another method for the synthesis of compound - **197** involves the reaction of the compound - **193** with malanonitrile to afford compound - **194**, 2-(2-bromobenzylidene)malononitrile and then the

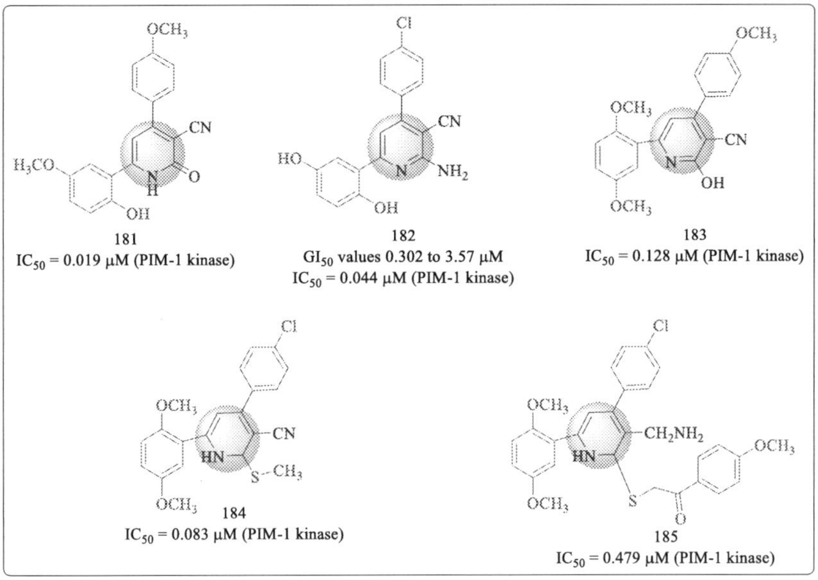

FIGURE 9.18 Most active molecules from pyridine and thieno[2,3-*b*] pyridine derivatives.

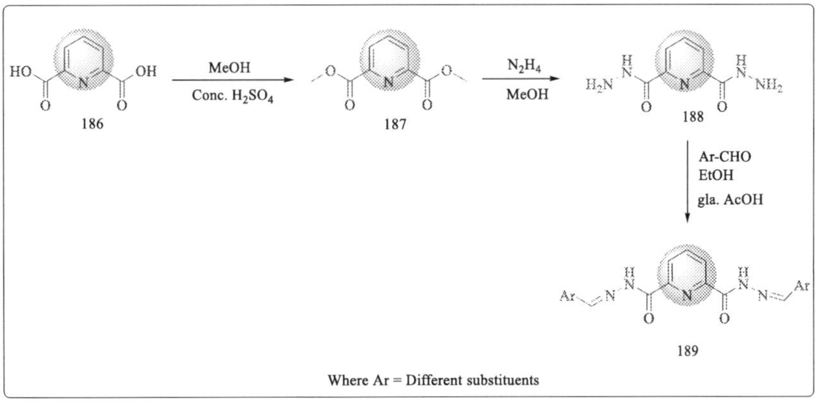

SCHEME 9.24 Synthetic route for 2,6-disubstituted pyridine hydrazones.

reaction of compound - **194** with the compound - **195** in the presence of piperidine formed compound - **197** (Scheme 9.25).

Compound - **197** on cyclization with ethyl chloroformate yielded compound - **199**, 7-(2-bromophenyl)-5-oxo-3,5-dihydro-[1,2,4]triazolo[1,5-a]pyridine-6,8-dicarbonitrile. The reaction of compound - **197** with acetic anhydride yielded compound - **200**, 7-(2-bromophenyl)-2-methyl-5-oxo-3,5-dihydro-[1,2,4]triazolo[1,5-a]pyridine-6,8-dicarbonitrile. Compound - **197** on reaction with benzaldehyde yielded compound - **201**, 7-(2-bromophenyl)-5-oxo-2-phenyl-1,2,3,5-tetrahydro-[1,2,4]triazolo[1,5-a]pyridine-6,8-dicarbonitrile, which on reaction with trifluoroacetic acid produced compound - **202**, [1,2,4]triazolo[1,5-a]pyridine derivative.

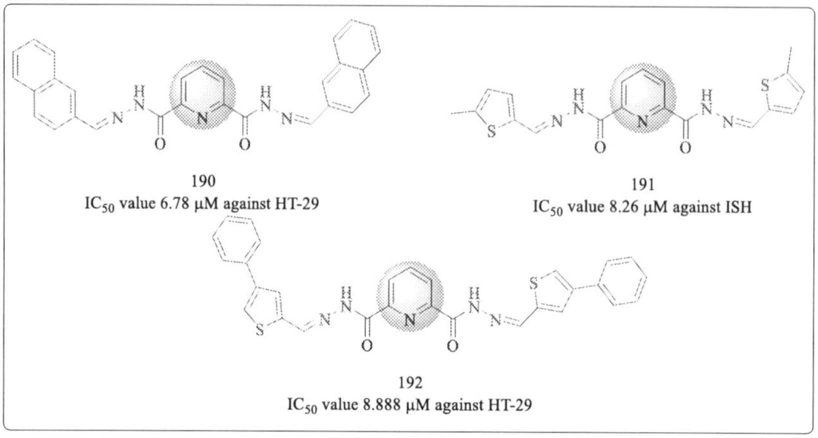

FIGURE 9.19 Most active molecules from 2,6-disubstituted pyridine hydrazones.

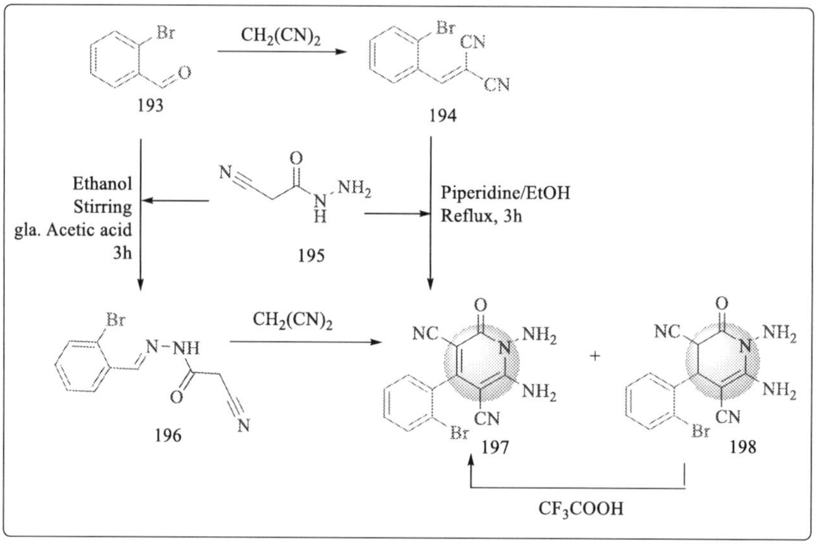

SCHEME 9.25 Synthetic route for pyridine-derived heterocycles.

Compound - **5** when reacted with sodium nitrite in acetic acid formed compound - **203**, 6-amino-4-(2-bromophenyl)-2-oxo-1,2-dihydropyridine-3,5-dicarbonitrile rather than compound - **204** (Scheme 9.26).

Compound - **196** on reaction with ethyl cyanoacetate in the presence of triethylamine produced compound - **205**, ethyl-1,2-diamino-4-(2-bromophenyl)-5-cyano-6-oxo-1,6-dihydropyridine-3-carboxylate rather than compound - **206** (Scheme 9.27). Compound - **196** on reaction with p-methoxybenzylidene malononitrile in the presence of triethylamine produced compound - **207**, 6-amino-1-[(2-bromobenzylidene)amino]-4-(4-methoxyphenyl)-2-oxo-1,2,3,4-tetrahydropyridine-3,5-dicarbonitrile.

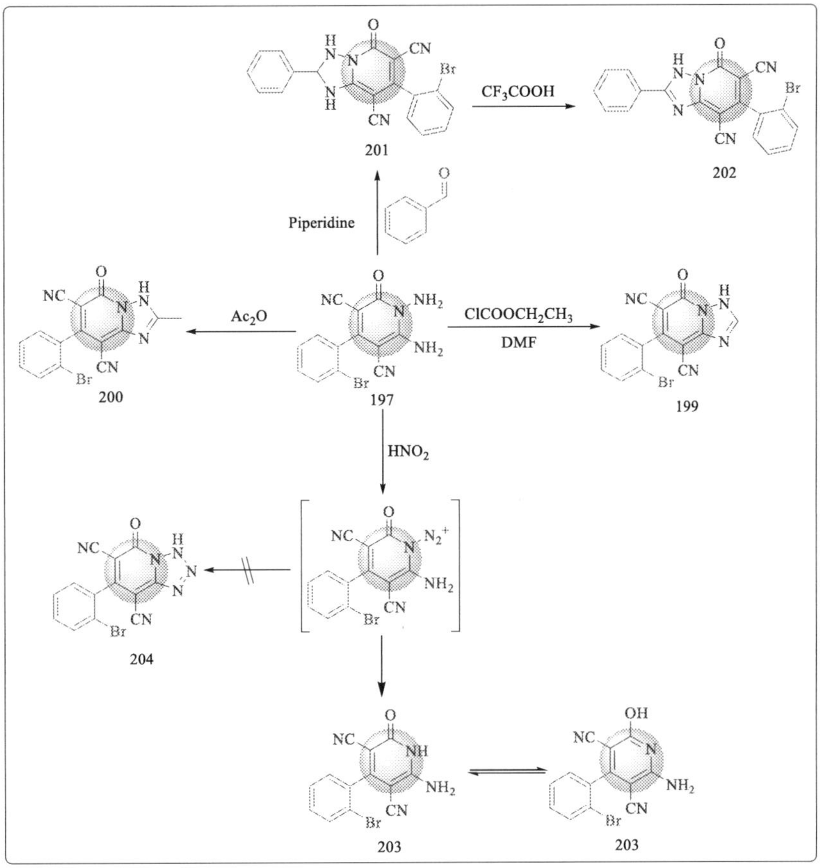

SCHEME 9.26 Synthetic route for pyridine-derived heterocycles.

The synthesized compounds were evaluated for anticancer activity against two human tumor cell lines, hepatocellular carcinoma (HepG2) and breast cancer (MCF-7) cell lines using 3-[4,5-dimethylthiazol-2-yl]-2,5-diphenyltetrazolium bromide colorimetric assay (MTT assay) and sorafenib and doxorubicin as standard drugs. Compounds **202, 201, 200**, and **207** were found most potent compounds with IC_{50} values of 4.25 ± 0.03, 6.08 ± 0.06 μM, 4.68 ± 0.06, 11.06 ± 0.09 μM, 4.34 ± 0.04, 10.29 ± 0.09 μM, 6.37 ± 0.06, and 12.83 ± 0.13 μM against HepG2 and MCF-7 cancer cell lines, respectively. Compounds **198, 200, 201, 202**, and **207** were 1.13, 3.74, 4.18, 3.64, and 2.81 times more toxic to HepG2 and 2.06, 1.58, 1.76, 2.54, and 1.40, times more toxic to MCF-7 breast cancer cells than in normal Vero cells, respectively (Fig. 9.20).

Raman et al. [38] prepared pyridine-based transition metal (II) complexes as potential anticancer agents. The Scheff base, **211** was prepared by condensation of pyridin-2-ylmethanamine, **209** and 2-hydroxy-4-nitrobenzaldehyde, **210**. The Schiff base then condensed with different metal(II)chlorides (Cu, Co, Ni, Zn) to form complexes **212** (Schemes 9.28, Scheme 9.29).

The synthesized compounds were evaluated for cytotoxicity against human cancer cell lines, HeLa (human cervical carcinoma), MCF-7 (human breast adenocarcinoma), Hep2

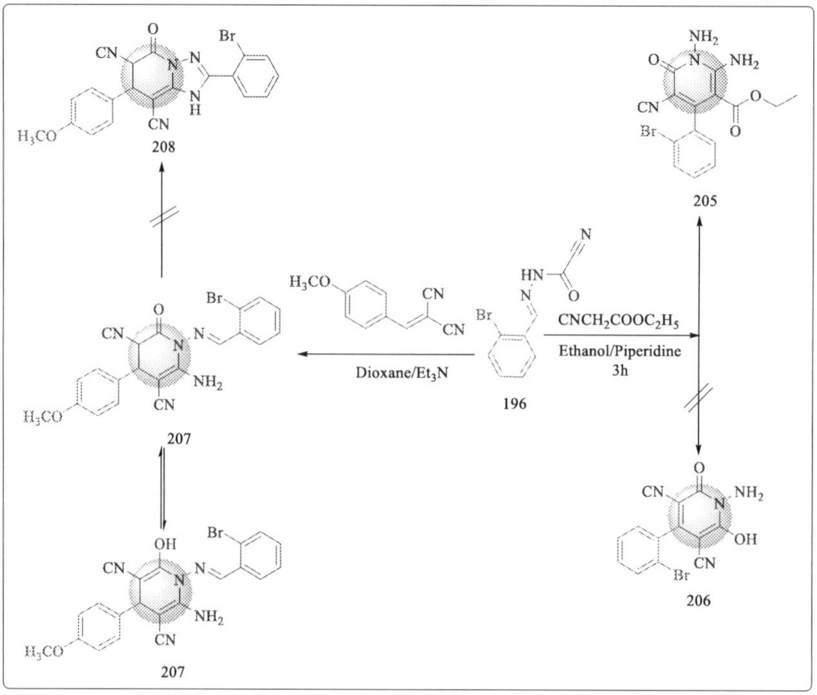

SCHEME 9.27 Synthetic route for pyridine-derived heterocycles.

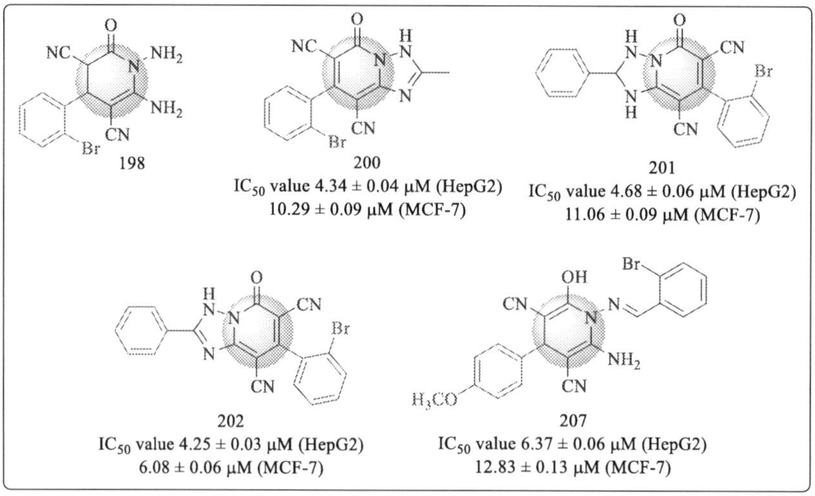

FIGURE 9.20 Most active molecules from pyridine-derived heterocycles.

(human laryngeal epithelial cancer) along with NHDF (normal human dermal fibroblasts) cell lines using cisplatin as positive control by MTT assay. All the synthesized compounds showed cytotoxic activity. Compound - **213** possessed better cytotoxicity with IC_{50} values of

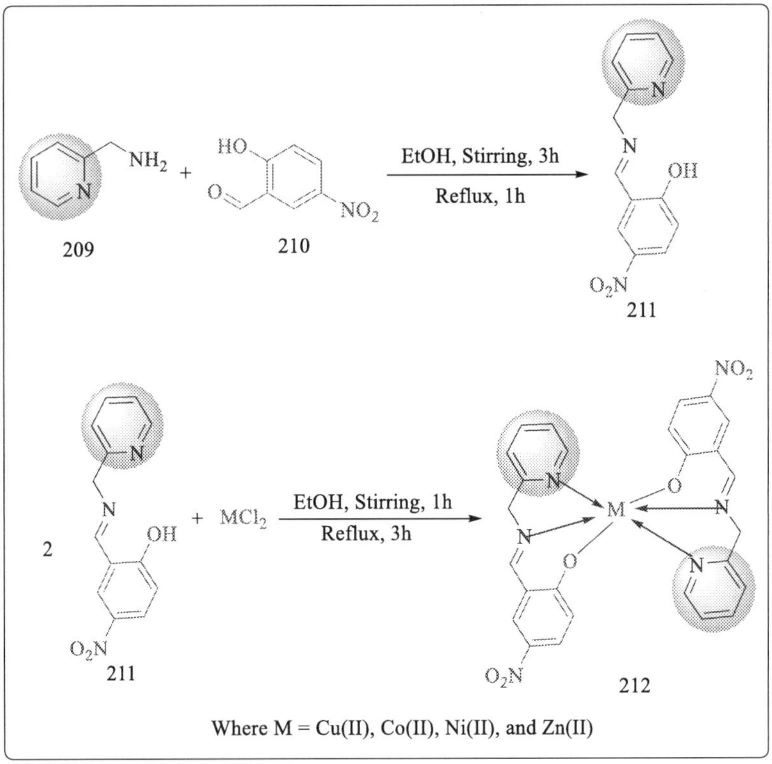

SCHEME 9.28 Synthetic route for pyridine-based transition metal (II) complexes.

44.02 μM, 66.12 μM, and 72.05 μM against HeLa, Hep2, and MCF-7 cancer cell lines, respectively (Fig. 9.21).

9.7 Synthesis and anti-inflammatory activity of some pyridine derivatives

Kamat et al. [25] prepared pyridine and thiazole-based hydrazides with promising anti-inflammatory activity (Scheme 9.9). The synthesized compounds were evaluated for their *in vitro* anti-inflammatory activity by inhibition of the protein (bovine albumin) denaturation method using the standard drug diclofenac sodium. The synthesized compound - **51** showed anti-inflammatory activity with inhibition of 32.63% in comparison to the standard drug (Fig. 9.22).

Bilavendran et al. [39] prepared pyrazolo-pyridine analogs as inflammation medications. Compound - **219** was synthesized by reaction of the compound - **217**, 1-methylpiperidin-4-one with different thiophene-carbaldehydes, **218**. Further reaction of compound - **219** with phenylhydrazine in using isopropyl alcohol as solvent produced compounds - **220** (Scheme 9.29).

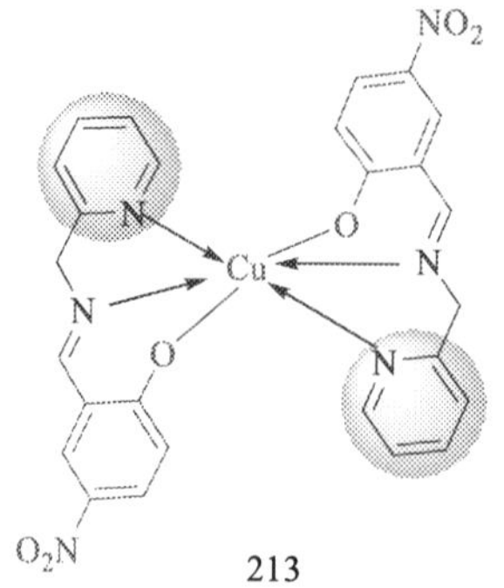

213
IC_{50} value 44.02 μM (HeLa)
66.12 μM (Hep-2)
72.05 μM (MCF-7)

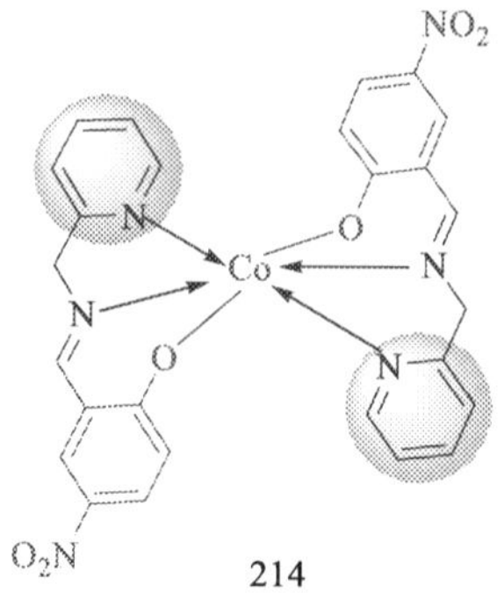

214
IC_{50} value 54.72 μM (HeLa)
48.19 μM (Hep-2)
76.11 μM (MCF-7)

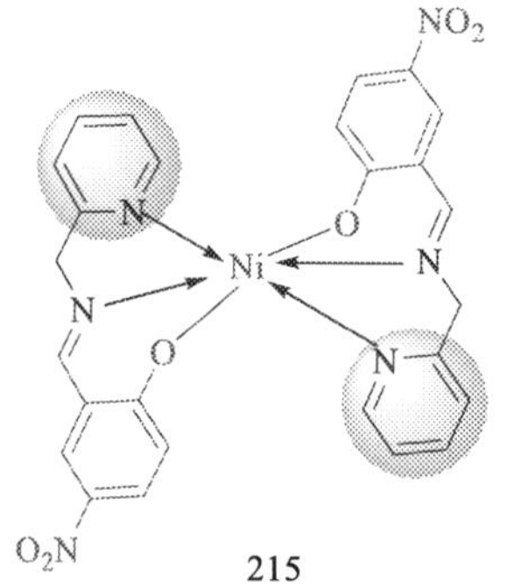

215
IC_{50} value 63.34 μM (HeLa)
53.39 μM (Hep-2)
49.75 μM (MCF-7)

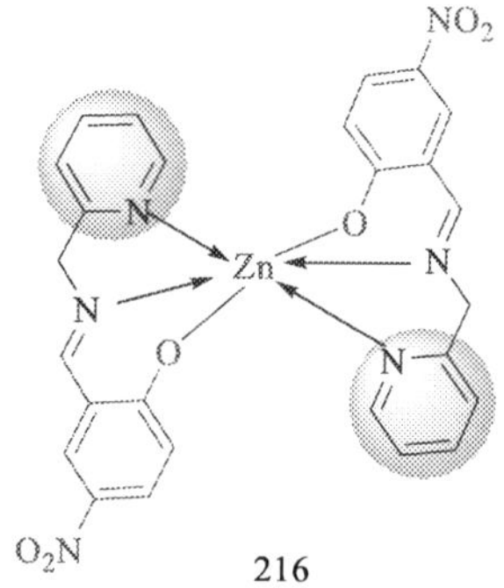

216
IC_{50} value 53.83 μM (HeLa)
42.09 μM (Hep-2)
50.23 μM (MCF-7)

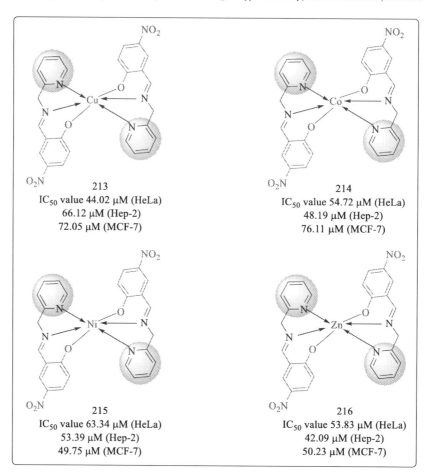

FIGURE 9.21 Most active molecules from pyridine-based transition metal (II) complexes.

The synthesized compounds - **221**, and **222** were screened for further *in vitro* anti-inflammatory activity and the results showed by relative percentage activity (RPA) and IC_{50} values using diclofenac as a standard drug. Both the compounds - **221** and **222** were most active with RPA and IC_{50} values 96.25 ± 2.15, 0.006 ± 0.001 μM and 91.75 ± 2.24, 0.008 ± 0.001 μM, respectively (Fig. 9.23).

Shringare et al. [40] prepared combretastatin-A4 analogues of pyridine derivatives as anti-inflammatory agents. Compound - **225**, 3-(3-hydroxy-4-methoxyphenyl)-1-(3,4,5-trimethoxyphenyl)prop-2-en-1-one was prepared by reaction of compound - **223**, 1-(3,4,5-trimethoxyphenyl)ethan-1-one and 3-hydroxy-4-methoxybenzaldehyde, **224**. Compound - **225**, on reaction with hydrazine hydrate formed compound - **226**, 2-methoxy-5-(3-(3,4,5-trimethoxyphenyl)-4,5-dihydro-1*H*-pyrazol-5-yl)phenol. Compound - **226** on reaction with 6-amino-1,3-dimethyluracil formed compound 5-(3-hydroxy-4-methoxyphenyl)-1,3-dimethyl-7-(3,4,5-trimethoxyphenyl)pyrido[2,3-*d*]pyrimidine-2,4(1*H*,3*H*)-dione, **227**. Compound - **226**

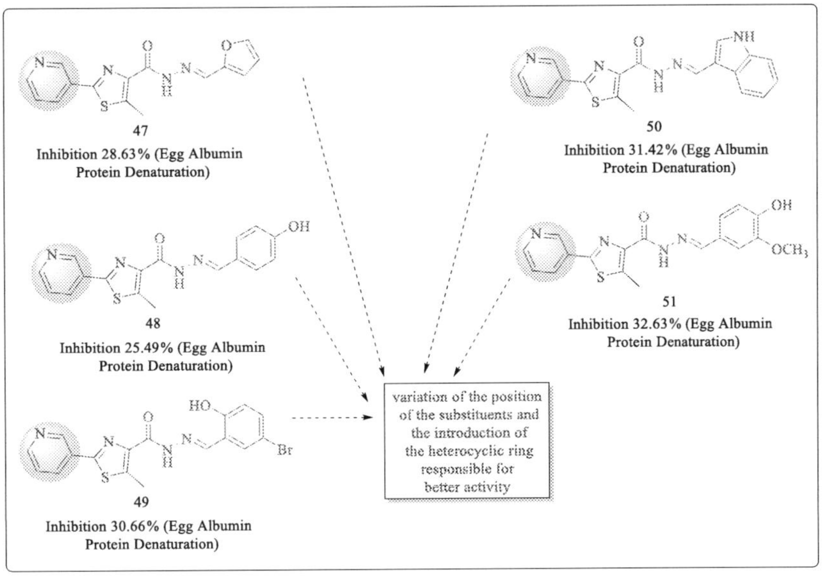

FIGURE 9.22 Most active molecules from pyridine and thiazole-based hydrazides.

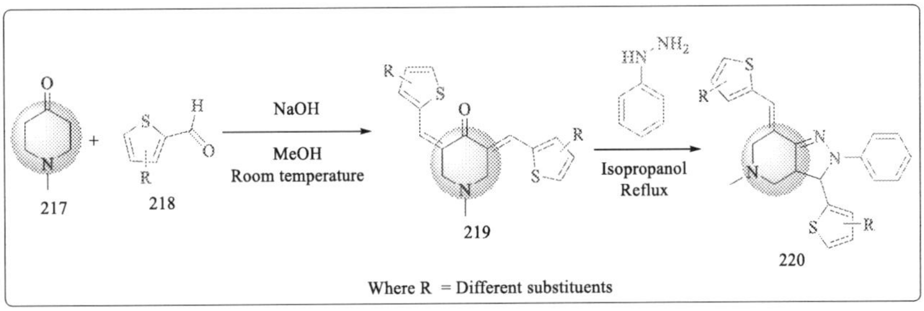

SCHEME 9.29 Synthetic route for pyrazolo-pyridine analogs.

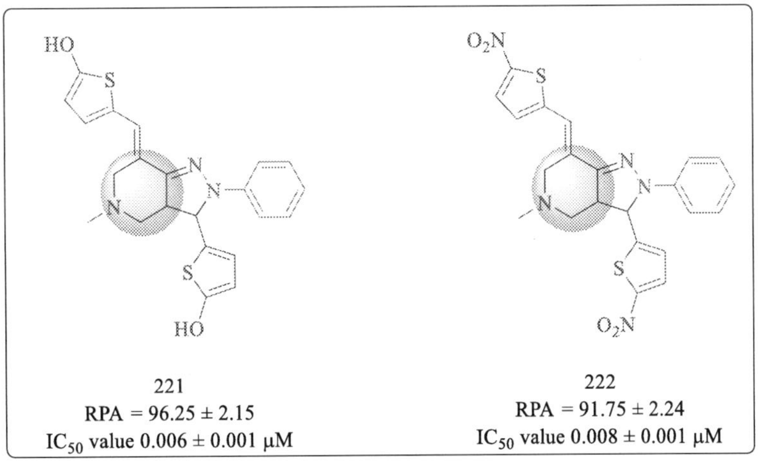

FIGURE 9.23 Most active molecules from pyrazolo-pyridine analogs.

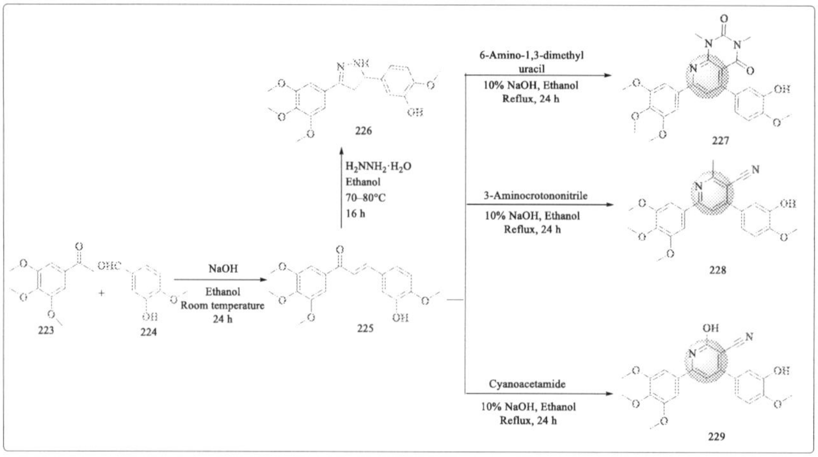

SCHEME 9.30 Synthetic route for combretastatin-A4 analogs of pyridine derivatives.

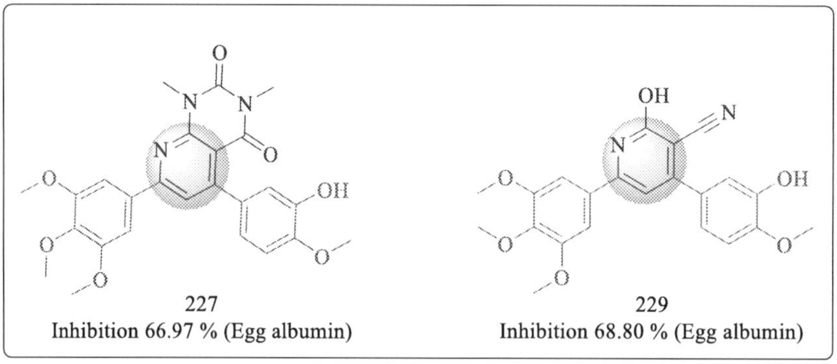

FIGURE 9.24 Most active molecules from combretastatin-A4 analogs of pyridine derivatives.

on reaction with acetoacetanilide formed compound 4-(3-hydroxy-4-methoxyphenyl)-2-methyl-6-(3,4,5-trimethoxyphenyl)nicotinonitrile, **228**. Compound - **226** on reaction with cyanoacetamide formed compound 2-hydroxy-4-(3-hydroxy-4-methoxyphenyl)-6-(3,4,5-trimethoxyphenyl)nicotinonitrile, **229** (Scheme 9.30).

The compounds were tested for anti-inflammatory efficacy *in vitro* using the protein denaturation of the egg albumin method and diclofenac sodium as a reference drug. Heat-induced albumin denaturation was effectively inhibited by Compounds **227** and **229** (66.97 and 68.80 percent) (Fig. 9.24).

9.8 Synthesis and antimalarial activity of some pyridine derivatives

Elshemy et al. [41] prepared pyridyl-indole derivatives as antimalarial agents. Compound - **230**, indole on reaction with cyanoacetic acid formed 3-cyanoacetyl indole, **231**. Compound - **231** on reaction with different aldehydes formed compound - **232**. Compound

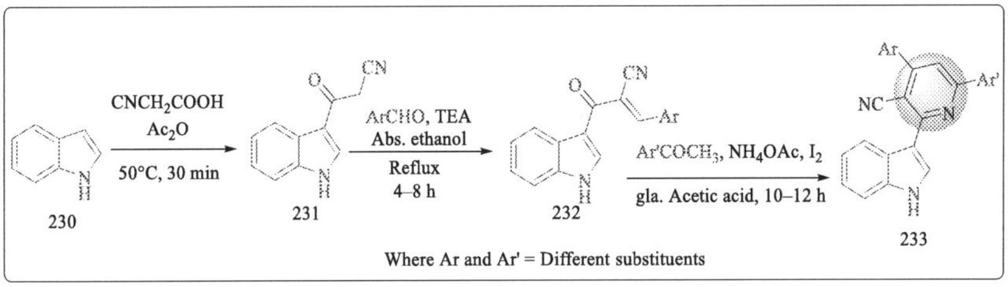

SCHEME 9.31 Synthetic route for pyridyl-indole derivatives.

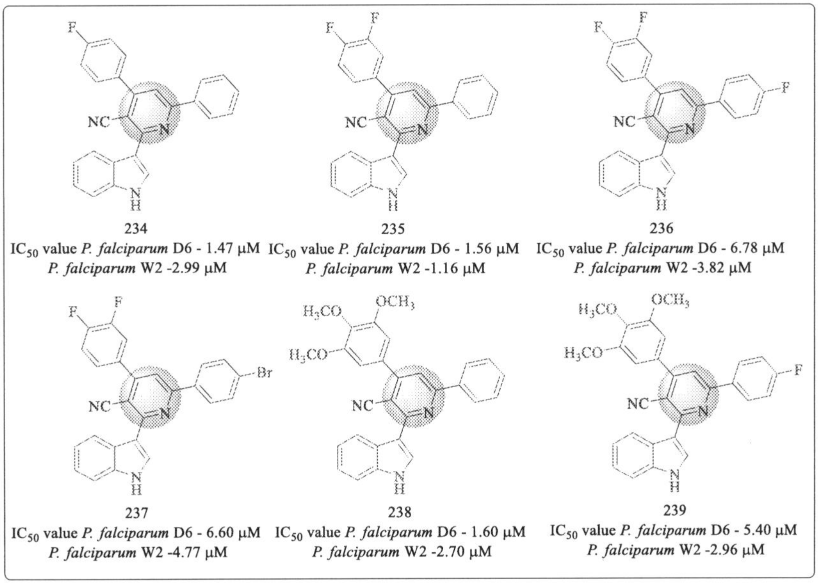

FIGURE 9.25 Most active molecules from pyridyl-indole derivatives.

- **232** on reaction with aromatic ketones and ammonium acetate in glacial acetic acid using I_2 as catalyst formed compounds - **233** (Scheme 9.31).

The synthesized compounds were evaluated for antimalarial activity against chloroquine-sensitive (D6) and chloroquine-resistant (W2) strains of *P. falciparum* using chloroquine and artemisinin as standard drugs. The synthesized Compounds **234–239** possessed promising antimalarial activity with IC_{50} values in the range of 1.47–6.60 μM and 1.16–4.77 μM against *P. falciparum* D6 and W2 strains, respectively (Fig. 9.25).

Karpina et al. [42] developed [1,2,4]triazolo[4,3-*a*]pyridine sulfonamides from 2-chloropyridine-3-sulfonyl chloride and 2-chloropyridine-5-sulfonyl chloride as potential antimalarial agents (Schemes 9.32 and 9.33).

The synthesized compounds were evaluated for *in vitro* antiprotozoal activity against Plasmodium falciparum 2/K chloroquine-resistant strain. The synthesized Compounds **251** and **252** possessed promising antimalarial activity with IC_{50} values 4.98 μM and 2.24 μM, respectively (Fig. 9.26).

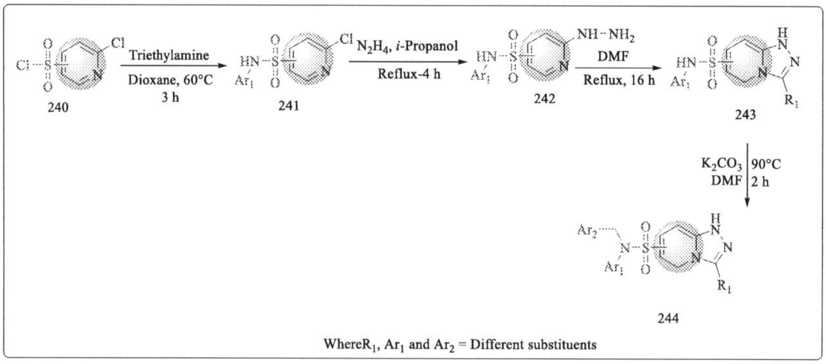

SCHEME 9.32 Synthetic route for [1,2,4]triazolo[4,3-*a*]pyridine sulfonamides.

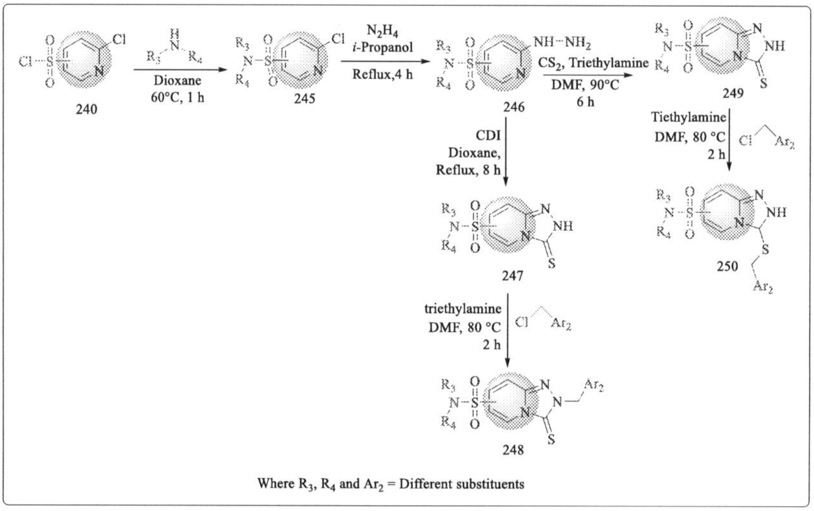

SCHEME 9.33 Synthetic route for [1,2,4]triazolo[4,3-*a*]pyridine sulfonamides.

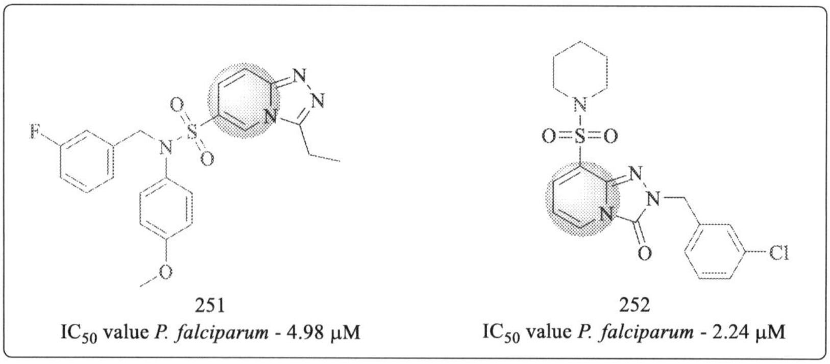

FIGURE 9.26 Most active molecules from [1,2,4]triazolo[4,3-*a*]pyridine sulfonamides.

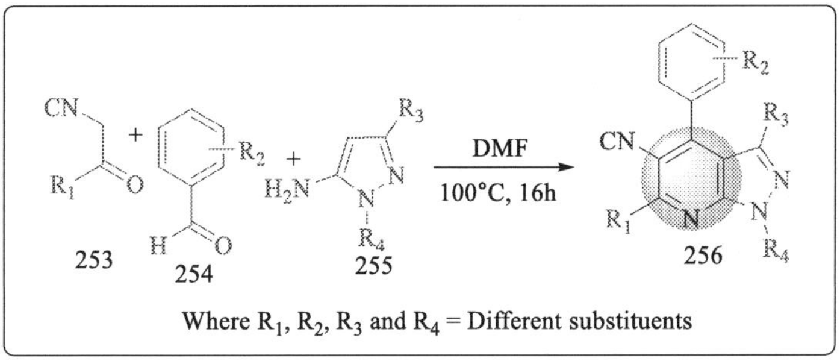

SCHEME 9.34 Synthetic route for pyrazolo[3,4-*b*]pyridines.

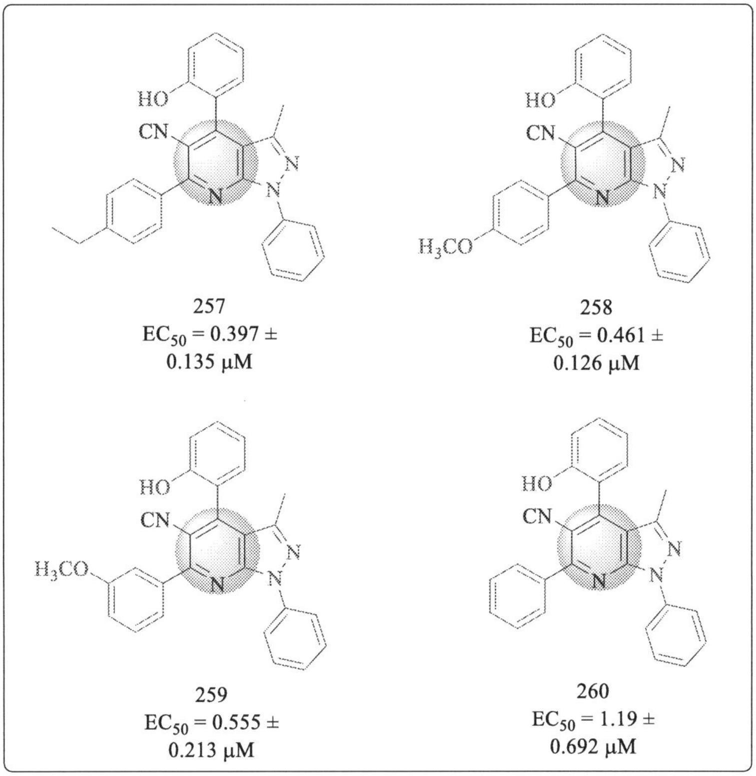

FIGURE 9.27 Most active molecules from pyrazolo[3,4-*b*]pyridines.

Eagon et al. [43] prepared pyrazolo[3,4-*b*]pyridines as antimalarial agents using one-pot, multicomponent synthetic procedures. Compounds - **256** were synthesized from cyanoacetophenones, **253**, benzaldehydes, **254** and 3-methyl-1-phenyl-1*H*-pyrazol-5-amine, **255** by heating at 100°C in DMF (Scheme 9.34).

The synthesized compounds were evaluated for antimalarial activity against chloroquine-sensitive *P. falciparum* 3D7 strain (MRA-102). The synthesized Compounds **257–260** possessed EC_{50} values in the range of 0.397–1.19 μM (Fig. 9.27).

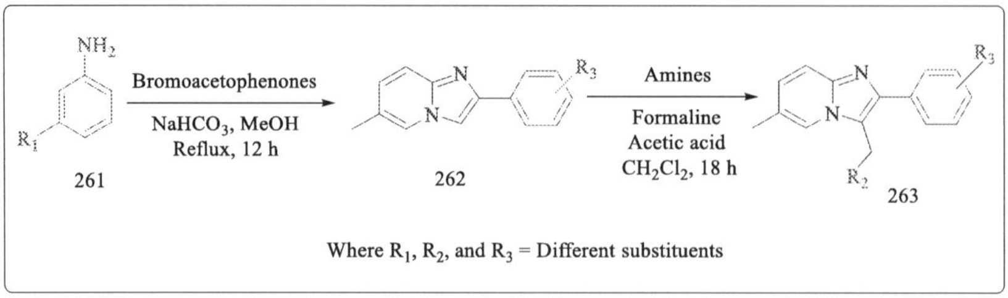

SCHEME 9.35 Synthetic route for imidazopyridine-based compounds.

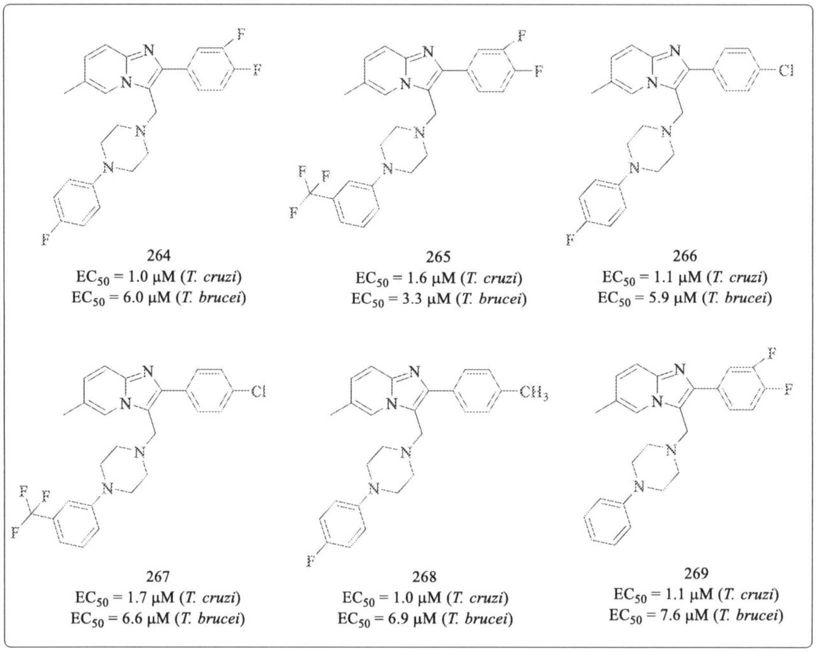

FIGURE 9.28 Most active molecules from imidazopyridine-based compounds.

9.9 Synthesis and anti-infective activity of some pyridine derivatives

Silva et al. prepared [44] imidazopyridine-based compounds as potential anti-infective agents against trypanosomiases. The compounds - **263** were synthesized from substituted amines, **261** as described in Scheme 9.35 by reacting it with bromoacetophenones. The intermediate - **262** were then treated with different amines to form compounds - **263**.

The synthesized derivatives were tested for *in vitro* anti-infective activity against *Trypanosomiases cruzi* and *Trypanosomiases brucei* and the activity is determined by EC_{50} and EC_{90}. The synthesized compounds possessed EC_{50} values in the range of 1.0 to 7.6 μM against *T. cruzi* and *T. brucei* (Fig. 9.28).

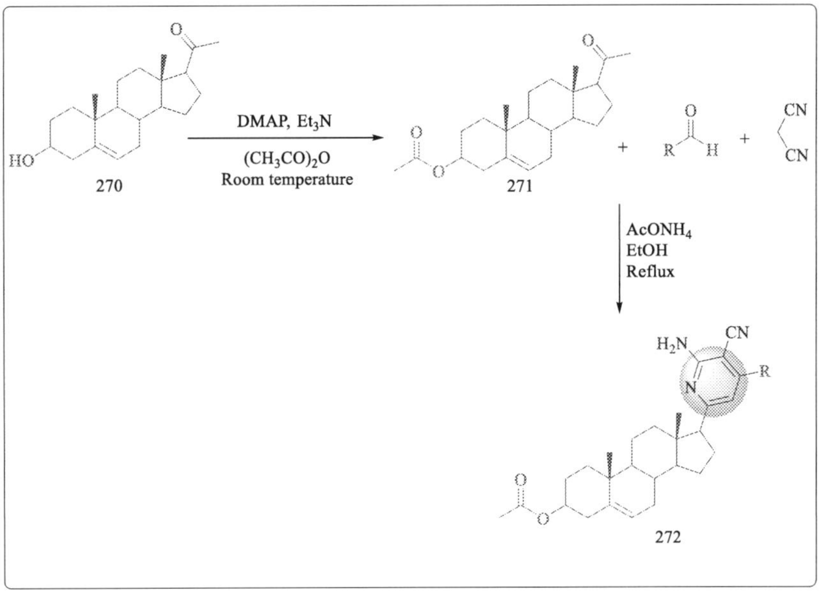

SCHEME 9.36 Synthetic route for steroidal pyridine compounds.

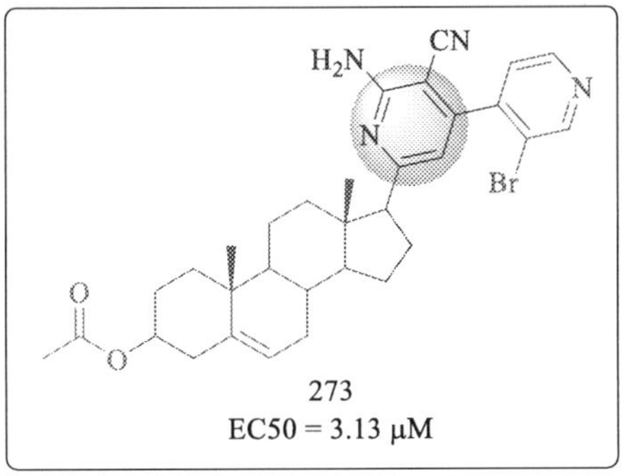

FIGURE 9.29 Most active molecules from steroidal pyridine compounds.

Wang et al. [45] prepared steroidal pyridine compounds as *in vitro* and *in vivo* antirespiratory syncytial virus agents. Compound - **272** was prepared using DMAP And ammonium acetate as Scheme 9.36.

The synthesized compounds were evaluated for antiviral activity against Respiratory syncytial virus (RSV) using ribavirin as a positive control (EC_{50} = 20.12 μM). The synthesized Compound - **273** possessed excellent antiviral activity with an EC_{50} value of 3.13 μM against Respiratory syncytial virus (RSV) (Fig. 9.29).

9.10 Conclusion

Traditionally regarded as a classical bioisosteric replacement, the substitution of a CH group with *N* atom in aromatic and heteroaromatic ring systems can have a wide range of impacts on molecular and physicochemical properties relevant to drug development.

In continuation to this, the efforts devoted toward the many applications of pyridine are highlighted in this chapter. The pyridine moiety has been demonstrated as an important chemical with numerous pharmacophore applications in this chapter. The biological actions of several substituted pyridines are significant. Anti-infective, anti-malarial, anti-cancer, anti-tubercular, anti-microbial, and anti-inflammatory effects of pyridine derivatives against a variety of biological targets and pyridine core-containing drugs with diverse biological or pharmacological activities were discovered and reviewed in this chapter. Furthermore, these strategies have been used to synthesize a number of useful drugs. We expect that this chapter will encourage the discovery of novel pharmacologically active pyridine compounds.

9.11 Graphical conclusion

Way Foreword,

The world is always searching for a potential lead in Medicinal Chemistry wherein the primary objective of new advancements in synthetic drug design is to offer rapid access to a variety of functionalized heterocyclic compounds that are critical for chemists to study. The Pyridine-based blend of potential lead provides an immense potential to expand the available drug-like chemical motifs and drive the development toward a more efficient designing offering targeted information in drug discovery. Additionally, the presence of pyridine nuclei in drug design creates robust synthetic pathways to study a variety of desired options that can help a chemist expedite the drug discovery process. Pyridine-containing compounds with unique heterocyclic designs enable various bonding strategies, which have a substantial impact on drug development. Future development of Pyridine-containing heterocycles has immense promise, potentially leading to a lead. The drug design concentrating on the bioisostere skeleton of known Pyridine molecular leads shall offer the opportunity of replacing an atom or a functional group, or a broadly similar, atom or active group, resulting in increased bioavailability and improved medication efficacy. Furthermore, the molecular docking of potential lead-in like anti-infective, antimalarial, anticancer, antitubercular, antimicrobial, and anti-inflammatory shall also offer the likely lead to the target. Considering the importance of pyridine-based heterocycles in the drug development process, the upcoming development shall focus on these molecules.

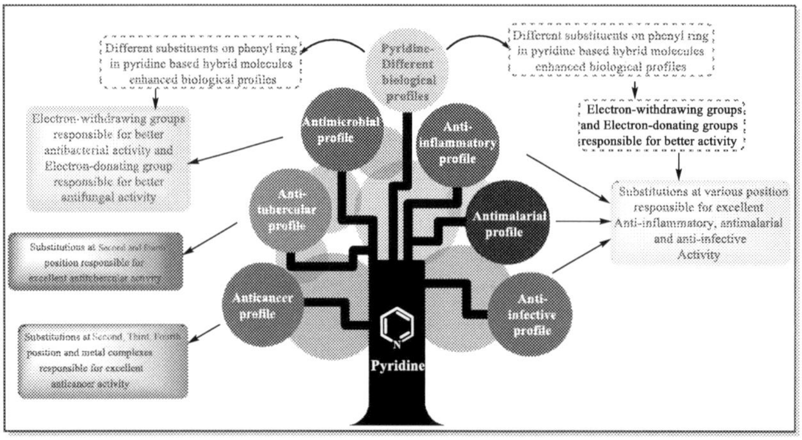

Acknowledgments

Prof. Nisheeth C. Desai is thankful to the University Grant Commission, New Delhi, for the financial support under BSR Faculty-Fellowship 2019 (F18-1/2011 (BSR)). Miss Jahnvi D. Monapara is grateful to the Department of Science and Technology, INSPIRE PROGRAM, for the award of INSPIRE Fellowship (IF180817). Miss Aratiba M. Jethawa is thankful to the SHODH—ScHeme Of Developing High quality research for providing research fellowship. The authors are also thankful to Priyanka Desai, founder of iScribblers, for the linguistic editing of the manuscript.

References

[1] T. Anderson, XIV.-on the constitution and properties of picoline, a new organic base from coal-tar, Trans. R. Soc. Edinburgh. 16 (1846) 123–136. https://doi.org/10.1017/S0080456800024984.

[2] M. Barczak, Synthesis and structure of pyridine-functionalized mesoporous SBA-15 organosilicas and their application for sorption of diclofenac, J. Solid State Chem. 258 (2018) 232–242. https://doi.org/10.1016/j.jssc.2017.10.006.

[3] X. Yu, X. Zhu, Y. Zhou, Q. Li, Z. Hu, T. Li, J. Tao, M. Dou, M. Zhang, Y. Shao, R. Sun, Discovery of N-Aryl-pyridine-4-ones as novel potential agrochemical fungicides and bactericides, J. Agric. Food Chem. 67 (2019) 13904–13913. https://doi.org/10.1021/acs.jafc.9b06296.

[4] X.L. Zhao, T. Jun, Y.N. Feng, H.F. Qian, W. Huang, Extremely high pH stability for a class of heterocyclic azo dyes having the common N2,N6-bis(3-methoxypropyl)pyridine-2,6-diamine coupling component, Dye. Pigment. 145 (2017) 315–323. https://doi.org/10.1016/j.dyepig.2017.06.030.

[5] H. Wang, L. Wei, C. Yang, J. Liu, J. Shen, A pyridine-Fe gel with an ultralow-loading Pt derivative as ORR catalyst in microbial fuel cells with long-term stability and high output voltage, Bioelectrochemistry 131 (2020) 107370. https://doi.org/10.1016/j.bioelechem.2019.107370.

[6] A.C. Reiersølmoen, D. Csókás, S. Øien-ØDegaard, A. Vanderkooy, A.K. Gupta, A.C.C. Carlsson, A. Orthaber, A. Fiksdahl, I. Pápai, M. Erdélyi, Catalytic activity of trans-bis(pyridine)gold complexes, J. Am. Chem. Soc. 142 (2020) 6439–6446. https://doi.org/10.1021/jacs.0c01941.

[7] H. Yang, F. Cai, Y. Luo, X. Ye, C. Zhang, S. Wu, The interphase and thermal conductivity of graphene oxide/butadiene-styrene-vinyl pyridine rubber composites: a combined molecular simulation and experimental study, Compos. Sci. Technol. 188 (2020) 107971. https://doi.org/10.1016/j.compscitech.2019.107971.

[8] L.D. Pennington, D.T. Moustakas, The necessary nitrogen atom: a versatile high-impact design element for multiparameter optimization, J. Med. Chem. 60 (2017) 3552–3579. https://doi.org/10.1021/acs.jmedchem.6b01807.

[9] N.C. Desai, B.Y. Patel, B.P. Dave, Synthesis and antimicrobial activity of novel quinoline derivatives bearing pyrazoline and pyridine analogues, Med. Chem. Res. 26 (2017) 109–119. https://doi.org/10.1007/s00044-016-1732-6.

[10] H. Wang, A. Wang, J. Gu, L. Fu, K. Lv, C. Ma, Z. Tao, B. Wang, M. Liu, H. Guo, Y. Lu, Synthesis and antitubercular evaluation of reduced lipophilic imidazo[1,2-a]pyridine-3-carboxamide derivatives, Eur. J. Med. Chem. 165 (2019) 11–17. https://doi.org/10.1016/j.ejmech.2018.12.071.

[11] J. Deng, P. Yu, Z. Zhang, J. Wang, J. Cai, N. Wu, H. Sun, H. Liang, F. Yang, Designing anticancer copper(II) complexes by optimizing 2-pyridine-thiosemicarbazone ligands, Eur. J. Med. Chem. 158 (2018) 442–452. https://doi.org/10.1016/j.ejmech.2018.09.020.

[12] R.A. Azzam, R.E. Elsayed, G.H. Elgemeie, Design and synthesis of a new class of pyridine-based N-sulfonamides exhibiting antiviral, antimicrobial, and enzyme inhibition characteristics, ACS Omega 5 (2020) 26182–26194. https://doi.org/10.1021/acsomega.0c03773.

[13] E.M Kirwen, T. Batra, C. Karthikeyan, G.S. Deora, V. Rathore, C. Mulakayala, N. Mulakayala, A.C. Nusbaum, J. Chen, H. Amawi, K. McIntosh, Sahabjada, N. Shivnath, D. Chowarsia, N. Sharma, M. Arshad, P. Trivedi, A.K. Tiwari, 2,3-Diaryl-3H-imidazo[4,5-b]pyridine derivatives as potential anticancer and anti-inflammatory agents, Acta Pharm. Sin. B. 7 (2017) 73–79. https://doi.org/10.1016/j.apsb.2016.05.003.

[14] Z.L. Ren, L. Su, S. Cai, W.T. Lu, Y. Qiao, P. He, M.W. Ding, Synthesis of polysubstituted pyridine derivatives via sequential AlCl3-catalyzed condensation/aza-wittig/isomerization reactions and a study of their antifungal activities, Asian J. Org. Chem. 8 (2019) 1394–1397. https://doi.org/10.1002/ajoc.201900291.

[15] Y. Kaddouri, F. Abrigach, E.B. Yousfi, M. El Kodadi, R. Touzani, New thiazole, pyridine and pyrazole derivatives as antioxidant candidates: synthesis, DFT calculations and molecular docking study, Heliyon 6 (2020), doi:10.1016/j.heliyon.2020.e03185.

[16] G. Komeili, F. Ghasemi, A.R. Rezvani, K. Ghasemi, F. Khadem Sameni, M. Hashemi, The effects of a new antidiabetic glycinium [(pyridine-2, 6-dicarboxylato) oxovanadate (V)] complex in high-fat diet of streptozotocin-induced diabetic rats, Arch. Physiol. Biochem. (2019) 1–7, doi:10.1080/13813455.2019.1663218.

[17] P.O. Miranda, B. Cubitt, N.T. Jacob, K.D. Janda, J.C. De La Torre, Mining a Kröhnke pyridine library for anti-arenavirus activity, ACS Infect. Dis. 4 (2018) 815–824. https://doi.org/10.1021/acsinfecdis.7b00236.

[18] E. Knoevenagel, A. Fries, Synthesen in der Pyridinreihe. Ueber eine Erweiterung der Hantzsch'schen Dihydropyridinsynthese, Ber. Dtsch. Chem. Ges. 31 (1898) 761–767. 10.1002/cber.189803101157.

[19] J.J. Li, E.J. Corey, Name reactions in heterocyclic chemistry, Name React. Heterocycl. Chem. (2005) 1–558, doi:10.1002/0471704156.

[20] F. Kroehnke, The specific synthesis of pyridines and oligopyridines, Synthesis 1 (1976) 1–24. 10.1055/s-1976-23941.

[21] M. Kurva, S.G. Pharande, A. Quezada-Soto, R. Gámez-Montaño, Ultrasound assisted green synthesis of bound type bis-heterocyclic carbazolyl imidazo[1,2-a]pyridines via Groebke-Blackburn-Bienayme reaction, Tetrahedron Lett. 59 (2018) 1596–1599. https://doi.org/10.1016/j.tetlet.2018.03.031.

[22] N. Kerru, L. Gummidi, S. Maddila, S.B. Jonnalagadda, Gadolinium oxide loaded zirconia and multi-component synthesis of novel dihydro-pyrazolo[3,4-d]pyridines under green conditions, Sustain. Chem. Pharm. 18 (2020) 100316. https://doi.org/10.1016/j.scp.2020.100316.

[23] M.A.A. Radwan, M.A. Alshubramy, M. Abdel-Motaal, B.A. Hemdan, D.S. El-Kady, Synthesis, molecular docking and antimicrobial activity of new fused pyrimidine and pyridine derivatives, Bioorg. Chem. 96 (2020) 103516. https://doi.org/10.1016/j.bioorg.2019.103516.

[24] N.C. Desai, K.A. Jadeja, D.J. Jadeja, V.M. Khedkar, P.C. Jha, Design, synthesis, antimicrobial evaluation, and molecular docking study of some 4-thiazolidinone derivatives containing pyridine and quinazoline moiety, Synth. Commun. 51 (2021) 952–963. https://doi.org/10.1080/00397911.2020.1861302.

[25] V. Kamat, R. Santosh, B. Poojary, S.P. Nayak, B.K. Kumar, M. Sankaranarayanan, S.K Faheem, D.A. Barretto, S.K. Vootla, Pyridine- and thiazole-based hydrazides with promising anti-inflammatory and antimicrobial activities along with their in silico studies, ACS Omega 5 (2020) 25228–25239. https://doi.org/10.1021/acsomega.0c03386.

[26] A.M.A. Hamid, H.A. El-Sayed, S.M. Mohammed, A.H. Moustafa, H.A. Morsy, Functionalization of 1,2,3-triazole to pyrimidine, pyridine, pyrazole, and isoxazole fluorophores with antimicrobial activity, Russ. J. Gen. Chem. 90 (2020) 476–482. https://doi.org/10.1134/S1070363220030226.

[27] T.H. Al-Tel, R.A. Al-Qawasmeh, Post groebke-blackburn multicomponent protocol: synthesis of new polyfunctional imidazo[1,2-a]pyridine and imidazo[1,2-a]pyrimidine derivatives as potential antimicrobial agents, Eur. J. Med. Chem. 45 (2010) 5848–5855. https://doi.org/10.1016/j.ejmech.2010.09.049.

[28] N.C. Desai, N.B. Bhatt, S.B. Joshi, K.A. Jadeja, V.M. Khedkar, Synthesis, antimicrobial activity and 3D-QSAR study of hybrid oxazine clubbed pyridine scaffolds, ChemistrySelect 4 (2019) 7541–7550. https://doi.org/10.1002/slct.201901391.

[29] S. Bodige, P. Ravula, K.C. Gulipalli, S. Endoori, P.K.R. Cherukumalli, J.N. N.S.C., N. Seelam, Design, synthesis, antitubercular and antibacterial activities of pyrrolo[3,2-b]pyridine-3-carboxamide linked 2-methoxypyridine derivatives and in silico docking studies, Synth. Commun. 49 (2019) 2219–2234. https://doi.org/10.1080/00397911.2019.1618874.

[30] A.N. Ambhore, S.S. Kamble, S.N. Kadam, R.D. Kamble, M.J. Hebade, S.V. Hese, M.V. Gaikwad, R.J. Meshram, R.N. Gacche, B.S. Dawane, Design, synthesis and in silico study of pyridine based 1,3,4-oxadiazole embedded hydrazinecarbothioamide derivatives as potent anti-tubercular agent, Comput. Biol. Chem. 80 (2019) 54–65. https://doi.org/10.1016/j.compbiolchem.2019.03.002.

[31] R. Reddyrajula, U.K. Dalimba, Structural modification of zolpidem led to potent antimicrobial activity in imidazo[1,2-: A] pyridine/pyrimidine-1,2,3-triazoles, New J. Chem. 43 (2019) 16281–16299. https://doi.org/10.1039/c9nj03462e.

[32] N.C. Desai, A. Trivedi, H. Somani, K.A. Jadeja, D. Vaja, L. Nawale, V.M. Khedkar, D. Sarkar, Synthesis, biological evaluation, and molecular docking study of pyridine clubbed 1,3,4-oxadiazoles as potential antituberculars, Synth. Commun. 48 (2018) 524–540. https://doi.org/10.1080/00397911.2017.1410892.

[33] N.C. Desai, H. Somani, A. Trivedi, K. Bhatt, L. Nawale, V.M. Khedkar, P.C. Jha, D. Sarkar, Synthesis, biological evaluation and molecular docking study of some novel indole and pyridine based 1,3,4-oxadiazole derivatives

as potential antitubercular agents, Bioorganic Med. Chem. Lett. 26 (2016) 1776–1783. https://doi.org/10.1016/j.bmcl.2016.02.043.

[34] H.R.M. Rashdan, I.A. Shehadi, A.H. Abdelmonsef, Synthesis, anticancer evaluation, computer-aided docking studies, and ADMET prediction of 1,2,3-triazolyl-pyridine hybrids as human aurora B kinase inhibitors, ACS Omega 6 (2021) 1445–1455. https://doi.org/10.1021/acsomega.0c05116.

[35] M.E. Abdelaziz, M.M.M. El-Miligy, S.M. Fahmy, M.A. Mahran, A.A. Hazzaa, Design, synthesis and docking study of pyridine and thieno[2,3-b] pyridine derivatives as anticancer PIM-1 kinase inhibitors, 2018. 10.1016/j.bioorg.2018.07.024.

[36] S. Şenkardeş, A. Türe, S. Ekrek, A.T. Durak, M. Abbak, Ö. Çevik, B. Kaşkatepe, İ. Küçükgüzel, G. Küçükgüzel, Novel 2,6-disubstituted pyridine hydrazones: synthesis, anticancer activity, docking studies and effects on caspase-3-mediated apoptosis, J. Mol. Struct. 1223 (2021) 1–9. https://doi.org/10.1016/j.molstruc.2020.128962.

[37] N.M. Saleh, A.A.H. Abdel-Rahman, A.M. Omar, M.M. Khalifa, K. El-Adl, Pyridine-derived VEGFR-2 inhibitors: Rational design, synthesis, anticancer evaluations, in silico ADMET profile, and molecular docking, Arch. Pharm. (Weinheim). 354 (2021), doi:10.1002/ardp.202100085.

[38] N. Raman, P.P. Utthra, T. Chellapandi, Insight into the in vitro anticancer screening, molecular docking and biological efficiency of pyridine-based transition metal(II) complexes, J. Coord. Chem. 73 (2020) 103–119. https://doi.org/10.1080/00958972.2020.1716218.

[39] J. Dennis Bilavendran, A. Manikandan, P. Thangarasu, K. Sivakumar, Synthesis and discovery of pyrazolo-pyridine analogs as inflammation medications through pro- and anti-inflammatory cytokine and COX-2 inhibition assessments, Bioorg. Chem. 94 (2020) 103484. https://doi.org/10.1016/j.bioorg.2019.103484.

[40] S.N. Shringare, H.V. Chavan, P.S. Bhale, S.B. Dongare, Y.B. Mule, S.B. Patil, B.P. Bandgar, Synthesis and pharmacological evaluation of combretastatin-A4 analogs of pyrazoline and pyridine derivatives as anticancer, anti-inflammatory and antioxidant agents, Med. Chem. Res. 27 (2018) 1226–1237. https://doi.org/10.1007/s00044-018-2142-8.

[41] H.A.H. Elshemy, M.A. Zaki, E.I. Mohamed, S.I. Khan, P.F. Lamie, A multicomponent reaction to design antimalarial pyridyl-indole derivatives: Synthesis, biological activities and molecular docking, Bioorg. Chem. 97 (2020) 103673. https://doi.org/10.1016/j.bioorg.2020.103673.

[42] V.R. Karpina, S.S. Kovalenko, S.M. Kovalenko, O.G. Drushlyak, N.D. Bunyatyan, V.A. Georgiyants, V.V. Ivanov, T. Langer, L. Maes, A novel series of [1, 2, 4] triazolo [4, 3-a] pyridine sulfonamides as potential antimalarial agents: in silico studies, synthesis and in vitro evaluation, Molecules 25 (2020) 4485. https://doi.org/10.3390/molecules25194485.

[43] S. Eagon, J.T. Hammill, M. Sigal, K.J. Ahn, J.E. Tryhorn, G. Koch, B. Belanger, C.A. Chaplan, L. Loop, A.S. Kashtanova, K. Yniguez, H. Lazaro, S.P. Wilkinson, A.L. Rice, M.O. Falade, R. Takahashi, K. Kim, A. Cheung, C. Dibernardo, J.J. Kimball, E.A. Winzeler, K. Eribez, N. Mittal, F.J. Gamo, B. Crespo, A. Churchyard, I. García-Barbazán, J. Baum, M.O. Anderson, B. Laleu, R.K. Guy, Synthesis and structure-activity relationship of dual-stage antimalarial pyrazolo[3,4- b]pyridines, J. Med. Chem. 63 (2020) 11902–11919. https://doi.org/10.1021/acs.jmedchem.0c01152.

[44] D.G. Silva, A. Junker, S.M.G. de Melo, F. Fumagalli, J.R. Gillespie, N. Molasky, F.S. Buckner, A. Matheeussen, G. Caljon, L. Maes, F.S. Emery, Synthesis and structure−activity relationships of imidazopyridine/pyrimidine- and furopyridine-based anti-infective agents against trypanosomiases, ChemMedChem 16 (2021) 966–975. https://doi.org/10.1002/cmdc.202000616.

[45] Z. Wang, D. Hou, J. Fang, L. Zhu, Y. Sun, Y. Tan, Z. Gu, L. Shan, Screening and pharmacodynamic evaluation of the antirespiratory syncytial virus activity of steroidal pyridine compounds in vitro and in vivo, J. Med. Virol. 93 (2021) 3428–3438. https://doi.org/10.1002/jmv.26604.

CHAPTER

10

Synthesis of pyridine derivatives using multicomponent reactions

Shah Imtiaz, Bhoomika Singh and Md. Musawwer Khan

Department of Chemistry, Aligarh Muslim University, Aligarh, India

10.1 Introduction

Nitrogen-containing heterocycles are of great importance and have occupied a prominent place in medicinal chemistry. These N-containing heterocycles have been used extensively as agrochemicals [1], pharmaceuticals [2], cosmetics [3], and have been found as a constituent in many naturally occurring organic compounds. These, N-heterocyclic architectures though look simple but, their synthesis is a challenging task. Multicomponent synthesis [4] has made the synthesis of the complex heterocycles much easier as compared to other synthetic methods. A large number of N-containing heterocyclic compounds of biological importance have been synthesized using one-pot multicomponent approach. The heterocyclic derivatives synthesized from multicomponent reactions (MCRs) include pyrrole derivatives [5], pyrimidine derivatives [6], phosphonate derivatives [7], pyridopyrimidine derivatives [8], pyridine derivatives [9,10], piperidine derivatives [11], pyridazine derivatives [12], quinazolinone derivatives [13], oxindole derivatives [14], pederin, and psymberin derivatives [15]. Few heterocyclic derivatives with biomedical significance are given in Fig. 10.1.

In this chapter, our main focus is the heterocyclic derivatives containing pyridine rings and their synthesis by MCRs. These derivatives are of high importance because of their biological applications and other applications in products of daily use. Patilhis and coworkers [16] have synthesized the novel pyrazolo[3,4-b][1,8]naphthyridine derivatives **I** using an efficient one-pot multicomponent approach. In this reaction 2-aminopyridine, 3-methyl-1-phenyl-2-pyrazolin-5-one and aromatic aldehydes were reacted in a single pot. The reagent used was 15 mol% of phosphorus oxide in refluxing ethanol. It was studied further that the obtained pyridine derivative acts as an anti-inflammatory agent. In another study, Ghorbani-Vaghei and his team reported the synthesis of a pyridine derivative, bis-dihydrotetrazolo[1,5-a]pyrimidine **II** [17]. In this reaction, diacetyl pyridine was reacted with 5-aminotetrazole

Recent Developments in the Synthesis and Applications of Pyridines.
DOI: https://doi.org/10.1016/B978-0-323-91221-1.00006-3

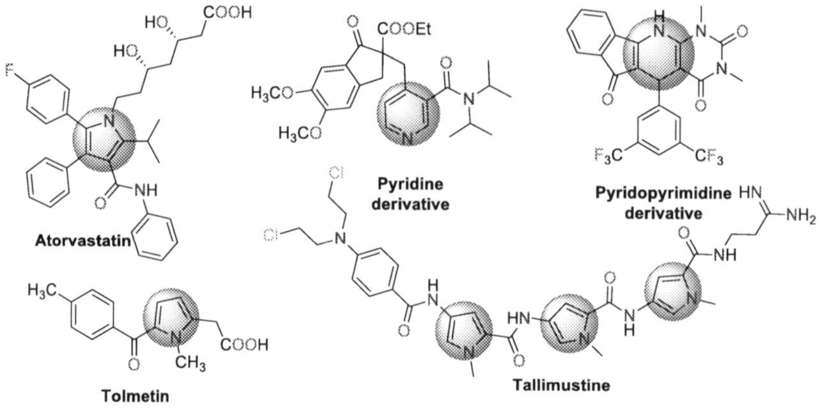

FIGURE 10.1 Few heterocyclic derivatives having biological significance synthesized by MCRs.

and two equivalents of aldehydes in a single pot in the presence of tetrabromobenzene-1,3-disulfonamide (TBBDA) at about 80°C. A series of pyrazolopyridine derivatives **V** have been synthesized via MCRs. These pyrimidine derivatives have shown inhibitory characters to protein kinases like ALK tyrosine kinase. A biologically important 3-sulfenylimidazo [1,2-a]pyridine derivatives **III** were reported by S. Kundu and B. Basu using a one-pot multicomponent approach [18]. The catalyst used was a green carbocatalyst of graphene oxide and NaI was used as an additive in this reaction. Further a calcium channel blocker, nifidipene, marketed as Procardia was synthesized using Hantzsch three component reactions [19].

10.2 Importance of pyridine and its derivatives

Pyridine ring is present in many organic compounds that have a lot of applications in pharmaceuticals, agrochemicals, and is also found in some vitamins. Few applications of pyridine are given below. Some pyridine derivatives synthesized by MCRS are given in Fig. 10.2 and few biologically active pyridine derivatives are given in Fig. 10.3.

(a) *Pyridine as precursor to important chemicals.*
Pyrithione-based fungicides are synthesized from pyridine [20]. Cetylpyridinium and laurylpyridinium have been used as an antiseptic in dental and oral care products are also obtained from pyridine (zincke reaction). Insecticide chlorpyrifos is obtained from chlorination of pyridine. The pyridine is used as a precursor in the synthesis of herbicides named paraquat (Scheme 10.1) and diquat [21].

(b) *Pyridine as a reagent in many reactions.*
Pyridine has a lone pair on the N-atom in it that gives it the Lewis base character. Therefore, it is used as a base in many reactions to abstract the acidic protons. It is also used to neutralize the excess of acid present in the reaction mixture. Because of its basic nature it is used as Karl Fischer reagent. Furthermore, it has a pleasant odor and can be used in place of those basic reagents having bad smell such as imidazole. It has also been used in the reagents required in the oxidation of alcohols such as PCC, PDC, and Collins reagents, etc. [22].

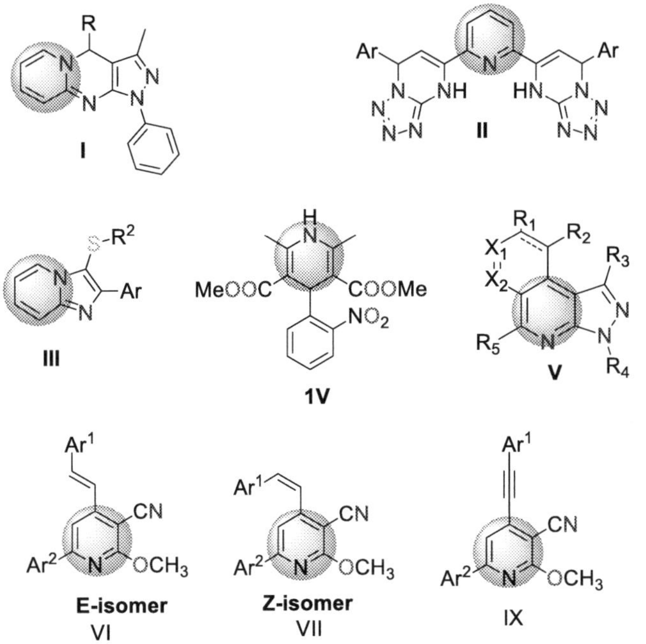

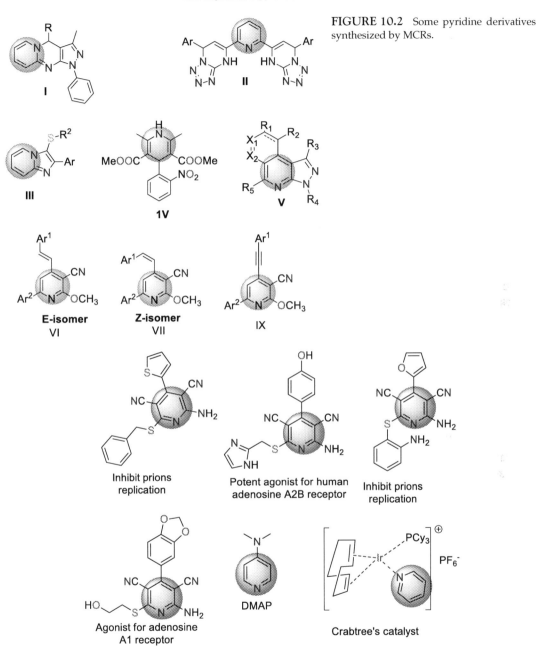

FIGURE 10.2 Some pyridine derivatives synthesized by MCRs.

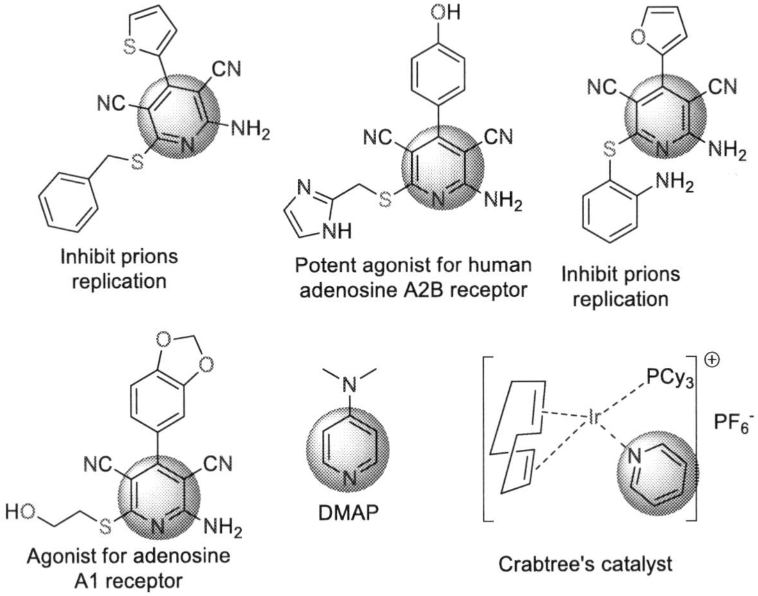

FIGURE 10.3 Some pyridine derivatives showing biological activity and used as reagents.

(c) *Pyridine as a solvent.*

Pyridine is basic, less reactive, and polar solvent. It is used in many condensation reactions such as Knoevenagel condensations [20]. Pyridine is generally used as a solvent in those reactions in which dehalogenation process is required. After dehalogenation pyridine gets

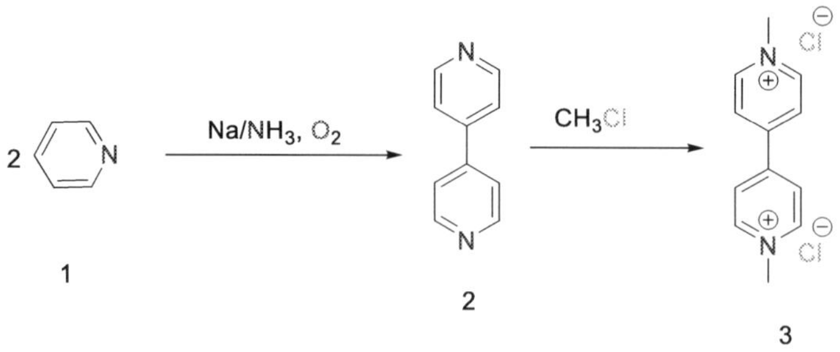

SCHEME 10.1 Synthesis of paraquat.

converted into pyridinium salt with a hydrogen atom attached to the N-atom of pyridine. Pyridine activates acid halides and acidic anhydrides of carboxylic acid in acylation and esterification reactions. Furthermore, pyridine derivatives such as 4-(1-pyrrolidinyl) pyridine and 4-dimethylaminopyridine (DMAP) are used to activate these reactions more efficiently as compared to pure pyridine. Pyridine has found its application in improving the network capacity of cotton in textile industries.

Pyridine derivatives have also found many applications in daily life. Many natural products of biological importance have shown pyridine as a structural subunit in them. In addition, these pyridine derivatives have found many applications in pharmaceuticals and medicines. The diseases like Parkinson's disease, asthma, hypoxia, cancer, epilepsy, and cardiovascular diseases are related to adenosine receptors [23]. Many derivatives of pyridines have shown inhibiting activity against adenosine receptors therefore, these derivatives can be used to treat above-mentioned diseases [24]. Pyridine derivatives can also be used to treat neurodegenerative diseases because of their ability to inhibit cholinesterases [25]. Moreover, pyridine derivatives are used as antibacterial, ant-HBV, anti-infective and antibiofilm agents [26,27]. Furthermore, these have been used to treat Creutzfeldt–Jacob disease because of their ability to inhibit prion replication [28]. Another derivative of pyridine, Crabtree's catalyst has been used as a homogenous catalyst in hydrogenation and hydrogen transfer reactions [29].

10.3 Classification of pyridines

10.3.1 Aromatic

10.3.1.1 Simple Pyridines

Pyridine (azabenzene) is a heterocyclic organic compound with chemical formula C_5H_5N that has similar structure to benzene with one –CH group replaced by N-atom. It is a colorless material, water miscible, weakly alkaline, and highly flammable. Pyridine was produced from coal tar. The structure of pyridine is given in Fig. 10.4. The derivatives of pyridine can be obtained from the substitution of different groups at positions 1–5. Sometimes the substituent can be added to N-atom of pyridine by the formation of a coordinate bond.

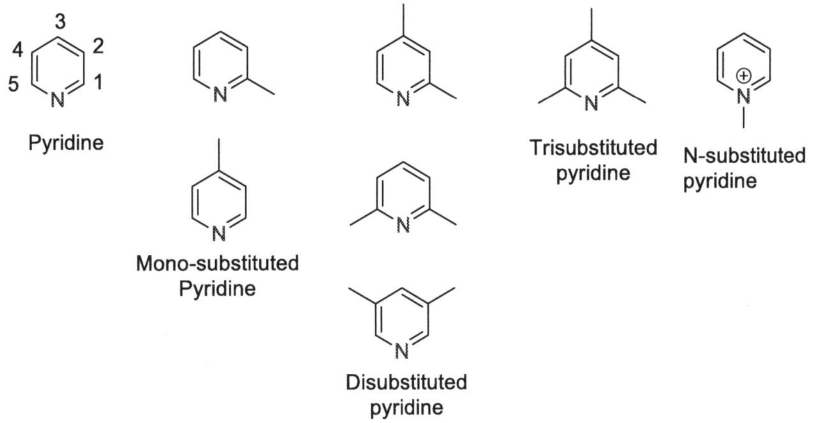

FIGURE 10.4 Possible substitution of simple pyridine to get different derivatives.

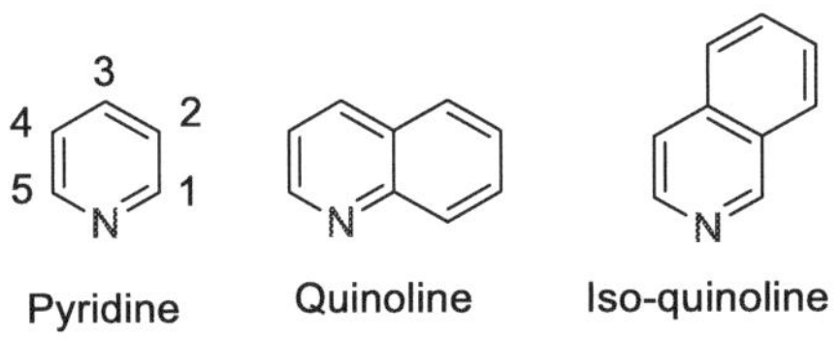

FIGURE 10.5 Chemical structure of benzofused pyridines.

10.3.1.2 *Fused pyridines*

10.3.1.2.1 Benzofused pyridines

In these types of pyridines the pyridine ring is fused with the benzene ring. The benzene ring can show the fusion at C1–C2 or C4–C5 and C2–C3 or C3–C4. If the fusion of benzene ring occurs at C1–C2 or C4–C5 the obtained structure is known as quinoline. Furthermore, if the fusion of benzene occurs at C2–C3 or C3–C4 of pyridine the structure obtained is named isoquinoline (Fig. 10.5).

10.3.1.2.2 Nonbenzofused pyridines

In these pyridines the nonbenzenoids aromatic rings such as pyrrole, pyridine itself, etc. are fused to pyridine ring. The fusion occurs in the same positions as shown in above section. Few examples possible from fusion of nonbenzenoids to pyridines are given in Fig. 10.6. Further substitution of these pyridines results in a vast variety of compounds of biological interest.

10.3.2 Nonaromatic

10.3.2.1 *Dihydropyridines*

Dihydropyridines are also a class of pyridines which are obtained from the hydrogenation of aromatic pyridine. The molecular formula of dihydropyridine is given as C_5H_7N. These pyridines are of two types 1,2-dihydropyridines and 1,4-dihydropyridines. The structures of

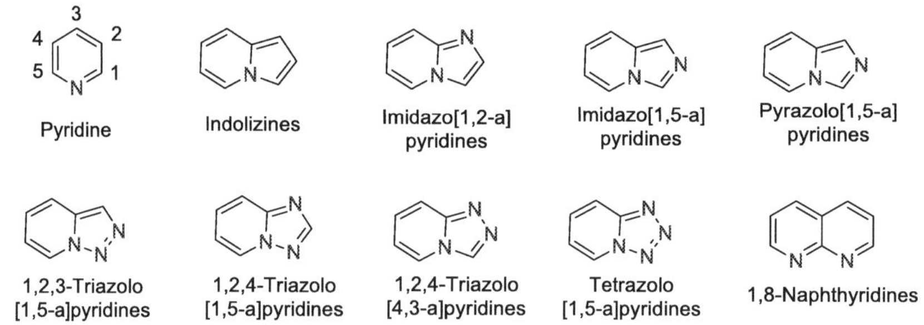

FIGURE 10.6 Chemical structure of nonbenzofused pyridines.

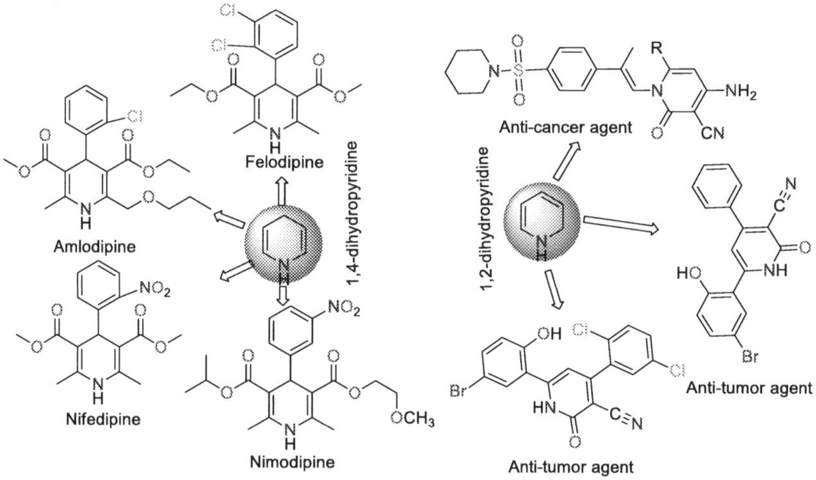

FIGURE 10.7 Some derivatives of dihydropyridines having biological importance.

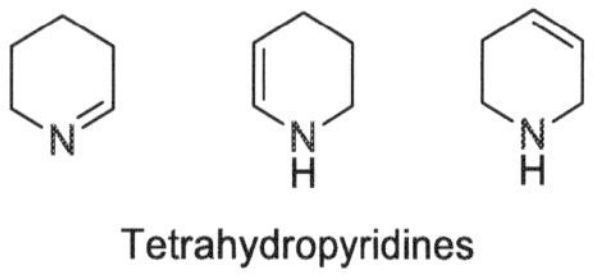

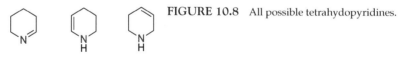

FIGURE 10.8 All possible tetrahydopyridines.

dihydropyridines are given in Fig. 10.7. These pyridines on further substitution with different groups result in a large number of derivatives having biological importance.

10.3.2.2 *Tetrahydropyridines*

Tetrahydropyridine (Fig. 10.8) having molecular formula C_4H_8N is also a type of pyridine with two double bonds reduced and having no aromaticity. The one double bond may be either present between the nitrogen atom or next carbon to nitrogen or it can be present between the two carbons next to the nitrogen atom as shown in figure below. Further substitutions of these pyridines with different substituents can form a library of compounds with different pharmaceutical properties. Moreover, these tetrahydropyridines are found in many important derivatives in fused form.

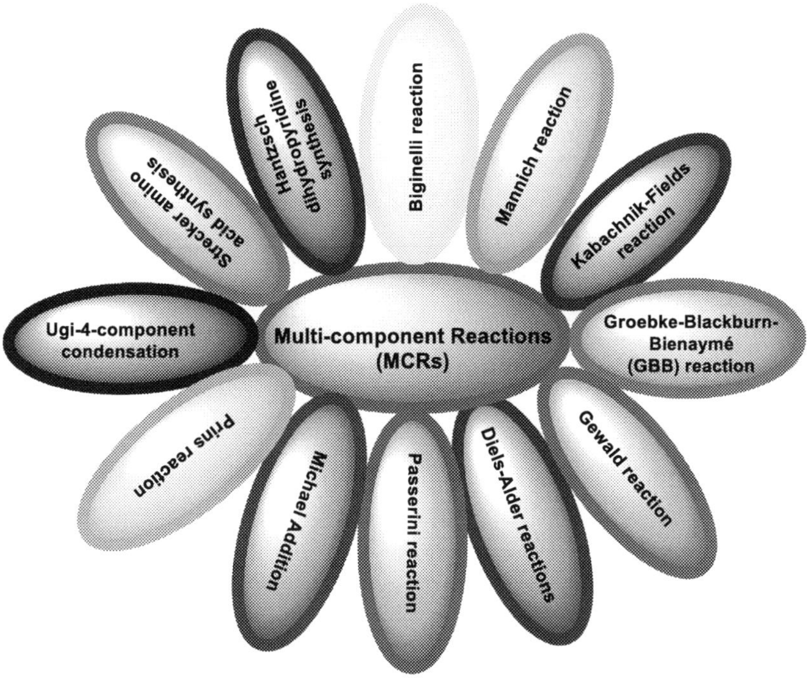

FIGURE 10.9 Some well-known multicomponent reactions.

10.4 Multicomponent reactions: its importance and green features

To achieve the synthesis of biologically active organic molecules and complex heterocyclic molecules, there is a need for efficient, economic, and environmentally benign methods. Multicomponent reactions (MCRs) have attracted the attention of many organic chemists because of their use in many organic transformations. These reactions show remarkable advantages over the multistep conventional bimolecular reactions. The concept of MCR is considered a perspective of green chemistry in which three or more than three reactant materials react in one-pot to provide the desired product selectively and also reduces the waste of reaction [30]. Some of the outstanding benefits of MCRs are high atom economy, short reaction times, operational simplicity, less by-products, good yields [31]. These reactions do not require the purification of intermediates and complex isolation. MCRs have found a new dimension in the production of new molecular frameworks and biologically active compounds to be used as potential drugs. These have been extensively used to synthesize heterocyclic compounds showing different biological activities such as anticancer activity, antimalarial, antiviral, antianxiety, antiparasitic activity, etc. [32]. Some of established MCRs include Strecker amino acid synthesis, Ugi-4-component condensation, Hantzsch dihydropyridine synthesis, Mannich reaction, Biginelli reaction, Kabachnik–Fields reaction, Gewald reaction, Groebke–Blackburn–Bienaymé (GBB) reaction, etc. (Fig. 10.9) [33].

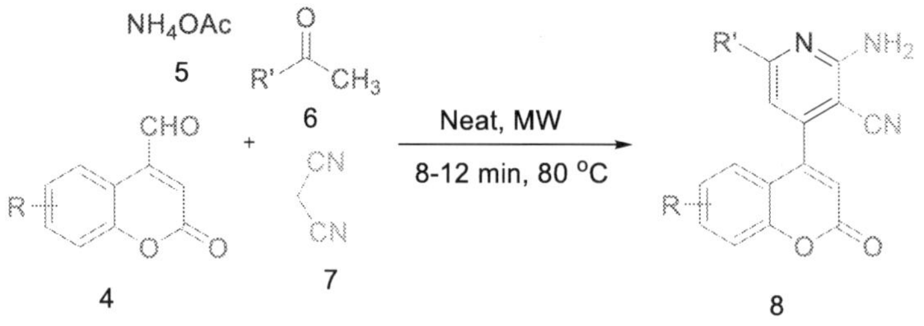

SCHEME 10.2 Synthesis of coumarinyl substituted pyridine.

In recent times MCRs in water have gained much attraction of chemistry because water is safe, environmentally benign, and cheap solvent [34]. Furthermore, organic products obtained from MCRs in aqueous medium can be easily isolated by simple phase separation. It has been observed in many studies related to MCRs that the rate of reaction enhances in aqueous medium as compared to other solvents. This significant enhancement in reaction rate can be attributed to the polarity of the solvent, hydrophobic packing of solvent, and H-bonding in solvent [35]. Aqueous solvents have shown a significant effect on reaction rate and selectivity in the organic transformation of few reactions such as aldol reactions, Diels–Alder reactions, and Michael additions [36,37].

10.5 Synthesis of simple pyridine derivatives using multicomponent reactions

10.5.1 Ammonium acetate as a source of nitrogen to pyridine ring

In 2017, Chougala et al. have reported the synthesis and anticancer activity of novel 2-amino-4-(2-oxo-2H-chromen-4-yl)-6-arylpyridine-3-carbonitrile derivatives **8** [38]. In this reaction four components, that is, 4-formylcoumarins **4**, ketones **6**, malononitrile **7**, and ammonium acetate **5** were reacted in a pot (Scheme 10.2). The coupling reaction between these compounds under neat microwave conditions resulted in desired pyridine derivatives.

Heravil et al. have synthesized polyfunctional derivatives of pyridines **10** and pyrans in an extremely efficient, one-pot multicomponent approach. In this synthesis a highly effective heterogeneous catalyst, nano-Fe_3O_4 was utilized that provides a large surface area for the reactants to get converted into the product. The catalyst can be recycled and used in an eco-friendly manner and can also be used for at least four times without any loss in catalytic activity. In this reaction an aldehyde **9**, ketone **6**, malononitrile **7**, and a ketone **6** was reacted to yield pyridine derivatives **10** in good yield in the presence of nano-Fe_3O_4 (Scheme 10.3) [39].

Thirumurugan and his coworkers developed a synthetic methodology having simple and eco-friendly approach to prepare indol-3-yl pyridinyl derivatives **14** through MCR [40]. In this synthesis a number of structurally diverse aldehydes **12**, 2-acetylfuran (or) 2-acetylpyridine **13** and 3-cyanoacetyl indole **11** were reacted in ammonium acetate **5** using one-pot four component approach under neat condition (Scheme 10.4). The versatility of this approach provides a valuable methodology for the synthesis of Kr€ohnke pyridines.

Mohammadi has developed a simple and efficient synthesis of 2,4,6-trisubstituted pyridines **16** utilizing microwave irradiation and solvent free conditions in the presence of γ-MnO_2

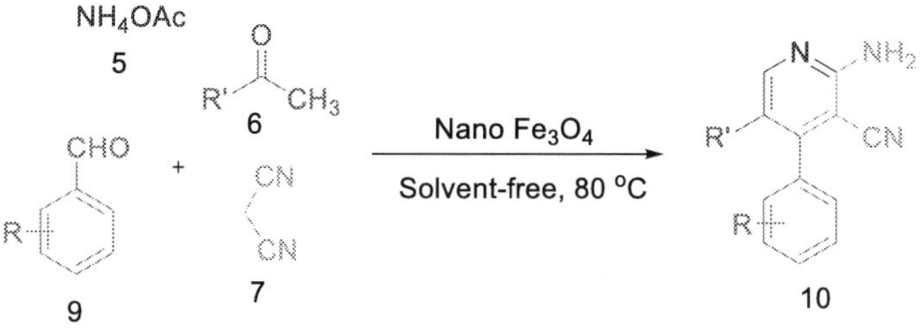

SCHEME 10.3 Synthesis of 2-amino-3-cyanopyridine derivatives.

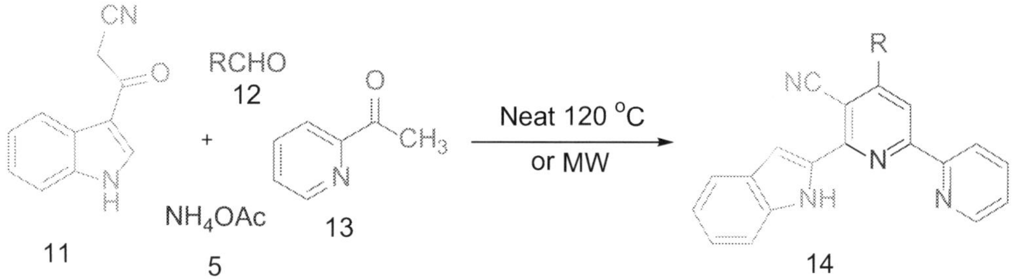

SCHEME 10.4 Synthesis of functionalized indol-3-yl pyridine derivatives.

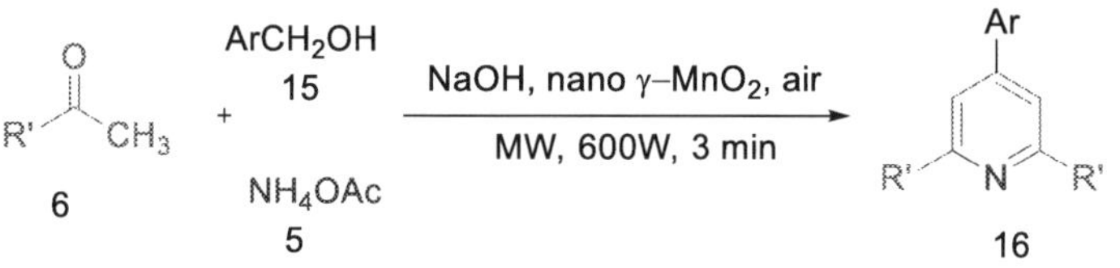

SCHEME 10.5 Synthesis of 2,4,6-trisubstituted pyridines using nano γ-MnO_2.

nanoparticles [41]. In this synthesis he reacted acetophenone **6**, benzyl alcohols **15**, and ammonium acetate **5** as a nitrogen source in a single pot using multicomponent approach (Scheme 10.5). The pyridine derivatives were obtained in excellent yields and short reaction times were required in this methodology. Furthermore, the use of simple chemicals, mild reaction conditions and easy availability of starting materials are the main advantages of this methodology.

Recently Reddy and his team have synthesized a new class of 2-amino-4-(3/2-(alkynyl)/3-(alkenyl)phenyl)-6-phenylnicotinonitriles **19** using one-pot multicomponent approach. In this multicomponent reaction the mixture of malononitrile **7**, bromo-benzaldehyde **17**, NH_4OAc **5**, acetophenone **18**, and a series of alkene/terminal alkynes was reacted (Scheme 10.6). The catalyst used in this reaction was Pd-catalyst in the presence of pyrrolidine in a mixture of H_2O–DME (1:4 ratio) [42]. The reaction was considered to proceed through Heck-type coupling wherein terminal olefins couple by showing stereoselectively. Furthermore, the reactions showed the exclusive formation of *E*-isomers.

In one of the researches, Edrisi et al. have synthesized a series of polysubstituted pyridines **22** by using ultrasound-assisted preparation a multicomponent one-pot approach. The products were obtained under milder and faster reaction conditions in water as a green medium in

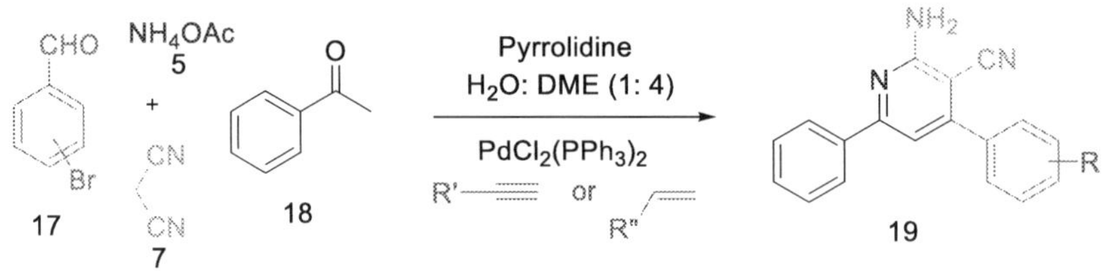

SCHEME 10.6 Synthesis of 2-amino-4-(3/2-(alkynyl)/3-(alkenyl)phenyl)-6-phenylnicotinenitriles via multicomponent reaction.

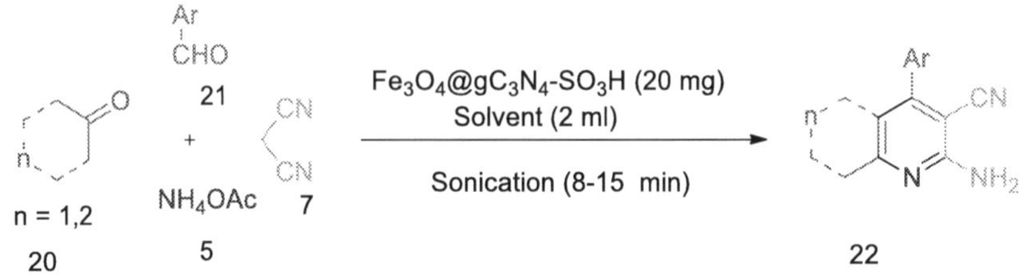

SCHEME 10.7 Ultrasound promoted greener syntheses of multisubstituted pyridine derivatives in water.

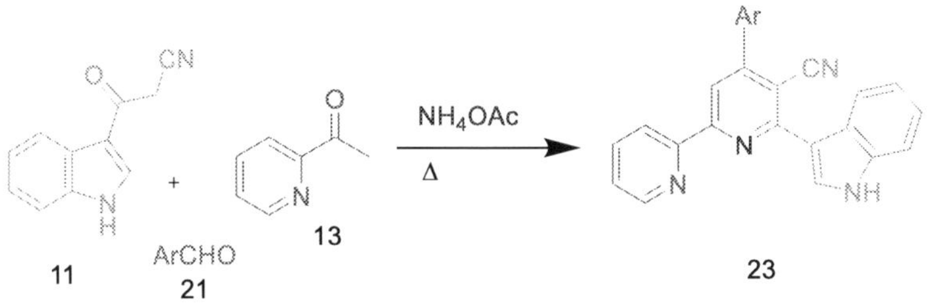

SCHEME 10.8 Synthesis of indole substituted bipyridines.

presence of polar and strongly acidic sulfonated magnetic graphitic carbon nitride (Fe_3O_4@g–C_3N_4–SO_3H) [43]. This sulfonated magnetic graphitic carbon nitride has unique chemical stability and electronic structure. It has drawn the attention of many researchers as a catalytic material because of its interesting sustainability and is a soft nanomaterial. Furthermore, in this reaction aromatic aldehydes **21** were reacted with cyclic and acyclic ketones **20**, ammonium acetate **5**, and malononitrile **7** to get desired pyridine derivatives **22** (Scheme 10.7).

Zahao et al. have synthesized indole-substituted bipyridine derivatives **23** via a one-pot and multicomponent approach [44]. In this synthesis the reaction between 3-(cyanoacetyl)indole **11**, an aldehyde **21**, 2-acetylpyridine **13**, and ammonium acetate **5** was done to get desired product (Scheme 10.8). The mechanism of formation was not clear, it was proposed that the reaction may proceed through a sequential procedure Knoevenagel condensation occurs followed by the Hantzsch condensation (Scheme 10.9). In the first step formation of two different intermediates **26** and **25** can be observed, first from condensation of 3-(cyanoacetyl)indole **11** and aldehyde **21** and second from 2-acetylpyridine **13** and ammonium acetate **5**. Then these two intermediates react with each other via Michael addition to give a material **27** which can further isomerize to an intermediate **28**. This resultant intermediate undergoes cyclization

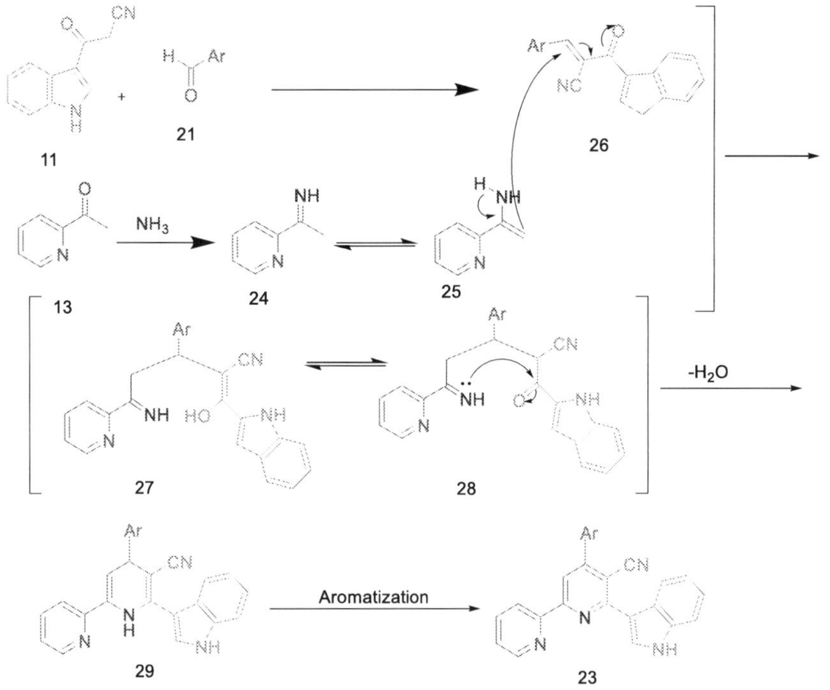

SCHEME 10.9 Proposed mechanism for synthesis of indole substituted bipyridines.

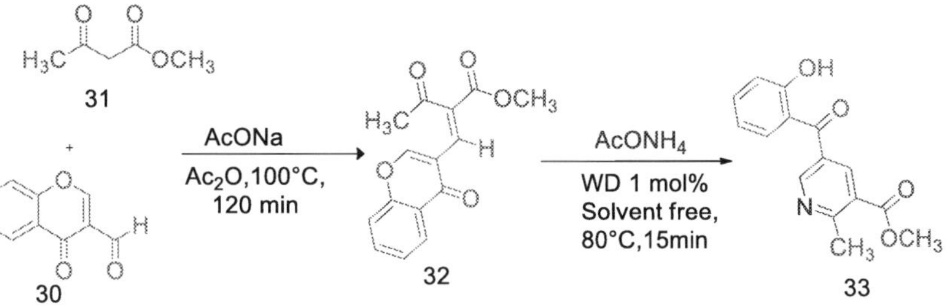

SCHEME 10.10 Solvent free synthesis of substituted pyridines in two steps.

reaction to give a dihydro type intermediate **29** which on aromatization gives desired pyridine containing heterocycles **23**.

Sanchez et al. in their research synthesized 2-amino-3, 5-dicarbonitrile-6-thio pyridine derivatives **33** in solvent free conditions via three-component condensation reaction [45]. In this reaction the 3-formylchromones **31** were taken as aldehydic component, ammonium acetate **5** as a nitrogen source, and a β-ester component **31** to get the desired product (Scheme 10.10). The Wells–Dawson heteropolyacids ($H_6P_2W_{18}O_{62}\cdot 24H_2O$) were used as catalysts in this multicomponent Hantzsch condensation reaction.

Pagadala and his team have synthesized pyridine derivatives **36** using one-pot multicomponent approach. In this multicomponent synthesis the coupling occurred by utilizing Au/MgO as an efficient reusable catalyst. They heated a mixture of ammonia, acetaldehyde **9**,

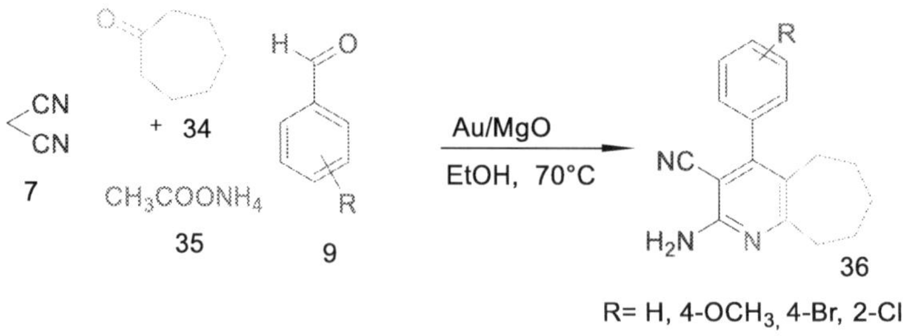

SCHEME 10.11 Four component synthesis of multisubstituted pyridine derivatives.

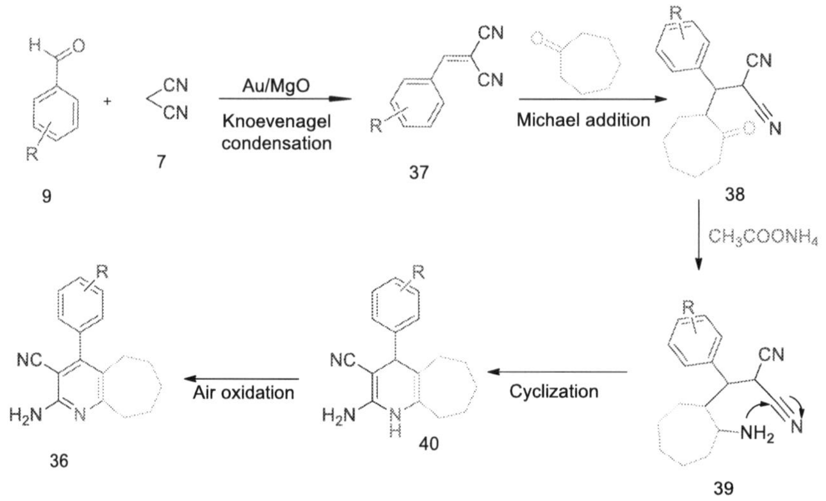

SCHEME 10.12 Plausible mechanism for synthesis of multisubstituted pyridine derivatives.

and ethyl 3-oxobutanoate **34** in solvent free conditions at 70°C (Scheme 10.11) [46]. The compound obtained from this MCR was identified as diethyl 2,4,6-trimethyl-1,4-dihydropyridine-3,5-dicarboxylate **36**. The catalyst used was recovered by facile workup and showing no loss of catalytic activity. Furthermore, a plausible mechanism was proposed for this reaction (Scheme 10.12). In the initial step arylidenemalononitrile is formed as an intermediate **37** via Knoevenagel Condensation reaction. The activated double bond of arylidene intermediate **37** undergoes Michael addition to ketone. The intermediate **37** then reacts with ammonium acetate to give an adduct **38** followed by intermolecular cyclization to get desired product **36**.

An efficient one-pot, multicomponent reaction of aldehydes **41**, NH_4OAc **5** and β-dicarbonyl compounds **42** was carried out in the absence of solvent (Method B) to afford target active 2,3,6-trisubstituted pyridine derivatives **43**. The same reaction was also performed using isopropanol as the solvent (Method A) [47] (Scheme 10.13). In both the methods same catalyst $K_5CoW_{12}O_{40}\cdot 3H_2O$ showing the reusable property was used to get desired product with good yield.

In one of the interesting research the synthesis of highly substituted pyridine derivatives **44** was reported by Mobinikhaledi and his group through one-pot multicomponent approach.

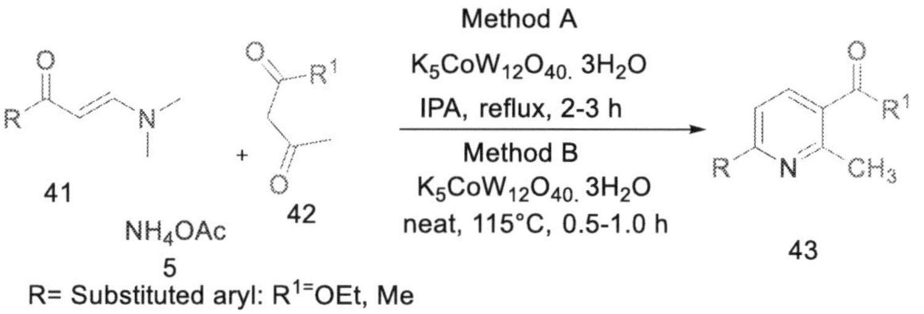

SCHEME 10.13 Synthesis of 2,3,6-trisubstituted pyridines using $K_5CoW_{12}O_{40}{\cdot}3H_2O$ catalyst.

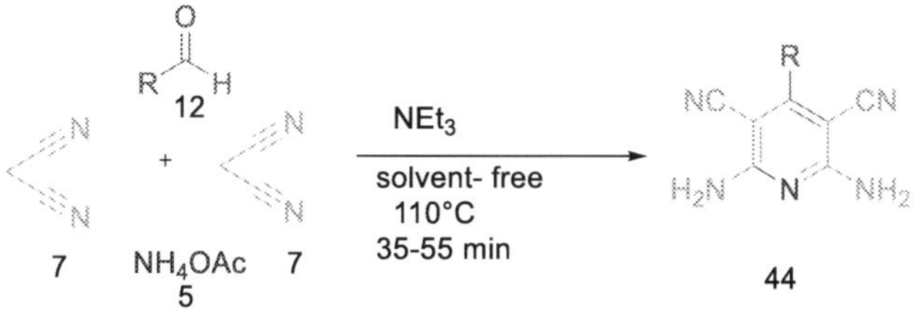

SCHEME 10.14 Multicomponent synthesis of pyridine derivative.

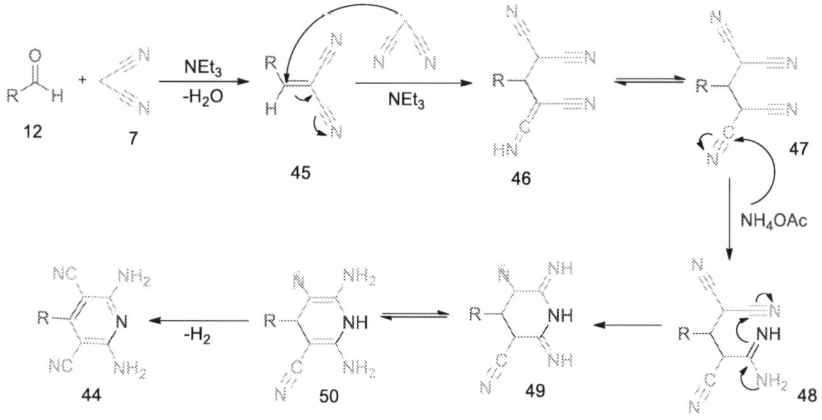

SCHEME 10.15 Suggested mechanism for multicomponent synthesis of pyridine derivative.

They reacted the mixture of malononitrile **7**, aldehydes **12**, and ammonium acetate **5** in the presence of triethylamine as a catalyst, in absence of the solvent (Scheme 10.14) [48]. The authors also proposed a plausible mechanism for this reaction (Scheme 10.15). The reaction was considered to be initiated by Knoevenagel condensation of malononitrile **7** and benzaldehyde **12** in the presence of triethylamine. This condensation step leads to the formation of an intermediate benzylidenemalononitrile. In the next step the Michael additions of another molecule of malononitrile to produced intermediate to give an adduct with four nitrile groups **47**. Then this adduct undergoes nucleophillic attack by ammonium acetate **5** on one of the nitrile group to give an intermediate **48** which on further cyclization followed by tautomerization and aromatization yields the desired product **44**.

There were a lot of reactions in which ammonium acetate was used as nitrogen source in pyridine ring synthesis. We have summarized a few of these reactions in Table 10.1 (three component MCRs) and Table 10.2 (four component MCRs).

TABLE 10.1 Synthesis of pyridine derivatives by one-pot three component reaction, utilizing NH_4OAc as source of nitrogen for pyridine ring.

S/No.	Component 1	Component 2	Component 3	Reagents	Product	Ref.
1	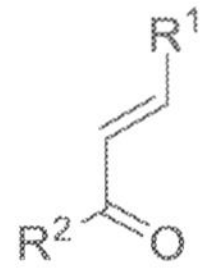	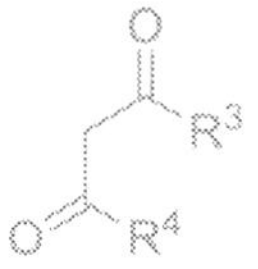	NH_4OAc	Molecular selves Activated carbon (50wt%) Toluene/AcOH reflux, 24 h	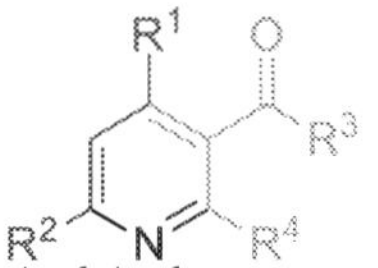 And And more product	[49]
2	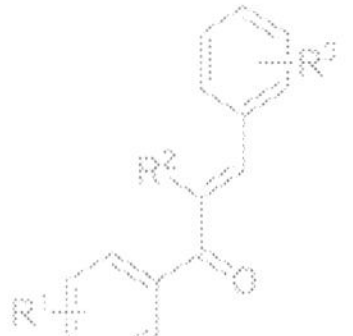	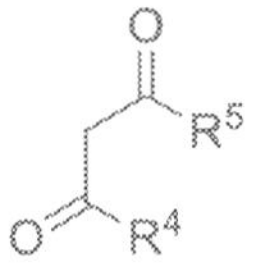	NH_4OAc	CAN (10mol%)EtOH, reflux, 15–30 h	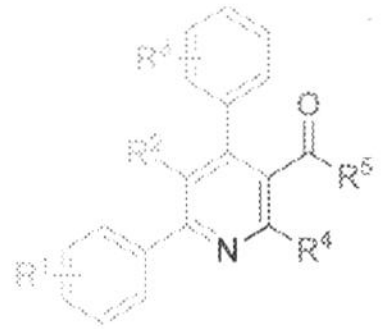	[50]
3	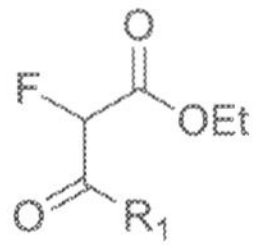	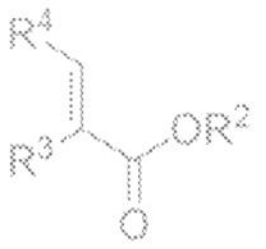	NH_4OAc	Cs_2CO_3(1eq.) MeCN,120°C, 18–24 h	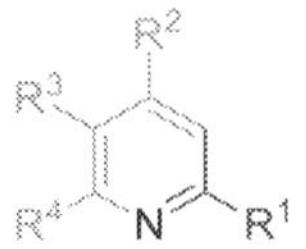	[51]
4	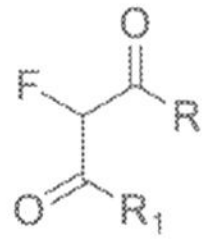	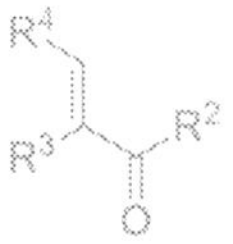	NH_4OAc	Cs_2CO_3(1eq.) 60°C, 4–12 h	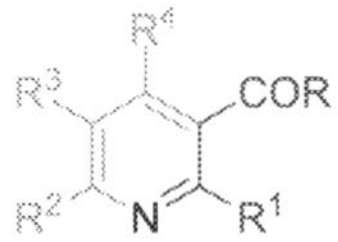	[51]
5	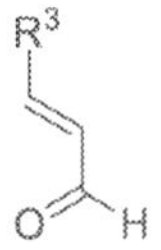	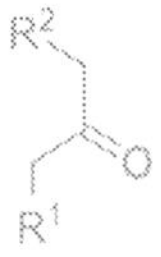	NH_4OAc	L-Proline (40 mol%)Toluene, rt, 10–20 h	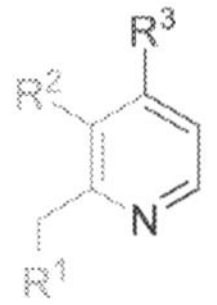	[52]
6			NH_4OAc	$NaHCO_3$ (1 eq.), Air. 1,4-dioxane, 90°C. 5 h	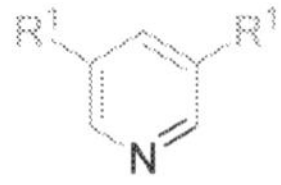	[53]
7	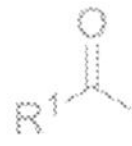	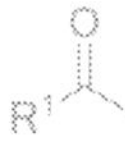	NH_4OAc	NH_4I (0.5 eq.)DMSO, 130°C, 14 h	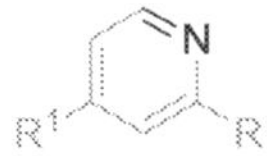	[54]
8	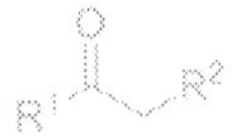	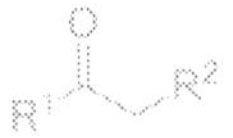	NH_4OAc	NH_4I (0.5 eq.)DMSO, 130°C, 14 h	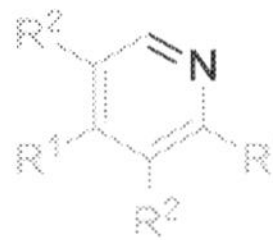 And one more product	[54]

(continued on next page)

TABLE 10.1 Synthesis of pyridine derivatives by one-pot three component reaction, utilizing NH_4OAc as source of nitrogen for pyridine ring—cont'd

S/No.	Component 1	Component 2	Component 3	Reagents	Product	Ref.
9	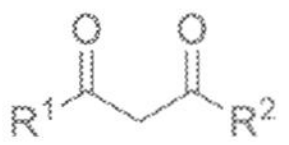	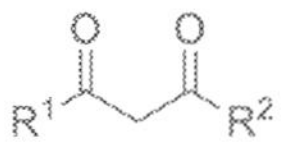	NH_4OAc	TFA (6 eq.)DMSO, 120°C, 20–48 h	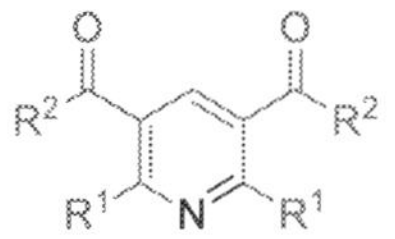	[55]
10	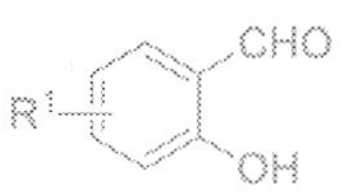		NH_4OAc	K_2CO_3 (1 eq.) DMF/EtOH, reflux 9 h	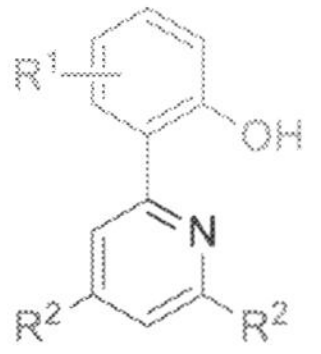	[56]

10.5.2 Amine group as a source of nitrogen to pyridine ring

In one of the studies by Lou and his coworkers, the highly substituted pyridinium salts **51** were synthesized by three component one-pot synthesis. In this reaction the amines **52** and alkynes **53** were reacted with vinyl aldehydes/ketones **51** in the presence of rhodium as a catalyst (Scheme 10.16). The reaction was assisted by Rh(III)-catalyzed vinylic C–H activation and found to proceed by in situ imine generation [62]. The authors have proposed the catalytic loop for the detailed mechanism of the reaction (Scheme 10.17). In this reaction α, β-unsaturated carbonyl compound **51** reacts with amine **52** to produce in situ α,β-unsaturated imine **55**. The imine **55** coordinated to Rh(III) complex followed by vinylic C–H cleavage that leads to the formation of five membered rhodacycle **56**. In the next step alkyne molecule **53** coordinates to rhodium metal in **56** followed by regioselective insertion resulting in the formation of seven-membered N containing rhodacycle **59**. This then affords pyridinium salt **54** and Rh(I) which again gets oxidized to Rh(III) in the presence of $Cu(OAc)_2 \cdot H_2O$ making Rh catalyst ready for the next cycle.

Huang et al. in 2017 synthesized pyridine derivatives **62** by utilizing amine as the source of nitrogen. They used multicomponent and one-pot approach to cyclize two equivalents of α,β-unsaturated carbonyl compounds **60** with primary amines **61** in the presence of TfOH as a catalyst (Scheme 10.18) [63]. The pyridines obtained by this method are symmetric in nature. The authors also depicted a plausible mechanism for this reaction. The α,β-unsaturated carbonyl compounds **60** undergo reverse aldol condensation reaction to the methyl ketone followed by addition to amine leading to imine adduct formation. Then this imine shows 1,4-additon to another molecule of α,β-unsaturated carbonyl compound to give an intermediate that further undergoes intermolecular nucleophillic addition reaction followed by aerobic oxidation to a pyridine derivate **62.**

Using the same strategy in 2016 Wan and his team have also synthesized pyridines derivatives **64** using the same catalyst in one-pot multicomponent approach [64]. They successfully cyclized two equivalents of enaminones/enaminoesters **63** with one equivalent of aldehyde **12** to afford a library of polysubstituted pyridines **64** in presence of triflic acid as a promoter catalyst (Scheme 10.19).

TABLE 10.2 Synthesis of pyridine derivatives by one-pot four component reaction, utilizing NH_4OAc as source of nitrogen for pyridine ring.

S/No.	Component 1	Component 2	Component 3	Reagents	Product	Ref.
1*	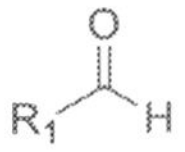	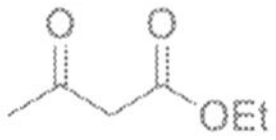	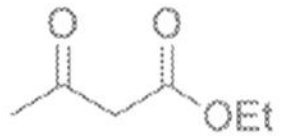	Triton X-100 (10mol%) $K_2S_2O_3$, H_2O, hv, 2–2.5 h	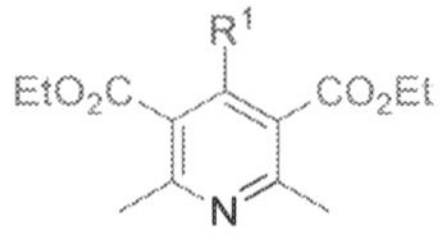	[57]
2*	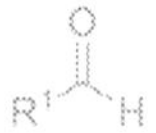	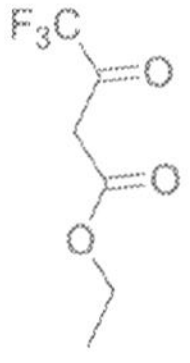	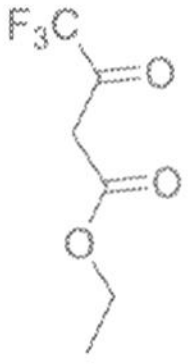	K_2CO_3(1eq.)Solvent-free, 100°C, 2–5 h	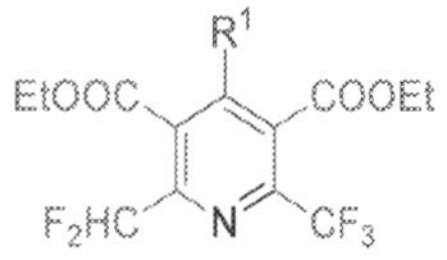	[58]
3*	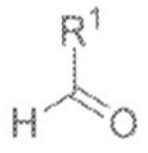	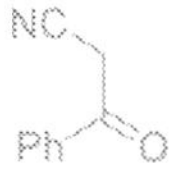	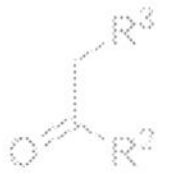	H2O, MW150–170°C, 18–60 min	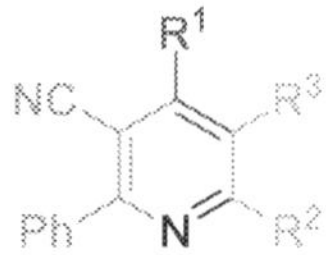	[59]
4*	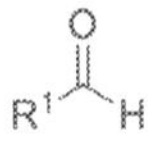		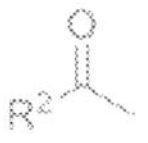	MW (120°C, 30min)	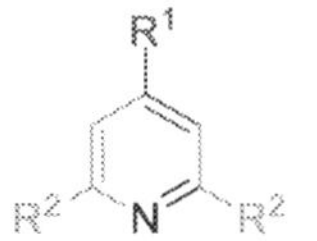	[60]
5*	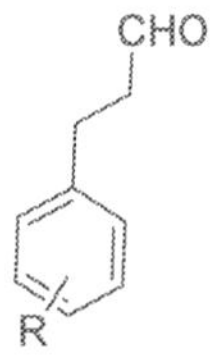 +	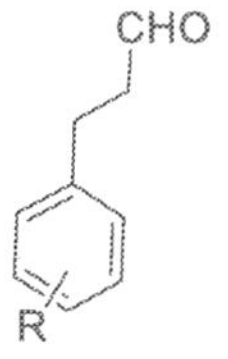 +	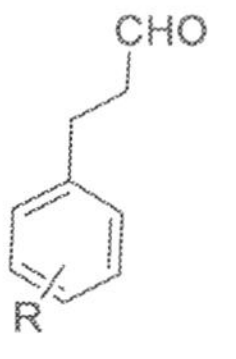 +	NaHCO3 (1 equiv.), Air 1,4-dioxane, 90°C, 5 h	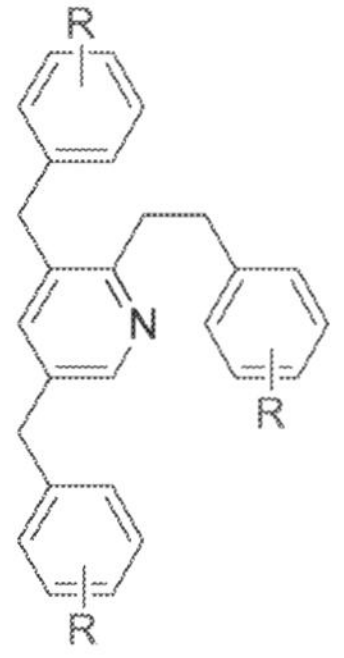	[61]

* The fourth component in all the above reactions is NH_4OAc.

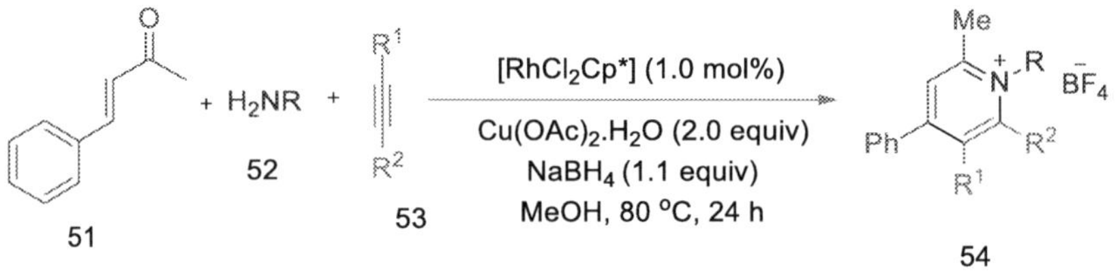

SCHEME 10.16 Rh-catalyzed synthesis pyridinium salts.

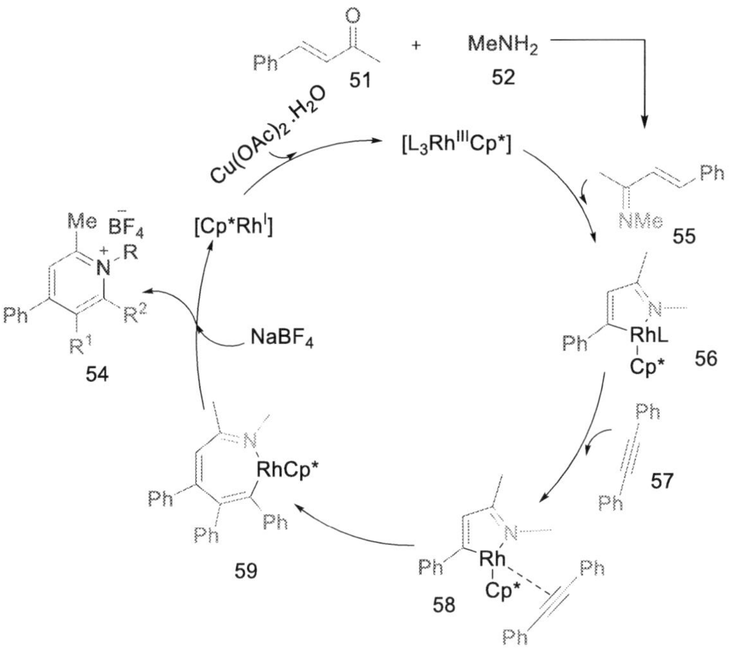

SCHEME 10.17 Proposed reaction mechanism.

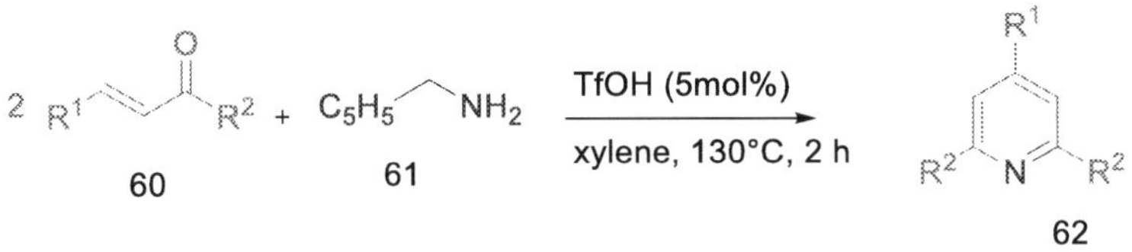

SCHEME 10.18 Triflic acid catalyzed synthesis of pyridines.

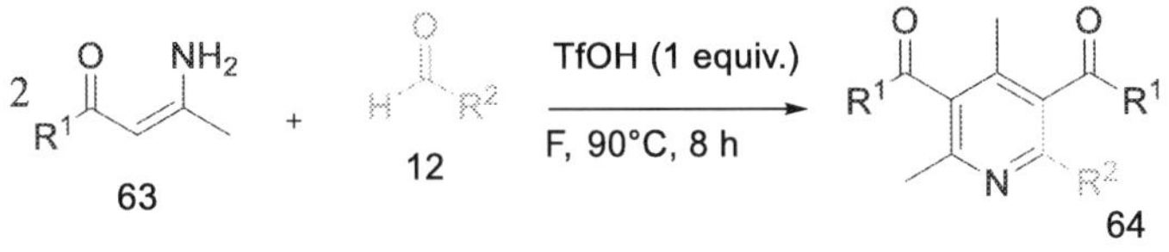

SCHEME 10.19 TfoH promoted three component reaction.

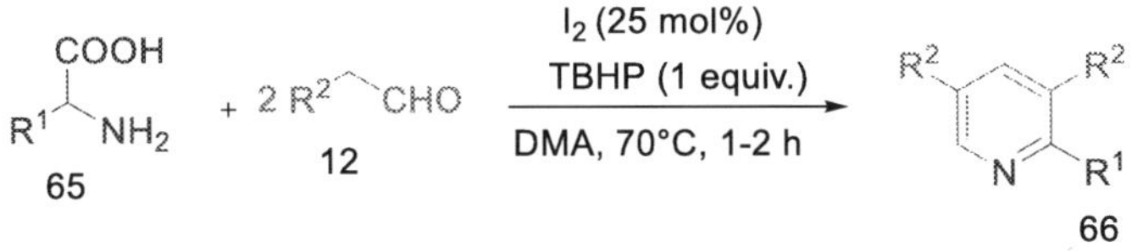

SCHEME 10.20 I_2/TBHP catalyzed synthesis of pyridines.

The same team of workers also synthesized the pyridines derivatives **66** from amino-acids **65** utilizing –NH_2 group as the source of nitrogen in the pyridine ring (Scheme 10.20). In this reaction they used multicomponent approach to synthesize pyridines **66** from two equivalents of aldehydes **12** and one equivalent of amino acid **65**. The catalyst used was iodine in the presence of one equivalent of TBHP. They studied many amino acids in this

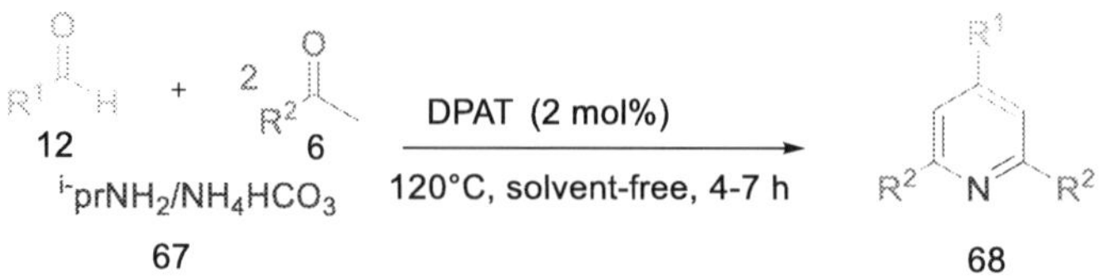

SCHEME 10.21 DPAT-catalyzed synthesis of symmetrical pyridines.

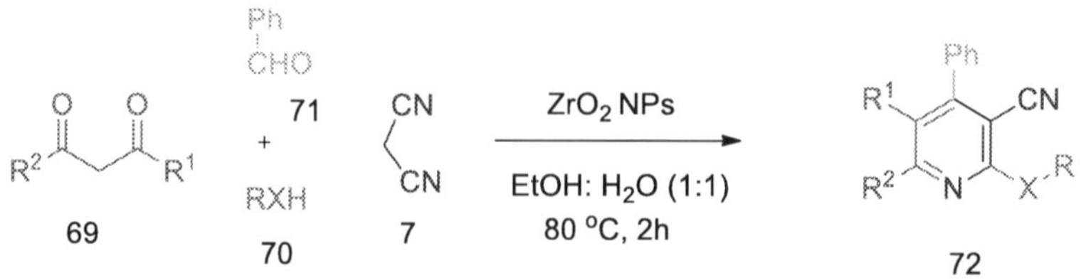

S/No.	R^1	R^2	X	Catalyst	Ref
1	$-CO_2Et$	-Me	O, NH	Nano-ZrO_2	67
2	-CN	$-NH_2$	S, O, NH	-	68
3	$-CO_2Et$	-Me	O, NH	$FeCl_3$	69
4	$-CO_2Et$	-Me	O, NH	$SnCl_2$.2H2O	70

SCHEME 10.22 Tetragonal nano-ZrO_2 catalyzed synthesis of polyfunctionalized pyridine derivatives.

reaction for synthesis. The amino acids used were aspartic acid (Asp), alanine (Ala), 5-methyl glutamine acid, isoleucine, glycine, leucine, nor-leucine, valine, phenylalanine, nor-valine, phenyl glycine, arginine (Arg), lysine (Lys), serine (Ser), and threonine (Thr) [65]. It was further revealed that the amino acids having active hydrogen are more effective in the production of pyridines.

Similarly, Li and Yu showed an efficient one-pot multicomponent synthesis of symmetrical pyridines **68**, in which amine **67** was used as the source of nitrogen to generate pyridine ring (Scheme 10.21). The reaction was performed in presence of diphenylammonium triflate (DPAT) as catalyst [66]. In this reaction methyl ketone **6**, an aldehyde **12** and mixture of an amine **67** and ammonium bicarbonate **67** were taken in one-pot under solvent free conditions to yield symmetric pyridine derivatives **68**.

10.5.3 Malononitrile as a source of nitrogen to pyridine ring

Saha et al. have reported synthesis of ethyl 4-hydroxy-2-(4-nitroaryl)-5-oxo-1-aryl-2,5-dihydro-1*H*-pyrrole-3-carboxylate **72** and ethyl 5-cyano-2-methyl-4-aryl-6-(arylamino) nicotinate by four component one-pot synthesis (Scheme 10.22) [67]. They reacted aryl

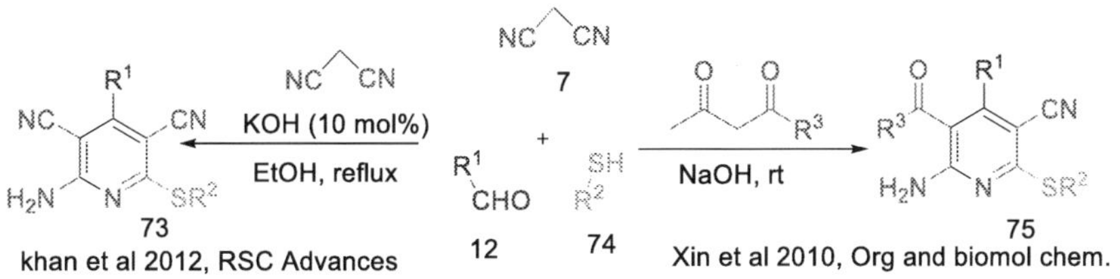

SCHEME 10.23 Synthesis of 2-amino-3,5-dicarbonitrile-6-thio-pyridines.

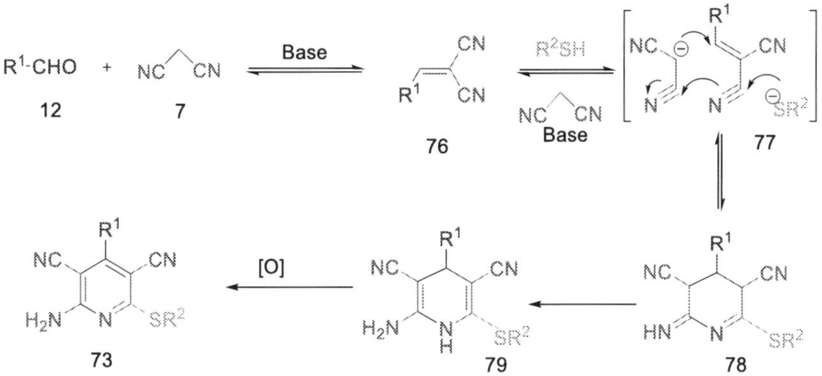

SCHEME 10.24 Proposed mechanism for the KOH-mediated synthesis of thio-pyridine derivatives.

amines **70**, aryl aldehydes **71**, dinitriles **7**, and esters **69** in the presence of tetragonal nano-ZrO_2 as a reusable catalyst. It was observed that Zr-catalyst can be recycled eight times. These poly substituted pyridines have also been synthesized via multicomponent approach from reaction mixture of one molecule of aldehyde **71**, two molecules of malononitrile **7**, and a nucleophile–XH (X = O, S, and NH) **70**. These reactions were considered to proceed via intermolecular cyclization and nucleophillic addition processes [68]. The same set of polysubstituted pyridines was also produced by He and his coworkers. They used the catalyst $FeCl_3$ in ethanol and pyridines were accessed through nucleophillic addition/intermolecular cyclization [69]. Furthermore, the same set of pyridine derivatives was synthesized by Reddy et al. using $SnCl_2 \cdot 2H_2O$ [70]. The limitation of both the methods established by He and Reddy was moderate product yields, long reaction time, and catalysts were not reusable.

In one of the researches, Khan and his coworkers have synthesized a library of highly functionalized pyridines **73** by efficient one-pot multicomponent synthesis (Scheme 10.23). The reaction mixture of malononitrile **7**, aldehydes **12**, and thiols **74** was taken that utilized potassium hydroxide (a bronsted base) as the catalyst to produce desired product [71]. They also proposed a plausible mechanism explaining the steps occurring in reaction (Scheme 10.24). In the first step, the Knoevenagel condensation occurs between aldehyde **12** and a molecule of malononitrile **7** to give Knoevenagel product. In the next step, another molecule of malononitrile **7** undergoes Michael addition followed by thiolate addition simultaneously.

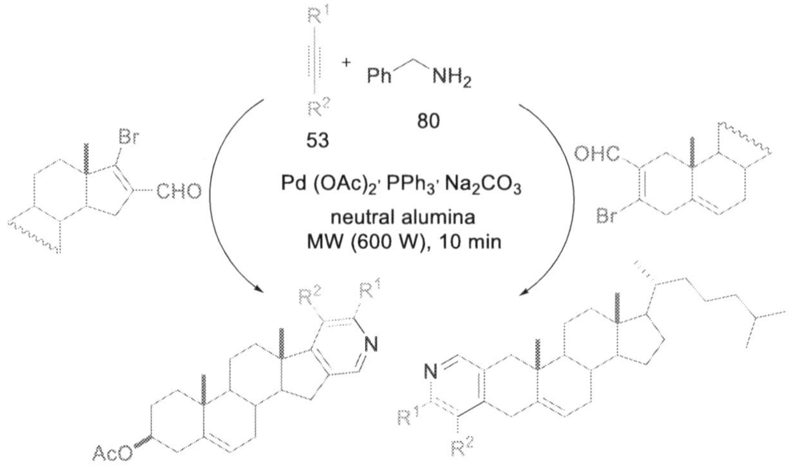

SCHEME 10.25 Synthesis of steroidal A- and D-ring fused 5,6-disubstituted pyridines.

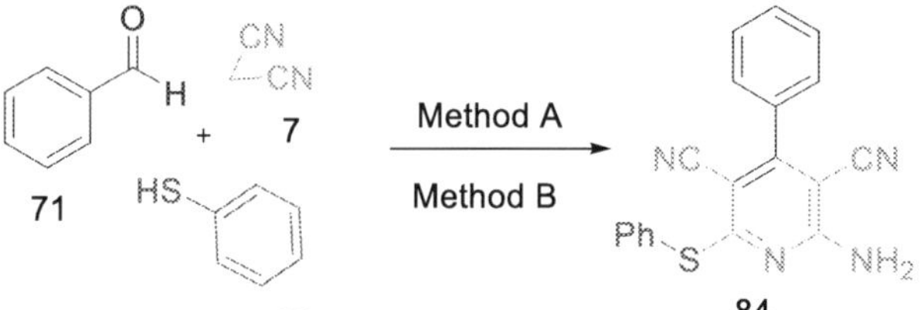

SCHEME 10.26 Synthesis of 2-amino-6-(aryithio)pyridines-3,5-dicarbonitriles. Method A: Thimmaiah et al., Zn–Cd MOF as catalyst. Method B: Gujar et al., NaCl as catalyst.

The adduct **77** obtained then undergoes cyclization to dihydropyridines **79** that on further oxidation leads to aromatization to yield pyridine derivatives **73**. Xin et al. have synthesized the same type of pyridine derivatives **75** utilizing diketone instead of one molecule of malononitrile **7** in the presence of NaOH as catalyst [71].

Shekarrao et al. have synthesized steroidal type pyridines in which 5, 6-disubstituted pyridines **81** were fused with ring A and ring D simultaneously by one-pot and multicomponent approach (Scheme 10.25). He further successfully synthesized nonsteroidal substituted pyridines **82** using the same strategy. Moreover, the reaction was performed under microwave irradiation using solvent-free conditions. The reaction between steroidal/nonsteroidal bhalovinylaldehyde, benzylamine **80**, and alkynes **53** was done using $Pd(OAc)_2$ as a catalyst to get desired product [72].

Thimmaiah et al. have been synthesized 2-amino-3,5-dicarbonitrile-6-thio-pyridines derivatives **84** by multicomponent synthesis. The reaction mixture of aldehydes **83**, malononitrile **7**, and thiophenols **83** was reacted in a reaction flask. The Zn(II) or Cd(II) metal organic framework was taken as a heterogeneous catalyst to get desired product (Scheme 10.26) [73]. Based on the same strategy, Gujar et al. have reported a facile and

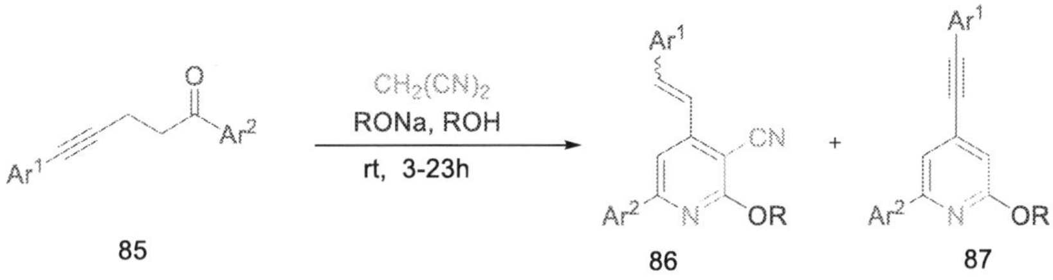

SCHEME 10.27 Reactions of 1,5-diarylpent-2-en-4-yn-1-ones with malononitrile and sodium alkoxides in the corresponding alcohols leading to pyridines.

convenient one-pot multicomponent synthesis of substituted pyridine derivatives **84** utilizing NaCl as catalyst [73]. He replaced the ketone by an aromatic aldehyde. The synthesis was performed in an aqueous medium under reflux and ultrasound irradiation. The formation of four new bonds was observed in this reaction. In this reaction the three components, that is, aldehydes, malononitrile, and thiophenol were condensed to build a pyridine skeleton in the presence of Lewis/Bronsted acids or bases.

Kuznetcova et al. synthesized a library of pyridine derivatives **87** using one-pot multicomponent approach. In this reaction the conjugated enynones, 1,5-diarylpent-2-en-4-yn-1-ones **85**, were reacted with malononitrile **7** and sodium alkoxide in presence of corresponding alcohols. Two types of products were obtained in reaction, first (E)-/(Z)-6-aryl-4-(2-arylethenyl)-2-alkoxypyridine-3-carbonitriles **86** and second 6-aryl-4-arylethynyl-2-alkoxy pyridines **87** as the minor product (Scheme 10.27). The plausible reaction mechanism for this reaction was also proposed (Scheme 10.28). In the initial step malononitrile anion adds to carbon–carbon double bond of enynone **85** by Michael type addition to produce anion **88**. The anion **88** then abstracts the proton from ROH to give compound **89**. The RO^- anion attacks **89** as a nucleophile onto the –CN group to form adducts **90**, followed by **91**, and then an enamine **92**. The enamine **92** shows cyclization by interaction between amino group and carbonyl carbon that result to the formation of an anion **93** containing N-type heterocycle. Then proton transfer occurs in anion **93** leading to formation of another N-centered anion **94**, that on elimination of –OH group produces compound **95**. Isomerization of double bonds in **95** may lead to structure **96**. Then **96** deprotonates to form propargy-allyl anion **97** which is the key intermediate to form both the products **86** and **87**. Furthermore, the formation of alkene and alkyne containing pyridine derivative is discussed in detail in scheme [74].

Zolfigol et al. in 2016 have reported a catalyst of mixed oxide $\{Fe_3O_4@SiO_2@(CH_2)_3$-Im$\}C(CN)_3$which was further used for the four-component synthesis of 2-amino-3-cyanopyridine derivatives (Scheme 10.29). The pyridine derivatives were synthesized by reacting aromatic aldehyde **9**, malononitrile **7**, acetophenone, and NH_4OAc **5** as a nitrogen source in pyridine ring formation. The reaction was completely solvent-free and refluxed at 100°C within short reaction times [75].

The mechanism for the use of catalyst in this reaction was proposed to understand the basics of the approach. The cationic site of the given catalyst activates the acetophenone and aromatic aldehydes **9** as electrophiles. At the same time the anionic site of catalyst with the acidic hydrogen of malononitrile **7** followed by nucleophillic addition. The reaction then moves via

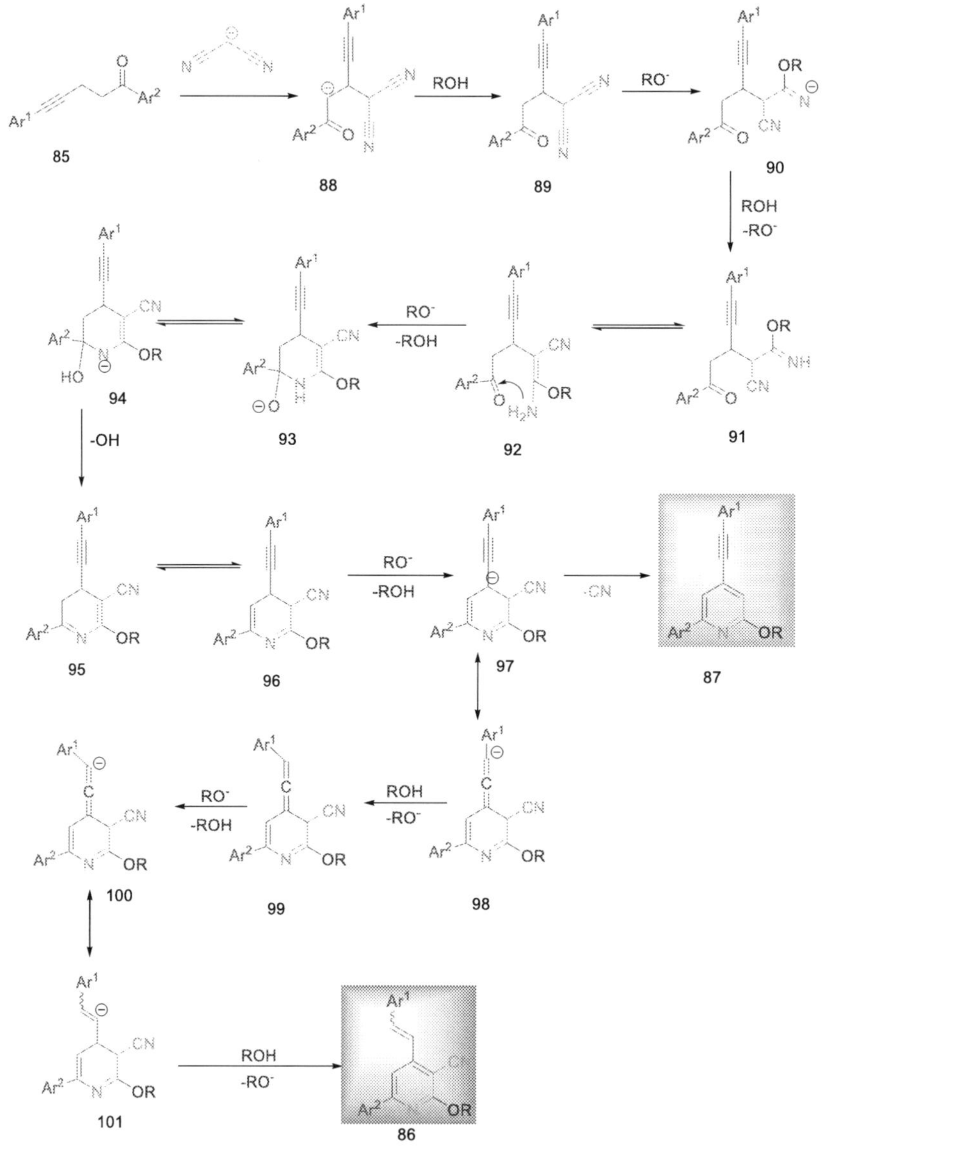

SCHEME 10.28 Plausible mechanism of the reaction of enynones with malononitrile and sodium alkoxides leading to pyridines.

alkylidenemalononitrile intermediate **104a** and enamine intermediates **104b** which results in the formation of another large sized intermediate **107**. This intermediate then follows few more steps including tautomerization, cyclization, anomeric oxidation, and hybrid transfer aromatization to yield desired products **103** and the plausible mechanism for the reaction is given in Scheme 10.30.

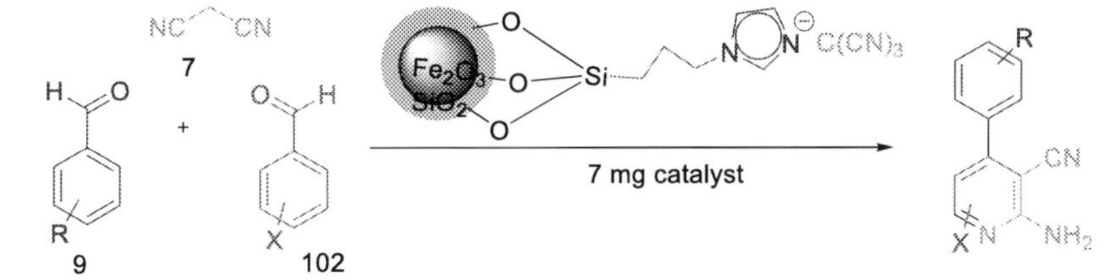

R= H, 4-OMe, 4-F/Cl/Br, 3/4-NO_2
X= H, 4-OMe, 4-OH, 4-C

SCHEME 10.29 Catalytic applicability of the (Fe_3O_4@SiO_2@$(CH_2)_3$lm) $C(CN)_3$ in one-pot synthesis of 2-amino-3-cyanopyridines.

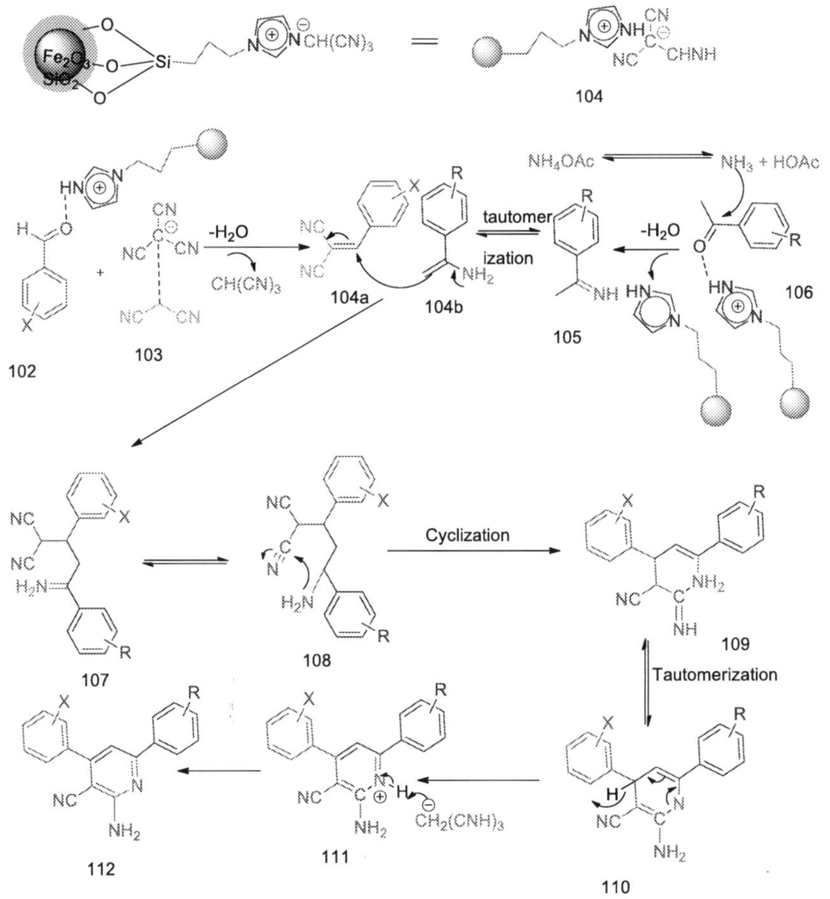

SCHEME 10.30 Plausible mechanism.

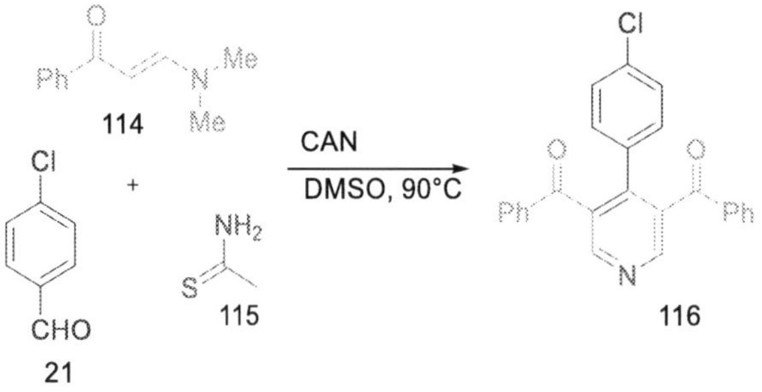

SCHEME 10.31 Synthesis of pyridines derivatives.

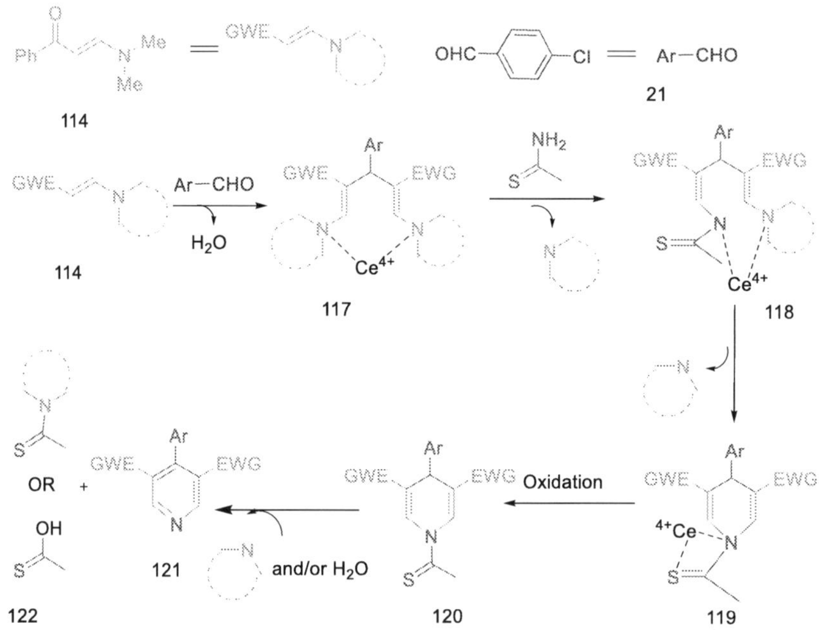

SCHEME 10.32 Plausible mechanism for synthesis of pyridines derivatives.

10.5.4 Miscellaneous

10.5.4.1 Thioacetamide as a nitrogen source for pyridine ring formation

Wan and his coworkers have been established a method for multicomponent synthesis of pyridine derivatives **116**. The ammonium source utilized was thioacetamide **115** which is efficient and cheap. It was reacted with electron-deficient enamines **114** or alkynes and aldehydes **21** and to get desired 3,4,5-trisubstituted pyridines **116** in moderate to good yields (Scheme 10.31) [76]. The authors have also proposed the mechanism for the reaction (Scheme 10.32). It was proposed that this reaction proceeds via domino transformation. The steps involved are enamine-based homo-condensation, thioacetamide-initiated trans amination, and cerium-promoted elimination–aromatization. In the first step the electron-deficient enamine **114** reacts with aldehyde to **21** give complex **117** after condensation. The Ce metal ion from CAN

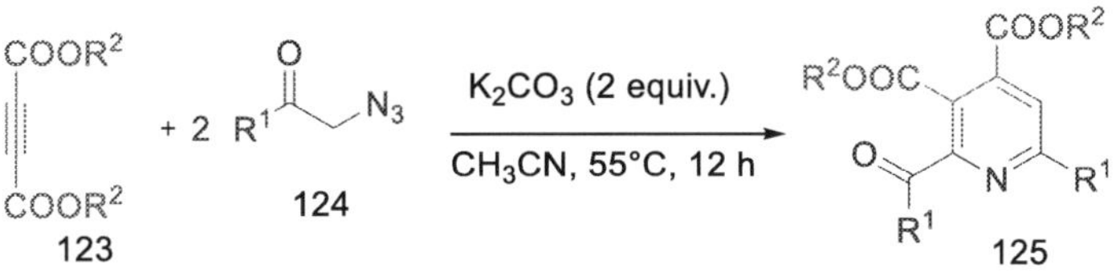

SCHEME 10.33 Synthesis of pyridines using alkynes as substrate.

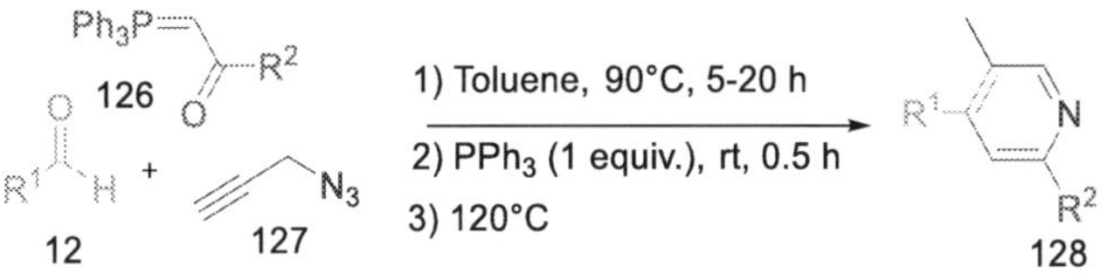

SCHEME 10.34 Three component synthesis of pyridines.

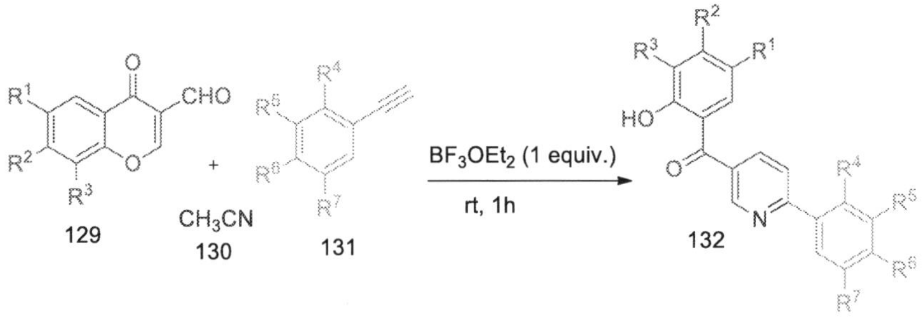

SCHEME 10.35 BF_3 Et_2O promoted synthesis of pyridines at room temperature.

transforms **117** to an intermediate **119** (a dihydropyridines) on reaction with thioacetamide **115**. The CAN then oxidizes **119** followed by aromatization to produce thioacetylpyridinium salt **120**. This thioacetyl cation is further removed by secondary amine or water already present in the reaction mixture to give pyridine derivatives **122**.

10.5.4.2 Azide group as a nitrogen source for pyridine ring formation

Yu et al. in 2013 synthesized the pyridines **125** by one-pot multicomponent approach in which azide was taken as the source of nitrogen to pyridine ring. In this reaction the mixture of one mole of dialkyl acetylene dicarboxylates, **123** and two moles of α-azidomethyl ketones **124** were reacted in the presence of K_2CO_3 as a reagent (Scheme 10.33) [77].

In another study by Li and Zhai in 2016, azides were used successfully as a nitrogen source in the pyridine ring of their derivatives **128**, synthesized by one-pot multicomponent approach. In this reaction, they reacted a mixture of propargyl azide **127**, aldehydes **12**, and phosphorus ylides **126** to get highly substituted derivatives of pyridines **128** in good yields (Scheme 10.34) [78].

10.5.4.3 Nitrile as a nitrogen source for pyridine ring formation

Lee et al. have used nitrile as the source of nitrogen to pyridine ring formation in multicomponent approach. In this reaction they have taken aryl acetylene **131**, formylchromones **129**, and acetonitrile **130** in reaction pot in the presence of one equivalent of boron trifluoride etherate to get the desired product **132** (Scheme 10.35) [79]. The proposed mechanism shows that reaction proceeds via [3+2+1] annulation and involves the steps such as Michael addition, nucleophilic addition, hydrolysis, ring opening, and elimination steps, etc.

10.6 Conclusion

Pyridines and their derivatives are of great importance in various fields especially biomedical and are widely distributed in nature. Thus, the synthesis of these pyridine derivatives by one-pot multicomponent reactions is an excellent approach as compared to other synthetic methods. We hope that this audit will help in research by further synthesizing novel pyridine derivatives.

Acknowledgments

The authors are highly thankful to Department of Chemistry, Aligarh Muslim University for providing basic facilities for research. Shah Imtiaz is also grateful to CSIR-New Delhi for providing fellowship.

Abbreviations

MCRs	multicomponent reactions
TBBDA	tetrabromobenzene-1,3-disulfonamide
ALK	anaplastic lymphoma kinase
PCC	pyridinium chlorochromate
PDC	pyridinium dichromate
DMAP	4-dimethylaminopyridine
GBB	Groebke–Blackburn–Bienaymé
DPAT	diphenylammonium triflate
DME	dimethoxyethane
TBHP	tert-butyl hydroperoxide
DMA	dimethylacetamide
DMF	dimethylformamide
TFA	trifluoroacetic acid
CAN	ceric (IV) ammonium nitrate
DMSO	dimethyl sulfoxide
IPA	isopropyl alcohol

References

[1] (a) L.F. Lee, EP patent 133612, 1985.; (b) Z. Song, X. Huang, W. Yi, W. Zhang, One-pot reactions for modular synthesis of polysubstituted and fused pyridines, Org. Lett. 18 (21) (2016) 5640–5643; (c) M.A. Cinelli, A. Morrell, T.S. Dexheimer, E.S. Scher, Y. Pommier, M. Cushman, Design, synthesis, and biological evaluation of 14-substituted aromathecins as topoisomerase I inhibitors, J. Med. Chem. 51 (15) (2008) 4609–4619; (d) Y. Nakao, H. Idei, K.S. Kanyiva, T. Hiyama, Direct alkenylation and alkylation of pyridone derivatives by Ni/$AlMe_3$ catalysis, J. Am. Chem. Soc. 131 (44) (2009) 15996–15997; (e) F. Bossert, H. Meyer, E. Wehinger, 4-Aryldihydropyridines, a new class of highly active calcium antagonists, Angew. Chem. Int. Ed. Engl. 20 (9) (1981) 762–769; (f) B. Loev, M.M. Goodman, K.M. Snader, R. Tedeschi, E. Macko, Hantzsch-type dihydropyridine hypotensive agents, J. Med. Chem. 17 (9) (1974) 956–965; (g) S. Cosconati, L. Marinelli, A. Lavecchia, E. Novellino, Characterizing the 1, 4-dihydropyridines binding interactions in the L-type Ca^{2+} channel: model construction and docking calculations, J. Med. Chem. 50 (7) (2007) 1504–1513; (h) R.A. Janis, D.J. Triggle, New developments in calcium ion channel antagonists, J. Med. Chem. 26 (6) (1983) 775–785; (i) A.H. Li, S. Moro, N. Forsyth, N. Melman, X.D. Ji, K.A. Jacobson, Synthesis, CoMFA analysis, and receptor docking of 3, 5-diacyl-2, 4-dialkylpyridine derivatives

as selective A3 adenosine receptor antagonists, J. Med. Chem. 42 (4) (1999) 706–721; (j) B.R. Henke, D.H. Drewry, S.A. Jones, E.L. Stewart, S.L. Weaver, RW. Wiethe, 2-Amino-4, 6-diarylpyridines as novel ligands for the estrogen receptor, Bioorg. Med. Chem. Lett. 11 (14) (2001) 1939–1942.

[2] (a) KN. Singh, Metal-free multicomponent reactions: a benign access to monocyclic six-membered N-heterocycles, Org. Biomol. Chem. 19 (12) (2021) 2622–2657; (b) J. Rivera-Utrilla, I. Bautista-Toledo, M.A. Ferro-Garcıa, C. Moreno-Castilla, Bioadsorption of Pb (II), Cd (II), and Cr (VI) on activated carbon from aqueous solutions, Carbon 41 (2) (2003) 323–330; (c) C.E. Garrett, K. Prasad, The art of meeting palladium specifications in active pharmaceutical ingredients produced by Pd-catalyzed reactions, Adv. Synth. Catal. 346 (8) (2004) 889–900; (d) C.L. Sun, ZJ. Shi, Transition-metal-free coupling reactions, Chem. Rev. 114 (18) (2014) 9219–9280.

[3] (a) KN. Singh, Metal-free multicomponent reactions: a benign access to monocyclic six-membered N-heterocycles, Org. Biomol. Chem. 19 (12) (2021) 2622–2657.

[4] (a) J.P. Wan, L. Gan, Y. Liu, Transition metal-catalyzed C–H bond functionalization in multicomponent reactions: a tool toward molecular diversity, Org. Biomol. Chem. 15 (43) (2017) 9031–9043; (b) J.P. Wan, L. Gan, Y. Liu, Transition metal-catalyzed C–H bond functionalization in multicomponent reactions: a tool toward molecular diversity, Org. Biomol. Chem. 15 (43) (2017) 9031–9043; (c) C.V. Galliford, KA. Scheidt, Catalytic multicomponent reactions for the synthesis of N-aryl trisubstituted pyrroles, J. Org. Chem. 72 (5) (2007) 1811–1813.

[5] V. Estevez, M. Villacampa, JC. Menendez, Multicomponent reactions for the synthesis of pyrroles, Chem. Soc. Rev. 39 (11) (2010) 4402–4421.

[6] H.A. Younus, M. Al-Rashida, A. Hameed, M. Uroos, U. Salar, S. Rana, KM. Khan, Multicomponent reactions (MCR) in medicinal chemistry: a patent review (2010-2020), Expert Opin. Ther. Pat. 31 (3) (2021) 267–289.

[7] M. Haji, Multicomponent reactions: a simple and efficient route to heterocyclic phosphonates, Beilstein J. Org. Chem. 12 (1) (2016) 1269–1301.

[8] H.A. Younus, M. Al-Rashida, A. Hameed, M. Uroos, U. Salar, S. Rana, K.M. Khan, Multicomponent reactions (MCR) in medicinal chemistry: a patent review (2010-2020), Expert Opin. Ther. Pat. 31 (3) (2021) 267–289.

[9] C.M. Hoock, A. Qadan, B. Terhaag, inventors; Teva GmbH, assignee. Multicomponent crystals made of ([2-amino-6-(4-fluoro-benzylamino)-pyridin-3-yl]-carbamic acid ethyl ester and an arylpropionic acid. United States patent US 8,962,847. 2015 Feb 24.

[10] F. Marsais, V. Levacher, C. Papamicael, P. Bohn, L. Peauger, V. Gembus, N. Le Fur, M.L. Dumartin-lepine, inventors; INSA (Institut National des Sciences Appliquees) de Rouen, VFP Therapies, Centre National de la Recherche Scientifique CNRS, assignee. Oxidisable pyridine derivatives, their preparation and use as anti-alzheimer agents. United States patent US 9,376,387. 2016 Jun 28.

[11] R. Blaszczyk, A. Gzik, B. Borek, M. Dziegielewski, K. Jedrzejczak, J. Nowicka, J. Chrzanowski, J. Brzezinska, A. Golebiowski, J. Olczak, Grzybowski MM, inventors. Dipeptide piperidine derivatives. United States patent US 10,851,099. 2020 Dec 1.

[12] M.D. Moldes, P.B. Pereira, T.C. Caamaño, M.D. Lago, N.V. Molares, D.V. Castelao, inventors; Universidade de Santiago de Compostela, Universidade de Vigo, assignee. Pyridazin-3 (2H)-one derivatives as monoamine oxidase selective isoform B inhibitors. United States patent US 10,253,000. 2019 Apr 9.

[13] H.A. Younus, M. Al-Rashida, A. Hameed, M. Uroos, U. Salar, S. Rana, KM. Khan, Multicomponent reactions (MCR) in medicinal chemistry: a patent review (2010-2020), Expert Opin. Ther. Pat. 31 (3) (2021) 267–289.

[14] H.A. Younus, M. Al-Rashida, A. Hameed, M. Uroos, U. Salar, S. Rana, KM. Khan, Multicomponent reactions (MCR) in medicinal chemistry: a patent review (2010-2020), Expert Opin. Ther. Pat. 31 (3) (2021) 267–289.

[15] P.E. Floreancig, B.W. Day, S. Wan, F. Wu, inventors; University of Pittsburgh, assignee. Pederin and psymberin agents. United States patent US 9,364,555. 2016 Jun 14.

[16] P.T. Patil, P.P. Warekar, K.T. Patil, S.S. Undare, D.K. Jamale, S.S. Vibhute, N.J. Valekar, G.B. Kolekar, M.B. Deshmukh, PV. Anbhule, A simple and efficient one-pot novel synthesis of pyrazolo [3, 4-b][1, 8] naphthyridine and pyrazolo [3, 4-d] pyrimido [1, 2-a] pyrimidine derivatives as anti-inflammatory agents, Res. Chem. Intermed. 44 (2) (2018) 1119–1130.

[17] S. Kundu, B. Basu, Graphene oxide (GO)-catalyzed multi-component reactions: green synthesis of library of pharmacophore 3-sulfenylimidazo [1, 2-a] pyridines, RSC Adv. 5 (2015) 50178–50185.

[18] B. Roy, D. Sengupta, B. Basu, Graphene oxide (GO)-catalyzed chemoselective thioacetalization of aldehydes under solvent-free conditions. Tetrahedron Lett. 55 (2014) 6596–6600.

[19] R. Vardanyan, V. Hruby, Synthesis of Essential Drugs, Elsevier, Amsterdam, The Netherlands, 2006.

[20] X. Hua, W. Liu, Y. Su, X. Liu, J. Liu, N. Liu, G. Wang, X. Jiao, X. Fan, C. Xue, Y. Liu, Studies on the novel pyridine sulfide containing SDH based heterocyclic amide fungicide, Pest Manage. Sci. 76 (7) (2020) 2368–2378.

[21] S. Schimizu, N. Watanabe, T. Kataoka, T. Shoji, N. Abe, S. Morishita, I. Ichimura, Ullmann's encycl. ind. chem. Pyridine and Pyridine Derivatives. Ullmann's Encyclopedia of Industrial Chemistry (2000) Weinheim, Germany : Wiley-VCH.
[22] G. Tojo, M.I. Fernandez, Oxidation of Alcohols to Aldehydes and Ketones: A Guide to Current Common Practice, Springer Science & Business Media, Netherlands, Springer U.S., 2006.
[23] L.C. Chang, J.K. von Frijtag Drabbe Künzel, T. Mulder-Krieger, R.F. Spanjersberg, S.F. Roerink, G. van den Hout, M.W. Beukers, J. Brussee, AP. IJzerman, A series of ligands displaying a remarkable agonistic–antagonistic profile at the adenosine A1 receptor, J. Med. Chem. 48 (6) (2005) 2045–2053.
[24] B.B. Fredholm, A.P. IJzerman, K.A. Jacobson, K.N. Klotz, J. Linden, International Union of Pharmacology. XXV. Nomenclature and classification of adenosine receptors, Pharmacol. Rev. 53 (4) (2001) 527–552.
[25] (a) B.C. May, J.A. Zorn, J. Witkop, J. Sherrill, A.C. Wallace, G. Legname, S.B. Prusiner, FE. Cohen, Structure–activity relationship study of prion inhibition by 2-aminopyridine-3, 5-dicarbonitrile-based compounds: parallel synthesis, bioactivity, and in vitro pharmacokinetics, J. Med. Chem. 50 (1) (2007) 65–73; (b) M.W. Beukers, L.C. Chang, J.K. von Frijtag Drabbe Künzel, T. Mulder-Krieger, R.F. Spanjersberg, J. Brussee, AP. IJzerman, New, non-adenosine, high-potency agonists for the human adenosine A2B receptor with an improved selectivity profile compared to the reference agonist N-ethylcarboxamidoadenosine, J. Med. Chem. 47 (15) (2004) 3707–3709; (c) L.C. Chang, J.K. von Frijtag Drabbe Künzel, T. Mulder-Krieger, R.F. Spanjersberg, S.F. Roerink, G. van den Hout, M.W. Beukers, J. Brussee, AP. IJzerman, A series of ligands displaying a remarkable agonistic– antagonistic profile at the adenosine A1 receptor, J. Med. Chem. 48 (6) (2005) 2045–2053; (d) A.S. Chavan, A.S. Kharat, M.R. Bhosle, RA. Mane, A convenient Baker yeast accelerated, one-pot synthesis of pentasubstituted thiopyridines, Synth. Commun. 47 (19) (2017) 1777–1782; (e) U. Rosentreter, T. Kraemer, A. Vaupel, W. Huebsch, N. Diedrichs, T. Krahn, K. Dembowsky, J.P. Stasch, PCT Int. Appl. WO 2002, 2002070520, A1.
[26] (a) A. Samadi, M. Chioua, M. Alvarez Perez, E. Soriano Santamaria, C. Valderas Cortina, J.L. Marco Contelles, S.C. De Los Rios, A. Romero Martinez, M. Villarroya Sanchez, M. Garcia Lopez, A. Garcia Garcia, Span. Patent ES 2011, 2365233 A1; (b) A. Samadi, J. Marco-Contelles, E. Soriano, M. Álvarez-Pérez, M. Chioua, A. Romero, L. González-Lafuente, L. Gandía, J.M. Roda, M.G. López, M. Villarroya.
[27] (a) H. Chen, W. Zhang, R. Tam, A.K. Raney, PCT Int. Appl. WO 2005, 2005058315, A1.; (b) B.B. Fredholm, A.P. IJzerman, K.A. Jacobson, K.N. Klotz, J. Linden.
[28] S.B. Levy, M.N. Alekshun, B.L. Podlogar, K. Ohemeng, A.K. Verma, T. Warchol, B. Bhatia, T. Bowser, M. Grier, P. US Appl. Publ. US 2005, 2005124678, A1.
[29] J.J. Verendel, O. Pamies, M. Dieguez, PG. Andersson, Asymmetric hydrogenation of olefins using chiral crabtree-type catalysts: scope and limitations, Chem. Rev. 114 (4) (2014) 2130–2169.
[30] R.W. Armstrong, A.P. Combs, P.A. Tempest, S.D. Brown, TA. Keating, Multiple-component condensation strategies for combinatorial library synthesis, Acc. Chem. Res. 29 (3) (1996) 123–131.
[31] M. Haji, Multicomponent reactions: a simple and efficient route to heterocyclic phosphonates, Beilstein J. Org. Chem. 12 (1) (2016) 1269–1301.
[32] S. Imtiaz, S. Banoo, α-Aminoazoles/azines: key reaction partners for multicomponent reactions, RSC Adv. 11 (19) (2021) 11083–11165.
[33] (a) N. Taha, Y. Sasson, M. Chidambaram, Phase transfer methodology for the synthesis of substituted stilbenes under Knoevenagel condensation condition, Appl. Catal. A 350 (2) (2008) 217–224; (b) D.A. Powell, RA. Batey, Lanthanide (III)-catalyzed multi-component aza-Diels–Alder reaction of aliphatic N-arylaldimines with cyclopentadiene, Tetrahedron Lett. 44 (41) (2003) 7569–7573; (c) A.A. Kudale, J. Kendall, D.O. Miller, J.L. Collins, GJ. Bodwell, Povarov reactions involving 3-aminocoumarins: synthesis of 1, 2, 3, 4-tetrahydropyrido [2, 3-c] coumarins and pyrido [2, 3-c] coumarins, J. Org. Chem. 73 (21) (2008) 8437–8447; (d) M.J. Thompson, B. Chen, Versatile assembly of 5-aminothiazoles based on the Ugi four-component coupling, Tetrahedron Lett. 49 (36) (2008) 5324–5327; (e) K.C. Majumdar, S. Samanta, B. Sinha, Recent developments in palladium-catalyzed formation of five-and six-membered fused heterocycles, Synthesis 44 (06) (2012) 817–847; (f) L. Fan, A.M. Adams, B. Ganem, Multicomponent reaction design: a one-pot route to substituted di-O-acylglyceric acid amides from α-diazoketones, Tetrahedron Lett. 49 (41) (2008) 5983–5985; (g) L.X. Shao, M.H. Qi, M. Shi, Lewis acid-catalyzed Prins-type reactions of methylenecyclopropylcarbinols with aldehydes and aldimines, Tetrahedron Lett. 49 (1) (2008) 165–168; (h) K. Gewald, E. Schinke, H. Bottcher, 2-Amino-thiophene aus methylenaktiven Nitrilen, Carbonylverbindungen und Schwefel, Chem. Ber. 99 (1966) 94–100.

[34] C.I. Herrerias, X. Yao, Z. Li, CJ. Li, Reactions of C– H bonds in water, Chem. Rev. 107 (6) (2007) 2546–2562.

[35] (a) D.C. Rideout, R. Breslow, Hydrophobic acceleration of Diels-Alder reactions, J. Am. Chem. Soc. 102 (26) (1980) 7816–7817; (b) J.F. Blake, W.L. Jorgensen, Solvent effects on a Diels-Alder reaction from computer simulations, J. Am. Chem. Soc. 113 (19) (1991) 7430–7432.

[36] (a) A. Lubineau, J. Augé Water as solvent in organic synthesis. Mod. Solvents Org. Synth. (1999) 1–39; (b) C.J. Li, T.H. Chan Organic Reactions in Aqueous Media. Wiley, Hoboken, New Jersey, U.S., 1997.

[37] C.J. Li, Organic reactions in aqueous media-with a focus on carbon-carbon bond formation, Chem. Rev. 93 (6) (1993) 2023–2035.

[38] B.M. Chougala, M. Holiyachi, N.S. Naik, L.A. Shastri, S. Dodamani, S. Jalalpure, S.R. Dixit, S.D. Joshi, V.A. Sunagar, Microwave synthesis of coumarinyl substituted pyridine derivatives as potent anticancer agents and molecular docking studies, Chem. Select. 2 (18) (2017) 5234–5242.

[39] M.M. Heravi, S.Y. Beheshtiha, M. Dehghani, N. Hosseintash, Using magnetic nanoparticles Fe_3O_4 as a reusable catalyst for the synthesis of pyran and pyridine derivatives via one-pot multicomponent reaction, J. Iran. Chem. Soc. 12 (11) (2015) 2075–2081.

[40] P. Thirumurugan, A. Nandakumar, D. Muralidharan, P.T. Perumal, Simple and convenient approach to the Kr€ ohnke pyridine type synthesis of functionalized indol-3-yl pyridine derivatives using 3-cyanoacetyl indole, J. Comb. Chem. 12 (1) (2010) 161–167.

[41] B. Mohammadi, Microwave assisted one-pot pseudo four-component synthesis of 2, 4, 6-trisubstituted pyridines using γ-MnO_2 nanoparticles, Monat. Chem.-Chem. Monthly 147 (11) (2016) 1939–1943.

[42] M.R. Bodireddy, N.G. Reddy, S.D. Kumar, Synthesis of alkynyl/alkenyl-substituted pyridine derivatives via heterocyclization and Pd-mediated Sonogashira/Heck coupling process in one-pot: a new MCR strategy, RSC Adv. 4 (33) (2014) 17196–17205.

[43] M. Edrisi, N. Azizi, Sulfonic acid-functionalized graphitic carbon nitride composite: a novel and reusable catalyst for the one-pot synthesis of polysubstituted pyridine in water under sonication, J. Iran. Chem. Soc. 17 (4) (2020) 901–910.

[44] K. Zhao, X.P. Xu, S.L. Zhu, D.Q. Shi, Y. Zhang, SJ. Ji, Facile and efficient synthesis of a new class of indole-substituted pyridine derivatives via one-pot multicomponent reactions, Synthesis 2009 (16) (2009) 2697–2708.

[45] L.M. Sanchez, Á.G. Sathicq, J.L. Jios, G.T. Baronetti, H.J. Thomas, GP. Romanelli, Solvent-free synthesis of functionalized pyridine derivatives using Wells-Dawson heteropolyacid as catalyst, Tetrahedron Lett. 52 (34) (2011) 4412–4416.

[46] R. Pagadala, S. Maddila, V. Moodley, W.E. van Zyl, SB. Jonnalagadda, An efficient method for the multicomponent synthesis of multisubstituted pyridines, a rapid procedure using Au/MgO as the catalyst, Tetrahedron Lett. 55 (29) (2014) 4006–4010.

[47] S. Kantevari, M.V. Chary, SV. Vuppalapati, A highly efficient regioselective one-pot synthesis of 2, 3, 6-trisubstituted pyridines and 2, 7, 7-trisubstituted tetrahydroquinolin-5-ones using $K_5CoW_{12}O_{40} \cdot 3H_2O$ as a heterogeneous recyclable catalyst, Tetrahedron 63 (52) (2007) 13024–13031.

[48] S.V. Bhaskaruni, S. Maddila, K.K. Gangu, S.B. Jonnalagadda, A review on multi-component green synthesis of N-containing heterocycles using mixed oxides as heterogeneous catalysts, Arab. J. Chem. 13 (1) (2020) 1142–1178.

[49] A.R. Longstreet, B.S. Campbell, B.F. Gupton, D.T. McQuade, Improved synthesis of mono-and disubstituted 2-halonicotinonitriles from alkylidene malononitriles, Org. Lett. 15 (20) (2013) 5298–5301.

[50] G. Tenti, M.T. Ramos, J.C. Menéndez, One-pot access to a library of structurally diverse nicotinamide derivatives via a three-component formal aza [3+ 3] cycloaddition, ACS Combinator. Sci. 14 (10) (2012) 551–557.

[51] Z. Song, X. Huang, W. Yi, W. Zhang, One-pot reactions for modular synthesis of polysubstituted and fused pyridines, Org. Lett. 18 (21) (2016) 5640–5643.

[52] H.D. Khanal, Y.R. Lee, Organocatalyzed oxidative N-annulation for diverse and polyfunctionalized pyridines, Chem. Commun. 51 (46) (2015) 9467–9470.

[53] L. Xiang, F. Zhang, B. Chen, X. Pang, X. Yang, G. Huang, R. Yan, An I_2-catalyzed oxidative cyclization for the synthesis of indolizines from aromatic/aliphatic olefins and α-picoline derivatives, RSC Adv. 5 (37) (2015) 29424–29427.

[54] X. Pan, Q. Liu, L. Chang, G. Yuan, An Efficient Synthesis of Polysubstituted Pyridines via C-H Oxidation and C-S Cleavage of Dimethyl Sulfoxide. Adv. Synth. Catal. 358 (2016) 218–225.

[55] S.Y. Khatavi, K. Kantharaju, Microwave accelerated synthesis of 2-oxo-2H-chromene-3-carboxylic acid using WELFSA, Curr. Microwave Chem. 5 (3) (2018) 206–214.

[56] X. Qing, T. Wang, F. Zhang, C. Wang, One-pot synthesis of 2, 4, 6-triarylpyridines from β-nitrostyrenes, substituted salicylic aldehydes and ammonium acetate, RSC Adv. 6 (98) (2016) 95957–95964.
[57] P.P. Ghosh, P. Mukherjee, AR. Das, Triton-X-100 catalyzed synthesis of 1, 4-dihydropyridines and their aromatization to pyridines and a new one pot synthesis of pyridines using visible light in aqueous media, RSC Adv. 3 (22) (2013) 8220–8226.
[58] L. Shen, S. Cao, J. Wu, H. Li, J. Zhang, M. Wu, X. Qian, K_2CO_3-assisted one-pot sequential synthesis of 2-trifluoromethyl-6-difluoromethylpyridine-3, 5-dicarboxylates under solvent-free conditions, Tetrahedron Lett. 51 (37) (2010) 4866–4869.
[59] B. Jiang, W.J. Hao, X. Wang, F. Shi, S.J. Tu Diversity-oriented synthesis of Kröhnke pyridines. J. Comb. Chem. 11 (5) (2009) 846–850
[60] G. Yin, Q. Liu, J. Ma, N. She, Solvent-and catalyst-free synthesis of new hydroxylated trisubstituted pyridines under microwave irradiation, Green Chem. 14 (6) (2012) 1796–1798.
[61] R. Yan, X. Zhou, M. Li, X. Li, X. Kang, X. Liu, X. Huo, G. Huang, Metal-free synthesis of substituted pyridines from aldehydes and NH_4OAc under air, RSC Adv. 4 (92) (2014) 50369–50372.
[62] C.Z. Luo, J. Jayakumar, P. Gandeepan, Y.C. Wu, C.H. Cheng. Rhodium (III)-Catalyzed vinylic C–H activation: a direct route toward pyridinium salts. Org. Lett. 2015;17(4):924-927.
[63] Z.Y. Mao, X.Y. Liao, H.S. Wang, C.G. Wang, K.B. Huang, YM. Pan, Acid-catalyzed tandem reaction for the synthesis of pyridine derivatives via C [double bond, length as m-dash] C/C (sp^3)–N bond cleavage of enones and primary amines, RSC Adv. 7 (22) (2017) 13123–13129.
[64] J.P. Wan, Y. Jing, C. Hu, S. Sheng, Metal-free synthesis of fully substituted pyridines via ring construction based on the domino reactions of enaminones and aldehydes, J. Org. Chem. 81 (15) (2016) 6826–6831.
[65] Q. Wang, C. Wan, Y. Gu, J. Zhang, L. Gao, Z. Wang, A metal-free decarboxylative cyclization from natural α-amino acids to construct pyridine derivatives, Green Chem. 13 (3) (2011) 578–581.
[66] J. Li, P. He, C. Yu, DPTA-catalyzed one-pot regioselective synthesis of polysubstituted pyridines and 1, 4-dihydropyridines, Tetrahedron 68 (22) (2012) 4138–4144.
[67] A. Saha, S. Payra, S. Banerjee, In-water facile synthesis of poly-substituted 6-arylamino pyridines and 2-pyrrolidone derivatives using tetragonal nano-ZrO_2 as reusable catalyst, RSC Adv. 6 (104) (2016) 101953–101959.
[68] (a) B.C. Ranu, R. Jana, S. Sowmiah, An improved procedure for the three-component synthesis of highly substituted pyridines using ionic liquid, J. Org. Chem. 72 (8) (2007) 3152–3154; (b) K. Guo, R. Mutter, W. Heal, T.R. Reddy, H. Cope, S. Pratt, M.J. Thompson, B. Chen, Synthesis and evaluation of a focused library of pyridine dicarbonitriles against prion disease, Eur. J. Med. Chem. 43 (1) (2008) 93–106; (c) M. Sridhar, B.C. Ramanaiah, C. Narsaiah, B. Mahesh, M. Kumaraswamy, K.K. Mallu, V.M. Ankathi, P.S. Rao, Novel $ZnCl_2$-catalyzed one-pot multicomponent synthesis of 2-amino-3, 5-dicarbonitrile-6-thio-pyridines, Tetrahedron Lett. 50 (27) (2009) 3897–3900.
[69] X. He, Y. Shang, Z. Yu, M. Fang, Y. Zhou, G. Han, F. Wu, $FeCl_3$-catalyzed four-component nucleophilic addition/intermolecular cyclization yielding polysubstituted pyridine derivatives, J. Org. Chem. 79 (18) (2014) 8882–8888.
[70] D.N. Reddy, K.B. Chandrasekhar, Y.S. Ganesh, B.S. Kumar, R. Adepu, M. Pal, $SnCl_2 \cdot 2H_2O$ as a precatalyst in MCR: synthesis of pyridine derivatives via a 4-component reaction in water, Tetrahedron Lett. 56 (31) (2015) 4586–4589.
[71] (a) M.N. Khan, S. Pal, T. Parvin, L.H. Choudhury, A simple and efficient method for the facile access of highly functionalized pyridines and their fluorescence property studies, RSC Adv. 2 (32) (2012) 12305–12314; (b) X. Xin, Y. Wang, S. Kumar, X. Liu, Y. Lin, D. Dong, Efficient one-pot synthesis of substituted pyridines through multicomponent reaction, Org. Biomol. Chem. 8 (13) (2010) 3078–3082.
[72] K. Shekarrao, D. Nath, P.P. Kaishap, S. Gogoi, R.C. Boruah, Palladium-catalyzed multi-component synthesis of steroidal A-and D-ring fused 5, 6-disubstituted pyridines under microwave irradiation, Steroids *78* (11) (2013) 1126–1133.
[73] (a) M. Thimmaiah, P. Li, S. Regati, B. Chen, J.C. Zhao, Multi-component synthesis of 2-amino-6-(alkylthio) pyridine-3, 5-dicarbonitriles using Zn (II) and Cd (II) metal–organic frameworks (MOFs) under solvent-free conditions, Tetrahedron Lett. 53 (36) (2012) 4870–4872; (b) J.B. Gujar, M.A. Chaudhari, D.S. Kawade, M.S. Shingare, Sodium chloride: a proficient additive for the synthesis of pyridine derivatives in aqueous medium, Tetrahedron Lett. 55 (50) (2014) 6939–6942.

[74] A.V. Kuznetcova, I.S. Odin, A.A. Golovanov, I.M. Grigorev, AV. Vasilyev, Multicomponent reaction of conjugated enynones with malononitrile and sodium alkoxides: complex reaction mechanism of the formation of pyridine derivatives, Tetrahedron 75 (33) (2019) 4516–4530.
[75] M.A. Zolfigol, M. Kiafar, M. Yarie, A.A. Taherpour, M. Saeidi-Rad, Experimental and theoretical studies of the nanostructured {Fe_3O_4@ SiO_2@(CH_2) 3Im} C $(CN)_3$ catalyst for 2-amino-3-cyanopyridine preparation via an anomeric based oxidation, RSC Adv. 6 (55) (2016) 50100–50111.
[76] J.P. Wan, Y. Zhou, K. Jiang, H. Ye, Thioacetamide as an ammonium source for multicomponent synthesis of pyridines from aldehydes and electron-deficient enamines or alkynes, Synthesis 46 (23) (2014) 3256–3262.
[77] J. Chen, H. Ni, W. Chen, G. Zhang, Y. Yu, A new strategy for facile synthesis of tetrasubstituted pyridine derivatives, Tetrahedron 69 (37) (2013) 8069–8073.
[78] H. Wei, Y. Li, K. Xiao, B. Cheng, H. Wang, L. Hu, H. Zhai, Synthesis of polysubstituted pyridines via a one-pot metal-free strategy, Org. Lett. 17 (24) (2015) 5974–5977.
[79] S. Sultana, S.M. Maezono, M.S. Akhtar, J.J. Shim, Y.J. Wee, S.H. Kim, Y.R. Lee, $BF_3\cdot OEt_2$-promoted annulation for substituted 2-arylpyridines as potent UV filters and antibacterial agents, Adv. Synth. Catal. 360 (4) (2018) 751–761.

CHAPTER

11

Recent green synthesis of pyridines and their fused systems catalyzed by nanocatalysts

Amira Elsayed Mahmoud Abdallah

Chemistry Department, Faculty of Science, Helwan University Ain, Helwan, Cairo, Egypt

11.1 Introduction

The most important six-membered heterocyclic ring system, pyridine was first discovered by the Scottish chemist Thomas Anderson in 1868. Due to the flammability properties of pyridine, Anderson added the prefix (*pyr*) is which related to the Greek: $\pi\tilde{\upsilon}\rho$ meaning *fire*. Then, added the suffix (*idine*) to indicate a nitrogen cyclic compound [1,2]. Many medicinal present drugs, pesticides, and natural products such as vitamins and alkaloids contain pyridine ring system. In addition, pyridines have several biological applications, such as, anticancer [3,4], antidiabetic [5], antimicrobial [6,7], antioxidant [8], and antiviral [9]. Due to the important characteristics of pyridine moiety, it can be used as solvent, catalyst, and base in many parts of organic chemical reactions. Recently, the green synthesis procedures of pendant and fused heterocyclic ring systems have great attention in the last years. The reasons for this importance to avoid any problems take place in the other methods such as, use expensive reagents, large amounts of solvent, long reaction time, and reduce the formation of biproduct via side reactions [10]. Green synthesis methods by using heterogeneous nanocatalysts introduce economic and environmental procedures which make the processes occur in a clean, simple way, inexpensive, short reaction times, high yields of the products, easy work-up, reusability of these catalysts for several times without any observed decreased in its catalytic activity [11].

This chapter will describe the most recent efficient methodology in the synthesis of pyridine derivatives during the last 6 years by using green synthesis methods in the presence of various types of nanocatalysts. The catalytic activity of the present nanocatalyst was evaluated for the formation of pyridine derivatives by using microwave, ultrasound radiation or other methods

Recent Developments in the Synthesis and Applications of Pyridines.
DOI: https://doi.org/10.1016/B978-0-323-91221-1.00011-7

as green methods. This chapter will be divided into two parts according to the type of pyridine ring, pendant, or fused systems, and each part divided into many subtitles parts according to the type of pendant substituted pyridines or fused pyridine ring systems and the types of nanocatalysts used sequentially.

11.2 Recent green synthesis of polysubstituted pyridines by using heterogeneous nanocatalyst systems

Due to the important applications of most of the compounds containing pyridine moiety, several research articles describe their synthesis especially focusing on green synthesis by using nanocatalyst system. The later eco-friendly green methods introduce pure pyridine derivatives in a short time, with excellent to high yields of the desired products. Therefore, green preparation of pyridines by using nanocatalyst is described through the last 6 years in a sequential manner.

11.2.1 Multicomponent synthesis of 2-amino-3-cyano pyridines by using various heterogeneous nanocatalysts

11.2.1.1 By using copper nanoparticles on charcoal (Cu/C) catalyst

Khalifeh and Ghamari (2016) reported a direct one-pot four-component coupling reaction to afford several 2-amino-3-cyanopyridine derivatives **5a–o**. The reaction was carried out between aldehydes (**1**), malononitrile (**2**), ketone (**3**) (aromatic or aliphatic), and ammonium acetate (**4**) in the presence of copper nanoparticles on charcoal (Cu/C) catalyst (Scheme 11.1) [12]. On the other hand, using different cyclic ketones like cyclohexanone (**6a**) and *p*-methyl cyclohexanone (**6b**) afforded the fused pyridine systems namely 2-amino-4-phenyl-5,6,7,8-tetrahydroquinoline-3-carbonitrile and 2-amino-6-methyl-4-phenyl-5,6,7,8-tetrahydroquinoline-3-carbonitrile, respectively **7a–b** (Scheme 11.2). The main advantage for this method was the recoverability of the catalyst from the reaction mixture by simple filtration and its reusability for several times (eight times) with no observed decrease in its catalytic activity.

11.2.1.2 Via using {Fe_3O_4@SiO_2@$(CH_2)_3$Im}$C(CN)_3$ nanocatalyst under solvent-free conditions

Zolfigol and his group (2016) described a novel one-pot green method for the preparation of 2-amino-4,6-diphenylnicotinonitriles in the presence of Fe_3O_4@SiO_2@$(CH_2)_3$Im}$C(CN)_3$ as a nanostructured heterogeneous catalyst with an ionic liquid tag under solvent-free and mild conditions [13]. The reaction involving condensation between aromatic aldehydes derivatives (**1**), malononitrile (**2**), substituted acetophenone (**3**), and ammonium acetate (**4**) to yield the corresponding desired products 2-amino-3-cyanopyridines **8a–q** (Scheme 11.3) in short times and good yields. The later desired products were synthesized earlier by other research groups with different methods [14–21] except only one new compound **8m**. The last step of the plausible mechanism in the synthetic pathway to obtain the desired products occurred via an anomeric-based oxidation mechanism.

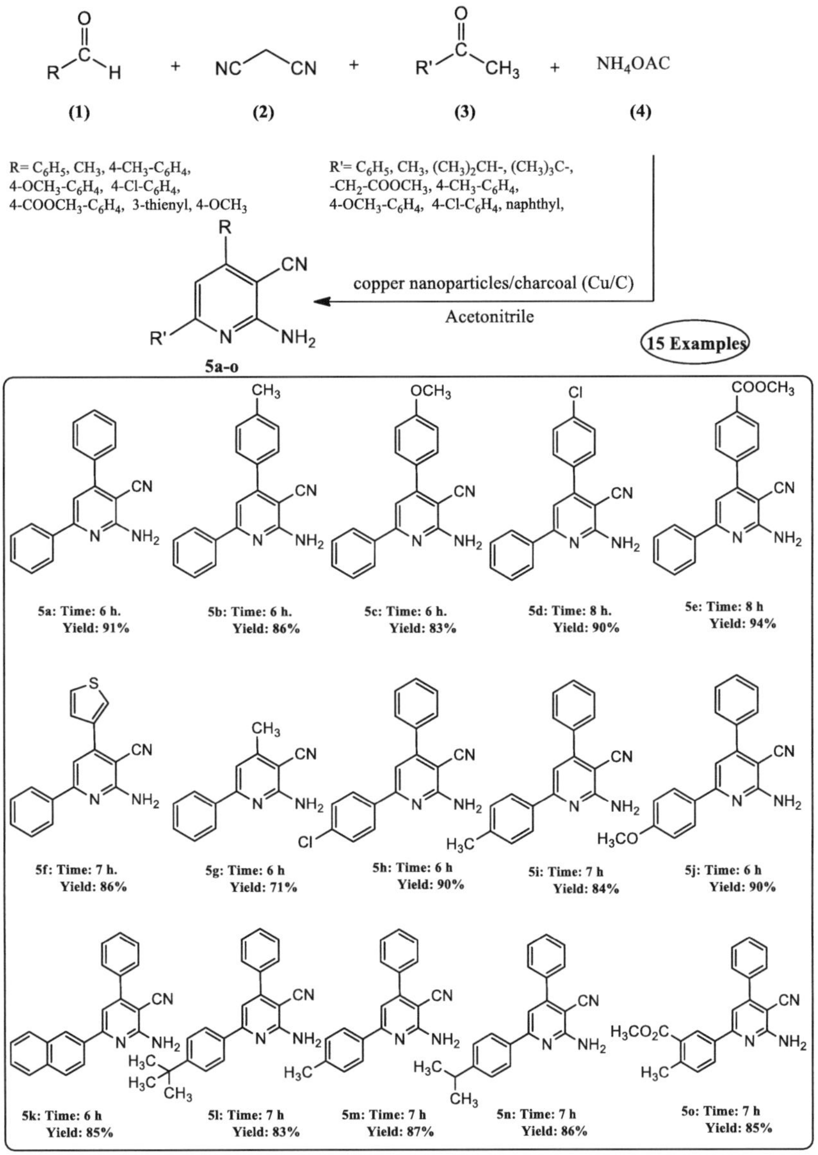

SCHEME 11.1 Copper nanoparticles on charcoal (Cu/C) catalyzed the synthesis of 2-amino-3-cyanopyridines.

11.2.1.3 *In aqueous medium via modified green biocatalyst vitamin B_3 (Fe_3O_4@Niacin) under microwave irradiation*

Super-paramagnetic nanoparticles of modified vitamin B_3 (Fe_3O_4@Niacin) were synthesized as a new green, efficient, and reusable heterogeneous biocatalyst by Afradi et al. in 2018 [22]. The catalytic application of the present new biocatalyst was evaluated for the synthesis of 2-amino-3-cyanopyridine derivatives **9a–o** via four-component condensation reaction of aryl

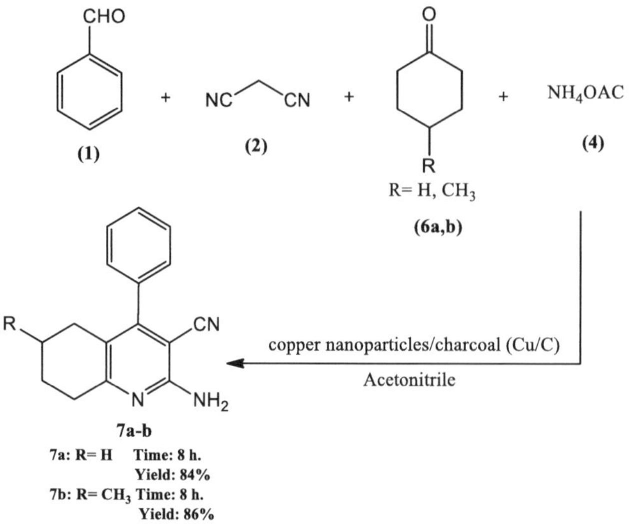

SCHEME 11.2 Synthesis of 2-amino-4-phenyl-5,6,7,8-tetrahydroquinoline-3-carbonitrile derivatives via copper nanoparticles on charcoal (Cu/C).

aldehydes (**1**), malononitrile (**2**), substituted ketones (**3**), and ammonium acetate (**4**) under microwave irradiation in water (Scheme 11.4). All the prepared compounds were synthesized and published previously [15,23], except five compounds **9b–e** and **9h**. The main advantages of this methodology were the very short reaction time (7–10 min), excellent yields and reusability of the catalyst at least six times without remarkable losing in its activity.

11.2.1.4 By using ($CoFe_2O_4@SiO_2–SO_3H$) solid acid nanocatalyst under microwave irradiation

A facile, efficient, and green method was described for the preparation of 2-amino-4,6-diarylnicotinonitrile in the presence of $CoFe_2@SiO_2–SO_3H$ as a recoverable catalyst under microwave irradiation and solvent-free conditions were reported by Hosseinzadeh and her group in 2018 [24]. The four-component condensation reaction occurred between aromatic aldehydes derivatives (**1**), malononitrile (**2**), acetophenone (**3**), and ammonium acetate (**4**) in the presence of $CoFe_2O_4@SiO_2–SO_3H$ for the synthesis of 2-amino-4,6-diarylnicotinonitriles **10a–m** (Scheme 11.5). All the prepared compounds were new, except two compounds only were synthesized earlier **10a, b** by other research group [21]. The significant features of this green methodology were short time of reaction, high yields, simple workup and recrystallization to obtain high pure compounds, and reusability of the catalyst at least five times without any loss in its catalytic activity.

11.2.1.5 Via a heterogeneous nanocatalyst (MNPs@AMTT/Cu(II) under solvent-free conditions

An efficient, simple, and facile one-pot, four-component reaction between aldehydes (**1**), malononitrile (**2**), ketone (**3**), and ammonium acetate (**4**) for the synthesis of 2-amino-3-cyanopyridine derivatives **11a–j** and their cycloalkanes **12a–o** via MNPs@AMTT/Cu(II) a novel reusable heterogeneous magnetic nanocatalyst system under solvent-free conditions was reported by Ji Zhou and coworkers in (2019) [25] (Schemes 11.6 and 7), respectively.

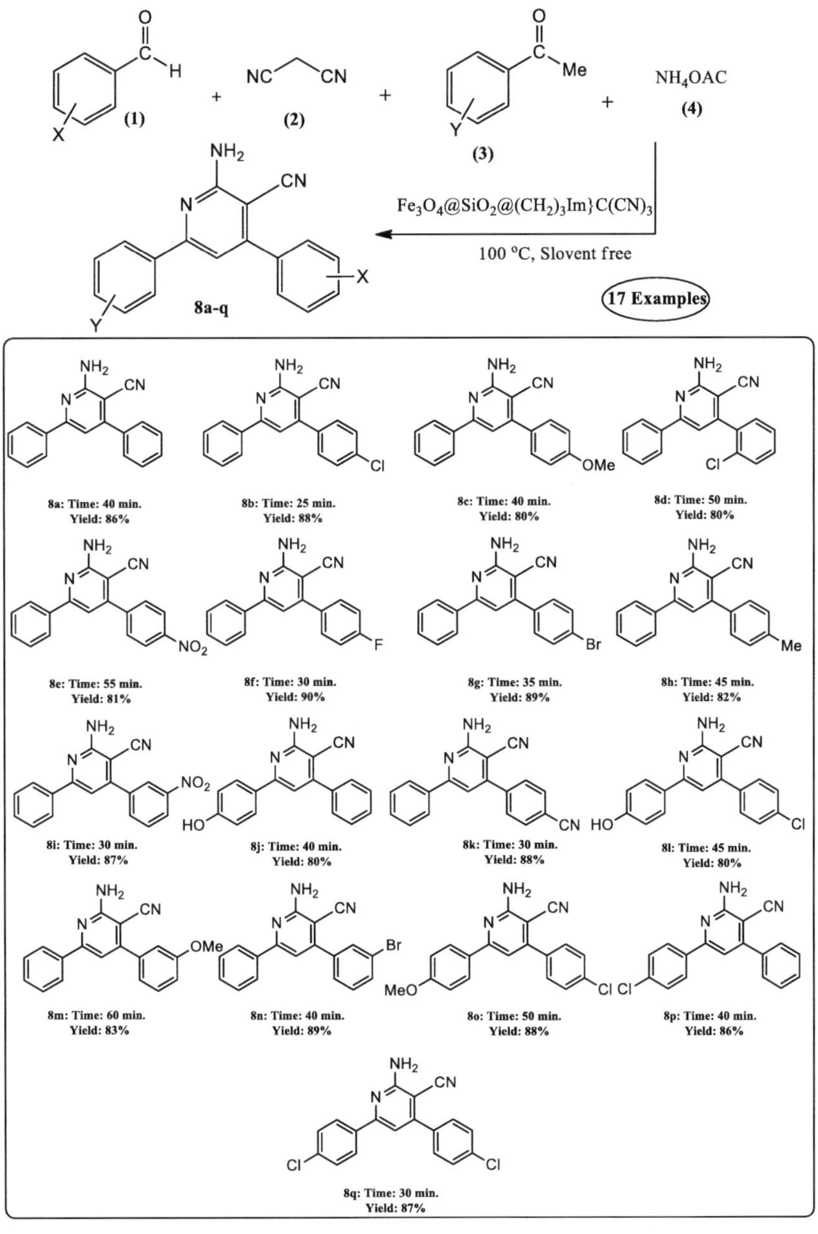

SCHEME 11.3 Catalytic application of the synthesis of 2-amino-3-cyanopyridines via $\{Fe_3O_4@SiO_2@(CH_2)_3Im\}C(CN)_3$ under solvent-free conditions.

The most important features of this green synthesis were the easy preparation of the catalyst, short reaction times, mild reaction conditions, excellent yields, and the catalyst is recoverable and can be reused several times at least six runs with good efficiency. Other published paper

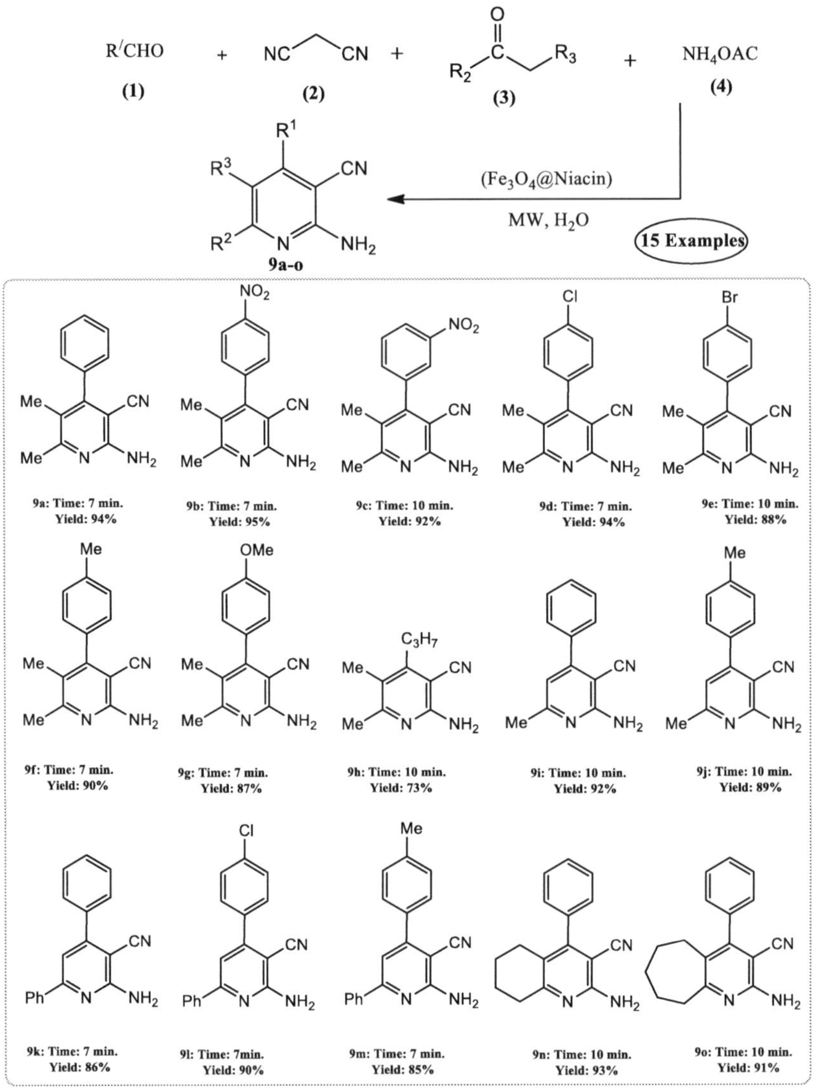

SCHEME 11.4 Synthesis of 2-amino-3-cyanopyridines using modified green biocatalyst vitamin B_3 (Fe_3O_4 @Niacin).

describes the synthesis of 2-amino-3-cyanopyridine derivatives and their cycloalkanes but by using a different methodology [15,20,26–34].

11.2.1.6 *By using a powerful magnetic nanoporous supported melamine (Fe_3O_4@MCM-41/melamine)*

In 2021, Zahra Haydari et al. synthesized 2-amino-3-cyanopyridines **8a–g** in water by using a novel magnetic mesoporous silica-supported melamine as a powerful nano-organocatalyst

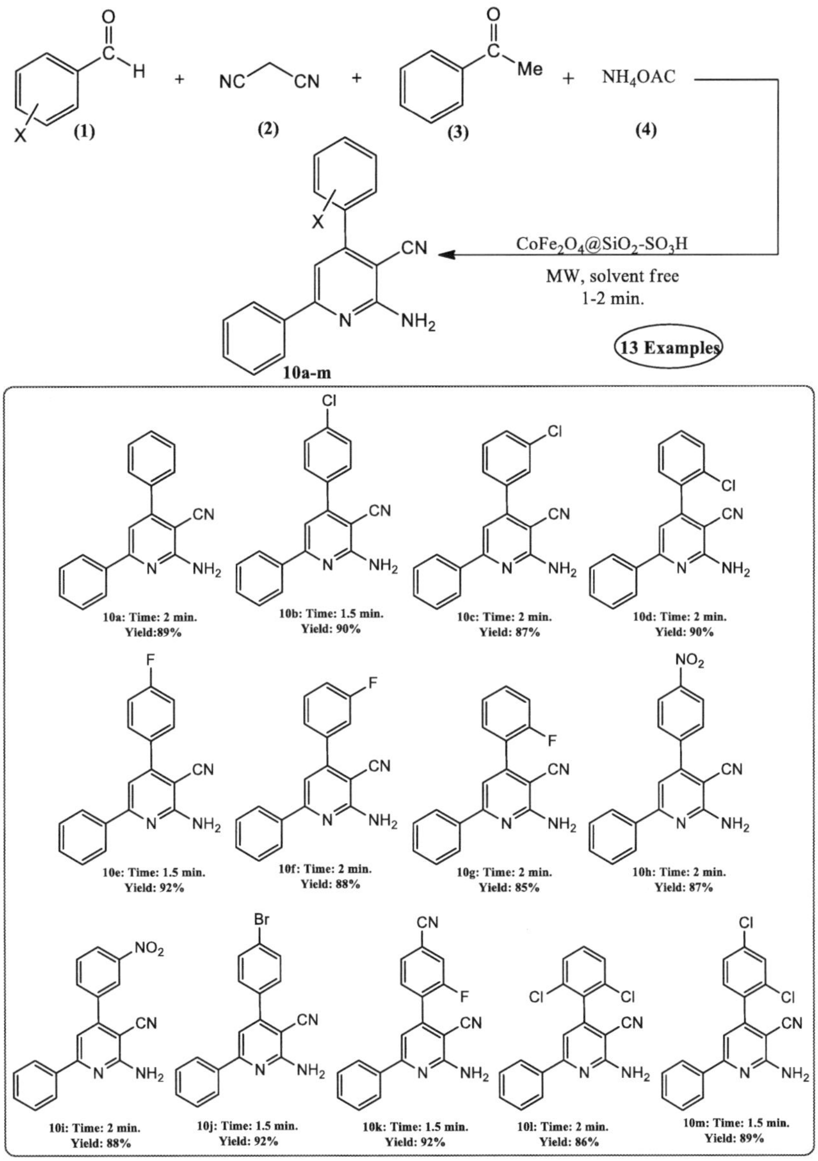

SCHEME 11.5 Catalytic activity of the ($CoFe_2O_4@SiO_2–SO_3H$) solid acid nanocatalyst for the synthesis of 2-amino-4,6-diarylnicotinonitrile derivatives.

(Fe_3O_4@MCM-41/melamine) under green ultrasonic conditions [35]. In this methodology, aldehydes (**1**), malononitrile (**2**), acetophenone (**3**), and ammonium acetate (**4**) were reacted in the presence of Fe_3O_4@MCM-41/melamine catalyst in H_2O as solvent under ultrasonic bath at 50°C for the appropriate time as required in each derivative to obtain the pure 2-amino-3-cyanopyridine derivatives **13a–g** (Scheme 11.8). Although, many published

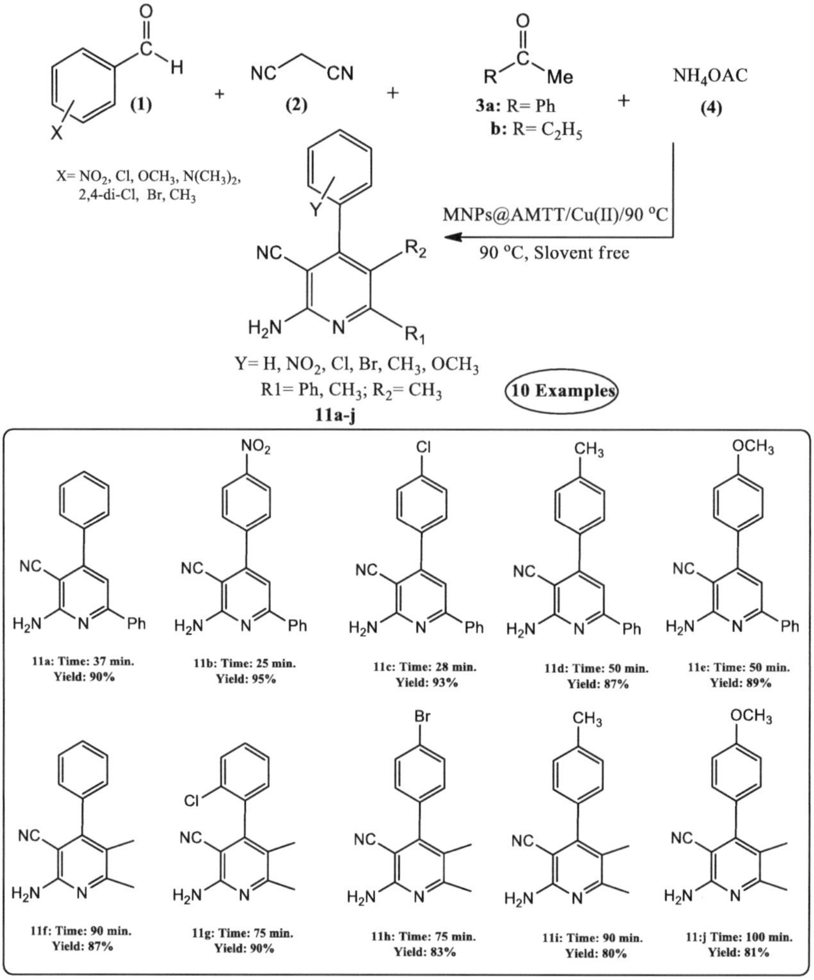

SCHEME 11.6 Synthesis of 2-amino-3-cyanopyridine derivatives **11a–j** catalyzed by (MNPs@AMTT/Cu(II) nanocatalyst.

research synthesized 2-amino-3-cyanopyridine derivatives by using different reaction conditions , the current study including several features than the others. This green procedure introduces many advantages such as the use of water as a green solvent, high stability, recoverability, reusability of the catalyst, and simple separation of the products.

11.2.1.7 *By using bifunctional ionic–liquid nanocatalyst (CaO@SiO$_2$@BAIL) under solvent-free conditions*

The synthesis of 2-amino-4,6-diarylpyridine-3-carbonitrile derivatives **15a–j** were described by Sameri and her groups (2021) through the reaction of arylaldehyde, arylketone, malononitrile, and ammonium acetate in the presence of CaO@SiO$_2$@BAIL as a catalyst with stirring at 80°C under solvent-free condition for the required time (Scheme 11.9) [36]. The

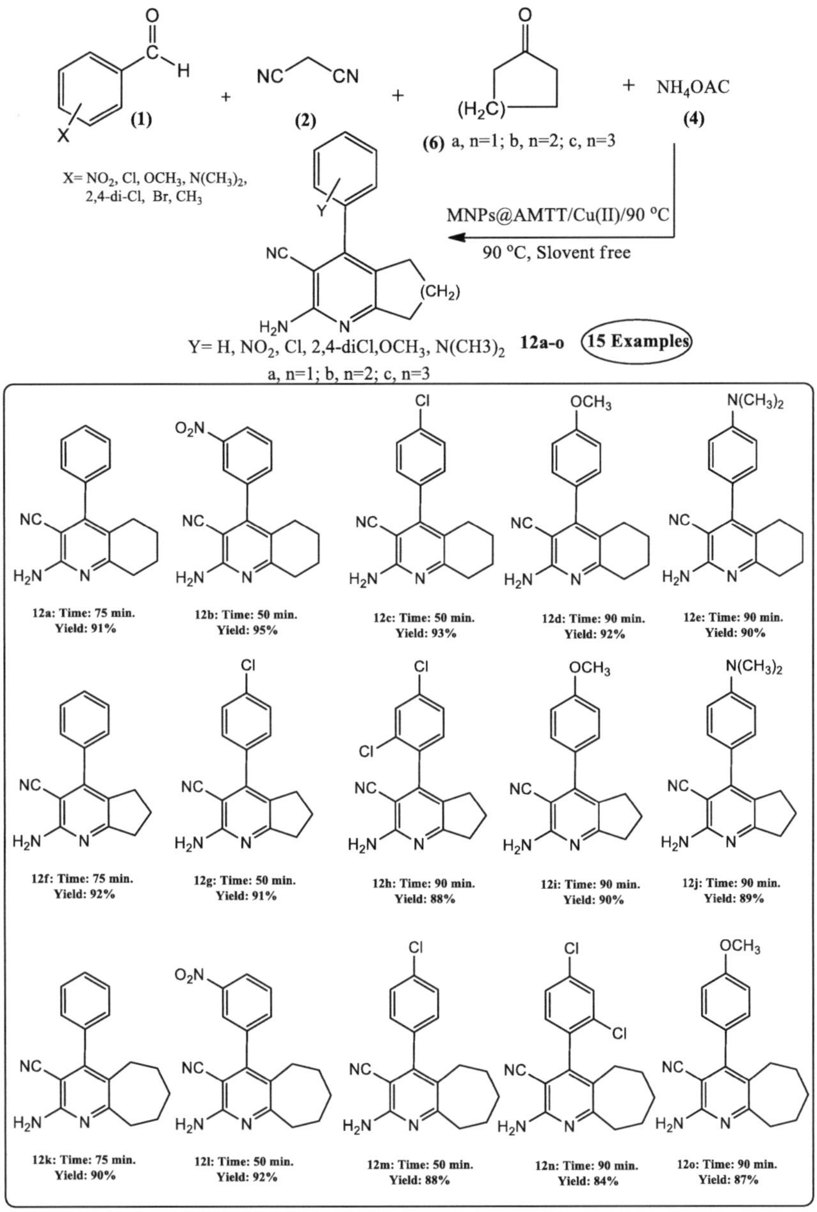

SCHEME 11.7 Synthesis of 2-amino-3-cyanopyridine cycloalkane derivatives **12a–o** catalyzed by (MNPs @AMTT/Cu(II) nanocatalyst.

presented green methods have several advantages such as fast, simple work up, high yields of products, low time of reaction, and the present catalyst was reused at least six times by keeping the same catalytic activity. Some compounds of 2-amino-4,6-diarylpyridine-3-carbonitrile derivatives **15a–j** were reported previously but using different methods [37–39].

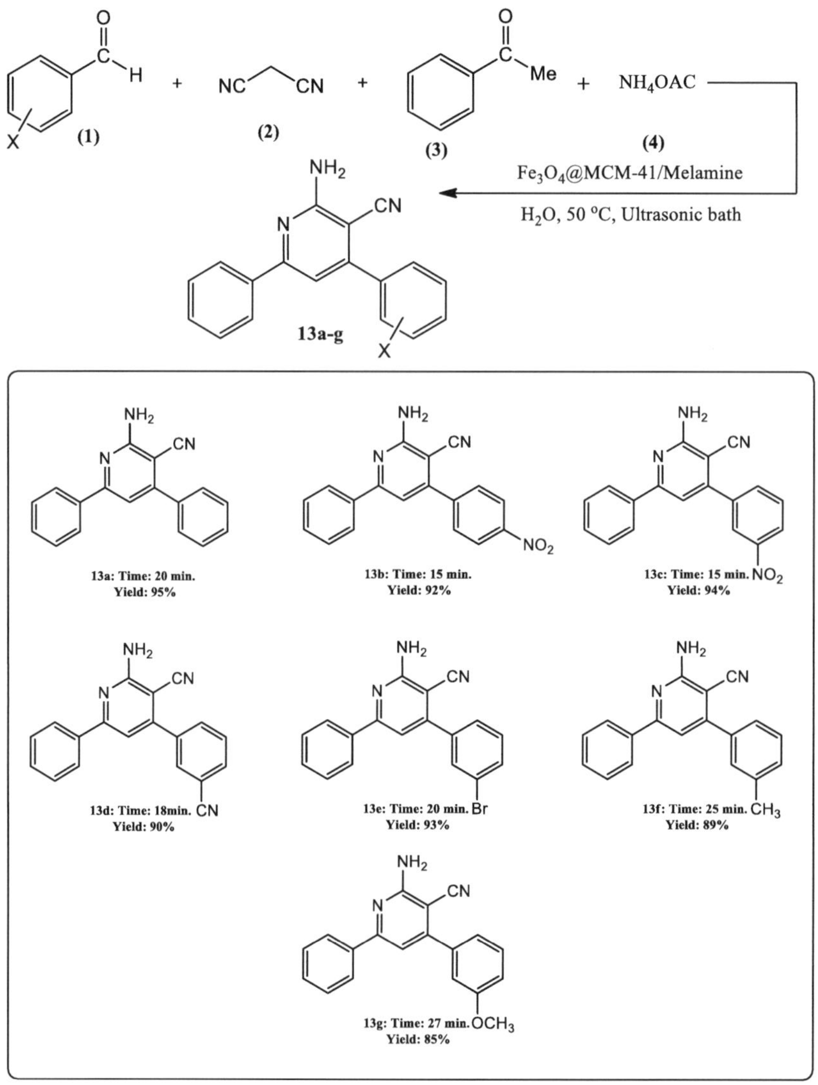

SCHEME 11.8 Synthesis of 2-amino-3-cyanopyridine derivatives **13a–g** catalyzed by Fe_3O_4@MCM-41.

11.2.2 Different heterogeneous nanocatalysts catalyzed the multicomponent synthesis of 1,4-dihydropyridines (1,4-DHPs)

11.2.2.1 By using monodisperse (PdRuNi@GO NPs) heterogeneous catalyst

1,4-Dihydropyridines (1,4-DHPs) were synthesized by Demirci and coworkers (2016) through one-pot, four-component condensation reactions of various aldehydes (**1**) with dimedone (**18**), ammonium acetate (**4**), and ethyl acetoacetate (**16**) at 70°C for 45 min in DMF with a catalytic amount of monodispersed PdRuNi nanoparticles furnished with graphene oxide (PdRuNi@GO NPs) as a stable and reusable nanocatalyst under mild conditions [40]. The

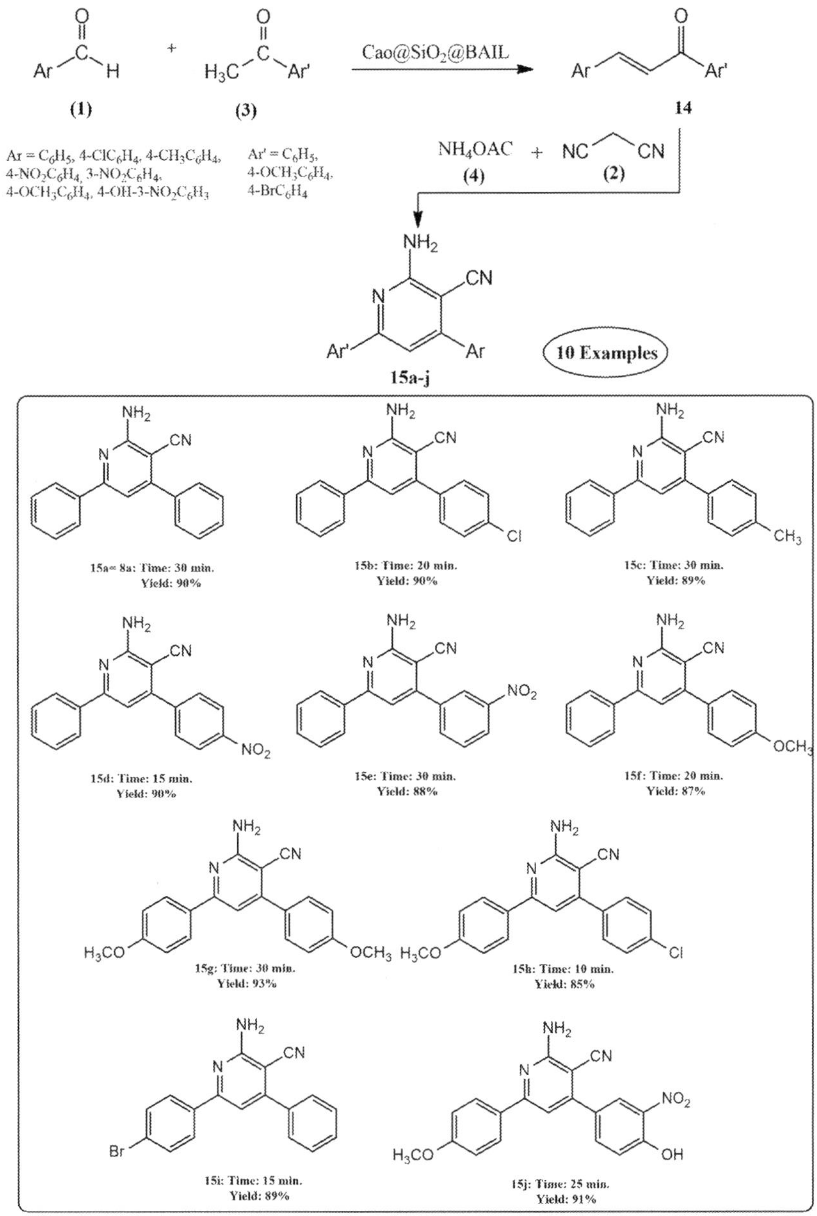

SCHEME 11.9 Synthesis of 2-amino-4,6-diarylpyridine-3-carbonitrile derivatives **12a–j** by using nanocatalyst ($CaO@SiO_2@BAIL$).

later novel catalyst provided this methodology with the highest yields and shortest reaction time. The products resulting from this method were 1,4-dihydropyridine-3,5-dicarboxylates **17a–f** (Scheme 11.10) and 1,4,5,6,7,8-hexahydroquinoline-3-carboxylate derivatives **19a–l** (Scheme 11.11).

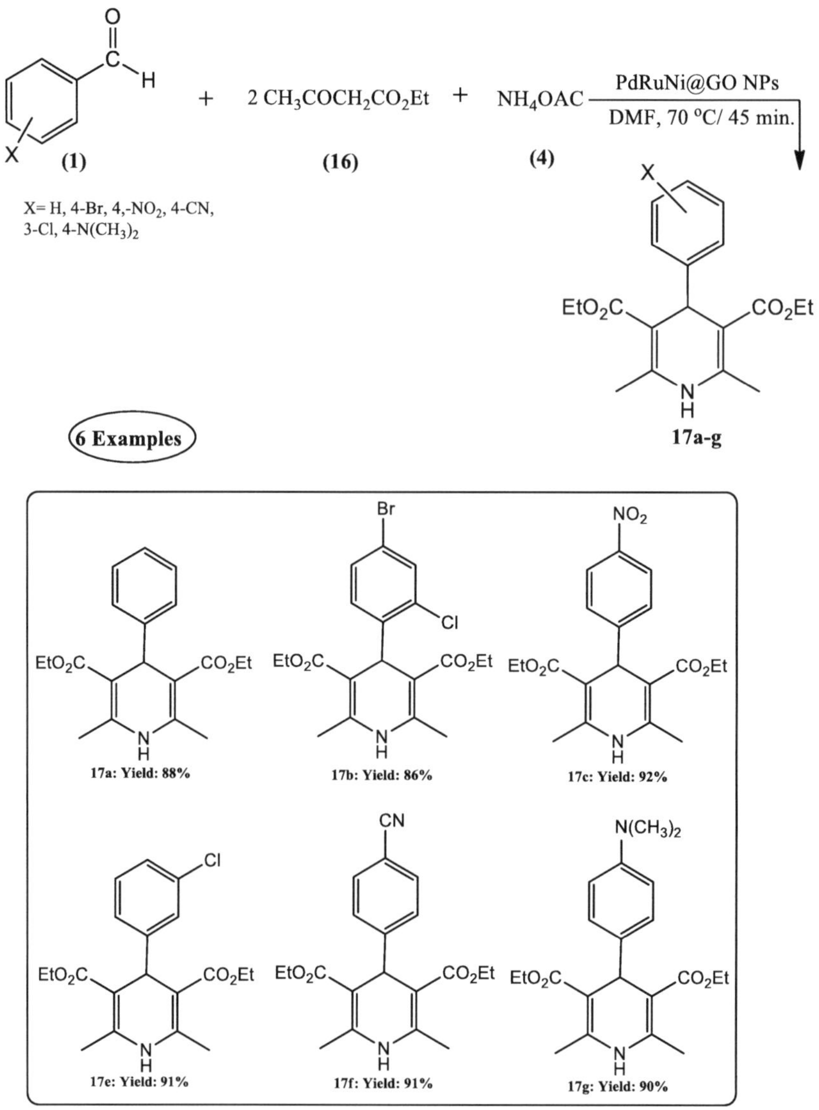

SCHEME 11.10 Synthesis of 1,4-dihydropyridine-3,5-dicarboxylates **17a–f** via (PdRuNi@GO NPs) nanocatalyst.

11.2.2.2 Via (ZnO NPs)

Reen et al. (2017) described a simple, efficient, and eco-friendly procedure using heterogenous catalyst (ZnO NPs) for the synthesis of 1,4-diaryl dihydropyridine derivatives **22a–j** [41]. The one-pot three-component reaction carried out between aldehydes (**1**), β-dicarbonyl compounds (**20**), and some of the substituted anilines (**21**) in the presence of a catalytic amount of ZnO NPs under neat condition at 80°C for a range of reaction times 80–135 min

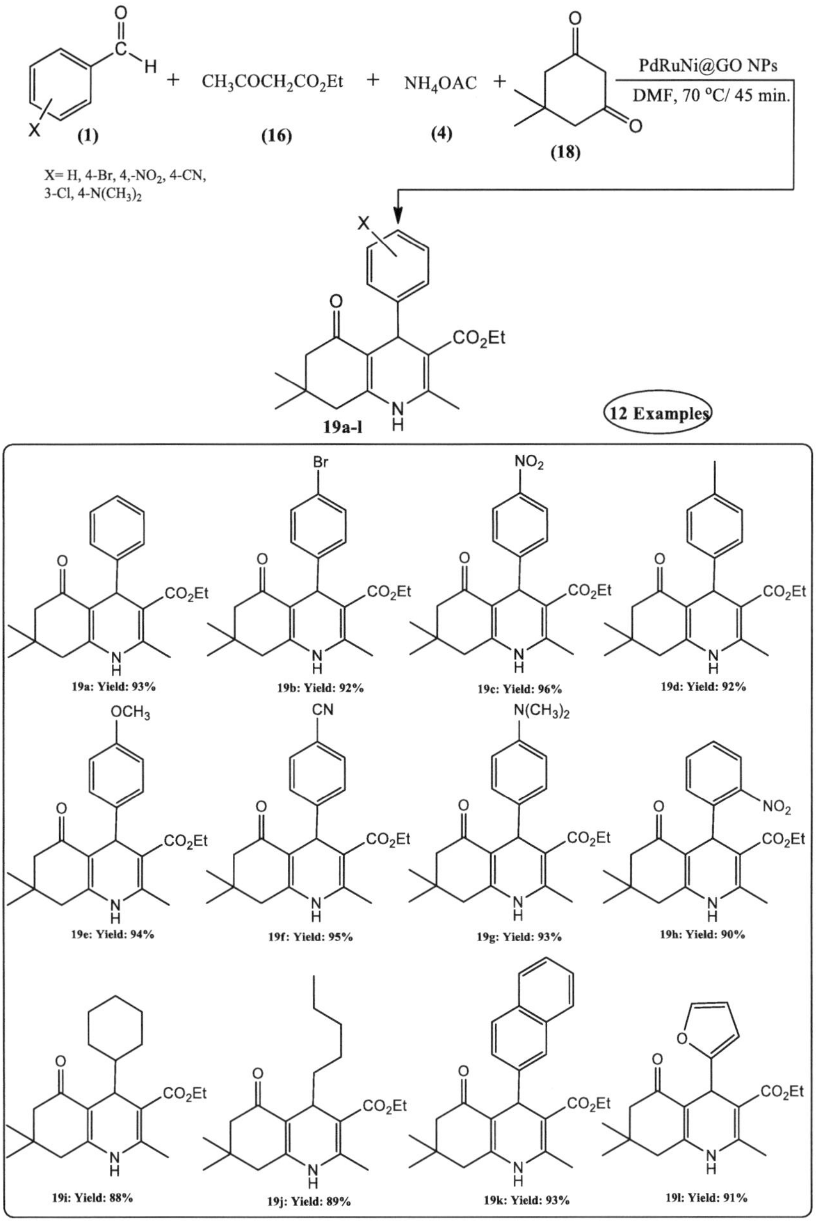

SCHEME 11.11 Synthesis of 1,4,5,6,7,8-hexahydroquinoline-3-carboxylate derivatives **19a–l** via (PdRuNi@GO NPs) nanocatalyst.

(Scheme 11.12). This methodology offers remarkable advantages as a green method in the high yields, short reaction time, easy work up, and neat conditions. In addition, the catalyst used can be easy recovery and also have good reusability.

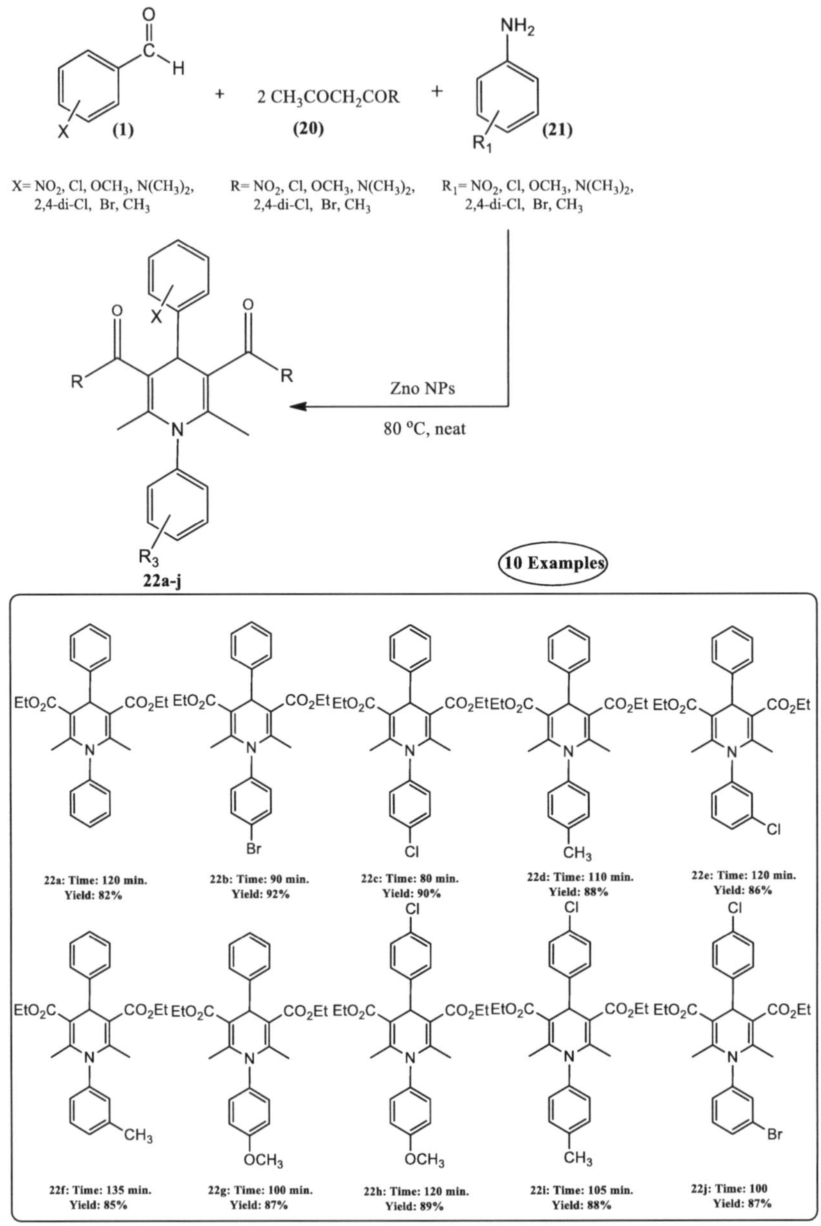

SCHEME 11.12 ZnO NPs catalyzed the synthesis of 1,4-diaryl dihydropyridines **22a–j**.

11.2.2.3 *By using anchored sulfonic acid on silica-layered nickel ferrite magnetic nanoparticles ($NiFe_2O_4@SiO_2@SO_3H$ MNPs)*

The recent green method which describes the synthesis of 1,4-DHPs under green solvent by using nanocatalyst was reported by Zeynizadeh et al. in **2019** [42]. Various 1,4-DHP derivatives

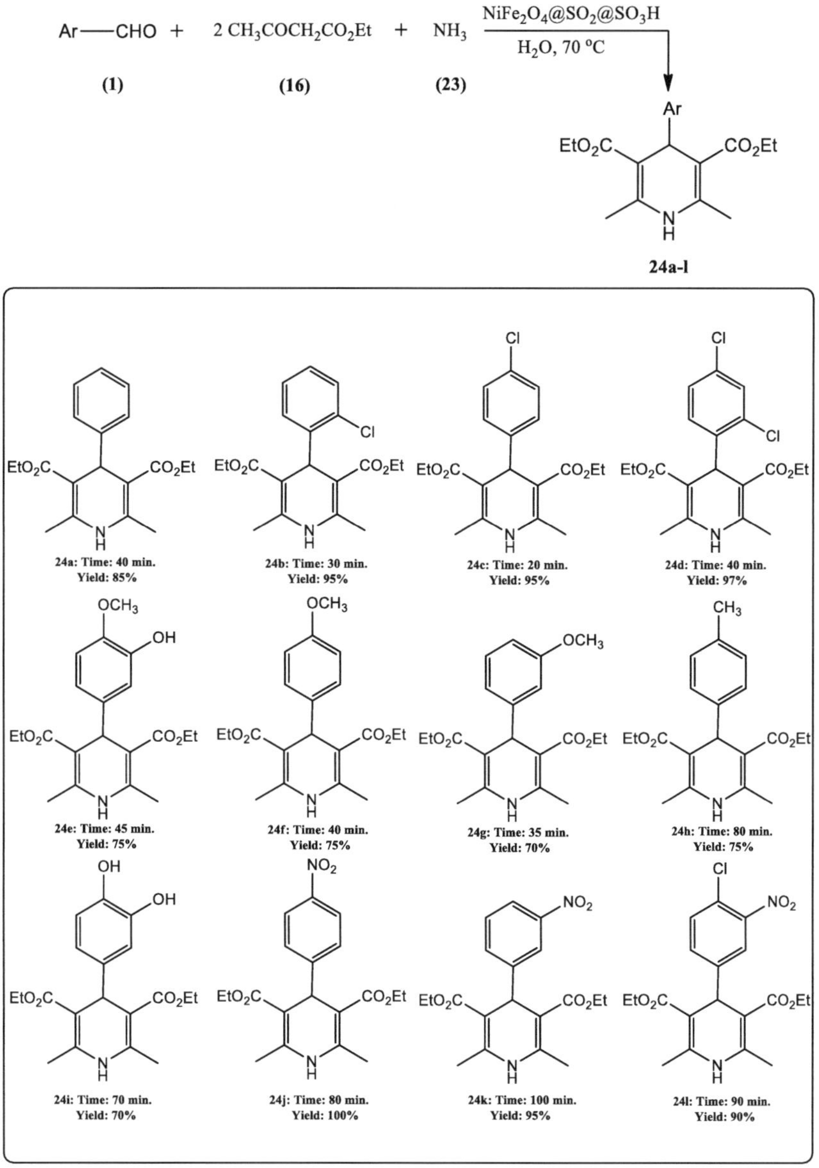

SCHEME 11.13 Synthesis of diethyl 2,6-dimethyl-4-aryl-1,4-dihydropyridine-3,5-dicarboxylate derivatives **24a–l** catalyzed by $NiFe_2O_4@SiO_2@SO_3H$ MNPs.

24a–l in high yields using an efficient magnetic solid acid nanocatalyst $NiFe_2O_4@SiO_2@SO_3H$ through Hantzsch synthesis (Scheme 11.13) were synthesized. The reaction proceeds via a one-pot three-component condensation reaction of aromatic aldehydes (**1**), (1,3-diketone) ethyl acetoacetate (**16**), and aqueous ammonia (**23**) in H_2O with stirring at 70°C for the appropriate required time. This current study provides notable advantages in terms of mild reaction conditions, using H_2O as a green environmental-friendly solvent, the stability and easy separation of the Ni-nanocatalyst and high to excellent yield of products within 10–100 min stirring time. Also, the reusability of $NiFe_2O_4@SiO_2@SO_3H$ MNPs was tested for seven successive cycles without the critical loss of catalytic activity.

11.2.3 Preparation of 2,4,6-triarylpyridine derivatives via various heterogeneous nanocatalysts

11.2.3.1 *By using copper (II) supported on magnetic chitosan (MCs–Cu(II)) as a green nanocatalyst under solvent-free and aerobic conditions*

A green strategy to prepare 2,4,6-triaryl pyridines catalyzed by a new Cu/magnetic chitosan (MCs–Cu(II)) nanocatalyst has been described by Shaabani and his group in **2016** [43]. The reaction of ketone and benzyl amine (**25**) under solvent-free conditions with added (MCs–Cu(II)) as a green nanocatalyst afforded symmetric and asymmetric 2,4,6-triaryl pyridines **26a–j, 27a–b** in good yields (Schemes 11.14 and 11.15). The reaction proceeds via C–N bond cleavage of benzyl amines under aerobic oxidation at 90°C with stirring for 8 h. The advantages of this catalytic system, Cu/magnetic chitosan was magnetic separability, stability, and recycled several times (5) without any loss in its catalytic activity.

11.2.3.2 *Via using $Fe_3O_4@TiO_2@O_2PO_2(CH_2)_2NHSO_3H$ catalyst*

In 2018, Zolfigol et al. have reported the greener one-pot MCR protocols in the presence of $Fe_3O_4@TiO_2@O_2PO_2(CH_2)_2NHSO_3H$ as a sulfonic acid-functionalized titanacoated magnetic nanoparticle catalyst under mild and solvent-free reaction conditions for the synthesis of 2,4,6-triarylpyridines **28a–n** (Scheme 11.16) and 1,8-dioxodecahydroacridines **29a–q** (Scheme 11.17) [44]. All the desired products were obtained in good to high yields with short reaction times furthermore, the excellent reusability of the used catalyst. All the resultant compounds of 2,4,6-triarylpyridines were prepared earlier in different catalytic reaction conditions [45–49]. Moreover, the produced compounds of 1,8-dioxodecahydroacridines, reported previously except compounds **29d** and **29I** were new products [50–57].

11.2.3.3 *By using bifunctional ionic–liquid nanocatalyst ($CaO@SiO_2@BAIL$) under solvent-free conditions*

Fatemeh Sameri and her coauthors (2021) reported a novel bifunctional ionic liquid-coated nanocatalyst ($CaO@SiO_2@BAIL$) [36]. By anchoring the 1-(3-(trimethoxysilyl)propyl)-1,3,5,7-tetraazaadamantan-1-ium chlorozincate (II) bifunctional ionic liquid (BAIL) onto the surface of silica-coated CaO nanoparticles, (CaO@SiO2@BAIL) nanocatalyst was synthesized. The activity of the later nanomaterial as a catalyst was tested for the efficient synthesis of a variety of pyridine derivatives within green conditions (Scheme 11.18). In this multicomponent one-pot synthesis of 2,4,6-triarylpyridine derivatives **30a–s**, arylaldehyde (**1**), arylketone (**3**), and CaO@SiO2@BAIL were added at room temperature under solvent-free condition. Add to the later reaction mixture after 10 min, the same equivalent of arylketone (**3**) and ammonium acetate (**4**). Some compounds of 2,4,6-triarylpyridine derivatives **30a–s** were reported previously but using different methods [58–62].

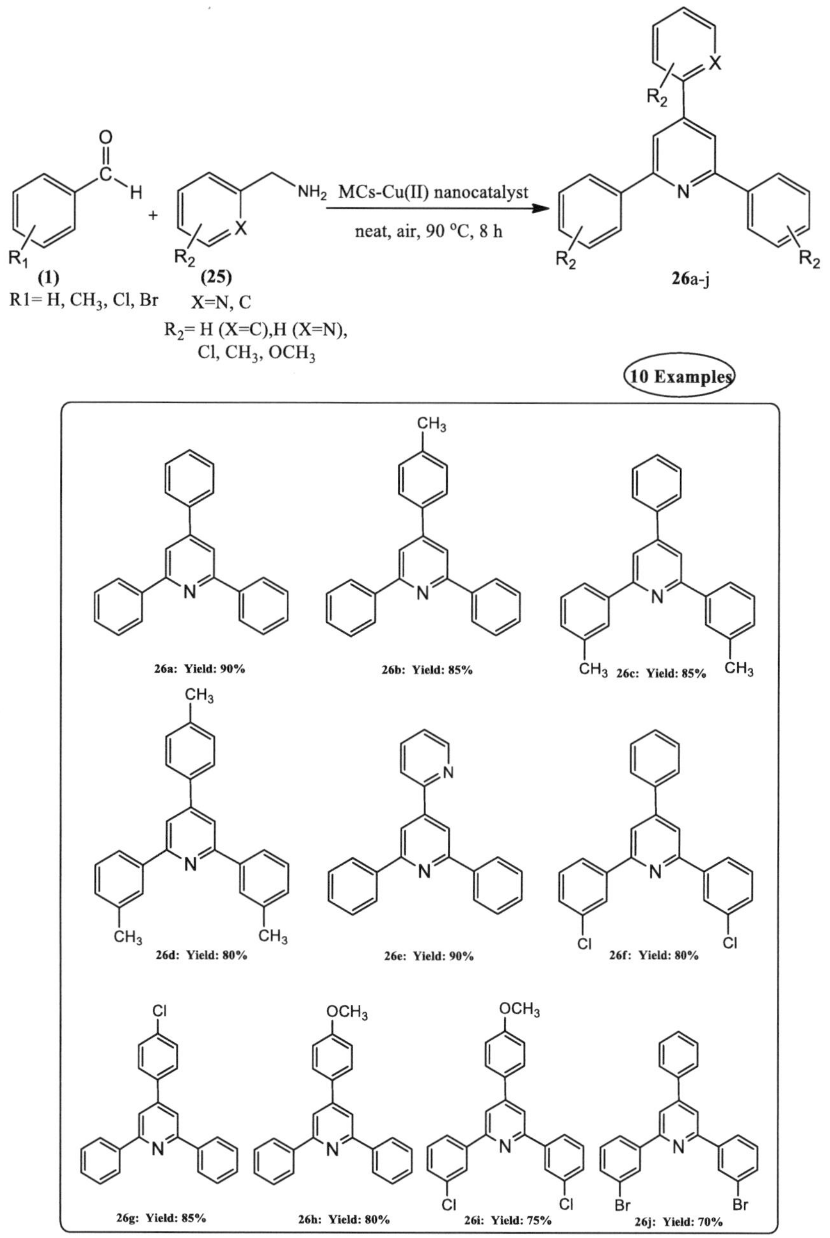

SCHEME 11.14 (MCs–Cu(II)) Nanocatalyst catalyzed the reaction synthesis of 2,4,6-triaryl pyridines **26a–j**.

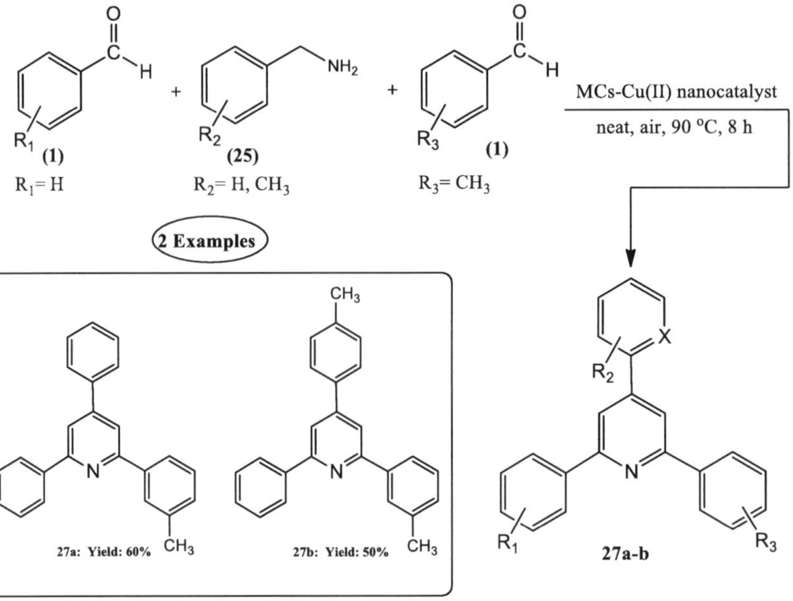

SCHEME 11.15 Preparation of the asymmetric 2,4,6-triaryl pyridines **27a–b** using the (MCs–Cu(II)) as a nanocatalyst.

11.2.4 Various heterogeneous nanocatalysts catalyzed the synthesis of pyridine dicarbonitriles

11.2.4.1 Via copper ferrite nanocatalyst

Copper ferrite nanocatalyst was used previously to prepare 1,4 dihydro pyridines through one-pot three multicomponent reaction [63]. On the other hand, a similar protocol for synthesizing pyridine dicarbonitriles was reported earlier using thiophenol with aromatic aldehydes and malononitrile in the presence of various catalysts [64–69]. Douglas et al. (2016) [70], reported the later reaction but using substituted phenolic derivatives instead of thiophenol to produce pyridine dicarbonitrile derivatives **32a–j** by the cyclocondensation reaction. The later products were synthesized in the presence of copper ferrite nanocatalyst through the reaction between aromatic aldehydes (**1**) and malononitrile (**2**) (1:2) in ethanol at 50°C with stirring for 15 min then a substituted phenol (**31**) was added to the reaction mixture which refluxed for appropriate time (Scheme 11.19). This method considers a green environmentally benign process, due to some notable advantages such as low time of reaction, high yielding of the desired products, catalyst recoverability, and reusability.

11.2.4.2 By using TiO_2 as efficient nanocatalysts

Shams-Najafi1 and his coworkers in 2019 described the catalytic effects of nano-TiO_2 prepared via ordinary or by a magnetized process, in the synthesis of pyridine dicarbonitrile derivatives (Scheme 11.20) [71]. The one-pot three multicomponent reaction of aryl aldehydes (**1**), malononitrile (**2**), 4-methyl thiophenol (**33**) in the presence of TiO_2 nanocatayst produced pyridine dicarbonitriles **34a–h**. The later products represented a higher range of yields and

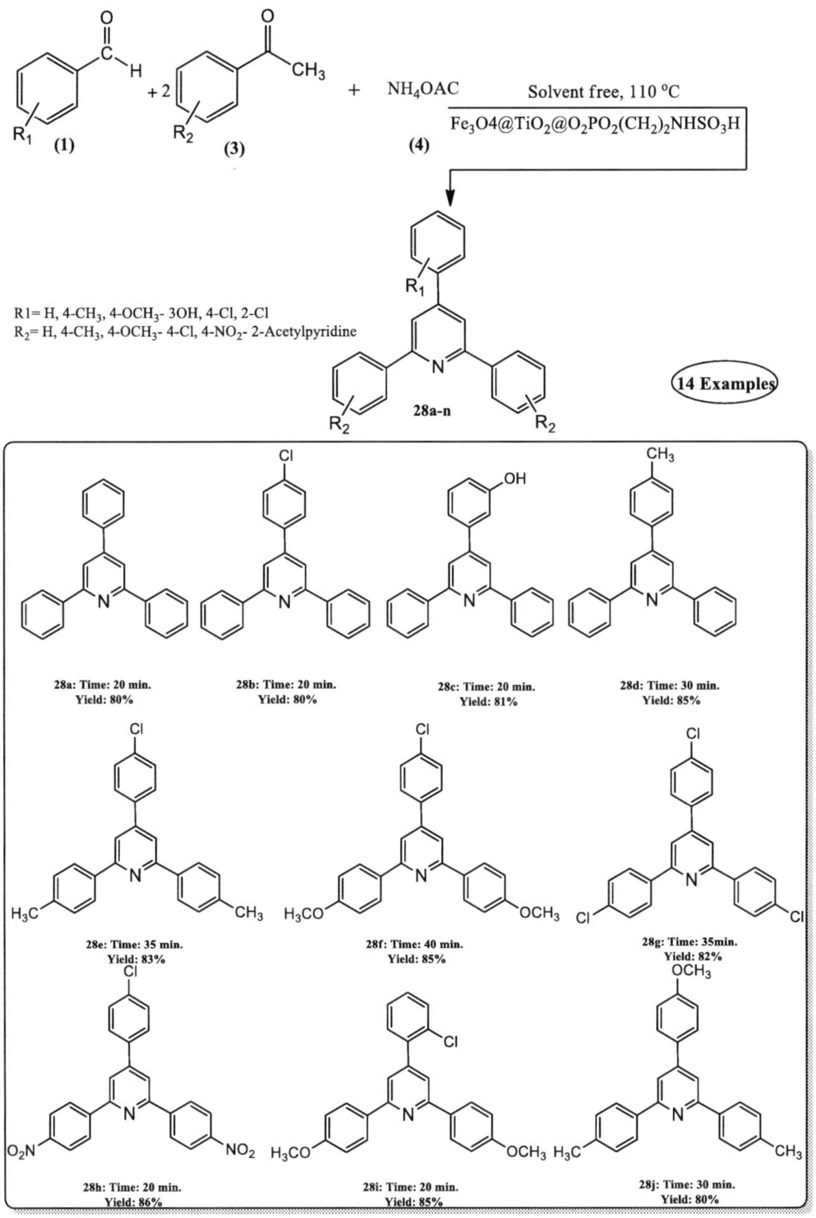

SCHEME 11.16 Synthesis of 2,4,6-triarylpyridines **28a–n** catalyzed by $Fe_3O_4@TiO_2@O_2PO_2(CH_2)_2NHSO_3H$.

shorter reaction times in nano-TiO_2 prepared by a magnetized process than the other ordinary process. These desired products of pyridine derivatives were synthesized earlier by different synthetic protocols in the presence of various catalyst and different reaction conditions [72–78]. Some remarkable advantages of this new catalytic method than the other available were short

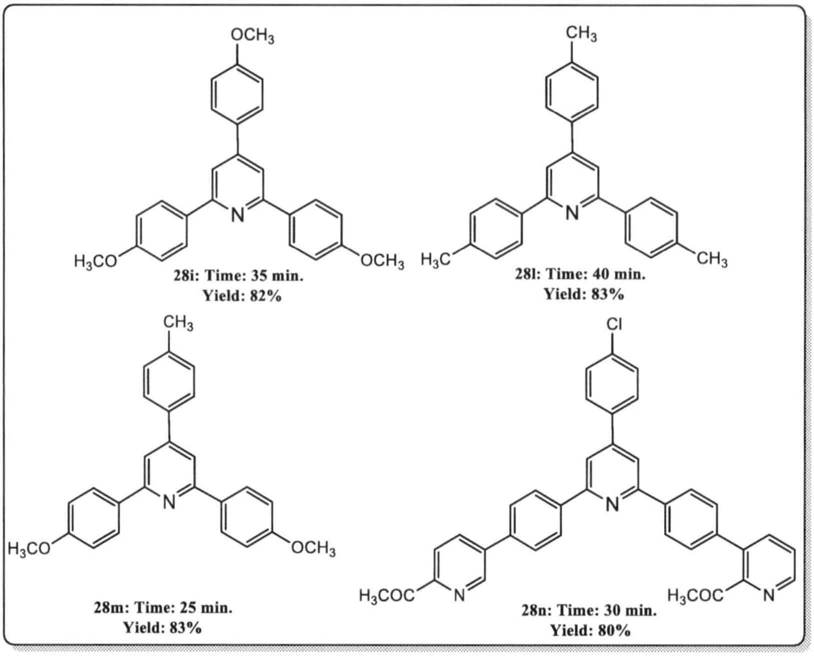

SCHEME 11.16, cont'd.

reaction time, high yields, and uses water as a green solvent, simple work up, inexpensive catalyst, stable, easily obtainable, removable from the reaction, and reused several times.

11.2.5 Nanocatalytic activity of Cu/NCNTs catalyst for the synthesis of pyridine derivatives

A new methodology using high temperature technique to synthesize green nanocatalyst Cu/NCNTs [copper nanoparticles (Cu) supported by *N*-doped carbon nanotubes (NCNTs)] was developed by Kour's group (2016) [79]. The catalytic activity of the Cu/NCNTs nanocatalyst was tested to prepare pyridine derivatives **35a–i** via four-component reaction of substituted aromatic aldehydes (**1**), aniline derivatives (**21**), malononitrile (**2**), and ethyl acetoacetate (**16**) in ethanol (Scheme 11.21). The synthesized compounds were reported previously by another research group [80], except for three new compounds **35c, 35i**, and **35j**.

11.3 Recent green synthesis of some selected fused heterocyclic-pyridine derivatives by using heterogeneous nanocatalyst systems

In this part some selected fused systems of pyridines such as pyrazolopyridines, pyridopyrimidines, thiazolopyridines, and imidazopyridines were synthesized by using environmentally green procedures in the presence of various nanocatalysts through the last 6 years in a sequential way.

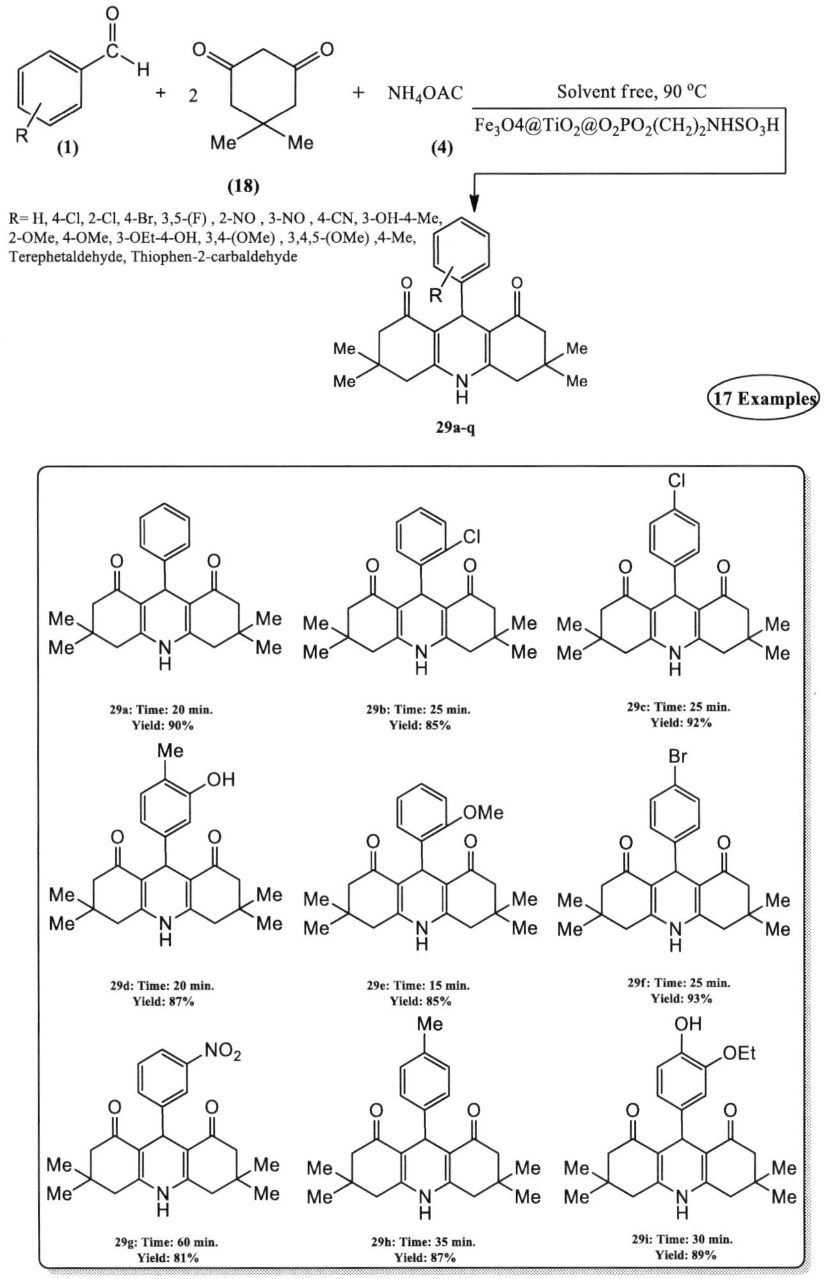

SCHEME 11.17 Preparation of 1,8-dioxodecahydroacridines by using $Fe_3O_4@TiO_2@O_2PO_2(CH_2)_2NHSO_3H$.

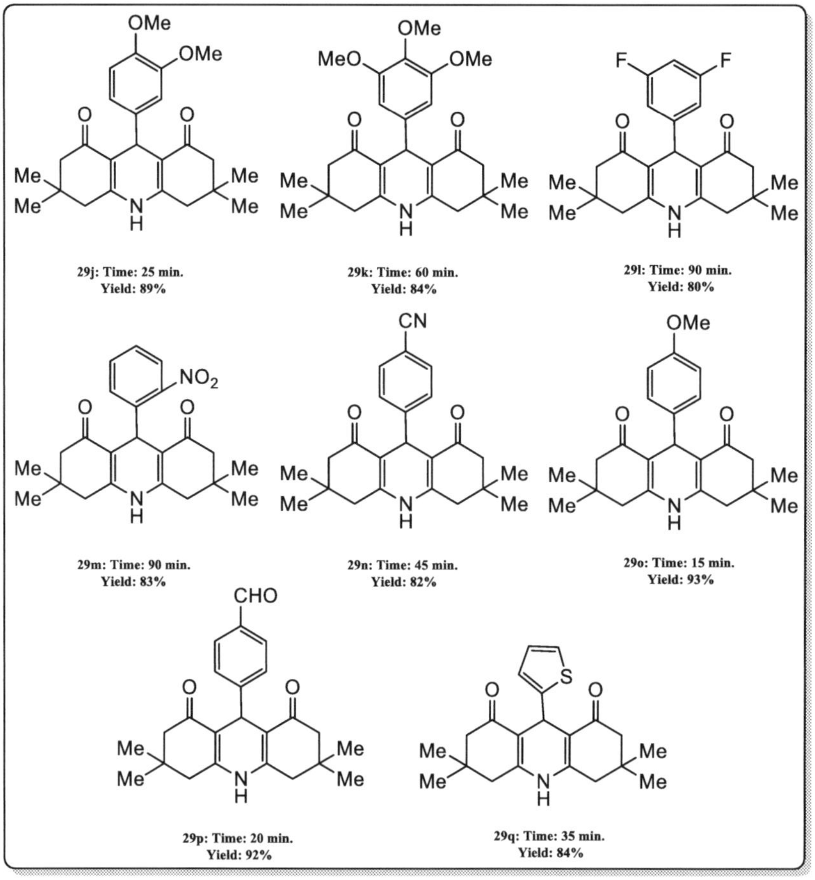

SCHEME 11.17, cont'd.

11.3.1 Recent green synthesis of pyrazolo[3,4-*b*]pyridine derivatives under various nanocatalysts

11.3.1.1 *Graphene oxide anchored sulfonic acid catalyst ($CoFe_3O_4$–GO–SO_3H) catalyzed the preparation of 3,6-di(pyridin-3-yl)-1H-pyrazolo[3,4-b]pyridine-5-carbonitriles under microwave irradiation*

A novel heterogeneous magnetically separable graphene oxide anchored sulfonic acid catalyst ($CoFe_3O_4$–GO–SO_3H) was successfully prepared and used for the synthesis of 3,6-di(pyridin-3-yl)-1*H*-pyrazolo[3,4-*b*]pyridine-5-carbonitrile derivatives **38a–o** by Zhang and his coworkers (2016) [81]. The reaction proceeds via one pot, three-component of 1-phenyl-3-(pyridin-3-yl)-1*H*-pyrazol-5-amine (**36**), 3-oxo-3-(pyridin-3-yl)propanenitrile (**37**), and substituted aldehydes (**1**) in ($CoFe_3O_4$–GO–SO_3H) catalyst and choline chloride (ChCl)/glycerol

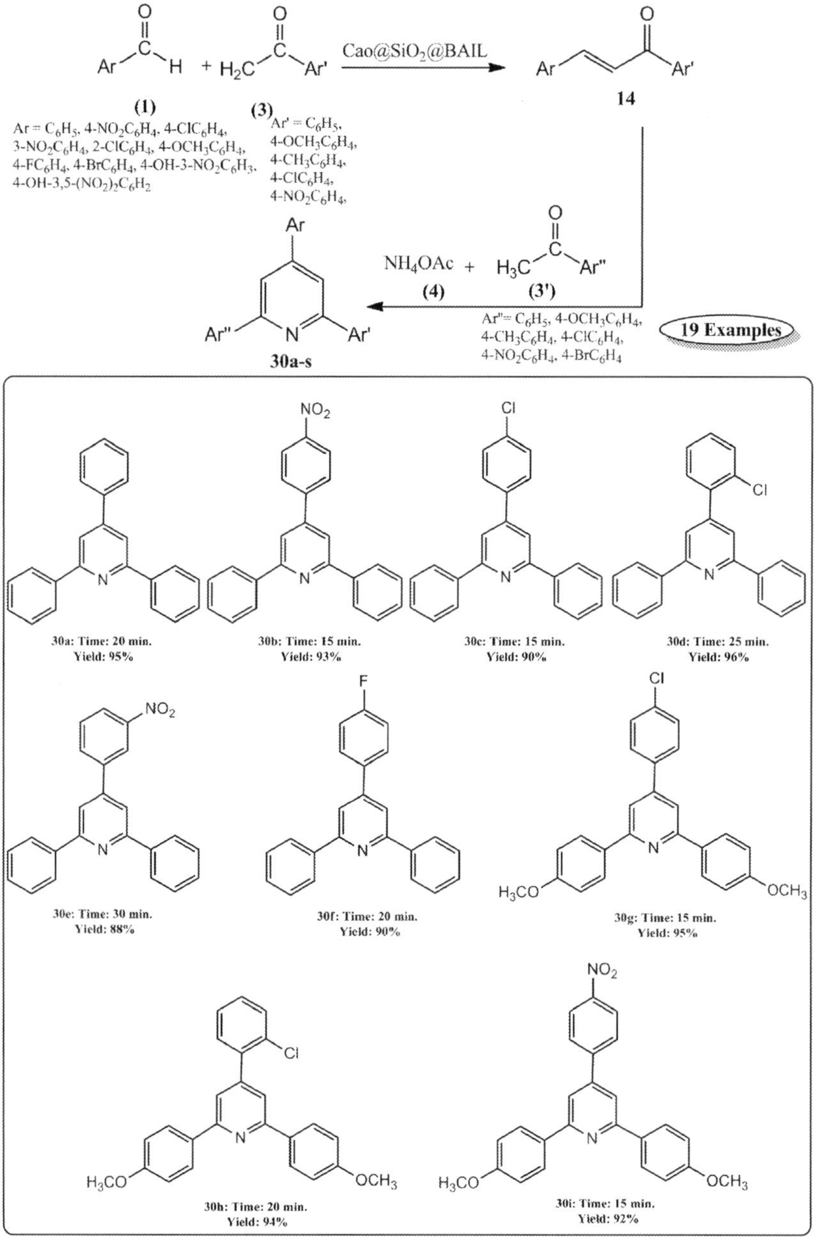

SCHEME 11.18 Synthesis of 2,4,6-triarylpyridine derivatives **30a–s** by using nanocatalyst (CaO@SiO_2@BAIL).

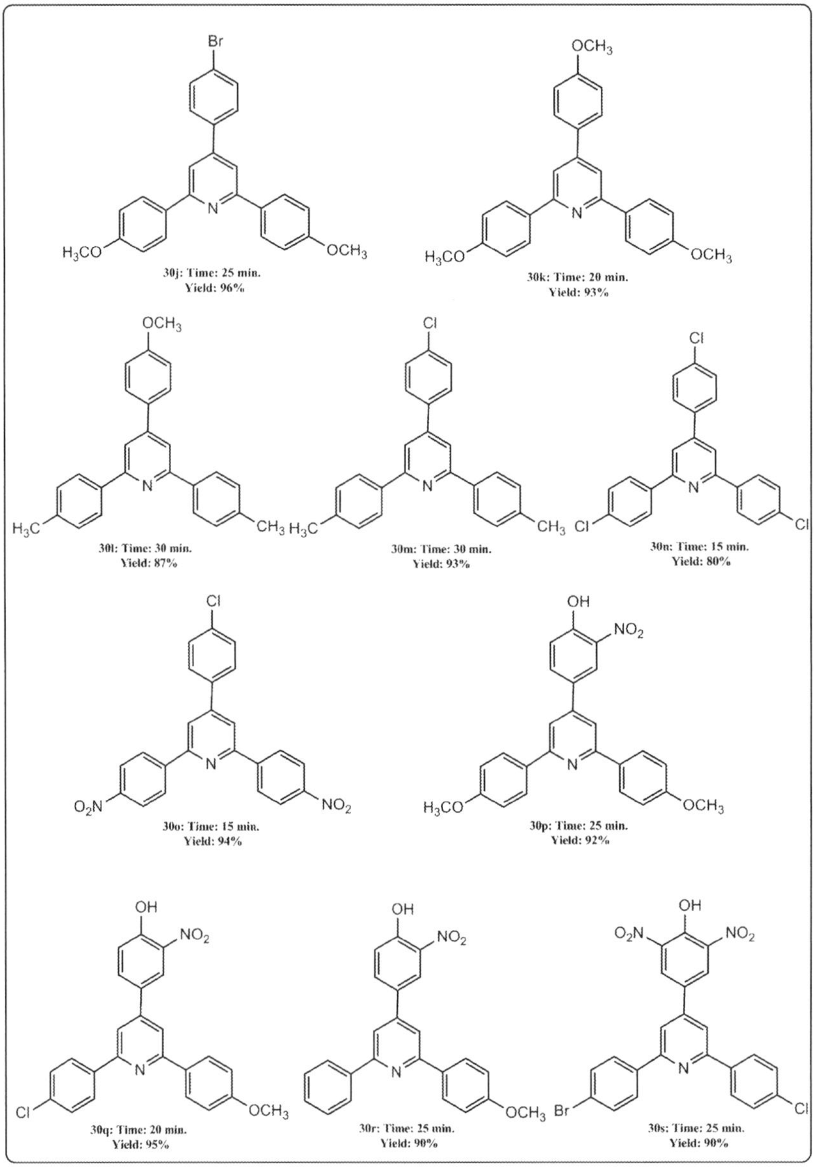

SCHEME 11.18, **cont'd.**

as a green solvent, the reaction mixture stirred under microwave irradiation at 80°C for an appropriate time (Scheme 11.22). The prepared catalyst was reused eight times by keeping its catalytic activity. All the synthesized products were new compounds except the compound **38d** [R = 4-MeOC_6H_4 in aldehydes] was prepared previously using a different method [82].

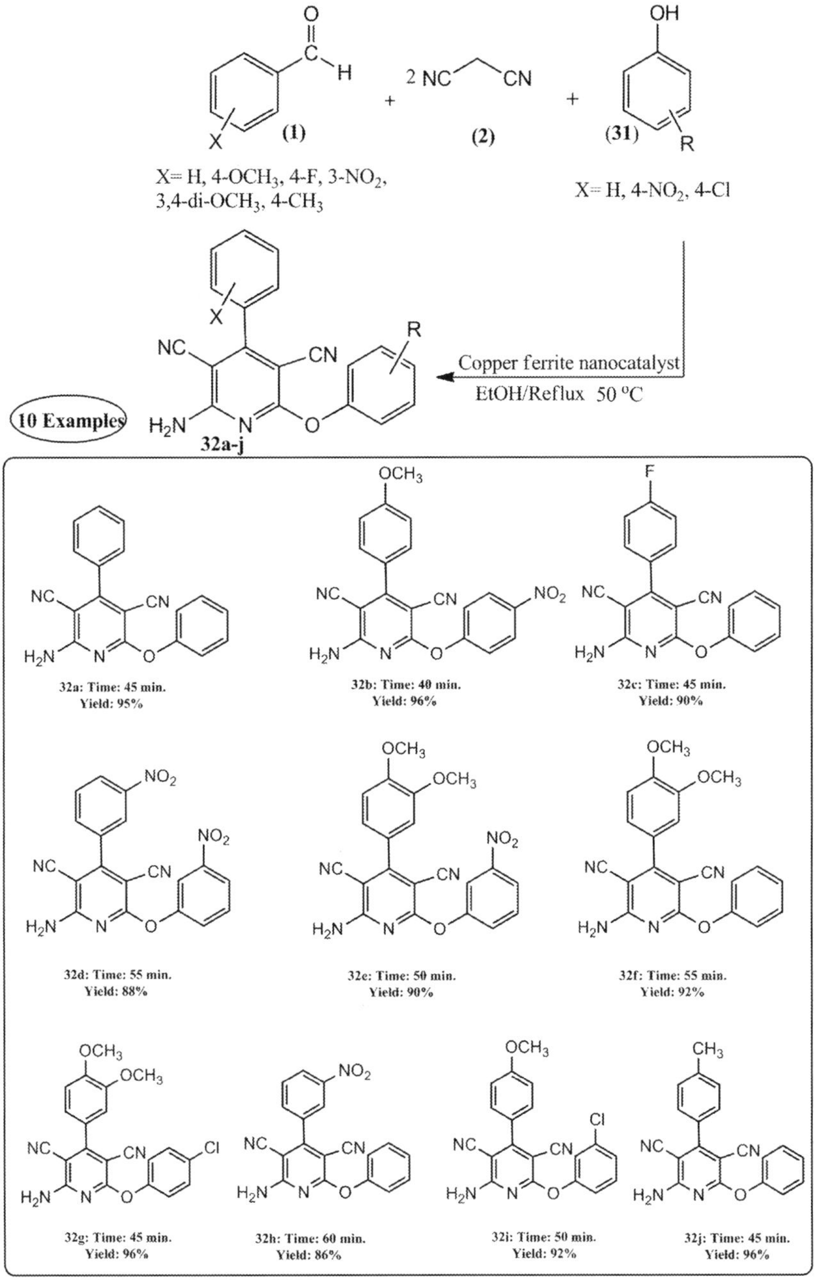

SCHEME 11.19 Copper ferrite nanocatalyst catalyzed the formation of pyridine dicarbonitriles.

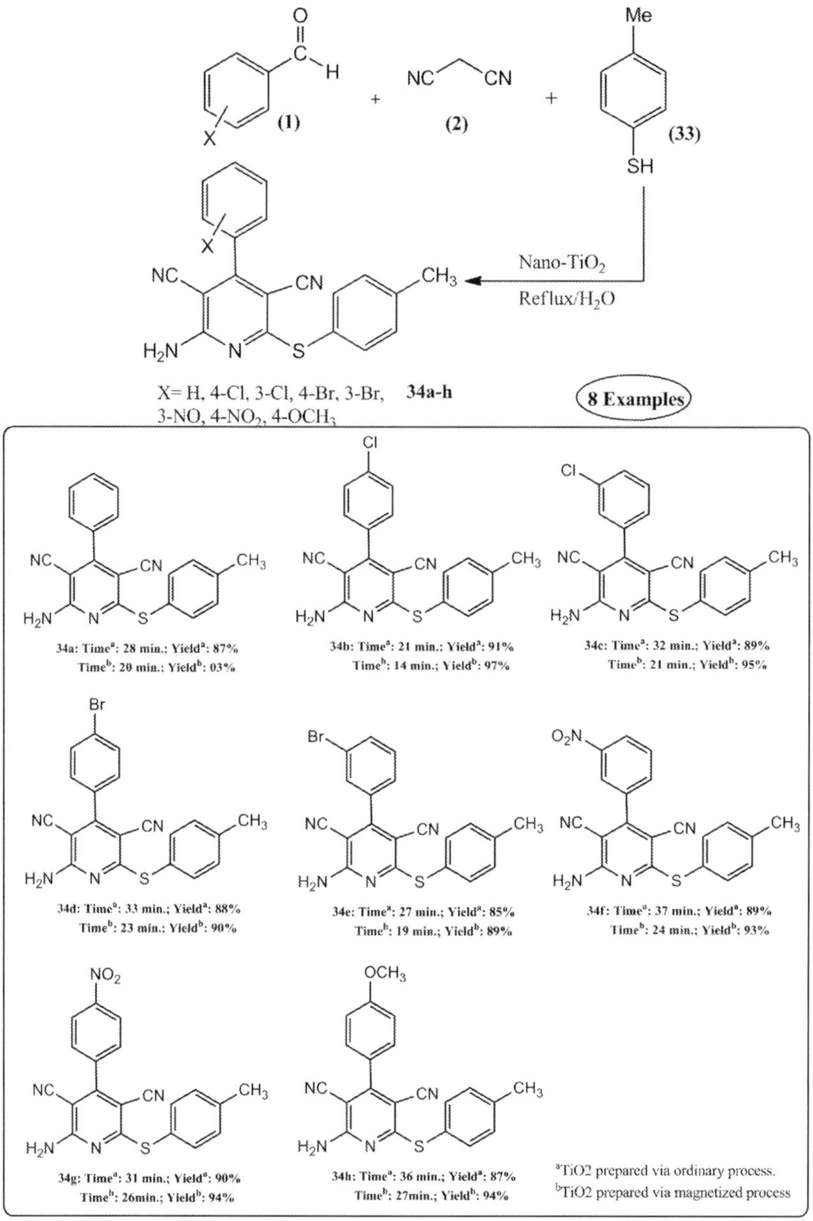

SCHEME 11.20 Synthesis of pyridine dicarbonitrile **34a–h** catalyzed by (nano-TiO_2).

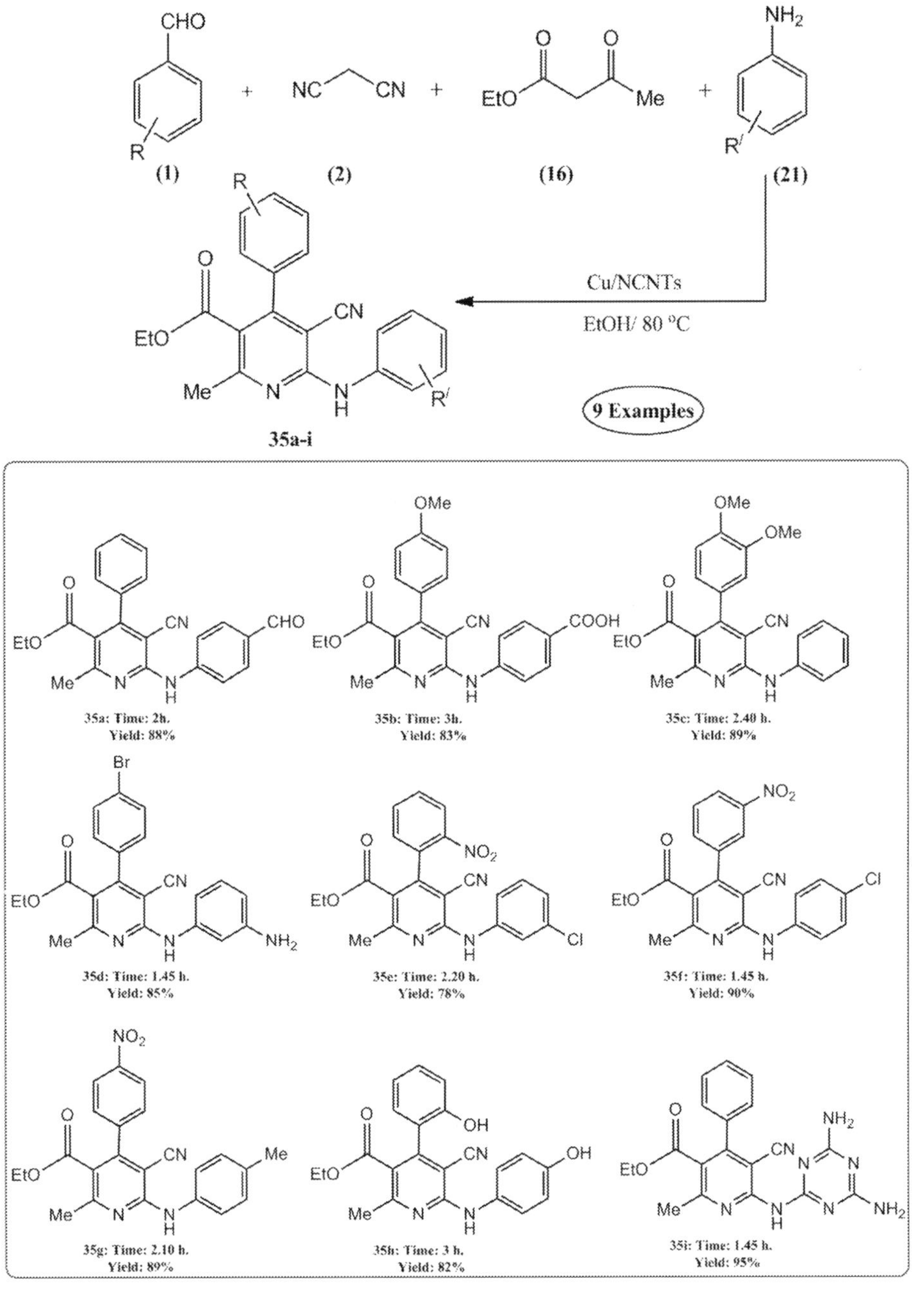

SCHEME 11.21 Four-component synthesis of pyridine derivatives **35a–i** via Cu/NCNTs nanocatalyst.

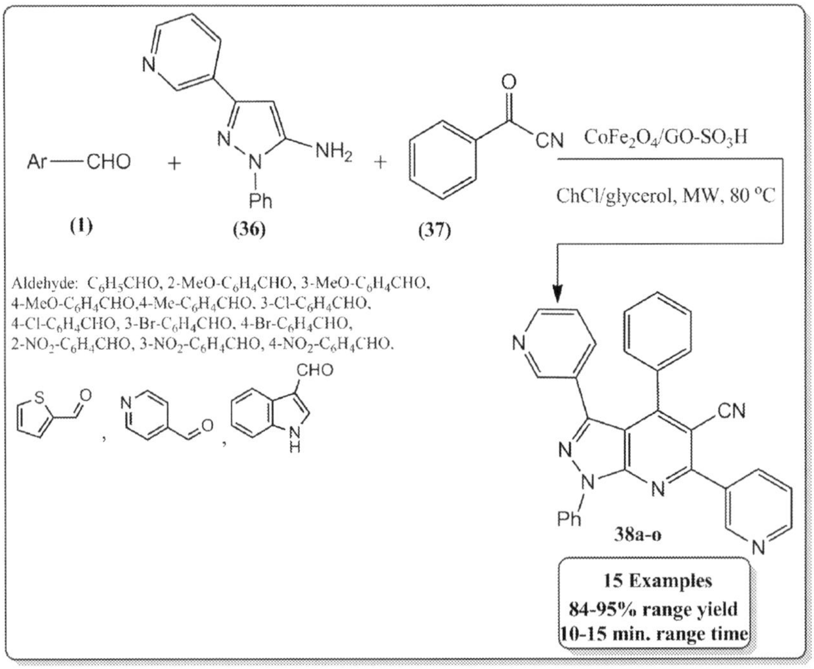

SCHEME 11.22 Use ($CoFe_3O_4$–GO–SO_3H) catalyst for the synthesis of 3,6-di(pyridin-3-yl)-1*H*-pyrazolo[3,4-*b*]pyridine-5-carbonitriles **38a–o**.

11.3.1.2 *Different green nanocatalysts catalyzed the synthesis of 3-methyl-4-aryl-2,4,5,7-tetrahydropyrazolo[3,4-b]pyridine-6-one derivatives*

11.3.1.2.1 By using nano-Fe_3O_4-supported organocatalyst based on 3,4-dihydroxypyridine (Fe_3O_4/Py)

A one-pot, three-component, strategy green preparation of 3-methyl-4-aryl-2,4,5,7-tetrahydropyrazolo[3,4-*b*]pyridine-6-ones **41a–o** in good to excellent yields have been reported by Pirhayati et al. in the year 2016 [83]. In this strategy, the first MNPs–pyridine (Fe_3O_4/Py) magnetically heterogeneous basic nanocatalyst was used in the condensation reaction of substituted aromatic aldehydes **1**, 5-methylpyrazol-3-amine **39**, and meldrum's acid **40** in ethanol at 80°C with stirring for about 1 h to afford the corresponding desired products **41a–o** (Scheme 11.23 Path A). The results indicated that pyridine-functionalized Fe_3O_4 nanoparticles were an efficient base catalyst which was reused five times without any remarkable decrease in its catalytic activity. Moreover, all reactions were completed in short times. The synthesized products in this paper were reported previously in different methods of reaction conditions and catalyst [84–91].

11.3.1.2.2 Via mesoporous SBA-15 silica phenylsulfonic acid (SBA-15-Ph-SO_3H) nanocatalyst

Veisi et al. in 2017 synthesized various fused systems of 2,4,5,7-tetrahydro pyrazolo[3,4-*b*]pyridine-6-one derivatives **41a–o** in high yields via one-pot three-component condensation

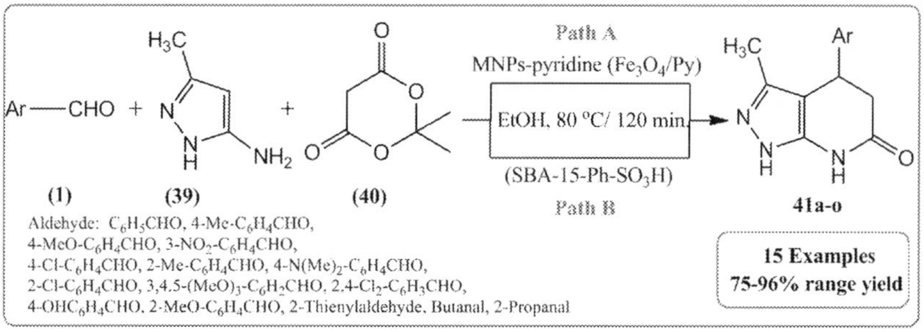

SCHEME 11.23 Application of MNPs–pyridine (Fe_3O_4/Py) [Path A] and (SBA-15-Ph-SO_3H) [Path B] nanocatalysts in the synthesis of 3-methyl-4-aryl-2,4,5,7-tetrahydropyrazolo[3,4-*b*]pyridine-6-ones **41a–o**.

reaction [92]. This environmentally benign protocol introduces a simple and efficient method by reacting a variety of aryl aldehydes (**1**), and 5-methylpyrazol-3-amine (**39**), and Meldrum's acid (**40**) in the presence of recyclable and heterogonous solid acid nanocatalyst (SBA-15-Ph-SO_3H) (Scheme 11.23 Path B). The reaction proceeds in ethanol at 80°C and stirred for 120 min furnished the desired systems in 75–96% rang yield.

11.3.1.3 *Preparation of 4-aryl-1,4,7,8-tetrahydro-3,5-dimethyldipyrazolo [3,4-b;4′,3′-e]pyridine derivatives*

11.3.1.3.1 Via an efficient catalyst (nano-$CuCr_2O_4$)

A novel green efficient methodology for the one-pot, four-component reaction of aryl aldehydes (**1**), ethyl acetoacetate (**16**), hydrazine hydrate (**42**), and ammonium acetate (**4**) in ethanol at room temperature in the presence of (nano-$CuCr_2O_4$) to obtain the 4-aryl-1,4,7,8-tetrahydro-3,5-dimethyldipyrazolo[3,4-b;4′,3′-e]pyridine derivatives **44a–k** was achieved by Shahbazi-Alavi et al. (2016) (Scheme 11.24) [93]. All the prepared dipyrazolopyridine derivatives about nine compounds were reported earlier by other research groups with different methods and reaction conditions systems [94,95] except the two compounds **44f** and **44k** were new compounds. The important features of this green process involved shorter time of reactions, good yields, use of ethanol as green solvent at room temperatures, simple practical method and the ability to recover the catalyst several times.

11.3.1.3.2 By using a heteropolyacid containing an ionic liquid-based organosilica (Fe_3O_4/KCC-1/IL/HPW) as heterogeneous catalysts

Sadeghzadeh in the same year (2016), prepared a novel heterogeneous catalysts heteropolyacid containing an ionic liquid-based ordered mesofibers magnetic organosilica (Fe_3O_4/KCC-1/IL/HPW) [96] and tested its catalytic activity in the synthesis of the earlier prepared 4-aryl-1,4,7,8-tetrahydro-3,5-dimethyldipyrazolo[3,4-b;4′,3′-*e*]pyridine derivatives **45a–i** (Scheme 11.24) [94,97]. The later nanocatalyst showed high catalytic activity in the preparation of the desired compounds. This high catalytic application was represented in the excellent yields of all the synthesized compounds (90–98%). The most important key advantage of this green procedure was the reusability of the present catalyst for several times (10 at least) with keeping its activity and selectivity. In addition, the use of water at room temperature with a short reaction time (30 min) confirmed this importance.

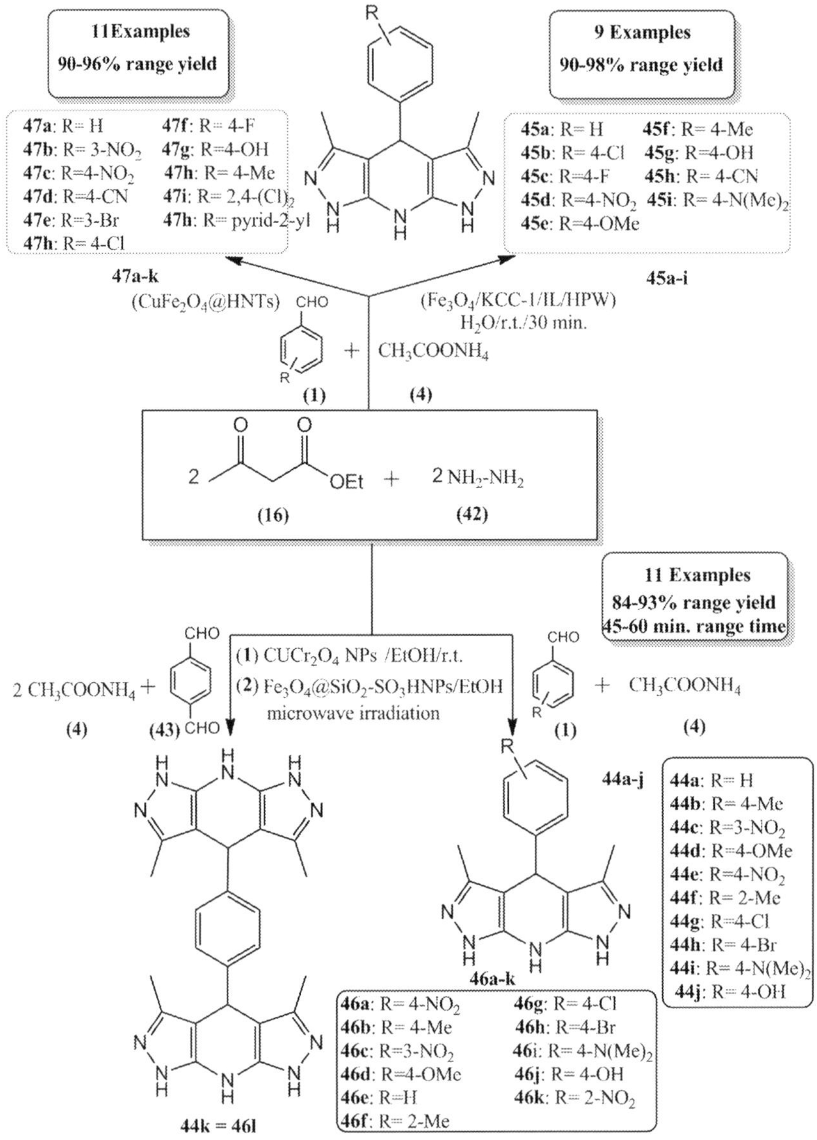

SCHEME 11.24 Synthesis of 4-aryl-1,4,7,8-tetrahydro-3,5-dimethyldipyrazolo[3,4-b;4′,3′-*e*]pyridine derivatives **44a–k, 45a–I, 46a–l**, and **47a–k** in the presence of (nano-$CuCr_2O_4$), Fe_3O_4/KCC-1/IL/HPW, Fe_3O_4@SiO_2–SO_3H, and $CuFe_2O_4$@HNTs, respectively.

11.3.1.3.3 By using Fe_3O_4@SiO_2–SO_3H under microwave irradiation

Safaei-Ghomi and Shahbazi-Alavi (**2017**) developed a green one-pot multicomponent reaction of ethyl acetoacetate (**16**), aldehydes (**1, 43**), hydrazine hydrate (**42**), and ammonium acetate (**4**) in the presence of newly prepared Fe_3O_4@SiO_2–SO_3H as nanocatalyst under microwave irradiation to obtain the 1,4,7,8-tetrahydrodipyrazolo[3,4-b;4′,3′-e]pyridines **46a–l**

(Scheme 11.24) [98]. All the later synthesized products were published earlier using different methods and reaction conditions except the compounds **46f**, and **46l** [R = 2-MeC_6H_4 and 4-CHO in aldehydes, respectively] were new compounds [94,99]. The important features of this green protocol were the excellent yields in short times, reusability of the catalyst (eight times) and use of microwave as green procedure.

11.3.1.3.4 Via mesoporous halloysite nanotubes modified by $CuFe_2O_4$ spinel ferrite nanoparticles ($CuFe_2O_4$@HNTs)

Maleki and his group (**2019**), described the first report on the design, synthesis and structure elucidation of a novel $CuFe_2O_4$@HNTs nanocomposite heterogeneous catalyst by modified the mesoporous halloysite nanotubes (HNTs) with $CuFe_2O_4$ nanoparticles [100]. The catalytic ability of the later prepared nanocomposite was tested as an efficient and recyclable heterogeneous nanocatalyst in the synthesis of the previously reported 1,4,7,8-tetrahydrodipyrazolo[3,4-b;4′,3′-*e*]pyridine derivatives by a one-pot multicomponent reaction [93,94,96,97] (Scheme 11.24). This procedure introduces important advantages as a green method due to some features, such as the excellent yields of the synthesized products (90–96%), green media where ethanol used as a solvent, the reaction occurred at room temperature under mild reaction conditions, the recoverability and reusability of the catalyst several times without remarkable loss in its activity.

11.3.1.4 *Synthesis of 4-aroyl-1,6-diaryl-3-methyl-1*H*-pyrazolo[3,4-*b*]pyridine-5-carbonitriles*

11.3.1.4.1 By using aluminum oxide nanocatalyst (nano-Al_2O_3)

Several novel 4-aroyl-1,6-diaryl-3-methyl-1*H*-pyrazolo[3,4-*b*]pyridine-5-carbonitrile derivatives **53a–p** in the presence of aluminum oxide nanocatalyst under reflux conditions were synthesized by Arlan and her group (2018) [101]. The eco-friendly, simple, and efficient methodology involved one-pot three-component reaction between 3-aryl-3-oxopropanenitriles (**48**), arylglyoxals (**49**), and 5-amino-1-aryl-3-methylpyrazoles (**50**) in H_2O–EtOH (1:1) using Al_2O_3 nanocatalyst, afforded pyrazolo[3,4-*b*]pyridine-5-carbonitrile derivatives **52a–p** in best yields (70–91%) (Scheme 11.25 Path A). The main advantages of this protocol were using of green solvents, high yields, the novelty of all the obtained products, easy prepare the used nanocatalyst and the simplicity of the workup procedure.

11.3.1.4.2 By using metal oxide silica based-metal bifunctional LDH (layered double hydroxide) (Fe_3O_4@SiO_2@Ni–Zn–Fe LDH) nanocatalyst

This green methodology reported the same earlier compounds **52a–p** prepared previously by the same research group but using a different substituted pyrazole and nanocatalyst by Arlan and her group (2020) [102]. Moreover, all the prepared compounds were published earlier using different method [103]. In this protocol, a novel (Fe_3O_4@SiO_2@Ni–Zn–Fe LDH) was used as a nanocatalyst for one-pot, four-component reaction of 3-aryl-3-oxopropanenitriles (**48**), 1-aryl-3-methyl-1*H*-pyrazol-5 (4*H*) one (**51**), arylglyoxals (**49**), and ammonium acetate (**4**) in EtOH/H_2O (1:1) under the reflux conditions to afford some pyrazolo[3,4-*b*]pyridine-5-carbonitrile derivatives **52a–p** (Scheme 11.25 Path B).

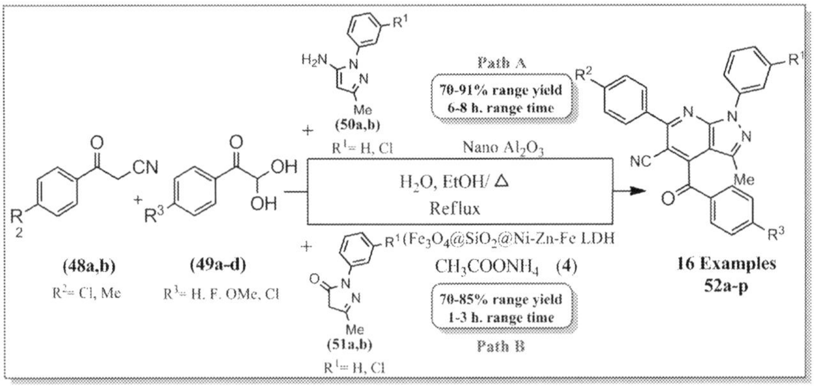

SCHEME 11.25 Use of nanocatalysts (nano-Al_2O_3) [Path A] and (Fe_3O_4@SiO_2@Ni–Zn–Fe LDH) [Path B] in the synthesis of 4-aroyl-1,6-diaryl-3-methyl-1*H*-pyrazolo[3,4-*b*]pyridine-5-carbonitriles **52a–p**.

11.3.1.5 One-pot synthesis of tetrahydrospiro[pyrazole-4,5′-pyrazolo[3,4-b] pyridine derivatives by novel magnetic nanoparticle supported ionic liquid (Fe_3O_4-GuHS)

Shojaei et al. (2019) reported a magnetically supported ionic liquid, guanidinium hydrogen sulfate (GuHS) on Fe_3O_4 nanoparticles as a novel nanocatalyst (Fe_3O_4–GuHS) and evaluated its activity for the synthesis a series of tetrahydrospiro[pyrazole-4,5′-pyrazolo[3,4-*b*]pyridine derivatives **54a–f** under solvent-free conditions [104]. The synthesized products were prepared by two different conditions, in ionic liquids (GuHS) and by the supported ionic liquids (Fe_3O_4–GuHS). The later method provided the best results, due to the many advantages present in this method, such as easy synthesis, separation and thermal stability of the nanocatalyst, the reusability of the catalyst for at least eight times without any loss of its activity and the simple separation and purification of the products. The procedure involves the one-pot reaction of 3-amino-1-phenyl-2-pyrazolin-5-one (**53**) with several substituted aryl aldehydes (**1**) and (Fe_3O_4-GuHS) in two drops DMSO and heated in an oil-bath at 60°C until obtained the desired products (Scheme 11.26).

11.3.2 Recent green synthesis of pyridopyrimidine derivatives by using different nanocatalysts

11.3.2.1 A triply green synthesis of pyrazolo[41,31:5,6]pyrido[2,3-d] pyrimidine-diones by using nano-ZnO in water

Triply novel green one-pot procedures were developed by Heravi and Daraie in 2016 [105] to synthesize pyrazolo[4′3′:5,6]pyrido[2,3-*d*]pyrimidine-diones via four-component in the presence of nano-ZnO, four-component in the presence of L-proline and five-component in the presence of nano-ZnO. The later synthesized products **58a–o** were obtained via the reaction between ethyl acetoacetate (**16**), hydrazine hydrate (**42**) or phenyl hydrazine (**55**), 1,3-dimethyl barbituric acid (**56**), substituted aryl aldehydes (**1**), and ammonium acetate (**4**) in catalytic amount of nano-ZnO or L-proline in water under reflux system (Scheme 11.27).

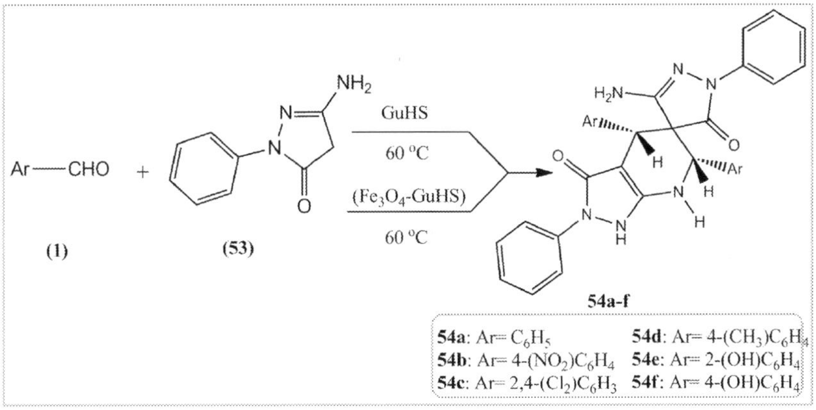

SCHEME 11.26 Magnetic nanoparticle supported ionic liquid (Fe_3O_4–GuHS) for the synthesis of tetrahydrospiro[pyrazole-4,5′-pyrazolo[3,4-*b*]pyridines **54a–f**.

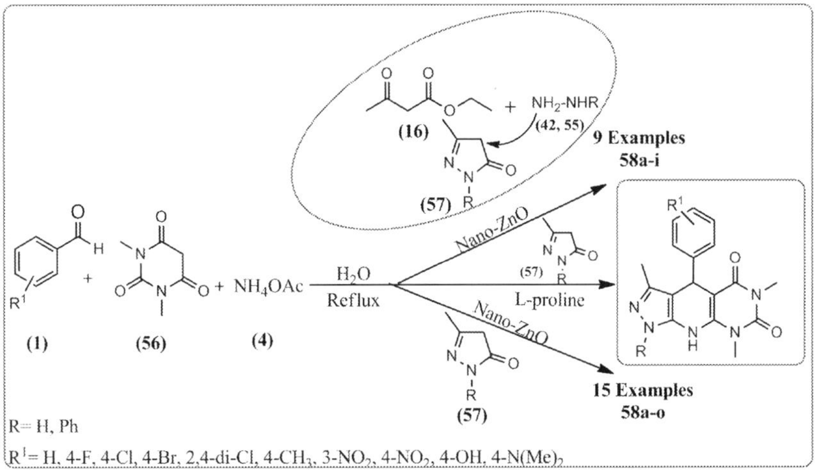

SCHEME 11.27 (Nano-ZnO) catalyzed the synthesis of pyrazolo[4′,3′:5,6]pyrido[2,3-*d*]pyrimidine-dione derivatives for four-components **58a–o** and five-component **58a–i** synthesis, respectively.

The important advantages of using nano-ZnO as a catalyst were in the easy work-up and high yields of products with short times. The prepared compounds were published earlier by other research group [106] except only two were new compounds **58b** and **58l**.

11.3.2.2 *Microwave-assisted preparation of pyrido[2,3-d:6,5-d′]dipyrimidine derivatives via magnetically $CuFe_2O_4$ nanoparticles as a catalyst in water*

Naeimi and his group (2017) described a convenient green strategy for the synthesis of 5-substituted-2,8-dithioxo-2,3,7,8,9,10-hexahydropyrido[2,3-*d*:6,5-*d*]dipyrimidine-4,6(1*H*,5*H*)-dione derivatives **60a–k** in one-pot, four-component condensation reactions

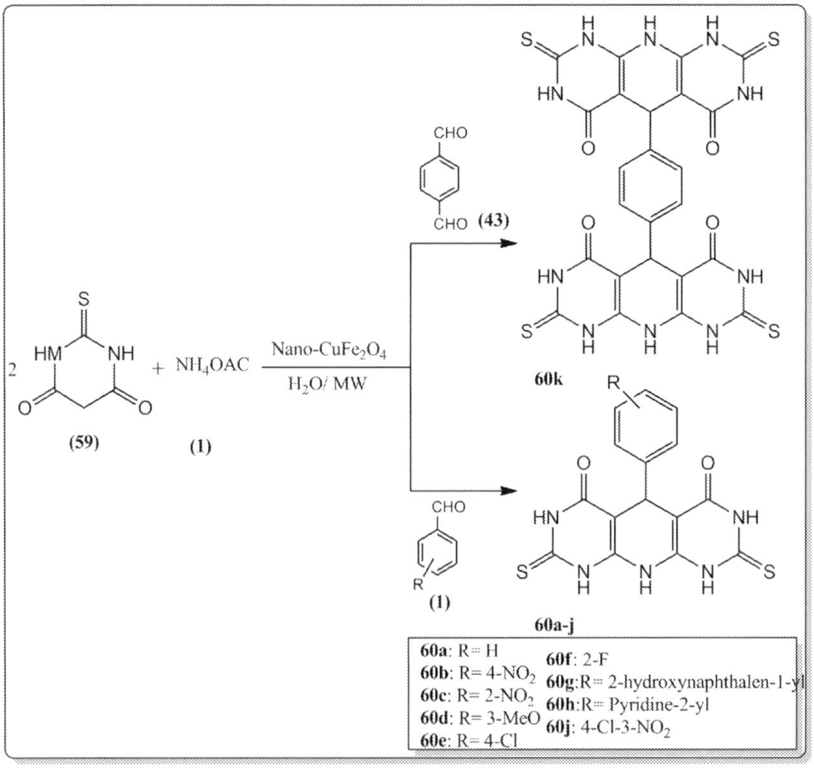

SCHEME 11.28 Catalytic application of synthesis pyrido[2,3-*d*:6,5-*d*′]dipyrimidine using (nano-$CuFe_2O_4$) **60a–k**.

of substituted aryl aldehydes (**1**), 2-thiobarbituric acid (**59**), and ammonium acetate (**4**) by using the green nanocatalyst (nano-$CuFe_2O_4$) and water as a green solvent under microwave irradiation (Scheme 11.28) [107]. Two compounds (**60a** and **60b**) from the prepared series **60a–k** were reported in the literature [108,109]. The notable features of this green protocol were the use of water green solvent, easy workup, high to excellent yields of the products, shorter times of reaction and the reusability of the catalyst.

11.3.2.3 *Novel nanomagnetic silica-bonded S-sulfonic acid [Fe_3O_4@SiO_2@$(CH_2)_3$S–SO_3H] catalyst for the synthesis of pyrido[2,3-d]pyrimidines under neat conditions*

Moradi et al. (2017) described an efficient method for the synthesis of pyrido[2,3-*d*]pyrimidines containing the 1,4-dihydropyridine moiety by using a novel nanocatalyst [Fe_3O_4@SiO_2@$(CH_2)_3$S–SO_3H] [110]. This procedure involving the reaction between substituted aryl aldehydes (**1**), 2,4-diamino-6-hydroxypyrimidine (**61**), and either or malononitrile (**2**), methyl or ethyl cyanoacetate (**16a, b**) in the presence of a catalytic amount of [Fe_3O_4@SiO_2@$(CH_2)_3$S–SO_3H], all the reaction mixture was stirred at 100°C under neat conditions to obtain the desired pyrido[2,3-*d*]pyrimidine derivatives **62a–c** and **63a–p**

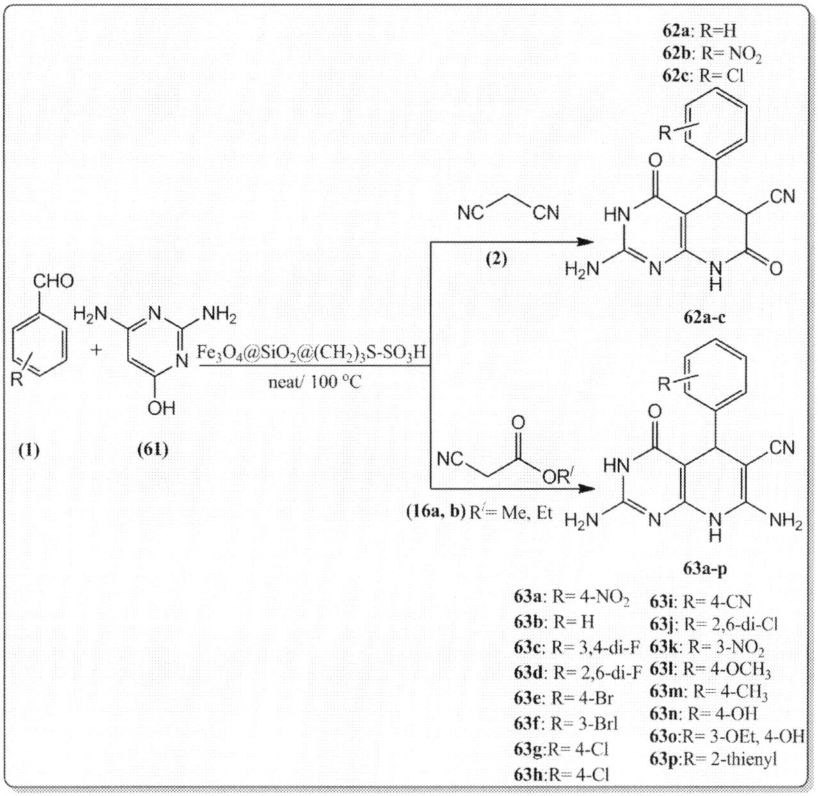

SCHEME 11.29 Synthesis of pyrido[2,3-*d*]pyrimidines **62a–c** and **63a–p** under neat conditions using [$Fe_3O_4@SiO_2@(CH_2)_3S–SO_3H$] nanocatalyst.

(and (Scheme 11.29). Some of the prepared compounds were published previously [111] **63a, 63e, 63g, 63h, 63k, 63m** and the other were new compounds **63b, 63c, 63d, 63f, 63i, 63j, 63l, 63n, 63o, 63p**.

11.3.2.4 Synthesis of new fused pyridines and 1,4-dihydropyridines with pyrazole and pyrimidine moieties by using $Fe_3O_4@Co(BDC)-NH_2$ catalyst under ultrasonic irradiation

A novel metal–organic framework $Fe_3O_4@Co(BDC)-NH_2$ as a new catalyst was prepared by Sepehrmansourie and his group in (2021) [112]. The later novel new catalyst was applied for the synthesis of new fused pyrazolo[4′,3′:5,6]pyrido[2,3-*d*]pyrimidine derivatives **64a–q**, the reaction carried out between the present catalyst, substituted aryl aldehydes (**1**), pyrimidine (1,3-dimethylpyrimidine- 2,4,6(1*H*,3*H*,5*H*)-trione or pyrimidine-2,4,6(1*H*,3*H*,5*H*)-trione) (**56**), and pyrazole-5-amine (3-methyl-1*H*-pyrazole-5-amine or 3-methyl-1-phenyl-1*H*-pyrazole-5-amine) derivatives (**39**) were mixed in dimethylsulfoxide (DMF) as solvent under ultrasonic irradiation (Scheme 11.30). Some prepared compounds were synthe-

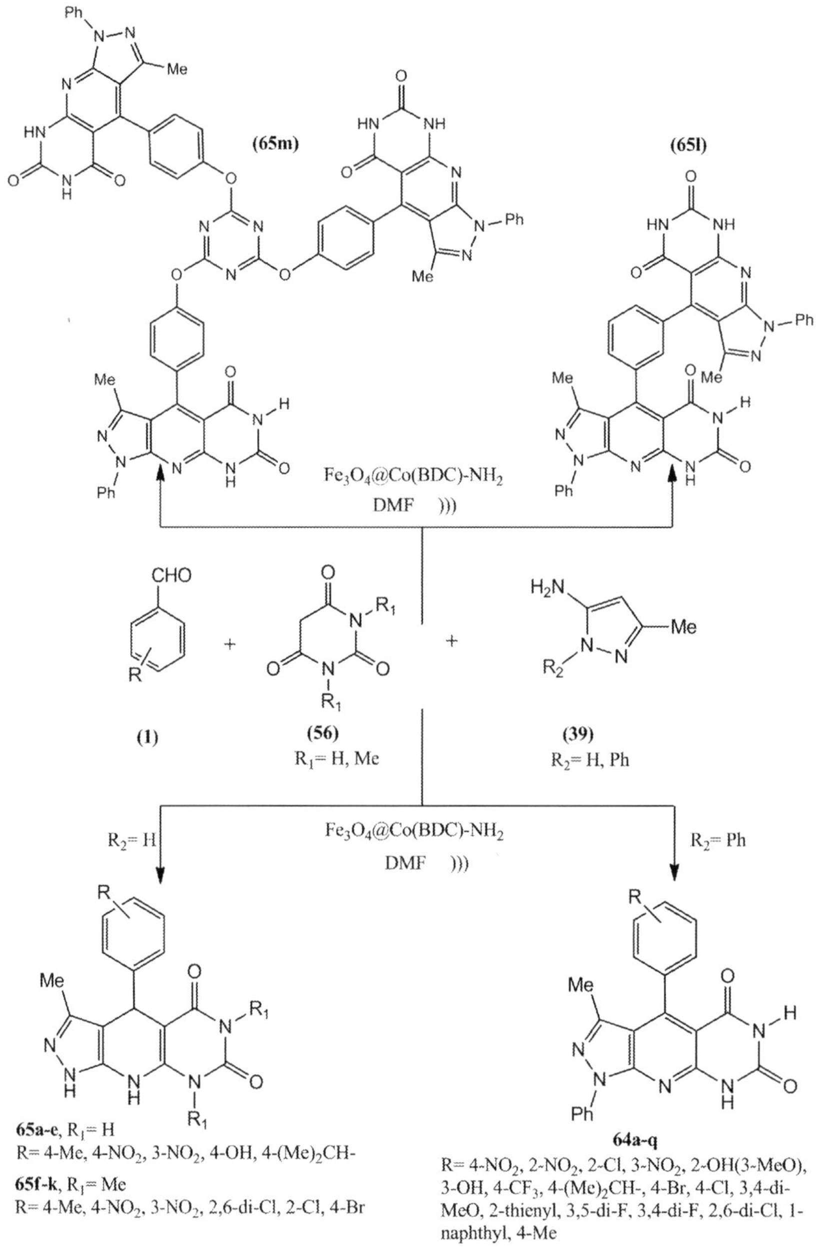

SCHEME 11.30 Fe_3O_4@Co(BDC)-NH_2 nanocatalyst catalyzed the synthesis of novel fused pyrazolo[4′,3′:5,6] pyrido[2,3-*d*]pyrimidine derivatives **64a–q, 65a–e, 65f–k, 65l**, and **65m**, respectively, were prepared.

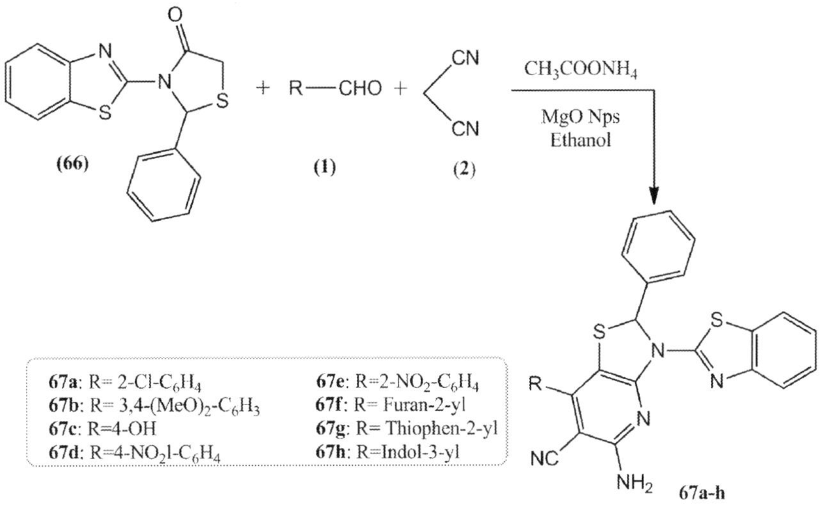

SCHEME 11.31 Synthesis of thiazolo[4,5-*b*]pyridine-6-carbonitrile derivatives by using MgO nanocatalyst.

sized through a cooperative vinylogous anomeric based oxidation **64a–q, 65a–e** and **65f–k**. Also, their bis and tris 3-methyl-1,4-diphenyl-1,8-dihydro-5*H*-pyrazolo[4′,3′:5,6]pyrido[2,3-*d*]pyrimidine-5,7(6*H*)-dione derivatives **65l** and **65m**, respectively, were prepared.

11.3.3 Eco-benign approach for the synthesis of thiazolo[4,5-b]pyridine-6-carbonitrile derivatives via MgO nanocatalyst

Gandhi and Agarwal in 2018 reported an environmentally friendly procedure for the synthesis of thiazolo[4,5-*b*]pyridine-6-carbonitrile derivatives in the presence of MgO heterogeneous base nanocatalyst [113]. The later prepared nanocatalyst MgO were used in multicomponent one-pot synthesis of thiazolo[4,5-*b*]pyridine-6-carbonitriles (**67a–h**) via the reaction between aromatic aldehydes (**1**), malononitrile (**2**), ammonium acetate (**4**), and 3-benzothiazol-2-yl-2-phenylthiazolidin-4-one (**66**) in ethanol (Scheme 11.31). The simple catalyst preparation, easily recoverable and reusable nanocatalyst, use of ethanol green solvent, and excellent product yields were the main features of this methodology.

11.3.4 Synthesis of 3-iminoaryl-imidazo[1,2-*a*]pyridines via manganese Schiff-base complex immobilized on chitosan-coated Fe_3O_4 (Fe_3O_4@CSBMn)

This new method for the preparation of chitosan-coated iron oxide nanoparticles (Fe_3O_4@CSBMn) was designed by Rakhtshah and Yaghoobi in 2019 [114]. The catalytic activity of the present prepared nanocatalyst was examined for the synthesis of some 3-iminoaryl-imidazo[1,2-*a*]pyridine derivatives (**70a–l**) through one-pot three-component reaction of various aldehyde derivatives (**1**) with trimethylsilyl cyanide (TMSCN) (**68**) and 2-aminopyridine (**69**) under solvent-free conditions (Scheme 11.32). The same derivatives were synthesized

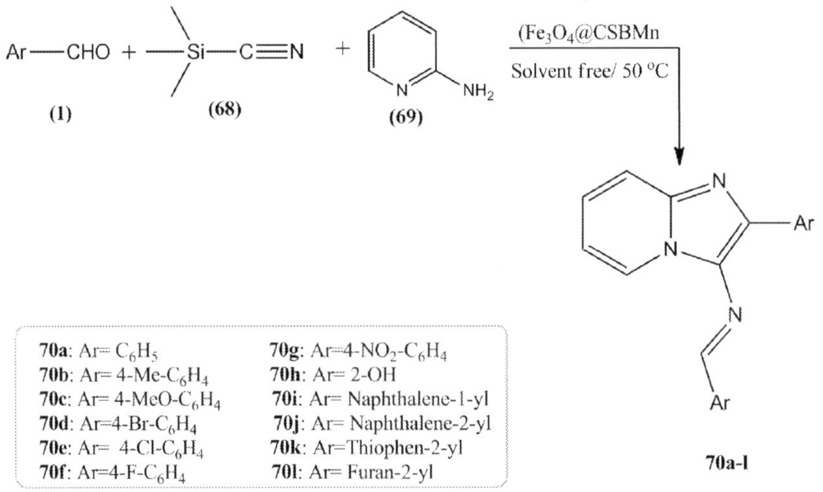

SCHEME 11.32 One-pot three-component synthesis of 3-iminoaryl-imidazo[1,2-*a*]pyridines using (Fe_3O_4 @CS-BMn) nanocatalyst.

earlier with trimethylsilyl cyanide (TMSCN) as a newer cyanide source on the catalytic system [115–117]. This promising novel nanocatalyst providing fast and short reaction times, high yields, and reusability for five times of the nanocatalyst.

11.4 Conclusion

This chapter describes the recent green synthesis of polyfunctional pyridine derivatives and their fused heterocyclic systems especially for pyrazolopyridines, pyridopyrimidine, thiazolopyridine, and imidazopyridine during the last 6 years (2016–2021). The later green methodologies occurred via multicomponent reaction in the presence of various types of nanocatalyst which providing many remarkable advantages of the reaction such as high yielding of the desired products, fast with short reaction time and the ability to use the catalyst for many times without significant loss in its catalytic activity.

References

[1] T. Anderson, "On the products of the destructive distillation of animal substances, Part II", Trans. R. Soc. Edinb. 20 (2) (1851) 247–260.
[2] T. Anderson, "Ueber die Producte der trocknen Destillation thierischer Materien" [On the products of dry distillation of animal matter], Ann. Chem. Pharm. (in German) 80 (1851) 44–65.
[3] D. Viradiya, S. Mirza, F. Shaikh, R. Kakadiya, A. Rathod, N. Jain, R. Rawal R, A. Shah, Design and synthesis of 1,4-dihydropyridine derivatives as anti-cancer agent, Anticancer Agents Med. Chem. 17 (7) (2017) 1003–1013.
[4] J. Zhang, J. Xi, R. He, R. Zhuang, L. Kong, L. Fu, Y. Zhao, C. Zhang, L. Zeng, J. Lu, R. Tao, Z. Liu, H. Zhu, S. Liu, Discovery of 3-(thiophen/thiazole-2-ylthio)pyridine derivatives as multitarget anticancer agents, Med. Chem. Res. 28 (2019) 1633–1647.

[5] J. Briede, M. Stivrina, B. Vigante, D. Stoldere, G. Duburs, Acute effect of antidiabetic 1,4-dihydropyridine compound cerebrocrast on cardiac function and glucose metabolism in the isolated, perfused normal rat heart, Cell Biochem. Funct. 26 (2) (2008) 238–245.
[6] Y.L. Murthy, A. Rajack, T.R. Moturu, J. Jeson babu, Ch. Praveen, K. Aruna Lakshmi, Design, solvent free synthesis, and antimicrobial evaluation of 1,4 dihydropyridines, Bioorg. Med. Chem. Lett. 22 (18) (2021) 6016–6023.
[7] H.S.A. Alnassar, M.H.E. Helal, A.A. Askar, D.M. Masoud, A.E.M. Abdallah, Pyridine azo disperse dye derivatives and their selenium nanoparticles (SeNPs): synthesis, fastness properties and antimicrobial evaluations, Int. J. Nanomed. 14 (2019) (2019) 7903–7918.
[8] E.R Kotb, H.A Soliman, E.M.H. Morsy, N.A.M. Abdelwahed, New Pyridine And Triazolopyridine Derivatives: Synthesis, Antimicrobial And Antioxidant Evaluation, Acta Pol. Pharm. 74 (3) (2017) 861–872.
[9] T.S. Ibrahim, H.A. Sayed, M.T. Khayat, A.M.M. Al-Mahmoudy, A.H. Moustafa, A.K.S. El-Deen, S.A.F. Rostom, S.S. Panda, Synthesis of nucleosides and non-nucleosides based 4,6-disubstituted-2-oxo-dihydropyridine-3-carbonitriles as antiviral agents, Med. Chem. 14 (8) (2018) 791–808.
[10] G. Elgemeie, M. Hamed, Microwave synthesis of guanine and purine analogs, Curr. Microwave Chem. 1 (2) (2014) 155–176.
[11] M.J. Ledoux, C. Pham-Huu, High specific surface area carbides of silicon and transition metals for catalysis, Catal. Today 15 (2) (1992) 263–284.
[12] R. Khalifeh, M. Ghamari, A multicomponent synthesis of 2-amino-3-cyanopyridine derivatives catalyzed by heterogeneous and recyclable copper nanoparticles on charcoal, J. Braz. Chem. Soc. 27 (4) (2016) 759–768.
[13] M.A. Zolfigol, M. Kiafar, M. Yarie, A.(A.) Taherpourb, M. Saeidi-Rada, Experimental and theoretical studies of the nanostructured {$Fe_3O_4@SiO_2@(CH_2)_3Im$}$C(CN)_3$ catalyst for 2-amino-3-cyanopyridine preparation via an anomeric based oxidation, RSC Adv. 6 (2016) 50100–50111.
[14] M.M. Heravi, S.Y.S. Beheshtiha, M. Dehghani, N. Hosseintash, Using magnetic nanoparticles Fe3O4 as a reusable catalyst for the synthesis of pyran and pyridine derivatives via one-pot multicomponent reaction., J. Iran. Chem. Soc. 12 (2015) 2075–2081.
[15] J. Tang, L. Wang, Y. Yao, L. Zhang, W. Wang, One-pot synthesis of 2-amino-3-cyanopyridine derivatives catalyzed by ytterbium perfluorooctanoate [$Yb(PFO)_3$], Tetrahedron Lett. 52 (4) (2011) 509–511.
[16] Q. Wu, Y. Zhang, S. Cui, Divergent syntheses of 2-aminonicotinonitriles and pyrazolines by copper-catalyzed cyclization of oxime ester, Org. Lett. 16 (2014) 1350–1353.
[17] C. Kurumurthy, R. Naresh Kumar, T. Yakaiah, P. Shanthan Rao, B. Narsaiah, Novel $Bu_4N{+}Br_2$-catalyzed one-pot multi-componentsynthesis of 2-amino nicotinonitriles in aqueous medium, Res. Chem. Intermed. 41 (2015) 3193–3199.
[18] R. Bodireddy Mohan, N.C. Gangi Reddy, S.D. Kumar, Synthesis of alkynyl/alkenyl-substituted pyridine derivatives via heterocyclization and Pd-mediated Sonogashira/Heck coupling process in one-pot: a new MCR strategy, RSC Adv. 4 (2014) 17196–17205.
[19] J. Safari, S.H. Banitab, S.D. Khalili, Ultrasound-promoted an efficient method for one-pot synthesis of 2-amino-4,6-diphenylnicotinonitriles in water: a rapid procedure without catalyst, Ultrason. Sonochem. 19 (2012) 1061–1069.
[20] S. Khaksar, M. Yaghoobi, A concise and versatile synthesis of 2-amino-3-cyanopyridine derivatives in 2, 2, 2-trifluoroethanol, J. Fluorine Chem. 142 (2012) 41–44.
[21] H.C. Shah, V.H. Shah, N.D. Desai, A novel strategy for the synthesis of 2-amino-4,6-diarylnicotinonitrile, Arkivoc (2009) 76–87.
[22] M. Afradi, S.A. Pour, M. Dolat, A. Yazdani-Elah-Abadi, Nanomagnetically modified vitamin B3 (Fe_3O_4@Niacin): an efficient and reusable green biocatalyst for microwave-assisted rapid synthesis of 2-amino-3-cyanopyridines in aqueous medium, Appl. Organomet. Chem. 32 (2018) e4103.
[23] S. Kambe, K. Saito, A. Sakurai, H. Midorikawa, Simple Method for the Preparation of 2-Amino-4-aryl-3-cyanopyridines by the Condensation of Malononitrile with Aromatic Aldehydes and Alkyl Ketones in the Presence of Ammonium Acetate., Synthesis 1980 (1980) 366–368.
[24] Z. Hosseinzadeh, A. Ramazani, H. Ahankar, K. Ślepokura, T. Lis, Synthesis of 2-amino-4,6-diarylnicotinonitrile in the presence of $CoFe_2O_4@SiO_2\text{-}SO_3H$ as a reusable solid acid nanocatalyst under microwave irradiation in solvent-freeconditions, Silicon 2018 (2018) 1–8.

[25] J. Zhou, D. Liu, F. Wu, Y. Xiong, F. R.Sheykhahmad, MNPs@AMTT/Cu(II): a heterogeneous and reusable magnetic nanocatalyst for the synthesis 2-amino-3-cyanopyridine and α,α'-bis (substituted-benzylidene) cycloalkanone derivatives under solvent-free conditions, Mater. Res. Express 6 (2019) 105086.

[26] M.M. Heravi, S.Y. Beheshtiha, M. Dehghani, N. Hosseintash, Using magnetic nanoparticles Fe_3O_4 as a reusable catalyst for the synthesis of pyran and pyridine derivatives via one-pot multicomponent reaction, J. Iran. Chem. Soc. 12 (2015) 2075–2081.

[27] R. Pagadala, S. Maddila, V. Moodley, W.E. van Zyl, S.B. Jonnalagadda, An efficient method for the multicomponent synthesis of multisubstituted pyeidines, a repid procedure using Au/MgO as the catalyst, Tetrahedron Lett. 55 (2014) 4006–4010.

[28] D. Khalili, Graphene oxide: a reusable and metal-free carbocatalyst for the one-pot synthesis of 2-amino-3-cyanopyridines in water, Tetrahedron Lett. 57 (2016) 1721–1723.

[29] Y. Wan, R. Yuan, F.R. Zhang, L.L. Pang, R. Ma, C.H. Yue, W. Lin, W. Yin, R.C. Bo, H. Wu, One-pot synthesis of N_2-substituted 2-amino-4-aryl-5,6,7,8-tetrahydroquinoline-3-carbonitrile in basic ionic liquid [bmim]OH, Synth. Commun. 41 (20) (2011) 2997–3015.

[30] A. Solhy, W. Amer, M. Karkouri, R. Tahir, A. El Bouari, A. Fihri, M. Bousmina, M. Zahouily, Bi-functional modified-phosphate catalyzed the synthesis of α-α'-(EE)-bis (benzylidene)-cycloalkanones: microwave versus conventional-heating, J. Mol. Catal. A: Chem. 336 (2011) 8–15.

[31] B. Krishnakumar, M. Swaminathan, Solvent free synthesis of quinoxalines, dipyridophenazines and chalcones under microwave irradiation with sulfated Degussa titania as a novel solid acid catalyst, J. Mol. Catal. A: Chem. 350 (2011) 16–25.

[32] A.M. Rahman, B.S. Jeong, D.H. Kim, J.K. Park, E.S. Lee, Y. Jahng, A facile synthesis of α, α'-bis (substituted-benzylidene)-cycloalkanones and substituted-benzylidene heteroaromatics: utility of NaOAc as a catalyst for aldol-type reaction, Tetrahedron 63 (2007) 2426–2431.

[33] J.T. Li, W.Z. Yang, G.F. Chen, T.S. Li, A facile synthesis of α,α'-bis(substituted benzylidene) cycloalkanones catalyzed by KF/Al_2O_3 under ultrasound irradiation, Synth. Commun. 33 (2003) 2619–2625.

[34] G. Sabitha, G.S. Reddy, K.B. Reddy, J.S. Yadav, Iodotrimethylsilane-Mediated Cross-Aldol Condensation: A Facile Synthesis of α,α'-Bis(substituted benzylidene)cycloalkanones, Synthesis 2 (2004) 263–266.

[35] Z. Haydari, D. Elhamifar, M. Shaker, M. Norouzi, Magnetic nanoporous MCM-41 supported melamine: a powerful nanocatalyst for synthesis of biologically active 2-amino-3-cyanopyridines, Appl. Surf. Sci. Adv. 5 (2021) 100096.

[36] F. Sameri, M.A. Bodaghifard, A. Mobinikhaledi, Ionic liquid-coated nanoparticles ($CaO@SiO_2@BAIL$): a bi-functional and environmentally benign catalyst for green synthesis of pyridine, pyrimidine, and pyrazoline derivatives, Polycycl. Aromat. Compd. (2021) 1903954.

[37] M.A. Zolfigol, M. Yarie, $Fe_3O_4@TiO_2@O_2PO_2(CH_2)NHSO_3H$ as a novel nanomagnetic catalyst: application to the preparation of 2-amino-4, 6 diphenylnicotinonitriles via anomeric-based oxidation, Appl. Organometal. Chem. 31 (5) (2017) e3598.

[38] S.S. Mansoor, K. Aswin, K. Logaiya, S. Sudhan, [Bmim] BF_4 ionic liquid: an efficient reaction medium for the one-pot multi-component synthesis of 2-amino-4, 6-diphenylpyridine-3-carbonitrile derivatives, J. Saudi Chem. Soc. 20 (5) (2016) 517–522.

[39] F. Shi, S. Tu, F. Fang, T. Li, One-pot synthesis of 2-amino-3-cyanopyridine derivatives under microwave irradiation without solvent, Arkivoc 1 (1) (2005) 137–142.

[40] T. Demirci, B. Çelik, Y. Yıldız, S. Eris¸, M. Arslan, F. Sen, B. Kilbas, One-pot synthesis of Hantzsch dihydropyridines using a highly efficient and stable PdRuNi@GO catalyst, RSC Adv. 6 (2016) 76948–76956.

[41] G.K. Reen, M. Ahuja, A. Kumar, R. Patidar, P. Sharma, ZnO nanoparticle-catalyzed multicomponent reaction for the synthesis of 1,4-diaryl dihydropyridines, Org. Prep. Proced. Int. 49 (3) (2017) 273–286.

[42] B. Zeynizadeh, S. Rahmani, E. Eghbali, Anchored sulfonic acid on silica-layered $NiFe_2O_4$: a magnetically reusable nanocatalyst for Hantzsch synthesis of 1,4-dihydropyridines, Polyhedron 168 (2019) 57–66.

[43] A. Shaabani, M.B. Boroujeni, M.S. Laeini, Copper(II) supported on magnetic chitosan: a green nanocatalyst for the synthesis of 2,4,6-triarylpyridines by C–N bond cleavage of benzylamines, RSC Adv. 6 (2016) 27706–27713.

[44] M.A. Zolfigol, F. Karimi, M. Yarie, M. Torabi, Catalytic application of sulfonic acid-functionalized titana-coated magnetic nanoparticles for the preparation of 1,8 dioxodecahydroacridines and 2,4,6-triarylpyridines via anomeric-based oxidation, Appl. Organomet. Chem. 32 (2018) e4063.

[45] M.A. Zolfigol, M. Safaiee, F. Afsharnadery, N. Bahrami-Nejad, S. Baghery, S. Salehzadeh, F. Maleki, Silica vanadic acid [SiO_2–VO $(OH)_2$] as an efficient heterogeneous catalyst for the synthesis of 1, 2-dihydro-1-aryl-3 H-naphth [1, 2-e][1, 3] oxazin-3-one and 2, 4, 6, RSC Adv. 5 (2015) 100546–100559.

[46] R. Karki, P. Thapa, M.J. Kang, T.C. Jeong, J.M. Nam, H.-L. Kim, Y. Na, W.J. Cho, Y. Kwon, Synthesis, topoisomerase I and II inhibitory activity, cytotoxicity, and structure–activity relationship study of hydroxylated 2, 4-diphenyl-6-aryl pyridines, Bioorg. Med. Chem. 18 (2010) 3066–3077.

[47] P.V. Shinde, V.B. Labade, J.B. Gujar, B.B. Shingate, M.S. Shingare, Bismuth triflate catalyzed solvent-free synthesis of 2, 4, 6-triaryl pyridines and an unexpected selective acetalization of tetrazolo [1, 5-a]-quinoline-4-carbaldehydes, Tetrahedron Lett. 53 (12) (2012) 1523.

[48] A.R. Moosavi-Zare, M.A. Zolfigol, S. Farahmand, A. Zare, A.R. Pourali, R. Ayazi-Nasrabadi, Synthesis of 2,4,6-triarylpyridines using $ZrOCl_2$ under solvent-free conditions, Synlett 25 (2014) 193–196.

[49] S. Tu, T. Li, F. Shi, Q. Wang, J. Zhang, J. Xu, X. Zhu, X. Zhang, S. Zhu, D. Shi, A convenient one-pot synthesis of 4′-aryl-2,2′:6′,2″-terpyridines and 2,4,6-triarylpyridines under microwave irradiation, Synthesis 18 (2005) 3045–3050.

[50] X. Fan, Y. Li, X. Zhang, G. Qu, J. Wang, An efficient and green preparetion of 9-arylacridine-1,8-dione derivatives, Heteroat. Chem. 18 (2007) 786–790.

[51] D.-Q. Shi, S.-N. Ni, F. Yang, J.W. Shi, G.L. Dou, X.Y. Li, X.S. Wang, An efficient synthesis of polyhydroacridine derivatives by the three-component reaction of aldehydes, amines and dimedone in ionic liquid, J. Heterocycl. Chem. 45 (3) (2008) 653–660.

[52] S.M. Vahdat, S. Khaksar, M. Akbari, S. Baghery, Sulfonated organic heteropolyacid salts as a highly efficient and green solid catalysts for the synthesis of 1,8-dioxo-decahydroacridine derivatives in water, Arab. J. Chem. 42 (7) (2014). https://doi.org/10.1016/j.arabjc.2014.10.026.

[53] S. Balalaie, F. Chadegani, F. Darviche, H.R. Bijanzadeh, One-pot synthesis of 1,8-dioxo-decahydroacridine derivatives in aqueous media, Chin. J. Chem. 27 (2009) 1953–1956.

[54] C.A. Navarro, C.A. Sierra, C. Ochoa-Puentes, Evaluation of sodium acetate trihydrate–urea DES as a benign reaction media for the Biginelli reaction. Unexpected synthesis of methylenebis (3-hydroxy-5, 5-dimethylcyclohex-2, RSC Adv. 6 (70) (2016) 65355 -63365.

[55] M.V. Reddy, Y.T. Jeong, Polystyrene-supported p-toluenesulfonic acid: a new, highly efficient, and recyclable catalyst for the synthesis of hydropyridine derivatives under solvent-free conditions, Synlett 23 (20) (2012) 2985 –1991.

[56] M.A. Zolfigol, N. Bahrami-Nejad, S. Baghery, A convenient method for the synthesis of 1,8-dioxodecahydroacridine derivatives using 1-methylimidazolium tricyanomethanide {[HMIM]C(CN)3} as a nanostructured molten salt catalyst, J. Mol. Liq. 218 (2016) 558–564.

[57] P. Mahesh, K. Guruswamy, B.S. Diwakar, B.R. Devi, Y.L.N. Murthy, P. Kollu, S.V.N. Pammi, Magnetically separable recyclable nano-ferrite catalyst for the synthesis of acridinediones and their derivatives under solvent-free conditions, Chem. Lett. 44 (2015) 1386–1388.

[58] H. Wang, W. Zhao, J. Du, F. Wei, Q. Chen, X. Wang, Cationic organotin cluster [t-$Bu_2Sn(OH)(H_2O)]_{22}$+2OTf-catalyzed one-pot three-component syntheses of 5 substituted 1H-tetrazoles and 2,4,6-triarylpyridines in water, Appl. Organomet. Chem. 33 (10) (2019) e5132.

[59] M. Adib, B. Mohammadi, S. Rahbari, P. Mirzaei, Reaction between guanidine hydrochloride and chalcones: an efficient solvent-free synthesis of 2,4,6 triarylpyridines under microwave irradiation, Chem. Lett. 37 (10) (2008) 1048–1049.

[60] T. Kobayashi, H. Kakiuchi, H. Kato, On the reaction of N-(diphenylphosphinyl)-1-phenylethanimine with aromatic aldehydes giving 4-aryl-2,6-diphenylpyridine derivatives, Bull. Chem. Soc. Jpn. 64 (2) (1991) 392–395.

[61] M. Wang, Z. Yang, Z. Song, Q. Wang, Three-component one-pot synthesis of 2,4,6-triarylpyridines without catalyst and solvent, J. Heterocycl. Chem. 52 (3) (2015) 907–910.

[62] M. Adib, H. Tahermansouri, S.A. Koloogani, B. Mohammadi, H.R. Bijanzadeh, Kr€ohnke pyridines: an efficient solvent-free synthesis of 2,4,6-triarylpyridines, Tetrahedron Lett. 47 (33) (2006) 5957–5960.

[63] K.V. Ivaturi, Y.L.N. Murthy, One-pot, three-component synthesis of 1, 4-dihydropyridines by using nano crystalline copper ferrite, Chem. Sci. Trans. 2 (1) (2013) 227–233.

[64] N.M. Evdokimov, I.V. Magedov, A.S. Kireev, A. Konienko, One-step, three-component synthesis of pyridines and 1,4-dihydropyridines with manifold medicinal utility, Org. Lett. 8 (5) (2006) 899–902.

[65] M. Sarita, G. Rina, ChemInform abstract: K_2CO_3-mediated, one-pot, multicomponent synthesis of medicinally potent pyridine and chromeno[2,3-b]pyridine scaffolds, Synth. Commun. 42 (15) (2012) 2229–2244.

[66] B.C. Ranu, R. Jana, S. Sowmiah, An improved procedure for the three-component synthesis of highly substituted pyridines using ionic liquid, J. Org. Chem. 72 (8) (2007) 3152–3154.

[67] K.M. Lakshmi, K. Mahendar, B. Suresh, One-pot, three-component synthesis of highly substituted pyridines and 1,4-dihydropyridines by using nanocrystalline magnesium oxide, J. Chem. Sci. 122 (1) (2010) 63–69.

[68] Y.I. Shaikh, A.A. Shaikh, G.M. Nazeruddin, Ammonia solution catalyzed one-pot synthesis of highly functionalized pyridine derivatives, J. Chem. Pharm. Res. 4 (11) (2012) 4953–4956.

[69] K. Guo, M.J. Thompson, B. Chen, Exploring catalyst and solvent effects in the multicomponent synthesis of pyridine-3,5-dicarbonitriles, J. Org. Chem. 74 (18) (2009) 6999–7006.

[70] S. Paul Douglas, B. Swathi, M. Ravi Kumar, Y. Ramesh, K. Jaya Rao, B. Satyanarayana, C.H. Pandu Naidu, Nano copper ferrite catalysed improved procedure for one-pot synthesis of poly substituted pyridine derivatives, Chem. Sci. Trans. 5 (2) (2016) 325–334.

[71] S.J. Shams-Najafi, M. Gholizadeh, A. Ahmadpour, TiO_2-nanoparticles as efficient catalysts for the synthesis of pyridine dicarbonitriles, J. Chin. Chem. Soc. 66 (2019) 1531–1536.

[72] N.M. Evdokimov, A.S. Kireev, A.A. Yakovenko, M.Y. Antipin, I.V. Magedov, A. Kornienko, One-step synthesis of heterocyclic privileged medicinal scaffolds by a multicomponent reaction of malononitrile with aldehydes and thiols, J. Org. Chem. 72 (9) (2007) 3443–3453.

[73] S.S. Takale, J. Patil, V. Padalkar, R. Pisal, A. Chaskar, O-iodoxybenzoic acid in water: optimized green alternative for multicomponent one-pot synthesis of 2-amino-3,5-dicarbonitrile-6-thiopyridines, J. Braz. Chem. Soc. 23 (2012) 966–969.

[74] S. Mishra, R. Ghosh, K_2CO_3-mediated, one-pot, multicomponent synthesis of medicinally potent pyridine and chromeno[2,3-b]pyridine scaffolds, Synth. Commun. 42 (15) (2012) 2229–2244.

[75] M.N. Khan, S. Pal, T. Parvin, L.H. Choudhury, A simple and efficient method for the facile access of highly functionalized pyridines and their fluorescence property studies, RSC Adv. 2 (32) (2012) 12305–12314.

[76] P.V. Shinde, V.B. Labade, B.B. Shingate, M.S. Shingare, Application of unmodified microporous molecular sieves for the synthesis of poly functionalized pyridine derivatives in water, J. Mol. Catal. A Chem. 336 (2011) 100–105.

[77] J. Safaei-Ghomi, M.A. Ghasemzadeh, CuI nanoparticles: a highly active and easily recyclable catalyst for the synthesis of 2-amino-3, 5-dicyano-6-sulfanyl pyridines, J. Sulfur Chem. 34 (3) (2013) 233–241.

[78] M. Lakshmi Kantam, K. Mahendar, S. Bhargava, One-pot, three-component synthesis of highly substituted pyridines and 1,4-dihydropyridines by using nanocrystalline magnesium oxide, J. Chem. Sci. 122 (2010) 63–69.

[79] G. Kour, M. Gupta, B. Vishwanathan, K. Thirunavukkarasu, (Cu/NCNTs): a new high temperature technique to prepare a recyclable nanocatalyst for four component pyridine derivative synthesis and nitroarenes reduction, New J. Chem. 40 (2016) 8535–8542.

[80] J. Xiong, Y. Wang, Q. Xue, X. Wu, Synthesis of highly stable dispersions of nanosized copper particles using L-ascorbic acid, Green Chem. 13 (2011) 900–904.

[81] M. Zhang, P. Liu, Y.-H. Liu, Z.-R. Shang, H.-C. Hu, Z.-H. Zhang, Magnetically separable graphene oxide anchored sulfonic acid: a novel, highly efficient and recyclable catalyst for one-pot synthesis of 3,6-di(pyridin-3-yl)-1H-pyrazolo[3,4-b]pyridine-5-carbonitriles in deep eutectic solvent under microwave irradiation, RSC Adv. 6 (2016) 106160–110617.

[82] M.A. El-borai, H.F. Rizk, M.F. Abd-Aal, I.Y. El-Deeb, Synthesis of pyrazolo [3, 4-b] pyridines under microwave irradiation in multi-component reactions and their antitumor and antimicrobial activities—Part 1, Eur. J. Med. Chem. 48 (2012) 92–96.

[83] M. Pirhayati, A. Kakanejadifard, H. Veisi, A new nano-Fe_3O_4-supported organocatalyst based on 3,4-dihydroxypyridine: an efficient heterogeneous nanocatalyst for one-pot synthesis of pyrazolo[3,4-*b*]pyridines and pyrano[2,3-*d*]pyrimidines, Appl. Organometal. Chem. 30 (2016) 1004–1008.

[84] X. Zhong, G. Dou, D. Wang, Polyethylene glycol (PEG-400): an efficient and recyclable reaction medium for the synthesis of pyrazolo[3,4-*b*]pyridin-6(7*H*)-one derivatives, Molecules 18 (11) (2013) 13139–13147.

[85] A. Azimi-Roshan, M. Mamaghani, N.O. Mahmoodi, F. Shirini, An efficient regioselective sonochemical synthesis of novel 4-aryl-3-methyl-4, 5-dihydro-1*H*-pyrazolo [3, 4-*b*] pyridin-6 (7*H*)-ones, Chin. Chem. Lett. 23 (4) (2012) 399–402.

[86] N. Ma, B. Jiang, G. Zhang, S.-J. Tu, W. Wever, G. Li, New multicomponent domino reactions (MDRs) in water: highly chemo-, regio- and stereoselective synthesis of spiro{[1,3]dioxanopyridine}-4,6-diones and pyrazolo[3,4-*b*]pyridines, Green Chem. 12 (2010) 1357–1361.

[87] M. Mamaghani, F. Shirini, N.O. Mahmoodi, A. Azimi-Roshan, H. Hashemlou, A green, efficient and recyclable Fe ± 3@ K10 catalyst for the synthesis of bioactive pyrazolo [3, 4-*b*] pyridin-6 (7H)-ones under "on water" conditions, J. Mol. Struct. 1051 (2013) 169–176.

[88] A. Stein, Advances in microporous and mesoporous solids—highlights of recent progress, Adv. Mater, 15 (10) (2003) 763–775.

[89] H. Veisi, D. Kordestani, A.R. Faraji, Palladium nanoparticles supported on an organosuperbase denderon-modified mesoporous SBA-15 as a heterogeneous catalyst in Heck coupling reaction, J. Porous Mater. 21 (2014) 141–148.

[90] D.K. Yadav, M.A. Quraishi, Choline chloride $ZnCl_2$: green, effective and reusable ionic liquid for synthesis of 7-amino-2, 4-dioxo-5-phenyl-2, 3, 4, 5-tetrahydro-1H-pyrano [2, 3-d] pyrimidine-6-carbonitrile derivative, J. Mater. Environ. Sci. 5 (2014) 1075–1078.

[91] J. Yu, H. Wang, Green synthesis of pyrano[2,3-*d*]-pyrimidine derivatives in ionic liquids, Synth. Commun. 35 (24) (2005) 3133–3140.

[92] H. Veisi, M. Hekmati, S. Hemmati, Mesoporous SBA-15 silica phenylsulfonic acid (SBA-15-Ph-SO3H) as efficient nanocatalyst for one-pot three-component synthesis of 3-methyl-4-aryl-2,4,5,7-tetrahydropyrazolo[3,4-b]pyridine-6-ones, J. Heterocyclic Chem. 54 (2017) 1630–1635.

[93] H. Shahbazi-Alavi, F. Eshteghal, S. Zahedi, S.H. Nazemzadeh, F. Alemi-Tameh, M. Tavazo, H. Basharnavaz, M.R. Lashkari, Nano-$CuCr_2O_4$: an efficient catalyst for a one-pot synthesis of tetrahydrodipyrazolopyridine, J. Chem. Res. 40 (2016) 361–363.

[94] K. Zhao, M. Lei, L. Ma, L. Hu, A facile protocol for the synthesis of 4-aryl-1,4,7,8-tetrahydro-3,5-dimethyldipyrazolo[3,4-b:4′,3′-e]pyridine derivatives by a Hantzsch-type reaction, Monatsh. Chem. 142 (2011) 1169–1173.

[95] N.G. Shabalala, R. Pagadala, S.B. Jonnalagadda, Ultrasonic-accelerated rapid protocol for the improved synthesis of pyrazoles, Ultrason. Sonochem. 27 (2015) 423–429.

[96] S.M. Sadeghzadeh, A heteropolyacid-based ionic liquid immobilized onto magnetic fibrous nano-silica as robust and recyclable heterogeneous catalysts for the synthesis of tetrahydrodipyrazolopyridines in water, RSC Adv. 6 (2016) 75973–75980.

[97] M. Dabiri, P. Salehi, M. Koohshari, Z. Hajizadeh, D.I. MaGee, An efficient synthesis of tetrahydropyrazolopyridine derivatives by a one-pot tandem multi-component reaction in a green media, Arkivoc 4 (2014) 204–214.

[98] J. Safaei-Ghomi, H. Shahbazi-Alavi, A exible one-pot synthesis of pyrazolopyridines catalyzed by $Fe_3O_4@SiO_2$-SO_3H nanocatalyst under microwave irradiation, Sci. Iran. C 24 (3) (2017) 1209–1219.

[99] J. Safaei-Ghomi, R. Sadeghzadeh, H. Shahbazi-Alavi, A pseudo six-component process for the synthesis of tetrahydrodipyrazolo pyridines using an ionic liquid immobilized on a $FeNi_3$ nanocatalyst, RSC Adv. 6 (40) (2016) 33676–33685.

[100] A. Maleki, Z. Hajizadeh, P. Salehi, Mesoporous halloysite nanotubes modified by $CuFe_2O_4$ spinel ferrite nanoparticles and study of its application as a novel and efficient heterogeneous catalyst in the synthesis of pyrazolopyridine derivatives, Sci. Rep. 9 (2019) 5552–5559.

[101] F.M. Arlan1, J. Khalafy, R. Maleki, One-pot three-component synthesis of a series of 4-aroyl-1,6-diaryl-3-methyl-1H-pyrazolo[3,4-b]pyridine-5-carbonitriles in the presence of aluminum oxide as a nanocatalyst, Chem. Heterocycl. Compd. 54 (1) (2018) 51–57.

[102] F.M. Arlana, R. Javahershenas, J. Khalafy, An efficient one-pot, four-component synthesis of a series of pyrazolo [3,4-b] pyridines in the presence of magnetic LDH as ananocatalyst, Asian J. Nanosci. Mater. 3 (2020) 238–250.

[103] F.L. Theiss, G.A. Ayoko, R.L. Frost, Synthesis of layered double hydroxides containing Mg^{2+}, Zn^{2+}, Ca^{2+} and Al^{3+} layer cations by co-precipitation methods—a review, Appl. Surf. Sci. 383 (2016) 200–213.

[104] R. Shojaei, M. Zahedifar, P. Mohammadi, K. Saidi, H. Sheibani, Novel magnetic nanoparticle supported ionic liquid as an efficient catalyst for the synthesis of spiro [pyrazole-pyrazolo[3,4-b]pyridine]dione derivatives under solvent free conditions, J. Mol. Struct. 1178 (2019) 401–407.

[105] M.M. Heravi, M. Daraie, A novel and efficient five-component synthesis of pyrazole based pyrido[2,3-d]pyrimidine-diones in water: a triply green synthesis, Molecules 21 (2016) 441–452.

[106] S.P. Satasia, P.N. Kalaria, D.K. Raval, Catalytic regioselective synthesis of pyrazole based pyrido[2,3-*d*]pyrimidine-diones and their biological evaluation, Org. Biomol. Chem. 12 (2014) 1751–1758.

[107] H. Naeimi, A. Didar, Z. Rashid, Microwave-assisted synthesis of pyrido-dipyrimidines using magnetically $CuFe_2O_4$ nanoparticles as an efficient, reusable, and powerful catalyst in water, J. Iran Chem. Soc. 14 (2017) 377–385.

[108] R.K. Rawal, R. Tripathi, S.B. Katti, C. Pannecouquec, E.D. Clercq, Synthesis and evaluation of 2-(2, 6-dihalophenyl)-3-pyrimidinyl-1, 3-thiazolidin-4-one analogues as anti-HIV-1 agents, Bioorg. Med. Chem. 15 (9) (2007) 3134–3142.

[109] S. Tao, F. Gao, X. Liu, O.T. Sørensen, Preparation and gas-sensing properties of $CuFe_2O_4$ at reduced temperature, Mater. Sci. Eng. 77 (2000) 172–176.

[110] S. Moradi, M.A. Zolfigol, M. Zarei, D.A. Alonso, A. Khoshnood, A. Tajally, An efficient catalytic method for the synthesis of pyrido[2,3-d]pyrimidines as biologically drug candidates by using novel magnetic nanoparticles as a reusable catalyst, Appl. Organomet. Chem. 32 (2018) e4043.

[111] B.X. Du, Y.L. Li, X.S. Wang, D.Q. Shi, Ionic liquid as an efficient and recyclable reaction medium for the synthesis of pyrido[2,3-*d*]pyrimidines, J. Heterocyclic Chem. 50 (2013) 534–538.

[112] H. Sepehrmansourie, M. Zarei, M.A. Zolfigol, S. Babaee, S. Rostamnia, Application of novel nanomagnetic metal–organic frameworks as a catalyst for the synthesis of new pyridines and 1,4-dihydropyridines via a cooperative vinylogous anomeric based oxidation, Sci. Rep. 11 (2021) 5279–5294.

[113] D. Gandhi, S. Agarwal, MgO NPs catalyzed eco-friendly reaction: a highly effective and green approach for the multicomponent one-pot synthesis of polysubstituted pyridines using 2-aminobenzothiazole, J. Heterocycl. Chem. 55 (2018) 2977–2984.

[114] J. Rakhtshah, F. Yaghoobi, Catalytic application of new manganese Schiff-base complex immobilized on chitosan-coated magnetic nanoparticles for onepot synthesis of 3-iminoaryl-imidazo[1,2-*a*]pyridines, Int. J. Biol. Macromol. 139 (2019) 904–916.

[115] M. Abdollahi-Alibeik, A. Rezaeipoor-Anari, BF3/MCM-41 as nano structured solid acid catalyst for the synthesis of 3-iminoaryl-imidazo[1,2-*a*]pyridines, Catal. Sci. Technol. 4 (2014) 1151–1159.

[116] S. Baghery, M.A. Zolfigol, R. Schirhagl, M. Hasani, Application of triphenylammonium tricyanomethanide as an efficient and recyclable nanostructured molten-salt catalyst for the synthesis of *N*-benzylidene-2-arylimidazo[1,2-*a*] pyridin-3-amines, Synlett 28 (2017) 1173–1176.

[117] R. Venkatesham, A. Manjula, B. Vittal Rao, Bromodimethylsulfonium) bromide-catalyzed one-pot three-component synthesis of imidazo[1,2-*a*]pyridines, J. Heterocycl. Chem. 48 (2011) 942–947.

CHAPTER

12

Pyridine ring as an important scaffold in anticancer drugs

Amr Elagamy[a], *Laila K. Elghoneimy*[a] *and Reem K. Arafa*[a,b]

[a] Drug Design and Discovery Lab, Helmy Institute for Medical Sciences, Zewail City of Science and Technology, Cairo, Egypt [b] Biomedical Sciences Program, University of Science and Technology, Zewail City of Science and Technology, Cairo, Egypt

12.1 Introduction

Cancer is an abnormal cell growth that has the potential to spread and invade other parts of the body in a process called metastasis [1]. As a result of the uncontrolled proliferation, generated tumors will cause several damages locally and/or systemically [2]. Cancer and chemotherapy-induced anemia (CIA) is an example of a local symptom that is developed due to hemorrhage, like in colon cancer, or due to the chemotherapy and radiotherapy that cause immunosuppression, which inhibits erythropoiesis [3]. Systemic symptoms could be exemplified by the decrease in the body mass index (BMI) in a symptom called cachexia [2,4]. In 2014, 20% of cancer deaths were reported to be due to cachexia [4]. Expectedly, statistics have shown that cancer is the second leading cause of death globally [5,6]. In 2020, the World Health Organization (WHO) reported that 10 million people have died from cancer worldwide. Moreover, in 2010, the annual economic cost of cancer was 1.16 trillion US dollars according to WHO. On this basis, we studied the role of pyridine-based scaffolds in treating cancer via affecting different biological pathways. Pyridine was first isolated from coal tar and bone oil by Anderson in 1846 [7]. Its cyclic nature was recognized by Dewar and Korner in 1869 [8]. Pyridine and its related derivatives are a privileged class of heterocyclic compounds found in various chemotherapeutic agents [9,10]. Pyridine is widely found in many natural products such as vitamins, alkaloids and coenzymes and forms the nucleus of around 253 FDA-approved drugs. Pyridine-based compounds are known to possess a remarkable biological and pharmacological activity including antimicrobial [11], antimalarial [12], anticancer [13], HIV inhibitors [14], antioxidant [15], and many others [16]. Pyridine

Recent Developments in the Synthesis and Applications of Pyridines.
DOI: https://doi.org/10.1016/B978-0-323-91221-1.00004-X

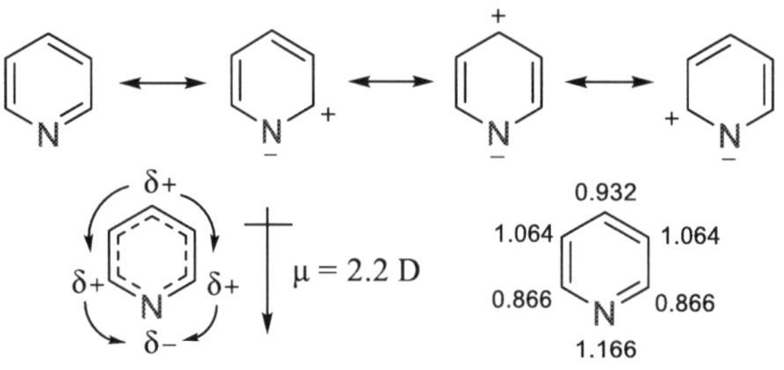

FIGURE 12.1 Resonance structures, dipole moment, and distortion of electronic distribution of pyridine [17].

moieties are frequently used in drugs due to their characteristics such as stability, water solubility, basicity, hydrogen bond-forming capacity, and small molecular size, which help to improve pharmacokinetic, pharmacological, and physicochemical properties. The effective therapeutic behavior of pyridines in drug discovery is mainly based on their mimicking nature. Pyridine ring acts as bioisostere of benzene, amines, amides, and other nitrogen-containing heterocycles. Pyridine is aromatic and possesses six delocalized electrons in the ring (Fig. 12.1). The nitrogen's lone electron pair of pyridine occupies sp^2 orbital parallel to the ring plane. Therefore, this electron pair does not participate in the aromatic π-system ring and more available for donation. In comparison to benzene, pyridine is an electron deficient and polar molecule due to the electron-withdrawing effect of the highly electronegative nitrogen atom. The electron density in pyridine ring at the C-2, C-4, and C-6 positions is less than that at the C-3 and C-5 positions. Therefore, pyridine has a dipole moment and a weaker resonant stabilization than benzene. The electrophilic substitution reactions preferably occur at the C-3 and C-5 positions, while the nucleophilic reactions such as alkylation, acylation, arylation, and amination occur at the C-2, C-4, and C-6 positions.

12.2 Pyridine-containing APIs in clinical use for treatment of cancer

12.2.1 Abemaciclib

Normally, cellular division is controlled by several regulatory enzymes, which prevent premature division and abnormal excessive proliferation. Retinoblastoma (RB1) is a negative regulatory enzyme that acts as a tumor suppressor via halting G1/S transition [18]. RB1 forms a complex with E2F that is needed by the dividing cells to enter the S phase [19]. Once RB1 is phosphorylated via CDK4/6, it will leave E2F free to encode for S phase genes and proceed with the cell cycling process [18]. Therefore, the dysregulation of CDK4/6-RB1 axis will result in cancerous tumors [20–23]. According to breast oncology, breast cancer is classified into three major subtypes; hormone receptor-positive and human epidermal growth factor receptor 2 negative breast cancer (HR+/HER2−), HER2-positive breast cancer, and triple-negative breast cancer [24]. It was found that 75% of stage IV breast cancer patients are of HR+/HER2− type [25–27]. In receptor positive breast cancer, estrogen shows excessive binding to estrogen receptor alpha (ERα) and hence induces the expression of cyclin D1 [28–31]. Cyclin D1 transcription activates CDK4/6 leading to hyperphosphorylation of RB1 [32,33]. Consequently, inhibiting CDK4/6 will arrest the cell cycle at the G1 phase and

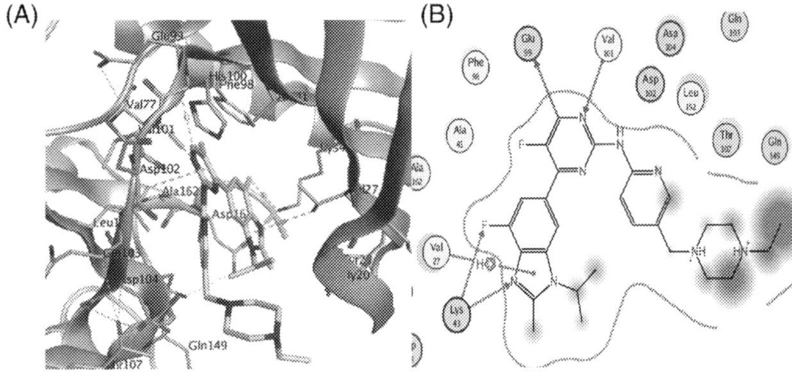

FIGURE 12.2 Binding interaction of abemaciclib with CDK6 (PDB ID: 5LS2); abemaciclib is shown as cyan sticks. (A) The 3D representation of the interactions. (B) The 2D interactions. Favorable bindings are shown as dashed lines. Nonpolar hydrogens are omitted for clarity.

interrupt malignant cells proliferation [24,34,35]. Abemaciclib, palbociclib, and ribociclib are FDA-approved third-generation CDK4/6 selective inhibitors [34,36]. They were approved to be used against HR+/HER2− metastatic breast cancer in combination with endocrine therapy [24,35]. Unlike palbociclib and ribociclib, abemaciclib, that was approved in 2017 and marketed as Verzenios [34,36], can be used alone as a monotherapy [24,35]. Generally, the three inhibitors share the same mechanism of action. However, they may show minor differences in terms of toxicity profile and dosing scheduling [37–39]. These differences are due to their different binding mode with CDK4/6 [36]. They bind to the inactive conformation of kinase by establishing two hydrogen bonds; one between the pyridine nitrogen and His100 residue and the other between the exocyclic NH of the side chain and Val101 [20,40]. Additional hydrogen bonding is formed between the carbonyl of palbociclib/ribociclib and Asp136 of the DGF motif [20,40]. Palbociclib and ribociclib have dimethylamide and acetyl functions causing steric features that account for their higher selectivity [20,36]. Whereas, the F atom in abemaciclib causes its burring inside the ATP pocket, therefore it may act against a wider range of kinases, such as HIPK, PIM, CAMK, and DYRK [20,36,41]. In this section, we will introduce the pharmacokinetics profile of abemaciclib along with its cocrystal structure with CDK6 (PDB ID: 5L2S) (Fig. 12.2). For palbociclib and ribociclib, they will be discussed later in the chapter according to their alphabetical order. Abemaciclib or Verzenios achieves a C_{max} of 249 ng/mL after 8 h of its oral administration. The elimination half-life of the drug is from 17 to 38 h. It has a distribution volume of 690.3 L. Verzenios metabolizes via CYP3A4 into M2 (*N*-desethylation), M20 (hydroxylation), and M18 (hydroxy-*N* desethylation). For the excretion, it is primarily excreted via feces with only 3% in urine [34,36].

In 2010, Eli Lilly & company described a synthetic approach to abemaciclib as in US20100160340 [36,42]. The heterocyclic scaffold **5** of abemaciclib was prepared from condensation of acetamide **2** and 4-bromo-2,6-difluoroaniline **3** followed by base mediated intramolecular cyclization. Then, it was converted to boronate ester **6**, which underwent Suzuki coupling with **7** to give **8**. Scaffold **12** was synthesized via reductive amination reaction between **9** and **10** followed by the replacement of bromine atom with an amino group. Buchwald

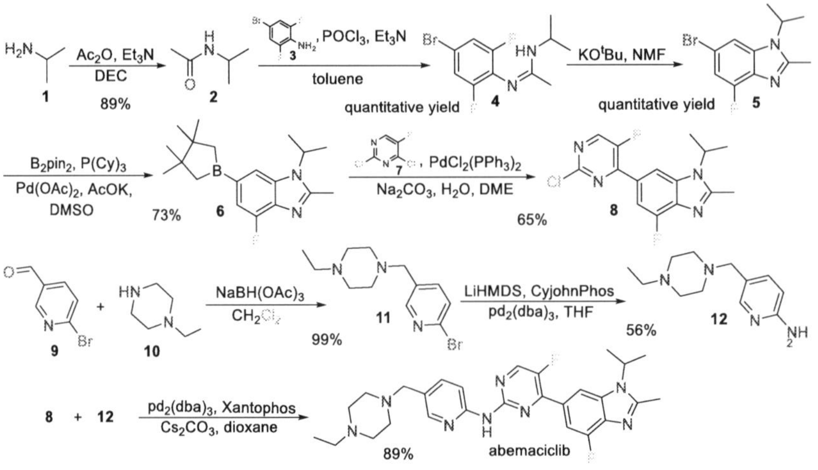

SCHEME 12.1 Synthesis of abemaciclib.

condensation between **8** and **12** provides abemaciclib (Scheme 12.1). In 2015, a modified synthetic strategy was reported by the same company [43]. In which, amination of compound **11** was conducted using liquid ammonia and copper oxide in place of LiHMDS. Recent publications have described improved routes to abemaciclib [7,44]. In addition, deuterated abemaciclib was patented as for the other third-generation inhibitors [45].

12.2.2 Abiraterone acetate

Metastatic castration-resistant prostate cancer (mCRPC) is a stage reached by prostate cancer patients after medical castration due to the continued production of androgens within prostate cancer cells [46]. Cytochrome P450 17A1, also known as CYP17A1 and cytochrome P450c17, is an enzyme found in the endoplasmic reticulum of placenta, adrenal glands, testis, and ovaries, which is responsible for the biosynthesis of sex steroids and glucocorticoid [47,48]. Abiraterone, also named abiraterone acetate or CB7630 and marketed under the brand names of Zytiga or Yonsa, selectively inhibits CYP17A1 leading to lowering androgens levels and inhibiting cancer cell proliferation [49]. Abiraterone acetate is a prodrug that is hydrolyzed by the action of esterase into abiraterone [50,51]. Generally, it is poorly absorbed by the body and reaches its C_{max} after 2 h of administration [49]. Zytiga is highly protein bound to α1-glycoprotein and albumin with Vd equals to 5630 L [49]. It is mainly metabolized into abiraterone sulfate (M45) and *N*-oxide abiraterone sulfate (M31) by sulfotransferase 2A1 (SULT2A1) and CYP3A4 [49,52]. The excretion was found to be through feces for the acetate form and 5% of the metabolites were found in urine [52]. The pyridine nitrogen forms irreversible covalent interaction with the heme iron and the 3β-hydroxyl group shows hydrogen bonding with Asn202, as shown in Fig. 12.3 [53]. Abiraterone acetate was approved in Feb 2018 for mCRPC in combination with prednisolone and is still under clinical investigation for breast cancer treatment.

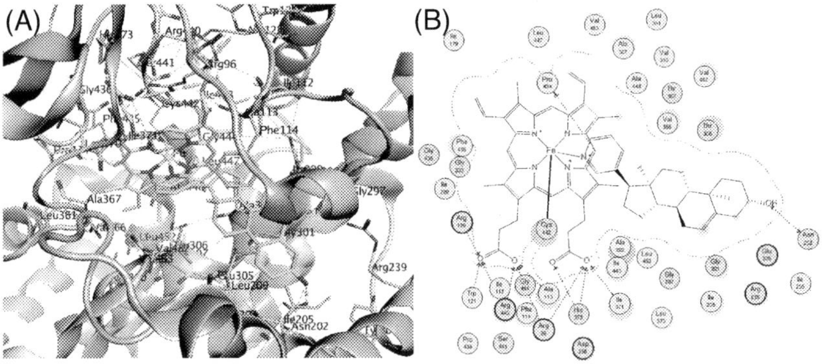

FIGURE 12.3 Binding interaction of abiraterone with CDK6 (PDB ID: 3RUK); abiraterone is shown as cyan sticks. (A) The 3D representation of the interactions. (B) The 2D interactions. Favorable bindings are shown as dashed lines. Nonpolar hydrogens are omitted for clarity.

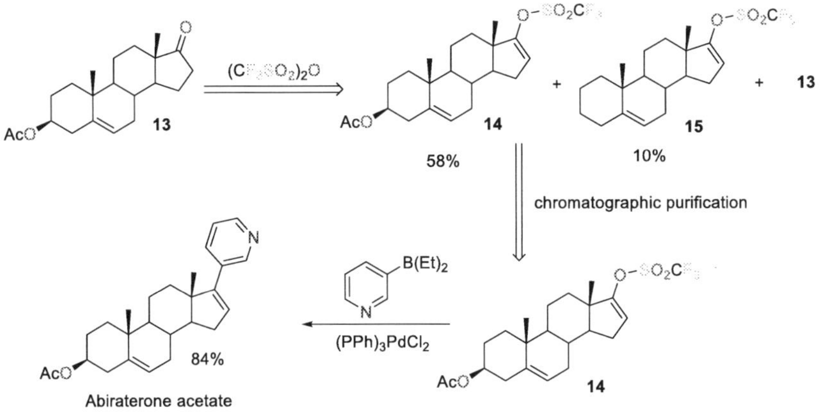

SCHEME 12.2 Synthesis of abiraterone acetate.

Abiraterone acetate is a synthetic androstane steroid. Specifically, it is an acetate ester derivative of androstadienol with a pyridine ring attached at the C-17 position. Abiraterone acetate has been disclosed for the first time in the international patent WO 93/20097 (Scheme 12.2) [54]. Practically, this method is not viable, mainly because of the use of expensive reagents, chromatographic separation is necessary, the difficulty to prepare enol triflate **14**, and the obtained yield is moderate. Later, alternative methods suitable for cross-coupling reaction were developed using vinyl iodide/bromide intermediate in lieu of compound **14** [55–57].

12.2.3 Acalabrutinib

Most of the non-Hodgkin lymphoma (NHL) cases are found to be originated from B-cell malignancies such as small lymphocytic lymphoma (SLL), chronic lymphocytic leukemia (CLL), and diffuse large B-cell lymphoma (DLBC) [58]. Bruton tyrosine kinase (BTK) is one

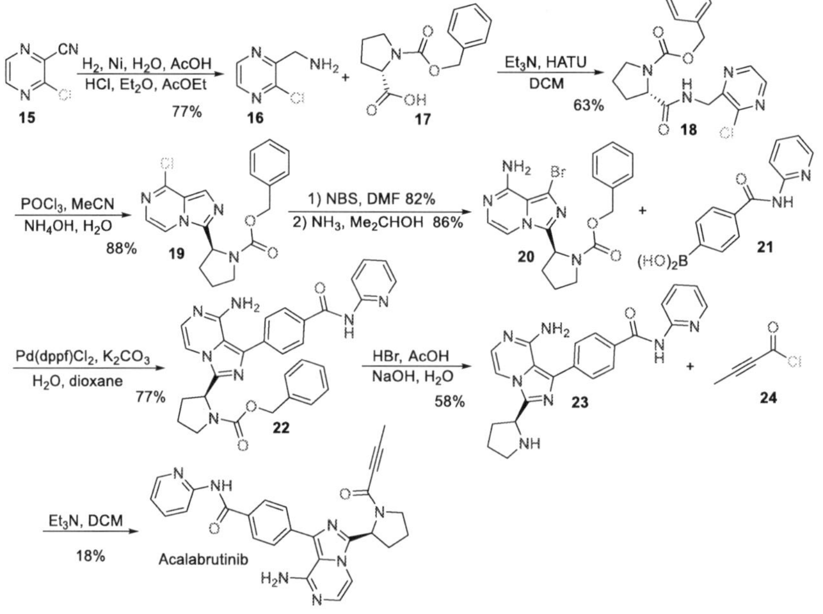

SCHEME 12.3 Synthesis of acalabrutinib.

of the vital components of B-cell receptor signaling axis, which is involved in the proliferation and survival of normal [59,60] and malignant [61] B-cells. As a consequence, targeting BTK is a promising strategy in treating hematological malignancies related to B-cells [62,63]. Acalabrutinib is a BTK second-generation inhibitor that was developed to overcome the adverse effects resulted from the off-target inhibition of other BTK inhibitors [64]. It binds irreversibly to Cys481 residue of the BTK via forming a covalent bond with 2-pyridylbenzamide and 2-butynamide moieties [63,65]. It was approved in 2017 for the treatment of relapsed or refractory (RR) mantle cell lymphoma (MCL) and in 2019 for chronic lymphocytic leukemia (CLL) [62]. On the 4th of October 2019, AstraZeneca marketed acalabrutinib under the brand name Calquence [66]. It is indicated either as a monotherapy in adults with RR CLL or as a combination in treatment-naïve (TN) chronic lymphocytic leukemia (CLL) [66]. Calquence shows rapid absorption and elimination after oral administration with T_{max} and half-life equal 0.9 h and 1 h, respectively. Regarding its elimination, it is metabolized by CYP3A and excreted mainly through feces. Moreover, renal impairments show no significant effect on its PK profile. Interestingly, Calquence absorption is affected negatively by proton pump inhibitors (PPIs) coadministration and positively by antacids and H2-receptor antagonists. Therefore, it is contraindicated with PPIs and recommended with H2-receptor antagonists and antacids [62]. Acalabrutinib is a second-generation Bruton's tyrosine inhibitor discovered and developed by Acerta pharma. The first approach for the synthesis of acalabrutinib was described as in 2013010868 starting with 2-chloro-3-cyanopyrazine precursor **15** (Scheme 12.3) [67]. This method suffered from skin irritation while handling **15** and difficulties in scale controlling

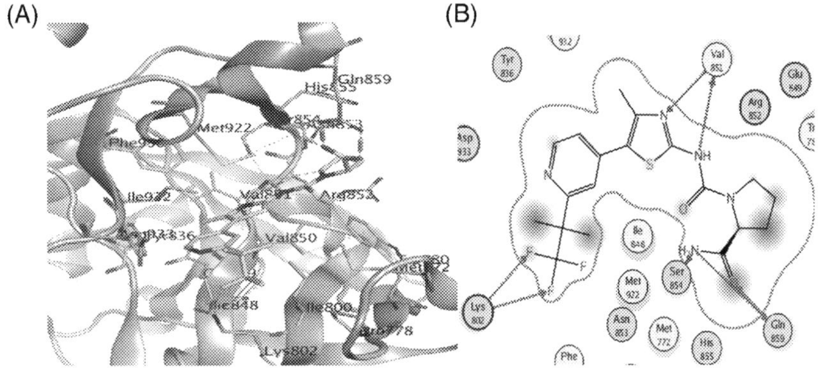

FIGURE 12.4 Binding interaction of alpelisib with CDK6 (PDB ID: 4JPS); alpelisib is shown as cyan sticks. (A) The 3D representation of the interactions. (B) The 2D interactions. Favorable bindings are shown as dashed lines. Nonpolar hydrogens are omitted for clarity.

the over reduction in the first step. Several alternatives synthetic routes were developed as potential alternatives on the commercial scale [68–71].

12.2.4 Alpelisib

Phosphoinositide 3 kinase (PI3K) is involved in regulating cellular proliferation, motility, adhesion and survival. PI3K is classified into three categories, Class I PI3K is further divided into IA and IB subtypes [72,73]. PI3K IA subclass consists of two subunits, catalytic (p110) and regulatory (p85). PIK3CA gene encodes for the catalytic (p110α) domain [74–76]. The phosphatidylinositol 3-kinase (PI3K)/AKT/mammalian target of rapamycin (mTOR) pathway is found to be dysregulated in cancer cells resulting in oncogenesis, cancer progression, and chemoresistance [77,78]. PIK3CA amplification or mutation will activate the axis [79]. Mutations in this pathway are frequently found among breast cancer patients with a percentage of around 40% leading to PI3K/AKT/mTOR axis activation [74–76,80,81]. Consequently, studies show that PIK3CA activation mutation is associated with bad prognosis and chemotherapy resistance in breast cancer patients [76,82–84]. Alpelisib (Piqray) is a phosphoinositide 3 kinase (PI3k) inhibitor that shows selectivity toward the alpha isoform [85,86]. It was granted the FDA approval recently in May 2019 to be used in combination with fulvestrant in patients prescribed with hormone receptor positive/HER-2 negative advanced or metastatic breast cancer with mutation in PIK3CA [85]. Clinically, solid tumor patients demonstrated sensitivity against Piqray monotherapy [80]. Regarding the PK parameters, alpelisib reaches a C_{max} of 1320 ± 912 ng/mL within around 2 h of oral administration. The half-life of alpelisib in plasma equals 13.7 ± 5.9 h and distributes with a volume of 838 ± 541 L [85,87]. Fig. 12.4 of the cocrystal structure of alpelisib with PI3K (PDB ID: 4JPS) reveals seven hydrogen bonding interactions with Lys802, Val851, Ser854, and Gln859 amino acid residues.

Alpelisib is an alpha-selective phosphatidylinositol 3-kinase inhibitor (PI3K) originally described in WO 2010/029082. In 2012, Novartis Pharma AG described a synthetic approach to alpelisib as in WO 2012/175522 (Scheme 12.4) [88]. The reaction of 4-methoxybut-3-en-2-one

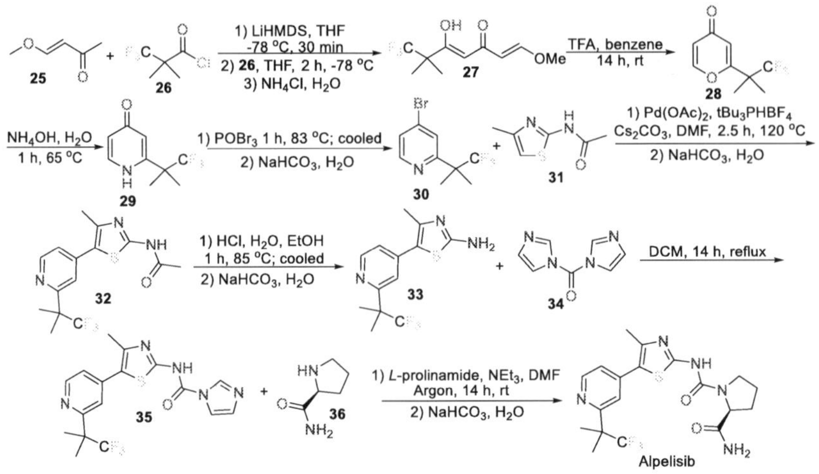

SCHEME 12.4 Synthesis of alpelisib.

25 with 3,3,3-trifluoro-2,2-dimethylpropanoyl chloride **26** in the presence of LiHMDS yields compound **27** which, upon treatment with TFA, provides **28**. The treatment of **28** with ammonium hydroxide followed by $POBr_3$ result in formation of 4-bromo-2-(1,1,1-trifluoro-2-methylpropan-2-yl)pyridine **30**. In the next step, coupling of **30** with *N*-(4-methylthiazol-2-yl)acetamide **31** yields **32**, which undergoes deprotection followed by reaction with di(1*H*-imidazol-1-yl)methanone **34** to give **35**. In the final step, alpelisib was obtained via reaction of **35** with (*S*)-pyrrolidine-2-carboxamide.

12.2.5 Apalutamide

Apalutamide (Erleada) is a nonsteroidal androgenic receptor (AR) antagonist that was approved by the United States (US) in February 2018 for the treatment of nonmetastatic castration-resistant prostate cancer (nmCRPC) [89,90]. Nonmetastatic castration-resistant prostate cancer (nmCRPC) is a stage reached by prostate cancer patients who still show elevated levels of prostate specific antigen (PSA) despite their subjection to androgen deprivation therapy (ADT) (hormone therapy) and low serum testosterone levels ($\leq$50 ng/mL) with no metastasis [91–95]. The main goal while treating those patients is to delay the time to metastasis and hence control the health-associated complications. Since prostate malignant cells will not be able to propagate and divide without androgens, inhibiting AR-signaling constitutes a promising strategy in the treatment protocol of nmCRPC [96]. Erleada blocks the androgenic effect by binding directly to the ligand binding site of AR. As a result of apalutamide blockage, AR nuclear translocation, DNA binding, and AR mediated transcription will be inhibited [89,90,97]. Therefore, it lengthens time to metastasis, metastasis-free survival (MFS) as well as progression-free survival (PFS) [98]. Apalutamide achieves its C_{max} within 2–3 h and then declines gradually with a $T_{1/2}$ of 3–4 days as it reaches a steady state [98,99]. At the steady state, Erleada distributes with a distribution volume of 276 L. For the elimination, it is metabolized

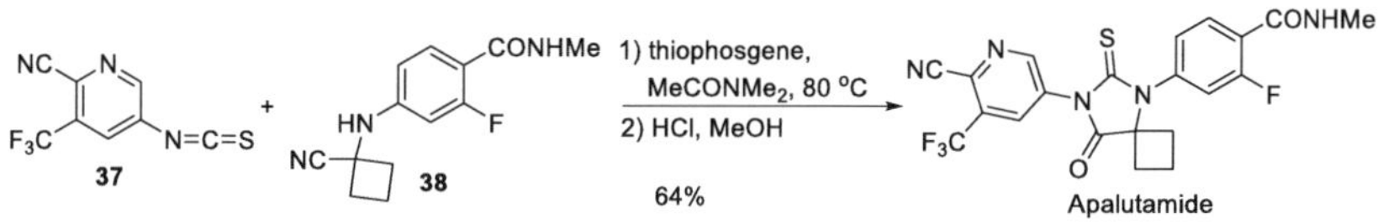

SCHEME 12.5 Final step to apalutamide disclosed by Sloan-kettering.

through CYP3A4 and CYP2C8 to its active form (*N*-desmethyl apalutamide) and excreted in the urine mainly with 24% in feces [90]. Apalutamide is a nonsteroidal antiandrogen drug discovered by Sawyers and Jung at UCLA in 2007. First-generation routes to apalutamide were reported by UCLA and Sloan-kettering [100–104]. This strategy discloses coupling reaction between suitably functionalized pyridine and cyanoaniline. The coupling of 5-isothiocyanato-3-(trifluoromethyl)picolinonitrile with 4-((1-cyanocyclobutyl)amino)-2-fluoro-*N*-methylbenzamide under microwave conditions provides 35–87% yield of apalutamide [100,101]. In 2008, a patent application from Sloan-kettering shows a modified procedure in which, microwave heating was avoided [103–105]. The reaction between **37** and **38** in the presence of thiophosgene in *N,N*-dimethylacetamide provides apalutamide in 64% yield on a 20 g scale (Scheme 12.5). Various routes were described in patent applications for the synthesis of apalutamide fragments **37** and **38** [106]. Aragon Pharmaceuticlas has patented two second-generation routes to apalutamide via late stage amide formation and C–N coupling [107].

12.2.6 Axitinib

Renal cell carcinoma (RCC) is a lethal and frequent disease that affected around 365,943 patients worldwide in 2015 [108]. Moreover, it was resulted in around 14,000 deaths in the United States in 2016 [109]. At the early localized stage of the kidney cancer, curative surgical operations are conducted to remove the localized cancerous tissues. However, systematic treatment is needed when the disease reaches the metastatic stage [108,110]. Tyrosine kinase inhibitors (TKI) targeting vascular endothelial growth factor receptors (VEGFR) are found to be effective in this context [108,111]. The renal cell carcinoma (RCC) is associated with mutations in a tumor suppressor gene called Von Hippel-Lindau (VHL), which will lead to excessive VEGFR transcription [112–116]. VEGFR plays a pivotal role in the pathogenesis of the kidney cancer by aiding in its progression, growth, and spread [114,117,118]. Axitinib is a second line VEGFR-1, -2, and -3 inhibitor that was approved for the treatment of advanced or metastatic renal carcinoma in January 2012 [108,111]. It inhibits VEGFR-1, -2, and -3 selectively by binding tightly to the deep pocket conformation of the kinase domain and stabilize its inactive conformation [119,120]. Fig. 12.5 (PDB ID: 4AG8) shows the interactions between the catalytic pocket and Inlyta in 2D and 3D. Axitinib (Inlyta) shows a C_{max} of 27.8 ng/mL after 2–4 h of administration [111,120]. Although, it is highly absorbed in the acidic PH, antacids show no significant effect on its absorption and hence Inlyta is not contraindicated with antacids [111,120]. Axitinib distributes with apparent volume of distribution of 160 L [111,120]. Foremost, it is metabolized by the liver enzymes CYP3A4/5 [111,120,121] and to a lesser extent by CYP1A2, 2C19, and UGT1A1 to the inactive metabolites glucuronide and

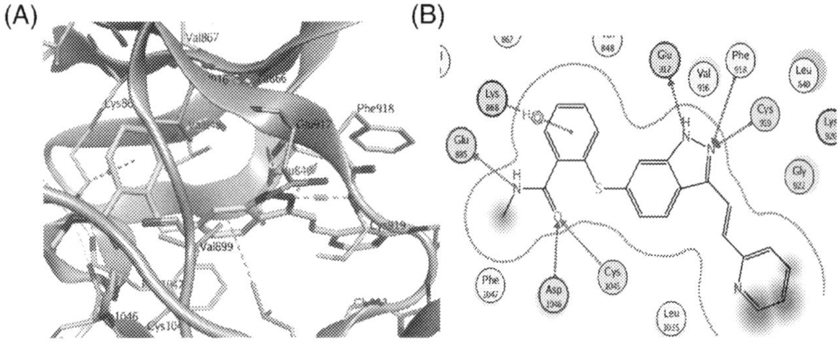

FIGURE 12.5 Binding interaction of axitinib with VEGFR (PDB ID: 4AG8); axitinib is shown as cyan sticks. (A) The 3D representation of the interactions. (B) The 2D interactions. Favorable bindings are shown as dashed lines. Nonpolar hydrogens are omitted for clarity.

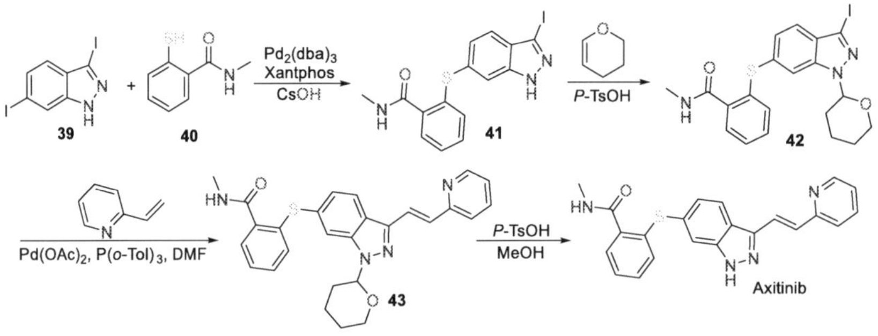

SCHEME 12.6 The first synthetic method to axitinib.

sulfoxide [120,122]. Then eliminated mainly in feces with less than 20% in urine, while the unchanged drug is found to be excreted in feces with less than 1% in urine [122]. The drug has a short half-life ranging between 2 and 6 h, which explains its dosing schedule (twice per day) [111,120].

Axitinib is a small molecule containing pyridine and indazole moieties developed by Pfizer. The first approach for the synthesis of axitinib was described as in US20060094881 (Agouron Pharmaceuticals) and Ep2163544 (Pfizer) using 3,6-diiodo-indazole as a starting material (Scheme 12.6) [123]. Iodo group at the C-6 position was substituted with mercapto group followed by protection of NH group using 3,4-dihydropyran. The obtained product **42** underwent heck reaction with 2-vinylpyridine and deprotection to afford axitinib. Although this method offers a short route to axitinib, the catalysts and reagents are relatively expensive, the total yield is low and requires further purification through the column. Another method was described as in W02006048745 (Pfizer) starting with 6-nitro-indazole. This method can apply to industrial production, however; it requires further purification through the column. In 2015, Sun's group developed an economic and practical protocol to obtain more than 300 g active pharmaceutical ingredients of axitinib via heck-type and C–S coupling reactions catalyzed by cupper iodide [124].

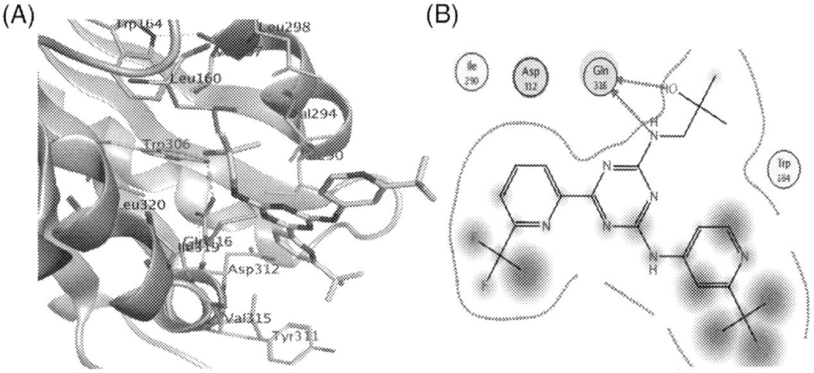

FIGURE 12.6 Binding interaction of enasidenib with mutated IDH-2 (PDB ID: 5I96); enasidenib is shown as cyan sticks. (A) The 3D representation of the interactions. (B) The 2D interactions. Favorable bindings are shown as dashed lines. Nonpolar hydrogens are omitted for clarity.

12.2.7 Enasidenib

Acute myeloid leukemia (AML) is a hematological malignancy that is widely spread among adults with an average age of 62 years old at the time of diagnosis [125,126]. This disease arises from the rapid proliferation of undifferentiated myeloid cells with a potential of recurrence [127,128]. Isocitrate dehydrogenase-1 and -2 (IDH1) and (IDH2) are among Krebs cycle enzymes that catalyze the conversion of isocitrate to alpha ketoglutarate. Several mutations were reported in IDH enzymes during cancer either solid tumors or hematological malignancies, which have an essential role in leukemogenesis [129–132]. The inhibition of IDH1 and IDH2 will aid in managing Relapsed/Refractory AML (R/R AML) [133]. Enasidenib and ivosidenib are FDA-approved drugs for treating R/R AML by selectively inhibit IDH2 and IDH1, respectively, with susceptible mutation [134]. Mutated IDH2 will hinder the differentiation of myeloid cells by hypermethylation of DNA and histones [135]. Enasidenib/Idhifa, which was approved in 2017 [126,136], reaches a C_{max} of 1.3 mcg/mL in 4 h with a $T_{1/2}$ of 137 h. It distributes among 55.8 L and metabolized by multiple cytochrome (CYP) and uridine 5′-diphospho-glucuronosyltransferase (UDP glucuronosyltransferase, UGT) enzymes. Eliminated mainly in feces (89%) and 11% in urine [134,136,137]. Fig. 12.6 represents the two- and three-dimensional interactions between Idhifa and the mutated form of IDH2 enzyme (PDB ID: 5I96).

Enasidenib is a small molecule containing one s-triazine and two pyridine rings developed by Agios Pharmaceuticals and then it was licensed to Celgene for further development. Enasidenib has been disclosed for the first time in the international patent WO2013102431 as shown in Scheme 12.7 [138].

12.2.8 Imatinib

Imatinib is a competitive pyridine-based inhibitor that binds to the ATP binding site and inhibits several tyrosine kinases (TK), such as platelet-derived growth factor receptors (PDGFR), stem-cell factor receptor (c-kit), and BCR-ABL fusion protein [139–142]. These

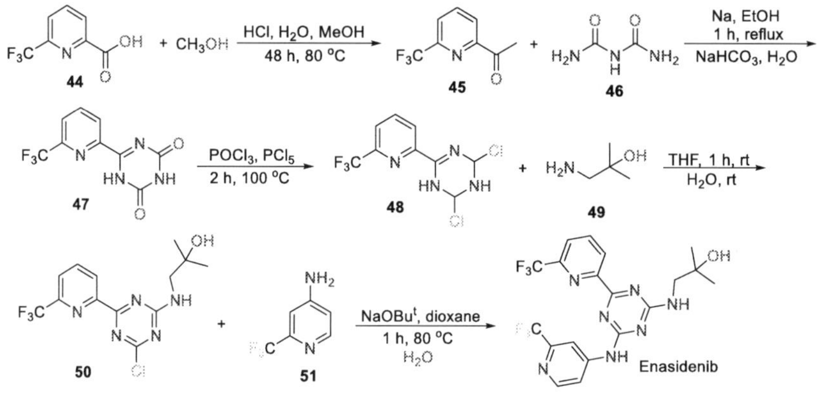

SCHEME 12.7 Synthesis of enasidenib.

kinases are involved in several transduction signaling pathways essential for cell growth, differentiation, and apoptosis [140]. As a result of mutations, these enzymes will be continuously active leading to malignancy [140]. The TK inhibitors will block the undesirable activation and its associated cascade and subsequently induce apoptosis before differentiation [143]. Imatinib (Gleevec or Glivec) was first approved in 2001 to be used against chronic myelogenous leukemia (CML) then received FDA approvals for a range of indications, such as Philadelphia chromosome positive acute lymphoblastic leukemia (Ph + ALL), myelodysplastic/myeloproliferative diseases, aggressive systemic mastocytosis, hypereosinophilic syndrome and/or chronic eosinophilic leukemia (CEL), dermatofibrosarcoma protuberans, and malignant gastrointestinal stromal tumors (GIST) [142,144,145]. CML is a fatal hematopoietic stem cell disorder that is characterized by a reciprocal translocation mutation between chromosomes 9 and 22 (BCR-ABL fusion protein) called Philadelphia chromosome (Ph) [142,146,147]. Although, imatinib shows selective inhibition toward BCR-ABL, it has inhibitory effects against c-kit and PDGFR [148,149]. Since c-kit and PDGFR are associated to a variety of cancers, such as GIST and hypereosinophilic syndrome, respectively, imatinib can be used in managing such disorders as well [150,151]. At the steady state, imatinib reaches a C_{max} of 2596.0 ± 786.7 ng/mL within a T_{max} of 3.3 ± 1.1 h. $T_{1/2\beta}$ equals 19.3 ± 4.4 h [145] and is distributed within 435 L [152]. It is mainly metabolized by hepatic enzymes CYP3A4 and CYP3A5 isoenzyme to *N*-demethylated piperazine derivative CGP 74588 [153] and excreted mainly in feces (68%) with 13% in urine [153–155]. $T_{1/2}$ of imatinib 18 h and for the derivative 40 h [156]. Fig. 12.7 represents the two- and three-dimensional interactions between imatinib and ABL kinase domain (PDB ID: 3HYY). The interactions depict a hydrogen bond between Met 318 and the N atom of pyridine.

In 1993, EP0564409 describes the first process for the synthesis of imatinib as an antitumor agent [157]. A synthetic route suitable for industrial production is described as in WO2004/108669 (Scheme 12.8) [158]. 2-Methyl-5-nitroaniline **52** reacts with cyanamide to give guanidine **53**, which undergoes cyclization reaction with 3-(dimethylamino)-1-(pyridin-3-yl)prop-2-en-1-one **56** to provide phenylamino-pyrimidine **57**. The reduction of nitro to amine followed by condensation with compound **59** affords imatinib. A similar synthetic route

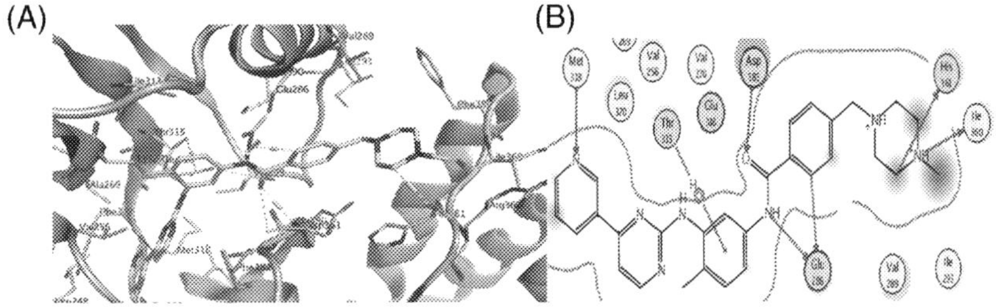

FIGURE 12.7 Binding interaction of imatinib with mutated IDH-2 (PDB ID: 3HYY). Imatinib is shown as cyan sticks. (A) The 3D representation of the interactions. (B) The 2D interactions. Favorable bindings are shown as dashed lines. Nonpolar hydrogens are omitted for clarity.

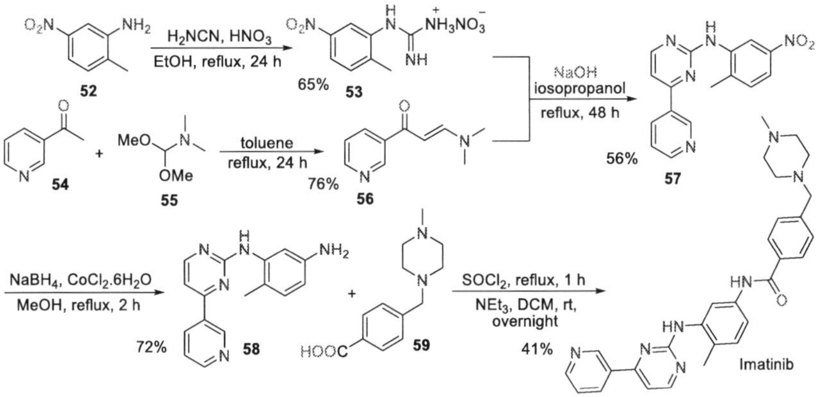

SCHEME 12.8 Synthesis of imatinib.

was described according to WO03/066613 in which the condensation reaction was conducted first and cyclization of pyrimidine chain takes place thereafter. These methods suffer from the use of virulent reagent cyanamide. Later other methods were developed avoiding the use of cyanamide [159,160]. To date, several methods have been reported for the synthesis of imatinib and its analogs [161–164].

12.2.9 Ivosidenib

As was discussed previously in Section 12.2.7, ivosidenib is a selective DH-1 inhibitor for treating AML. Ivosidenib or Tibsovo got its approval on July 20, 2018 [165,166]. Mutated IDH enzymes will further convert alpha ketoglutarate to D-2-hydroxyglutarate (2-HG). The building up of D-2-hydroxyglutarate (2-HG) in the cells will contribute to oncogenesis [167,168]. The peak plasma concentration of Tibsovo is 6551 ng/mL that was achieved in approximately 3 h. The Vd of enasidenib is 234 L with a half-life of 93 h. It is found to be metabolized by CYP3A4 and execrated in feces (77%) and urine (17%) [165,169,170]. The synthetic approach to ivosidenib includes the synthesis of phenyl glycine intermediate **64**, which was obtained

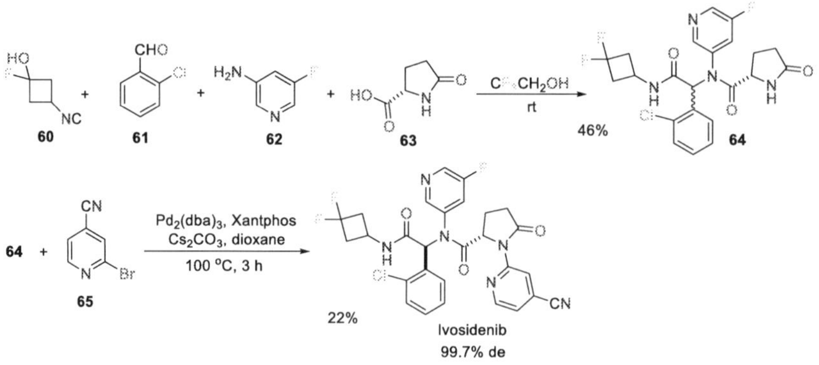

SCHEME 12.9 Synthesis of ivosidenib.

as a mixture of two diastereoisomers via Ugi four-component reaction. A Buchwald–Hartwig coupling between **64** and 2-bromoisonicotinonitrile **65** affords ivosidenib (Scheme 12.9) [171].

12.2.10 Neratinib

Neratinib or Nerlynx was approved on the 17th of July 2017 to be used for extended adjuvant therapy of early-stage human epidermal growth factor receptor 2 (HER2) positive breast cancer [172,173]. Neratinib is a pan-HER kinase inhibitor that binds irreversibly to the ATP-kinase domain of HER2 inhibiting its proliferation and autophosphorylation leading to a blockage in the downstream phosphorylation of AKT and MAPK [174–176]. Nerlynx achieved the maximum peak concentration within 3–6.5 h [173], whereas eliminated with a $T_{1/2}$ of 14 h (240 mg) [177]. The drug distributes extensively among tissues and binds reversibly to albumin [178,179]. Since it is metabolized by hepatic CYP3A, attention should be paid while administering neratinib with inducers or inhibitors [180]. Neratinib is a 4-anilinoquinoline-3-carbonitrile derivative. It was discovered and initially developed by Wyeth. The development up to Phase III in breast cancer was achieved via Pfizer then it was licensed to Puma Biotechnology for global development and commercialization. Various synthetic approaches have been reported for the synthesis of neratinib that vary in the sequence of introducing side chains at the C-4 and C-6 positions [171]. Scheme 12.10 shows substituted quinoline **66** coupled with 3-chloro-4-(2-pyridinylmethoxy)-aniline **67** at the C-4 position, deacetylated, and reacted with (*E*)-4-(dimethylamino) but-2-enoyl chloride hydrochloride **70** at the C-6 position to give neratinib. On the other hand, compound **66** could firstly react with **70** at the C-6 position followed by introduction of aniline derivative **67** at the C-4 position to achieve neratinib (Scheme 12.10).

12.2.11 Nilotinib

As was discussed, ABL-BCR is a protein kinase oncogene that characterizes CML [181]. During CML treatment, mutations may develop in different amino acid residues of the

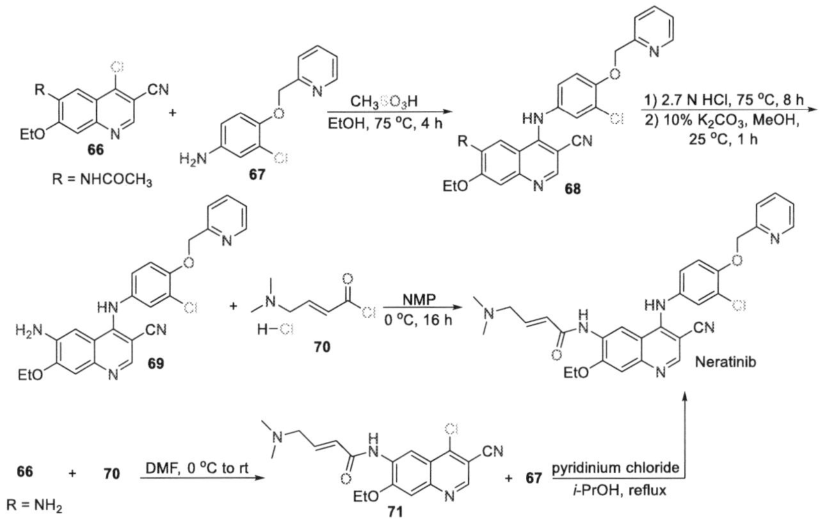

SCHEME 12.10 Synthesis of neratinib.

ABL-BCR leading to resistance and treatment failure. Nilotinib or Tasigna is a second-generation ABL-BCR inhibitor used when patients are intolerant to imatinib therapy [182]. It acts against several mutations, however it has no inhibitory activity against the gate-keeper mutation T315I [183,184] and some others, such as V299L, L248R, and E255K [185]. Moreover, it depicts higher potency than imatinib by 20 folds against the wild ABL-BCR type since it has more lipophilic structure that aids in its selectivity and binding potency [181,183,186]. Tasigna gained the approval in 2007 to be used in patients with CML who received a prior treatment with imatinib and show resistance or intolerance [187,188], while in 2010 it was approved to be used as a fist line therapy [189,190]. Speaking about the drug pharmacokinetics, it has a C_{max} of 3.6 μM at steady state after administering a dose of 400 mg twice per day. The peak plasma concentration is achieved within 3 h and the half-life is shown to be 15 h [181]. The cocrystallized structure of nilotinib with the kinase domain of ABL (PDB ID: 3CS9) reveals multiple hydrogen bindings with two H-pi stackings (Fig. 12.8). Pyridine nitrogen participates by one hydrogen bonding with Met 318 amino acid residue.

Nilotinib has been developed by Novartis based on the structure of the Abl-imatinib complex in order to address imatinib resistance and intolerance [191]. It is almost 20 orders of magnitude more potent than imatinib and being active against all current resistant mutants excluding T315I [192]. The Initial Novartis discovery route is described in Scheme 12.11 [193]. Ethyl 3-amino-4-methylbenzoate **72** reacts with cyanamide to provide guanidine **73**, which undergoes condensation with enone to give 2-aminopyrimidine **75**. In the further step, compound **75** undergoes hydrolysis followed by coupling with substituted aniline **77** to yield nilotinib.

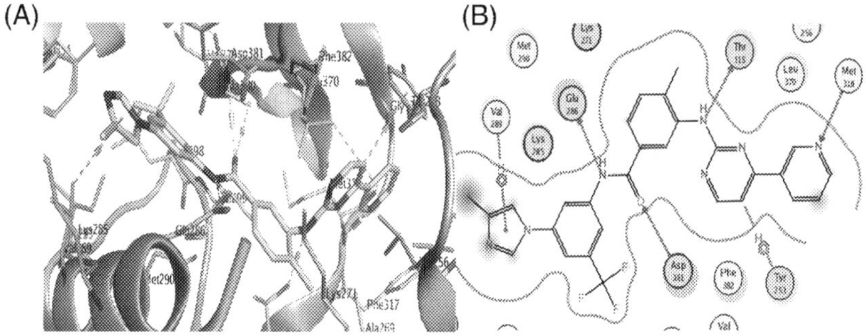

FIGURE 12.8 Binding interaction of nilotinib with mutated IDH-2 (PDB ID: 3CS9). Nilotinib is shown as cyan sticks. (A) The 3D representation of the interactions. (B) The 2D interactions. Favorable bindings are shown as dashed lines. Nonpolar hydrogens are omitted for clarity.

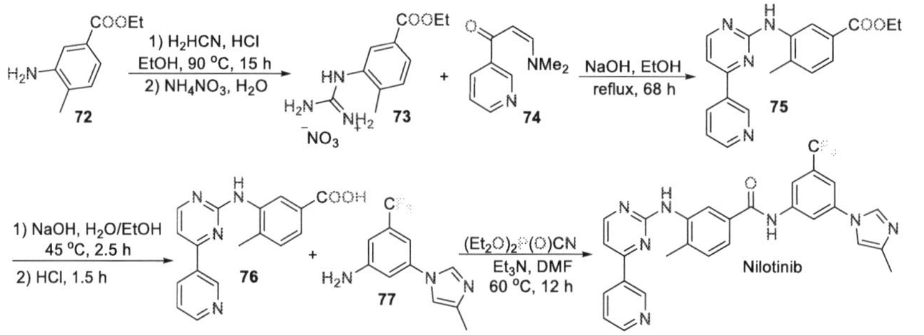

SCHEME 12.11 Synthesis of nilotinib.

12.2.12 Palbociclib

In 2015, palbociclib (Ibrance) gained its approval as the first FDA-approved third-generation CDK4/6 inhibitor [34,36]. It has the same MOA of abemaciclib, discussed previously. The nitrogen of the pyridine ring of palbociclib is essential for the drug selectivity toward CDK4 [36]. The cocrystallized structure with CDK6 (PDB ID: 5L2I) shows reproducible interactions to what has been discussed previously in Section 12.2.1 (Fig. 12.9). The drug reaches C_{max} equals 97 ng/mL within 6–12 h of administration and is eliminated with a $T_{1/2}$ of 24–34 h. It distributes with a volume of 2583 L and metabolizes via CYP3A4 and SULT2A1 into palbociclib-glucuronide. Around 74% of palbociclib is excreted in feces and 18% in urine [34,36].

In 2005, Pfizer reported the first synthetic approach to palbociclib. This method encountered various difficulties including the use of expensive and unstable reagents, harsh conditions, low yield, and 11-step sequences is required [194]. Recently, a novel route for the synthesis of palbociclib was described using an economic starting material and involving an eight-step reaction. This route offered good yield, low cost, less hazards, and readily controllable reaction conditions (Scheme 12.12) [195].

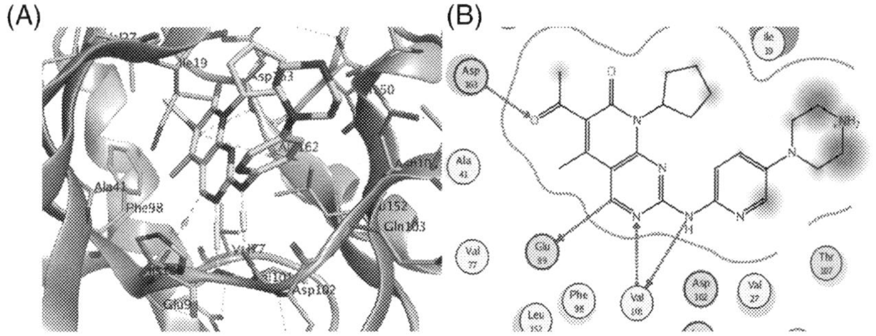

FIGURE 12.9 Binding interaction of palbociclib with CDK6 (PDB ID: 5LSI); palbociclib is shown as cyan sticks. (A) The 3D representation of the interactions. (B) The 2D interactions. Favorable bindings are shown as dashed lines. Nonpolar hydrogens are omitted for clarity.

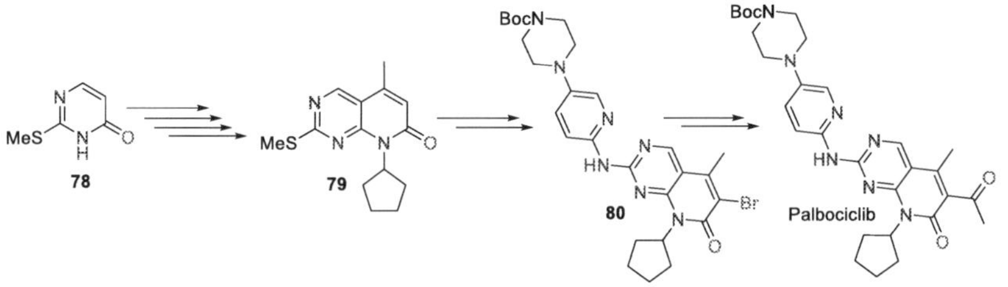

SCHEME 12.12 Synthesis of palbociclib.

12.2.13 Pexidartinib

Tenosynovial giant cell tumor (TGCT) is a rare tumor that grows in the synovium, bursa, and tendon sheath [196,197]. TGCT affects the joints and hence impairs the physical activity, as well as the mobility [198]. The reason behind this tumor is the overly expressed colony stimulating factor 1 (CSF-1) that arises from the translocation mutation of chromosome 1p13 to 2q35 [199,200]. Colony stimulating factor 1 receptor (CSF1-R) is a member of platelet-derived growth factor (PDGF) receptor family, which is found to be expressed on the surface of macrophages and tumor cells [201]. Therefore, once activated it will aid in the differentiation, proliferation, and survival of tumor associated macrophages (TAM) and tumor cells [202,203]. On this basis, inhibiting CSF-1/CSF-1R pathway is a promising therapeutic strategy for TGCT treatment. Earlier, TGCT was primarily treated via surgical resection, however, this is not possible with the nodular pattern of the disease [204–206]. In August 2019, the FDA-approved pexidartinib or PLX3397 as the first systemic treatment against TGCT [198,207]. Pexidartinib (Turalio) is a selective CSF1-R inhibitor that binds to the juxtamembrane domain (JMD) region and stabilizes the autoinhibited state of the receptor. Therefore, prevents the binding of CSF-1 and ATP [198]. Pexidartinib has also shown an inhibitory effect against KIT proto-oncogene receptor tyrosine kinase (cKIT), FMS-like tyrosine kinase 3 (FLT3) harboring an internal tandem duplication (ITD) mutation (FLT3) and platelet-derived growth factor receptor-β (PDGFR-β) [198,207]. Fig. 12.10 shows the interaction pattern of pexidartinib with

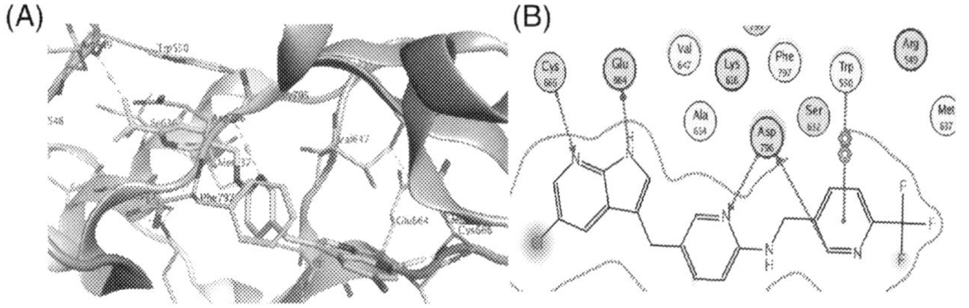

FIGURE 12.10 Binding interaction of pexidartinib with CDK6 (PDB ID: 4R7H); pexidartinib is shown as cyan sticks. (A) The 3D representation of the interactions. (B) The 2D interactions. Favorable bindings are shown as dashed lines. Nonpolar hydrogens are omitted for clarity.

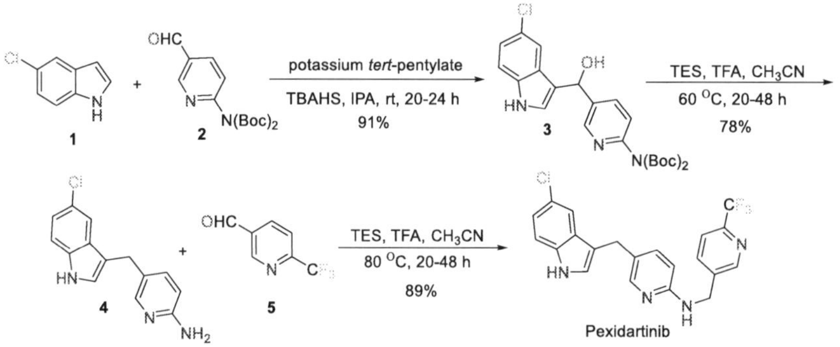

SCHEME 12.13 Synthesis of pexidartinib.

CSF-1R (PDB code: 4R7H), where the pyridine ring appears to participate in the main binding with Trp550 amino acid residue via a π–π stacking interaction [208]. Additionally, the pyridine nitrogen shows hydrogen binding with Asp796 [208]. Regarding the drug pharmacokinetics, the reported T_{max} (time to reach the peak concentration), after oral administration, is 2.5 h and the apparent volume of distribution (Vd) is 187 L [207]. To be eliminated, it is first metabolized into *N*-glucuronide via cytochrome P450 3A4 (CYP 3A4) oxidation and then glucuronidation by UDP glucuronosyltransferase family 1 member A4 (UGT1A4) [207]. Afterward, it will be excreted in feces with a half-life ($T_{1/2}$) equals to 26.6 h [207].

Pexidartinib is a selective tyrosine kinase inhibitor that inhibits the binding of colony stimulating factor-1 (CSF1) to colony stimulating factor-1 receptor (CSF1R). The synthesis of pexidartinib was described by Daiichi Sankyo Company in 2016 (Scheme 12.13) [209]. Compound **3** was obtained in 91% yield via Friedel–Crafts benzylation mediated by TBAHS, which upon dehydroxylation and Boc deprotection gives 78% yield of **4**. A reductive amination reaction between **4** and **5** provides the target product in 89% yield. Although, this route is much concise and provides 63% overall yield on hundred grams scale, the starting materials are considered expensive and significantly increased the cost. In 2019, Liu and coworkers described a new five-step synthetic approach to pexidartinib in 49% overall yield and 99.2% purity using readily available materials and reagents [210].

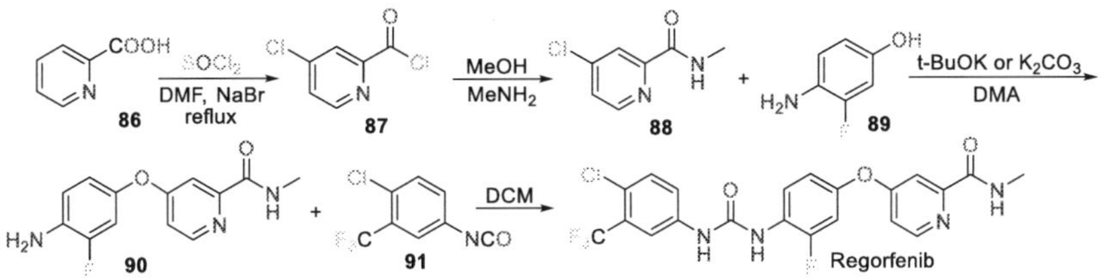

SCHEME 12.14 Traditional synthetic approach to regorafenib.

12.2.14 Regorafenib

Regorafenib (Stivarga) is a multikinase inhibitor that was FDA-approved to be used in several types of cancer [211,212]. In 2012, Stivarga has been approved for metastatic colorectal cancer (mCRC) and by 2013 it was approved to be used against gastrointestinal stromal tumors (GIST) following imatinib and sunitinib. Since 2017 the FDA approved it against hepatocellular carcinoma (HCC) after sorafenib [212]. Regorafenib acts against broad types of cancer due to its unique mechanism of action [211]. Regorafenib inhibits different kinases, which are essential for tumor immunity, maintenance, and progression. Inhibiting these kinases will hinder tumor angiogenesis and disrupt its immunity [211,212]. The tumor immunity is disrupted by inhibiting colony stimulating factor-1 receptor (CSF-1) [213–215] that is required for the survival and differentiation of the tumor associated macrophages (TAMs) involved in several types of cancer including GIST, HCC, and mCRC [216,217]. Stivarga affects tumor angiogenesis via targeting kinases, such as vascular endothelial growth factor receptors (VEGFRs) 1–3, fibroblast growth factor receptors (FGFRs) 1–2, platelet-derived growth factor receptor (PDGFR), and epidermal growth factor-like domains 2 (TIE2), proliferation through KIT, RAF, RET, and others and for the immunity and tumor microenvironment via VEGFR2, VEGFR3, and PDGFR [212,213,215,218,219]. Moreover, it has inhibitory activity against DDR2, TrkA, Eph2A, BRAF, BRAFV600E, SAPK2, PTK5, and Abl [212,220–222]. It metabolizes into pharmacologically active metabolites M-2 (*N*-oxide) and M-5 (*N*-oxide and *N*-desmethyl) through the action of CYP3A4 and UGT1A9. Therefore, it has a long elimination half-life of 28 and 25 h for eliminating regorafenib and M2, respectively, and 51 h for M5 elimination. This drug has a main excretion route through feces (71%) and 19% via urine [212]. Several synthetic pathways have been reported for the preparation of regorafenib (Scheme 12.14) [223–225]. These methods have industrial limitations due to impurity generation, low yield, isolation of intermediates, and stringent reaction requirements. In addition, the desired intermediate **90** was obtained with drawbacks such as the use of expensive materials, inert gas requirement, formation of impurities, and using column chromatography. In 2015, an improved and high-yielding protocol was developed by Gong and coworkers to overcome these limitations. This protocol describes the synthesis of 4-(4-amino-3-fluorophenoxy)-*N*-methylpicolinamide **90** as a key intermediate via *O*-alkylation, nitration, and reduction reactions [226].

12.2.15 Ribociclib

Ribociclib or Kisqali got its FDA approval in 2017 [34,36]. Similar to palbociclib and abemaciclib, ribociclib is a third-generation selective CDK4/6 inhibitor. It exhibits similar

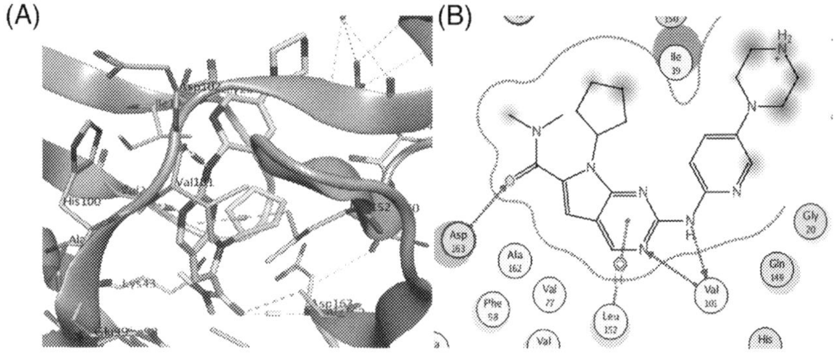

FIGURE 12.11 Binding interaction of ribociclib with CDK6 (PDB ID: 5LST); ribociclib is shown as cyan sticks. (A) The 3D representation of the interactions. (B) The 2D interactions. Favorable bindings are shown as dashed lines. Nonpolar hydrogens are omitted for clarity.

interactions that were explained in Section 12.2.1 (PDB ID: 5L2T) (Fig. 12.11). After 1–4 h of oral administration, its concentration reaches 1680 ng/mL. It distributes with a Vd equals to 1090 L and metabolized into M13 (*N*-hydroxylation), M4 (*N*-demethylation), and M1 (secondary glucuronide) via CYP3A4. Around 69% of ribociclib is excreted in feces and 23% in urine [34,36].

Ribociclib is a pyrrolopyrimidine compound bearing similar side chains as palbociclib. The required key intermediate for the synthesis of ribociclib is described by Novartis as in WO2010020675 [227]. Pyrrolopyrimidine intermediate **100** was obtained via Sonogashira coupling followed by cyclization reaction mediated by TBAF as shown in Scheme 12.15. Then, it was condensed by means of Buchwald reaction with 2-aminopyridine derivative **101**. In the final step, deprotection occurs to provide ribociclib. In 2011, Novartis has reported a more convenient patent incorporating a lower number of synthetic steps [228].

12.2.16 Selpercatinib

Selpercatinib is a selective tyrosine kinase receptor (RET) inhibitor [229]. In various cancerous tumors, RET is abnormally active acting as an oncogene [230]. Nonsmall cell lung cancer (NSCLC) and thyroid cancer are driven by RET fusions. While, medullary thyroid cancer (MTC) is driven by activating RET mutations [230]. In May 2020, selpercatinib (Retevmo) was first approved in the United States to treat RET fusion-positive NSCLC, RET fusion-positive thyroid cancer, and MTC [231]. Fig. 12.12 depicts the binding mode of selpercatinib with RET (PDB ID: 7JU6). The pyridine ring shows H-π stacking with Val738 amino acid residue.

Retevmo has a C_{max} equals to 2980 ng/mL, when reaches the steady state. It achieves the peak plasma concentration within 2 h and is eliminated with a $T_{1/2}$ of 32 h. Like many others, the drug is primarily metabolized by CYP3A4. Around 24% and 68% of the metabolized drug was found in urine and feces, respectively. For the unchanged form, 14% was detected in feces and 12% in urine [232].

Selpercatinib has been developed by Loxo Oncology, Inc. (a wholly owned subsidiary of Eli Lilly & Company). A medicinal chemistry approach to selpercatinib was patented by Charles

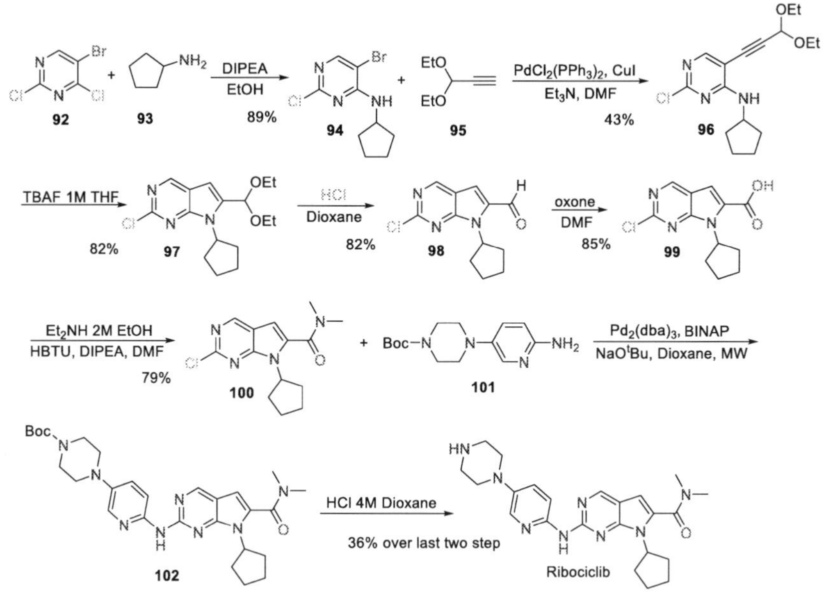

SCHEME 12.15 Synthesis of ribociclib.

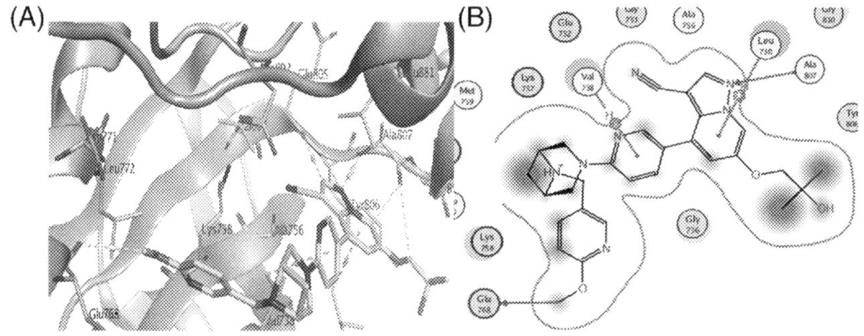

FIGURE 12.12 Binding interaction of selpercatinib with RET (PDB ID: 7JU6). Selpercatinib is shown as cyan sticks. (A) The 3D representation of the interactions. (B) The 2D interactions. Favorable bindings are shown as dashed lines. Nonpolar hydrogens are omitted for clarity.

Todd Eary et al. of Loxo Oncology, Inc. employing Suzuki coupling reaction as a key step as shown in Scheme 12.16 [233].

12.2.17 Sonidegib

Generally, basal cell carcinoma (BCC), which accounts for 80% of skin cancers [234], is manageable by curative surgery or radiation. However, in some cases, it may progress to an advanced stage that is irresponsive to the usual treatment regimens [235]. Therefore, Hedgehog-GLI (HH) pathway inhibitors show effectiveness in such cases [235,236].

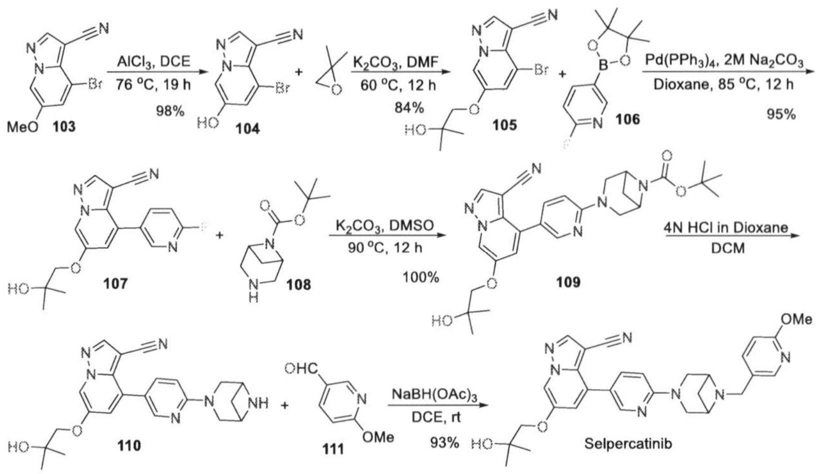

SCHEME 12.16 Synthesis of selpercatinib.

Hedgehog-GLI (HH) axis is normally silenced [237] and associated with cellular differentiation and proliferation [238,239]. However, it is found to be abnormally active during various types of cancer [237]. This activation will stimulate glioma-associated oncogene (GLI) family that favors tumor growth [237]. Smoothened (SMO) controls the correct signaling of HH axis [240,241], therefore, it is considered a promising therapeutic target to control the tumor among several kinds of cancer, including BCC [235,236,242]. Sonidegib (Odomzo), also known as erismodegib, is the second approved drug, in July 2015, to be used against locally advanced BCC by acting as SMO antagonist [240]. Odomzo is absorbed rapidly while fasting with a T_{max} of 2–4 h [243,244]. It is distributed with a volume of 9170 L [235] and eliminated with a $T_{1/2}$ of 28 days [245,246]. Sonidegib is metabolized into the oxide form via 3A4 isoform and eliminated in feces (93%) [244]. Sonidegib was first described by Novartis [247]. Various synthetic pathways were established to prepare sonidegib. However, these processes have drawbacks such as the use of unsustainable catalyst and hazardous organic solvents [248–251]. In 2019, Lipshutz and his group developed an industrial approach to sonidegib using a commercially available and inexpensive reagents, and avoiding column chromatography. In this protocol, sonidegib was obtained through five steps and three pots approach in which ppm level of palladium catalyst was used and all steps were carried under aqueous micellar conditions (Scheme 12.17) [252].

12.2.18 Sorafenib

Sorafenib or Nexavar is a multikinase inhibitor that functions uniquely by several modes of action. It inhibits VEGFRs, PDGFR, Kit, FMS-related tyrosine kinase 3 (FLT-3) [253], and oncogenic rearranged during transfection (RET) kinase [254] and degrades antiapoptotic myeloid cell leukemia 1 (Mcl-1) protein [255]. Furthermore, Nexavar inhibits RAS-Raf-MEK-ERK pathway that silence tumor suppressor genes and aid in tumor invasion and angiogenesis [255,256]. Therefore, sorafenib has the ability to cease cancer proliferation and angiogenesis

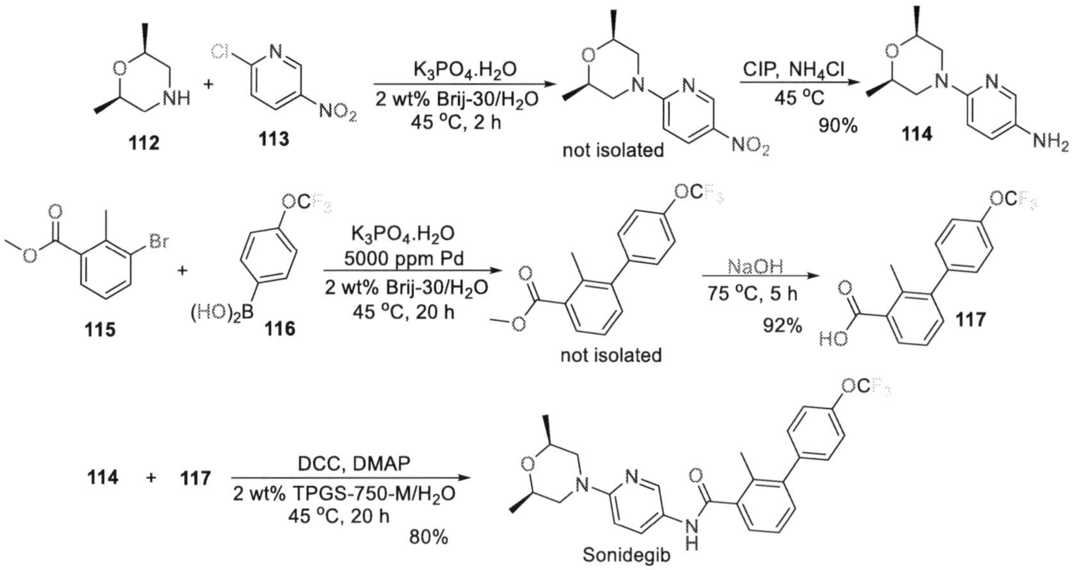

SCHEME 12.17 Large-scale synthesis of sonidegib.

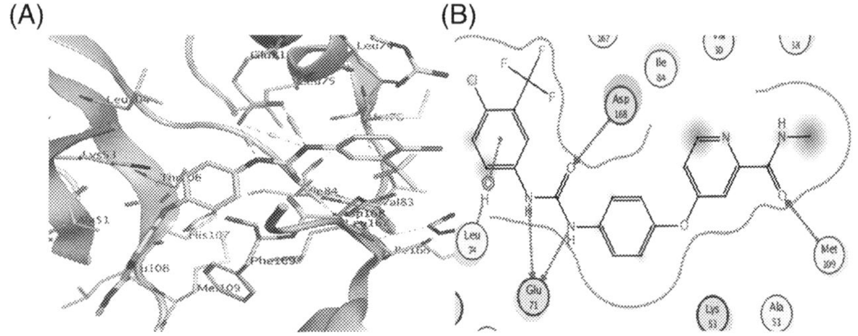

FIGURE 12.13 Binding interaction of sorafenib with P38 MAP kinase (PDB ID: 3GCS). Sorafenib is shown as cyan sticks. (A) The 3D representation of the interactions. (B) The 2D interactions. Favorable bindings are shown as dashed lines. Nonpolar hydrogens are omitted for clarity.

and promote cancer apoptosis [253,257–259]. It was approved to be used in the treatment of advanced RCC, unresectable HCC, and differentiated thyroid cancer (DTC) refractory to radioactive iodine in 2005, 2007, and 2013, respectively [260]. In RCC, a tumor suppressor gene called Von Hippel-Lindau is found to be inactive leading to VEGF overexpression [261], hence sorafenib is a promising therapeutic agent during RCC [261,262]. On the same basis, in HCC, Raf-MEK-ERK axis and VEGF are responsible for the pathogenesis so using Nexavar will control tumor progression [259,263,264]. During DTC, VEGF is overly expressed, RAS is mutated and BRAF V600E mutation is present, which will result in tumor progression and recurrence. These pathogenic targets are targeted by sorafenib [265,266]. Fig. 12.13 illustrates

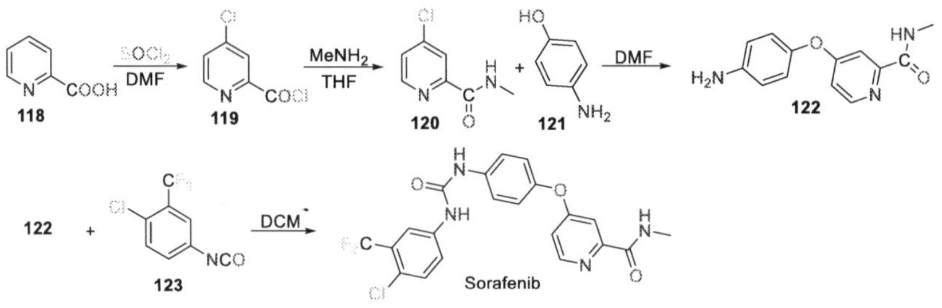

SCHEME 12.18 Classical synthetic approach to sorafenib.

the interactions of sorafenib in the kinase domain of the MAP (PDB ID: 3GCS). After oral administration, Nexavar reaches the C_{max} and the steady state within around 3 h and 7 days, respectively, and eliminated with a half-life ranging from 25 to 48 h [262,267]. It undergoes oxidative and glucuronidation metabolism pathways via the action of CYP3A4 and UGT1A9 [268]. For the excretion, the parent drug was found in feces only [267], however, the glucuronidated metabolites were excreted with 77% in feces and 19% in urine [268].

Sorafenib, a molecule possessing a structure of diaryl urea, was developed by the Bayer and Onyx companies [269] WO0041698 and WO0042012 describe a medicinal chemistry protocol for the synthesis of sorafenib (Scheme 12.18) [270]. In this approach, picolinic acid **118** reacts with thionyl chloride to form acid chloride salt **119**, which on treatment with methylamine provides 4-chloro-*N*-methylpicolinamide **120**. The obtained carboxamide **120** undergoes reaction with 4-aminophenol **121** to give 4-(4-aminophenoxy)-*N*-methylpicolinamide **122**. Subsequent reaction of **122** with 1-chloro-4-isocyanato-2-(trifluoromethyl)benzene **123** yields sorafenib. The drawbacks of this method are due to the requirement of chromatographic purification and low yields at every step. In another reported method sorafenib was obtained via direct coupling of **122** with 4-chloro-3-(trifluoromethyl)aniline in the presence of 1,1′-carbonyldiimidazole [271]. This method also affords product with poor yield. Bankston and his group have developed an improved approach to sorafenib in 63% yield without involving chromatographic purification [272]. Nowadays, several methods have been reported for the synthesis of sorafenib analogs [273–278].

12.2.19 Vismodegib

As previously discussed in Section 12.2.17, BCC can be controlled via Hh-pathway blockage. Vismodegib (Erivedge) is another Hh-axis inhibitor that was the first-in-class to be approved against locally advanced BCC (laBCC), which is irresponsive to radiation and/or surgery and metastatic BCC (mBCC) [279,280]. It was granted the FDA approval in January 2012 to be the first in class to be approved as Hh inhibitor [281,282]. Fig. 12.14 represents the 2D and 3D binding modes of vismodegib with SMO (PDB ID: 5L7I). For the PK parameters, the drug has a low volume of distribution between 16.4 and 26.6 L [279]. It is metabolized through oxidation, glucuronidation, and an uncommon cleavage of the pyridine moiety by CYP2C9 and excreted by 82% in feces and 4.4% in urine [279,283–285].

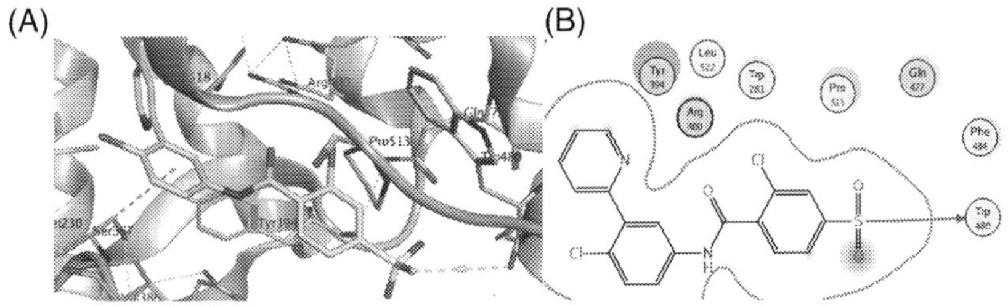

FIGURE 12.14 Binding interaction of vismodegib with SMO (PDB ID: 5L7I). Vismodegib is shown as cyan sticks. (A) The 3D representation of the interactions. (B) The 2D interactions. Favorable bindings are shown as dashed lines. Nonpolar hydrogens are omitted for clarity.

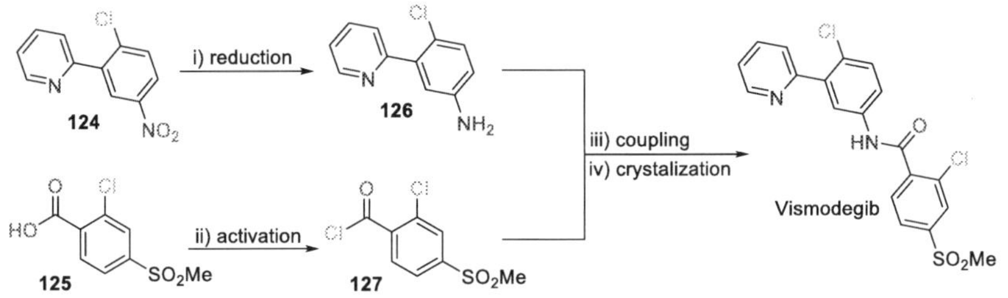

SCHEME 12.19 General approach to vismodegib.

Vismodegib has been developed and launched by Roche and Genentech, under license from Curis. It was first disclosed in Patent No. WO06/028959. Vismodegib was prepared using two key starting materials, 2-(2-chloro-5-nitrophenyl)pyridine **124** and 2-chloro-4-(methylsulfonyl)benzoic acid **125** via a four-step pathway as shown in Scheme 12.19 [286].

References

[1] Y. Suhail, et al., Systems biology of cancer metastasis, Cell Syst. 9 (2) (2019) 109–127.
[2] K. Fearon, et al., Definition and classification of cancer cachexia: an international consensus, Lancet Oncol. 12 (5) (2011) 489–495.
[3] H. Abdel-Razeq, H. Hashem, Recent update in the pathogenesis and treatment of chemotherapy and cancer induced anemia, Crit. Rev. Oncol. Hematol. 145 (2020) 102837.
[4] S. Yang, et al., A narrative review of cancer-related fatigue (CRF) and its possible pathogenesis, Cells 8 (7) (2019) 738.
[5] M.P. Heron, Deaths: Leading Causes for 2017, 68, 6th ed., Natl Vital Stat Rep, 2019, pp. 1–77.
[6] W.H.O.J.G. Organization, Global Health Estimates 2015: Deaths by Cause, Age, Sex, by Country and by Region, 2000-2016, 2018.
[7] M.P. Carroll, et al., Development of an improved route for the synthesis of an abemaciclib intermediate, Org. Process Res. Dev. 23 (11) (2019) 2549–2555.
[8] A. Chaubey, S.N. Pandeya, Pyridine: a versatile nucleus in pahramceutical field, Asian J. Pharm. Clin. Res. 4 (2011) 5–8.
[9] A.K. Elansary, et al., Synthesis and anticancer activity of some novel fused pyridine ring system, Arch. Pharm. Res. 35 (11) (2012) 1909–1917.
[10] S. Kumar, et al., Pyridine: potential for biological activities, J. Chronother. Drug Deliv. 2 (2011) 71–78.

[11] R.W. Friesen, et al., 2-Pyridinyl-3-(4-methylsulfonyl)phenylpyridines: selective and orally active cyclooxygenase-2 inhibitors, Bioorg. Med. Chem. Lett. 8 (19) (1998) 2777–2782.

[12] J. Xue, et al., Antimalarial and structural studies of pyridine-containing inhibitors of 1-deoxyxylulose-5-phosphate reductoisomerase, ACS Med. Chem. Lett. 4 (2) (2013) 278–282.

[13] J. Easmon, et al., Synthesis, structure-activity relationships, and antitumor studies of 2-benzoxazolyl hydrazones derived from alpha-(N)-acyl heteroaromatics, J. Med. Chem. 49 (21) (2006) 6343–6350.

[14] M. Stevens, et al., Inhibition of human immunodeficiency virus by a new class of pyridine oxide derivatives, Antimicrob. Agents Chemother. 47 (9) (2003) 2951–2957.

[15] M.V.D. Almeida, et al., Synthesis and antimicrobial activity of pyridine derivatives substituted at C-2 and C-6 positions, Lett. Drug Des. Discov. 4 (2007) 149–153.

[16] A.A. Altaf, et al., A review on the medicinal importance of pyridine derivatives, J. Drug Des. Med. Chem 1 (2015) 1–11.

[17] M.L.P. Reddya, K.S. Bejoymohandas, Evolution of 2, 3-bipyridine class of cyclometalating ligands as efficient phosphorescent iridium(III) emitters for applications in organic light emitting diodes, JPPC 29 (2016) 29–47.

[18] R.A. Weinberg, The retinoblastoma protein and cell cycle control, Cell 81 (3) (1995) 323–330.

[19] J.Y. Wang, E.S. Knudsen, P.J. Welch, The retinoblastoma tumor suppressor protein, Adv. Cancer. Res. 64 (1994) 25–85.

[20] S. Goel, et al., CDK4/6 inhibition in cancer: beyond cell cycle arrest, Trends Cell Biol. 28 (11) (2018) 911–925.

[21] E. Hamilton, J.R. Infante, Targeting CDK4/6 in patients with cancer, Cancer Treat. Rev. 45 (2016) 129–138.

[22] M.E. Klein, et al., CDK4/6 inhibitors: the mechanism of action may not be as simple as once thought, Cancer Cell 34 (1) (2018) 9–20.

[23] B. O'Leary, R.S. Finn, N.C. Turner, Treating cancer with selective CDK4/6 inhibitors, Nat. Rev. Clin. Oncol. 13 (7) (2016) 417–430.

[24] L.M. Spring, et al., Cyclin-dependent kinase 4 and 6 inhibitors for hormone receptor-positive breast cancer: past, present, and future, Lancet 395 (10226) (2020) 817–827.

[25] N. Howlader, et al., US incidence of breast cancer subtypes defined by joint hormone receptor and HER2 status, J. Natl. Cancer Inst. 106 (5) (2014) dju055 1–8.

[26] D.J. Lobbezoo, et al., Prognosis of metastatic breast cancer subtypes: the hormone receptor/HER2-positive subtype is associated with the most favorable outcome, Breast Cancer Res. Treat. 141 (3) (2013) 507–514.

[27] C.A. Parise, et al., Breast cancer subtypes as defined by the estrogen receptor (ER), progesterone receptor (PR), and the human epidermal growth factor receptor 2 (HER2) among women with invasive breast cancer in California, 1999-2004, Breast J. 15 (6) (2009) 593–602.

[28] L. Altucci, et al., Estrogen induces early and timed activation of cyclin-dependent kinases 4, 5, and 6 and increases cyclin messenger ribonucleic acid expression in rat uterus, Endocrinology 138 (3) (1997) 978–984.

[29] D. Geum, et al., Estrogen-induced cyclin D1 and D3 gene expressions during mouse uterine cell proliferation in vivo: differential induction mechanism of cyclin D1 and D3, Mol. Reprod. Dev. 46 (4) (1997) 450–458.

[30] C. Thangavel, et al., Therapeutically activating RB: reestablishing cell cycle control in endocrine therapy-resistant breast cancer, Endocr. Relat. Cancer 18 (3) (2011) 333–345.

[31] C.A. Lange, D. Yee, Killing the second messenger: targeting loss of cell cycle control in endocrine-resistant breast cancer, Endocr. Relat. Cancer 18 (4) (2011) C19–C24.

[32] L. Ding, et al., The roles of cyclin-dependent kinases in cell-cycle progression and therapeutic strategies in human breast cancer, Int. J. Mol. Sci. 21 (6) (2020) 1960 1–28.

[33] S. Parylo, et al., Role of cyclin-dependent kinase 4/6 inhibitors in the current and future eras of cancer treatment, J. Oncol. Pharm. Pract. 25 (1) (2019) 110–129.

[34] C.L. Braal, et al., Inhibiting CDK4/6 in breast cancer with palbociclib, ribociclib, and abemaciclib: similarities and differences, Drugs 81 (3) (2021) 317–331.

[35] L.M. Spring, et al., CDK 4/6 inhibitors in breast cancer: current controversies and future directions, Curr. Oncol. Rep. 21 (3) (2019) 25.

[36] M. Poratti, G. Marzaro, Third-generation CDK inhibitors: a review on the synthesis and binding modes of Palbociclib, Ribociclib and Abemaciclib, Eur. J. Med. Chem. 172 (2019) 143–153.

[37] A. Patnaik, et al., Efficacy and safety of abemaciclib, an inhibitor of CDK4 and CDK6, for patients with breast cancer, non-small cell lung cancer, and other solid tumors, Cancer Discov. 6 (7) (2016) 740–753.

[38] N. Vidula, H.S. Rugo, Cyclin-dependent kinase 4/6 inhibitors for the treatment of breast cancer: a review of preclinical and clinical data, Clin. Breast Cancer 16 (1) (2016) 8–17.
[39] K.A. Lee, S.T. Shepherd, S.R. Johnston, Abemaciclib, a potent cyclin-dependent kinase 4 and 6 inhibitor, for treatment of ER-positive metastatic breast cancer, Future Oncol. 15 (29) (2019) 3309–3326.
[40] E.S. Kim, L.J. Scott, Palbociclib: a review in HR-positive, HER2-negative, advanced or metastatic breast cancer, Target Oncol. 12 (3) (2017) 373–383.
[41] P. Chen, et al., Spectrum and degree of CDK drug interactions predicts clinical performance, Mol. Cancer Ther. 15 (10) (2016) 2273–2281.
[42] D.A. Coates, et al., *Protein kinase inhibitors*. US20100160340, 2010.
[43] E.M. Chan, *Combination therapy for cancer.* WO2015130540, 2015.
[44] M.O. Frederick, et al., Development of a Leuckart–Wallach reaction in flow for the synthesis of abemaciclib, Org. Process Res. Dev 21 (9) (2017) 1447–1451.
[45] D. Hou, *Deuterate d N-(5-((4-ethylpiperazin-1-yl)methyl) pyridin-2-yl)-5-fluoro-4-(4- fluoro-1-isopropyl-2-methyl-1H-benzo[d]imidazol-6-yl)pyridmidine-2-amine.* WO201808120 2018.
[46] H.I. Scher, C.L. Sawyers, Biology of progressive, castration-resistant prostate cancer: directed therapies targeting the androgen-receptor signaling axis, J. Clin. Oncol. 23 (32) (2005) 8253–8261.
[47] S. Dharia, et al., Colocalization of P450c17 and cytochrome b5 in androgen-synthesizing tissues of the human, Biol. Reprod. 71 (1) (2004) 83–88.
[48] M. Katagiri, N. Kagawa, M.R. Waterman, The role of cytochrome b5 in the biosynthesis of androgens by human P450c17, Arch. Biochem. Biophys. 317 (2) (1995) 343–347.
[49] G.E. Benoist, et al., Pharmacokinetic aspects of the two novel oral drugs used for metastatic castration-resistant prostate cancer: abiraterone acetate and enzalutamide, Clin. Pharmacokinet. 55 (11) (2016) 1369–1380.
[50] A. O'Donnell, et al., Hormonal impact of the 17alpha-hydroxylase/C(17,20)-lyase inhibitor abiraterone acetate (CB7630) in patients with prostate cancer, Br. J. Cancer 90 (12) (2004) 2317–2325.
[51] C.J. Ryan, et al., Phase I clinical trial of the CYP17 inhibitor abiraterone acetate demonstrating clinical activity in patients with castration-resistant prostate cancer who received prior ketoconazole therapy, J. Clin. Oncol. 28 (9) (2010) 1481–1488.
[52] M. Acharya, et al., A phase I, open-label, single-dose, mass balance study of 14C-labeled abiraterone acetate in healthy male subjects, Xenobiotica 43 (4) (2013) 379–389.
[53] N.M. DeVore, E.E. Scott, Structures of cytochrome P450 17A1 with prostate cancer drugs abiraterone and TOK-001, Nature 482 (7383) (2012) 116–119.
[54] G.A. Potter, et al., Novel steroidal inhibitors of human cytochrome P45017 alpha (17 alpha-hydroxylase-C17,20-lyase): potential agents for the treatment of prostatic cancer, J. Med. Chem. 38 (13) (1995) 2463–2471.
[55] G.A. Potter, I.R. Hardcastle, M. Jarman, A convenient, large-scale synthesis of abiraterone acetate [3p-acetoxy-1 7-(3-pyridyl)androsta-5,1 6-diene], a potential new drug for the treatment of prostate cancer, Org. Prep. Proc. Int. 29 (1997) 123–134.
[56] A.N. Balaev, A.V. Gromyko, V.E. Fedorov, Four-step synthesis of abiraterone acetate from dehydroepiandrosterone, Pharm. Chem. J. 50 (2016) 36–38.
[57] M.K. Madhra, et al., Improved procedure for preparation of abiraterone acetate, Org. Process Res. Dev. 18 (2014) 555–558.
[58] M. Al-Hamadani, et al., Non-Hodgkin lymphoma subtype distribution, geodemographic patterns, and survival in the US: a longitudinal analysis of the National Cancer Data Base from 1998 to 2011, Am. J. Hematol. 90 (9) (2015) 790–795.
[59] S. Tsukada, et al., Deficient expression of a B cell cytoplasmic tyrosine kinase in human X-linked agammaglobulinemia, Cell 72 (2) (1993) 279–290.
[60] D. Vetrie, et al., The gene involved in X-linked agammaglobulinaemia is a member of the src family of protein-tyrosine kinases, Nature 361 (6409) (1993) 226–233.
[61] M. Gururajan, C.D. Jennings, S. Bondada, Cutting edge: constitutive B cell receptor signaling is critical for basal growth of B lymphoma, J. Immunol. 176 (10) (2006) 5715–5719.
[62] H.A. Abbas, W.G. Wierda, Acalabrutinib: a selective Bruton tyrosine kinase inhibitor for the treatment of B-cell malignancies, Front. Oncol. 11 (2021) 668162.
[63] B. Fakhri, C. Andreadis, The role of acalabrutinib in adults with chronic lymphocytic leukemia, Ther. Adv. Hematol. 12 (2021) 2040620721990553.

[64] T. Barf, et al., Acalabrutinib (ACP-196): a covalent Bruton tyrosine kinase inhibitor with a differentiated selectivity and in vivo potency profile, J. Pharmacol. Exp. Ther. 363 (2) (2017) 240–252.
[65] J. Wu, M. Zhang, D. Liu, Acalabrutinib (ACP-196): a selective second-generation BTK inhibitor, J. Hematol. Oncol. 9 (2016) 21.
[66] J. Delgado, et al., EMA review of acalabrutinib for the treatment of adult patients with chronic lymphocytic leukemia, Oncologist 26 (3) (2021) 242–249.
[67] T.A. Barf, et al., 4-Imidazopyridazin-1-yl-benzamides and 4-imidazotriazin-1-yl-benzamides as BTK-inhibitors, Google Patents, Google Patents, 2013.
[68] X.C. Wu, Yijie, Y. Chen, M. Wang, Method for Preparing Acalabrutinib as Anti-Leukemia Drug Using 4-Aldehyde-N-2-Pyridine-Benzamide, Assignee Anqing CHICO Pharmaceutical Co., Ltd, 2018 CN 109020977.
[69] A.R. Hamdy, Wayne, R. Izumi, B. Lannutti, T. Covey, R. Ulrich, D. Johnson, T. Barf, A. Kaptein, Therapeutic Combinations of a BTK Inhibitor, a PI3K Inhibitor, a JAK-2 Inhibitor, a PD-1 Inhibitor and/or a PD-L1 Inhibitor, Assignee Acerta Pharma B.V., Neth, 2016 WO 2016024228.
[70] P.L. Wang, Pixu,, X. Gu, Y. Ge, Z. Wang, F. Gao, Q. Du, *Processes for the preparation of acalabrutinib.*WO 2019090269, 2019.
[71] X.X.S. Miracpharma), *Process for the preparation of acalabrutinib.*CN 107056786 2017.
[72] B. Verret, et al., Efficacy of PI3K inhibitors in advanced breast cancer, Ann. Oncol. 30 (Suppl_10) (2019) x12–x20.
[73] B. Vanhaesebroeck, et al., The emerging mechanisms of isoform-specific PI3K signalling, Nat. Rev. Mol. Cell Biol. 11 (5) (2010) 329–341.
[74] L. Angus, et al., The genomic landscape of metastatic breast cancer highlights changes in mutation and signature frequencies, Nat. Genet. 51 (10) (2019) 1450–1458.
[75] O. Martínez-Sáez, et al., Frequency and spectrum of PIK3CA somatic mutations in breast cancer, Breast Cancer Res. 22 (1) (2020) 45.
[76] F. Mosele, et al., Outcome and molecular landscape of patients with PIK3CA-mutated metastatic breast cancer, Ann. Oncol. 31 (3) (2020) 377–386.
[77] R. Dienstmann, et al., Picking the point of inhibition: a comparative review of PI3K/AKT/mTOR pathway inhibitors, Mol. Cancer Ther. 13 (5) (2014) 1021–1031.
[78] P. Liu, et al., Targeting the phosphoinositide 3-kinase pathway in cancer, Nat. Rev. Drug Discov. 8 (8) (2009) 627–644.
[79] J.A. Engelman, Targeting PI3K signalling in cancer: opportunities, challenges and limitations, Nat. Rev. Cancer 9 (8) (2009) 550–562.
[80] D. Juric, et al., Phosphatidylinositol 3-kinase α-selective inhibition with alpelisib (BYL719) in PIK3CA-altered solid tumors: results from the first-in-human study, J. Clin. Oncol. 36 (13) (2018) 1291–1299.
[81] B. Pereira, et al., The somatic mutation profiles of 2,433 breast cancers refines their genomic and transcriptomic landscapes, Nat. Commun. 7 (2016) 11479.
[82] J. Baselga, et al., Biomarker analyses in CLEOPATRA: a phase III, placebo-controlled study of pertuzumab in human epidermal growth factor receptor 2-positive, first-line metastatic breast cancer, J. Clin. Oncol. 32 (33) (2014) 3753–3761.
[83] J.D. Jensen, et al., PIK3CA mutations, PTEN, and pHER2 expression and impact on outcome in HER2-positive early-stage breast cancer patients treated with adjuvant chemotherapy and trastuzumab, Ann. Oncol. 23 (8) (2012) 2034–2042.
[84] N. Sobhani, et al., The prognostic value of PI3K mutational status in breast cancer: a meta-analysis, J. Cell. Biochem. 119 (6) (2018) 4287–4292.
[85] A. Markham, Alpelisib: first global approval, Drugs 79 (11) (2019) 1249–1253.
[86] C. Fritsch, et al., Characterization of the novel and specific PI3Kα inhibitor NVP-BYL719 and development of the patient stratification strategy for clinical trials, Mol. Cancer Ther. 13 (5) (2014) 1117–1129.
[87] A. James, et al., Absorption, distribution, metabolism, and excretion of ,(14)C]BYL719 (alpelisib) in healthy male volunteers, Cancer Chemother. Pharmacol. 76 (4) (2015) 751–760.
[88] I.S. Gallou, C. Gauer, and F. Stowasser, *Polymorphs of (s)-pyrrolidine-1, 2-dicarboxylic acid 2-amide 1-({4-methyl-5-,2-(2, 2, 2-trifluoro-1, 1-dimethyl-ethyl)-pyridin-4-yl]-thiazol-2-yl}-amide*. 2015, Google Patents.
[89] N.J. Clegg, et al., ARN-509: a novel antiandrogen for prostate cancer treatment, Cancer Res. 72 (6) (2012) 1494–1503.
[90] Z.T. Al-Salama, Apalutamide: first global approval, Drugs 78 (6) (2018) 699–705.

[91] J. Esther, et al., Management of nonmetastatic castration-resistant prostate cancer: recent advances and future direction, Curr. Treat. Options Oncol. 20 (2) (2019) 14.

[92] T. Karantanos, et al., Understanding the mechanisms of androgen deprivation resistance in prostate cancer at the molecular level, Eur. Urol. 67 (3) (2015) 470–479.

[93] F. Rozet, et al., Non-metastatic castrate-resistant prostate cancer: a call for improved guidance on clinical management, World J. Urol. 34 (11) (2016) 1505–1513.

[94] H.I. Scher, et al., Trial design and objectives for castration-resistant prostate cancer: updated recommendations from the prostate cancer clinical trials working group 3, J. Clin. Oncol. 34 (12) (2016) 1402–1418.

[95] S. Magnan, et al., Intermittent vs continuous androgen deprivation therapy for prostate cancer: a systematic review and meta-analysis, JAMA Oncol. 1 (9) (2015) 1261–1269.

[96] J. El-Amm, J.B. Aragon-Ching, The current landscape of treatment in non-metastatic castration-resistant prostate cancer, Clin. Med. Insights Oncol. 13 (2019) 1179554919833927.

[97] H.T. Borno, E.J. Small, Apalutamide and its use in the treatment of prostate cancer, Future Oncol. 15 (6) (2019) 591–599.

[98] Z.T. Al-Salama, Apalutamide: a review in non-metastatic castration-resistant prostate cancer, Drugs 79 (14) (2019) 1591–1598.

[99] D.E. Rathkopf, H.I. Scher, Apalutamide for the treatment of prostate cancer, Expert Rev. Anticancer Ther. 18 (9) (2018) 823–836.

[100] M.E. Jung, et al., *Androgen receptor modulator for the treatment of prostate cancer and androgen receptor-associated diseases*. Int. Patent Appl. WO 2007/126765 A2, 2007.

[101] M.E. Jung, et al., *Androgen receptor modulator for the treatment of prostate cancer and androgen receptor-associated diseases*.U.S. Patent 8445507 B2, 2013.

[102] O. Ouerfelli, et al., *Synthesis of thiohydantoins*.Int. Patent Appl. WO 2008/119015 A2, 2008.

[103] O. Ouerfelli, et al., *Synthesis of thiohydantoins*.U.S. Patent Appl. 2010/0190991, 2010.

[104] O. Ouerfelli, et al., *Synthesis of thiohydantoins*.U.S. Patent 9926291 B2, 2018.

[105] O. Ouerfelli, et al., *Synthesis of thiohydantoins*. 2008, Int. Patent Appl. WO 2008/119015 A2.

[106] Hughes, D.L.J.O.P.R. and Development, Review of synthetic routes and crystalline forms of the antiandrogen oncology drugs enzalutamide, apalutamide, and darolutamide, Org. Process Res. Dev. 24 (3) (2020) 347–362.

[107] C.B. Haim, et al., *Processes for the preparation of a diarylthiohydantoin compound*, in *Google Patents*. 2016.

[108] Y. Umeyama, Y. Shibasaki, H. Akaza, Axitinib in metastatic renal cell carcinoma: beyond the second-line setting, Future Oncol. 13 (21) (2017) 1839–1852.

[109] R.L. Siegel, K.D. Miller, A. Jemal, Cancer statistics, 2016, CA Cancer J. Clin. 66 (1) (2016) 7–30.

[110] B.I. Rini, et al., Active surveillance in metastatic renal-cell carcinoma: a prospective, phase 2 trial, Lancet Oncol. 17 (9) (2016) 1317–1324.

[111] A. Bellesoeur, et al., Axitinib in the treatment of renal cell carcinoma: design, development, and place in therapy, Drug Des. Dev. Ther 11 (2017) 2801–2811.

[112] J. Brieger, et al., Inverse regulation of vascular endothelial growth factor and VHL tumor suppressor gene in sporadic renal cell carcinomas is correlated with vascular growth: an in vivo study on 29 tumors, J. Mol. Med. (Berl.) 77 (6) (1999) 505–510.

[113] D.I. Gabrilovich, et al., Antibodies to vascular endothelial growth factor enhance the efficacy of cancer immunotherapy by improving endogenous dendritic cell function, Clin. Cancer Res. 5 (10) (1999) 2963–2970.

[114] B.I. Rini, E.J. Small, Biology and clinical development of vascular endothelial growth factor-targeted therapy in renal cell carcinoma, J. Clin. Oncol. 23 (5) (2005) 1028–1043.

[115] A.R. Schoenfeld, et al., The von Hippel-Lindau tumor suppressor gene protects cells from UV-mediated apoptosis, Oncogene 19 (51) (2000) 5851–5857.

[116] L. Albiges, et al., Vascular endothelial growth factor-targeted therapies in advanced renal cell carcinoma, Hematol. Oncol. Clin. North Am. 25 (4) (2011) 813–833.

[117] L.M. Ellis, D.J. Hicklin, VEGF-targeted therapy: mechanisms of anti-tumour activity, Nat. Rev. Cancer 8 (8) (2008) 579–591.

[118] N. Ferrara, H.P. Gerber, J. LeCouter, The biology of VEGF and its receptors, Nat. Med. 9 (6) (2003) 669–676.

[119] D.D. Hu-Lowe, et al., Nonclinical antiangiogenesis and antitumor activities of axitinib (AG-013736), an oral, potent, and selective inhibitor of vascular endothelial growth factor receptor tyrosine kinases 1, 2, 3, Clin. Cancer Res. 14 (22) (2008) 7272–7283.

[120] M. Gross-Goupil, et al., Axitinib: a review of its safety and efficacy in the treatment of adults with advanced renal cell carcinoma, Clin. Med. Insights Oncol. 7 (2013) 269–277.
[121] B. Escudier, M. Gore, Axitinib for the management of metastatic renal cell carcinoma, Drugs R. D. 11 (2) (2011) 113–126.
[122] H.S. Rugo, et al., Phase I trial of the oral antiangiogenesis agent AG-013736 in patients with advanced solid tumors: pharmacokinetic and clinical results, J. Clin. Oncol. 23 (24) (2005) 5474–5483.
[123] Brigitte Leigh EwanickiMethods of preparing indazole compounds, 2006 US2006/0094881.
[124] L.-H. Zhai, et al., Effective laboratory-scale preparation of axitinib by two CuI-catalyzed coupling reactions, Org. Process Res. Dev. 19 (7) (2015) 849–857.
[125] X. Song, et al., Incidence, survival, and risk factors for adults with acute myeloid leukemia not otherwise specified and acute myeloid leukemia with recurrent genetic abnormalities: analysis of the Surveillance, Epidemiology, and End Results (SEER) database, 2001-2013, Acta Haematol. 139 (2) (2018) 115–127.
[126] R.A. Myers, et al., Enasidenib: an oral IDH2 inhibitor for the treatment of acute myeloid leukemia, J. Adv. Pract. Oncol. 9 (4) (2018) 435–440.
[127] F. Thol, et al., How I treat refractory and early relapsed acute myeloid leukemia, Blood 126 (3) (2015) 319–327.
[128] R.F. Schlenk, et al., Relapsed/refractory acute myeloid leukemia: any progress? Curr. Opin. Oncol. 29 (6) (2017) 467–473.
[129] M.F. Amary, et al., IDH1 and IDH2 mutations are frequent events in central chondrosarcoma and central and periosteal chondromas but not in other mesenchymal tumours, J. Pathol. 224 (3) (2011) 334–343.
[130] M. Arai, et al., Frequent IDH1/2 mutations in intracranial chondrosarcoma: a possible diagnostic clue for its differentiation from chordoma, Brain Tumor Pathol. 29 (4) (2012) 201–206.
[131] D.R. Borger, et al., Frequent mutation of isocitrate dehydrogenase (IDH)1 and IDH2 in cholangiocarcinoma identified through broad-based tumor genotyping, Oncologist 17 (1) (2012) 72–79.
[132] L. Goyal, et al., Prognosis and clinicopathologic features of patients with advanced stage isocitrate dehydrogenase (IDH) mutant and IDH wild-type intrahepatic cholangiocarcinoma, Oncologist 20 (9) (2015) 1019–1027.
[133] A. Tefferi, et al., IDH mutations in primary myelofibrosis predict leukemic transformation and shortened survival: clinical evidence for leukemogenic collaboration with JAK2V617F, Leukemia 26 (3) (2012) 475–480.
[134] M.I. Del Principe, et al., An evaluation of enasidenib for the treatment of acute myeloid leukemia, Expert Opin. Pharmacother. 20 (16) (2019) 1935–1942.
[135] S. Nassereddine, et al., The role of mutant IDH1 and IDH2 inhibitors in the treatment of acute myeloid leukemia, Ann. Hematol. 96 (12) (2017) 1983–1991.
[136] J. Dugan, D. Pollyea, Enasidenib for the treatment of acute myeloid leukemia, Expert Rev. Clin. Pharmacol. 11 (8) (2018) 755–760.
[137] E.M. Stein, et al., Enasidenib in mutant IDH2 relapsed or refractory acute myeloid leukemia, Blood 130 (6) (2017) 722–731.
[138] G.D. Cianchetta, Byron,, J. Popovici-Muller, FG. Salituro, JO. Saunders, J. Travins, S. Yan, T. Guo, Li. Zhang, *Aryltriazinediamine Compounds as IDH2 Inhibitors and Their Preparation and Use in the Treatment of Cancer.* Assignee Agios Pharmaceuticals, Inc., USA: p. WO 2013102431.
[139] D.G. Savage, K.H. Antman, Imatinib mesylate—a new oral targeted therapy, N. Engl. J. Med. 346 (9) (2002) 683–693.
[140] K.S. Kolibaba, B.J. Druker, Protein tyrosine kinases and cancer, Biochim. Biophys. Acta 1333 (3) (1997) F217–F248.
[141] P. le Coutre, et al., In vivo eradication of human BCR/ABL-positive leukemia cells with an ABL kinase inhibitor, J. Natl. Cancer Inst. 91 (2) (1999) 163–168.
[142] C.F. Waller, Imatinib mesylate, Recent Results Cancer Res. 184 (2010) 3–20.
[143] C. Gambacorti-Passerini, et al., Inhibition of the ABL kinase activity blocks the proliferation of BCR/ABL+ leukemic cells and induces apoptosis, Blood Cells Mol. Dis. 23 (3) (1997) 380–394.
[144] J.P. Flynn, V. Gerriets, *Imatinib*, in *StatPearls*, StatPearls PublishingStatPearls Publishing LLC., Treasure Island, FL, 2021 Copyright © 2021.
[145] B. Peng, P. Lloyd, H. Schran, Clinical pharmacokinetics of imatinib, Clin. Pharmacokinet. 44 (9) (2005) 879–894.
[146] H. Bower, et al., Life expectancy of patients with chronic myeloid leukemia approaches the life expectancy of the general population, J. Clin. Oncol. 34 (24) (2016) 2851–2857.

[147] K. Sasaki, et al., Relative survival in patients with chronic-phase chronic myeloid leukaemia in the tyrosine-kinase inhibitor era: analysis of patient data from six prospective clinical trials, Lancet Haematol. 2 (5) (2015) e186–e193.
[148] M.C. Heinrich, et al., Inhibition of KIT tyrosine kinase activity: a novel molecular approach to the treatment of KIT-positive malignancies, J. Clin. Oncol. 20 (6) (2002) 1692–1703.
[149] E. Buchdunger, et al., Abl protein-tyrosine kinase inhibitor STI571 inhibits in vitro signal transduction mediated by c-kit and platelet-derived growth factor receptors, J. Pharmacol. Exp. Ther. 295 (1) (2000) 139–145.
[150] J.V. Jovanovic, et al., Low-dose imatinib mesylate leads to rapid induction of major molecular responses and achievement of complete molecular remission in FIP1L1-PDGFRA-positive chronic eosinophilic leukemia, Blood 109 (11) (2007) 4635–4640.
[151] M.C. Heinrich, et al., Kinase mutations and imatinib response in patients with metastatic gastrointestinal stromal tumor, J. Clin. Oncol. 21 (23) (2003) 4342–4349.
[152] B. Peng, et al., Absolute bioavailability of imatinib (Glivec) orally versus intravenous infusion, J. Clin. Pharmacol. 44 (2) (2004) 158–162.
[153] M.H. Cohen, et al., Approval summary for imatinib mesylate capsules in the treatment of chronic myelogenous leukemia, Clin. Cancer Res. 8 (5) (2002) 935–942.
[154] H.P. Gschwind, et al., Metabolism and disposition of imatinib mesylate in healthy volunteers, Drug Metab. Dispos. 33 (10) (2005) 1503–1512.
[155] D.E. Baker, Imatinib mesylate, Rev. Gastroenterol. Disord. 2 (2) (2002) 75–86.
[156] Z. Nikolova, et al., Bioequivalence, safety, and tolerability of imatinib tablets compared with capsules, Cancer Chemother. Pharmacol. 53 (5) (2004) 433–438.
[157] A.A. Kamath, et al., US 2011/0306763 A1.
[158] K.E. Kil, et al., Synthesis and positron emission tomography studies of carbon-11-labeled imatinib (Gleevec), Nucl. Med. Biol. 34 (2) (2007) 153–163.
[159] D. Loiseleur, et al., WO Patent 2003066613, 2003.
[160] Y. Liu, et al., A facile total synthesis of imatinib base and its analogues, Org. Process Res. Dev. 12 (2008) 490.
[161] J. Kang, et al., Synthesis of imatinib, a tyrosine kinase inhibitor, labeled with carbon-14, J. Labelled Comp. Radiopharm. 63 (4) (2020) 174–182.
[162] K.C. Nicolaou, et al., Synthesis and biopharmaceutical evaluation of imatinib analogues featuring unusual structural motifs, ChemMedChem 11 (1) (2016) 31–37.
[163] I.M. Fawzy, et al., Design, synthesis and 3D QSAR based pharmacophore study of novel imatinib analogs as antitumor-apoptotic agents, Future Med. Chem. 10 (12) (2018) 1421–1433.
[164] H. Liu, et al., A novel synthesis of imatinib and its intermediates, Monat. Chem.-Chem. Monthly 141 (8) (2010) 907–911.
[165] C.D. DiNardo, et al., Durable remissions with ivosidenib in IDH1-mutated relapsed or refractory AML, N. Engl. J. Med. 378 (25) (2018) 2386–2398.
[166] G.J. Roboz, et al., Ivosidenib induces deep durable remissions in patients with newly diagnosed IDH1-mutant acute myeloid leukemia, Blood 135 (7) (2020) 463–471.
[167] L. Dang, K. Yen, E.C. Attar, IDH mutations in cancer and progress toward development of targeted therapeutics, Ann. Oncol. 27 (4) (2016) 599–608.
[168] J. Mondesir, et al., IDH1 and IDH2 mutations as novel therapeutic targets: current perspectives, J. Blood Med. 7 (2016) 171–180.
[169] S. Dhillon, Ivosidenib: first global approval, Drugs 78 (14) (2018) 1509–1516.
[170] D. Dai, et al., Effect of itraconazole, food, and ethnic origin on the pharmacokinetics of ivosidenib in healthy subjects, Eur. J. Clin. Pharmacol. 75 (8) (2019) 1099–1108.
[171] J. Popovici-Muller, et al., Discovery of AG-120 (ivosidenib): a first-in-class mutant IDH1 inhibitor for the treatment of IDH1 mutant cancers, ACS Med. Chem. Lett. 9 (4) (2018) 300–305.
[172] A. Nasrazadani, A. Brufsky, Neratinib: the emergence of a new player in the management of HER2+ breast cancer brain metastasis, Future Oncol. 16 (7) (2020) 247–254.
[173] I. Echavarria, et al., Neratinib for the treatment of HER2-positive early stage breast cancer, Expert Rev. Anticancer Ther. 17 (8) (2017) 669–679.

[174] D.M. Collins, et al., Preclinical characteristics of the irreversible pan-HER kinase inhibitor neratinib compared with lapatinib: implications for the treatment of HER2-positive and HER2-mutated breast cancer, Cancers (Basel) 11 (6) (2019) 737 1–27.
[175] S.K. Rabindran, et al., Antitumor activity of HKI-272, an orally active, irreversible inhibitor of the HER-2 tyrosine kinase, Cancer Res. 64 (11) (2004) 3958–3965.
[176] H.R. Tsou, et al., Optimization of 6,7-disubstituted-4-(arylamino)quinoline-3-carbonitriles as orally active, irreversible inhibitors of human epidermal growth factor receptor-2 kinase activity, J. Med. Chem. 48 (4) (2005) 1107–1131.
[177] K.K. Wong, et al., A phase I study with neratinib (HKI-272), an irreversible pan ErbB receptor tyrosine kinase inhibitor, in patients with solid tumors, Clin. Cancer Res. 15 (7) (2009) 2552–2558.
[178] A. Chandrasekaran, et al., Reversible covalent binding of neratinib to human serum albumin in vitro, Drug Metab. Lett. 4 (4) (2010) 220–227.
[179] J. Wang, et al., Characterization of HKI-272 covalent binding to human serum albumin, Drug Metab. Dispos. 38 (7) (2010) 1083–1093.
[180] R. Abbas, et al., Pharmacokinetics of oral neratinib during co-administration of ketoconazole in healthy subjects, Br. J. Clin. Pharmacol. 71 (4) (2011) 522–527.
[181] T. Sacha, G. Saglio, Nilotinib in the treatment of chronic myeloid leukemia, Future Oncol. 15 (9) (2019) 953–965.
[182] E. Weisberg, et al., AMN107 (nilotinib): a novel and selective inhibitor of BCR-ABL, Br. J. Cancer 94 (12) (2006) 1765–1769.
[183] E. Weisberg, et al., Characterization of AMN107, a selective inhibitor of native and mutant Bcr-Abl, Cancer Cell 7 (2) (2005) 129–141.
[184] T. O'Hare, et al., In vitro activity of Bcr-Abl inhibitors AMN107 and BMS-354825 against clinically relevant imatinib-resistant Abl kinase domain mutants, Cancer Res. 65 (11) (2005) 4500–4505.
[185] G. Rosti, et al., Dasatinib and nilotinib in imatinib-resistant Philadelphia-positive chronic myelogenous leukemia: a 'head-to-head comparison', Leuk. Lymphoma 51 (4) (2010) 583–591.
[186] M. Golemovic, et al., AMN107, a novel aminopyrimidine inhibitor of Bcr-Abl, has in vitro activity against imatinib-resistant chronic myeloid leukemia, Clin. Cancer Res. 11 (13) (2005) 4941–4947.
[187] H. Kantarjian, et al., Nilotinib in imatinib-resistant CML and Philadelphia chromosome-positive ALL, N. Engl. J. Med. 354 (24) (2006) 2542–2551.
[188] P. le Coutre, et al., Nilotinib (formerly AMN107), a highly selective BCR-ABL tyrosine kinase inhibitor, is active in patients with imatinib-resistant or -intolerant accelerated-phase chronic myelogenous leukemia, Blood 111 (4) (2008) 1834–1839.
[189] A. Hochhaus, et al., Long-term benefits and risks of frontline nilotinib vs imatinib for chronic myeloid leukemia in chronic phase: 5-year update of the randomized ENESTnd trial, Leukemia 30 (5) (2016) 1044–1054.
[190] G. Saglio, et al., Nilotinib versus imatinib for newly diagnosed chronic myeloid leukemia, N. Engl. J. Med. 362 (24) (2010) 2251–2259.
[191] R. Capdeville, et al., Glivec (STI571, imatinib), a rationally developed, targeted anticancer drug, Nat. Rev. Drug Discov. 1 (7) (2002) 493–502.
[192] B.J. Druker, Circumventing resistance to kinase-inhibitor therapy, N. Engl. J. Med. 354 (24) (2006) 2594–2596.
[193] B.J. Deadman, et al., The synthesis of Bcr-Abl inhibiting anticancer pharmaceutical agents imatinib, nilotinib and dasatinib, Org. Biomol. Chem. 11 (11) (2013) 1766–1800.
[194] P.L. Toogood, et al., Discovery of a potent and selective inhibitor of cyclin-dependent kinase 4/6, J. Med. Chem. 48 (7) (2005) 2388–2406.
[195] S.t. Li, et al., A new route for the synthesis of Palbociclib, Chem. Pap. 73 (2019) 3043–3051.
[196] F. Gouin, T. Noailles, Localized and diffuse forms of tenosynovial giant cell tumor (formerly giant cell tumor of the tendon sheath and pigmented villonodular synovitis), Orthop. Traumatol. Surg. Res. 103 (1s) (2017) S91–S97.
[197] S.K. Shetty, et al., Chondroid tenosynovial giant cell tumor of temporomandibular joint, Ann. Maxillofac. Surg. 8 (2) (2018) 327–329.
[198] B. Benner, et al., Pexidartinib, a novel small molecule CSF-1R inhibitor in use for tenosynovial giant cell tumor: a systematic review of pre-clinical and clinical development, Drug Des. Dev. Ther. 14 (2020) 1693–1704.
[199] S.P. Granowitz, J. D'Antonio, H.L. Mankin, The pathogenesis and long-term end results of pigmented villonodular synovitis, Clin. Orthop. Relat. Res. (114) (1976) 335–351.

[200] R.B. West, et al., A landscape effect in tenosynovial giant-cell tumor from activation of CSF1 expression by a translocation in a minority of tumor cells, Proc. Natl. Acad. Sci. U S A, 103 (3) (2006) 690–695.
[201] J.S. Cupp, et al., Translocation and expression of CSF1 in pigmented villonodular synovitis, tenosynovial giant cell tumor, rheumatoid arthritis and other reactive synovitides, Am. J. Surg. Pathol. 31 (6) (2007) 970–976.
[202] M. Chittezhath, et al., Molecular profiling reveals a tumor-promoting phenotype of monocytes and macrophages in human cancer progression, Immunity 41 (5) (2014) 815–829.
[203] R. Ostuni, et al., Macrophages and cancer: from mechanisms to therapeutic implications, Trends Immunol. 36 (4) (2015) 229–239.
[204] M. Brahmi, A. Vinceneux, P.A. Cassier, Current systemic treatment options for tenosynovial giant cell tumor/pigmented villonodular synovitis: targeting the CSF1/CSF1R axis, Curr. Treat. Options Oncol. 17 (2) (2016) 10.
[205] N. Giustini, et al., Tenosynovial giant cell tumor: case report of a patient effectively treated with pexidartinib (PLX3397) and review of the literature, Clin. Sarcoma Res. 8 (2018) 14.
[206] W.D. Tap, et al., Pexidartinib versus placebo for advanced tenosynovial giant cell tumour (ENLIVEN): a randomised phase 3 trial, Lancet 394 (10197) (2019) 478–487.
[207] S. Monestime, D. Lazaridis, Pexidartinib (TURALIO^{TM}): the first FDA-indicated systemic treatment for tenosynovial giant cell tumor, Drugs R. D. 20 (3) (2020) 189–195.
[208] W.D. Tap, et al., Structure-guided blockade of CSF1R kinase in tenosynovial giant-cell tumor, N. Engl. J. Med. 373 (5) (2015) 428–437.
[209] D. Chen, et al., Exploratory process development of pexidartinib through the tandem Tsuji–Trost reaction and heck coupling, Synthesis 51 (12) (2019) 2564–2571.
[210] P.N. Ibrahim., M. Jin., and S. Matsuura., *Synthesis of 1H-pyrrolo,2,3-B]pyridine derivatives that modulate kinases.* WO2016179412, 2017.
[211] A. Grothey, et al., Evolving role of regorafenib for the treatment of advanced cancers, Cancer Treat. Rev. 86 (2020) 101993.
[212] T.J. Ettrich, T. Seufferlein, Regorafenib, Recent Results Cancer Res. 211 (2018) 45–56.
[213] L. Abou-Elkacem, et al., Regorafenib inhibits growth, angiogenesis, and metastasis in a highly aggressive, orthotopic colon cancer model, Mol. Cancer Ther. 12 (7) (2013) 1322–1331.
[214] M.A. Cannarile, et al., Colony-stimulating factor 1 receptor (CSF1R) inhibitors in cancer therapy, J. Immunother. Cancer 5 (1) (2017) 53.
[215] D. Zopf, et al., Pharmacologic activity and pharmacokinetics of metabolites of regorafenib in preclinical models, Cancer Med. 5 (11) (2016) 3176–3185.
[216] W. Hu, et al., Tumor-associated macrophages in cancers, Clin. Transl. Oncol. 18 (3) (2016) 251–258.
[217] S.M. Pollack, et al., Emerging targeted and immune-based therapies in sarcoma, J. Clin. Oncol. 36 (2) (2018) 125–135.
[218] R. Schmieder, et al., Regorafenib (BAY 73-4506): antitumor and antimetastatic activities in preclinical models of colorectal cancer, Int. J. Cancer 135 (6) (2014) 1487–1496.
[219] S.M. Wilhelm, et al., Regorafenib (BAY 73-4506): a new oral multikinase inhibitor of angiogenic, stromal and oncogenic receptor tyrosine kinases with potent preclinical antitumor activity, Int. J. Cancer 129 (1) (2011) 245–255.
[220] S. Dhillon, Regorafenib: a review in metastatic colorectal cancer, Drugs 78 (11) (2018) 1133–1144.
[221] M. Mohammadi, H. Gelderblom, Systemic therapy of advanced/metastatic gastrointestinal stromal tumors: an update on progress beyond imatinib, sunitinib, and regorafenib, Expert Opin. Investig. Drugs 30 (2) (2021) 143–152.
[222] M. Papadimitriou, C.A. Papadimitriou, Antiangiogenic tyrosine kinase inhibitors in metastatic colorectal cancer: focusing on regorafenib, Anticancer Res. 41 (2) (2021) 567–582.
[223] J. Dumas, et al., WO2005009961 2005.
[224] M.L. Gers and R. Gehring, WO2006034796 2006.
[225] O. Christensen and I. Kuss, WO2012012404 2012.
[226] L.-M. Wang, et al., An efficient and high-yielding protocol for the production of Regorafenib via a new synthetic strategy, Res. Chem. Intermed. 42 (4) (2016) 3209–3218.
[227] B. Gilbert, et al., *Pyrrolopyrimidine compounds as CDK inhibitors*, in *Google Patents*2014.
[228] C.T. Brain., et al., *Pyrrolopyrimidine compounds as inhibitors of CDK4/6*.WO2011101409 2011.

[229] D. Bradford, et al., FDA approval summary: selpercatinib for the treatment of lung and thyroid cancers with RET gene mutations or fusions, Clin. Cancer Res. 27 (8) (2021) 2130–2135.

[230] V. Subbiah, et al., State-of-the-art strategies for targeting RET-dependent cancers, J. Clin. Oncol. 38 (11) (2020) 1209–1221.

[231] K. Sidhom, P.O. Obi, A. Saleem, A review of exosomal isolation methods: is size exclusion chromatography the best option? Int. J. Mol. Sci. 21 (18) (2020) 6466 1–19.

[232] A. Markham, Selpercatinib: first approval, Drugs 80 (11) (2020) 1119–1124.

[233] J. Rayadurgam, et al., Palladium catalyzed C–C and C–N bond forming reactions: an update on the synthesis of pharmaceuticals from 2015–2020, Org. Chem. Front. 8 (2) (2021) 384–414.

[234] J.A.C. Verkouteren, et al., Epidemiology of basal cell carcinoma: scholarly review, Br. J. Dermatol. 177 (2) (2017) 359–372.

[235] G. Brancaccio, et al., Sonidegib for the treatment of advanced basal cell carcinoma, Front. Oncol. 10 (2020) 582866.

[236] O. Sanmartín, et al., Sonidegib in the treatment of locally advanced basal cell carcinoma, Actas Dermosifiliogr (Engl. Ed.) 112 (4) (2021) 295–301.

[237] R.L. Carpenter, H.W. Lo, Hedgehog pathway and GLI1 isoforms in human cancer, Discov. Med. 13 (69) (2012) 105–113.

[238] Z. Choudhry, et al., Sonic hedgehog signalling pathway: a complex network, Ann. Neurosci. 21 (1) (2014) 28–31.

[239] P.W. Ingham, Y. Nakano, C. Seger, Mechanisms and functions of hedgehog signalling across the metazoa, Nat. Rev. Genet. 12 (6) (2011) 393–406.

[240] C.B. Burness, Sonidegib: first global approval, Drugs 75 (13) (2015) 1559–1566.

[241] C. Wang, et al., Structure of the human smoothened receptor bound to an antitumour agent, Nature 497 (7449) (2013) 338–343.

[242] E.H. Epstein, Basal cell carcinomas: attack of the hedgehog, Nat. Rev. Cancer 8 (10) (2008) 743–754.

[243] J. Rodon, et al., A phase I, multicenter, open-label, first-in-human, dose-escalation study of the oral smoothened inhibitor Sonidegib (LDE225) in patients with advanced solid tumors, Clin. Cancer Res. 20 (7) (2014) 1900–1909.

[244] M. Zollinger, et al., Absorption, distribution, metabolism, and excretion (ADME) of ^{1}C-sonidegib (LDE225) in healthy volunteers, Cancer Chemother. Pharmacol. 74 (1) (2014) 63–75.

[245] V. Goel, et al., Population pharmacokinetics of sonidegib (LDE225), an oral inhibitor of hedgehog pathway signaling, in healthy subjects and in patients with advanced solid tumors, Cancer Chemother. Pharmacol. 77 (4) (2016) 745–755.

[246] S. Pan, et al., Discovery of NVP-LDE225, a potent and selective smoothened antagonist, ACS Med. Chem. Lett. 1 (3) (2010) 130–134.

[247] B. Hu, et al., An efficient synthesis of Erismodegib, J. Chem. Res. 38 (1) (2014) 18–20.

[248] W. Gao, et al., WO2008/154259A1, 2007.

[249] S. Pan, et al., Discovery of NVP-LDE225, a potent and selective smoothened antagonist, ACS Med. Chem. Lett. 1 (2010) 130–134.

[250] A. Fritze, K. Corcelle, and M.E. Grubesa, WO2011009852A2, 2011.

[251] J.S. Bajwa, et al., WO2010033481A1, 2010.

[252] B.S. Takale, et al., An environmentally responsible 3-pot, 5-step synthesis of the antitumor agent sonidegib using ppm levels of Pd catalysis in water, Green Chem. 21 (23) (2019) 6258–6262.

[253] S.M. Wilhelm, et al., BAY 43-9006 exhibits broad spectrum oral antitumor activity and targets the RAF/MEK/ERK pathway and receptor tyrosine kinases involved in tumor progression and angiogenesis, Cancer Res. 64 (19) (2004) 7099–7109.

[254] F. Carlomagno, et al., BAY 43-9006 inhibition of oncogenic RET mutants, J. Natl. Cancer Inst. 98 (5) (2006) 326–334.

[255] C. Yu, et al., The role of Mcl-1 downregulation in the proapoptotic activity of the multikinase inhibitor BAY 43-9006, Oncogene 24 (46) (2005) 6861–6869.

[256] G. Salvatore, et al., BRAF is a therapeutic target in aggressive thyroid carcinoma, Clin. Cancer Res. 12 (5) (2006) 1623–1629.

[257] Y.S. Chang, et al., Sorafenib (BAY 43-9006) inhibits tumor growth and vascularization and induces tumor apoptosis and hypoxia in RCC xenograft models, Cancer Chemother. Pharmacol. 59 (5) (2007) 561–574.

[258] S. Kim, et al., Sorafenib inhibits the angiogenesis and growth of orthotopic anaplastic thyroid carcinoma xenografts in nude mice, Mol. Cancer Ther. 6 (6) (2007) 1785–1792.

[259] L. Liu, et al., Sorafenib blocks the RAF/MEK/ERK pathway, inhibits tumor angiogenesis, and induces tumor cell apoptosis in hepatocellular carcinoma model PLC/PRF/5, Cancer Res. 66 (24) (2006) 11851–11858.
[260] B. Escudier, F. Worden, M. Kudo, Sorafenib: key lessons from over 10 years of experience, Expert Rev. Anticancer Ther. 19 (2) (2019) 177–189.
[261] R. Sawhney, F. Kabbinavar, Angiogenesis and angiogenic inhibitors in renal cell carcinoma, Curr. Urol. Rep. 9 (1) (2008) 26–33.
[262] D. Strumberg, et al., Phase I clinical and pharmacokinetic study of the Novel Raf kinase and vascular endothelial growth factor receptor inhibitor BAY 43-9006 in patients with advanced refractory solid tumors, J. Clin. Oncol. 23 (5) (2005) 965–972.
[263] M. Fernández, et al., Angiogenesis in liver disease, J. Hepatol. 50 (3) (2009) 604–620.
[264] Y. Ito, et al., Activation of mitogen-activated protein kinases/extracellular signal-regulated kinases in human hepatocellular carcinoma, Hepatology 27 (4) (1998) 951–958.
[265] F. Pitoia, F. Jerkovich, Selective use of sorafenib in the treatment of thyroid cancer, Drug Des. Dev. Ther. 10 (2016) 1119–1131.
[266] R.M. Tuttle, et al., Serum vascular endothelial growth factor levels are elevated in metastatic differentiated thyroid cancer but not increased by short-term TSH stimulation, J. Clin. Endocrinol. Metab. 87 (4) (2002) 1737–1742.
[267] G.M. Keating, Sorafenib: a review in hepatocellular carcinoma, Target Oncol. 12 (2) (2017) 243–253.
[268] C. Lathia, et al., Lack of effect of ketoconazole-mediated CYP3A inhibition on sorafenib clinical pharmacokinetics, Cancer Chemother. Pharmacol. 57 (5) (2006) 685–692.
[269] P.R. Muddasani, V.C. Nannapaneni, Novel process for the preparation of Sorafenib, WO 2009/054004 A2, 2009.
[270] C. Randrup Hansen, et al., Effects and side effects of using Sorafenib and Sunitinib in the treatment of metastatic renal cell carcinoma, Int. J. Mol. Sci. 18 (2) (2017) 461.
[271] L. Zhang, et al., Convenient synthesis of sorafenib and its derivatives, Synth. Commun. 41 (21) (2011) 3140–3146.
[272] D. Bankston, et al., A scaleable synthesis of BAY 43-9006: a potent Raf kinase inhibitor for the treatment of cancer, Org. Process Res. Dev. 6 (6) (2002) 777–781.
[273] M. Wang, et al., Synthesis, biological evaluation and docking studies of Sorafenib derivatives N-(3-fluoro-4-(pyridin-4-yloxy)phenyl)-4(5)-phenylpicolinamides, Med. Chem. 13 (2) (2017) 176–185.
[274] W. Wang, et al., Synthesis, activity and docking studies of phenylpyrimidine-carboxamide Sorafenib derivatives, Bioorg. Med. Chem. 24 (23) (2016) 6166–6173.
[275] M. Khandan, et al., Synthesis and cytotoxic evaluation of some novel quinoxalinedione diarylamide sorafenib analogues, Res. Pharm. Sci. 13 (2) (2018) 168–176.
[276] R.M. Sbenati, et al., Design, synthesis, biological evaluation, and modeling studies of novel conformationally-restricted analogues of sorafenib as selective kinase-inhibitory antiproliferative agents against hepatocellular carcinoma cells, Eur. J. Med. Chem. 210 (2021) 113081.
[277] M. Wang, et al., Design, synthesis and antitumor activity of Novel Sorafenib derivatives bearing pyrazole scaffold, Bioorg. Med. Chem. 25 (20) (2017) 5754–5763.
[278] M. Wang, et al., Design, synthesis and activity of novel sorafenib analogues bearing chalcone unit, Bioorg. Med. Chem. Lett. 26 (22) (2016) 5450–5454.
[279] J.E. Frampton, N. Basset-Séguin, Vismodegib: a review in advanced basal cell carcinoma, Drugs 78 (11) (2018) 1145–1156.
[280] G.M. Keating, Vismodegib: in locally advanced or metastatic basal cell carcinoma, Drugs 72 (11) (2012) 1535–1541.
[281] F. Meiss, H. Andrlová, R. Zeiser, Vismodegib, Recent Results Cancer Res. 211 (2018) 125–139.
[282] F. Meiss, R. Zeiser, Vismodegib, Recent Results Cancer Res. 201 (2014) 405–417.
[283] R.A. Graham, et al., Pharmacokinetics of hedgehog pathway inhibitor vismodegib (GDC-0449) in patients with locally advanced or metastatic solid tumors: the role of alpha-1-acid glycoprotein binding, Clin. Cancer Res. 17 (8) (2011) 2512–2520.
[284] P.M. Lorusso, et al., Pharmacokinetic dose-scheduling study of hedgehog pathway inhibitor vismodegib (GDC-0449) in patients with locally advanced or metastatic solid tumors, Clin. Cancer Res. 17 (17) (2011) 5774–5782.
[285] T. Lu, et al., Semi-mechanism-based population pharmacokinetic modeling of the hedgehog pathway inhibitor vismodegib, CPT Pharmacom. Syst. Pharmacol. 4 (11) (2015) 680–689.
[286] R. Angelaud, et al., Manufacturing development and genotoxic impurity control strategy of the hedgehog pathway inhibitor vismodegib. Org. Process Res. Dev. 20 (8) (2016) 1509-1519.

CHAPTER

13

Recent developments in the synthesis of pyridine analogues as a potent anti-Alzheimer's therapeutic leads

Aluru Rammohan[a], Baki Vijaya Bhaskar[b] and Grigory V. Zyryanov[a,c]

[a]Department Organic and Biomolecular Chemistry, Ural Federal University, 19 Mira, Ekaterinburg 620002, Russian Federation [b]Department of Biochemisrty, University of Nebraska, Lincoln, Nebraska, United States [c]I. Ya. Postovskiy Institute of Organic Synthesis, UB of the RAS, Yekaterinburg, Russian Federation

13.1 Introduction

Alzheimer's disease (AD) is a type of neurological disorder that mainly affects memory loss and recorded 60–80% dementia cases. According to the WHO, 2016 estimates that more than 47.5 million people will be living with dementia, with 75.6 million infected by the end of 2030 [1,2]. The brains of AD patients undergo significant changes over time because of damage in brain nerve cells related to regions like learning, thinking, and cognitive functions (memory) and subsequent experiences memory loss and speech damage [3,4]. As the disease progression neurons in the brain parts are started to damage and hence bodily functions such as walking, and swallowing are also affected [4]. In this situation, they start to face difficulties even in common activities such as trouble to recalling earlier conversations, names or events and it is termed as apathy. Advanced symptoms include communication impairment, disorientation, confusion, poor judgment, behavioral changes, and finally difficulty in speaking, swallowing, and walking (Fig. 13.1). Therefore, AD patients are become bed-bound and need help around the clock and can eventually lead to malignancy. The hallmark pathologies of Alzheimer's disease are the accumulation of the protein fragment β-amyloid (plaques) outside neurons in the brain and twisted strands of the protein tau (tangles) inside neurons [5,6]. Also, these

Recent Developments in the Synthesis and Applications of Pyridines.
DOI: https://doi.org/10.1016/B978-0-323-91221-1.00009-9

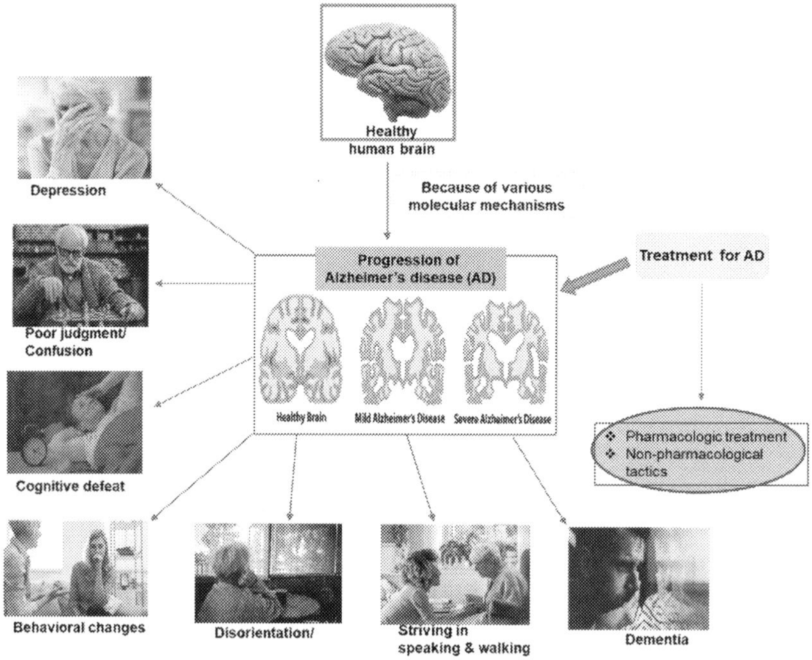

FIGURE 13.1 Pathogenesis of AD is associated symptoms and their therapeutic approaches.

changes are accompanied by the death of neurons and damage to brain tissue. Alzheimer's is a slowly progressive brain disease and associated with several disorders like cerebrovascular problems, frontotemporal lobar degeneration (FTLD), Lewy body disease (DLB), hippocampal sclerosis (HS), and Parkinson's disease (PD) [7].

Further, the pathogenesis of AD is associated with *various molecular mechanisms* as follows,

1. *β-Amyloid (Aβ) hypothesis:* Amyloid precursor protein (APP) is a transmembrane protein that the key protein undergoes misguided proteolytic cleavage by secretase enzymes and produces a huge amount of β-amyloid and then leads to the formation of plaques between the neuronal cells [5,6]. Ultimately, these plaques can trigger an array of cascade that can lead to synaptic damage and neuron loss and are found to be a key hallmark of AD.
2. *Tau hypothesis:* The tau protein plays a major role in the polymerization and stability of microtubules [8]. Tau protein is highly expressed in the mammalian brain and predominantly localized in the axon to perform transportations. Tau proteins undergo several post-translation modifications such as phosphorylation, acetylation, glycosylation, nitration, glycation, methylation, ubiquitination, sumoylation, truncation, and prolyl isomerization [9]. In contrast, hyperphosphorylated tau proteins accumulated in the paired helical filaments of neurofibrillary tangles are identified within the brains of patients with AD as well as in the cerebrospinal fluid (CSF). Therefore, the hyperphosphorylated tau proteins in the CSF proposed to be biomarkers for the identification of AD.
3. *Cholinergic hypothesis*: In the cholinergic hypothesis, the choline acetyltransferase synthesizes neurotransmitter, that is, acetylcholine (*ACh*), which is playing a prominent role in

the regulation of cognitive functions, memory, and learning in the brain [10]. In contrast, the catalase (CAT) levels were found to be low in AD-affected brains and thus, the *ACh* role in AD brain came to light for better understanding. Furthermore, reduced CAT levels in hippocampal and neocortical regions were observed in which neurodegeneration occurs during AD. Apart from this, hurdles have been identified with cholinergic system as numerous cellular events like choline transport, expression of nicotinic and muscarinic, releasing of *ACh*, and axonal transport [11]. Though the researcher thought replacement of cholinergic therapy would be the primary remedy for AD.

4. *Neuroinflammation:* Neuroinflammation is defined as an inflammatory event(s) in nervous tissue triggered by viral or bacterial invasion, toxic substances, and trauma. The version of neuroinflammation-triggered AD has been enriched during past 20 years, and major cells were revealed, such as neuronal cells, microglia, and astrocytes, which are involved in series of inflammatory cascade in the central nervous system [12,13]. Glial cells or neuroglial cells are non-neuronal cells which associate with brain matter by forming microarchitecture and work with astrocytes. In general, microglia is in the resting state in normal or healthy brains, while stimuli trigger it switched to classical (M1) or alternative state (M2) [14]. The M1 state microglia secrets proinflammatory cytokines such as tumor necrosis factor-α (TNF-α), interleukin-6 (IL-6), chemokines, ROS, interleukin-1β (IL- 1β), and acute phase proteins whereas activated M2 state microglia released anti-inflammatory cytokines to downregulate the effects of neuroinflammation. In contrast, the activated M1 microglia surrounded Aβ-plaques and secretes proinflammatory response to clear Aβ from the AD brains, however, during this event key pathways activated such as activation of complement system, cytokines release and occur phagocytosis which exhibits detrimental impact on adjoining neuronal cells by disrupting synaptic plasticity, enhancing tauopathies, and hindrance the microglial clearance of Aβ [14,15]. Ultimately, this event can cause severe neuron damage in AD.

Treatment of Alzheimer disease and AD's dementia: The current treatment strategies of AD were broadly classifieds into two groups.

1. *Pharmacologic treatment:* To date, no specific vigorous pharmacological treatment is available for the treatment of AD's dementia. However, US FDA-approved five drugs (Fig. 13.2) such as rivastigmine, galantamine, donepezil, memantine, and memantine in combination with donepezil for the primary treatment of AD [7]. The stated clinical drugs can instantly enhance the cognitive function by elevating the neurotransmitters in the brain. Despite the fact, the impact of permitted drugs is different from person to person and exerts lower residence time. Moreover, currently there is no specific drug(s) accessible for the treatment of behavioral and psychiatric symptoms of moderate/advanced Alzheimer's dementia. Whereas the drug memantine dysfunction numerous receptors from excess stimulation in the brain that could lead to nerve damage [16]. Therefore, considering all the circumstances of AD treatment several other approaches were also considered as crucial drug targets for the treatment of AD.

Consequently, the current pharmacological therapies include (a) cholinergic inhibitors, (b) *N*-methyl D-aspartate (NMDA) receptor antagonism, (c) amyloid-targeted therapy (β,γ-secretase inhibitors & Aβ immunotherapy), (d) tau-targeted therapy, (e) antioxidant therapy, and (f) metal chelators, etc. [7,17].

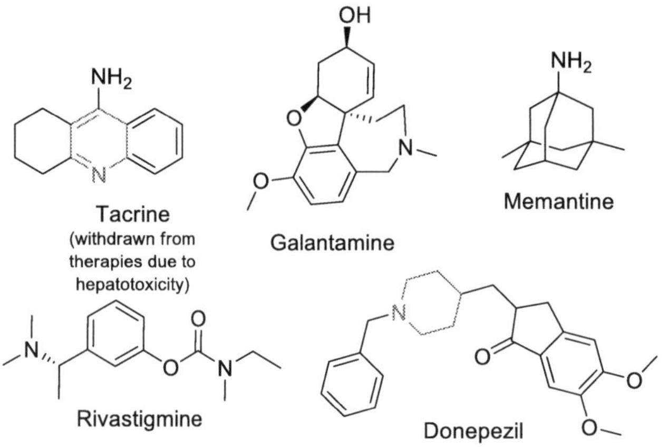

FIGURE 13.2 FDA-approved currently available clinically persuasive key Alzheimer's disease drugs.

2. *Nonpharmacologic therapy:* Most importantly nonpharmacologic therapies are involved in the improvement of cognitive function, life quality, and to ensure perfection in daily chores as well as decrease behavioral symptoms such as depression, apathy, wandering or confusing (about their location), sleep disturbances, agitation, and aggression [18,19]. A recent review and analysis of nonpharmacologic treatments for agitation and aggression in people with dementia concluded that nonpharmacologic interventions seemed to be more effective than pharmacologic interventions for reducing aggression and agitation [19]. Nonpharmacologic therapies include computerized memory training, listening to favorite music to stir recall, and using special lighting to lessen sleep disorders.

13.2 Role of pyridine in drug discovery

Pyridine is an imperative nitrogenous six-membered heteroaromatic basic compound with molecular formula C_6H_5N that resembles with benzene structure. Pyridine is relatively electron-deficient and hence less susceptible to electrophilic substitution reactions than benzene [20,21]. While, it is readily involved in nucleophilic substitutions at C-2 and C-4 positions and is familiar for its metalation reactions over organometallic bases [21,22]. The basic nature of pyridine is due to its readily accessible free lone-pair of nitrogen as like in tertiary amines for the formation of pyridine-N-oxide and pyridinium salts. Another important aspect of pyridine-based scaffolds is connected with the possibility of enhancing their solubility; drug lipophilicity, and stability via the converting of pyridine nitrogen atom(s) to pyridinium quaternary salt(s) or N-oxide(s) [23]. Also, pyridine is an important key template for the existence of several natural molecules such as nicotinic acid (vitamin B3), β-picoline, quinoline, isoqunoline, trigonelline, and coenzymes (e.g., NADP) etc. [24–26]. Pyridine and pyridine derivatives such as piperidine, pyrimidine, quinolone, isoquinoline, pyridone, bipyridine, nicotine derivatives, and heteroring-fused pyridine (e.g., imidazo[1,2-a]pyridine) analogues (Fig. 13.3) have innumerable applications in pharmaceutical, agricultural, and material science sectors [27–29]. Principally, among all the nitrogen-based heterocycles pyridines and their analogues are holding a leading role in drug developments because they were renowned as

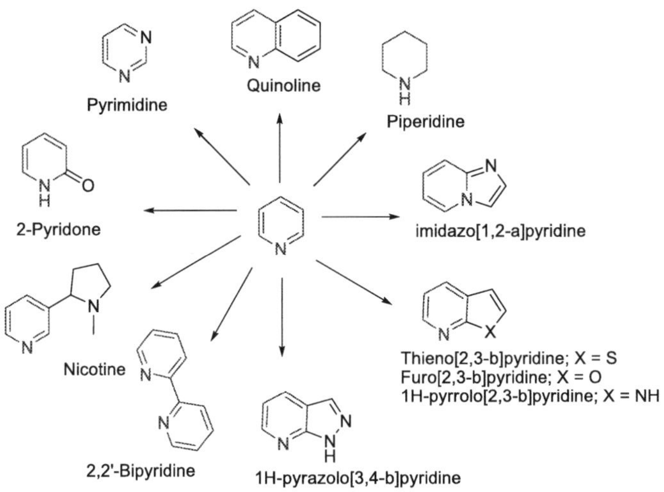

FIGURE 13.3 Few important pyridine equivalents that have leading role in the drug development.

antimicrobial, anticancer, antioxidant, antidiabetic, insecticidal, and neuroprotective agents [27,30].

In contrast, the synthesis of pyridine analogues was very convenient than other six-membered heterocycles by using renowned synthesis methods such as Hantzsch synthesis, Emil Knoevenagel process, Aleksei Chichibabin, Bonnemann cyclization, Gattermann–Skita synthesis, and Ciamician–Dennstedt rearrangement [31,32]. The industrial-scale production of pyridine analogues became easier after the invention of so called "Chichibabin reaction (Chichibabin synthesis)" by the famous Russian chemist Aleksei Chichibabin through inexpensive staring materials and a more efficient route that is still practiced [33]. At present, a great number of organometallic catalysts are available to develop efficient approaches to the diverse pyridine scaffolds without tedious procedures [34]. Accordingly, the recent methodologies involving transmetalation, cross-coupling, ring-closing metathesis (RCM), aza-Diels-Alder 6π-electrocyclization reactions are the eminent synthesis procedures to afford the quantifiable functionalized pyridines derivatives [34,35].

13.3 Pyridine analogues as Alzheimer's disease (AD) drug agents

Pyridine/pyridine analogues have been acknowledged as promising therapeutic agents for the treatment of Alzheimer's disease as well as other neurodegenerative disorders [36,37]. For instance, among the FDA-approved clinical drugs the Donepezil (piperdine) and Tacrine (amino-pyridine) prominent pyridine analogues were valuable in the treatment of AD [38,39]. These approved pyridine analogues have played a key role in the treatment of mild to moderate AD patients through blocking the active site of plaque deposition [40]. Further, several long-term studies have been revealed that pyridine analogues were showed significant activities against the pathogenesis of AD by regulating the β-amyloid plaque (AB), cholinesterase's (*ChEs*), α,β,γ-secretase, tau phosphorylation, and inflammatory toxicity, etc. [40,41]. Thus, it expands the importance of pyridine analogues for the treatment of AD and hence further

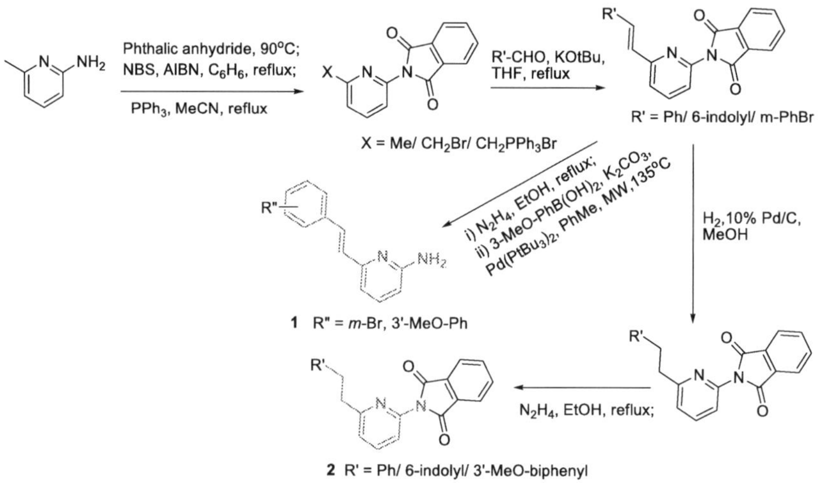

SCHEME 13.1 Synthesis of 2-aminopyridine analogues as BACE-1 inhibitors.

studies are recommended for the development of potent anti-Alzheimer drugs. Therefore, the present section describes the role of pyridine analogues and their recent developments in the management of AD in the following ladders.

13.3.1 Simple pyridine analogues

Recently, the advances of pyridines both as small molecules and scaffolds of traditional natural products have been presenting remarkable tactics against the pathogenies of Alzheimer's disease (AD) through regulating the metabolic dysfunctions like accumulation of β-amyloid, cholinesterase (ChEs) secretion, tau phosphorylation, and other neurodegenerative disorders [42,43]. Likewise, simple pyridine derivatives like aminopyridine analogues [44,45], 2-methylpyridine derivatives [46], pyridine sulfonamides [47], pyridine-2-yl-urea [48], and pyridine-2-carboxaldehyde derivatives [49] were also found to be significant key molecules in the treatment of AD. Congreve et al. [44] have proposed a series of 2-aminopyridine analogues as prominent aspartyl protease enzyme β-secretase (BACE-1) inhibitors which are promising for treating AD. In the primary synthesis attempt they have proposed (*E*)-6-styrylpyridin-2-amine derivatives **1** and their dihydro derivatives **2** through sequential region-selective reactions of 6-methylpyridin-2-amine (Scheme 13.1) with 0.02–2 mM BACE1 inhibitory concentrations (IC_{50}) and a ligand efficacy (LE) greater than 0.25. Further, the synthetic attempts occasioned more potent BACE1 inhibitors based on 3-substituted 2-aminopyridine derivatives **3** and **4** (Scheme 13.2) with IC_{50} 0.69–4.2 and 24–100 μM, and, respectively, they together enhanced LE ~ 0.23–0.26. These structure-based strategies revealed that the aminopyridine showed substantial interactions with the active site of BACE1 and beneficial as an inhibitor(s) of β-secretase.

Further, in his study Singh et al. [45] evaluated that N-(pyridine-4-yl)-hydrazinecarboxamide derivatives **5** (Scheme 13.3) showed important acetylcholinesterase (*AChE*) inhibition and cognitive enhancing tactics through regulating the scopolamine-induced amnesia.

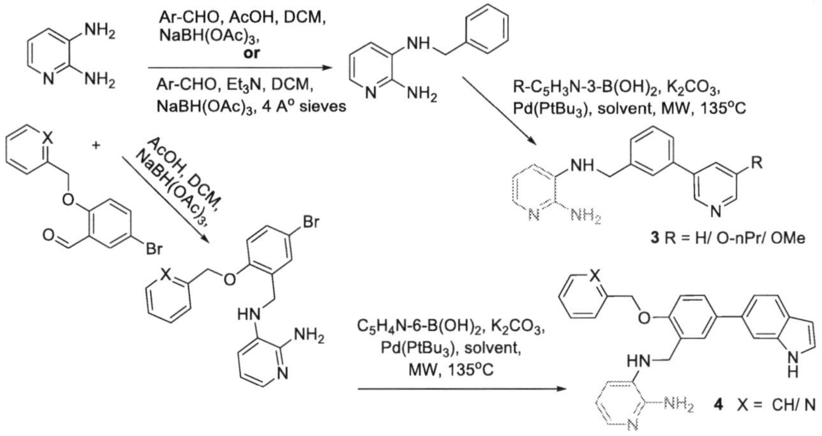

SCHEME 13.2 Synthesis of 3-substituted 2-aminopyridine derivatives as BACE-1 inhibitors.

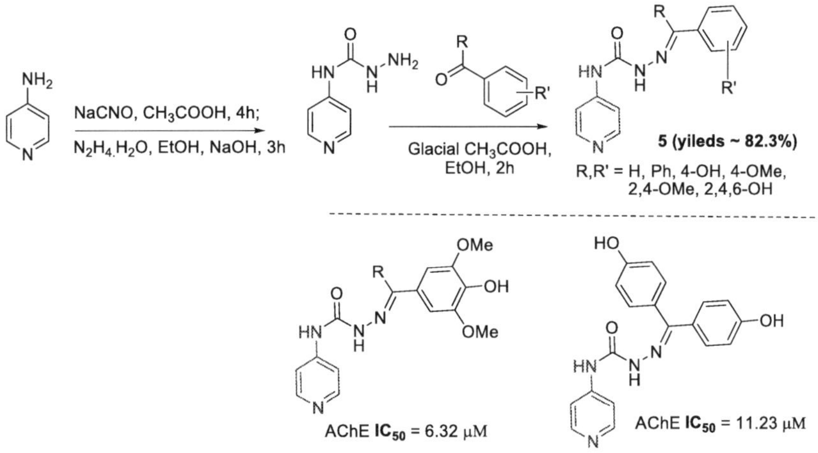

SCHEME 13.3 Synthesis of N-(pyridine-4-yl)hydrazinecarboxamide derivatives as *AChE* inhibitors.

Likewise, Chen et al. [46] proposed a series of 2-methylpyridine biarylamide analogues **6** (Scheme 13.4) for treating Alzheimer's disease (AD). Based on the structural analysis and molecular docking the above mentioned studies revealed that these 2-methylpyridine analogues were useful for treating AD through inhibiting the γ-secretase activity as well as dropping the Aβ42 levels in the cerebrospinal fluid (CSF) with the enlightening the cell potency.

The diabetes mellitus (DM) is a complex syndrome peculiarly observed with Alzheimer's disease (AD) because of impaired insulin levels and glucose hemostasis showing abnormal neuronal and synaptic functions in the brain [47,50]. Therefore, several researchers have focused on the development and evaluation of therapeutic agents for AD-associated DM. In this context, Riaz et al. [47] by means of facile sulfonation procedure in acceptable to

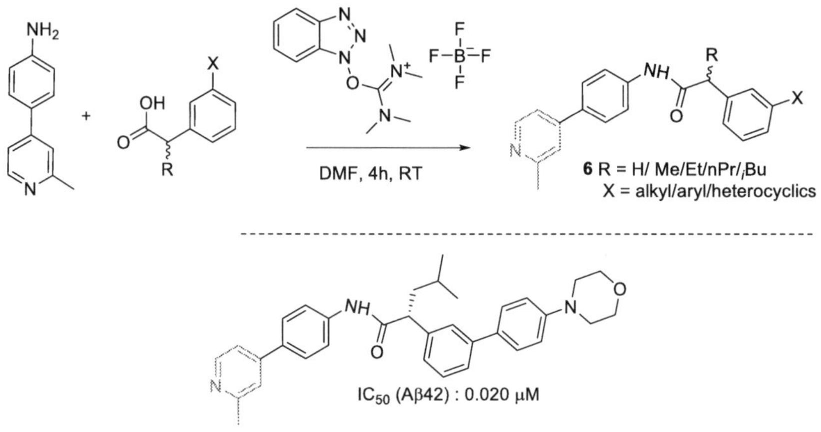

SCHEME 13.4 Synthesis of 2-methylpyridine biarylamide analogues as γ-secretase modulators (GSMs).

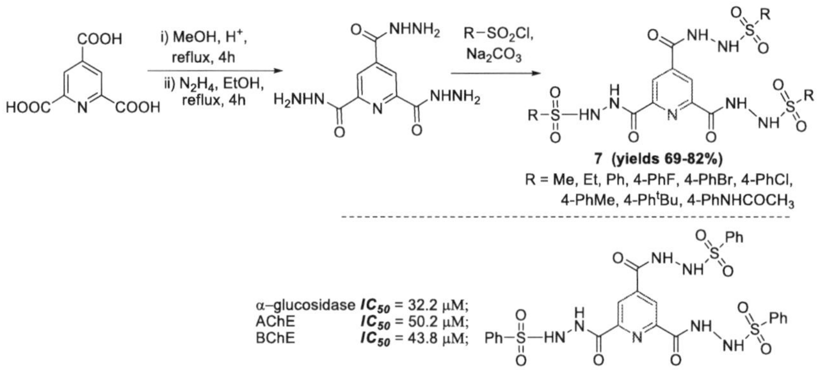

SCHEME 13.5 Synthesis of pyridine 2,4,6-tricarbohydrazide analogues as inhibitors of α-glucosidase and cholinesterase (*ChEs*).

good yields 69–82% developed a series of novel pyridine 2,4,6-tricarbohydrazide analogues **7** (Scheme 13.5) as a significant inhibitors of α-glucosidase and cholinesterase (*ChEs*). Further, the biological screening and molecular docking studies revealed that these pyridine tris-sulfonamides were displayed dual enzyme inhibition, and, thus, they have great potential for the development of future lead agents against diabetes-associated neurodegenerative disorders.

Likewise, Elkamhawy et al. [48] reported a new series of *N*-2-(3-benzyloxypyridyl)urea derivatives **8–10** (Scheme 13.6) with a valuable potency against β-amyloid (Aβ)-induced mitochondrial membrane depolarization. The biological evaluation and in silico studies suggested that the stated pyridyl-urea derivatives were promising lead compounds in the development of potential clinical drug candidates for the treatment Alzheimer disease through deteriorating the Aβ-induced oxidative cell signaling and mitochondrial dysfunctions. Another study [49] stated that the target compound, (E)-*N'*-(pyridin-2-ylmethylene)isonicotinohydrazide **11**

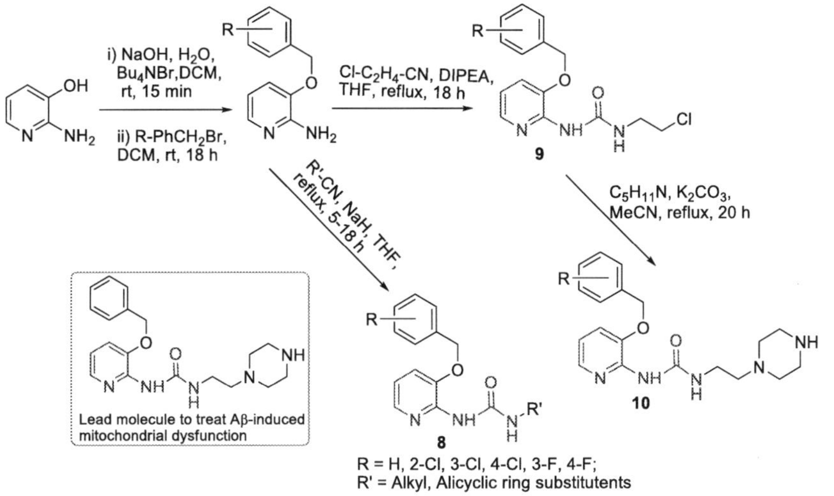

SCHEME 13.6 Synthesis of *N*-2-(3-benzyloxypyridyl)urea derivatives as significant inhibitors of Aβ-induced mitochondrial dysfunction.

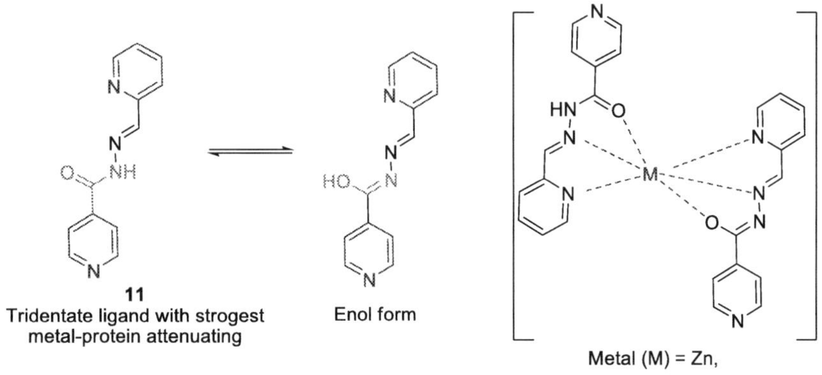

FIGURE 13.4 The important (E)-*N*′-(pyridin-2-ylmethylene)isonicotinohydrazide with potential anti-Alzheimer's activity through metal–protein attenuating property.

(Fig. 13.4), has promising anti-AD tactics by metal–protein attenuating property and decline the Aβ-toxicity and hence, valuable for treating this neurodegenerative disease.

13.3.2 Bipyridine analogues

Bipyridines are another important key molecules that have impressive applications across diverse areas like pharmaceutical chemistry [51,52], material science [53–55], and functionalization of complex organic molecules [56,57]. Moreover, bipyridines have been accredited as a functional metal-chelating scaffolds and were potential for treating Alzheimer's disease and other neurodegenerative disorders [58,59]. Recently, Ji et al. [60] demonstrated

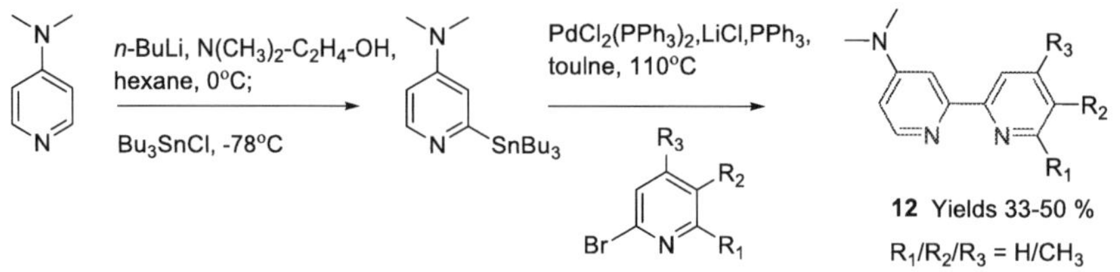

SCHEME 13.7 Development of bipyridine analogues with latent inhibition of metal-induced Aβ-plaque.

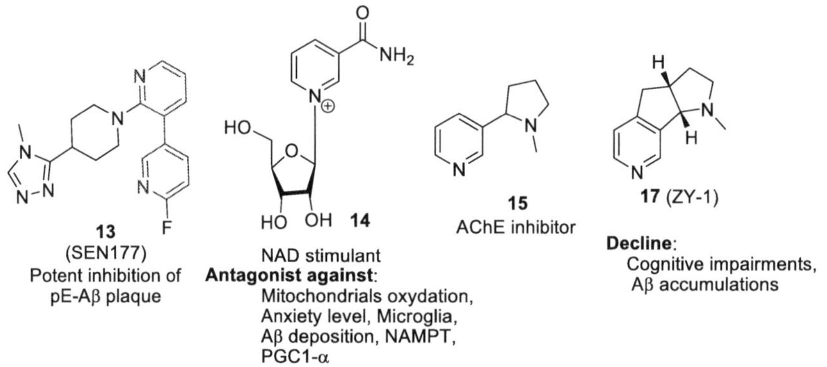

FIGURE 13.5 Few pyridine derivatives such as a novel 3,3′-bipyridine analogue 13 termed as human glutaminyl cyclase (hQC, SEN177), a naturally occurring pyridine nucleoside, that is, nicotinamide riboside 14 and nicotine hybrid 15 and 17 were showed potent restraint of pathogenesis of AD through multiple assets and can be considered as credible drug lead for neurodegenerative diseases.

interesting bipyridine derivatives **12**, which were obtained in acceptable yields of 33–50% through transition metal-catalyzed approach. Further, the stated bipyridines bearing electron-donating groups like diethylamino and methyl with proper substitutions (Scheme 13.7) showed potential inhibition of metal-induced amyloid-β (Aβ) accumulations as well as metal chelating properties, and both of these properties/activities are potentially usable in obvious strategies in the treatment/prevention of Alzheimer's disease. Further, another study [61] revealed that 6′-fluoro-2-(4-(4-methyl-4*H*-1,2,4-triazol-3-yl)piperidin-1-yl)-3,3′-bipyridine **13** (termed as SEN177, Fig. 13.5) inhibition of human glutaminyl cyclase (*hQC*) which play a key role in the pathogenesis of AD through the concomitant accumulation of *N*-pyroglutamic acid peptides (pE-Aβ).

13.3.3 Nicotine analogues

Nicotine, a naturally occurring pyrrolidine–pyridine chiral alkaloid, has a pharmacologically great prominence as a therapeutic psychoactive drug [62,63]. Moreover, nicotine and its analogues have been revealed as potential stimulants with anxiolytic efficacy to regulate the cognitive impairments in Alzheimer's disease [64]. For instance, nicotinamide riboside **14** (*syn.* vitamin B3, Fig. 13.5), a naturally occurring pyridine nucleoside, was reported [65] as a pioneering drug candidate stimulating the synthesis of nicotinamide adenine dinucleotide (NAD^+) in the brain. Moreover, nicotinamide riboside is a key neuroprotective molecule in

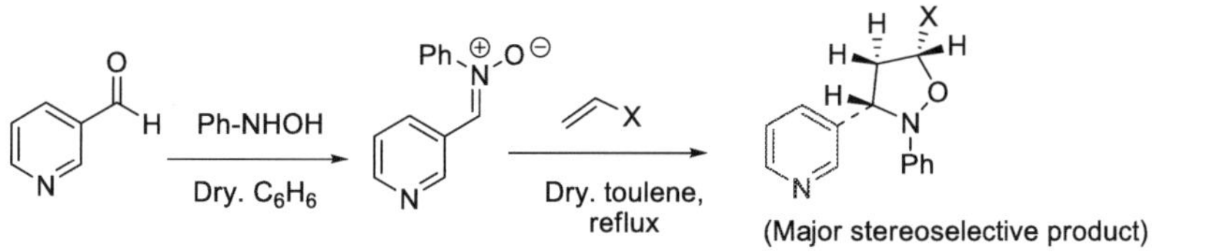

SCHEME 13.8 Development of regio- and stereoselective nicotine derivatives 16 through 1,3-dipolar cycloaddition method.

the stimulation of peroxisome proliferator-activated receptor gamma coactivator 1 (PGC1-α) that augments mitochondrial biogenesis through various oxygen detoxification paths [65,66]. Accordingly, endogenous nicotinamide riboside has great prominence in the treatment of AD through multiple assets like NAD stimulant, regulation of cognitive function, anxiety level, microglia, Aβ-deposition, serum nicotinamide phosphoribosyl transferase (NAMPT), etc. [66].

Likewise, Khurana et al. [67] assessed the feasible nootropic potential and acetylcholine nicotinic agonist activities of nicotine analogues **15** through structure-based theoretical study. Further, another comprehensive study [68] has summarized the positive thoughts on the nicotine analogues as drug targets of acetylcholine receptors (*nAChEs*) to treat AD and other neuronal diseases. Later, Sing et al. [69] have reported a facile regio- and stereoselective synthesis of nicotine derivatives **16** through 1,3-dipolar cycloaddition procedure as described in Scheme 13.8. Moreover, another nicotine analogue **17** (ZY-1, Fig. 13.5) was found to be promising lead compound to decline the cognitive impairments, acetyl choline receptor ($\alpha4\beta2$-nAChRs) agonists, and inhibits Aβ accumulations [70]. Thus, the study revealed that the nicotine analogues [69–71] are beneficial for the treatment of pathogenesis of cognitive decline, choline-associated AD, oxidative stress mechanisms, and other neurodegenerative disorders.

13.3.4 Chromanone and *iso*-chromanone cohesive pyridine analogues

The naturally existing oxygen heterocycles have a stimulating role in medicinal chemistry because still they show great prominence in drug discovery [72,73]. Particularly, chromanone and *iso*-chromanone analogues display a significant role in treating several diseases including Alzheimer's disease [73,74]. Alipour et al. [75] have developed a synthesis of new class chalcone analogues **18** (Scheme 13.9) through condensation of two pharmacologically important chromophores, namely coumarin(s) and pyridine derivatives, followed by quaternization with various aryl halides. Further, bioassay optimization studies endorsed that the coumarin–pyridine analogues **18** were of significant inhibiting activity of both acetylcholinesterase (*AChE*) and butyrylcholinesterase (*BuChE*) progression. Further, another study [76] also discloses the dihydropyrano[3,2-*c*]chromene cohesive pyridinium scaffolds **19**, obtained through conventional nanocatalyzed reaction pathways as described in Scheme 13.10, as potent cholinesterase inhibitors.

Likewise, Xu and his coworkers [77,78] had proposed a two series of donepezil analogues **20** and **21** through simple condensation of 4-isochromanone and pyridine-4-carboxyaldehyde at room temperature followed by quaternization with suitable aryl bromides as depicted in

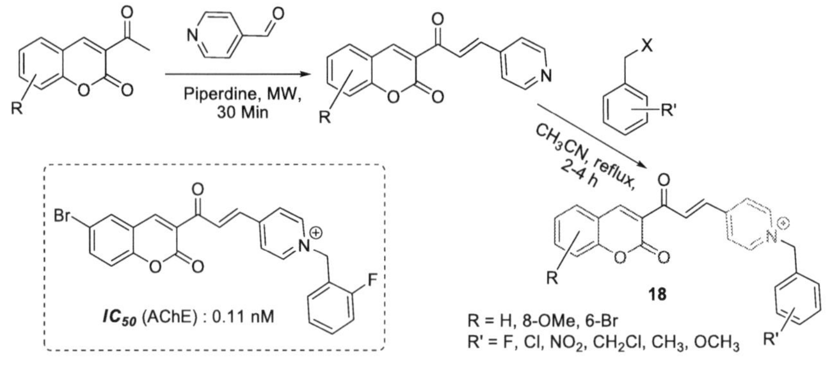

SCHEME 13.9 Synthesis of coumarin–pyridine chalcone analogues with compelling inhibitors of cholinesterase's (*AChE* and *BuChE*).

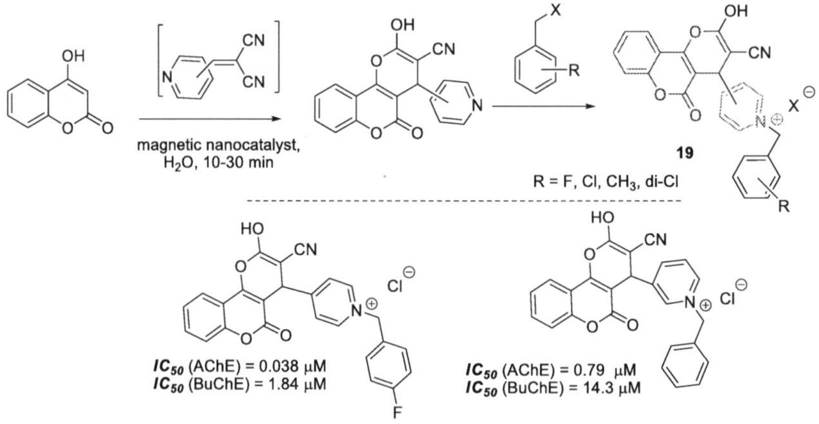

SCHEME 13.10 Synthesis of dihydropyrano[3,2-c]chromenes cohesive pyridium scaffolds as potent cholinesterase inhibitors.

Scheme 13.11. Further, in vitro and in silico studies explicated the pharmacodynamic and pharmacokinetic properties of the target analogues as potent cholinesterase (*AChE* & *BuChE*) inhibitors, which are beneficial in terms of future creating promising therapeutic lead agents for treating AD.

13.3.5 Benzofuran–pyridine analogues

The privileged scaffold benzofuran is a dexterous basic heterocyclic stencil that exist in various natural products and therapeutic drugs such as aurones, coumestans, diptoindonesin G, malibatol A, cannabifuran, etc. [79]. Moreover, benzofuran analogues have been revealed as important AD therapeutic lead templates owing to their effectiveness against various routes of neurotoxicity [80]. Considering the role of benzofuran and pyridine toward AD pathogenesis, Baharloo and his coworkers [81,82] have reported two novel benzofuran-based groups of molecules depicted in Scheme 13.12. The preliminary synthetic attempt of Baharloo et al. [81] has accomplished diverse pyridine integrated aurone analogues **22**

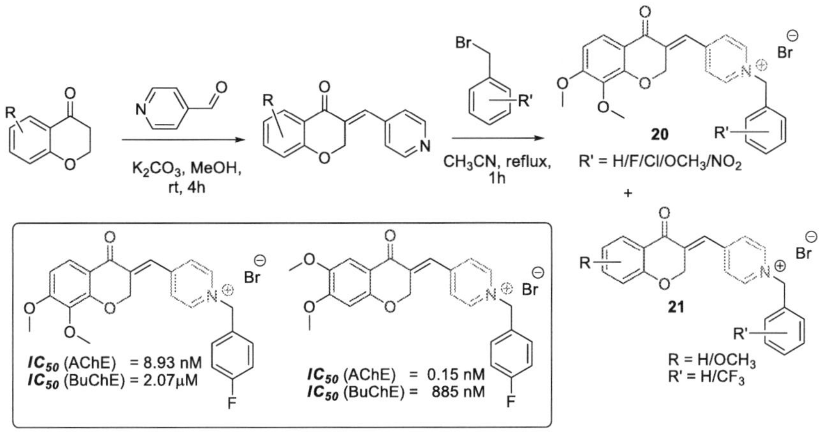

SCHEME 13.11 Synthesis of two series of donepezil analogues with potent cholinesterase (*AChE* & *BuChE*) inhibitors.

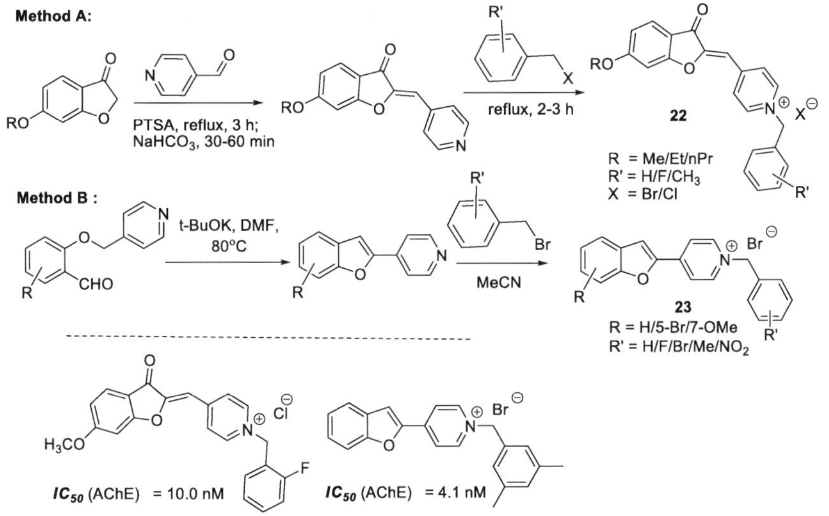

SCHEME 13.12 Synthesis of benzofuran cohesive pyridine analogues as significant inhibitors of acetylcholinesterase (*AChE*).

(Scheme 13.12, Method A) by successive condensation of benzopyran-3(2*H*)-one with pyridine-4-carboxyaldehyde followed by quaternization paths. Later, the same group of authors [82] developed another group of 4-(benzofuran-2-yl)pyridine derivative **23** (Scheme 13.12, Method B) through the cyclization of 2-(pyridin-4-yl)benzaldehyde derivatives in strong basic condition (*t*-BuOK/DMF) followed by *N*-benzylation with various benzyl bromides. Further, the biological eminence studies disclosed that the synthesized benzofuran–pyridine analogues **22** and **23** were found to be significant inhibitors of acetylcholinesterase (*AChE*) enzyme-induced AD.

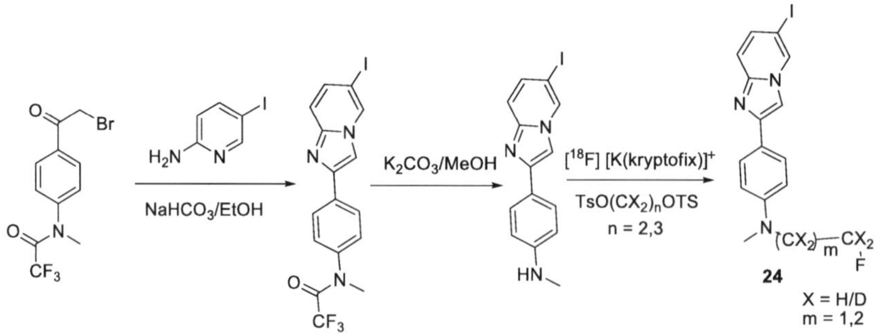

SCHEME 13.13 Synthesis of radioactive F^{18}-labeled imidazo[1,2-a]pyridine analogues as a promising β-amyloid imaging agents.

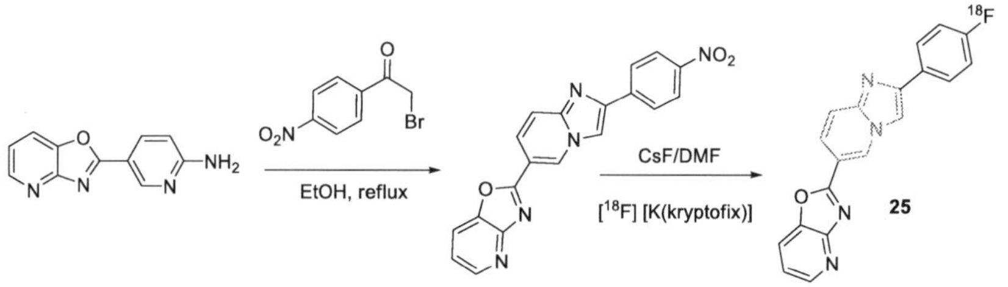

SCHEME 13.14 Synthesis of radioactive F^{18}-labeled imidazo[1,2-a]pyridine analogues as a promising PET tracer for β-amyloid imaging.

13.3.6 Imidazole–pyridine analogues

Among the nitrogen heterocyclic motifs "imidazole" occupies a unique substantial role for the drug development because of its wide range of biological activities like antimicrobial, antidiabetic, antioxidant, antiviral, anti-inflammatory, anticancer, and neuroprotective tactics [83]. Particularly, imidazole-fused pyridines were found to be key molecules for the treatment of various chronic diseases including cancers and Alzheimer's disease [84,85]. Earlier efforts of Cai et al. [86] have demonstrated the synthesis of radioactive F^{18}-labeled imidazo[1,2-*a*]-pyridine analogues **24** for the detection of Aβ-plaque through the positron emission tomography (PET) substantial groups as depicted in Scheme 13.13. Further, the in vivo and structural studies of substantial F^{18}-laballed imidazole analogues were displayed pleasing pharmacokinetics and affinities for imaging Aβ-plaque. Recently, Singh et al. [87] also developed a unique 2[-2-(4-flurophenyl)imidazo[1,2-*a*]pyrine-6-yl]oxazolo[4,5-b]pyridine **25** as a promising β-amyloid imaging agent through conventional reaction procedure with an outstanding yield of 92% (Scheme 13.14). Therefore, the illustrated studies are compassionate stencils for the development of potent radioactive metabolites for treating AD. Further, Nirogi et al. [88] reported various imidazo[1,5-*a*]pyridine derivatives (**26**–**29**) having dissimilar pharmacophores like 4-aminomethylpiperidine **26**, bicyclic aminopiperdine **27**, morpholine **28**, and 1,3,5-oxadiazole **29** as shown in Scheme 13.15. The in vitro studies of projected compounds **26–29** having various substituents, such as alicyclic ring of various sizes, and alkyl

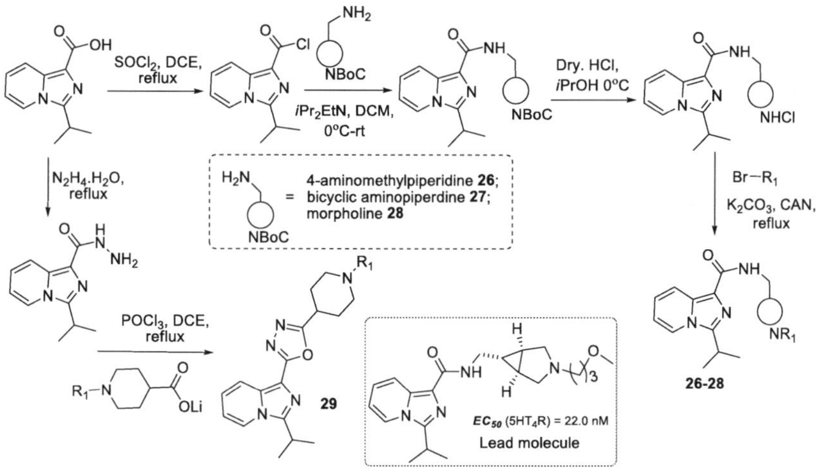

SCHEME 13.15 Synthesis of imidazo[1,5-a]pyridine derivatives as a lead molecules to treat pathogenesis of neurodegenerative disorders.

with various lengths and geometry, were carried to study their activity as partial agonists for 5-hydroxytryptamine-4 receptor (HT_4R) that known to play a key role in maintaining cognitive function. According to the authors, the stated imidiazo[1,5-*a*]pyridine analogues were showed acceptable efficacy as antagonists for brain penetrant 5-HT4R, optimal structure activity relation (SAR), and ADME properties, and, therefore, these compounds are beneficial as lead molecules to treat pathogenesis of neurodegenerative disorders.

Recently, Haghighijoo et al. [89] have also proposed a new class imidazo[1,2-*a*]-pyridines **30** with triazole pharmacophore obtained through Cu(II) mediated click reaction protocol (Scheme 13.16). Further, on optimization of biological activities and structure activity relation (SAR) analysis revealed that the *N*-cyclohexylimidazo[1,2-*a*]pyridine bearing triazole motifs were beneficial for the treating multiple pathogenesis of AD including β-site amyloid precursor protein cleaving enzyme 1 (BACE1), *BuChE* triggered Aβ-plaque, metal imbalance and oxidative-stress induced neurotoxicity, etc. Moreover, the inhibitory assessments and pharmacokinetics profiles suggested that the β-chloropyridine analogues **30** were potent future lead/drug candidates for managing the multiple syndrome, that is, AD.

13.3.7 Pyrazole–pyridine analogues

Pyrazole, a five-membered structural isomer of imidazole, is another imperative heterocyclic pharmacophore that has proficiency as a chelating ligand in numerous metal complexes for electron-transfer reactions, organic synthesis and chemiluminescence applications [90]. Moreover, various pyrazolo[3,4-*b*]pyridine analogues [91] such as cartazolate, etazolate, tracazolate, riociguat, etc. have been practicing to treat various neurological disorders. Barreiro et al. [92] synthesized new class tacrine analogues (**31–33**) by integrating pharmacologically significant pyrazole and pyridine chromophores as depicted in Scheme 13.17. As a primary effort the target pyrazolo[4,3-*d*]pyridine derivatives **31** & **32** were achieved through Lewis

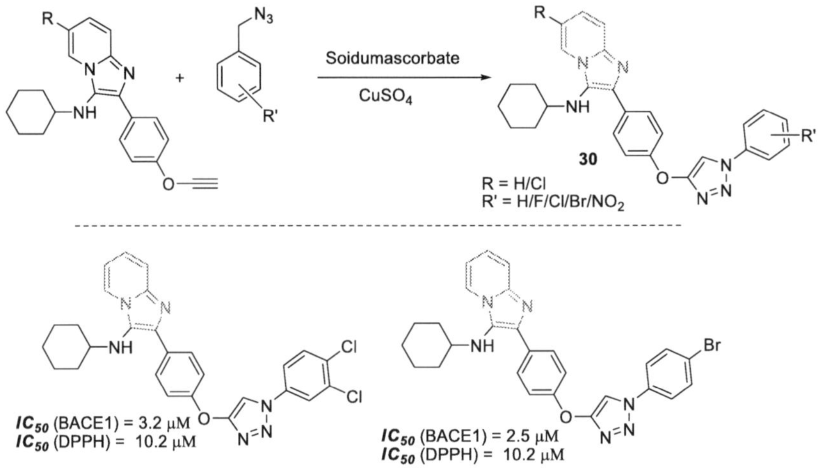

SCHEME 13.16 Synthesis of 1,2,3-triazole cohesive *N*-cyclohexylimidazo[1,2-a]pyridine as a drug candidates for managing the multiple syndrome AD.

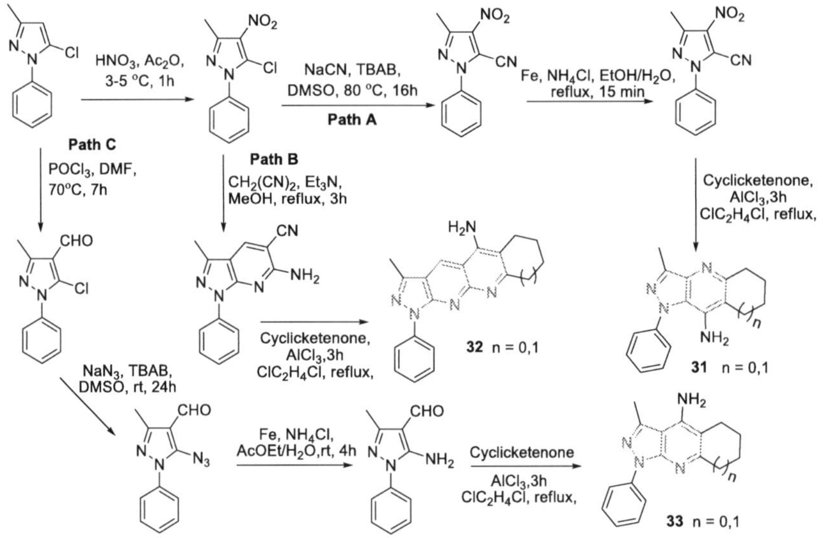

SCHEME 13.17 Synthesis of pyrazolo[4,3-d]pyridine and pyrazolo[4,3-b][1,8] naphthyridine derivatives as cholinesterase inhibitors.

acid-catalyzed cyclization of *O*-amino, nitrile derivatives of 1-phenyl -1*H*-pyrazole with cyclopentanone/cyclohexanone (Scheme 13.17; Path A & B). Similarly, the cyclization of *O*-amino, nitrile derivatives of 1-phenyl-1*H*-pyrazole[3,4-*b*]-pyridine with respective alicyclic compounds occasioned the pyrazolo[4,3-*b*][1,8] naphthyridine derivatives **33** (Scheme 13.17; Path C). Further, in vivo and molecular dynamic studies revealed that naphthyridine

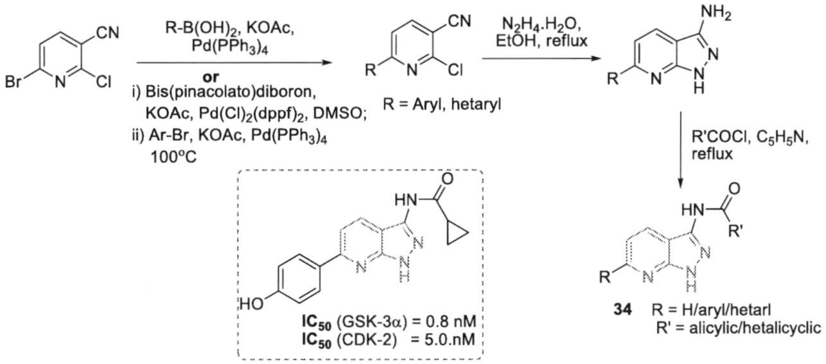

SCHEME 13.18 Synthesis of pyrazolo[3,4-b]pyridines as dual inhibitors of GSK-3 and cyclin dependent kinase-2 (CDK-2)

analogues **33** exhibited substantial total cholinesterase inhibition (IC_{50} 23.2 ± 5.3 and 6.39 ± 0.86 μM) with tolerable selective index (SI).

Similarly, Witherington et al. [93,94] developed a novel series of pyrazolo[3,4-*b*]pyridines **34** as potent inhibitors of glycogen synthase kinase 3 (GSK-3), and these compounds were prominent to treat several physiological disorders including AD. The synthesis of pyrazole–pyridine analogues **34** were achieved in outstanding yields through the cyclization of 3-carbonitrile-pyridin-2-one derivative followed by *N*-acylation of pyrazolo[3,4-*b*]pyridine amines as prescribed in Scheme 13.18. Further, optimization of biological studies resolved that the stated compounds **34** were showed dual inhibition of GSK-3 and cyclin dependent kinase-2 (CDK-2) because of the homology of two kinases. Therefore, these dual-kinase inhibitors are beneficial for treating multiple syndromes like AD, diabetes and cancers, etc.

Likewise, Umar et al. [91] have developed several unique pyrazolo[3,4-*b*]pyridine amides **35** comprising diverse pharmacophores such as triazole, quinoline, piperazine, piperidine, morpholine, etc. (Scheme 13.19). Further in vitro and in silico optimization studies revealed that the synthesized amide analogues **35** were found to be potent inhibitors of cholinesterase (*AChE, BuChE*), Aβ-plaque, oxidative-stress and metal-induced neurotoxicity. Therefore, from the foregoing results, it is assessed that the pyrazolo–pyridine analogues are beneficial future drug agents for managing multiple pathogenesis of AD.

13.3.8 Triazole–pyridine analogues

Triazole nucleus encompassing scaffolds can be achieved certainly, for instance, by click reaction, and thus triazole-bearing hybrids have great prominence in medicinal chemistry owing to diverse bioactivities [95,96]. Moreover, triazole pharmacophore has been revealed as a significant template for the development of Alzheimer's drug leads [97]. Bulut et al. [98] have developed a series of 2,5-di(4*H*-1,2,4-triazol-3-yl)pyridine derivatives **36** in outstanding yields 85–95% through sequential reactions of pyridine-2,5-dicarboxylic acid as described in Scheme 13.20. Further, in vitro bioactivity studies shown that the stated 1,2,4-triazole-3-

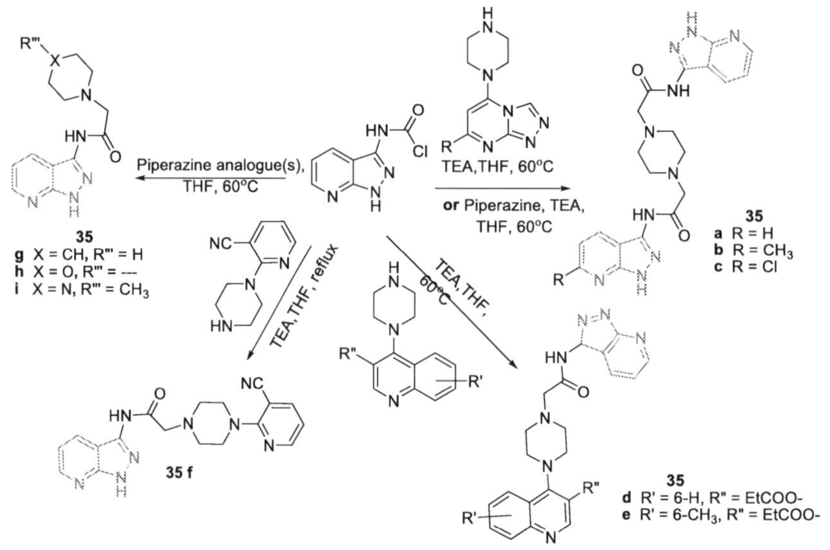

SCHEME 13.19 Synthesis of pyrazolo[3,4-b]pyridine amides 36 comprising diverse pharmacophores as drug agents for multiple syndrome AD.

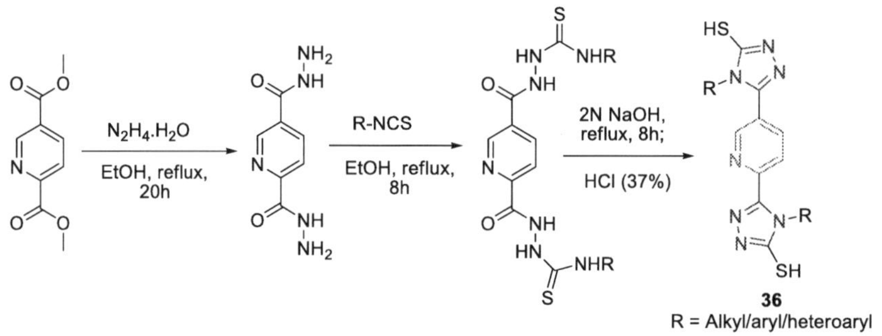

SCHEME 13.20 Synthesis of 1,2,4-triazole-3-thione-pyridines as a carbonic anhydrase (CA) I, II, and AChE ameliorative AD drug targets.

thione-pyridines **36** were demonstrating moderate inhibition of carbonic anhydrase (CA) I, II, and *AChE*, and potent antioxidant profiles.

Likewise, Zribi et al. [99] proposed a novel triazolopyridopyrimidines **37** through a one-pot multicomponent reaction of 2-(5-phenyl-4*H*-1,2,4-triazol-3-yl)acetonitrile with malonitrile and appropriate aldehydes followed by cyclization with formamide as shown in Scheme 13.21. The bioactivity assessment and molecular modeling studies concluded that 6-amino-2,5-diphenyl-[1,2,4]triazolo-[1′,5′:1,6]-pyrido[2,3-*d*]pyrimidine-4-carbonitrile hybrids **37** have substantial *AChE* inhibitory and radical reducing potency and, therefore, can be considered as a promising structural lead for future AD drug developments.

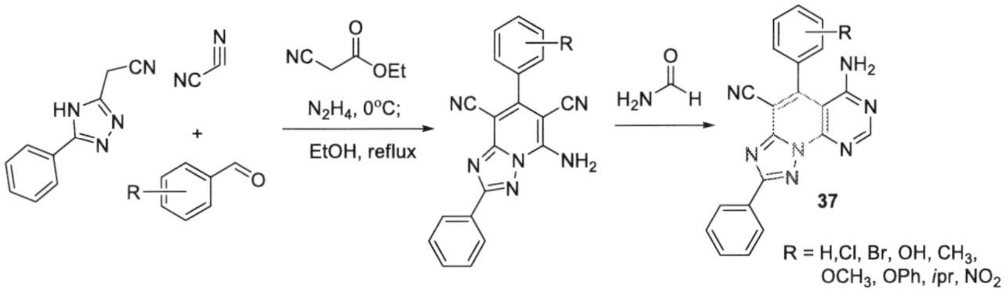

SCHEME 13.21 Synthesis of 6-amino-2,5-diphenyl-[1,2,4]triazolo-[1′,5′:1,6]pyrido[2,3-*d*]pyrimidine-4-carbonitrile analogues as promising structural lead for AD drug developments.

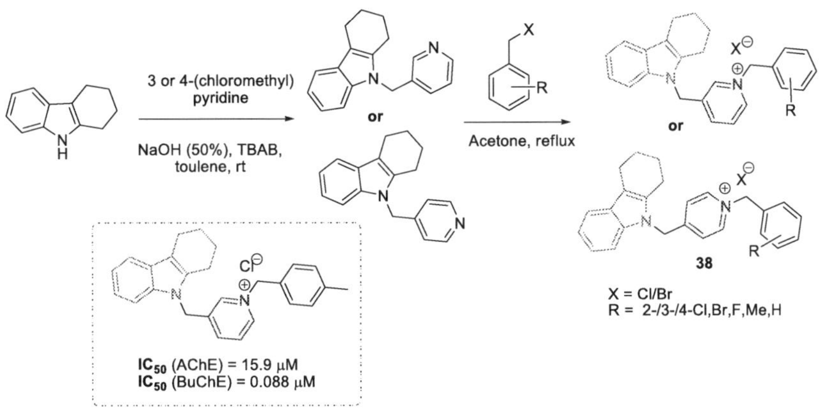

SCHEME 13.22 Synthesis of (2,3,4,9-tetrahydro-1*H*-carbazol-9(2*H*)-yl)methyl)pyridinium moieties were significant inhibitors of cholinesterase's.

13.3.9 Carbazole–pyridine analogues

Carbazole is another pharmacologically important privileged nitrogen scaffold that has shown a promising role in drug discovery like anticancer leads, antiviral agents and potent antitubercular drugs, etc. [100–102]. Further, some reports have accredited that carbazole derivatives play an important role in neuroprotection tactics. For instance, Ghobadian et al. [103] have developed a new series of (2,3,4,9-tetrahydro-1*H*-carbazol-9(2*H*)-yl)methyl)pyridinium derivatives **38** through a phase transfer catalyst, that is, tetrabutylammonium bromide (TBAB)-mediated N-alkylation of carbazole with 3- or 4-chloromethyl pyridine followed by N-quaternization of pyridine residue with various benzyl halides as shown in Scheme 13.22. Further, in vitro and in silico optimization biological studies revealed that tetrahydro-1*H*-carbazole allied benzyl pyridine moieties were significant inhibitors of *AChE*-induced Aβ-peptide aggregation as well as new potent selective *BuChE* inhibitors with therapeutic potential for the treatment of AD.

Likewise, Ni et al. [104] have also reported novel tadalafil hybrids **39** through Pictet–Spengler reaction of tryptophan derivative with *p*-methoxybenzaldehyde followed by acylation with 2-chloroacetyl chloride and cyclization with various piperidin-4-yl-ethane amines

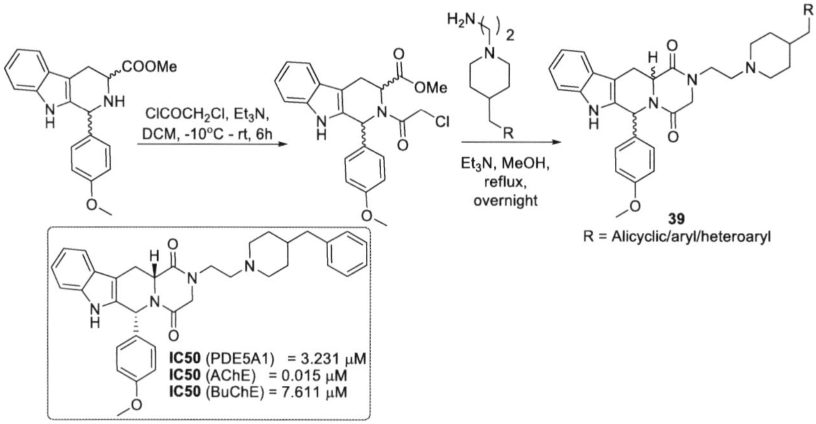

SCHEME 13.23 Synthesis of tadalafil analogues as selective inhibitors of AChE and phosphodiesterase 5 for the treating multiple pathogenesis of AD.

as depicted in Scheme 13.23. The stated tadalafil derivatives **39** were showed worthy selective inhibition of *AChE* ($IC_{50} < 1$ μM) and phosphodiesterase 5 (PDE5, IC_{50} 0.05–3.23 μM)) activities according to in vitro model studies. Moreover, the pharmacophore analysis of established tadalafil analogues **39** exhibited acceptable drug-likeness and BBB permeability and hence these drug candidates are beneficial for the treating multiple pathogenesis of AD and neurodegenerative diseases.

13.3.10 Tacrine analogues

Tacrine was one of the FDA-approved skillful clinical drugs that inhibits the pathogenesis of Alzheimer's disease [105,106]. Among all the current practicing cholinesterase inhibitors tacrine has the foremost potential and also auspicious pyridine-template (i.e., tacrine) against the other pathogenesis paths of neurodegenerative disorders [106]. Therefore, the researchers have been focused for the development of new class potent tacrine-based scaffolds for treating the multiple pathogenesis of AD. Initially, Fernandez-Bachiller et al. [107] have proposed a series of 8-hydroxyquinoline integrated tacrine analogues **40** with significant yields through Mannich aminomethylation by the reaction with paraformaldehyde under reflux (Scheme 13.24). Further, the biological assessment studies revealed that the projected tacrine hybrids **40** were beneficial as anti-Alzheimer agents acting via inhibiting the cholinesterase's activity (*ChEs*) and decline the mitochondrial oxidation, metal toxicity through neuroprotective mechanisms.

Later, Fernandez-Bachiller and coworkers [108] also developed a series of innovative tacrine-4*H*-chromen-4-one hybrids **41** through (benzotriazol-1-yloxy)tris(dimethylamino)-phosphonium hexafluorophosphate (i.e., BOP reagent) catalyzed peptide-bond synthesis between tacrine-amines and 4-oxo-4*H*-chromene-2-carboxylic acids as revealed in Scheme 13.25. As conjectured the developed tacrine-flavonoid amides **41** showed stimulating radical absorbance strategies that were comparable with the standards trolox and ascorbic

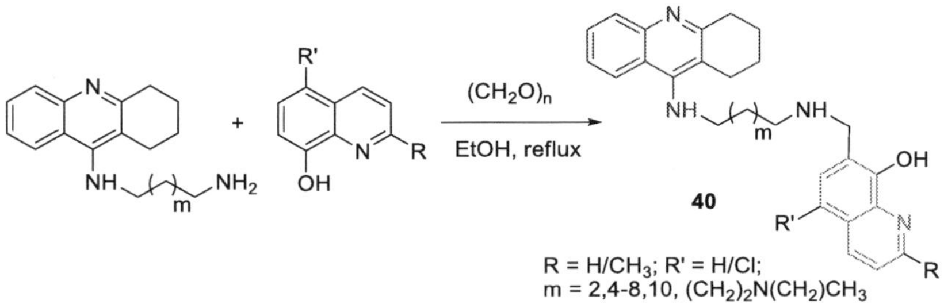

SCHEME 13.24 Synthesis of 8-hydroxyquinoline integrated tacrine analogues as anti-Alzheimer agents.

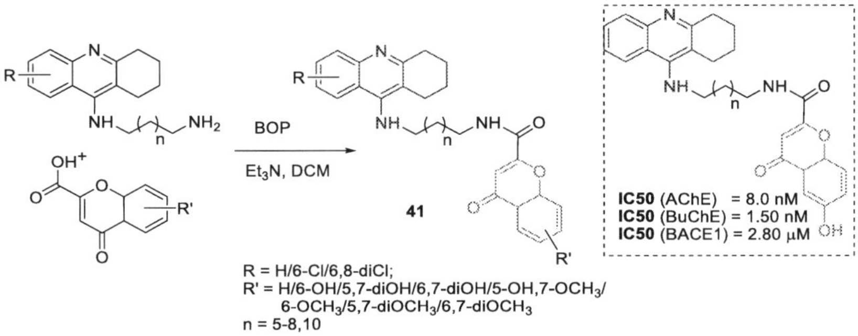

SCHEME 13.25 Synthesis of tacrine–flavonoid hybrids as a multiple drug targets of AD.

acid. Moreover, the in vitro biological studies revealed that the dual-chromophore analogues **41** were showed greater cholinesterase (*AChE* & *BuChE*) inhibition potency than tacrine and good blood–brain barrier penetration strategies. Therefore, the investigated tacrine–flavonoid hybrids are beneficial for the management of multiple pathogenesis of AD.

Najafi et al. [109] have established a sequence of triazole cohesive tacrine analogues **42** via click reaction of propargylated tacrine with in situ generated aryl azides as shown in Scheme 13.26. The optimized tacrine compounds **42** were showed significant inhibitory activities through binding the catalytic site (CS) and peripheral anionic site (PAS) of both cholinesterase (*AChE* & *BuChE*) enzymes. Moreover, the in vitro studies and molecular modeling kinetic studies suggest that the tacrine–1,2,3-triazole hybrids are beneficial for enlightening the cognitive condition by regulating the role of scopolamine.

Likewise, De Los Ríos and Marco-Contelles [110] have developed a new series of 5-amino-2-methyl-6,7,8,9-tetrahydro-benzo[*b*][1,8]naphthyridine-3-carboxylate esters **43** through condensation of 6-amino-5-cyano-2-methylnicotinates with respective alicyclic ketones followed by cyclization as described in Scheme 13.27. Further, in vivo and structure-guided pharmacophore modeling studies revealed that the new pyrido-tacrines **43** were showed interesting biological profiles and beneficial as anti-Alzheimer's drug leads by targeting multiple pathogenesis such as cholinesterase activity, voltage-gated calcium channels (VGCC), hepatotoxicity, and scopolamine-induced cognitive impairments, etc. Similarly, Saeedi et al. [111] proposed a new class of tacrine analogues, that is, thieno[2,3-*b*]pyridine amines **44** through Lewis

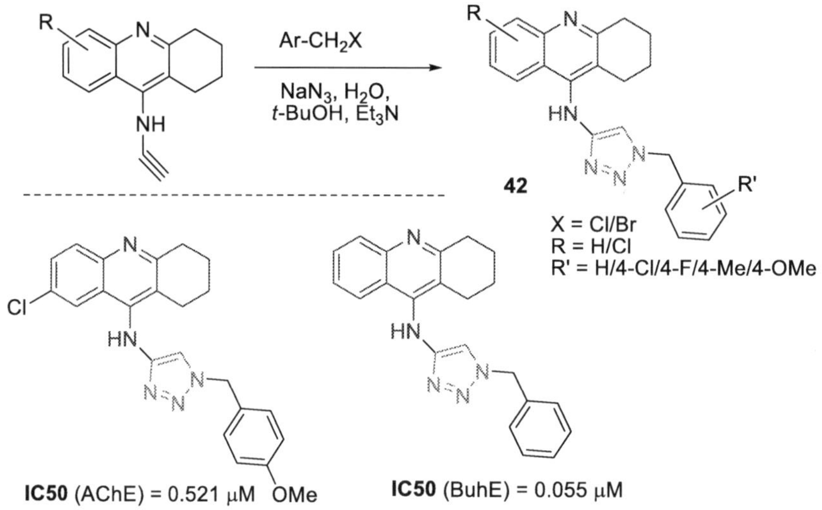

SCHEME 13.26 Synthesis of click reaction-mediated tacrine–1,2,3-triazole hybrids as beneficial cognitive condition enhancer.

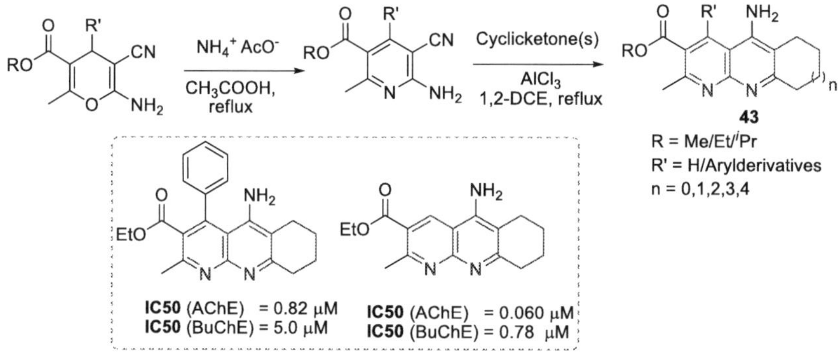

SCHEME 13.27 Synthesis of tetrahydrobenzo[*b*][1,8]naphthyridine-3-carboxylate esters as anti-AD to treat scopolamine-induced cognitive impairments.

acid catalyzed condensation of 2-aminothiophene-3-carbonitile derivatives with cyclic ketones in dichloroethane (DCE) at reflux (Scheme 13.28). The stated tacrine hybrids **44** were found to be substantial inhibitors of *AChE* prime Aβ-neurotoxicity and β-secretase (BACE1) induced cleavage of amyloid protein precursor (APP) activities.

Likewise, Chalupova et al. [112] have also described an interesting tacrine-indole carbamate derivative **45** through BOP reagent catalyzed condensation of assorted tacrine derivatives and *L*-tryptophan as shown in Scheme 13.29. Further, the accompanied bioactivity and in silico analysis disclosed that the stated tacrine–tryptophan hybrids **45** can regulate the cholinesterase induced Aβ-plaque and neuronal nitric oxide synthase owing to good BBB penetration and procognitive potential. Recently, Ceschi et al. [113] have synthesized

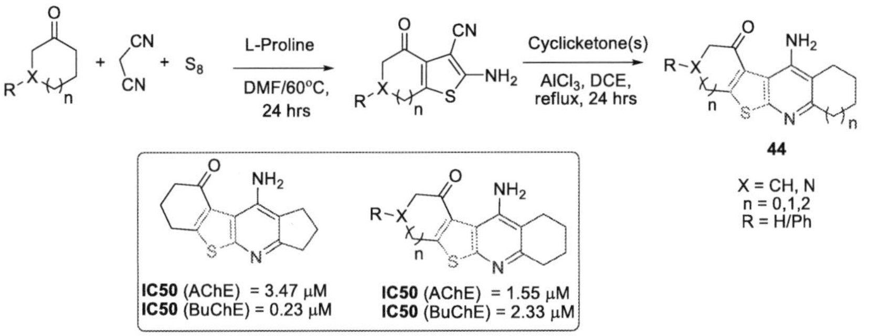

SCHEME 13.28 Synthesis of thieno[2,3-*b*]pyridine amines substantial inhibitors of *AChE* prime Aβ-neurotoxicity and β-secretase (BACE1).

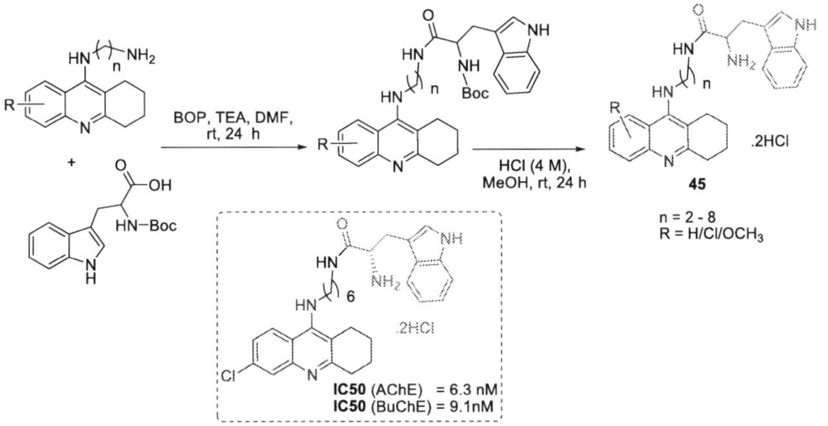

SCHEME 13.29 Synthesis of tacrine–tryptophan hybrids as anti-AD drug leads by decking multiple pathogenesis.

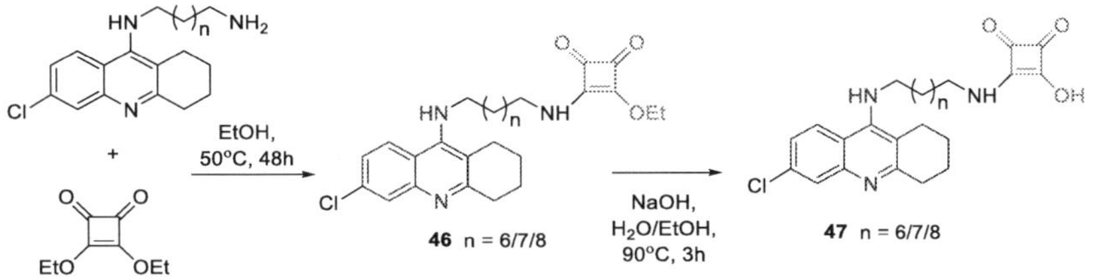

SCHEME 13.30 Synthesis of tacrine–monosquaramides and tacrine–squaramic acid derivatives as cholinesterase inhibitors.

stimulating tacrine bridged monosquaramides **46** and tacrine-squaramic acid derivatives **47** with outstanding yields of 71–99% through facile condensation of tacrine amines with diethoxysquarate in ethanol (Scheme 13.30). The in vitro and molecular modeling studies disclosed that the squaramic derivatives **46** were exhibited potent dual-binding inhibition at *nM* concentrations through interacting with catalytic active cite (CS) and peripheral anionic site (PAS) of *AChE* cavity. Hence, tacrine–squaramic derivatives **47** are valuable to managing the cholinesterase-induced Alzheimer's disease as like clinical drug agent tacrine. Moreover,

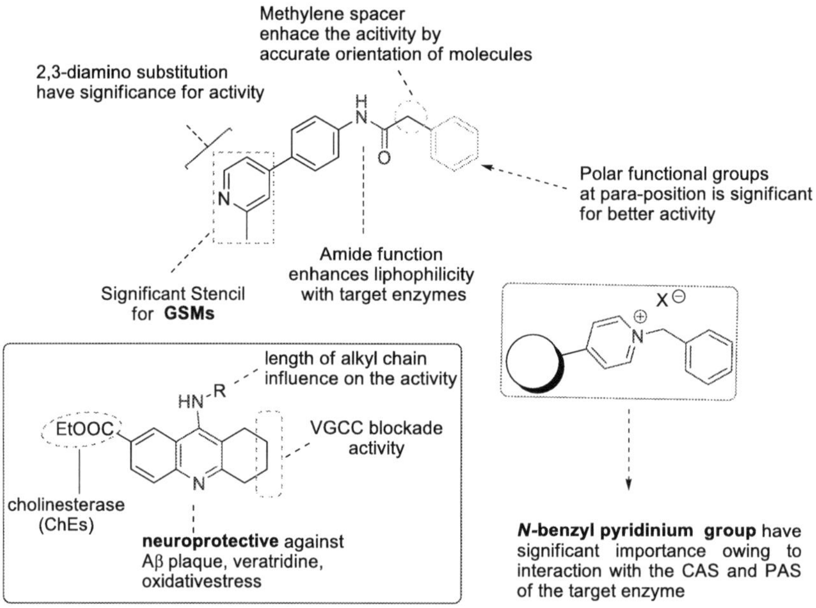

FIGURE 13.6 The essential key notes for skeletal modification assessed by SARs for the optimization of neurodegenerative disease.

some of the exceeding stated tacrine analogues have been stated as fewer hepatotoxic than the parental tacrine drug and hence optimistically the stated new class tacrine-based analogues might be valuable as future anti-Alzheimer's leads to treat the multiple pathogenesis of neurodegenerative disorders.

13.4 Structure–activity relationship studies (SARs)

Currently, the importance of SARs for the design and discovery of promising new therapeutic drug leads is growing exponentially. Since, SARs can be helpful to improve certain pharmacokinetic properties such as absorption, bioavailability, distribution, metabolism, excretion etc., and elimination of mutagenic functional groups before drug(s) was developed [114–116]. Moreover, the structure-based designed drug agents have greater properties like solubility, metabolic stability, better therapeutic potential, and less toxic effects [116]. Therefore, considering the role of SAR studies, we have summarized the key structural assignments (Fig. 13.6) from the reported pyridine-drug developments for AD that may be beneficial for future drug design and innovation.

The SAR lead optimization study by Chen et al. [46] demonstrated noteworthy substitutional pattern for improved activities against Alzheimer disease. Thus, the 2-methylpyridine function has greater potential for creating γ-secretase modulators (GSMs) than other nitrogen heterocycles. The introduction of carbamide/sulfonamide groups have significant strategy for lowering $A\beta42$ plaque in the cerebrospinal fluid (CSF) through exhibiting high lipophilicity and passive permeability. Furthermore, the methylene carbon (without substitution) bridge between amide and aromatic ring enhances the orientation and selectivity of molecule with target enzymes (Scheme 13.4). Likewise, the terminal aromatic ring with polar substitutes (amide, acetyl, carboxylic, methoxy and amino groups, etc.) at para-position (Fig. 13.6) has an influence on the enrichment of potential activity against pathogenesis of Aβ plaque. Another AD drug lead optimization study [44] identified the role of substitutional pattern in the pyridine ring as a factor influencing compound potency as β-secretase 1 (BACE1) inhibitor(s). Certainly, 2,3-diaminopyridine analogues (Scheme 13.2) have charged bidentate interactions with catalytic aspartates (Asp32 and Asp228) in the active site of enzyme BACE1 with considerable micromolar potency and hence amino-substituted pyridines or bipyridines are good eminence drug leads for AD. Furthermore, Ji et al. [60] stretch that the constructive role of bidentate ligands such as bipyridines with appropriate probe groups (Scheme 13.7) was valuable to treat the Aβ-plaque, metal-neurotoxicity (e.g., Cu, Zn) as well as metal-associated biosystems.

Further, Foroumadi and coworkers [81,82] have demonstrated two groups of novel benzofuran-derived benzyl pyridinium halides (Scheme 13.12) as compelling cholinesterase inhibitors. The potent anti-*AChE* activities might be due to the planar aurone ring π–π stacking interaction(s) and quaternary nitrogen in the pyridine ring system π-cation interaction with the targeted enzyme *AChE* [117,118]. Therefore, further optimization studies of these sort of analogues are necessary to select the best candidates to treat AD. Likewise, assorted studies [75–78] have reported with benzyl pyridinium chromophore group (Fig. 13.6) integrated natural entities (Schemes 13.9–13.12) with remarkable activities against the pathogenesis of AD. Wang et al. [77] study revealed that the role of *N*-benzyl pyridinium moiety is in enhancing the *AChE*-inhibitory activity by interacting with the catalytic activity site (CAS) and peripheral anionic site (PAS) of the target enzyme. In addition, according to the molecular modeling studies and ADME properties the quaternary nitrogen scaffolds have better solubilities and improved pharmacokinetics as of likelihood lipophilic interactions. Therefore, the incidence of *N*-benzyl pyridinium group is an additional structural feature in the AD drug discovery for boosted activity.

Likewise, a comprehensive study [105] has summarized momentous structural importance in the development of various tacrine analogues and their absolute role in the AD drug discovery. The constructive study disclosed that the 6,8-positions of tacrine moiety have imperative for the enhanced pharmacological activities against AD. Also, amino moieties at 9-position of tacrine, the length of alkyl chain of the 9-amino substituent (Fig. 13.6) have been of significant influence on the activity and structural modifications. Further, the integration of natural compounds into tacrine nucleus (Schemes 13.24, 13.25, 13.29, and 13.30) was a significant skeletal modification to provide a combined therapeutic activity for neurodegenerative diseases by targeting multiple paths. For instance, De Los Ríos and Marco-Contelles [110] have reported that a novel series of tetrahydrobenzo[*b*][1,8]naphthyridine-3-carboxylate esters (Scheme 13.27), obtained through skeletal modification of tacrine core, as multitargeted AD drug agents. Further, the stated naphthyridine–tacrine analogues presented stimulating anti-Alzheimer's drug profiles through pointing multiple pathogenesis like cholinesterase's, VGCC channels, hepatotoxicity, and cognitive impairments, etc. Therefore, the précised skeletal assessments and SARs may be supportive for the development of future AD drug lead agents.

13.5 Clinical approaches

The prevalence of Alzheimer disease worldwide is huge, with an estimated 5.8 million people being treated for AD in the United States alone [119]. So far there are no promising drugs for the treatment of this multiple pathogenesis of AD, and all the currently available drugs have been developed only to reduce the severity of diseases. Furthermore, some drugs have also been reported to be associated with an incidence of clinical lethal effects, for example, some cases of hepatotoxicity caused by FDA-approved drug tacrine were known [120]. Therefore, leading clinical trials are necessary to treat the multisyndrome AD and to overcome the impending clinical tactics. According to the National Institute of Aging (NIA) data more than 250 active clinical trials have yet to be registered [121]. In addition, the current clinical works more effectively as it considers various strategies such as prevention, treatment, diagnostic, and enhances the quality of life. Recently, Cummings et al. [119] have summarized the progression in the clinical trials for the treatment of Alzheimer's disease. According to the Cummings study approximately 28 conducts are in phase III and 74 are in phase II trials while 24 are phase I. Furthermore, Hung and Fu [122] have also abridged the downsides for the failures of AD drug clinical trials. There are still trivial promising trials existing as per the current data and hence, new therapies and targets are urgently needed for the treatment of AD.

In this concern, pyridine scaffolds are also auspicious therapeutic leads and some of the important pyridine analogues have been registered for clinical trials (Fig. 13.7). The phase-I clinical trial (NCT01429740) of pyridine analogue, PF-05180999 was showed prominent brain-penetrating phosphodiesterase (PDE2A) inhibition [123]. Further, an imidazole[1,2-*a*]pyridine analogue was found to be a potent inhibitor cholesterol 24-hydralyse (CYP46A1) in its trial WO2014061676 that was beneficial for treating AD [124]. Likewise, another clinical candidate TAK-915 showed phosphodiesterase 2A inhibition in its preclinical studies which was valuable to ameliorate the cognitive impairments [125]. Another preclinical bipyridine hybrid, soticlestat has also been found to be a novel inhibitor of cholesterol 24-hydralyse and associated mechanism with a potential of neural hyperexcitation [126]. Bitopertin (RO4917838) was a pyridine derivative that showed persistent, predominant negative symptoms in patients of Schizophrenia in its phase-III trials (NCT01192906) [127]. Even though very tiny pre-/primary clinical trials have registered yet the precise studies reported on pyridine scaffolds were showed impending strategy for the development of new anti-AD drug agents. Hence, further gigantic clinical studies of pyridine scaffolds are needed to achieve imminent clinical agents for multisyndrome AD.

13.6 Summary

Alzheimer's disease is a complex syndrome involving several pathophysiology paths including β-amyloid (Aβ) deposition, metal-ion dyshomeostatis, oxidative-stress, tau protein phosphorylation, and neurotransmitter dysfunction, etc. Although the specific basis of AD is not yet known, the most common criteria are cognitive impairments followed by memory loss and dementia. Over the past decade, several accelerated attempts have been made to develop

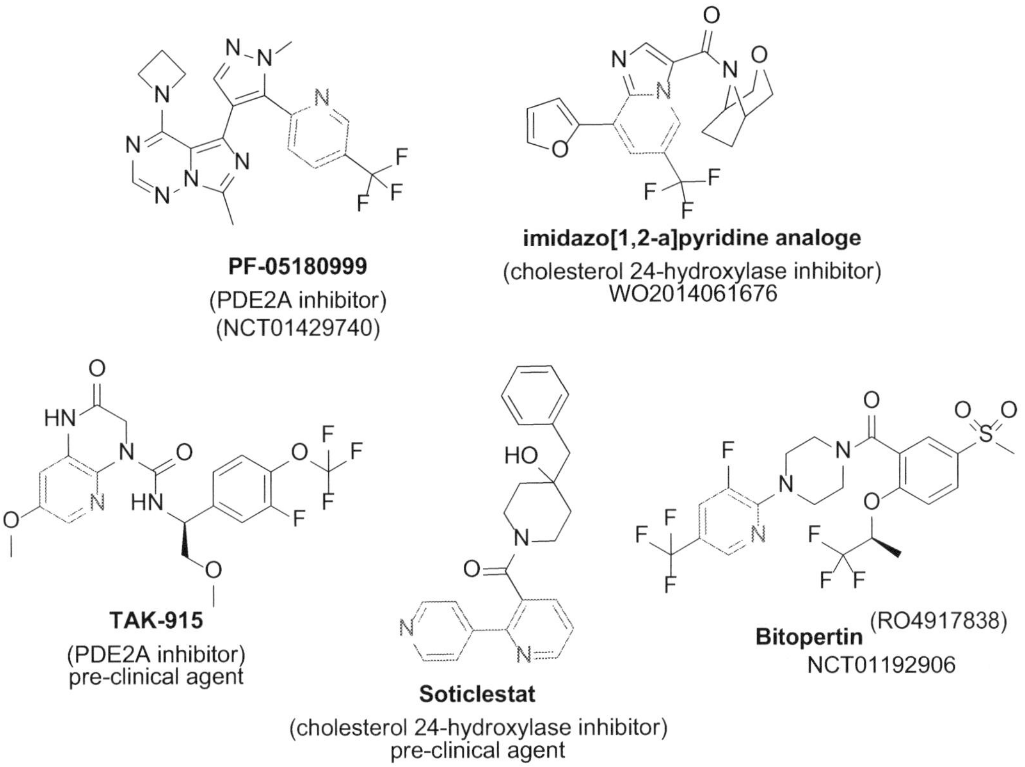

FIGURE 13.7 Some of the imperative pre- and postclinical candidates of pyridine analogues.

new strategies and new drug agents for the prevention/treatment of AD. So far, very few drug agents have been compiled for clinical trials and some of them have been found to be the lowest clinical campaigns. In this scenario, diverse pyridine hybrids have been revealed as auspicious therapeutic leads for the treatment of AD as well as other neurodegenerative disorders. Some aminopyridines (**4,5**) displayed important interactions with the active site of BACE1 which are valuable to develop inhibitors of β-secretase. Further, bipyridine analogues (**13,14**) have been found to be potent inhibitors of metal-induced β-amyloid (Aβ)-plaque and human glutaminyl cyclase (hQC) which play a key role in the amelioration of AD pathogenesis. Moreover, some of the natural compounds integrated pyridine scaffolds (**19–22, 42**) have shown promising strategy against oxidative stress, cholinesterase, and Aβ-plaque, etc., together with fewer hepatotoxic than exiting clinical AD drugs. Likewise, some tacrine hybrids (**44–48**) have been found to be skillful molecules to targeting multiple pathogenesis of AD such as cholinesterase activity, voltage-gated calcium channels (VGCC), hepatotoxicity, and scopolamine-induced cognitive impairments. Considering the above précised impending tactics of pyridine scaffolds further massive studies are needed to achieve more efficient clinical drugs and to evolving new therapies for the Alzheimer disease.

Acknowledgments

The authors A.R. and G.V.Z. were thankful to the Ministry of Science and Higher Education of the Russian Federation (Grant no. 075-15-2020-777) for supporting this work.

References

[1] B.S. Schoenberg, Epidemiology of dementia, Neurol. Clin. 4 (1986) 447–458.
[2] World Health OrganizationConsultation on the Development of the Global Dementia Observatory, World Health Organization, Geneva, 2017 5-6 July 2016: meeting report. (No. WHO/MSD/MER/17.4), WHO.
[3] C.G. Lyketsos, M.C. Carrillo, J.M. Ryan, A.S. Khachaturian, P. Trzepacz, J. Amatniek, et al., Neuropsychiatric symptoms in Alzheimer's disease, Alzheimers Dement. 7 (2011) 532–539.
[4] S.T. Fujimoto, L. Longhi, K.E. Saatman, T.K. McIntosh, Motor and cognitive function evaluation following experimental traumatic brain injury, Neurosci. Biobehav. Rev. 28 (2004) 365–378.
[5] R.H. Takahashi, C.G. Almeida, P.F. Kearney, F. Yu, M.T. Lin, T.A. Milner, et al., Oligomerization of Alzheimer's β-amyloid within processes and synapses of cultured neurons and brain, J. Neurosci. 24 (2004) 3592–3599.
[6] F. Kametani, M. Hasegawa, Reconsideration of amyloid hypothesis and tau hypothesis in Alzheimer's disease, Front. Neurosci. 30 (2018) 25.
[7] L. Blaikie, G. Kay, P.K. Lin, Current and emerging therapeutic targets of alzheimer's disease for the design of multi-target directed ligands, Med. Chem. Comm. 10 (2019) 2052–2072.
[8] H. Kadavath, R.V. Hofele, J. Biernat, S. Kumar, K. Tepper, H. Urlaub, et al., Tau stabilizes microtubules by binding at the interface between tubulin heterodimers, Proc. Natl. Acad. Sci. 112 (2015) 7501–7506.
[9] S. Muralidar, S.V. Ambi, S. Sekaran, D. Thirumalai, B. Palaniappan, Role of tau protein in Alzheimer's disease: the prime pathological player, Int. J. Biol. Macromol. 163 (2020) 1599–1617.
[10] E.K. Perry, P.H. Gibson, G. Blessed, R.H. Perry, B.E. Tomlinson, Neurotransmitter enzyme abnormalities in senile dementia: choline acetyltransferase and glutamic acid decarboxylase activities in necropsy brain tissue, J. Neurol. Sci. 34 (1977) 247–265.
[11] T.A. Slotkin, Cholinergic systems in brain development and disruption by neurotoxicants: nicotine, environmental tobacco smoke, organophosphates, Toxicol. Appl. Pharmacol. 198 (2004) 132–151.
[12] G. Cisbani, A. Koppel, D. Knezevic, I. Suridjan, R. Mizrahi, R.P. Bazinet, Peripheral cytokine and fatty acid associations with neuroinflammation in AD and aMCI patients: An exploratory study, Brain Behav. Immun. 87 (2020) 679–688.
[13] V. Calsolaro, P. Edison, Neuroinflammation in Alzheimer's disease: current evidence and future directions, Alzheimers Dement. 12 (2016) 719–732.
[14] Y. Tang, W. Le, Differential roles of M1 and M2 microglia in neurodegenerative diseases, Mol. Neurobiol. 53 (2016) 1181–1194.
[15] Z.Q. Sun, J.F. Liu, W. Luo, C.H. Wong, K.F. So, Y. Hu, et al., *Lycium barbarum* extract promotes M2 polarization and reduces oligomeric amyloid-β-induced inflammatory reactions in microglial cells, Neural Regen. Res. 17 (2022) 203.
[16] C.G. Parsons, A. Stöffler, W. Danysz, Memantine: a NMDA receptor antagonist that improves memory by restoration of homeostasis in the glutamatergic system-too little activation is bad, too much is even worse, Neuropharmacology 53 (2007) 699–723.
[17] S. Bais, R. Kumari, Y. Prashar, A review on current strategies and future perspective in respect to Alzheimer's disease treatment, Curr. Res. Neurosci. 6 (2016) 1–5.
[18] C. Zucchella, E. Sinforiani, S. Tamburin, A. Federico, E. Mantovani, S. Bernini, et al., The multidisciplinary approach to Alzheimer's disease and dementia. A narrative review of non-pharmacological treatment, Front. Neurol. 9 (2018) 1058.
[19] J.A. Watt, Z. Goodarzi, A.A. Veroniki, V. Nincic, P.A. Khan, M. Ghassemi, et al., Comparative efficacy of interventions for aggressive and agitated behaviors in dementia: a systematic review and network meta-analysis, Ann. Intern. Med. 171 (2019) 633–642.
[20] R.A. Abramovitch (Ed.), Pyridine and Its Derivatives, Supplement, Part 2, Wiley, USA, New York, 2009.
[21] M. Schlosser, R. Ruzziconi, Nucleophilic substitutions of nitroarenes and pyridines: new insight and new applications, Synthesis 13 (2010) 2111–2123.

[22] G. Bertuzzi, L. Bernardi, M. Fochi, Nucleophilic dearomatization of activated pyridines, Catalysts 8 (2018) 632.

[23] Y. Hamada, Role of Pyridines in Medicinal Chemistry and Design of BACE1 Inhibitors Possessing a Pyridine Scaffold, In: P.P. Pandey (Ed.), London: Intech Open. 2018, pp. 9-26.

[24] A.A. Altaf, A. Shahzad, Z. Gul, N. Rasool, A. Badshah, B. Lal, E. Khan, A review on the medicinal importance of pyridine derivatives, J. Drug Des. Med. Chem. 1 (2015) 1–11.

[25] P. Kiuru, J. Yli-Kauhaluoma, Pyridine and its derivatives, in: K. Majumdar, S.K. Chattopadhyay (Eds.), Heterocycles in Natural Product Synthesis, Wiley-VCH Verlag GmbH & Co. KGaA, Weinheim, Germany, 2011, pp. 267–297.

[26] P.M. Dewick, Medicinal Natural Products: A Biosynthetic Approach, 2nd ed., Wiley, New York, 2002, pp. 311–315.

[27] M. Baumann, I.R. Baxendale, An overview of the synthetic routes to the best selling drugs containing 6-membered heterocycles, Beilstein J. Org. Chem. 9 (2013) 2265–2319.

[28] A.M.K. El-Dean, A.A. Abd-Ella, R. Hassanien, M.E. El-Sayed, S.A.A. Abdel-Raheem, Design, synthesis, characterization, and insecticidal bioefficacy screening of some new pyridine derivatives, ACS Omega 4 (2019) 8406–8412.

[29] G. Volpi, R. Rabezzana, Imidazo[1, 5-*a*]pyridine derivatives: useful, luminescent and versatile scaffolds for different applications, New J. Chem. 45 (2021) 5737–5743.

[30] M.A. Chiacchio, D. Iannazzo, R. Romeo, S.V. Giofrè, L. Legnani, Pyridine and pyrimidine derivatives as privileged scaffolds in biologically active agents, Curr. Med. Chem. 26 (2019) 7166–7195.

[31] D.W. Hopper, A.L. Crombie, J.J. Clemens, S. Kwon, Six-membered ring systems: pyridine and benzo derivatives, Prog. Heterocycl. Chem. 21 (2019) 330–374.

[32] A.R. Katritzky, C.A. Ramsden, E.F.V. Scriven, R.J.K. Taylor, Six-membered rings with one heteroatom, and their fused carbocyclic derivatives, Comprehens. Heterocyclic Chem. III 7 (2008) 1–1066.

[33] R.L. Frank, R.P. Seven, I.V. Pyridines., A study of the Chichibabin synthesis, J. Am. Chem. Soc. 71 (1949) 2629–2635.

[34] M.D. Hill, Recent strategies for the synthesis of pyridine derivatives, Chemistry 16 (2010) 12052–12062.

[35] B. Heller, M. Hapke, The fascinating construction of pyridine ring systems by transition metal-catalysed [2+2+2] cycloaddition reactions, Chem. Soc. Rev. 36 (2007) 1085–1094.

[36] J. Kumar, A. Gill, M. Shaikh, A. Singh, A. Shandilya, E. Jameel, et al., Pyrimidine-triazolopyrimidine and pyrimidine-pyridine hybrids as potential acetylcholinesterase inhibitors for Alzheimer's disease, Chem. Select. 3 (2018) 736–747.

[37] A. Hiremathad, L. Piemontese, Heterocyclic compounds as key structures for the interaction with old and new targets in Alzheimer's disease therapy, Neural Regen. Res. 12 (2017) 1256.

[38] Z. Liu, A. Zhang, H. Sun, Y. Han, L. Kong, X. Wang, Two decades of new drug discovery and development for Alzheimer's disease, RSC Adv. 7 (2017) 6046–6058.

[39] I. Grabowska, H. Radecka, A. Burza, J. Radecki, M. Kaliszan, R. Kaliszan, Association constants of pyridine and piperidine alkaloids to amyloid β peptide determined by electrochemical impedance spectroscopy, Curr. Alzheimer Res. 7 (2010) 165–172.

[40] Y. Yang, M. Cui, Radiolabeled bioactive benzoheterocycles for imaging β-amyloid plaques in Alzheimer's disease, Eur. J. Med. Chem. 87 (2014) 703–721.

[41] N. Guzior, A. Wieckowska, D. Panek, B. Malawska, Recent development of multifunctional agents as potential drug candidates for the treatment of Alzheimer's disease, Curr. Med. Chem. 22 (2015) 373–404.

[42] V. Patel, X. Zhang, N.A. Tautiva, A.N. Nyabera, O.O. Owa, M. Baidya, H.C. Sung, P.S. Taunk, S. Abdollahi, S. Charles, R.A. Gonnella, N. Gadi, K.T. Duong, J.N. Fawver, C. Ran, T.O. Jalonen, I.V.J. Murray, Small molecules and Alzheimer's disease: misfolding, metabolism and imaging, Curr. Alzheimer Res. 12 (2015) 445.

[43] X. Wu, H. Cai, L. Pan, G. Cui, F. Qin, Y. Li, Z. Cai, Small molecule natural products and Alzheimer's disease, Curr. Top. Med. Chem. 19 (2019) 187–204.

[44] M. Congreve, D. Aharony, J. Albert, O. Callaghan, J. Campbell, R.A. Carr, et al., Application of fragment screening by X-ray crystallography to the discovery of aminopyridines as inhibitors of β-secretase, J. Med. Chem. 50 (2007) 1124–1132.

[45] S.K. Singh, S.K. Sinha, M.K. Shirsat, Design, synthesis and evaluation of 4-aminopyridine analogues as cholinesterase inhibitors for management of Alzheimer's diseases, Indian J. Pharm. Educ. Res. 52 (2018) 644–654.

[46] J.J. Chen, W. Qian, K. Biswas, et al., Discovery of 2-methylpyridine-based biaryl amides as γ-secretase modulators for the treatment of Alzheimer's disease, Bioorg. Med. Chem. Lett. 23 (2013) 6447–6454.
[47] S. Riaz, I.U. Khan, M. Bajda, M. Ashraf, A. Shaukat, T.U. Rehman, et al., Pyridine sulfonamide as a small key organic molecule for the potential treatment of type-II diabetes mellitus and Alzheimer's disease: *in vitro* studies against yeast α-glucosidase, acetylcholinesterase and butyrylcholinesterase, Bioorg. Chem. 63 (2015) 64–71.
[48] A. Elkamhawy, J.E. Park, A.H. Hassan, H. Ra, A.N. Pae, J. Lee, et al., Discovery of 1-(3-(benzyloxy) pyridin-2-yl)-3-(2-(piperazin-1-yl) ethyl) urea: a new modulator for amyloid beta-induced mitochondrial dysfunction, Eur. J. Med. Chem. 128 (2017) 56–69.
[49] D.S. Cukierman, E. Accardo, R.G. Gomes, A. De Falco, M.C. Miotto, M.C. Freitas, et al., Aroylhydrazones constitute a promising class of 'metal-protein attenuating compounds' for the treatment of Alzheimer's disease: a proof of concept based on the study of the interactions between zinc (II) and pyridine-2-carboxaldehyde isonicotinoylhydrazone, J. Biol. Inorg. Chem. 23 (2018) 1227–1241.
[50] W.Q. Zhao, M. Townsend, Insulin resistance and amyloidogenesis as common molecular foundation for type 2 diabetes and Alzheimer's disease, Biochim. Biophys. Acta Mol. Basis Dis. 1792 (2009) 482–496.
[51] S. Gordon, M. Kittleson, Drugs used in the management of heart disease and cardiac arrhythmias, in: J.E. Maddison, S.W. Page, D.B. Church (Eds.), Small Animal Clinical Pharmacology, Elsevier Health Sciences, Philadelphia, PA 19103-2899, USA, 2008, p. 380. 5.
[52] B.M. Kelly-Basetti, D.J. Cundy, S.M. Pereira, W.H. Sasse, G.P. Savage, G.W. Simpson, Synthesis and fungicidal activity of 2, 2′-bipyridine derivatives, Bioorg. Med. Chem. Lett. 5 (1995) 2989–2992.
[53] A.P. Krinochkin, D.S. Kopchuk, G.A. Kim, V.A. Shevyrin, T.A. Tseitler, S. Santra, et al., Synthesis and luminescent properties of functionalized bipyridyl based Eu complexes, Chem. Select. 5 (2020) 9180–9183.
[54] J.Y. Tsao, J.D. Sai, C.I. Yang, Azide-bridged Cu (ii), Mn (ii) and Co (ii) coordination polymers constructed with a bifunctional ligand of 6-(1 H-tetrazol-5-yl)-2, 2′-bipyridine, Dalton Trans. 45 (2016) 3388–3397.
[55] B.L. Elbert, A.J. Farley, T.W. Gorman, T.C. Johnson, C. Genicot, B. Lallemand, C–H cyanation of 6-ring N-containing heteroaromatics, Chem. Chem. Eur. J. 23 (2017) 14733.
[56] G.V. Zyryanov, D.S. Kopchuk, I.S. Kovalev, S. Santra S, M. Rahman M, A.F. Khasanov, et al., Rational synthetic methods in creating promising (hetero) aromatic molecules and materials, Mendel. Commun. 30 (2020) 537–554.
[57] P. Hayoz, A. Von Zelewsky, New versatile optically active bipyridines as building blocks for helicating and caging ligands, Tetrahedron Lett. 33 (1992) 5165–5168.
[58] D.S. Kopchuk, A.P. Krinochkin, D.N. Kozhevnikov, P.A. Slepukhin, Novel neutral lanthanide complexes of 5-aryl-2,2′-bipyridine-6′-carboxylic acids with improved photophysical properties, Polyhedron 118 (2016) 30–36.
[59] M.A. Santos, K. Chand, S. Chaves, Recent progress in multifunctional metal chelators as potential drugs for Alzheimer's disease, Coord. Chem. Rev. 327 (2016) 287–303.
[60] Y. Ji, H.J. Lee, M. Kim, G. Nam, S.J. Lee, J. Cho, et al., Strategic design of 2,2′-bipyridine derivatives to modulate metal–amyloid-β aggregation, Inorg. Chem. 56 (2017) 6695–6705.
[61] C. Pozzi, F. Di Pisa, M. Benvenuti, S. Mangani, The structure of the human glutaminyl cyclase–SEN177 complex indicates routes for developing new potent inhibitors as possible agents for the treatment of neurological disorders, J. Biolog. Inorg. Chem. 23 (2018) 1219–1226.
[62] D. O'Hagan, Pyrrole, pyridine and alkaloids, Nat. Prod. Rep. 17 (2000) 435–446.
[63] M. Hadjiconstantinou, N.H. Neff, Nicotine and endogenous opioids: neurochemical and pharmacological evidence, Neuropharmacy 60 (2011) 1209–1220.
[64] A.H. Rezvani, E.D. Levin, Cognitive effects of nicotine, Biol. Psychiatry 49 (2001) 258–267.
[65] Y. Chi, A.A. Sauve, Nicotinamide riboside, a trace nutrient in foods, is a vitamin B3 with effects on energy metabolism and neuroprotection, Curr. Opin. Clin. Nutr. Metab. Care. 16 (2013) 657–661.
[66] X. Xie, Y. Gao, M. Zeng, Y. Wang, T.F. Wei, Y.B. Lu, et al., Nicotinamide ribose ameliorates cognitive impairment of aged and Alzheimer's disease model mice, Metab. Brain Dis. 34 (2019) 353–366.
[67] N. Khurana, M.P. Ishar, A. Gajbhiye, R.K. Goel, PASS assisted prediction and pharmacological evaluation of novel nicotinic analogs for nootropic activity in mice, Eur. J. Pharmacol. 662 (2011) 22–30.
[68] M.W. Holladay, M.J. Dart, J.K. Lynch, Neuronal nicotinic acetylcholine receptors as targets for drug discovery, J. Med. Chem. 40 (1997) 4169–4194.
[69] G. Singh, M.P. Ishar, N.K. Girdhar, L. Singh, Investigations on region- and stereoselectivities in cycloadditions involving α-(3-pyridyl)-N-phenylnitrone: development of an efficient route to novel nicotine analogs, J. Heterocycl. Chem. 42 (2005) 1047–1054.

[70] H. Nie, Z. Wang, W. Zhao, J. Lu, C. Zhang, K. Lok, et al., New nicotinic analogue ZY-1 enhances cognitive functions in a transgenic mice model of Alzheimer's disease, Neurosci. Lett. 537 (2013) 29–34.

[71] L. Hritcu, R. Ionita, D.E. Motei, C. Babii, M. Stefan, M. Mihasan, Nicotine versus 6-hydroxy-l-nicotine against chlorisondamine induced memory impairment and oxidative stress in the rat hippocampus, Biomed. Pharmacother. 86 (2017) 102–108.

[72] M.A. Brimble, J.S. Gibson, J. Sperry, Pyrans and their Benzo Derivatives: Synthesis, The University of Auckland, Auckland, New Zealand, 2008, pp. 425–675. Elsevier.

[73] S. Kamboj, R. Singh, Chromanone-A prerogative therapeutic scaffold: an overview, Arab. J. Sci. Eng. 47 (2022) 75–111.

[74] N. Jiang, J. Ding, J. Liu, X. Sun, Z. Zhang, Z. Mo, et al., Novel chromanone-dithiocarbamate hybrids as multifunctional AChE inhibitors with β-amyloid anti-aggregation properties for the treatment of Alzheimer's disease, Bioorg. Chem. 89 (2019) 103027.

[75] M. Alipour, M. Khoobi, A. Foroumadi, H. Nadri, A. Moradi, A. Sakhteman, et al., Novel coumarin derivatives bearing *N*-benzyl pyridinium moiety: potent and dual binding site acetylcholinesterase inhibitors, Bioorg. Med. Chem. 20 (2012) 7214–7222.

[76] M. Khoobi, M. Alipour, A. Sakhteman, H. Nadri, A. Moradi, M. Ghandi, et al., Design, synthesis, biological evaluation and docking study of 5-oxo-4, 5-dihydropyrano [3,2-*c*] chromene derivatives as acetylcholinesterase and butyryl-cholinesterase inhibitors, Eur. J. Med. Chem. 68 (2013) 260–269.

[77] C. Wang, Z. Wu, H. Cai, S. Xu, J. Liu, J. Jiang, et al., Design, synthesis, biological evaluation and docking study of 4-isochromanone hybrids bearing *N*-benzyl pyridinium moiety as dual binding site acetylcholinesterase inhibitors, Bioorg. Med. Chem. Lett. 25 (2015) 5212–5216.

[78] J. Wang, C. Wang, Z. Wu, X. Li, S. Xu, J. Liu, et al., Design, synthesis, biological evaluation, and docking study of 4-isochromanone hybrids bearing *N*-benzyl pyridinium moiety as dual binding site acetylcholinesterase inhibitors (part II), Chem. Biol. Drug Des. 91 (2018) 756–762.

[79] R. Zhu, J. Wei, Z. Shi, Benzofuran synthesis via copper-mediated oxidative annulation of phenols and unactivated internal alkynes, Chem. Sci. 4 (2013) 3706.

[80] S. Rizzo, A. Tarozzi, M. Bartolini, G. Da Costa, A. Bisi, S. Gobbi, et al., 2-Arylbenzofuran-based molecules as multipotent Alzheimer's disease modifying agents, Eur. J. Med. Chem. 58 (2012) 519–532.

[81] H. Nadri, M. Pirali-Hamedani, M. Shekarchi, M. Abdollahi, V. Sheibani, et al., Design, synthesis and anticholinesterase activity of a novel series of 1-benzyl-4-((6-alkoxy-3-oxobenzofuran-2 (3H)-ylidene) methyl) pyridinium derivatives, Bioorg. Med. Chem. 18 (2010) 6360–6366.

[82] F. Baharloo, M.H. Moslemin, H. Nadri, A. Asadipour, M. Mahdavi, S. Emami, et al., Benzofuran-derived benzylpyridinium bromides as potent acetylcholinesterase inhibitors, Eur. J. Med. Chem. 93 (2015) 196–201.

[83] A. Siwach, P.K. Verma, Synthesis and therapeutic potential of imidazole containing compounds, BMC Chem 15 (2021) 1–69.

[84] M. Lawson, J. Rodrigo, B. Baratte, T. Robert, C. Delehouzé, O. Lozach, et al., Synthesis, biological evaluation and molecular modeling studies of imidazo[1,2-*a*] pyridines derivatives as protein kinase inhibitors, Eur. J. Med. Chem. 123 (2016) 105.

[85] Z.P. Zhuang, M.P. Kung, A. Wilson, C.W. Lee, K. Plössl, C. Hou, et al., Structure– activity relationship of imidazo[1,2-a] pyridines as ligands for detecting β-amyloid plaques in the brain, J. Med. Chem. 46 (2003) 237–243.

[86] L. Cai, F.T. Chin, V.W. Pike, H. Toyama, J.S. Liow, S.S. Zoghbi, et al., Synthesis and evaluation of two 18F-labeled 6-iodo-2-(4′-N, N-dimethylamino) phenylimidazo [1, 2-a] pyridine derivatives as prospective radioligands for β-amyloid in Alzheimer's disease, J. Med. Chem. 47 (2004) 2208–2218.

[87] S. Singh, S. Singh, A.K. Tiwari, R.K. Sharma, R. Mathur, A. Kaul, et al., A novel 18F labelled imidazo-oxazolopyridine derivative as β-amyloid imaging agent: synthesis and preliminary evaluation, Asian J. Chem. 30 (2018) 183–190.

[88] R. Nirogi, A.R. Mohammed, A.K. Shinde, N. Bogaraju, S.R. Gagginapalli, S.R. Ravella, et al., Synthesis and SAR of imidazo [1, 5-a] pyridine derivatives as 5-HT4 receptor partial agonists for the treatment of cognitive disorders associated with Alzheimer's disease, Eur. J. Med. Chem. 103 (2015) 289–301.

[89] Z. Haghighijoo, S. Akrami, M. Saeedi, A. Zonouzi, A. Iraji, B. Larijani, et al., *N*-cyclohexylimidazo [1,2-*a*] pyridine derivatives as multi-target-directed ligands for treatment of Alzheimer's disease, Bioorg. Chem. 103 (2020) 104146.

[90] J.A. McCleverty, Comprehensive Coordination Chemistry II, Elsevier, Amsterdam, 2003.
[91] T. Umar, S. Shalini, M.K. Raza, S. Gusain, J. Kumar, P. Seth, et al., A multifunctional therapeutic approach: synthesis, biological evaluation, crystal structure and molecular docking of diversified 1*H*-pyrazolo [3,4-*b*] pyridine derivatives against Alzheimer's disease, Eur. J. Med. Chem. 175 (2019) 2–19.
[92] E.J. Barreiro, C.A. Camara, H. Verli, L. Brazil-Más, N.G. Castro, W.M. Cintra, et al., Design, synthesis, and pharmacological profile of novel fused pyrazolo [4,3-*d*] pyridine and pyrazolo [3,4-*b*][1, 8] naphthyridine isosteres: a new class of potent and selective acetylcholinesterase inhibitors, J. Med. Chem. 46 (2003) 1144–1152.
[93] J. Witherington, V. Bordas, A. Gaiba, A. Naylor, A.D. Rawlings, B.P. Slingsby, et al., 6-Heteroaryl-pyrazolo [3, 4-*b*] pyridines: potent and selective inhibitors of glycogen synthase kinase-3 (GSK-3), Bioorg. Med. Chem. Lett. 13 (2003) 3059–3062.
[94] J. Witherington, V. Bordas, S.L. Garland, D.M. Hickey, R.J. Ife, J. Liddle, 5-Aryl-pyrazolo [3,4-*b*] pyridines: potent inhibitors of glycogen synthase kinase-3 (GSK-3), Bioorg. Med. Chem. Lett. 13 (2003) 1577–1580.
[95] K. Bozorov, J. Zhao, H.A. Aisa, 1,2,3-Triazole-containing hybrids as leads in medicinal chemistry: a recent overview, Bioorg. Med. Chem. 27 (2019) 3511–3531.
[96] Z. Xu, 1,2,3-Triazole-containing hybrids with potential antibacterial activity against methicillin-resistant *Staphylococcus aureus* (MRSA), Eur. J. Med. Chem. 206 (2020) 112686.
[97] M. Xu, Y. Peng, L. Zhu, S. Wang, J. Ji, K.P. Rakesh, Triazole derivatives as inhibitors of Alzheimer's disease: current developments and structure-activity relationships, Eur. J. Med. Chem. 180 (2019) 656–672.
[98] N. Bulut, U.M. Kocyigit, I.H. Gecibesler, T. Dastan, H. Karci, P. Taslimi, et al., Synthesis of some novel pyridine compounds containing bis-1, 2, 4-triazole/thio-semicarbazide moiety and investigation of their antioxidant properties, carbonic anhydrase, and acetylcholinesterase enzymes inhibition profiles, J. Biochem. Mol. Toxicol. 32 (2018) 22006.
[99] L. Zribi, I. Pachòn-Angona, O.M. Bautista-Aguilera, D. Diez-Iriepa, J. Marco-Contelles, L. Ismaili, Triazolopyridopyrimidine: a new scaffold for dual-target small molecules for Alzheimer's disease therapy, Molecules 25 (2020) 3190.
[100] A. Głuszyńska, Biological potential of carbazole derivatives, Eur. J. Med. Chem. 94 (2015) 405–426.
[101] A. Caruso, J. Ceramella, D. Iacopetta, C. Saturnino, M.V. Mauro, R. Bruno, et al., Carbazole derivatives as antiviral agents: an overview, Molecules 24 (2019) 1912.
[102] P.Y. Wang, H.S. Fang, W.B. Shao, J. Zhou, Z. Chen, B.A. Song, et al., Synthesis and biological evaluation of pyridinium-functionalized carbazole derivatives as promising antibacterial agents, Bioorg. Med. Chem. Lett. 27 (2017) 4294–4297.
[103] R. Ghobadian, M. Mahdavi, H. Nadri, A. Moradi, N. Edraki, T. Akbarzadeh, et al., Novel tetrahydrocarbazole benzyl pyridine hybrids as potent and selective butryl cholinesterase inhibitors with neuroprotective and β-secretase inhibition activities, Eur. J. Med. Chem. 155 (2018) 49–60.
[104] W. Ni, H. Wang, X. Li, X. Zheng, M. Wang, J. Zhang, et al., Novel tadalafil derivatives ameliorates scopolamine-induced cognitive impairment in mice via inhibition of acetylcholinesterase (AChE) and phosphodiesterase 5 (PDE5), ACS Chem. Neurosci. 9 (2018) 1625–1636.
[105] B. Sameem, M. Saeedi, M. Mahdavi, A. Shafiee, A review on tacrine-based scaffolds as multi-target drugs (MTDLs) for Alzheimer's disease, Eur. J. Med. Chem. 128 (2017) 332–345.
[106] J. Grutzendler, J.C. Morris, Cholinesterase inhibitors for Alzheimer's disease, Drugs 61 (2001) 41–52.
[107] M.I. Fernández-Bachiller, C. Pérez, G.C. González-Munoz, S. Conde, M.G. López, M. Villarroya, et al., Novel tacrine–8-hydroxyquinoline hybrids as multifunctional agents for the treatment of Alzheimer's disease, with neuroprotective, cholinergic, antioxidant, and copper-complexing properties, J. Med. Chem. 53 (2010) 4927–4937.
[108] M.I. Fernández-Bachiller, C. Pérez, L. Monjas, J. Rademann, M.I. Rodríguez-Franco, New tacrine–4-Oxo-4*H*-chromene hybrids as multifunctional agents for the treatment of alzheimer's disease, with cholinergic, antioxidant, and β-amyloid-reducing properties, J. Med. Chem. 55 (2012) 1303–1317.
[109] Z. Najafi, M. Mahdavi, M. Saeedi, E. Karimpour-Razkenari, R. Asatouri, F. Vafadarnejad, et al., Novel tacrine-1, 2,3-triazole hybrids: *in vitro, in vivo* biological evaluation and docking study of cholinesterase inhibitors, Eur. J. Med. Chem. 125 (2017) 1200–1212.
[110] C. De Los Ríos, J. Marco-Contelles, Tacrines for Alzheimer's disease therapy. III. The PyridoTacrines, Eur. J. Med. Chem. 166 (2019) 381–389.

[111] M. Saeedi, M. Safavi, E. Allahabadi, A. Rastegari, R. Hariri, S. Jafari, et al., Thieno [2,3-*b*] pyridine amines: synthesis and evaluation of tacrine analogs against biological activities related to Alzheimer's disease, Arch. Pharm. 353 (2020) 2000101.

[112] K. Chalupova, J. Korabecny, M. Bartolini, B. Monti, D. Lamba, R. Caliandro, et al., Novel tacrine-tryptophan hybrids: multi-target directed ligands as potential treatment for Alzheimer's disease, Eur. J. Med. Chem. 168 (2019) 491–514.

[113] M.A. Ceschi, R.M. Pilotti, J.P. Lopes, H. Dapont, J.B. da Rocha, B.A. Afolabi, et al., An expedient synthesis of tacrine-squaric hybrids as potent, selective and dual-binding cholinesterase inhibitors, J. Braz. Chem. Soc. 31 (2020) 857–866.

[114] R. Guha, On exploring structure–activity relationships. *In silico* models for drug discovery. Methods Mol. Biol. 993 2013, 81–94.

[115] A. Rammohan, J.S. Reddy, G. Sravya, C.N. Rao, G.V. Zyryanov, Chalcone synthesis, properties and medicinal applications: a review, Environ. Chem. Lett. 18 (2020) 433–458.

[116] S. Wang, G. Dong, C. Sheng, Structural simplification: an efficient strategy in lead optimization, Acta Pharm. Sin. B 9 (2019) 880–901.

[117] R. Sheng, Y. Xu, C. Hu, J. Zhang, X. Lin, J. Li, et al., Design, synthesis and AChE inhibitory activity of indanone and aurone derivatives, Eur. J. Med. Chem. 44 (2009) 7–17.

[118] H. Sugimoto, Y. Yamanish, Y. Iimura, Y. Kawakami, Donepezil hydrochloride (E2020) and other acetylcholinesterase inhibitors, Curr. Med. Chem. 7 (2000) 303–339.

[119] J. Cummings, G. Lee, K. Zhong, J. Fonseca, K. Taghva, Alzheimer's disease drug development pipeline: 2021, Alzheimers Dement. 7 (2021) 12179.

[120] W. Summers, A. Koehler, G. Marsh, K. Tachiki, A. Kling, Long-term hepatotoxicity of tacrine, Lancet 1 (1989) 729.

[121] N.I.A., Fund Active, Alzheimer's: Related Dementias Clinical Trials and Studies, National Institute on Aging, BetheBethesda, MD 20892, USA, 2020, p. 15.

[122] S.Y. Hung, W.M. Fu, Drug candidates in clinical trials for Alzheimer's disease, J. Biomed. Sci. 24 (2017) 1–2.

[123] A.B. Forster, P. Abeywickrema, J. Bunda, C.D. Cox, T.D. Cabalu, M. Egbertson, The identification of a novel lead class for phosphodiesterase 2 inhibition by fragment-based drug design, Bioorg. Med. Chem. Lett. 27 (2017) 5167–5171.

[124] Y. Uto, Imidazo[1,2-*a*]pyridines as cholesterol 24-hydroxylase (CYP46A1) inhibitors: a patent evaluation (WO2014061676), Expert Opin. Ther. Pat. 25 (2015) 373–377.

[125] M. Nakashima, H. Imada, E. Shiraishi, Y. Ito, N. Suzuki, M. Miyamoto, et al., Phosphodiesterase 2A inhibitor TAK-915 ameliorates cognitive impairments and social withdrawal in N-methyl-D-aspartate receptor antagonist-induced rat models of schizophrenia, J. Pharm. Exper. Ther. 365 (2018) 179–188.

[126] T. Nishi, S. Kondo, M. Miyamoto, S. Watanabe, S. Hasegawa, S. Kondo S, et al., Soticlestat, a novel cholesterol 24-hydroxylase inhibitor shows a therapeutic potential for neural hyperexcitation in mice, Sci. Rep. 10 (2020) 17081.

[127] J. Neef, D.S. Palacios, Progress in mechanistically novel treatments for schizophrenia, RSC Med. Chem. 12 (2021) 1459–1475.

CHAPTER

14

Pyridine-based probes and chemosensors

*Pawan Kumar, Bindu Syal and Princy Gupta**

Department of Chemistry and Chemical Sciences, Central University of Jammu, Jammu, J&K, India

14.1 Introduction

A molecule that interacts with a specific analyte to produce a noticeable change is called a sensor. Interactions between the medium and the sensor lead to a change in optical properties which can be read, interpreted, and used for the creation of optical sensors. The terms *probe, sensor*, and *chemosensor* are used almost interchangeably by the supramolecular chemists to recognize the same (supra) molecular entity, a system capable of identifying analytes which upon interrogation with a technique of choice, gives a different signal from its free or unbound state. Based on IUPAC classification [1], optical sensors can be sectioned according to the type of optical properties, which have been applied for sensing:

(a) Absorbance caused by the absorptivity of the analyte itself or by a reaction with some suitable indicator;
(b) Reflectance measured using an immobilized indicator;
(c) Luminescence measured by the intensity of light emitted by a chemical reaction in the receptor system;
(d) Fluorescence measured as the positive emission effect caused by irradiation;
(e) Refractive index measured as the result of a change in solution composition;
(f) Optothermal effect measured by the thermal effect caused by light absorption;
(g) Light scattering caused by particles of definite size present in the sample;

Sensors offer a communication interface between humans and the world. The principle of a chemosensor design involves three major processes: (i) to separate analytes, (ii) to capture a particular analyte from a complex mixture, and (iii) to output a signal from chemosensor-analyte complex [2]. From analytical and supramolecular chemistry points of view, chemosensors are one of the substantial topics in chemistry [3–9] and chemists have been working

* Corresponding author

DOI: https://doi.org/10.1016/B978-0-323-91221-1.00013-0

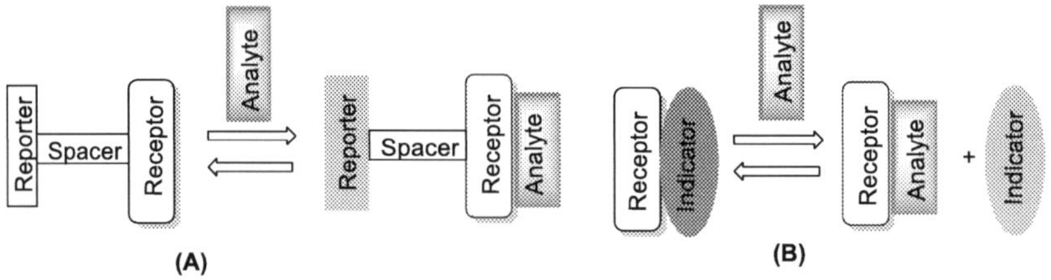

FIGURE 14.1 Illustration of the main approaches for the design of colorimetric chemical sensors: (A) receptor–spacer–reporter (RSR) and (B) indicator displacement assay (IDA).

on developing synthetic chemical sensors to detect various analytes due to their worth in environmental protection and biomedical analysis.

Colorimetric sensors are optical sensors that establish a distinguishable color change upon reaction with the analyte. The color change is usually detected by the naked eye and the change in intensity can be determined using special instrumentation which falls within the visible range. The design of colorimetric chemical sensors is based on the concept of assembly/disassembly which takes place between different molecules resulting in the emergence of an optical signal. It is worthwhile to mention that one of the desired properties of colorimetric sensors is their ability to recuperate and therefore chemical interactions involved in the design of chemical sensors are selective but reversible. There are two main methods for the design of colorimetric chemical sensors: (i) sensing based on direct binding of an analyte and (ii) sensing based on competitive binding of an analyte (Fig. 14.1) [10]. The receptor–spacer–receptor scheme is a classical representative of the direct binding approach which ages back to Emil Fisher's lock and key principle. In disparity, sensors developed using indicator displacement assay (IDA) scheme starts from a receptor–indicator assembly, employs competition between an analyte and the indicator, and signal change results from the displacement of indicator. The binding constants of the analyte to receptor should be comparable to that of the indicator to the receptor for the development of IDA-based sensors with desired sensitivity [11].

Colorimetric chemosensors for metals are becoming gradually predominant in the arena of inorganic and organometallic chemistry and allow for on-site analysis of samples without the use of costly and complicated instrumentation. The origin of a colorimetric sensor for heavy metal detection is that a color change will be detected in the presence of the metal species of interest and the change is preferably noticeable by the naked eye but sometimes can also be observed using a simplistic spectrophotometer [12]. All the different types of colorimetric sensors capable of detecting metal species have one common thing and that is the metal ion forms complex which results in color change. The color change is caused by the electron density on the chromophore being altered and many N-, S-, and O-containing organic molecules are good receptors due to their empty d-orbitals and lone pairs of electrons [13].

Fluorescent chemosensors are compounds which incorporate a binding site, a fluorophore and a mechanism for communication between these two sites and give fluorescence as the output signal [14]. The indicators are defined as fluorescent chemodosimeters if the binding sites are irreversible. It is pertinent to mention that these definitions as well as the term fluorescent probe have been used interchangeably and equivocally over the past decades. These sensors have unveiled tremendous advantages compared to traditional analytical methods like mass

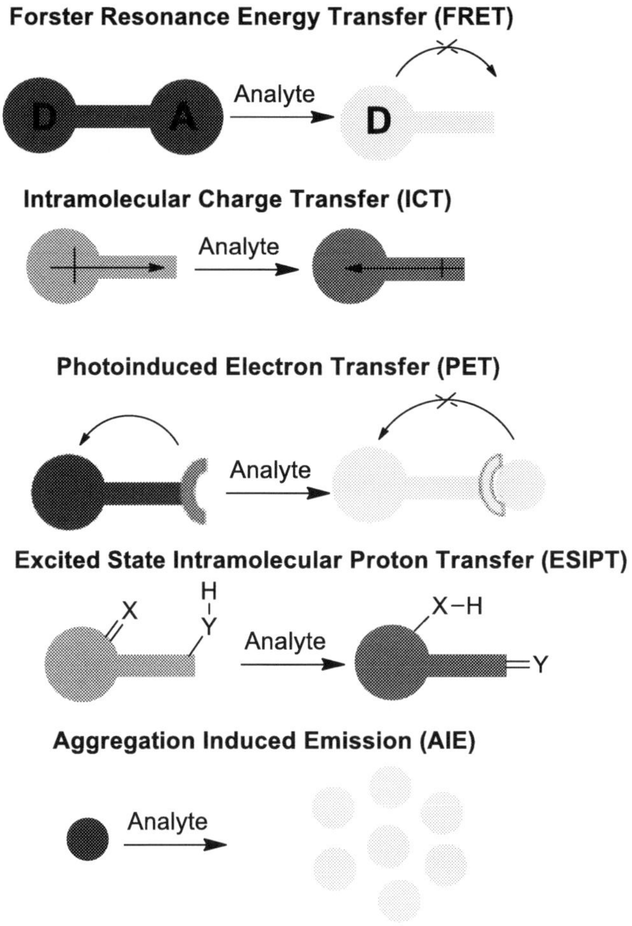

FIGURE 14.2 Schematic representation of the fluorescent mechanisms.

spectrometry [15], Raman spectroscopy [16], and ion mobility spectroscopy [17] because of high sensitivity, selectivity, fast measurement, low cost, and occasionally naked eye detection of fluorescence signals.

The design of fluorescent chemosensors is based on a system which consists of receptor and fluorophore as the key components. The receptor interacts with analyte through coordination, ionic interaction, hydrogen bonding or chemical reaction, and subsequently the fluorophore changes its photophysical property when interactions between receptor and analyte happen. The photophysical mechanisms usually used for analyte detection by fluorescent chemosensors include Förster resonance energy transfer (FRET) [18], intramolecular charge transfer (ICT) [19], photoinduced electron transfer (PET) [20], excited state intramolecular proton transfer (ESIPT) [21], aggregation-induced emission (AIE) [22], or involve dual/triple sensing mechanisms (Fig. 14.2) and the number of approaches is still increasing. The performance of a fluorescent chemosensor depends on selectivity, sensitivity while characteristics such as response time, recovery time, reversibility, overall lifetime cost are also taken into account for real applications [23]. Moreover, ease of use, high accuracy, and acceptable reproducibility of the results are also anticipated from the sensors which are expected from any other analytical tool [24]. The room of applicability of fluorescent chemosensors is extended to embrace numerous biologically important analytes particularly biologically or environmentally important cations, anions, small neutral molecules as well biomacromolecules have been advanced along with progression in microscopic imaging technologies [25].

Receptor moiety being one of the key components of chemosensors is vital for achieving optimal detection performance. In the earlier reports, different types of interaction between analyte and receptors have been used in chemosensors such as Lewis acid–base interaction [**26**], collision quenching [27], formation/cleavage of chemical bond [28,29]. Out of these, Lewis acid–base interaction is the most commonly used approach due to its rapid response. Electron donating systems containing nitrogen, sulfur, or phosphorus are the receptors used and aromatic N-heterocycle receptors especially pyridine and its derivatives occupy a prominent position which have been efficaciously introduced into either the main chains or side groups of chemosensors.

Pyridines have higher nitrogen contents and are capable of donating delocalized electrons to the system they belong to and this is due to the fact that nitrogen atom possesses greater number of nonbonding valence electrons. The presence of pyridine moiety in a molecule enriches the molecule with electron density and compatibly improves its stability and binding capability. It is familiar that bipyridine, terpyridine, and its derivatives show great abilities to coordinate a large number of metal ions and range could be extended from alkali metal ion selective systems to transition metal ion selective systems. Moreover, pyridine and its derivatives are reactive for metal complex-forming reactions such as N-oxidation, N-protonation, and quaternization with alkyl halide which can alter their optical and electrical properties [30].

This chapter aims to provide an overview of colorimetric and fluorescent chemosensors based on pyridine scaffold for ions and metals of biological interest with their characteristic rational designs as reported in the literature during the last 5–6 years. We hope this chapter instigates the development of novel and powerful colorimetric and fluorescent chemosensors for a wide range of applications.

14.2 Pyridine-based colorimetric chemosensors

It is well known that colorimetric chemosensors enable on-site monitoring of analyte by color change without the need for expensive equipment, time-consuming sample preparation, or trained operators [31]. Among various colorimetric chemosensors, pyridine-based chemosensors have acquired huge attention because of their increased detection perceptivity for sensing a wide range of environmental contaminants and bioactive chemicals [32].

14.2.1 Pyridine–pyrazole chemosensors

A novel chemosensor **1** (Fig. 14.3) consisting of pyridine and pyrazole moiety was developed by the reaction between substituted imidazole hydrochloride and pyridine-2-carboxaldehyde [33] which exhibited an excellent colorimetric response for the successive

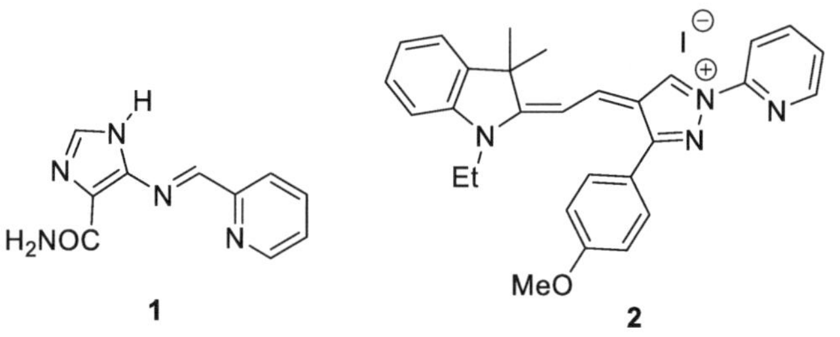

FIGURE 14.3 Pyridine–pyrazole-based chemosensors.

recognition of Ni^{2+} and CN^{-} ions. ICT transitions occurred from metal to **1** on adding Ni^{2+} to the aqueous solution of **1** that resulted in a color change from colorless to yellowish color of **1**-Ni^{2+} complex. The reaction of CN^{-} ions with **1**-Ni^{2+} brought an apparent color modification from yellow to colorless.

Another chemosensor hemicyanine-functionalized *N*-(2-pyridyl) pyrazole **2** (Fig. 14.3) was reported for CN^{-} detection [34], which is basically a donor–π–acceptor dye (dark yellow) in which donor aryl group is present in the pyrazole ring and the acceptor group is $C{=}N^{+}$. **2** exhibited two absorption bands, one at around 300 nm and the other in the range 400–450 nm due to π–π^* transition and ICT from the pyrazole N atom to the $C{=}N^{+}$. On enhancing the donation ability of the aryl ring of the pyrazole, the ICT band of the probe underwent a red shift. Due to the nucleophilic attack of the cyanide ions to the probe, the ICT band at 420 nm disappeared significantly and the band at 300 nm gradually decreased in intensity which resulted in a visible color change from dark yellow to colorless.

14.2.2 Pyridine–phenol chemosensors

Kim et al. [35] designed a chemosensor **3** (Fig. 14.4) composed of phenol and pyridine moiety for the sequential sensing of Cu^{2+} and CN^{-} ions. When **3** was treated with Cu^{2+} ions, it showed an absorption maximum at 436 nm along with the reduction in the intensity of absorption peak at 385 nm, and a noticeable color variation from colorless to yellow was observed. The exposure of this complex to CN^{-} ions induced a blue shift of the band at 436 nm and released the sensor into the solution and the color of the solution again changed from yellow to colorless.

Interestingly, another work regarding pyridine-phenol chemosensor **4** (Fig. 14.4) was reported by Wang et al. [36]. The prepared sensor contained the OH group and the N atom of the pyridine unit as chelating sites. A perceptible color transition from colorless to yellow could be easily noticed upon the addition of Cu^{2+} ions to **4**, attributed to ligand to metal charge transfer from the heteroatoms of the pyridine and hydroxyl groups to the Cu^{2+} ions.

Suzuki–Miyaura cross-coupling reaction between 2-(5-bromo-2-pyridylazo)-5-(diethylamino) phenol and tetraphenylethene boron derivative resulted in the formation of a pyridine-phenol chemosensor **5** (Fig. 14.4) which helped in the colorimetric recognition of Cu^{2+} and ${UO_2}_{+}{}^{2+}$ ions by varying the solvent mixture ratio [37]. Different colors were shown by the sensor in solid form (red color) and in dilute solutions (yellow color). The reaction of Cu^{2+} with **5** in DMSO/H_2O mixture changed the pale-yellow solution of **5** to purple. When various ions (Fe^{3+}, Co^{2+}, Ni^{2+}, UO_2^{2+}) were added into the solution of **5** in DMSO/H_2O

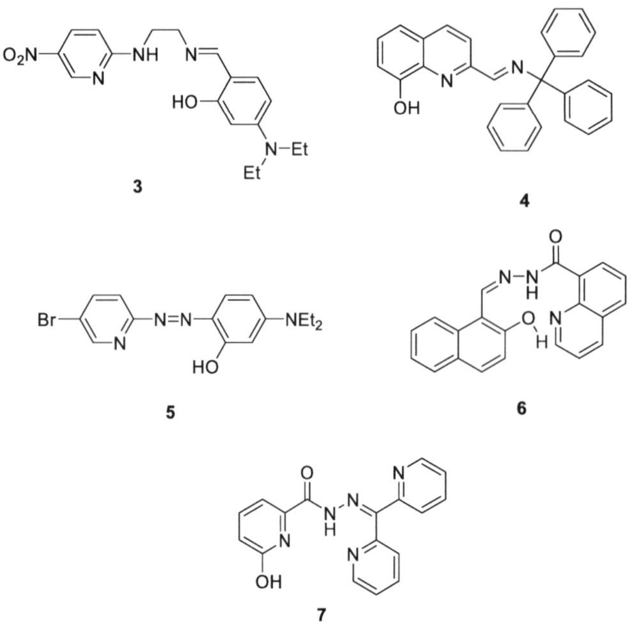

FIGURE 14.4 Pyridine–phenol chemosensors.

containing 40% water, a visible blue color was observed only in the case of uranyl cation whereas the other ions displayed a purple color.

Li and coworkers fabricated a colorimetric probe **6** (Fig. 14.4) via condensation reaction between 2-hydroxy-1-naphthylaldehyde and 8-quinolylhydrazine in ethanol solution [38]. The reaction of F^- with **6** led to the disappearance of the absorption peak at 369 nm and the appearance of a new band at 464 nm along with a shoulder peak at 443 nm compared to other anions in which only a little or no spectral change was observed. Meanwhile, due to the formation of N–H···F and O–H···F hydrogen bonds, there was a significant change in color from colorless to yellow.

A multiple-target chemosensor, *N′*-(di(pyridine-2-yl) methylene-6-hydroxypicolinohydride **7** (Fig. 14.4) was synthesized by Kim et al. [39]. Formation of **7**-Fe^{2+} complex on the reaction of **7** and Fe^{2+} prompted a color change from colorless to blue over other ions (Zn^{2+}, Cd^{2+}) that caused no color change. On the other hand, the coordination of Co^{2+} with **7** turned the solution yellow due to the formation of **7**-Co^{2+} ions (2:1) complex.

Pyridine–phenol containing dye **8** (Fig. 14.5) has been synthesized which served as the colorimetric chemosensor for the recognition of Cu^{2+} ions and cysteine [40]. When Cu^{2+} was bound to **8** in an aqueous solution at pH greater than 7, it was found that the new band appeared with an absorption maximum at 508 nm accompanied by the reduction in the absorption intensity at 412 nm and the yellow-colored solution of the sensor turned bright red. The prepared **8**-Cu^{2+} was then used for the sensing of cysteine (an amino acid). The exposure of the complex to cysteine caused the displacement reaction between the complex and cysteine leading to the binding of the cysteine to the Cu^{2+} and discharge of **8** into the solution which converted the bright red color of the solution to yellow.

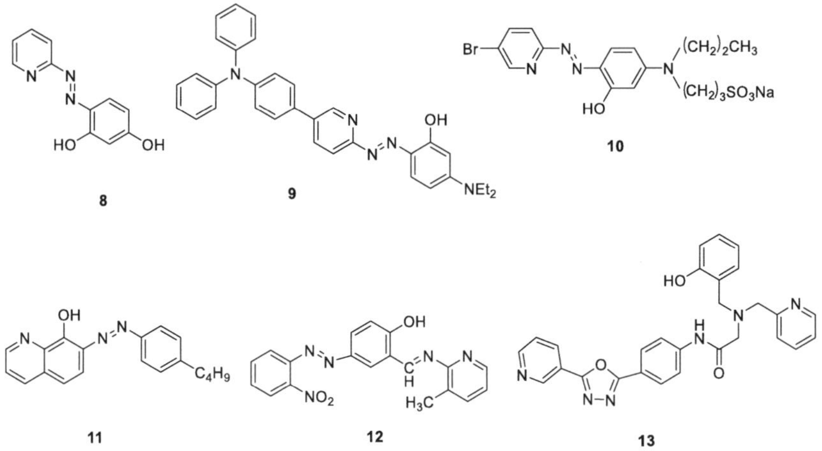

FIGURE 14.5 Pyridine–phenol chemosensors.

Uranium has been extensively employed in nuclear power stations for the production of nuclear fuel used to generate electricity and the migration of the radiations into the environment pollute it leading to severe health problems and even death [41]. Therefore, colorimetric sensor **9** (Fig. 14.5) was synthesized by adding triphenylamine to 2-(5-bromo-2-pyridylazo)-5-(diethylamino)phenol (Br-PADAP) for the monitoring of uranyl ions [42]. The solution of **9** was orange in color corresponding to absorption maxima at 480 nm and the absorption maxima shifted to 602.5 nm after the addition of UO_2^{2+} which turned the color of the solution blue.

Palladium complex of pyridine phenol substituted derivative **10** (Fig. 14.5) was employed by Morikawa et al. [43] for the detection of CN^- ions in the blood. A yellow-colored dye **10** on chelation with palladium ions using pyridinic nitrogen atoms formed a blue-colored complex. The addition of cyanide ions to the complex resulted in a color change from blue to yellow due to its competitive binding with palladium.

Interestingly, impregnating the probe **11** (Fig. 14.5) into the mesoporous silica and polymer monoliths resulted in a solid-state chemosensor [44] that was used for the sensing of Hg^{2+} ions. Due to metal to ligand charge transfer from Hg^{2+} to **11** on complexation, a distinct and noticeable color change from initial light-orange to deep red was seen.

Although trivalent metal ions (Cr^{3+}, Fe^{3+}, Al^{3+}) have their own biological and environmental significance, they are highly toxic in high doses [45]. Therefore, a Schiff-base probe **12** (Fig. 14.5) was employed to detect these ions colorimetrically [46]. The reaction of **12** with Cr^{3+}, Fe^{3+}, Al^{3+} in 4:1 DMF/H_2O produced corresponding complexes with 2:1 ligand–metal binding stoichiometric ratios. These reactions were accompanied by a remarkable change in color from yellow to colorless for Al^{3+}, brown for Fe^{3+}, and green for Cr^{3+}. Apart from detecting trivalent metal ions, probe **12** also detected HSO_4^- and the strong interaction between the Schiff-base sensor and HSO_4^- turned the yellow solution of **12** colorless.

Recently, Panunzi et al. [47] prepared a highly water-soluble and flexible multidentate ligand **13** for sensing Co^{3+}, Cu^{2+}, Fe^{3+}, V^{3+} (Fig. 14.5), which exhibited an absorption band

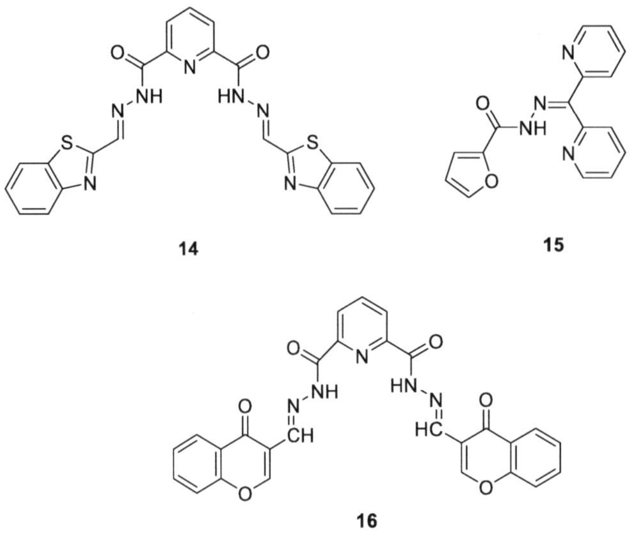

FIGURE 14.6 Pyridine–carbohydrazide chemosensors.

at 308 nm attributed to intramolecular $\pi-\pi^*$ transition and its solution was colorless. In the presence of first-row transition metal ions such as V^{3+}, Fe^{3+}, and Cu^{2+} in an acidic medium (pH = 6.25), a new peak was observed at 506 nm, 599 nm, and at 418 nm, respectively, in addition to a peak at 308 nm. Meanwhile, the colorless solution of the sensor changed to pink for V^{3+}, blue for Fe^{3+}, and yellow for Cu^{2+}. When other first-row transition metals were added, no color change was observed. At pH 8.25, the sensor displayed high selectivity for Co^{3+}, Cu^{2+}, and Fe^{3+}, and **13** exhibited an absorption maximum corresponding to broad shoulder at 430 nm, 490 nm, and 581 nm upon the addition of Fe^{3+}, Co^{2+}, and Cu^{2+}, respectively, which was accompanied by a color change from colorless to reddish for Fe^{3+}, peach-yellow for Co^{2+}, and lime-green for Cu^{2+}.

14.2.3 Pyridine–carbohydrazide chemosensors

Cu^{2+} ions and anions adenosine monophosphate anions (AMP^{2-}), F^-, AcO^- having biological importance could be detected by a chemosensor **14** (Fig. 14.6) containing carbohydrazide and pyridine moiety as receptor sites [48]. The sensor displayed immediate colorless to pale yellow color change on the addition of Cu^{2+}. On the other hand, **14** underwent a color change to light yellow (AcO^-, F^-) and to intense yellow (AMP^{2-}).

Harrison and coworkers [49] reported a colorimetric sensor **15** (Fig. 14.6) composed of pyridine and carbohydrazide moiety obtained by the reaction of 2-furoic hydrazide with di-2-pyridylketone using acid as a catalyst. When Fe^{2+} was bound to **15**, it created enhanced conjugation in the sensor which decreased the energy gap between frontier molecular orbitals and shifted the absorbance to a longer wavelength accompanied by a color change from colorless to blue. The sensor **15** developed a yellow color on complexation with Cu^{2+} ions and remained colorless in the case of other metal ions.

A colorimetric chemosensor based on pyridine–dicarbohydrazide **16** (Fig. 14.6) was synthesized by Khanmohammadi et al. [50] for the naked-eye detection of inorganic anions

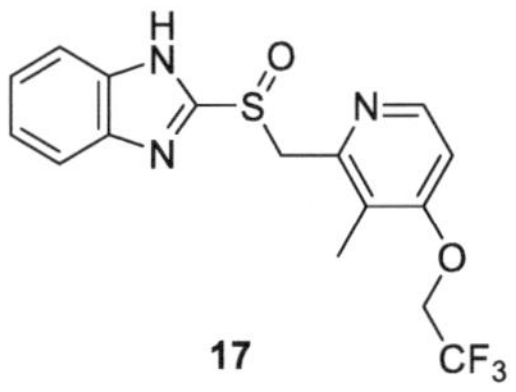

17

FIGURE 14.7 Lansopyrazole chemosensor.

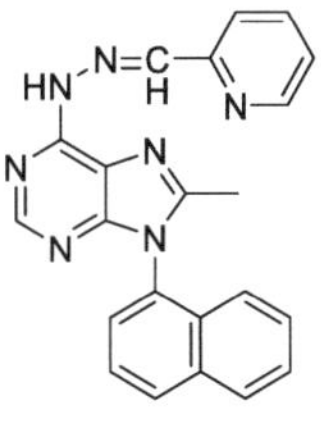

18

FIGURE 14.8 Naphthyl–pyridine chemosensor.

(CN^-) and cations (Zn^{2+}, Cu^{2+}) in an aqueous medium using chromone and pyridine–dicarbohydrazide which showed two absorption bands at 275 and 310 nm corresponding to p/p* transitions. On adding cyanide anions to receptor solution, the intensity of the absorbance peak at 310 nm reduced significantly and a new band at 430 nm emerged due to intramolecular charge transfer which resulted in a color change from colorless to yellow.

14.2.4 Lansopyrazole chemosensors

Carbonate ions are widely used in various industries for the production of paper, pulp, portable cement, rubber, lime, ceramic glazes, and lithium-ion rechargeable batteries and besides this, CO_3^{2-} in high concentration can cause severe diarrhea, abdominal pain, flatulence, and even death. Ziarani et al. employed benzimidazole-scaffold, Lansopyrazole **17** (Fig. 14.7) for the visible detection of CO_3^{2-} ions [51]. Upon introduction of carbonate ions to **17**, a bathochromic shift was observed in the solution of receptor which induced a color change from colorless to yellow.

14.2.5 Naphthyl–pyridine chemosensors

Cupric ions and sulfide anions could be detected by various sensors [52,53]. Lu et al. designed a purine derivative **18** (Fig. 14.8) for the visible monitoring of Cu^{2+} and S^{2-} ions [54]. Treatment of Cu^{2+} ions with **18** shifted the absorption band from 344 to 436 nm and induced color change from colorless to yellow due to ICT process. Interestingly, the prepared complex could detect S^{2-} via a color change from yellow ($\lambda_{max} = 436$ nm) to colorless ($\lambda_{max} = 344$ nm).

Recently, bidentate imine derivative **19** comprising of pyridine and naphthyl units has been synthesized following the pathway outlined in Scheme 14.1 [55]. The binding of the metal ions with **19** through nitrogen atom of the azomethine group and the oxygen atoms of the hydroxyl group developed a new complex which underwent intramolecular charge transfer from donor hydroxyl group to the metal ions and changed yellow-colored solution of **19** to a deep green

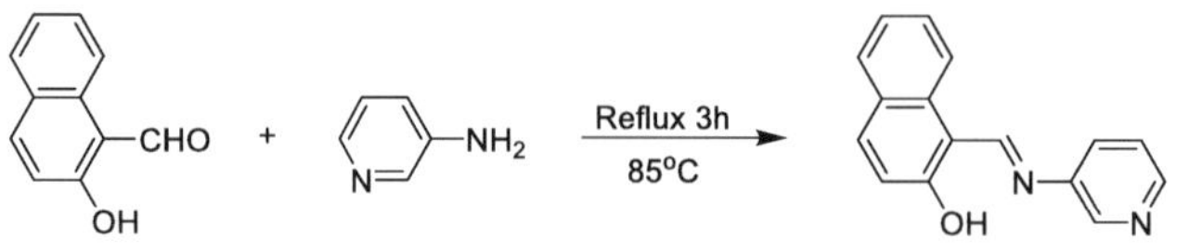

19

SCHEME 14.1 Synthesis of naphthyl–pyridine-based chemosensors.

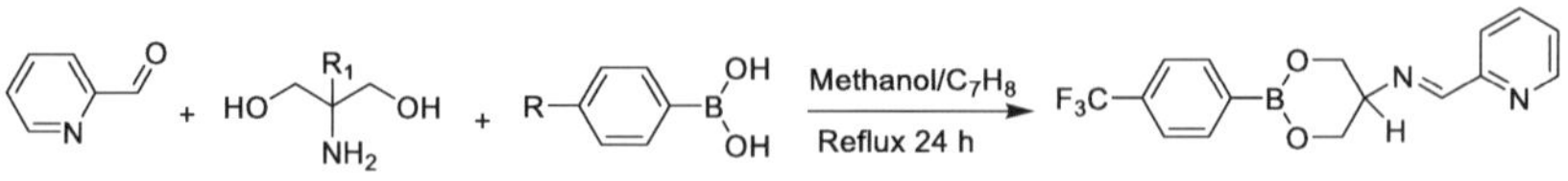

20

SCHEME 14.2 Synthesis of pyridine–boron chemosensors.

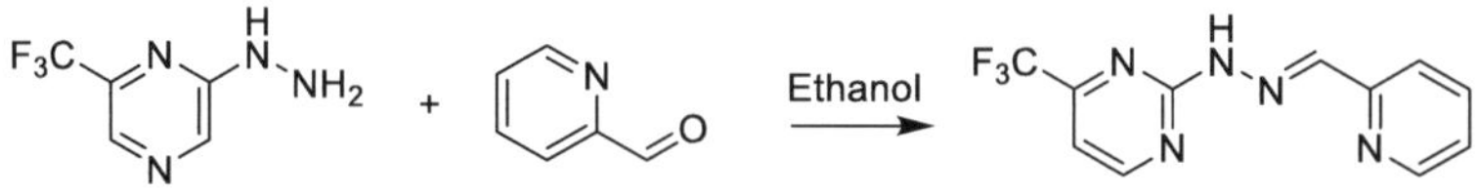

21

SCHEME 14.3 Synthesis of hydrazinyl–pyridine-based chemosensor.

for VO^{2+}, orange for Pd^{2+}, deep yellow for Mn^{2+} and Zn^{2+}, deep brown for Fe^{2+} and Cr^{3+}, dark green for Co^{2+}, green for Ni^{2+}, and green–yellow for Cu^{2+}.

14.2.6 Pyridine–boron chemosensors

Recently, Portillo and coworkers [56] developed an imine-boronic ester functionalized with pyridine moiety **20** (Scheme 14.2) that could selectively detect Fe^{2+}, Co^{2+}, Cu^{2+} ions. The interaction of the receptor with Fe^{2+}, Co^{2+}, Cu^{2+} ions led to the emergence of two new absorption peaks at 360 and 566 nm for Fe^{2+}, one peak at 360 nm and 462 nm in case of Co^{2+} and Cu^{2+} due to charge transfer from ligand to metal which changed the colorless solution of the sensor to purple, orange, and green, respectively.

14.2.7 Hydrazinyl–pyridine chemosensors

Jung et al. synthesized a selective and multifunctional chemosensor **21** (Scheme 14.3) based on hydrazinyl–pyridine components for the simultaneous sensing of soft metal ions [57]. The soft coordination environment of **21** due to the presence of three nitrogen atoms enabled it to bind to the soft metal ions in a nearly perfect aqueous solution and it recognized Fe^{2+}, Co^{2+}, and Cu^{2+} ions via color change from colorless to pink (Fe^{3+}) and yellow (Co^{2+}, Cu^{2+}).

Remarkably, another hydrazinyl pyridine-based receptors **22** and **23** could visually detect anions [58]. Receptors **22** and **23** (Fig. 14.9) contained furan and thiophene ring as a heterocyclic moiety respectively along with the presence of a nitro group at one of the carbon atoms adjacent to the heteroatom (O, S). On the treatment of **22** and **23** with various anions, **22** reacted only with fluoride ions and induced pale yellow to aqua color change, whereas

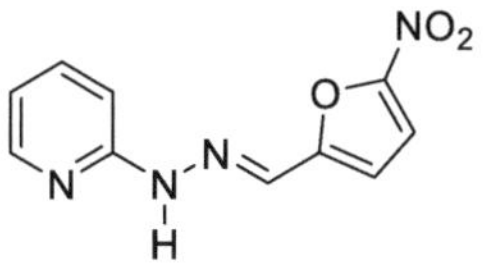

22

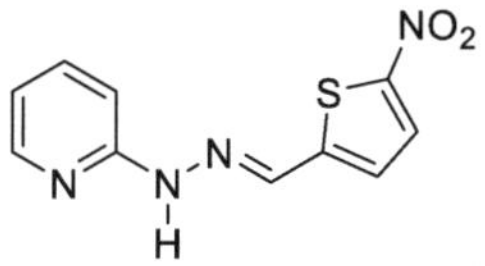

23

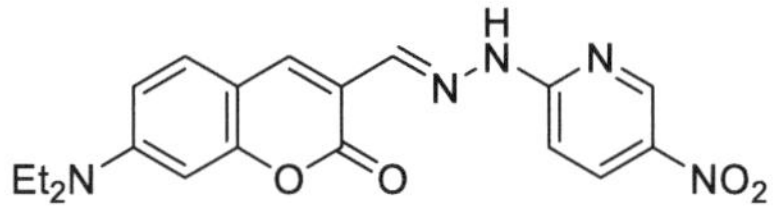

24

FIGURE 14.9 Hydrazinyl–pyridine chemosensors.

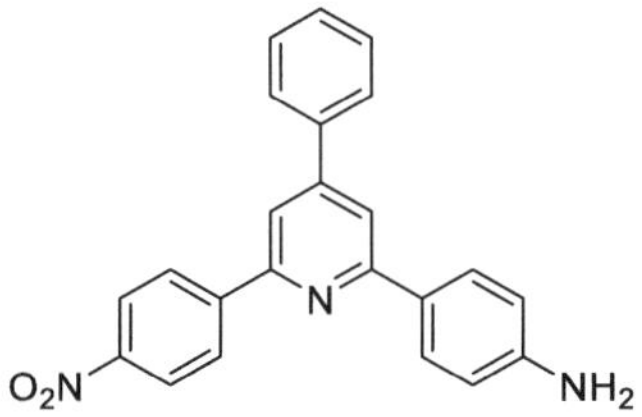

25

FIGURE 14.10 Triarylpyridine chemosensor.

23 discriminated between F^-, CO_3^{2-}, and AcO^- via a color change from pale yellow to aqua (F^-, CO_3^{2-}) and green (AcO^-).

Heavy transition metals such as Cu^{2+} and Ni^{2+} ions in water and human blood serum could be detected by a dual-responsive colorimetric probe **24** (Fig. 14.9) [59]. The absorption band of the probe at 538 nm amplified dramatically accompanied by a light yellow to red color change after adding Cu^{2+} and the complexation of Ni^{2+} with the probe **24** shifted the absorption peak to 494 nm and converted the light yellow color of **24** to purple. The color change was due to the chelation between the probe and the heavy transition metal ions followed by the ICT process.

14.2.8 Triarylpyridine chemosensors

Human exposure to formaldehyde can lead to respiratory ailments, irritation in the eyes, nose, and throat [60,61] and it could be detected by triarylpyridine chemosensor **25** [62]. Before adding formaldehyde, the solution of substituted triarylpyridine **25** (Fig. 14.10) was colorless ($\lambda_{max} = 400$ nm) and the compound formed after adding formaldehyde adsorbed blue–purple color light ($\lambda_{max} = 430$ nm) and displayed a complementary yellow color.

14.2.9 Pyridine–Schiff base chemosensors

Schiff base receptors **26** and **27** have been reported for the detection of Fe^{3+} ions [63]. The receptors showed an absorption band at 200 nm and it was colorless and an increase in the

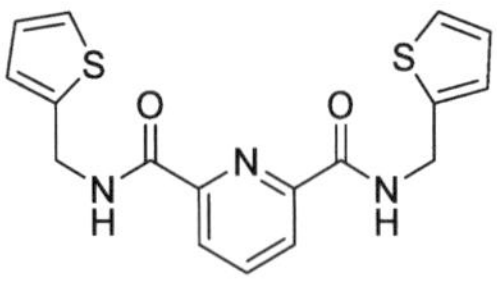

26

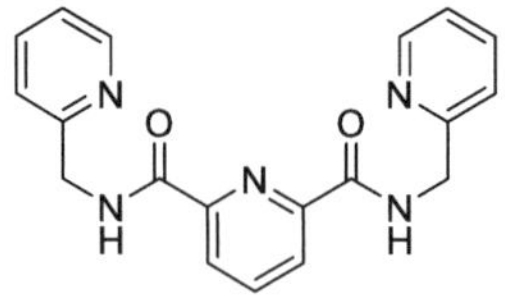

27

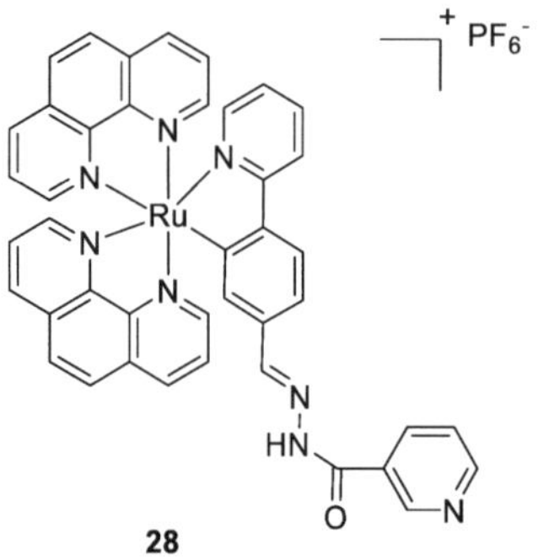

28

FIGURE 14.11 Pyridine–Schiff base chemosensors.

29

SCHEME 14.4 Pyridine–thiourea chemosensor.

amount of Fe^{3+} in the methanolic solution of **26** or **27** (Fig. 14.11) resulted in the development of two absorption bands at 270 and 360 nm which was visualized by a color change from colorless to bright yellow.

Another Schiff base complex **28** (Fig. 14.11) containing ruthenium metal and functional group nicotinohydrazide could selectively detect Cu^{2+} and Fe^{2+} ions in acetonitrile [64]. The probe displayed two absorption bands in the UV region, one at 326 nm due to π–π^* transition in the **28** and the other at 486 nm due to charge transfer transition from metal to **28**. When Cu^{2+} was added to the solution of the complex, the band at 486 nm completely diminished along with the reduction in the band intensity at 386 nm followed by an evident color variation from pink to yellow.

14.2.10 Pyridine–thiourea chemosensors

Sadaphal et al. [65] developed a chemosensor **29** (Scheme 14.4) by reacting 2-aminopyridine and phenyl isothiocyanate in the absence of solvent at ambient temperature. The solution of **29** was colorless because it had no visible light adsorption which on chelation with Cu^{2+} undertook a color transition from colorless to yellowish-green owing to the absorption maxima at 393 nm.

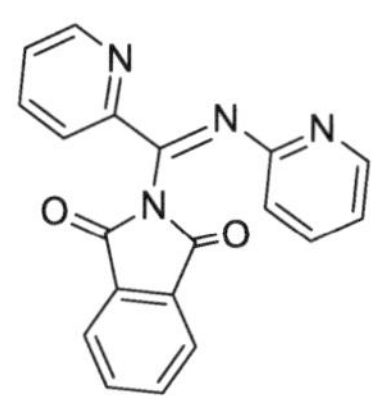

30

FIGURE 14.12 Pyridine–phthalimide chemosensor.

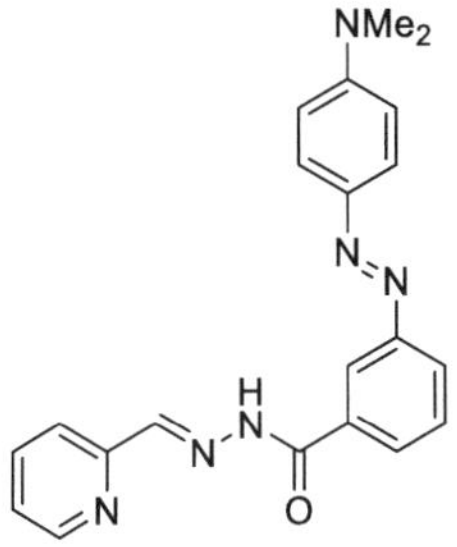

31

FIGURE 14.13 Pyridine–hydrazone chemosensor.

14.2.11 Pyridine–phthalimide chemosensors

Sahoo and coworkers [66] have reported a new substituted isoindoline-1,3-dione chemosensor **30** (Fig. 14.12) which exhibited two strong absorption bands at 288 and 352 nm due to π–π^* transitions. After the addition of Cu^{2+} ions, the absorption band at 288 nm underwent a blue shift to 277 nm and the band at 352 nm disappeared and there was an emergence of a new peak in the visible region at 434 nm which induced a distinguishable color variation from colorless to bright yellow.

14.2.12 Pyridine–hydrazone chemosensors

A colorimetric probe **31** (Fig. 14.13) containing tridentate chelating units has been prepared from *m*-methyl red dye for the recognition of Cu^{2+} ions in an aqueous medium [67]. The chelation of Cu^{2+} ions with **31** followed by ICT and metal to ligand charge transfer (MLCT) brought a visible color change of the solution from light yellow to red.

14.2.13 Dipicolinimidamide chemosensors

A new receptor based on dipicolinimidamide **32** (Fig. 14.14) has been fabricated by Patil and coworkers [68], which displayed an absorption band at 279 nm. The reaction of Cu^{2+} ions with **32** shifted the absorption peak to 290 nm and caused the color alteration from colorless to yellow. The ICT transitions from the nitrogen atom of the pyridine and amidine group to the copper ion were accountable for the red shift and color change.

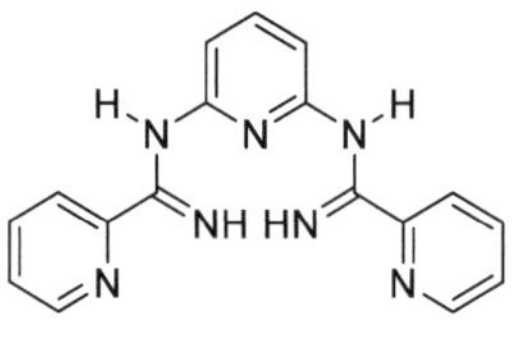

32

FIGURE 14.14 Dipicolinimidamide-based chemosensor.

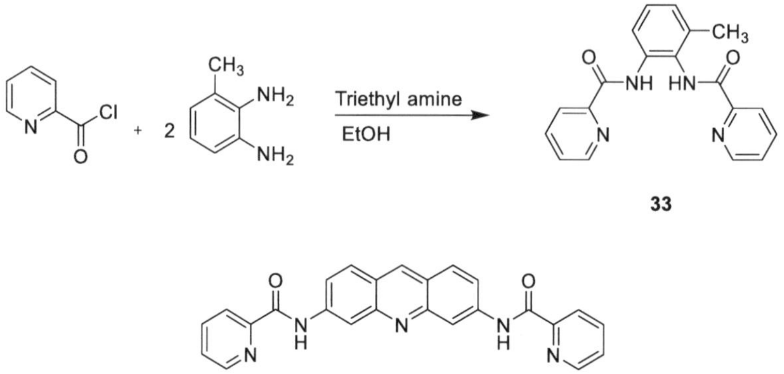

33

34

SCHEME 14.5 Chemical structure of pyridine–carboxamide derivatives.

14.3 Pyridine-based fluorescent chemosensors

14.3.1 Pyridine–carboxamide chemosensors

Faridbod and coworkers [69] introduced a fluorescent probe **33** prepared by the reaction of 3-methylbenzene-1,2-diamine and pyridine-2-carbonyl chloride, and used it for the detection of lutetium ions (Lu^{3+}) (Scheme 14.5). On the interaction of Lu^{3+} with **33**, the fluorescent emission was observed with a blue shift from 462 to 450 nm in ethanol/water solution at room temperature due to strong binding of lutetium ions with pyridine chemosensor and results indicated that the Lu^{3+} ions selectivity was much higher than that of common metal ions.

The fluorescent sensing probes for the recognition and imaging in the biological systems have proved to be an imperative approach due to their simplicity, easy operation and sensitivity in the detection process. For instance, when central fragment 3-methyl phenyl in **33** was replaced with acridine [**70**], compound **34** was synthesized and has been explored as a selective chemosensing agent for Cu^{2+} which might be due to the participation of nitrogen of pyridine and amide with Cu^{2+} leading to stronger chelation with compound **34** (Scheme 14.5). Moreover, the prepared compound was found to be a potential sensor for the recognition of Cu^{2+} in living cells with little cytotoxicity and decent imaging characteristics.

On the other hand, pyridine-2,6-dicarboxamide-based scaffolds are widely used for the discerning detection of various analytes owing to their outstanding features such as generation of a pincer cavity on binding with metals, chelation stabilization of metals in a higher coordination state, and the opportunity to introduce diverse groups/fluorophores [71,72].

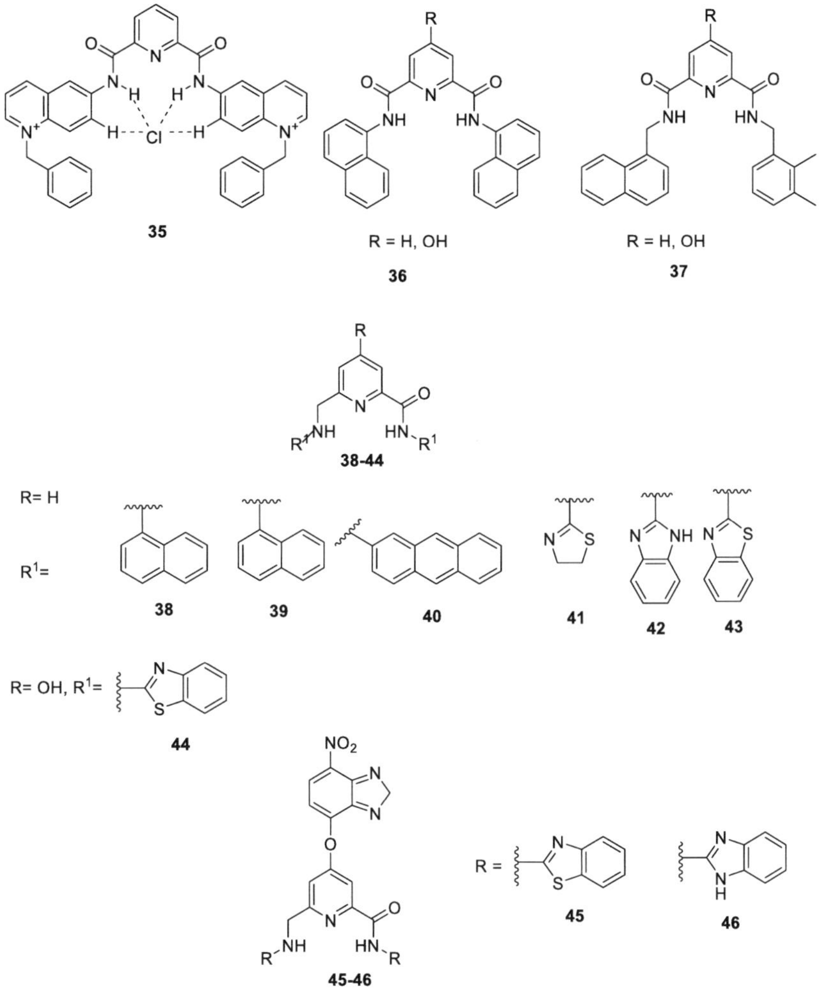

FIGURE 14.15 Pyridine-2,6-dicarboxamide scaffolds as fluorescent probes.

However, detection of a particular analyte depends on various factors like solvent, pH, fluorophore, etc. Sensing of chloride with different systems has been attained through hydrogen binding interactions despite the fact that water molecules, halides, and other pseudohalides efficiently compete for the binding sites. To overcome this problem, a highly sensitive water-soluble dicationic chloride sensor [73] was developed based on a pyridine-2,6-dicarboxamide compound **35** with suitable cleft for hydrogen bonding that resulted in blue fluorescence in an aqueous solution having a detection limit of 33 μmol/L. Only Cl^- formed a strong complex with receptors over other ions such as F^-, I^-, $CH_3CO_2^-$, NO_3^-, SO_4^{2-} (Fig. 14.15). As revealed

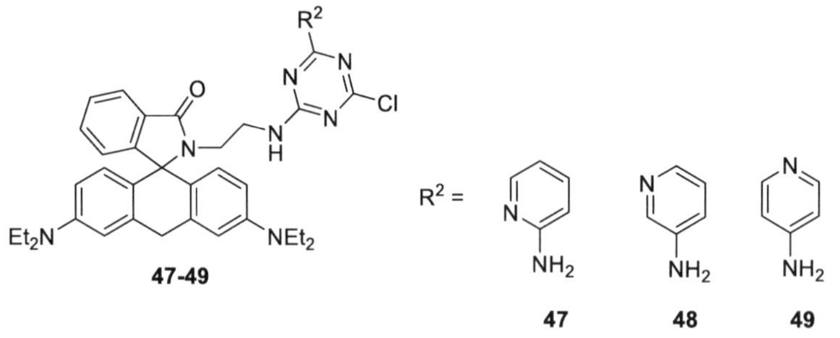

FIGURE 14.16 Pyridine–rhodamine chemosensors.

in the crystal structure calculations, the size of the cavity radius of **35** was 2.12 Å, which matched very well with H··· Cl^- distance of 2.22 Å resulting in strong binding interaction.

The selective detection of Pd^{2+} ions has also been achieved using pyridine-2,6-dicarboxamide scaffolds **36** and **37** (Fig. 14.15) in an aqueous medium [74]. Inclusion of Pd^{2+} to these compounds, when OH group is at the para position to pyridine ring displayed fluorescence quenching with red-shift from 395 to 406 nm at pH 7.2 but fluorescence intensity was considerably reduced.

Gupta and coworkers [75] reported the synthesis of various pyridine-2,6-dicarboxamide-based chemosensors (compounds **38–44**) for the selective detection of S^- and H_2S (Fig. 14.15). Compound **43** has shown good detection for S^- as well as H_2S and fluorescent emission at pH 7.2 was observed under UV light with the naked eye. On the other hand, hard donor sites have also been generated for the chemosensing of metal cations that prefer hard binding sites. The representative examples are compounds **45** and **46** that were applied for the fluorescent detection of Al^{3+} and Fe^{3+} at pH 7.2, utilized hard donor sites O and N for the selective sensing of Al^{3+} and Fe^{3+}. It was found that Fe^{3+} was coordinated to the pincer cavity whereas Al^{3+} was detected by 4-nitrobenzaoxadizole moiety [76].

14.3.2 Pyridine–rhodamine chemosensors

Rhodamine dye and its derivatives-based fluorescent probes are widely used for the recognition of different metal ions as well as intracellular imaging due to their unique features such as high extinction coefficient, photostability, high quantum yields, long emission wavelength, low detection limit, and compatibility in an aqueous medium [77–80]. Rhodamine is a fluorophore that acts as an acceptor in the generation of the fluorescence resonance energy transfer process.

Yan's group developed three pyridine–rhodamine derivatives **47–49** (Fig. 14.16) as fluorescent probes to recognize Fe^{3+}ions [81]. The addition of Fe^{3+} induced a significant bathochromic shift in absorption at 562 nm with emission band at 582 nm attributed to the ring-opening form of spirolactam and color change from colorless to pink only in the solution of **47,** in which 2-aminopyridine had correct spatial orientation for bonding (Scheme 14.6). Additionally, probe **47** has also been used for intracellular imaging in *HL-7702* cells for the recognition of Fe^{3+} ions.

Cao et al. [82] described a ratiometric fluorescent probe **50** that exhibited fluorescence change upon interaction with Hg^{2+} ions in EtOH/H_2O (2:8) system. The fluorescence

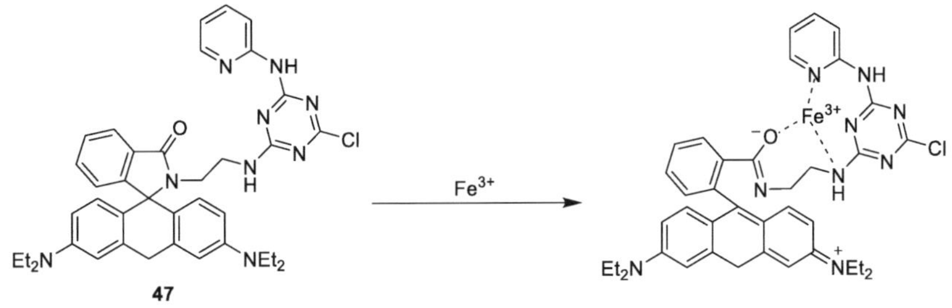

SCHEME 14.6 Interaction of Fe^{3+} with the pyridine–rhodamine chemosensor.

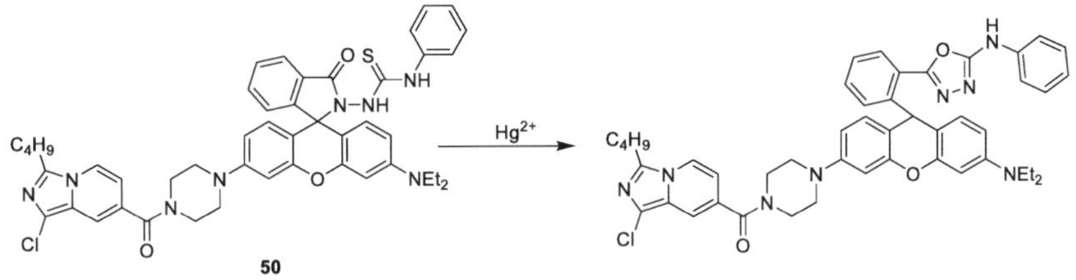

SCHEME 14.7 Sensor for sensing of Hg^{2+} ions.

spectrum showed a dramatic increase at 584 nm due to the ring-opening reaction of thiosemicarbazides to form 1,3,4-oxadiazoles promoted by the addition of Hg^{2+} ions (Scheme 14.7). In another approach, a highly potential dual-channel probe **51** was developed for Fe^{3+} and Hg^{2+} sensing by Liu et al. [83] where the detection was recorded within 2 s. When the fluorescence spectra were recorded in aqueous ethanol (1:1) at pH 7.4, a fluorescent band of 1,8-naphthalimide at 516 nm was decreased and fluorescent emission intensity at 578 nm was significantly enhanced over other metal ions that was attributed to the PET process. The ring-opening form of spirolactam was the possible associated mechanism upon chelation of Fe^{3+}/Hg^{2+} ions (Scheme 14.8).

Kan and coworkers [84] synthesized pyridine-3-sulfonyl chloride-based reversible fluorescent probe **52** for selective detection of Al^{3+} in aqueous ethanol $EtOH/H_2O$ (1:2) (Scheme 14.9). The addition of Al^{3+} to **52** increased the fluorescent intensity, and the solution turned pink from colorless which was visible even with the naked eye. Contrary to the previous reports [83], detection at physiological pH of 7.2 was ideal and at low pH values, the spironolactone ring was not stable which produced strong fluorescence and at higher pH, the fluorescence intensity was not significantly observed. However, the probe has efficiently been utilized for the detection of Al^{3+} not only in natural water but also in living cells as well as in plant tissues.

In another approach, the scope of these rhodamine–pyridine-based probes was expanded to a rhodamine–ferrocene conjugated pyridine **53** fluorescent system that was potentially utilized for the recognition of Hg^{2+} in water and in living cells [85]. Upon addition of Hg^{2+} to **53**, significant fluorescence enhancement at 590 nm with a color change to pink was observed. Additionally, spirolactam ring-opened with the binding of Hg^{2+} to chemosensor (Scheme 14.9), and the key role of pyridine ring was the generation of recognition center.

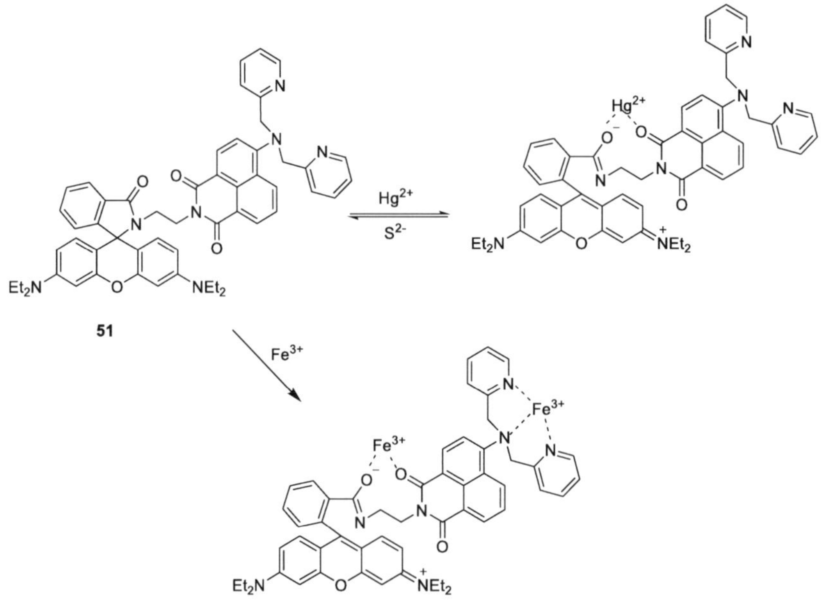

SCHEME 14.8 Mechanism of detection of Fe^{3+} and Hg^{2+} ions.

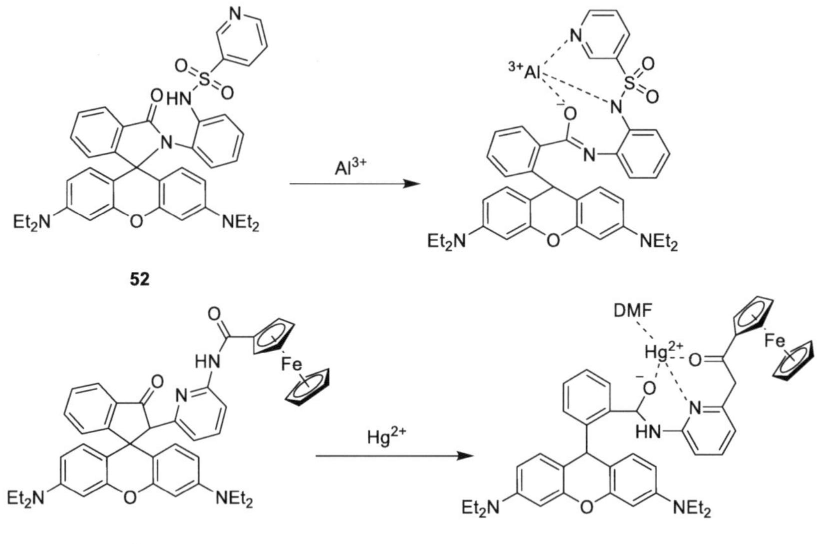

SCHEME 14.9 Proposed mechanism for the detection of Al^{3+} and Hg^{2+} ions.

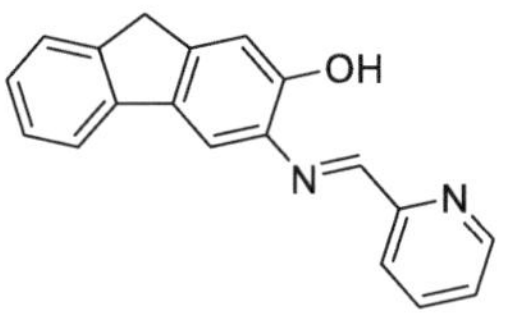

54

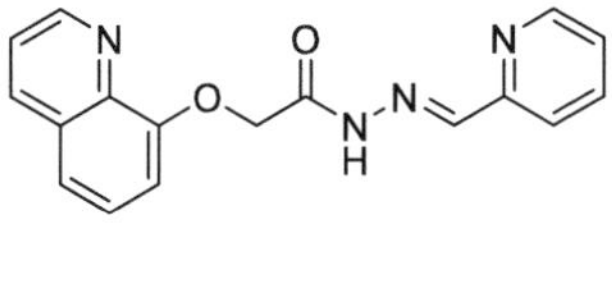

55

FIGURE 14.17 Chemical structure of pyridine–Schiff base chemosensors.

14.3.3 Pyridine–Schiff base chemosensors

Schiff base, an organic compound containing azomethine moiety synthesized by the condensation of aldehyde/ketone with a primary amine, is widely used as a ligand in various fields such as intermediates in organic synthesis, catalysis, chemosensors, food industry, polymers, etc. Schiff base has a strong affinity toward metal ions resulting in a wide range of applications such as antioxidative, antitumor, electronic, fluorescence, and photophysical properties [86–90]. Some of the recently reported pyridine–Schiff base fluorescent chemosensors have been discussed in this section.

A fluorine-bearing Schiff base–pyridine conjugated compound **54** was developed as a turn-on fluorescent sensor for Zn^{2+} and Cd^{2+} by Roy et al. [91]. It was found that the compound selectively reacted with Zn^{2+} and Cd^{2+} over other metal ions in CH_3CN/H_2O system at pH 7.4 with fluorescence emission centered around 430 nm inducing bathochromic shift at 485 nm and 495 nm in case of Zn^{2+} and Cd^{2+}, respectively. Spectroscopic studies indicated the coordination of metal through O atom of OH and N atom of imine group followed by deprotonation (Fig. 14.17). Compound **54** was also found to be applicable for intracellular imaging of Zn^{2+} and Cd^{2+} ions.

Karaoglu and coworkers [92] designed a metal displacement sensor **55** based on 8-hydroxyquinoline conjugated pyridine and utilized it for the detection of Fe^{2+} in an aqueous media. Introduction of Cu^{2+} to compound **55** resulted in quenching of fluorescence emission at 400 nm attributed to the five coordinated complex formation but noteworthy fluorescence enhancement was observed on addition of Fe^{2+} to **55**-Cu^{2+} complex, suggesting that **55**-Cu^{2+} complex served as the displacement probe (Fig. 14.17).

Hydrazine (N_2H_4) is a carcinogenic agent, widely used in different chemical industries as a strong reduced agent due to its good H_2O solubility, but its consumption brings serious health problems. A fluorescent sensor based on natural product derivatives **56** was developed for the recognition of hydrazine with prominent selectivity in living cells [93]. Fluorescence emission intensity was induced at 442 nm in phosphate buffer solution at pH 7.4 as a result of the reaction of hydrazine with compound **56**, but at lower pH, no reaction was observed (Scheme 14.10). Furthermore, the practical application of **56** was also explored for bioimaging of N_2H_4 in living *HeLa* cells.

Choudhury's group [94] had reported a copolymeric probe **57** based on tryptophan fused pyridine units for selective detection of Cu^{2+} and Hg^{2+} in aqueous media as well as for intracellular imaging of *U-87* neuroblastoma cells. The recognition mechanism was involved in the complex formation of metal via nitrogen atom of pyridine and imine group (Fig. 14.18) with a significant change in color from yellow to green and the binding mechanism was rationalized based on ^{1}H NMR, DFT, and IR studies.

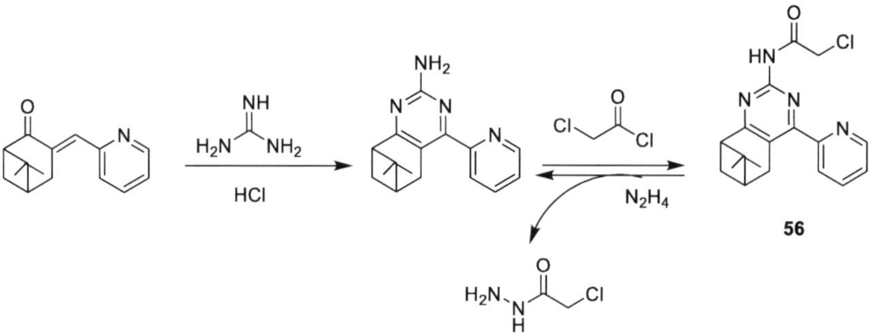

56

SCHEME 14.10 Proposed mechanism for sensing of hydrazine.

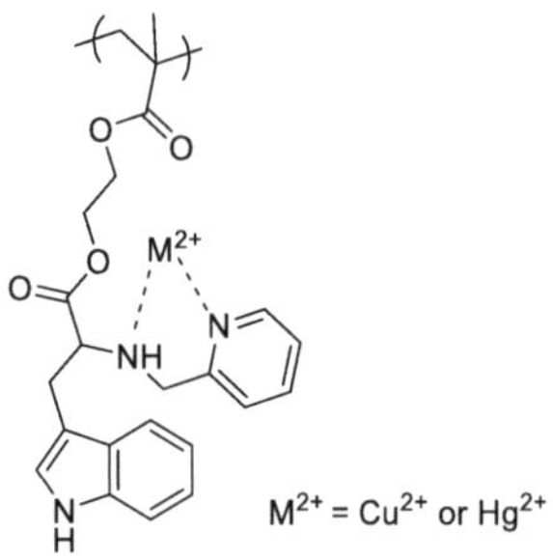

57

FIGURE 14.18 Structure of pyridine–imine chemosensor showing interaction with M^{2+} ions.

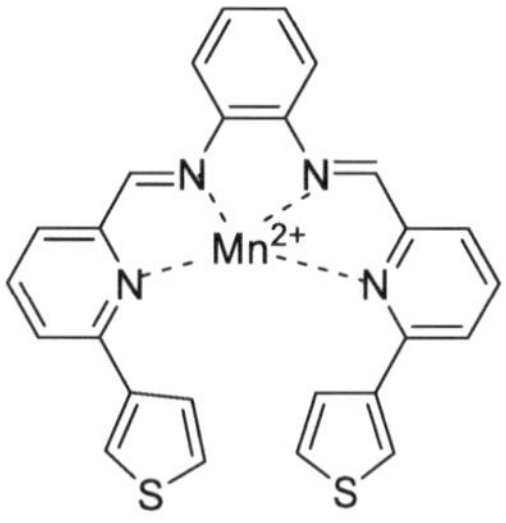

58

FIGURE 14.19 Structure of pyridine–Schiff base chemosensor.

Singh's group [95] devised a novel pyridine–Schiff base fluorescent chemoselective sensor **58** for the detection of Mn^{2+} ions which on interaction with probe **58** resulted in a bathochromic shift from 365 to 385 nm and fluorescence emission intensity was found at acidic pH (3–7). DFT calculations showed the distorted square planar geometry and it was also found that the coordination ability of N atom of imine was higher than that of N atom of pyridine (Fig. 14.19).

The detection of thiols in biological system was achieved by utilizing a fluorescent probe **59** bearing a phenol substituent linked to the Schiff base–pyridine system [96]. Compound **59** exhibited selective detection of glutathione (GSH) in the presence of cysteine in an aqueous solution, resulting in a decrease in fluorescence emission at 488 nm as a result of intramolecular

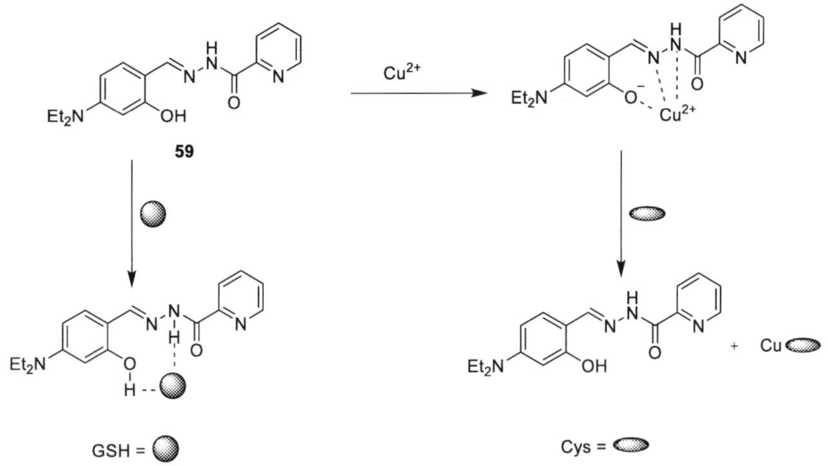

SCHEME 14.11 Proposed mechanism for sensing of glutathione and cystine.

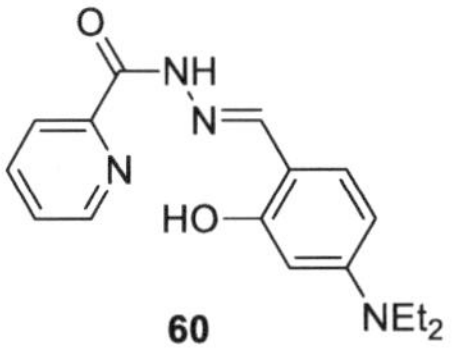

FIGURE 14.20 Structure of pyridine–Schiff base chemosensor.

hydrogen bonding. On the other hand, complex of **59** with Cu^{2+} ions was found selective for cysteine (Cys) in phosphate buffer saline (PBS) solution and fluorescent imaging in living cells (Scheme 14.11).

Wong and coworkers [97] proposed a fluorescent biosensor **60** that showed potential selectivity for sensing Al^{3+}/Zn^{2+} ions in living cells. Pyridine–Schiff base derivative **60** showed specific interaction for Al^{3+} and Zn^{2+} at pH 7.4, an associated fluorescence enhancement was observed at 504 and 575 nm for Al^{3+} and Zn^{2+}, respectively. Under acidic conditions, the dissociation of **60**-Al^{3+} complex was detected due to protonation of the probe and in basic pH conditions, the presence of OH groups were able to capture Al^{3+} resulted in hydrolyzed complexes (Fig. 14.20).

The recognition of Al^{3+} ions was also achieved using similar kinds of probes **61** and **62** in an aqueous system [98]. Inclusion of Al^{3+} to **61** and **62** induced the fluorescence emission at 483 and 467 nm with color change from colorless to aquamarine over other competitive ions like Co^{2+} and Fe^{3+} in DMF/H_2O (1:9) solvent at pH 7. There occurred coordination of O and both N atoms of imine as well as pyridine with Al^{3+} (Scheme 14.12) as suggested by the ^1H NMR studies.

Numerous efforts have been made to design and synthesize fluorescent biosensors based on Schiff-base to be employed in biological systems. Roy's group [99] utilized two such sensors **63** and **64** for pH dependent recognition of normal and cancer cells. At low pH, both the probes showed little fluorescence at 530 nm that increased with increase in pH, a dramatic increase

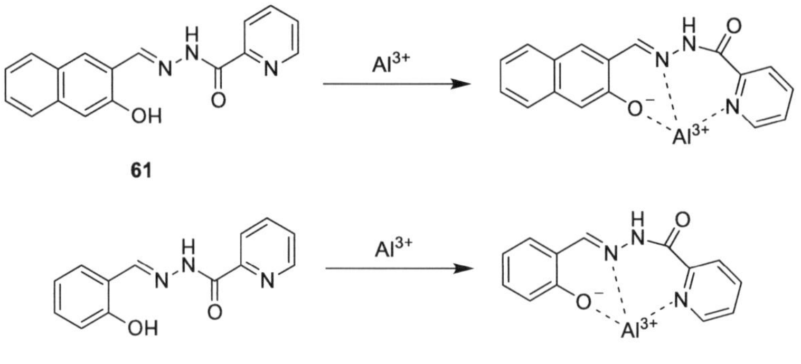

SCHEME 14.12 Structure and sensing mechanism of pyridine–Schiff base chemosensors.

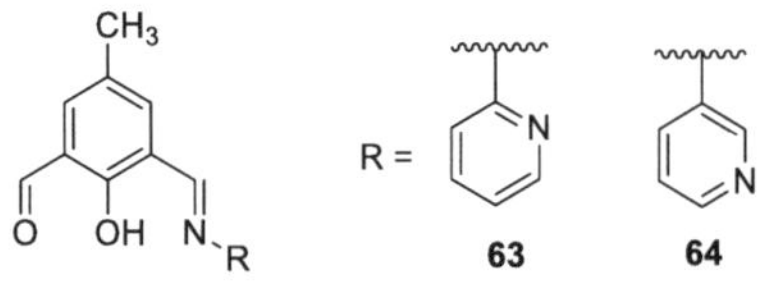

FIGURE 14.21 Chemical structure of pyridine–Schiff base chemosensors.

was observed at the higher pH (4–10) range. The solution of probes **63** and **64** exhibited a change in color from colorless to yellow (in acidic pH) and yellowish-green (in alkaline pH) observed by the naked eye. It was suggested that cancer cells have a pH range of 5.5–7.0, so the proposed method could potentially be applied in a biological system for selective recognition. Moreover, there was no considerable effect of quinoline or pyridine fragment of probes **63** and **64** on fluorescent properties (Fig 14.21).

14.3.4 Bi/terpyridine chemosensors

Wang et al. [100] have explored the pyridine-linked naphthalimide derivative **65** that enabled efficient discrimination between Zn^{2+} and Cd^{2+} ions via amide tautomerization. The fluorescence intensity was significantly increased on adding Cd^{2+} which led to amide tautomerization in aqueous ethanol under a wide pH range (4–9) over other metal ions. The bright yellow and cyan emission fluorescence was observed for the complex of **65** with Cd^{2+} and Zn^{2+}, respectively. It was experimentally found that the compound was permeable to the cell membrane and its stability over a wide pH range enabled it to be used for sensing Cd^{2+} ions in living systems (Scheme 14.13).

The detection of Hg^{2+} and Fe^{3+} ions was achieved using three dipicolylamine derivatives bearing dipyridine, **66, 67,** and **68** by Ajavakom and coworkers [101]. Compound **68** showed substantial fluorescence enhancement upon addition of Hg^{2+} in aqueous medium while that of Fe^{3+} in acetonitrile was attributed to PET inhibition in the probe. Metal ions were coordinated to a bidentate cavity of dipyridine as demonstrated by the ^1H NMR experiments (Fig. 14.22).

A bipyridine-based fluorescent system expended to porphyrin-linked bipyridine sensor **69** for the detection of Cd^{2+} was introduced by Shen et al. [102] in which blue shift emission

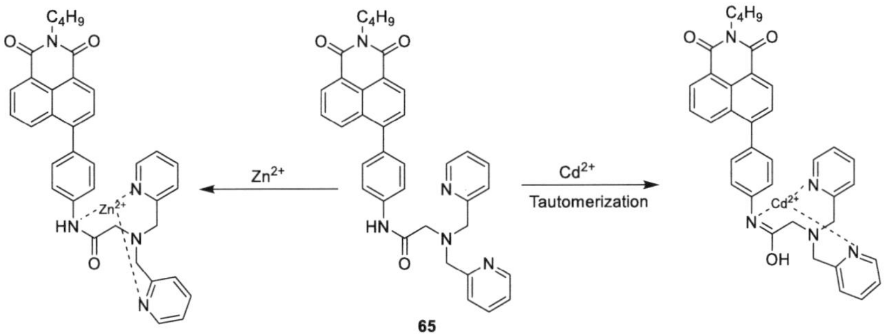

SCHEME 14.13 Structure of bipyridine derivative and its binding with Zn^{2+} and Cd^{2+} ions.

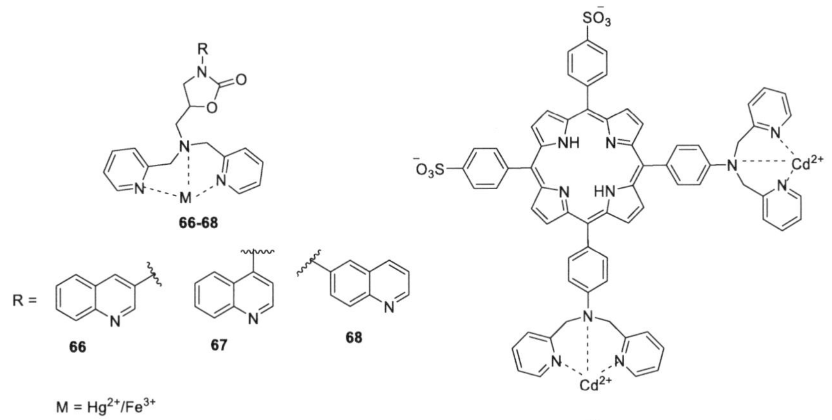

FIGURE 14.22 Structure of bipyridine-based probes and their interaction with metal ions.

was observed on the interaction of Cd^{2+} with the sensor due to ICT process. ^{1}H and ^{13}C NMR analysis were performed to illustrate the binding behavior and studies indicated that Cd^{2+} ions coordinated to bipyridine cavity led to a weakened electron-donating ability of dipyridine to a porphyrin system. The solubility of **69** in water enabled the system to be utilized for intracellular imaging under physiological pH conditions (Fig. 14.22).

In 2020, Asha's group [103] reported a novel graphene oxide bearing dipyridine moiety-based green fluorescent probe **70**, the addition of Zr^{4+} ions to which induced fluorescence enhancement response that was attributed to the FRET mechanism involved during the complex formation. Fluorescence spectra were recorded in an aqueous solution at pH 7; no complex formation was observed at lower or higher pH because of protonation of pyridinic nitrogen at a lower pH range resulted in blocking of the binding sites for Zr^{4+} ions (Fig. 14.23).

A dipyridine appended to pyrene moiety-based probe **71** showed selective detection for Cu^{2+} ions in an aqueous system as well as in living cell under physiological pH (4–9) [104]. The fluorescence emission intensity at 466 nm was significantly enhanced in presence of Cu^{2+} ions over other tested metal ions (Fig. 14.23).

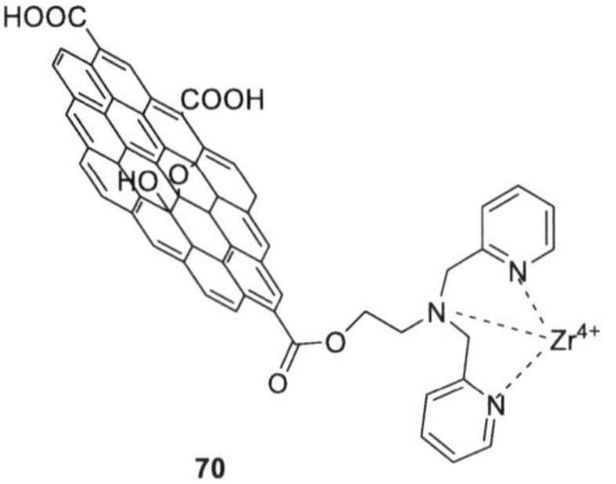

70

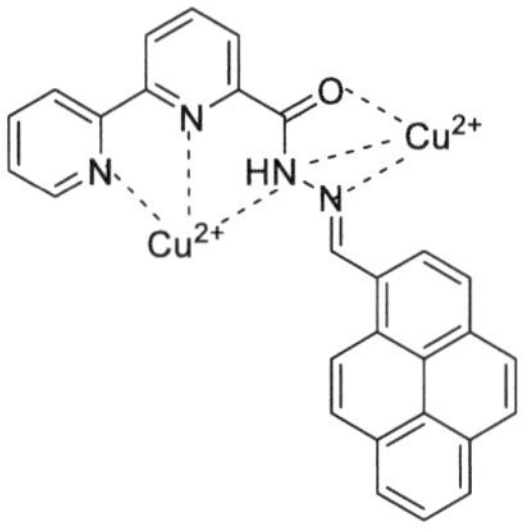

71

FIGURE 14.23 Structure of bipyridine-based probes for sensing of Zr^{4+} and Cu^{2+} ions.

Yang et al. [105] synthesized a terpyridine bearing copolymeric fluorescent probe **72** for the selective detection of Mn^{2+} and Ni^{2+}. Solution of **72** in THF resulted in complete fluorescence quenching when Mn^{2+} and Ni^{2+} ions were introduced due to strong chelation of terpyridine to these transition metal ions (Fig. 14.24).

Patra's group [106] reported terpyridine-based highly selective turn-on fluorescent probe **73** for the detection of Cd^{2+} ions (Fig. 14.24). The reaction of probe with Cd^{2+} showed significant fluorescence enhancement and an increase in quantum yield due to metal chelation enhanced fluorescence (MCHEF) effect. The chemosensor competently distinguished Cd^{2+} and Zn^{2+} based on the change in color of the solution. The blue color of the solution changed to emissive bright greenish for Cd^{2+} and nonemissive pale yellow for Zn^{2+} ions, both visible under UV radiance by the naked eye.

Lu and coworkers [107] developed a turn-on terpyridine containing a cationic poly(perylene diimide) derivative **74** for selective detection of Fe^{3+} ions in pure water and in living systems. The addition of Fe^{3+} ions to compound **74** induced emission at 548 and 591 nm due to PET process. Moreover, in the presence of a reducing agent like $Na_2S_2O_3$, the emission decreased instantly which was attributed to the reduction of Fe^{3+} to Fe^{2+}, showing that the probe could potentially be utilized to monitor Fe^{3+}/Fe^{2+} transitions (Fig. 14.24).

Another polymeric fluorescent probe **75** having terpyridine moiety used for Fe^{3+} ion sensing was reported by Yang et al. [108] in 2016. The complete quenching in fluorescence by Fe^{3+} resulted due to PET mechanism in THF solution. Adding CN^- to **75**-Fe^{3+} complex brought a prominent increase in fluorescence intensity due to strong coordination between CN^--**75**-Fe^{3+} complex and demonstrated the protuberant sensitivity and selectivity of the chemosensor (Fig. 14.24).

Shen and group have explored the selective detection of Ga^{3+} and Cr^{3+} ions in vitro and in vivo systems [109]. Adding Ga^{3+} and Cr^{3+} ions to 4-chloro-7-nitrobenzo-2-oxa-1,3-diazole-based derivative **76,** led to fluorescence enhancement due to PET process from donor to acceptor. The detection of these metals was assessed exogenously in cancer cells as well as intracellular imaging in Zebrafish (Fig. 14.24).

The use of nitro-based compounds in explosive materials may have both short-term or long-term exposure to soil and water system which ultimately results in serious health problems. So, there is a need for the detection of such toxic components by using rapid and selective sensor methods with ease of operation. Patra et al. [110] have reported terpyridine-based conjugated probes **77–80** for the sensitive sensing of nitroaromatic compounds in solution and in vapors visible by the naked eye (Fig. 14.25). The addition of nitroaromatic compounds like

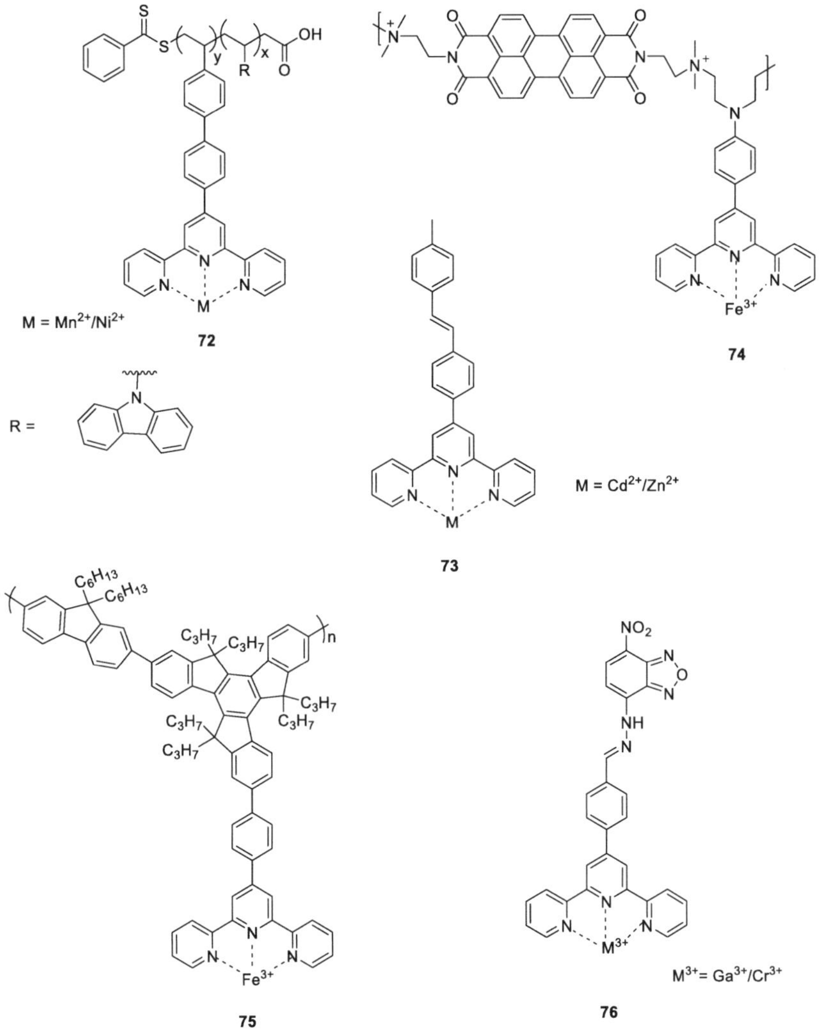

FIGURE 14.24 Chemical structure of terpyridine-based probes.

nitrotoluene, dinitrotoluene, nitrobenzene, hydroxy nitrobenzene, nitro benzoic acid, etc., to the probes **77–79** induced blue fluorescence while **80** displayed green emission. The quenching process was attributed to PET mechanism from the electron-rich probe to electron-deficient nitro compounds upon complex formation as illustrated by ^{1}H NMR titrations.

14.3.5 Pyridine-BODIPY (4,4-difluoro-4-borato-3a,4a-diaza-s-indacene) chemosensors

In 2015, pyridine-BODIPY-based conjugated fluorescent chemosensor, **81** was used for the recognition of Au^{3+} ions by Emrullahoglu and coworkers [111]. Compound **81** showed low

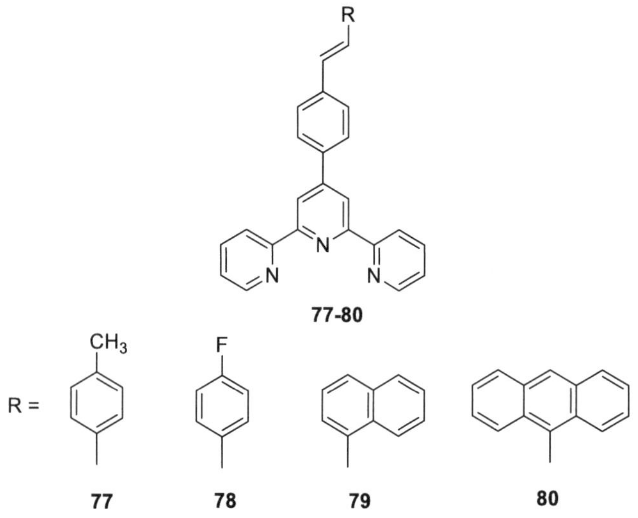

FIGURE 14.25 Chemical structure of probes for the detection of nitroaromatic compounds.

fluorescence at 591 nm but momentous fluorescence increase was detected on adding Au^{3+} with a blue shift from 591 to 577 nm at pH 7.2. Au^{3+} coordinated to the N atom of pyridine which was confirmed by ^{1}H NMR data (Fig. 14.26) and the reduction in fluorescence intensity on the addition of CN^{-} verified the reversibility of probe.

Kim's group [112] reported ethynylpyridine-linked E-BODIPY sensor **82** for the fluorescent detection of Zn^{2+} ions. Fluorescence intensity increased on chelation of Zn^{2+} ions to probe with blue shift (at pH 6). Tetrahedral geometry was predicted in which both N atoms of pyridyl groups were involved in coordination with zinc metal ions (Fig. 14.26).

A novel tetrapyridine-based fluorescent probe **83** was explored by Bharadwaj et al. [113] for the detection of either Hg^{2+} or Cd^{2+} ions in an aqueous medium and it has shown fluorescence enhancement when added to the solution of **83**. However, Cu^{2+} ions were seen competing due to good binding affinity toward picolyl amine substituent of the probe although fluorescence was relatively lower than that of Hg^{2+} or Cd^{2+} ions (Fig. 14.26).

Turan and coworkers [114] have also reported the fluorescence detection of Zn^{2+} ions by using two probes **84** and **85** in biological systems. On adding Zn^{2+} to probe **84**, an instant decline in fluorescence intensity with red shift was observed. **84**-Zn^{2+} system was potentially employed in the biological system for the recognition of cancer cells (Fig. 14.26).

Xu et al. [115] developed a dual responsive turn-off fluorescent probe **86** which exhibited fluorescence quenching in methanol at 589 nm in presence of Cu^{2+} or Ni^{2+} that might be due to the transfer of excited electrons from the probe to unoccupied 3d orbital of metal ions. The detection limit for both the metals was found as low as $0.1 \mu M$ and the binding affinity between probe and Cu^{2+}/Ni^{2+} was very strong that resulted in stable complex formation (Fig. 14.26).

14.3.6 Pyridine–pyrazole chemosensors

A novel pyridine–pyrazole probe **87** was developed by Portilla et al. [116] for the detection of CN^{-} ions that displayed fluorescence emission with a bathochromic shift at 490 nm in

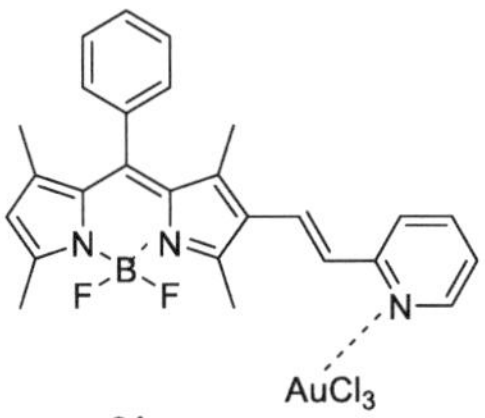

81

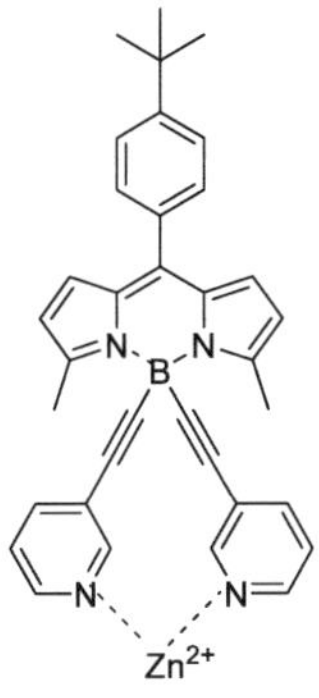

82

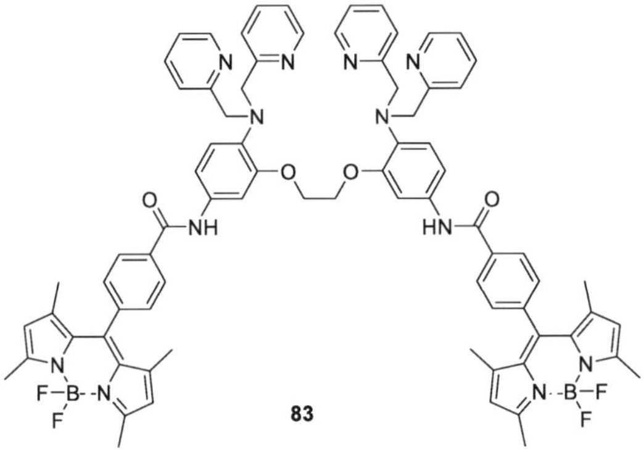

83

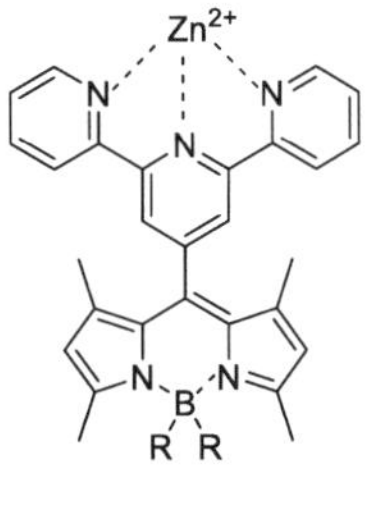

R = Methoxytriethyleneglycol

84

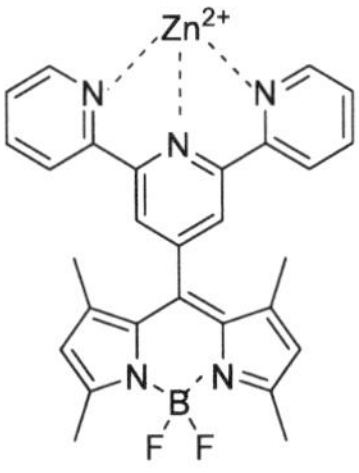

85

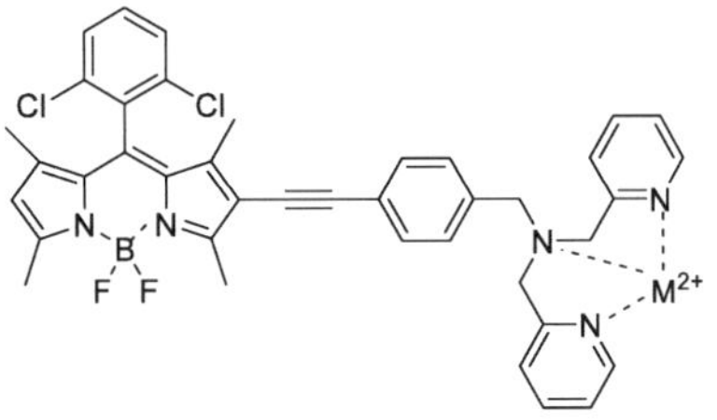

$M^{2+} = Cu^{2+}/Ni^{2+}$

86

FIGURE 14.26 Molecular structures of pyridine–BODIPY-based chemosensors.

an aqueous medium. The emission was attributed to ICT process, due to donor–π–acceptor (D–π–A) nature of CN^- ions and Michael addition of CN^- to C=C double bond of **87** was associated mechanism as unveiled by ^1H NMR spectra (Scheme 14.14).

A coumarin-linked pyrazole-pyridine probe **88** showed ON–OFF–ON behavior for the selective detection of fluoride (F^-) ions in DMSO, as developed by Seferoglu et al. [117]. The fluorescence emission centered at 482 nm was due to tautomerization of **88** and the sensing mechanism revealed that the pyrazole ring acted as an electron source as well as a hydrogen

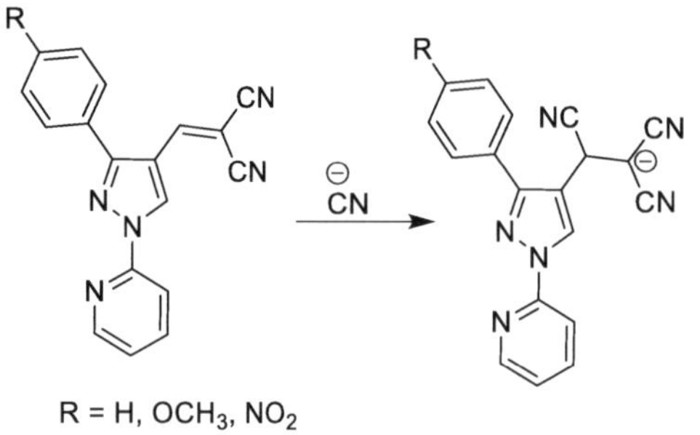

87

SCHEME 14.14 Proposed mechanism of sensing of cyanide ions.

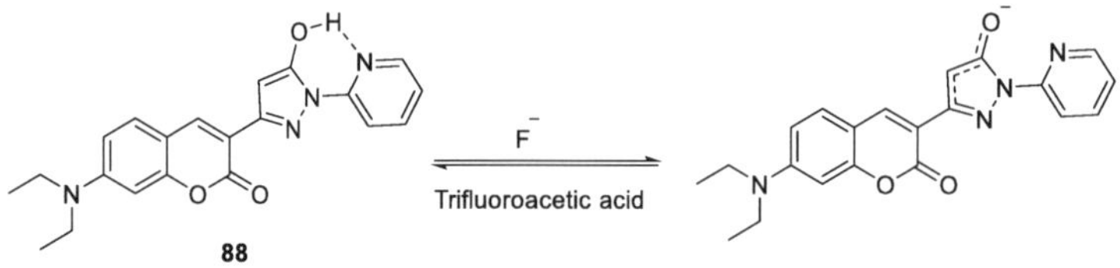

88

SCHEME 14.15 Sensing of fluoride ions.

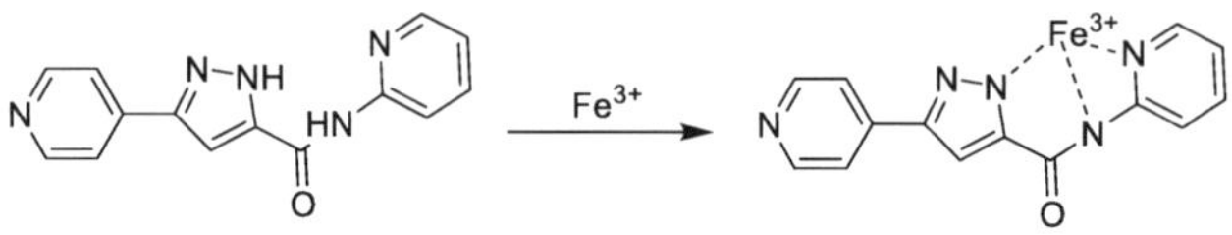

89

SCHEME 14.16 Sensing mechanism of Fe^{3+} ions.

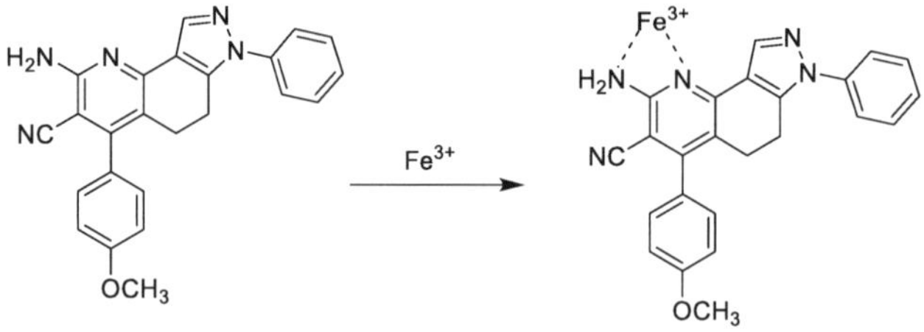

90

SCHEME 14.17 Sensing mechanism of Fe^{3+} ions.

bond donor whereas the tautomer was stabilized by pyridine. The key role of coumarin substituent was to accept the electrons via PET process (Scheme 14.15).

Madhu's group [118] explored the selective detection of Fe^{3+} ions using pyridine-based probe **89** in DMSO/H_2O (9:1) solution (Scheme 14.16). The fluorescence emission involved ICT from pyridine moiety to Fe^{3+} environment. In complex formation, N atoms of both pyridine and amide units offered the binding site to metal ions as indicated by ^1H NMR and FT-IR spectra.

The detection of Fe^{3+} ion in drinking water was also accomplished by the use of pyrazolo-quinoline derivative **90** in DMSO solvent that exhibited blue shift in fluorescence emission [119]. ICT from the methoxy group of the phenyl ring to the electron-withdrawing cyanide group of pyrazolo-quinoline moiety was ascribed to fluorescence emission (Scheme 14.17).

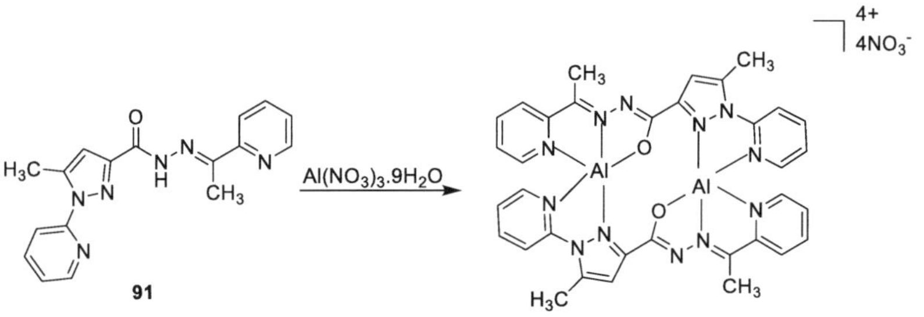

SCHEME 14.18 Structure of pyridine–pyrazole probe and its complex with Al^{3+} ions.

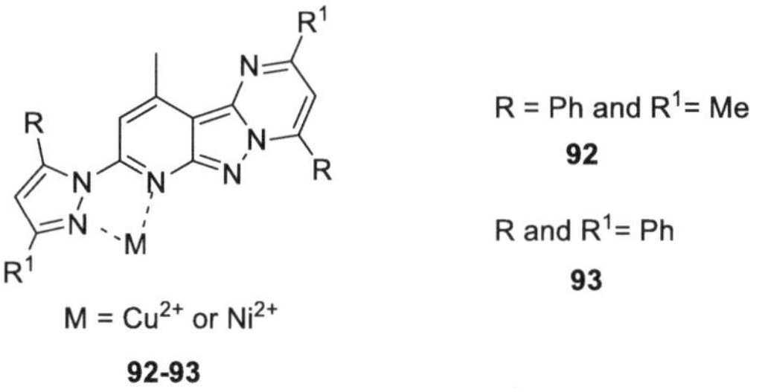

FIGURE 14.27 Molecular structure of pyridine–pyrazole probes for the detection of Cu^{2+} and Ni^{2+} ions.

A similar type of pyridine–pyrazole derivative **91** showed sensing of Al^{3+} ions due to the CHEF process with the detection limit of 1.2 nM in refluxing ethanol [120]. The fluorescence enhancement of **91** upon coordination of Al^{3+} was the result of PET that occurred from the nitrogen of imino group to the π-conjugated system of the probe. Moreover, the sensing mechanism at physiological pH enabled its applicability in biological systems (Scheme 14.18).

A pyrazolopyrimidine-based derivative **92** was reported as a fluorescent sensor for the recognition of Cu^{2+} and Ni^{2+} [121], in which ICT process led to the fluorescence quenching at 491 nm because of the paramagnetic nature of these metal ions. The detection limit was calculated as 0.043 and 0.038 μM for Cu^{2+} and Ni^{2+}, respectively. Cu^{2+} ions were found to be in a distorted trigonal bipyramidal geometry on binding to both N atoms of pyrazole and pyridine with appropriate ligands, whereas Ni^{2+} ions adopted octahedral environment (Fig. 14.27).

Probe **93** was explored for the detection of Cu^{2+} and Ni^{2+} ions, the significant fluorescence emission was observed at 494 nm upon complex formation in methanol solution with a detection limit below 8.7 nM for Cu^{2+} ions while for Ni^{2+} it was 8.9 nM [122]. The predicted binding stoichiometry between **93** and metal ions was 2:1 based on Job's plot, whereas stoichiometry was 1:1 in case of probe **92** (Fig. 14.27).

Portilla and coworkers [123] developed compound **94** that exhibited blue emission due to twisted intramolecular charge transfer (TICT) in the detection mechanism of Cu^{2+} ions. Ligand to metal charge transfer was involved during the sensing mechanism that was also responsible for inhibiting the ICT, so the TICT process occurred (Scheme 14.19).

The selective sensing of Al^{3+} ions was reported by Mondal et al. [124] using probe **95** based on CHEF mechanism. Al^{3+} coordinated to a bis-bidentate cavity of **95** via two O atoms and

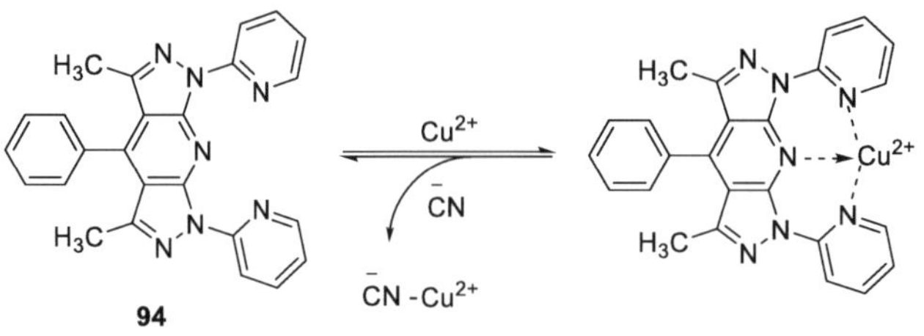

SCHEME 14.19 Proposed mechanism of Cu^{2+} ions detection.

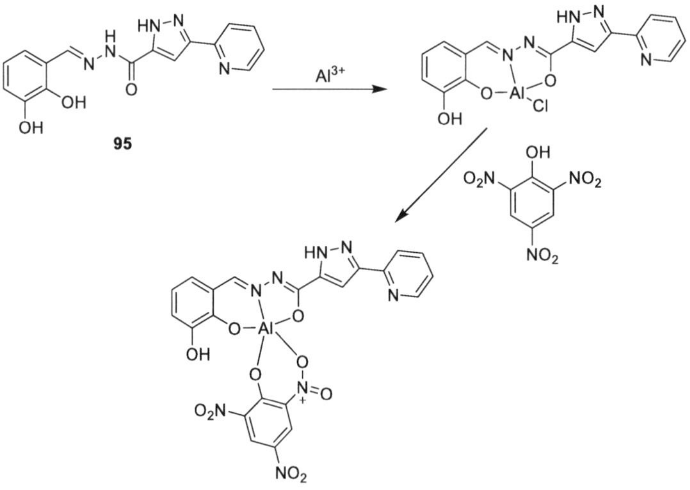

SCHEME 14.20 Proposed sensing mechanism of Al^{3+} ions and picric acid.

N atom that resulted in stable complex formation. **95**-Al^{3+} complex was used as a fluorescent probe for the selective sensing of picric acid and bioimaging of human breast cancer cells was also rationalized (Scheme 14.20).

14.3.7 Pyridine–imidazole chemosensors

Badiei and coworkers [**125**] have reported silica (SBA-15) functionalized 2,6-bis(2-benzimidazolyl) pyridine probe, **96** for the selective detection of Hg^{2+} ions in aqueous media. Addition of Hg^{2+} ions to **96** induced fluorescence quenching at pH range of 6–8 which was attributed to LMCT from probe **96** to Hg^{2+} ions. Under acidic pH, reaction of Hg^{2+} with probe was not observed because of protonation of heteroatom of probe due to the presence of hydronium ions, while **96**-Hg^{2+} complex dissociated at pH higher than 8 (Scheme 14.21).

Benzimidazole-based derivatives **97** and **98** were reported by He et al. [126] for the selective sensing of Cu^{2+} ions. Compound **97** exhibited selective fluorescence quenching with red shift in wavelength at 341 and 356 nm in the presence of Cu^{2+} ions while for **98**, the emission band occurred at 461 nm but did not showed any selectivity. Compound **97** was also used for the recognition of Cu^{2+} ions in living *HepG2 cells* (Fig. 14.28).

Mondal's group [127] designed a pyridine–imidazole-based probe **99** for sensing of Hg^{2+} ions, and sensing application was extended to the imaging of cancer cells. Hg^{2+} ions quenched the fluorescence intensity and showed red shift at 485 nm over other metal ions while the

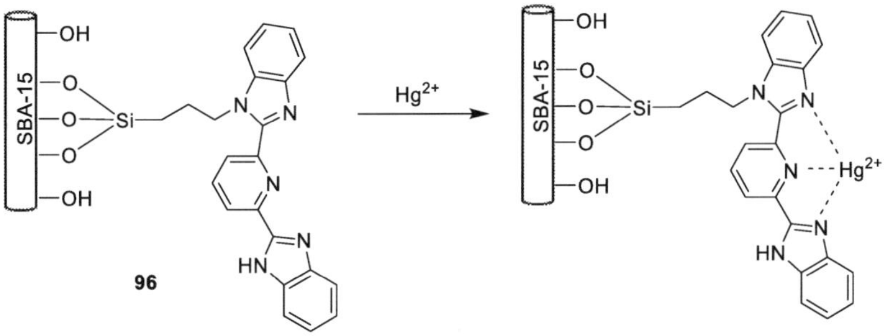

SCHEME 14.21 Possible sensing mechanism for the detection of Hg^{2+} ions.

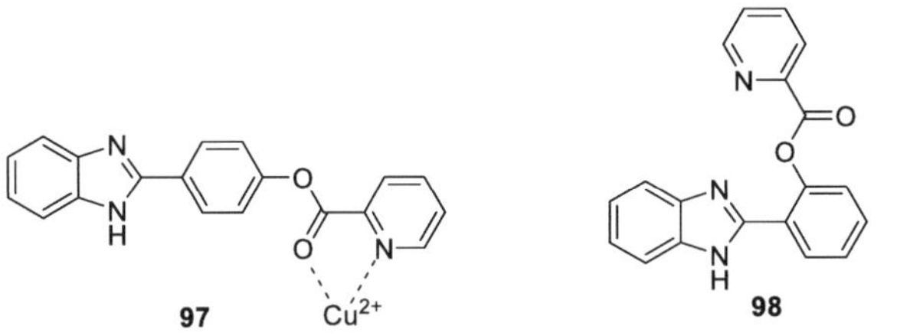

FIGURE 14.28 Chemical structure of pyridine–imidazole chemosensors.

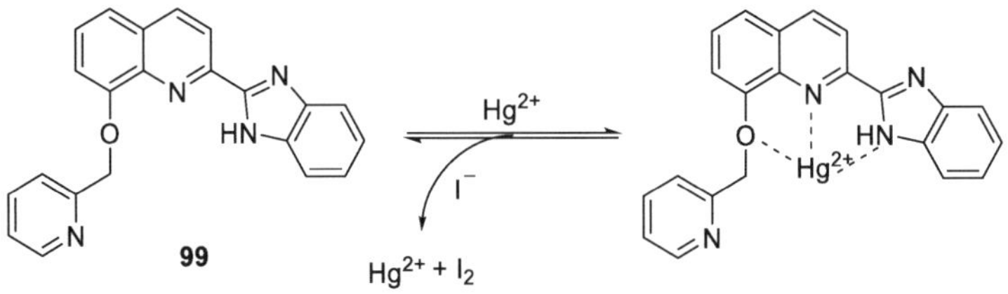

SCHEME 14.22 Proposed sensing mechanism of Hg^{2+} and I^- ions.

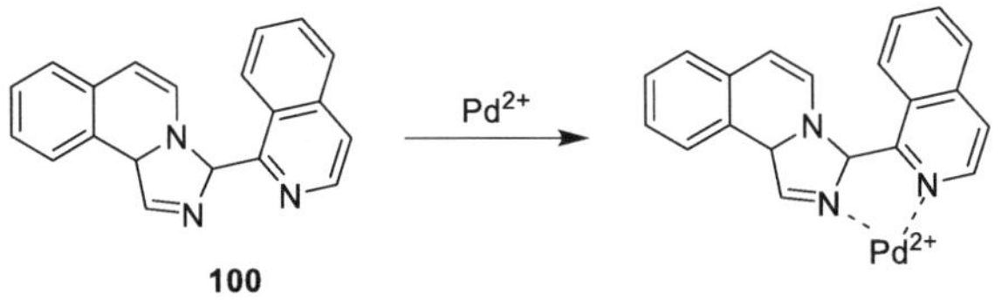

SCHEME 14.23 Sensing of Pd^{2+} ions using pyridine–imidazole chemosensor.

emission intensity reverted when I^- ions were added to the solution of the **99**-Hg^{2+} complex because of HgI_2 formation, so the **99**-Hg^{2+} complex could also be used for the sensing of iodide ions selectively (Scheme 14.22).

Manivannan's group [128] reported selective detection of Pd^{2+} ions based on compound **100** as a reusable probe. Addition of Pd^{2+} ions to probe **100** displayed fluorescence emission with change in color from sky blue to yellow that was visible by the naked eye at pH 7.4 using CH_3OH/4-(2-hydroxyethyl)-1-piperazineethanesulfonic acid (HEPES) buffer. A bidentate coordination site from probe **100** with appropriate ligands formed square planar geometry around Pd^{2+} ions (Scheme 14.23).

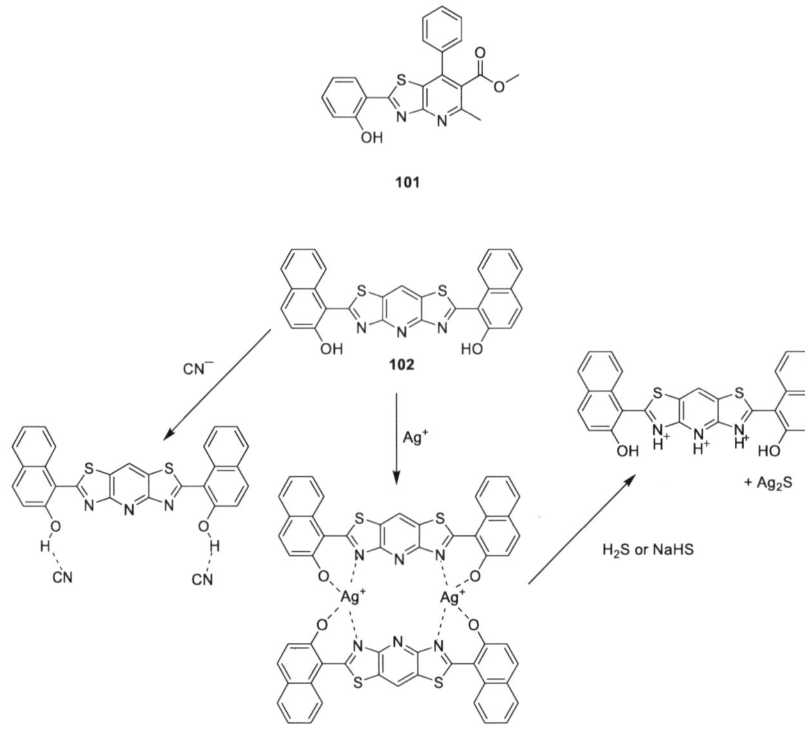

SCHEME 14.24 Chemical structure of pyridine–thiazole chemosensors.

14.3.8 Pyridine–thiazole chemosensors

In 2018, Huang's group [129] developed a pyridine–thiazole-based fluorescent probe **101** for the selective sensing of Zn^{2+} ions which showed red shifted fluorescence enhancement at 510 nm (at pH 7.4) with the detection limit of 3.48×10^{-7} M over other metal ions. Probe **101** was also utilized for the detection of Zn^{2+} ions in living cells (Scheme 14.24).

Darabi and coworkers [130] reported bisthiazolopyridine linked naphthol-based derivative **102** for the detection of Ag^+ and CN^- ions. Chemosensor **102** exhibited fluorescence quenching with red-shift in wavelength attributed to d^{10} configuration of Ag^+ leading to quenching via electron transfer. On the other hand, a new fluorescence band appeared when CN^- ions were added but the original band of **102** was reduced. **102**-Ag^+ complex on adding S^- and HS^- resulted in blue-shift in fluorescence (Scheme 14.24).

14.3.9 Pyridine–hydrazone chemosensors

Sensors **103–105** for Zn^{2+} ions were developed based on naphthaldehyde-2-pyridinehydrazone [131], in which probe **103** exhibited fluorescence enhancement at 484 nm on reaction with Zn^{2+} ions at pH 7 in aqueous solution due to CHEF process. Although the

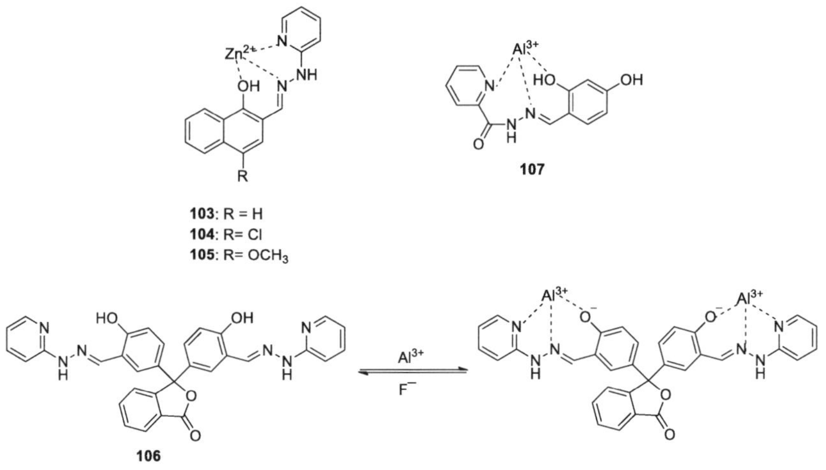

SCHEME 14.25 Structure of pyridine–hydrazone chemosensors.

electron-rich character of the naphthol ring induced red-shift in fluorescence at 490 nm in case of **103** and 510 nm for **104** and **105** respectively but the enhancement was not significant in case of **104** and **105** as in case of **103** (Scheme 14.25).

Wang's group [132] reported fluorescent probe **106** based on phenolphthalein bearing pyridine–hydrazone moieties. Reaction of Al^{3+} with **106** induced fluorescence at 465 nm in aqueous solution, further **106**-Al^{3+} complex exhibited fluorescence quenching with good selectivity on interaction with F^- ions. Fluorescence enhancement induced by Al^{3+} was the result of CHEF process and rigidity of the probe inhibited C=N isomerization after complex formation. Additionally, the probe was also utilized for the chronological recognition of Al^{3+} and F^- ions in human cells (Scheme 14.25).

Peng and coworkers [133] have also reported the detection of Al^{3+} ions based on pyridine acyl hydrazone derivative **107** that exhibited significant fluorescence attributed to PET process, as well as ESIPT process. Nitrogen atom of pyridine, imine, and oxygen atom of the OH group were involved in binding with Al^{3+} as revealed by the ^1H NMR spectra (Scheme 14.25).

14.3.10 Pyridine–coumarin chemosensors

Pyridine–coumarin derivative **108** was reported by Zhang's group [134] for the chronological detection of Fe^{3+} and PO_4^{3-} ions over pH range of 4–8 in an aqueous solution. Addition of Fe^{3+} ions to **108** encouraged fluorescence quenching with change in color from colorless to yellow and addition of PO_4^{3-} ions caused fluorescence recovery with color change from colorless to blue. Probe was also found effective for the recognition of Fe^{3+} ions in human B lymphocyte tumor cells (Scheme 14.26).

Recently, recognition of Cu^{2+} ions was reported using probe **109** as a turn-off sensor in aqueous media by Ngororabanga et al. [135] that exhibited fluorescence quenching at 455 nm due to PET from probe **109** to metal ions. DFT calculations showed that the presence of OH

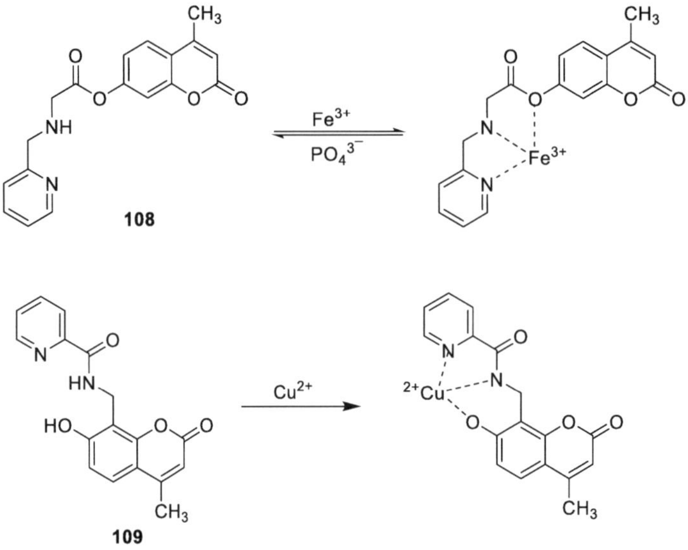

SCHEME 14.26 Chemical structure of pyridine–coumarin chemosensors.

group also contributed to fluorescence enhancement via electron transfer to the carbonyl group. During complex formation, O atom of OH group, N atom of pyridine, and amide group of sensor provided the binding sites that interacted with copper ions and resulted in a stable complex formation (Scheme 14.26).

14.3.11 Pyridine-based other chemosensors

The selective detection of Cr^{3+} ions was explored by Kim and coworkers [136] using a ratiometric fluorescent sensor **110** in aqueous media and *HeLa* cells which showed three fluorescent peaks centered at 385, 402, and at 475 nm as the interaction of Cr^{3+} ions with **110** caused quenching of peak at 385 nm and augmentation at 475 nm. The fluorescence enhancement was ascribed to the increased π–π stacking as a result of complex formation that pulled both pyrene fragments closer to each other (Fig. 14.29).

Nandha Kumar and group [137] reported chalcone-derived chemosensor **111** that induced fluorescence on complexation with Pb^{2+} ions, attributed to PET process and the presence of other competitive metal ions did not interfere with Pb^{2+} ions. IR analysis confirmed the involvement.of N atom of pyridine and carbonyl group with Pb^{2+} ions in complex formation (Fig. 14.29).

For the detection of La^{3+}, a polydentate ligand-based derivative **112** was developed that showed fluorescence enhancement (~60-fold) along with redshift (65 nm) in wavelength due to PET process from ligand to metal [138]. The complex formation due to interaction of metal ions with chemosensor reduced the electron density of **112**, raised planarity and increased π conjugation content contributed toward the red-shift fluorescence. It was also streamlined that metal to ligand ideal stoichiometric ratio was 1:3, lower or higher stoichiometric ratio resulted in poor fluorescence (Fig. 14.29).

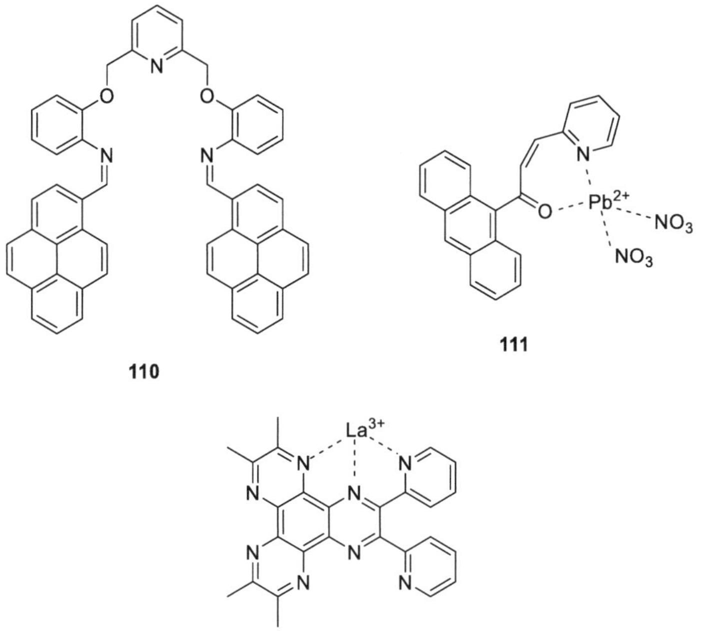

FIGURE 14.29 Chemical structure of pyridine-based derivatives for sensing of Cr^{3+}, Pb^{2+}, and La^{3+} ions.

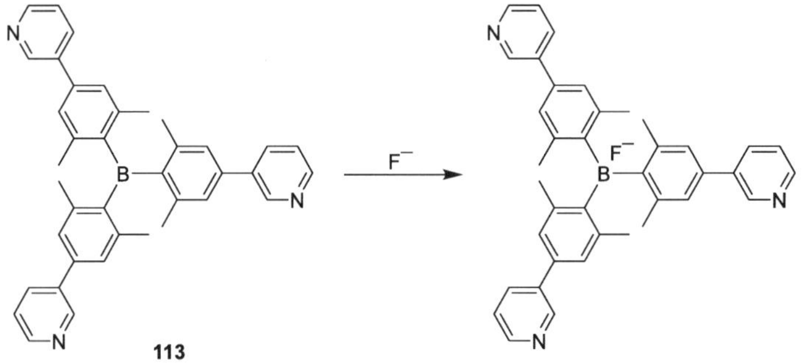

SCHEME 14.27 Proposed sensing mechanism of HF.

Yoon's group [139] described the sensing of HF based on triaryl borane derivative **113** that displayed blue shift in fluorescence quenching at 410 nm. On coordination, the vacant p-orbital of central boron was occupied by HF that resulted in perturbation of ICT as well as $\pi-\pi$ interaction in probe **113**. However, the strong affinity of HF with **113** was due to the increase in Lewis acidity via conjugation of boron to pyridyl moieties (Scheme 14.27).

A diphenylpyridine derivative **114** as a turn-on sensor was developed by Wang and group [140] for Ag^+ ions sensing in aqueous media. Upon addition of Ag^+ ions to probe **114**, fluorescence intensity was enhanced to about 12-fold at 349 nm. The luminescence quantum yield was also increased on coordination of Ag^+ ions to two ligands that resulted in intraligand charge transfer process. However, the planar geometry around Ag^+ ions was not observed due to steric hindrance created by phenyl rings (Scheme 14.28).

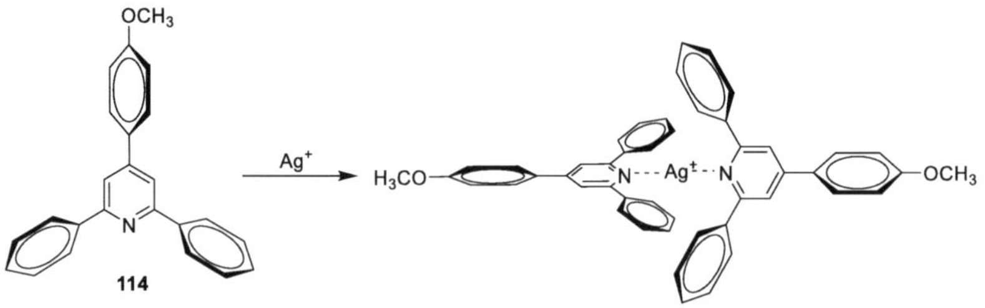

SCHEME 14.28 Binding of Ag^+ ions with diphenylpyridine derivative.

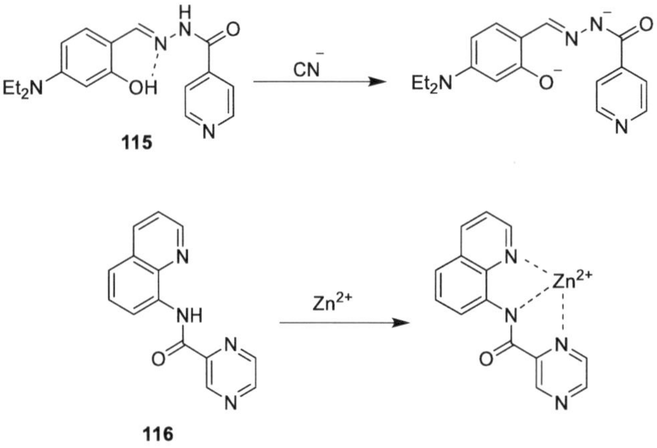

SCHEME 14.29 Proposed sensing mechanism of CN^- and Zn^{2+} ions.

14.4 Pyridine-based dual-mode chemosensors

14.4.1 Pyridine–carboxamide chemosensors

In 2016, Hu's group [141] explored the sensing of CN^- in food samples using compound **115** in an aqueous solution. The pale yellow colored solution **115** displayed fluorescence emission at 552 nm, the interaction of CN^- ions induced enhancement in emission intensity about 14-fold with eminent color change to yellow. Intramolecular H-bonding accompanied by the deprotonation was the sensing mechanism operated during the interaction of CN^- ions with compound **115** and addition of H^+ to **115**-CN^- complex caused a reduction in emission intensity, hence H^+ could be used to establish the reversibility of the compound (Scheme 14.29).

Amirnasr and coworkers [142] reported fluorescent compound **116** that exhibited fluorescence enhancement at 467 nm upon interaction with Zn^{2+} ions in CH_3CN solvent. CHEF mechanism involved in the complexation of metal ions with compound led to increased fluorescence response. Interestingly, an instant change in color from colorless to yellow was detected when Cu^{2+} (100 μM) was added to the solution of **116**, and could be used as a colorimetric probe for Cu^{2+} ions sensing selectively over other interfering cations (Scheme 14.29).

Similarly, compound **117** has also been reported for the sensing of Zn^{2+} ions fluorometrically [143]. Addition of Zn^{2+} ions (half-equivalent) to **117** induced remarkable fluorescence

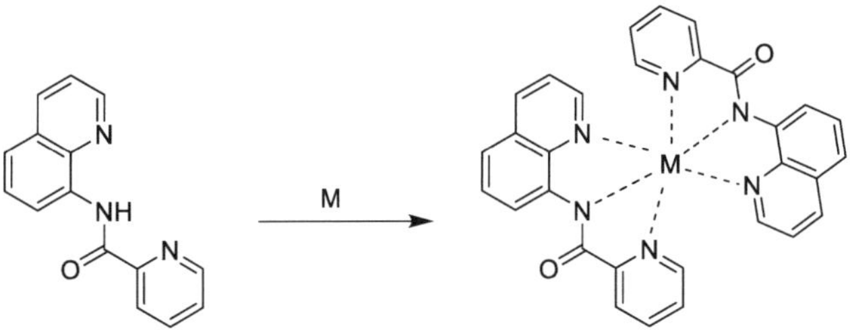

117 M = Zn^{2+},Co^{2+}

SCHEME 14.30 Chemical structure of pyridine–carboxamide chemosensors.

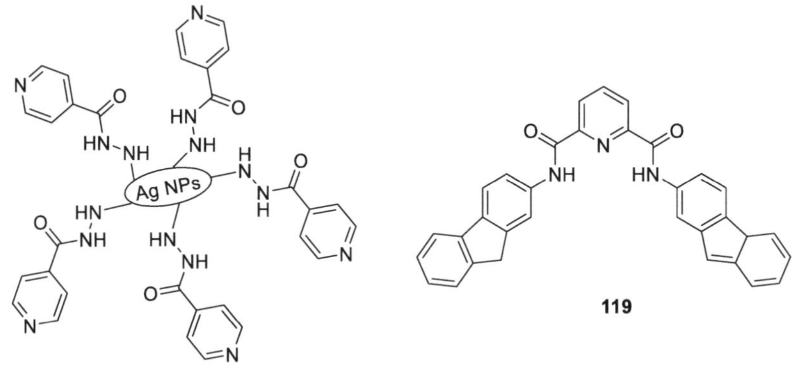

119

118

FIGURE 14.30 Chemical structure of pyridine–carboxamide chemosensors.

enhancement of about 120-fold in CH_3CN solvent at 481 nm ascribed to MCHEF process. In the colorimetric investigation of **117** with different cations, the absorption band at 401 nm increased steadily upon reaction with Co^{2+} ions and the appearance of yellow color consequently indicated that charge transfer process might be involved (Scheme 14.30).

Sakthivel and group [**144**] synthesized functionalized silver nanoparticles-based sensor **118** for Hg^{2+} ions sensing. The absorption band at 396 nm gradually decreased and the yellow color turned to colorless as a result of reaction of **118** with Hg^{2+} ions. The fluorescence emission band at 380 and 398 nm declined with the appearance of a new band located at 412 nm, but no change in emission intensity was observed except for Hg^{2+} ions which quenched the emission. The binding mode of the sensor was rationalized based on FT-IR spectra; there was no change at 1637 cm^{-1}, a stretching band of C=O, but the C=N characteristic band at 1385 cm^{-1} significantly decreased as a result of complex formation which confirmed the coordination of sensor by pyridine nitrogen atom (Fig. 14.30).

Dicarboxamide derivative **119** was developed as a fluorescent and colorimetric cation probe that displayed the sensing of Cu^{2+} and Pb^{2+} ions [145]. Compound **119** showed absorption bands at 268 and 316 nm that were attributed to π–π^* and n–π^* transitions of aromatic rings. Exposure of Cu^{2+} and Pb^{2+} ions caused bathochromic as well as hypochromic shift of the second band and the fluorescence intensity at 480 and 531 nm was declined due to

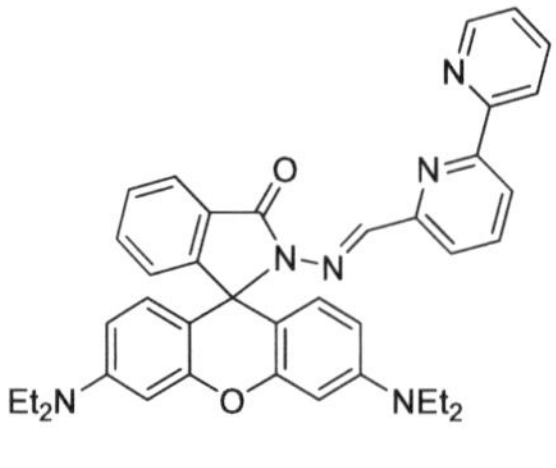

120

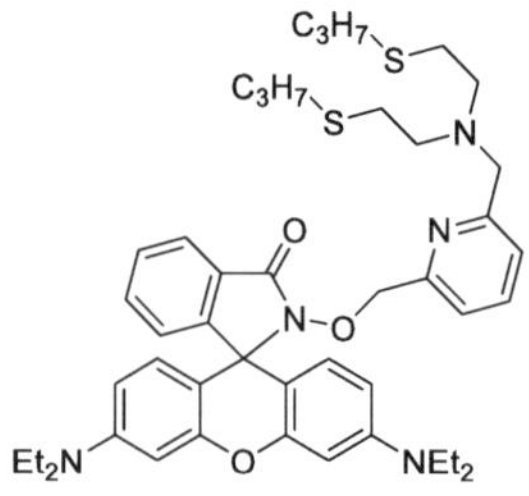

121

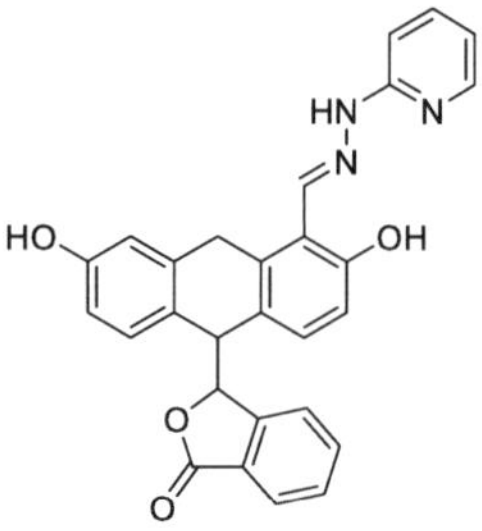

122

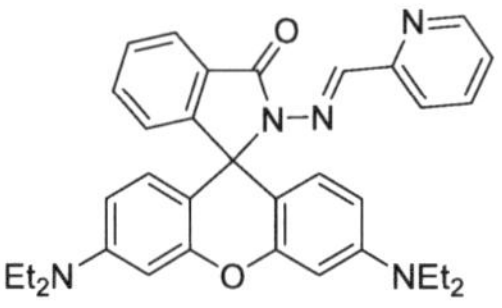

123

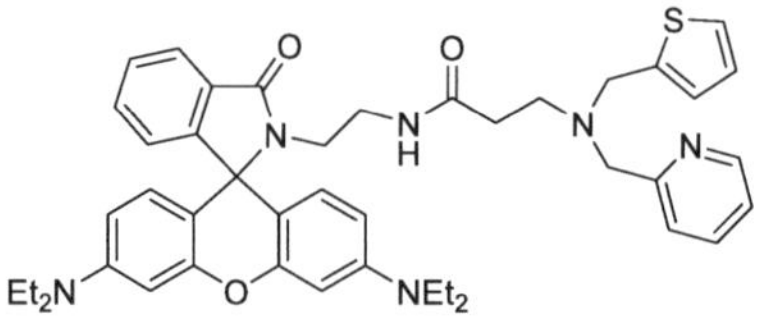

124

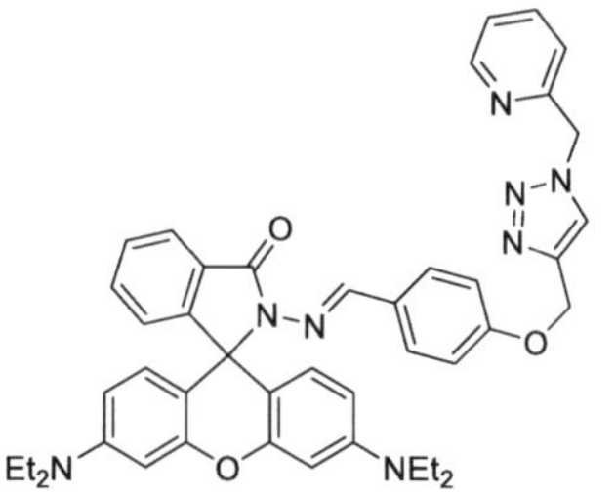

125

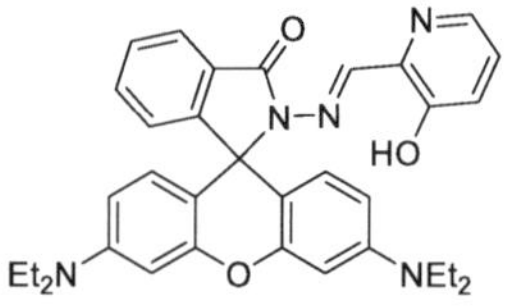

126

FIGURE 14.31 Chemical structures of pyridine–rhodamine-based scaffolds for sensing.

chelation-enhanced fluorescence quenching (CHEQ) effect and the spin–orbital interaction induced by Cu^{2+} ions resulted in yellow color. Whereas in case of Pb^{2+} complex, the reverse PET was responsible for the quenching process. Pb^{2+} ions coordinated with amide oxygen of the **119** (2:1 stoichiometry) while Cu^{2+} coordinated to the N_3 pincer cavity (1:1 stoichiometry) (Fig. 14.30).

14.4.2 Pyridine–rhodamine chemosensors

A bipyridine-linked rhodamine derivative **120** was used for the sensing of Hg^{2+} ions at physiological pH [146]. On coordination of Hg^{2+} to **120**, color change from colorless to pink induced absorption band at 568 nm and also the fluorescence emission at 584 nm increased 18-fold along with the appearance of red color. Further, **120**-Hg^{2+} complex was utilized for the sensing of anions, but only the presence of I^- and S^{2-} led to significant fluorescence quenching (Fig. 14.31).

Xu's group [147] reported probe **121** that involved the ring-opening mechanism of spirolactam in the sensing of metal ions. Colorless solution of **121** displayed absorption peak at 562 nm and a shoulder at around 520 nm; a dramatic absorption enhancement of about 251-fold was observed accompanied by the visible purple color after the addition of Cu^{2+} ions. However, the red-shift in fluorescence enhancement around 259-fold at 582 nm was detected upon selective coordination of Hg^{2+} with **121** as a consequence of the PET mechanism (Fig. 14.31).

Dual-responsive Al^{3+} and Cu^{2+} sensor **122** was developed and used for the colorimetric and fluorescent detection by Hou and coworkers [148]. When Cu^{2+} ions were added to **122**, absorption band at 492 nm gradually weakened with blue shift and pink color turned to orange–yellow, while in case of Al^{3+} ions pale pink to yellow color change was observed with the weakening of absorption intensity at 344 nm and the appearance of a new red-shifted peak at 382 nm. Moreover, fluorescence enhancement was pH-dependent, Cu^{2+} ions were detected at higher pH (7–9), but Al^{3+} ions coordinated in the pH range of 3.8–4.2 and fluorescence enhancement was attributed to CHEF process (Fig. 14.31).

Recognition of Cd^{2+} was achieved using probe **123** based on FRET process reported by Stalin et al. [149]. Interaction of Cd^{2+} ions with **123** caused opening of spirolactam ring that displayed color change from colorless to a deep magenta accompanied by new absorption band at 560 nm. Also, orange fluorescence was induced at 590 nm with 20-fold enhancement and Cd^{2+} ions adopted a near planar geometry as predicted by DFT calculations (Fig. 14.31).

Lu's group [150] designed probe **124** that could be potentially used for sensing Pd^{2+} ions both in an aqueous as well as in living system. Colorless solution of **124** turned pink along with fluorescence upon selective coordination with Pd^{2+} ions over other competitive metal ions above pH 4 and PET led to a significant fluorescence response (Fig. 14.31).

A multidentate conjugated compound **125** was applied for sensing of Sn^{2+} as reported by Xu and coworkers [151] where the colorless solution of **125** changed to pink and a new absorption peak centered at 560 nm emerged. Likewise, fluorescence increased at 587 nm with orange color and the involvement of N and O atoms of the probe with Sn^{2+} formed a stable five- or six-membered chelate ring (Fig. 14.31).

A similar concept was applied for the detection of both anions and cations using compound **126** by Hu et al. [152]. Among various coexisting anions, only CN^- ions selectively induced the change in color from colorless to pale yellow with emergence of a new absorption band at 478 nm and the previously observed band at 562 nm vanished. Fluorescence spectra displayed color change to pale yellow from pale purple with enhanced fluorescence at 498 nm and addition of Cu^{2+} caused fluorescence quenching at 486 nm with a concomitant purple color appearance (Fig. 14.31).

Novel pyridine–rhodamine B-based compounds **127** and **128** were developed and used for the ferric ion (Fe^{3+}) detection in living cells based on "OFF–ON" mechanism [153]. Sensing behavior toward various cations was evaluated but except Fe^{3+} ions, no other metal displayed a good spectral response. Addition of Fe^{3+} ions to sensor significantly induced absorption centered at 560 nm and fluorescence enhancement at 582 nm with gradual appearance of pink color in aqueous ethanol solvent at physiological conditions. Fe^{3+} ions were coordinated through nitrogen and carbonyl oxygen to the compound but pyridine's nitrogen was not involved in binding as revealed by the NMR analysis. Moreover, both the compounds were potentially used for imaging Fe^{3+} ions in human breast adenocarcinoma (*MCF-7*) cells (Fig. 14.32).

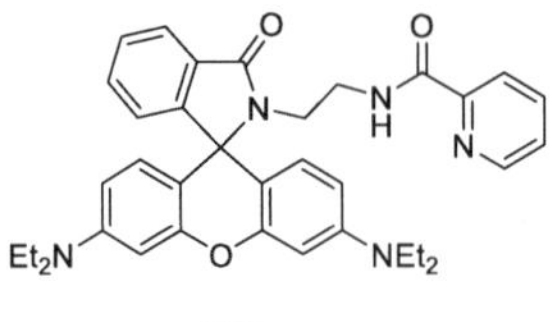

127

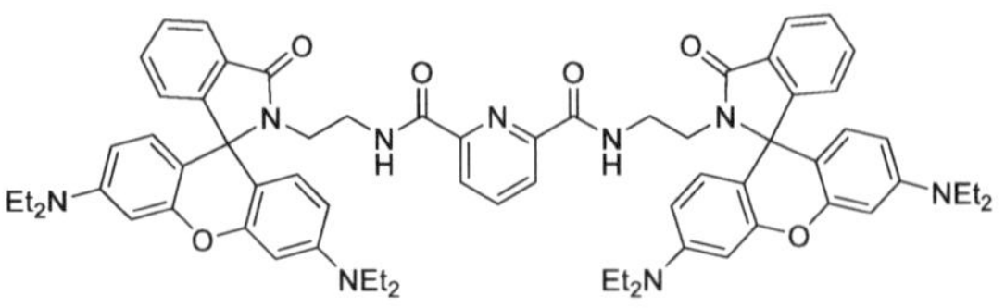

128

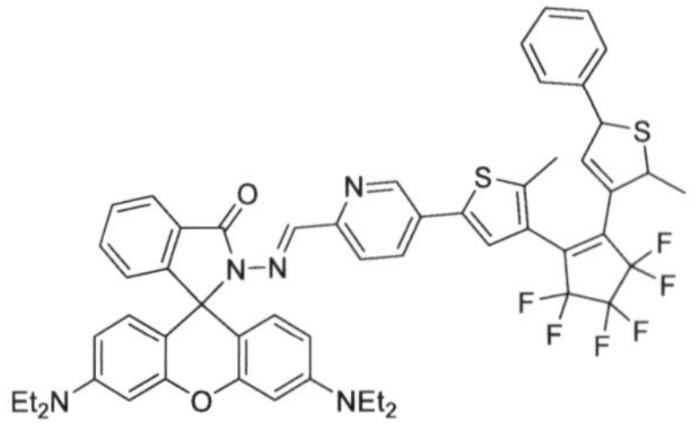

129

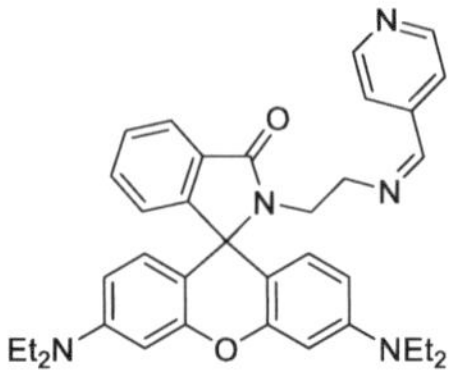

130

FIGURE 14.32 Structures of pyridine–rhodamine-based scaffolds.

Liu's group [154] explored solvent-dependent probe **129** for the recognition of Cu^{2+} and Hg^{2+} ions. Introduction of Hg^{2+} ions to **129** caused fluorescence enhancement (34-fold) centered at 617 nm in DMS solvent and dark color changed to orange–red while colorimetric response was changed only on exposure to Cu^{2+} ions and a new absorption band centered at 555 nm was detected in THF solvent with the associated pink color. These results demonstrated that **129** could be used for dual-sensing of Hg^{2+} and Cu^{2+} ions (Fig. 14.32).

Wang et al. [155] reported the same compound **128** that exhibited fluorescence emission at 585 nm, the emission steadily intensified on coordination with Fe^{3+} ions and colorless solution of **128** exhibited absorption at 560 nm. An increase in intensity with a pink color appearance indicated the formation of a new complex. On testing the sensing property for imaging Fe^{3+} ions in human hepatocarcinoma cells, bright red fluorescence was developed showing that the prepared compound could be potentially employed for imaging in biological systems (Fig. 14.32).

Sun's [156] group also explored the detection of Fe^{3+} ions using novel sensor **130** that displayed turn-on fluorescence enhancement (20-fold) at 592 nm and exhibited absorptions at 289 and 316 nm with a very weak band at 559 nm; coordination of Fe^{3+} ions intensified the absorption band with the appearance of a pale pink color. It was revealed that PET might have caused the enhancement of fluorescence and absorption bands (Fig. 14.32).

Interestingly, a similar kind of compound **131** was developed by Yuan et al. [157] and explored for the detection of glyphosate pesticide residues in soil and vegetables based on displacement strategy. Compound **131** displayed green fluorescence quenching (at 539 nm) response toward Cu^{2+} ions, due to ICT process inhibition while the absorption band at 354 and 530 nm vanished with a concomitant appearance of red–purple color. Further, exposure of glyphosate to **131**-Cu^{2+} complex induced yellow–green fluorescence intensity at 539 nm about 20-fold indicated the displacement reaction due to strong binding affinity of glyphosate toward the Cu^{2+} ions (Scheme 14.31).

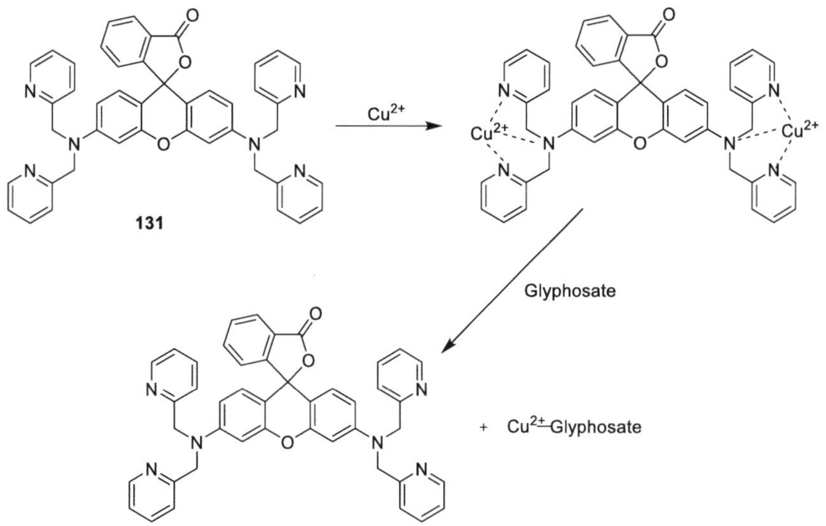

SCHEME 14.31 Illustration of sensing mechanism of glyphosate.

14.4.3 Pyridine–Schiff base chemosensors

Pyridine–Schiff base derivative **132** was explored for the selective sensing of Al^{3+} ions both by colorimetric and fluorescent techniques [158]. Addition of Al^{3+} to **132** caused emergence of yellow–green color from colorless solution with substantial fluorescence enhancement at 454 nm in aqueous solution as a result of ICT inhibition in the sensor. Pyridine and naphthalene rings of **132** displayed two absorptions centered at 328 and 373 nm because of π–π^* transitions and a red-shift from 373 to 415 nm was observed after coordination of Al^{3+} ions to **132**. Also, associated yellow–green color was the result of the deprotonation of the phenolic OH groups (Fig. 14.33).

Lu's group [159] reported colorimetric Cu^{2+} ions sensing as well as fluorometric Fe^{3+} ions detection by using compound **133**. Colorless solution of **133** exhibited two absorption bands at 285 and 352 nm; addition of Fe^{3+} ions boosted the absorption intensity, and Cu^{2+} ions caused red-shift from 352 to 373 nm with the appearance of orange and pink color, respectively. Fluorescence quenching at 370 nm was observed by the interaction of Fe^{3+} ions with compound **133** which could be expected due to the paramagnetic nature of Fe^{3+} ions, whereas the fluorescence response was not significantly affected by the presence of Cu^{2+} ions (Fig. 14.33).

Compound **134** displayed fluorescent detection of Al^{3+} and the obvious colorimetric response to OH ions [160]; interaction of Al^{3+} ions to **134** induced fluorescence at 534 nm possibly due to CHEF effect with bright green color over other coexisting metal ions. A redshift in the absorption from 554 to 662 nm with change in color from violet to blue was seen in the presence of OH ions (Fig. 14.33).

Cui's group [161] reported compound **135** that exhibited red-shift fluorescence emission enhancement (106-fold) at 506 nm on coordination with Cd^{2+} ions in methanol. There was no change in fluorescence of **135** in the presence of other cations and NMR analysis revealed

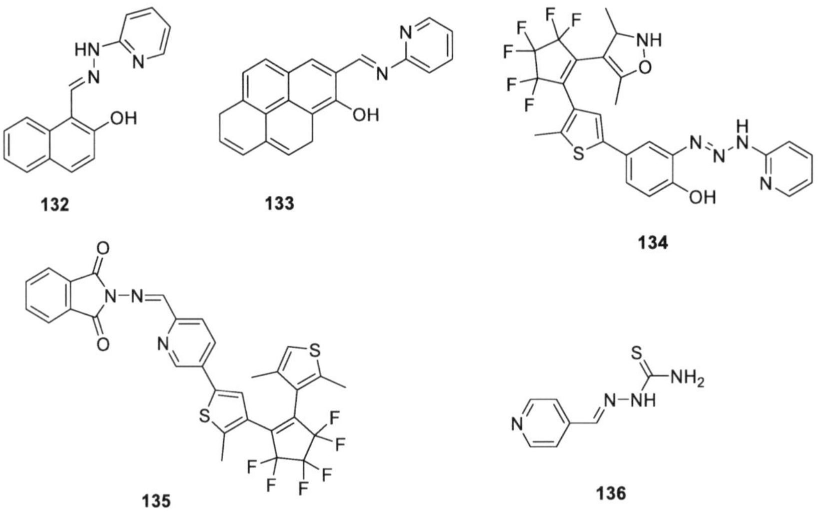

FIGURE 14.33 Chemical structures of pyridine–Schiff base chemosensors.

that N atom of Schiff base, pyridine, and O atom of carbonyl group provided binding sites for metal ions. On the other hand, absorption band of colorless solution of **135** at 344 nm gradually vanished accompanied by a new band located at 412 nm with yellow color upon reaction with Cu^{2+} selectively. These results signified that the prepared compound could be utilized for fluorometric as well as colorimetric detection of respective metal ions (Fig. 14.33).

Pyridine–Schiff base compound **136** was reported as a dual-mode sensing probe by Patra and coworkers [162], which exhibited absorption band at 322 nm which was ascribed to the π–π^* transition of the pyridine ring. Absorption band was red-shifted from 365 to 340 nm after the addition of Cu^{2+} and Ag^{+} and the colorless solution turned yellow due to LMCT mechanism. Additionally, Ag^{+} ions significantly enhanced the blue fluorescence emission at 416 nm that might be due to the chelation enhancement fluorescence effect (Fig. 14.33).

14.4.4 Pyridine–BODIPOY chemosensors

In 2015, He and group [163] developed a compound **137** to be used as a Cu^{2+} ion probe that exhibited a remarkable change in color from pink to blue in CH_3CN. Strong absorption maxima centered at 603 nm with a weak band at 400 nm was attributed to transition in BODIPY chromophore; addition of Cu^{2+} ions cause absorption band to redshift from 603 to 608 nm with associated blue color which indicated the formation of new complex. Also, the fluorescence emission intensity at 617 nm decreased with bright red color that might be due to energy transfer from BODIPY to metal ions. However, introduction of EDTA or S^{2-} to **137**-Cu^{2+} ions could restore the fluorescence intensity due to their strong binding affinity (Fig. 14.34).

The detection of Zn^{2+} ions utilizing compound **138** was reported by Wang et al. [164] interaction of which with Zn^{2+} ions showed a bathochromic shift in the absorption band located at 525–552 nm accompanied by the appearance of purple color from pale yellow. Salmon pink

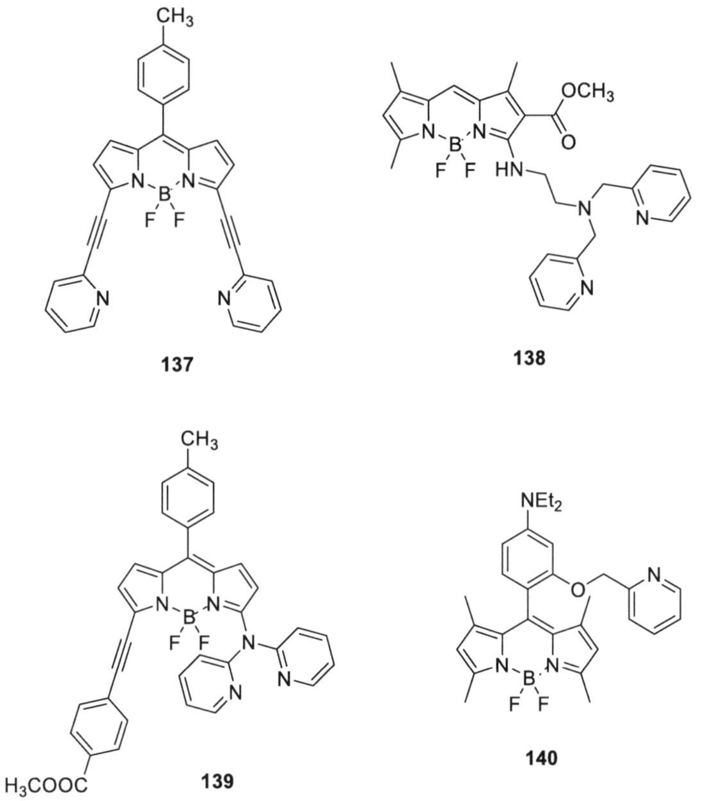

FIGURE 14.34 Chemical structure of pyridine–BODIPOY-based compounds.

color fluorescence emission band was also red-shifted and found at 580 nm indicating the complex formation (Fig. 14.34).

Qiu and coworkers [165] developed compound **139** which shows a significant colorimetric and fluorescent change for simultaneous monitoring of Cu^{2+}, Hg^{2+}, and Pb^{2+} ions. **139** displayed absorption bands at 324 and at 595 nm in CH_3CN attributed to ICT process (Fig. 14.34).

Adhikari's group [166] established compound **140** for the recognition of aspartic and glutamic acid in an aqueous system and in living cells which displayed a weak emission band at 527 nm due to PET process. Also, addition of either acid to the solution of **140** led to a decline in the absorption bands, and a blue shift in absorption from 514 to 403 nm with the color change from pink to colorless, whereas other amino acids did not produced any spectral change. The prepared compound could be effectively used for intracellular imaging in different cells such as *MDA-MB-468, HEK-293T, and HeLa* cells (Fig. 14.34).

14.4.5 Pyridine–imidazole chemosensors

Razavi and coworkers [167] reported compound **141** for colorimetric sensing Fe^{2+}/Fe^{3+} ions and fluorescent sensing of Zn^{2+} ions. When Fe^{2+} and Fe^{3+} ions were added to the colorless

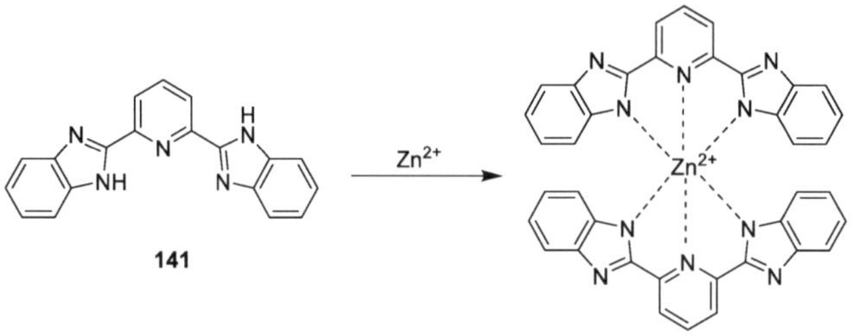

SCHEME 14.32 Illustration of the proposed binding mechanism of Zn^{2+}.

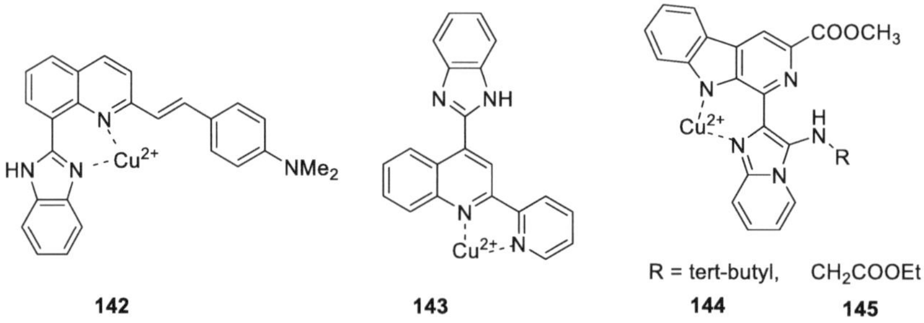

FIGURE 14.35 Molecular structures of pyridine–imidazole sensors.

solution of **141**, a remarkable color change to purple and yellow were observed due to strong and weak MLCT, respectively. After the addition of these ions, new band evolved at 566 nm for Fe^{2+} and a weaker band centered at 400 nm led to the color change. **141** exhibited fluorescence emission at 376 and 431 nm; addition of Zn^{2+} ions caused red-shift of the band at 376 nm as well as quenching at 431 nm due to the formation of a new complex (Scheme 14.32).

Compound **142** as a dual-mode Cu^{2+} ions sensor was described by Dan et al. [168] that exhibited absorption bands at 342 and 412 nm and on addition of Cu^{2+} ions, absorption intensity gradually decreased with red-shifted band at 550 nm and change in color from yellow to purple. Furthermore, interaction of Cu^{2+} ions to **142** caused quenching in fluorescence at 610 nm (pH range 4–9). It was revealed that N atom of both pyridine and imidazole cooperatively participated in binding to Cu^{2+} ions (Fig. 14.35).

Similarly, Shankarling's group [169] also reported detection of Cu^{2+} ions by employing compound **143**, colorless solution of which turned yellow on introducing Cu^{2+} ions with redshift in absorption peak from 332 to 345 nm. Paramagnetic nature of Cu^{2+} ions induced fluorescence quenching at 438 nm indicating the formation of complex and it was found that fluorescence quenching was the consequence of electron transfer from Cu^{2+} ions to chemosensor (Fig. 14.35).

Another efficient Cu^{2+} ions sensors, **144** and **145** were reported by Shrivastava and coworkers [170], which exhibited absorption maxima at 286 and 386 nm; interaction of Cu^{2+} ions caused red-sift in absorption at 386–424 nm, and visible color change to yellow in both the compounds. The intense fluorescence emission centered at 425 nm was significantly enhanced due to CHEF as a result of stable complex formation (Fig. 14.35).

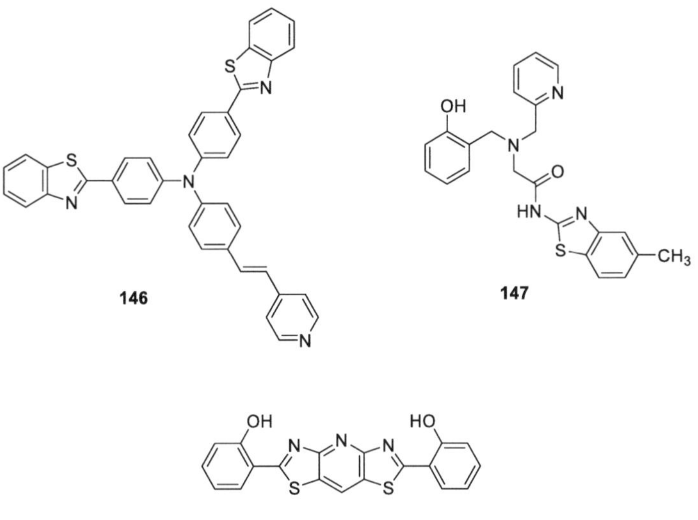

148

FIGURE 14.36 Structure of pyridine–thiazole chemosensors.

14.4.6 Pyridine–thiazole chemosensors

Siva's group [171] developed compound **146** that displayed visible color change from colorless to reddish-orange color on complexation with Fe^{3+} ions and a new absorption peak originated at 460 nm, whereas other divalent metals were unable to produce any significant changes. The fluorescence emission centered at 480 nm was selectively quenched by Fe^{3+} ions. Proton acceptor nature of pyridine in **146** was responsible for making it an appropriate candidate for acidic pH sensing because after protonation electron density on Natom of pyridine increased that led to stronger ICT effect. Furthermore, utilization of **146** was extended for sensing nitroaromatics; addition of picric acid induced a remarkable increase in absorption peak at 400 nm and a strong fluorescence quenching at 500 nm (Fig. 14.36).

Caruso and coworkers [172] described compound **147** for multitarget detection, which exhibited colorimetric sensing of Fe^{2+} and Fe^{3+} ions and fluorescent sensing of Zn^{2+} and Cd^{2+} ions. The colorless solution of **147** turned pale violet and violet on reaction with Fe^{2+} and Fe^{3+} ions. **147** displayed absorption band at 278 nm ascribed to intramolecular $\pi-\pi^*$ charge-transfer transitions and interaction of Fe^{2+}/Fe^{3+} reduced intensity at 278 nm and increased intensity at 307 nm. The red-shifted fluorescence enhancement at 348–370 nm for Zn^{2+} ions and 370 nm for Cd^{2+} ions were detected, probably ascribed to CHEF mechanism involved during the complex formation (Fig. 14.36).

Detection of CN^- ions via deprotonation mechanism was reported by Darabi and group [173] using compound **148** as a dual-mode probe which displayed visible change from colorless to yellow toward CN^- ions in $MeOH/H_2O$ (9:1) solvent. Addition of CN^- ions decreased the absorption bands accompanied by a new band located at 438 nm and red-shift signified the formation of a new product. Fluorescence of **148** at 420 nm drastically decreased on addition of CN^- ions and new fluorescence maxima at 515 nm with a concomitant green color appearance. The spectral change was attributed to the disruption of the intramolecular H-bond in **148** due to deprotonation of phenolic OH by CN^- ions (Fig. 14.36).

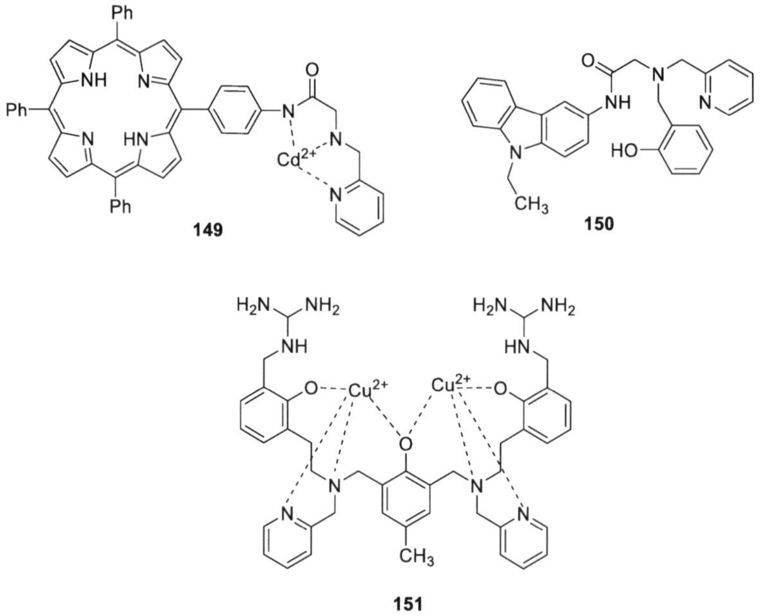

FIGURE 14.37 Chemical structures of pyridine–amine derivatives.

14.4.7 Pyridine–amine chemosensors

A pyridine porphyrin derivative **149** was developed for the selective sensing of Cd^{2+} ions in the moderate pH range [174], pink-colored solution of which turned to yellow–green immediately after addition of Cd^{2+} ions. Introduction of Cd^{2+} ions to **149** caused absorption band to decrease at 415 nm and increase in blue-shifted peak at 645 nm. Also, the fluorescence enhancement was observed at 606 nm, but the intensity at 653 nm decreased gradually as a result of internal charge transfer process (Fig. 14.37).

Kim's group [175] designed compound **150** that showed significant spectral change toward Fe^{2+}/Fe^{3+} and Cu^{2+} ions in an aqueous solution. Addition of Fe^{2+}/Fe^{3+} to **150** enhanced the absorbance intensity at 375 nm accompanied by appearance of bright brown color from colorless solution, whereas addition of Cu^{2+} ions decreased the absorbance at 365 and 242 nm, and a new absorption band originated at 337 nm with yellow color which might be due to charge-transfer from ligand to metal on complex formation (Fig. 14.37).

Chen and coworkers [176] reported detection of pyrophosphate based on compound **151** as a di-nuclear Cu^{2+} complex sensor that showed a strong binding affinity toward pyrophosphate. Reaction of pyrophosphate with blue-colored solution of **151** increased the absorption intensity at 443 nm, but reduced at 685 nm accompanied by yellow color. The fluorescence quenching at 465 nm with the dark color was seen showing the formation of a new product. Additionally, it was found that **151** could be potentially employed to monitor the hydrolysis reactions of pyrophosphate (Fig. 14.37).

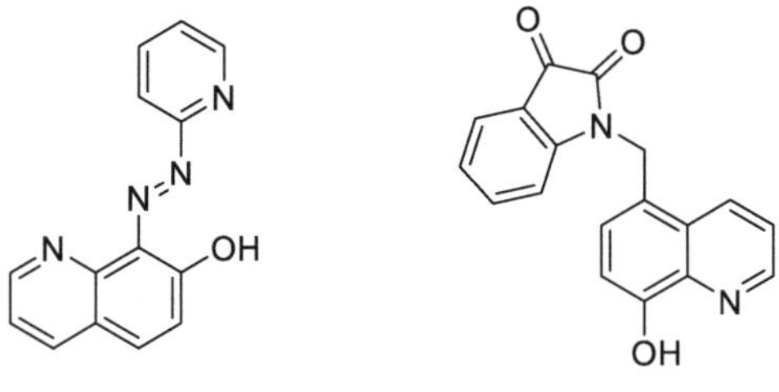

152 153

FIGURE 14.38 Structure of pyridine-based chemosensors for sensing of Al^{3+} ions.

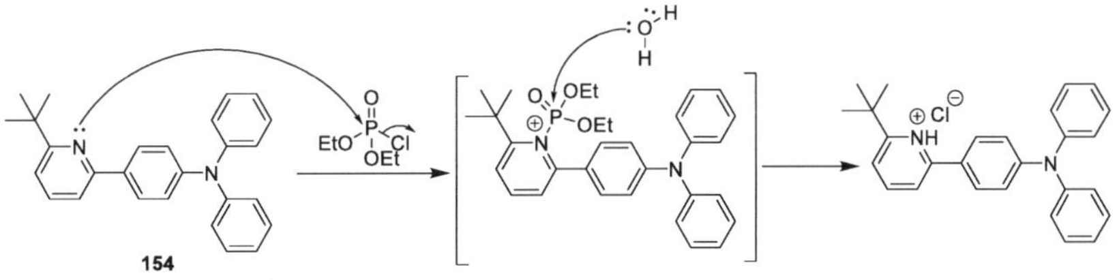

SCHEME 14.33 Illustration of sensing mechanism of DCP.

14.4.8 Pyridine-based some other chemosensors

In 2015, Gupta and coworkers [177] reported compound **152** based on pyridine azo derivative which on reaction with Al^{3+} ions produced dark red colored complex from the colorless solution and induced red-shift in fluorescence from 521 to 569 nm with a pinkish-red color appearance (Fig. 14.38).

Lashgari's group [178] reported fluorescent sensing of Al^{3+} ions using compound **153** (Fig. 14.38). Among various tested metal ions, only Al^{3+} ions induced substantial fluorescence enhancement (15-fold) at 486 nm after coordination with **153**. Absorption spectra displayed new bands at 371, 460, and 580 nm when Fe^{2+} and Fe^{3+} ions were added, consequently the yellow color of the sensor changed to black which was ascribed to MLCT upon complex formation.

Yao's group [179] reported a similar kind of pyridine-based compound **154** which could be used for real-time monitoring of diethyl chlorophosphate (DCP), a toxic chemical agent. Exposure of DCP vapors to solution of **154** induced red shift in fluorescence emission centered at 410–522 nm with 98% quenching; and a change in color from sky-blue to yellowish-green. The intensity of band at 345 nm reduced after addition of DCP accompanied by the appearance of new peaks at 294 and 410 nm with yellowish-green in color. Also, sensing of DCP was based on the cascade type of mechanism as revealed in the NMR analysis (Scheme 14.33).

Sensing of nitrate anions via electrostatic interaction was reported by Tian et al. [180] based on compound **155**. Among different nitrates of Co^{2+}, Ni^{2+}, Zn^{2+}, and Cd^{2+}, nickel nitrate displayed a strong affinity toward sensor and **155**-$Ni(NO_3)_2$ exhibited absorption band at 470 and 570 nm; addition of an organic solvent decreased the absorption at 470 nm accompanied by increase at 570 nm (Fig. 14.39).

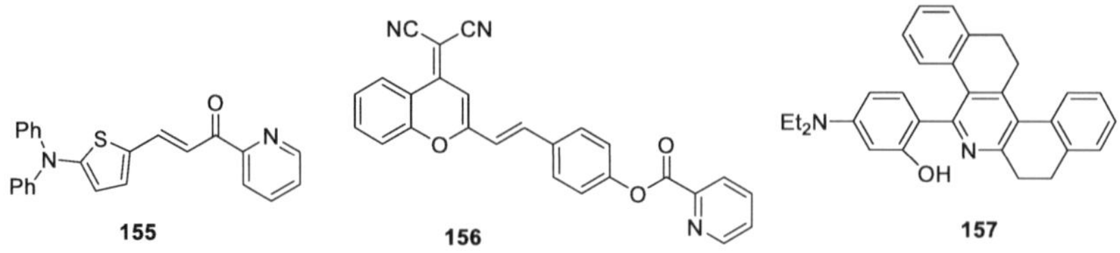

FIGURE 14.39 Chemical structure of pyridine-based chemosensors.

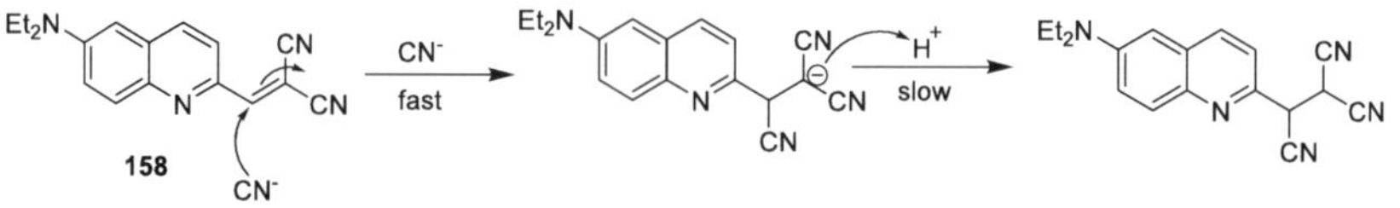

SCHEME 14.34 The proposed sensing mechanism of CN^-.

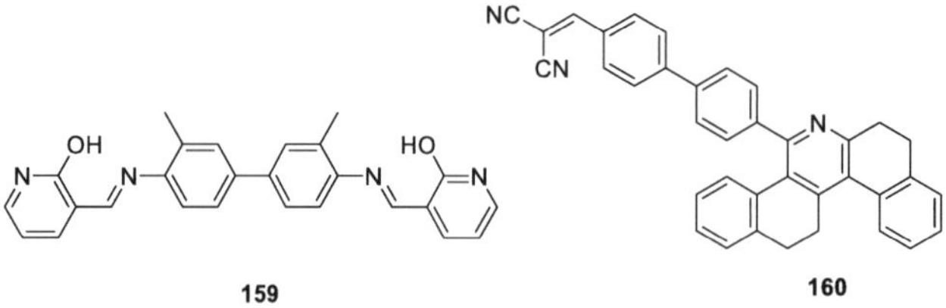

FIGURE 14.40 Structure of pyridine-based chemosensors.

Gu and coworkers [181] reported compound **156** as Cu^{2+} ions near IR fluorescent and colorimetric probe that exhibited near IR fluorescence at 676 nm on interaction with Cu^{2+} ions with concomitant visible color change from yellow to purple. Sensing of Cu^{2+} ions was based on intramolecular charge transfer mechanism in which compound **156** underwent hydrolysis promoted by Cu^{2+} ions which resulted in strong electron donor phenolate anion that increased ICT process and shifted the emission to the near IR region (Fig. 14.39).

Sundaramurthy's group [182] described TICT process for the sensing of aniline using **157** that was facilitated by $BF_3{\bullet}OEt_2$. Reaction of aniline with **157** caused a decline in absorption at 333 nm with related change in color from colorless to yellow (Fig. 14.39).

Song's group [183] described detection of CN^- ions via Michael addition both in solution and in gas phase based on diethylaminoquinoline derivatives **158,** addition of CN^- ions to which caused a significant visible color change from red to colorless within 10 s and quenching in fluorescence at 614 nm with emergence of new blue-shift in fluorescence at 494 nm due to ICT process, whereas other anions did not showed any significant change. A two-step sensing mechanism was demonstrated, in which the vinyl β-carbon of sensor underwent nucleophilic addition of CN^- followed by slow protonation of resultant carbanion to yield the final product (Scheme 14.34).

Hu and coworkers [184] reported compound **159** for sensing of CN^- ions via deprotonation mechanism. Reaction of CN^- ions with the yellow-colored solution of **159** turned the solution colorless along with the appearance of an absorption peak at 310 nm. Cyanide ion abstracted proton from OH of the sensor that led to ICT in senor as revealed by NMR analysis (Fig. 14.40).

Recently, recognition of biogenic primary amine was described by Sathiyanarayanan's group [185] based on intramolecular approach using compound **160** that showed high selectivity in acetonitrile. **160** displayed absorption maxima at 396 nm, addition of dopamine (DA) and ethylenediamine (EDA) to **160** caused the appearance of a new absorption band at 368 nm and the intensity at 396 nm reduced significantly, on further additions new bands around 279–326 nm and 260–321 nm were observed indicating the formation of a new product. The fluorescence emission was also decreased with a blue shift from 594 to 414 nm on interaction with DA and EDA with related change in color from orange to sky blue for DA and blue for EDA, respectively (Fig. 14.40).

14.5 Conclusions and future prospects

Incessant progress has been made in the past decades in the field of chemosensors and it is growing into a vibrant and active research field. This prodigious growth can partly be credited to the ground-breaking research of Professor Anthony W. Czarnik's and Professor A. Prasanna de Silva who have inspired innumerable researchers through their influential contributions to the field of chemosensors and molecular logic.

It may appear that most of the problems in chemosensors research have already been solved but we will always need "new chemosensors" for hitherto unknown analytes as an existing chemosensor may work for one application but may fall short of the obligatory selectivity or sensitivity required for use in a specific practical application. So, we will continue to need increasing number of chemosensors whether in the form of new receptors or modification of the existing systems and it is envisaged that many of the "old" chemosensors can be repurposed for use in as yet unknown applications. Presently, countless work directed toward the development of chemosensors has been published however, there are still several challenges that need to be addressed to improve the performance of chemosensors such as long-term stability, biocompatibility, cost, and so on.

Numerous authors have mentioned the difficulty of attaining high specificity as the superiority of design and synthetic efforts required to produce highly selective and sensitive chemosensors are the key features in this field. Moreover, most supramolecular chemosensors are cross reactive, they respond to a variety of analytes although this may seem counterproductive but it is the same strategy that nature uses in the recognition of molecules.

To sum up, it has been very well said the past set us going in the precise direction, the present-day provided us with the challenges that need to be unraveled and the future gives us prodigious hope that our lives will be improved by chemosensors.

Acknowledgments

The authors wish to thank all the authors whose work has been reported in this chapter.

Reference

[1] A. Hulanicki, S. Glab, F. Ingman, Chemical sensors: definitions and classification, Pure Appl. Chem. 63 (1991) 1247–1250.

[2] G. Fukuhara, Analytical supramolecular chemistry: Colorimetric and fluorimetric chemosensors, J. Photochem. Photobiol. C 42 (2020) 10034.
[3] B. Wang, E. V. Anslyn (Eds), Chemosensors: principles, strategies, and applications Wiley, Hoboken, 2011.
[4] T.W. Bell, N.M. Hext, Supramolecular optical chemosensors for organic analytes, Chem. Soc. Rev. 33 (2004) 589–598.
[5] J. Wu, W. Liu, J. Ge, H. Zhang, P. Wang, New sensing mechanisms for design of fluorescent chemosensors emerging in recent years, Chem. Soc. Rev. 40 (2011) 3483–3495.
[6] L. He, B. Dong, Y. Liu, W. Lin, Fluorescent chemosensors manipulated by dual/triple interplaying sensing mechanisms, Chem. Soc. Rev. 45 (2016) 6449–6461.
[7] N. Kwon, Y. Hu, J. Yoon, Fluorescent chemosensors for various analytes including reactive oxygen species, biothiol, metal ions, and toxic gases, ACS Omega 3 (2018) 13731–13751.
[8] V. Schroeder, S. Savagatrup, M. He, S. Lin, T.M. Swager, Carbon nanotube chemical sensors, Chem. Rev. 119 (2019) 599–663.
[9] M. Mayer, A.J. Baeumner, A megatrend challenging analytical chemistry: biosensor and chemosensor concepts ready for the internet of things, Chem. Rev. 119 (2019) 7996–8027.
[10] I. I. Ebralidze, N. O. Laschuk, J. Poisson, O. V. Zenkina, Book chapter from nanomaterials design for sensing applications faculty of science, university of ontario institute of technology, Oshawa, ON, Canada.
[11] L. You, D.J. Zha, E.V. Anslyn, Recent advances in supramolecular analytical chemistry using optical sensing, Chem. Rev. 115 (2015) 7840–7892.
[12] H. Zhu, J. Fan, B. Wang, X. Peng, Fluorescent, MRI, and colorimetric chemical sensors for the first-row d-block metal ions, Chem. Soc. Rev. 44 (2015) 4337–4366.
[13] H. Sharma, N. Kaur, A. Singh, A. Kuwar, N. Singh, Optical Chemosensor for water sample analysis, J. Mater. Chem. 4 (2016) 5154–5194.
[14] A.W. Czarnik, Chemical communication in water using fluorescent chemosensors, Acc. Chem. Res. 27 (1994) 302–308.
[15] A. Popov, H. Chen, O.N. Kharybin, E.N. Nikolaev, R.G. Cooks, Detection of explosives on solid surfaces by thermal desorption and ambient ion/molecule reactions, Chem. Commun. (2005) 1953–1955.
[16] J.M. Sylvia, J.A. Janni, J.D. Klein, K.M. Spencer, Surface-enhanced raman detection of 2,4-dinitrotoluene impurity vapor as a marker to locate landmines, Anal. Chem. 72 (2000) 5834–5840.
[17] (a) E. Wallis, T.M. Griffin, N. Popkie, M.A. Eagan, R.F. McAtee, D. Vrazel, J.Mc Kinly, Instrument response measurements of ion mobility spectrometers in situ: maintaining optimal system performance of fielded systems, Proc. SPIE-Int. Soc. Opt. Eng. 5795 (2005) 54–64; (b) G.A. Eiceman, J.A. Stone, Ion mobility spectrometers in national defence, Anal. Chem. 76 (2004) 390A–397A.
[18] L. Wu, C. Huang, B.P. Emery, A.C. Sedgwick, S.D. Bull, -P.He X, H. Tian, J. Yoon, J.L. Sessler, T.D. James, Forster resonance energy transfer (FRET)-based small-molecule sensors and imaging agents, Chem. Soc. Rev. 49 (2020) 5110–5139.
[19] A.P. Silva, H.Q.N. de, Gunaratne, T. Gunnlaugsson, A.J.M. Huxley, C.P. McCoy, J.T. Rademacher, T.E. Rice, Signaling recognition events with fluorescent sensors and switches, Chem. Rev. 97 (1997) 1515–1566.
[20] B. Daly, J. Ling, A.P. de Silva, Current developments in fluorescent PET (photoinduced electron transfer) sensors and switches, Chem. Soc. Rev. 44 (2015) 4203–4211.
[21] A.C. Sedgwick, L. Wu, H-H. Han, S.D. Bull, X-P. He, T.D. James, J.L. Sessler, B.Z. Tang, H. Tian, J. Yoon, Excited-state intramolecular proton-transfer (ESIPT) based fluorescence sensors and imaging agents, Chem. Soc. Rev. 47 (2018) 8842–8880.
[22] R.T.K. Kwok, C.W.T. Leung, J.W.Y. Lam, B.Z. Tang, Biosensing by luminogens with aggregation-induced emission characteristics, Chem. Soc. Rev. 44 (2015) 4228–4238.
[23] (a) S.H. Park, N. Kwon, J.H. Lee, J. Yoon, I. Shin, Synthetic ratiometric fluorescent probes for detection of ions, Chem. Soc. Rev. 49 (2020) 143–179; (b) D. Wu, L.Y. Chen, Q.L. Xu, X.Q. Chen, J. Yoon, Design principles, sensing mechanisms, and applications of highly specific fluorescent probes for $HOCl/OCl^-$, Acc. Chem. Res. 52 (2019) 2158–2168.
[24] B.R. Eggins, Chemical sensors and biosensors, Wiley, West Sussex, 2008, p. 28.
[25] X. Tian, L.C. Murfin, L. Wu, S.E. Lewis, T.D. James, Fluorescent small organic probes for biosensing, Chem. Sci. 12 (2021) 3406–3426.

[26] (a) S. Khatua, M. Schmittel, A single molecular light-up sensor for quantification of Hg^{2+} and Ag^{+} in aqueous medium: high selectivity toward Hg^{2+} over Ag^{+} in a mixture, Org. Lett. 15 (2013) 4422–4425; (b) O. Kotova, S. Comby, T. Gunnlaugsson, Sensing of biologically relevant d-metal ions using a Eu(III)-cyclen based luminescent displacement assay in aqueous pH 7.4 buffered solution, Chem. Commun. 47 (2011) 6810–6812; (c) Z.H. Lin, S.J. Ou, C.Y. Duan, B.G. Zhang, Z.P. Bai, Naked-eye detection of fluoride ion in water: a remarkably selective easy-to-prepare test paper, Chem. Commun. (2006) 624–626.

[27] (a)C. Y. K. Chan, J. W. Y. Lam, C. Deng, X. J. Chen, K. S. Wong, B. Z. Tang, Synthesis, light emission, explosive detection, fluorescent photopatterning, and optical limiting of disubstituted polyacetylenes carrying tetraphenylethene luminogens, macromolecules 48 (2015) 1038–1047;; (b) M. Wang, G.X. Zhang, D.Q. Zhang, D.B. Zhu, B.Z. Tang.

[28] (a) S.Y. Lim, K.H. Hong, D. Kim, H. Kwon, H.J. Kim, Tunable heptamethine–azo dye conjugate as an NIR fluorescent probe for the selective detection of mitochondrial glutathione over cysteine and homocysteine, J. Am. Chem. Soc. 136 (2014) 7018–7025; (b) Y.K. Yue, F.J. Huo, P. Yue, X.M. Meng, J.C. Salamanca, J.O. Escobedo, R.M. Strongin, C.X. Yin, In situ lysosomal cysteine-specific targeting and imaging during dexamethasone-induced apoptosis, Anal. Chem. 90 (2018) 7018–7024; (c) H. Zhang, L.Z. Xu, W.Q. Chen, J. Huang, C.S. Huang, J.R. Sheng, X.Z. Song, A lysosome-targetable fluorescent probe for simultaneously sensing Cys/Hcy, GSH, and H_2S from different signal patterns, ACS Sens 12 (2018) 2513–2517.

[29] (a) S.Y. Li, L.H. Liu, H. Cheng, B. Li, W.X. Qiu, X.Z. Zhang, A dual-FRET-based fluorescence probe for the sequential detection of MMP-2 and caspase-3, Chem. Commun. 51 (2015) 14520–14523; (b) K.Y. Tan, C.Y. Li, Y.F. Li, J.J. Fei, B. Yang, Y.J. Fu, F. Li, Real-time monitoring ATP in mitochondrion of living cells: a specific fluorescent probe for ATP by dual recognition sites, Anal. Chem. 89 (2017) 1749–1756; (c) J. Yin, Y. Kwon, D. Kim, D. Lee, G. Kim, Y. Hu, J.H. Ryu, J. Yoon, Cyanine-based fluorescent probe for highly selective detection of glutathione in cell cultures and live mouse tissues, J. Am. Chem. Soc. 136 (2014) 5351–5358.

[30] T. Yamamoto, T. Maruyama, Z-H. Zhou, T. Ito, T. Fukuda, Y. Yoneda, F. Begum, T. Ikeda, S. Sasaki, pi. conjugated poly(pyridine-2,5-diyl), poly(2,2′-bipyridine-5,5′-diyl), and their alkyl derivatives. preparation, linear structure, function as a ligand to form their transition metal complexes, catalytic reactions, n-type electrically conducting properties, optical properties, and alignment on substrates, J. Am. Chem. Soc. 116 (1994) 4832–4845.

[31] G.S. Lucka, E. Nowak, J. Kawchuk, M. Hoofar, H. Najjaran, Portable device for the detection of colorimetric assays, R. Soc. Open Sci. 4 (2017) 171025.

[32] S. Upadhyay, A. Singh, R. Sinha, S. Omer, K. Negi, Colorimetric chemosensors for d-metal ions: a review in the past, present and future prospect, J. Mol. Struct. 1193 (2019) 89–102.

[33] J.H. Kang, S.Y. Lee, H.M. Ahn, C. Kim, A novel colorimetric chemosensor for the sequential detection of Ni^{2+} and CN^{-} in aqueous solution, Sens. Actuators B Chem. 242 (2017) 25–34.

[34] L.M. Garzon, J. Portilla, Synthesis of novel D–π–A dyes for colorimetric cyanide sensing based on hemicyanine–functionalized N-(2-pyridyl) pyrazoles, Eur. J. Org. Chem. (2019) 7079–7088.

[35] J.H. Kang, S.Y. Lee, H.M. Ahn, C. Kim, Sequential detection of copper (II) and cyanide by a simple colorimetric chemosensor, Inorg. Chem. Commun. 74 (2016) 62–65.

[36] D. Wang, X.J. Zheng, A colorimetric chemosensor for Cu (II) ion in aqueous medium, Inorg. Chem. Commun. 84 (2017) 178–181.

[37] J. Wen, S. Li, Z. Huang, W. Li, X. Wang, Colorimetric detection of Cu^{2+} and ${UO_2}^{2+}$ by mixed solvent effect, Dyes pigm 152 (2018) 67–74 G6.

[38] Z. Li, S. Wang, L. Xiao, X. Li, X. Shao, X. Jing, X. Peng, L. Ren, An efficient colorimetric probe for fluoride ion based on Schiff base, Inorg. Chim. Acta. 476 (2018) 7–11.

[39] H. Kim, V. Seo, Y. Youn, H. Lee, M. Yang, C. Kim, Determination of Fe^{2+} and Co^{2+} by a multiple-target colorimetric chemosensor with low detection limit in aqueous solution, ChemistrySelect 4 (2019) 1199–1204.

[40] H. Tavallali, G.D. Rad, M.A. Karimi, E. Rahimy, A novel dye-based colorimetric chemosensors for sequential detection of Cu^{2+} and cysteine in aqueous solution, Anal. Biochem. 583 (2019) 113376.

[41] Q. Yue, J. He, L. Stamford, A. Azapagic, Nuclear power in China: an analysis of the current and near-future uranium flows, Energy Technol 5 (2017) 681–691.

[42] X. Wu, Y. Mao, D. Wang, Q. Huang, Q. Yin, M. Zheng, Q. Hu, H. Wang, Designing a colorimetric sensor containing nitrogen and oxygen atoms for uranyl ions identification: Chromatic mechanism, binding feature and on-site application, Sens. Actuators B Chem 307 (2020) 127681.

[43] Y. Morikawa, K. Nishiwaki, S. Suzuki, N. Yasaka, Y. Okada, I. Nakanishi, A new chemosensor for cyanide in blood based on the Pd complex of 2-(5-bromo-2-pyridylazo)-5-[Nn-propyl-N-(3-sulfopropyl) amino] phenol, Analyst 145 (2020) 7759–7764.

[44] T. Madhesan, A.M. Mohan, Porous silica and polymer monolith architectures as solid-state optical chemosensors for Hg^{2+} ions, Anal. Bioanal. Chem. 412 (2020) 7357–7370.

[45] H.J. Jang, J.H. Kang, D. Yun, C. Kim, A multi-responsive naphthalimide-based "turn-on" fluorescent chemosensor for sensitive detection of trivalent cations Ga^{3+}, Al^{3+} and Cr^{3+}, J Fluoresc 28 (2018) 785–794.

[46] M. Orojloo, S. Amani, Colorimetric detection of pollutant trivalent cations and HSO_4^- in aqueous media using a new Schiff-base probe: An experimental and DFT studies, Polycycl. Aromat. Compd. 41 (2021) 33–46.

[47] R. Diana, U. Caruso, L.D. Costanzo, F.S. Gentile, B. Panunzi, Colorimetric recognition of multiple first-row transition metals: A single water-soluble chemosensor in acidic and basic conditions, Dyes Pigm 184 (2021) 108832.

[48] R. Kumar, H. Jain, P. Gahlyan, A. Joshi, C.N. Ramachandran, A highly sensitive pyridine-dicarbohydrazide based chemosensor for colorimetric recognition of Cu^{2+}, AMP^{2-}, F^- and AcO^- ions, New J. Chem. 42 (2018) 8567–8576.

[49] M. Yang, J.B. Chae, C. Kim, R.G. Harrison, A visible chemosensor based on carbohydrazide for Fe (II), Co (II) and Cu (II) in aqueous solution, Photochem. Photobiol. Sci. 18 (2019) 1249–1258.

[50] K. Rezaeian, H. Khanmohammadi, A. Talebbaigy, Detection of CN^-, Cu^{2+} and Zn^{2+} ions using a new chromone-based colorimetric chemosensor: half-adder and integrated circuits, Anal. Methods 12 (2020) 1759–1766.

[51] M. Darroudi, G.M. Ziarani, S. Bahar, J.B. Ghasemi, A. Badiei, Lansoprazole-based colorimetric chemosensor for efficient binding and sensing of carbonate ion: spectroscopy and DFT studies, Front. chem. 8 (2020) 626472.

[52] H. Wang, D.L. Shi, J. Li, H.Y. Tang, J. Li, Y. Guo, A facile fluorescent probe with a large Stokes shift for sequentially detecting copper and sulfide in 100% aqueous solution and imaging them in living cells, Sens. Actuators B Chem. 256 (2018) 600–608.

[53] Y. Liu, Y. Ding, J. Huang, X. Zhang, T. Fang, Y. Zhang, X. Zheng, X. Yang, A benzothiazole-based fluorescent probe for selective detection of H_2S in living cells and mouse hippocampal tissues, Dyes Pigm 138 (2017) 112–118.

[54] X. Jin, H. Chen, W. Zhang, B. Wang, W. Shen, H. Lu, A novel purine derivative-based colorimetric chemosensor for sequential detection of copper ion and sulfide anion, Appl. Organomet. Chem. 32 (2018) 4577.

[55] L.HA. Rahman, A.M.A. Dief, A. Mawgoud, A.A H., Novel di-and tri-azomethine compounds as chemo sensors for the detection of various metal ions, Int. J. Nanomater. Chem. 5 (2019) 1–17.

[56] P.S. Portillo, A.H. Sirio, C.G. Alcanter, P.G. Lacroix, V. Agarwal, R. Santillan, V. Barba, Colorimetric metal ion (II) sensors based on imine boronic esters functionalized with pyridine, Dyes Pigm 186 (2021) 108991.

[57] J.M. Jung, S.Y. Lee, C. Kim, A novel colorimetric chemosensor for multiple target metal ions Fe^{2+}, Co^{2+}, and Cu^{2+} in a near-perfect aqueous solution: Experimental and theoretical studies, Sens. Actuators B Chem. 251 (2017) 291–301.

[58] A. Singh, M. Mohan, D.R. Trivedi, Chemosensor based on hydrazinyl pyridine for selective detection of F^- ion in organic media and CO_3^{2-} ions in aqueous media: design, synthesis, characterization and practical application, ChemistrySelect 4 (2019) 14120–14131.

[59] G. Yin, J. Yao, S. Hong, Y. Zhang, Z. Xiao, T. Yu, H. Li, P. Yin, A dual-responsive colorimetric probe for the detection of Cu^{2+} and Ni^{2+} species in real water samples and human serum, Analyst 144 (2019) 6962–6967.

[60] T. Salthammer, The formaldehyde dilemma, Int. J. Hyg. Environ. Health. 218 (2015) 433–436.

[61] J. Rovira, N. Roig, M. Nadal, M. Schuhmacher, J.L. Domingo, Human health risks of formaldehyde indoor levels: An issue of concern, J. Environ. Sci. Heal. A 51 (2016) 357–363.

[62] D. Ovianto, F.A.R. Maruf, N.N. Fadilah, I.B.A.R. Sugiharta, B. Purwono, Study on the colorimetric properties of 2, 4, 6-triarylpyridine derivative compound for imaging Formaldehyde, J. Phys. Conf. Ser. 909 (2017) 012079.

[63] M.H. Kao, C.F. Wan, A.T. Wu, A selective colorimetric chemosensor for Fe^{3+}, J. Lumin. 32 (2017) 1561–1566.

[64] X. Xia, D. Zhang, C. Fan, S. Pu, Naked-eye detection of Cu (II) and Fe (III) based on a Schiff Base Ruthenium complex with nicotinohydrazide, Appl. Organomet. Chem. 34 (2020) 5841.

[65] Y.R. Sadaphal, S.S. Gholap, A highly selective colorimetric chemosensor for copper (II) based on N-phenyl-N′-(pyridin-2-yl) thiourea (HPyPT), Sens. Actuators B Chem. 253 (2017) 173–179.

[66] P. Patil, S. Sehlangia, A. Patil, C. Pradeep, S.K. Sahoo, U. Patil, A new phthalimide based chemosensor for selective spectrophotometric detection of Cu (II) from aqueous medium, Spectrochim. Acta A 220 (2019) 117129.

[67] Y. Zhang, Y.T. Wang, X.X. Kang, M. Ge, H.Y. Feng, J. Han, D.H. Wang, D.Z. Zhao, Azobenzene disperse dye-based colorimetric probe for naked eye detection of Cu^{2+} in aqueous media: spectral properties, theoretical insights, and applications, J. Photochem. Photobiol. A Chem. 356 (2018) 652–660.

[68] P.A. Patil, S. Sehlangia, C.P. Pradeep, Dipicolinimidamide functionalized chromogenic chemosensor for recognition of Cu^{2+} ions and its applications, Sens. Int. 2 (2021) 100075.

[69] F. Faridbod, M. Sedaghat, M. Hosseini, M.R. Ganjali, M. Khoobi, A. Shafiee, P. Norouzi, Turn-on fluorescent chemosensor for determination of lutetium ion, Spectrochim. Acta A 137 (2015) 1231–1234.

[70] Q. Dai, H. Liu, C. Gao, W. Li, C. Zhu, C. Lin, Y. Tan, Z. Yuan, Y. Jiang, A one-step synthesized acridine-based fluorescent chemosensor for selective detection of copper (II) ions and living cell imaging, New J. Chem. 42 (2018) 613–618.

[71] A. Rajput, R. Mukherjee, Coordination chemistry with pyridine/pyrazine amide ligands. Some noteworthy results, Coord. Chem. Rev. 257 (2013) 350–368.

[72] P. Kumar, R. Gupta, The wonderful world of pyridine-2,6-dicarboxamide based scaffolds, Dalton Trans 45 (2016) 18769–18783.

[73] I.J.B. Rodriguez, D.M. Otero, J.B. Flores, A.K. Yatsimirsky, A.D Gonzalez, Sensitive water-soluble fluorescent chemosensor for chloride based on a bisquinolinium pyridine-dicarboxamide compound, Sens. Actuators B 221 (2015) 1348–1355.

[74] P. Kumar, V. Kumar, R. Gupta, Selective fluorescent turn-off sensing of Pd2+ ion: applications as paper strips, polystyrene films, and in cell imaging, RSC Adv 7 (2017) 7734–7741.

[75] P. Kumar, V. Kumar, S. Pandey, R. Gupta, Detection of sulfide ion and gaseous H_2S using a series of pyridine-2,6-dicarboxamide based scaffolds, Dalton Trans 47 (2018) 9536–9545.

[76] P.Kumar Sudheer, V. Kumar, R. Gupta, Detection of Al^{3+} and Fe^{3+} ions by nitrobenzoxadiazole bearing pyridine-2,6-dicarboxamiade based chemosensors: Effect of solvents on detection, New J. Chem. 44 (2020) 13285–13294.

[77] A.K. Mahapatra, S.K. Manna, D. Mandal, C.D. Mukhopadhyay, Highly Sensitive and Selective Rhodamine-Based "Off–On" Reversible Chemosensor for Tin (Sn^{4+}) and Imaging in Living Cells, Inorg. Chem. 52 (2013) 10825–10834.

[78] S-L. Shen, X-P. Chen, X-F. Zhang, J-Y. Miao, B-X. Zhao, A rhodamine B-based lysosomal pH probe, J. Mater. Chem. B 3 (2015) 919–925.

[79] M. Homma, Y. Takei, A. Murata, T. Inoue, S. Takeoka, A ratiometric fluorescent molecular probe for visualization of mitochondrial temperature in living cell, Chem. Commun. 51 (2015) 6194–6197.

[80] M. Ozdemir, A selective fluorescent ''turn-on'' sensor for recognition of Zn^{2+} in aqueous media, Spectrochim. Acta A 161 (2016) 115–121.

[81] F. Yan, T. Zheng, D. Shi, Y. Zou, Y. Wang, M. Fu, L. Chen, W. Fu, Rhodamine-aminopyridine based fluorescent sensors for Fe^{3+} in water: Synthesis, quantum chemical interpretation and living cell application, Sens. Actuators, B Chem. 215 (2015) 598–606.

[82] Y. Ge, X. Xing, A. Liu, R. Ji, S. Shen, X. Cao, A novel imidazo[1,5-*a*]pyridine-rhodamine FRET system as an efficient ratiometric fluorescent probe for Hg^{2+} in living cells, Dyes Pigm 146 (2017) 136–142.

[83] J. Liu, Y. Qian, A novel naphthalimide-rhodamine dye: Intramolecular fluorescence resonance energy transfer and ratiometric chemodosimeter for Hg^{2+} and Fe^{3+}, Dyes Pigm 136 (2017) 782–790.

[84] C. Kan, X. Shao, L. Wu, Y. Zhang, X. Bao, J. Zhu, A novel "OFF–ON–OFF" fluorescence chemosensor for hypersensitive detection and bioimaging of Al(III) in living organisms and natural water environment, J. Photochem. Photobiol. A 398 (2020) 112618.

[85] Y-S. Guo, M. Zhao, Q. Wang, Y-Q. Chen, D-S. Guo, New pyridine-bridged ferrocene–rhodamine receptor for the multifeature detection of Hg^{2+} in water and living cells, ACS Omega 5 (2020) 17672–17678.

[86] S.B. Roy, J. Mondal, A.R.K. Bukhsh, K.K. Rajak, A novel fluorene based "turn on" fluorescent sensor for the determination of zinc and cadmium: experimental and theoretical studies along with live cell imaging, New. J. Chem. 40 (2016) 9593–9608.

[87] K. Karaoglu, H.T. Akcay, I. Yilmaz, Detection of Fe^{2+} in acetonitrile/water mixture by new 8-hydroxyquinolin based sensor through metal displacement mechanism, J. Mol. Struct. 1133 (2017) 492–498.

[88] M. Li, J. He, Z. Wang, Q. Jiang, H. Yang, J. Song, Y. Yang, X. Xu, S. Wang, A novel nopinone-based turn-on fluorescent probe for hydrazine in living cells with high selectivity, Ind. Eng. Chem. Res. 58 (2019) 22754–22762.

[89] N. Choudhury, B. Ruidas, B. Sah, K. Srikanth, C.D. Mukhopadhyay, P. De, Multifunctional tryptophan-based fluorescent polymeric probes for sensing, bioimaging and removal of Cu^{2+} and Hg^{2+} Ions, Polym. Chem. 11 (2020) 2015–2026.

[90] P. Raj, A. Singh, A. Singh, N. Singh, Syntheses and photophysical properties of Schiff-Base Ni (II) Complexes: Application for sustainable antibacterial activity and cytotoxicity, ACS Sustain. Chem. Eng. 5 (2017) 6070–6080.
[91] S. Shirase, K. Shinohara, H. Tsurugi, K. Mashima, Oxidation of alcohols to carbonyl compounds catalyzed by oxo-bridged dinuclear cerium complexes with pentadentate Schiff-base ligands under dioxygen atmosphere, ACS Catal 8 (2018) 6939–6947.
[92] K. Karaoglu, H.T. Akcay, I. Yilmaz, Detection of Fe^{2+} in acetonitrile/water mixture by new 8- hydroxyquinolin based sensor through metal displacement mechanism, J. Mol. Struct. 1133 (2017) 492–498.
[93] M.J. Reimann, D.R. Salmon, J.T. Horton, E.C. Gier, L.R. Jefferies, Water-soluble sulfonate Schiff-base ligands as fluorescent detectors for metal ions in drinking water and biological systems, ACS Omega 4 (2019) 2874–2882.
[94] D. Basak, V.J. Leusen, T. Gupta, P. Koogerler, V. Bertolasi, D. Ray, Unusually distorted pseudo-octahedral coordination environment around Co II from thioether Schiff-base ligands in dinuclear [CoLn] (Ln = La, Gd, Tb, Dy, Ho) complexes: synthesis, structure, and understanding of magnetic behavior, Inorg. Chem. 59 (2020) 2387–2405.
[95] N. Roy, A. Dutta, P. Mondal, P.C. Paul, T.S. Singh, A new turn-on fluorescent chemosensor based on sensitive Schiff base for Mn^{2+} ion, J. Lumin. 165 (2015) 167–173.
[96] X. Yu, K. Wang, D. Cao, Z. Liu, R. Guan, Q. Wu, Y. Xu, Y. Sun, X. Zhao, A diethylamino pyridine formyl Schiff base as selective recognition chemosensor for biological thiols, Sens. Actuators B Chem. 250 (2017) 132–138.
[97] H. Liu, T. Liu, J. Li, Y. Zhang, J. Li, J. Song, J. Qua, A simple Schiff base as dual-responsive fluorescent sensor for bioimaging recognition of Zn^{2+} and Al^{3+} in living cells, J. Mater. Chem. 6 (2018) 5435–5442.
[98] H. Peng, Y. Han, N. Lin, H. Liu, Two pyridine-derived Schiff-bases as turn-on fluorescent sensor for detection of aluminium ion, Opt. Mater. 95 (2019) 109210.
[99] T. Dhawa, A. Hazra, A. Barma, K. Pal, P. Karmakar, P. Roy, 4-Methyl-2,6-diformylphenol based biocompatible chemosensors for pH: discrimination between normal cells and cancer cells, RSC Adv 10 (2020) 15501–15513.
[100] Y. Zhang, X. Chen, J. Liu, G. Gao, X. Zhang, S. Hou, H. Wang, A highly selective and sensitive fluorescent chemosensor for distinguishing cadmium(II) from zinc(II) based on amide tautomerization, New J. Chem. 42 (2018) 19245–19251.
[101] W. Paisuwan, P. Rashatasakhon, V. Ruangpornvisuti, M. Sukwattanasinitt, A. Ajavakom, Dipicolylamino quinoline derivative as novel dual fluorescent detecting system for Hg^{2+} and Fe^{3+}, Sens. Bio-Sensing Res. 24 (2019) 100283.
[102] W-B. Huang, W. Gu, H-X. Huang, J-B. Wang, W-X. Shen, Y-Y. Lv, J. Shen, A porphyrin-based fluorescent probe for optical detection of toxic Cd^{2+} ion in aqueous solution and living cells, Dyes Pigm 143 (2017) 427–435.
[103] J.B. Asha, P. Suresh, Covalently modified graphene oxide as highly fluorescent and sustainable carbonaceous chemosensor for selective detection of zirconium ion in complete aqueous medium, ACS Sustain. Chem. Eng. 8 (2020) 14301–14311.
[104] S.M. Hossain, V. Prakash, P. Mamidi, S. Chattopadhyay, A.K. Singh, Pyrene-appended bipyridine hydrazone ligand as a turn-on sensor for Cu^{2+} and its bioimaging application, RSC Adv 10 (2020) 3646–3658.
[105] R-S. Juang, P-C. Yang, H-W. Wen, Lin C-Y, S-C. Lee, T-W. Chang, Synthesis and chemosensory properties of terpyridine-containing diblock polycarbazole through RAFT polymerization, React. Funct. Polym. 93 (2015) 130–137.
[106] A. Sil, A. Maity, D. Giri, S.K. Patra, A phenylene-vinylene terpyridine conjugate fluorescent probe for distinguishing Cd^{2+} from Zn^{2+} with high sensitivity and selectivity, Sens. Actuators B Chem. 226 (2016) 403–411.
[107] L. Jin, C. Liu, N. An, Q. Zhang, J. Wang, L. Zhao, Y. Lu, Fluorescence turn-on detection of Fe^{3+} in pure water based on a cationic poly(perylene diimide) derivative, RSC Adv 6 (2016) 58394–58400.
[108] P-C. Yang, H-W. Wen, C-W. Huang, Y-N. Zhu, Synthesis and chemosensory properties of two-arm truxene-functionalized conjugated polyfluorene containing terpyridine moiety, RSC Adv 6 (2016) 87680–87689.
[109] X. He, C. Wu, Y. Qian, Y. Li, L. Zhang, F. Ding, H. Chen, J. Shen, Highly sensitive and selective light-up fluorescent probe for monitoring gallium and chromium ions in vitro and in vivo, Analyst 144 (2019) 3807–3816.
[110] A. Sil, D. Giri, S.K. Patra, Arylene-vinylene terpyridine conjugates: Highly sensitive, reusable and simple fluorescent probes for the detection of nitroaromatics, J. Mater. Chem. C 5 (2017) 11100–11110.
[111] M. Ucuncu, E. Karakus, M. Emrullahoglu, A BODIPY/pyridine conjugate for reversible fluorescence detection of gold(III) ions, New J. Chem. 39 (2015) 8337–8341.

[112] I. Roy, J-Y. Shin, D. Shetty, J.K. Khedkar, J.H. Park, K. Kim, E-Bodipy fluorescent chemosensor for Zn^{2+} ion, J. Photochem. Photobiol. A 331 (2016) 233–239.
[113] S.B. Maity, S. Banerjee, K. Sunwoo, J.S. Kim, P.K. Bharadwaj, A Fluorescent chemosensor for Hg^{2+} and Cd^{2+} ions in aqueous medium under physiological pH and its applications in imaging living Cells, Inorg. Chem. 54 (2015) 3929–3936.
[114] I.S. Turan, G. Gunaydin, S. Ayan, E.U. Akkaya, Molecular demultiplexer as a terminator automaton, Nat. Commun. 8 (2018) 805.
[115] Y. Song, J. Tao, Y. Wang, Z. Cai, X. Fang, S. Wang, H. Xu, A novel dual-responsive fluorescent probe for the detection of copper(II) and nickel(II) based on BODIPY derivatives, Inorg. Chim. Acta 516 (2021) 120099.
[116] J.O. Hernandez, J. Portilla, Synthesis of dicyanovinyl-substituted 1(2-pyridyl) pyrazoles: Design of a fluorescent chemosensor for selective recognition of cyanide, J. Org. Chem. 82 (2017) 13376–13385.
[117] M. Alkis, D. Pekyilmaz, E. Yalcin, B. Aydiner, Y. Dede, Z Seferoglu, H-bond stabilization of a tautomeric coumarin-pyrazole-pyridine triad generates a PET driven, reversible and reusable fluorescent chemosensor for anion detection, Dyes Pigm 141 (2017) 493—500.
[118] P. Madhu, P. Sivakumar, A novel pyridine-pyrazole based selective "turn-off" fluorescent chemosensor for Fe(III) ions, J. Photochem. Photobiol. A 371 (2019) 341–348.
[119] A. Shylaja, S.S. Roja, R.V. Priya, R.R. Kumar, Four-Component Domino Synthesis of Pyrazolo[3,4h]quinoline-3-carbonitriles: "Turn-Off" Fluorescent Chemosensor for Fe^{3+} Ions, J. Org. Chem. 83 (2018) 14084–14090.
[120] B. Naskar, K. Das, R.R. Mondal, D.K. Maiti, A. Requena, J.P.C. Carrasco, C. Prodhan, K. Chaudhuri, S. Goswami, A new fluorescence turn-on chemosensor for nanomolar detection of Al^{3+} constructed from pyridine-pyrazole system, New. J. Chem. 42 (2018) 2933–2941.
[121] Y-Q. Gu, W-Y. Shen, Y. Zhou, S-F. Chen, Y. Mi, B-F. Long, D.J. Young, F-L. Hu, A pyrazolopyrimidine based fluorescent probe for the detection of Cu^{2+} and Ni^{2+} and its application in living cells, Spectrochim. Acta A 209 (2019) 141–149.
[122] Y-Q. Gu, W-Y. Shen, Y. Mi, Y-F. Jing, J-M. Yuan, P. Yu, X-M.Zhu.F-L. Hu, Dual-response detection of Ni^{2+} and Cu^{2+} ions by a pyrazolopyrimidine-based fluorescent sensor and the application of this sensor in bioimaging, RSC Adv 9 (2019) 35671–35676.
[123] M. Garcia, I. Romero, J. Portilla, Synthesis of fluorescent 1,7-Dipyridyl-bis-pyrazolo[3,4*b*:4′,3′*e*]pyridines: Design of reversible chemosensors for nanomolar detection of Cu^{2+}, ACS Omega 4 (2019) 6757—6768.
[124] S. Saha, A. De, A. Ghosh, A. Ghosh, K. Bera, K.S. Das, S. Akhtar, N.C. Maiti, A.K. Das, B.B. Das, R. Mondal, Pyridine-pyrazole based Al(III) 'turn on' sensor for MCF7 cancer cell imaging and detection of picric acid, RSC Adv 11 (2021) 10094–10109.
[125] A. Badiei, B.C. Razavi, H. Goldooz, G.M. Ziarani, F. Faridbod, M.R. Ganjali, A Novel fluorescent chemosensor assembled with 2,6-bis(2-benzimidazolyl)pyridine-functionalized nanoporous silica-type SBA-15 for recognition of Hg^{2+} ion in aqueous media, Int. J. Environ. Res. 12 (2018) 109–115.
[126] Y. He, Q. Bing, Y. Wei, H. Zhang, G. Wang, A new benzimidazole-based selective and sensitive 'on–off' fluorescence chemosensor for Cu^{2+} ions and application in cellular bioimaging, Luminescence 34 (2019) 153–161.
[127] S. Gharami, K. Aich, P. Ghosh, L. Patra, N. Murmu, T.K. Mondal, A fluorescent "ON–OFF–ON" switch for the selective and sequential detection of Hg^{2+} and I^- with applications in imaging using human AGS gastric cancer cells, Dalton Trans 49 (49) (2020) 187–195.
[128] S. Mahata, A. Bhattacharya, J.P. Kumar, B.B. Mandal, V. Manivannan, Naked-eye detection of Pd^{2+} ion using a highly selective fluorescent heterocyclic probe by "turn-off" response and in-vitro live cell imaging, J. Photochem. Photobiol. A 394 (2020) 112441.
[129] J. Gong, Y-H. Li, C-J Zhang, J. Huang, Q. Sun, A thiazolo[4,5-*b*]pyridine-based fluorescent probe for detection of zinc ions and application for in vitro and in vivo bioimaging, Talanta 185 (2018) 396–404.
[130] M. Kargar, H.R. Darabi, A. Sharifi, A. Mostashari, A new chromogenic and fluorescence chemosensor based on naphthol–bisthiazolopyridine hybrid: Fast response and selective detection of multiple targets, silver, cyanide, sulfide, hydrogen sulfide ions and gaseous H_2S, Analyst 145 (2020) 2319–2330.
[131] Y. Liu, Y. Li, Q. Feng, N. Li, K. Li, H. Hou, B. Zhang, Turn-on' fluorescent chemosensors based on naphthaldehyde-2-pyridinehydrazone compounds for the detection of zinc ion in water at neutral pH, Luminescence 33 (2017) 29–33.
[132] K-Y. Kong, L-J. Hou, X-Q. Shao, S-M. Shuang, Y. Wang, C. Dong, A phenolphthalein-based fluorescent probe for the sequential sensing of Al^{3+} and F^- ions in aqueous medium and live cells, Spectrochim. Acta A 208 (2019) 131–139.

[133] H. Peng, X. Peng, J. Huang, A. Huang, S. Xu, J. Zhou, S. Huang, X. Cai, Synthesis and crystal structure of a novel pyridine acylhydrazone derivative as a "turn on" fluorescent probe for Al^{3+}, J. Mol. Struct. 1212 (2020) 128138.
[134] R. Liu, X. Tang, Y. Wang, J. Han, H. Zhang, C. Li, W. Zhang, L. Ni, H. Li, A fluorescent chemosensor for relay recognition of Fe^{3+} and PO_4^{3-} in aqueous solution and its applications, Tetrahedron 73 (2017) 5229–5238.
[135] J.M.V. Ngororabanga, C. Moyo, E. Hosten, N. Mama, Z.R. Tshent, A novel coumarin-based ligand: a turn-off and highly selective fluorescent chemosensor for Cu^{2+} in water, Anal. Methods 11 (2019) 3857–3865.
[136] L. Rasheed, M. Yousuf, S. Youn, T. Yoon, K-Y. Kim, Y-K. Seo, G. Shi, M. Saleh, J-H. Hur, K.S. Kim, Turn-On ratiometric fluorescent probe for selective discrimination of Cr^{3+} from Fe^{3+} in aqueous media for living cell imaging, Chem. Eur. J. 21 (2015) 16349–16353.
[137] J. Prabhu, K. Velmurugan, R. Nandhakumar, Development of fluorescent Lead II sensor based on an anthracene derived chalcone, Spectrochim. Acta A 144 (2015) 23–28.
[138] Q. Zhao, X-M. Liu, H-R. Li, Y-H. Zhang, X-H. Bu, High-performance fluorescence sensing of lanthanum ions (La^{3+}) by a polydentate pyridyl-based quinoxaline derivative, Dalton Trans 45 (2016) 10836–10841.
[139] Y. Hu, M. Kye, J.Y. Jung, Y-B. Lim, J. Yoon, Design for a small molecule based chemosensor containing boron and pyridine moieties to detect HF, Sens. Actuators, B Chem. 255 (2018) 2621–2627.
[140] Y. Zhang, D. Wang, C. Sun, H. Feng, D. Zhao, Y. Bi, A simple 2,6-diphenylpyridine-based fluorescence "turn-on" chemosensor for Ag^{+} with a high luminescence quantum yield, Dyes Pigm 141 (2017) 202–208.
[141] J-H. Hu, Y. Sun, J. Qi, P-X. Pei, Q. Lin, Y-M. Zhang, A colorimetric and "turn-on" fluorimetric chemosensor for selective detection of cyanide and its application in food sample, RSC Adv 6 (2016) 100401–100406.
[142] M. Sohrabi, M. Amirnasr, H. Farrokhpour, S. Meghdadi, A single chemosensor with combined ionophore/fluorophore moieties acting as a fluorescent "Off-On" Zn^{2+} sensor and a colorimetric sensor for^{2+}: Experimental, logic gate behavior and TD-DFT calculations, Sens. Actuators, B Chem. 250 (2017) 647–658.
[143] M. Sohrabi, M. Amirnasr, S. Meghdadi, M. Lutz, M.B. Torbati, M. Farrokhpour, A highly selective fluorescence turn-on chemosensor for Zn^{2+}, its application in live cell Imaging, and a colorimetric sensor for Co^{2+}: Experimental and TD-DFT calculations, New J. Chem. 42 (2018) 12595–12606.
[144] P. Sakthivel, K. Sekar, A sensitive isoniazid capped silver nanoparticles-selective colorimetric fluorescent sensor for Hg^{2+} ions in aqueous medium, J Fluoresc 30 (2020) 91–101.
[145] H. Rahimi, R. Hosseinzadeh, M. Tajbakhsh, A new and efficient pyridine-2,6-dicarboxamide-based fluorescent and colorimetric chemosensor for sensitive and selective recognition of Pb^{2+} and Cu^{2+}, J. Photochem. Photobiol. A 407 (2021) 113049.
[146] X. Huang, Z. Lu, Z. Wang, C. Fan, W. Fan, X. Shi, H. Zhang, M. Pei, A colorimetric and turn-on fluorescent chemosensor for selectively sensing Hg^{2+} and its resultant complex for fast detection of I^{-} over S^{2-}, Dyes Pigm 128 (2016) 33–40.
[147] M. Li, Y. Sun, L. Dong, Q-C. Feng, H. Xu, S-Q. Zang, T.C.W. Mak, Colorimetric recognition of Cu^{2+} and fluorescent detection of Hg^{2+} in aqueous media by a dual chemosensor derived from rhodamine B dye with a NS_2 receptor, Sens. Actuators, B Chem. 226 (2016) 332–341.
[148] L. Hou, J. Feng, Y. Wang, C. Dong, S. Shuang, Y. Wang, Single fluorescein-based probe for selective colorimetric and fluorometric dual sensing of Al^{3+} and Cu^{2+} Sens, Actuators, B Chem 247 (2017) 451–460.
[149] M. Maniyazagan, R. Mariadasse, J. Jeyakanthan, N.K. Lokanath, S. Naveen, K. Premkumar, P. Muthuraja, P. Manisankar, T. Stalin, Rhodamine based "turn–on" molecular switch FRET–sensor for cadmium and sulfide ions and live cell imaging study, Sens. Actuators, B Chem. 238 (2017) 565–577.
[150] H. Chen, X. Jin, W. Zhang, H. Lu, W. Shen, A new rhodamine B-based 'off-on' colorimetric chemosensor for Pd^{2+} and its imaging in living cells, Inorganica Chim. Acta 482 (2018) 122–129.
[151] Z. Yan, G. Wei, S. Guang, M. Xu, X. Ren, R. Wu, G. Zhao, F. Ke, H. Xu, A multidentate ligand chromophore with rhodamine-triazole-pyridine units and its acting mechanism for dual-mode visual sensing trace Sn^{2+}, Dyes Pigm 159 (2018) 542–550.
[152] C. Long, J.H. Hu, Q-Q. Fu, P-W. Ni, A new colorimetric and fluorescent probe based on Rhodamine B hydrazone derivatives for cyanide and Cu^{2+} in aqueous media and its application in real life, Spectrochim. Acta A 219 (2019) 297–306.
[153] F. Song, C. Yang, H. Liu, Z. Gao, J. Zhu, X. Bao, C. Kan, Dual-binding pyridine and rhodamine B conjugate derivatives as fluorescent chemosensors for Ferric ion in aqueous media and living cells, Analyst 144 (2019) 3094–3102.

[154] H. Zhao, H. Ding, H. Kang, C. Fan, G. Liu, S. Pu, A solvent-dependent chemosensor for fluorimetric detection of Hg^{2+} and colorimetric detection of Cu^{2+} based on a new diarylethene with a rhodamine B unit, RSC Adv 9 (2019) 42155–44216.

[155] X. Wang, T. Li, A novel "off-on" rhodamine-based colorimetric and fluorescent chemosensor based on hydrolysis driven by aqueous medium for the detection of Fe^{3+} Spectrochim, Acta A 229 (2020) 117951.

[156] M. Shellaiah, N. Thirumalaivasan, B. Aazaad, K. Awasthi, K.W. Sun, S-P. Wu, M.C. Lin, N. Ohta, Novel rhodamine probe for colorimetric and fluorescent detection of Fe^{3+} ions in aqueous media with cellular imaging, Spectrochim. Acta A 242 (2020) 118757.

[157] J. Guan, J. Yang, Y. Zhang, X. Zhang, H. Deng, J. Xu, J. Wang, M-S. Yuan, Employing a fluorescent and colorimetric picolyl-functionalized rhodamine for the detection of glyphosate pesticide, Talanta 224 (2021) 121834.

[158] F. Yu, L.J. Hou, L.Y. Qin, J.B. Chao, Y. Wang, W.J. Jin, A new colorimetric and turn-on fluorescent chemosensor for Al^{3+} in aqueous medium and its application in live-cell imaging, J Photochem. Photobiol. A 315 (2016) 8–13.

[159] Y. Guo, L. Wang, J. Zhuo, B. Xu, X. Li, J. Zhang, Z. Zhang, H. Chi, Y. Dong, G. Lu, A pyrene-based dual chemosensor for colorimetric detection of Cu2+ and fluorescent detection of Fe^{3+}, Tetrahedron Lett 58 (2017) 3951–3956.

[160] S.Y. Li, D.B. Zhang, J.Y. Wang, R.M. Lu, C.H. Zheng, S.Z. Pu, A novel diarylethene-hydrazinopyridine-based probe for fluorescent detection of aluminum ion and naked-eye detection of hydroxide ion, Sens. Actuators, B Chem. 245 (2017) 263–272.

[161] H. Liu, S. Cui, F. Shi, S. Pu, A diarylethene based multi-functional sensor for fluorescent detection of Cd^{2+} and colorimetric detection of Cu^{2+}, Dyes Pigm 161 (2019) 34–43.

[162] M. Sahu, A.K. Manna, K. Rout, J. Mondal, G.K. Patra, A diarylethene based multi-functional sensor for fluorescent detection of Cd^{2+} and colorimetric detection of Cu^{2+}, Inorganica Chim. Acta 508 (2020) 119633.

[163] L. Huang, J. Zhang, X. Yu, Y. Ma, T. Huang, X. Shen, H. Qiu, X. He, X. Yin, A Cu^{2+} selective fluorescent chemosensor based on BODIPY with two pyridine ligands and logic gate Spectrochim, Acta A 145 (2015) 25–32.

[164] S. Xia, J. Shen, J. Wang, H. Wang, M. Fang, H. Zhou, M. Tanasova, Ratiometric fluorescent and colorimetric BODIPY-based sensor for zinc Ions in solution and living cells, Sens. Actuators, B Chem. 258 (2018) 1279–1286.

[165] Z. Gu, H. Cheng, X. Shen, T. He, K. Jiang, H. Qiu, Q. Zhang, S. Yin, A BODIPY derivative for colorimetric fluorescence sensing of Hg^{2+}, Pb^{2+} and Cu^{2+} ions and its application in logic gates, Spectrochim. Acta A 203 (2018) 315–323.

[166] S. Guria, A. Ghosh, K. Manna, A. Pal, A. Adhikary, S. Adhikari, Rapid detection of aspartic acid and glutamic acid in water by BODIPY-based fluorescent probe: Live-cell imaging and DFT studies, Dyes Pigm 168 (2019) 111–122.

[167] B.V. Razavi, A Badiei, N. Lashgari, G.M. Ziarani, 2,6-bis(2-benzimidazolyl)pyridine fluorescent red-shifted sensor for recognition of Zinc(II) and a calorimetric sensor for iron Ions, J Fluoresc 26 (2016) 1723–1728.

[168] L. Liu, F. Dan, W. Liu, X. Lu, Y. Han, S. Xiao, H. Lan, A high-contrast colorimetric and fluorescent probe for Cu^{2+} based on benzimidazole-quinoline, Sens. Actuators B Chem 247 (2017) 445–450.

[169] P.A. More, G.S. Shankarling, Reversible 'turn off' fluorescence response of Cu^{2+} ions towards 2-pyridyl quinoline based chemosensor with visible colour change, Sens. Actuators, B Chem. 241 (2017) 552–559.

[170] S. Swami, D. Behera, A. Agarwala, V.P. Verma, R. Shrivastava, β-Carboline-imidazopyridine hybrid: A selective and sensitive optical sensor for sensing of copper and fluoride ions, New J. Chem. 42 (2018) 10317–10326.

[171] A.J. Beneto, A. Siva, Benzothiazole, pyridine functionalized triphenylamine based fluorophore for solid state fluorescence switching, Fe^{3+} and picric acid sensing, Sens. Actuators B Chem. 242 (2017) 535–544.

[172] R. Diana, U. Caruso, S. Concilio, S. Piotto, A. Tuzi, B. Panunzi, A real-time tripodal colorimetric/fluorescence sensor for multiple target metal ions, Dyes Pigm 155 (2018) 249–257.

[173] N. Assadollahnejad, M. Kargar, H.R. Darabi, N. Abouali, S. Jamshidi, A. Sharifi, A. Aghapoor, H. Sayahi, A new ratiometric, colorimetric and "turn-on" fluorescent chemosensor for detection of cyanide ion based on phenol–bisthiazolopyridine hybrid, New J. Chem. 43 (2019) 13001–13009.

[174] Y. Lv, L. Wu, W. Shen, J. Wang, G. Xuan, X. Sun, A porphyrin-based chemosensor for colorimetric and fluorometric detection of cadmium(II) with high selectivity, J. Porphyr. Phthalocyanines 19 (2015) 1–6.

[175] G.J. Park, G.R. You, Y.W. Choi, C. Kim, A naked-eye chemosensor for simultaneous detection of iron and copper ions and its copper complex for colorimetric/fluorescent sensing of cyanide, Sens. Actuators B Chem. 229 (2016) 257–271.

[176] Z. Li, J. Qiang, J. Li, T. Wei, Z. Zhang, Y. Chen, Q. Kong, F. Wang, X. Chen, Di-nuclear Cu(II) complex bearing guanidinium arms for sensing pyrophosphate and monitoring the activity of PPase, Dyes Pigm 163 (2019) 243–248.

[177] V.K. Gupta, S.K. Shoora, L.K. Kumawat, A.K. Jain, A highly selective colorimetric and turn-on fluorescent chemosensor based on 1-(2-pyridylazo)-2-naphthol for the detection of aluminium(III) ions, Sens. Actuators B Chem. 209 (2015) 15–24.

[178] N. Lashgari, A. Badiei, G.M. Ziarani, A Fluorescent Sensor for Al(III) and Colorimetric Sensor for Fe(III) and Fe(II) Based on a Novel 8-Hydroxyquinoline Derivative, J. Fluoresc. 26 (2016) 1885–1894.

[179] J. Yao, Y. Fu, W. Xu, T. Fan, Y. Gao, Q. He, D. Zhu, H. Cao, J. Cheng, A concise and efficient fluorescent probe via an intromolecular charge transfer for the chemical warfare agent mimic diethylchlorophosphate vapor detection, Anal. Chem. 88 (2016) 2497–2501.

[180] J. Tan, X. Wang, Q. Zhang, H. Zhou, J. Yang, J. Wu, Y. Tian. X. Zhang, Chalcone based ion-pair recognition towards nitrates and the application for the colorimetric and fluorescence turn-on determination of water content in organic solvents, Sens. Actuators B Chem. 260 (2018) 727–735.

[181] B. Gu, L. Huang, Z. Xu, Z. Tan, M. Hu, Z. Yang, Y. Chen, C. Peng, W. Xiao, D. Yu, H. Li, A reaction-based, colorimetric and near-infrared fluorescent probe for Cu2+ and its applications, Sens. Actuators B Chem 273 (2018) 118–125.

[182] U. Balijapalli, S. Manickam, K. Thirumoorthy, K.N. Sundaramurthy, K.I. Sathiyanarayanan, Tetrahydrodibenzo[*a*,*i*]phenanthridin-5-yl)phenol as a fluorescent probe for the detection of aniline, J. Org. Chem. 84 (2019) 11513–11523.

[183] L. Zhong, H. Li, S-L. Wang, Q-H. Song, The sensing property of charge-transfer chemosensors tuned by acceptors for colorimetric and fluorometric detection of CN^-/HCN in solutions and in gas phase, Sens. Actuators B Chem. 266 (2018) 703–709.

[184] P-W. Ni, Y. Yao, Q-Q. Fu, C. Long, G-H. Hu, Highly selective calorimetric and fluoremetric chemosensor for CN^- based on *o*-tolidine, J. Appl. Spectrosc. 87 (2020) 832–838.

[185] M. Saravanakumar, B. Umamahesh, R. Selvakumar, J. Dhanapal, S.K.A. Kumar, K.L. Sathiyanarayanan, A colorimetric and ratiometric fluorescent sensor for biogenic primary amines based on dicyanovinyl substituted phenanthridine conjugated probe, Dyes Pigm 178 (2020) 108346.

CHAPTER

15

Recent advances in catalytic synthesis of pyridine derivatives

Morteza Torabi, Meysam Yarie, Saeed Baghery and Mohammad Ali Zolfigol

Department of Organic Chemistry, Faculty of Chemistry, Bu-Ali Sina University, Hamedan, Iran

15.1 Introduction

Pyridine skeletons have fascinated extensive consideration since they not only are common nucleus in drug molecules and natural products but also show noteworthy roles in coordination chemistry, functional materials, and organic catalysis [1,2]. Pyridines are substantial resources in supramolecular chemistry because of their p-stacking facility and outstanding thermal stability and employed in high-performance organic light-emitting devices (OLED) [3]. Especially, β-(pyridin-2-yl) ketones were known as main structures for various pyridines with umami flavor (Scheme 15.1), which were broadly used as superb replacements to monosodium glutamate in the food chemistry [4]. Also, compound B3PyPB (Scheme 15.1), as one of the pyridine-based π-conjugated mediums, was applied in OLED [5].

Pyridines are traditionally generated from aldehydes and amines through multicomponent reactions [6–8] or transition-metal-catalyzed cycloaddition methods [9–11]. These compounds have been used in varied fields such as materials and surfaces, polymer science, supramolecular structures, and in catalysis [12,13]. Pyridines display an appropriate role in the biologically active natural materials containing oxidative-reductive NADP-NADPH coenzymes, nicotine, and vitamin B_6. Also, pyridines and its derivatives have been produced for several usages such as acetylcholinesterase inhibitor, anti-inflammatory, anti-depressant, anti-cancer and HIV protease inhibitor [14–16]. Fascinatingly, several marketed drug agents have this special core in their structures [17]. Some of the most important of them are MRZ"8676 (mGluR5 receptor antagonist) [18], Streptonigrin (anticancer) [19], Lavendamycin (anticancer) [20], and Bay 60"5521 (CETP inhibitors) [21] which have pyridine nucleus as essential moiety (Scheme 15.2). In recent years, using the capability of review articles, researchers who are interested in this field, summarized the methods of synthesizing these valuable compounds, each from

Recent Developments in the Synthesis and Applications of Pyridines.
DOI: https://doi.org/10.1016/B978-0-323-91221-1.00002-6

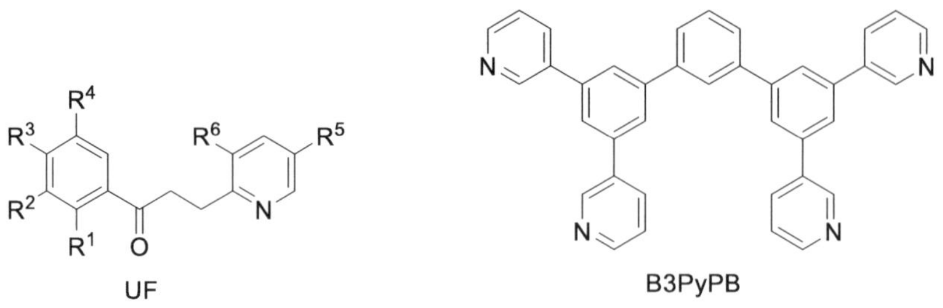

SCHEME 15.1 Structures of umami flavor (UF) and B3PyPB as pyridine-based compounds.

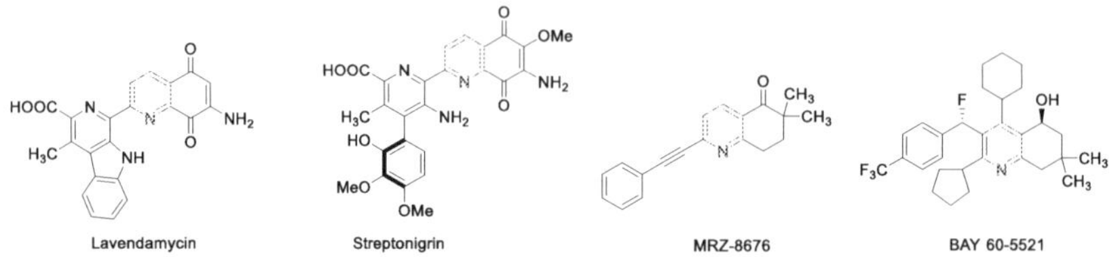

SCHEME 15.2 Some typical bioactive molecules with pyridine core.

a different perspective [22–25]. Here, in the light of the above-mentioned properties and applications of pyridine and its derivatives, we decided to collect the recent advances in the catalytic synthesis of pyridines.

The concept of catalyst which was first introduced by Berzelious in 1836 [24], increasingly attracted the attention of scientists. These materials which cover the important domain of chemistry play a pivotal role in a variety of chemicals, biological systems and fuels. The broad applications of catalysts in academic and industrial institutions are undeniable [25–28]. According to the green chemistry definition, a catalyst is a beneficial tool to convert basic chemicals into valuable products [29]. Generally, catalytic processes are divided into homogeneous and heterogeneous catalysis. In homogeneous systems, catalyst and reactants are in the same phase and due to this attribute, the interactions between catalyst and reactants become more and more effective and catalytic sites are accessible. This leads to lowering the activation energies for chemical reactions. Other advantages of homogeneous catalysts are including chemoselectivity, enantioselectivity, regioselectivity, high turnover number (TON) and high turnover frequency (TOF) [30–33]. Homogeneous catalysis can be divided into simple acids or bases [34], transition metal catalysis [35], metal complexes [36], organocatalysts [37], ionic liquids [38], etc. Homogeneous catalysis covers many of catalytic reactions such as multicomponent reactions [39], carbonylation [40], oxidation [35], reduction [41], and hydrogenation [42].

The importance of heterogeneous catalysis as the center of catalytically chemical reactions is obvious and palpable. In heterogeneous catalysis, catalyst and reactants are in different phase. These functional materials due to having active sites, reacted with reactants that drive the reaction to form the desired product [43–46]. The production of 90% of valuable chemical compounds is assisted by heterogeneous catalytic processes [47]. The two key merits of heterogeneous catalysts are easy separation and recycling and high durability which attract the attention of many researchers [45]. Heterogeneous catalysts are including of metal

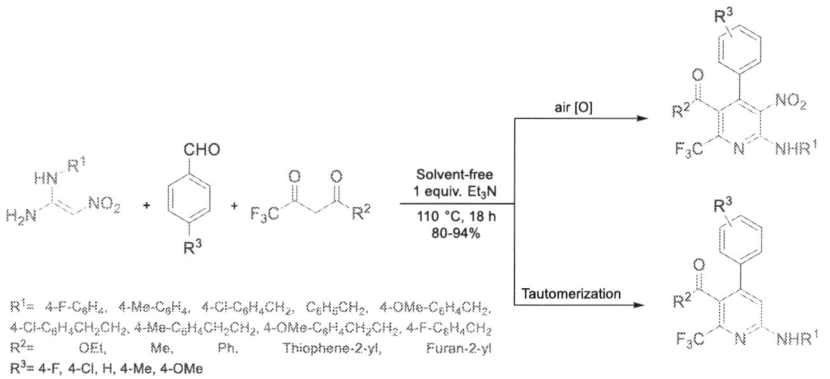

SCHEME 15.3 Synthesis of fluorinated 2-aminopyridine compounds 4 and 5 by using 1.0 equivalent Et_3N.

oxides (Fe_3O_4, ZnO, V_2O_5, Al_2O_3, etc.) [24, 48], silica-based catalysts [49], polymeric catalysts [50], metal–organic frameworks [51], covalent organic frameworks [52], etc. Applications of heterogeneous catalysts are seen in synthesis of gases, oxidation, reduction, decomposition, hydrogenation, and multicomponent reactions [53,54].

15.2 Basic catalysts

Using Et_3N as a base, the reaction of different types of 1,1-enediamines (EDAMs), various aldehydes, and 1,3-dicarbonyl compounds through the Knoevenagel, Michael, and cyclization reactions, leads to the formation of fluorinated 2-aminopyridine derivatives via a facile and benign process (Scheme 15.3) [55]. The obtained results indicated that EDAMs with the electron-withdrawing groups (–F, –Cl) produced higher yields than EDAMs with the electron-donating groups (–Me, –OMe). The reactions took 18 h under solvent-free conditions at 110°C in 1.0 equivalent Et_3N. The 2-aminopyridine products were attained in moderate yields and high conversion rate (80–94%). Nevertheless, the ratio of products is about 1:1.

The suggested mechanism of the synthesis of fluorinated 2-aminopyridines is displayed in Scheme 15.4 [55]. Initially, 1,3-dicarbonly compounds were tautomerized to their enol form. Then, the reaction of generated enol form with aldehyde derivatives and a dehydration process leads to the formation of Knoevenagel adduct (**I**). In the next step, Knoevenagel adduct (**I**) reacted with EDAMs via Michael addition promoted by the base to afford intermediate **(II)**. After that, intermediate **(III)** was made by imine–enamine tautomerism, intramolecular cyclization, and dehydration reaction, respectively. The fluorinated 2-aminopyridine was synthesized via oxidation reaction of intermediate **(III)**, which share an analogous concept with literature reports [56, 57]. All at once, the intermediate **(III)** can produce the desired molecules by a tautomerization and HNO_2 releasing or air oxidation.

The synthesis of a series of chromeno[2,3-*b*]pyridine derivatives has been reported under microwave irradiation in 5 mL propanol and two drop TEA at 150°C for appropriate reaction times (5–30 min) and good yields (80–97%) via one-pot reaction between substituted salicylaldehydes, substituted phenols, and malononitrile (Scheme 15.5). Also, it is demonstrated

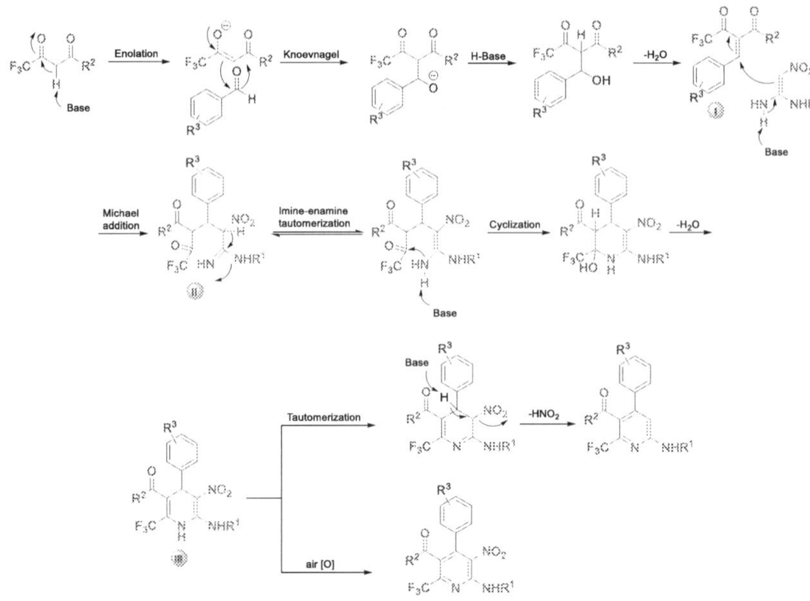

SCHEME 15.4 The proposed mechanism for the synthesis of fluorinated 2-aminopyridine compounds 4 and 5.

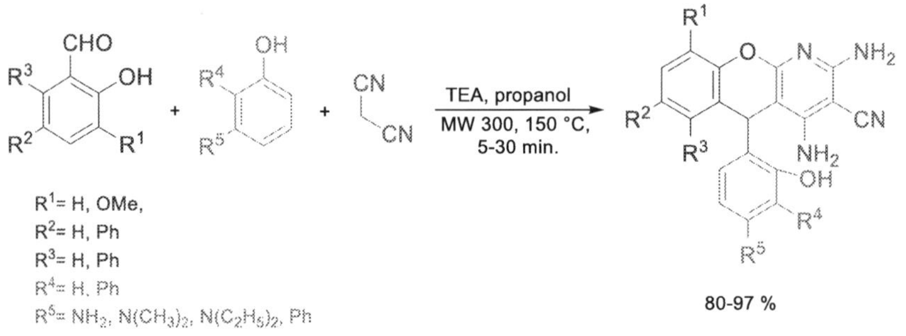

SCHEME 15.5 Synthesis of chromeno[2,3-*b*]pyridines in the presence of TEA.

that the prepared chromeno[2,3-*b*]pyridines inhibited the generation of NO in both human and porcine chondrocytes in vitro [58].

In 2017, diethylamine Dess–Martin periodinane was used both as catalyst and oxidant for the consecutive and one-pot synthesis of 2-amino-3,5-dicarbonitrile-6-sulfanylpyridines as medicinally important structures through a pseudo-four component reaction between 2,6-disubstituted benzaldehyde derivatives, malononitrile, and thiols [59]. Significant advantages of the presented work include mild reaction conditions, easy isolation, and purification processes and excellent yields at reasonable times (1.5–2.5 h) (Scheme 15.6).

As depicted in Scheme 15.7, the three-component reaction of conjugated enynones, 1,5-diarylpent-2-en-4-yn-1-ones, malononitrile, and sodium alkoxides in the corresponding alcohols for 3–23 h at room temperature leads to the formation of (*E*)-/(*Z*)-6-aryl-4-(2-arylethenyl)-2-alkoxypyridine-3-carbonitrile derivatives, as the major products (ca. yields up to 40–80%)

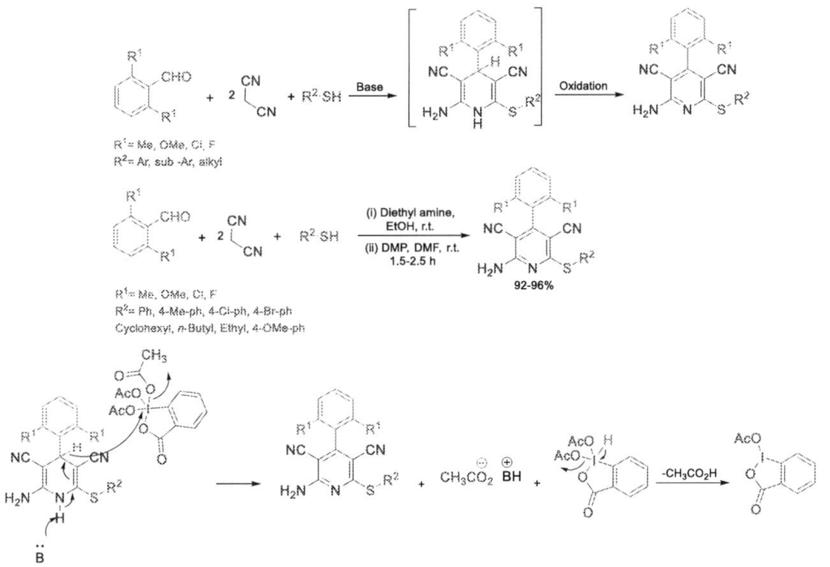

SCHEME 15.6 Two step and successive synthesis of pyridine 3,5-dicarbonitriles by diethylamine Dess-Martin periodinane.

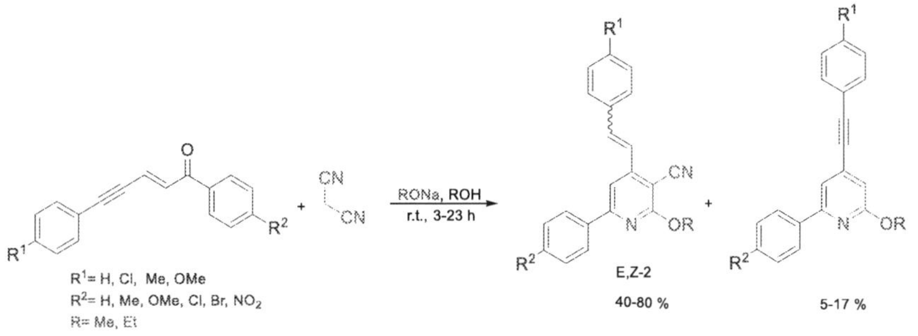

SCHEME 15.7 Synthesis of pyridine derivatives by using conjugated enynones, 1,5-diarylpent-2-en-4-yn-1-ones.

and 6-aryl-4-arylethynyl-2-alkoxypyridine derivatives, as the minor products (yields of 5–17%) [60]. Also, the plausible reaction mechanism was depicted in Scheme 15.8.

A mild and efficient synthesis of pyridines and pyridine-2-one derivatives by using bromoacetic acid is reported by applying a DMAP-promoted in situ activation approach [61]. In this method, simply available bromoacetic acid has been used as a 2C synthon to undergoes [2 + 4] cycloaddition with various acyclic and cyclic 1-azadienes (Scheme 15.9).

K_2CO_3 promoted a clean and efficient method for the synthesis of 2-amino-3,5-dicarbonitrile-6-thio-pyridine derivatives through a three-component reaction procedure between aldehydes, malononitrile, and thiophenol derivatives in PEG-400 as a green solvent (Scheme 15.10) [62]. The advantages of this process are good isolated yields (82–92%), short reaction times (60–120 min), easy operational manner, varied precursor scope, and recyclability of the solvent.

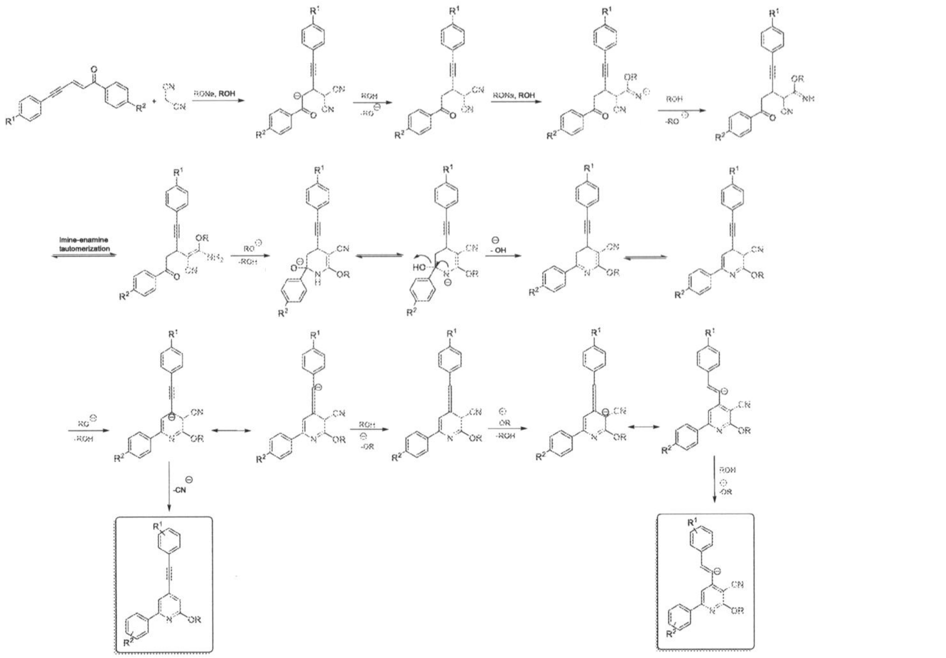

SCHEME 15.8 Suggested mechanism for the synthesis of pyridines by using conjugated enynones, 1,5-diarylpent-2-en-4-yn-1-ones.

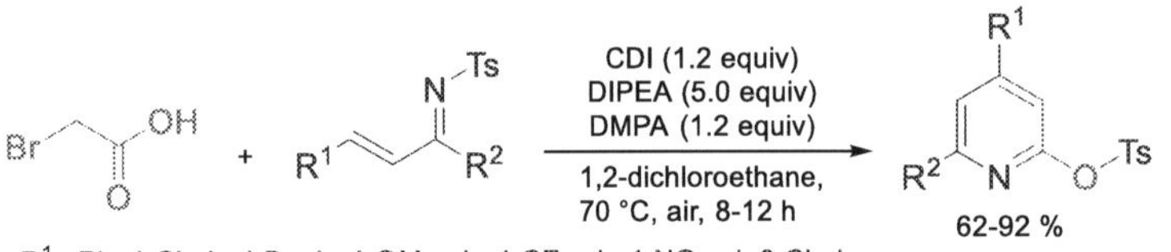

R^1= Ph, 4-Cl-ph, 4-Br-ph, 4-OMe-ph, 4-CF_3-ph, 4-NO_2-ph,3-Cl-ph, 3-OMe-ph, 2-Cl-ph, 2-OMe-ph, Naphthalene-1-yl, Furan-2-yl
R^2= Phenyl, 4-OMe-ph, 4-Br-ph, 4-NO_2-ph, 3-NO_2-ph, Naphthalene-2-yl, Furan-2-yl, Isobutyl
CDI = N,N'-Carbonyldiimidazole
DIPEA = N,N'-diisopropylethylamine
DMAP = 4-dimethylaminopyridine

SCHEME 15.9 2,4,6-Trisubstituted pyridines synthesized via the reaction between bromoacetic acid and 1-azadienes.

The synthesis of substituted pyridines with suitable yields (31–95%) was reported through modified Guareschi–Thorpe cyclization of β-keto esters and 2-cyanoacetamide using DBU as a base [63]. The chlorination of DBU salts of pyridones with $POCl_3$ by using a quaternary ammonium salt under standard atmospheric reflux conditions as opposed to the usual pressure equipment led to appropriate yields of substituted 2,6-dichloropyridines (Scheme 15.11). The cyclization is common for various alkyl, fluoro alkyl, and aryl β-keto esters. Remarkably, the yields for aryl β-keto esters are lower. The use of excess cyanoacetamide and DBU afforded higher yields of related products.

A benign and efficient three-component reaction for the synthesis of 1*H*-pyrazolo[3,4-*b*]pyridines is reported by Hill using Et_3N in DMF at 90°C [64]. In this synthetic method,

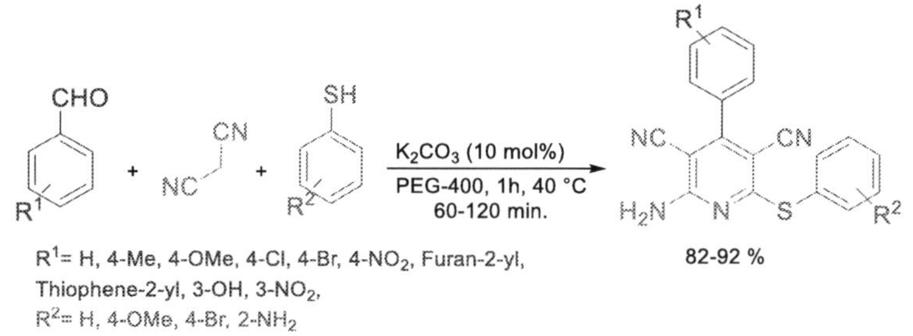

SCHEME 15.10 Synthesis of 2-amino-3,5-dicarbonitrile-6-thiopyridines by using K_2CO_3/PEG-400 medium.

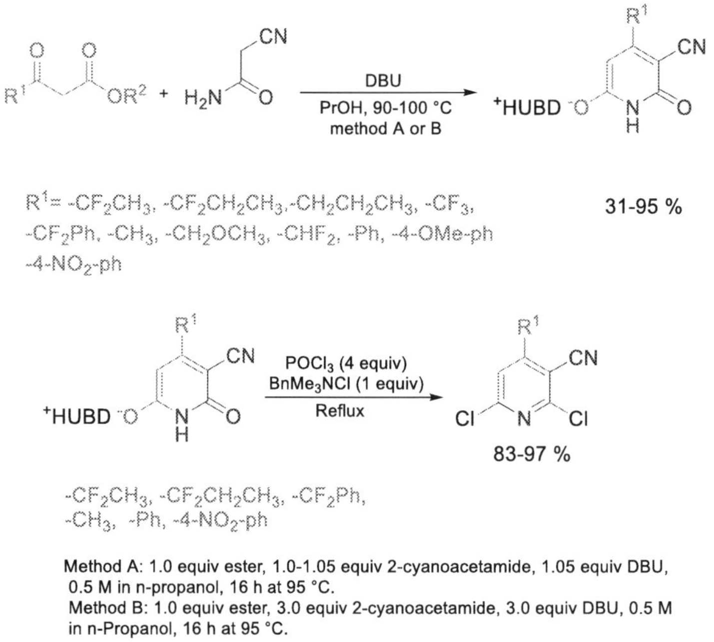

SCHEME 15.11 Guareschi–Thorpe cyclization for the synthesis of substituted pyridines.

aldehyde derivatives, 3-oxopropanenitrile and 1*H*-pyrazol-5-amine are applied as reactants to access related products in appropriate yields. This transformation proceeded through a Knoevenagel condensation and following, selective dehydrate cyclization to give related products (Scheme 15.12).

Vinamidinium (1,5-diazapentadienium) salts as significant three-carbon building blocks have been broadly applied for the construction of substituted benzenes and heterocyclic compounds [65]. Xu et al. [66] reported a facile and convenient method for the synthesis of 3-chloro-α-carbolines using indolinones and the 3-chlorovinamidinium salt (Scheme 15.13). This technique displayed advantages such as simply accessible reactants, appropriate yields, and informal workup.

A possible mechanism for the synthesis of 3-chloro-α-carbolines was suggested in Scheme 15.14. At first, in the presence of Et_3N, 3-chlorovinamidinium salt reacts with 2-indolinones to provide related intermediate (**I**), then activation of this intermediate with $POCl_3$, followed by ring-closure and isomerization under high-temperature conditions process leads to the formation of desired molecules.

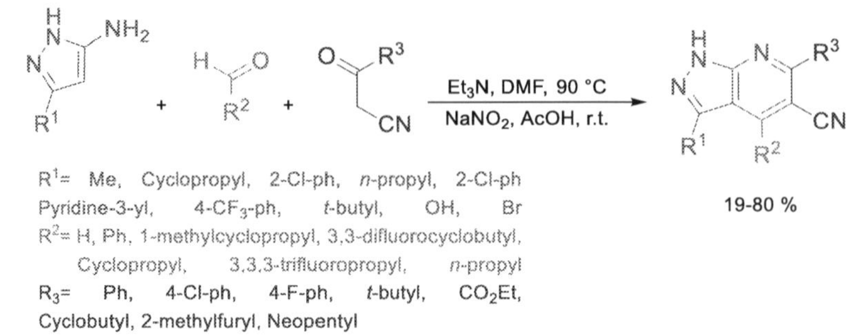

SCHEME 15.12 Synthesis of 1*H*-pyrazolo[3,4-*b*]pyridines in three-component synthesis of aldehydes, 3-oxopropanenitrile, and 1*H*-pyrazol-5-amine.

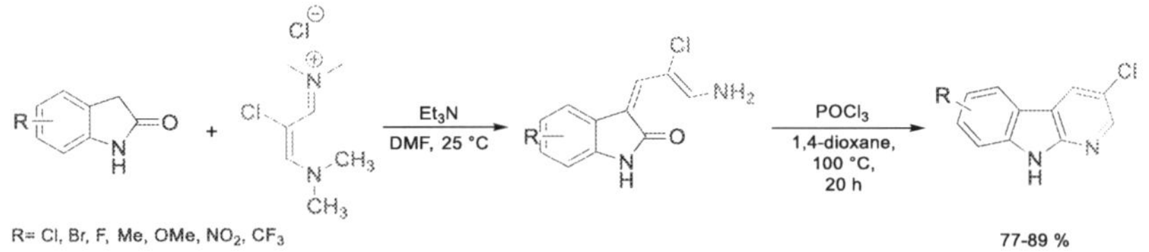

SCHEME 15.13 Synthesis of 3-chloro-α-carbolines from 3-chlorovinamidinium salt with 2-indolinones.

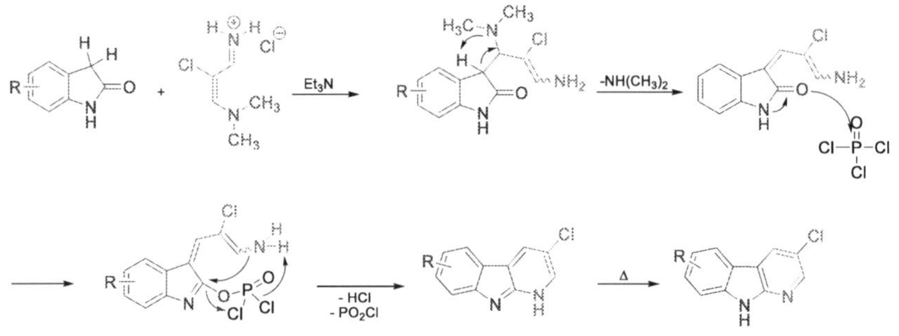

SCHEME 15.14 Probable mechanism for the synthesis of 3-chloro-α-carbolines.

Hanan et al. [67] reported a one-pot protocol for the synthesis of various terpyridine ligands (Scheme 15.15). Also, to confirm the metal ion coordination ability of the prepared terpyridine ligands, three metal complexes [Fe (II), Ni (II), Cu (II)] were synthesized. In terms of the solubility in an organic solvent, it is reported that the di-*tert*-butylterpyridine ligands possess an improved solubility in various organic solvents in which their homologues without the *tert*-butyl group are only sparingly soluble or insoluble. Practically, any organic solvent except hexanes (and other aliphatic solvents) can be applied to dissolve these ligands with the *tert*-butyl moieties.

Wei and Li [68] developed a facile method for the construction of substituted pyridine derivatives via a tandem condensation/alkyne isomerization/6π-3-azatriene electrocyclization processes. The present protocol was applied to generates several pyridines in appropriate yields (Scheme 15.16). Also, unsaturated ketones were applied in this procedure,

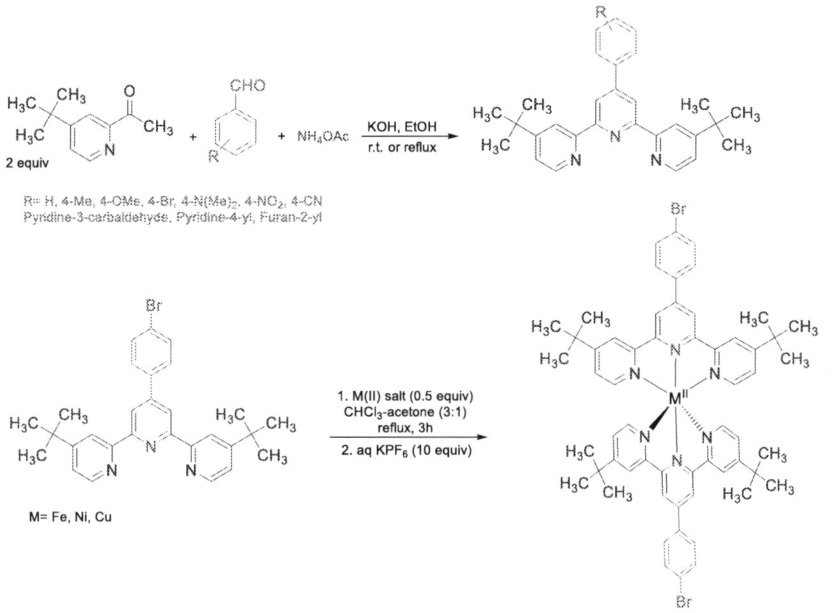

SCHEME 15.15 Synthesis of 4′-aryl- and 4′-heteroaryl-4,4″-di-*tert*-butyl-2,2′:6′,2″-terpyridine and their corresponding metal complexes.

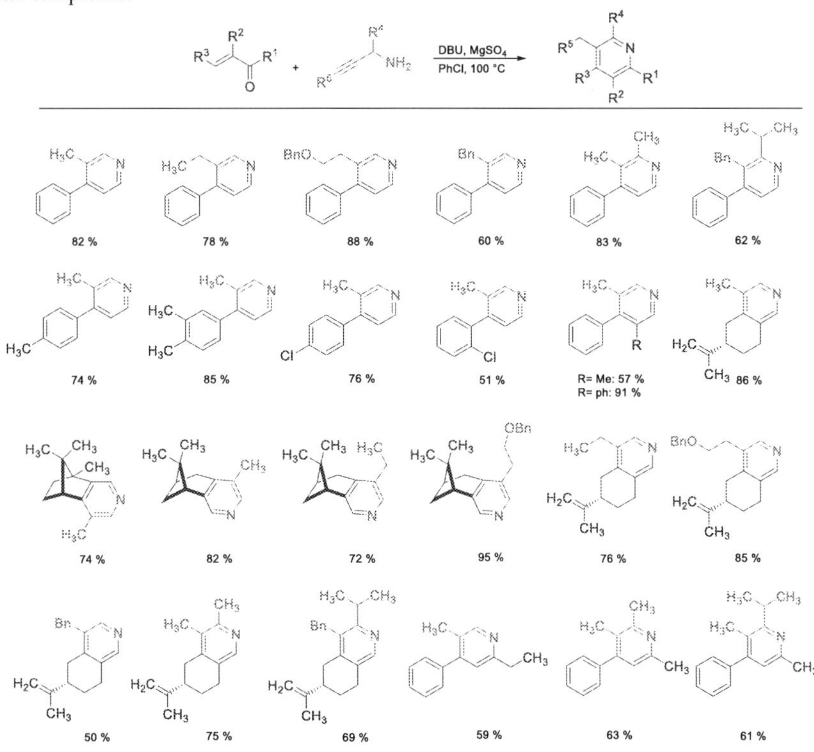

SCHEME 15.16 Synthesis of substituted pyridines via a tandem condensation/alkyne isomerization/6π-3-azatriene electrocyclization system.

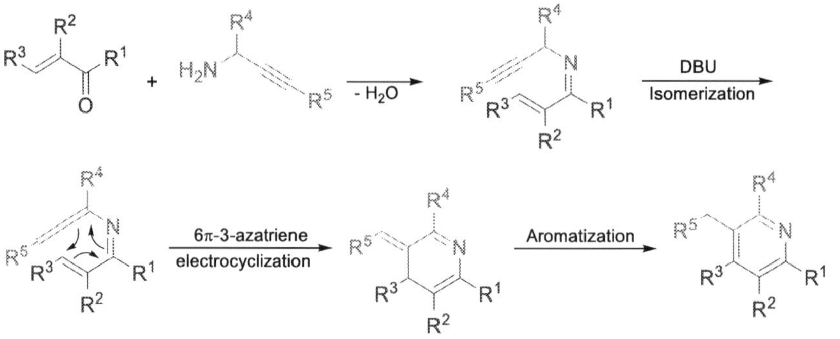

SCHEME 15.17 Plausible mechanism for the synthesis of substituted pyridines.

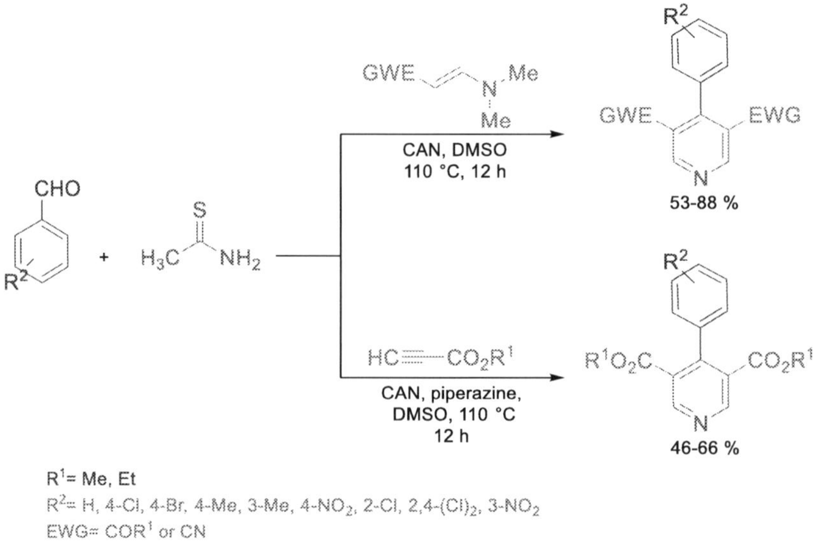

SCHEME 15.18 Synthesis of pyridines by using electron-deficient enamine and alkyne.

but at a much higher reaction temperature (150°C). In a plausible mechanism sequential step including imine formation, isomerization, electrocyclization, and aromatization processes lead to the formation of desired molecules (Scheme 15.17).

By using aldehyde derivatives and electron-deficient enamines or alkynes as the key building blocks, Wan et al. have attained a direct one-step synthetic manner to pyridine derivatives by using thioacetamide as nitrogen source and CAN as a catalyst (Scheme 15.18). The reported approach has the advantages of a facile and benign one-step process and a much wider tolerance to several reactants. Their reaction consequently affords a versatile pathway for the synthesis of numerous pyridines [69–71].

In 2021, Shabalin and coworkers reported a regiocontrolled synthetic route for the preparation of 2,4,6-triarylpyridine derivatives by utilizing C-vinylation of methyl ketone derivatives using alkynones, and then cyclization of in situ produced unsaturated 1,5-dicarbonyls by applying ammonium acetate. In addition to high regioselectivity, the presented method has

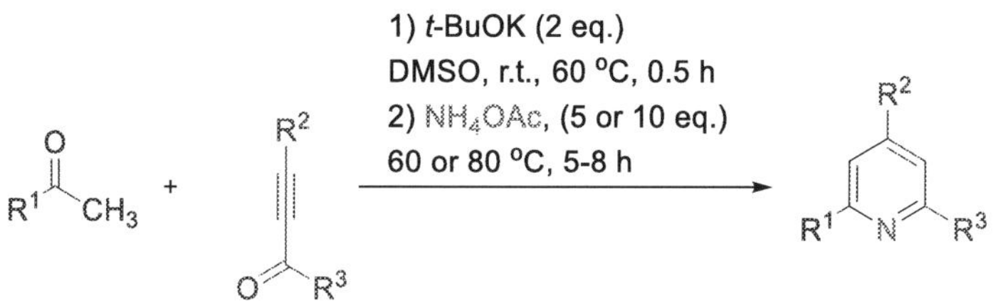

SCHEME 15.19 Regiocontrolled synthetic route to 2,4,6-triarylpyridine derivatives.

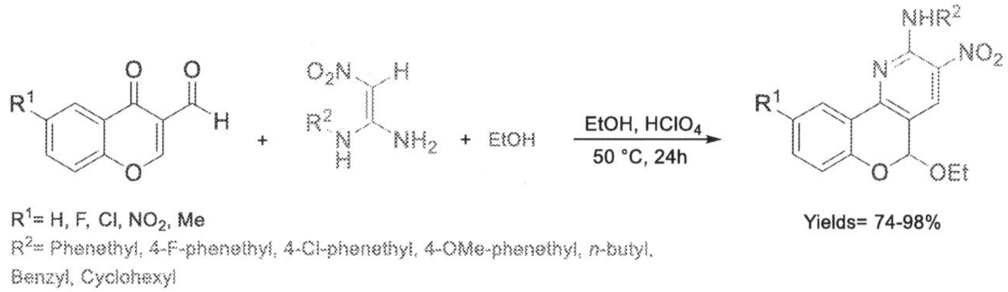

SCHEME 15.20 One-pot synthesis 5*H*-chromeno[4,3-*b*]pyridines in the presence of $HClO_4$.

several merits such as one-pot and high yielding protocol, transition metal-free catalysis, and available starting materials (Scheme 15.19) [72].

15.3 Acidic catalysts

The one-pot synthesis 5*H*-chromeno[4,3-*b*]pyridines was reported via a benign and facile cascade reaction between 3-formylchromones, various 1,1-enediamines (EDAMs), and numerous alcohols or amines by using one drop of $HClO_4$ as a catalyst in ethanol or acetone as a solvent [73]. This process is especially suitable and efficient protocol for the synthesis of 5*H*-chromeno[4,3-*b*]pyridine derivatives in good yields (74–92%) and reasonable times (Scheme 15.20). This method has several advantages such as the convenience of operational work, using green solvent and simplicity of purification by washing the crude products [group-assisted purification (GAP) chemistry] with ethanol.

Cinchomeronic dinitrile derivatives were attained as products from the reaction between the accessible and reactive adducts of tetracyanoethylene and ketones (4-oxoalkane-1,1,2,2-tetracarbo-nitriles) with hydrogen chloride in 69–98% yields (Scheme 15.21) [74]. The reaction progressed effortlessly in anhydrous propan-2-ol, which has been initially saturated by a fivefold excess of dry hydrogen chloride. Nevertheless, this process had a category of demerits, containing the requirement of complex equipment, constant monitoring of the amount of hydrogen chloride in the reaction mixture, and the preparation of 2-chloropropane as a reaction by-product. Also, acidic conditions were applied for the synthesis of 2-iodocycloalka[*b*]pyridine-3,4-dicarbonitrile derivatives [75,76].

Kondratov et al. reported a facile and straightforward protocol for the synthesis of pyridine carboxylate fused with five- or six-membered hetero-aromatic rings using acidic conditions [77]. The technique is based on the Combes-type condensation of the low molecular-weight β-alkoxyvinyl glyoxylates as *CCC* bis-electrophiles with hetero-aromatic amines as *NCC* binucleophiles. In most studies, β-alkoxyvinyl glyoxylates without extra substituent at the

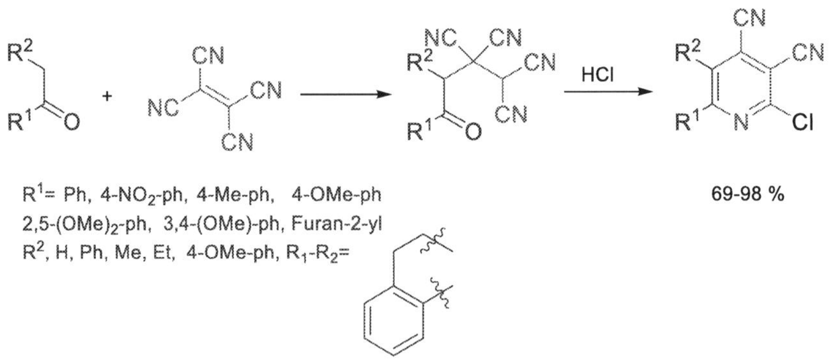

SCHEME 15.21 Synthesis of cinchomeronic dinitriles.

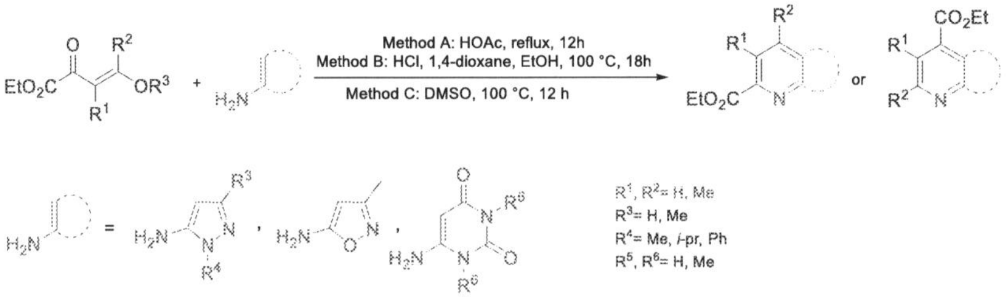

SCHEME 15.22 Synthesis of fused pyridine α- and γ-carboxylates **12** and **13**.

β position led to the resulting α-pyridine carboxylates (67–87% yield). In the state of β-methyl-substituted derivative, γ-pyridine carboxylates were achieved in 84–99% yield (Scheme 15.22). It was establishing that region-selectivity of the condensation could be strongly modified by altering conditions, for example, solvents and acidic additives (HOAc, DMSO, or HCl-1,4-dioxane).

2,4,6-Trisubstituted pyridine derivatives were efficiently synthesized through a one-pot acid-catalyzed tandem reaction in the absence of any metallic reagents or further oxidants [78]. This reaction includes a C=C bond cleavage of enones through a "masked" reverse Aldol reaction, and C(sp^3)–N bond cleavage of primary amines to give nitrogen sources for the assembly of pyridines in suitable yields in 2 h with outstanding functional group tolerance (Scheme 15.23). Using air as an oxidant, applying accessible reactants as a nitrogen source, and experimentally appropriate catalytic method are some of the advantages of this strategy.

Khan et al. [79] have established a versatile and convenient procedure for the synthesis of pyrido[2,3-*c*]coumarins via an intramolecular Povarov reaction of 2-(propargyloxy)benzaldehydes and 3-aminocoumarins in the presence of a catalytic amount of triflic acid as catalyst in acetonitrile. No co-oxidant was requisite for the aromatization of the related products. This technique is facile and needs a short reaction time, and the products are effortlessly isolated without column chromatography or aqueous workup (Scheme 15.24).

Khanal and Lee [80] reported a useful and mild approach for the synthesis of 2,3,4-trisubstituted pyridines using an organocatalyzed three-component reaction. Various pyridines are produced from the reaction between ketones with α,β-unsaturated aldehydes and ammonium acetate in the presence of L-proline as acidic organocatalyst (Scheme 15.25).

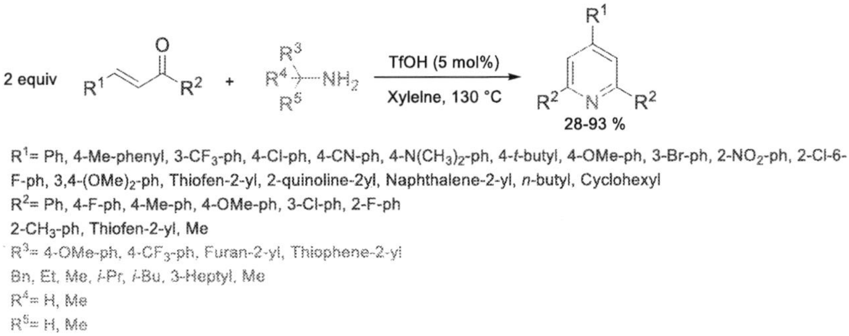

SCHEME 15.23 Synthesis of 2,4,6-trisubstituted pyridines via consecutive reverse Aldol reaction catalyzed by TfOH.

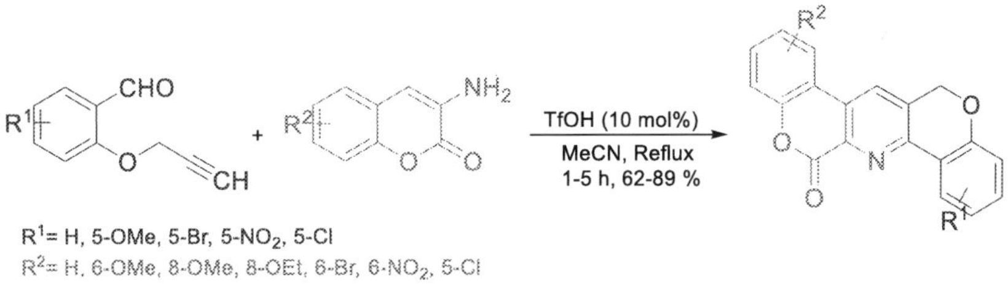

SCHEME 15.24 Synthesis of fused pyrido[2,3-*c*]coumarin catalyzed by TfOH.

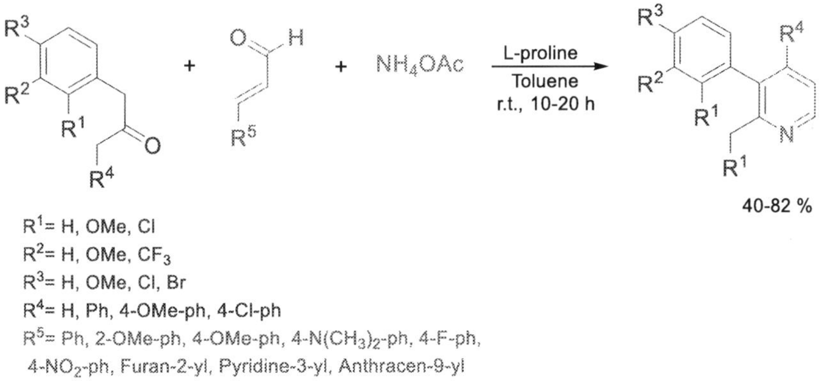

SCHEME 15.25 Synthesis of numerous pyridines catalyzed by *L*-proline.

The reported procedure leads to fast *N*-annulation via C–C and C–N bond formation in a single operation, which avoids the synthesis of other functional groups such as oximes, imines, or azides. The furnished compounds are employed for the evaluation of antibacterial activities and as fluorescence sensors for Cu^{2+} ions.

Also, L-proline is applied for the synthesis of other types of pyridine derivatives having a cyclic ring or a heteroatom on the pyridine tag by using cyclic ketone or 1-(*p*-tolylthio)propan-2-one as the reactants. For instance, the reaction of ketones with α,β-unsaturated aldehydes and NH_4OAc at room temperature for 12 h afforded the related 2,3,4-trisubstituted pyridine

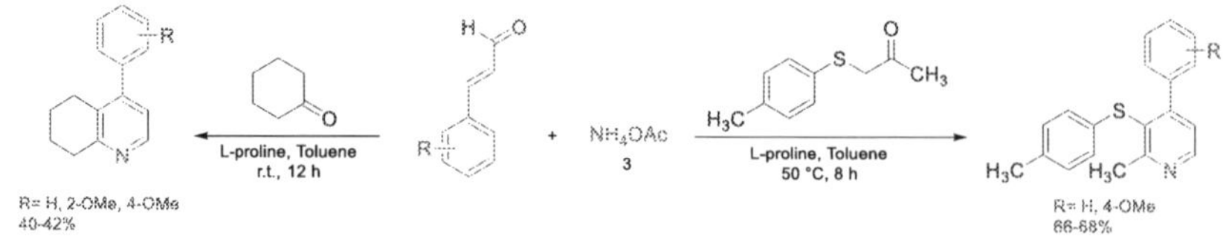

SCHEME 15.26 Synthesis of pyridines in the presence of L-proline.

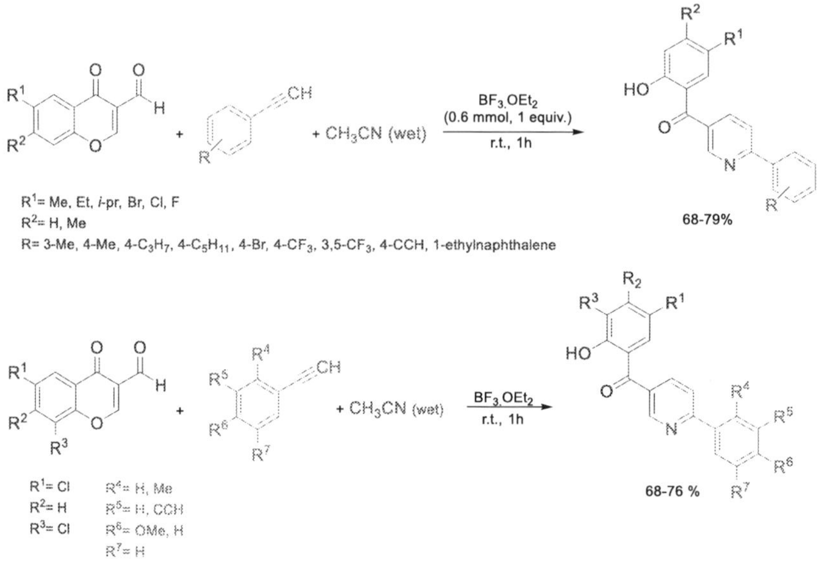

SCHEME 15.27 One-pot synthesis of pyridines via [3 + 2 + 1] annulation by using $BF_3{\cdot}OEt_2$.

derivatives with 40–42% yields (Scheme 15.26) [80]. In another work, *N*-protected amino acids were applied as precursor for the synthesis of pyridine derivatives [81].

A mild and facile $BF_3{\cdot}OEt_2$-mediated approach was studied for the preparation of 2-phenylpyridine derivatives bearing benzophenone tags from simply accessible 3-formylchromones and phenylacetylenes in wet acetonitrile (as nitrogen source) [82]. The reported one-pot process progresses via [3 + 2 + 1] annulation by using cascade nucleophilic addition, hydrolysis, Michael-type addition, ring-opening, and elimination reactions (Scheme 15.27). The produced molecules may have uses as UV filters and display potent antibacterial activities.

Also, $BF_3{\cdot}OEt_2$ was applied in a single-step and benign process to achieving substituted pyridines in suitable yields (61–92%) with a wide range of functional group tolerance. It is worthy to mention that this Lewis acid-promoted [3 + 3] annulation approach between 3-ethoxycyclobutanones and enamines or amidines could be simply scaled up without loss of efficacy (Scheme 15.28) [83].

In another investigation, HBF_4 was used as an oxidizing catalyst for the synthesis of 2-amino-4,6-diphenylnicotinonitrile under mild reaction conditions. Also, theoretical investigations supported the suggested anomeric-based oxidation (ABO) mechanism for the synthesis of target molecule [84]. The theoretical obtained data display that the intermediate

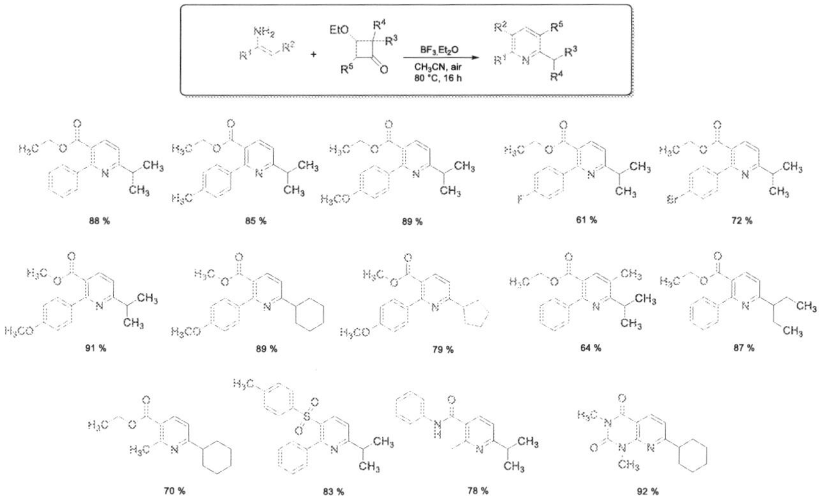

SCHEME 15.28 Synthesis of various pyridines in the presence of $BF_3{\cdot}OEt_2$.

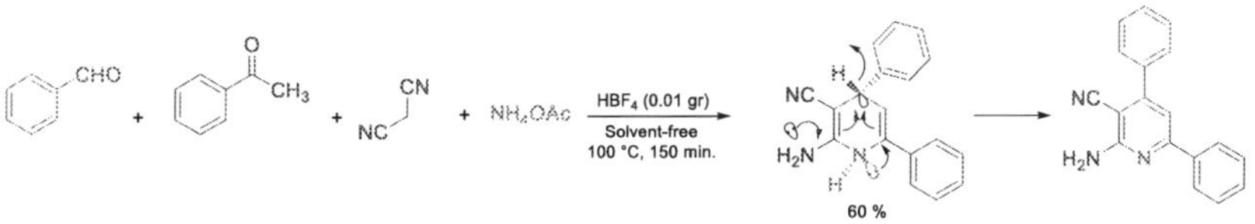

SCHEME 15.29 Synthesis of 2-amino-4,6-diphenylnicotinonitrile with HBF_4 as catalyst.

isomers with 5R- and 5S-chiral positions possess appropriate skeletons for the aromatization via anomeric-based oxidation in the final step of the mechanistic route (Scheme 15.29).

15.4 Ionic liquids and molten salts

Shi et al. [85] have studied a facile and mild one-pot three-component reaction of aldehyde, acyl acetonitrile and electron-rich amino heterocycles (such as 5-aminopyrazole or 6-aminopyrimidine-2,4-dione), for the preparation of pyrazolo[3,4-*b*]pyridines and pyrido[2,3-*d*]pyrimidine derivatives in ionic liquid [bmim]Br (2 mL) as a catalytically active solvent (Scheme 15.30).

A one-pot three-component reaction strategy has been reported for the construction of 2-amino-4-aryl-6-(arylamino)pyridine-3,5-dicarbonitrile derivatives in the presence of triethylammonium trinitromethanide [TEATNM] and triethylammonium tricyanomethanide [TEATCM] as ionic liquid catalysts in short reaction times (5–35 min) and high yields (86–97%) via *an* anomeric-based oxidation mechanistic pathway (Scheme 15.31) [86]. Both of the experimental and theoretical investigations have supported the anomeric-based oxidation for the final step of the synthetic mechanism for the described approaches.

An appropriate suggested mechanism for the synthesis of 2-amino-4-aryl-6-(arylamino)pyridine-3,5-dicarbonitriles **4** in the presence of [TEATCM] or [TEATNM] is

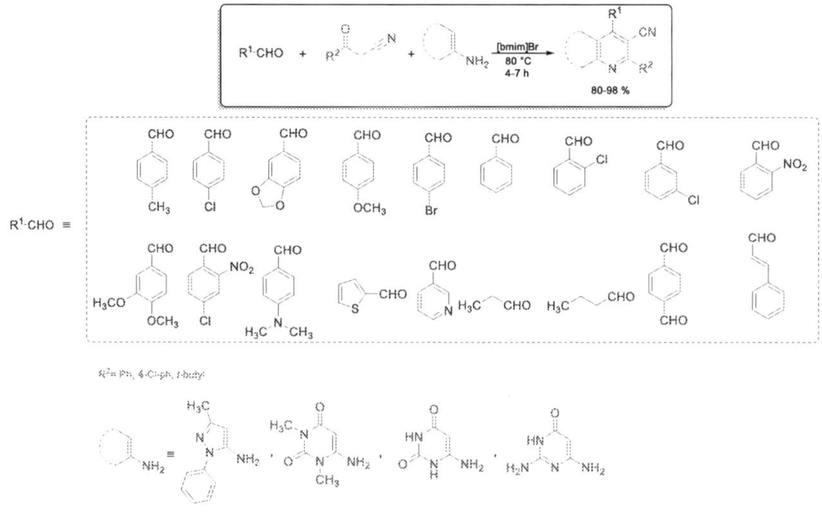

SCHEME 15.30 Synthesis of fused pyridine derivatives in [bmim]Br.

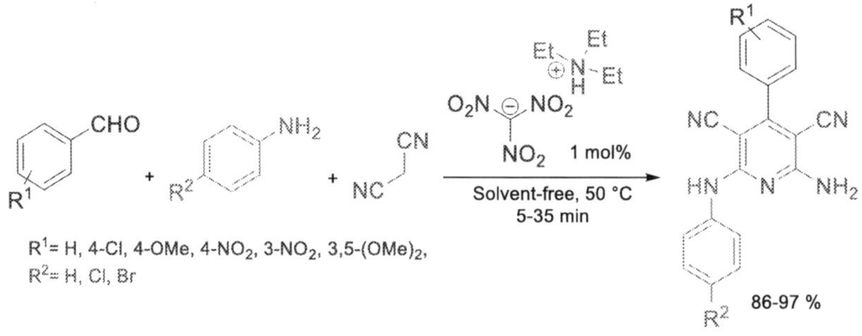

SCHEME 15.31 Synthesis of 2-amino-4-aryl-6-(arylamino)pyridine-3,5-dicarbonitriles 4 catalyzed by [TEATNM] and [TEATCM].

depicted in Scheme 15.32 [87]. Initially, the reaction happens via a Knoevenagel condensation between activated malononitrile **3** and aldehyde **1**. At that point, the consequent Michael-type addition of the second molecule of activated malononitrile **3** to the Knoevenagel adduct **5** causes the synthesis of intermediate **6**. In the next step, dihydropyridine **7** was formed via nucleophilic attack of anilines **2** to cyanide group of intermediate **6** and then cyclization. Following, dihydropyridine **7** tautomerized to dihydropyridine **8**. Lastly, dihydropyridine **8** via anomeric-based oxidation [87] was converted to its corresponding pyridines **4**. Conversion of intermediate **8** to pyridines **4** might be occurred by using unusual hydride transfer and releasing of molecular hydrogen (H_2). The C–H bond is weakened via stereoelectronic interactions [88] of the nitrogen lone pairs with the antibonding orbital of C–H ($\sigma^*_{C\text{-}H}$ orbital) which can be broken by reaction with a proton to give molecular hydrogen. The major reason for ABO is the driving force of aromatization which will be supported via stereo electronic and/or anomeric effect. A theoretical study has been also used for approving the

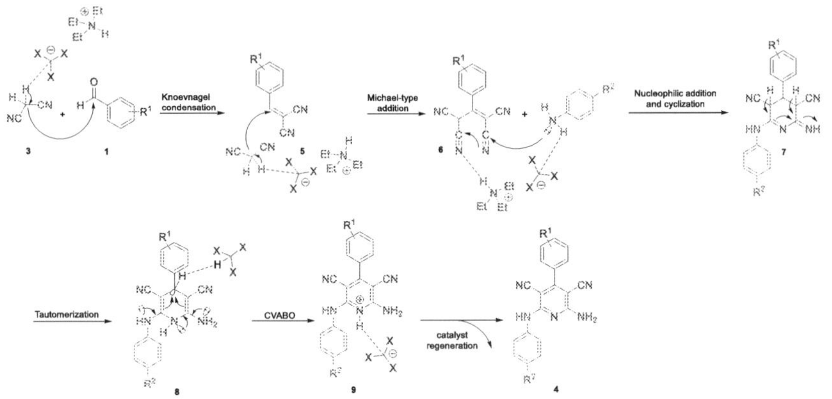

SCHEME 15.32 The possible mechanism for the synthesis of 2-amino-4-aryl-6-(arylamino)pyridine-3,5-dicarbonitriles 4 by using [TEATCM] or [TEATNM] as catalysts.

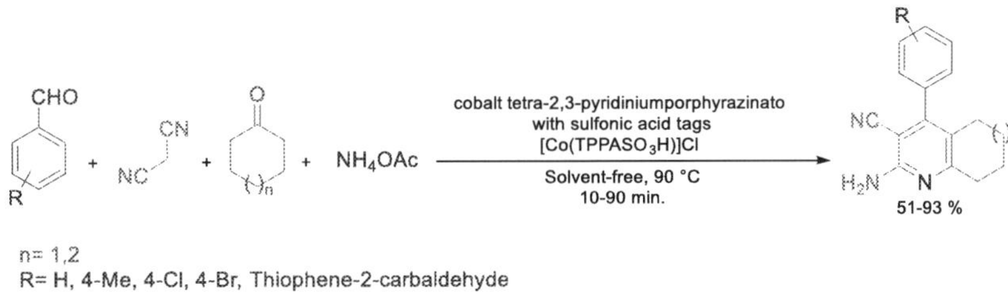

SCHEME 15.33 Synthesis of 2-amino-3-cyanopyridines in the presence of $[Co(TPPASO_3H)]Cl$.

anomeric-based oxidation mechanism which was proposed for the final step in the synthesis of pyridines **4**.

$[Co(TPPASO_3H)]Cl$ efficiently catalyzed the synthesis of 2-amino-3-cyanopyridine derivatives (Scheme 15.33). This synthesis progressed via a cooperative vinylogous anomeric-based oxidation mechanism. In this study, 2-amino-3-cyanopyridines were synthesized in good yields (51–93%) and short reaction times (10–90 min). This process has main advantages, for example, avoidance of column chromatography purification, broad scope of substrates, operational ease, ready available, and thermally stable catalyst [89].

A quinoline-based dendrimer-like ionic liquid was synthesized and its catalytic behavior investigated for the construction of pyridine derivatives bearing sulfonamide tags via a cooperative vinylogous anomeric-based oxidation mechanism (Scheme 15.34). These target compounds were obtained in appropriate yields (62–88%) at short reaction times (15–45 min) [90].

$\{[1,4\text{-pyrazine-}NO_2][C(NO_2)_3]_2\}$ as a versatile and green nanostructured molten salt was synthesized and used as a catalyst for the benign preparation of 2-amino-3,5-dicarbonitrile-6-sulfanylpyridines through the one-pot three-component condensation reaction of numerous aldehydes, malononitrile, and benzyl mercaptan at room temperature under solvent-free conditions (Scheme 15.35). The reported approach has advantages such as effortlessness of separation, appropriate yields (83–93%), and short reaction times (20–40 min). Also,

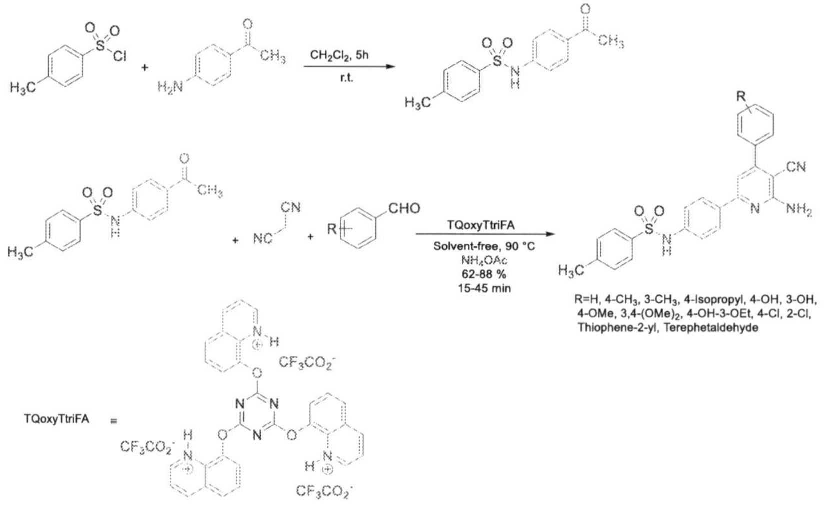

SCHEME 15.34 Synthesis of pyridines with sulfonamide tag by using TQoxyTtriFA.

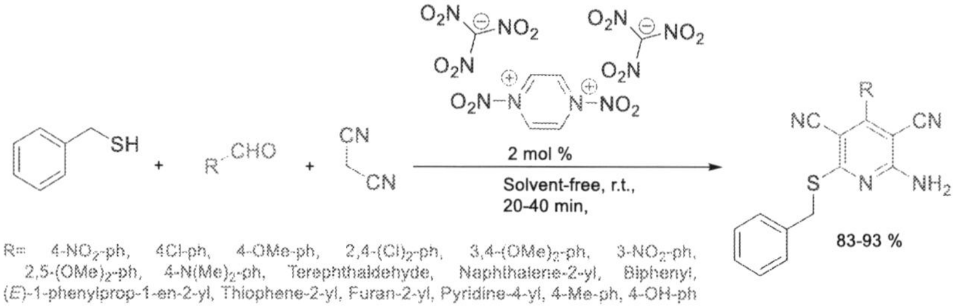

SCHEME 15.35 Synthesis of 2-amino-3,5-dicarbonitrile-6-sulfanylpyridines via {[1,4-pyrazine-NO_2][$C(NO_2)_3$]$_2$} as a nanostructured molten salt catalyst.

Zolfigol et al. [91] proposed that the final step synthesis of the targeted pyridine derivatives proceeds via anomeric-based oxidation pathway (ABO).

15.5 Transition metal catalysis

15.5.1 Cu catalysis

Kumar and Kapur [92] have established an effective and benign copper-mediated synthetic method to pyridine derivatives through a reductive cleavage of isoxazoles (Scheme 15.36). 3,5-Diaryl isoxazoles were used as masked enaminones, and dimethyl sulfoxide was applied as the one-carbon substitute producing an active methylene group in the course of the reaction to form two C–C bonds. The uncommon reactivity of 4-vinyl isoxazoles was identified under the reaction conditions subsequent in tri-substituted pyridines (53–83%). Deuterium labeling and other control experiments were performed to suggest probable ways for this

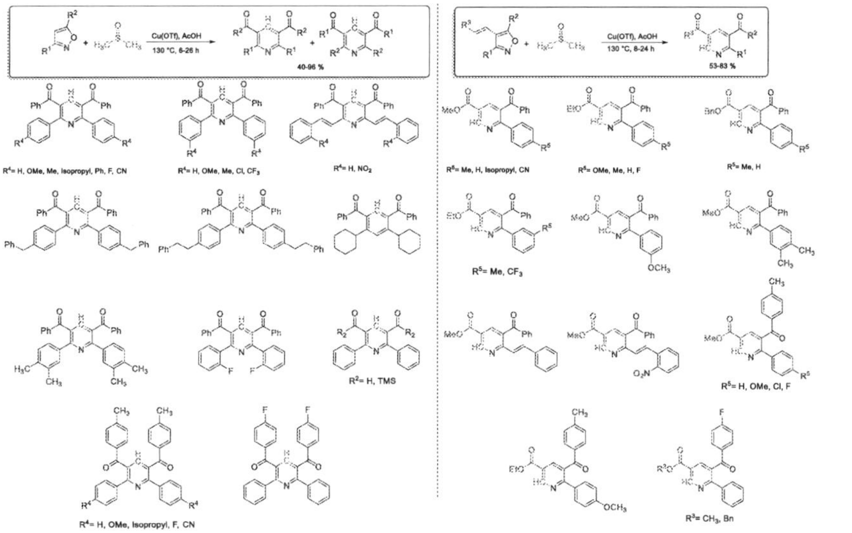

SCHEME 15.36 Synthesis of tetra-substituted pyridines nicotinate derivatives.

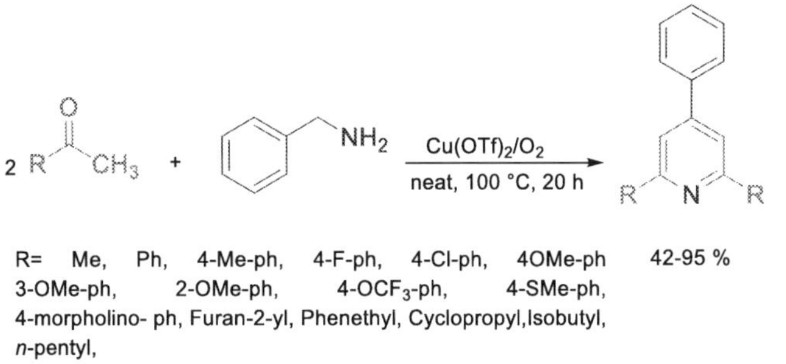

SCHEME 15.37 Synthesis of 2,4,6-trisubstituted pyridines via C–N bond cleavage by using $Cu(OTf)_2$.

transformation. This protocol affords an efficient and expeditious route for the synthesis of tetra-substituted pyridine derivatives (40–96%).

Also, a benign and facile copper-catalyzed C–N bond cleavage of aromatic methylamines was reported for the synthesis of pyridine derivatives [93]. With neat conditions and efficient operation, the fragment-assembling approach gives a wide range of 2,4,6-trisubstituted pyridines in appropriate yields (42–95%) from simple accessible reactants (Scheme 15.37).

Chen et al. [94] reported a benign and efficient one-pot synthetic manner of 2,4,6-triphenylpyridines in the presence of $Cu(OTf)_2$ through oxidative sp^3 CH coupling of acetophenones with toluene derivatives. This reaction showed a suitable functional group tolerance for preparation of 2,4,6-triphenylpyridine derivatives in appropriate yields at 100–120°C for 14 h (Scheme 15.38).

To extend the variety of reactants for pyridines, they attempted to application propiophenones and methyl heterocycles as reactants for this transformation (Scheme 15.39). Propiophenones and 2 methylfuran gave the related pyridines in good yields.

In 2018, Zheng and coworkers reported a benign, facile and eco-friendly green technique for the preparation of aromatic β-carbolines by applying $CuBr_2$ as a catalyst for oxidation

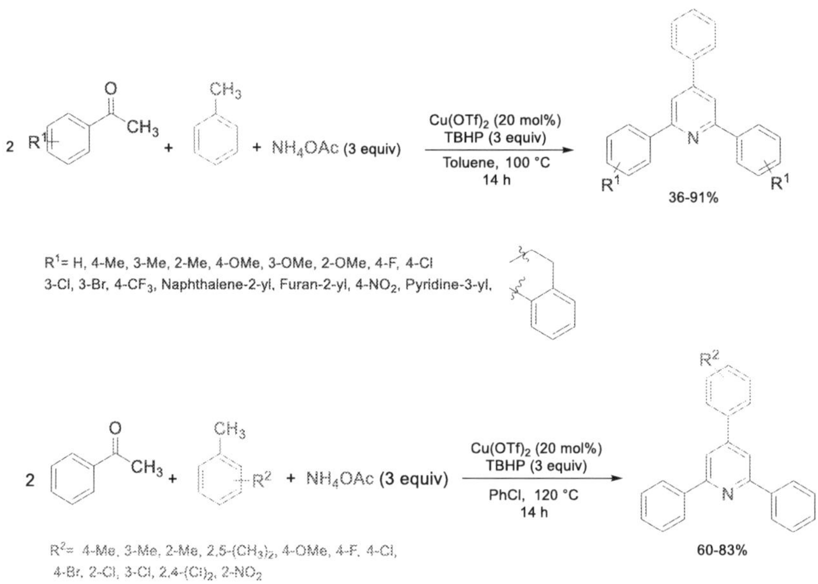

SCHEME 15.38 Synthesis of pyridines via copper-catalyzed $Cu(OTf)_2$ oxidative sp^3 CH coupling.

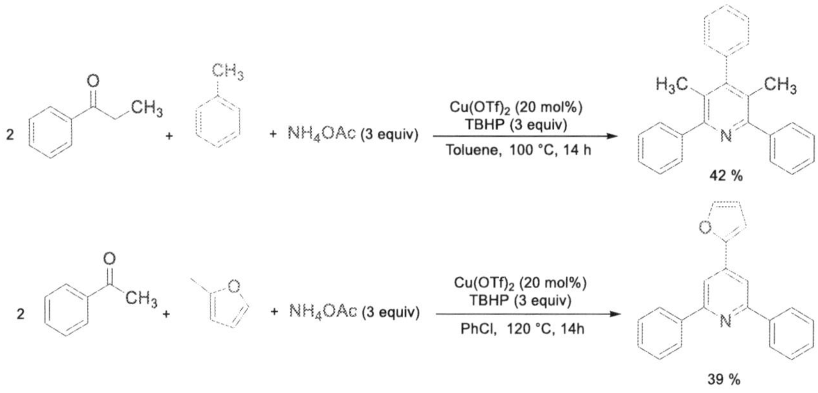

SCHEME 15.39 Synthesis of pyridines by using $Cu(OTf)_2$ catalyst.

of 1,2,3,4-tetrahydro-β-carbolines (THβCs) by air (O_2) [95]. The main merits of the reported methods are including appropriate tolerance of functional groups, high yielding, simple experimental procedure. Furthermore, this method was effectively used for the total synthesis of β-carboline alkaloids perlolyrine and flazin (Scheme 15.40).

In the presence of a catalytic amount of CuCl, arylpropynyloxy-benzonitrile derivatives were applied as the precursor for the synthesis of fused pyridines in appropriate yields (32–78%) by Novák et al. [96]. When the reactions were carried out in ethyl acetate, full conversion was reached in just 1 h (Scheme 15.41). It performed that the steric effect of the diaryliodonium salts had a key effect on the medium. When the functional group was positioned at the *meta*-

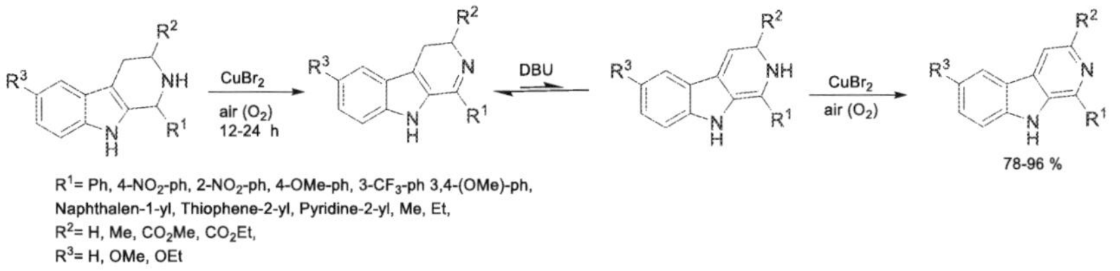

SCHEME 15.40 $CuBr_2$-catalyzed oxidation of several substituted THβCs 1 to β-carbolines 2 via air oxidant.

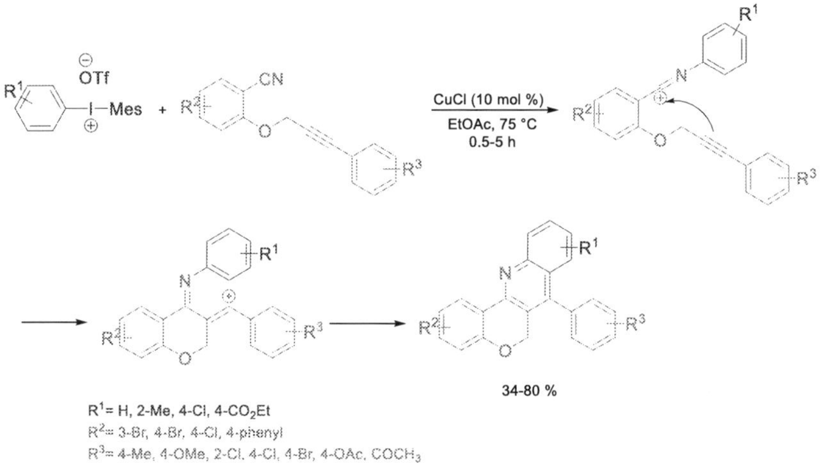

SCHEME 15.41 Synthesis of fused pyridines via cascade annulation by arylpropynyloxy-benzonitriles.

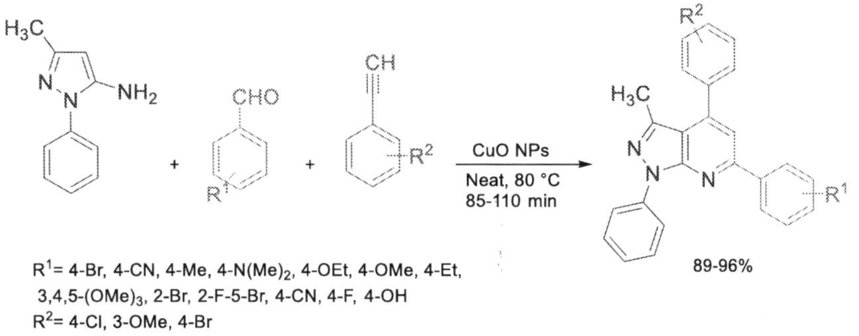

SCHEME 15.42 Synthesis of phenyl-1*H*-pyrazolo[3,4b]pyridines catalyzed by CuONPs.

or *para*-position of the phenyl group of the iodonium salt, the reaction was fast and the related product was achieved in good yield. But when the *ortho*-position was employed, for instance, by a methyl substituent, the reaction became slow and low-yielding; also, only 8% conversion was identified in the presence of *ortho*-halo substituents.

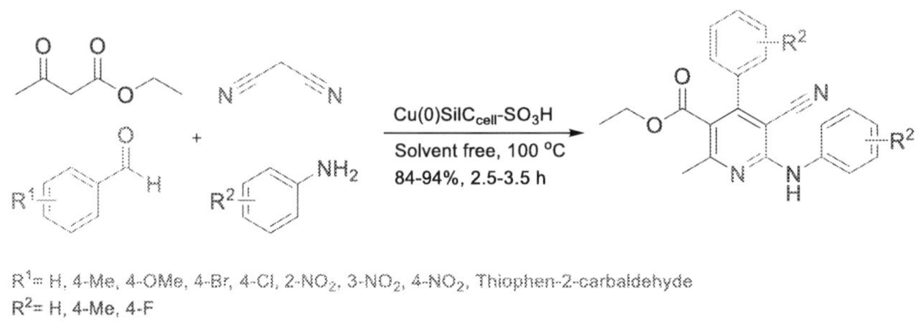

SCHEME 15.43 One-pot synthesis of polysubstituted pyridine derivatives in the presence of Cu(0)–SilC$_{cell}$–SO$_3$H.

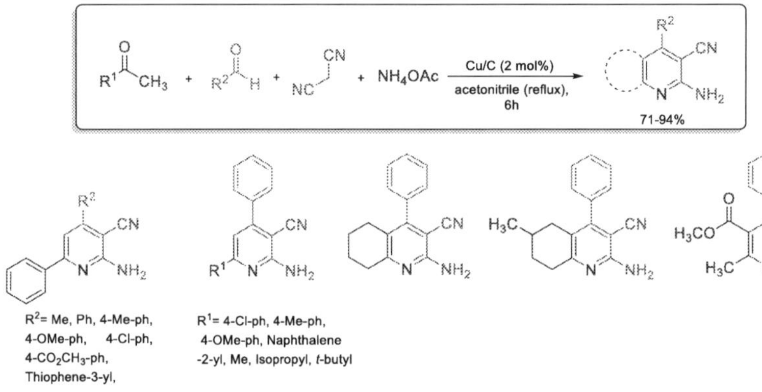

SCHEME 15.44 Synthesis of 2-amino-3-cyanopyridine derivatives using copper nanoparticles on charcoal as catalyst.

CuO nanoparticles catalyzed the facile synthesis of phenyl-1*H*-pyrazolo[3,4-*b*]pyridine derivatives through C–H bond activation subsequently C–C bond formation via intramolecular cyclization under solvent-free conditions at 80°C for the first time (Scheme 15.42) [97]. This method has the advantages of steric and functional group tolerance, high atom economy efficacy, high yields (89–96%) at relatively short reaction times (85–110 min).

Cu(0)–SilC$_{cell}$–SO$_3$H efficiently catalyzed a one-pot synthesis of polysubstituted pyridine derivatives by utilizing arylaldehydes, aniline derivatives, malononitrile, and NH$_4$OAc under solvent-free conditions. High catalytic activity, thermal stability, simple recovering, and reusing ability of the catalyst are the main merits of the presented procedure (Scheme 15.43) [98].

Catalytic potential of copper nanoparticles was investigated in the synthesis of 2-amino-3-cyanopyridine derivatives through a multicomponent reaction pathway between several aromatic and aliphatic ketones, malononitrile, ammonium acetate, and a variety of aromatic and aliphatic aldehydes utilizing 2 mol% of copper nanoparticles on charcoal (Cu/C) as efficient heterogeneous catalyst [99]. Durability and recyclability of catalyst and performing the reaction on large scale are notable. The reaction was performed in the presence of different protic and aprotic solvents and other catalysts such as Cu(OAc)$_2$ and CuI. The best conditions and also the structures of the products are illustrated in Scheme 15.44.

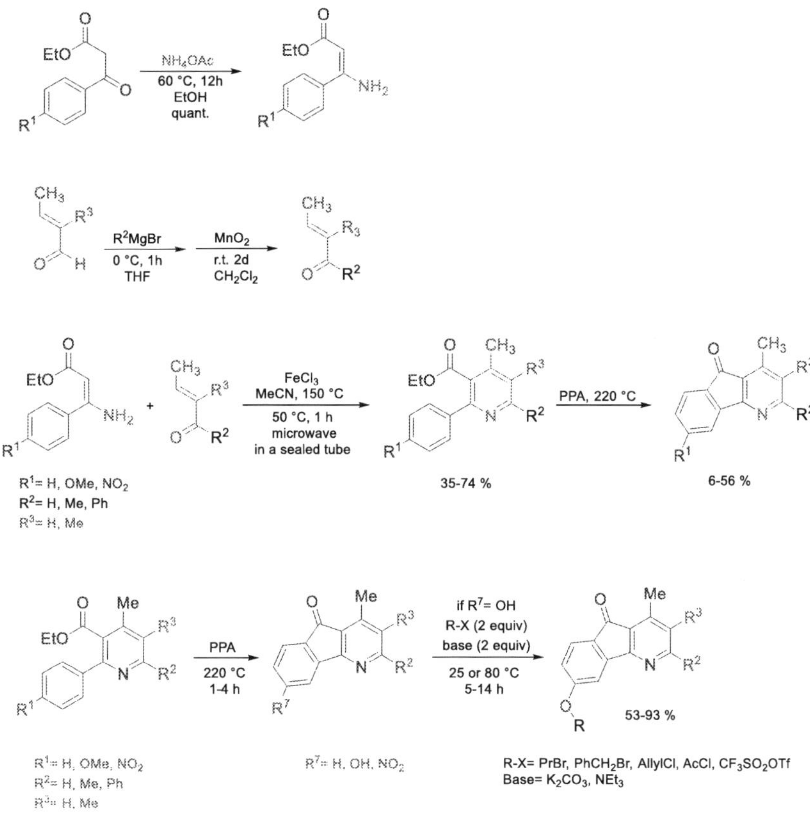

SCHEME 15.45 Synthesis of onychines by using nicotinates **4** and PPA.

15.5.2 Fe catalysis

The condensation of an enamino ester with an α,β-unsaturated carbonyl compound by applying $FeCl_3$, then, intramolecular Friedel–Crafts reaction, emerges as a versatile and efficient technique for preparation onychines [100]. This technique simplifies the modification of the onychine structure only by varying the beginning enamino esters and enones, while the electron-withdrawing group on the 2-phenyl group avoided the second cyclization step (Scheme 15.45). Also, hydrolysis of a methoxy group was established to happen through the cyclization when 2-(4-methoxyphenyl)nicotinate was heated in polyphosphoric acid (PPA). The hydroxy group was effortlessly modified upon treatment with electrophiles for example alkyl halides, acetyl chloride, and triflic anhydride, which therefore gave various onychine derivatives. This technique will be valuable for the investigation of the biological activity of onychine and its derivatives.

$FeCl_3$ as catalyst controlled cyclization and selective N–O bond cleavage of *N*-vinyl-α,β-unsaturated nitrones under benign reaction conditions for the synthesis of pyridine derivatives in good yields (23–82%) (Scheme 15.46) [101]. For this goal, the reaction mixture was stirred at 25°C for 5 min and then heated at 80°C for 12–15 h.

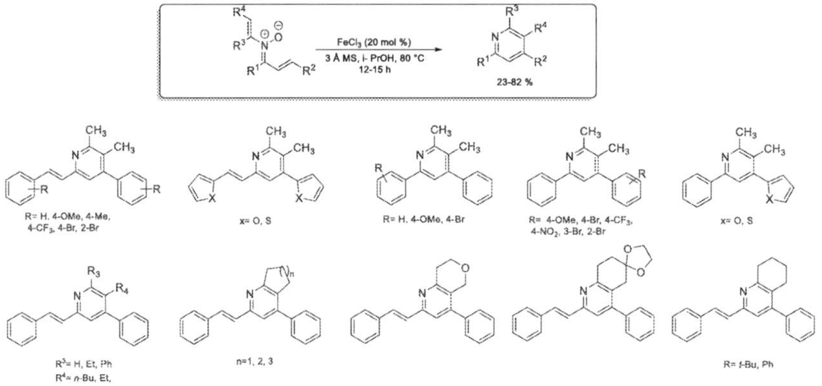

SCHEME 15.46 Synthesis of pyridines 2 by using $FeCl_3$ as catalyst.

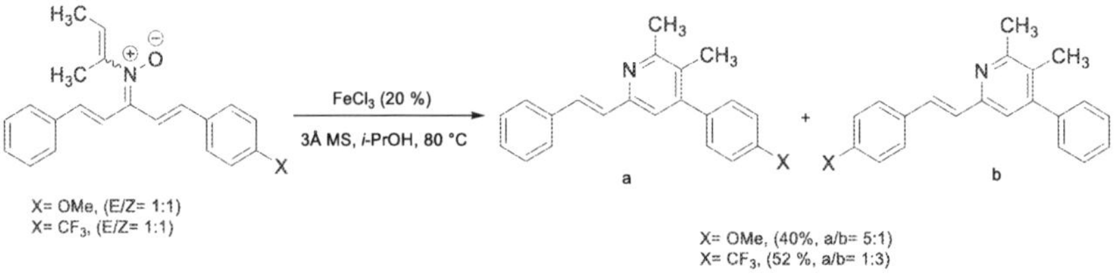

SCHEME 15.47 Regioselectivity studies for the synthesis of pyridines in the presence of $FeCl_3$.

Also, the regioselectivity of the reported protocol for the synthesis of targeted pyridines was investigated (Scheme 15.47). For the synthesis of pyridines, when 1:1 E/Z ratio of nitrone with a methoxy group was exposed to 20 mol% of $FeCl_3$ in *i*-PrOH at 80°C, the product was afforded in 40% yield and the regioselectivity of cyclization was detected in 5:1 ratio. Nevertheless, 1:1 E/Z ratio of nitrone bearing trifluoromethyl group generated the desired product in 52% yield and the regioselectivity of cyclization was detected in 1:3 ratio.

In another work, $FeCl_3$ catalyzed facile synthesis of substituted pyridines through the cyclization of ketoxime acetates and aldehyde derivatives [102]. This protocol presents a suitable functional group tolerance to afford 2,4,6-triarylsubstituted symmetrical pyridines in good yields (64–95%) without any additive. Also, a gram-scale reaction system was performed to establish the scaled-up applicability of this synthetic process (Scheme 15.48).

Pan et al. [103] have basically studied a category of alkyne-tethered oximes, and used them into a unique iron-catalyzed radical relay technique for the convenient assembly of a varied range of fused pyridine derivatives. This process displays wide substrate scope and appropriate functional group tolerance, and can be employed for a few biologically active molecules. Furthermore, the fused pyridines can be variously functionalized via numerous simple transformations, such as cyclization, C–H alkylation and click reaction. DFT calculation investigations show that the reactions include cascade 1,5-hydrogen atom transfer, 5-exo-dig radical addition and cyclization routes. In this study, fused pyridines were synthesized by several substrates and conditions in the reaction between oximes and maleimides or cinnamonitrile or cyano-substituted alkenes in appropriate yields (Scheme 15.49).

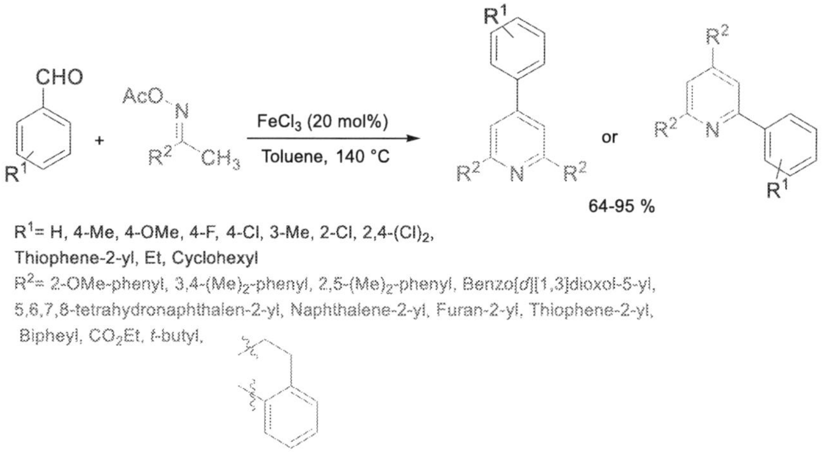

SCHEME 15.48 Fe-catalyzed facile synthesis of pyridines from ketoxime acetates.

Also, a twofold reaction between **(I)** and *m*-phenylenedimaleimide simply afforded the fused pyridine **(II)** in 55% yield (Scheme 15.50). This product mainly underlines the power of the present radical relay reaction, as most of the predictable approaches would not allow its synthesis with such great effortlessness.

In another study, Pan et al. [103] consequently evaluated the agreement of this method to some biologically active compounds (Scheme 15.51). The estrone- and D-galactose-derived reactants were exposed to this method, therefore delivering the target product in good yields. Meclizine, an antihistamine that is employed for motion sickness, was converted into the related oxime, which then undertook the reaction to give 61% yield. Also, starting from erlotinib, an EGF receptor tyrosine kinase inhibitor, the related fused pyridine was achieved in appropriate yield.

15.5.3 Pd catalysis

Li et al. reported a Pd-catalyzed addition of organoboron reagents to dinitrile derivatives for the preparation of 2,6-diarylpyridines in appropriate yields (Scheme 15.52) [104].

Selvakumar et al. [105] have investigated a useful one-pot synthesis of 2,3-disubstituted quinolines from (het)aryl-substituted Morita–Baylis–Hillman (MBH) adducts by using palladium-catalyzed Heck reaction and cyclization (Scheme 15.53). The isolable α-benzyl β-keto esters can be converted into the related quinolones under the optimized conditions in appropriate yields (67–83%).

15.5.4 Ru catalysis

Severin et al. [106] have studied the cyclotrimerization of 1-triazenes for the synthesis of substituted pyridyl triazene derivatives. These results meaningfully develop the synthetic efficacy of triazenes and highlight their compatibility with transition metal catalysts, chiefly

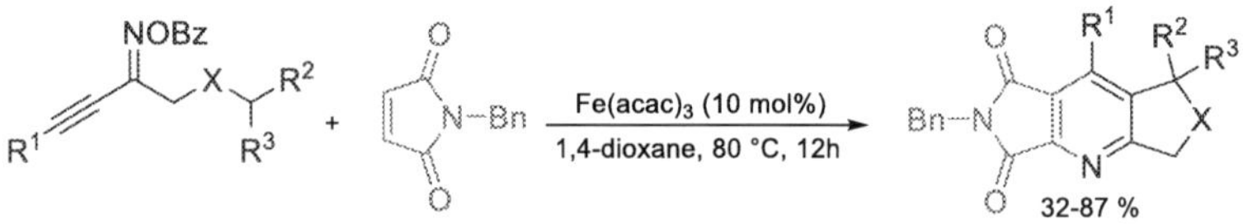

R_1= Phenyl, 4-OMe-phenyl, 4-Me-phenyl, 4-N(Me)$_2$-phenyl, 4-TMS-phenyl, 4-Cl-phenyl
4-Br-phenyl, 4-CF$_3$-phenyl, 4-CN-phenyl, 4-CO$_2$Me-phenyl, 3-Me-phenyl, 3-Me-phenyl,
3-F-phenyl, 3-Cl-phenyl, 2-Me-phenyl, 2F-phenyl, 3,5-(F)$_2$-phenyl, 4-CCH-phenyl,
Naphthalene-2-yl, Benzofuran-2-yl, Pyridine-3-yl, Thiophene-3-yl
R_2= Me
R_3= Me, n-butyl
x= C, O

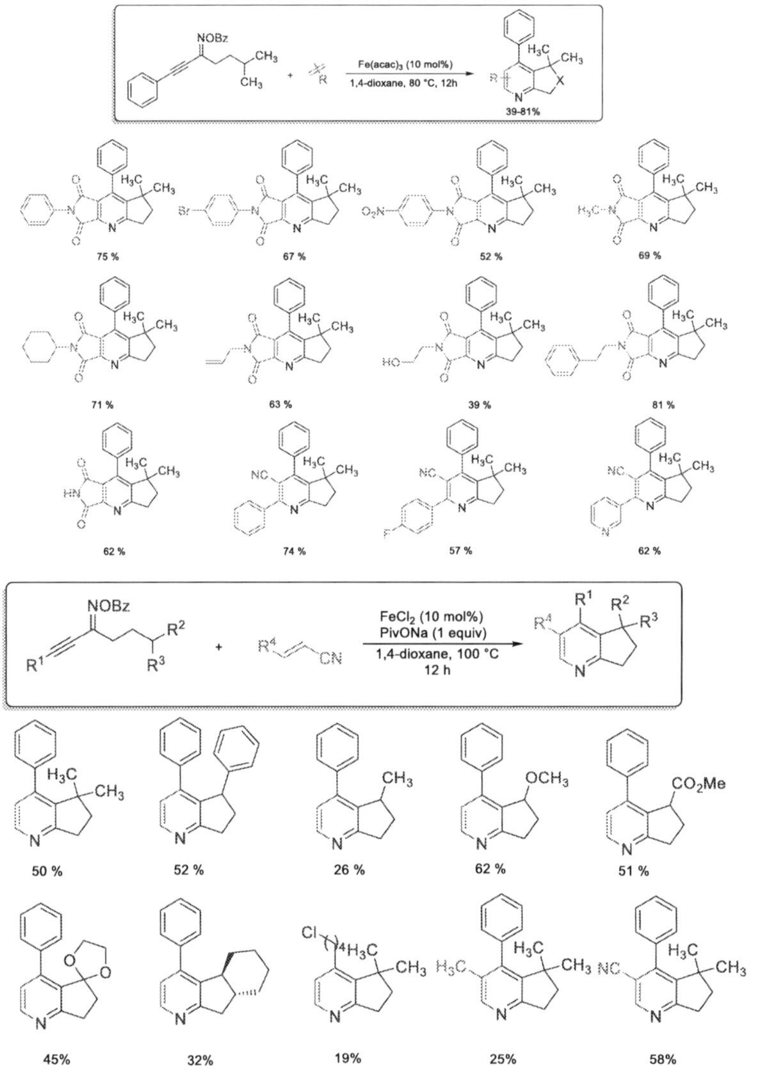

SCHEME 15.49 Synthesis of fused pyridines by using several substrates and various reaction conditions.

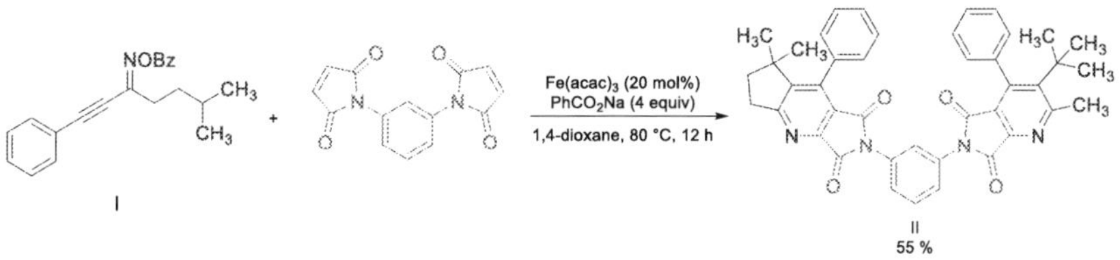

SCHEME 15.50 Synthesis of bis-fused pyridines catalyzed by Fe(acac)$_3$.

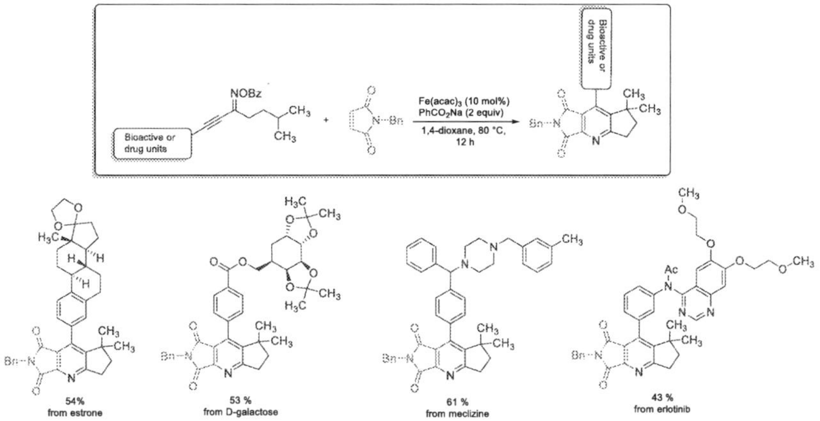

SCHEME 15.51 Application of biologically active compounds for the synthesis of fused pyridines catalyzed by Fe(acac)$_3$.

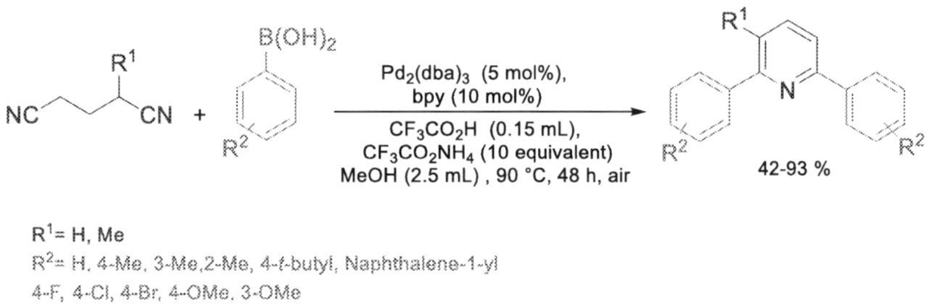

SCHEME 15.52 Synthesis of pyridines by using Pd$_2$(dba)$_3$.

ruthenium. The Cp*Ru-catalyzed [2 + 2 + 2] cyclotrimerization progresses well with triazenes (Scheme 15.54). In general, the reactions are described by a distinct regioselectivity contained by the triazenyl group and especially yield the sterically more hindered product. This generates the process very valuable for different functionality influences on these bicyclic motifs. Furthermore, such triazenes involve similarly well in intramolecular transformation to give various useful heterocycles. Studies of the unknown coordination chemistry of triazenes displayed not only a rare Cp*RuCl(η^2-alkyne) complex but as well uncommon additional complexes. These types may be related to catalyst deactivation and can afford some control to design even better performing catalysts.

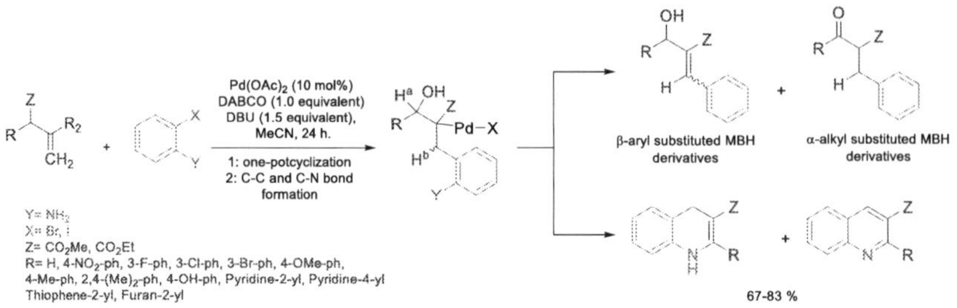

SCHEME 15.53 Synthesis of 2,3-disubstituted quinolines from (het)aryl-substituted Morita–Baylis–Hillman (MBH) adducts.

A facile method for achieving 2-aminopyridines is reported by Yi and coworkers, in the presence of a Ru complex through [2 + 2 + 2] cycloaddition of a category of 1,6- and 1,7-diynes with cyanamides [107]. The catalytic performances of the $Cp^*Ru(CH_3CN)_3PF_6$ were studied with neither additives nor ligands using a wide range of substrates under benign conditions for the synthesis of a broad range of 2-aminopyridines with isolated yields up to 99% and good region selectivity's up to >99/1 (Scheme 15.55).

Tran and coworkers reported a mild and versatile synthetic strategy for the preparation of selenopyridines via a [2 + 2 + 2] cycloaddition of α,ω-diynes and selenocyanates in the presence of ($[Cp^*Ru(CH_3CN)_3]PF_6$) [108]. This benign and efficient method lets access to various selenopyridine derivatives with suitable yields and good region-selectivities, in the presence of dichloromethane or dichloroethane as solvents at 50°C or 80°C (Scheme 15.56).

Deshmukh and Bhanage [109] have promoted a microwave-assisted facile approach for the synthesis of fused pyridines by annulation reaction of internal alkynes through C–H/N–N activation. *N*-Cbz hydrazones were used as directing group for redox-neutral Ru-catalyzed transformation (Scheme 15.57). Moreover, the reported method works capably without any additives in addition to the external oxidants. This procedure is appropriate to various *N*-Cbz hydrazones bearing electron-releasing and electron-withdrawing substituents for the synthesis of isoquinoline derivatives in suitable yields (34–93%). Applying nonvolatile and biodegradable solvent is the other advantage of this method.

A benign and facile technique for the construction of 2-triazolyl thio-/selenopyridines with suitable yields in the presence of a ruthenium (II) complex through one-pot [3 + 2]/[2 + 2 + 2] cycloaddition reactions of azides, 1-alkynyl thio-/selenocyanates, and 1,6-diynes is studied by Bhatt and coworkers (Scheme 15.58). This atom-economical catalytic protocol represents an efficient and useful approach for the preparation of various cycloadducts with good regioselectivities. The process was additional prolonged to the synthesis of 3,3′-bis(triazolyl thio-/seleno)-2,2′-bipyridines by the reaction of tetraynes with 1-alkynyl thio-/selenocyanates in the presence of aryl/alkyl azides [110].

15.5.5 Ag catalysis

Nizami and Hua [111] have investigated the annulation of propargyl amine or *N*-methyl propargyl amine with electron-deficient alkynes by using silver salts as catalysts and $K_2S_2O_8$ or DBU as additives to progress the chemo selective [4 + 2] cycloaddition giving pyridines

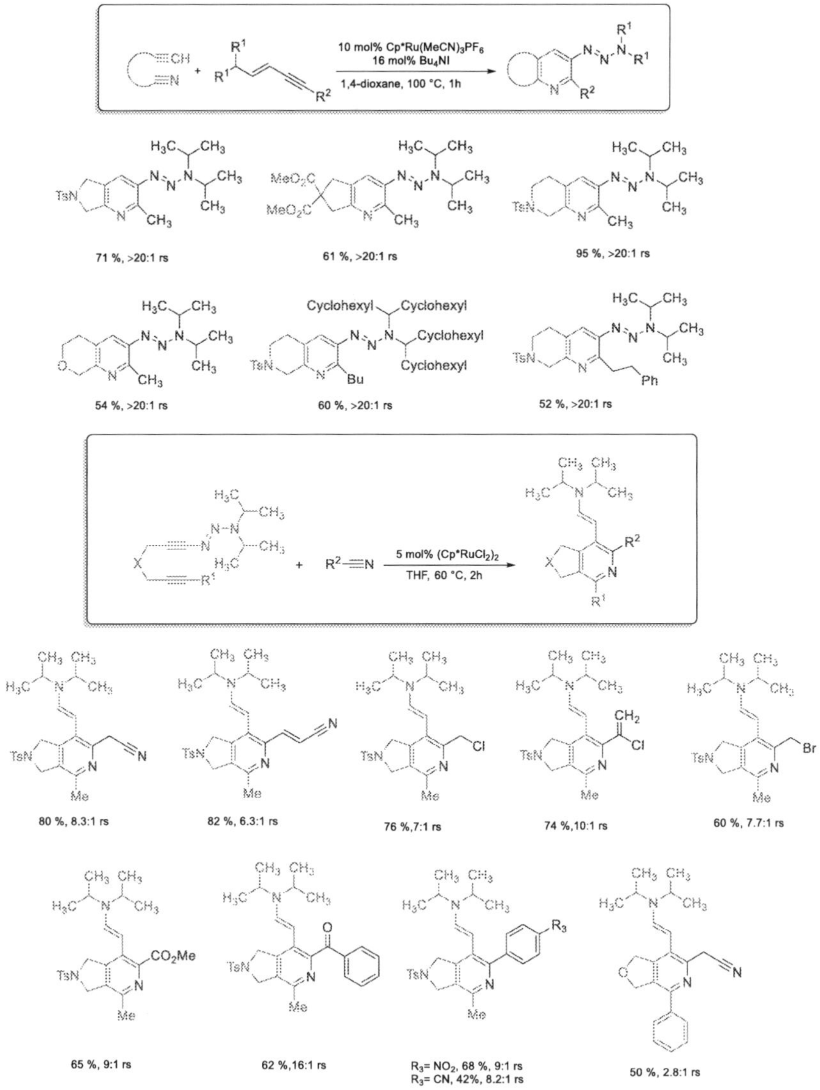

SCHEME 15.54 Synthesis of 3-triazenyl pyridines (**8**) and synthesis of pyridyl triazenes (**10**) from nitriles by using Cp*Ru-catalyzed.

(Scheme 15.59). The current methods have the advantages of good atom economy and appropriate chemo selectivity with the application of readily accessible substrates.

The suggested mechanism for the current silver-catalyzed chemo selective [4 + 2] annulation reaction between propargyl amines with alkynes is displayed in Scheme 15.60. It includes the Michael addition of propargyl amine to α,β-unsaturated carbonyl compound, and the carbo-cyclization synthesizing six-membered nitrogen-heterocyclic intermediates to then chemo selectively afford pyridines via further aromatization under several conditions. The

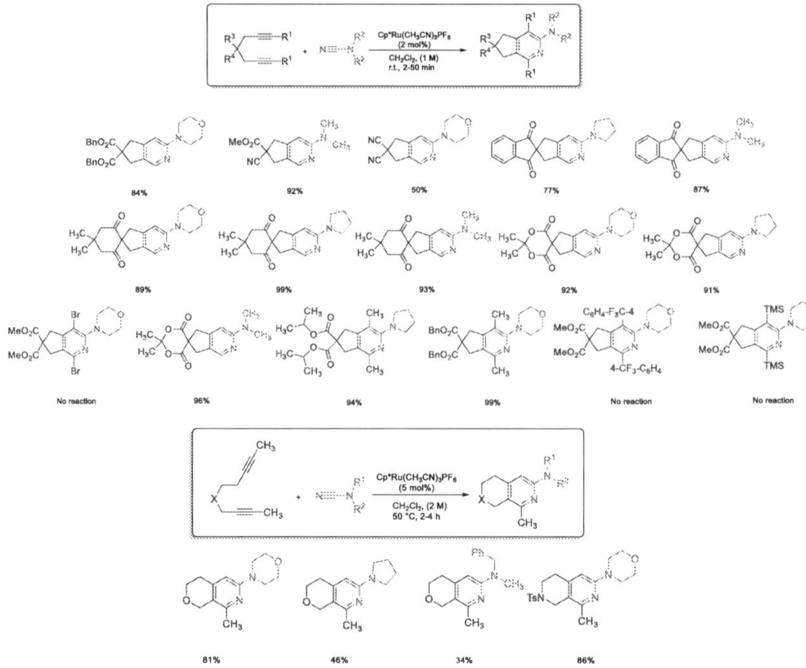

SCHEME 15.55 Ruthenium-catalyzed [2 +2 + 2] cycloaddition reaction of symmetrical and unsymmetrical diynes with cyanamides.

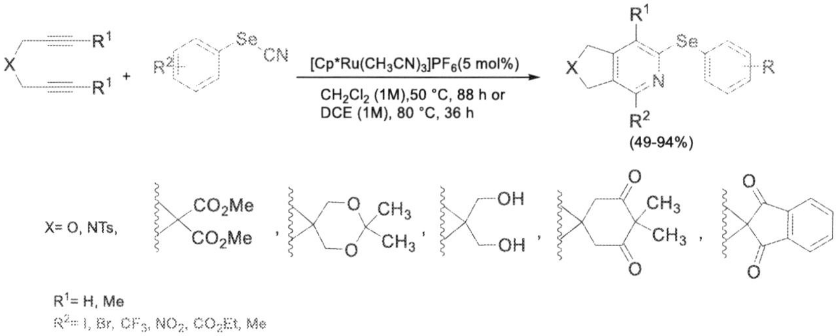

SCHEME 15.56 Synthesis of selenopyridines via a [2 + 2 + 2] cycloaddition of α,ω-diynes and selenocyanates.

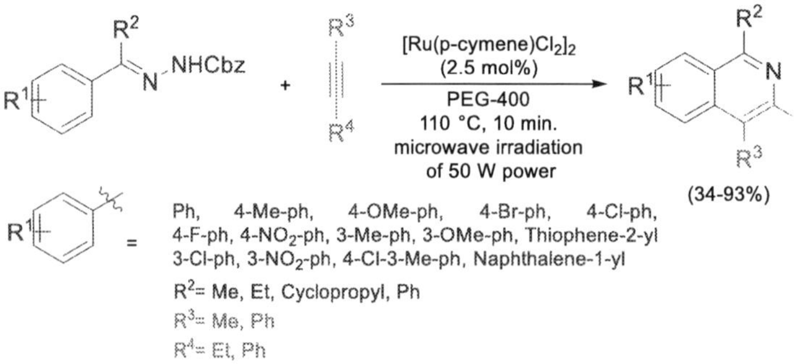

SCHEME 15.57 Synthesis of isoquinolines catalyzed by $[Ru(p\text{-cymene})Cl_2]_2$.

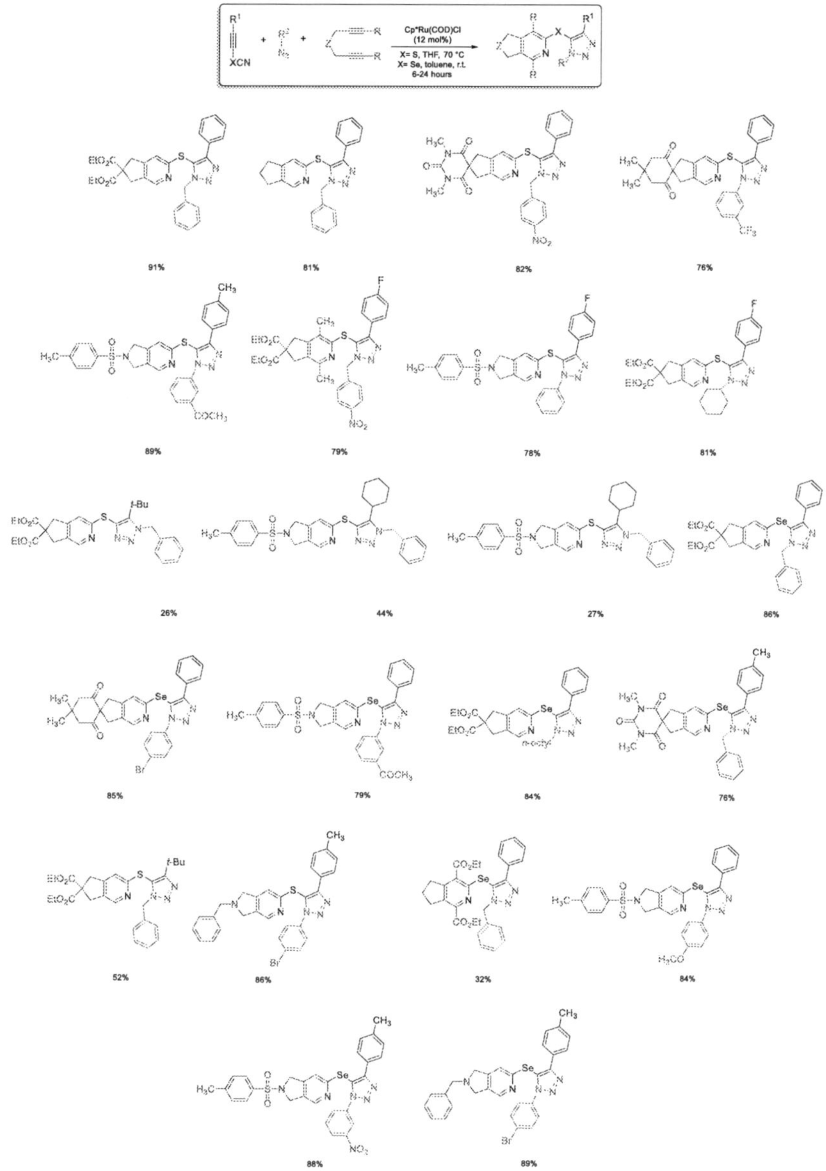

SCHEME 15.58 Synthesis of 2-triazolyl thio-/selenopyridines and 3,3′-bis(triazolyl thio-/seleno)-2,2′-bipyridines catalyzed by Cp*Ru(COD)Cl.

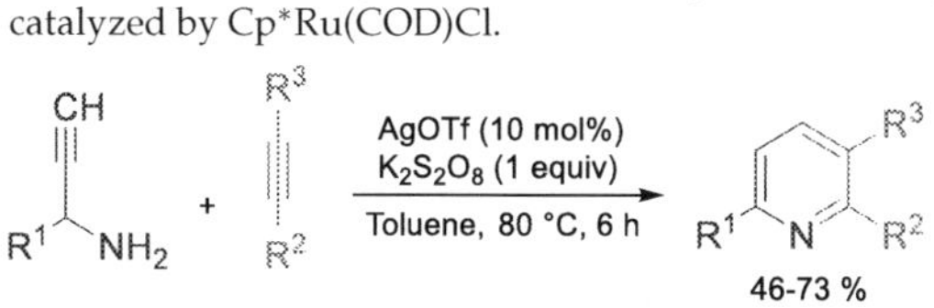

R^1= H, Me
R^2= H, Et, Ph
R^3= COMe

SCHEME 15.59 Chemo selective [4 + 2] cycloaddition synthesis of pyridines.

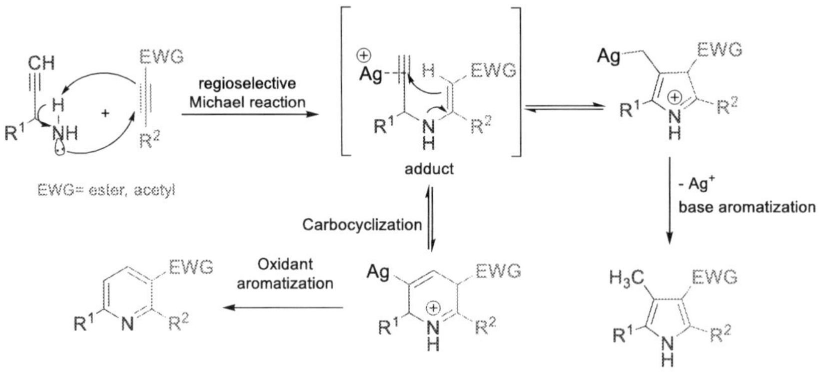

SCHEME 15.60 Suggested reaction mechanism for the [4 + 2] cycloaddition synthesis of pyridines.

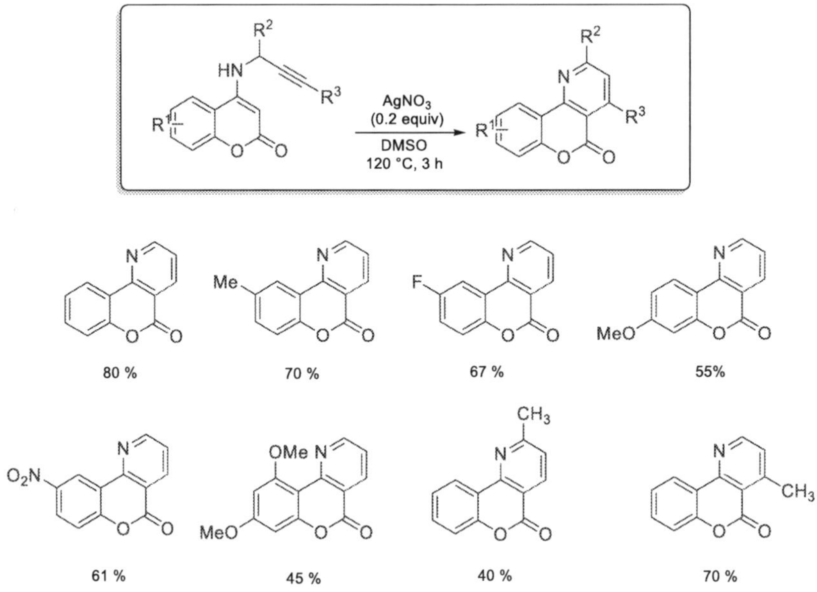

SCHEME 15.61 Synthesis of pyrido[3,2-*c*]coumarins catalyzed by $AgNO_3$.

first step of Michael addition happening with good regioselectivity to produce *N*-propargylic β-enaminones or β-enamino esters is the important step to give the pyridines and pyrroles bearing electron-withdrawing group at 3-position [112].

Yoon and Han [113] reported the synthesis of the pyrido[3,2-*c*]coumarin derivatives through a versatile and benign process by using $AgNO_3$ via a cyclo-isomerization of 4-(propynylamino)coumarins. This catalytic route provides pyrido[3,2-*c*]coumarins bearing varied substituents on benzene or pyridine ring. As displayed in Scheme 15.61, *N*-coumarinyl propargylamines having electron-releasing or electron-withdrawing groups on the coumarin applied and afforded the desired pyrido[3,2-*c*]coumarins in moderate to good yields (45–70%).

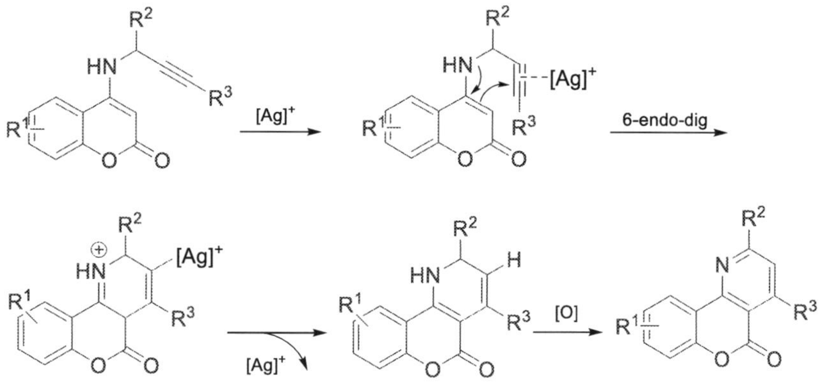

SCHEME 15.62 Possible mechanism for the synthesis of pyrido[3,2-c]coumarins.

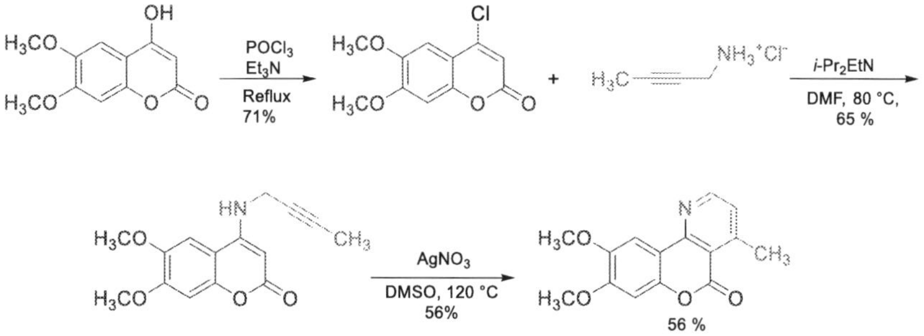

SCHEME 15.63 Synthesis of polynemoraline C (1) by using $AgNO_3$ as a catalyst.

A possible mechanism for the synthesis of pyrido[3,2-*c*]coumarins is showed in Scheme 15.62. The reaction is thought to progress with the initial coordination of the alkyne tag of the molecules to the metal catalyst. Once activated, the nucleophilic carbon atom of the enamine attacks the electrophilic alkyne bond of the structure to produce a 6-membered ring via a 6-*endo-dig* cyclization process. Protodemetalation and oxidation provide the desired pyrido[3,2-*c*]coumarins.

In another study, chlorination of the dimethoxy-4-hydroxycoumarin [114] in the presence of $POCl_3$ gives 4-chlorocoumarin, which was subjected to nucleophilic aromatic substitution with 2-butynylamine hydrochloride [115], to afford *N*-coumarinyl butynylamine. Finally, cyclo-isomerization process in the presence of $AgNO_3$ provided the related polyneomarline (Scheme 15.63).

15.5.6 Rh catalysis

Cheng et al. [116] reported the synthesis of 3,4-disubstituted isoquinoline derivatives through a rhodium-catalyzed annulation of benzylamine with diazo compounds (such as diazoacetoacetate esters and 2-diazoacetoacetone) in appropriate yields with high regioselectivity. Significantly, in addition to the methyl group, other groups such asethyl, *n*-propyl and *i*-propyl can be located on the 3-position of isoquinoline ring and leads to moderate

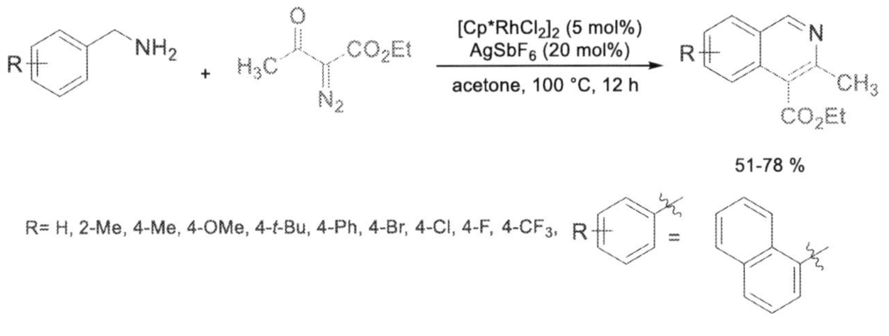

SCHEME 15.64 Rhodium-catalyzed synthesis of isoquinoline by using benzylamine with α-diazo ketone.

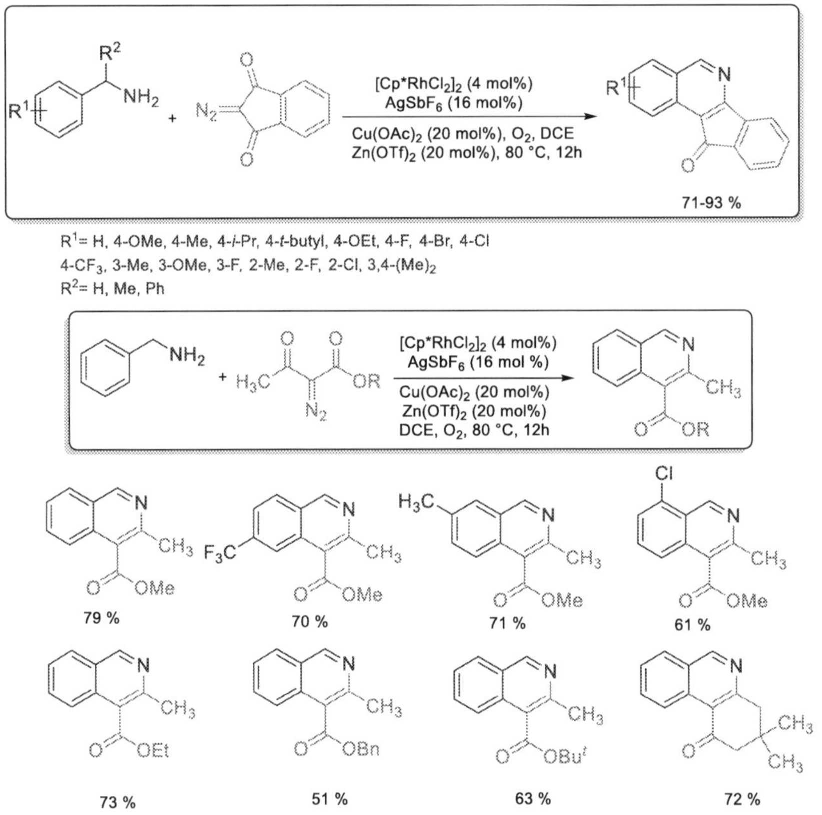

SCHEME 15.65 Synthesis of fused pyridines catalyzed by Rh/Cu.

reaction yields. Nevertheless, 2-diazobenzoylacetate was unsuccessful to work under the process (Scheme 15.64).

Wang and Li [117] reported an Rh/Cu cocatalyzed synthetic manner as a benign strategy for the preparation of fused pyridines. They applied benzylamine derivatives as building blocks and O_2 as environmentally friendly terminal oxidant (Scheme 15.65). Benzylamine was used

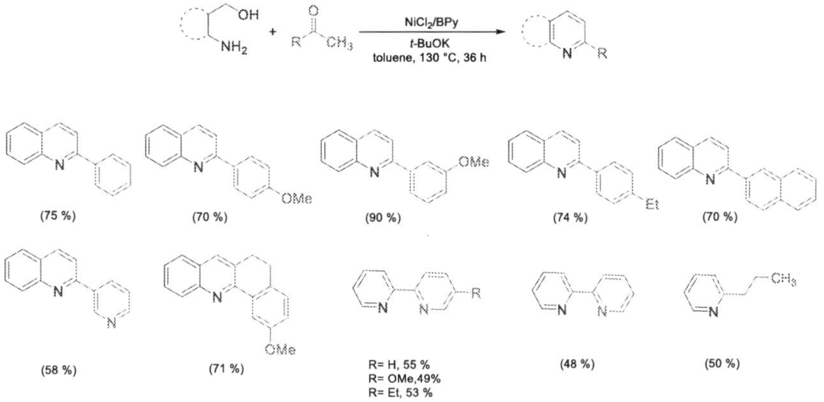

SCHEME 15.66 Synthesis of C-2-substituted pyridines.

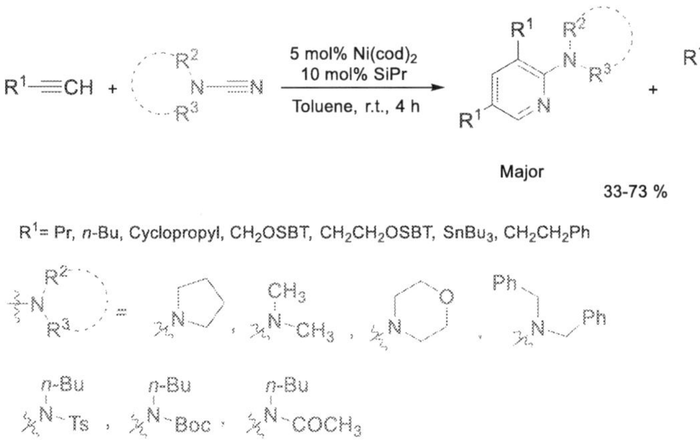

SCHEME 15.67 Ni/SIPr-catalyzed [2 + 2 + 2] cycloaddition of **2a** or **2b** and various terminal alkynes **1**.

as an arene source in C–H activation and coupling with various types of diazo compounds for the synthesis of related structures.

15.5.7 Ni catalysis

The reaction of 3-amino-1-propanol with acetophenone derivatives leads to the synthesis of C-2-substituted pyridines in reasonable reaction yield in the presence of nitrogen-ligated nickel catalyst (Scheme 15.66). Remarkably, 2-acetylpyridine capably transformed into the bipyridine ligand. Additionally, the application of more challenging 2-pentanone yielded the desired 2-*n*-propyl pyridine in 71% yield (Scheme 15.66) [118].

Louie et al. [119] applied nickel(cod)$_2$ as a catalyst for the [2 + 2 + 2] cycloaddition of terminal alkynes and cyanamides for the preparation of 2-aminopyridine molecules in appropriate yields. Remarkably, the major products are the 3,5-disubstituted regioisomer, which is distinctive in the field of pyridine cycloaddition chemistry (Scheme 15.67). The process is benign

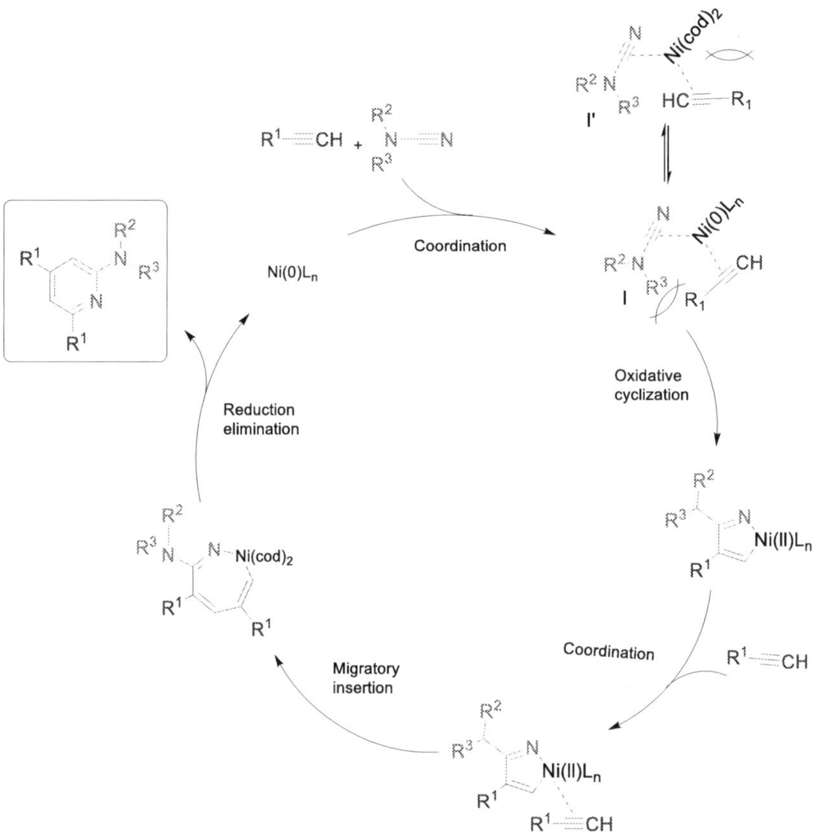

SCHEME 15.68 Suggested mechanism of Ni/SIPr-catalyzed [2 + 2 + 2] cycloaddition for the synthesis of 2-aminopyridine.

and facile with alkyl–alkyne reactants and tolerates some functional groups.

The suggested mechanistic way for the synthesis of target molecules is included the coordination of one alkyne and one cyanamide molecule to the metal center, oxidative cyclization, coordination with another alkyne molecule, migratory insertion, and finally reductive elimination of target molecules [120]. The regioselectivity of the process can be attributed to the coordination types **I** compared with **I′**, which is estimated to be less stable for the presence of larger steric repulsion between the alkyne substituents and the large NHC ligand (Scheme 15.68).

15.5.8 Sn catalysis

Khalafy et al. reported an efficient and appropriate pathway for the synthesis of pyrazolo[5,4-*b*]quinolines via reaction of 5-amino-3-(arylamino)-1*H*-pyrazole-4-carbonitrile derivatives with cyclohexane-1,3-dione or dimedone in the presence of a catalytic amount of $SnCl_2{\cdot}2H_2O$ under solvent-free conditions [121]. The advantages of this procedure contain the ease of isolation of the related products and the low cost of the catalyst (Scheme 15.69).

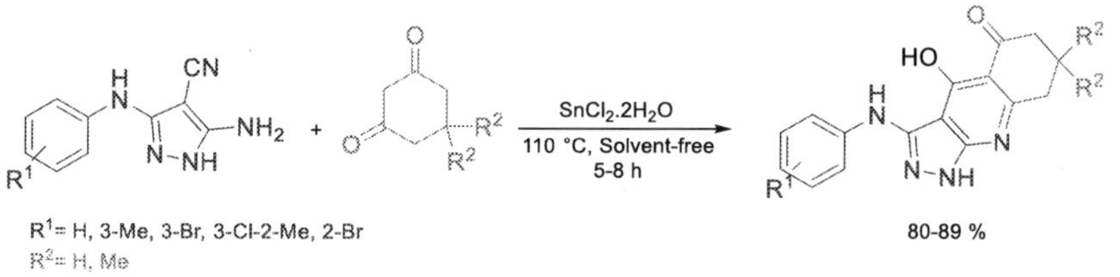

SCHEME 15.69 Synthesis of tetrahydropyrazolo[5,4-*b*]quinolones catalyzed by $SnCl_2{\cdot}2H_2O$.

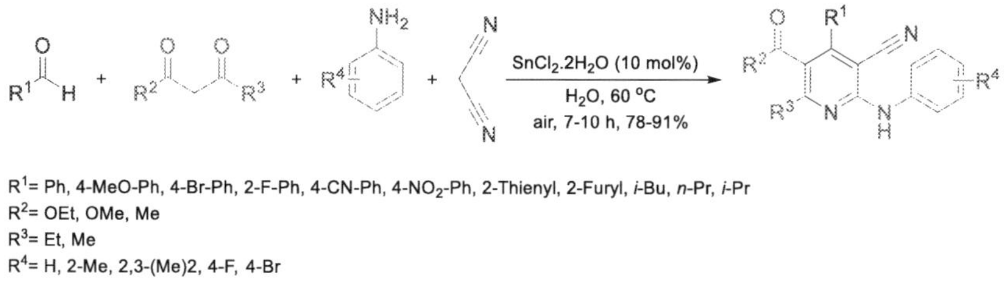

SCHEME 15.70 Sn-catalyzed preparation of polysubstituted pyridine derivatives.

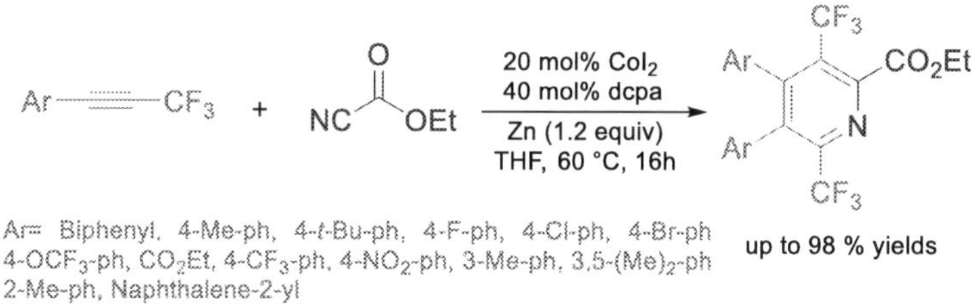

SCHEME 15.71 Synthesis of pyridine derivatives by using the CoI_2/dcype/Zn catalyst medium.

Several polysubstituted pyridine derivatives were prepared through the reaction of different alkyl/(hetero)aryl aldehydes, β-keto esters (or 1,3-diketones), aniline derivatives, and malononitrile in the presence of $SnCl_2{\cdot}2H_2O$ as a convenient precatalyst. The main merits of the reported protocol are using an inexpensive catalytic system, aqueous media of the reaction, wide substrate window with reasonable operation yields (Scheme 15.70) [122].

15.5.9 Co catalysis

Kawatsura et al. [123] investigated the intermolecular [2 + 2 + 2] cycloaddition of aryl- and trifluoromethyl-substituted internal alkyne and ethyl cyanoformate, for the preparation of functionalized pyridine derivatives by applying the CoI_2/dcype/Zn as catalyst medium. The reaction progressed with good regioselectivity and two trifluoromethyl-substituted ethyl picolinates were achieved as a single regioisomer (Scheme 15.71).

The catalytic application of synthesized chiral indenylcobalt(I) complex was investigated in asymmetric [2 + 2 + 2] cycloaddition reaction between naphthyldiynes and nitrile derivatives [124]. The catalytic performance of the synthesized structures was investigated in asymmetric cyclization reactions of triyne derivatives and also, diynes with nitriles to generate chiral triaryls and heterobiaryls, respectively (Table 15.1).

TABLE 15.1 Reaction between diyne (**12**) and benzonitrile (**13**) in the presence of complex 18 under various conditions.

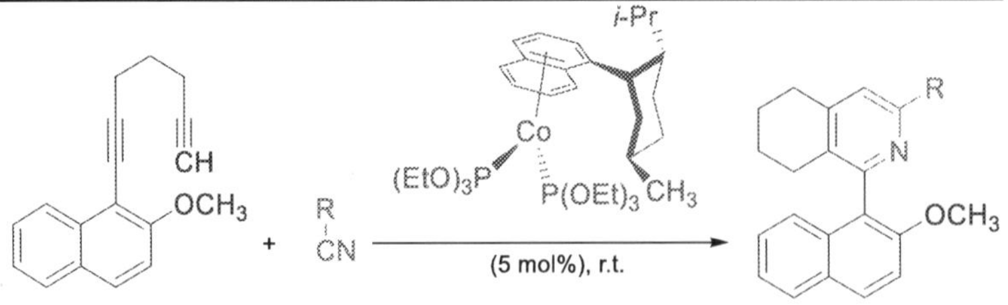

R= Me, 2-F-6-OMe-phenyl, 3,4,5-(OMe)$_3$

Entry	Solvent	Temp. (°C)	Time (h)	Yield (%)	Selectivity
1	THF	50	70	20	66% ee(R)
2	Toluene	80	24	54	27% ee(R)
3	THF	-20. hv	24	2	-
4	THF	0. hv	24	86	64% ee(R)

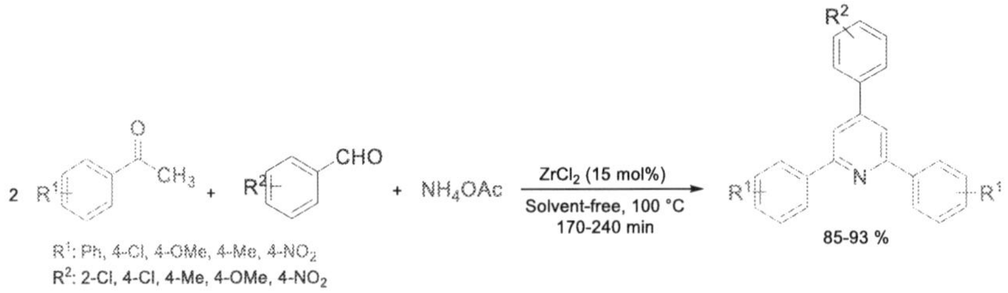

SCHEME 15.72 Synthesis of 2,4,6-triarylpyridines by using catalytic amount of $ZrOCl_2$.

15.5.10 Zr catalysis

Moosavi-Zare et al. applied oxozirconium(IV) chloride as a catalyst for the benign and facile preparation of 2,4,6-triarylpyridines (Kröhnke pyridines) via a one-pot three-component condensation reaction between aldehydes, acetophenone derivatives, and ammonium acetate (as nitrogen source) under solvent-free conditions [125]. The advantages of this method over other approaches include higher yields of related products, recyclability and reusability of catalyst, and cleaner reaction system, producing it an attractive method for the synthesis of Kröhnke pyridines (Scheme 15.72). Numerous aromatic aldehydes containing electron-donating substituents, electron-withdrawing substituents and halogens were reacted with various acetophenones and ammonium acetate to provide the desired molecules in relatively short reaction times.

15.5.11 Mn catalysis

In 2020, Chai and coworkers reported the synthesis and application of a cheap, phosphine-free, and bidentate manganese (I) complex as a catalyst for the preparation of pyridine derivatives by using secondary alcohols and amino alcohol derivatives as starting materials (Scheme 15.73) [126].

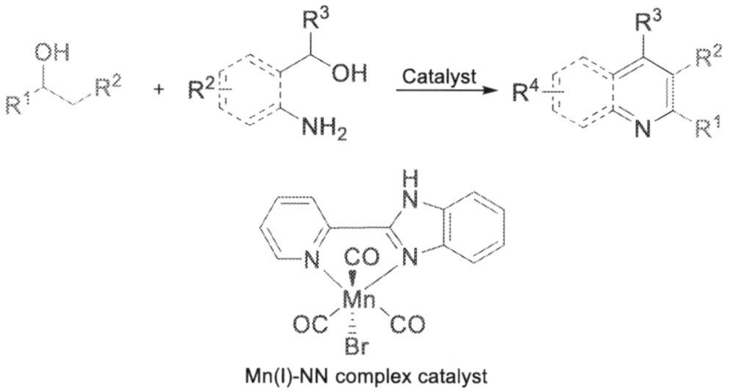

SCHEME 15.73 Synthesis of pyridine derivatives using Mn(I)-NN complex catalyst.

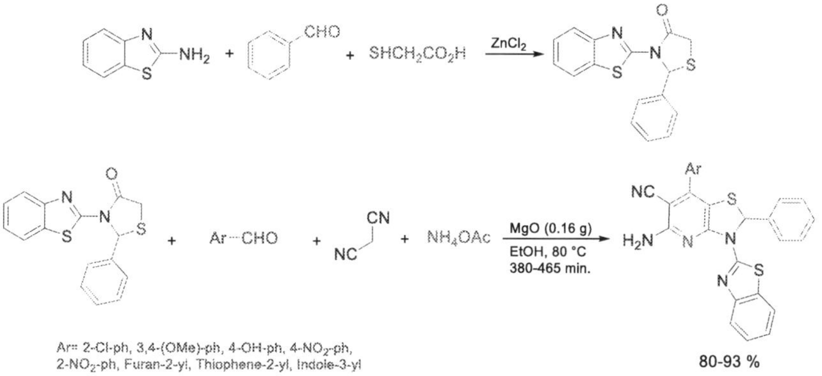

SCHEME 15.74 General procedure for the synthesis of thiazolo[4,5-b]pyridine-6-carbonitrile using MgO.

15.5.12 Mg catalysis

An efficient and environmentally friendly strategy was promoted by Agarwal and coworkers for the synthesis of thiazolo[4,5-*b*]pyridine-6-carbonitriles via condensation reaction following Michael addition reaction between synthesized 3-benzothiazol-2-yl-2-phenyl-thiazolidin-4-one, aromatic aldehydes, malononitrile and ammonium acetate using MgO NPs as green, low-cost, forceful, and recoverable nanocatalyst. In this protocol, all of the multicomponent reactions were performed in mild conditions and all of the products were prepared with a good yield (80–93%) (Scheme 15.74) [127].

15.5.13 Al catalysis

In 2018, catalytic application of aluminum oxide (Al_2O_3) was investigated for the synthesis of a series of 4-aroyl-1,6-diaryl-3-methyl-1*H*-pyrazolo[3,4-*b*]pyridine-5-carbonitrile derivatives in a one-pot three-component reaction manner between arylglyoxals, 3-aryl-3-oxopropanenitriles, and 5-amino-1-aryl-3-methylpyrazoles. In this study, the products have an excellent yield (70–91%) and the reactions were performed under green conditions (Scheme 15.75) [128]. Moreover, in a comparative study, the reaction was performed in the presence of other catalysts such as concentrated HCl, *p*-TSA, $ZnCl_2$, AcOH, and L-proline that the yield of the product was lower than catalytic reaction with Al_2O_3.

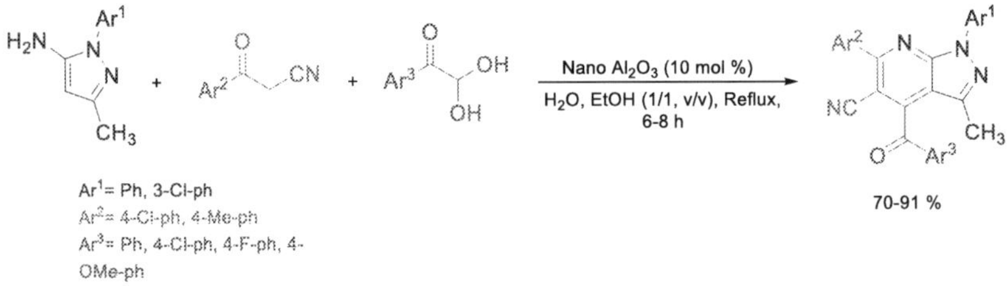

SCHEME 15.75 Green synthesis of 4-aroyl-1,6-diaryl-3-methyl-1*H*-pyrazolo[3,4-*b*]pyridine-5-carbonitrile derivatives with Al_2O_3 as catalyst.

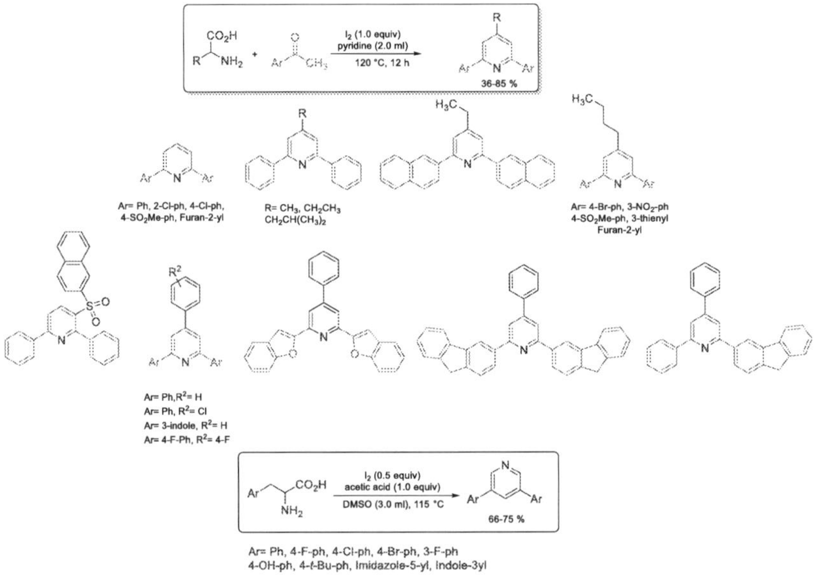

SCHEME 15.76 Selective synthesis of 2,6-disubstituted, 2,4,6-trisubstituted and 3,5-disubstituted pyridines by using catalytic amounts of I_2.

15.5.14 I_2 catalysis

In an interesting report, 2,6-disubstituted, 2,4,6-trisubstituted, and 3,5-disubstituted pyridines were synthesized according to the catabolism and reconstruction performances of natural amino acids by using molecular iodine as a catalyst in a chemoselective manner [129]. In this process, molecular iodine acts as a tandem catalyst and initiates the decarboxylation-deamination of the applied amino acids (Scheme 15.76). It is important that the reaction of glycine with 1-phenyl-2-(phenylsulfonyl)ethan-1-one under the optimized reaction conditions led to the unexpected cleavage of one of the two sulfonyl benzene groups and leads to the formation of 2,3,6-trisubstituted pyridine as a major product in 74% yield. These transformations involved a catabolism and reconstruction reaction model containing a unique I_2-mediated oxidative cleavage of unreactive C–N bonds. Besides the inherent importance of this biomimetic pathway, the key merit of this process is avoiding of toxic aldehydes as starting material and also, harsh reaction conditions.

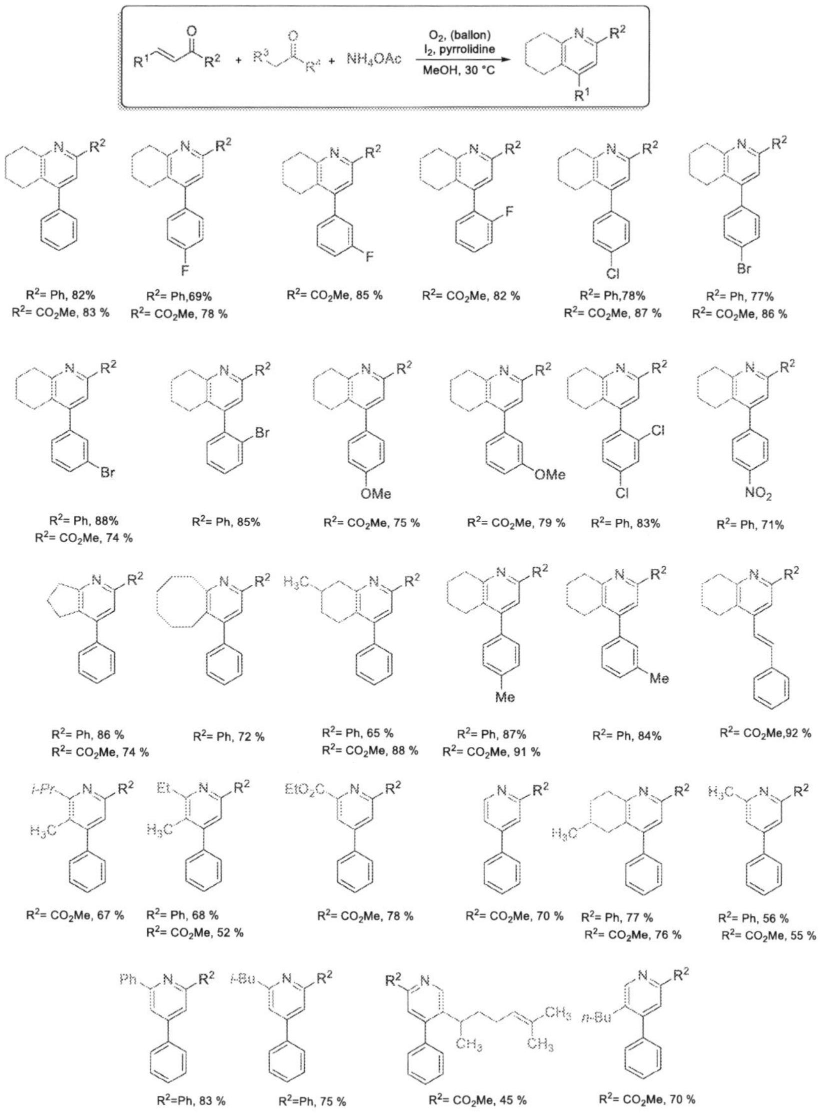

SCHEME 15.77 Iodine-catalyzed aerobic oxidative synthesis of polyfunctionalized pyridines.

Iodine was applied as a catalyst in an aerobic oxidative formal [4 + 2] annulation for the preparation of pyridine derivatives via a green and benign reaction medium [130]. In this reaction the catalytic amounts of molecular iodine in combination with oxygen were used. Numerous ketones and aldehydes are subjected to the reaction with various chalcones and β,γ-unsaturated α-ketoesters in this approach (Scheme 15.77).

To know the role of iodine in this aerobic oxidative prescribed [4 + 2] annulation, Zhu et al. [130] reacted chalcone, cyclohexanone, and ammonium acetate under the reaction procedure. Also, TEMPO (2.0 equiv.) was applied as a radical scavenger. In this testing, the authors

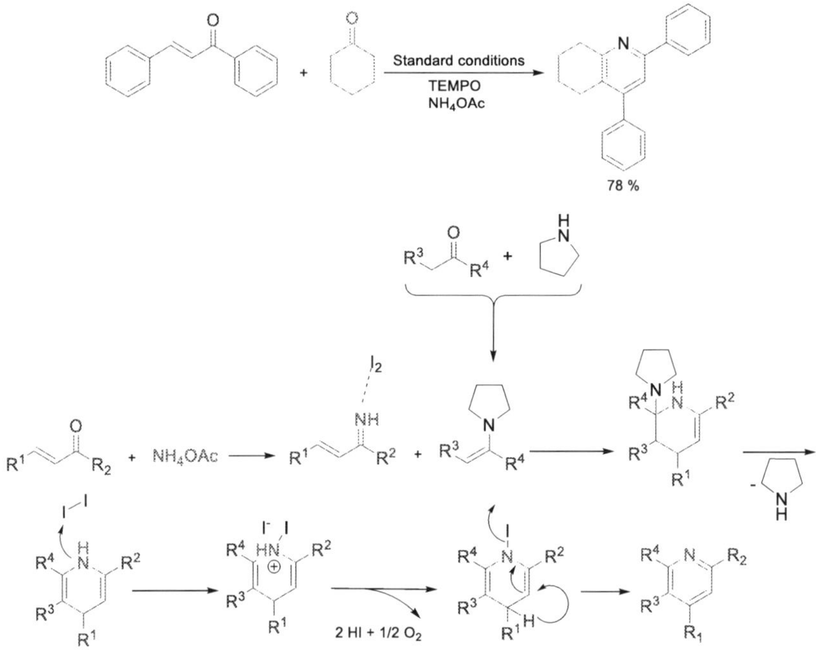

SCHEME 15.78 Possible mechanism for the aerobic oxidative prescribed [4 + 2] annulation synthesis of pyridines.

observed the synthesis of desired molecules in 78% yield and no TEMPO-bound intermediate was distinguished (Scheme 15.78). Consequently, they think that a radical-based reaction mechanism can be excepted. First, the reaction of ketone with amine affords an enamine that can undertake [4 + 2] annulation with chalcone by using I_2. The corresponding intermediate A provides intermediate B by eliminating one molecular amine. Then, intermediate B reacts with I_2 to synthesize C which can generate D via removal of one molecular HI. Lastly, **3** is attained by the removal of further molecular HI. Especially, two molecular HI produced in situ are oxidized by oxygen to regenerate iodine to complete the I_2/I^- catalytic cycle.

Iodine–dimethyl sulfoxide (I_2-DMSO) system was applied for the synthesis of various fused pyridine derivatives via a one-pot and benign consecutive oxidative cross coupling then, intramolecular cyclization of pyridoimidazole arylamines with carbonyl compounds (Scheme 15.79). Facile reaction conditions, metal- and base-free catalyst, selective product design and suitable yields (that without needing prior C(sp^2)–H activation) are some of the advantages of this method [131].

A versatile and simple strategy was reported for the metal-free preparation of 2,4,6-trisubstituted pyridines using aryl methyl ketones and amine derivatives through cascade condensation–cyclization–aromatization method in the presence of I_2 (Scheme 15.80). The reaction was accomplished in good yield by using 1 mol % I_2 acting as an initiator to generate HI in situ as the actual catalyst, subsequently aerobic oxidation under neat heating conditions. This efficient method displays numerous advantages for example no use of any metal catalyst or additional oxidant, wide scope of substrates and high functional group compatibility, great potential for large-scale synthesis, and no tedious extractive workup [132].

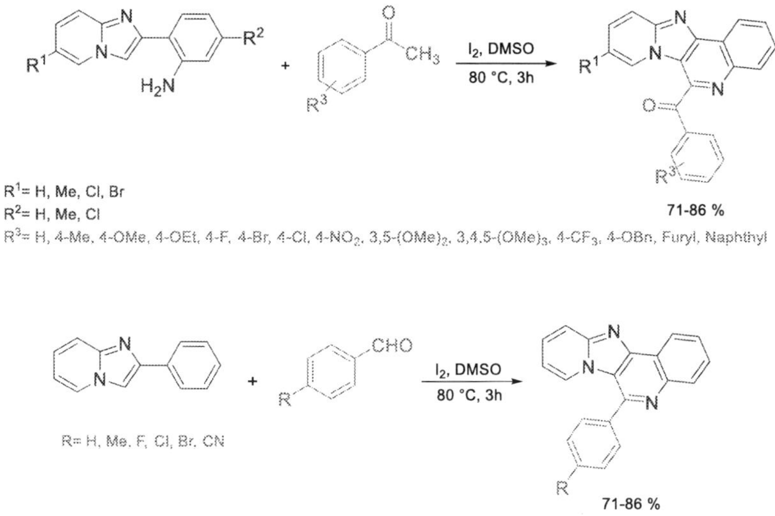

SCHEME 15.79 Synthesis of pyrido-fused imidazo[4,5-*c*]quinolines catalyzed by I_2-DMSO.

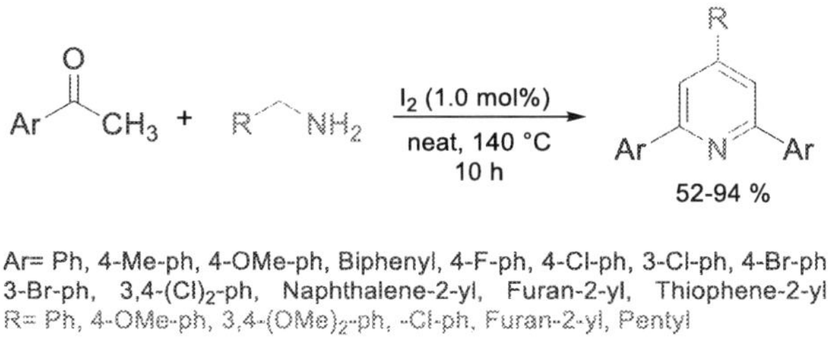

SCHEME 15.80 Synthesis of 2,4,6-trisubstituted pyridines catalyzed by I_2 under neat heating.

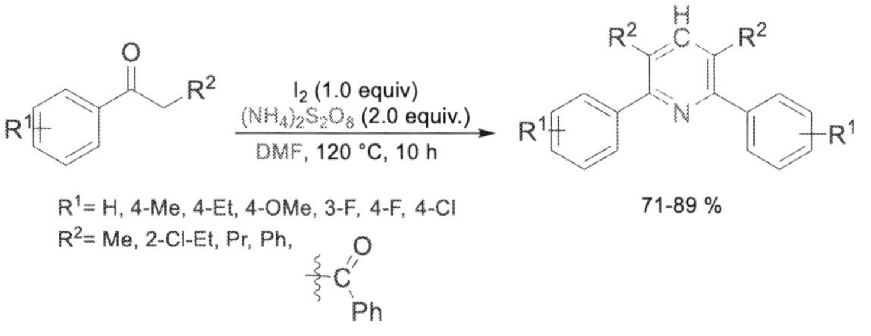

SCHEME 15.81 Synthesis of pyridines by using DMF as the sources of carbon in the presence of I_2 and $(NH_4)_2S_2O_8$.

In another work, iodine catalyzed [2 + 2 + 1 + 1]-cycloaddition reaction for the synthesis of symmetrical tetra-substituted pyridines in appropriate reaction yields by using ammonium persulfate-mediated cyclization of α-substituted arones with DMF (Scheme 15.81). In this reaction medium, both DMF and $(NH_4)_2S_2O_8$ play a significant dual role, DMF is employed not only as a benign reaction system, but also as the source of carbon C4, and $(NH_4)_2S_2O_8$ appears both as an oxidant and nitrogen source for the construction of target pyridines [133].

To have a deeper insight into the mechanism for the synthesis of desired pyridine, numerous parallel experiments were conducted (Scheme 15.82). First, to determine the C4 source, a

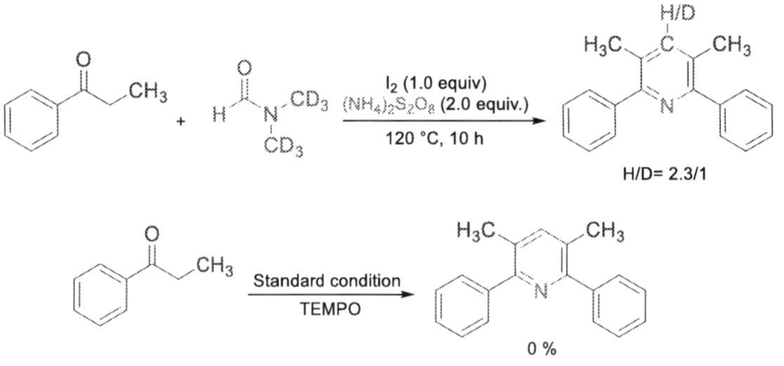

SCHEME 15.82 Studies of the reaction mechanism via synthesis of 4-deuterated pyridine 2a–d_1.

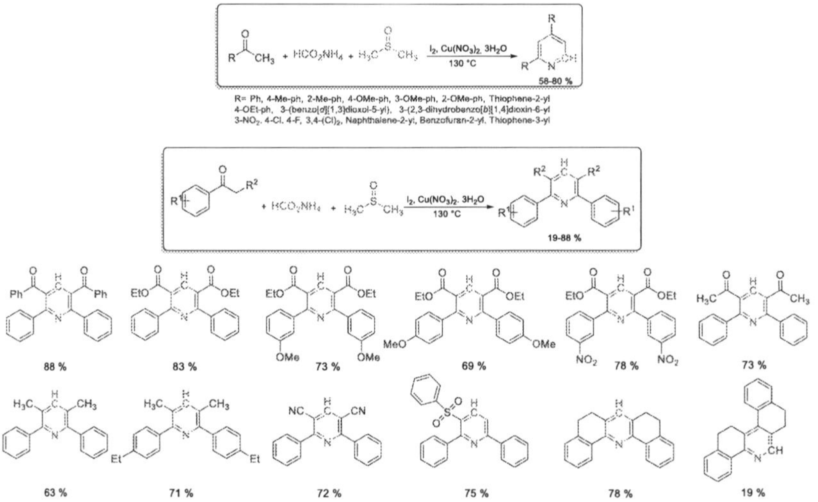

SCHEME 15.83 Synthesis of pyridines via C_{sp3}–H oxidation and C–S cleavage of dimethyl sulfoxide.

deuterium labeled experiment was performed by using *N,N*-dimethyl-d6-formamide (DMF-d_6) as the solvent. It is establishing that this transformation becomes a little slower but gives the expense of 4-deuterated pyridine 2a–d_1 as the only product (Scheme 15.82, Eq. (1)). The resulting data proved that the C4 carbon part of the pyridine ring derives from the methyl group rather than the aldehyde group of DMF. Nevertheless, this conversion is completely inhibited when the radical scavenger TEMPO is applied (Scheme 15.2, Eq. (2)), which proposed that a radical way may include in the mechanism.

An efficient and versatile cleavage of the C–S bond and C_{sp3}–H oxidation of dimethyl sulfoxide (DMSO) was reported for the synthesis of substituted pyridines from methyl ketones [134]. In this method, the formic acid was produced as a coproduct from ammonium formate, which acted as a significant catalyst for the reaction. Remarkably, this transformation showed a wide substrate scope toward various ketones to afford the related pyridines in appropriate yields (Scheme 15.83). This technique displays a valuable approach for capturing in situ-generated methylene intermediates by using DMSO as a source of methylene units.

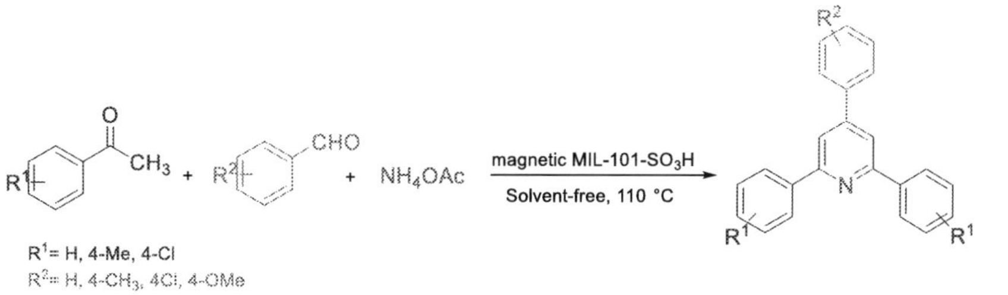

SCHEME 15.84 Synthesis of 2,4,6-triaryl pyridines in the presence of magnetic MIL-101–SO_3H nanocatalyst.

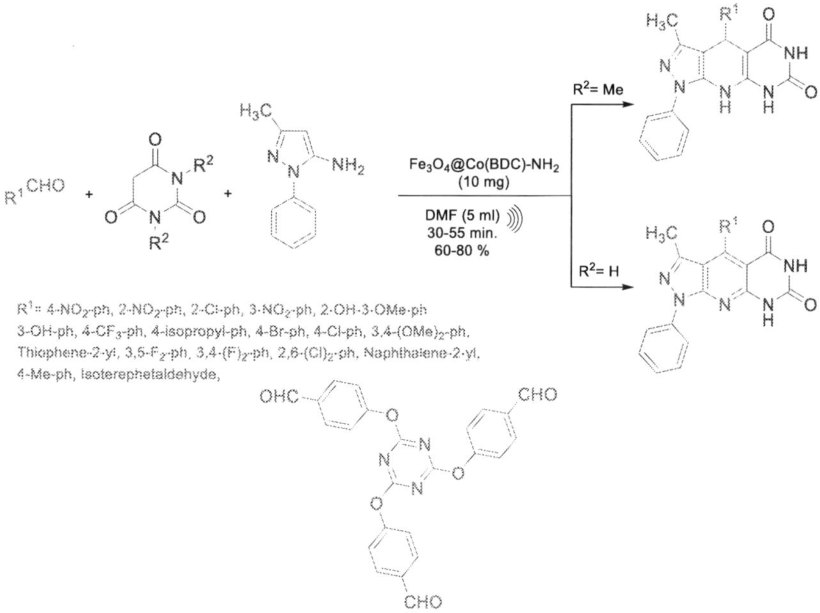

SCHEME 15.85 Synthesis of fused pyridines in the presence of Fe_3O_4@Co(BDC)-NH_2 as a magnetic metal–organic frameworks catalyst.

15.6 Heterogeneous catalytic systems

15.6.1 MOF-based heterogeneous catalysts

The magnetic MIL-101–SO_3H nanocatalyst was effectively applied for the preparation of the 2,4,6-triaryl pyridine derivatives via the reaction between acetophenone, benzaldehyde, and ammonium acetate under solvent-free conditions in appropriate reaction times [135]. The reaction was carried out by using a 30 mg nanocatalyst under solvent-free conditions at 110°C (Scheme 15.84).

Zolfigol et al. [136] synthesized magnetic metal–organic frameworks based on Fe_3O_4 with high surface area. Then, the prepared magnetic porous catalyst was employed for the synthesis of fused pyridine systems with pyrazole and pyrimidine tags as under ultrasonic irradiation (Scheme 15.85). The noteworthy advantages of this process are versatile, easy workup,

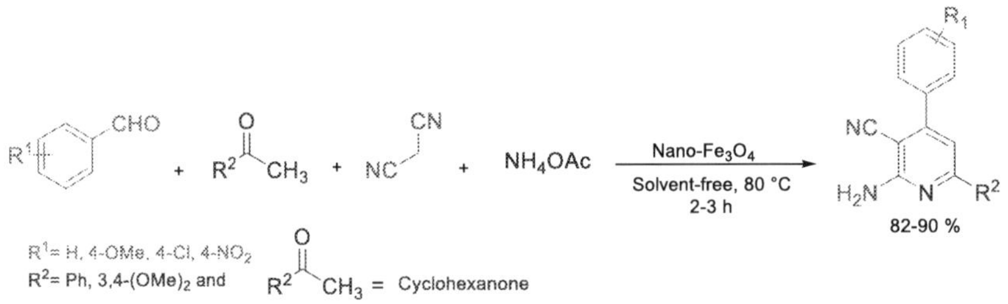

SCHEME 15.86 Synthesis of 2-amino-3-cyanopyridines by using nano Fe_3O_4.

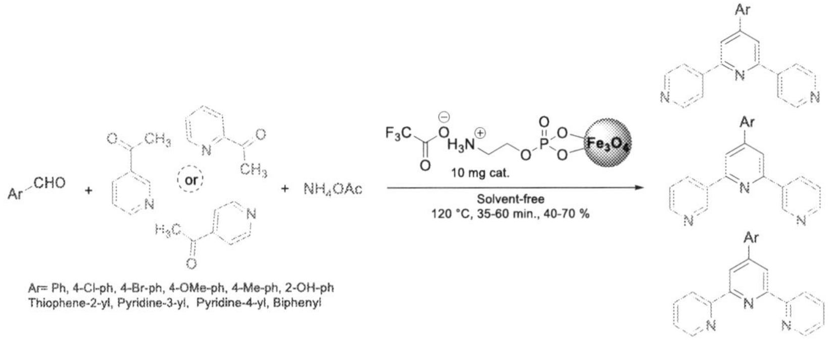

SCHEME 15.87 Synthesis of terpyridines by using $Fe_3O_4@O_2PO_2(CH_2)_2NH_3^+CF_3CO_2^-$.

high yields (60–80%), short reaction times (30–55 min), and also, high thermal stability and reusability of NMMOFs catalyst.

15.6.2 Nanomagnetic catalysis

An efficient and benign procedure is reported for the synthesis of 2-amino-3-cyanopyridine derivatives by using Fe_3O_4 nanoparticles as an appropriate heterogeneous catalyst via one-pot four-component cyclo-condensation reaction between aldehydes, ketones, ammonium acetate, and malononitrile under solvent-free condition at 80°C in good yields and short reaction times [137]. The most significant advantage of this process is the suitable separation of the commercially available nano-Fe_3O_4 catalyst from the reaction mixture and its reusability in numerous runs, without substantial loss of activity (Scheme 15.86).

$Fe_3O_4@O_2PO_2(CH_2)_2NH_3^+CF_3CO_2^-$ as an efficient nanomagnetic catalyst was applied for the three-component synthesis of teryridines using acetylpyridines, aldehydes, and ammonium acetate as starting materials under solvent-free conditions (Scheme 15.87). The desired terpyridine derivatives were attained with satisfactory yields (40–70%) and short reaction times (35–60 min) through a cooperative vinylogous anomeric-based oxidation pathway [138].

In another work, Zolfigol et al. [139] reported $Fe_3O_4@O_2PO_2(CH_2)_2NH_3^+CF_3CO_2^-$ as a benign and reusable ionically tagged nanomagnetic catalyst at the synthesis 2-amino-6-(2-oxo-2*H*-chromen-3-yl)-4-arylnicotinonitriles through the reaction between aldehydes, 3-acetylcoumarins, malononitrile, and ammonium acetate as a source of nitrogen under solvent-free condition at 70°C with high yields (Scheme 15.88). Experimental data have con-

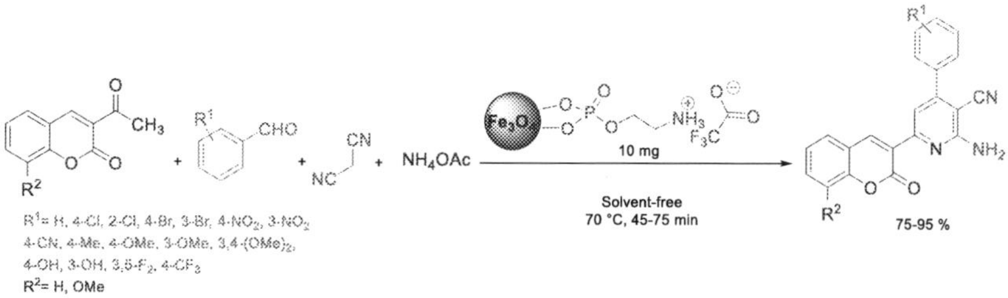

SCHEME 15.88 Synthesis 2-amino-6-(2-oxo-2*H*-chromen-3-yl)-4-arylnicotinonitriles catalyzed by $Fe_3O_4@O_2PO_2(CH_2)_2NH_3^+CF_3CO_2^-$ MNPs.

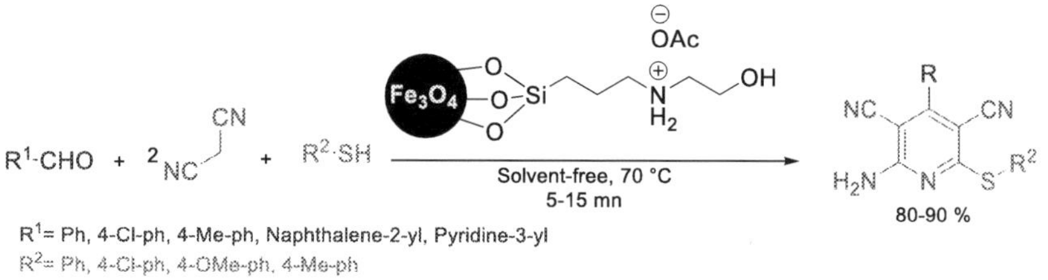

SCHEME 15.89 Synthesis of 2-amino-3,5-dicarbonitrile-6-thio-pyridines by using Fe_3O_4–2-HEAA.

firmed that the final step of the probable mechanism progressed via a vinylogous anomeric-based oxidation. Described catalyst displays appropriate potential of recycling and reusing at this four-component reaction. Simplification and informal work-up, mild reaction conditions, and short reaction times are the main advantages of this presented technique.

Sobhani et al. [140] have studied an efficient and eco-friendly process for the synthesis of uniform Fe_3O_4 magnetic nanoparticles via a coprecipitation process by using 2-hydroxyethylammonium acetate (2-HEAA) as a cost-effective ionic liquid. This is the first investigation for the synthesis of iron oxide via functionalized ionic liquids. The synthesized Fe_3O_4–2-HEAA was then successfully employed as a magnetically recyclable heterogeneous catalyst with dual mode of activation for the one-pot synthesis of 2-amino-3,5-dicarbonitrile-6-thio-pyridines via the reaction between aldehydes, malononitrile, and thiols in appropriate yields (80–90%). The catalyst was effortlessly separated from the reaction media by an external magnet and reused five times without loss in activity. Using a cost-effective and reusable catalyst with appropriate TON and TOF, being agreeable for both aromatic and aliphatic aldehydes/thiols and short reaction times (5–15 min) are the other advantages of this process (Scheme 15.89).

Recently, Zolfigol et al. [141] reported the synthesis of triarylpyridines with sulfonate or sulfonamide moieties using novel magnetic nanoparticles with pyridinium bridges namely $Fe_3O_4@SiO_2@$PCLH-TFA via a cooperative vinylogous anomeric-based oxidation. With a good performance from recovering and reusing points of view, the catalyst demonstrated itself in tested reactions. Moreover, UV–visible spectroscopy was used for the investigation of kinetics behavior of reactions (Scheme 15.90).

Another heterogeneous nanomagnetic catalyst having morpholine tags was synthesized by Zolfigol's group [142]. The catalytic performance of the prepared nanomagnetic catalyst was investigated in four-component synthesis of 2-amino-4,6-diphenylnicotinonitriles via

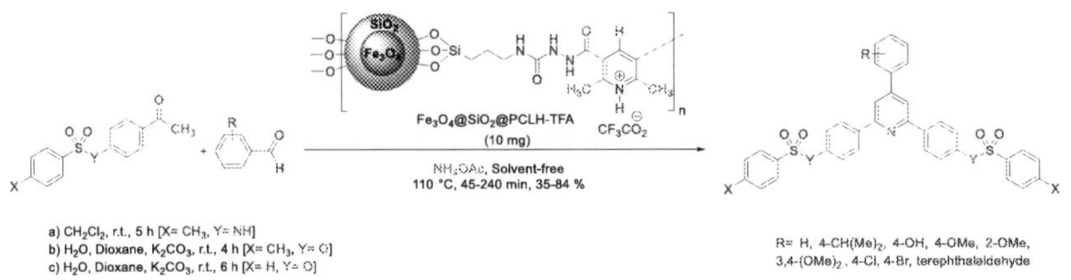

SCHEME 15.90 Synthesis of triarylpyridines with sulfonate or sulfonamide moieties in the presence of $Fe_3O_4@SiO_2@PCLH-TFA$ as catalyst.

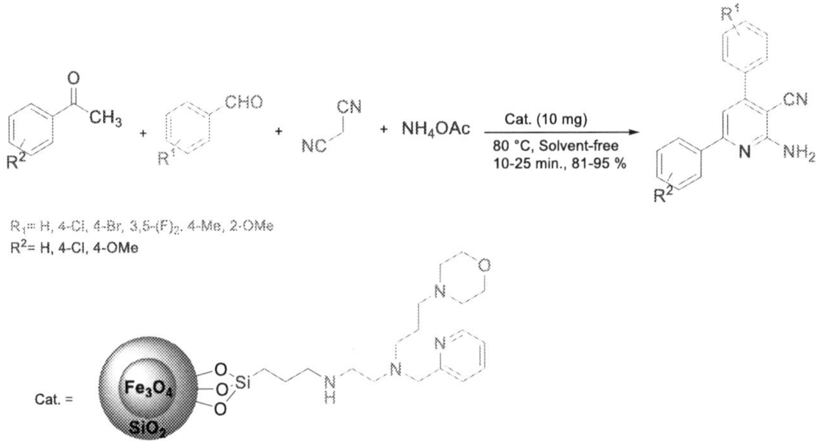

SCHEME 15.91 Synthesis of 2-amino-4,6-diphenylnicotinonitriles by using facile magnetic nanoparticles with morpholine tags as multirole catalyst.

anomeric-based-oxidation mechanistic pathway under benign reaction conditions with appropriate yields in short reaction times (Scheme 15.91).

$Fe_3O_4@SiO_2@(CH_2)_3$–urea–thiourea as a retrievable nanomagnetic hydrogen-bond and was prepared and its catalytic application was studied for the multicomponent synthesis of bipyridine-5-carbonitriles through a cooperative vinylogous anomeric-based oxidation (CVABO) mechanism under solvent-free conditions (Scheme 15.92). The applied catalyst can easily be recovered and reused four times without considerable destruction of its catalytic effectiveness [143].

Zolfigol group reported the synthesis of $Fe_3O_4@SiO_2@(CH_2)_3$–urea–benzimidazole sulfonic acid as an efficient reusable biological urea-based nanomagnetic catalyst for the four-component synthesis of 2-amino-3-cyano pyridines via a vinylogous anomeric-based oxidation mechanistic pathway in short reaction times and good yields (Scheme 15.93). This catalyst was recycled in the presence of an external magnet and reused for four further runs. Also, $Fe_3O_4@SiO_2@(CH_2)_3Im\}C(CN)_3$ as a nanostructured catalyst with ionic liquid tags was applied for the synthesis of 2-amino-3-cyano pyridine derivatives [144].

Also, the same team, applied $Fe_3O_4@SiO_2@(CH_2)_3$–urea–benzimidazole sulfonic acid as a catalyst for the synthesis of coumarin-linked nicotinonitrile derivatives through a cooperative vinylogous anomeric-based oxidation mechanistic way (Scheme 15.94). Titled compounds

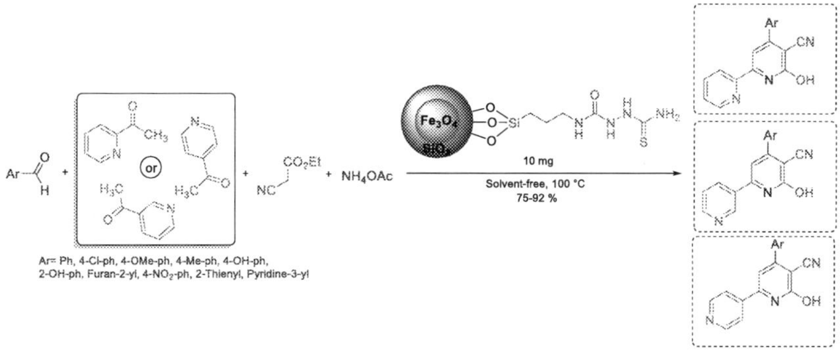

SCHEME 15.92 Synthesis of bipyridine-5-carbonitriles in the presence of $Fe_3O_4@SiO_2@(CH_2)_3$–urea–thiourea.

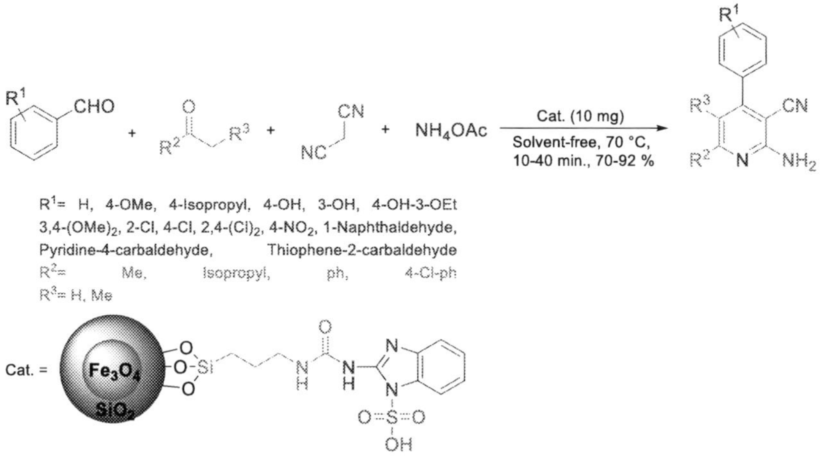

SCHEME 15.93 Synthesis of 2-amino-3-cyano pyridines catalyzed by $Fe_3O_4@SiO_2@(CH_2)_3$–urea–benzimidazole sulfonic acid.

prepared easily with appropriate yields and short reaction times under solvent-free conditions at 80°C. In an extension of the typical anomeric effect, this investigation offered vinyl heteroatoms can donate electron density from their lone pairs into the accepting σ^* orbital via an intervening π orbitals [145].

In another investigation [146] an ionically tagged magnetic nanoparticles possessing urea linkers, namely, $Fe_3O_4@SiO_2@(CH_2)_3$–urea–thiazole sulfonic acid chloride was synthesized and employed as an efficient catalyst for the synthesis of 2-aryl-quinoline-4-carboxylic acids from the reaction between aldehydes, pyruvic acid, and 1-naphthylamine through an anomeric-based oxidation mechanistic way under in the absence of any solvent (Scheme 15.95). The titled products were synthesized in appropriate yields and short reaction times. The described catalyst demonstrates elegant recovery and reusing potential in the investigated multicomponent reactions.

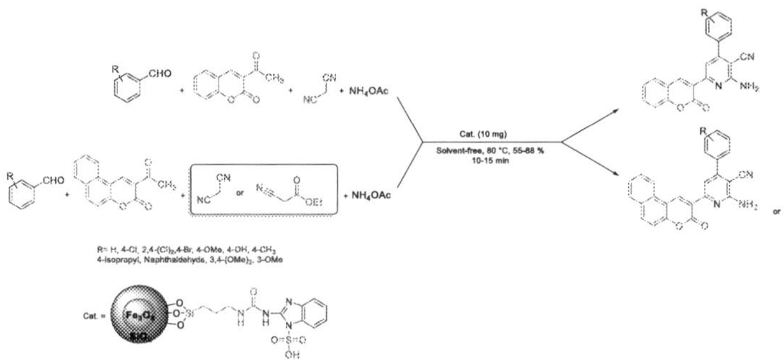

SCHEME 15.94 Synthesis of coumarin-linked nicotinonitriles in the presence of $Fe_3O_4@SiO_2@(CH_2)_3$–urea–benzimidazole with sulfonic acid moieties.

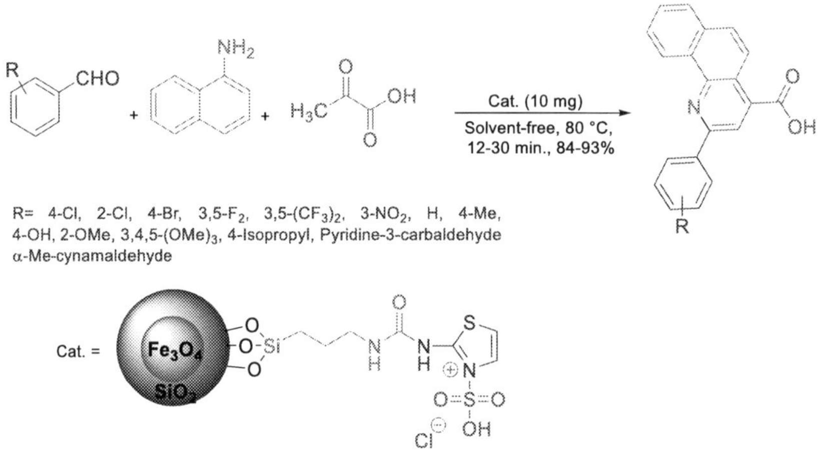

SCHEME 15.95 Synthesis of 2-aryl-quinoline-4-carboxylic acids catalyzed by $Fe_3O_4@SiO_2@(CH_2)_3$–urea–thiazole sulfonic acid chloride.

A nanomagnetic catalyst with $Cl[DABCO\text{-}NO_2]C(NO_2)_3$ tags was designed and synthesized by Zolfigol and coworkers [147]. The prepared nanomagnetic catalyst was employed as an expedient and recyclable catalyst for the one-pot three-component synthesis of pyrazolo[3,4-*b*]-pyridines via a condensation reaction between 3-methyl-1-phenyl-1*H*-pyrazol-5-amine or 3-methyl-1*H*-pyrazol-5-amine, aldehyde derivatives, and malononitrile for the synthesis of pyrazolo[3,4-*b*]-pyridines under neat conditions at 100°C with appropriate yields and good reaction times (Scheme 15.96). Informal work-up and reusability of catalyst are the main advantages of this approach. Lastly, an anomeric-based oxidation was proposed for the final step of the synthesis of desired compounds.

Also, Zolfigol et al. [148] reported the design and synthesis of an efficient nanomagnetic catalyst possessing phosphate linkers namely $Fe_3O_4@SiO_2@(CH_2)_3NH(CH_2)_2O_2P(OH)_2$ for the synthesis of benzo-[*h*]quinoline-4-carboxylic acids through a three-component reaction of aldehydes, naphthylamine, and pyruvic acid under solvent–solvent condition at 80°C with appropriate yields (Scheme 15.97). The main advantages of this study are the eco-friendly and green reaction conditions, simple separation and purification of related products, short

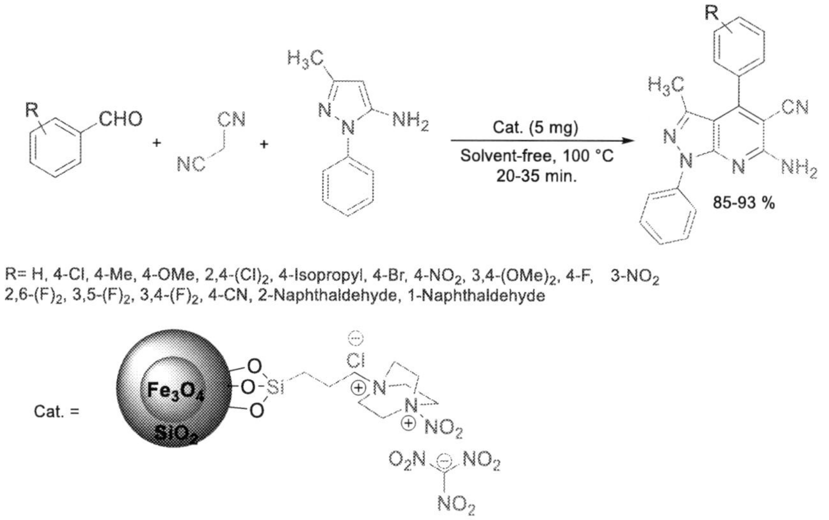

SCHEME 15.96 Synthesis of pyrazolo[3,4-b]pyridines catalyzed by $Fe_2O_3@SiO_2(CH_2)_3$–Cl[DABCO-NO_2] $C(NO_2)_3$ as a versatile nanomagnetic catalyst.

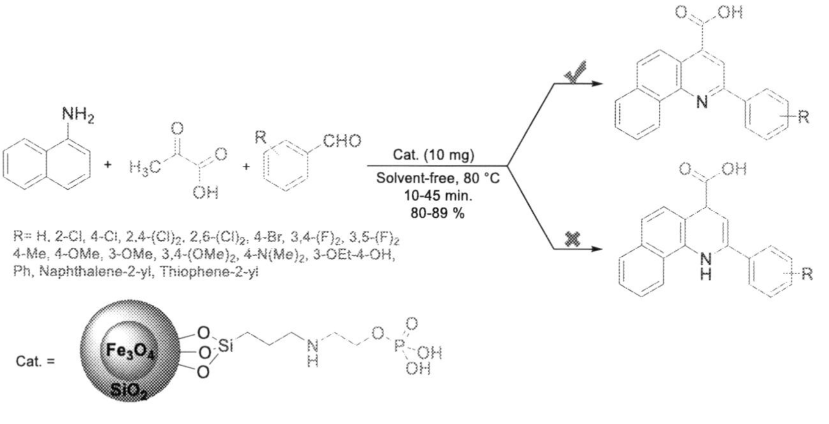

SCHEME 15.97 Synthesis of benzo-[*h*]quinoline-4-carboxylic acids catalyzed by $Fe_3O_4@SiO_2@(CH_2)_3NH(CH_2)_2$ $O_2P(OH)_2$.

reaction times, and easy separation and reusability of the catalyst. In the probable mechanism for the described synthesis, a cooperative anomeric-based oxidation was proposed for the final step.

Maleki et al. [149] report a facile, benign, and one-pot four-component process for the synthesis of highly substituted pyridines via the reaction of numerous aldehydes, malononitrile, cyclic ketones, and ammonium acetate by using a catalytic amount of cellulose-based magnetic composite nanocatalyst (Fe_3O_4/cellulose) at room temperature using ethanol as solvent (Scheme 15.98). The Fe_3O_4/cellulose nanocatalyst can be recycled effortlessly and reused various times without substantial loss of catalytic activity.

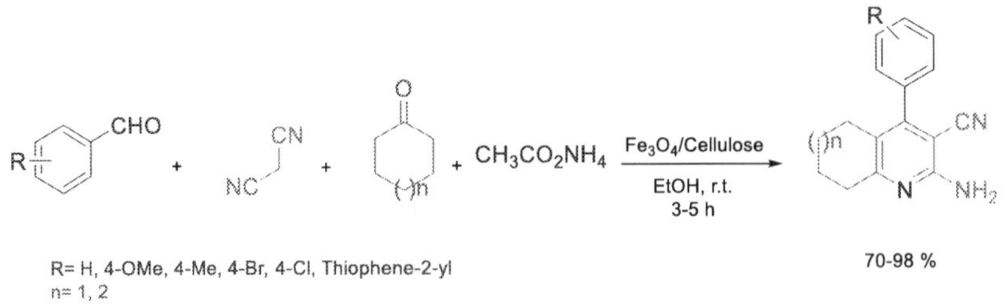

SCHEME 15.98 Nano Fe_3O_4/cellulose catalyzed synthesis of highly substituted pyridines.

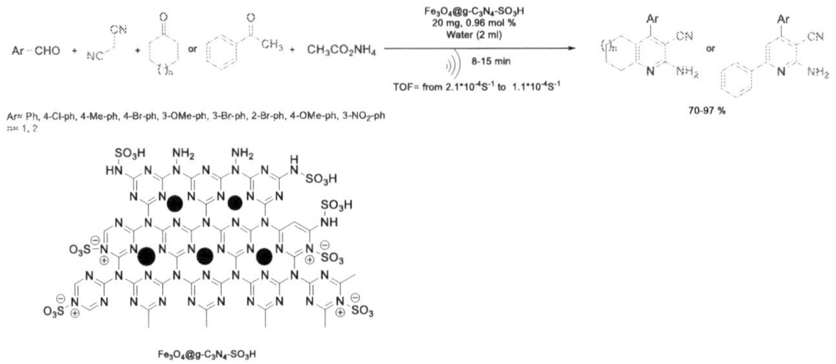

SCHEME 15.99 Synthesis of pyridines catalyzed by recyclable Fe_3O_4@g-C_3N_4–SO_3H.

A benign, efficient and effortlessly recyclable Fe_3O_4@g–C_3N_4–SO_3H as an influential, easy-to-use catalyst for the synthesis of pyridines under ultrasound irradiation has been investigated (Scheme 15.99). This one-pot four-component reaction affords the pyridine derivatives in suitable yields (70–97%) and short reaction times (8–15 min) [150]. In view of environmental attentions, this method suggests a straightforward and expedient route to produce biologically active pyridines in a green route.

A magnetic separable graphene oxide anchored sulfonic acid tags was used as a green and benign catalyst for the synthesis of 3,6-di(pyridin-3-yl)-1*H*-pyrazolo[3,4-*b*]pyridine-5-carbonitriles through one-pot three-component reaction between 1-phenyl-3-(pyridin-3-yl)-1*H*-pyrazol-5-amine, 3-oxo-3(pyridin-3-yl)propanenitrile and aldehyde derivatives in choline chloride/glycerol under microwave irradiation (Scheme 15.100) [151].

In 2019, highly substituted thiopyridine derivatives were prepared by using Fe_3O_4 supported Co (II) macrocyclic Schiff base ligand as a heterogeneous and reusable catalyst. In this method, thiophenol was condensed with aldehyde derivatives and malononitrile at 100°C under solvent-free conditions (Scheme 15.101) [152].

A straightforward synthetic protocol has been developed by Gajaganti and coworkers for the preparation of symmetrical pyridine derivatives in the presence of magnetic Fe_3O_4 nanoparticles. In this three-component reaction method, methyl arene derivatives reacted with acetophenone and ammonium acetate to provide desired molecules with good to excellent operational yields (Scheme 15.102) [153]. Also, in 2020, Fe_3O_4@GOTfOH/Ag/St-PEG-AcA has been applied as a recoverable catalyst for the synthesis of 2,4,6-triarylpyridines [154].

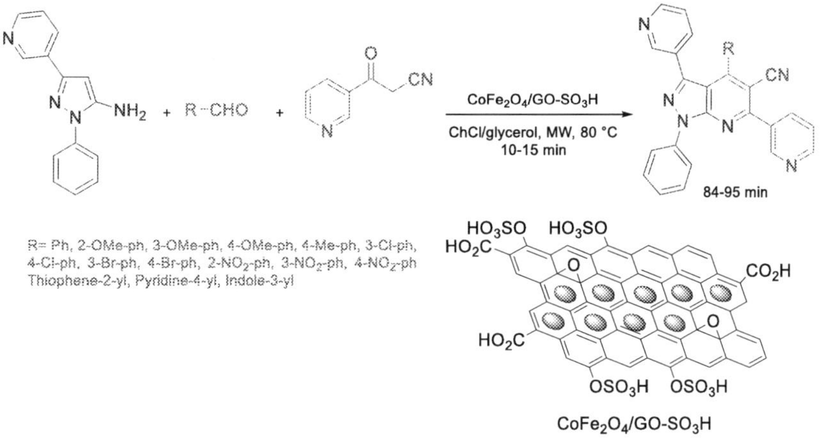

SCHEME 15.100 Synthesis of 3,6-di(pyridin-3-yl)-1*H*-pyrazolo[3,4-*b*]pyridine-5-carbonitriles by using $CoFe_2O_4$–GO–SO_3H.

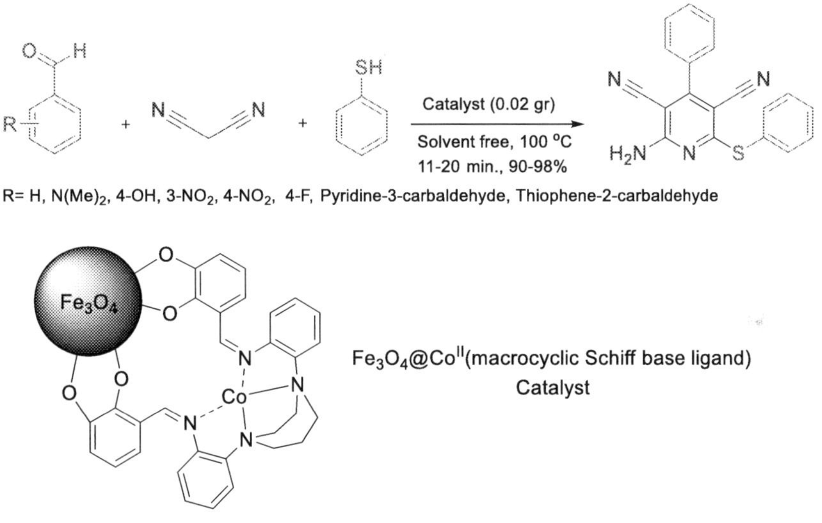

SCHEME 15.101 Fe_3O_4 supported Co (II) macrocyclic Schiff base ligand promoted the synthesis of thiopyridine derivatives.

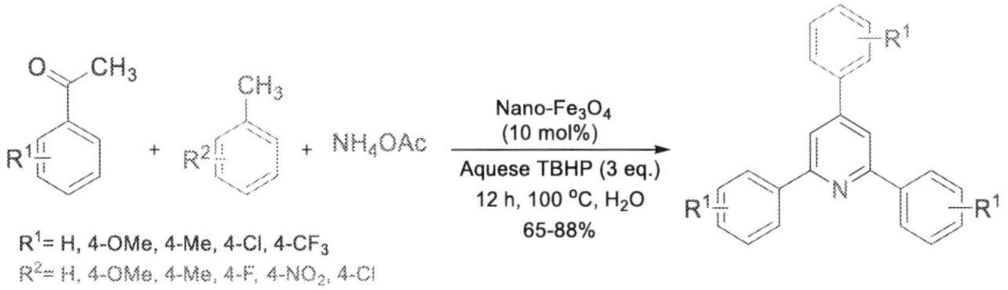

SCHEME 15.102 Preparation of symmetrical pyridines in the presence of magnetic Fe_3O_4 nanoparticles.

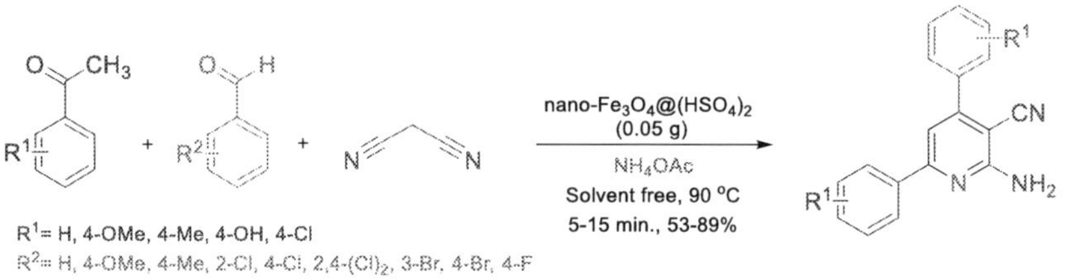

SCHEME 15.103 Nano-Fe_3O_4@$(HSO_4)_2$ prepared and applied as a solid acid catalyst for the synthesis of 2-amino-3-cyanopyridines.

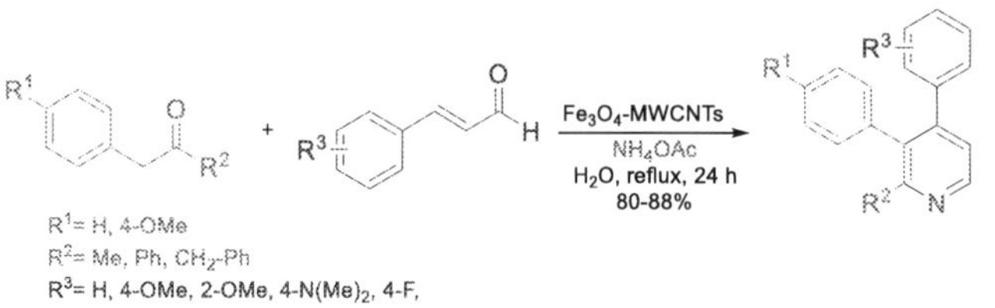

SCHEME 15.104 Fe_3O_4-multiwalled carbon nanotube nanoparticles promoted the synthesis of polyfunctionalized pyridine derivatives.

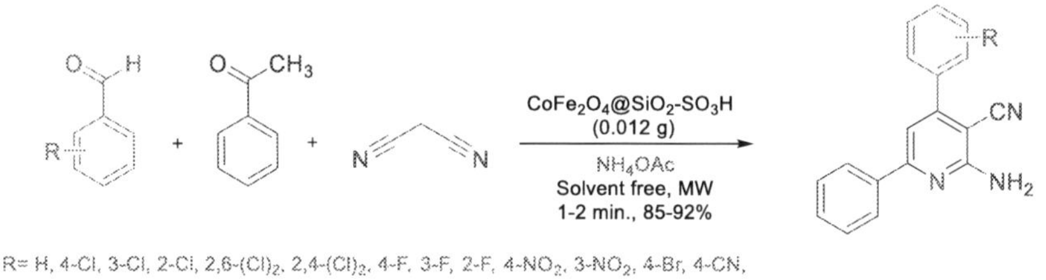

SCHEME 15.105 Catalytic activity of $CoFe_2O_4$@SiO_2–SO_3H for the preparation of 2-amino-4,6-diarylnicotinonitriles.

In another work, in a catalytic route by using nano-Fe_3O_4@$(HSO_4)_2$ as a solid acid catalyst, 2-amino-3-cyanopyridine derivatives were prepared through a condensation reaction between arylaldehyde derivatives, acetophenone, malononitrile, and NH_4OAc as nitrogen source (Scheme 15.103) [155]. In another work, $SrFe_{12}O_{19}$ efficiently catalyzed the preparation of these versatile molecules [156].

In 2019, Basavegowda and coworkers reported the synthesis and characterization of Fe_3O_4-multiwalled carbon nanotube nanoparticles. Then, they applied the resulting structure as a heterogeneous nanocatalyst for the preparation of polyfunctionalized pyridine derivatives in water. Benign reaction conditions, excellent yields, and easy recovering and reusing of the catalyst are the key features of the presented synthetic manner (Scheme 15.104) [157].

After synthesis and characterization by using several methods, $CoFe_2O_4$@SiO_2–SO_3H represents high capability as a reusable heterogenous solid acid catalyst for the preparation of 2-amino-4,6-diarylnicotinonitrile derivatives under microwave irradiation. The investigated procedure represents short reaction time, simple practical method, excellent reaction yields as significant features (Scheme 15.105) [158]. Also, APSMI@Fe_3O_4 magnetic nanocatalyst [159] and poly *N,N*-dimethylaniline-formaldehyde supported on silica-coated magnetic nanoparticles [160] were synthesized, characterized, and applied as a catalyst for the preparation of 2-amino-3-cayano pyridines.

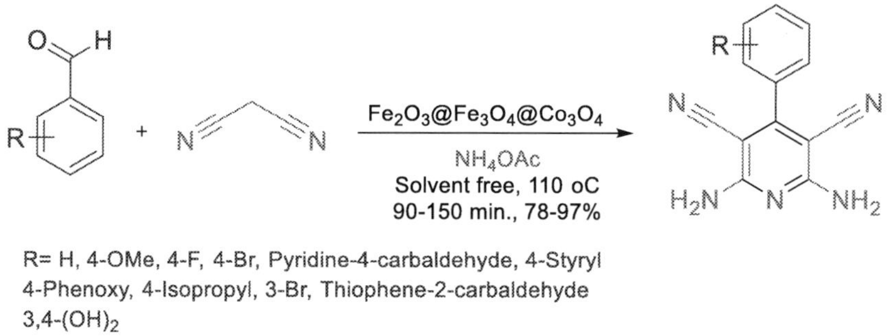

SCHEME 15.106 $Fe_2O_3@Fe_3O_4@Co_3O_4$ composite applied as catalyst for the synthesis of polysubstituted pyridines.

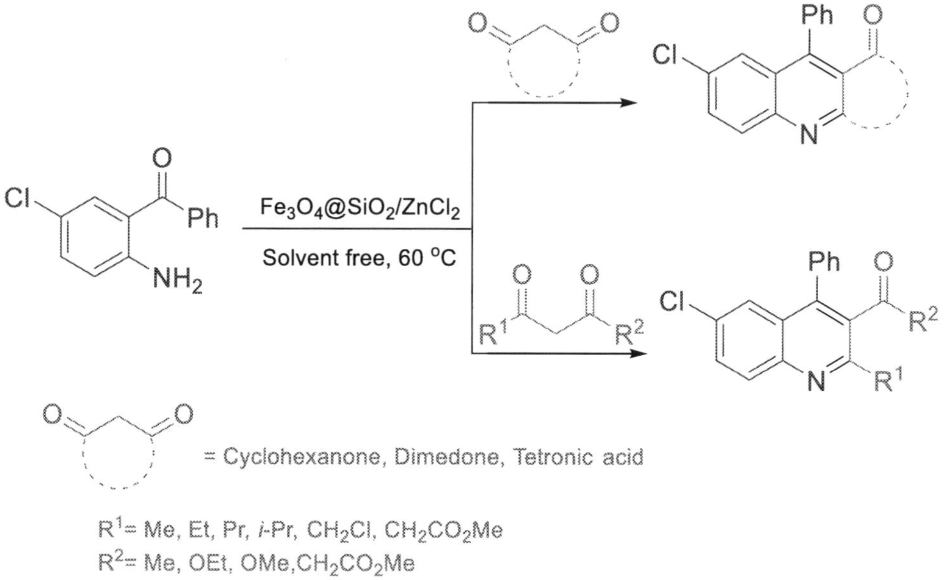

SCHEME 15.107 $Fe_3O_4@SiO_2/ZnCl_2$ promoted the synthesis of pyridine derivatives.

Maleki and coworkers reported the synthesis, characterization, and catalytic performance of the $Fe_2O_3@Fe_3O_4@Co_3O_4$ composite for the one-pot preparation of polysubstituted pyridine derivatives. The reported protocol provides the desired molecules through an easy work-up with high yields (Scheme 15.106) [161].

After supporting of $ZnCl_2$ on silica-coated magnetic nanoparticles, $Fe_3O_4@SiO_2/ZnCl_2$ was prepared and applied as a recoverable catalyst for the preparation of pyridine derivatives through Friedländer synthetic manner. In comparison with $ZnCl_2$ as homogenous catalyst, $Fe_3O_4@SiO_2/ZnCl_2$ shows higher catalytic activity with a high ability of recycling and reusing (Scheme 15.107) [162].

Chromeno[2,3-b]pyridines have tremendous incorporation in medicinal chemistry. In 2015, $Fe_3O_4@SiO_2–NH_2$ core–shell nanocomposite was introduced as a green and robust catalyst for

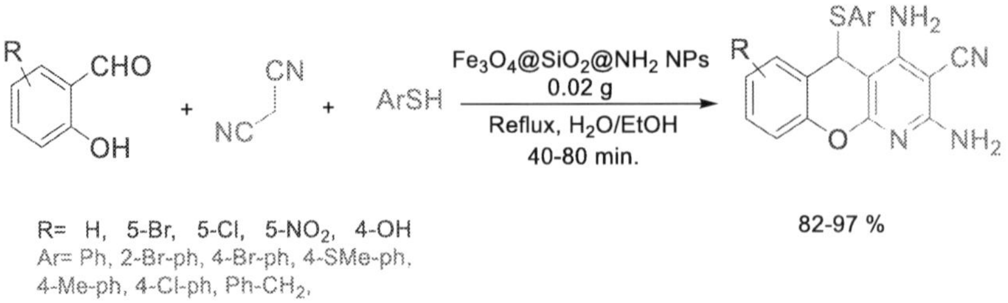

SCHEME 15.108 General procedure for the synthesis of highly substituted chromeno[2,3-*b*]pyridines in the presence of $Fe_3O_4@SiO_2–NH_2$ as catalyst.

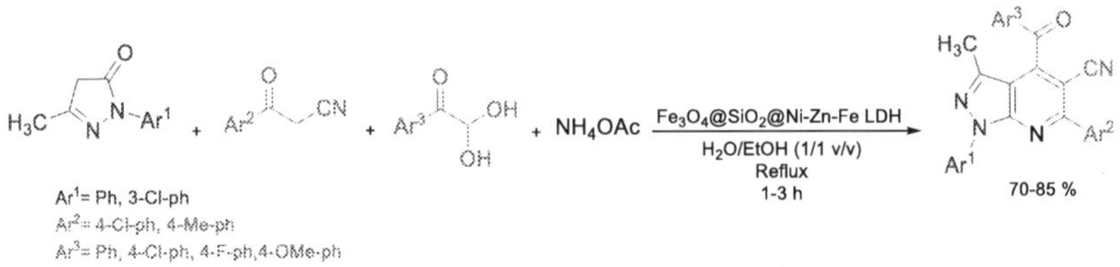

SCHEME 15.109 Synthesis of pyrazolo [3,4-*b*] pyridines using $Fe_3O_4@SiO_2@Ni–Zn–Fe$ LDH as catalyst.

the synthesis of highly substituted chromeno[2,3-b]pyridines via three-component reactions of salicylaldehydes, thiol derivatives, and malononitrile under refluxing of aqueous ethanol. This method has several benefits including straightforward, high yields (82–97%), short reaction times (40–80 min), little catalyst loading, and recovery and reusability of the catalyst (Scheme 15.108) [163].

In 2020, pyrazolo [3,4-*b*] pyridines were synthesized through a one-pot, four-component reaction method between 3-aryl-3-oxopropanenitriles, 1-aryl-3-methyl-1*H*-pyrazol-5 (4*H*) one, arylglyoxals, and ammonium acetate using $Fe_3O_4@SiO_2@Ni–Zn–Fe$ LDH (layered double hydroxide) as an efficient catalyst in aqueous ethanol as a green reaction media. This system has several benefits such as mild reaction conditions, high yields (70–85%), and easy workup (Scheme 15.109) [164].

In 2017, Yazdani-Elah-Abadi and coworkers reported vitamin B_3 functionalized superparamagnetic nanoparticles ($Fe_3O_4@Niacin$) as a green and drastic heterogeneous biocatalyst for the synthesis of 2-amino-3-cyanopyridine derivatives using multicomponent reaction strategy between ketones, malononitrile, ammonium acetate, and aromatic aldehydes bearing electron-withdrawing and electron-releasing substituents under microwave irradiation in water as a green solvent. High yields (73–95%), appropriate reaction times (7–10 min), and the use of environmentally friendly solvent and catalyst are important advantages of this approach (Scheme 15.110) [165].

The use of $Fe_3O_4@SiO_2$ decorated L-proline as a magnetically heterogeneous catalyst was studied by Maleki and coworker for the synthesis of 2,4,6-triarylpyridine derivatives via a one-pot, multicomponent reaction of substituted acetophenone, aryl aldehydes with electron-withdrawing and electron-releasing substituents and ammonium acetate as nitrogen source. Here, all of pyridine derivatives was synthesized under solvent-free conditions, low temperature (60°C), and have an excellent yield (75–94%) (Scheme 15.111). The recoverability

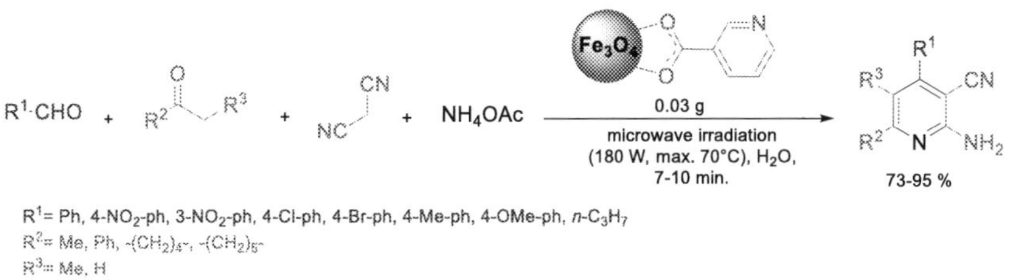

SCHEME 15.110 A straightforward procedure for the synthesis of 2-amino-3-cyanopyridine derivatives in the presence of Fe_3O_4@Niacin as catalyst under microwave irradiation.

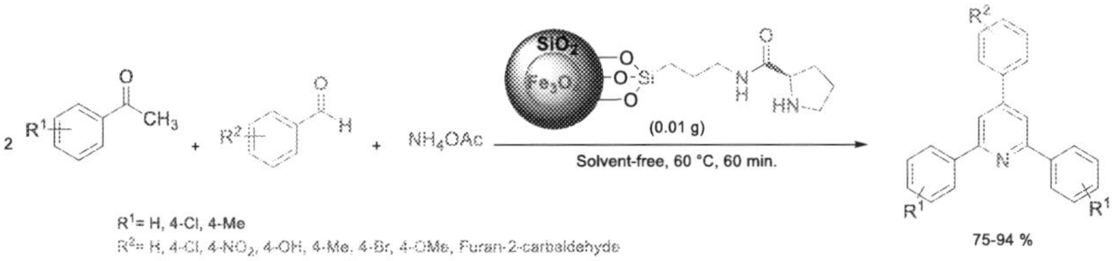

SCHEME 15.111 Synthesis of 2,4,6-triarylpyridines by Fe_3O_4@SiO_2 decorated L-proline as catalyst.

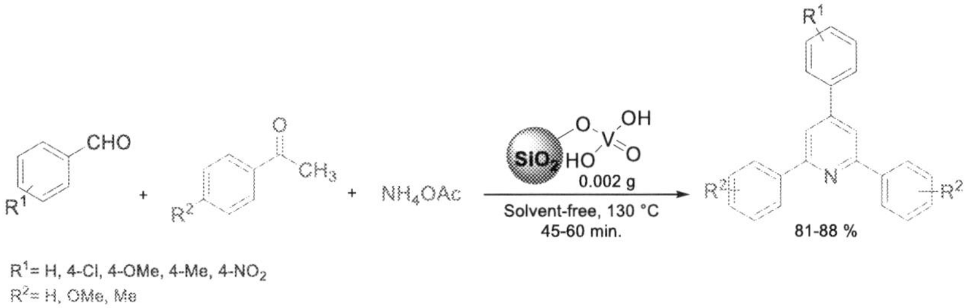

SCHEME 15.112 Synthesis of 2,4,6-triarylpyridine catalyzed by [SiO_2–$VO(OH)_2$].

of catalyst was investigated and recycled several times without any significant reduction in activity [166].

15.6.3 Silica-supported catalysts

Silica-bonded vanadic acid [SiO_2–$VO(OH)_2$] (SVA) catalyzed the facile and mild synthesis of 2,4,6-triarylpyridine via the condensation reaction between ammonium acetate, aldehydes, and numerous acetophenones in appropriate yields (81–88%) and short reaction times (45–60 min) [167]. The main advantages of the described process are cleaner reaction profile, reusability of the [SiO_2–$VO(OH)_2$] catalyst which generates it in close influence with the green chemistry disciplines (Scheme 15.112).

Thiel and Hapke [168] report the first solid-supported CpCo(I)-catalyst for the synthesis of pyridines. For the investigation of catalytic property of this catalyst, the reaction between 1,7-heptadiyne (**9**) and benzonitrile (**10**) was studied. Upon discovering that the sol–gel-supported catalysts SG-(8) and SGL-(8) did not show any catalytic activity toward the [2 + 2 + 2] cycloaddition reaction under thermal or microwave conditions, **8** was supported on silica

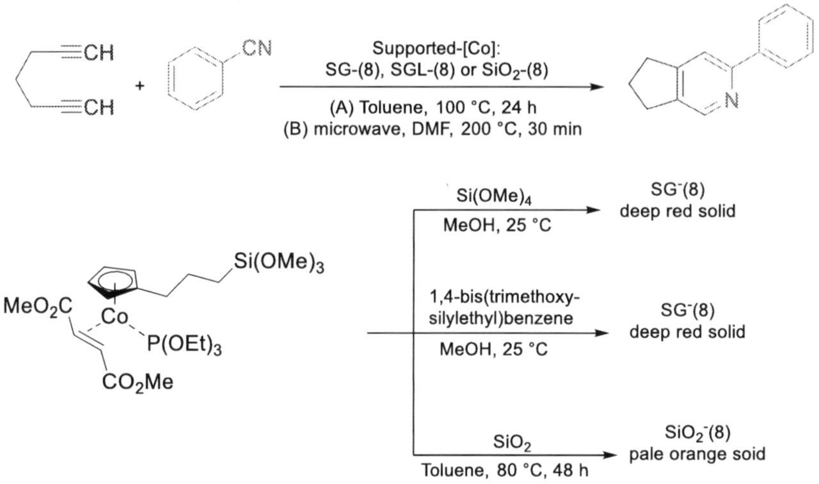

SCHEME 15.113 Catalytic activity of solid phase-supported complex **8** for the synthesis of pyridine.

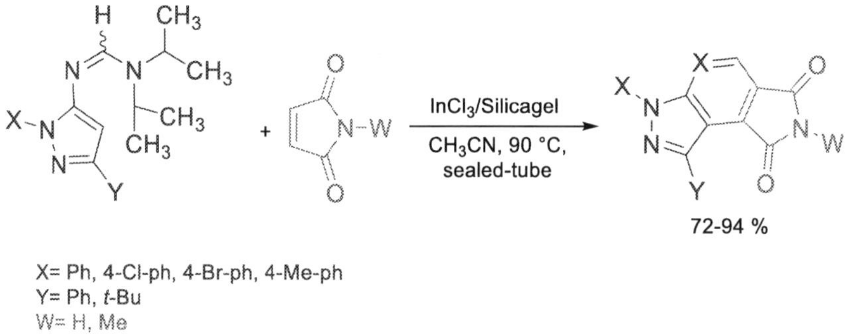

SCHEME 15.114 Synthesis of pyrazolo[3,4-*b*]pyrrolo[3,4-*d*]pyridines via Aza-Diels–Alder reaction catalyzed by $InCl_3/SiO_2$.

instead. SiO_2-(**8**) confirmed capable catalytic activity in the cyclization reaction (21% yield under thermal conditions, 47% under microwave conditions) and was effortlessly removable from the reaction mixture by filtration. Whereas the sol–gel-entrapped catalyst formed no cycloaddition product, the silica-bound analogue led to the preparation of the pyridine derivative, representing the first solid-supported CpCo(I)-complex to catalyze the cyclotrimerization of a diyne and a nitrile (Scheme 15.113).

Lin et al. [169] have effectively established an indium (III) chloride/silica gel (20% wt.) catalyzed Aza-Diels–Alder reaction of *N*,*N*-diisopropylamidinyl pyrazoylimines with maleimide or *N*-methylmaleimide. Pyrazolo[3,4-*b*]pyrrolo[3,4-*d*]pyridines were attained in appropriate yields (72–94%). Additionally, they confirmed the evidence that Lewis acidic indium (III) chloride/silica gel have more catalytic activity than Brønsted–Lowery acid in the Aza-Diels–Alder reaction (Scheme 15.114).

In another investigation, the authors reported the synthesis and catalytic application of (MCM-41)-pyridine nanopowder at the preparation of polysubstituted pyridine derivatives

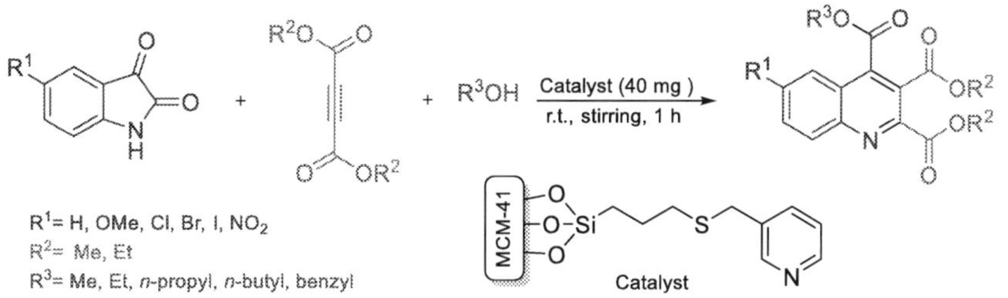

SCHEME 15.115 Catalytic synthesis of polysubstituted pyridine derivatives.

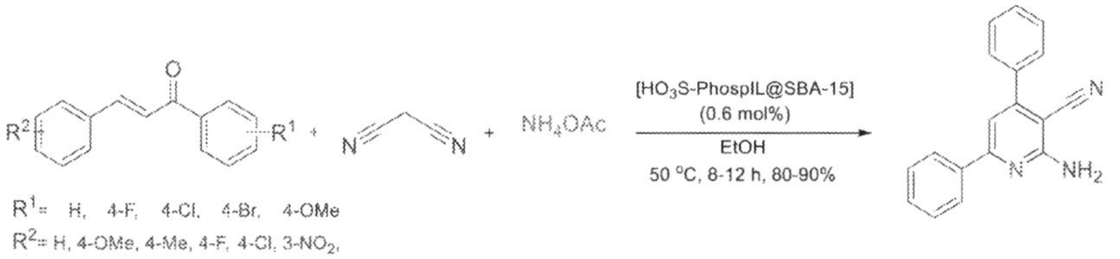

SCHEME 15.116 [HO3S-PhospIL@SBA-15] applied as catalyst for the preparation of 2-amino-3-cyano-4,6-diarylpyridines.

using isatin derivatives and dialkylacetylenedicarboxylates via a domino approach. This synthetic manner is privileged by benign reaction conditions, simplicity of target molecules separation and purification protocol, recovering and reusing capability of the catalyst and reasonable obtained yields (Scheme 15.115) [170].

[HO3S-PhospIL@SBA-15] was applied as an efficient and recyclable catalyst for the construction of 2-amino-3-cyano-4,6-diarylpyridine derivatives by using different chalcones, malononitrile, and ammonium acetate under mild conditions in EtOH and green catalyst (Scheme 15.116) [171].

In another work, Rahmani and coworkers reported the synthesis of 2-amino-3-cyano-pyridine derivatives by using propylphosphonium hydrogen carbonate ionic liquid supported on nanosilica as catalyst under solvent-free reaction conditions (Scheme 15.117) [172].

Lin and coworkers developed an efficient one-pot method for the synthesis of coumarin-fused pyrazolo[3,4-*b*]pyridine derivatives by utilizing silica sulfuric acid (SSA) as catalyst under microwave irradiation. It is worthy to mention that, the authors applied alcohol derivatives as one of the starting materials in their practical procedure. The key elements in this report are moderate-to-good yields, relatively short reactions times, and applying eco-compatible solvent and catalyst (Scheme 15.118) [173].

15.7 Organocatalysis

In another study, trityl chloride (Ph_3CCl) was used as a catalyst for presenting a convenient and versatile approach for the synthesis of 2,4,6-triarylpyridines through a one-pot pseudo four component condensation reaction between aldehydes, acetophenones, and ammonium acetate under solvent-free conditions (Scheme 15.119). Mechanistically, it is fascinating that in situ generation of trityl carbocation from trityl chloride (Ph_3C^+) assists the reaction. The

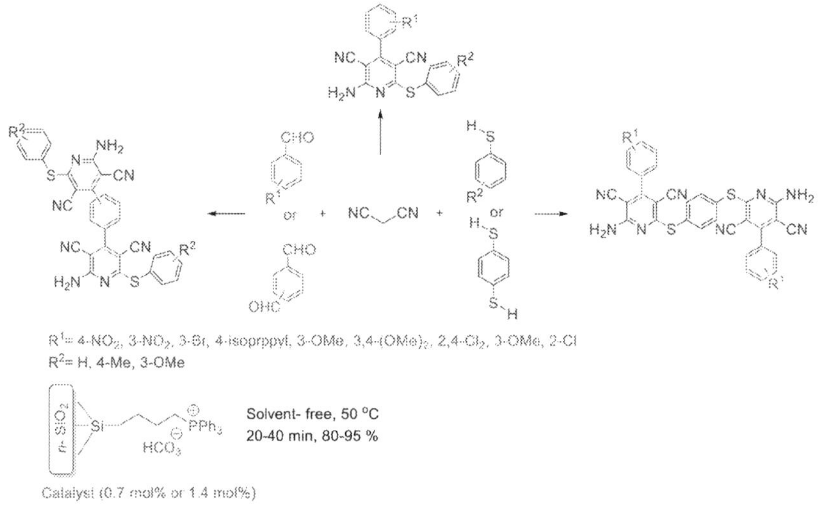

SCHEME 15.117 PPHC–nSiO$_2$ applied as catalyst for the synthesis of 2-amino-3-cyano-pyridines.

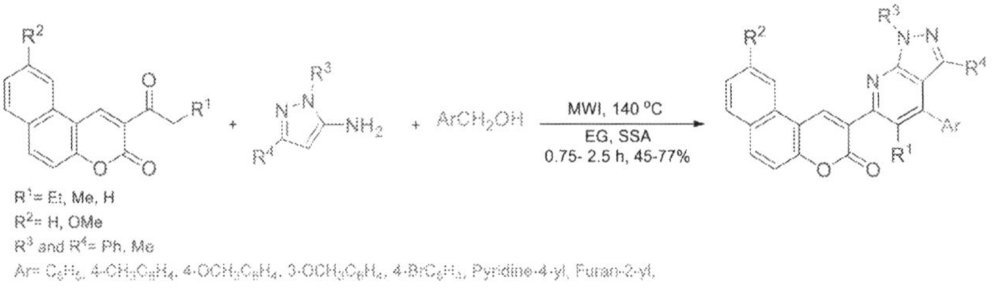

SCHEME 15.118 Synthesis of coumarin-fused pyrazolo[3,4-*b*]pyridine in the presence of SSA.

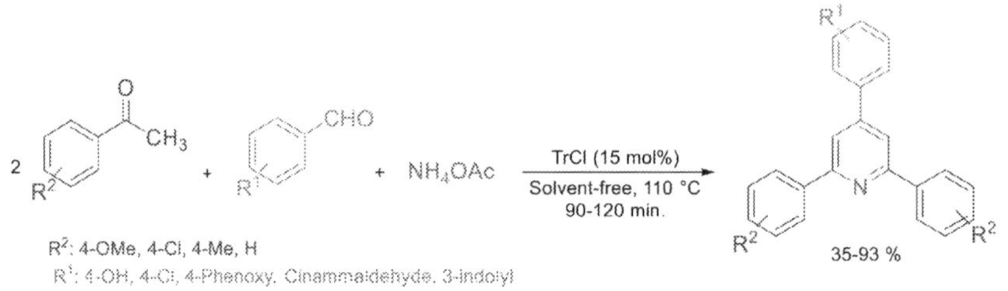

SCHEME 15.119 Synthesis of 2,4,6-triarylpyridines and 3-(2,6-diarylpyridin-4-yl)-1*H*-indoles by using TrCl as a natural catalyst.

hopeful points of this catalytic method are simplification, effectiveness, good yields (35–93%), short reaction times (90–120 min), and appropriate compliance with the green chemistry processes [174].

Chitosan as a heterogeneous organocatalyst applied in the Guareschi–Thorpe type method for the synthesis of alicyclic[*b*]-fused pyridines [175]. This approach is operationally appropriate and displayed a broad variety of functional group tolerance and substrate compatibility,

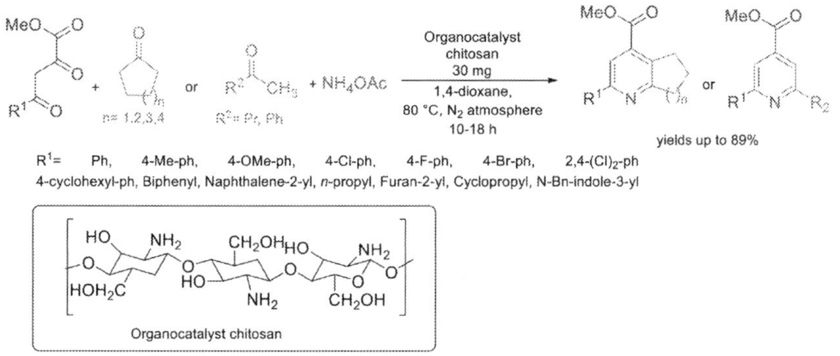

SCHEME 15.120 Synthesis of alicyclic[*b*]-fused pyridines by using chitosan as a catalyst via Guareschi–Thorpe type approach.

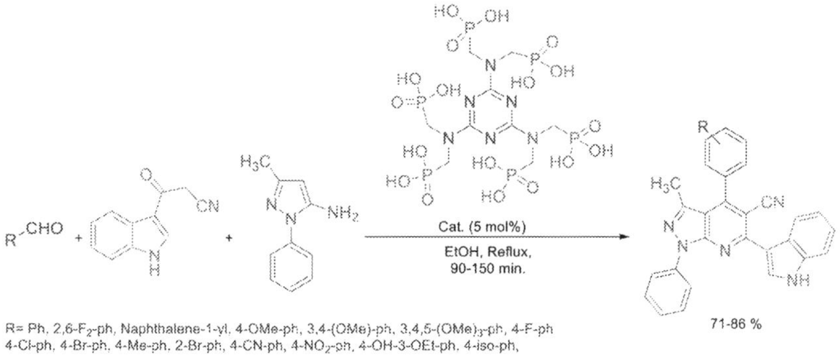

SCHEME 15.121 MHMHPA catalyzed synthesis of (3′-indolyl)pyrazolo[3,4-b]pyridines.

low catalyst loading, and catalyst reusing capability. Using optimized reaction conditions, numerous substituted diketo-esters, various carbonyl compounds, and ammonium acetate were reacted with each other and provided the desired alicyclic[*b*]-fused pyridines (Scheme 15.120).

15.7.1 Solid acids

Melamine hexakis(methylene)hexakis(phosphonic acid) (MHMHPA) as a heterogeneous catalyzed one-pot reaction between cyanoacetylindole, 3-methyl-1-phenyl-1*H*-pyrazol-5-amine, and aldehydes (Scheme 15.121). (3′-Indolyl)pyrazolo[3,4-*b*]pyridines were synthesized in appropriate yields (71–86%) and suitable reaction times (90–150 min) in refluxing ethanol [176].

15.7.2 Carbon nanotube-supported catalysts

Preparing procedure of *N*-doped carbon nanotubes (NCNTs) modified with copper nanoparticles via high temperature technique and its catalytic application for the synthesis of pyridine derivatives was reported by Gupta and coworkers. Although copper nanoparticles

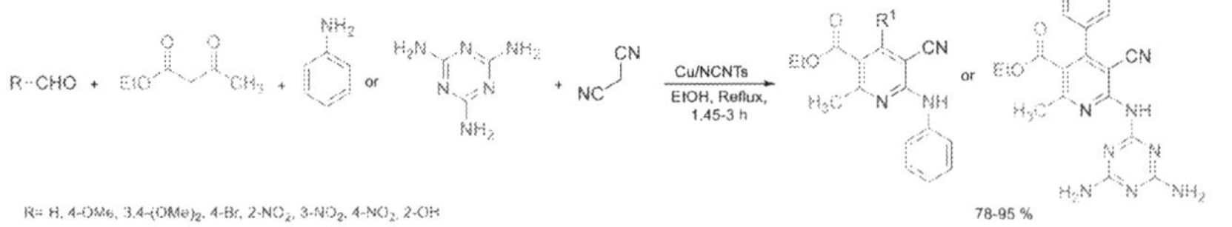

SCHEME 15.122 Experimental procedure for the synthesis of pyridine derivatives with Cu/NCNTs as catalyst.

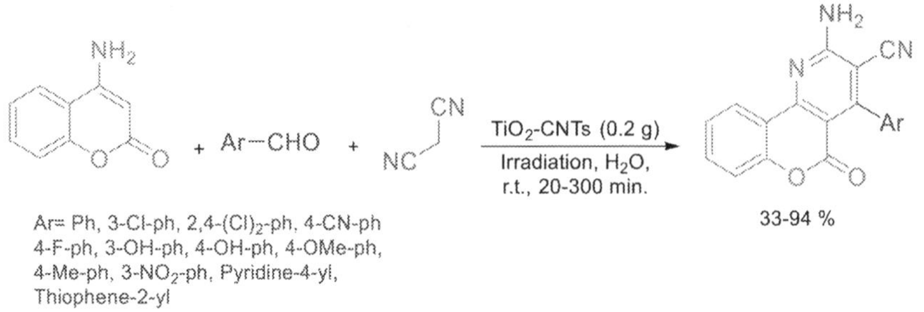

SCHEME 15.123 Synthesis of chromeno[*b*]pyridine derivatives in the presence of TiO_2–CNTs.

are not very stable under normal conditions, they find significant stability using NCNTs as a supporting system. This approach has considerable shorter times of reaction (1.45–3 h), and good yields (78–95%) in comparison of related previous works and also, ethyl 6-(4,6-diamino-1,3,5-triazin-2-yl amino)-5-cyano-2-methyl-4-phenylpyridine-3-carboxylate was synthesized for the first time (Scheme 15.122) [177].

Abdolmohammadi and coworkers introduced a straightforward and efficient method for the catalytic synthesis of chromeno[*b*]pyridine derivatives via three-component reaction between 4-aminocoumarin, aromatic aldehydes, and malononitrile using carbon nanotubes decorated TiO_2 nanoparticles (TiO_2-CNTs) as privileged heterogeneous catalyst under irradiation conditions. The main advantages of this protocol are including the use of H_2O as green solvent, short reaction times (20–300 min), easy work-up procedure and good yields (33–94%) (Scheme 15.123) [178].

15.7.3 Miscellaneous

Safaiee et al. [179] reported chitosan-supported oxo-vanadium (5 mg) as a green and recyclable biopolymer catalyst for the synthesis of 2,4,6-triarylpyridines via three-component reaction between aldehydes, ketones, and ammonium acetate under solvent-free condition at 130°C in good yields (53–88%) and short reaction times (55–75 min) (Scheme 15.124).

A composite cocrystalline zeolite HZSM-5/11(78) was synthesized and studied in the conversion of glycerol with ammonia to pyridine bases (pyridine, 2-methylpyridine, 3-methylpyridine) [180]. The HZSM-5/11(78) displayed appropriate catalytic activity compared to other zeolites with similar Si/Al ratios for example HZSM-5(80), HZSM-11(80) and the physical mixture of HZSM-5(80) and HZSM-11(80). The results showed that the proper catalytic activity of HZSM-5/11(78) was linked to the higher surface area and coexistence of a suitable ratio of Lewis and Brønsted tags. The optimum conditions for synthesizing pyridine

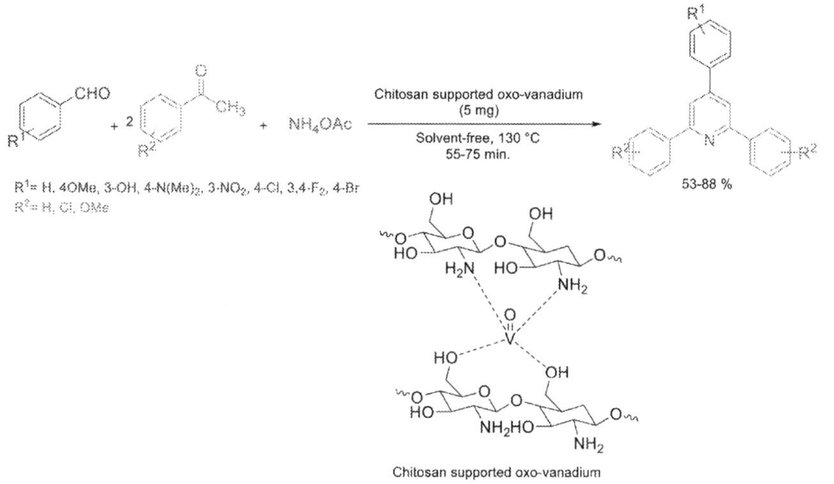

SCHEME 15.124 Synthesis of 2,4,6-triarylpyridines catalyzed by chitosan-supported oxo-vanadium (ChVO).

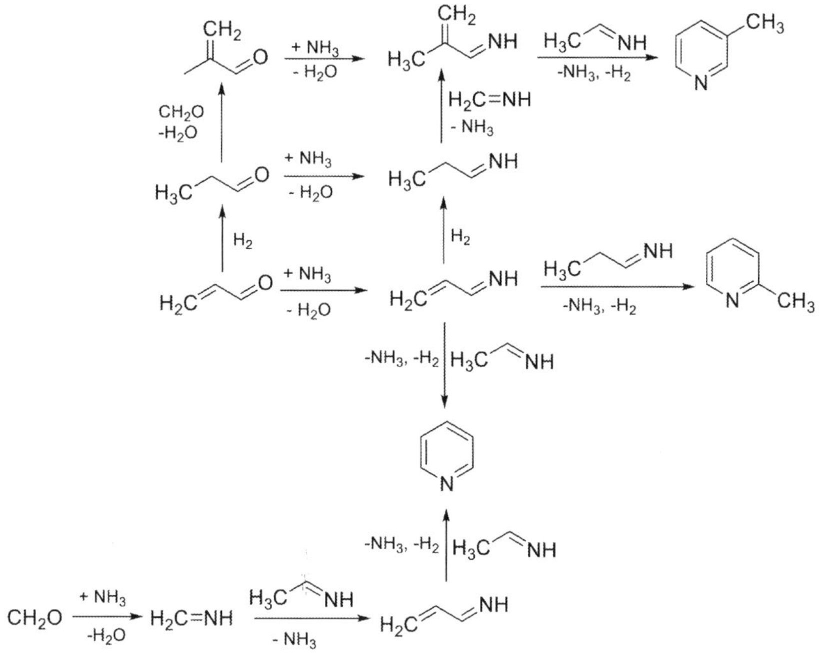

SCHEME 15.125 Synthesis of pyridine bases by using cocrystalline zeolite HZSM-5/11 (**78**).

bases from glycerol with ammonia over this catalyst were investigated, containing a reaction temperature of 520°C, 0.1 MPa pressure with a molar ratio of ammonia to glycerol of 12:1, and a GHSV of 300 h^{-1} (Scheme 15.125).

A graphene oxide–TiO_2 composite (GO–TiO_2) has been synthesized as an efficient, benign, and recyclable catalyst for the synthesis of pyridines via the reaction between aldehydes,

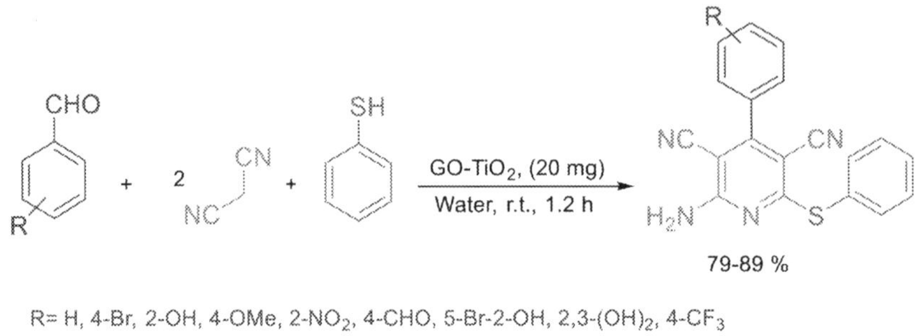

SCHEME 15.126 Synthesis of pyridine derivatives catalyzed by GO–TiO_2.

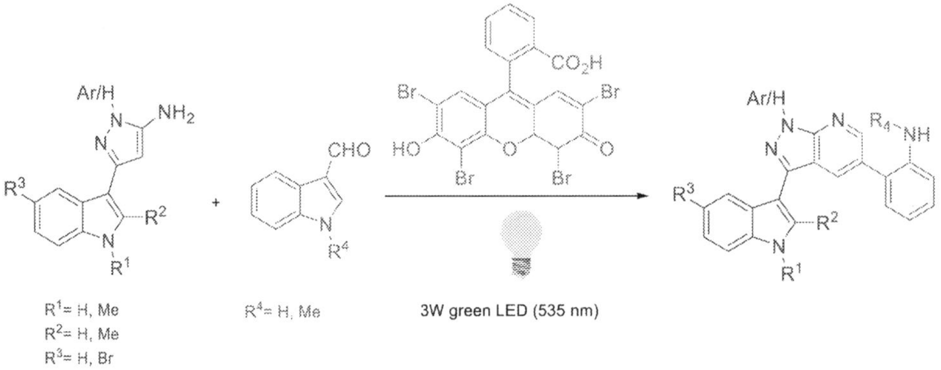

SCHEME 15.127 Synthesis of heterobiaryl-pyrazolo[3,4-*b*]pyridines by using dual Lewis acid and oxidative photocatalysis.

thiophenol, and malononitrile in short reaction times (60–120 min) and high isolated yields (79–89%) in aqueous medium at room temperature (Scheme 15.126). This catalyst is inexpensive, effortlessly recycled after the reaction, and reused up to five times without any important loss in catalytic activity [181].

It is reported that the mixture of Eosin Y and $Yb(OTf)_3$ efficiently catalyzes the electrocyclization and indole-ring opening under photocatalytic conditions at room temperature (Scheme 15.127). As compared to the existing approaches, this catalytic system represents a facile and mild pathway for the synthesis of heterobiaryl-pyrazolo[3,4-*b*]pyridine derivatives in appropriate yields [182].

Zhang et al. [183] have studied a facile, benign, and useful process for the preparation of 2,4-diamino-5-(1*H*-indol-3-yl)-5*H*-chromeno[2,3-*b*]pyridine-3-carbonitriles through one-pot pseudo four-component reaction between salicylaldehydes, malononitrile, and indole under visible light irradiation (Scheme 15.128). The reaction progresses in aqueous ethyl lactate under catalyst-free conditions to provide the probable products in suitable yields (70–93%) with a wide variety of salicylaldehydes and indole. This method occurs the necessities of green chemistry and opens a path for the improvement of more viable multicomponent reactions for synthesizing biologically and synthetically significant molecules under visible light irradiation. Also, this process is appropriate to the synthesis of compound A [MK^{-2} inhibitor].

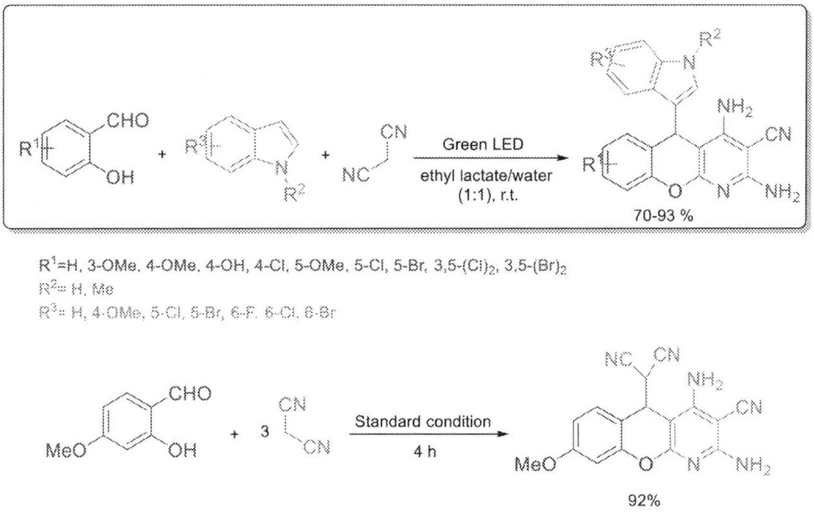

SCHEME 15.128 Synthesis of 2,4-diamino-5-(1*H*-indol-3-yl)-5*H*-chromeno[2,3-*b*]pyridine-3-carbonitriles via visible light initiated and synthesis of MK^{-2} inhibitor (A) via visible light initiated.

When 2-hydroxy-4-methoxy benzaldehyde reacted with three equivalents of malononitrile under similar reaction conditions, MK^{-2} inhibitor was achieved in 92% yield.

Stratakis et al. [184] reported the catalytic capability of supported Au nanoparticles on TiO_2, in oxidative cyclization reaction between conjugated allenones or allenyl esters and propargylamines or homopropargylamine, producing substituted 3-keto pyridines (60–82%; 12–32 h) or 4-picolines (55–85%; 12–36 h) (Scheme 15.129). The identified cyclization pathway is different from that happening under homogeneous ionic Au-catalysis. This catalytic medium is simply accessible and the procedure is purely heterogeneous, allowing its easy recycling and reusing. The enaminones are synthesized in situ upon mixing a conjugated allenone or allenyl ester with the alkynylamine, therefore the pyridine-producing transformation is typically a one-pot manner.

In 2019, pyridine dicarbonitrile derivatives were synthesized by Gholizadeh and coworkers using TiO_2-nanoparticles as heterogeneous catalysts. This system was prepared in an ordinary and magnetized process. The two mentioned protocols have a good result for the synthesis of pyridine dicarbonitrile derivatives. Nevertheless, the synthesis of nano-TiO_2 through a magnetized process, due to good performance as Lewis acid, revealed better catalytic activity (Scheme 15.130). This system has better yields of products and shorter reaction times. Moreover, magnetized nano-TiO_2 is inexpensive, storable, and sustainable [185].

In another work, titanium silicate was applied as a solid acid catalyst in the solvent-free and one-pot manner to 2,4,6-triaryl pyridine derivatives. This method demonstrates good recovering and reusing capability for the catalyst (Scheme 15.131) [186].

In 2019, Yang and coworkers present chemoselective catalytic synthesis for the preparation of 2,4,6-trisubstituted pyridine derivatives by utilizing merrifield resin-supported quinone as a recoverable heterogeneous biomimetic catalyst. Good catalytic performance, chemoselectivity of the protocol, and mild reaction conditions are the key advantages of the method (Scheme 15.132) [187].

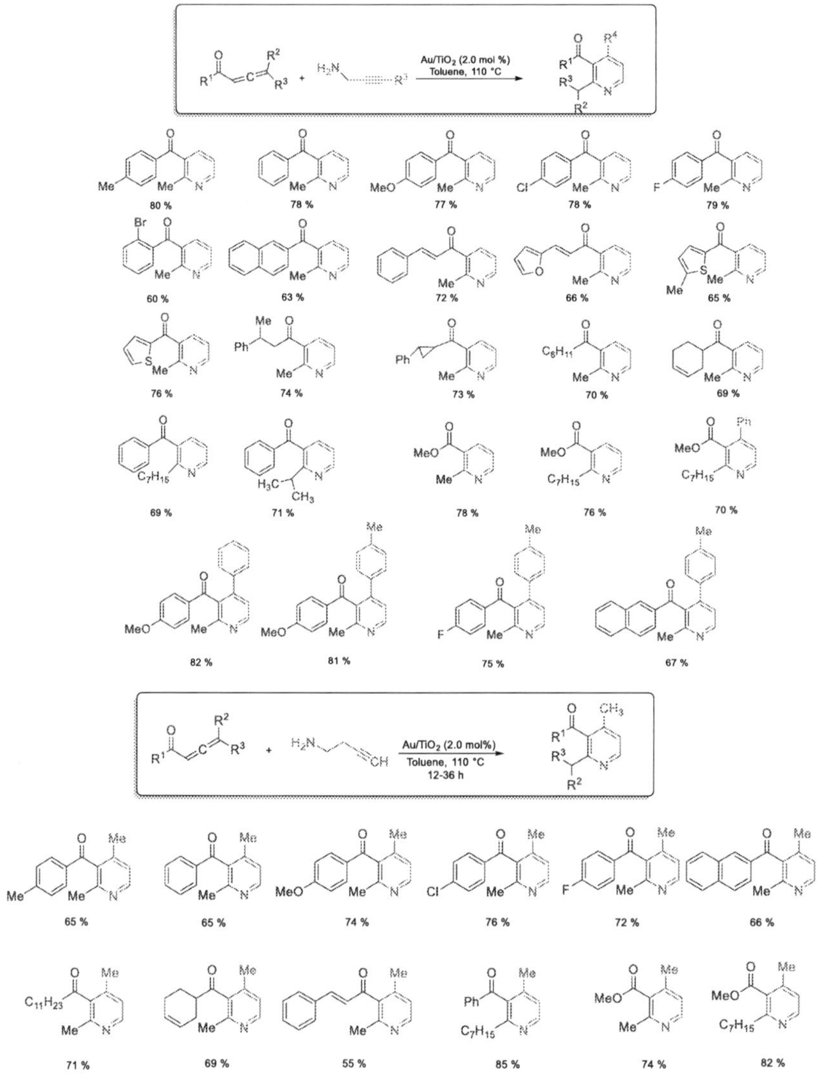

SCHEME 15.129 Synthesis of 3-keto pyridines and 4-picolines in the presence of Au/TiO_2.

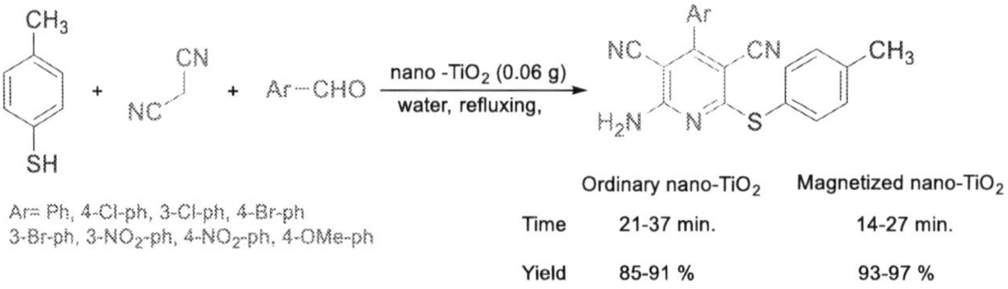

SCHEME 15.130 Synthesis of pyridine dicarbonitrile derivatives using two types of nano-TiO_2 as catalyst.

SCHEME 15.131 Titanium silicate promoted the synthesis of 2,4,6-triaryl pyridines.

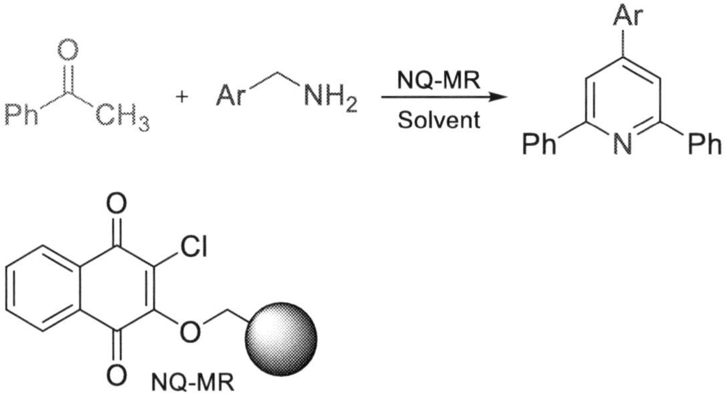

SCHEME 15.132 Chemoselective synthesis of 2,4,6-trisubstituted pyridine derivatives.

15.8 Conclusion

Among the nitrogen containing heterocyclic molecules, pyridine derivatives have a unique position. This special position refers to their various applications in diverse chemical domains such as coordination chemistry, supramolecular structures, polymers, catalysis, and also, their undeniable biological activities. Therefore, the synthesis and evaluation of the applications of these versatile molecules are an excellent target for academic and industrial chemists. In this way, we tried to collect the varied catalytic routes including homogeneous and heterogeneous catalytic systems, which were recently applied for the synthesis of pyridine derivatives in this chapter. However, many efforts have been made in past decades for the synthesis of these versatile compounds and it is very difficult to collect all of these methods in a single chapter or a review paper, herein we have tried to collect some of the newest methods [188]. We hope we have done our duty for these valuable compounds.

Acknowledgments

Authors thank the Bu-Ali Sina University for financial support to our research group.

References

[1] D.F. Fischer, R. Sarpong, Total synthesis of (+)-complanadine a using an iridium-catalyzed pyridine C-H functionalization, J. Am. Chem. Soc. 132 (17) (2010) 5926–5927.

[2] E. Belhadj, A. El-Ghayoury, M. Mazari, M. Salle, The parent tetrathiafulvalene-terpyridine dyad: synthesis and metal binding properties, Tetrahedron Lett. 54 (24) (2013) 3051–3054.

[3] Y. Wang, A. Daniher, A. De Klerk, C. Winkel, Pyridine derivatives with umami flavour. US 20120121783, May 17 (2012).

[4] H. Sasabe, J. Kido, Multifunctional materials in high-performance OLEDs: challenges for solid-state lighting, J. Chem. Mater. 23 (3) (2011) 621–630.

[5] Z.Y. Li, X.Q. Huang, F. Chen, C. Zhang, X.Y. Wang, N. Jiao, Cu-catalyzed concise synthesis of pyridines and 2-(1*H*)-pyridones from acetaldehydes and simple nitrogen donors, Org. Lett. 17 (3) (2015) 584–587.

[6] H.D. Khanal, Y.R. Lee, Organocatalyzed oxidative *N*-annulation for diverse and polyfunctionalized pyridines, Chem. Commun. 51 (46) (2015) 9467–9470.

[7] L. Shen, S. Cao, J. Wu, J. Zhang, H. Li, N. Liu, X. Qian, A revisit to the Hantzsch reaction: unexpected products beyond 1, 4-dihydropyridines, Green Chem. 11 (9) (2009) 1414–1420.

[8] A.V. Gulevich, A.S. Dudnik, N. Chernyak, V. Gevorgyan, Transition metal-mediated synthesis of monocyclic aromatic heterocycles, Chem. Rev. 113 (5) (2013) 3084–3213.

[9] G. Zeni, R.C. Larock, Synthesis of heterocycles via palladium-catalyzed oxidative addition, Chem. Rev. 106 (11) (2006) 4644–4680.

[10] J.A. Varela, C. Saa, Construction of pyridine rings by metal-mediated [2+ 2+ 2] cycloaddition, Chem. Rev. 103 (9) (2003) 3787–3802.

[11] T.J.J. Muller, U.H.F. Bunz, Functional Organic Materials: Syntheses, Strategies and Applications, Wiley-VCH, Weinheim, 2007 Eds.

[12] J.M. Lehn, Supramolecular Chemistry—Concepts and Perspectives, VCH, Weinheim, 1995.

[13] M. Baumann, I.R. Baxendale, An overview of the synthetic routes to the best selling drugs containing 6-membered heterocycles, Belistein J. Org. Chem. 9 (1) (2013) 2265–2319.

[14] R.B. Lacerda, C.K.F. de Lima, L.L. da Silva, N.C. Romeiro, A.L.P. Miranda, E.J. Barreiro, C.A.M. Fraga, Discovery of novel analgesic and anti"inflammatory 3-arylamine-imidazo[1,2-a]pyridine symbiotic prototypes, Bioorg. Med. Chem. 17 (1) (2009) 74–84.

[15] B. Vacher, B. Bonnaud, P. Funes, N. Jubault, W. Koek, M.B. Assie, C. Cosi, M. Kleven, Novel derivatives of 2"pyridinemethylamine as selective, potent, and orally active agonists at 5-HT1A receptors, J. Med. Chem. 42 (9) (1999) 1648–1660.

[16] A.E. Goetz, N.K. Garg, Regioselective reactions of 3,4-pyridynes enabled by the aryne distortion model, Nat. Chem. 5 (1) (2013) 54–60.

[17] J.P. Rocher, B. Bonnet, C. Bolea, R. Lutjens, E. Le Poul, S. Poli, M. Epping Jordan, A.S. Bessis, B. Ludwig, V. Mutel, mGluR5 negative allosteric modulators overview: a medicinal chemistry approach towards a series of novel therapeutic agents, Curr. Top. Med. Chem. 11 (6) (2011) 680–695.

[18] M.M. Harding, G.V. Long, C.L. Brown, Solution conformation of the antitumor drug streptonigrin, J. Med. Chem. 36 (21) (1993) 3056–3060.

[19] M. Hassani, W. Cai, D.C. Holley, J.P. Lineswala, B.R. Maharjan, G.R. Ebrahimian, H. Seradj, M.G. Stocksdale, F. Mohammadi, C.C. Marvin, J.M. Gerdes, H.D. Beall, M. Behforouz, Novel lavendamycin analogues as antitumor agents: synthesis, in vitro cytotoxicity, structure metabolism, and computational molecular modeling studies with NAD(P)H: quinone oxidoreductase 1, J. Med. Chem. 48 (24) (2005) 7733–7749.

[20] N.B. Mantlo, A. Escribano, Update on the discovery and development of cholesteryl ester transfer protein inhibitors for reducing residual cardiovascular risk, J. Med. Chem. 57 (1) (2014) 1–17.

[21] F.M. Arlan, A.P. Marjani, R. Javahershenas, J. Khalafy, Recent developments in the synthesis of polysubstituted pyridines via multicomponent reactions using nanocatalysts, New J. Chem. 45 (28) (2021) 12328–12345.

[22] G.M. Ziarani, Z. Kheilkordi, F. Mohajer, A. Badiei, R. Luque, Magnetically recoverable catalysts for the preparation of pyridine derivatives: an overview, RSC Adv. 11 (28) (2021) 17456–17477.

[23] C. Allais, J.M. Grassot, J. Rodriguez, T. Constantieux, Metal-free multicomponent syntheses of pyridines, Chem. Rev. 114 (21) (2014) 10829–10868.

[24] J.C. Vedrine, Heterogeneous catalysis on metal oxides, Catalysts 7 (11) (2017) 341.

[25] F. Dumeignil, J.F. Paul, S. Paul, Heterogeneous catalysis with renewed attention: principles, theories, and concepts, J. Chem. Educ. 94 (6) (2017) 675–689.
[26] X. Cui, W. Li, P. Ryabchuk, K. Junge, M. Beller, Bridging homogeneous and heterogeneous catalysis by heterogeneous single-metal-site catalysts, Nat. Catal. 1 (6) (2018) 385–397.
[27] M. Foscato, V.R. Jensen, Automated in silico design of homogeneous catalysts, ACS Catal. 10 (3) (2020) 2354–2377.
[28] K. Sordakis, C. Tang, L.K. Vogt, H. Junge, P.J. Dyson, M. Beller, G. Laurenczy, Homogeneous catalysis for sustainable hydrogen storage in formic acid and alcohols, Chem. Rev. 118 (2) (2018) 372–433.
[29] A. Kokel, C. Schafer, Application of green chemistry in homogeneous catalysis, Green Chemistry, An Inclusive Approach, Edited by Béla Török, Timothy Dransfield, Elsevier, 2018, pp. 375–414.
[30] S.K. Singh, A.W. Savoy, Ionic liquids synthesis and applications: an overview, J. Mol. Liq. 297 (2020) 112038–112061.
[31] V.S. Shende, V.B. Saptal, B.M. Bhanage, Recent advances utilized in the recycling of homogeneous catalysis, Chem. Rec. 19 (9) (2019) 2022–2043.
[32] D.J. Cole-Hamilton, Homogeneous catalysis-new approaches to catalyst separation, recovery, and recycling, Science 299 (5613) (2003) 1702–1706.
[33] S. Bhaduri, D. Mukesh, Homogeneous Catalysis: Mechanisms, Industrial Applications, second Ed., John Wiley & Sons, Hoboken, New Jersey, 2014.
[34] F. Polo-Garzon, Z. Wu, Acid-base catalysis over perovskites: a review, J. Mater. Chem. A 6 (7) (2018) 2877–2894.
[35] R.H. Crabtree, Homogeneous transition metal catalysis of acceptorless dehydrogenative alcohol oxidation: applications in hydrogen storage and to heterocycle synthesis, Chem. Rev. 117 (13) (2017) 9228–9246.
[36] A. Mukherjee, D. Milstein, Homogeneous catalysis by cobalt and manganese pincer complexes, ACS Catal. 8 (12) (2018) 11435–11469.
[37] P. Kisszékelyi, S. Nagy, Z. Fehér, P. Huszthy, J. Kupai, Membrane-supported recovery of homogeneous organocatalysts: a review, Chemistry 2 (3) (2020) 742–758.
[38] K. Dong, X. Liu, H. Dong, X. Zhang, S. Zhang, Multiscale studies on ionic liquids, Chem. Rev. 117 (10) (2017) 6636–6695.
[39] L.R. Melo, W.A. Silva, Ionic liquid in multicomponent reactions: a brief review, Curr. Green Chem. 3 (2) (2016) 120–132.
[40] L. Wang, W. Sun, C. Liu, Recent advances in homogeneous carbonylation using CO_2 as Co surrogate, Chin. J. Chem. 36 (4) (2018) 353–362.
[41] M.L. Pegis, C.F. Wise, D.J. Martin, J.M. Mayer, Oxygen reduction by homogeneous molecular catalysts and electrocatalysts, Chem. Rev. 118 (5) (2018) 2340–2391.
[42] A.N. Kim, B.M. Stoltz, Recent advances in homogeneous catalysts for the asymmetric hydrogenation of heteroarenes, ACS Catal. 10 (23) (2020) 13834–13851.
[43] C.M. Friend, B. Xu, Heterogeneous catalysis: a central science for a sustainable future, Acc. Chem. Res. 50 (3) (2017) 517–521.
[44] R. Schlögl, Heterogeneous catalysis, Angew. Chem. Int. Ed. Engl. 54 (11) (2015) 3465–3520.
[45] A. Corma, Heterogeneous catalysis: understanding for designing, and designing for applications, Angew. Chem. Int. Ed. Engl. 55 (21) (2016) 6112–6113.
[46] R.A. Van Santen, Modern Heterogeneous Catalysis: An Introduction, Wiley-VCH Verlag GmbH & Co. KGaA, Boschstr. 12, 69469 Weinheim, Germany, 2017.
[47] G. Rothenberg, Catalysis: Concepts and Green Applications, Wiley-VCH Verlag GmbH & Co. KGaA, Weinheim, 2017.
[48] L. Wu, A. Mendoza-Garcia, Q. Li, S. Sun, Organic phase syntheses of magnetic nanoparticles and their applications, Chem. Rev. 116 (18) (2016) 10473–10512.
[49] P. Veerakumar, P. Thanasekaran, K.L. Lu, S.B. Liu, S. Rajagopal, Functionalized silica matrices and palladium: a versatile heterogeneous catalyst for Suzuki, Heck, and Sonogashira reactions, ACS Sustain. Chem. Eng. 5 (8) (2017) 6357–6376.
[50] Y.B. Zhou, Z.P. Zhan, Conjugated microporous polymers for heterogeneous catalysis, Chem. Asian J. 13 (1) (2018) 9–19.
[51] M. Liu, J. Wu, H. Hou, Metal-organic framework (MOF)-based materials as heterogeneous catalysts for C-H bond activation, Chem. Eur. J. 25 (12) (2019) 2935–2948.

[52] H. Hu, Q. Yan, R. Ge, Y. Gao, Covalent organic frameworks as heterogeneous catalysts, Chin. J. Catal. 39 (7) (2018) 1167–1179.
[53] I. Fechete, Y. Wang, J.C. Vedrine, The past, present and future of heterogeneous catalysis, Catal. Today 189 (1) (2012) 2–27.
[54] M.A. Ghasemzadeh, B. Mirhosseini-Eshkevari, M. Tavakoli, F. Zamani, Metal-organic frameworks: advanced tools for multicomponent reactions, Green Chem. 22 (21) (2020) 7265–7300.
[55] X.X. Du, Q.X. Zi, Y.M. Wu, Y. Jin, J. Lin, S.J. Yan, An environmentally benign multi-component reaction: regioselective synthesis of fluorinated 2-aminopyridines using diverse properties of the nitro group, Green Chem. 21 (6) (2019) 1505–1516.
[56] J. Chen, D. Chang, F. Xiao, G.J. Deng, Four-component quinazoline synthesis from simple anilines, aromatic aldehydes and ammonium iodide under metal-free conditions, Green Chem. 20 (24) (2018) 5459–5463.
[57] M.L. Deb, P.J. Borpatra, P.K. Baruah, A one-pot catalyst/external oxidant/solvent-free cascade approach to pyrimidines via a 1, 5-hydride transfer, Green Chem. 21 (1) (2019) 69–74.
[58] S.T. Chung, W.H. Huang, C.K. Huang, F.C. Liu, R.Y. Huang, C.C. Wu, A.R. Lee, Synthesis and anti-inflammatory activities of 4*H*-chromene and chromeno [2, 3-*b*] pyridine derivatives, Res. Chem. Intermed. 42 (2) (2016) 1195–1215.
[59] R.V. Kupwade, S.S. Khot, M.A. Kulkarni, U.V. Desai, P.P. Wadgaonkar, Diethylamine Dess-Martin periodinane: an efficient catalyst–oxidant combination in a sequential, one-pot synthesis of difficult to access 2-amino-3,5-dicarbonitrile-6-sulfanylpyridines at ambient temperature, RSC Adv. 7 (62) (2017) 38877–38883.
[60] A.V. Kuznetcova, I.S. Odin, A.A. Golovanov, I.M. Grigorev, A.V. Vasilyev, Multicomponent reaction of conjugated enynones with malononitrile and sodium alkoxides: complex reaction mechanism of the formation of pyridine derivatives, Tetrahedron 75 (33) (2019) 4516–4530.
[61] L. Wang, G. Zhu, W. Tang, T. Lu, D. Du, DMAP-promoted in situ activation of bromoacetic acid as a 2-carbon synthon for facile synthesis of pyridines and fused pyridin-2-ones, Tetrahedron 72 (41) (2016) 6510–6517.
[62] M. Kidwai, R. Chauhan, K_2CO_3 catalyzed green and rapid access to 2-amino-3,5-dicarbonitrile-6-thio-pyridines, J. Iran. Chem. Soc. 11 (4) (2014) 1005–1013.
[63] M.C. Eriksson, X. Zeng, J. Xu, D.C. Reeves, C.A. Busacca, V. Farina, C.H. Senanayake, The Guareschi-Thorpe cyclization revisited—an efficient synthesis of substituted 2,6-dihydroxypyridines and 2,6-dichloropyridines, Synlett 29 (11) (2018) 1455–1460.
[64] M.D. Hill, A multicomponent approach to highly substituted 1*H*-pyrazolo [3, 4-*b*] pyridines, Synthesis 48 (14) (2016) 2201–2204.
[65] D. Lloyd, H. McNab, Vinamidines and vinamidinium salts—examples of stabilized push-pull alkenes, Angew. Chem. Int. Ed. Engl. 15 (8) (1976) 459–468.
[66] S. Tian, Y. Mao, Y. Jiang, G. Xu, The application of vinamidinium salt to the synthesis of 3-chloro-α-carbolines, Synlett 29 (07) (2018) 949–953.
[67] B. Laramee-Milette, T. Auvray, S. Nguyen, S. Tremblay, C. Lachance-Brais, M. Donguy, V. Taylor, D. Deschenes, G.S. Hanan, Simple solubilization of the traditional 2,2′:6′, 2″-terpyridine ligand in organic solvents by substitution with 4,4″-di-tert-butyl groups, Synthesis 47 (24) (2015) 3849–3858.
[68] H. Wei, Y. Li, Quick access to pyridines through 6π-3-azatriene electrocyclization: concise total synthesis of suaveoline alkaloids, Synlett 30 (14) (2019) 1615–1620.
[69] J.P. Wan, Y. Zhou, K. Jiang, H. Ye, Thioacetamide as an ammonium source for multicomponent synthesis of pyridines from aldehydes and electron-deficient enamines or alkynes, Synthesis 46 (23) (2014) 3256–3262.
[70] J.P. Wan, Y. Lin, K. Hu, Y. Liu, Secondary amine-initiated three-component synthesis of 3, 4-dihydropyrimidinones and thiones involving alkynes, aldehydes and thiourea/urea, Beilstein J. Org. Chem. 10 (1) (2014) 287–292.
[71] J. Wan, Y. Zhou, Y. Liu, Z. Fang, C. Wen, Multicomponent reactions for diverse synthesis of *N*-substituted and nh 1, 4-dihydropyridines, Chin. J. Chem. 32 (3) (2014) 219–226.
[72] D.A. Shabalin, M.Y. Dvorko, E.Y. Schmidt, B.A. Trofimov, Regiocontrolled synthesis of 2, 4,6-triarylpyridines from methyl ketones, electron-deficient acetylenes and ammonium acetate, Org. Biomol. Chem. 19 (12) (2021) 2703–2715.
[73] C.H. Zhang, R. Huang, X.M. Hu, J. Lin, S.J. Yan, Three-component site-selective synthesis of highly substituted 5 *H*-chromeno-[4, 3-*b*] pyridines, J. Org. Chem. 83 (9) (2018) 4981–4989.
[74] O.V. Ershov, M.Y. Ievlev, M.Y. Belikov, K.V. Lipin, A.I. Naydenova, V.A. Tafeenko, Synthesis and solid-state fluorescence of aryl substituted 2-halogenocinchomeronic dinitriles, RSC Adv. 6 (85) (2016) 82227–82232.

[75] O.V. Ershov, V.N. Maksimova, K.V. Lipin, M.Y. Belikov, M.Y. Ievlev, V.A. Tafeenko, O.E. Nasakin, Regiospecific synthesis of gem-dinitro derivatives of 2-halogenocycloalka [*b*] pyridine-3, 4-dicarbonitriles, Tetrahedron 71 (39) (2015) 7445–7450.

[76] M.Y. Belikov, M.Y. Ievlev, O.V. Ershov, K.V. Lipin, S.A. Legotin, O.E. Nasakin, Synthesis of photochromic 5,6-diaryl-2-chloropyridine-3,4-dicarbonitriles from 3,4-diaryl-4-oxobutane-1, 1, 2, 2-tetracarbonitriles, Russ. J. Org. Chem. 50 (9) (2014) 1372–1374.

[77] O.O. Stepaniuk, T.V. Rudenko, B.V. Vashchenko, V.O. Matvienko, I.S. Kondratov, A.A. Tolmachev, O.O. Grygorenko, Synthesis of fused pyridine carboxylates by reaction of β-alkoxyvinyl glyoxylates with amino heterocycles, Synthesis 52 (13) (2020) 1915–1926.

[78] Z.Y. Mao, X.Y. Liao, H.S. Wang, C.G. Wang, K.B. Huang, Y.M. Pan, Acid-catalyzed tandem reaction for the synthesis of pyridine derivatives via C [double bond, length as m-dash] C/C (sp^3)-*N* bond cleavage of enones and primary amines, RSC Adv. 7 (22) (2017) 13123–13129.

[79] M. Belal, D.K. Das, A.T. Khan, Synthesis of pyrido [2, 3-C] coumarin derivatives by an intramolecular povarov reaction, Synthesis 47 (08) (2015) 1109–1116.

[80] H.D. Khanal, Y.R. Lee, Organocatalyzed oxidative *N*-annulation for diverse and polyfunctionalized pyridines, Chem. Commun. 51 (46) (2015) 9467–9470.

[81] B. Prek, J. Bezenšek, B. Stanovnik, Synthesis of pyridines with an amino acid residue by [2+ 2] cycloadditions of electron-poor acetylenes on enaminone systems derived from *N*-Boc protected amino acids, Tetrahedron 73 (35) (2017) 5260–5267.

[82] S. Sultana, S.M.B. Maezono, M.S. Akhtar, J.J. Shim, Y.J. Wee, S.H. Kim, Y.R. Lee, $BF_3 \cdot OEt_2$-promoted annulation for substituted 2-arylpyridines as potent UV filters and antibacterial agents, Adv. Synth. Catal. 360 (4) (2018) 751–761.

[83] Y. Zhou, Z. Tang, Q. Song, Lewis acid-mediated [3+ 3] annulation for the construction of substituted pyrimidine and pyridine derivatives, Adv. Synth. Catal. 359 (2017) 952–958.

[84] M.A. Zolfigol, M. Kiafar, M. Yarie, A.A. Taherpour, T. Fellowes, A.N. Hancok, A. Yari, A convenient method for preparation of 2-amino-4, 6-diphenylnicotinonitrile using HBF_4 as an efficient catalyst via an anomeric-based oxidation: a joint experimental and theoretical study, J. Mol. Struct. 1137 (2017) 674–680.

[85] Z. Huang, Y. Hu, Y. Zhou, D. Shi, Efficient one-pot three-component synthesis of fused pyridine derivatives in ionic liquid, ACS Comb. Sci. 13 (1) (2011) 45–49.

[86] S. Baghery, M.A. Zolfigol, F. Maleki, [TEATNM] and [TEATCM] as novel catalysts for the synthesis of pyridine-3,5-dicarbonitriles via anomeric-based oxidation, New J. Chem. 41 (17) (2017) 9276–9290.

[87] (a) M. Yarie, Catalytic anomeric-based oxidation, Iran. J. Catal. 7 (2017) 85–88; (b) M. Yarie, Catalytic vinylogous anomeric-based oxidation (Part I), Iran. J. Catal. 10 (2020) 79–83.

[88] (a) I.V. Alabugin, L. Kuhn, M.G. Medvedev, NV. Krivoshchapov, VA. Vil', I.A. Yaremenko, P. Mehaffy, M. Yarie, A.O. Terent'ev, M.A. Zolfigol, Stereoelectronic power of oxygen in control of chemical reactivity: the anomeric effect is not alone, Chem. Soc. Rev. 50 (2021) 10253–10345 Advance Article, doi:10.1039/D1CS00386K; (b) I.V. Alabugin, L. Kuhn, N.V. Krivoshchapov, P. Mehaffy, M.G. Medvedev, Anomeric effect, hyperconjugation and electrostatics: lessons from complexity in a classic stereoelectronic phenomenon, Chem. Soc. Rev. 50 (2021) 10212–10252 Advance Article, doi:10.1039/D1CS00564B.

[89] M. Dashteh, M.A. Zolfigol, A. Khazaei, S. Baghery, M. Yarie, S. Makhdoomi, M. Safaiee, Synthesis of cobalt tetra-2, 3-pyridiniumporphyrazinato with sulfonic acid tags as an efficient catalyst and its application for the synthesis of bicyclic ortho-aminocarbonitriles, cyclohexa-1,3-dienamines and 2-amino-3-cyanopyridines, RSC Adv. 10 (46) (2020) 27824–27834.

[90] M. Torabi, M. Yarie, M.A. Zolfigol, S. Rouhani, S. Azizi, T.O. Olomola, M. Maaza, T.A. Msagati, Synthesis of new pyridines with sulfonamide moiety via a cooperative vinylogous anomeric-based oxidation mechanism in the presence of a novel quinoline-based dendrimer-like ionic liquid, RSC Adv. 11 (5) (2021) 3143–3152.

[91] M.A. Zolfigol, M. Safaiee, B. Ebrahimghasri, S. Baghery, S. Alaie, M. Kiafar, A. Taherpour, Y. Bayat, A. Asgari, Application of novel nanostructured dinitropyrazine molten salt catalyst for the synthesis of sulfanylpyridines via anomeric-based oxidation, J. Iran. Chem. Soc. 14 (9) (2017) 1839–1852.

[92] P. Kumar, M. Kapur, Unusual reactivity of 4-vinyl isoxazoles in the copper-mediated synthesis of pyridines, employing dmso as a one-carbon surrogate, Org. Lett. 22 (15) (2020) 5855–5860.

[93] H. Huang, X. Ji, W. Wu, L. Huang, H. Jiang, Copper-catalyzed formal C-N bond cleavage of aromatic methylamines: assembly of pyridine derivatives, J. Org. Chem. 78 (8) (2013) 3774–3782.

[94] J. Han, X. Guo, Y. Liu, Y. Fu, R. Yan, B. Chen, One-pot synthesis of benzene and pyridine derivatives via copper-catalyzed coupling reactions, Adv. Synth. Catal. 359 (15) (2017) 2676–2681.
[95] B. Zheng, T.H. Trieu, T.Z. Meng, X. Lu, J. Dong, Q. Zhang, X.X. Shi, Cu-catalyzed mild and efficient oxidation of THβCs using air: application in practical total syntheses of perlolyrine and flazin, RSC Adv. 8 (13) (2018) 6834–6839.
[96] K. Aradi, P. Bombicz, Z. Novak, Modular copper-catalyzed synthesis of chromeno [4, 3-*b*] quinolines with the utilization of diaryliodonium salts, J. Org. Chem. 81 (3) (2016) 920–931.
[97] M.V. Reddy, Y.T. Jeong, Copper (ii) oxide nanoparticles as a highly active and reusable heterogeneous catalyst for the construction of phenyl-1*H*-pyrazolo [3, 4-*b*] pyridine derivatives under solvent-free conditions, RSC Adv. 6 (105) (2016) 103838–103842.
[98] M. Bhardwaj, M. Kour, S. Paul, Cu (0) onto sulfonic acid functionalized silica/carbon composites as bifunctional heterogeneous catalysts for the synthesis of polysubstituted pyridines and nitriles under benign reaction media, RSC Adv. 6 (101) (2016) 99604–99614.
[99] R. Khalifeh, M. Ghamari, A multicomponent synthesis of 2-amino-3-cyanopyridine derivatives catalyzed by heterogeneous and recyclable copper nanoparticles on charcoal, J. Braz. Chem. Soc. 27 (4) (2016) 759–768.
[100] M. Arita, S. Yokoyama, H. Asahara, N. Nishiwaki, Facile synthesis of onychines, Synthesis 51 (09) (2019) 2007–2013.
[101] C.H. Chen, Q.Y. Wu, C. Wei, C. Liang, G.F. Su, D.L. Mo, Iron (III)-catalyzed selective N-O bond cleavage to prepare tetrasubstituted pyridines and 3, 5-disubstituted isoxazolines from N-vinyl-α, β-unsaturated ketonitrones, Green Chem. 20 (12) (2018) 2722–2729.
[102] Y. Yi, M.N. Zhao, Z.H. Ren, Y.Y. Wang, Z.H. Guan, Synthesis of symmetrical pyridines by iron-catalyzed cyclization of ketoxime acetates and aldehydes, Green Chem. 19 (4) (2017) 1023–1027.
[103] F. Du, S.J. Li, K. Jiang, R. Zeng, X.C. Pan, Y. Lan, Y.C. Chen, Y. Wei, Iron-catalyzed radical relay enabling the modular synthesis of fused pyridines from alkyne-tethered oximes and alkenes, Angew. Chem. 132 (52) (2020) 23963–23970.
[104] L. Qi, R. Li, X. Yao, Q. Zhen, P. Ye, Y. Shao, J. Chen, Syntheses of pyrroles, pyridines, and ketonitriles via catalytic carbopalladation of dinitriles, J. Org. Chem. 85 (2) (2019) 1097–1108.
[105] K. Selvakumar, K.A.P. Lingam, R.V.L. Varma, V. Vijayabaskar, Controlled and efficient synthesis of quinoline derivatives from Morita–Baylis–Hillman adducts by palladium-catalyzed heck reaction and cyclization, Synlett 26 (05) (2015) 646–650.
[106] J.F. Tan, C.T. Bormann, F.G. Perrin, F.M. Chadwick, K. Severin, N. Cramer, Divergent synthesis of densely substituted arenes and pyridines via cyclotrimerization reactions of alkynyl triazenes, J. Am. Chem. Soc. 141 (26) (2019) 10372–10383.
[107] F. Ye, F. Boukattaya, M. Haddad, V. Ratovelomanana-Vidal, V. Michelet, Synthesis of 2-aminopyridines via ruthenium-catalyzed [2+ 2+ 2] cycloaddition of 1,6-and 1,7-diynes with cyanamides: scope and limitations, New J. Chem. 42 (5) (2018) 3222–3235.
[108] C. Tran, M. Haddad, V. Ratovelomanana-Vidal, Ruthenium-catalyzed [2+2+2] cycloaddition of α, ω-diynes and selenocyanates: an entry to selenopyridine derivatives, Synthesis 51 (12) (2019) 2532–2541.
[109] D.S. Deshmukh, B.M. Bhanage, Ruthenium-catalyzed annulation of *N*-Cbz hydrazones via C-H/N-N bond activation for the rapid synthesis of isoquinolines, Synthesis 51 (12) (2019) 2506–2514.
[110] D. Bhatt, P.R. Singh, P. Kalaramna, K. Kumar, A. Goswami, An atom-economical approach to 2-triazolyl thio-/seleno pyridines via ruthenium-catalyzed one-pot [3+2]/[2+2+2] cycloadditions, Adv. Synth. Catal. 361 (23) (2019) 5483–5489.
[111] T.A. Nizami, R. Hua, Silver-catalyzed chemoselective annulation of propargyl amines with alkynes for access to pyridines and pyrroles, Tetrahedron 73 (42) (2017) 6080–6084.
[112] S. Cacchi, G. Fabrizi, E. Filisti, N-propargylic β-enaminones: common intermediates for the synthesis of polysubstituted pyrroles and pyridines, Org. Lett. 10 (13) (2008) 2629–2632.
[113] J.A. Yoon, Y.T. Han, Efficient synthesis of pyrido [3, 2-C] coumarins via silver nitrate catalyzed cycloisomerization and application to the first synthesis of polyneomarline c, Synthesis 51 (24) (2019) 4611–4618.
[114] M.L. Rao, A. Kumar, Pd-catalyzed chemo-selective mono-arylations and bis-arylations of functionalized 4-chlorocoumarins with triarylbismuths as threefold arylating reagents, Tetrahedron 70 (39) (2014) 6995–7005.
[115] R. Nishizawa, T. Nishiyama, K. Hisaichi, K. Hirai, H. Habashita, Y. Takaoka, H. Tada, K. Sagawa, S. Shibayama, K. Maeda, H. Mitsuya, H. Nakai, D. Fukushima, M. Toda, Discovery of orally available spirodiketopiperazine-based CCR5 antagonists, Bioorg. Med. Chem. 18 (14) (2010) 5208–5223.

[116] H. Chu, P. Xue, J.T. Yu, J. Cheng, Rhodium-catalyzed annulation of primary benzylamine with α-diazo ketone toward isoquinoline, J. Org. Chem. 81 (17) (2016) 8009–8013.
[117] Q. Wang, X. Li, Rhodium/copper-cocatalyzed annulation of benzylamines with diazo compounds: access to fused isoquinolines, Org. Chem. Front. 3 (9) (2016) 1159–1162.
[118] K. Singh, M. Vellakkaran, D. Banerjee, A nitrogen-ligated nickel-catalyst enables selective intermolecular cyclisation of β-and γ-amino alcohols with ketones: access to five and six-membered *N*-heterocycles, Green Chem. 20 (10) (2018) 2250–2256.
[119] Y. Zhong, N.A. Spahn, R.M. Stolley, M.H. Nguyen, J. Louie, 3, 5-Disubstituted 2-aminopyridines via nickel-catalyzed cycloaddition of terminal alkynes and cyanamides, Synlett 26 (03) (2015) 307–312.
[120] R.M. Stolley, H.A. Duong, J. Louie, Mechanistic evaluation of the Ni (IPr) 2-catalyzed cycloaddition of alkynes and nitriles to afford pyridines: evidence for the formation of a key η1-Ni $(IPr)_2(RCN)$ intermediate, Organometallics 32 (17) (2013) 4952–4960.
[121] A.P. Marjani, J. Khalafy, F. Salami, M. Mohammadlou, Tin (II) chloride catalyzed synthesis of new pyrazolo [5, 4-b] quinolines under solvent-free conditions, Synthesis 47 (11) (2015) 1656–1660.
[122] D.N.K. Reddy, K.B. Chandrasekhar, Y.S.S. Ganesh, B.S. Kumar, R. Adepu, M. Pal, $SnCl_2 \cdot 2H_2O$ as a precatalyst in MCR: synthesis of pyridine derivatives via a 4-component reaction in water, Tetrahedron Lett. 56 (31) (2015) 4586–4589.
[123] T. Ishikawa, T. Sonehara, S. Murakami, M. Minakawa, M. Kawatsura, Synthesis of trifluoromethyl-substituted ethyl picolinate derivatives by the cobalt-catalyzed regioselective intermolecular [2+ 2+ 2] cycloaddition, Synlett 27 (13) (2016) 2029–2033.
[124] P. Jungk, T. Taufer, I. Thiel, M. Hapke, Synthesis of chiral indenylcobalt (I) complexes and their evaluation in asymmetric [2+2+2] cycloaddition reactions, Synthesis 48 (13) (2016) 2026–2035.
[125] A.R. Moosavi-Zare, M.A. Zolfigol, S. Farahmand, A. Zare, A.R. Pourali, R. Ayazi-Nasrabadi, Synthesis of 2,4,6-triarylpyridines using $ZrOCl_2$ under solvent-free conditions, Synlett 25 (02) (2014) 193–196.
[126] H. Chai, W. Tan, Y. Lu, G. Zhang, J. Ma, Sustainable synthesis of quinolines (pyridines) catalyzed by a cheap metal Mn (I)-NN complex catalyst, Appl. Organomet. Chem. 34 (8) (2020) e5685.
[127] D. Gandhi, S. Agarwal, MgO NPs catalyzed eco-friendly reaction: a highly effective and green approach for the multicomponent one-pot synthesis of polysubstituted pyridines using 2-aminobenzothiazole, J. Heterocycl. Chem. 55 (12) (2018) 2977–2984.
[128] F.M. Arlan, J. Khalafy, R. Maleki, One-pot three-component synthesis of a series of 4-aroyl-1,6-diaryl-3-methyl-1*H*-pyrazolo [3, 4-*b*] pyridine-5-carbonitriles in the presence of aluminum oxide as a nanocatalyst, Chem. Heterocycl. Compd. 54 (1) (2018) 51–57.
[129] J.C. Xiang, M. Wang, Y. Cheng, A.X. Wu, Molecular iodine-mediated chemoselective synthesis of multisubstituted pyridines through catabolism and reconstruction behavior of natural amino acids, Org. Lett. 18 (1) (2016) 24–27.
[130] C. Zhu, B. Bi, Y. Ding, T. Zhang, Q.Y. Chen, Iodine-catalyzed aerobic oxidative formal [4+2] annulation for the construction of polyfunctionalized pyridines, Tetrahedron 71 (49) (2015) 9251–9257.
[131] A. Kale, C. Bingi, N.C. Ragi, P. Sripadi, P.R. Tadikamalla, K. Atmakur, Synthesis of pyrido-fused imidazo [4, 5-c] quinolines by I2-DMSO promoted oxidative cross coupling and intramolecular cyclization, Synthesis 49 (07) (2017) 1603–1612.
[132] H. Xu, J.C. Zeng, F.J. Wang, Z. Zhang, Metal-free synthesis of 2,4,6-trisubstituted pyridines via iodine-initiated reaction of methyl aryl ketones with amines under neat heating, Synthesis 49 (08) (2017) 1879–1883.
[133] W. Liu, H. Tan, C. Chen, Y. Pan, A method to access symmetrical tetrasubstituted pyridines via iodine and ammonium persulfate mediated [2+ 2+ 1+ 1]-cycloaddition reaction, Adv. Synth. Catal. 359 (9) (2017) 1594–1598.
[134] X. Wu, J. Zhang, S. Liu, Q. Gao, A. Wu, An efficient synthesis of polysubstituted pyridines via c-h oxidation and c-s cleavage of dimethyl sulfoxide, Adv. Synth. Catal. 358 (2) (2016) 218–225.
[135] M.B. Boroujeni, A. Hashemzadeh, M.T. Faroughi, A.S haabani, M.M. Amini, Magnetic MIL-101-SO_3H: a highly efficient bifunctional nanocatalyst for the synthesis of 1,3,5-triarylbenzenes and 2,4,6-triaryl pyridines, RSC Adv. 6 (102) (2016) 100195–100202.
[136] H. Sepehrmansourie, M. Zarei, M.A. Zolfigol, S. Babaee, S. Rostamnia, Application of novel nanomagnetic metal–organic frameworks as a catalyst for the synthesis of new pyridines and 1, 4-dihydropyridines via a cooperative vinylogous anomeric-based oxidation, Sci. Rep. 11 (1) (2021) 1–15.

[137] M.M. Heravi, S.Y.S. Beheshtiha, M. Dehghani, N. Hosseintash, Using magnetic nanoparticles Fe_3O_4 as a reusable catalyst for the synthesis of pyran and pyridine derivatives via one-pot multicomponent reaction, J. Iran. Chem. Soc. 12 (11) (2015) 2075–2081.

[138] F. Karimi, M. Yarie, M.A. Zolfigol, A convenient method for synthesis of terpyridines via a cooperative vinylogous anomeric-based oxidation, RSC Adv. 10 (43) (2020) 25828–25835.

[139] F. Karimi, M.A. Zolfigol, M. Yarie, A novel and reusable ionically tagged nanomagnetic catalyst: application for the preparation of 2-amino-6-(2-oxo-2*H*-chromen-3-yl)-4-arylnicotinonitriles via vinylogous anomeric-based oxidation, Mol. Catal. 463 (2019) 20–29.

[140] S. Sobhani, F. Nasseri, F. Zarifi, Unique role of 2-hydroxyethylammonium acetate as an ionic liquid in the synthesis of Fe_3O_4 magnetic nanoparticles and preparation of pyridine derivatives in the presence of a new magnetically recyclable heterogeneous catalyst, J. Iran. Chem. Soc. 15 (12) (2018) 2721–2732.

[141] M. Torabi, M.A. Zolfigol, M. Yarie, B. Notash, S. Azizian, M.M. Azandaryani, Synthesis of triarylpyridines with sulfonate and sulfonamide moieties via a cooperative vinylogous anomeric-based oxidation, Sci. Rep. 11 (1) (2021) 1–19.

[142] S. Kalhor, M. Yarie, M. Rezaeivala, M.A. Zolfigol, Novel magnetic nanoparticles with morpholine tags as multirole catalyst for synthesis of hexahydroquinolines and 2-amino-4,6-diphenylnicotinonitriles through vinylogous anomeric-based oxidation, Res. Chem. Intermed. 45 (6) (2019) 3453–3480.

[143] F. Karimi, M. Yarie, M.A. Zolfigol, Fe_3O_4@SiO_2@$(CH_2)_3$-urea-thiourea: a novel hydrogen-bonding and reusable catalyst for the construction of bipyridine-5-carbonitriles via a cooperative vinylogous anomeric-based oxidation, Mol. Catal. 497 (2020) 111201.

[144] [a] M. Torabi, M. Yarie, M.A. Zolfigol, Synthesis of a novel and reusable biological urea based acidic nanomagnetic catalyst: application for the synthesis of 2-amino-3-cyano pyridines via cooperative vinylogous anomeric-based oxidation, Appl. Organomet. Chem. 33 (6) (2019) e4933; [b] M.A. Zolfigol, M. Kiafar, M. Yarie, A.A. Taherpour, M. Saeidi-Rad, Experimental and theoretical studies of the nanostructured {Fe_3O_4@SiO_2@$(CH_2)_3$Im}$C(CN)_3$ catalyst for 2-amino-3-cyanopyridine preparation via an anomeric-based oxidation, RSC Adv. 6 (55) (2016) 50100–50111.

[145] M. Torabi, M. Yarie, F. Karimi, M.A. Zolfigol, Catalytic synthesis of coumarin-linked nicotinonitrile derivatives via a cooperative vinylogous anomeric-based oxidation, Res. Chem. Intermed. 46 (12) (2020) 5361–5376.

[146] P. Ghasemi, M. Yarie, M.A. Zolfigol, A.A. Taherpour, M. Torabi, Ionically tagged magnetic nanoparticles with urea linkers: application for preparation of 2-aryl-quinoline-4-carboxylic acids via an anomeric-based oxidation mechanism, ACS Omega 5 (7) (2020) 3207–3217.

[147] J. Afsar, M.A. Zolfigol, A. Khazaei, D.A. Alonso, A. Khoshnood, Y. Bayat, A. Asgari, Synthesis and application of a novel nanomagnetic catalyst with cl[DABCO-NO_2] $C(NO_2)_3$ tags in the preparation of pyrazolo [3, 4-*b*] pyridines via anomeric-based oxidation, Res. Chem. Intermed. 44 (12) (2018) 7595–7618.

[148] F. Karimi, M. Yarie, M.A. Zolfigol, Synthesis and characterization of Fe_3O_4@SiO_2@$(CH_2)_3NH(CH_2)_2O_2P(OH)_2$ and its catalytic application in the synthesis of benzo-[*h*] quinoline-4-carboxylic acids via a cooperative anomeric-based oxidation mechanism, Mol. Catal. 489 (2020) 110924.

[149] A. Maleki, A.A. Jafari, S. Yousefi, V. Eskandarpour, An efficient protocol for the one-pot multicomponent synthesis of polysubstituted pyridines by using a biopolymer-based magnetic nanocomposite, C. R. Chim. 18 (12) (2015) 1307–1312.

[150] M. Edrisi, N. Azizi, Sulfonic acid-functionalized graphitic carbon nitride composite: a novel and reusable catalyst for the one-pot synthesis of polysubstituted pyridine in water under sonication, J. Iran. Chem. Soc. 17 (4) (2020) 901–910.

[151] M. Zhang, P. Liu, Y.H. Liu, Z.R. Shang, H.C. Hu, Z.H. Zhang, Magnetically separable graphene oxide anchored sulfonic acid: a novel, highly efficient and recyclable catalyst for one-pot synthesis of 3, 6-di(pyridin-3-yl)-1*H*-pyrazolo [3,4-*b*] pyridine-5-carbonitriles in deep eutectic solvent under microwave irradiation, RSC Adv. 6 (108) (2016) 106160–106170.

[152] H. Ebrahimiasl, D. Azarifar, J. Rakhtshah, H. Keypour, M. Mahmoudabadi, Application of novel and reusable Fe_3O_4@CoII (macrocyclic schiff base ligand) for multicomponent reactions of highly substituted thiopyridine and 4*H*-chromene derivatives, Appl. Organomet. Chem. 34 (9) (2020) e5769.

[153] S. Gajaganti, D. Kumar, S. Singh, V. Srivastava, B.K. Allam, A new avenue to the synthesis of symmetrically substituted pyridines catalyzed by magnetic nano-Fe_3O_4: methyl arenes as sustainable surrogates of aryl aldehydes, Chem. Select. 4 (31) (2019) 9241–9246.

[154] S. Forouzandehdel, M. Meskini, M.R. Rami, Design and application of (Fe_3O_4)-GOTfOH based agnps doped starch/PEG-poly (acrylic acid) nanocomposite as the magnetic nanocatalyst and the wound dress, J. Mol. Struct. 1214 (2020) 128142.
[155] T. Akbarpoor, A. Khazaei, J.Y. Seyf, N. Sarmasti, M.M. Gilan, One-pot synthesis of 2-amino-3-cyanopyridines and hexahydroquinolines using eggshell-based nano-magnetic solid acid catalyst via anomeric-based oxidation, Res. Chem. Intermed. 46 (2) (2020) 1539–1554.
[156] G.M. Ziarani, S. Bahar, A. Badiei, The green synthesis of 2-amino-3-cyanopyridines using $SrFe_{12}O_{19}$ magnetic nanoparticles as efficient catalyst and their application in complexation with Hg^{2+} ions, J. Iran. Chem. Soc. 16 (2) (2019) 365–372.
[157] N. Basavegowda, K. Mishra, Y.R. Lee, Fe_3O_4-decorated MWCNTs as an efficient and sustainable heterogeneous nanocatalyst for the synthesis of polyfunctionalised pyridines in water, Mater. Technol. 34 (9) (2019) 558–569.
[158] Z. Hosseinzadeh, A. Ramazani, H. Ahankar, K. Slepokura, T. Lis, Synthesis of 2-amino-4, 6-diarylnicotinonitrile in the presence of $CoFe_2O_4$@SiO_2-SO_3H as a reusable solid acid nanocatalyst under microwave irradiation in solvent-free conditions, Silicon 11 (4) (2019) 2169–2176.
[159] M. Ashouri, H. Kefayati, S. Shariati, Synthesis, characterization, and catalytic application of Fe_3O_4-Si-$[CH_2]_3$-N=CH-aryl for the efficient synthesis of novel poly-substituted pyridines, J. Chin. Chem. Soc. 66 (4) (2019) 355–362.
[160] S. Asadbegi, M.A. Bodaghifard, A. Mobinikhaledi, Poly *N,N*-dimethylaniline-formaldehyde supported on silica-coated magnetic nanoparticles: a novel and retrievable catalyst for green synthesis of 2-amino-3-cyanopyridines, Res. Chem. Intermed. 46 (3) (2020) 1629–1643.
[161] B. Maleki, H. Natheghi, R. Tayebee, H. Alinezhad, A. Amiri, S.A. Hossieni, S.M.M. Nouri, Synthesis and characterization of nanorod magnetic Co-Fe mixed oxides and its catalytic behavior towards one-pot synthesis of polysubstituted pyridine derivatives, Polycycl. Aromat. Compd. 40 (3) (2020) 633–643.
[162] E. Soleimani, M. Naderi Namivandi, H. Sepahvand, $ZnCl_2$ supported on Fe_3O_4@SiO_2 core-shell nanocatalyst for the synthesis of quinolines via Friedländer synthesis under solvent-free condition, Appl. Organomet. Chem. 31 (2) (2017) e3566.
[163] M.A. Ghasemzadeh, M.H. Abdollahi-Basir, M. Babaei, Fe_3O_4@SiO_2–NH_2 core-shell nanocomposite as an efficient and green catalyst for the multi-component synthesis of highly substituted chromeno [2, 3-*b*] pyridines in aqueous ethanol media, Green Chem. Lett. Rev. 8 (3-4) (2015) 40–49.
[164] F. Majidi Arlan, R. Javahershenas, J. Khalafy, An efficient one-pot, four-component synthesis of a series of pyrazolo [3, 4-*b*] pyridines in the presence of magnetic ldh as a nanocatalyst, Asian J. Nanosci. Mater. 3 (3) (2020) 238–250.
[165] M. Afradi, S.A. Pour, M. Dolat, A. Yazdani-Elah-Abadi, Nanomagnetically modified vitamin B_3 (Fe_3O_4@ niacin): an efficient and reusable green biocatalyst for microwave-assisted rapid synthesis of 2-amino-3-cyanopyridines in aqueous medium, Appl. Organomet. Chem. 32 (2) (2018) e4103.
[166] A. Maleki, R. Firouzi-Haji, l-Proline functionalized magnetic nanoparticles: a novel magnetically reusable nanocatalyst for one-pot synthesis of 2,4,6-triarylpyridines, Sci. Rep. 8 (1) (2018) 1–8.
[167] M.A. Zolfigol, M. Safaiee, F. Afsharnadery, N. Bahrami-Nejad, S. Baghery, S.S alehzadeh, F. Maleki, Silica vanadic acid [SiO_2-$VO(OH)_2$] as an efficient heterogeneous catalyst for the synthesis of 1, 2-dihydro-1-aryl-3 *H*-naphth [1,2-*e*][1,3] oxazin-3-one and 2,4,6-triarylpyridine derivatives via anomeric-based oxidation, RSC Adv. 5 (122) (2015) 100546–100559.
[168] I. Thiel, M. Hapke, The first solid-supported cp′ Co (I)-catalyst for the synthesis of pyridines, J. Mol. Catal. A. Chem. 383 (2014) 153–158.
[169] W.P. Yen, P.L. Liu, N. Uramaru, H.Y. Lin, F.F. Wong, Indium (III) chloride/silica gel catalyzed synthesis of pyrazolo [3,4-*b*] pyrrolo[3,4-*d*] pyridines, Tetrahedron 71 (46) (2015) 8798–8803.
[170] K. Mal, S. Chatterjee, A. Bhaumik, C. Mukhopadhyay, Mesoporous MCM-41 silica supported pyridine nanoparticle: a highly efficient, recyclable catalyst for expeditious synthesis of quinoline derivatives through domino approach, Chem. Select. 4 (5) (2019) 1776–1784.
[171] L. Ma'Mani, E. Hajihosseini, M. Saeedi, M. Mahdavi, A. Asadipour, L. Firoozpour, A. Shafiee, A. Foroumadi, Sulfonic acid supported phosphonium based ionic liquid functionalized SBA-15 for the synthesis of 2-amino-3-cyano-4, 6-diarylpyridines, Synth. React. Inorg. Met. Org. Chem. 46 (2) (2016) 306–310.
[172] F. Rahmani, I. Mohammadpoor-Baltork, A.R. Khosropour, M. Moghadam, S. Tangestaninejad, V. Mirkhani, Propyl phosphonium hydrogen carbonate ionic liquid supported on nano-silica as a reusable catalyst for the

efficient multicomponent synthesis of fully substituted pyridines and bis-pyridines, RSC Adv. 5 (50) (2015) 39978–39991.

[173] W. Lin, C. Zhuang, X. Hu, J. Zhang, J. Wang, Alcohol participates in the synthesis of functionalized coumarin-fused pyrazolo [3,4-*b*] pyridine from a one-pot three-component reaction, Molecules 24 (15) (2019) 2835–2851.

[174] A.R. Moosavi-Zare, M.A. Zolfigol, Z. Rezanejad, Trityl chloride promoted the synthesis of 3-(2, 6-diarylpyridin-4-yl)-1*H*-indoles and 2, 4, 6-triarylpyridines by in situ generation of trityl carbocation and anomeric-based oxidation in neutral media, Can. J. Chem. 94 (7) (2016) 626–630.

[175] P.K. Jaiswal, V. Sharma, M. Mathur, S. Chaudhary, Organocatalytic modified Guareschi-Thorpe type regioselective synthesis: a unified direct access to 5, 6, 7, 8-tetrahydroquinolines and other alicyclic [*b*]-fused pyridines, Org. Lett. 20 (19) (2018) 6059–6063.

[176] J. Afsar, M.A. Zolfigol, A. Khazaei, M. Zarei, Y. Gu, D.A. Alonso, A. Khoshnood, Synthesis and application of melamine-based nano catalyst with phosphonic acid tags in the synthesis of (3′-indolyl) pyrazolo [3, 4-*b*] pyridines via vinylogous anomeric-based oxidation, Mol. Catal. 482 (2020) 110666–110678.

[177] G. Kour, M. Gupta, B. Vishwanathan, K. Thirunavukkarasu, Cu/NCNTs): a new high temperature technique to prepare a recyclable nanocatalyst for four component pyridine derivative synthesis and nitroarenes reduction, New J. Chem. 40 (10) (2016) 8535–8542.

[178] S. Abdolmohammadi, B. Mirza, E. Vessally, Immobilized tio_2 nanoparticles on carbon nanotubes: an efficient heterogeneous catalyst for the synthesis of chromeno [*b*] pyridine derivatives under ultrasonic irradiation, RSC Adv. 9 (71) (2019) 41868–41876.

[179] M. Safaiee, B. Ebrahimghasri, M.A. Zolfigol, S. Baghery, A. Khoshnood, D.A. Alonso, Synthesis and application of chitosan supported vanadium oxo in the synthesis of 1,4-dihydropyridines and 2,4,6-triarylpyridines via anomeric-based oxidation, New J. Chem. 42 (15) (2018) 12539–12548.

[180] Y. Zhang, X. Zhai, H. Zhang, J. Zhao, Enhanced selectivity in the conversion of glycerol to pyridine bases over HZSM-5/11 intergrowth zeolite, RSC Adv. 7 (38) (2017) 23647–23656.

[181] S. Kumari, A. Shekhar, D.D. Pathak, Graphene oxide–TiO_2 composite: an efficient heterogeneous catalyst for the green synthesis of pyrazoles and pyridines, New J. Chem. 40 (6) (2016) 5053–5060.

[182] S. Singh, P. Chauhan, M. Ravi, P.P. Yadav, Eosin Y-Yb $(OTf)_3$ catalyzed visible light mediated electrocyclization/indole ring opening towards the synthesis of heterobiaryl-pyrazolo [3, 4-B] pyridines, New J. Chem. 42 (9) (2018) 6617–6620.

[183] M. Zhang, M.N. Chen, Z.H. Zhang, Visible light-initiated catalyst-free one-pot, multicomponent construction of 5-substituted indole chromeno [2,3-*b*] pyridines, Adv. Synth. Catal. 361 (22) (2019) 5182–5190.

[184] M. Fragkiadakis, M. Kidonakis, L. Zorba, M. Stratakis, Synthesis of 3-keto pyridines from the conjugated allenone-alkynylamine oxidative cyclization catalyzed by supported Au nanoparticles, Adv. Synth. Catal. 362 (4) (2020) 964–968.

[185] S.J. Shams-Najafi, M. Gholizadeh, A. Ahmadpour, TiO_2-nanoparticles as efficient catalysts for the synthesis of pyridine dicarbonitriles, J. Chin. Chem. Soc. 66 (11) (2019) 1531–1536.

[186] S.P. Gadekar, M.K. Lande, Solid acid catalyst TS-1 zeolite-assisted solvent-free one-pot synthesis of poly-substituted 2,4,6-triaryl-pyridines, Res. Chem. Intermed. 44 (5) (2018) 3267–3278.

[187] Q. Yang, Y. Zhang, W. Zeng, Z.C. Duan, X. Sang, D. Wang, Merrifield resin-supported quinone as an efficient biomimetic catalyst for metal-free, base-free, chemoselective synthesis of 2,4,6-trisubstituted pyridines, Green Chem. 21 (20) (2019) 5683–5690.

[188] (a) S. Gupta, A. Maji, D. Panja, M. Halder, S. Kundu, CuO NPs catalyzed synthesis of quinolines, pyridines, and pyrroles via dehydrogenative coupling strategy. J. Catal. 413 (2022) 1017–1027. (b) S. R. Ambati, J. L. Patel, K. Chandrakar, U. Sarkar, S. Penta, S. Banerjee, R. S. Varma, One-pot, three-component synthesis of novel coumarinyl-pyrazolo [3, 4-b] pyridine-3-carboxylate derivatives using [AcMIm] FeCl4 as recyclable catalyst. J. Mol. Struct. 1268 (2022) 133623–133624. (c) M. A. Shalaby, H. M. Al-Matar, A. M. Fahim, S. A. Rizk, A new approach to chromeno [4, 3-b] pyridine: Synthesis, X-ray, spectral investigations, hirshfeld surface analysis, and computational studies. J. Phys. Chem. Solids 170 (2022) 110933–110947. (d) X. Li, W. Cai, Y. Huang, One-pot Synthesis of 2, 3, 6-trisubstituted pyridines by phosphine-catalyzed annulation of γ-vinyl allenoates with enamino esters followed by DDQ-promoted oxidative aromatization. Adv. Synth. Catal. 364 (11) (2022) 1879–1883. (e) N. Zarei, M. Torabi, M. Yarie, M. A. Zolfigol, Novel urea-functionalized magnetic nanoparticles as a heterogeneous hydrogen bonding catalyst for the synthesis of new 2-hydroxy pyridines. Polycycl. Aromat. Compd. 4 (2022)

1263–1280. (f) A. Maji, S. Gupta, M. Maji, S. Kundu, Well-defined phosphine-free manganese (II)-complex-catalyzed synthesis of quinolines, pyrroles, and pyridines. J. Org. Chem. 87 (13) (2022), 8351–8367. (g) P. Qian, L. Xu, W. Wang, L. Zhang, L. Tang, J. Liu, L. Sheng, Electrochemical synthesis of dipyrazolo/dipyrimidine-fused pyridines via oxidative domino cyclization of C (sp3)-H bonds. Org. Chem. Front. 9 (6) (2022) 1662–1667. (h) B. Nie, W. Wu, C. Jin, Q. Ren, J. Zhang, Y. Zhang, H. Jiang, Pd (II)-catalyzed synthesis of alicyclic [b]-fused pyridines via C (sp^2)H activation of α, β-unsaturated N-acetyl hydrazones with vinyl azides. J. Org. Chem. 87 (1) (2021) 159–171. (i) A. Rakshit, H. N. Dhara, A. K. Sahoo, T. Alam, B. K. Patel, Pd (II)-catalyzed synthesis of furo [2, 3-b] pyridines from β-ketodinitriles and alkynes via cyclization and NH/C annulation. Org. Lett. 24 (20) (2022) 3741–3746. (j) R. F. Barghash, W. M. Eldehna, M. Kovalová, V. Vojáčková, V. Kryštof, H. A. Abdel-Aziz, One-pot three-component synthesis of novel pyrazolo [3, 4-b] pyridines as potent antileukemic agents. Eur. J. Med. Chem. 227 (2022) 113952. (k) K. M. Al-Zaydi, M. Al-Boqami, N. M. Elnagdi, Green synthesis of dihydropyrimidines and pyridines utilizing biginelli reaction. Polycycl. Aromat. Compd. (2021)1-12. (l) K. Li, Y. Lv, Z. Lu, X. Yun, S. Yan, An environmentally benign multi-component reaction: Highly regioselective synthesis of functionalized 2-(diarylphosphoryl)-1, 2-dihydro-pyridine derivatives. Green Synth. Catal. 3 (1) (2022) 59–68. (m) A. Negi, S. I. Mirallai, S. Konda, P. V. Murphy, An improved method for synthesis of non-symmetric triarylpyridines. Tetrahedron 121 (2022) 132930. (n) M. Torabi, M. Yarie, M. A. Zolfigol, S. Azizian, Y. Gu, A magnetic porous organic polymer: catalytic application in the synthesis of hybrid pyridines with indole, triazole and sulfonamide moieties. RSC Adv. 12 (14) (2022) 8804–8814. (o) M. Torabi, M. A. Zolfigol, M. Yarie, Y. Gu, Application of ammonium acetate as a dual rule reagent-catalyst in synthesis of new symmetrical terpyridines. Mol. Catal. 516 (2021) 111959–111967. (p) H. Sepehrmansourie, M. Zarei, M. A. Zolfigol, S. Babaee, S. Azizian, S. Rostamnia, Catalytic synthesis of new pyrazolo [3, 4-b] pyridine via a cooperative vinylogous anomeric-based oxidation. Sci. Rep. 12 (1) (2022) 14145.

CHAPTER

16

Pyridine as a potent antimicrobial agent and its recent discoveries

Nitish Kumar, Harmandeep Kaur, Anchal Khanna, Komalpreet Kaur, Jatinder Vir Singh, Sarabjit Kaur, Preet Mohinder Singh Bedi and Balbir Singh

Department of Pharmaceutical Sciences, Guru Nanak Dev University, Amritsar, Punjab, India

16.1 Introduction

Microorganism, a class which is residing on earth prior to the evolution of humans. These either exist as a single cell or live in by forming colonies. These microbes were invisible till the scientific study of these microorganisms was initiated and the first observation was made by Antonie van Leeuwenhoek in 1670 [1]. Later, with advancement in technology, these microbes were classified into two classes, that is, prokaryotes (archaea and bacteria) and eukaryotes [2]. These microbes exist in every type of habitat on earth that includes very hot or cold conditions too [2].

These microbes are very useful to humans as they help in digesting the waste of sewage; the production of dairy products, biofuels, antibiotics, and many more. These are also present in every mammal where they assist the host's system to digest the ingested food by consuming the food and converting it into required products. In addition to this, these microbes have also been misused as bioweapons to attack their opposition country [3]. It is saying that "everything within limits is safe and not harmful to humans or the world." Therefore, when any microorganism crosses its safe limits, it starts harming the host which leads to degradation of the health of the host. Microbes have been affecting humans for a very long time and have caused a lot of casualties and have become a challenge to the healthcare system [4]. These increased casualties led to the discovery of antimicrobials.

Antimicrobials are the class of drugs which either kills the microbes or retards the growth of the microbes. The first antimicrobial was discovered in 1940 and was found to save many lives

DOI: https://doi.org/10.1016/B978-0-323-91221-1.00003-8

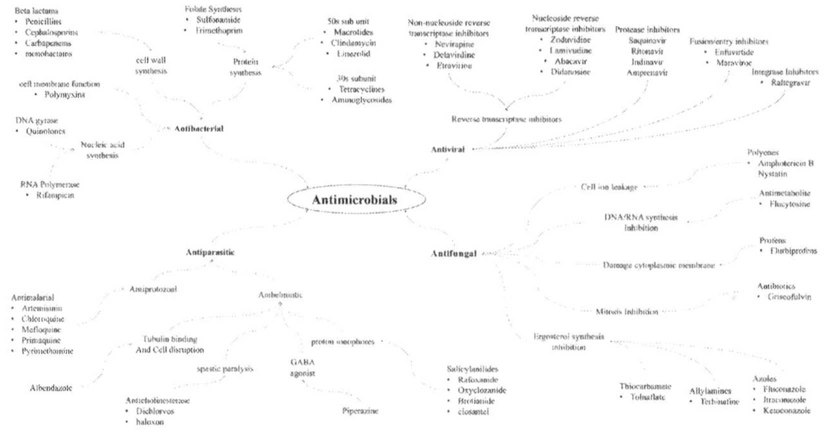

FIGURE 16.1 Flow chart of antimicrobials that are currently in use in daily clinic practice, with their brief mechanism of action.

till now. These antimicrobials are derived from synthetic (e.g., levofloxacin), semisynthetic or from natural (e.g., Penicillin) sources and every category of antibiotics has its own pathway or mechanism of action as shown in Fig. 16.1 [5].

Later, due to misuse or improper use of antibiotics created a new problem that led to the development of resistance to these developed antimicrobials [6]. This resistance has led to 0.7 million deaths every year which can reach up to 10 million by 2050 if uncontrolled [7]. This development of resistance and increase in deaths led researchers to search and develop more potent or efficient antimicrobials that have the power to mitigate this resistance problem. During the research, the researcher either works by modifying lead or by searching for a new antimicrobial nucleus.

Among all the nuclei in chemistry, heterocyclic compounds have a wide range of biological profiles. Among these compounds pyridine has played a vital role in enumerating various structures of antimicrobials and has gained a lot of attention too [8]. Pyridine is a six-membered heterocyclic compound where there is a bioisostere of one carbon of benzene with nitrogen which might be responsible for diverse biological profiles [9]. This Nitrogen atom also results in weaker resonance stabilization and development of dipole moment of 2.2 debyes. [10]. Moreover, crystals of pyridine are orthorhombic in shape and physically have a diamagnetic character. This physical and chemical nature of pyridine makes it easier for a researcher to modify pyridine in their desirable way. Moreover, pyridine is a least abundant compound in nature and has also been found to be present in leaves and roots of marshmallow and belladonna herbs. But pyridine is present as part of the main nucleus in a few classes of phytoconstituents such as alkaloids [11].

So, considering the importance of pyridine, we have discussed and classified heterocyclic compounds that have been discovered since 2016 with pyridine as a substituent or as the main nucleus which may help researchers to develop more or highly specific antimicrobial so that there will be the least chance of production of antimicrobial resistance.

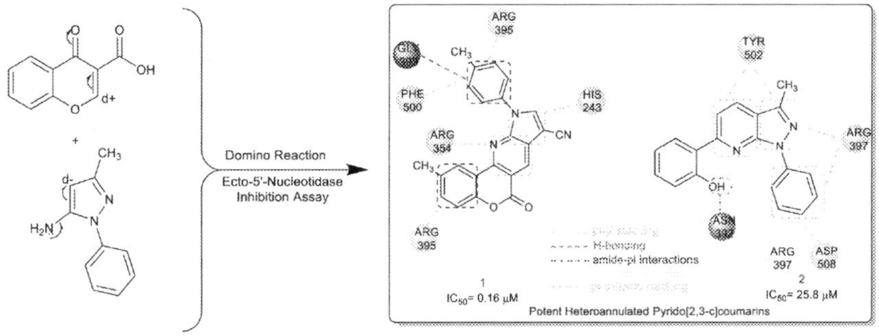

FIGURE 16.2 Heteroannulated pyrido[2,3-c] coumarin derivatives as inhibitors of both human and rat e5′NT.

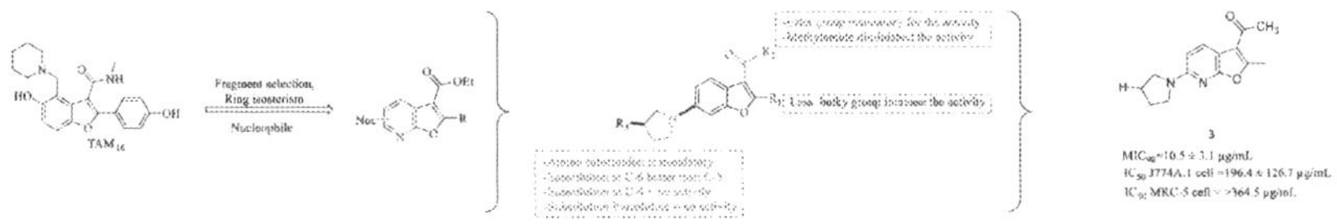

FIGURE 16.3 Furo[2,3-b] pyridine derivatives through C–H activation approach as *Mycobacterium tuberculosis* inhibitors.

16.2 Pyridine as antimicrobial agents

16.2.1 Compounds containing oxygen in the heterocyclic nucleus

Miliutina et al. (2017) synthesized a series of heteroannulated pyrido[2,3-c] coumarin derivatives by domino reactions of chromone-3-carboxylic acid derivatives with amino-substituted heterocycles and exhibited strong fluorescence as shown in Fig. 16.2. The synthesized compounds were further investigated for their inhibitory potential against both human and rat ecto-5′-nucleotidase (e5′NT). Almost all the compounds were active against human e5′NT except isoxazole derivatives. Structural activity relationship studies suggest that aromatic substitution at the 1-position of the pyrrole moiety and electron-donating group on the coumarin enhances activity. Compound **1** came out to be hit among the series against human e5′NT with an IC_{50} value of 0.16 μM. Also, compound **2** came out to be most potent against rat e5′NT. So, docking of the potent compounds **1** and **2** was carried out to study significant binding interactions with h-e5′ NT and r-e5′NT, respectively [12].

In an attempt to overcome the problem of antimicrobial resistance, Fumagalli et al. (2019) synthesized a series of furo[2,3-b] pyridine derivatives through the C–H activation approach as *Mycobacterium tuberculosis* inhibitors as shown in Fig. 16.3. All compounds were screened against three mycobacterial strains (H37Rv, J774A.1, MRC-5) and compound **3** showed potency with an $MIC_{90} < 10$ μg/mL. Cytotoxicity assay of the most promising compounds was carried out against MRC-5 cell line and J774A.1 and were found to be nontoxic. Antimicrobial resistance of potent compounds was further investigated against resistant clinical isolates and compound **3** was active against multidrug-resistant tuberculosis. Low molecular weight, low MIC_{90} against mycobacterial strains and drug-resistant clinical isolates of *M. tuberculosis* and a high SI of compound **3** makes it a lead-like compound and for use in fragment-based drug designing process [13].

16.2.2 Compounds containing nitrogen in the heterocyclic nucleus

Lu et al. (2017) designed a series of pyrazolo[1,5-a]pyridine-3-carboxamide hybrids which were synthesized and evaluated for their antitubercular activity as shown in Fig. 16.4. Among all the synthesized compounds, compound **4** was found to be the most potent compound. The antitubercular activity of compound **4** was further assessed using various methods such as in vivo cost-efficient mouse model infected with the selectable marker-free auto luminescent *M. tuberculosis* H37Ra strain. Dose-dependent activity was exhibited by compound **4** with an IC_{50} of range 0.003–0.014 μg/mL. Thus, concluded that compound **4** can be further used for the development of new antitubercular drugs [14].

Gilish Jose et al. (2017) designed and biologically evaluated new series of Mannich bases, obtained via one-pot three-component condensation of pyrrolo[3,2-c]pyridine scaffold (**6a–c**) with secondary amines as shown in Fig. 16.5. Upon evaluation via in vitro antimicrobial activities, few compounds showed good gram-positive antibacterial activity, that is, against *S. aureus, B. flexus, C. sporogenes*, and *S. mutans*. These active compounds were then screened against *Mycobacterium tuberculosis* H37Rv (ATCC 27294) using MABA. Among all the tested compounds, compound **5** was found to be the most potent antimycobacterial compound against *M. tuberculosis* and possesses a low cytotoxicity against the HEK-293T cell line [15].

Sunheekang et al. (2017) designed and synthesized a series of novel fused ring analogues of Q203, that is, imidazo[1,2-a]pyridine-3-carboxamide (IPA) analogue as potent antitubercular agents as shown in Fig. 16.6. In order to reduce the length of the side chain and log*P* values, the linearly extended side chain of Q203 was replaced with a set of shorter fused rings. In comparison with Q203, some of the synthesized analogues showed remarkable antitubercular activities without a long side chain. Out of all the molecules, compound **6** exhibited excellent orally available pharmacokinetic properties, high drug exposure level and long half-life, which makes this compound a promising antitubercular molecule that showed approximately the same potency and ADME properties as that of compound Q203. At last antitubercular activity was performed and the compound **6** possessed MIC value of 0.009 μg/mL (extracellular), <0.001 μg/mL (intracellular) against *M. tuberculosis* strain [16].

Watanabe et al. (2017) synthesized a series of compounds containing cyanomethylated tetrahydropyridine nucleus and evaluated them for their hepatitis C virus inhibition as shown in Fig. 16.7. Among them, compound **7** was found to be potent and exhibits optimum anti-HCV activity with the IC_{50} value of 10.85 ± 1.45 μM and suggested that these nitrile-containing compounds can be further optimized for efficiency [17].

Lv et al. (2017) designed and synthesized a series of novel imidazo [1,2-a] pyridine-3-carboxamide derivatives as potent antimycobacterial agents against various strains of *M. tuberculosis* as shown in Fig. 16.8. Among all the synthesized compounds, compounds **8** and **9** were found to be most potent against *M. tuberculosis* H37Rv (ATCC, 27294) strains of *M. tuberculosis*, drug-sensitive/resistant multidrug-resistant *M. tuberculosis* (MDR-MTB)11168 strain and multidrug-resistant *M. tuberculosis*9160 strain. Also, the compound **10** was found to exhibit good aqueous solubility as compared to the other synthesized compounds [18].

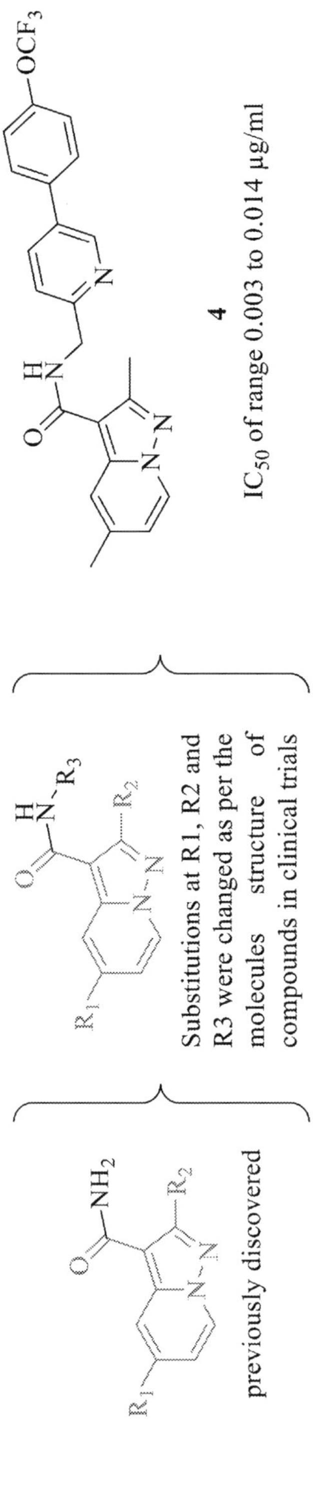

FIGURE 16.4 Pyrazolo[1,5-a]pyridine-3-carboxamide hybrids as antitubercular agents.

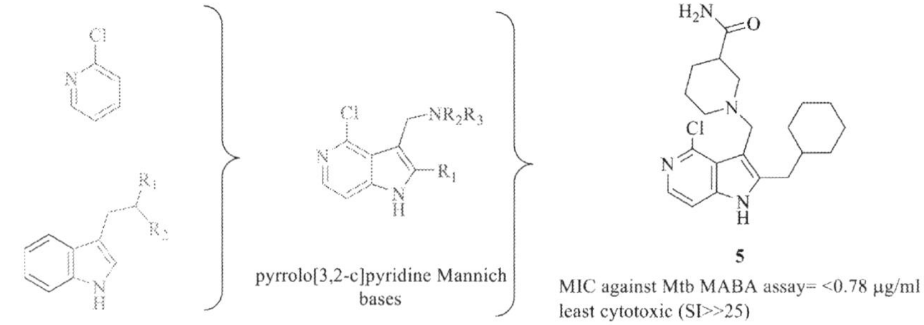

FIGURE 16.5 Pyrrolo[3,2-c]pyridine Mannich bases as antimicrobial agents.

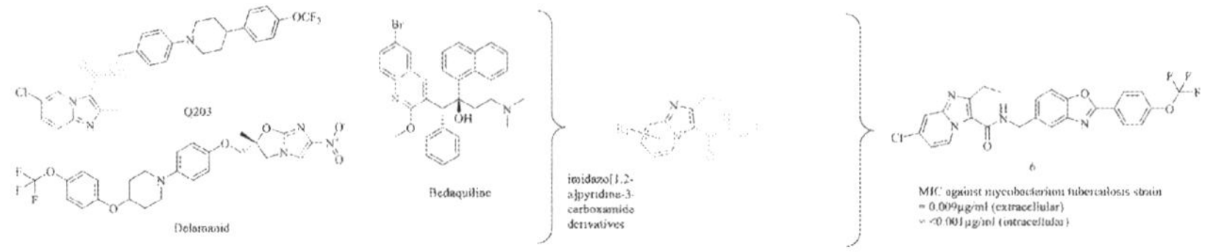

FIGURE 16.6 Imidazo[1,2-a]pyridine-3-carboxamide as antitubercular agents.

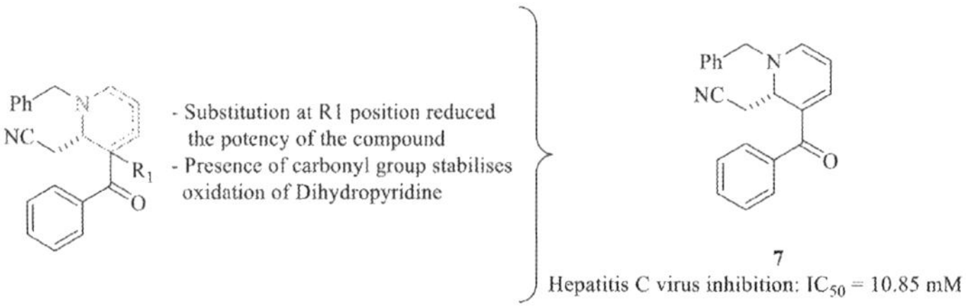

FIGURE 16.7 Cyanomethylated tetrahydropyridines as hepatitis C virus inhibitors.

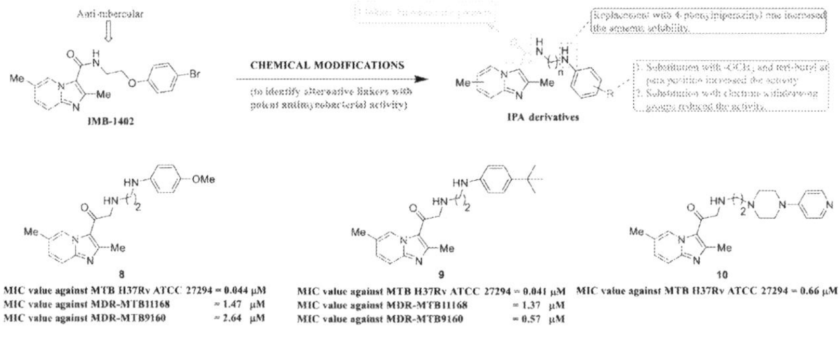

FIGURE 16.8 Novel imidazo [1,2-a] pyridine-3-carboxamide (IPA) derivatives as potent antimycobacterial agents.

Sajja et al. (2017) reported novel series of benzo [6,7]cyclohepta[1,2-b]pyridine-1,2,3-triazole hybrids as antimycobacterial agents and were evaluated against *M. tuberculosis*H37Rv as shown in Fig. 16.9. All compounds were found to be active with MICs ranging between 1.56 and 25 μg/mL. Compounds **11** and **12** were found to be potent with a MIC value of 1.56 μg/mL, taking isoniazid, rifampicin, ethambutol as standard antitubercular drugs.

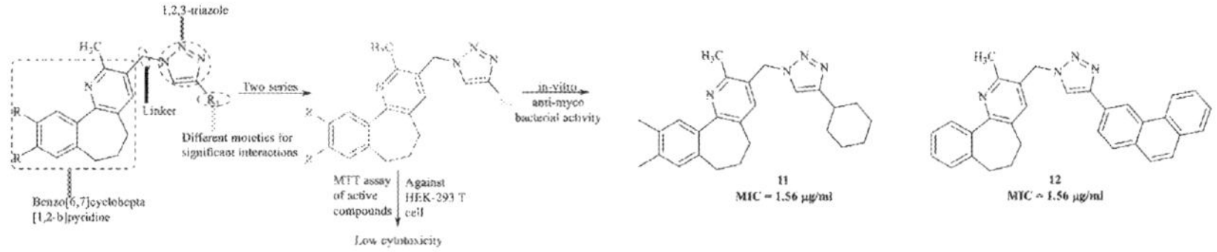

FIGURE 16.9 Benzo [6,7] cyclohepta[1,2-b]pyridine-1,2,3-triazole hybrids as antimycobacterial agents.

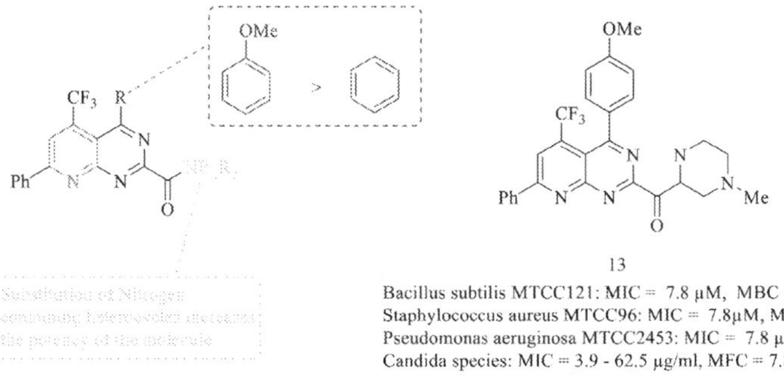

Bacillus subtilis MTCC121: MIC = 7.8 μM, MBC = 15.6 μg/ml
Staphylococcus aureus MTCC96: MIC = 7.8μM, MBC = 15.6 μg/ml
Pseudomonas aeruginosa MTCC2453: MIC = 7.8 μM, MBC = 15.6 μg/ml
Candida species: MIC = 3.9 - 62.5 μg/ml, MFC = 7.8 - 62.5 μg/ml
Biofilm inhibition Assay:
Bacillus subtilis MTCC121: IC_{50} = 2.7 ± 0.13 μg/ml
Pseudomonas aeruginosa MTCC2453: IC_{50} = 2.7 ± 0.16 μg/ml

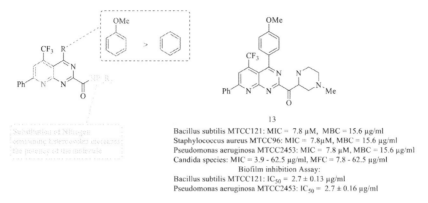

FIGURE 16.10 Novel pyrido[2,3-d] pyrimidine derivatives as antimicrobial agents.

MTT assay was carried out to evaluate the safety profile of active derivatives (MIC < 12.5 mg/mL) against HEK-293 T cells and were found to be nontoxic [19].

Veeraswamy et al. (2018) synthesized a novel pyrido[2,3-d] pyrimidine derivatives and evaluated them for their antimicrobial activity as shown in Fig. 16.10. Among them, compound **13** was the most potent against *Bacillus subtilis* MTCC121, *Staphylococcus aureus* MTCC96, *Pseudomonas aeruginosa* MTCC2453 for antibacterial activity. Further, a biofilm inhibition assay was carried out which shows compound **13** possesses excellent activity against *B. subtilis* MTCC121, *Pseudomonas aeruginosa* MTCC2453. Moreover, compound **13** was active against *Candida* species and showed minimum fungicidal concentration (MFC) between 7.8 and 62.5 μg/mL. Structure–activity pattern suggests that substitution of N-methylpyrazinyl on the carbonyl linkage of pyrimidine nucleus in the structure was most suitable for an antibacterial activity [20].

El-Gohary and Shaaban (2018) designed and synthesized novel pyrazolopyridine analogs which were further screened for their antimicrobial activity against various microbes as shown in Fig. 16.11. Among all the compounds screened, the 5-carboxylic acid analogs were found to possess broad-spectrum antimicrobial activity, with more efficacy for gram-positive bacteria. These compounds also showed remarkable activity over *S. aureus*, *E. coli*, and *C. albicans*. The compounds were assessed for prognosis of Lipinski's rule and Veber's norms using Molinspiration software. By these parameters, it was found that all the analogs were feasible for oral absorption. PreADMET software was used to predict carcinogenic effects if any. Most of the compounds were noncarcinogenic. Out of all compounds, compound **14** was found to be active against the majority of the strains. They were found to be antimicrobial, antiquorum, and antitumor [21].

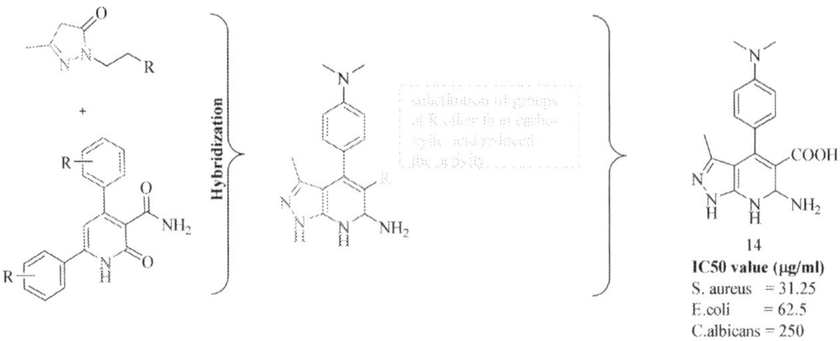

FIGURE 16.11 Pyrazolopyridine analogs as antimicrobial agent.

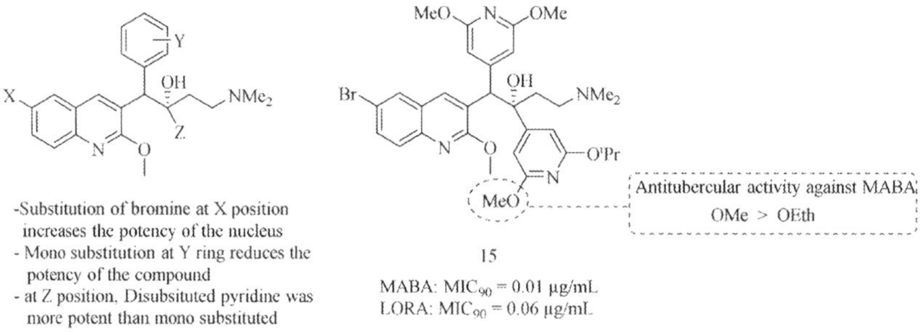

FIGURE 16.12 C-pyridyl analogues of the drug bedaquiline for tuberculosis.

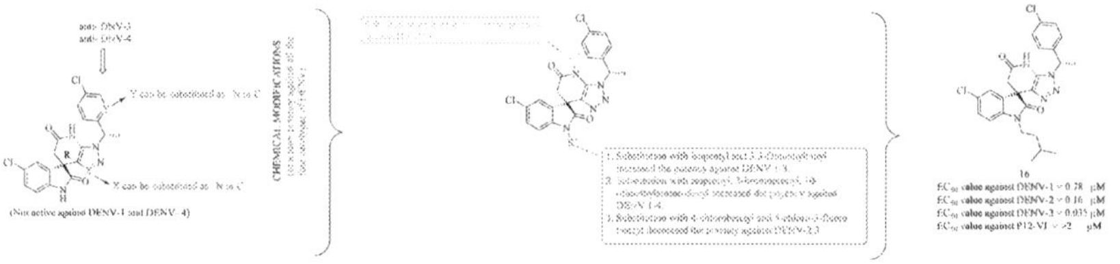

FIGURE 16.13 4,6-Dihydrospiro[[1,2,3]triazolo[4,5-b]pyridine-7,3′-indoline]-2′,5(3H)-dione analogues as potent NS4B inhibitors.

Blaser et al. (2019) developed structure–activity relationships for unit C pyridyl analogues of the drug bedaquiline and evaluated for tuberculosis as shown in Fig. 16.12. Among them compound **15** was the most potent against *M. tuberculosis*. Structure–activity pattern suggests that substitution of 3-methoxy next to the nitrogen atom of the pyridine ring in the structure was more suitable than 3-ethoxy against tuberculosis [22].

Xu et al. (2019) designed and synthesized a series of substituted 4,6-dihydrospiro[[1,2, 3]triazolo[4,5-b]pyridine-7,3′-indoline]-2′,5(3H)-dione analogues as potent NS4B inhibitors against DENV 1–4 serotypes of dengue virus as shown in Fig. 16.13. Among all the synthesized compounds, compound **16** was found to be most potent against DENV 1–3, respectively. On further evaluation, the efficacy of the compound **16** has also been revealed against resistant viruses P12-VI with an EC_{50} value of 2.0 μM. Lastly, the compound **16** demonstrated excellent in vivo pharmacokinetic properties and satisfactory efficacy in the A129 mouse model [23].

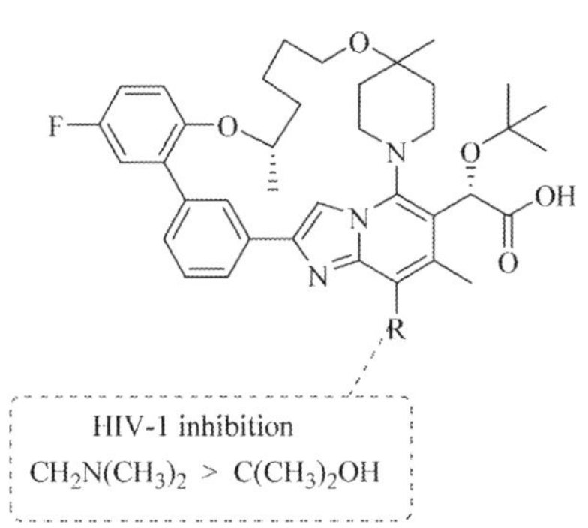

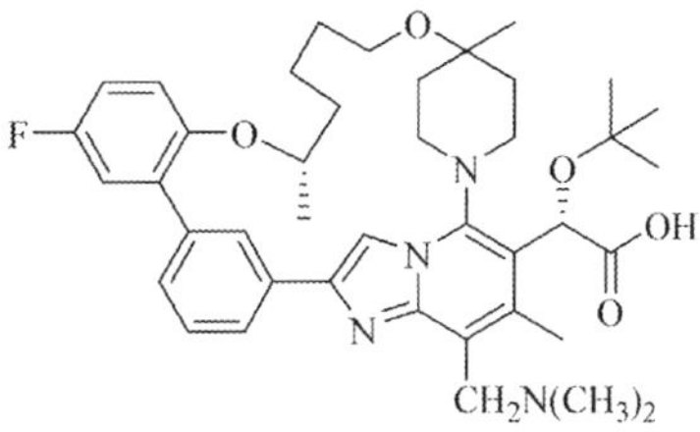

17
HIV-1 inhibition: EC_{50} = 4.4 nM
PXR transactivation: EC_{50} > 50 mM

FIGURE 16.14 Macrocyclic allosteric inhibitors of HIV-1 integrase.

Sivaprakasam et al. (2020) synthesized a series of macrocyclic allosteric inhibitors of HIV-1 integrase as shown in Fig. 16.14. Among them, compound **17** was the most potent against HIV-1 with the EC_{50} value of 4.4 ± 0.5 nM along with PXR transactivation with EC_{50} value >50 μM. Structure–activity pattern suggests that substitution of dimethylaminomethyl on the pyridine nucleus in the structure was most suitable for HIV-1 inhibition [24].

Singh et al. (2020) synthesized the series of functionalized pyridazine derivatives (**18–27**) by reacting tetrazine compound with various functional units by single-step iEDDA approach and studied for their antimicrobial activity, that is, against *E. coli, S. aureus, P. aeruginosa, K. pneumonia, B. subtilis* and *R. solani, S. rolfasii, F. oxysproum, A. niger* strains as shown in Fig. 16.15. Results of this study revealed that all compounds exhibited nonidentical behavior against different strains. The zone of inhibition (mm) of compounds **19, 20, 24, 26** and fungal inhibition (%) of compounds **20, 25, 24** came out to be 17–35 mm and 55–87% respectively as compared to other synthesized compounds. From in vitro results, it was concluded that compound **24** showed significant antimicrobial activity against bacteria and fungus as well [25].

16.2.3 Compounds containing more than one hetero atom in heterocyclic ring

Wang et al. (2016) synthesized a series of pyridinium-tailored 2,5-substituted-1,3,4-oxadiazole thioether/sulfoxide/sulfone derivatives and evaluated for their antibacterial activity as shown in Fig. 16.16. Among them, compound **28** was found to be the most potent. Structure–activity pattern suggests that increasing chain length between sulfur and pyridinium nucleus in the structure enhanced the antibacterial activity [26]

Keeping in view the vital role of indole, pyridine, and 1,2,3 oxadiazole moieties in pharmaceuticals, Desai et al. (2016) designed and amalgamated series of indole and pyridine based 1,2,4-oxadiazole derivatives and screened for their in vitro antitubercular, antiproliferative, and antibacterial activity as shown in Fig. 16.17. The antitubercular activity revealed that compounds **29, 30, 31,** and **32** exhibited excellent potency against *M. bovis* in both active and dormant strains with MIC values ranging from 0.94 to 5.17 μg/mL. Synthesized compounds were further evaluated for antiproliferative activity against three human cancer cell lines: HeLa, A5A9, PANC-1 and were found to be nontoxic. Cytotoxic assay reported that compounds **29, 30,** and **31** represented selectivity index greater than 10 which formed these

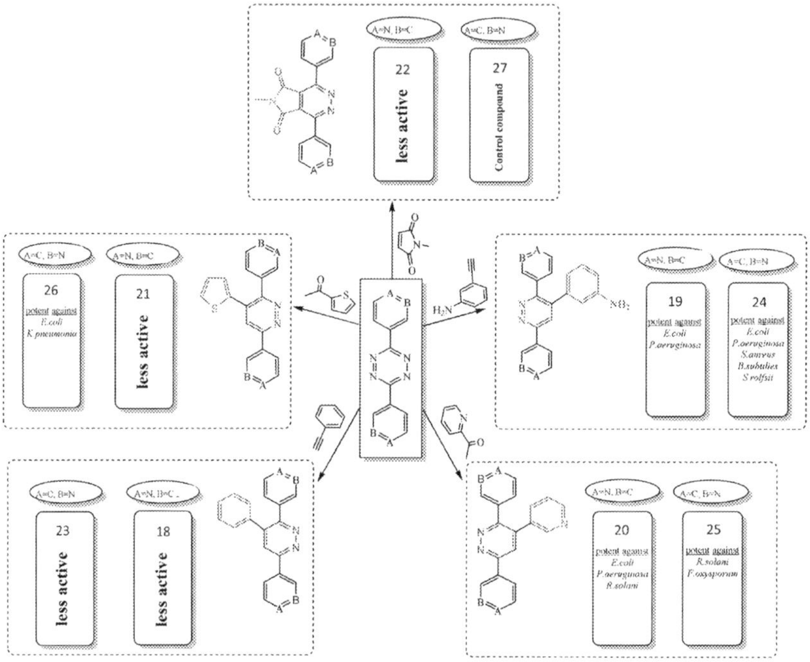

FIGURE 16.15 Functionalized tetrazine derivatives as antimicrobial agents.

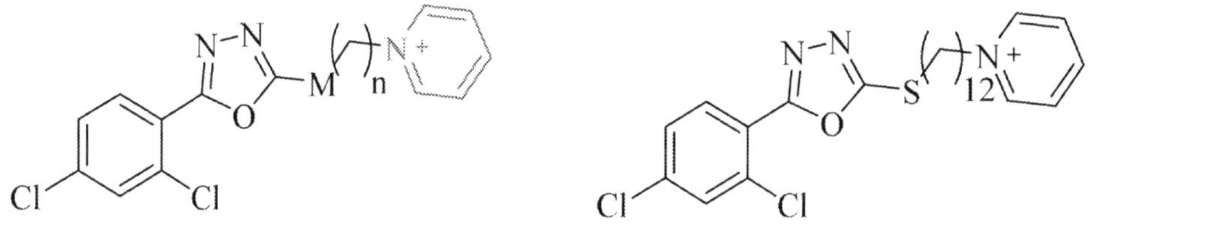

28

- n greater than 4 is necessary for optimal efficacy

Xoo: EC50 = 0.54 ± 0.03 ug/ml
Xac: EC50 = 1.62 ± 0.20 ug/ml
R. soolanacearum: EC 50 = 0.75± 0.03 ug/ml

FIGURE 16.16 Pyridinium-tailored 2,5-substituted-1,3,4-oxadiazole thioether/sulfoxide/sulfone derivatives as antibacterial agents.

compounds to be assured antitubercular compounds. Further, the active compounds were also docked into the active site of the mycobacterial enoyl reductase (InhA) enzyme [27].

Picconi et al. (2017) designed and synthesized a series of novel pyridyl nitrofuranylisoxazolines and screened them against multiple drug-resistant *Staphylococcus* strains as shown in Fig. 16.18. All the compounds were screened through the initial *Staphylococcus* strains panel and resultant potent compounds were screened through an extended *Staphylococcus* strains

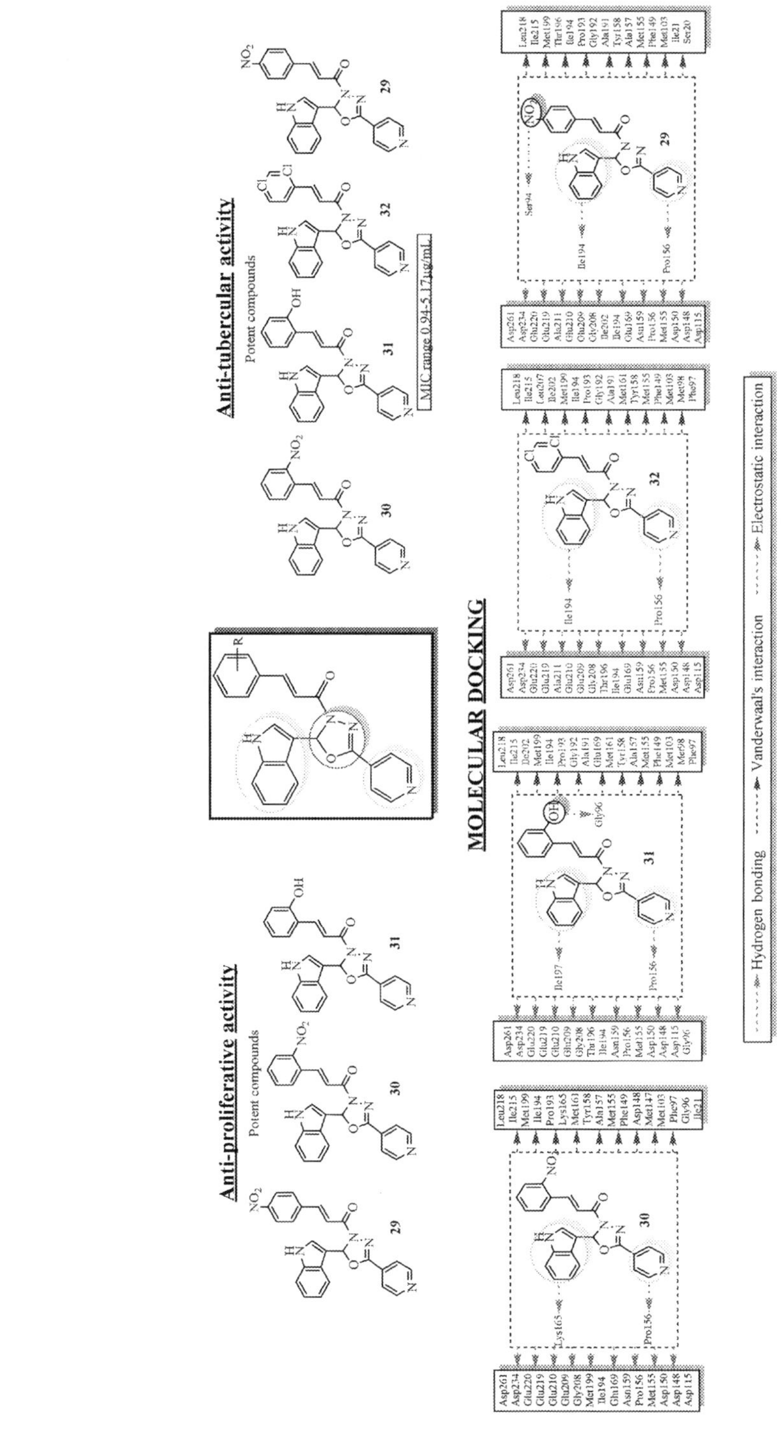

FIGURE 16.17 Indole and pyridine based 1,2,4-oxadiazole derivatives.

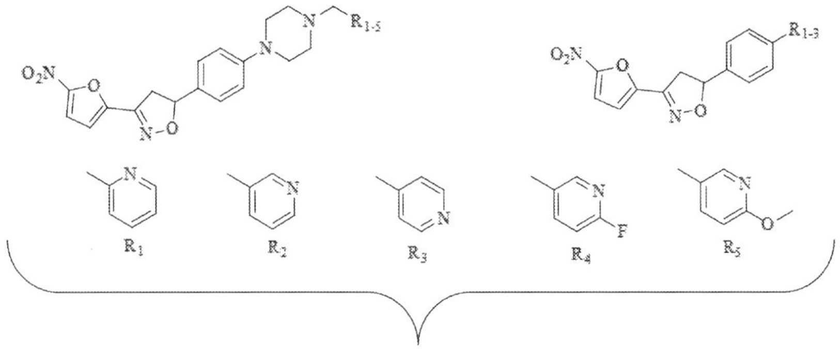

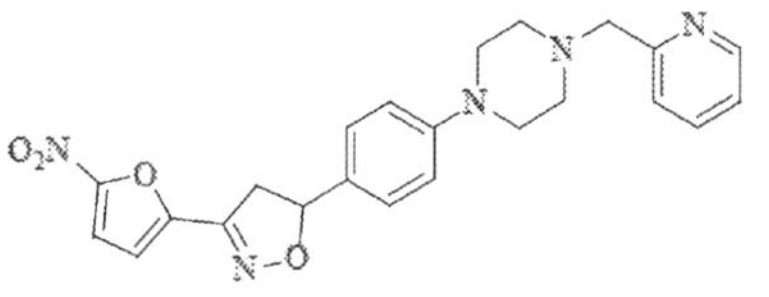

33

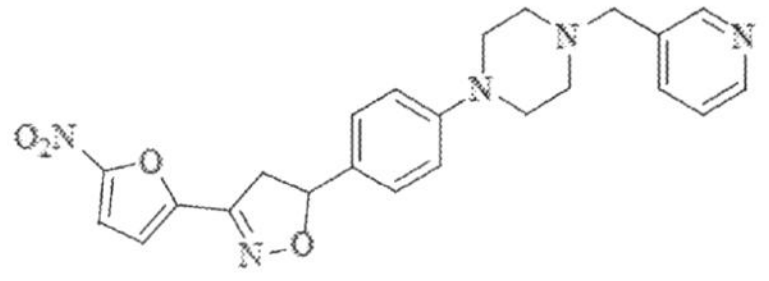

34

MIC values (μg/ml) Bacteria	33	34
S. epidermidis ATCC 35984	8–16	4–8
S. aureus A1988	16	8
S. aureus BGW541	8–16	8
MRSA NCTC 12493	8	4–8
S. aureus 13142	8–16	8–16
S. aureus 6571 PVL	16–32	16
S. aureus 8325 PVL	16	16

FIGURE 16.18 Pyridyl nitrofuranylisoxazolines derivatives active against multiple drug-resistant *Staphylococcus* strains.

panel. Results revealed that compounds with piperazine linker between pyridyl group and isoxazaline ring were found to be the most active. Compounds **33** and **34** were found to be potent in the initial *Staphylococcus* strains panel with MIC in the range of 4–32 μg/mL. These potent compounds on screening through extended *Staphylococcus* strains panel showed great activity with MIC's range 4–16 μg/mL. Hence it was concluded that these compounds are novel, noncytotoxic, and potent [28].

Patel et al. (2019) designed, synthesized, and evaluated a series of 1,3,4-thiadiazole derivatives and were further evaluated for their antimycobacterial activity against *M. tuberculosis* and resistance to multidrug-resistant tuberculosis strains as shown in Fig. 16.19. The results revealed that compound **35** has demonstrated the significant inhibitory activity with MIC of 9.87 mM against *M. tuberculosis*-H37Rv and multidrug-resistant TB, respectively, as compared to the isoniazid and rifampin. Structure–activity relationship reveals that the electron-withdrawing group on aliphatic side chain at second position of 1,3,4-thiadiazole has a diminishing effect on the antimycobacterial and multidrug-resistant inhibitory activity as compared to the unsubstituted aliphatic side chain. Upon checking their cytotoxic parameters, it was found that these compounds were not cytotoxic in nature [29].

Wang et al. (2019) designed and synthesized a series of less lipophilic Q203 imidazo[1,2-a] pyridine-3-carboxamide derivatives containing alkaline fused ring moieties as potent antimycobacterial agents against *M. tuberculosis*H37Rv strain and other drug-sensitive and

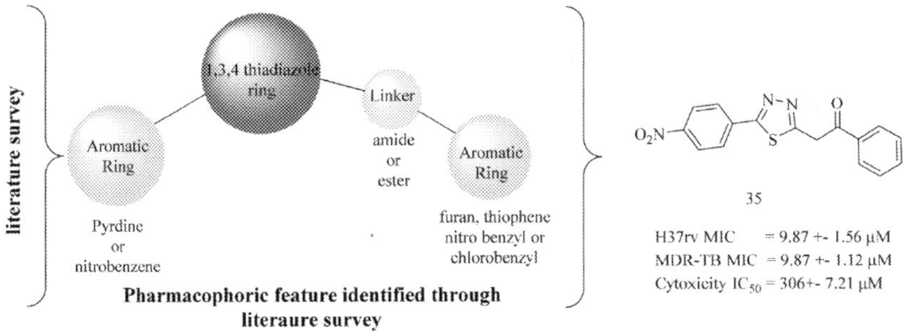

FIGURE 16.19 1,3,4-Thiadiazole derivatives as antimycobacterial agents.

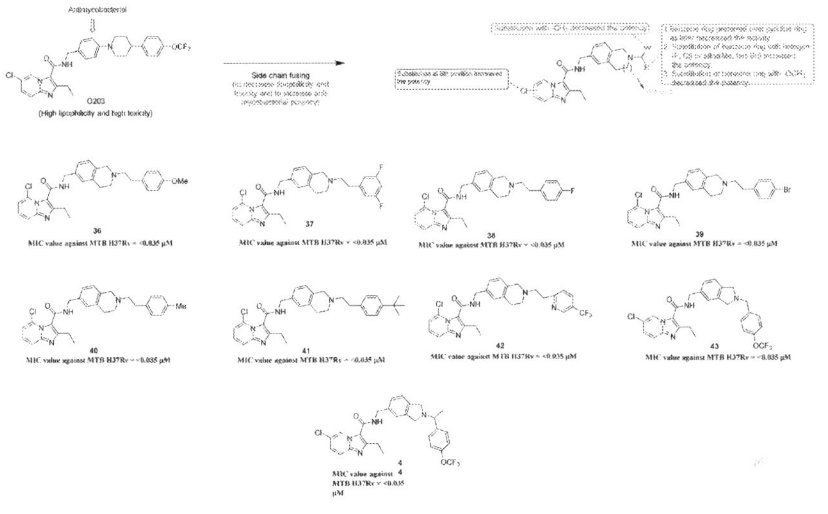

FIGURE 16.20 Q203 imidazo[1,2-a] pyridine-3-carboxamide derivatives containing alkaline fused ring moieties as potent antimycobacterial agents.

drug-resistant strains of *M. tuberculosis* as shown in Fig. 16.20. Among all the synthesized compounds, **36, 37, 38, 39, 40, 41, 42, 43**, and **44** were found to be most potent, each with the MIC value of < 0.035 μM against *M. tuberculosis* H37Rv strain of *M. tuberculosis*. This activity was found to be equipotent to the Q203 and PBTZ169. On further evaluation, it has been revealed that compound **42** displayed higher aqueous solubility and absorption in plasma and also exhibited a satisfactory safety level as compared to Q203. Lastly, the compounds **36, 37, 38, 39, 40, 41, 42, 43**, and **44** were reported to have lower cytotoxicity toward mammalian Vero cells [30].

To handle the expanding frequency of drug-resistant antifungal infections, Sun et al. (2019) designed and synthesized a series of compounds containing amide-pyridine scaffold as antifungal compounds using fragment-based drug discovery approach and tested for their in vitro antifungal activity against *C. albicans, C. glabrata, C. krusei*, and *C. tropicalis and Aspergillus fumigatus* strains as shown in Fig. 16.21. Results of the antifungal activity indicated that the compounds **45** showed potency against *C. tropicalis* with MIC value of 0.125 μg/mL and 46

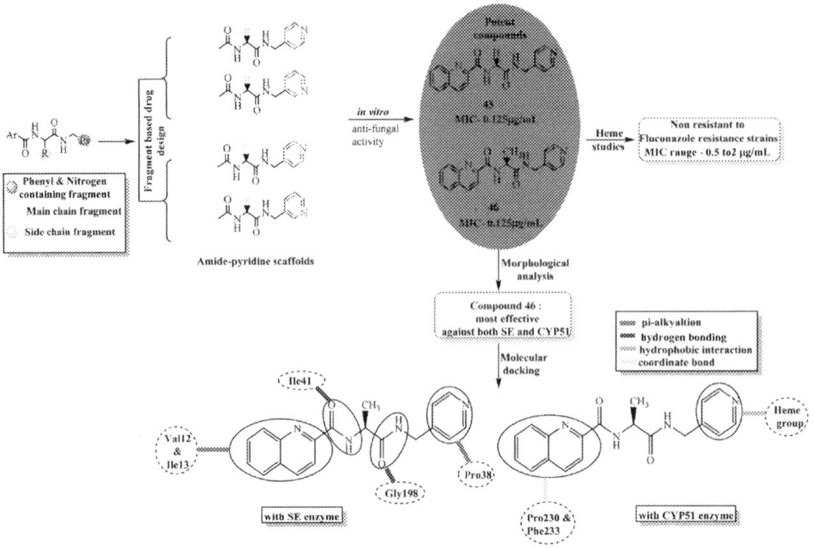

FIGURE 16.21 Compounds containing amide-pyridine scaffold as antifungal.

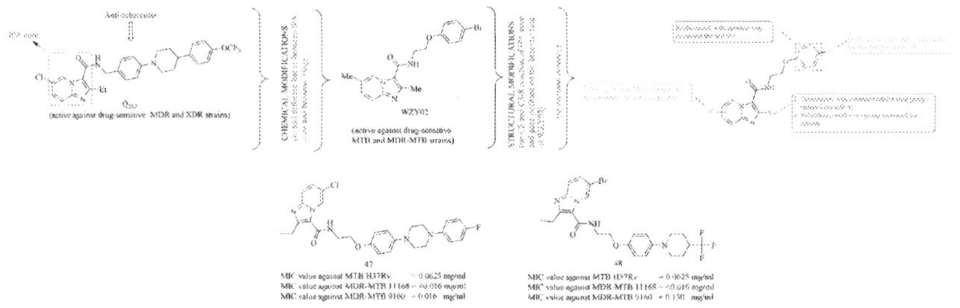

FIGURE 16.22 N-(2-phenoxy) ethyl imidazo[1,2-a] pyridine-3-carboxamides (IPAs) derivatives as potent antitubercular agents.

exhibited potency against *C. albicans* and *C. glabrata* with MIC value of 0.125 μg/mL. Then the potent compounds were further assessed with fluconazole-resistant strains of *C. albicans* and came out to be nonresistant with MIC values ranging 0.5–2 μg/mL. Morphological analysis revealed that compound **46** is the most effective against both SE and CYP51 activity. Further, heme studies were also performed to investigate the binding properties of **46** which exposed its strong attractions with heme-specific binding sites. In addition to it, molecular docking was also performed to examine the interactions between **46** and the dual target enzyme (SE and CYP51) [31].

Wang et al. (2019) designed and synthesized a series of N-(2-phenoxy) ethyl imidazo[1,2-a] pyridine-3-carboxamides derivatives as potent antitubercular agents against *M. tuberculosis as* shown in Fig. 16.22. Among all the synthesized compounds, **47** and **48** were found to be most potent with MIC values of 0.625 μg/mL each against drug-sensitive *M. tuberculosis*H37Rv strain, < 0.016 μg/mL each against multidrug-resistant *M. tuberculosis*11168 and finally 0.160 μg/mL and 0.130 μg/mL respectively against multidrug-resistant *M. tuberculosis*9160

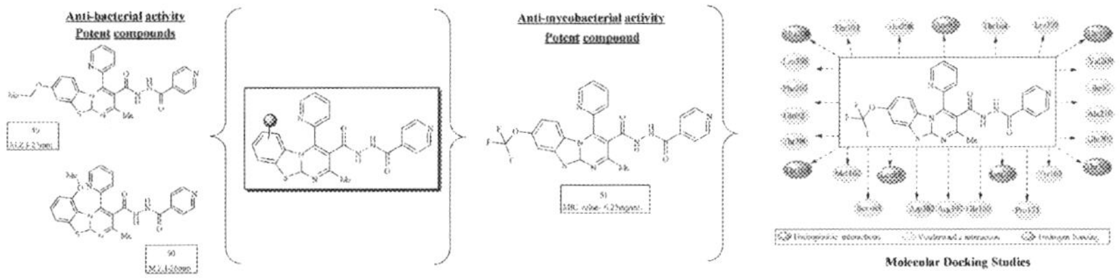

FIGURE 16.23 Isoniazid containing 4H-pyrimido[2,1-b]benzothiazoles derivatives as antibacterial agents.

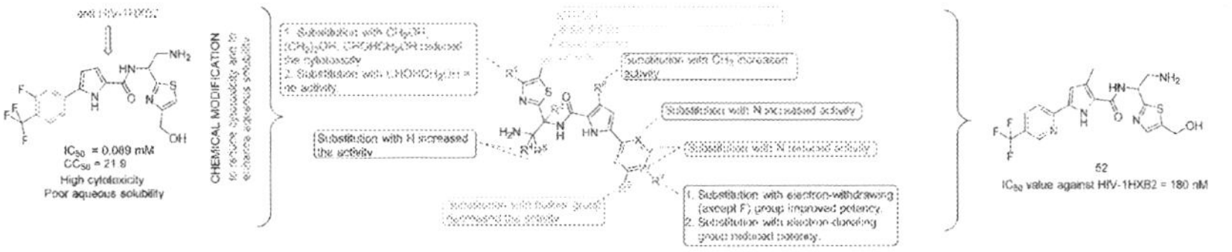

FIGURE 16.24 Novel pyridine-based gp 120 entry antagonists as potent anti-HIV-1 agents.

strain. On further evaluation, it has been revealed that the compounds **47** and **48** displayed high C_{max}, $AUC_{0-\infty}$, and long half-life ($T_{1/2}$). Lastly, compounds **47** and **48** showed very low oral acute lethal toxicity [32].

Bhoi et al. (2019) synthesized a series of novel isoniazid containing 4H-pyrimido[2,1-b] benzothiazoles derivatives and assessed for their in vitro antibacterial activity upon *E. aerogens, E. coli, M. luteus,* and *B. cereus* strains using streptomycin as the standard drug as shown in Fig. 16.23. Results of antibacterial activity highlighted the compounds **49** (M.Z.I-25mm) and 50 (M.Z.I-26mm) with the highest antibacterial potency against *E. aerogens* and *B. cereus*, respectively. Synthesized compounds were also studied for their in vitro antimycobacterial activity and concluded that compound **51** was most active with MIC value of 6.25 μg/mL. Further molecular docking was also conducted of compound **51** with the crystal structure of *M. tuberculosis* fatty acyl CoA synthetase enzyme [33].

Curreli et al. (2020) designed and synthesized a series of 48 novel pyridine-based gp120 Entry antagonists as potent anti-HIV-1 agents. Among all the synthesized compounds, **52** was found to be most potent with the mean IC_{50} value of 0.18 ± 0.008 μM against 50 HIV-1 Env-pseudo typed viruses as shown in Fig. 16.24. On further evaluation, it has been revealed that compound **52** inhibited WT HIV-1_{NL4-3} with an IC_{50} value of 2.1 μM. Furthermore, metabolic stability studies on human liver microsomes suggested that compound **52** has the highest metabolic stability among other compounds. Also, compound **52** was not found to inhibit CYP1A2, CYP2B6, and CYP3A enzymes at >50 μM dose level and CYP450 enzyme at 100 μM dose. Lastly, compound **52** was found to be noncytotoxic toward CD4-negative and CCR5-positive Cf2TH-CCR5 cells and also showed excellent ADME properties [34].

Elangovan et al. (2021) designed and synthesized (E)-4-((pyridine-4-ylmethylene)amino)-N-(pyrimidin-2-yl)benzenesulfonamide derivatives and investigated their antimicrobial activity as shown in Fig. 16.25. The compounds were initially screened by using a theoretical method, that is, density functional theory with B3LYP/6311 ++ G (d,p) basic set and then respective FTIR were computed. Out of all the compounds, compound **53**, which possessed a dock score of −7.6 kcal/mol, was screened for antibacterial and antifungal activity by using disc diffusion method. Evaluating the above-mentioned parameters and analyzing the zone

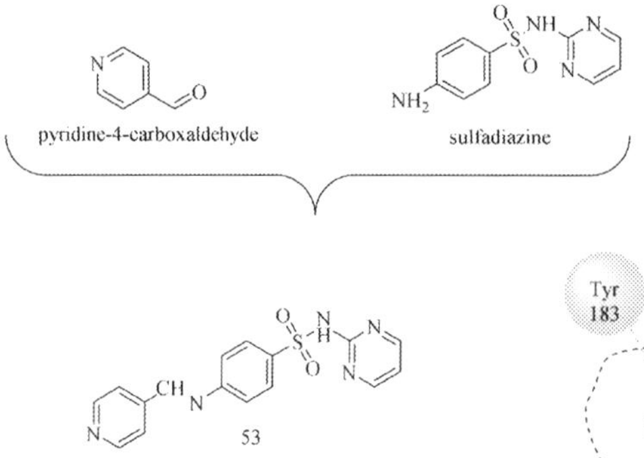

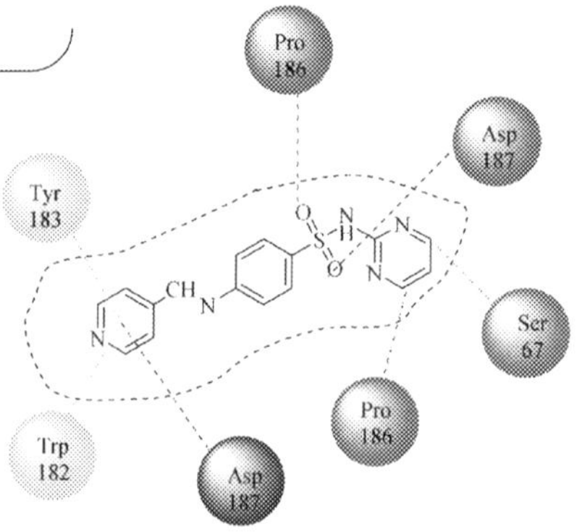

2D interaction diagram of potent compound 53

Microorganisms	Zone of inhibition (mm/ml)			
	25 μl	50 μl	75 μl	100 μl
Staphylococcus aureus	16	18	20	24
Bacillus subtilis	20	22	26	28
E.coli	18	20	22	25
Klebsiella pneumonia	20	22	25	27
Candida albicans	22	25	27	30
Aspergillus niger	18	20	22	24

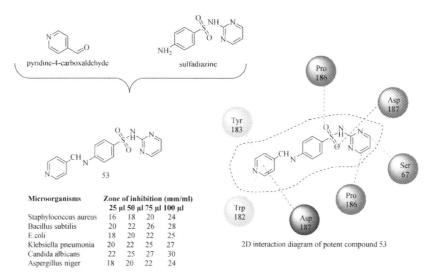

FIGURE 16.25 (E)-4-((pyridine-4-ylmethylene)amino)-N-(pyrimidin-2-yl)benzene-sulfonamide derivatives as antimicrobial agents.

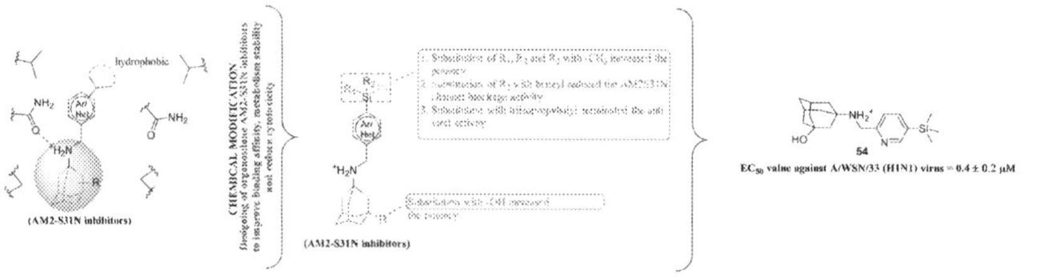

FIGURE 16.26 Organosilanes as potent antiviral agents against multidrug-resistant influenza A virus.

of inhibition, compound **53** could be designated as a novel potent antibacterial and antifungal agent [35].

16.2.4 Organometallic

Hu et al. (2017) designed and synthesized a series of organosilanes as potent antiviral agents against multidrug-resistant influenza A virus as shown in Fig. 16.26. Among all the synthesized compounds, compound **54** was found to be most potent with an EC_{50} value of 0.4 ± 0.2 μM against A/WSN/33 (H1N1) virus. On further evaluation, it has been revealed that compound **54** exhibited potent antiviral activity by blocking AM2-S31N channel. Furthermore, compound **54** was also found to be potent against oseltamivir-sensitive influenza A virus strains such as A/California/07/2009 (H1N1), A/Switzerland/9715293/2013 (H3N2) and A/Denmark/524/2009 (H1N1), oseltamivir-resistant, amantadine-resistant, and virus strains resistant from both the drugs. Lastly, the molecular docking studies of the compound **54** in the AM2-S31N channel demonstrated that trimethylsilyl-substituted benzene ring of **54** binds to the AM2-S31N channel by facing toward the N-terminus of the channel, the ammonium ion

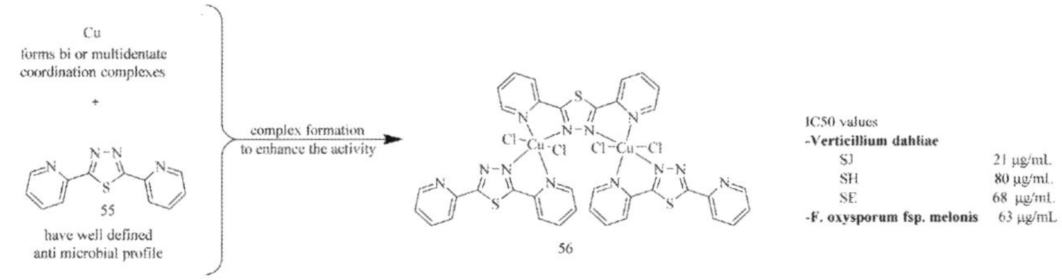

FIGURE 16.27 2,5-Bis(pyridine-2-yl)-1,3,4-thiadiazole as antimicrobial agent.

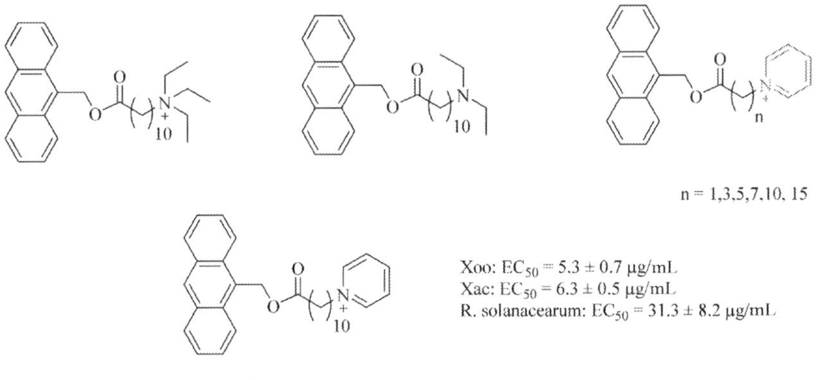

57

FIGURE 16.28 Pyridinium-tailored aromatic amphiphiles and evaluated them for their antibacterial activity.

formed hydrogen bond with one of the N31 carbonyl side chain and the adamantane cage fits in a space created by G34 [36].

Laachir et al. (2020) synthesized polymer based on 2,5-bis(pyridine-2-yl)-1,3,4-thiadiazole (**55**) as bridging ligand, noted **56**, has been synthesized by reaction of compound **55** with copper salt ($CuCl_{22}H_2O$) and characterized by single-crystal X-ray diffraction as shown in Fig. 16.27. These novel 1-D copper coordination polymer compounds exhibited antimicrobial activity against many fungal phytopathogens such as *Verticilliumdahliae* and *Fusarium oxysporum* sp. The copper complex **56** was obtained by reaction between 2,5-bis(pyridine-2-yl)-1,3,4-thiadiazole (**55**) and copper chloride. This compound was assessed against mycelial growth of fungal pathogens that affect several crops of economic interest. The compound at a dose of **50** μg/mL was found to inhibit 60% of the growth of the SJ strain of *Verticillium dahlia*. At higher doses, that is, at 100 and 200 μg/mL, growth inhibition reached up to 90% of many fungal phytopathogens [37].

16.2.5 Miscellaneous

Wang et al. (2016) synthesized a series of pyridinium-tailored aromatic amphiphiles and evaluated them for their antibacterial activity as shown in Fig. 16.28. Among them compound **57** was found to be the most potent against Xoo, *R. solanacearum*, Xac [38].

Sun et al. (2017) repurposed an inhibitor of ribosomal biogenesis and evaluated it for antifungal activity as shown in Fig. 16.29. It was found that compound **58** was found to be

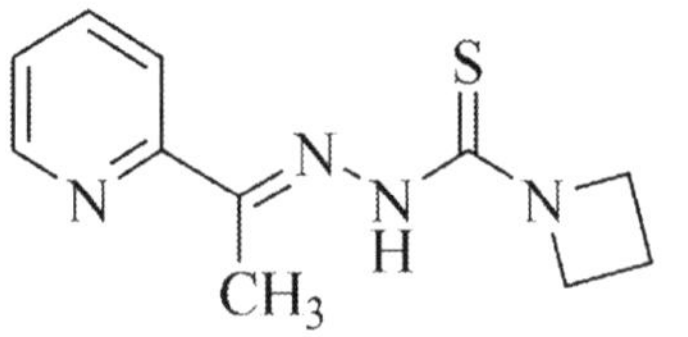

58

Candida albicans (SC5314): $MIC_{50} = 0.1\ \mu g/mL$
$MIC_{80} = 0.125\ \mu g/mL$, $MIC_{100} = 0.5\ \mu g/mL$
Candida parapsilosis: $MIC_{50} = 0.1\ \mu g/mL$
Candida lusitaniae : $MIC_{50} = 0.1\ \mu g/mL$
Candida apicola: $MIC_{50} = 0.1\ \mu g/mL$
Cryptococcus neoformans JEC-21: $MIC_{50} = 0.1\ \mu g/mL$

FIGURE 16.29 Repurposed pyridine containing inhibitor as antifungal.

the active against *C. albicans* (SC5314), *C. parapsilosis, C. lusitaniae, C. apicola*, and *C. neoformans* JEC-21 for antifungal activity [39].

The outbreak of visceral leishmaniasis infection is exacerbating day by day and is creating a threat to our healthcare system. Therefore, to develop efficacious compounds to inhibit the growth of this parasite, Thomson et al. in 2018 screened a library of ~ 900 compounds with bicyclic nitroimidazoles nucleus against L. don. amastigotes to get a hit molecule as shown in Fig. 16.30. Among these 900 compounds, only 248 compounds (28%) showed inhibition greater than 50% at a dose of 10 ug/mL and 89 compounds (36%) at a dose of 3.3 ug/mL. Later the author sorted or refined the library by eliminating the prescreened compounds, compounds with metabolic or solubility issues. This finally gave them 42 hits which were again screened at CDRI to get the lead compound **59**, which was possessing a lower log*D* value, optimum solubility, and weak CYP3A4 activity. Further, this compound **59** was screened on various strains of VL available worldwide and results depicted that the compound was possessing an optimum potency to inhibit the growth of this parasite and concluded that the bicyclic nitroimidazoles nucleus can be used for the treatment of this parasite infection [40].

Due to its innumerable therapeutic potential, Li et al. (2019) generated a novel series of scaffolds of conformationally rigid nonketolide versions of erythromycin as antibacterial agents as shown in Fig. 16.31. A total of 30 compounds were evaluated for bactericidal activity against a panel of erythromycin-resistant and erythromycin-susceptible clinical isolates and ATCC stains, including gram-positive bacteria: *S. aureus, S. pneumoniae, S. pyogenes*, and *S. epidermidis* and gram-negative bacteria: *H. influenzae* and *M. catarrhalis*. Compound **63a** was the most potent among the series. Further, **63a** was docked into *D. radiodurans* 50S ribosomal subunit, which suggested pi–pi stacking interactions. In addition to this, in vivo pharmacokinetic profile of **63a** was evaluated in terms of half-life, drug–plasma concentration, mean residence time, and cytochrome P450 inhibition pharmacokinetic testing in male SD rats via oral administration suggested **63a** exhibited both longer half-life and higher plasma concentration than telithromycin. PK properties of **63a** were better than telithromycin such as sevenfold higher AUC, sevenfold slower systemic plasma clearance, 2.4-fold longer $t_{1/2}$, 1.4-fold longer mean residence time, fivefold higher C_{max}, and 0.4-fold shorter T_{max} [41].

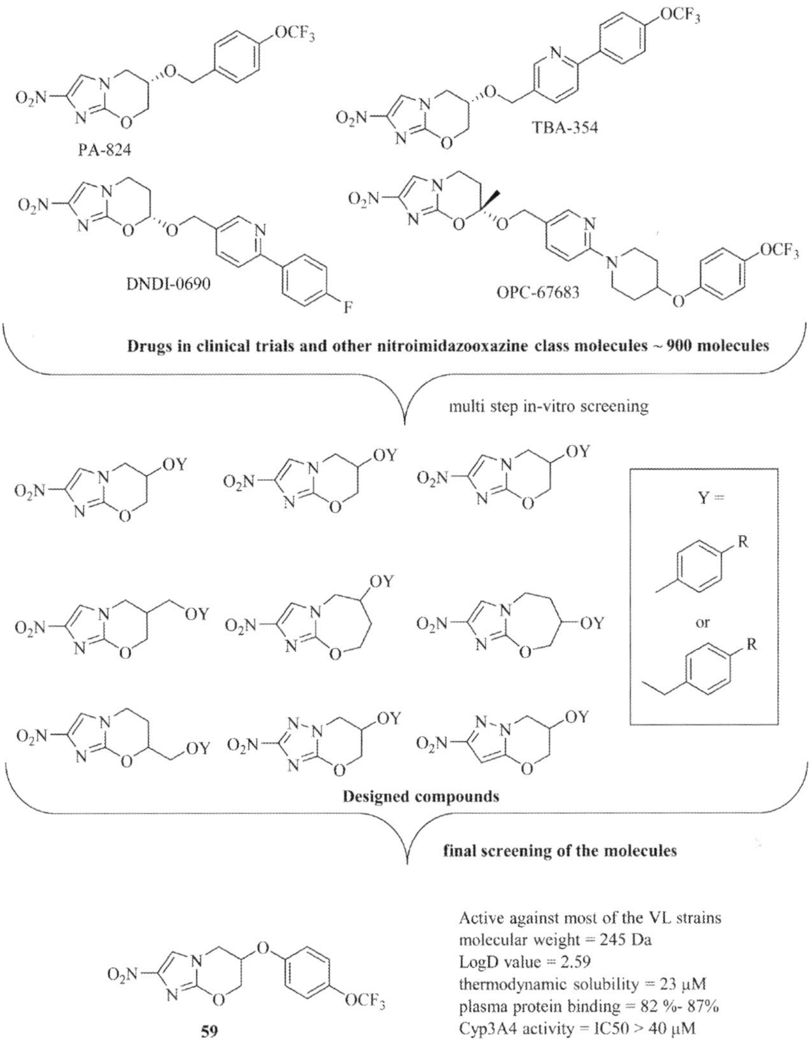

FIGURE 16.30 Bicyclic nitroimidazoles nucleus as antileishmanial agent.

16.3 Pharmacophoric features

In order to enlighten the core features that a molecule or an antimicrobial should have, we have used the structure of all the antimicrobial agents mentioned above, to develop a hypothetical pharmacophore model which will help researchers to design their molecules more efficiently. DS Biovia Discovery studio was used which gave us four models that are providing us information regarding the position of various pharmacophoric features such as hydrogen acceptor, hydrogen donor, or hydrophobic as shown in Fig. 16.32 [42–44].

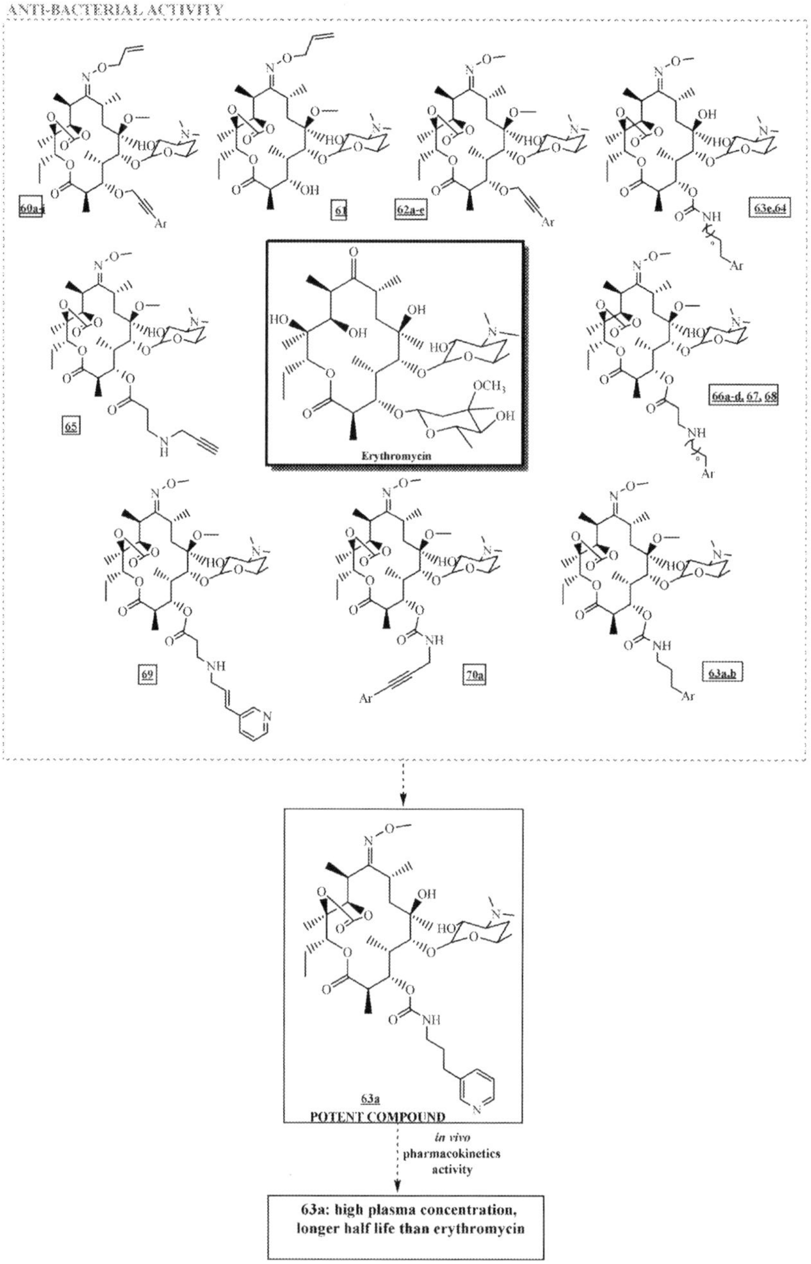

FIGURE 16.31 Scaffolds of conformationally rigid nonketolide versions of erythromycin as antibacterial agents.

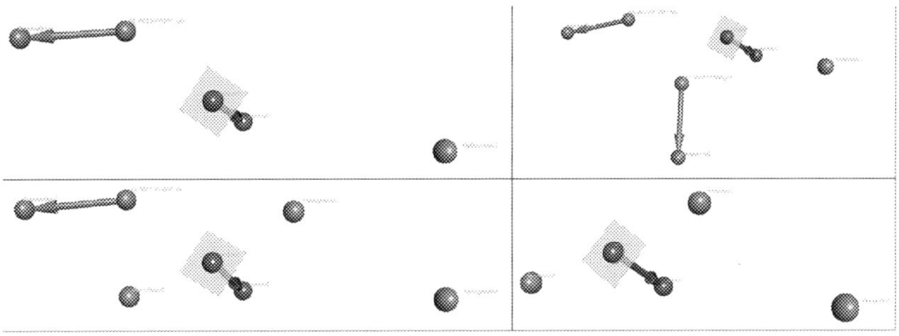

FIGURE 16.32 Proposed pharmacophore hypothetical model to develop or design antimicrobials.

16.4 Conclusion

As we all know that humans and other species are evolving with time to adapt to the changes in the environment. So, this adaptation leads to the development of antimicrobial resistance. Therefore, to mitigate this problem, continuous research is in progress and providing new nuclei or molecules as antimicrobial agents. Among these, pyridine has been found to possess diverse biological profiles and due to this various researchers have used this nucleus to design various antimicrobial agents. As discussed above, researchers have used pyridine as a substituent such as novel imidazo [1,2-a] pyridine-3-carboxamide derivatives by LV et al. (2017). Some have incorporated in their main nucleus benzo [6,7] cyclohepta[1,2-b]pyridine-1,2,3-triazole hybrids by Sajja et al. (2017) or used pyridine as linker such as pyrazolo[1,5-a]pyridine-3-carboxamide hybrids by Lu et al. (2017). This makes pyridine an important nucleus in designing and developing new antimicrobials. At last, we have developed a pharmacophore hypothetical model that will guide researchers to design their molecules as potent antimicrobial agents.

16.5 Authors contribution

Nitish Kumar: Conceptualization and writing of original draft; **Harmandeep Kaur, Anchal Khanna**: Data curation, formal analysis; **Komalpreet Kaur, Jatinder Vir Singh**: Writing, review, and editing; **Sarabjit Kaur, Preet Mohinder Singh Bedi**, and **Balbir Singh**: Conceptualization, supervision, and critical screening of the manuscript.

References

[1] H. Gest, The discovery of microorganisms by Robert Hooke and Antoni Van Leeuwenhoek, fellows of the Royal Society, Notes Rec. R. Soc. Lond. 58 (2004) 187–201. https://doi.org/10.1098/RSNR.2004.0055.

[2] G.-.C. Nina, P. Ana, O. Aharon, Strategies of adaptation of microorganisms of the three domains of life to high salt concentrations, FEMS Microbiol. Rev. 42 (2018) 353–375. https://doi.org/10.1093/FEMSRE/FUY009.

[3] B.Y. Vincent, The role of the microbiome in human health and disease: an introduction for clinicians, BMJ (Clin. Res. Ed.) 356 (2017). https://doi.org/10.1136/BMJ.J831.

[4] J.P. Narain, Public health challenges in India: seizing the opportunities, Indian J. Commun. Med. 41 (2016) 85. https://doi.org/10.4103/0970-0218.177507.

[5] K.D. Tripathi, Essentials of medical pharmacology. JP Medical Ltd, 2013.

[6] L.S.J. Roope, R.D. Smith, K.B. Pouwels, J. Buchanan, L. Abel, P. Eibich, et al., The challenge of antimicrobial resistance: what economics can contribute, Science 364 (2019) 6435. https://doi.org/10.1126/SCIENCE.AAU4679. eaau4679.

[7] Antimicrobial resistance n.d. https://www.who.int/news-room/fact-sheets/detail/antimicrobial-resistance (accessed August 3, 2021).

[8] A.C. Maria, I. Daniela, R. Roberto, V.G. Salvatore, L. Laura, Pyridine and pyrimidine derivatives as privileged scaffolds in biologically active agents, Curr. Med. Chem. 26 (2019) 7166–7195. https://doi.org/10.2174/0929867325666180904125400.

[9] D.L. Comins, S. O'Connor, R.S. Al-Awar, Pyridines and their benzo derivatives: reactivity at the ring, Comprehens. Heterocycl. Chem. III 7 (2008) 41–99. https://doi.org/10.1016/B978-008044992-0.00602-7.

[10] D.R. Lide, W.M. Haynes (Eds.), Physical Constants of Inorganic Compounds - Hydrogen Peroxide. CRC Handbook of Chemistry and Physics. Vol. 9. Boca Raton, FL: CRC press, 90th ed., 2010, pp. 4–67.

[11] Y. Aizenberg-Gershtein, I. Izhaki, R. Santhanam, P. Kumar, I.T. Baldwin, M. Halpern, Pyridine-type alkaloid composition affects bacterial community composition of floral nectar, Sci. Rep. 5 (2015) 1–11. https://doi.org/10.1038/srep11536.

[12] M. Miliutina, J. Janke, E. Chirkina, S. Hassan, S.A. Ejaz, S.U. Khan, et al., Domino reactions of chromone-3-carboxylic acids with aminoheterocycles: synthesis of heteroannulated pyrido[2,3-c]coumarins and their optical and biological activity, Eur. J. Org. Chem. (2017) 7148–7159. https://doi.org/10.1002/ejoc.201701276.

[13] F. Fumagalli, S.M.G. de Melo, C.M. Ribeiro, M.C. Solcia, F.R. Pavan, F. da Silva Emery, Exploiting the furo[2,3-b]pyridine core against multidrug-resistant *Mycobacterium tuberculosis*, Bioorg. Med. Chem. Lett. 29 (2019) 974–977. https://doi.org/10.1016/j.bmcl.2019.02.019.

[14] X. Lu, J. Tang, S. Cui, B. Wan, S.G. Franzblauc, T. Zhang, et al., Pyrazolo[1,5-a]pyridine-3-carboxamide hybrids: design, synthesis and evaluation of anti-tubercular activity, Eur. J. Med. Chem. 125 (2017) 41–48. https://doi.org/10.1016/j.ejmech.2016.09.030.

[15] G. Jose, T.H. Suresha Kumara, H.B.V. Sowmya, D. Sriram, T.N. Guru Row, A.A. Hosamani, et al., Synthesis, molecular docking, antimycobacterial and antimicrobial evaluation of new pyrrolo[3,2-c]pyridineMannich bases, Eur. J. Med. Chem. 131 (2017) 275–288. https://doi.org/10.1016/j.ejmech.2017.03.015.

[16] S. Kang, Y.M. Kim, H. Jeon, S. Park, M.J. Seo, S. Lee, et al., Synthesis and structure-activity relationships of novel fused ring analogues of Q203 as antitubercular agents, Eur. J. Med. Chem. 136 (2017) 420–427. https://doi.org/10.1016/j.ejmech.2017.05.021.

[17] R. Watanabe, H. Mizoguchi, H. Oikawa, H. Ohashi, K. Watashi, H. Oguri, Stereo-controlled synthesis of functionalized tetrahydropyridines based on the cyanomethylation of 1,6-dihydropyridines and generation of anti-hepatitis C virus agents, Bioorg. Med. Chem. 25 (2017) 2851–2855. https://doi.org/10.1016/j.bmc.2017.03.011.

[18] K. Lv, L. Li, B. Wang, M. Liu, B. Wang, W. Shen, et al., Design, synthesis and antimycobacterial activity of novel imidazo[1,2-a]pyridine-3-carboxamide derivatives, Eur. J. Med. Chem. 137 (2017) 117–125. https://doi.org/10.1016/j.ejmech.2017.05.044.

[19] Y. Sajja, S. Vanguru, H.R. Vulupala, R. Bantu, P. Yogeswari, D. Sriram, et al., Design, synthesis and in vitro anti-tuberculosis activity of benzo[6,7]cyclohepta[1,2-b]pyridine-1,2,3-triazole derivatives, Bioorg. Med. Chem. Lett. 27 (2017) 5119–5121. https://doi.org/10.1016/j.bmcl.2017.10.071.

[20] B. Veeraswamy, D. Madhu, G. Jitender Dev, Y. Poornachandra, G. Shravan Kumar, C. Ganesh Kumar, et al., Studies on synthesis of novel pyrido[2,3-d]pyrimidine derivatives, evaluation of their antimicrobial activity and molecular docking, Bioorg. Med. Chem. Lett. 28 (2018) 1670–1675. https://doi.org/10.1016/j.bmcl.2018.03.022.

[21] N.S. El-Gohary, MI. Shaaban, New pyrazolopyridine analogs: synthesis, antimicrobial, antiquorum-sensing and antitumor screening, Eur. J. Med. Chem. 152 (2018) 126–136. https://doi.org/10.1016/j.ejmech.2018.04.025.

[22] A. Blaser, H.S. Sutherland, A.S.T. Tong, P.J. Choi, D. Conole, S.G. Franzblau, et al., Structure-activity relationships for unit C pyridyl analogues of the tuberculosis drug bedaquiline, Bioorg. Med. Chem. 27 (2019) 1283–1291. https://doi.org/10.1016/j.bmc.2019.02.025.

[23] J. Xu, X. Xie, N. Ye, J. Zou, H. Chen, M.A. White, et al., Design, synthesis, and biological evaluation of substituted 4,6-dihydrospiro[[1,2,3]triazolo[4,5- b]pyridine-7,3′-indoline]-2′,5(3 H)-dione analogues as potent NS4B inhibitors for the treatment of dengue virus infection, J. Med. Chem. 62 (2019) 7941–7960. https://doi.org/10.1021/acs.jmedchem.9b00698.

[24] P. Sivaprakasam, Z. Wang, N.A. Meanwell, J.A. Khan, D.R. Langley, S.R. Johnson, et al., Structure-based amelioration of PXR transactivation in a novel series of macrocyclic allosteric inhibitors of HIV-1 integrase, Bioorg. Med. Chem. Lett. 30 (22) (2020) 127531. https://doi.org/10.1016/j.bmcl.2020.127531.

[25] B. Singh, R. Bhatia, B. Pani, D. Gupta, Synthesis, crystal structures and biological evaluation of new pyridazine derivatives, J. Mol. Struct. 1200 (2020) 127084. https://doi.org/10.1016/j.molstruc.2019.127084.

[26] P.Y. Wang, L. Zhou, J. Zhou, Z.B. Wu, W. Xue, B.A. Song, et al., Synthesis and antibacterial activity of pyridinium-tailored 2,5-substituted-1,3,4-oxadiazole thioether/sulfoxide/sulfone derivatives, Bioorg. Med. Chem. Lett. 26 (2016) 1214–1217. https://doi.org/10.1016/j.bmcl.2016.01.029.

[27] N.C. Desai, H. Somani, A. Trivedi, K. Bhatt, L. Nawale, V.M. Khedkar, et al., Synthesis, biological evaluation and molecular docking study of some novel indole and pyridine based 1,3,4-oxadiazole derivatives as potential antitubercular agents, Bioorg. Med. Chem. Lett. 26 (2016) 1776–1783. https://doi.org/10.1016/j.bmcl.2016.02.043.

[28] P. Picconi, P. Prabaharan, J.L. Auer, S. Sandiford, F. Cascio, M. Chowdhury, et al., Novel pyridyl nitrofuranylisoxazolines show antibacterial activity against multiple drug resistant Staphylococcus species, Bioorg. Med. Chem. 25 (2017) 3971–3979. https://doi.org/10.1016/j.bmc.2017.05.037.

[29] H. Patel, H. Jadhav, I. Ansari, R. Pawara, S. Surana, Pyridine and nitro-phenyl linked 1,3,4-thiadiazoles as MDR-TB inhibitors, Eur. J. Med. Chem. 167 (2019) 1–9. https://doi.org/10.1016/j.ejmech.2019.01.073.

[30] A. Wang, H. Wang, Y. Geng, L. Fu, J. Gu, B. Wang, et al., Design, synthesis and antimycobacterial activity of less lipophilic Q203 derivatives containing alkaline fused ring moieties, Bioorg. Med. Chem. 27 (2019) 813–821. https://doi.org/10.1016/j.bmc.2019.01.022.

[31] B. Sun, Y. Dong, K. Lei, J. Wang, L. Zhao, M. Liu, Design, synthesis and biological evaluation of amide-pyridine derivatives as novel dual-target (SE, CYP51)antifungal inhibitors, Bioorg. Med. Chem. 27 (2019) 2427–2437. https://doi.org/10.1016/j.bmc.2019.02.009.

[32] A. Wang, K. Lv, L. Li, H. Liu, Z. Tao, B. Wang, et al., Design, synthesis and biological activity of N-(2-phenoxy)ethylimidazo[1,2-a]pyridine-3-carboxamides as new antitubercular agents, Eur. J. Med. Chem. 178 (2019) 715–725. https://doi.org/10.1016/j.ejmech.2019.06.038.

[33] M.N. Bhoi, M.A. Borad, D.J. Jethava, P.T. Acharya, E.A. Pithawala, C.N. Patel, et al., Synthesis, biological evaluation and computational study of novel isoniazid containing 4H-Pyrimido[2,1-b]benzothiazoles derivatives, Eur. J. Med. Chem. 177 (2019) 12–31. https://doi.org/10.1016/j.ejmech.2019.05.028.

[34] F. Curreli, S. Ahmed, S.M. Benedict Victor, I.R. Iusupov, D.S. Belov, P.O. Markov, et al., Preclinical optimization of gp120 entry antagonists as anti-HIV-1 agents with improved cytotoxicity and ADME properties through rational design, synthesis, and antiviral evaluation, J. Med. Chem. 63 (2020) 1724–1749. https://doi.org/10.1021/acs.jmedchem.9b02149.

[35] N. Elangovan, S. Sowrirajan, K.P. Manoj, AM. Kumar, Synthesis, structural investigation, computational study, antimicrobial activity and molecular docking studies of novel synthesized (E)-4-((pyridine-4-ylmethylene)amino)-N-(pyrimidin-2-yl)benzenesulfonamide from pyridine-4-carboxaldehyde and sulfadiazine, J. Mol. Struct. 1241 (2021) 130544. https://doi.org/10.1016/j.molstruc.2021.130544.

[36] Y. Hu, Y. Wang, F. Li, C. Ma, J. Wang, Design and expeditious synthesis of organosilanes as potent antivirals targeting multidrug-resistant influenza A viruses, Eur. J. Med. Chem. 135 (2017) 70–76. https://doi.org/10.1016/j.ejmech.2017.04.038.

[37] A. Laachir, S. Guesmi, E.M. Ketatni, M. Saadi, L. elAmmari, S. Esserti, et al., Novel 1-D copper(II) coordination polymer based on 2,5-bis(pyridine-2-yl)-1,3,4-thiadiazole as bridging ligand: synthesis, crystal structure, Hirshfeld surface analysis, spectroscopic characterizations and biological assessment, J. Mol. Struct. 1218 (2020) 128533. https://doi.org/10.1016/j.molstruc.2020.128533.

[38] P. Wang, M. Gao, L. Zhou, Z. Wu, D. Hu, J. Hu, et al., Synthesis and antibacterial activity of pyridinium-tailored aromatic amphiphiles, Bioorg. Med. Chem. Lett. 26 (2016) 1136–1139. https://doi.org/10.1016/j.bmcl.2016.01.053.

[39] N. Sun, D. Li, Y. Zhang, K. Killeen, W. Groutas, R. Calderone, Repurposing an inhibitor of ribosomal biogenesis with broad anti-fungal activity, Sci. Rep. 7 (2017) 1–9. https://doi.org/10.1038/s41598-017-17147-x.

[40] A.M. Thompson, P.D. O'Connor, A.J. Marshall, A. Blaser, V. Yardley, L. Maes, et al., Development of (6 R)-2-nitro-6-[4-(trifluoromethoxy)phenoxy]-6,7-dihydro-5 H-imidazo[2,1-b] [1,3]oxazine (DNDI-8219): a new lead for visceral Leishmaniasis, J. Med. Chem. 61 (2018) 2329–2352. https://doi.org/10.1021/acs.jmedchem.7b01581.

[41] X.-.M. Li, W. Lv, S.-.Y. Guo, Y.-.X. Li, B.-.Z. Fan, M. Cushman, et al., Synthesis and structure-bactericidal activity relationships of non-ketolides: 9-oxime clarithromycin 11,12-cyclic carbonate featured with three-to eight-atom-length spacers at 3-OH, Eur. J. Med. Chem. 171 (2019) 235–254. https://doi.org/10.1016/j.ejmech.2019.03.037.

[42] Dassault Systèmes BIOVIADiscovery Studio Modeling Environment, 2020.

[43] N. Kumar, A. Singh, H.K. Gulati, K. Bhagat, K. Kaur, J. Kaur, et al., Phytoconstituents from ten natural herbs as potent inhibitors of main protease enzyme of SARS-COV-2: in silico study, Phytomed. Plus 1 (2021) 100083. https://doi.org/10.1016/j.phyplu.2021.100083.
[44] N. Kumar, H.K. Gulati, A. Sharma, S. Heer, A.K. Jassal, L. Arora, et al., Most recent strategies targeting estrogen receptor alpha for the treatment of breast cancer, Mol. Divers. 1 (2020) 3. https://doi.org/10.1007/s11030-020-10133-y.

CHAPTER

17

Synthesis of pyridine derivatives for diverse biological activity profiles: A review

Tejeswara Rao Allaka[a] *and Naresh Kumar Katari*[b]

[a]Centre for Chemical Sciences and Technology, Institute of Science and Technology, Jawaharlal Nehru Technological University Hyderabad, Hyderabad, Telangana, India
[b]Department of Chemistry, School of Science, GITAM Deemed to be University, Hyderabad, Telangana, India

17.1 Introduction

Medicinal chemistry is an important part of science, which encompasses discovery, identification, development of therapeutics, and interpretation of their mechanism of action in the body at the molecular level. Besides this, medicinal chemistry also involves the synthesis, characterization, and isolation of compounds that can be used in medicine for the prevention, treatment, and cure of diseases, providing a chemical basis for the interdisciplinary field of therapeutics. In this approach, many synthetic organic compounds have been developed as effective therapeutic medicinal agents. Among various heterocycles, pyridine is an important heterocyclic system that shows diverse biological properties. It is rather used as a precursor for the commercial synthesis of interesting pharmaceuticals and agrochemicals.

The determination of pyridine scaffolds is linked to a characteristic experiment carried out by Anderson in 1846, who was indeed studied for the pyrolysis of bones and was able to isolate picoline as the first known pyridine [1]. It has found many applications in diverse chemical domains and in coordination chemistry [2], terpyridines [3], bipyridines [4], mono pyridines [5], and is used to chelate metallic ions as N-donor ligands, affording efficient organometallic catalysts [6]. Pyridines are also involving in polymers [7], supramolecular structures [8], materials and surfaces [9], and also in organocatalysis, as represented by the important applications of DMAP and its derivations [10] (Fig. 17.1). Many other bioactive

DOI: https://doi.org/10.1016/B978-0-323-91221-1.00005-1

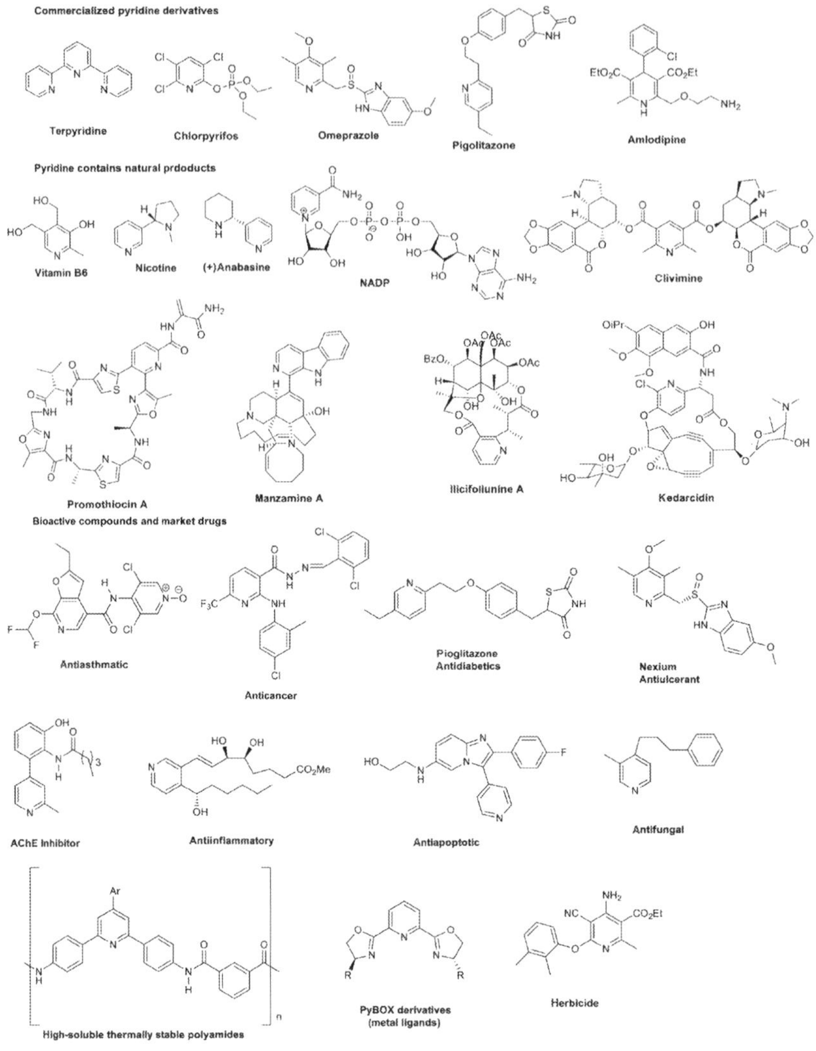

FIGURE 17.1 Representative pyridine scaffolds and their chemical domain of applications.

enlarged pyridines have been synthesized with different interesting effects like antiasthmatic ones [11], anti-inflammatory [12], inhibiting acetylcholinesterase (AChE) [13], inhibiting HIV protease [14], antidepressant [15], hypertension [16], or hypotension [17], preventing [18], and exploited in agrochemistry [19], insecticide [20], herbicide [21], and antifungal properties [22]. Thus, this nucleus constitutes a major scaffold to create antiviral or anticancer drugs [23]. The present review will constitute aggregate to the state of the art by providing a comprehensive compilation of synthetic applications involving specifocally metal-free multicomponent reactions and their biological efforts.

17.2 Synthetic strategy

Hantzsch approach is one of the most popular methods for the synthesis of pyridine **2** via two steps, which involves oxidation of 1,4-dihydropyridines **1** (DHPs), previously formed through a one-pot pseudo four-component reaction of two equivalent of a 1,3-dicarbonyl compound, a source of ammonia and an aldehyde [24]. The industrial formation of pyridine is derived from Chichibabin reaction involves condensation of ketones, aldehydes, α,β-unsaturated carbonyl compounds with ammonia and its derivatives. In particular, unsubstituted pyridine **3** is derived from acetaldehyde and formaldehyde through Knovenagel condensation. In the 1970s Katritzky and Sammes discussed the synthesis of 1-pyridinio-4-pyridone cations through the condensation of chelidonic acid with 1-aminopyridine salts **4**. These N–N activated pyridinium scaffolds were found to be reactive toward KCN addition, affording the cyanated pyridines **5, 6** with excellent yield and this reaction was regulate to be selective for the 2-position of the pyridine **5** [25]. However, 1-aminopyridinium salts also readily undergo a selective combination of cyanide to the 4-position of the pyridinium ring. Moreover, 4-cyanopyridines are unstable and spontaneously react in a [3 + 2]-cycloaddition with another equivalent of the 1-aminopyridinium salt to give 2-phenyl-[1,2,4]triazolo[1,5-a] pyridine **7** [26]. Within the past decade, three-component synthesis of 2-alkoxypyridines from sodium alkoxide, malononitrile, and Michael acceptor have been the area of intensive research reports. For example, in 2004 Goda et al. describe the use of chalcone and sodium ethoxide or sodium methoxide at rt (Scheme 17.1) [27].

Pyridines can be prevailed by Staudinger reaction, Wittig reaction, and Aza-Wittig reaction. Synthesis of polysubstituted pyridines from phousylides, aldehydes, in the presence of propargyl azide (**i**) [28]. Moreover, it has been transformation provides 2-aryl pyridines by the reaction of 1,3-diamino propane and acetophenones which is catalyzed by copper (**ii**) [29]. Also, 2-substituted pyridines were obtained by the addition of pyridine-N-oxides to Grignard reagents catalyzed by THF at rt in the presence of acetic anhydride under reflux conditions to provide 2-substituted pyridines (**iii**) [30]. Also Katritzky et al. reported a [3 + 2 + 1] synthesis of 2,4,6-tri and 2,3,4,6-tetrasubstituted pyridine compounds with excellent yield. This method depends on readily available α-benzo triazolyl ketone and α, β-unsaturated ketone substituents when heated with NH_4OAc in CH_3COOH (**iv**) [31]. Synthesis of 3-hydroxyl pyridine derivatives derived from olefin metathesis in the presence of DBU, DMF (**v**) [32].

Additionally, 2-Azabicyclo[3.3.1]non-2-en-1ol (**vi**) was synthesised via the reaction between vinyl azides and monocyclic cyclopropanol in the presence of catalytic amounts of Mn(acac)3 [33]. Another method for providing pyridine targets is a well-planned base KtOBu promoted reaction of aryl ethyl amines with alkyne-ones gives enaminones under metal-free conditions (**vii**) [34]. Dilman et al. proposed coupling of silylenol ethers with α,α-difluoro-β-iodoketones using fac-Ir $(ppy)_3$ under blue LED irradiation can be produced the substituted 3-fluoropyridines with subsequent one-pot condensation with ammonium acetate (**viii**) [35]. Craig and Henry described novel variation of well-established [5 + 1] methodology by condensation of 1,5-dicarbonyl with ammonia [36]. Both 2,4-di- and 2,4,6-trisubstituted pyridines with aryl, alkyl, and ester substituents were described from 1,6-dienes by ozonolysis, subsequent condensation with ammonia, and elimination of sulfinic acid (**ix**) [37] and to obtain a wide

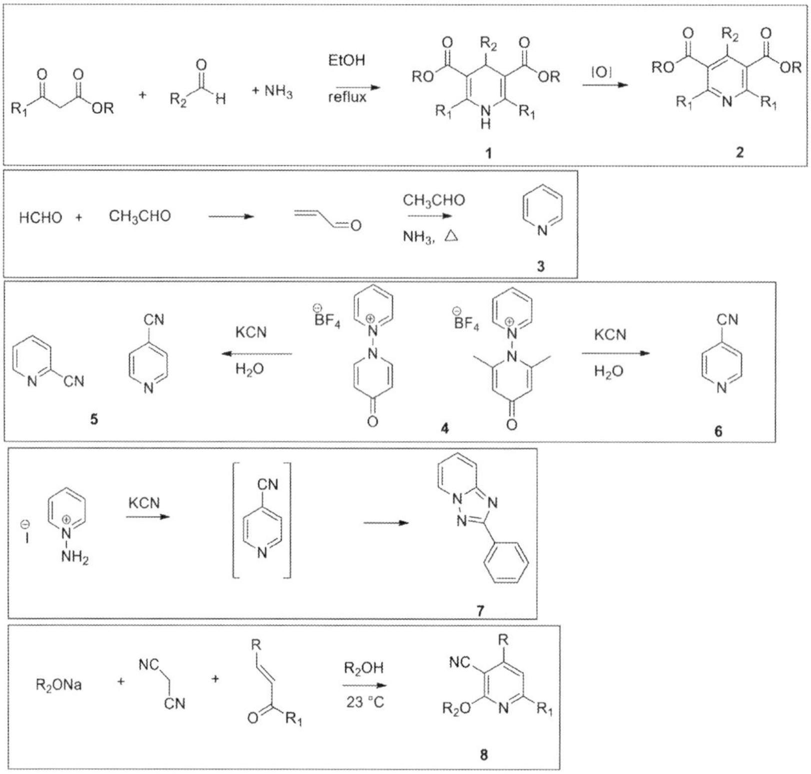

SCHEME 17.1 Sufficiently well-developed synthetic approaches of pyridines.

range of polysubstituted dihydro pyridine through lithiation/isomerization/intramolecular carbolithiation (**x**) [38]. By using metal-free protocol a wide variety of pyridine scaffolds has been prepared with good yields. Although this process cannot be caused by triethylamine or iodine, mechanistic experiments suggested a radical pathway (**xi**) (Fig. 17.2) [39].

Michael-addition reactions described up to now implied 1,3-dicarbonyl compound, an ammonia source, and a Michael acceptor connection with dimethylamino substituent on position 3. AlSaleh et al. reported from methyl acetoacetate or acetyl acetone reacting with acetic acid, enaminoketone, and ammonium acetate under refluxing conditions to produce **9** (Scheme 17.2) [40]. Bergman, Lewis, and Ellman discussed an elegant $[RhCl(coe)_2]_2$-catalyzed direct alkylation of 3,3-dimethyl-1-butene with 2-substituted pyridines in the presence of PCy_3, providing the examples of pyridines **10** as shown in Scheme 17.3 [41]. Moreover, Moore et al. reported in the direct acylation of pyridine in the presence of terminal alkenes, triruthenium dodecylcarbonyl as a catalyst and carbon monoxide atmosphere (Scheme 17.4) [42].

Hiyama and coworkers [43] described the methodology by activating pyridine in situ through the addition of a catalytic quantity of a mild Lewis acid (Scheme 17.5). The addition of $AlMe_3$, $ZnPh_2$, or $ZnMe_2$ was adequate to initiate 2-alkenyl pyridines and to avoid

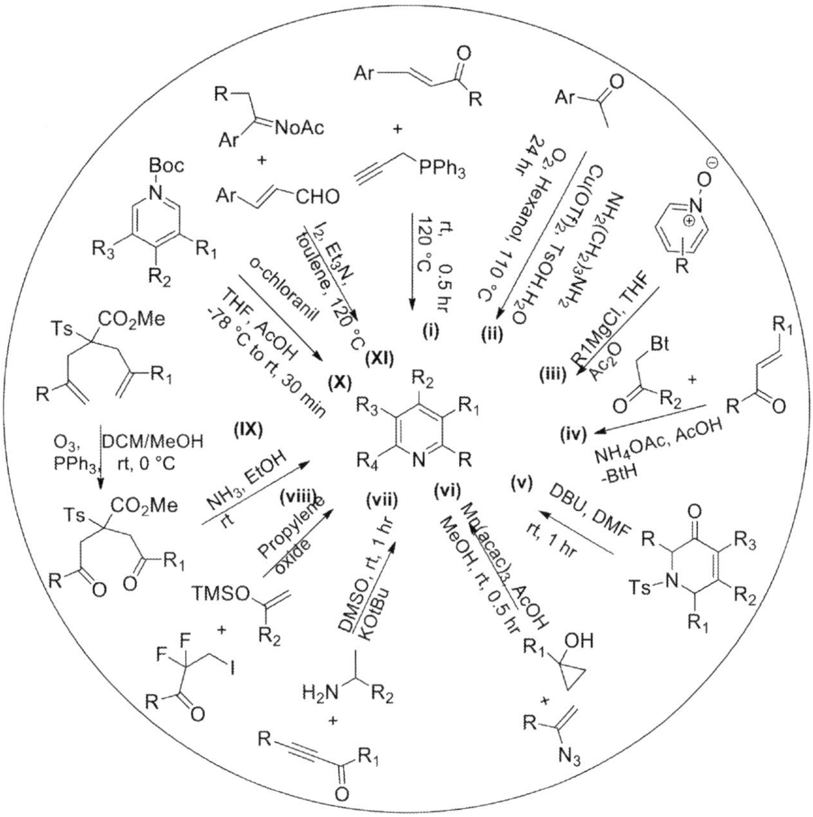

FIGURE 17.2 Design of different pyridine templates.

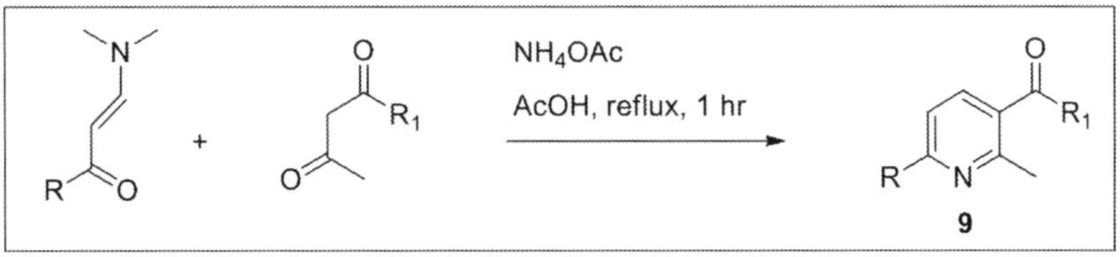

SCHEME 17.2 Synthesis of pyridine derivatives by using acetyl acetone.

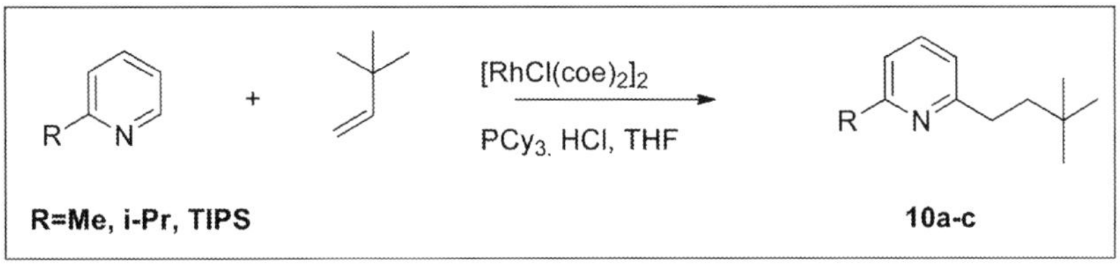

SCHEME 17.3 Preparation of pyridine scaffolds via Rh (II) catalyst.

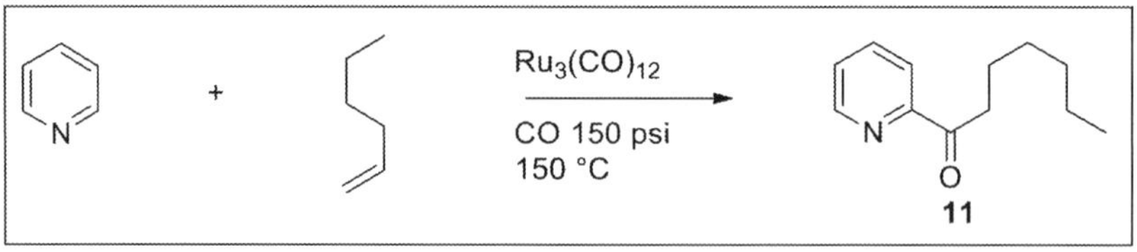

SCHEME 17.4 Preparation of pyridine scaffolds via Ru catalyst.

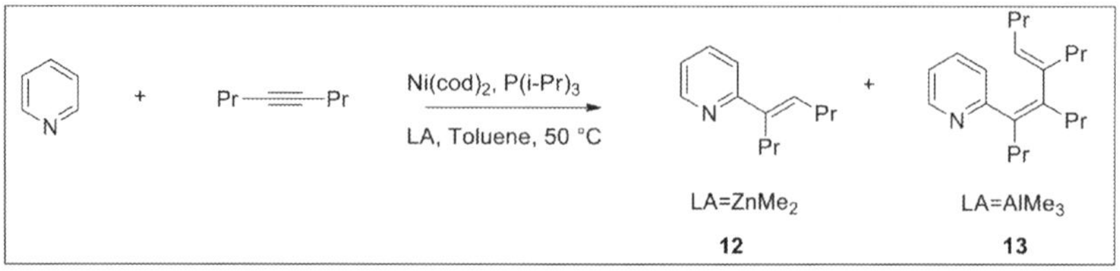

SCHEME 17.5 Preparation of pyridine scaffolds via Ru catalyst.

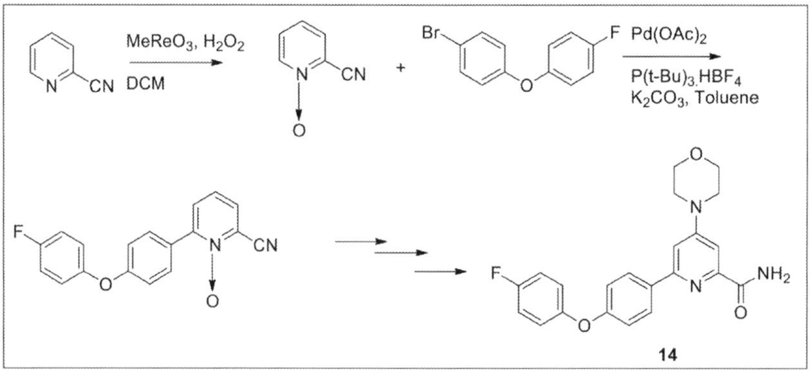

SCHEME 17.6 Synthesis of pyridine scaffolds linked to morpholine.

2,6-dialkenylation. The Zn-based Lewis acids produced monoalkenylated product **12**, where the aluminum reagents appoint the bis-alkenylated **13**. One of the most advantages of pyridine N-oxides from arylation products is activated pyridinium derivatives. The preparation of 2,6-bisarylated pyridine species of aryl triflates via pyridine N-oxide was a key intermediate for biologically active molecule **14** known to reveals antimicrobial and antimalarial activity (Scheme 17.6) [44]. Whereas Charette and coworkers [45] developed the synthesis of di-astereoselective addition of chiral pyridinium N-imidate salts to Grignard reagents. Highly regioselective and diastereoselective additions were carried through when enantiopure (S)-valinol amide was working as a chiral auxiliary. Bidentate chelation of the auxiliary with the nucleophiles directed to organometallic reagents at C-2 position of the corresponding pyridinium salt **15** with highest regioselective and diastereoselectivities (Scheme 17.7). Further investigation, Charette et al. [46] derived the regioselective reaction of Grignard reagents to 3-substituted pyridinium N-imidate salts. The reaction of MeMgBr or PhMgBr to 3-methoxypyridine gives excellent regioselective way, exclusively at the more hindered at C-2 position to obtained dihydropyridines **17, 18**. From this reaction, we have noticed a

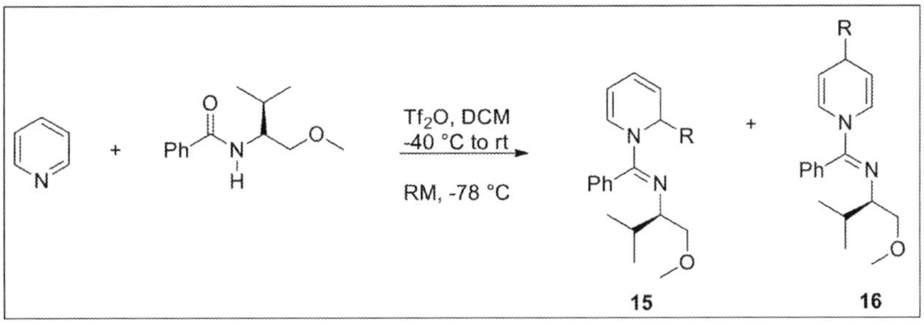

SCHEME 17.7 Synthesis of stereoselective pyridine derivatives.

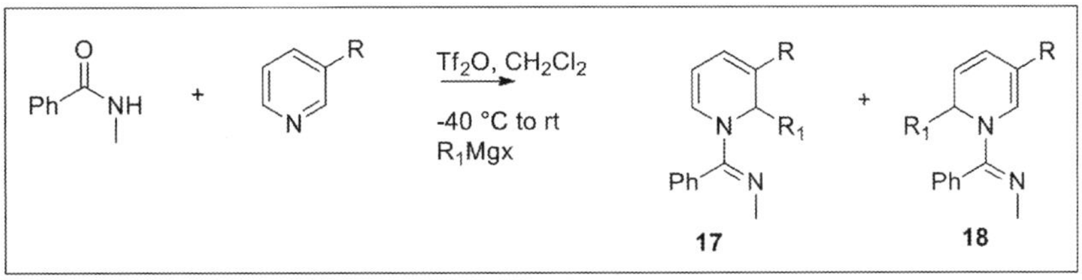

SCHEME 17.8 Synthesis of dihydropyridine derivatives.

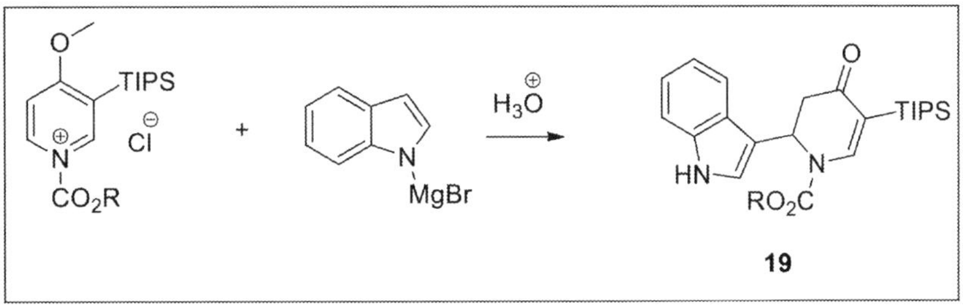

SCHEME 17.9 Preparation of Indole core with pyridines.

lower selectivity in the combination of PhMgBr or MeMgBr to 3-bromopyridine imidate salts, 3-chloro pyridine and 3-picoline to providing dihydropyridines (Scheme 17.8).

Further investigation on vinyl, alkyl, aryl, Grignard reagents, and other nucleophiles have been examined and numerous pyrrolyl, indolyl Grignard reagents were used for obtaining moderate diastereoselectivity (Scheme 17.9) [47]. Moreover, Comins et al. execute a robust methodology for the preparation of dihydropyridones with excellent stereoselectivity. This method involves the diastereoselective adduction of Grignard reagents to chiral N-acyl pyridinium salts from 4-methoxy-3-(triisopropylsilyl)pyridine and chloroformates by hydrolysis of 4-methoxy-1,2-dihydropyridine as shown in Scheme 17.10 [48]. The wide range application of this technology has led to many alkaloid natural products, such as (±)-Lasubine-II (Scheme 17.11) [49]. This approach to the synthesis of alkaloid bicyclic cores, presuming the addition of a functionalized Grignard nucleophiles that will subsequently permit cyclization has been widely exploited.

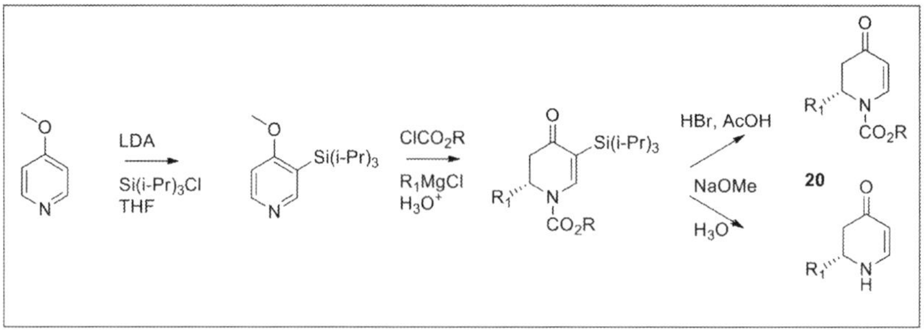

SCHEME 17.10 Preparation of chiral N-acyl pyridinium salts.

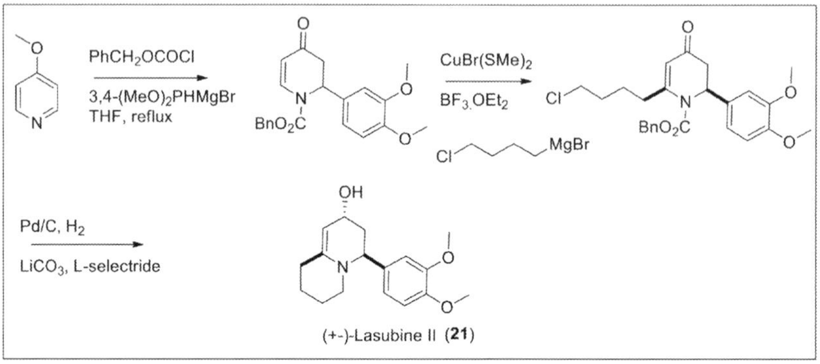

SCHEME 17.11 Synthesis of (±)-Lasubine-II alkaloid by using 4-methoxy pyridine.

17.3 Biological activity profiles

17.3.1 Antimicrobial activities

Di-acylhydrazine **22** and acyl(arylsulfonyl)hydrazines **23** belong to pyridine scaffolds and influence exciting antibacterial sensitivity against microorganisms *E. coli* and *S. albus* as compared to streptomycin sulfate. These compounds were also tested for their antibacterial activity against *C. dactylon, T. procumbens, E. crusgalli, C. rotundus, C. argentia, E. hirta,* and *E. indica*. Antifungal activity of the same target pyridines against *A. teniussiama* and *A. niger* [50]. Compounds **24–29** reveal the antimicrobial activity with the maximum zone of inhibition was observed against different microorganisms *S. aureus, B. subtitlis, S. typhi,* and *E. coli*. Pyridine derivative **26** showed good activity against *E. coli* and analogue **27** exhibits remarkable efficiency active with *S. aureus*, while the represented compounds **28** and **29** are active against *S. typhi* and target **29** have shown moderate activity against *B. subtilis* compared with standard drug [51]. Whereas Ali and Salem reported the antibacterial activity of the compound **30** to be nearly equivalent to Amphotericin B against fungal strain *Fusarium oxysporum* with MIC 0.98 μg/mL. Among the tested pyridines **31, 32, 33,** and **34** have showed remarkable antibacterial activities against different microorganisms (MIC = 0.58–2.1 μg/mL) in comparison to the standard ciprofloxacin (MIC = 0.65–1.9 μg/mL) [52]. Gomha and coauthors describe for antimicrobial activities of the synthesized pyridine derivatives were screened and the results

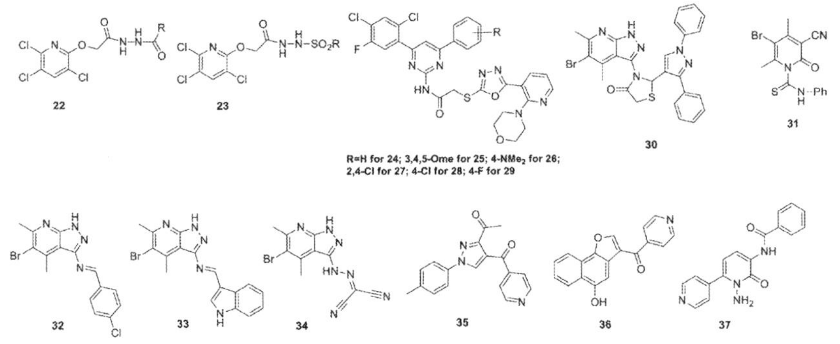

FIGURE 17.3 Antimicrobial properties of some of the pyridine derivatives.

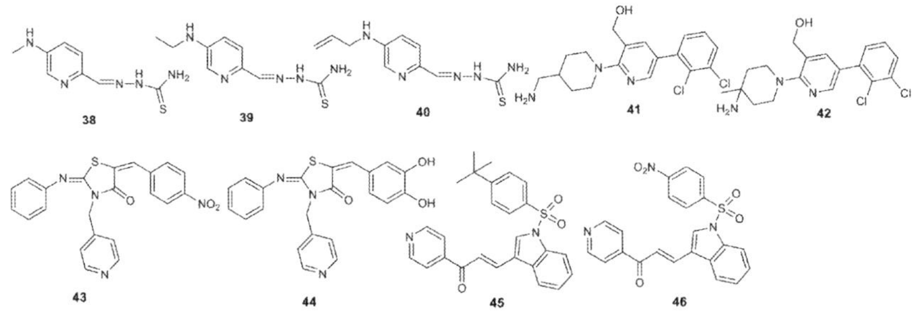

FIGURE 17.4 Anticancer activity of pyridine derivatives against SHP2 and CAIX inhibition.

showed that compounds **35, 36**, and **37** have good antibacterial activities against *S. aureus* with the zone of inhibition 5, 6, 7 mm compared with Cefepime reference drug (8 mm) (Fig. 17.3) [53].

17.3.2 Anticancer activity

The pyridine bearing compounds, 5-(methylamino)pyridine hydrazones **38**, 5-(ethylamino) pyridine hydrazone **39** and 5-(allylamino)pyridine-2-carboxaldehyde thiosemicarbazones **40** have shown a potent inhibition of ribonucleotide reductase with IC_{50} 1.3, 1.0, and 1.4 μM, respectively [54]. From literature a variety of cancers including liver, breast, and gastric cancers are mainly associated with SHP2[PS1]. Therefore, the developments of SHP2 inhibitors have enhanced extensive attention and targets (2-(4-(aminomethyl) piperidin-1-yl)-5-(2,3-dichlorophenyl)pyridin-3-yl)methanol **41**, (2-(4-amino-4-methylpiperidin-1-yl)-5-(2,3-dichlorophenyl)pyridin-3-yl)methanol **42** were most potent and highly selective SHP2 inhibitor with IC_{50} values 1.36, 1.41 μM, respectively [55]. Also pyridine-thiazolidinone contains nitro substitution on benzene ring **43** exhibited good activity against CAIX inhibitor with $IC_{50} = 1.61$ μM. From SAR studies pyridine scaffold with dihydroxyl groups on benzene **44** reveals an excellent effect on the behavior of CAIX inhibitor with $IC_{50} = 1.84$ μM. Furthermore, compounds **45** and **46** show inhibitory activity against CAIX ($IC_{50} = 0.15$ μM, 0.15 μM). Moreover, they exhibited predominant antiproliferative activity and produce apoptosis in the MCF-7 cancer cell line with % of inhibition 41.74%, 21.57%, respectively (Fig. 17.4) [56].

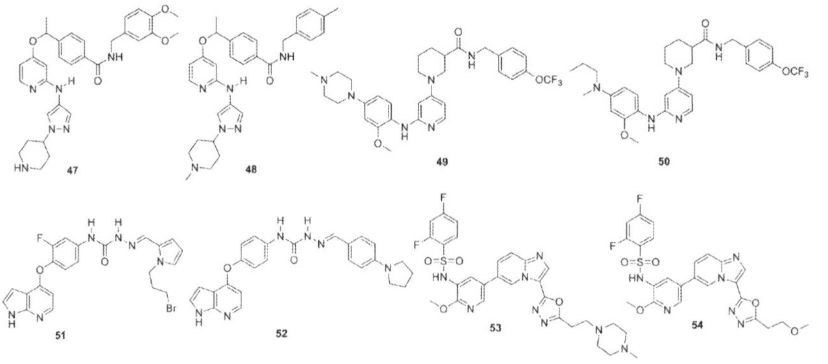

FIGURE 17.5 Cytotoxicity of pyridine compounds are active against with different cell lines.

ROS (proto-oncogene tyrosine protein kinase) is an enzyme encoded in humans by the ROS 1 gene and it has structural similarity with ALK (lymphoma kinase protein). Wang et al. [57] prepared a novel 4-(1-phenylethoxy)-2-amino-pyridine compounds that have been prepared and evaluated as ROS1 inhibitors. Among the tested compounds **47** and **48** reveal an antiproliferative effect against HCC78 cell line with IC_{50} 8.1 μM and 65.3 μM. In 2019, Liu et al. [58] prepared a novel series of 4-(1-phenylethoxy)-2-amino-pyridine scaffolds as anticrizotinib resistant ROS1/ALK inhibitors. Compounds **49, 50** showed more potent anticancer activity against ROS1G2032R harboring Ba/F_3 cell line with IC_{50} 104.7 nM, 643.5 nM compared crizotinib, whereas antiproliferative activity against cancer cell lines H3122 and HCC78 with IC_{50} 6.45 μM, 10.71 μM, respectively. In 2018 Wang et al. [59] studies led to the detection of novel and potent series of pyridine derivatives as c-Met inhibitors carrying hydrazone moiety. Many of the tested compounds reveal the highest anticancer activity against HepG2, A549, PC-3, and MCF-7 cancer cell lines and analogues **51, 52** exhibited perfect cytotoxicity against c-Met kinase with IC_{50} 0.506, 0.907 μM, respectively. In addition, final pyridine scaffold **53** exhibited excellent anticancer activity against HCT116, PI3Kα cell lines with IC_{50} 0.01, 0.20 μM and **54** displayed the strongest inhibitory activity against HCT116, PI3Kα cell lines with IC_{50} values 0.26, 0.17 μM, respectively (Fig. 17.5) [60].

Zhu et al. [61] discovered a novel potent 3,4-disubstituted-1H-pyrrolo[2,3-b] pyridines were introduced for use as irreversible inhibitors of mutant EGFR L858R/T790M through the different substituents by using Buchwald–Hartwig cross or Suzuki coupling reactions. Compounds **55, 56** exhibited excellent activity against T790M/EGFR-L858R enzyme assay with IC_{50} 0.001, 0.027 μM, respectively. HER2 (human epidermal growth factor receptor 2) also belongs to EGFR family and a gene that has breast cancer connection. Whereas compound **57** reveals that the most outstanding activity inhibitors against EGFR, HER-2 kinase with $IC_{50} = 0.09$ μM, 0.2μM, respectively. Moreover, Knudsen and coauthors designed and synthesized new imidazo[4,5-b]pyridine targets as cyclin-dependent kinase 9. Especially the pyridine derivatives **58, 59**, and **60** were the most active against CDK9 enzyme assay belongs to IC_{50} values range 0.50–1.002 μM. Furthermore, pyridine scaffold **61** undergo profiling activity against other kinases as PI3Kβ, PI3Kα, PI3Kδ, PI3Kγ, mTOR9 with IC_{50} 13, 4.2, 50, 64, and 78 μM, respectively [62,63]. In 2019 Wanga et al. prepared novel 3-substituted-1H-pyrrolo[2,3-b]pyridine derivatives and evaluated biological activities against maternal embryonic leucine

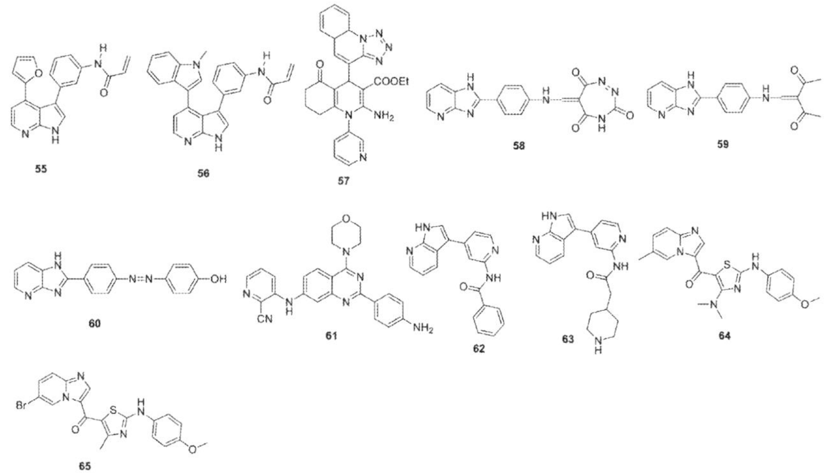

FIGURE 17.6 Anticancer properties of pyridine scaffolds with different cell lines.

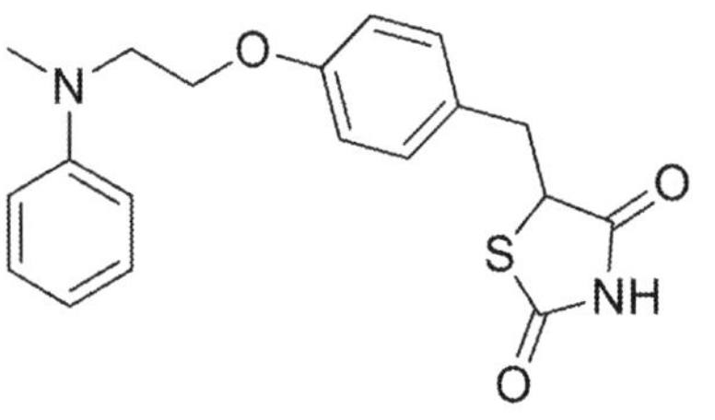

Rosiglitazone
antidiabetic
66

FIGURE 17.7 Antidiabetic drug contains pyridine scaffold.

zipper kinase (MELK) and compound **62** demonstrated moderate potency against MELK [IC_{50} 122–558 μM] [64]. Compound **63** displayed strong antiproliferative effects on cell lines MDA-MB-231, MCF-7, and A549 with IC_{50} 0.109–0.245 μM, respectively. Whereas Kamala et al. [65] recently established novel series of imidazoyl-pyridines linked to thiophene and thiazole motif via keto spacer. Compounds **64, 65** exhibited an absolute NF-YB activity against NF-KB reporter with IC_{50} 6.5 ± 0.6, 184.58 ± 1.47 μM, respectively (Fig. 17.6).

17.3.3 Antidiabetic activity

The pyridine scaffolds having thiazolidinones **66** reveal antidiabetic activity by GOD-POD method. The results show that some of these derivatives exhibited very effective antidiabetic activities, few of these pyridine derivatives manifest appreciable antidiabetic activities and finding to lead the development of novel class of antidiabetic agents in the coming time (Fig. 17.7) [66].

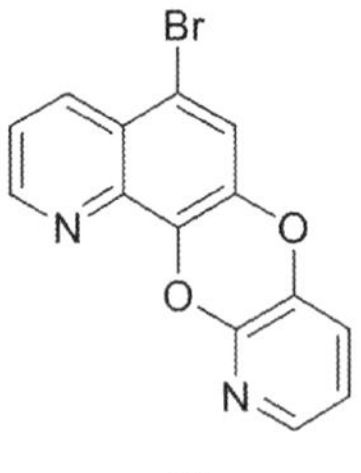

67

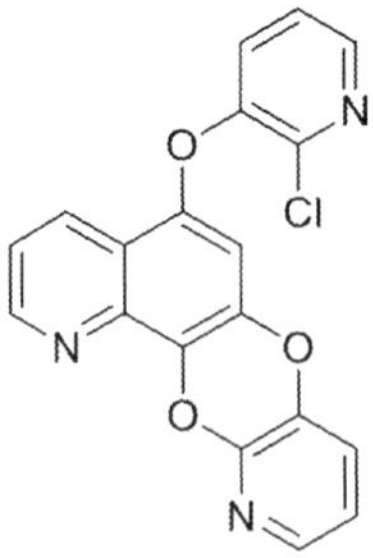

68

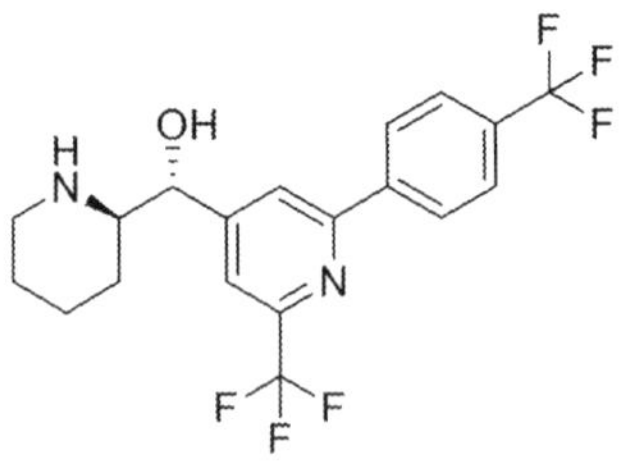

69

FIGURE 17.8 Antimalarial targets.

17.3.4 Antimalarial agents

Some of the pyridine quinoline hybrid molecules **67** and **68** were chosen for their antimalarial activity against *Plasmodium falciparum* strain with the active site of chloroquine-susceptible and the results of these molecules exhibited moderate to good antimalarial activity. These templates can be used for designing new antimalarial agents as improved their activities and also shows active against HPIA (heme polymerization inhibition) activities [67]. In addition, Pandya et al. [68] prepared for the synthesis of (*R*-2*R*)-piperidin-2-yl{2-(trifluoromethyl)-6-[4-(trifluoromethyl)substitutedphenyl]pyridin-4-yl}methanol (Enpirolin) derivatives used as antimalarial drugs as shown in Fig. 17.8.

17.3.5 Antituberculosis

Jones and coteam have been prepared for the pyridine-4-carbohydrazide (Isoniazide **70**) and it is the first-line medication in the prevention and treatment of antituberculosis [69]. Moreover, the synthesized compound 2-ethylpyridine-4-carbothioamide **71** is an antibiotic used in the treatment of tuberculosis [70]. Danac et al. [71] synthesized fused bipyridine heterocycles **72** and evaluated them for their antimycobacterial activities and nonreplicating, extracellular and intracellular organisms. Further, the end of 2017 have been prepared new drug 6-chloro-2-ethyl-N-(4-(4-(4-(trifluoromethoxy)phenyl)piperidin-1-yl) benzyl)imidazo[1,2-a]pyridine-3-carboxamide **73** in phase II clinical trials at is an imidazo pyridine derivative and it also shows good activity against MDR-TB and XDR-TB [72]. Dhumal et al. [73] were design and synthesis of novel 2-substituted pyridine linked to thiazolyl-5-aryl-1,3,4-oxadiazoles and evaluated for their antitubercular activity. Among them, two pyridine analogs **74, 75** showed potent activity against *M. bovis* BCG at concentrations <3 μg/mL. Velezheva et al. [74] have been synthesized novel indole scaffolds including pyridine nuclei and most active compound **76** showed MICs value equal to INH (0.05–2 μg/mL) against *M. tuberculosis $H_{37}Rv$* (Fig. 17.9).

In addition novel 2-substituted thiazolidinyl isonicotinamide derivatives, 3,4-disubstituted thiazolylidene isonicotinohydrazide derivatives and pyrrolyl isonicotinamide derivatives were synthesized by Elhakeem et al. [75]. These compounds were assessed for their in vitro antitubercular activity and minimum inhibitory concentration of compound **77** was found to be most active against *M. tuberculosis* H_{37}Ra 7131 (9.77 μg/mL). Further investigation by

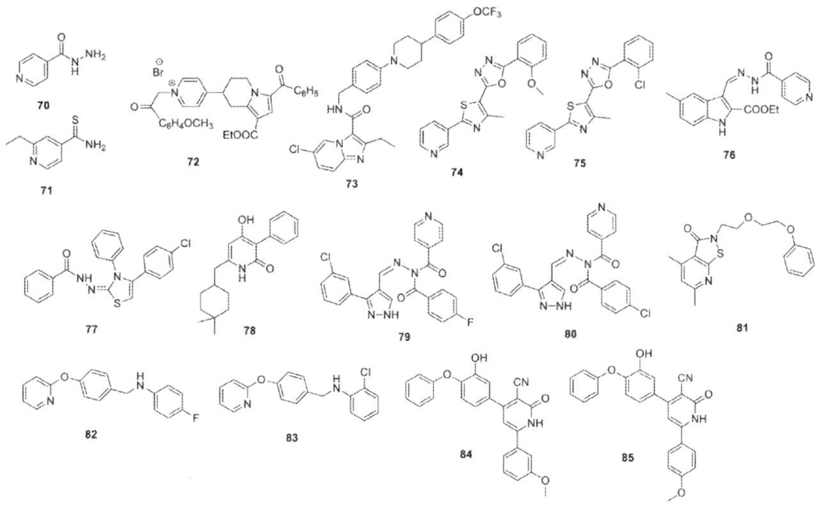

FIGURE 17.9 Antimycobacterial activities against different *Mycobacterium tuberculosis* (M.tb).

Shuyi Ng et al. [76] involved the synthesis of 4-hydroxy-2-pyridones and their anti-TB studies against *M. tuberculosis* $H_{37}Rv$. Compound **78** displayed potent activity against both drug sensitive and multidrug resistant TB clinical isolates and also showed favorable oral pharmacokinetic properties. A new series of isonicotinohydrazide based on pyrazole derivatives **79, 80** was prepared and evaluated their *M. tuberculosis* $H_{37}Rv$ (MTB). For the synthesized isothiazolopyridine derivatives assayed against *M. tuberculosis* $H_{37}Rv$ and 4,6-dimethyl-2-(2-(2-phenoxyethoxy) ethyl)isothiazolo[5,4-b]pyridin-3(2H)-one **81** is the most active against *M. tuberculosis* $H_{37}Rv$ with MIC 6.25 μg/mL [77]. In 2013, Malinka et al. [78] were prepared phenyl substituted-N-(4-(pyridin-2-yloxy)benzyl)anilines and showed better anti-TB activity with MIC range from 19 μM to 25 μM. The potent active compound was found to be 4-fluoro substituted derivative **82** with MIC 19 μM, whereas **83** also displayed good activity with MIC of 24 μM at 20 μg/mL and all the final compounds exhibited activity equal to or better than TCL in 7H9-based medium. Finally, among the tested compounds in GAST-Fe medium, pyridine analogues **84, 85** exhibited the maximum potency of 15, 20 μM at the second week of incubation when compared to TCL (MIC 30 μM) [79] (Fig. 17.9).

17.3.6 Anti-inflammatory activity

A group of imidazolyl-pyridine targets has been synthesized and from these compounds, **86** exhibit good anti-inflammatory activities [80]. Whereas below represented compounds **87–90** showed potent analgesic activity comparable to the standard drug Pentazocine (Fig. 17.10) [81,82].

17.3.7 Antihypertensive agents

Wang and coteam [83] used (RS)-3-ethyl-5-methyl-2-[(2-aminoethoxy) methyl]-4-(2-chlorophenyl)-6-methyl-1,4-dihydropyridine-3,5-dicarboxylate **91** as antihypertensive in the

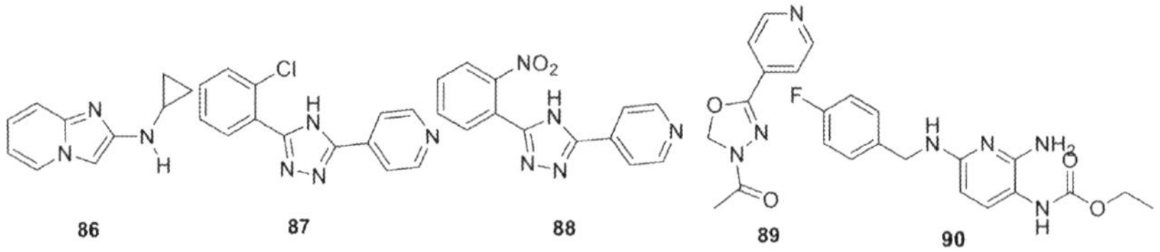
86 87 88 89 90

FIGURE 17.10 Anti-inflammatory activity profiles of some of the pyridine derivatives.

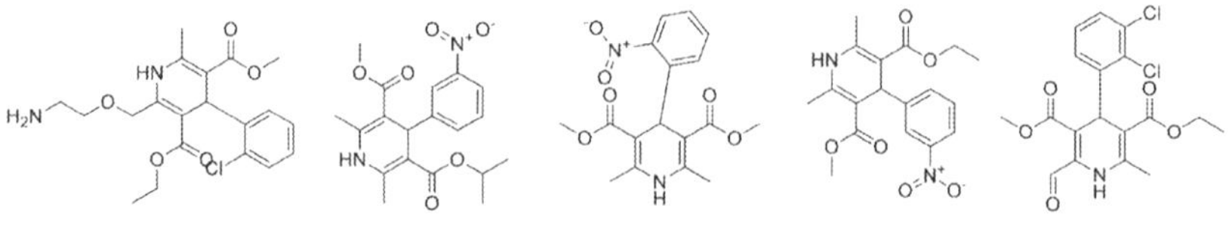

FIGURE 17.11 Antihypertensive agents.

treatment of angina pectoris. Whereas nilvadipine is a first-generation CCB (calcium channel blocker) used for the care of hypertension and compound methyl propan-2-yl 2,6-dimethyl-4-(3-nitrophenyl)-1,4-dihydropyridine-3,5-dicarboxylate for used to chronic major cerebral artery occlusion [84]. In addition, dimethyl-2,6-dimethyl-4-(2-nitrophenyl)-1,4-dihydropyridine-3,5-dicarboxylate is also first-generation calcium channel blocker having peripheral, coronary artery dilating properties and is used for the treatment of hypertension and angina [85]. Whereas the compound (RS)-ethyl methyl 2,6-dimethyl-4-(3-nitrophenyl)-1,4-dihydropyridine-3,5-dicarboxylate, 5-ethyl-3-methyl-4-(2,3-dichlorophenyl)-2-formyl-6-methyl-1,4-dihydropyridine-3,5-dicarboxylates is used for the treatment of primary (essential) hypertension to control high blood pressure (Fig. 17.11) [86,87].

17.3.8 Antiamoebic agents

Some of the ligands which have been originated from acetyl pyridines have shown antiamoebic activity, however, their complexes with ruthenium (II) have shown drastically increased antiamoebic activity. Complex **73** showed the highest antiamoebic activity compared with commercially available metronidazole. Binuclear complex of vanadium containing 2-acetylpyridine shows an excellent ameobocidal activity with IC_{50} 1.68–0.40 μM than standard metromidazole with IC_{50} 1.81 μM but ligand itself have no activity against amoeba (Fig. 17.12) [88].

17.3.9 Antiarhythmatic activity

Dokki and coauthors discuss the pyrimidine and pyridine scaffolds [PS2] can be easily combined with thiophene core. The resulted compounds were tested for their antiarrhythmic activities and exhibited the highest antiarrhythmic activity better than lidocaine and procaine amide as standard antiarrhythmic compounds [89].

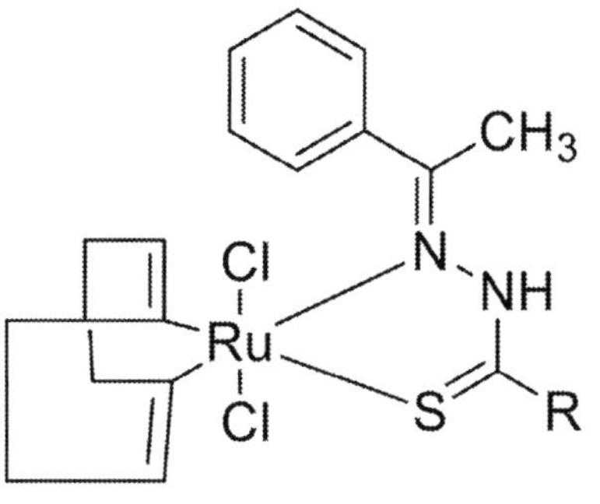

73

FIGURE 17.12 Antiamoebic activity.

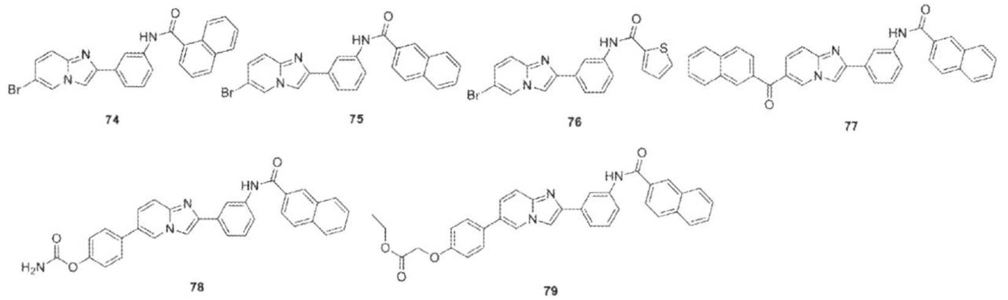

FIGURE 17.13 Enzymatic inhibitory studies of pyridine scaffolds.

17.3.10 Enzyme inhibition

Pyridine ring containing benzoimidazole derivatives shows gastric H^+/K^+-ATPase inhibitory activity. The series of imidazolyl-pyridine derivatives **74–79** have been prepared and discuss the ability to inhibition of cholesterol acyltransferase (acyl-CoA) (Fig. 17.13) [90].

17.4 Conclusion

It should be clear from the above facts, pyridine core has been introduced into drug discovery programs for diverse motivations. Very few compounds are designed to be an important part of the pharmacophore, admiringly contributing to ligand binding. Most of the research and development of anticancer drugs have been steady on molecules where the pyridine ring is present at the center part, contrasting various currently pyridine-containing drugs. It should be clear that the emerging complexity of the genetical environment that these processes are only beginning to be understood, and that multiple targets are being identified for the discovery of novel agents that have therapeutic activity placed in inflection of deviating transmitted processes. In addition with bioavailability of several probable targets such as ALK, ROS1, SHP2, CAIX HER2, L858R/T790M, EGFR, MELK, HPIA, *M. tuberculosis*, cholesterol acyl transferase along with good ability, there is immobile a wide range of potential pyridines from these features.

Acknowledgment

T.R. Allaka is thankful to CCST, IST, JNTUH University, Hyderabad for providing the facilities.

References

[1] T. Anderson, Uber Picolin: eine neue Basis aus dem Steinkohlen—Theerbl, Ann. Chem. Pharm. 60 (1846) 86–103.
[2] Y.T. Chan, C.N. Moorefield, M. Soler, G.R. Newkome, Unexpected isolation of a pentameric metallomacrocycle from the Fe(II)-mediated complexation of 120 degrees juxtaposed 2,2′:6′,2″-terpyridine ligands, Chemistry (Weinheim An Der Bergstrasse, Germany) 16 (2010) 1768–1771.
[3] J. Husson, M. Knorr, 2,2′:6′,2?-Terpyridines functionalized with thienyl substituents: synthesis and applications, J. Heterocycl. Chem. 49 (2012) 453–478.
[4] J. Bunzen, T. Bruhn, G. Bringmann, A. Lutzen, Synthesis and Helicate Formation of a New Family of BINOL-Based Bis (bipyridine) Ligands, J. Am. Chem. Soc. 131 (10) (2009) 3621–3630.
[5] D. Bora, B. Deb, A.L. Fuller, A.M.Z. Slawin, J.D Woollins, D.K. Dutta, Dicarbonyliridium(I) complexes of pyridine ester ligands and their reactivity towards various electrophiles, Inorganica Chim. Acta 363 (7) (2010) 1539–1546.
[6] S. Lin, X. Lu, Cationic Pd(II)/Bipyridine-Catalyzed Conjugate Addition of Arylboronic Acids to ß,ß-Disubstituted Enones: Construction of Quaternary Carbon Centers, Org. Lett. 12 (2010) 2536–2539.
[7] N.G. Kang, M. Changez, J.S. Lee, Living Anionic Polymerization of the Amphiphilic Monomer 2-(4-Vinylphenyl)pyridine, Macromolecules 40 (24) (2007) 8553–8559.
[8] (a) J.M. Lehn, Supramolecular Chemistry Concepts and Perspectives, VCH, Weinheim, 1995; (b) U.S. Schubert, C. Eschbaumer, Macromolecules Containing Bipyridine and Terpyridine Metal Complexes: Towards Metallo-supramolecular Polymers, 2002 Angew. Chem. Int. Ed. Chemie., 41 2892.
[9] (a)Functional Organic Materials, Muller, T.J.J., Bunz, U.H.F., Eds., Wiley-VCH: Weinheim, 2007.; (b) R. Makiura, S. Motoyama, Y. Umemura, H. Yamanaka, O. Sakata, H. Kitagawa.
[10] R. Murugan, E.F.V. Scriven, Applications of Dialkylaminopyridine (DMAP) Catalysts in Organic Synthesis, Aldrichimica Acta 36 (1) (2003) 21.
[11] G.M. Buckley, N. Cooper, R.J. Davenport, H.J. Dyke, F.P. Galleway, L. Gowers, A.F. Haughan, H.J. Kendall, C. Lowe, J.G. Montana, J. Oxford, J.C. Peake, C.L. Picken, M.D. Richard, V. Sabin, A. Sharpe, J.B.H. Warneck, 8-Methoxyquinoline-5-carboxamides as PDE4 inhibitors: a potential treatment for asthma, Bioorg. Med. Chem. Lett. 12 (12) (2002) 1613–1615.
[12] (a) C.D. Duffy, P. Maderna, C. McCarthy, C.E. Loscher, C. Godson, P.J. Guiry, Synthesis and biological evaluation of pyridine-containing lipoxin A4 analogues, Chem Med Chem 5 (4) (2010) 517–522; (b) R.B. Lacerda, C.K.F. de Lima, L.L. da Silva, N.C. Romeiro, A.L.P. Miranda, E.J. Barreiro, C.A.M. Fraga, Discovery of novel analgesic and anti-inflammatory 3-arylamine-imidazo[1,2-a]pyridine symbiotic prototypes, Bioorg. Med. Chem. 17 (1) (2009) 74–84.
[13] O.H. David, Pyrrole, pyrrolidine, pyridine, piperidine and tropane alkaloids, Nat. Prod. Rep. 17 (2000) 435–446.
[14] Vania A.F.F.M. Santos, Luis O. Regasini, Claudio R. Nogueira, Gabriela D. Passerini, Isabel Martinez, Vanderlan S. Bolzani, Marcia A.S. Graminha, Regina M.B. Cicarelli, Maysa Furlan, Antiprotozoal Sesquiterpene Pyridine Alkaloids from Maytenus ilicifolia, J. Nat. Prod. 75 (2012) 991–995.
[15] James A. Bull, James J. Mousseau, Guillaume Pelletier, Andre´ B. Charette, Synthesis of Pyridine and Dihydropyridine Derivatives by Regio- and Stereoselective Addition to N-Activated Pyridines, Chem. Rev. 112 (2012) 2642–2713.
[16] Z.J. Song, M. Zhao, R. Desmond, P. Devine, D.M. Tschaen, R. Tillyer, L. Frey, R. Heid, F. Xu, B. Foster, J. Li, R. Reamer, R. Volante, U.H. Dolling, P.J. Reider, S. Okada, Y. Kato, E. Mano, Practical Asymmetric Synthesis of an Endothelin Receptor Antagonist, J. Org. Chem. 64 (26) (1999) 9658–9667.
[17] Abe. Y, Kayakiri. H, Satoh. S, Inoue. T, Sawada. Y, Inamura. N, Asano. M, Aramori. I, Hatori. C, Sawai. H, Oku. T, Tanaka H, A novel class of orally active non-peptide bradykinin B2 receptor antagonists. 3. Discovering bioisosteres of the imidazo[1,2-a]pyridine moiety, J. Med. Chem. 41 (21) (1998) 4062–4079.
[18] S. Follot, J.C. Debouzy, D. Crouzier, C. Enguehard Gueiffier, A. Gueiffier, F. Nachon, B. Lefebvre, F. Fauvelle, Physicochemical properties and membrane interactions of anti-apoptotic derivatives 2-(4-fluorophenyl)-3-(pyridin-4-yl)imidazo [1,2-a]pyridine depending on the hydroxyalkylamino side chain length and conformation: An NMR and ESR study, Eur. J. Med. Chem. 44 (9) (2009) 3509–3518.

[19] G. Matolcsy, In Pesticide Chemistry, Elsevier, Amsterdam, 1998, p. 427.
[20] Ming Wei Zhang, Rui Feng Zhang, Fang Xuan Zhang, Rui Hai Liu, Phenolic Profiles and Antioxidant Activity of Black Rice Bran of Different Commercially Available Varieties, J. Agric. Food Chem. 58 (13) (2010) 7580–7587.
[21] N. Nayak, J. Ramprasad, U. Dalimba, Design, Synthesis, and Biological Evaluation of New 8-Trifluoromethylquinoline Containing Pyrazole-3-carboxamide Derivatives, J. Heterocycl. Chem. 54 (2017) 171–182.
[22] Ali Tarik El-Sayed, Synthesis of some novel pyrazolo[3,4-b]pyridine and pyrazolo[3,4-d]pyrimidine derivatives bearing 5,6-diphenyl-1,2,4-triazine moiety as potential antimicrobial agents, Eur. J. Med. Chem. 44 (2009) 4385–4392.
[23] (a) M. Heller, U.S. Schubert, Syntheses of Functionalized 2,2′:6′,2″-Terpyridines, Eur. J. Org. Chem. (2003) 947–961; (b) G.D. Henry, De Novo Synthesis of Substituted Pyridines, Tetrahedron 60 (2004) 6043–6061.
[24] D.M. Stout, A.I. Meyers, Recent advances in the chemistry of dihydropyridines, Chem. Rev. 82 (2) (1982) 223–243.
[25] A.R. Katritzky, Z. Zakaria, E. Lunt, P.G. Jones, O. Kennard, Photocyclisation of 1,2-diarylpyridinium cations and the photobis-cyclisation of 1,2,6-triarylpyridinium cations. X-Ray crystal structure of 9-phenyl-2,10b-diazadibenzo[fg,op]naphthacenium perchlorate, J. Chem. Soc. Chem. Commun. 6 (1979) 268–269.
[26] T. Okamoto, M. Hirobe, Y. Tamai, E. Yabe, Reaction of n-aminopyridinium derivatives. 3. Synthesis of s-triazolo [1,5-a]-pyridine ring, Chem. Pharm. Bull. 14 (5) (1966) 506–512.
[27] F.E. Goda, Abdel Aziz, A.M. A., O.A.Synthesis Attef, Antimicrobial Activity and Conformational Analysis of Novel Substituted Pyridines: BF3-Promoted Reaction of Hydrazine with 2-Alkoxy Pyridines, Bioorg. Med. Chem. 12 (8) (2004) 1845–1852.
[28] M. Aldenderfer, N.M. Craig, R.J. Speak, R.P. Filcoff, Four-thousand-year-old gold artifacts from the Lake Titicaca basin, southern Peru, Proc. Natl. Acad. Sci. U.S.A. 105 (13) (2008) 5002–5005.
[29] L.Y. Xi, R.Y. Zhang, S. Liang, S.Y. Chen, X.Q. Yu, Copper-Catalyzed Aerobic Synthesis of 2-Arylpyridines from Acetophenones and 1,3-Diaminopropane, Org. Lett. 16 (20) (2014) 5269–5271.
[30] H. Andersson, F. Almqvist, R. Olsson, Synthesis of 2-Substituted Pyridines via a Regiospecific Alkylation, Alkynylation, and Arylation of Pyridine N-Oxides, Org. Lett. 9 (2007) 1335–1337.
[31] Alan R. Katritzky, Ashraf A.A. Abdel-Fattah, Dmytro O. Tymoshenko, Samy A. Essawy, A Novel and Efficient 2,4,6-Trisubstituted Pyridine Ring Synthesis via alpha-Benzotriazolyl Ketones, Synthesis 12 (1999) 2114–2118.
[32] K. Yoshida, F. Kawagoe, K. Hayashi, S. Horiuchi, T. Imamoto, A. Yanagisawa, Synthesis of 3-Hydroxypyridines Using Ruthenium-Catalyzed Ring-Closing Olefin Metathesis, Org. Lett. 11 (2009) 515–518.
[33] H. Huang, J. Cai, L. Tang, Z. Wang, F. Li, G.J. Deng, Metal-Free Assembly of Polysubstituted Pyridines from Oximes and Acroleins, Org. Chem. 81 (4) (2016) 1499–1505.
[34] J. Shen, D. Cai, C. Kuai, Y. Liu, M. Wei, G. Cheng, X. Cui, Base-Promoted ß-C(sp3)–H Functionalization of Enaminones: An Approach to Poly substituted Pyridines, J. Org. Chem. 80 (13) (2015) 6584–6589.
[35] S.I. Scherbinina, O.V. Fedorov, V.V. Levin, V.A. Kokorekin, M.I. Struchkova, A.D. Dilman, Synthesis of 3-Fluoropyridines via Photoredox-Mediated Coupling of a,a-Difluoro-ß-iodoketones with Silyl Enol Ethers, J. Org. Chem. 82 (2017) 12967–12974.
[36] D. Craig, G.D. Henry, Sulfone-mediated synthesis of poly substituted pyridines, Tetrahedron Lett. 46 (2005) 2559–2562.
[37] Y.F. Wang, K.K. Toh, S. Chiba, K. Narasaka, Mn(III)-Catalyzed Synthesis of Pyrroles from Vinyl Azides and 1,3-Dicarbonyl Compounds, Org. Lett. 10 (21) (2008) 5019–5022.
[38] W. Gati, M.M. Rammah, M.B. Rammah, F. Couty, G.De Novo Evano, Synthesis of 1,4-Dihydropyridines and Pyridines, J. Am. Chem. Soc. 134 (2012) 9078–9081.
[39] I.N. Maruyama, Mechanisms of activation of receptor tyrosine kinases: monomers or dimers, Cells (2014) 304–330.
[40] B. Al-Saleh, M.M. Abdelkhalik, A.M. Eltoukhy, M.H. Elnagdi, Enaminones in heterocyclic synthesis: A new regioselective synthesis of 2,3,6-trisubstituted pyridines, 6-substituted-3-aroylpyridines and 1,3,5-triaroylbenzenes, J. Heterocycl. Chem. 39 (5) (2002) 1035–1038.
[41] J.C. Lewis, R.G. Bergman, J.A. Ellman, Rh (I)-catalyzed alkylation of quinolines and pyridines via C-H bond activation, J. Am. Chem. Soc. 129 (17) (2007) 5332–5333.
[42] G. Kamil, S. Bengu, S. Dalibor, Site-Specific Phenylation of Pyridine Catalyzed by Phosphido-Bridged Ruthenium Dimer Complexes: A Prototype for C-H Arylation of Electron-Deficient Heteroarenes, J. Am. Chem. Soc. 127 (11) (2005) 3648–3649.

[43] Y. Nakao, K.S. Kanyiva, T. Hiyama, A Strategy for C-H Activation of Pyridines: Direct C-2 Selective Alkenylation of Pyridines by Nickel/Lewis Acid Catalysis, J. Am. Chem. Soc. 130 (8) (2008) 2448–2449.
[44] D.J. Schipper, M. El-Salfiti, C.J. Whipp, K. Fagnou, Direct arylation of azine N-oxides with aryl triflates, Tetrahedron 65 (26) (2009) 4977–4983.
[45] A.B. Charette, M. Grenon, A. Lemire, M. Pourashraf, J. Martel, Practical and Highly Regio- and Stereoselective Synthesis of 2-Substituted Dihydropyridines and Piperidines: Application to the Synthesis of (-)-Coniine, J. Am. Chem. Soc. 123 (47) (2001) 11829–11830.
[46] A. Lemire, M. Grenon, M. Pourashraf, A.B. Charette, Nucleophilic Addition to 3-Substituted Pyridinium Salts: Expedient Syntheses of (-)-L-733,061 and (-)-CP-99,994, Org. Lett. 6 (20) (2004) 3517–3520.
[47] J.T. Kuethe., D.L Comins, Addition of Indolyl and Pyrrolyl Grignard Reagents to 1-Acylpyridinium Salts, J. Org. Chem. 69 (8) (2004) 2863–2866.
[48] D.L. Comins, S.P. Joseph, R.R. Goehring, Asymmetric Synthesis of 2-Alkyl(Aryl)-2,3-dihydro-4-pyridones by Addition of Grignard Reagents to Chiral 1-Acyl-4-methoxypyridinium Salts, J. Am. Chem. Soc. 116 (11) (1994) 4719–4728.
[49] S.D. Dharmpal, C.O. Allan, Efficient route to the synthesis of C-2, C-3 substituted 4-piperidones, Tetrahedron Letters 32 (30) (1991) 3643–3646.
[50] V. Chavan, S. Sonawane, M. Shingare, B. Karale, Synthesis, characterization, and biological activities of some 3,5,6-trichloropyridine derivatives, Chem. Heterocycl. Comp. 42 (2006) 625–630.
[51] T.A. Naik, K.H. Chikhalia, Studies on synthesis of pyrimidine derivatives and their pharmacological evaluation, Eur. J. Chem. 41 (2007) 60.
[52] S.S. Marwa, A.M.A. Mohamed, Novel Pyrazolo[3,4-b]pyridine Derivatives: Synthesis, Characterization, Antimicrobial and Antiproliferative Profile, Biol. Pharm. Bull. 39 (4) (2016) 473–483.
[53] M.A. Fathy, M.G. Sobhi, H.A. Aly, M. Peter, A.S. Mohsen, A Facile Synthesis and Drug Design of Some New Heterocyclic Compounds Incorporating Pyridine Moiety and Their Antimicrobial Evaluation, Lett. Drug Des. Discov. 14 (7) (2017) 752–762.
[54] L. Mao-Chin, L. Tai-Shun, G.C. Joseph, H.C. Ann, C.S. Alan, Synthesis and Biological Activity of 3- and 5-Amino Derivatives of Pyridine-2-carboxaldehyde Thiosemicarbazone, J. Med. Chem. 39 (13) (1996) 2586–2593.
[55] L. Wen-Shan, Y. Bing, W. Rui-Rui, L. Wei-Ya, M. Yang Chun, Z. Liang, D. Shan, M. Ying, W Run Ling, Design, synthesis and biological evaluation of pyridine derivatives as selective SHP2 inhibitors, Bioorg. Chem. 100 (2020) 103875.
[56] M. Esraa Ali, S.M.I. Nasser, H. Mohamed, R Hanan, Medicinal attributes of pyridine scaffold as anticancer targeting agents, Future J. Pharm. Sci. 7 (1) (2021) 24.
[57] Y. Tian, T. Zhang, L. Long, Z. Li, S. Wan, G. Wang, Y. Yu, J. Hou, X. Wu, J. Zhang, Design, synthesis, biological evaluation and molecular modeling of novel 2-amino-4-(1-phenylethoxy) pyridine derivatives as potential ROS1 inhibitors, Eur. J. Med. Chem. 143 (2018) 182–199.
[58] W. Chen, X. Guo, C. Zhang, D. Ke, G. Zhang, Y. Yu, Discovery of 2-aminopyridines bearing a pyridone moiety as potent ALK inhibitors to overcome the crizotinib-resistant mutants, Eur. J. Med. Chem. 183 (2019) 111734.
[59] W. Wang, S. Xu, Y. Duan, X. Liu, X. Li, C. Wang, B. Zhao, P. Zheng, W. Zhu, Synthesis and bioevaluation and doking study of 1H-pyrrolo[2,3-b]pyridine derivatives bearing aromatic hydrazone moiety as c-Met inhibitors, Eur. J. Med. Chem. 145 (2018) 315–327.
[60] X. Siyu, D. Alexej, J. Han, M. Sako, A. Fatemeh, C. Simon, Z. Peng, L. Xinyong, Inhibitors of SARS-CoV-2 Entry: Current and Future Opportunities, J. Med. Chem. 63 (21) (2020) 12256–12274.
[61] C.B. Sangani, J.A. Makawana, Y.T. Duan, S.B. Teraiya, N.J. Thumar, H.L. Zhu, Design, synthesis and molecular modeling of biquinoline-pyridine hybrids as a new class of potential EGFR and HER-2 kinase inhibitors, Bioorg. Med. Chem. Lett. 24 (18) (2014) 4472–4476.
[62] U. Asghar, A.K. Witkiewicz, N.C. Turner, E.S. Knudsen, The history and future of targeting cyclin-dependent kinases in cancer therapy, Nat. Rev. Drug. Discov. 14 (2) (2015) 130–146.
[63] W. Peng, Z. Tu, Z. Long, Q. Liu, G. Lu, Discovery of 2-(2-aminopyrimidin-5-yl)-4-morpholino-N-(pyridin-3-yl)quinazolin-7-amines as novel PI3K/mTOR inhibitors and anticancer agents, Eur. J. Med. Chem. 108 (2016) 644–654.
[64] R. Wang, Y. Chen, B. Yang, S. Yu, X. Zhao, C. Zhang, C. Hao, D. Zhao, M. Cheng, Design, synthesis, biological evaluation and molecular modeling of novel 1H-pyrrolo[2,3-b]pyridine derivatives as potential anti-tumor agents, Bioorg. Chem. 94 (2019) 103474.

[65] K.K. Vasu, C.S. Digwal, A.N. Pandya, D.H. Pandya, J.A. Sharma, S. Pateet, Imidazo[1,2-a]pyridines linked with thiazoles/thiophene motif through keto spacer as potential cytotoxic agents and NF-?B inhibitors, Bioorg. Med. Chem. Lett. 27 (2017) 5463–5466.
[66] S.D. Firke, B.M. Firake, R.Y. Chaudhari, V.R. Patil, Synthetic and Pharmacological Evaluation of Some Pyridine Containing Thiazolidinones, Asian. J. Research. Chem. 2 (2) (2009) 157–161.
[67] B. Narayan Acharya, D. Thavaselvam, M. Parshad Kaushik, Synthesis and antimalarial evaluation of novel pyridine quinoline hybrids, Med. Chem. Res. 17 (8) (2008) 487–494.
[68] S.N.A Pandya., Test Book of Medicinal Chemistry Vol. 1 (2009) 610–611.
[69] S.J. David, The Health Care Experiments at Many Farms: The Navajo, Tuberculosis, and the Limits of Modern Medicine, 1952-1962, Bull. Hist. Medic. 76 (4) (2002) 749–790.
[70] T.A. Vannelli, A. Dykman, P.R.O.D Montellano, The anti-tuberculosis drug ethionamide is activated by a flavoprotein monooxygenase, J. Biol. Chem. 277 (15) (2002) 12824–12829.
[71] R. Danac, I.I. Mangalagiu, Antimycobacterial activity of nitrogen heterocycles derivatives: bipyridine derivatives. Part III, Eur. J. Med. Chem. 74 (2014) 664–670.
[72] Newtbdrugs 2019. www.newtbdrugs.com. Available from: https://www.newtbdrugs.org/pipeline/compound/telacebec-q203. Accesed on 15th April, 2019.
[73] S.T. Dhumala, A.R. Deshmukha, M.R. Bhoslea, V.M. Khedkar, L.U. Nawaleb, D. Sarkar, R.A. Mane, Synthesis and antitubercular activity of new 1, 3, 4-oxadiazoles bearing pyridyl and thiazolyl scaffolds, Bioorg. Med. Chem. Lett. 26 (15) (2016) 3646–3651.
[74] V. Velezheva, P. Brennan, P. Ivanov, A. Kornienko, S. Lyubimov, K. Kazarian, B. Nikonenko, K. Majorov, A. Apt, Synthesis and anti-tuberculosis activity of indole-pyridine derived hydrazides, hydrazide-hydrazones, and thiosemicarbazones, Bioorg. Med. Chem. Lett. 26 (3) (2016) 978–985.
[75] M.A. Elhakeem, A.T. Taher, S.M. Abuel Maaty, Synthesis and anti-mycobacterial evaluation of some new isonicotinyl hydrazide analogues, Bull. Fac. Pharm. Cairo Univ. 53 (2015) 45–52.
[76] P.S. Ng, U.H. Manjunatha, S.P.S. Rao, L.R. Camacho, Ma.N. Ling, M. Herve, C.G. Noble, A. Goh, S. Peukert, T.T. Diagana, P.W. Smith, R.R. Kondreddi, Structure Activity Relationships of 4-Hydroxy-2-Pyridones: A Novel Class of Antituberculosis Agents, Eur. J. Med. Chem. 106 (2015) 144–156.
[77] N. Nayak, J. Ramprasad, U. Dalimba, New INH-pyrazole analogs: Design, synthesis and evaluation of antitubercular and antibacterial activity, Bioorg. Med. Chem. Lett. 25 (33) (2015) 5540–5545.
[78] W. Malinka, P. Swiatek, M. Sliwinska, A. Szponar Gamian, Z. Karczmarzyk, Andrzej Fruzinski, A. Synthesis of novel isothiazolopyridines and their in vitro evaluation against Mycobacterium and Propionibacteriumacnes, Bioorg. Med. Chem. 21 (17) (2013) 5282–5291.
[79] R. Verma, H.I. Boshoff, K. Arora, I. Bairy, M. Tiwari, B.G. Varadaraj, G.G. Shenoy, Synthesis, evaluation, molecular docking, and molecular dynamics studies of novel N-(4-[pyridin-2-yloxy]benzyl)arylamine derivatives as potential antitubercular agents, Drug. Dev. Res. 81 (2020) 315–328.
[80] S.M. Sondhi, M. Dinodia, A. Kumar, Synthesis, anti-inflammatory and analgesic activity evaluation of some amidine and hydrazone derivatives, Bioorg. Med. Chem. 14 (13) (2006) 4657–4663.
[81] G.B. Nigade, P. Chavan, M.N. Deodhar, Synthesis and analgesic activity of new pyridine-based heterocyclic derivatives, Med. Chem. Res. 21 (1) (2010) 27–37.
[82] M.D. Swedberg, H.E. Shannon, B. Nickel, S.R. Goldberg, Pharmacological mechanisms of action of flupirtine: a novel, centrally acting, nonopioid analgesic evaluated by its discriminative effects in the rat, J. Pharmacol. Exp. Ther. 246 (3) (1988) 1067–1074.
[83] J.G. Wang, A combined role of calcium channel blockers and angiotensin receptor blockers in stroke prevention, Vasc. Health Risk Manage. 5 (2009) 593–605.
[84] S.N.A Pandya, Test Book of Medicinal Chemistry, Vol 1, Synthetic and Biochemical Approach, 2009, pp. 43–44.
[85] S.N.A Pandya, Test Book of Medicinal Chemistry, Vol-1, Synthetic and Biochemical Approach, 2009, pp. 400–401.
[86] M.A. Siddiqui, G.L. Plosker, Fixed-dose combination enalapril/nitrendipine a review of its use in mild-to-moderate hypertension, Drugs 64 (2004) 1135–1148.
[87] P. Prathima, S.P. Sethy, T. Sameena, K. Shailaja, Pyridine and Its Biological Activity: A Review, Asian J. Res. Chem. 6 (10) (2013) 888–899.

[88] N. Bharti, M.R. Maurya, F. Naqvi, A. Azam, Synthesis and antiamoebic activity of new cyclooctadiene ruthenium (II) complexes with 2-acetylpyridine and benzimidazole derivatives, Bioorg. Med. Chem. Lett. 10 (20) (2000) 2243–2245.
[89] A.G.E.S. Amr, N.A.S. Abdel-Hafez, S.F. Mohamed, M.M. Abdalla, Synthesis, reactions, and antiarrhythmic activities of some novel pyrimidines and pyridines fused with thiophene moiety, Tur. J. Chem. 33 (3) (2009) 421–432.
[90] S.U. Jo, S.G. Gang, S.S. Kim, H.G. Jeon, J.G. Choe, E.G. Yeom, Synthesis and SAR of Benzimidazole Derivatives Containing Oxycyclic Pyridine as a Gastric H+/K+-ATPase Inhibitors, Bull. Korean Chem. Soc. 22 (11) (2001) 1217–1223.

Index

Page numbers followed by "*f*" and "*t*" indicate, figures and tables respectively.

A

Abemaciclib synthesis, 376, 378*f*
Abipyridine-based fluorescent system, 466
Abiraterone acetate, 379
Acalabrutinib, 379
Acetylcholine esterase, 160, 164
AChE inhibitor, 87
Acidic catalysts, 513
Acquired immune deficiency syndrome (AIDS), 1
Acute myeloid leukemia (AML), 385
Adenosine, 104
Aldose reductase inhibitors, 231
Alkylation, 163
Alpelisib, 381
Alzheimer disease, 86, 190, 221, 411
  pathogenesis, 412*f*
Amide carbonyl group bioisosteres, 30
Amide–pyridines, 238
Aminoimidazoles, 95
Amino oxazoline xanthenes, 170
Amyloid precursor protein (APP), 36, 412
Amyloid-specific imaging agents, 91
Anaplastic lymphoma kinase (ALK), 216
Anomeric-based oxidation mechanism, 332
Antiamoebic agents, 618
Antiarhythmatic activity, 618
Antibiotics, 14
  beta lactam inhibitors, 15*f*
  mechanism of action, 16*f*
Anticancer activity, 215, 613
Antidiabetic activity, 615
Antihypertensive agents, 617, 618*f*
Anti-infective agents, 9
Antiinflammatory activity, 235, 617
Antimalarial agents, 616*f*
Antimicrobial agents, 11
  cllassification of cell wall, 13*f*
  types, 12*f*
Antimicrobials, 581
  resistance, 9
Antioxidants, 234
  activities, 234
Antitubercular therapy, 36
Antituberculosis, 616
Antituberculosis activity, 240
Apalutamide, 382
Artemether, 149
Artesunate, 149
Axitinib, 383, 384
Azaindolizinone derivative, 169

B

BACE1 inhibitors, 198*f*
Baeyer pyridine synthesis, 254, 257*f*
Basal cell carcinoma (BCC), 395
Benzimidazole
  based derivatives, 474
  pyridine compounds, 241
Benzofuran-based N-benzylpyridinium derivatives, 87
Benzofuranone based derivatives, 191
Benzofuran–pyridine analogues, 422
Benzofused pyridines, 303
Biginelli reaction, 305
Bioisosteres, 25, 28
Bipyridines, 419
  based fluorescent system, 466
Body mass index (BMI), 375

C

Carbanions, 50
Carbazole, 429
Carbonate ions, 453
Carbonic anhydrase inhibition, 220
Catalytic anionic site (CAS), 160
Chemotherapy-induced anemia (CIA), 375
Chloroquine, 149
Cholinergic hypothesis, 412
Cholinesterase
  enzymes, 222
  inhibition activity, 221
  inhibitors, 86, 193*f*, 194*f*, 195*f*
Chromanone, 421
Cinchomeronic dinitrile derivatives, 513
Colorimetric chemosensors, 446, 452
Colorimetric sensors, 446
Copper ferrite nanocatalyst, 348

Coronavirus disease 2019 (COVID-19), 1
Coumarin-linked pyrazole-pyridine probe, 471
Creutzfeldt–Jakob disease, 11
Crystal field theory (CFT), 45
Cupric ions, 453
Cytotoxicity images of compounds, 129*f*

D
Dementia, 159
Diabetes mellitus, 417
Dicarboxamide derivative, 481
Dihydropyridines, 227, 303
Dipeptide bioisosteres, 30
Dipicolinimidamide
  based chemosensor, 458*f*
  chemosensors, 457
DNA gyrase, 20
Drugs containing pyridine moiety, 190*f*

E
Emil Knoevenagel process, 415
Enasidenib, 385, 386*f*
Enzyme inhibition, 619
Erlotinib, 152*t*
Ethyl-2-pyridinyl acetate, 178

F
FDA-approved drugs, 191*f*
Fibroblast growth factors (FGFs), 219
Fluorescent chemosensors, 446
Fluorescent sensing probes, 458
Förster resonance energy transfer (FRET), 447
Frontier molecular orbitals (FMO), 139
Furopyridinediones, 226
Furopyridines, 83

G
Gattermann–Skita synthesis, 415
Gemcitabine, 152*t*
Gewald reaction, 305
Glutamate receptors, 37
Glycogen synthase kinase-3 (GSK-3), 201
Green synthesis methods, 331
Grignard nucleophiles, 611
Grignard reagents, 611
Grimm's classification, 25
Grimm's hydride displacement law, 25
Groebke–Blackburn–Bienaymé (GBB) reaction, 305

H
Hantzsch approach, 607
Hantzsch dihydropyridine synthesis, 305*f*
Hantzsch pyridine synthesis, 72*f*, 254, 257*f*
Hantzsch synthesis methods, 415
Heat-resisting materials, 59
Heavy transition metals, 455
Heterocyclic organic compounds, 302, 582
Heterocyclic polymers, 52
Heterogeneous catalysis, 504
High-temperature electrolyte membranes, 59
Hippocampus, 160
Histamine, 36
Hückel aromaticity rule, 207
Human immunodeficiency virus (HIV), 1
Human pathogens
  process of development, 17*f*
  stages, 7*f*
  types of, 6*f*, 10*f*
Hydrazine, 463
Hydrazinyl pyridine, 242
Hydrazinyl–pyridine chemosensors, 454

I
Imatinib, 385
Imidazole–pyridine analogues, 424
Imidazopyridine, 78
  synthesis, 79*f*
Imidazopyridines, 212, 216, 223
Infectious diseases, 1
Infectious fungal diseases, 236
Influenza B-Mass virus, 64
International Diabetes Federation, 151
Ion-exchange resins, 59
Ionic liquids, 517
Isoniazid, 211
Ivosidenib, 387

K
Kabachnik–Fields reaction, 305
Katrizky pyridine synthesis, 256, 258*f*
Ketone carbonyl bioisosteres, 30
Kröhnke pyridine synthesis, 255, 258*f*

L
Lansopyrazole chemosensors, 453*f*, 453
Long-lasting endocrinal metabolic disease, 225

M
Mannich reaction, 305
MAO-B inhibitors, 101
Matrix metalloproteinases (MMPs), 36
Mefloquine, 149
Mepyramine, 36
Metal chelators, 178
Metastatic castration-resistant prostate cancer (mCRPC), 378
Michael-addition reactions, 608
Microorganism, 581

Microplate alamar blue assay method (MABA method), 266
MOF-based heterogeneous catalysts, 547
Monoalkylated bipyridyl, 52
Multicomponent
reactions, 165, 305*f*
synthesis, 299
Multiple-target chemosensor, 450
Multitarget directed ligands (MTDLs), 107
Mycobacterium tuberculosis, 24

N
Nanocatalyst catalyzed the reaction synthesis, 347*f*
Naphthyl–pyridine chemosensors, 453*f*, 453
Neratinib, 388
Nerlynx, 388
Neuroinflammation, 413
Neuroprotective agents, 102
Nevirapine, 210
Nexavar, 396
Nexium, 211
Niacin, 209
Nicaraven, 210
Nicotine, 208, 420
Nicotinic acetylcholine receptor (nAChRs) ligands, 106
Nicotinonitrile coumarin hybrids, 224
Nilotinib, 388, 389
synthesis, 390*f*
Nitrate anions, 491
Nitro-based compounds, 468
Nitrogen-containing heterocycles, 299
Nonbenzofused pyridines, 303, 304*f*
Nonpharmacologic therapy, 414
Nonsteroidal anti-inflammatory drugs (NSAID), 235

O
One-pot procedures, 362
Organometallic, 596
Oxidation, 163

P
Palbociclib, 390
synthesis, 391*f*
Peptide isosteres, 30
Pexidartinib, 391, 392
synthesis, 392*f*
Phosphate bioisosteres, 30
Phosphodiesterases, 106
Phosphoinositide 3 kinase (PI3K), 381
Photoinduced electron transfer (PET), 447
Photoresists, 59
Pictet–Spengler reaction, 429
Pioglitazone, 210
Platelet-derived growth factor receptors (PDGFR), 385
Plausible mechanism, 321*f*
Polycyclic pyridine compounds, 216
Polyol pathway, 231
Positron emission tomography (PET), 91, 92
PreADMET software, 587
Primaquine, 149
Pyrazolo-pyridine analogs, 292, 425
Pyrazolopyridines, 81, 216, 219, 350
Pyrazolopyrimidine-based derivative, 473
Pyridine, 14, 43, 44, 69, 87, 144, 163, 164, 207, 214, 253, 301, 324, 375, 414, 503, 607
alkaloids, 223
Alzheimer's disease therapy, 159
amine chemosensors, 490
amine derivative, 178
antimicrobial agents, 583
BACE1 inhibitors, 97*f*
bioisosteres, 34, 37
bioisosterism, 25
BODIPOY chemosensors, 486
bonding, 46*f*
boron chemosensors, 454
carbohydrazide chemosensors, 452
carboxamide chemosensors, 458, 480
chalcones, 228
chemical properties, 49
chemical structure, 44*f*
cholinesterase inhibitors, 90*f*
colorimetric chemosensors, 448
compounds, 14, 43, 375
coumarin chemosensors, 477
derivatives, 11, 20, 64*f*, 189, 190, 279*f*, 292, 302
BACE1 inhibitor, 196
dicarbonitriles, 230
fused rings, 183
grafted chitosan derivatives, 239
history, 44
hydrazine carbothioamide derivatives, 220
hydrazine carbothioamides libraries, 220
hydrazone chemosensors, 457*f*, 457, 476
imidazole chemosensors, 474, 487
ion exchange resin, 58*f*
ligand properties, 45
liquid alkyl, 48
malaria treatment, 147
marketed drugs, 209*f*
metal, 47*f*, 48*f*
microwave-assisted synthesis, 259
moieties, 14, 20*f*
molecular modeling and computational simulation, 182
monomers, 56*f*
mono-substituted, 44
neuroprotective agents., 104*f*

N-oxide compounds, 242
one-pot synthesis, 212
pharmacological activity, 47
phenol chemosensors, 449, 451*f*
phenol containing dye, 450
phthalimide chemosensors, 457
physical properties, 48
physicochemical characteristics, 49*t*
polymers, 54, 58
pyrazole chemosensors, 448, 470
resonance, 255*f*
rhodamine chemosensors, 460*f*, 460, 482
ring, 145, 619
scaffolds, 14, 71*f*, 605, 615*f*
schiff base chemosensors, 455, 463, 464*f*
schiff base derivative, 485
structure, 208*f*, 255*f*
sulfides, 238
summary, 145*f*, 166
synthesis, 70
synthetic route, 279*f*
thiazole chemosensors, 476, 489
thiazolidinone derivatives, 221
thiourea chemosensors, 456
toxicological manifestations, 182
transition metal compounds, 46
urethanes, 57
Pyridinium
ionic liquids, 76
salts, 76
Pyridinyl aminohydantoins, 95
Pyridopyrimidines, 227
Pyridotacrines, 103
Pyridoxine, 208
Pyridyl benzimidazole derivatives, 233
Pyrimidine-pyridine hybrids, 165, 236
Pyrimidine-triazolopyrimidine, 165
Pyrithione-based fungicides, 300
Pyrrolo aminopyridine derivatives, 181
Pyrrolopyridines, 218

Q
Quantum-chemical calculation methods, 139
Quinoline-based dendrimer-like ionic liquid, 519

R
Reactive oxygen species (ROS), 234
Receptor moiety, 448
Redox polymers, 60
Regorafenib, 393
Renal cell carcinoma (RCC), 383
Rhodamine dye, 460
Ribociclib, 393, 394
Ruthenium complexes, 199

S
SARS coronavirus (SARS-CoV) infection, 241
Secretase inhibitors, 95
Selpercatinib, 394
Single photon emission computed tomography (SPECT), 94
Singulair, 211
Sonidegib, 395
Sorafenib, 396, 398
Structure–activity relationship (SAR), 198*f*
Structure–activity relationship studies (SARs), 434
Substituted chelidamic acids, 173
Sulfadoxine, 149
Super-paramagnetic nanoparticles, 333
Suzuki–Miyaura cross-coupling reaction, 449

T
Tacrine, 164, 430
Tau aggregation inhibitors, 98
Tau hypothesis, 412
T-cell lymphoma, 218
Tenatoprazole, 212
Tenosynovial giant cell tumor (TGCT), 391
Tetrahydropyridine, 304
Thiazole-based hydrazides, 285
Thienopyridines, 218
Thiophene substituted pyridine derivatives, 128*f*
Thiourea isosteres, 31
Tillman regent, 60
Transition metal catalysis, 520
Triarylpyridine chemosensor, 455*f*
Triarylpyridine chemosensors, 455
Triazole, 427

U
Ultrasound-assisted green synthesis, 256
Uranium, 451
Urea, 31
Urease, 232
inhibitory activity, 232

V
Vinamidinium, 509
Vinorelbine, 152*t*
Vinyl pyridine
based polymers, 53
DVB copolymers, 55
monomers, 52
Vismodegib, 398

Z
Zn-based Lewis acids, 608

Printed in the United States
by Baker & Taylor Publisher Services